SOLUTIONS MANUAL

JOSEPH TOPICH
Virginia Commonwealth University

RUTH TOPICH
Virginia Commonwealth University

GENERAL CHEMISTRY
ATOMS FIRST

McMURRY • FAY

Prentice Hall

New York   Boston   San Francisco
London   Toronto   Sydney   Tokyo   Singapore   Madrid
Mexico City   Munich   Paris   Cape Town   Hong Kong   Montreal

Project Editor: Jennifer Hart
Acquisitions Editor: Terry Haugen
Editor in Chief, Chemistry and Geosciences: Nicole Folchetti
Marketing Manager: Erin Gardner
Managing Editor, Chemistry and Geosciences: Gina M. Cheselka
Project Manager: Wendy A. Perez
Operations Specialist: Amanda A. Smith
Supplement Cover Manager: Paul Gourhan
Supplement Cover Designer: Tina Krivoshein
Cover Photo Credit: Jim Cummins/Corbis

Printed in the United States of America

10 9 8 7 6 5 4 3 2

ISBN-13:   978-0-321-56398-9
ISBN-10:   0-321-56398-0

Prentice Hall
is an imprint of

www.pearsonhighered.com

Table of Contents

Preface

Chemistry is the study of the composition and properties of matter along with the changes that matter undergoes. The principles of chemistry are learned through a collection of different experiences. You read about them in your textbook. You hear about them from your chemistry instructor. You see them first hand in the laboratory, and probably most importantly, you use chemistry principles to solve chemistry problems.

Problem solving is one key to your success in chemistry! *GENERAL CHEMISTRY Atoms First* by McMurry and Fay contains thousands of problems that you can work. You want to develop a problem solving strategy. Read each problem carefully. List the information contained in the problem. Understand what the problem is asking. Use your knowledge of chemistry principles to identify connections between the information in the problem and the solution that you are seeking. Set up and attempt to solve the problem. Look at your answer. Is it reasonable? Are the units correct? Then, and only then, check your answer with the answer in the *Solutions Manual*.

The *Solutions Manual* to accompany *GENERAL CHEMISTRY Atoms First* contains the solutions to all in-chapter, key concept and end-of-chapter problems.

I have worked to ensure that the solutions in this manual are as error free as possible. Solutions have been double-checked and in many cases triple-checked. Small differences in numerical answers between those of students and those in the *Solutions Manual* may result because of rounding and/or significant figure differences. It should also be noted that there is, in many cases, more than one acceptable setup for a problem.

I would like to thank John McMurry and Robert Fay for the opportunity to contribute to their *GENERAL CHEMISTRY Atoms First* package. I also want to thank them for their helpful comments as I worked on this solutions manual. I also want to acknowledge and thank the entire Prentice Hall staff. Finally, I want to thank in a very special way my wife, Ruth, and our daughter, Judy, for their constant encouragement and support as I worked on this project.

Joseph Topich
Department of Chemistry
Virginia Commonwealth University

1 Chemistry: Matter and Measurement

1.1 (a) Cd (b) Sb (c) Am

1.2 (a) silver (b) rhodium (c) rhenium (d) cesium (e) argon (f) arsenic

1.3 (a) Ti, metal (b) Te, semimetal (c) Se, nonmetal
(d) Sc, metal (e) At, semimetal (f) Ar, nonmetal

1.4 The three "coinage metals" are copper (Cu), silver (Ag), and gold (Au).

1.5 (a) The decimal point must be shifted ten places to the right so the exponent is –10. The result is 3.72×10^{-10} m.
(b) The decimal point must be shifted eleven places to the left so the exponent is 11. The result is 1.5×10^{11} m.

1.6 (a) microgram (b) decimeter (c) picosecond
(d) kiloampere (e) millimole

1.7 $^{\circ}C = \dfrac{5}{9} \times (^{\circ}F - 32) = \dfrac{5}{9} \times (98.6 - 32) = 37.0\,^{\circ}C$

$K = \,^{\circ}C + 273.15 = 37.0 + 273.15 = 310.2\,K$

1.8 (a) $K = \,^{\circ}C + 273.15 = -78 + 273.15 = 195.15\,K = 195\,K$

(b) $^{\circ}F = (\dfrac{9}{5} \times \,^{\circ}C) + 32 = (\dfrac{9}{5} \times 158) + 32 = 316.4\,^{\circ}F = 316\,^{\circ}F$

(c) $^{\circ}C = K - 273.15 = 375 - 273.15 = 101.85\,^{\circ}C = 102\,^{\circ}C$

$^{\circ}F = (\dfrac{9}{5} \times \,^{\circ}C) + 32 = (\dfrac{9}{5} \times 101.85) + 32 = 215.33\,^{\circ}F = 215\,^{\circ}F$

1.9 $d = \dfrac{m}{V} = \dfrac{27.43\ g}{12.40\ cm^3} = 2.212\ g/cm^3$

1.10 $volume = 9.37\ g \times \dfrac{1\ mL}{1.483\ g} = 6.32\ mL$

1.11 $E_K = \dfrac{1}{2}mv^2 = \dfrac{1}{2} \times 1070\ kg \times (28.3\ m/s)^2 = 428{,}476\ \dfrac{kg \cdot m^2}{s^2} = 428{,}000\ J$

$428{,}000\ J \times \dfrac{1\ kJ}{1000\ J} = 428\ kJ$

1.12 (a) $500 \text{ Cal} \times \dfrac{1000 \text{ cal}}{1 \text{ Cal}} \times \dfrac{4.184 \text{ J}}{1 \text{ cal}} \times \dfrac{1 \text{ kJ}}{1000 \text{ J}} = 2092 \text{ kJ} = 2000 \text{ kJ}$

(b) 100 watts = 100 J/s

$\text{time} = 2092 \text{ kJ} \times \dfrac{1000 \text{ J}}{1 \text{ kJ}} \times \dfrac{1 \text{ s}}{100 \text{ J}} \times \dfrac{1 \text{ min}}{60 \text{ s}} \times \dfrac{1 \text{ hr}}{60 \text{ min}} = 5.811 \text{ hr} = 5.8 \text{ hr}$

1.13 The actual mass of the bottle and the acetone = 38.0015 g + 0.7791 g = 38.7806 g. The measured values are 38.7798 g, 38.7795 g, and 38.7801 g. These values are both close to each other and close to the actual mass. Therefore the results are both precise and accurate.

1.14 (a) 76.600 kJ has 5 significant figures because zeros at the end of a number and after the decimal point are always significant.
(b) $4.502\,00 \times 10^3$ g has 6 significant figures because zeros in the middle of a number are significant and zeros at the end of a number and after the decimal point are always significant.
(c) 3000 nm has 1, 2, 3, or 4 significant figures because zeros at the end of a number and before the decimal point may or may not be significant.
(d) 0.003 00 mL has 3 significant figures because zeros at the beginning of a number are not significant and zeros at the end of a number and after the decimal point are always significant.
(e) 18 students has an infinite number of significant figures because this is an exact number.
(f) 3×10^{-5} g has 1 significant figure.
(g) 47.60 mL has 4 significant figures because a zero at the end of a number and after the decimal point is always significant.
(h) 2070 mi has 3 or 4 significant figures because a zero in the middle of a number is significant and a zero at the end of a number and before the decimal point may or may not be significant.

1.15 (a) Because the digit to be dropped (the second 4) is less than 5, round down. The result is 3.774 L.
(b) Because the digit to be dropped (0) is less than 5, round down. The result is 255 K.
(c) Because the digit to be dropped is equal to 5 with nothing following, round down. The result is 55.26 kg.
(d) Because the digit to be dropped (1) is less than 5, round down. The first zero is significant because it is in the middle of the number. The second zero is significant because a zero at the end of a number and after the decimal point is always significant. The result is 906.40 kJ.

1.16 (a)
$$\begin{array}{r} 24.567 \text{ g} \\ + \ 0.044\,78 \text{ g} \\ \hline 24.611\,78 \text{ g} \end{array}$$
This result should be expressed with 3 decimal places. Because the digit to be dropped (7) is greater than 5, round up. The result is 24.612 g (5 significant figures).

(b)   4.6742 g / 0.003 71 L = 1259.89 g/L
0.003 71 has only 3 significant figures so the result of the division should have only 3 significant figures. Because the digit to be dropped (first 9) is greater than 5, round up. The result is 1260 g/L (3 significant figures), or 1.26×10^3 g/L.

(c)   0.378   mL   This result should be expressed with 1 decimal place.
      + 42.3   mL   Because the digit to be dropped (9) is greater than 5, round
      – 1.5833   mL   up. The result is 41.1 mL (3 significant figures).
        41.0947 mL

1.17   The level of the liquid in the thermometer is just past halfway between the 32 °C and 33 °C marks on the thermometer. The temperature is 32.6°C (3 significant figures).

1.18   (a) Calculation: $^\circ F = (\dfrac{9}{5} \times {}^\circ C) + 32 = (\dfrac{9}{5} \times 1064) + 32 = 1947\ ^\circ F$

Ballpark estimate: $^\circ F \approx 2 \times {}^\circ C$ if $^\circ C$ is large. The melting point of gold $\approx 2000\ ^\circ F$.

(b) r = d/2 = 3×10^{-6} m = 3×10^{-4} cm; h = 2×10^{-6} m = 2×10^{-4} cm
Calculation: volume = $\pi r^2 h$ = $(3.1416)(3 \times 10^{-4}\ cm)^2(2 \times 10^{-4}\ cm)$ = $6 \times 10^{-11}\ cm^3$
Ballpark estimate: volume = $\pi r^2 h \approx 3r^2 h \approx 3(3 \times 10^{-4}\ cm)^2(2 \times 10^{-4}\ cm) \approx 5 \times 10^{-11}\ cm^3$

1.19   1 carat = 200 mg = 200×10^{-3} g = 0.200 g

Mass of Hope Diamond in grams = 44.4 carats $\times\ \dfrac{0.200\ g}{1\ carat}$ = 8.88 g

1 ounce = 28.35 g

Mass of Hope Diamond in ounces = 8.88 g $\times\ \dfrac{1\ ounce}{28.35\ g}$ = 0.313 ounces

1.20   Volume of Hope Diamond = 8.88 g $\times \dfrac{1\ cm^3}{3.52\ g}$ = 2.52 cm^3

C atoms in Hope Diamond = 8.88 g $\times \dfrac{5.014 \times 10^{21}\ C\ atoms}{0.1000\ g}$ = 4.45×10^{23} C atoms

1.21   mass of salt = 155 lb $\times \dfrac{453.6\ g}{1\ lb} \times \dfrac{1\ kg}{1000\ g} \times \dfrac{4\ g}{1\ kg}$ = 281.2 g or 300 g

Key Concept Problems

1.22

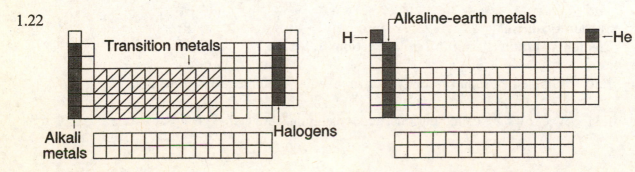

1.23

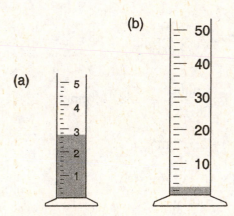

metals   nonmetals

1.24   red – gas; blue – 42;
        green – lithium, sodium, potassium or rubidium are possible answers

1.25   The element is americium (Am) with atomic number = 95. It is in the actinide series.

1.26   (a) Darts are clustered together (good precision) but are away from the bullseye (poor accuracy).
        (b) Darts are clustered together (good precision) and hit the bullseye (good accuracy).
        (c) Darts are scattered (poor precision) and are away from the bullseye (poor accuracy).

1.27   (a) 34.2 mL (3 significant figures)   (b) 2.68 cm (3 significant figures)

1.28

The 5 mL graduated cylinder is marked every 0.2 mL and can be read to ± 0.02 mL. The 50 mL graduated cylinder is marked every 2 mL and can only be read to ± 0.2 mL. The 5 mL graduated cylinder will give more accurate measurements.

1.29   A liquid that is less dense than another will float on top of it. The most dense liquid is mercury, and it is at the bottom of the cylinder. Because water is less dense than mercury but more dense than vegetable oil, it is the middle liquid in the cylinder. Vegetable oil is the least dense of the three liquids and is the top liquid in the cylinder.

Section Problems
Elements and the Periodic Table (Sections 1.2–1.4)

1.30   114 elements are presently known. About 90 elements occur naturally.

1.31   The rows are called periods, and the columns are called groups.

1.32   There are 18 groups in the periodic table. They are labeled as follows:
1A, 2A, 3B, 4B, 5B, 6B, 7B, 8B (3 groups), 1B, 2B, 3A, 4A, 5A, 6A, 7A, 8A

1.33   Elements within a group have similar chemical properties.

1.34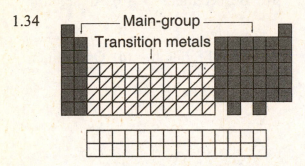

1.35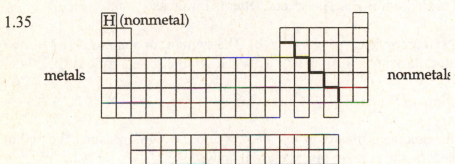

1.36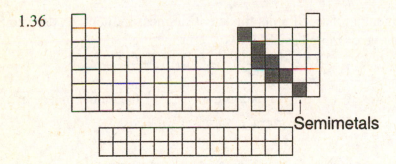

A semimetal is an element with properties that fall between those of metals and nonmetals.

1.37   (a) The alkali metals are shiny, soft, low-melting metals that react rapidly with water to form products that are alkaline.
(b) The noble gases are gases of very low reactivity.
(c) The halogens are nonmetallic and corrosive. They are found in nature only in combination with other elements.

1.38   Li, Na, K, Rb, and Cs

1.39   Be, Mg, Ca, Sr, and Ba

1.40   F, Cl, Br, and I

1.41   He, Ne, Ar, Kr, Xe, and Rn

1.42   (a) gadolinium, Gd   (b) germanium, Ge   (c) technetium, Tc   (d) arsenic, As

1.43   (a) cadmium, Cd   (b) iridium, Ir   (c) beryllium, Be   (d) tungsten, W

1.44   (a) Te, tellurium   (b) Re, rhenium   (c) Be, beryllium   (d) Ar, argon
        (e) Pu, plutonium

1.45   (a) B, boron   (b) Rh, rhodium   (c) Cf, californium   (d) Os, osmium
        (e) Ga, gallium

1.46   (a) Tin is Sn, Ti is titanium.
        (b) Manganese is Mn, Mg is magnesium.
        (c) Potassium is K, Po is polonium.
        (d) The symbol for helium is He. The second letter is lowercase.

1.47   (a) The symbol for carbon is C.   (b) The symbol for sodium is Na.
        (c) The symbol for nitrogen is N.   (d) The symbol for chlorine is Cl.

Units and Significant Figures (Sections 1.5–1.13)

1.48   Mass measures the amount of matter in an object, whereas weight measures the pull of gravity on an object by the earth or other celestial body.

1.49   There are only seven fundamental (base) SI units for scientific measurement. A derived SI unit is some combination of two or more base SI units.
        Base SI unit: Mass, kg;   Derived SI unit: Density, kg/m^3

1.50   (a) kilogram, kg   (b) meter, m   (c) kelvin, K   (d) cubic meter, m^3
        (e) joule, $(kg \cdot m^2) / s^2$ (f) kg/m^3 or g/cm^3

1.51   (a) kilo, k   (b) micro, μ (c) giga, G   (d) pico, p   (e) centi, c

1.52   A Celsius degree is larger than a Fahrenheit degree by a factor of $\dfrac{9}{5}$.

1.53   A kelvin and Celsius degree are the same size.

1.54   The volume of a cubic decimeter (dm^3) and a liter (L) are the same.

1.55   The volume of a cubic centimeter (cm^3) and a milliliter (mL) are the same.

1.56   (a) and (b) are exact numbers because they are both definitions.
        (c) and (d) are not exact numbers because they result from measurements.

1.57   4.8673 g   The result should contain only 1 decimal place. Because the digit to
        $\underline{-\ 4.8\ \ \ \ \ \ g}$ be dropped (6) is greater than 5, round up. The result is 0.1 g.
           0.0673 g

1.58   cL is centiliter (10^{-2} L)

1.59   (a) deciliter (10^{-1} L)   (b) decimeter (10^{-1} m)   (c) micrometer (10^{-6} m)
         (d) nanoliter (10^{-9} L)   (e) megajoule (10^{6} J)

1.60   1 mg = 1×10^{-3} g and 1 pg = 1×10^{-12} g

$$\frac{1 \times 10^{-3}\,\text{g}}{1\ \text{mg}} \times \frac{1\ \text{pg}}{1 \times 10^{-12}\,\text{g}} = 1 \times 10^{9}\ \text{pg/mg}$$

35 ng = 35×10^{-9} g

$$\frac{35 \times 10^{-9}\,\text{g}}{35\ \text{ng}} \times \frac{1\ \text{pg}}{1 \times 10^{-12}\,\text{g}} = 3.5 \times 10^{4}\ \text{pg/35 ng}$$

1.61   1 μL = 10^{-6} L $\dfrac{1\ \mu\text{L}}{10^{-6}\,\text{L}} = 10^{6}\ \mu\text{L/L}$

20 mL = 20×10^{-3} L $\dfrac{20 \times 10^{-3}\,\text{L}}{20\ \text{mL}} \times \dfrac{1\ \mu\text{L}}{10^{-6}\,\text{L}} = 2 \times 10^{4}\ \mu\text{L/20 mL}$

1.62   (a)   5 pm = 5×10^{-12} m

$$5 \times 10^{-12}\ \text{m} \times \frac{100\ \text{cm}}{1\ \text{m}} = 5 \times 10^{-10}\ \text{cm}$$

$$5 \times 10^{-12}\ \text{m} \times \frac{1\ \text{nm}}{1 \times 10^{-9}\,\text{m}} = 5 \times 10^{-3}\ \text{nm}$$

(b) $8.5\ \text{cm}^3 \times \left(\dfrac{1\ \text{m}}{100\ \text{cm}}\right)^3 = 8.5 \times 10^{-6}\ \text{m}^3$

$8.5\ \text{cm}^3 \times \left(\dfrac{10\ \text{mm}}{1\ \text{cm}}\right)^3 = 8.5 \times 10^{3}\ \text{mm}^3$

(c) $65.2\ \text{mg} \times \dfrac{1 \times 10^{-3}\,\text{g}}{1\ \text{mg}} = 0.0652\ \text{g}$

$65.2\ \text{mg} \times \dfrac{1 \times 10^{-3}\,\text{g}}{1\ \text{mg}} \times \dfrac{1\ \text{pg}}{1 \times 10^{-12}\,\text{g}} = 6.52 \times 10^{10}\ \text{pg}$

1.63   (a) A liter is just slightly larger than a quart.
         (b) A mile is about twice as long as a kilometer.
         (c) An ounce is about 30 times larger than a gram.
         (d) An inch is about 2.5 times larger than a centimeter.

1.64   (a) 35.0445 g has 6 significant figures because zeros in the middle of a number are significant.
         (b) 59.0001 cm has 6 significant figures because zeros in the middle of a number are significant.

(c) 0.030 03 kg has 4 significant figures because zeros at the beginning of a number are not significant and zeros in the middle of a number are significant.

(d) 0.004 50 m has 3 significant figures because zeros at the beginning of a number are not significant and zeros at the end of a number and after the decimal point are always significant.

(e) 67,000 m^2 has 2, 3, 4, or 5 significant figures because zeros at the end of a number and before the decimal point may or may not be significant.

(f) 3.8200 x 10^3 L has 5 significant figures because zeros at the end of a number and after the decimal point are always significant.

1.65   (a) $130.95 is an exact number and has an infinite number of significant figures.

(b) 2000.003 has 7 significant figures because zeros in the middle of a number are significant.

(c) The measured quantity, 5 ft 3 in., has 2 significant figures. The 5 ft is certain and the 3 in. is an estimate.

(d) 510 J has 2 or 3 significant figures because zeros at the end of a number and before the decimal point may or may not be significant.

(e) 5.10 x 10^2 J has 3 significant figures because zeros at the end of a number and after the decimal point are always significant.

(f) 10 students is a count, therefore 10 is an exact number with an infinite number of significant figures.

1.66   To convert 3,666,500 m^3 to scientific notation, move the decimal point 6 places to the left and include an exponent of 10^6. The result is 3.6665 x 10^6 m^3.

1.67   Because the digit to be dropped (3) is less than 5, round down. The result to 4 significant figures is 7926 mi or 7.926 x 10^3 mi.

Because the digit to be dropped (2) is less than 5, round down. The result to 2 significant figures is 7900 mi or 7.9 x 10^3 mi.

1.68   (a) To convert 453.32 mg to scientific notation, move the decimal point 2 places to the left and include an exponent of 10^2. The result is 4.5332 x 10^2 mg.

(b) To convert 0.000 042 1 mL to scientific notation, move the decimal point 5 places to the right and include an exponent of 10^{-5}. The result is 4.21 x 10^{-5} mL.

(c) To convert 667,000 g to scientific notation, move the decimal point 5 places to the left and include an exponent of 10^5. The result is 6.67 x 10^5 g.

1.69   (a) Because the exponent is a negative 3, move the decimal point 3 places to the left to get 0.003 221 mm.

(b) Because the exponent is a positive 5, move the decimal point 5 places to the right to get 894,000 m.

(c) Because the exponent is a negative 12, move the decimal point 12 places to the left to get 0.000 000 000 001 350 82 m^3.

(d) Because the exponent is a positive 2, move the decimal point 2 places to the right to get 641.00 km.

1.70   (a) Because the digit to be dropped (0) is less than 5, round down. The result is 3.567 x 10^4 or 35,670 m (4 significant figures).
Because the digit to be dropped (the second 6) is greater than 5, round up. The result is 35,670.1 m (6 significant figures).
(b) Because the digit to be dropped is 5 with nonzero digits following, round up. The result is 69 g (2 significant figures).
Because the digit to be dropped (0) is less than 5, round down. The result is 68.5 g (3 significant figures).
(c) Because the digit to be dropped is 5 with nothing following, round down. The result is 4.99 x 10^3 cm (3 significant figures).
(d) Because the digit to be dropped is 5 with nothing following, round down. The result is 2.3098 x 10^{-4} kg (5 significant figures).

1.71   (a) Because the digit to be dropped (1) is less than 5, round down. The result is 7.000 kg.
(b) Because the digit to be dropped is 5 with nothing following, round down. The result is 1.60 km.
(c) Because the digit to be dropped (1) is less than 5, round down. The result is 13.2 g/cm^3.
(d) Because the digit to be dropped (1) is less than 5, round down. The result is 2,300,000. or 2.300 000 x 10^6.

1.72   (a) 4.884 x 2.05 = 10.012
The result should contain only 3 significant figures because 2.05 contains 3 significant figures (the smaller number of significant figures of the two). Because the digit to be dropped (1) is less than 5, round down. The result is 10.0.
(b) 94.61 / 3.7 = 25.57
The result should contain only 2 significant figures because 3.7 contains 2 significant figures (the smaller number of significant figures of the two). Because the digit to be dropped (second 5) is 5 with nonzero digits following, round up. The result is 26.
(c) 3.7 / 94.61 = 0.0391
The result should contain only 2 significant figures because 3.7 contains 2 significant figures (the smaller number of significant figures of the two). Because the digit to be dropped (1) is less than 5, round down. The result is 0.039.

(d)
```
    5502.3
      24
+     0.01
   5526.31
```
This result should be expressed with no decimal places. Because the digit to be dropped (3) is less than 5, round down. The result is 5526.

(e)
```
    86.3
+   1.42
-   0.09
   87.63
```
This result should be expressed with only 1 decimal place. Because the digit to be dropped (3) is less than 5, round down. The result is 87.6.

(f) 5.7 x 2.31 = 13.167
The result should contain only 2 significant figures because 5.7 contains 2 significant figures (the smaller number of significant figures of the two). Because the digit to be dropped (second 1) is less than 5, round down. The result is 13.

1.73 (a) $\dfrac{3.41 - 0.23}{5.233} \times 0.205 = \dfrac{3.18}{5.233} \times 0.205 = 0.12457 = 0.125$

Complete the subtraction first. The result has 2 decimal places and 3 significant figures. The result of the multiplication and division must have 3 significant figures. Because the digit to be dropped is 5 with nonzero digits following, round up.

(b) $\dfrac{5.556 \times 2.3}{4.223 - 0.08} = \dfrac{5.556 \times 2.3}{4.143} = 3.08 = 3.1$

Complete the subtraction first. The result of the subtraction should have 2 decimal places and 3 significant figures (an extra digit is being carried until the calculation is completed). The result of the multiplication and division must have 2 significant figures. Because the digit to be dropped (8) is greater than 5, round up.

Unit Conversions (Section 1.14)

1.74 (a) $0.25 \text{ lb} \times \dfrac{453.59 \text{ g}}{1 \text{ lb}} = 113.4 \text{ g} = 110 \text{ g}$

(b) $1454 \text{ ft} \times \dfrac{12 \text{ in.}}{1 \text{ ft}} \times \dfrac{2.54 \text{ cm}}{1 \text{ in.}} \times \dfrac{1 \text{ m}}{100 \text{ cm}} = 443.2 \text{ m}$

(c) $2{,}941{,}526 \text{ mi}^2 \times \left(\dfrac{1.6093 \text{ km}}{1 \text{ mi}}\right)^2 \times \left(\dfrac{1000 \text{ m}}{1 \text{ km}}\right)^2 = 7.6181 \times 10^{12} \text{ m}^2$

1.75 (a) $5.4 \text{ in.} \times \dfrac{2.54 \text{ cm}}{1 \text{ in.}} \times \dfrac{1 \text{ m}}{100 \text{ cm}} = 0.14 \text{ m}$

(b) $66.31 \text{ lb} \times \dfrac{1 \text{ kg}}{2.2046 \text{ lb}} = 30.08 \text{ kg}$

(c) $0.5521 \text{ gal} \times \dfrac{3.7854 \text{ L}}{1 \text{ gal}} \times \dfrac{1 \times 10^{-3} \text{ m}^3}{1 \text{ L}} = 2.090 \times 10^{-3} \text{ m}^3$

(d) $65 \dfrac{\text{mi}}{\text{h}} \times \dfrac{1.6093 \text{ km}}{1 \text{ mi}} \times \dfrac{1000 \text{ m}}{1 \text{ km}} \times \dfrac{1 \text{ h}}{60 \text{ min}} \times \dfrac{1 \text{ min}}{60 \text{ s}} = 29 \dfrac{\text{m}}{\text{s}}$

(e) $978.3 \text{ yd}^3 \times \left(\dfrac{1 \text{ m}}{1.0936 \text{ yd}}\right)^3 = 748.0 \text{ m}^3$

(f) $2.380 \text{ mi}^2 \times \left(\dfrac{1.6093 \text{ km}}{1 \text{ mi}}\right)^2 \times \left(\dfrac{1000 \text{ m}}{1 \text{ km}}\right)^2 = 6.164 \times 10^6 \text{ m}^2$

1.76 (a) $1 \text{ acre-ft} \times \dfrac{1 \text{ mi}^2}{640 \text{ acres}} \times \left(\dfrac{5280 \text{ ft}}{1 \text{ mi}}\right)^2 = 43{,}560 \text{ ft}^3$

(b) $116 \text{ mi}^3 \times \left(\dfrac{5280 \text{ ft}}{1 \text{ mi}}\right)^3 \times \dfrac{1 \text{ acref-ft}}{43{,}560 \text{ ft}^3} = 3.92 \times 10^8 \text{ acre-ft}$

1.77 (a) $18.6 \text{ hands} \times \dfrac{1/3 \text{ ft}}{1 \text{ hand}} \times \dfrac{12 \text{ in.}}{1 \text{ ft}} \times \dfrac{2.54 \text{ cm}}{1 \text{ in.}} = 189 \text{ cm}$

(b) $(6 \times 2.5 \times 15) \text{ hands}^3 \times \left(\dfrac{1/3 \text{ ft}}{1 \text{ hand}}\right)^3 \times \left(\dfrac{12 \text{ in.}}{1 \text{ ft}}\right)^3 \times \left(\dfrac{2.54 \text{ cm}}{1 \text{ in.}}\right)^3 \times \left(\dfrac{1 \text{ m}}{100 \text{ cm}}\right)^3 = 0.2 \text{ m}^3$

1.78 (a) $\dfrac{200 \text{ mg}}{100 \text{ mL}} \times \dfrac{1000 \text{ mL}}{1 \text{ L}} = 2000 \text{ mg/L}$

(b) $\dfrac{200 \text{ mg}}{100 \text{ mL}} \times \dfrac{1 \times 10^{-3} \text{ g}}{1 \text{ mg}} \times \dfrac{1 \text{ } \mu g}{1 \times 10^{-6} \text{ g}} = 2000 \text{ } \mu g/\text{mL}$

(c) $\dfrac{200 \text{ mg}}{100 \text{ mL}} \times \dfrac{1 \times 10^{-3} \text{ g}}{1 \text{ mg}} \times \dfrac{1000 \text{ mL}}{1 \text{ L}} = 2 \text{ g/L}$

(d) $\dfrac{200 \text{ mg}}{100 \text{ mL}} \times \dfrac{1 \times 10^{-3} \text{ g}}{1 \text{ mg}} \times \dfrac{1000 \text{ mL}}{1 \text{ L}} \times \dfrac{1 \text{ ng}}{1 \times 10^{-9} \text{ g}} \times \dfrac{1 \times 10^{-6} \text{ L}}{1 \text{ } \mu L} = 2000 \text{ ng/}\mu L$

(e) $2 \text{ g/L} \times 5 \text{ L} = 10 \text{ g}$

1.79 $8.65 \text{ stones} \times \dfrac{14 \text{ lb}}{1 \text{ stone}} = 121 \text{ lb}$

1.80 $55 \dfrac{\text{mi}}{\text{h}} \times \dfrac{5280 \text{ ft}}{1 \text{ mi}} \times \dfrac{12 \text{ in.}}{1 \text{ ft}} \times \dfrac{2.54 \text{ cm}}{1 \text{ in.}} \times \dfrac{1 \text{ h}}{3600 \text{ s}} \times \dfrac{2.5 \times 10^{-4} \text{ s}}{1 \text{ shake}} = 0.61 \dfrac{\text{cm}}{\text{shake}}$

1.81 $160 \text{ lb} \times \dfrac{1 \text{ kg}}{2.2046 \text{ lb}} = 72.6 \text{ kg}$

$72.6 \text{ kg} \times \dfrac{20 \text{ } \mu g}{1 \text{ kg}} \times \dfrac{1 \text{ mg}}{1 \times 10^{3} \text{ } \mu g} = 1.452 \text{ mg} = 1.5 \text{ mg}$

Temperature (Section 1.8)

1.82 $^\circ F = (\dfrac{9}{5} \times {}^\circ C) + 32$

$^\circ F = (\dfrac{9}{5} \times 39.9 ^\circ C) + 32 = 103.8 \text{ } ^\circ F$ (goat)

$^\circ F = (\dfrac{9}{5} \times 22.2 ^\circ C) + 32 = 72.0 \text{ } ^\circ F$ (Australian spiny anteater)

1.83 For Hg: mp is $\left[\dfrac{9}{5} \times (-38.87)\right] + 32 = -37.97 \text{ } ^\circ F$

For Br$_2$: mp is $\left[\dfrac{9}{5} \times (-7.2)\right] + 32 = 19.0 \text{ } ^\circ F$

For Cs: mp is $\left[\dfrac{9}{5} \times (28.40)\right] + 32 = 83.12 \text{ } ^\circ F$

For Ga: mp is $\left[\dfrac{9}{5} \times (29.78)\right] + 32 = 85.60 \text{ } ^\circ F$

1.84 $^\circ C = \dfrac{5}{9} \times (^\circ F - 32) = \dfrac{5}{9} \times (6192 - 32) = 3422 \text{ } ^\circ C$

$K = {}^\circ C + 273.15 = 3422 + 273.15 = 3695.15 \text{ K or } 3695 \text{ K}$

1.85 $°F = (\frac{9}{5} \times °C) + 32 = (\frac{9}{5} \times 175) + 32 = 347\ °F$

1.86   Ethanol boiling point   78.5 °C   173.3 °F   200 °E
         Ethanol melting point   –117.3 °C   –179.1 °F   0 °E

(a) $\dfrac{200\ °E}{[78.5\ °C - (-117.3\ °C)]} = \dfrac{200\ °E}{195.8\ °C} = 1.021\ °E/°C$

(b) $\dfrac{200\ °E}{[173.3\ °F - (-179.1\ °F)]} = \dfrac{200\ °E}{352.4\ °F} = 0.5675\ °E/°F$

(c) $°E = \dfrac{200}{195.8} \times (°C + 117.3)$

H_2O melting point = 0°C; $°E = \dfrac{200}{195.8} \times (0 + 117.3) = 119.8\ °E$

H_2O boiling point = 100°C; $°E = \dfrac{200}{195.8} \times (100 + 117.3) = 222.0\ °E$

(d) $°E = \dfrac{200}{352.4} \times (°F + 179.1) = \dfrac{200}{352.4} \times (98.6 + 179.1) = 157.6\ °E$

(e) $°F = \left(°E \times \dfrac{352.4}{200} \right) - 179.1 = \left(130 \times \dfrac{352.4}{200} \right) - 179.1 = 50.0\ °F$

Because the outside temperature is 50.0°F, I would wear a sweater or light jacket.

1.87 NH_3 boiling point   –33.4 °C   –28.1 °F   100 °A
         NH_3 melting point   –77.7 °C   –107.9 °F   0° A

(a) $\dfrac{100\ °A}{[-33.4 - (-77.7\ °C)]} = \dfrac{100\ °A}{44.3\ °C} = 2.26\ °A/°C$

(b) $\dfrac{100\ °A}{[-28.1 - (-107.9\ °F)]} = \dfrac{100\ °A}{79.8\ °F} = 1.25\ °A/°F$

(c) $°A = \dfrac{100}{44.3} \times (°C + 77.7)$

H_2O melting point = 0°C; $°A = \dfrac{100}{44.3} \times (0 + 77.7) = 175\ °A$

H_2O boiling point = 100°C; $°A = \dfrac{100}{44.3} \times (100 + 77.7) = 401\ °A$

(d) $°A = \dfrac{100}{79.8} \times (°F + 107.9) = \dfrac{100}{79.8} \times (98.6 + 107.9) = 259\ °A$

Density (Section 1.10)

1.88 $250 \text{ mg} \times \dfrac{1 \times 10^{-3} \text{ g}}{1 \text{ mg}} = 0.25 \text{ g}; \quad V = 0.25 \text{ g} \times \dfrac{1 \text{ cm}^3}{1.40 \text{ g}} = 0.18 \text{ cm}^3$

$500 \text{ lb} \times \dfrac{453.59 \text{ g}}{1 \text{ lb}} = 226{,}795 \text{ g}; \quad V = 226{,}795 \text{ g} \times \dfrac{1 \text{ cm}^3}{1.40 \text{ g}} = 161{,}996 \text{ cm}^3 = 162{,}000 \text{ cm}^3$

1.89   For H_2: $V = 1.0078 \text{ g} \times \dfrac{1 \text{ L}}{0.0899 \text{ g}} = 11.2 \text{ L}$

For Cl_2: $V = 35.45 \text{ g} \times \dfrac{1 \text{ L}}{3.214 \text{ g}} = 11.03 \text{ L}$

1.90 $d = \dfrac{m}{V} = \dfrac{220.9 \text{ g}}{(0.50 \times 1.55 \times 25.00) \text{ cm}^3} = 11.4 \dfrac{\text{g}}{\text{cm}^3} = 11 \dfrac{\text{g}}{\text{cm}^3}$

1.91 $d = 2.40 \text{ mm} = 0.240 \text{ cm}, \ r = d/2 = 0.120 \text{ cm}, \text{ and } V = \pi r^2 h$

$d = \dfrac{m}{V} = \dfrac{0.3624 \text{ g}}{(3.1416)(0.120 \text{ cm})^2 (15.0 \text{ cm})} = 0.534 \text{ g/cm}^3$

1.92 $d = \dfrac{m}{V} = \dfrac{8.763 \text{ g}}{(28.76 - 25.00) \text{ mL}} = \dfrac{8.763 \text{ g}}{3.76 \text{ mL}} = 2.331 \dfrac{\text{g}}{\text{cm}^3} = 2.33 \dfrac{\text{g}}{\text{cm}^3}$

1.93   The explosion was caused by a chemical property. Na reacts violently with H_2O.

Energy (Section 1.11)

1.94   Car: $E_K = \frac{1}{2}(1400 \text{ kg}) \left(\dfrac{115 \times 10^3 \text{ m}}{3600 \text{ s}} \right)^2 = 7.1 \times 10^5 \text{ J}$

Truck: $E_K = \frac{1}{2}(12{,}000 \text{ kg}) \left(\dfrac{38 \times 10^3 \text{ m}}{3600 \text{ s}} \right)^2 = 6.7 \times 10^5 \text{ J}$

The car has more kinetic energy.

1.95   Heat $= q = 7.1 \times 10^5 \text{ J}$ (from Problem 1.94)
$q = (\text{specific heat}) \times m \times \Delta T$

$m = \dfrac{q}{(\text{specific heat}) \times \Delta T} = \dfrac{7.1 \times 10^5 \text{ J}}{\left(4.18 \dfrac{\text{J}}{\text{g} \cdot {}^{\circ}\text{C}} \right)(50\,{}^{\circ}\text{C} - 20\,{}^{\circ}\text{C})} = 5.7 \times 10^3 \text{ g of water}$

1.96 $1 \text{ oz} = 28.35 \text{ g}$

$\text{energy} = 0.450 \text{ oz} \times \dfrac{28.35 \text{ g}}{1 \text{ oz}} \times \dfrac{2498 \text{ kJ}}{45.0 \text{ g}} \times \dfrac{1 \text{ kcal}}{4.184 \text{ kJ}} = 169 \text{ kcal}$

1.97 $g\ Na = \dfrac{1.00\ g\ Na}{17.9\ kJ} \times \dfrac{4.184\ kJ}{1\ kcal} \times 171\ kcal = 40.0\ g\ Na$

$g\ Cl = \dfrac{1.54\ g\ Cl}{1.00\ g\ Na} \times 40.0\ g\ Na = 61.6\ g\ Cl$

Chapter Problems

1.98   (a) selenium, Se   (b) rhenium, Re   (c) cobalt, Co (d) rhodium, Rh

1.99   (a) Element 117 is a halogen because it would be found directly below At in group 7A.
(b) Element 119
(c) Element 115 would be found directly below Bi and would be a metal. Element 117 might have the properties of a semimetal.
(d) Element 119, at the bottom of group 1A, would likely be a soft, shiny, very reactive metal forming a +1 cation.

1.100 NaCl melting point = 1074 K
$^{\circ}C = K - 273.15 = 1074 - 273.15 = 800.85\ ^{\circ}C = 801\ ^{\circ}C$

$^{\circ}F = (\dfrac{9}{5} \times\ ^{\circ}C) + 32 = (\dfrac{9}{5} \times 800.85) + 32 = 1473.53\ ^{\circ}F = 1474\ ^{\circ}F$

NaCl boiling point = 1686 K
$^{\circ}C = K - 273.15 = 1686 - 273.15 = 1412.85\ ^{\circ}C = 1413\ ^{\circ}C$

$^{\circ}F = (\dfrac{9}{5} \times\ ^{\circ}C) + 32 = (\dfrac{9}{5} \times 1412.85) + 32 = 2575.13\ ^{\circ}F = 2575\ ^{\circ}F$

1.101 $^{\circ}F = \left[\dfrac{9}{5} \times\ ^{\circ}C\right] + 32 = \left[\dfrac{9}{5} \times (-38.9)\right] + 32 = -38.0\ ^{\circ}F$

1.102 $V = 112.5\ g \times \dfrac{1\ mL}{1.4832\ g} = 75.85\ mL$

1.103 $15.28\ \dfrac{lb}{gal} \times \dfrac{453.59\ g}{1\ lb} \times \dfrac{1\ gal}{3.7854\ L} \times \dfrac{1\ L}{1000\ mL} = 1.831\ g/mL$

1.104 $V = 8.728 \times 10^{10}\ lb \times \dfrac{453.59\ g}{1\ lb} \times \dfrac{1\ mL}{1.831\ g} \times \dfrac{1\ L}{1000\ mL} = 2.162 \times 10^{10}\ L$

1.105 $0.22\ in. \times \dfrac{2.54\ cm}{1\ in.} \times \dfrac{10\ mm}{1\ cm} = 5.6\ mm$

1.106 (a) density $= \dfrac{1 \text{ lb}}{1 \text{ pint}} \times \dfrac{8 \text{ pints}}{1 \text{ gal}} \times \dfrac{1 \text{ gal}}{3.7854 \text{ L}} \times \dfrac{453.59 \text{ g}}{1 \text{ lb}} \times \dfrac{1 \text{ L}}{1000 \text{ mL}} = 0.95861 \text{ g/mL}$

(b) area in m^2 =

$1 \text{ acre} \times \dfrac{1 \text{ mi}^2}{640 \text{ acres}} \times \left(\dfrac{5280 \text{ ft}}{1 \text{ mi}}\right)^2 \times \left(\dfrac{12 \text{ in.}}{1 \text{ ft}}\right)^2 \times \left(\dfrac{2.54 \text{ cm}}{1 \text{ in.}}\right)^2 \times \left(\dfrac{1 \text{ m}}{100 \text{ cm}}\right)^2 = 4047 \text{ m}^2$

(c) mass of wood =

$1 \text{ cord} \times \dfrac{128 \text{ ft}^3}{1 \text{ cord}} \times \left(\dfrac{12 \text{ in.}}{1 \text{ ft}}\right)^3 \times \left(\dfrac{2.54 \text{ cm}}{1 \text{ in.}}\right)^3 \times \dfrac{0.40 \text{ g}}{1 \text{ cm}^3} \times \dfrac{1 \text{ kg}}{1000 \text{ g}} = 1450 \text{ kg} = 1400 \text{ kg}$

(d) mass of oil =

$1 \text{ barrel} \times \dfrac{42 \text{ gal}}{1 \text{ barrel}} \times \dfrac{3.7854 \text{ L}}{1 \text{ gal}} \times \dfrac{1000 \text{ mL}}{1 \text{ L}} \times \dfrac{0.85 \text{ g}}{1 \text{ mL}} \times \dfrac{1 \text{ kg}}{1000 \text{ g}} = 135.1 \text{ kg} = 140 \text{ kg}$

(e) fat Calories =

$0.5 \text{ gal} \times \dfrac{32 \text{ servings}}{1 \text{ gal}} \times \dfrac{165 \text{ Calories}}{1 \text{ serving}} \times \dfrac{30.0 \text{ Cal from fat}}{100 \text{ Cal total}} = 792 \text{ Cal from fat}$

1.107 amount of chocolate =

$2.0 \text{ cups coffee} \times \dfrac{105 \text{ mg caffeine}}{1 \text{ cup coffee}} \times \dfrac{1.0 \text{ ounce chocolate}}{15 \text{ mg caffeine}} = 14 \text{ ounces of chocolate}$

14 ounces of chocolate is just under 1 pound.

1.108 (a) number of Hershey's Kisses =

$2.0 \text{ lb} \times \dfrac{453.59 \text{ g}}{1 \text{ lb}} \times \dfrac{1 \text{ serving}}{41 \text{ g}} \times \dfrac{9 \text{ kisses}}{1 \text{ serving}} = 199 \text{ kisses} = 200 \text{ kisses}$

(b) Hershey's Kiss volume $= \dfrac{41 \text{ g}}{1 \text{ serving}} \times \dfrac{1 \text{ serving}}{9 \text{ kisses}} \times \dfrac{1 \text{ mL}}{1.4 \text{ g}} = 3.254 \text{ mL} = 3.3 \text{ mL}$

(c) Calories/Hershey's Kiss $= \dfrac{230 \text{ Cal}}{1 \text{ serving}} \times \dfrac{1 \text{ serving}}{9 \text{ kisses}} = 25.55 \text{ Cal/kiss} = 26 \text{ Cal/kiss}$

(d) % fat Calories =

$\dfrac{13 \text{ g fat}}{1 \text{ serving}} \times \dfrac{9 \text{ Cal from fat}}{1 \text{ g fat}} \times \dfrac{1 \text{ serving}}{230 \text{ Cal total}} \times 100\% = 51\% \text{ Calories from fat}$

1.109 Let Y equal volume of vinegar and $(422.8 \text{ cm}^3 - Y)$ equal the volume of oil.

Mass = volume x density

$397.8 \text{ g} = (Y \times 1.006 \text{ g/cm}^3) + [(422.8 \text{ cm}^3 - Y) \times 0.918 \text{ g/cm}^3]$

$397.8 \text{ g} = (1.006 \text{ g/cm}^3)Y + 388.1 \text{ g} - (0.918 \text{ g/cm}^3)Y$

$397.8 \text{ g} - 388.1 \text{ g} = (1.006 \text{ g/cm}^3)Y - (0.918 \text{ g/cm}^3)Y$

$9.7 \text{ g} = (0.088 \text{ g/cm}^3)Y$

$Y = \text{vinegar volume} = \dfrac{9.7 \text{ g}}{0.088 \text{ g/cm}^3} = 110 \text{ cm}^3$

oil volume $= (422.8 \text{ cm}^3 - Y) = (422.8 \text{ cm}^3 - 110 \text{ cm}^3) = 313 \text{ cm}^3$

1.110 $^\circ C = \dfrac{5}{9} \times (^\circ F - 32);$ Set $^\circ C = ^\circ F$: $^\circ C = \dfrac{5}{9} \times (^\circ C - 32)$

Solve for $^\circ C$: $^\circ C \times \dfrac{9}{5} = ^\circ C - 32$

$(^\circ C \times \dfrac{9}{5}) - ^\circ C = -32$

$^\circ C \times \dfrac{4}{5} = -32$

$^\circ C = \dfrac{5}{4}(-32) = -40\,^\circ C$

The Celsius and Fahrenheit scales "cross" at $-40\,^\circ C$ $(-40\,^\circ F)$.

1.111   Cork: volume = 1.30 cm x 5.50 cm x 3.00 cm = 21.45 cm^3

mass $= 21.45$ cm^3 x $\dfrac{0.235 \text{ g}}{1 \text{ cm}^3} = 5.041$ g

Lead: volume $= (1.15 \text{ cm})^3 = 1.521$ cm^3

mass $= 1.521$ cm^3 x $\dfrac{11.35 \text{ g}}{1 \text{ cm}^3} = 17.26$ g

total mass = 5.041 g + 17.26 g = 22.30 g
total volume = 21.45 cm^3 + 1.521 cm^3 = 22.97 cm^3

average density $= \dfrac{22.30 \text{ g}}{22.97 \text{ cm}^3} = 0.971$ g/cm^3 so the cork and lead will float.

1.112   Convert 8 min, 25 s to s.   8 min x $\dfrac{60 \text{ s}}{1 \text{ min}} + 25$ s = 505 s

Convert 293.2 K to $^\circ F$:

$293.2 - 273.15 = 20.05\,^\circ C$ and $^\circ F = (\dfrac{9}{5} \times 20.05) + 32 = 68.09\,^\circ F$

Final temperature $= 68.09\,^\circ F + 505$ s x $\dfrac{3.0\,^\circ F}{60 \text{ s}} = 93.34\,^\circ F$

$^\circ C = \dfrac{5}{9} \times (93.34 - 32) = 34.1\,^\circ C$

1.113   Ethyl alcohol density $= \dfrac{19.7325 \text{ g}}{25.00 \text{ mL}} = 0.7893$ g/mL

total mass = metal mass + ethyl alcohol mass = 38.4704 g
ethyl alcohol mass = total mass − metal mass = 38.4704 g − 25.0920 g = 13.3784 g

ethyl alcohol volume = 13.3784 g x $\dfrac{1 \text{ mL}}{0.7893 \text{ g}} = 16.95$ mL

metal volume = total volume − ethyl alcohol volume = 25.00 mL − 16.95 mL = 8.05 mL

metal density $= \dfrac{25.0920 \text{ g}}{8.05 \text{ mL}} = 3.12$ g/mL

Chapter 1 – Chemistry: Matter and Measurement

1.114 Average brass density = $(0.670)(8.92\ \text{g/cm}^3) + (0.330)(7.14\ \text{g/cm}^3) = 8.333\ \text{g/cm}^3$

length = 1.62 in. x $\dfrac{2.54\ \text{cm}}{1\ \text{in.}}$ = 4.115 cm

diameter = 0.514 in. x $\dfrac{2.54\ \text{cm}}{1\ \text{in.}}$ = 1.306 cm

volume = $\pi r^2 h$ = $(3.1416)[(1.306\ \text{cm})/2]^2(4.115\ \text{cm})$ = 5.512 cm^3

mass = 5.512 cm^3 x $\dfrac{8.333\ \text{g}}{1\ \text{cm}^3}$ = 45.9 g

1.115 35 sv = 35 x 10^9 $\dfrac{\text{m}^3}{\text{s}}$

(a) gulf stream flow = $\left(35 \times 10^9\ \dfrac{\text{m}^3}{\text{s}}\right)\left(\dfrac{100\ \text{cm}}{1\ \text{m}}\right)^3\left(\dfrac{1\ \text{mL}}{1\ \text{cm}^3}\right)\left(\dfrac{60\ \text{s}}{1\ \text{min}}\right)$ = 2.1 x 10^{18} mL/min

(b) mass of H$_2$O = $\left(2.1 \times 10^{18}\ \dfrac{\text{mL}}{\text{min}}\right)\left(\dfrac{60\ \text{min}}{1\ \text{h}}\right)(24\ \text{h})\left(\dfrac{1.025\ \text{g}}{1\ \text{mL}}\right)$ = 3.1 x 10^{21} g = 3.1 x 10^{18} kg

(c) time = $\left(1.0 \times 10^{15}\ \text{L}\right)\left(\dfrac{1000\ \text{mL}}{1\ \text{L}}\right)\left(\dfrac{1\ \text{min}}{2.1 \times 10^{18}\ \text{mL}}\right)$ = 0.48 min

1.116 (a) Gallium is a metal.

(b) Indium, which is right under gallium in the periodic table, should have similar chemical properties.

(c) Ga density = $\dfrac{0.2133\ \text{lb}}{1\ \text{in.}^3}$ x $\dfrac{453.59\ \text{g}}{1\ \text{lb}}$ x $\dfrac{1\ \text{in.}^3}{(2.54\ \text{cm})^3}$ = 5.904 g/cm^3

(d) Ga boiling point   2204 °C   1000 °G
     Ga melting point   29.78 °C   0 °G

$\dfrac{1000\ °G - 0\ °G}{2204\ °C - 29.78\ °C} = \dfrac{1000\ °G}{2174.22\ °C} = 0.4599\ °G/°C$

°G = 0.4599 x (°C – 29.78)
°G = 0.4599 x (801 – 29.78) = 355 °G
The melting point of sodium chloride (NaCl) on the gallium scale is 355 °G.

1.117 1 knot = 1 nautical mile per hour
Historically, 1 knot = 47 ft 3 in. per 28 s

(a) 1 knot = $\dfrac{\left(47\ \text{ft} \times \dfrac{12\ \text{in.}}{1\ \text{ft}} + 3\ \text{in.}\right)}{28\ \text{s}}$ = 20.25 in/s

Convert knots in in./s to knots in ft/hr.

$$1 \text{ knot} = 20.25 \text{ in./s} \times \frac{1 \text{ ft}}{12 \text{ in.}} \times \frac{60 \text{ s}}{1 \text{ min}} \times \frac{60 \text{ min}}{1 \text{ hr}} = 6075 \text{ ft/hr}$$

Therefore, 1 nautical mile = 6075 ft

$$1 \text{ nautical mile in meters} = 6075 \text{ ft} \times \frac{12 \text{ in.}}{1 \text{ ft}} \times \frac{2.54 \text{ cm}}{1 \text{ in.}} \times \frac{1 \text{ m}}{100 \text{ cm}} = 1851.66 \text{ m} = 1852 \text{ m}$$

(b) $\text{speed in knots} = 48 \text{ mi/hr} \times \dfrac{5280 \text{ ft}}{1 \text{ mi}} \times \dfrac{1 \text{ nautical mi}}{6075 \text{ ft}} = 41.7 \text{ knots} = 42 \text{ knots}$

(c) $\text{depth in leagues} = 35,798 \text{ ft} \times \dfrac{1 \text{ nautical mi}}{6075 \text{ ft}} \times \dfrac{1 \text{ league}}{3 \text{ nautical mi}} = 1.964 \text{ leagues}$

(d) $\left| \dfrac{1851.66 \text{ m} - 1852 \text{ m}}{1851.66 \text{ m}} \right| \times 100 = 0.0184 \%$

The current definition of the nautical mile is 0.0184 % larger than the original definition.

Convert the current definition of the nautical mile in meters to feet.

$$1 \text{ nautical mile in feet} = 1852 \text{ m} \times \frac{100 \text{ cm}}{1 \text{ m}} \times \frac{1 \text{ in.}}{2.54 \text{ cm}} \times \frac{1 \text{ ft}}{12 \text{ in.}} = 6076.115 \text{ ft}$$

$$\left| \frac{5280 \text{ ft} - 6076.115 \text{ ft}}{5280 \text{ ft}} \right| \times 100 = 15.1 \%$$

The current definition of the nautical mile is 15.1 % larger than the statute mile.

2   The Structure and Stability of Atoms

2.1   First, find the S:O ratio in each compound.
Substance A: S:O mass ratio = (6.00 g S) / (5.99 g O) = 1.00
Substance B: S:O mass ratio = (8.60 g S) / (12.88 g O) = 0.668

$$\frac{\text{S:O mass ratio in substance A}}{\text{S:O mass ratio in substance B}} = \frac{1.00}{0.668} = 1.50 = \frac{3}{2}$$

2.2 $0.005 \text{ mm} \times \dfrac{1 \text{ cm}}{10 \text{ mm}} \times \dfrac{1 \text{ Au atom}}{2.9 \times 10^{-8} \text{ cm}} = 2 \times 10^4 \text{ Au atoms}$

2.3 $1 \times 10^{19} \text{ C atoms} \times \dfrac{1.5 \times 10^{-10} \text{ m}}{\text{C atom}} \times \dfrac{1 \text{ km}}{1000 \text{ m}} \times \dfrac{1 \text{ time}}{40{,}075 \text{ km}} = 37.4 \text{ times} \approx 40 \text{ times}$

2.4 $^{75}_{34}\text{Se}$ has 34 protons, 34 electrons, and (75 − 34) = 41 neutrons.

2.5 $^{35}_{17}\text{Cl}$ has (35 − 17) = 18 neutrons. $^{37}_{17}\text{Cl}$ has (37 − 17) = 20 neutrons.

2.6   The element with 47 protons is Ag. The mass number is the sum of the protons and the neutrons, 47 + 62 = 109. The isotope symbol is $^{109}_{47}\text{Ag}$.

2.7   atomic mass = (0.6915 × 62.93 amu) + (0.3085 × 64.93 amu) = 63.55 amu

2.8 $2.15 \text{ g} \times \dfrac{1 \text{ amu}}{1.6605 \times 10^{-24} \text{ g}} \times \dfrac{1 \text{ Cu}}{63.55 \text{ amu}} = 2.04 \times 10^{22} \text{ Cu atoms}$

2.9   (a) $\text{g Ti} = 1.505 \text{ mol Ti} \times \dfrac{47.867 \text{ g Ti}}{1 \text{ mol Ti}} = 72.04 \text{ g Ti}$

       (b) $\text{g Na} = 0.337 \text{ mol Na} \times \dfrac{22.989\ 770 \text{ g Na}}{1 \text{ mol Na}} = 7.75 \text{ g Na}$

       (c) $\text{g U} = 2.583 \text{ mol U} \times \dfrac{238.028\ 91 \text{ g U}}{1 \text{ mol U}} = 614.8 \text{ g U}$

2.10   (a) $\text{mol Ti} = 11.51 \text{ g Ti} \times \dfrac{1 \text{ mol Ti}}{47.867 \text{ g Ti}} = 0.2405 \text{ mol Ti}$

       (b) $\text{mol Na} = 29.127 \text{ g Na} \times \dfrac{1 \text{ mol Na}}{22.989\ 770 \text{ g Na}} = 1.2670 \text{ mol Na}$

(c) $\text{mol U} = 1.477 \text{ kg} \times \dfrac{1000 \text{ g}}{1 \text{ kg}} \times \dfrac{1 \text{ mol U}}{238.028\ 91 \text{ g U}} = 6.205 \text{ mol U}$

2.11   (a) In beta emission, the mass number is unchanged, and the atomic number increases by one. $^{106}_{44}\text{Ru} \rightarrow\ ^{0}_{-1}\text{e} +\ ^{106}_{45}\text{Rh}$

(b) In alpha emission, the mass number decreases by four, and the atomic number decreases by two. $^{189}_{83}\text{Bi} \rightarrow\ ^{4}_{2}\text{He} +\ ^{185}_{81}\text{Tl}$

(c) In electron capture, the mass number is unchanged, and the atomic number decreases by one. $^{204}_{84}\text{Po} +\ ^{0}_{-1}\text{e} \rightarrow\ ^{204}_{83}\text{Bi}$

2.12   The mass number decreases by four, and the atomic number decreases by two. This is characteristic of alpha emission. $^{214}_{90}\text{Th} \rightarrow\ ^{210}_{88}\text{Ra} +\ ^{4}_{2}\text{He}$

2.13 $^{148}_{69}\text{Tm}$ decays to $^{148}_{68}\text{Er}$ by either positron emission or electron capture.

2.14   (a) ^{199}Au has a higher neutron/proton ratio and decays by beta emission. ^{173}Au has a lower neutron/proton ratio and decays by alpha emission.

(b) ^{196}Pb has a lower neutron/proton ratio and decays by positron emission. ^{206}Pb is nonradioactive.

2.15   H and He

Key Concept Problems

2.16   Drawing (a) represents a collection of SO_2 units. Drawing (d) represents a mixture of S atoms and O_2 units.

2.17   To obey the law of mass conservation, the correct drawing must have the same number of red and yellow spheres as in drawing (a). The correct drawing is (d).

2.18   Figures (b) and (d) illustrate the law of multiple proportions. The ⬭⬭⚫/⬭⚫ mass ratio is 2.

2.19   A Na atom has 11 protons and 11 electrons [drawing (b)].
A Ca^{2+} ion has 20 protons and 18 electrons [drawing (c)].
A F^- ion has 9 protons and 10 electrons [drawing (a)].

2.20   The isotope contains 8 neutrons and 6 protons. The isotope symbol is $^{14}_{6}\text{C}$.

$^{14}_{6}\text{C}$ would decay by beta emission because the n/p ratio is high.

2.21   The shorter arrow pointing right is for beta emission. The longer arrow pointing left is for alpha emission.

$$A = {}^{147+94}_{94}X = {}^{241}_{94}Pu$$

$$B = {}^{146+95}_{95}X = {}^{241}_{95}Am$$

$$C = {}^{144+93}_{93}X = {}^{237}_{93}Np$$

$$D = {}^{142+91}_{91}X = {}^{233}_{91}Pa$$

$$E = {}^{141+92}_{92}X = {}^{233}_{92}U$$

Section Problems
Atomic Theory (Sections 2.1–2.6)

2.22   The law of mass conservation in terms of Dalton's atomic theory states that chemical reactions only rearrange the way that atoms are combined; the atoms themselves are not changed.
The law of definite proportions in terms of Dalton's atomic theory states that the chemical combination of elements to make different substances occurs when atoms join together in small, whole-number ratios.

2.23   The law of multiple proportions states that if two elements combine in different ways to form different substances, the mass ratios are small, whole-number multiples of each other. This is very similar to Dalton's statement that the chemical combination of elements to make different substances occurs when atoms join together in small, whole-number ratios.

2.24   First, find the C:H ratio in each compound.
Benzene: C:H mass ratio = (4.61 g C) / (0.39 g H) = 12
Ethane: C:H mass ratio (4.00 g C) / (1.00 g H) = 4.00
Ethylene: C:H mass ratio = (4.29 g C) / (0.71 g H) = 6.0

$$\frac{\text{C:H mass ratio in benzene}}{\text{C:H mass ratio in ethane}} = \frac{12}{4.00} = \frac{3}{1}$$

$$\frac{\text{C:H mass ratio in benzene}}{\text{C:H mass ratio in ethylene}} = \frac{12}{6.0} = \frac{2}{1}$$

$$\frac{\text{C:H mass ratio in ethylene}}{\text{C:H mass ratio in ethane}} = \frac{6.0}{4.00} = \frac{3}{2}$$

2.25   First, find the C:O ratio in each compound.
Carbon suboxide: C:O mass ratio = (1.32 g C) / (1.18 g O) = 1.12
Carbon dioxide: C:O mass ratio = (12.00 g C) / (32.00 g O) = 0.375

$$\frac{\text{C:O mass ratio in carbon suboxide}}{\text{C:O mass ratio in carbon dioxide}} = \frac{1.12}{0.375} = \frac{3}{1}$$

2.26   (a) For benzene:

$$4.61 \text{ g} \times \frac{1 \text{ amu}}{1.6605 \times 10^{-24} \text{ g}} \times \frac{1 \text{ C atom}}{12.011 \text{ amu}} = 2.31 \times 10^{23} \text{ C atoms}$$

$$0.39 \text{ g} \times \frac{1 \text{ amu}}{1.6605 \times 10^{-24} \text{ g}} \times \frac{1 \text{ H atom}}{1.008 \text{ amu}} = 2.3 \times 10^{23} \text{ H atoms}$$

$$\frac{C}{H} = \frac{2.31 \times 10^{23} \text{ C atoms}}{2.3 \times 10^{23} \text{ H atoms}} = \frac{1 \text{ C}}{1 \text{ H}} \qquad \text{A possible formula for benzene is CH.}$$

For ethane:

$$4.00 \text{ g} \times \frac{1 \text{ amu}}{1.6605 \times 10^{-24} \text{ g}} \times \frac{1 \text{ C atom}}{12.011 \text{ amu}} = 2.01 \times 10^{23} \text{ C atoms}$$

$$1.00 \text{ g} \times \frac{1 \text{ amu}}{1.6605 \times 10^{-24} \text{ g}} \times \frac{1 \text{ H atom}}{1.008 \text{ amu}} = 5.97 \times 10^{23} \text{ H atoms}$$

$$\frac{C}{H} = \frac{2.01 \times 10^{23} \text{ C atoms}}{5.97 \times 10^{23} \text{ H atoms}} = \frac{1 \text{ C}}{3 \text{ H}} \qquad \text{A possible formula for ethane is } CH_3.$$

For ethylene:

$$4.29 \text{ g} \times \frac{1 \text{ amu}}{1.6605 \times 10^{-24} \text{ g}} \times \frac{1 \text{ C atom}}{12.011 \text{ amu}} = 2.15 \times 10^{23} \text{ C atoms}$$

$$0.71 \text{ g} \times \frac{1 \text{ amu}}{1.6605 \times 10^{-24} \text{ g}} \times \frac{1 \text{ H atom}}{1.008 \text{ amu}} = 4.2 \times 10^{23} \text{ H atoms}$$

$$\frac{C}{H} = \frac{2.15 \times 10^{23} \text{ C atoms}}{4.2 \times 10^{23} \text{ H atoms}} = \frac{1 \text{ C}}{2 \text{ H}} \qquad \text{A possible formula for ethylene is } CH_2.$$

(b) The results in part (a) give the smallest whole-number ratio of C to H for benzene, ethane, and ethylene, and these ratios are consistent with their modern formulas.

2.27 $$1.32 \text{ g} \times \frac{1 \text{ amu}}{1.6605 \times 10^{-24} \text{ g}} \times \frac{1 \text{ C atom}}{12.011 \text{ amu}} = 6.62 \times 10^{22} \text{ C atoms}$$

$$1.18 \text{ g} \times \frac{1 \text{ amu}}{1.6605 \times 10^{-24} \text{ g}} \times \frac{1 \text{ O atom}}{15.9994 \text{ amu}} = 4.44 \times 10^{22} \text{ O atoms}$$

$$\frac{C}{O} = \frac{6.62 \times 10^{22} \text{ C atoms}}{4.44 \times 10^{22} \text{ O atoms}} = \frac{1.5 \text{ C}}{1 \text{ O}};$$

therefore the formula for carbon suboxide is $C_{1.5}O$, or C_3O_2.

2.28   The mass of 6.02×10^{23} atoms is its atomic mass expressed in grams. If the atomic mass of an element is X, then 6.02×10^{23} atoms of this element weighs X grams.

2.29   The mass of 6.02×10^{23} atoms is its atomic mass expressed in grams. If the mass of 6.02×10^{23} atoms of element Y is 83.80 g, then the atomic mass of Y is 83.80. Y is Kr.

2.30 $mass = \dfrac{x \ g}{6.02 \ x \ 10^{23} \ atoms} \ x \ 3.17 \ x \ 10^{20} \ atoms = (x) \ x \ 5.27 \ x \ 10^{-4} \ g$

2.31 $atomic \ mass \ in \ g = \dfrac{0.815 \ g}{4.61 \ x \ 10^{21} \ atoms} \ x \ 6.02 \ x \ 10^{23} \ atoms = 106 \ g; \ Z = Pd$

2.32   Assume a 1.00 g sample of the binary compound of zinc and sulfur.

$0.671 \ x \ 1.00 \ g = 0.671 \ g \ Zn;$ $0.329 \ x \ 1.00 \ g = 0.329 \ g \ S$

$0.671 \ g \ x \ \dfrac{1 \ amu}{1.6605 \ x \ 10^{-24} \ g} \ x \ \dfrac{1 \ Zn \ atom}{65.39 \ amu} = 6.18 \ x \ 10^{21} \ Zn \ atoms$

$0.329 \ g \ x \ \dfrac{1 \ amu}{1.6605 \ x \ 10^{-24} \ g} \ x \ \dfrac{1 \ S \ atom}{32.066 \ amu} = 6.18 \ x \ 10^{21} \ S \ atoms$

$\dfrac{Zn}{S} = \dfrac{6.18 \ x \ 10^{21} \ Zn \ atoms}{6.18 \ x \ 10^{21} \ S \ atoms} = \dfrac{1}{1}$

2.33   Assume a 1.000 g sample of one of the binary compounds.

$0.3104 \ x \ 1.000 \ g = 0.3104 \ g \ Ti;$ $0.6896 \ x \ 1.000 \ g = 0.6896 \ g \ Cl$

$0.3104 \ g \ x \ \dfrac{1 \ amu}{1.6605 \ x \ 10^{-24} \ g} \ x \ \dfrac{1 \ Ti \ atom}{47.88 \ amu} = 3.90 \ x \ 10^{21} \ Ti \ atoms$

$0.6896 \ g \ x \ \dfrac{1 \ amu}{1.6605 \ x \ 10^{-24} \ g} \ x \ \dfrac{1 \ Cl \ atom}{35.453 \ amu} = 1.17 \ x \ 10^{22} \ Cl \ atoms$

$\dfrac{Cl}{Ti} = \dfrac{1.17 \ x \ 10^{22}}{3.90 \ x \ 10^{21}} = \dfrac{3}{1}$

Assume a 1.000 g sample of the other binary compound.

$0.2524 \ x \ 1.000 \ g = 0.2524 \ g \ Ti;$ $0.7476 \ x \ 1.000 \ g = 0.7476 \ g \ Cl$

$0.2524 \ g \ x \ \dfrac{1 \ amu}{1.6605 \ x \ 10^{-24} \ g} \ x \ \dfrac{1 \ Ti \ atom}{47.88 \ amu} = 3.17 \ x \ 10^{21} \ Ti \ atoms$

$0.7476 \ g \ x \ \dfrac{1 \ amu}{1.6605 \ x \ 10^{-24} \ g} \ x \ \dfrac{1 \ Cl \ atom}{35.453 \ amu} = 1.27 \ x \ 10^{22} \ Cl \ atoms$

$\dfrac{Cl}{Ti} = \dfrac{1.27 \ x \ 10^{22}}{3.17 \ x \ 10^{21}} = \dfrac{4}{1}$

Elements and Atoms (Sections 2.3–2.6)

2.34   The atomic number is equal to the number of protons.
The mass number is equal to the sum of the number of protons and the number of neutrons.

2.35   The atomic number is equal to the number of protons.
The atomic mass is the weighted average mass (in amu) of the various isotopes for a particular element.

2.36   Atoms of the same element that have different numbers of neutrons are called isotopes.

2.37   The mass number is equal to the sum of the number of protons and the number of neutrons for a particular isotope.

For $^{14}_{6}C$, mass number = 6 protons + 8 neutrons = 14.

For $^{14}_{7}N$, mass number = 7 protons + 7 neutrons = 14.

2.38   The subscript giving the atomic number of an atom is often left off of an isotope symbol because one can readily look up the atomic number in the periodic table.

2.39   Te has isotopes with more neutrons than the isotopes of I.

2.40   (a) carbon, C   (b) argon, Ar   (c) vanadium, V

2.41 $^{137}_{55}Cs$

2.42   (a) $^{220}_{86}Rn$ (b) $^{210}_{84}Po$ (c) $^{197}_{79}Au$

2.43   (a) $^{140}_{58}Ce$ (b) $^{60}_{27}Co$

2.44   (a) $^{15}_{7}N$, 7 protons, 7 electrons, (15 – 7) = 8 neutrons

(b) $^{60}_{27}Co$, 27 protons, 27 electrons, (60 – 27) = 33 neutrons

(c) $^{131}_{53}I$, 53 protons, 53 electrons, (131 – 53) = 78 neutrons

(d) $^{142}_{58}Ce$, 58 protons, 58 electrons, (142 – 58) = 84 neutrons

2.45   (a) ^{27}Al, 13 protons and (27 – 13) = 14 neutrons

(b) ^{32}S, 16 protons and (32 – 16) = 16 neutrons

(c) ^{64}Zn, 30 protons and (64 – 30) = 34 neutrons

(d) ^{207}Pb, 82 protons and (207 – 82) = 125 neutrons

2.46   (a) $^{24}_{12}Mg$, magnesium   (b) $^{58}_{28}Ni$, nickel

(c) $^{104}_{46}Pd$, palladium   (d) $^{183}_{74}W$, tungsten

2.47   (a) $^{202}_{80}Hg$, mercury   (b) $^{195}_{78}Pt$, platinum

(c) $^{184}_{76}Os$, osmium   (d) $^{209}_{83}Bi$, bismuth

2.48 $(0.199 \times 10.0129 \text{ amu}) + (0.801 \times 11.009\,31 \text{ amu}) = 10.8 \text{ amu for B}$

2.49 $(0.5184 \times 106.9051 \text{ amu}) + (0.4816 \times 108.9048 \text{ amu}) = 107.9 \text{ amu for Ag}$

2.50 $24.305 \text{ amu} = (0.7899 \times 23.985 \text{ amu}) + (0.1000 \times 24.986 \text{ amu}) + (0.1101 \times Z)$
Solve for Z. $Z = 25.982 \text{ amu for } ^{26}\text{Mg}$.

2.51   The total abundance of all three isotopes must be 100.00%. The natural abundance of ^{29}Si is
4.68%. The natural abundance of ^{28}Si and ^{30}Si together must be 100.00% – 4.68% = 95.32%.
Let Y be the natural abundance of ^{28}Si and [95.32 – Y] the natural abundance of ^{30}Si.
$28.0855 \text{ amu} = (0.0468 \times 28.9765 \text{ amu}) + (Y \times 27.9769 \text{ amu}) + ([0.9532 - Y] \times 29.9738 \text{ amu})$

Solve for Y. $Y = \dfrac{-1.842}{-1.997} = 0.9224$

^{28}Si natural abundance = 92.24% ^{30}Si natural abundance = 95.32 – 92.24 = 3.08%

Nuclear Reactions and Radioactivity (Sections 2.7–2.8)

2.52   Positron emission is the conversion of a proton in the nucleus into a neutron plus an
ejected positron.
Electron capture is the process in which a proton in the nucleus captures an inner-shell
electron and is thereby converted into a neutron.

2.53   An alpha particle $\left(^{4}_{2}\text{He}^{2+}\right)$ is a helium nucleus. The He atom has two electrons and is neutral.

2.54   In beta emission a neutron is converted to a proton and the atomic number increases. In
positron emission a proton is converted to a neutron and the atomic number decreases.

2.55   "Neutron rich" nuclides emit beta particles to decrease the number of neutrons and
increase the number of protons in the nucleus.

"Neutron poor" nuclides decrease the number of protons and increase the n/p ratio by
either alpha emission, positron emission, or electron capture.

2.56   (a) $^{126}_{50}\text{Sn} \rightarrow {}^{0}_{-1}\text{e} + {}^{126}_{51}\text{Sb}$ (b) $^{210}_{88}\text{Ra} \rightarrow {}^{4}_{2}\text{He} + {}^{206}_{86}\text{Rn}$
(c) $^{77}_{37}\text{Rb} \rightarrow {}^{0}_{1}\text{e} + {}^{77}_{36}\text{Kr}$ (d) $^{76}_{36}\text{Kr} + {}^{0}_{-1}\text{e} \rightarrow {}^{76}_{35}\text{Br}$

2.57   (a) $^{90}_{38}\text{Sr} \rightarrow {}^{0}_{-1}\text{e} + {}^{90}_{39}\text{Y}$ (b) $^{247}_{100}\text{Fm} \rightarrow {}^{4}_{2}\text{He} + {}^{243}_{98}\text{Cf}$
(c) $^{49}_{25}\text{Mn} \rightarrow {}^{0}_{1}\text{e} + {}^{49}_{24}\text{Cr}$ (d) $^{37}_{18}\text{Ar} + {}^{0}_{-1}\text{e} \rightarrow {}^{37}_{17}\text{Cl}$

2.58   (a) $^{188}_{80}\text{Hg} \rightarrow {}^{188}_{79}\text{Au} + {}^{0}_{1}\text{e}$ (b) $^{218}_{85}\text{At} \rightarrow {}^{214}_{83}\text{Bi} + {}^{4}_{2}\text{He}$
(c) $^{234}_{90}\text{Th} \rightarrow {}^{234}_{91}\text{Pa} + {}^{0}_{-1}\text{e}$

2.59   (a) $^{24}_{11}\text{Na} \rightarrow {}^{24}_{12}\text{Mg} + {}^{0}_{-1}\text{e}$ (b) $^{135}_{60}\text{Nd} \rightarrow {}^{135}_{59}\text{Pr} + {}^{0}_{1}\text{e}$

       (c) $^{170}_{78}\text{Pt} \rightarrow {}^{166}_{76}\text{Os} + {}^{4}_{2}\text{He}$

2.60   (a) $^{162}_{75}\text{Re} \rightarrow {}^{158}_{73}\text{Ta} + {}^{4}_{2}\text{He}$ (b) $^{138}_{62}\text{Sm} + {}^{0}_{-1}\text{e} \rightarrow {}^{138}_{61}\text{Pm}$

       (c) $^{188}_{74}\text{W} \rightarrow {}^{188}_{75}\text{Re} + {}^{0}_{-1}\text{e}$ (d) $^{165}_{73}\text{Ta} \rightarrow {}^{165}_{72}\text{Hf} + {}^{0}_{1}\text{e}$

2.61   (a) $^{157}_{63}\text{Eu} \rightarrow {}^{157}_{64}\text{Gd} + {}^{0}_{-1}\text{e}$ (b) $^{126}_{56}\text{Ba} + {}^{0}_{-1}\text{e} \rightarrow {}^{126}_{55}\text{Cs}$

       (c) $^{146}_{62}\text{Sm} \rightarrow {}^{142}_{60}\text{Nd} + {}^{4}_{2}\text{He}$ (d) $^{125}_{56}\text{Ba} \rightarrow {}^{125}_{55}\text{Cs} + {}^{0}_{1}\text{e}$

2.62 ^{160}W is neutron poor and decays by alpha emission. ^{185}W is neutron rich and decays by beta emission.

2.63 $^{136}_{53}\text{I}$ is neutron rich and decays by beta emission.

       $^{122}_{53}\text{I}$ is neutron poor and decays by positron emission.

2.64 $^{241}_{95}\text{Am} \rightarrow {}^{237}_{93}\text{Np} + {}^{4}_{2}\text{He}$

       $^{237}_{93}\text{Np} \rightarrow {}^{233}_{91}\text{Pa} + {}^{4}_{2}\text{He}$

       $^{233}_{91}\text{Pa} \rightarrow {}^{233}_{92}\text{U} + {}^{0}_{-1}\text{e}$

       $^{233}_{92}\text{U} \rightarrow {}^{229}_{90}\text{Th} + {}^{4}_{2}\text{He}$

       $^{229}_{90}\text{Th} \rightarrow {}^{225}_{88}\text{Ra} + {}^{4}_{2}\text{He}$

       $^{225}_{88}\text{Ra} \rightarrow {}^{225}_{89}\text{Ac} + {}^{0}_{-1}\text{e}$

       $^{225}_{89}\text{Ac} \rightarrow {}^{221}_{87}\text{Fr} + {}^{4}_{2}\text{He}$

       $^{221}_{87}\text{Fr} \rightarrow {}^{217}_{85}\text{At} + {}^{4}_{2}\text{He}$

       $^{217}_{85}\text{At} \rightarrow {}^{213}_{83}\text{Bi} + {}^{4}_{2}\text{He}$

       $^{213}_{83}\text{Bi} \rightarrow {}^{213}_{84}\text{Po} + {}^{0}_{-1}\text{e}$

       $^{213}_{84}\text{Po} \rightarrow {}^{209}_{82}\text{Pb} + {}^{4}_{2}\text{He}$

       $^{209}_{82}\text{Pb} \rightarrow {}^{209}_{83}\text{Bi} + {}^{0}_{-1}\text{e}$

2.65 $^{222}_{86}\text{Rn} \rightarrow {}^{218}_{84}\text{Po} + {}^{4}_{2}\text{He}$

       $^{218}_{84}\text{Po} \rightarrow {}^{214}_{82}\text{Pb} + {}^{4}_{2}\text{He}$

       $^{214}_{82}\text{Pb} \rightarrow {}^{210}_{80}\text{Hg} + {}^{4}_{2}\text{He}$

       $^{210}_{80}\text{Hg} \rightarrow {}^{210}_{81}\text{Tl} + {}^{0}_{-1}\text{e}$

       $^{210}_{81}\text{Tl} \rightarrow {}^{210}_{82}\text{Pb} + {}^{0}_{-1}\text{e}$

2.66　Each alpha emission decreases the mass number by four and the atomic number by two. Each beta emission increases the atomic number by one.

$$^{232}_{90}\text{Th} \rightarrow {}^{208}_{82}\text{Pb}$$

Number of α emissions $= \dfrac{\text{Th mass number} - \text{Pb mass number}}{4}$

$$= \dfrac{232 - 208}{4} = 6 \text{ α emissions}$$

The atomic number decreases by 12 as a result of 6 alpha emissions. The resulting atomic number is $(90 - 12) = 78$.

Number of β emissions = Pb atomic number $- 78 = 82 - 78 = 4$ β emissions

2.67　Each alpha emission decreases the mass number by four and the atomic number by two. Each beta emission increases the atomic number by one.

$$^{235}_{92}\text{U} \rightarrow {}^{207}_{82}\text{Pb}$$

Number of α emissions $= \dfrac{\text{U mass number} - \text{Pb mass number}}{4}$

$$= \dfrac{235 - 207}{4} = 7 \text{ α emissions}$$

The atomic number decreases by 14 as a result of 7 alpha emissions. The resulting atomic number is $(92 - 14) = 78$.

Number of β emissions = Pb atomic number $- 78 = 82 - 78 = 4$ β emissions

Chapter Problems

2.68　atomic mass $= (0.205 \times 69.924 \text{ amu}) + (0.274 \times 71.922 \text{ amu})$
$+ (0.078 \times 72.923 \text{ amu}) + (0.365 \times 73.921 \text{ amu})$
$+ (0.078 \times 75.921 \text{ amu}) = 72.6 \text{ amu}$

2.69　Deuterium is ^2H and deuterium fluoride is ^2HF.
^2H has 1 proton, 1 neutron, and 1 electron.
F has 9 protons, 10 neutrons, and 9 electrons.
^2HF has 10 protons, 11 neutrons, and 10 electrons.
Chemically, ^2HF is like HF and is a weak acid.

2.70　For NH_3, $(2.34 \text{ g N})\left(\dfrac{3 \times 1.0079 \text{ amu H}}{14.0067 \text{ amu N}}\right) = 0.505 \text{ g H}$

For N_2H_4, $(2.34 \text{ g N})\left(\dfrac{4 \times 1.0079 \text{ amu H}}{2 \times 14.0067 \text{ amu N}}\right) = 0.337 \text{ g H}$

2.71　$\text{g N} = (1.575 \text{ g H})\left(\dfrac{3.670 \text{ g N}}{0.5275 \text{ g H}}\right) = 10.96 \text{ g N}$

From Problem 2.70:

$$\text{for NH}_3, \quad \frac{g\,N}{g\,H} = \frac{2.34\,g\,N}{0.505\,g\,H} = 4.63$$

$$\text{for N}_2\text{H}_4, \quad \frac{g\,N}{g\,H} = \frac{2.34\,g\,N}{0.337\,g\,H} = 6.94$$

$$\text{for compound X,} \quad \frac{g\,N}{g\,H} = \frac{10.96\,g\,N}{1.575\,g\,H} = 6.96; \qquad \text{X is N}_2\text{H}_4$$

2.72   (a) I   (b) Kr

2.73 $^1\text{H}^{35}\text{Cl}$ has 18 protons, 18 neutrons, and 18 electrons.
       $^1\text{H}^{37}\text{Cl}$ has 18 protons, 20 neutrons, and 18 electrons.
       $^2\text{H}^{35}\text{Cl}$ has 18 protons, 19 neutrons, and 18 electrons.
       $^2\text{H}^{37}\text{Cl}$ has 18 protons, 21 neutrons, and 18 electrons.
       $^3\text{H}^{35}\text{Cl}$ has 18 protons, 20 neutrons, and 18 electrons.
       $^3\text{H}^{37}\text{Cl}$ has 18 protons, 22 neutrons, and 18 electrons.

2.74 $\dfrac{12.0000\,\text{amu}}{15.9994\,\text{amu}} = \dfrac{X}{16.0000\,\text{amu}}; \quad X = 12.0005$ amu for ^{12}C prior to 1961.

2.75 $\dfrac{39.9626\,\text{amu}}{15.9994\,\text{amu}} = \dfrac{X}{16.0000\,\text{amu}}; \quad X = 39.9641$ amu for ^{40}Ca prior to 1961.

2.76   Molecular mass = (8 x 12.011 amu) + (9 x 1.0079 amu) + (1 x 14.0067 amu)
                            + (2 x 15.9994 amu) = 151.165 amu

2.77 $\text{mass \% C} = \dfrac{8\ \text{x}\ 12.011}{151.165}\ \text{x}\ 100 = 63.565\%$

       $\text{mass \% H} = \dfrac{9\ \text{x}\ 1.0079}{151.165}\ \text{x}\ 100 = 6.0008\%$

       $\text{mass \% N} = \dfrac{14.0067}{151.165}\ \text{x}\ 100 = 9.2658\%$

       $\text{mass \% O} = \dfrac{2\ \text{x}\ 15.9994}{151.165}\ \text{x}\ 100 = 21.168\%$

2.78 $^{100}_{43}\text{Tc} \rightarrow\ ^{0}_{1}\text{e} +\ ^{100}_{42}\text{Mo}$ (positron emission)

       $^{100}_{43}\text{Tc} +\ ^{0}_{-1}\text{e} \rightarrow\ ^{100}_{42}\text{Mo}$ (electron capture)

2.79 α emission: $^{226}_{89}\text{Ac} \rightarrow\ ^{222}_{87}\text{Fr} +\ ^{4}_{2}\text{He}$

       β emission: $^{226}_{89}\text{Ac} \rightarrow\ ^{226}_{90}\text{Th} +\ ^{0}_{-1}\text{e}$

       electron capture: $^{226}_{89}\text{Ac} +\ ^{0}_{-1}\text{e} \rightarrow\ ^{226}_{88}\text{Ra}$

3 Periodicity and the Electronic Structure of Atoms

3.1 Gamma ray $\quad \nu = \dfrac{c}{\lambda} = \dfrac{3.00 \times 10^8 \text{ m/s}}{3.56 \times 10^{-11} \text{ m}} = 8.43 \times 10^{18} \text{ s}^{-1} = 8.43 \times 10^{18} \text{ Hz}$

Radar wave $\quad \nu = \dfrac{c}{\lambda} = \dfrac{3.00 \times 10^8 \text{ m/s}}{10.3 \times 10^{-2} \text{ m}} = 2.91 \times 10^9 \text{ s}^{-1} = 2.91 \times 10^9 \text{ Hz}$

3.2 $\quad \nu = 102.5 \text{ MHz} = 102.5 \times 10^6 \text{ Hz} = 102.5 \times 10^6 \text{ s}^{-1}$

$\lambda = \dfrac{c}{\nu} = \dfrac{3.00 \times 10^8 \text{ m/s}}{102.5 \times 10^6 \text{ s}^{-1}} = 2.93 \text{ m}$

$\nu = 9.55 \times 10^{17} \text{ Hz} = 9.55 \times 10^{17} \text{ s}^{-1}$

$\lambda = \dfrac{c}{\nu} = \dfrac{3.00 \times 10^8 \text{ m/s}}{9.55 \times 10^{17} \text{ s}^{-1}} = 3.14 \times 10^{-10} \text{ m}$

3.3 The wave with the shorter wavelength (b) has the higher frequency. The wave with the larger amplitude (b) represents the more intense beam of light. The wave with the shorter wavelength (b) represents blue light. The wave with the longer wavelength (a) represents red light.

3.4 $\quad m = 2; \; R_\infty = 1.097 \times 10^{-2} \text{ nm}^{-1}$

$\dfrac{1}{\lambda} = R_\infty \left[\dfrac{1}{m^2} - \dfrac{1}{n^2} \right]; \quad \dfrac{1}{\lambda} = R_\infty \left[\dfrac{1}{2^2} - \dfrac{1}{7^2} \right]; \quad \dfrac{1}{\lambda} = 2.519 \times 10^{-3} \text{ nm}^{-1}; \quad \lambda = 397.0 \text{ nm}$

3.5 $\quad m = 3; \; R_\infty = 1.097 \times 10^{-2} \text{ nm}^{-1}$

$\dfrac{1}{\lambda} = R_\infty \left[\dfrac{1}{m^2} - \dfrac{1}{n^2} \right]; \quad \dfrac{1}{\lambda} = R_\infty \left[\dfrac{1}{3^2} - \dfrac{1}{4^2} \right]; \quad \dfrac{1}{\lambda} = 5.333 \times 10^{-4} \text{ nm}^{-1}; \quad \lambda = 1875 \text{ nm}$

3.6 $\quad m = 3; \; R_\infty = 1.097 \times 10^{-2} \text{ nm}^{-1}$

$\dfrac{1}{\lambda} = R_\infty \left[\dfrac{1}{m^2} - \dfrac{1}{n^2} \right]; \quad \dfrac{1}{\lambda} = R_\infty \left[\dfrac{1}{3^2} - \dfrac{1}{\infty^2} \right]; \quad \dfrac{1}{\lambda} = 1.219 \times 10^{-3} \text{ nm}^{-1}; \quad \lambda = 820.4 \text{ nm}$

3.7 $\quad \lambda = 91.2 \text{ nm} = 91.2 \times 10^{-9} \text{ m}$

$\nu = \dfrac{c}{\lambda} = \dfrac{3.00 \times 10^8 \text{ m/s}}{91.2 \times 10^{-9} \text{ m}} = 3.29 \times 10^{15} \text{ s}^{-1}$

$E = h\nu = (6.626 \times 10^{-34} \text{ J·s})(3.29 \times 10^{15} \text{ s}^{-1}) = 2.18 \times 10^{-18} \text{ J/photon}$

$E = (2.18 \times 10^{-18} \text{ J/photon})(6.022 \times 10^{23} \text{ photons/mol}) = 1.31 \times 10^6 \text{ J/mol} = 1310 \text{ kJ/mol}$

3.8   IR, $\lambda = 1.55 \times 10^{-6}$ m

$$E = \frac{hc}{\lambda} = (6.626 \times 10^{-34} \text{ J} \cdot \text{s})\left(\frac{3.00 \times 10^8 \text{ m/s}}{1.55 \times 10^{-6} \text{ m}}\right)(6.022 \times 10^{23} / \text{mol})$$

$E = 7.72 \times 10^4$ J/mol $= 77.2$ kJ/mol

UV, $\lambda = 250$ nm $= 250 \times 10^{-9}$ m

$$E = \frac{hc}{\lambda} = (6.626 \times 10^{-34} \text{ J} \cdot \text{s})\left(\frac{3.00 \times 10^8 \text{ m/s}}{250 \times 10^{-9} \text{ m}}\right)(6.022 \times 10^{23} / \text{mol})$$

$E = 4.79 \times 10^5$ J/mol $= 479$ kJ/mol

X ray, $\lambda = 5.49$ nm $= 5.49 \times 10^{-9}$ m

$$E = \frac{hc}{\lambda} = (6.626 \times 10^{-34} \text{ J} \cdot \text{s})\left(\frac{3.00 \times 10^8 \text{ m/s}}{5.49 \times 10^{-9} \text{ m}}\right)(6.022 \times 10^{23} / \text{mol})$$

$E = 2.18 \times 10^7$ J/mol $= 2.18 \times 10^4$ kJ/mol

3.9 $\lambda = \dfrac{h}{mv} = \dfrac{6.626 \times 10^{-34} \text{ kg m}^2 \text{ s}^{-1}}{(1150 \text{ kg})(24.6 \text{ m/s})} = 2.34 \times 10^{-38}$ m

This wavelength is shorter than the diameter of an atom.

3.10

n	l	m_l		Orbital	No. of Orbitals
5	0	0		5s	1
	1	−1, 0, +1		5p	3
	2	−2, −1, 0, +1, +2		5d	5
	3	−3, −2, −1, 0, +1, +2, +3		5f	7
	4	−4, −3, −2, −1, 0, +1, +2, +3, +4		5g	9

There are 25 possible orbitals in the fifth shell.

3.11   (a) 2p   (b) 4f   (c) 3d

3.12   (a) 3s orbital: $n = 3$, $l = 0$, $m_l = 0$
      (b) 2p orbital: $n = 2$, $l = 1$, $m_l = -1, 0, +1$
      (c) 4d orbital: $n = 4$, $l = 2$, $m_l = -2, -1, 0, +1, +2$

3.13   The g orbitals have four nodal planes.

3.14   The figure represents a d orbital, $n = 4$ and $l = 2$.

3.15 $m = 1$, $n = \infty$; $R_\infty = 1.097 \times 10^{-2}$ nm^{-1}

$$\frac{1}{\lambda} = R_\infty\left[\frac{1}{m^2} - \frac{1}{n^2}\right]; \quad \frac{1}{\lambda} = R_\infty\left[\frac{1}{1^2} - \frac{1}{\infty^2}\right]; \quad \frac{1}{\lambda} = R_\infty\left[\frac{1}{1}\right] = 1.097 \times 10^{-2} \text{ nm}^{-1}; \lambda = 91.2 \text{ nm}$$

$$E = (6.626 \times 10^{-34} \text{ J} \cdot \text{s})\left(\frac{3.00 \times 10^8 \text{ m/s}}{91.2 \times 10^{-9} \text{ m}}\right)(6.022 \times 10^{23}/\text{mol})$$

$$E = 1.31 \times 10^6 \text{ J/mol} = 1.31 \times 10^3 \text{ kJ/mol}$$

3.16   Cr, Cu, Nb, Mo, Ru, Rh, Pd, Ag, La, Ce, Gd, Pt, Au, Ac, Th, Pa, U, Np, Cm, Ds, Rg

3.17   (a) Ti, $1s^2 2s^2 2p^6 3s^2 3p^6 4s^2 3d^2$ or $[Ar] 4s^2 3d^2$

[Ar] ⇅   ↑ ↑ __ __ __
     4s   3d

(b) Zn, $1s^2 2s^2 2p^6 3s^2 3p^6 4s^2 3d^{10}$ or $[Ar] 4s^2 3d^{10}$

[Ar] ⇅   ⇅ ⇅ ⇅ ⇅ ⇅
     4s   3d

(c) Sn, $1s^2 2s^2 2p^6 3s^2 3p^6 4s^2 3d^{10} 4p^6 5s^2 4d^{10} 5p^2$ or $[Kr] 5s^2 4d^{10} 5p^2$

[Kr] ⇅   ⇅ ⇅ ⇅ ⇅ ⇅   ↑ ↑ __
     5s   4d   5p

(d) Pb, $[Xe] 6s^2 4f^{14} 5d^{10} 6p^2$

3.18   For Na^+, $1s^2 2s^2 2p^6$;   for Cl^-, $1s^2 2s^2 2p^6 3s^2 3p^6$

3.19   The ground-state electron configuration contains 28 electrons. The atom is Ni.

3.20   (a) Ba; atoms get larger as you go down a group.
(b) Hf; atoms get smaller as you go across a period.
(c) Sn; atoms get larger as you go down a group.
(d) Lu; atoms get smaller as you go across a period.

3.21   Excited mercury atoms in a fluorescent bulb emit photons, some in the visible but most in the ultraviolet region. Visible photons contribute to light we can see; ultraviolet photons are invisible to our eyes. To utilize this ultraviolet energy, fluorescent bulbs are coated on the inside with a phosphor that absorbs ultraviolet photons and re-emits the energy as visible light.

Key Concept Problems

3.22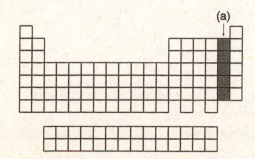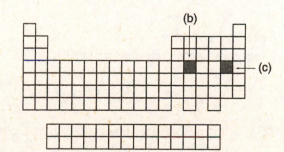

3.23

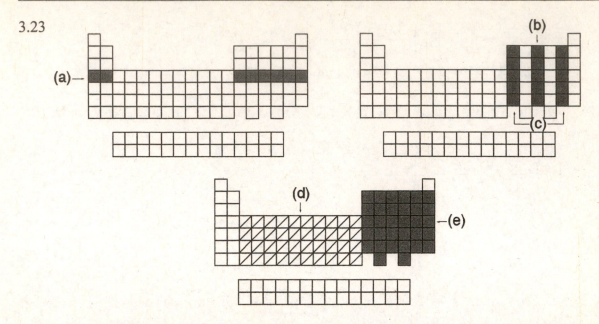

3.24 The green element, molybdenum, has an anomalous electron configuration. Its predicted electron configuration is [Ar] $5s^2 4d^4$. Its anomalous electron configuration is [Ar] $5s^1 4d^5$ because of the resulting half-filled d-orbitals.

3.25 The wave with the larger amplitude (a) has the greater intensity. The wave with the shorter wavelength (a) has the higher energy radiation. The wave with the shorter wavelength (a) represents yellow light. The wave with the longer wavelength (b) represents infrared radiation.

3.26 [Ar] $4s^2 3d^{10} 4p^1$ is Ga.

3.27 There are 34 total electrons in the atom, so there are also 34 protons in the nucleus. The atom is selenium (Se)

Se,   [Ar]   ⇅   ⇅ ⇅ ⇅ ⇅ ⇅   ⇅ ↑ ↑
           4s   3d   4p

3.28 Ca and Br are in the same period, with Br to the far right of Ca. Ca is larger than Br. Sr is directly below Ca in the same group, and is larger than Ca. The result is
Sr (215 pm) > Ca (197 pm) > Br (114 pm)

3.29 (a) $3p_y$ n = 3, l = 1   (b) $4d_{z^2}$ n = 4, l = 2

Section Problems
Electromagnetic Energy (Section 3.1)

3.30 Violet has the higher frequency and energy. Red has the higher wavelength.

3.31 Ultraviolet light has the higher frequency and the greater energy. Infrared light has the longer wavelength.

3.32 $\lambda = \dfrac{c}{v} = \dfrac{3.00 \times 10^8 \text{ m/s}}{5.5 \times 10^{15} \text{ s}^{-1}} = 5.5 \times 10^{-8} \text{ m}$

3.33 $\lambda = \dfrac{c}{v} = \dfrac{3.00 \times 10^8 \text{ m/s}}{4.33 \times 10^{-3} \text{ m}} = 6.93 \times 10^{10} \text{ s}^{-1} = 6.93 \times 10^{10} \text{ Hz}$

3.34 (a) $v = 99.5 \text{ MHz} = 99.5 \times 10^6 \text{ s}^{-1}$
$E = hv = (6.626 \times 10^{-34} \text{ J·s})(99.5 \times 10^6 \text{ s}^{-1})(6.022 \times 10^{23} \text{ /mol})$
$E = 3.97 \times 10^{-2} \text{ J/mol} = 3.97 \times 10^{-5} \text{ kJ/mol}$
$v = 1150 \text{ kHz} = 1150 \times 10^3 \text{ s}^{-1}$
$E = hv = (6.626 \times 10^{-34} \text{ J·s})(1150 \times 10^3 \text{ s}^{-1})(6.022 \times 10^{23} \text{ /mol})$
$E = 4.589 \times 10^{-4} \text{ J/mol} = 4.589 \times 10^{-7} \text{ kJ/mol}$
The FM radio wave (99.5 MHz) has the higher energy.
(b) $\lambda = 3.44 \times 10^{-9} \text{ m}$

$E = \dfrac{hc}{\lambda} = (6.626 \times 10^{-34} \text{ J·s})\left(\dfrac{3.00 \times 10^8 \text{ m/s}}{3.44 \times 10^{-9} \text{ m}}\right)(6.022 \times 10^{23} \text{ /mol})$

$E = 3.48 \times 10^7 \text{ J/mol} = 3.48 \times 10^4 \text{ kJ/mol}$
$\lambda = 6.71 \times 10^{-2} \text{ m}$

$E = \dfrac{hc}{\lambda} = (6.626 \times 10^{-34} \text{ J·s})\left(\dfrac{3.00 \times 10^8 \text{ m/s}}{6.71 \times 10^{-2} \text{ m}}\right)(6.022 \times 10^{23} \text{ /mol})$

$E = 1.78 \text{ J/mol} = 1.78 \times 10^{-3} \text{ kJ/mol}$
The X ray ($\lambda = 3.44 \times 10^{-9}$ m) has the higher energy.

3.35 $v = 400 \text{ MHz} = 400 \times 10^6 \text{ s}^{-1}$
$E = (6.626 \times 10^{-34} \text{ J·s})(400 \times 10^6 \text{ s}^{-1})(6.02 \times 10^{23}\text{/mol}) = 0.160 \text{ J/mol} = 1.60 \times 10^{-4} \text{ kJ/mol}$

3.36 (a) $E = 90.5 \text{ kJ/mol} \times \dfrac{1000 \text{ J}}{1 \text{ kJ}} \times \dfrac{1 \text{ mol}}{6.02 \times 10^{23}} = 1.50 \times 10^{-19} \text{ J}$

$v = \dfrac{E}{h} = \dfrac{1.50 \times 10^{-19} \text{ J}}{6.626 \times 10^{-34} \text{ J·s}} = 2.27 \times 10^{14} \text{ s}^{-1}$

$\lambda = \dfrac{c}{v} = \dfrac{3.00 \times 10^8 \text{ m/s}}{2.27 \times 10^{14} \text{ s}^{-1}} = 1.32 \times 10^{-6} \text{ m} = 1320 \times 10^{-9} \text{ m} = 1320 \text{ nm, near IR}$

(b) $E = 8.05 \times 10^{-4} \text{ kJ/mol} \times \dfrac{1000 \text{ J}}{1 \text{ kJ}} \times \dfrac{1 \text{ mol}}{6.02 \times 10^{23}} = 1.34 \times 10^{-24} \text{ J}$

$v = \dfrac{E}{h} = \dfrac{1.34 \times 10^{-24} \text{ J}}{6.626 \times 10^{-34} \text{ J·s}} = 2.02 \times 10^9 \text{ s}^{-1}$

$\lambda = \dfrac{c}{v} = \dfrac{3.00 \times 10^8 \text{ m/s}}{2.02 \times 10^9 \text{ s}^{-1}} = 0.149 \text{ m, radio wave}$

(c) $E = 1.83 \times 10^3$ kJ/mol $\times \dfrac{1000 \text{ J}}{1 \text{ kJ}} \times \dfrac{1 \text{ mol}}{6.02 \times 10^{23}} = 3.04 \times 10^{-18}$ J

$v = \dfrac{E}{h} = \dfrac{3.04 \times 10^{-18} \text{ J}}{6.626 \times 10^{-34} \text{ J·s}} = 4.59 \times 10^{15} \text{ s}^{-1}$

$\lambda = \dfrac{c}{v} = \dfrac{3.00 \times 10^8 \text{ m/s}}{4.59 \times 10^{15} \text{ s}^{-1}} = 6.54 \times 10^{-8} \text{ m} = 65.4 \times 10^{-9} \text{ m} = 65.4 \text{ nm, UV}$

3.37　(a) $E = hv = (6.626 \times 10^{-34} \text{ J·s})(5.97 \times 10^{19} \text{ s}^{-1})\left(\dfrac{1 \text{ kJ}}{1000 \text{ J}}\right)(6.022 \times 10^{23}/\text{mol})$

　　　　$E = 2.38 \times 10^7$ kJ/mol

　　　(b) $E = hv = (6.626 \times 10^{-34} \text{ J·s})(1.26 \times 10^6 \text{ s}^{-1})\left(\dfrac{1 \text{ kJ}}{1000 \text{ J}}\right)(6.022 \times 10^{23}/\text{mol})$

　　　　$E = 5.03 \times 10^{-7}$ kJ/mol

　　　(c) $E = hv = (6.626 \times 10^{-34} \text{ J·s})\left(\dfrac{3.00 \times 10^8 \text{ m/s}}{2.57 \times 10^2 \text{ m}}\right)\left(\dfrac{1 \text{ kJ}}{1000 \text{ J}}\right)(6.022 \times 10^{23}/\text{mol})$

　　　　$E = 4.66 \times 10^{-7}$ kJ/mol

3.38　$\lambda = \dfrac{h}{mv} = \dfrac{6.626 \times 10^{-34} \text{ kg m}^2 \text{ s}^{-1}}{(9.11 \times 10^{-31} \text{ kg})(0.99 \times 3.00 \times 10^8 \text{ m/s})} = 2.45 \times 10^{-12} \text{ m, } \gamma \text{ ray}$

3.39　$\lambda = \dfrac{h}{mv} = \dfrac{6.626 \times 10^{-34} \text{ kg m}^2 \text{ s}^{-1}}{(1.673 \times 10^{-27} \text{ kg})(0.25 \times 3.00 \times 10^8 \text{ m/s})} = 5.28 \times 10^{-15} \text{ m, } \gamma \text{ ray}$

3.40　$156 \text{ km/h} = 156 \times 10^3 \text{ m}/3600 \text{ s} = 43.3 \text{ m/s}; \ 145 \text{ g} = 0.145 \text{ kg}$

　　　$\lambda = \dfrac{h}{mv} = \dfrac{6.626 \times 10^{-34} \text{ kg m}^2 \text{ s}^{-1}}{(0.145 \text{ kg})(43.3 \text{ m/s})} = 1.06 \times 10^{-34} \text{ m}$

3.41　$1.55 \text{ mg} = 1.55 \times 10^{-3} \text{ g} = 1.55 \times 10^{-6} \text{ kg}$

　　　$\lambda = \dfrac{h}{mv} = \dfrac{6.626 \times 10^{-34} \text{ kg m}^2 \text{ s}^{-1}}{(1.55 \times 10^{-6} \text{ kg})(1.38 \text{ m/s})} = 3.10 \times 10^{-28} \text{ m}$

3.42　$145 \text{ g} = 0.145 \text{ kg}; \ 0.500 \text{ nm} = 0.500 \times 10^{-9} \text{ m}$

　　　$v = \dfrac{h}{m\lambda} = \dfrac{6.626 \times 10^{-34} \text{ kg m}^2 \text{ s}^{-1}}{(0.145 \text{ kg})(0.500 \times 10^{-9} \text{ m})} = 9.14 \times 10^{-24} \text{ m/s}$

3.43　$750 \text{ nm} = 750 \times 10^{-9} \text{ m}$

　　　$v = \dfrac{h}{m\lambda} = \dfrac{6.626 \times 10^{-34} \text{ kg m}^2 \text{ s}^{-1}}{(9.11 \times 10^{-31} \text{ kg})(750 \times 10^{-9} \text{ m})} = 970 \text{ m/s}$

Chapter 3 – Periodicity and the Electronic Structure of Atoms

Atomic Spectra (Section 3.2)

3.44   For n = 3; $\lambda = 656.3$ nm $= 656.3 \times 10^{-9}$ m

$$E = \frac{hc}{\lambda} = (6.626 \times 10^{-34} \text{ J·s})\left(\frac{2.998 \times 10^8 \text{ m/s}}{656.3 \times 10^{-9} \text{ m}}\right)\left(\frac{1 \text{ kJ}}{1000 \text{ J}}\right)(6.022 \times 10^{23}/\text{mol})$$

$E = 182.3$ kJ/mol

For n = 4; $\lambda = 486.1$ nm $= 486.1 \times 10^{-9}$ m

$$E = \frac{hc}{\lambda} = (6.626 \times 10^{-34} \text{ J·s})\left(\frac{2.998 \times 10^8 \text{ m/s}}{486.1 \times 10^{-9} \text{ m}}\right)\left(\frac{1 \text{ kJ}}{1000 \text{ J}}\right)(6.022 \times 10^{23}/\text{mol})$$

$E = 246.1$ kJ/mol

For n = 5; $\lambda = 434.0$ nm $= 434.0 \times 10^{-9}$ m

$$E = \frac{hc}{\lambda} = (6.626 \times 10^{-34} \text{ J·s})\left(\frac{2.998 \times 10^8 \text{ m/s}}{434.0 \times 10^{-9} \text{ m}}\right)\left(\frac{1 \text{ kJ}}{1000 \text{ J}}\right)(6.022 \times 10^{23}/\text{mol})$$

$E = 275.6$ kJ/mol

3.45 $m = 2, n = \infty$; $R_\infty = 1.097 \times 10^{-2}$ nm^{-1}

$$\frac{1}{\lambda} = R_\infty\left[\frac{1}{m^2} - \frac{1}{n^2}\right]; \quad \frac{1}{\lambda} = R_\infty\left[\frac{1}{2^2} - \frac{1}{\infty^2}\right]; \quad \frac{1}{\lambda} = R_\infty\left[\frac{1}{4}\right] = 2.74 \times 10^{-3} \text{ nm}^{-1}; \lambda = 364.6 \text{ nm}$$

3.46   From problem 3.45, for $n = \infty$, $\lambda = 364.6$ nm $= 364.6 \times 10^{-9}$ m

$$E = \frac{hc}{\lambda} = (6.626 \times 10^{-34} \text{ J·s})\left(\frac{2.998 \times 10^8 \text{ m/s}}{364.6 \times 10^{-9} \text{ m}}\right)\left(\frac{1 \text{ kJ}}{1000 \text{ J}}\right)(6.022 \times 10^{23}/\text{mol})$$

$E = 328.1$ kJ/mol

3.47 $m = 4, n = 5$; $R_\infty = 1.097 \times 10^{-2}$ nm^{-1}

$$\frac{1}{\lambda} = R_\infty\left[\frac{1}{m^2} - \frac{1}{n^2}\right]; \quad \frac{1}{\lambda} = R_\infty\left[\frac{1}{4^2} - \frac{1}{5^2}\right] = 2.468 \times 10^{-4} \text{ nm}^{-1}; \lambda = 4051 \text{ nm}$$

$$E = \frac{hc}{\lambda} = (6.626 \times 10^{-34} \text{ J·s})\left(\frac{2.998 \times 10^8 \text{ m/s}}{4051 \times 10^{-9} \text{ m}}\right)\left(\frac{1 \text{ kJ}}{1000 \text{ J}}\right)(6.022 \times 10^{23}/\text{mol})$$

$E = 29.55$ kJ/mol, IR

$m = 4, n = 6$; $R_\infty = 1.097 \times 10^{-2}$ nm^{-1}

$$\frac{1}{\lambda} = R_\infty\left[\frac{1}{m^2} - \frac{1}{n^2}\right]; \quad \frac{1}{\lambda} = R_\infty\left[\frac{1}{4^2} - \frac{1}{6^2}\right] = 3.809 \times 10^{-4} \text{ nm}^{-1}; \lambda = 2625 \text{ nm}$$

$$E = \frac{hc}{\lambda} = (6.626 \times 10^{-34} \text{ J·s})\left(\frac{2.998 \times 10^8 \text{ m/s}}{2625 \times 10^{-9} \text{ m}}\right)\left(\frac{1 \text{ kJ}}{1000 \text{ J}}\right)(6.022 \times 10^{23}/\text{mol})$$

$E = 45.60$ kJ/mol, IR

3.48 $\lambda = 330$ nm $= 330 \times 10^{-9}$ m

$E = \dfrac{hc}{\lambda} = (6.626 \times 10^{-34} \text{ J·s})\left(\dfrac{3.00 \times 10^8 \text{ m/s}}{330 \times 10^{-9} \text{ m}}\right)\left(\dfrac{1 \text{ kJ}}{1000 \text{ J}}\right)(6.022 \times 10^{23}/\text{mol})$

$E = 363$ kJ/mol

3.49   795 nm $= 795 \times 10^{-9}$ m

$E = \dfrac{hc}{\lambda} = (6.626 \times 10^{-34} \text{ J·s})\left(\dfrac{3.00 \times 10^8 \text{ m/s}}{795 \times 10^{-9} \text{ m}}\right)\left(\dfrac{1 \text{ kJ}}{1000 \text{ J}}\right)(6.022 \times 10^{23}/\text{mol})$

$E = 151$ kJ/mol

Orbitals and Quantum Numbers (Sections 3.5–3.10)

3.50   n is the principal quantum number. The size and energy level of an orbital depends on n. l is the angular-momentum quantum number. l defines the three-dimensional shape of an orbital. m_l is the magnetic quantum number. m_l defines the spatial orientation of an orbital. m_s is the spin quantum number. m_s indicates the spin of the electron and can have either of two values, $+\frac{1}{2}$ or $-\frac{1}{2}$.

3.51   The Heisenberg uncertainty principle states that one can never know both the position and the velocity of an electron beyond a certain level of precision. This means we cannot think of electrons circling the nucleus in specific orbital paths, but we can think of electrons as being found in certain three-dimensional regions of space around the nucleus, called orbitals.

3.52   The probability of finding the electron drops off rapidly as distance from the nucleus increases, although it never drops to zero, even at large distances. As a result, there is no definite boundary or size for an orbital. However, we usually imagine the boundary surface of an orbital enclosing the volume where an electron spends 95% of its time.

3.53   A 4s orbital has three nodal surfaces.

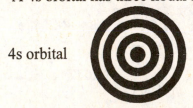

4s orbital

nodes are white

regions of maximum electron probability are black

3.54   Part of the electron-nucleus attraction is canceled by the electron-electron repulsion, an effect we describe by saying that the electrons are shielded from the nucleus by the other electrons. The net nuclear charge actually felt by an electron is called the effective nuclear charge, Z_{eff}, and is often substantially lower than the actual nuclear charge, Z_{actual}.

$$Z_{eff} = Z_{actual} - \text{electron shielding}$$

3.55   Electron shielding gives rise to energy differences among 3s, 3p, and 3d orbitals in multielectron atoms because of the differences in orbital shape. For example, the 3s orbital is spherical and has a large probability density near the nucleus, while the 3p orbital is dumbbell shaped with a node at the nucleus. An electron in a 3s orbital can penetrate closer to the nucleus than an electron in a 3p orbital can and feels less of a shielding effect from other electrons. Generally, for any given value of the principal quantum number n, a lower value of l corresponds to a higher value of Z_{eff} and to a lower energy for the orbital.

3.56   (a) 4s $n = 4$; $l = 0$; $m_l = 0$; $m_s = \pm\frac{1}{2}$
       (b) 3p $n = 3$; $l = 1$; $m_l = -1, 0, +1$; $m_s = \pm\frac{1}{2}$
       (c) 5f $n = 5$; $l = 3$; $m_l = -3, -2, -1, 0, +1, +2, +3$; $m_s = \pm\frac{1}{2}$
       (d) 5d $n = 5$; $l = 2$; $m_l = -2, -1, 0, +1, +2$; $m_s = \pm\frac{1}{2}$

3.57   (a) 3s   (b) 2p   (c) 4f   (d) 4d

3.58   (a) is not allowed because for $l = 0$, $m_l = 0$ only.
       (b) is allowed.
       (c) is not allowed because for $n = 4$, $l = 0, 1, 2,$ or 3 only.

3.59   Co $1s^2\, 2s^2\, 2p^6\, 3s^2\, 3p^6\, 4s^2\, 3d^7$
       (a) is not allowed because for $l = 0$, $m_l = 0$ only.
       (b) is not allowed because $n = 4$ and $l = 2$ is for a 4d orbital.
       (c) is allowed because $n = 3$ and $l = 1$ is for a 3p orbital.

3.60   For $n = 5$, the maximum number of electrons will occur when the 5g orbital is filled:
       [Rn] $7s^2\, 5f^{14}\, 6d^{10}\, 7p^6\, 8s^2\, 5g^{18} = 138$ electrons

3.61 $n = 4$, $l = 0$ is a 4s orbital. The electron configuration is $1s^2\, 2s^2\, 2p^6\, 3s^2\, 3p^6\, 4s^2$. The number of electrons is 20.

3.62 $0.68 \text{ g} = 0.68 \times 10^{-3} \text{ kg}$

$$(\Delta x)(\Delta mv) \geq \frac{h}{4\pi}; \quad \Delta x \geq \frac{h}{4\pi(\Delta mv)} = \frac{6.626 \times 10^{-34} \text{ kg m}^2 \text{ s}^{-1}}{4\pi(0.68 \times 10^{-3} \text{ kg})(0.1 \text{ m/s})} = 8 \times 10^{-31} \text{ m}$$

3.63 $4.0026 \text{ amu} \times \dfrac{1.660\,540 \times 10^{-27} \text{ kg}}{1 \text{ amu}} = 6.6465 \times 10^{-27} \text{ kg}; \quad (\Delta x)(\Delta mv) \geq \dfrac{h}{4\pi}$

$$\Delta x \geq \frac{h}{4\pi(\Delta mv)} = \frac{6.626 \times 10^{-34} \text{ kg m}^2 \text{ s}^{-1}}{4\pi(6.6465 \times 10^{-27} \text{ kg})(0.01 \times 1.36 \times 10^3 \text{ m/s})} = 5.833 \times 10^{-10} \text{ m}$$

Electron Configurations (Sections 3.10–3.13)

3.64   The number of elements in successive periods of the periodic table increases by the progression 2, 8, 18, 32 because the principal quantum number n increases by 1 from one period to the next. As the principal quantum number increases, the number of orbitals in a shell increases. The progression of elements parallels the number of electrons in a particular shell.

3.65   The n and l quantum numbers determine the energy level of an orbital in a multielectron atom.

3.66   (a) 5d   (b) 4s   (c) 6s

3.67   (a) 2p < 3p < 5s < 4d   (b) 2s < 4s < 3d < 4p   (c) 3d < 4p < 5p < 6s

3.68   (a) 3d after 4s   (b) 4p after 3d   (c) 6d after 5f   (d) 6s after 5p

3.69   (a) 3s before 3p   (b) 3d before 4p   (c) 6s before 4f   (d) 4f before 5d

3.70   (a) Ti, Z = 22 $1s^2 2s^2 2p^6 3s^2 3p^6 4s^2 3d^2$
      (b) Ru, Z = 44 $1s^2 2s^2 2p^6 3s^2 3p^6 4s^2 3d^{10} 4p^6 5s^2 4d^6$
      (c) Sn, Z = 50 $1s^2 2s^2 2p^6 3s^2 3p^6 4s^2 3d^{10} 4p^6 5s^2 4d^{10} 5p^2$
      (d) Sr, Z = 38 $1s^2 2s^2 2p^6 3s^2 3p^6 4s^2 3d^{10} 4p^6 5s^2$
      (e) Se, Z = 34 $1s^2 2s^2 2p^6 3s^2 3p^6 4s^2 3d^{10} 4p^4$

3.71   (a) Z = 55, Cs   [Xe] $6s^1$ (b) Z = 40, Zr   [Kr] $5s^2 4d^2$
      (c) Z = 80, Hg   [Xe] $6s^2 4f^{14} 5d^{10}$ (d) Z = 62, Sm   [Xe] $6s^2 4f^6$

3.72   (a) Rb, Z = 37   [Kr] ↑
                        5s

      (b) W, Z = 74   [Xe] ↑↓   ↑↓ ↑↓ ↑↓ ↑↓ ↑↓ ↑↓ ↑↓   ↑ ↑ ↑ ↑ _
                        6s   4f   5d

      (c) Ge, Z = 32   [Ar] ↑↓   ↑↓ ↑↓ ↑↓ ↑↓ ↑↓   ↑ ↑ _
                        4s   3d   4p

      (d) Zr, Z = 40   [Kr] ↑↓   ↑ ↑ _ _ _
                        5s   4d

3.73   (a) Z = 25, Mn   [Ar] ↑↓   ↑ ↑ ↑ ↑ ↑
                        4s   3d

      (b) Z = 56, Ba   [Xe] ↑↓
                        6s

      (c) Z = 28, Ni   [Ar] ↑↓   ↑↓ ↑↓ ↑↓ ↑ ↑
                        4s   3d

      (d) Z = 47, Ag   [Kr] ↑   ↑↓ ↑↓ ↑↓ ↑↓ ↑↓
                        5s   4d

3.74   4s > 4d > 4f

3.75   K < Ca < Se < Kr

3.76   Z = 116   [Rn] $7s^2 5f^{14} 6d^{10} 7p^4$

3.77   Z = 119   [Rn] $7s^2 5f^{14} 6d^{10} 7p^6 8s^1$

3.78 (a) O $1s^2 2s^2 2p^4$ $\underline{\uparrow\downarrow}\;\underline{\uparrow}\;\underline{\uparrow}$ 2 unpaired e^-
                                                                2p

      (b) Si $1s^2 2s^2 2p^6 3s^2 3p^2$ $\underline{\uparrow}\;\underline{\uparrow}\;\underline{\;\;}$ 2 unpaired e^-
                                                                3p

      (c) K   [Ar] $4s^1$ 1 unpaired e^-

      (d) As   [Ar] $4s^2 3d^{10} 4p^3$ $\underline{\uparrow}\;\underline{\uparrow}\;\underline{\uparrow}$ 3 unpaired e^-
                                                                4p

3.79 (a) Z = 31, Ga   (b) Z = 46, Pd

3.80 Order of orbital filling:
      1s→2s→2p→3s→3p→4s→3d→4p→5s→4d→5p→6s→4f→5d→6p→7s→5f→6d→7p→8s→5g
      Z = 121

3.81 A g orbital would begin filling at atomic number = 121 (see 5.82). There are nine g
      orbitals that can each hold two electrons. The first element to have a filled g orbital
      would be atomic number = 138.

Atomic Radii and Periodic Properties (Section 3.14)

3.82 Atomic radii increase down a group because the electron shells are farther away from the
      nucleus.

3.83 Across a period, the effective nuclear charge increases, causing a decrease in atomic radii.

3.84 F < O < S

3.85 (a) K, lower in group 1A
      (b) Ta, lower in group 5B
      (c) V, farther to the left in same period
      (d) Ba, four periods lower and only one group to the right

3.86 Mg has a higher ionization energy than Na because Mg has a higher Z_{eff} and a smaller size.

3.87 F has a higher electron affinity than C because of a higher effective nuclear charge and
      room in the valence shell for the additional electron. In addition, F^- achieves a noble gas
      electron configuration.

Chapter Problems

3.88 m = 2; R_∞ = 1.097 x 10^{-2} nm^{-1}

$$\frac{1}{\lambda} = R_\infty\left[\frac{1}{m^2} - \frac{1}{n^2}\right]; \; \frac{1}{\lambda} = R_\infty\left[\frac{1}{2^2} - \frac{1}{6^2}\right] = 2.438 \times 10^{-3}\ nm^{-1}$$

      λ = 410.2 nm = 410.2 x 10^{-9} m

$$E = \frac{hc}{\lambda} = (6.626 \times 10^{-34} \text{ J·s})\left(\frac{2.998 \times 10^8 \text{ m/s}}{410.2 \times 10^{-9} \text{ m}}\right)\left(\frac{1 \text{ kJ}}{1000 \text{ J}}\right)(6.022 \times 10^{23}/\text{mol})$$

$$E = 291.6 \text{ kJ/mol}$$

3.89 $m = 5$; $R_\infty = 1.097 \times 10^{-2} \text{ nm}^{-1}$

$$\frac{1}{\lambda} = R_\infty\left[\frac{1}{m^2} - \frac{1}{n^2}\right]$$

$n = 6$, $\dfrac{1}{\lambda} = R_\infty\left[\dfrac{1}{5^2} - \dfrac{1}{6^2}\right] = 1.341 \times 10^{-4} \text{ nm}^{-1}$; $\lambda = 7458 \text{ nm} = 7458 \times 10^{-9} \text{ m}$

$$E = \frac{hc}{\lambda} = (6.626 \times 10^{-34} \text{ J·s})\left(\frac{2.998 \times 10^8 \text{ m/s}}{7458 \times 10^{-9} \text{ m}}\right)\left(\frac{1 \text{ kJ}}{1000 \text{ J}}\right)(6.022 \times 10^{23}/\text{mol})$$

$$E = 16.04 \text{ kJ/mol}$$

$n = 7$, $\dfrac{1}{\lambda} = R_\infty\left[\dfrac{1}{5^2} - \dfrac{1}{7^2}\right] = 2.149 \times 10^{-4} \text{ nm}^{-1}$; $\lambda = 4653 \text{ nm} = 4653 \times 10^{-9} \text{ m}$

$$E = \frac{hc}{\lambda} = (6.626 \times 10^{-34} \text{ J·s})\left(\frac{2.998 \times 10^8 \text{ m/s}}{4653 \times 10^{-9} \text{ m}}\right)\left(\frac{1 \text{ kJ}}{1000 \text{ J}}\right)(6.022 \times 10^{23}/\text{mol})$$

$$E = 25.71 \text{ kJ/mol}$$

The lines in this series are in the infrared region of the electromagnetic spectrum.

3.90 $m = 5$, $n = \infty$; $R_\infty = 1.097 \times 10^{-2} \text{ nm}^{-1}$

$$\frac{1}{\lambda} = R_\infty\left[\frac{1}{5^2} - \frac{1}{\infty^2}\right] = R_\infty\left[\frac{1}{25}\right] = 4.388 \times 10^{-4} \text{ nm}^{-1};\ \lambda = 2279 \text{ nm}$$

3.91   (a) $E = \left(142 \dfrac{\text{kJ}}{\text{mol}}\right)\left(\dfrac{1000 \text{ J}}{1 \text{ kJ}}\right)\left(\dfrac{1 \text{ mol}}{6.02 \times 10^{23}}\right) = 2.36 \times 10^{-19} \text{ J}$

$E = \dfrac{hc}{\lambda}$, $\lambda = \dfrac{hc}{E} = \dfrac{(6.626 \times 10^{-34} \text{ J·s})(3.00 \times 10^8 \text{ m/s})}{2.36 \times 10^{-19} \text{ J}}$

$\lambda = 8.42 \times 10^{-7} \text{ m}$ (infrared)

(b) $E = \left(4.55 \times 10^{-2} \dfrac{\text{kJ}}{\text{mol}}\right)\left(\dfrac{1000 \text{ J}}{1 \text{ kJ}}\right)\left(\dfrac{1 \text{ mol}}{6.02 \times 10^{23}}\right) = 7.56 \times 10^{-23} \text{ J}$

$E = \dfrac{hc}{\lambda}$, $\lambda = \dfrac{hc}{E} = \dfrac{(6.626 \times 10^{-34} \text{ J·s})(3.00 \times 10^8 \text{ m/s})}{7.56 \times 10^{-23} \text{ J}}$

$\lambda = 2.63 \times 10^{-3} \text{ m}$ (microwave)

(c) $E = \left(4.81 \times 10^4 \dfrac{\text{kJ}}{\text{mol}}\right)\left(\dfrac{1000 \text{ J}}{1 \text{ kJ}}\right)\left(\dfrac{1 \text{ mol}}{6.02 \times 10^{23}}\right) = 7.99 \times 10^{-17} \text{ J}$

$E = \dfrac{hc}{\lambda}$, $\lambda = \dfrac{hc}{E} = \dfrac{(6.626 \times 10^{-34} \text{ J·s})(3.00 \times 10^8 \text{ m/s})}{7.99 \times 10^{-17} \text{ J}}$

$\lambda = 2.49 \times 10^{-9} \text{ m}$ (X ray)

3.92   (a) $E = h\nu = (6.626 \times 10^{-34} \text{ J·s})(3.79 \times 10^{11} \text{ s}^{-1})\left(\dfrac{1 \text{ kJ}}{1000 \text{ J}}\right)(6.022 \times 10^{23}/\text{mol}) = 0.151 \text{ kJ/mol}$

(b) $E = h\nu = (6.626 \times 10^{-34} \text{ J·s})(5.45 \times 10^{4} \text{ s}^{-1})\left(\dfrac{1 \text{ kJ}}{1000 \text{ J}}\right)(6.022 \times 10^{23}/\text{mol}) = 2.17 \times 10^{-8} \text{ kJ/mol}$

(c) $E = h\nu = (6.626 \times 10^{-34} \text{ J·s})\left(\dfrac{3.00 \times 10^{8} \text{ m/s}}{4.11 \times 10^{-5} \text{ m}}\right)\left(\dfrac{1 \text{ kJ}}{1000 \text{ J}}\right)(6.022 \times 10^{23}/\text{mol}) = 2.91 \text{ kJ/mol}$

3.93 $\nu = 9{,}192{,}631{,}770 \text{ s}^{-1} = 9.19263 \times 10^{9} \text{ s}^{-1}$

$E = h\nu = (6.626 \times 10^{-34} \text{ J·s})(9.19263 \times 10^{9} \text{ s}^{-1})\left(\dfrac{1 \text{ kJ}}{1000 \text{ J}}\right)(6.022 \times 10^{23}/\text{mol}) = 3.668 \times 10^{-3} \text{ kJ/mol}$

3.94   (a) Ra   [Rn] $7s^2$ [Rn] ↑↓
                                          7s

(b) Sc   [Ar] $4s^2 3d^1$ [Ar] ↑↓   ↑ _ _ _ _
                                          4s   3d

(c) Lr   [Rn] $7s^2 5f^{14} 6d^1$ [Rn] ↑↓   ↑↓ ↑↓ ↑↓ ↑↓ ↑↓ ↑↓ ↑↓   ↑ _ _ _ _
                                          7s   5f   6d

(d) B   [He] $2s^2 2p^1$ [He] ↑↓   ↑ _ _
                                          2s   2p

(e) Te   [Kr] $5s^2 4d^{10} 5p^4$ [Kr] ↑↓   ↑↓ ↑↓ ↑↓ ↑↓ ↑↓   ↑↓ ↑ ↑
                                          5s   4d   5p

3.95   (a) row 1 $n = 1$, $l = 0$ 1s   2 elements
                          $l = 1$ 1p   6 elements
                          $l = 2$ 1d   10 elements

           row 2 $n = 2$, $l = 0$ 2s   2 elements
                          $l = 1$ 2p   6 elements
                          $l = 2$ 2d   10 elements
                          $l = 3$ 2f   14 elements
There would be 50 elements in the first two rows.
(b) There would be 18 elements in the first row [see (a) above]. The fifth element in the second row would have atomic number = 23.
(c) Z = 12   ↑↓   ↑↓ ↑↓ ↑↓   ↑ ↑ ↑ ↑ _
                1s   1p   1d

3.96 $206.5 \text{ kJ} = 206.5 \times 10^3 \text{ J}$; $E = \dfrac{206.5 \times 10^3 \text{ J}}{1 \text{ mol}} \times \dfrac{1 \text{ mol}}{6.022 \times 10^{23}} = 3.429 \times 10^{-19} \text{ J}$

$E = \dfrac{hc}{\lambda}, \quad \lambda = \dfrac{hc}{E} = \dfrac{(6.626 \times 10^{-34} \text{ J·s})(3.00 \times 10^{8} \text{ m/s})}{3.429 \times 10^{-19} \text{ J}} = 5.797 \times 10^{-7} \text{ m} = 580. \text{ nm}$

3.97 780 nm is at the red end of the visible region of the electromagnetic spectrum.
780 nm = 780 x 10^{-9} m

$$E = \frac{hc}{\lambda} = (6.626 \text{ x } 10^{-34} \text{ J} \cdot \text{s})\left(\frac{3.00 \text{ x } 10^8 \text{ m/s}}{780 \text{ x } 10^{-9} \text{ m}}\right)\left(\frac{1 \text{ kJ}}{1000 \text{ J}}\right)(6.022 \text{ x } 10^{23}/\text{mol}) = 153 \text{ kJ/mol}$$

3.98 (a) Sr, Z = 38   [Kr] ⇅
                                              5s

(b) Cd, Z = 48   [Kr] ⇅   ⇅ ⇅ ⇅ ⇅ ⇅
                                              5s   4d

(c) Z = 22, Ti   [Ar] ⇅   ↑ ↑ _ _ _
                                              4s   3d

(d) Z = 34, Se   [Ar] ⇅   ⇅ ⇅ ⇅ ⇅ ⇅   ⇅ ↑ ↑
                                              4s   3d   4p

3.99 La ([Xe] $6s^2 5d^1$) is directly below Y ([Kr] $5s^2 4d^1$) in the periodic table. Both have similar valence electron configurations, but for La the valence electrons are one shell farther out leading to its larger radius.

Although Hf ([Xe] $6s^2 4f^{14} 5d^2$) is directly below Zr ([Kr] $5s^2 4d^2$) in the periodic table, Zr and Hf have almost identical atomic radii because the 4f electrons in Hf are not effective in shielding the valence electrons. The valence electrons in Hf are drawn in closer to the nucleus by the higher Z_{eff}.

3.100 For K, $Z_{eff} = \sqrt{\dfrac{(418.8 \text{ kJ/mol})(4^2)}{1312 \text{ kJ/mol}}} = 2.26$

For Kr, $Z_{eff} = \sqrt{\dfrac{(1350.7 \text{ kJ/mol})(4^2)}{1312 \text{ kJ/mol}}} = 4.06$

3.101 75 W = 75 J/s; 550 nm = 550 x 10^{-9} m;   (0.05)(75 J/s) = 3.75 J/s

$$E = \frac{hc}{\lambda} = (6.626 \text{ x } 10^{-34} \text{ J} \cdot \text{s})\left(\frac{3.00 \text{ x } 10^8 \text{ m/s}}{550 \text{ x } 10^{-9} \text{ m}}\right) = 3.61 \text{ x } 10^{-19} \text{ J/photon}$$

$$\text{number of photons} = \frac{3.75 \text{ J/s}}{3.61 \text{ x } 10^{-19} \text{ J/photon}} = 1.0 \text{ x } 10^{19} \text{ photons/s}$$

3.102 q = (350 g)(4.184 J/g·°C)(95 °C − 20 °C) = 109,830 J
λ = 15.0 cm = 15.0 x 10^{-2} m

$$E = (6.626 \text{ x } 10^{-34} \text{ J} \cdot \text{s})\left(\frac{3.00 \text{ x } 10^8 \text{ m/s}}{15.0 \text{ x } 10^{-2} \text{ m}}\right) = 1.33 \text{ x } 10^{-24} \text{ J/photon}$$

$$\text{number of photons} = \frac{109,830 \text{ J}}{1.33 \text{ x } 10^{-24} \text{ J/photon}} = 8.3 \text{ x } 10^{28} \text{ photons}$$

3.103 $E = \left(310 \dfrac{kJ}{mol}\right)\left(\dfrac{1000\,J}{1\,kJ}\right)\left(\dfrac{1\,mol}{6.022 \times 10^{23}}\right) = 5.15 \times 10^{-19}\,J$

$E = \dfrac{hc}{\lambda}, \quad \lambda = \dfrac{hc}{E} = \dfrac{(6.626 \times 10^{-34}\,J \cdot s)(3.00 \times 10^{8}\,m/s)}{5.15 \times 10^{-19}\,J} = 3.86 \times 10^{-7}\,m = 386\,nm$

3.104 $48.2\,nm = 48.2 \times 10^{-9}\,m$

$E(photon) = 6.626 \times 10^{-34}\,J \cdot s \times \dfrac{3.00 \times 10^{8}\,m/s}{48.2 \times 10^{-9}\,m} \times \dfrac{1\,kJ}{1000\,J} \times \dfrac{6.022 \times 10^{23}}{mol} = 2.48 \times 10^{3}\,kJ/mol$

$E_K = E(electron) = \tfrac{1}{2}(9.109 \times 10^{-31}\,kg)(2.371 \times 10^{6}\,m/s)^{2}\left(\dfrac{1\,kJ}{1000\,J}\right)\left(\dfrac{6.022 \times 10^{23}}{mol}\right)$

$E_K = 1.54 \times 10^{3}\,kJ/mol$

$E(photon) = E_i + E_K; \quad E_i = E(photon) - E_K = (2.48 \times 10^{3}) - (1.54 \times 10^{3}) = 940\,kJ/mol$

3.105 Charge on electron $= 1.602 \times 10^{-19}\,C;\quad 1\,V \cdot C = 1\,J = 1\,kg\,m^2/s^2$
 (a) $E_K = (30{,}000\,V)(1.602 \times 10^{-19}\,C) = 4.806 \times 10^{-15}\,J$

$E_K = \tfrac{1}{2}mv^2; \quad v = \sqrt{\dfrac{2\,E_K}{m}} = \sqrt{\dfrac{2 \times 4.806 \times 10^{-15}\,kg\,m^2/s^2}{9.109 \times 10^{-31}\,kg}} = 1.03 \times 10^{8}\,m/s$

$\lambda = \dfrac{h}{mv} = \dfrac{6.626 \times 10^{-34}\,kg\,m^2/s}{(9.109 \times 10^{-31}\,kg)(1.03 \times 10^{8}\,m/s)} = 7.06 \times 10^{-12}\,m$

(b) $E = \dfrac{hc}{\lambda} = (6.626 \times 10^{-34}\,J \cdot s)\left(\dfrac{3.00 \times 10^{8}\,m/s}{1.54 \times 10^{-10}\,m}\right) = 1.29 \times 10^{-15}\,J/photon$

3.106 Substitute the equation for the orbit radius, r, into the equation for the energy level, E, to
get $E = \dfrac{-Ze^2}{2\left(\dfrac{n^2 a_o}{Z}\right)} = \dfrac{-Z^2 e^2}{2 a_o n^2}$

Let E_1 be the energy of an electron in a lower orbit and E_2 the energy of an electron in a higher orbit. The difference between the two energy levels is

$\Delta E = E_2 - E_1 = \dfrac{-Z^2 e^2}{2 a_o n_2^2} - \dfrac{-Z^2 e^2}{2 a_o n_1^2} = \dfrac{-Z^2 e^2}{2 a_o n_2^2} + \dfrac{Z^2 e^2}{2 a_o n_1^2} = \dfrac{Z^2 e^2}{2 a_o n_1^2} - \dfrac{Z^2 e^2}{2 a_o n_2^2}$

$\Delta E = \dfrac{Z^2 e^2}{2 a_o}\left[\dfrac{1}{n_1^2} - \dfrac{1}{n_2^2}\right]$

Because Z, e, and a_o are constants, this equation shows that ΔE is proportional to

$\left[\dfrac{1}{n_1^2} - \dfrac{1}{n_2^2}\right]$ where n_1 and n_2 are integers with $n_2 > n_1$.

This is similar to the Balmer-Rydberg equation where $1/\lambda$ or ν for the emission spectra of atoms is proportional to $\left[\dfrac{1}{m^2} - \dfrac{1}{n^2}\right]$ where m and n are integers with n > m.

3.107 (a) $\underset{1s}{\underline{\uparrow\downarrow}}$ $\underset{2s}{\underline{\uparrow\downarrow}}$ $\underset{2p}{\underline{\uparrow\downarrow}\ \underline{\uparrow\downarrow}\ \underline{\uparrow\downarrow}}$ $\underset{3s}{\underline{\uparrow\downarrow}}$ $\underset{3p}{\underline{\uparrow\downarrow}\ \underline{\uparrow\circ}\ \underline{\uparrow\circ}}$ $\underset{4s}{\underline{\ \ }}$

Two partially filled orbitals.
(b) The element in the 3rd column and 4th row under these new rules would have an atomic number of 30 and be in the s-block.

3.108 (a) 3d, $n = 3$, $l = 2$
(b) 2p, $n = 2$, $l = 1$, $m_l = -1, 0, +1$
3p, $n = 3$, $l = 1$, $m_l = -1, 0, +1$
3d, $n = 3$, $l = 2$, $m_l = -2, -1, 0, +1, +2$
(c) N, $1s^2 2s^2 2p^3$ so the 3s, 3p, and 3d orbitals are empty.
(d) C, $1s^2 2s^2 2p^2$ so the 1s and 2s orbitals are filled.
(e) Be, $1s^2 2s^2$ so the 2s orbital contains the outermost electrons.
(f) 2p and 3p ($\underline{\uparrow}\ \underline{\uparrow}\ \underline{\ }$) and 3d ($\underline{\uparrow}\ \underline{\uparrow}\ \underline{\ }\ \underline{\ }\ \underline{\ }$).

3.109 $\lambda = 1.03 \times 10^{-7}$ m $= 103 \times 10^{-9}$ m $= 103$ nm

$$\frac{1}{\lambda} = R_\infty\left[\frac{1}{m^2} - \frac{1}{n^2}\right], \quad R_\infty = 1.097 \times 10^{-2} \text{ nm}^{-1}$$

$$\frac{1}{103 \text{ nm}} = (1.097 \times 10^{-2} \text{ nm}^{-1})\left[\frac{1}{1^2} - \frac{1}{n^2}\right], \text{ solve for n.}$$

$$\frac{(1/103 \text{ nm})}{(1.097 \times 10^{-2} \text{ nm}^{-1})} - 1 = -\frac{1}{n^2}$$

$$\frac{1}{n^2} = 0.115; \quad n^2 = \frac{1}{0.115}; \quad n = \sqrt{\frac{1}{0.115}} = 2.95$$

The electron jumps to the third shell.

3.110 (a) $E = h\nu$; $\quad \nu = \dfrac{E}{h} = \dfrac{7.21 \times 10^{-19} \text{ J}}{6.626 \times 10^{-34} \text{ J·s}} = 1.09 \times 10^{15} \text{ s}^{-1}$

(b) $E(\text{photon}) = E_i + E_K$; from (a), $E_i = 7.21 \times 10^{-19}$ J

$E(\text{photon}) = \dfrac{hc}{\lambda} = (6.626 \times 10^{-34} \text{ J·s})\left(\dfrac{3.00 \times 10^8 \text{ m/s}}{2.50 \times 10^{-7} \text{ m}}\right) = 7.95 \times 10^{-19}$ J

$E_K = E(\text{photon}) - E_i = (7.95 \times 10^{-19} \text{ J}) - (7.21 \times 10^{-19} \text{ J}) = 7.4 \times 10^{-20}$ J

Calculate the electron velocity from the kinetic energy, E_K.

$E_K = 7.4 \times 10^{-20}$ J $= 7.4 \times 10^{-20}$ kg·m²/s² $= \frac{1}{2}mv^2 = \frac{1}{2}(9.109 \times 10^{-31} \text{ kg})v^2$

$$v = \sqrt{\frac{2 \times (7.4 \times 10^{-20}\ kg \cdot m^2/s^2)}{9.109 \times 10^{-31}\ kg}} = 4.0 \times 10^5\ m/s$$

$$deBroglie\ wavelength = \frac{h}{mv} = \frac{6.626 \times 10^{-34}\ kg \cdot m^2/s}{(9.109 \times 10^{-31}\ kg)(4.0 \times 10^5\ m/s)} = 1.8 \times 10^{-9}\ m = 1.8\ nm$$

3.111 (a) $E = h\nu$; $\nu = \dfrac{E}{h} = \dfrac{4.70 \times 10^{-16}\ J}{6.626 \times 10^{-34}\ J \cdot s} = 7.09 \times 10^{17}\ s^{-1}$

(b) $\lambda = \dfrac{c}{\nu} = \dfrac{3.00 \times 10^8\ m/s}{7.09 \times 10^{17}\ s^{-1}} = 4.23 \times 10^{-10}\ m = 0.423 \times 10^{-9}\ m = 0.423\ nm$

(c) $\lambda = \dfrac{h}{mv}$; $v = \dfrac{h}{m\lambda} = \dfrac{6.626 \times 10^{-34}\ kg \cdot m^2/s}{(9.11 \times 10^{-31}\ kg)(4.23 \times 10^{-10}\ m)} = 1.72 \times 10^6\ m/s$

(d) $E_K = \dfrac{mv^2}{2} = \dfrac{(9.11 \times 10^{-31}\ kg)(1.72 \times 10^6\ m/s)^2}{2} = 1.35 \times 10^{-18}\ kg \cdot m^2/s^2 = 1.35 \times 10^{-18}\ J$

3.112 (a) 5f subshell: $n = 5$, $l = 3$, $m_l = -3, -2, -1, 0, +1, +2, +3$

3d subshell: $n = 3$, $l = 2$, $m_l = -2, -1, 0, +1, +2$

(b) In the H atom the subshells in a particular energy level are all degenerate, i.e., all have the same energy. Therefore, you only need to consider the principal quantum number, n, to calculate the wavelength emitted for an electron that drops from the 5f to the 3d subshell.

$m = 3$, $n = 5$; $R_\infty = 1.097 \times 10^{-2}\ nm^{-1}$

$$\frac{1}{\lambda} = R_\infty \left[\frac{1}{m^2} - \frac{1}{n^2} \right]; \quad \frac{1}{\lambda} = R_\infty \left[\frac{1}{3^2} - \frac{1}{5^2} \right]; \quad \frac{1}{\lambda} = 7.801 \times 10^{-4}\ nm^{-1}; \quad \lambda = 1282\ nm$$

(c) $m = 3$, $n = \infty$; $R_\infty = 1.097 \times 10^{-2}\ nm^{-1}$

$$\frac{1}{\lambda} = R_\infty \left[\frac{1}{m^2} - \frac{1}{n^2} \right]; \quad \frac{1}{\lambda} = R_\infty \left[\frac{1}{3^2} - \frac{1}{\infty^2} \right]; \quad \frac{1}{\lambda} = R_\infty \left[\frac{1}{3^2} \right] = 1.219 \times 10^{-3}\ nm^{-1}; \quad \lambda = 820.4\ nm$$

$$E = (6.626 \times 10^{-34}\ J \cdot s) \left(\frac{3.00 \times 10^8\ m/s}{820.4 \times 10^{-9}\ m} \right) (6.022 \times 10^{23}/mol) = 1.46 \times 10^5\ J/mol = 146\ kJ/mol$$

3.113 (a) [Kr] $5s^2\ 4d^{10}\ 5p^6$ (b) [Kr] $5s^2\ 4d^{10}\ 5p^5\ 6s^1$

(c) Both Xe* and Cs have a single electron in the 6s orbital with similar effective nuclear charges. Therefore the 6s electrons in both cases are held with similar strengths and require almost the same energy to remove.

4 Ionic Bonds and Some Main-Group Chemistry

4.1   (a) LiBr is composed of a metal (Li) and nonmetal (Br) and is ionic.
(b) $SiCl_4$ is composed of only nonmetals and is molecular.
(c) BF_3 is composed of only nonmetals and is molecular.
(d) CaO is composed of a metal (Ca) and nonmetal (O) and is ionic.

4.2   Figure (a) most likely represents an ionic compound because there are no discrete molecules, only a regular array of two different chemical species (ions). Figure (b) most likely represents a molecular compound because discrete molecules are present.

4.3   (a) Ra^{2+} [Rn]   (b) Y^{3+} [Kr]   (c) Ti^{4+} [Ar]   (d) N^{3-} [Ne]
Each ion has the ground-state electron configuration of the noble gas closest to it in the periodic table.

4.4   The neutral atom contains 30 e^- and is Zn. The ion is Zn^{2+}.

4.5   (a) O^{2-}; decrease in effective nuclear charge and an increase in electron-electron repulsions lead to the larger anion.
(b) S; atoms get larger as you go down a group.
(c) Fe; in Fe^{3+} electrons are removed from a larger valence shell and there is an increase in effective nuclear charge leading to the smaller cation.
(d) H^-; decrease in effective nuclear charge and an increase in electron-electron repulsions lead to the larger anion.

4.6 K^+ is smaller than neutral K because the ion has one less electron. K^+ and Cl^- are isoelectronic, but K^+ is smaller than Cl^- because of its higher effective nuclear charge. K is larger than Cl^- because K has one additional electron and that electron begins the next shell (period). K^+, r = 133 pm; Cl^-, r = 184 pm; K, r = 227 pm

4.7   (a) Br   (b) S   (c) Se   (d) Ne

4.8   (a) Be $1s^2 2s^2$;   N $1s^2 2s^2 2p^3$
Be would have the larger third ionization energy because this electron would come from the 1s orbital.
(b) Ga   [Ar] $4s^2 3d^{10} 4p^1$;   Ge   [Ar] $4s^2 3d^{10} 4p^2$
Ga would have the larger fourth ionization energy because this electron would come from the 3d orbitals. Ge would be losing a 4s electron.

4.9   (b) Cl has the highest E_{i1} and smallest E_{i4}.

4.10   Ca (red) would have the largest third ionization energy of the three because the electron being removed is from a filled valence shell. For Al (green) and Kr (blue), the electron being removed is from a partially filled valence shell. The third ionization energy for Kr would be larger than that for Al because the electron being removed from Kr is coming out of a 4p orbital while the electron being removed from Al makes Al isoelectronic with Ne. In addition, Z_{eff} is larger for Kr than for Al. The ease of losing its third electron is Al < Kr < Ca.

4.11   Cr   [Ar] $4s^1 3d^5$;   Mn   [Ar] $4s^2 3d^5$;   Fe   [Ar] $4s^2 3d^6$
Cr can accept an electron into a 4s orbital. The 4s orbital is lower in energy than a 3d orbital. Both Mn and Fe accept the added electron into a 3d orbital that contains an electron, but Mn has a lower value of Z_{eff}. Therefore, Mn has a less negative E_{ea} than either Cr or Fe.

4.12   The least favorable E_{ea} is for Kr (red) because it is a noble gas with a filled set of 4p orbitals. The most favorable E_{ea} is for Ge (blue) because the 4p orbitals would become half filled. In addition, Z_{eff} is larger for Ge than it is for K (green).

4.13   (a) Rb would lose one electron and adopt the Kr noble gas configuration.
(b) Ba would lose two electrons and adopt the Xe noble gas configuration.
(c) Ga would lose three electrons and adopt an Ar-like noble gas configuration (note that Ga^{3+} has ten 3d electrons in addition to the two 3s and six 3p electrons).
(d) F would gain one electron and adopt the Ne noble gas configuration.

4.14   Group 6A elements will gain 2 electrons.

4.15
$$K(s) \rightarrow K(g) \qquad\qquad +89.2 \text{ kJ/mol}$$
$$K(g) \rightarrow K^+(g) + e^- \qquad +418.8 \text{ kJ/mol}$$
$$\tfrac{1}{2}\,[F_2(g) \rightarrow 2\,F(g)] \qquad +79 \text{ kJ/mol}$$
$$F(g) + e^- \rightarrow F^-(g) \qquad -328 \text{ kJ/mol}$$
$$K^+(g) + F^-(g) \rightarrow KF(s) \qquad \underline{-821 \text{ kJ/mol}}$$
$$\text{Sum} = -562 \text{ kJ/mol for } K(s) + \tfrac{1}{2}\,F_2(g) \rightarrow KF(s)$$

4.16   (a) KCl has the higher lattice energy because of the smaller K^+.
(b) CaF_2 has the higher lattice energy because of the smaller Ca^{2+}.
(c) CaO has the higher lattice energy because of the higher charge on both the cation and anion.

4.17   The anions are larger than the cations. Cl^- is larger than O^{2-} because it is below it in the periodic table. Therefore, (a) is NaCl and (b) is MgO. Because of the higher ion charge and shorter cation–anion distance, MgO has the larger lattice energy.

4.18   (a) CsF, cesium fluoride   (b) K_2O, potassium oxide   (c) CuO, copper(II) oxide
(d) BaS, barium sulfide   (e) $BeBr_2$, beryllium bromide

4.19   (a) vanadium(III) chloride, VCl_3 (b) manganese(IV) oxide, MnO_2
(c) copper(II) sulfide, CuS   (d) aluminum oxide, Al_2O_3

4.20 red – potassium sulfide, K_2S; green – strontium iodide, SrI_2; blue – gallium oxide, Ga_2O_3

4.21 (a) $Ca(ClO)_2$, calcium hypochlorite
(b) $Ag_2S_2O_3$, silver(I) thiosulfate or silver thiosulfate
(c) NaH_2PO_4, sodium dihydrogen phosphate   (d) $Sn(NO_3)_2$, tin(II) nitrate
(e) $Pb(CH_3CO_2)_4$, lead(IV) acetate   (f) $(NH_4)_2SO_4$, ammonium sulfate

4.22 (a) lithium phosphate, Li_3PO_4 (b) magnesium hydrogen sulfate, $Mg(HSO_4)_2$
(c) manganese(II) nitrate, $Mn(NO_3)_2$ (d) chromium(III) sulfate, $Cr_2(SO_4)_3$

4.23 Drawing 1 represents ionic compounds with one cation and two anions. Only (c) $CaCl_2$ is consistent with drawing 1.
Drawing 2 represents ionic compounds with one cation and one anion. Both (a) LiBr and (b) $NaNO_2$ are consistent with drawing 2.

4.24 (a) O^{2-} (b) O_2^{2-} (c) O_2^-

4.25 (a) $2\,Cs(s) + 2\,H_2O(l) \rightarrow 2\,Cs^+(aq) + 2\,OH^-(aq) + H_2(g)$
(b) $Rb(s) + O_2(g) \rightarrow RbO_2(s)$
(c) $2\,K(s) + 2\,NH_3(g) \rightarrow 2\,KNH_2(s) + H_2(g)$

4.26 (a) $Be(s) + Br_2(l) \rightarrow BeBr_2(s)$
(b) $Sr(s) + 2\,H_2O(l) \rightarrow Sr(OH)_2(aq) + H_2(g)$
(c) $2\,Mg(s) + O_2(g) \rightarrow 2\,MgO(s)$

4.27 $Mg(s) + S(s) \rightarrow MgS(s)$;   In MgS, the oxidation number of S is –2.

4.28 Only about 10% of current world salt production comes from evaporation of seawater. Most salt is obtained by mining the vast deposits of <u>halite</u>, or <u>rock salt</u>, formed by evaporation of ancient inland seas. These salt beds can be up to hundreds of meters thick and may occur anywhere from a few meters to thousands of meters below Earth's surface.

Key Concept Problems

4.29 (a) $MgSO_4$ (b) Li_2CO_3 (c) $FeCl_2$ (d) $Ca_3(PO_4)_2$

4.30 (a) shows an extended array, which represents an ionic compound.
(b) shows discrete units, which represent a covalent compound.

4.31

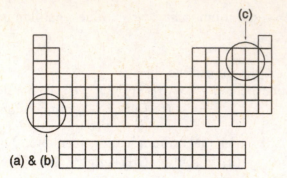

4.32

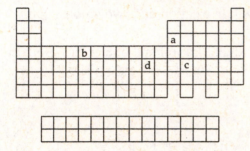

(a) Al^{3+}
(b) Cr^{3+}
(c) Sn^{2+}
(d) Ag^+

4.33 The first sphere gets larger on going from reactant to product. This is consistent with it being a nonmetal gaining an electron and becoming an anion. The second sphere gets smaller on going from reactant to product. This is consistent with it being a metal losing an electron and becoming a cation.

4.34 (a) I_2 (b) Na   (c) NaCl   (d) Cl_2

4.35 (c) has the largest lattice energy because the charges are closest together.
(a) has the smallest lattice energy because the charges are farthest apart.

4.36 Green, CBr_4;   Blue, SrF_2;   Red, PbS or PbS_2

4.37

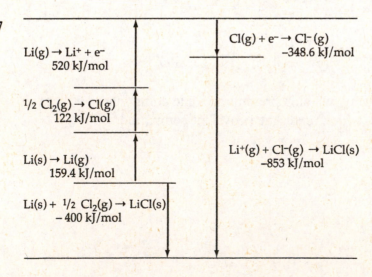

Section Problems
Molecules, Ions, Ionization Energy, and Electron Affinity (Sections 4.1–4.6)

4.38 A covalent bond results when two atoms share several (usually two) of their electrons. An ionic bond results from a complete transfer of one or more electrons from one atom to another. The C–H bonds in methane (CH_4) are covalent bonds. The bond in NaCl (Na^+Cl^-) is an ionic bond.

4.39 A molecule is the unit of matter that results when two or more atoms are joined by covalent bonds. An ion results when an atom gains or loses electrons. CH_4 is a methane molecule. Na^+ is the sodium cation.

4.40 (a) Be^{2+}, 4 protons and 2 electrons   (b) Rb^+, 37 protons and 36 electrons
      (c) Se^{2-}, 34 protons and 36 electrons   (d) Au^{3+}, 79 protons and 76 electrons

4.41 (a) A 2+ cation that has 36 electrons must have 38 protons. X = Sr.
      (b) A 1– anion that has 36 electrons must have 35 protons. X = Br.

4.42 (a) La^{3+},   [Xe]   (b) Ag^+,   [Kr] $4d^{10}$ (c) Sn^{2+},   [Kr] $5s^2 4d^{10}$

4.43 (a) Se^{2-},   [Kr]   (b) N^{3-},   [Ne]

4.44 Cr^{2+} [Ar] $3d^4$ $\uparrow \;\; \uparrow \;\; \uparrow \;\; \uparrow \;\; \_$
                                   3d

      Fe^{2+} [Ar] $3d^6$ $\uparrow\downarrow \;\; \uparrow \;\; \uparrow \;\; \uparrow \;\; \uparrow$
                                   3d

4.45 Z = 30, Zn

4.46 The largest E_{i1} are found in Group 8A because of the largest values of Z_{eff}.
The smallest E_{i1} are found in Group 1A because of the smallest values of Z_{eff}.

4.47 Fr would have the smallest ionization energy, and He would have the largest.

4.48 (a) K   [Ar] $4s^1$ Ca   [Ar] $4s^2$
Ca has the smaller second ionization energy because it is easier to remove the second 4s valence electron in Ca than it is to remove the second electron in K from the filled 3p orbitals.
(b) Ca   [Ar] $4s^2$ Ga   [Ar] $4s^2 3d^{10} 4p^1$
Ca has the larger third ionization energy because it is more difficult to remove the third electron in Ca from the filled 3p orbitals than it is to remove the third electron (second 4s valence electron) from Ga.

4.49 Sn has a smaller fourth ionization energy than Sb because of a smaller Z_{eff}.
Br has a larger sixth ionization energy than Se because of a larger Z_{eff}.

4.50   (a) $1s^2 2s^2 2p^6 3s^2 3p^3$ is P   (b) $1s^2 2s^2 2p^6 3s^2 3p^6$ is Ar   (c) $1s^2 2s^2 2p^6 3s^2 3p^6 4s^2$ is Ca
       Ar has the highest E_{i2}. Ar has a higher Z_{eff} than P. The 4s electrons in Ca are easier to remove than any 3p electrons.
       Ar has the lowest E_{i7}. It is difficult to remove 3p electrons from Ca, and it is difficult to remove 2p electrons from P.

4.51   The atom in the third row with the lowest E_{i4} is the 4A element, Si. $1s^2 2s^2 2p^6 3s^2 3p^2$

4.52   Using Figure 4.6 as a reference:

	Lowest E_{i1}	Highest E_{i1}
(a)	K	Li
(b)	B	Cl
(c)	Ca	Cl

4.53   (a) Group 2A   (b) Group 6A

4.54   The relationship between the electron affinity of a univalent cation and the ionization energy of the neutral atom is that they have the same magnitude but opposite signs.

4.55   The relationship between the ionization energy of a univalent anion and the electron affinity of the neutral atom is that they have the same magnitude but opposite signs.

4.56 Na^+ has a more negative electron affinity than either Na or Cl because of its positive charge.

4.57   Br would have a more negative electron affinity than Br^- because Br^- has no room in its valence shell for an additional electron.

4.58   Energy is usually released when an electron is added to a neutral atom but absorbed when an electron is removed from a neutral atom because of the positive Z_{eff}.

4.59 E_{i1} increases steadily across the periodic table from Group 1A to Group 8A because electrons are being removed from the same shell and Z_{eff} is increasing. The electron affinity increases irregularly from 1A to 7A and then falls dramatically for Group 8A because the additional electron goes into the next higher shell.

Ionic Bonds and Lattice Energy (Sections 4.8–4.9)

4.60 $MgCl_2$ > LiCl > KCl > KBr

4.61 $AlBr_3$ > CaO > $MgBr_2$ > LiBr

4.62 $Li \rightarrow Li^+ + e^-$ +520 kJ/mol
       $Br + e^- \rightarrow Br^-$ <u>−325 kJ/mol</u>
                                 +195 kJ/mol

4.63   The total energy = (376 kJ/mol) + (–349 kJ/mol) = +27 kJ/mol, which is unfavorable because it is positive.

4.64
$Li(s) \rightarrow Li(g)$	+159.4 kJ/mol
$Li(s) \rightarrow Li(g) + e^-$	+520 kJ/mol
$\frac{1}{2} [Br_2(l) \rightarrow Br_2(g)]$	+15.4 kJ/mol
$\frac{1}{2} [Br_2(g) \rightarrow 2\ Br(g)]$	+112 kJ/mol
$Br(g) + e^- \rightarrow Br^-(g)$	–325 kJ/mol
$Li^+(g) + Br^-(g) \rightarrow LiBr(s)$	–807 kJ/mol

Sum = –325 kJ/mol for $Li(s) + \frac{1}{2} Br_2(l) \rightarrow LiBr(s)$

4.65   (a)
$Li(s) \rightarrow Li(g)$	+159.4 kJ/mol
$Li(g) \rightarrow Li^+(g) + e^-$	+520 kJ/mol
$\frac{1}{2}[F_2(g) \rightarrow 2\ F(g)]$	+79 kJ/mol
$F(g) + e^- \rightarrow F^-(g)$	–328 kJ/mol
$Li^+(g) + F^-(g) \rightarrow LiF(s)$	–1036 kJ/mol

Sum = –606 kJ/mol for $Li(s) + \frac{1}{2} F_2(g) \rightarrow LiF(s)$

(b)
$Ca(s) \rightarrow Ca(g)$	+178.2 kJ/mol
$Ca(g) \rightarrow Ca^+(g) + e^-$	+589.8 kJ/mol
$Ca^+(g) \rightarrow Ca^{2+}(g) + e^-$	+1145 kJ/mol
$F_2(g) \rightarrow 2\ F(g)$	+158 kJ/mol
$2[F(g) + e^- \rightarrow F^-(g)]$	2(–328) kJ/mol
$Ca^{2+}(g) + 2\ F^- \rightarrow CaF_2(s)$	–2630 kJ/mol

Sum = –1215 kJ/mol for $Ca(s) + F_2(g) \rightarrow CaF_2(s)$

4.66
$Na(s) \rightarrow Na(g)$	+107.3 kJ/mol
$Na(g) \rightarrow Na^+(g) + e^-$	+495.8 kJ/mol
$\frac{1}{2} [H_2(g) \rightarrow 2\ H(g)]$	½(+435.9) kJ/mol
$H(g) + e^- \rightarrow H^-(g)$	–72.8 kJ/mol
$Na^+(g) + H^-(g) \rightarrow NaH(s)$	–U

Sum = –60 kJ/mol for $Na(s) + \frac{1}{2} H_2(g) \rightarrow NaH(s)$

–U = –60 – 107.3 – 495.8 – 435.9/2 + 72.8 = –808 kJ/mol; U = 808 kJ/mol

4.67
$Ca(s) \rightarrow Ca(g)$	+178.2 kJ/mol
$Ca(g) \rightarrow Ca^+(g) + e^-$	+589.8 kJ/mol
$Ca^+(g) \rightarrow Ca^{2+}(g) + e^-$	+1145 kJ/mol
$H_2(g) \rightarrow 2\ H(g)$	+435.9 kJ/mol
$2[H(g) + e^- \rightarrow H^-(g)]$	2(–72.8) kJ/mol
$Ca^{2+}(g) + 2\ H^-(g) \rightarrow CaH_2(s)$	–U

Sum = –186.2 kJ/mol for $Ca(s) + H_2(g) \rightarrow CaH_2(s)$

–U = –186.2 – 178.2 – 589.8 – 1145 – 435.9 + 2(72.8) = –2390 kJ/mol; U = 2390 kJ/mol

4.68 $Cs(s) \rightarrow Cs(g)$ +76.1 kJ/mol
        $Cs(g) \rightarrow Cs^+(g) + e^-$ +375.7 kJ/mol
        $\frac{1}{2}[F_2(g) \rightarrow 2 F(g)]$ +79   kJ/mol
        $F(g) + e^- \rightarrow F^-(g)$ −328 kJ/mol
        $Cs^+(g) + F^-(g) \rightarrow CsF(s)$ −740 kJ/mol

        Sum = −537   kJ/mol for $Cs(s) + \frac{1}{2} F_2(g) \rightarrow CsF(s)$

4.69 $Cs(s) \rightarrow Cs(g)$ +76.1 kJ/mol
        $Cs(g) \rightarrow Cs^+(g) + e^-$ +375.7 kJ/mol
        $Cs^+(g) \rightarrow Cs^{2+}(g) + e^-$ +2422 kJ/mol
        $F_2(g) \rightarrow 2 F(g)$ +158 kJ/mol
        $2[F(g) + e^- \rightarrow F^-(g)]$ 2(−328) kJ/mol
        $Cs^{2+}(g) + 2 F^-(g) \rightarrow CsF_2(s)$ −2347 kJ/mol

        Sum = +29   kJ/mol for $Cs(s) + F_2(g) \rightarrow CsF_2(s)$

The overall reaction absorbs 29 kJ/mol.
In the reaction of cesium with fluorine, CsF will form because the overall energy for the formation of CsF is negative, whereas it is positive for CsF_2.

4.70 $Ca(s) \rightarrow Ca(g)$ +178.2 kJ/mol
        $Ca(g) \rightarrow Ca^+(g) + e^-$ +589.8 kJ/mol
        $\frac{1}{2}[Cl_2(g) \rightarrow 2 Cl(g)]$ +121.5 kJ/mol
        $Cl(g) + e^- \rightarrow Cl^-(g)$ −348.6 kJ/mol
        $Ca^+(g) + Cl^-(g) \rightarrow CaCl(s)$ −717 kJ/mol

        Sum = −176   kJ/mol for $Ca(s) + \frac{1}{2} Cl_2(g) \rightarrow CaCl(s)$

4.71 $Ca(s) \rightarrow Ca(g) + e^-$ +178.2 kJ/mol
        $Ca(g) \rightarrow Ca^+(g) + e^-$ +589.8 kJ/mol
        $Ca^+(g) \rightarrow Ca^{2+}(g)$ +1145 kJ/mol
        $Cl_2(g) \rightarrow 2 Cl(g)$ +243 kJ/mol
        $2[Cl(g) + e^- \rightarrow Cl^-(g)]$ 2(−348.6) kJ/mol
        $Ca^{2+}(g) + 2 Cl^-(g) \rightarrow CaCl_2(s)$ −2258 kJ/mol

        Sum = −799   kJ/mol for $Ca(s) + Cl_2(g) \rightarrow CaCl_2(s)$

In the reaction of calcium with chlorine, $CaCl_2$ will form because the overall energy for the formation of $CaCl_2$ is much more negative than for the formation of CaCl.

4.72

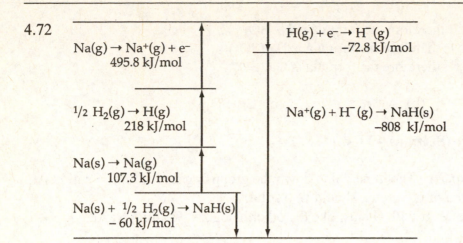

4.73

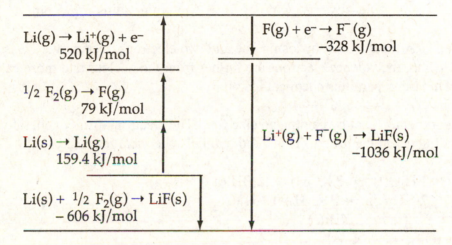

Naming Ionic Compounds (Section 4.10)

4.74 (a) KCl (b) $SnBr_2$ (c) CaO (d) $BaCl_2$ (e) AlH_3

4.75 (a) $Ca(CH_3CO_2)_2$ (b) $Fe(CN)_2$ (c) $Na_2Cr_2O_7$ (d) $Cr_2(SO_4)_3$ (e) $Hg(ClO_4)_2$

4.76 (a) barium ion (b) cesium ion (c) vanadium(III) ion
(d) hydrogen carbonate ion (e) ammonium ion (f) nickel(II) ion
(g) nitrite ion (h) chlorite ion (i) manganese(II) ion (j) perchlorate ion

4.77 (a) Zn^{2+} (b) Fe^{3+} (c) Ti^{4+} (d) Sn^{2+} (e) Hg_2^{2+} (f) Mn^{4+} (g) K^+ (h) Cu^{2+}

4.78 (a) zinc(II) cyanide (b) iron(III) nitrite (c) titanium(IV) sulfate
(d) tin(II) phosphate (e) mercury(I) sulfide (f) manganese(IV) oxide
(g) potassium periodate (h) copper(II) acetate

4.79 (a) magnesium sulfite (b) cobalt(II) nitrite (c) manganese(II) hydrogen carbonate
(d) zinc(II) chromate (e) barium sulfate (f) potassium permanganate
(g) aluminum sulfate (h) lithium chlorate

4.80 (a) Na^+ and SO_4^{2-}; therefore the formula is Na_2SO_4
(b) Ba^{2+} and PO_4^{3-}; therefore the formula is $Ba_3(PO_4)_2$
(c) Ga^{3+} and SO_4^{2-}; therefore the formula is $Ga_2(SO_4)_3$

4.81 (a) Na_2O_2 (b) $AlBr_3$ (c) $Cr_2(SO_4)_3$

Main-Group Chemistry (Sections 4.11–4.14)

4.82 (a) At is in Group 7A. The trend going down the group is gas → liquid → solid. At, being at the bottom of the group, should be a solid.
(b) At is likely to react with Na just like the other halogens, yielding NaAt.

4.83 Predicted for Fr: melting point ≈ 23 °C   boiling point ≈ 650 °C
density ≈ 2 g/cm^3 atomic radius ≈ 275 pm

4.84 Group 1A and 2A metals react by losing one and two electrons, respectively. As you go down each group, the valence electrons are farther from the nucleus and more easily removed. This trend parallels chemical reactivity.

4.85 Group 7A nonmetals react by gaining an electron. The electron affinity generally decreases going down the group. This trend parallels chemical reactivity.

4.86 (a) $2 K(s) + 2 H_2O(l) \rightarrow 2 K^+(aq) + 2 OH^-(aq) + H_2(g)$
(b) $2 K(s) + 2 NH_3(g) \rightarrow 2 KNH_2(s) + H_2(g)$
(c) $2 K(s) + Br_2(l) \rightarrow 2 KBr(s)$
(d) $K(s) + O_2(g) \rightarrow KO_2(s)$

4.87 (a) $Ca(s) + 2 H_2O(l) \rightarrow Ca^{2+}(aq) + 2 OH^-(aq) + H_2(g)$
(b) $Ca(s) + He(g) \rightarrow$ N. R.
(c) $Ca(s) + Br_2(l) \rightarrow CaBr_2(s)$
(d) $2 Ca(s) + O_2(g) \rightarrow 2 CaO(s)$

4.88 (a) $Cl_2(g) + H_2(g) \rightarrow 2 HCl(g)$
(b) $Cl_2(g) + Ar(g) \rightarrow$ N. R.
(c) $Cl_2(g) + 2 Rb(s) \rightarrow 2 RbCl(s)$

4.89 $2 Mg(s) + O_2(g) \rightarrow 2 MgO(s)$
$MgO(s) + H_2O(l) \rightarrow Mg(OH)_2(aq)$

Chapter Problems

4.90 Cu^{2+} has fewer electrons and a larger effective nuclear charge; therefore it has the smaller ionic radius.

4.91 $S^{2-} > Ca^{2+} > Sc^{3+} > Ti^{4+}$, Z_{eff} increases on going from S^{2-} to Ti^{4+}.

4.92

$Mg(s) \rightarrow Mg(g)$	+147.7 kJ/mol
$Mg(g) \rightarrow Mg^+(g) + e^-$	+737.7 kJ/mol
$\frac{1}{2} F_2(g) \rightarrow F(g)$	+79 kJ/mol
$F(g) + e^- \rightarrow F^-(g)$	−328 kJ/mol
$Mg^+(g) + F^-(g) \rightarrow MgF(s)$	−930 kJ/mol

Sum = −294   kJ/mol for $Mg(s) + \frac{1}{2} F_2(g) \rightarrow MgF(s)$

$Mg(s) \rightarrow Mg(g)$	+147.7 kJ/mol
$Mg(g) \rightarrow Mg^+(g) + e^-$	+737.7 kJ/mol
$Mg^+(g) \rightarrow Mg^{2+}(g) + e^-$	+1450.7 kJ/mol
$F_2(g) \rightarrow 2 F(g)$	+158 kJ/mol
$2[F(g) + e^- \rightarrow F^-(g)]$	2(−328) kJ/mol
$Mg^{2+}(g) + 2 F^-(g) \rightarrow MgF_2(s)$	−2952 kJ/mol

Sum = −1114   kJ/mol for $Mg(s) + F_2(g) \rightarrow MgF_2(s)$

In the reaction of magnesium with fluorine, MgF_2 will form because the overall energy for the formation of MgF_2 is much more negative than for the formation of MgF.

4.93

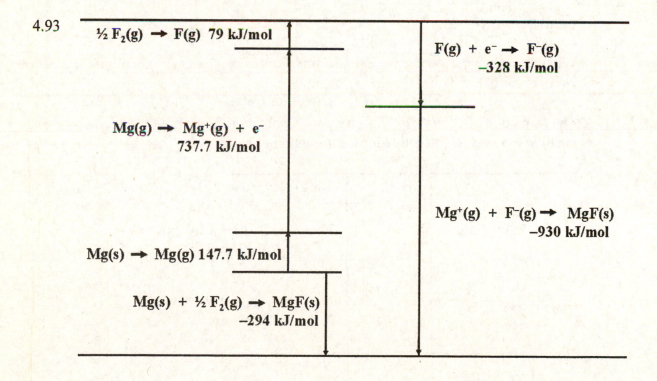

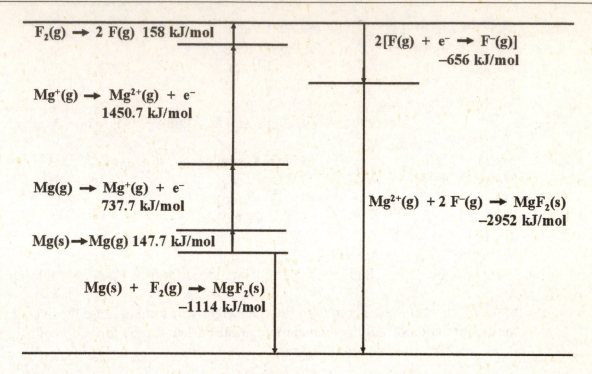

$F_2(g) \rightarrow 2 \ F(g) \ \ 158 \ kJ/mol$

$2[F(g) + e^- \rightarrow F^-(g)]$
$-656 \ kJ/mol$

$Mg^+(g) \rightarrow Mg^{2+}(g) + e^-$
$1450.7 \ kJ/mol$

$Mg(g) \rightarrow Mg^+(g) + e^-$
$737.7 \ kJ/mol$

$Mg^{2+}(g) + 2 \ F^-(g) \rightarrow MgF_2(s)$
$-2952 \ kJ/mol$

$Mg(s) \rightarrow Mg(g) \ \ 147.7 \ kJ/mol$

$Mg(s) + F_2(g) \rightarrow MgF_2(s)$
$-1114 \ kJ/mol$

4.94
$Na(s) \rightarrow Na(g)$ $\qquad +107.3 \ kJ/mol$
$Na(g) + e^- \rightarrow Na^-(g)$ $\qquad -52.9 \ kJ/mol$
$\frac{1}{2}[Cl_2(g) \rightarrow 2 \ Cl(g)]$ $\qquad +122 \quad kJ/mol$
$Cl(g) \rightarrow Cl^+(g) + e^-$ $\qquad +1251 \quad kJ/mol$
$Na^-(g) + Cl^+(g) \rightarrow ClNa(s)$ $\qquad \underline{-787 \quad kJ/mol}$
$\qquad\qquad$ Sum $= +640 \quad$ kJ/mol for $Na(s) + \frac{1}{2} \ Cl_2(g) \rightarrow Cl^+Na^-(s)$

The formation of Cl^+Na^- from its elements is not favored because the net energy change is positive whereas it is negative for the formation of Na^+Cl^-.

4.95

$Cl(g) \rightarrow Cl^+(g) + e^-$
$1251 \ kJ/mol$

$Na(g) + e^- \rightarrow Na^-(g)$
$-52.9 \ kJ/mol$

$Na^-(g) + Cl^+(g) \rightarrow ClNa(s)$
$-787 \ kJ/mol$

$\frac{1}{2} \ Cl_2(g) \rightarrow Cl(g)$
$122 \ kJ/mol$

$Na(s) + \frac{1}{2} \ Cl_2(g) \rightarrow ClNa(s)$
$640 \ kJ/mol$

$Na(s) \rightarrow Na(g)$
$107.3 \ kJ/mol$

4.96 (a) $2 \ Li(s) + 2 \ H_2O(l) \rightarrow 2 \ Li^+(aq) + 2 \ OH^-(aq) + H_2(g)$
$\qquad$ (b) $2 \ Li(s) + 2 \ NH_3(g) \rightarrow 2 \ LiNH_2(s) + H_2(g)$
$\qquad$ (c) $2 \ Li(s) + Br_2(l) \rightarrow 2 \ LiBr(s)$
$\qquad$ (d) $6 \ Li(s) + N_2(g) \rightarrow 2 \ Li_3N(s)$
$\qquad$ (e) $4 \ Li(s) + O_2(g) \rightarrow 2 \ Li_2O(s)$

4.97 (a) $F_2(g) + H_2(g) \rightarrow 2\,HF(g)$
(b) $F_2(g) + 2\,Na(s) \rightarrow 2\,NaF(s)$
(c) $F_2(g) + Sr(s) \rightarrow SrF_2(s)$

4.98 (a) Mg^{2+} and Cl^-, $MgCl_2$, magnesium chloride   (b) Ca^{2+} and O^{2-}, CaO, calcium oxide
(c) Li^+ and N^{3-}, Li_3N, lithium nitride   (d) Al^{3+} and O^{2-}, Al_2O_3, aluminum oxide

4.99 (a) Because X reacts by losing electrons, it is likely to be a metal.
(b) Because Y reacts by gaining electrons, it is likely to be a nonmetal.
(c) X_2Y_3
(d) X is likely to be in group 3A and Y is likely to be in group 6A.

4.100 When moving diagonally down and right on the periodic table, the increase in atomic radius caused by going to a larger shell is offset by a decrease caused by a higher Z_{eff}. Thus, there is little net change in the charge density.

4.101 TeO_4^{2-}, tellurate; TeO_3^{2-}, tellurite.
TeO_4^{2-} and TeO_3^{2-} are analogous to SO_4^{2-} and SO_3^{2-}.

4.102

$Mg(s) \rightarrow Mg(g)$	$+147.7$ kJ/mol
$Mg(g) \rightarrow Mg^+(g) + e^-$	$+738$ kJ/mol
$Mg^+(g) \rightarrow Mg^{2+}(g) + e^-$	$+1451$ kJ/mol
$\frac{1}{2}[O_2(g) \rightarrow 2\,O(g)]$	$+249.2$ kJ/mol
$O(g) + e^- \rightarrow O^-(g)$	-141.0 kJ/mol
$O^-(g) + e^- \rightarrow O^{2-}(g)$	E_{ea2}
$Mg^{2+}(g) + O^{2-}(g) \rightarrow MgO(s)$	-3791 kJ/mol
$Mg(s) + \frac{1}{2}O_2(g) \rightarrow MgO(s)$	-601.7 kJ/mol

$147.7 + 738 + 1451 + 249.2 - 141.0 + E_{ea2} - 3791 = -601.7$
$E_{ea2} = -147.7 - 738 - 1451 - 249.2 + 141.0 + 3791 - 601.7 = +744$ kJ/mol
Because E_{ea2} is positive, O^{2-} is not stable in the gas phase. It is stable in MgO because of the large lattice energy that results from the $+2$ and -2 charge of the ions and their small size.

4.103 (a) (i) Ra because it is farthest down (7th period) in the periodic table.
(ii) In because it is farthest down (5th period) in the periodic table.
(b) (i) Tl and Po are farthest down (6th period) but Tl is larger because it is to the left of Po and thus has the smaller ionization energy.
(ii) Cs and Bi are farthest down (6th period) but Cs is larger because it is to the left of Bi and thus has the smaller ionization energy.

4.104 (a) The more negative the E_{ea}, the greater the tendency of the atom to accept an electron, and the more stable the anion that results. Be, N, O, and F are all second row elements. F has the most negative E_{ea} of the group because the anion that forms, F^-, has a complete octet of electrons and its nucleus has the highest effective nuclear charge.
(b) Se^{2-} and Rb^+ are below O^{2-} and F^- in the periodic table and are the larger of the four.

Se^{2-} and Rb^+ are isoelectronic, but Rb^+ has the higher effective nuclear charge so it is smaller. Therefore Se^{2-} is the largest of the four ions.

4.105

$Ca(s) \rightarrow Ca(g)$		+178 kJ/mol
$Ca(g) \rightarrow Ca^+(g)$		+590 kJ/mol
$Ca^+(g) \rightarrow Ca^{2+}(g)$		+1145 kJ/mol
$2\,C(s) \rightarrow 2\,C(g)$		2(+717 kJ/mol)
$2\,C(g) \rightarrow C_2(g)$		–614 kJ/mol
$C_2(g) \rightarrow C_2^-(g)$		–315 kJ/mol
$C_2^-(g) \rightarrow C_2^{2-}(g)$		+410 kJ/mol
$Ca^{2+}(g) + C_2^{2-}(g) \rightarrow CaC_2(s)$		–U
$Ca(s) + 2\,C(s) \rightarrow CaC_2(s)$		–60 kJ/mol

$-U = -60 - 178 - 590 - 1145 - 2(717) + 614 + 315 - 410 = -2888$ kJ/mol
$U = 2888$ kJ/mol

4.106

$Cr(s) \rightarrow Cr(g)$		+397 kJ/mol
$Cr(g) \rightarrow Cr^+(g)$		+652 kJ/mol
$Cr^+(g) \rightarrow Cr^{2+}(g)$		+1588 kJ/mol
$Cr^{2+}(g) \rightarrow Cr^{3+}(g)$		+2882 kJ/mol
$\frac{1}{2}(I_2(s) \rightarrow I_2(g))$		+62/2 kJ/mol
$\frac{1}{2}(I_2(g) \rightarrow 2\,I(g))$		+151/2 kJ/mol
$I(g) + e^- \rightarrow I^-(g)$		–295 kJ/mol
$Cl_2(g) \rightarrow 2\,Cl(g)$		+243 kJ/mol
$2(Cl(g) + e^- \rightarrow Cl^-(g))$		2(–349) kJ/mol
$Cr^{3+}(g) + 2\,Cl^-(g) + I^-(g) \rightarrow CrCl_2I(s)$		–U
$Cr(s) + Cl_2(g) + \frac{1}{2}\,I_2(g) \rightarrow CrCl_2I(s)$		–420 kJ/mol

$-U = -420 - 397 - 652 - 1588 - 2882 - 62/2 - 151/2 + 295 - 243 + 2(349) = -5295.5$ kJ/mol
$U = 5295$ kJ/mol

Multi-Concept Problems

4.107 (a) $E = (703\text{ kJ/mol})(1000\text{ J/1 kJ})/(6.022 \times 10^{23}\text{ photons/mol}) = 1.17 \times 10^{-18}$ J/photon

$E = \dfrac{hc}{\lambda}$

$\lambda = \dfrac{hc}{E} = \dfrac{(6.626 \times 10^{-34}\text{ J·s})(3.00 \times 10^8\text{ m/s})}{1.17 \times 10^{-18}\text{ J}} = 1.70 \times 10^{-7}\text{ m} = 170 \times 10^{-9}\text{ m} = 170\text{ nm}$

(b) Bi   [Xe] $6s^2\,4f^{14}\,5d^{10}\,6p^3$
     Bi^+ [Xe] $6s^2\,4f^{14}\,5d^{10}\,6p^2$

(c) $n = 6,\ l = 1$

(d) Element 115 would be directly below Bi in the periodic table. The valence electron is farther from the nucleus and less strongly held than in Bi. The ionization energy for element 115 would be less than that for Bi.

4.108 (a) Fe   [Ar] $4s^2 3d^6$
        Fe^{2+} [Ar] $3d^6$
        Fe^{3+} [Ar] $3d^5$

(b) A 3d electron is removed on going from Fe^{2+} to Fe^{3+}. For the 3d electron, n = 3 and l = 2.

(c) E(J/photon) = 2952 kJ/mol x $\dfrac{1 \text{ mol photons}}{6.022 \times 10^{23} \text{ photons}}$ x $\dfrac{1000 \text{ J}}{1 \text{ kJ}}$ = 4.90 x 10^{-18} J/photon

$$E = \frac{hc}{\lambda}$$

$$\lambda = \frac{hc}{E} = \frac{(6.626 \times 10^{-34} \text{ J} \cdot \text{s})(3.00 \times 10^8 \text{ m/s})}{4.90 \times 10^{-18} \text{ J}} = 4.06 \times 10^{-8} \text{ m} = 40.6 \times 10^{-9} \text{ m} = 40.6 \text{ nm}$$

(d) Ru is directly below Fe in the periodic table and the two metals have similar electron configurations. The electron removed from Ru to go from Ru^{2+} to Ru^{3+} is a 4d electron. The electron with the higher principal quantum number, n = 4, is farther from the nucleus, less tightly held, and requires less energy to remove.

4.109 (a) 58.4 nm = 58.4 x 10^{-9} m

$$E(\text{photon}) = 6.626 \times 10^{-34} \text{ J} \cdot \text{s} \times \frac{3.00 \times 10^8 \text{ m/s}}{58.4 \times 10^{-9} \text{ m}} \times \frac{1 \text{ kJ}}{1000 \text{ J}} \times \frac{6.022 \times 10^{23}}{\text{mol}} = 2050 \text{ kJ/mol}$$

$$E_K = E(\text{electron}) = \tfrac{1}{2}(9.109 \times 10^{-31} \text{ kg})(2.450 \times 10^6 \text{ m/s})^2 \left(\frac{1 \text{ kJ}}{1000 \text{ J}} \right) \left(\frac{6.022 \times 10^{23}}{\text{mol}} \right)$$

E_K = 1646 kJ/mol
E(photon) = E_i + E_K; E_i = E(photon) – E_K = 2050 – 1646 = 404 kJ/mol

(b) 142 nm = 142 x 10^{-9} m

$$E(\text{photon}) = 6.626 \times 10^{-34} \text{ J} \cdot \text{s} \times \frac{3.00 \times 10^8 \text{ m/s}}{142 \times 10^{-9} \text{ m}} \times \frac{1 \text{ kJ}}{1000 \text{ J}} \times \frac{6.022 \times 10^{23}}{\text{mol}} = 843 \text{ kJ/mol}$$

$$E_K = E(\text{electron}) = \tfrac{1}{2}(9.109 \times 10^{-31} \text{ kg})(1.240 \times 10^6 \text{ m/s})^2 \left(\frac{1 \text{ kJ}}{1000 \text{ J}} \right) \left(\frac{6.022 \times 10^{23}}{\text{mol}} \right)$$

E_K = 422 kJ/mol
E(photon) = E_i + E_K; E_i = E(photon) – E_K = 843 – 422 = 421 kJ/mol

5

Covalent Bonds and Molecular Structure

5.1

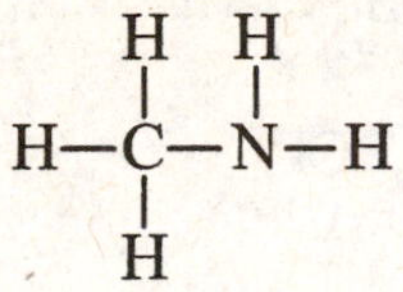

$$\begin{array}{ccc} & H & H \\ & | & | \\ H - & C - N - H \\ & | & \\ & H & \end{array}$$

5.2 $C_5H_{11}NO_2S$

5.3 $C_9H_{13}NO_3$

5.4   (a) $SiCl_4$ chlorine $EN = 3.0$
                         silicon $\underline{EN = 1.8}$
                                       $\Delta EN = 1.2$ The Si–Cl bond is polar covalent.

      (b) $CsBr$ bromine $EN = 2.8$
                        cesium $\underline{EN = 0.7}$
                                         $\Delta EN = 2.1$ The Cs^+Br^- bond is ionic.

      (c) $FeBr_3$ bromine $EN = 2.8$
                        iron $\underline{EN = 1.8}$
                                         $\Delta EN = 1.0$ The Fe–Br bond is polar covalent.

      (d) CH_4 carbon $EN = 2.5$
                        hydrogen $\underline{EN = 2.1}$
                                         $\Delta EN = 0.4$ The C–H bond is polar covalent.

5.5   (a) CCl_4 chlorine $EN = 3.0$
                        carbon $\underline{EN = 2.5}$
                                         $\Delta EN = 0.5$

      (b) $BaCl_2$ chlorine $EN = 3.0$
                        barium $\underline{EN = 0.9}$
                                         $\Delta EN = 2.1$

      (c) $TiCl_3$ chlorine $EN = 3.0$
                        titanium $\underline{EN = 1.5}$
                                         $\Delta EN = 1.5$

      (d) Cl_2O oxygen $EN = 3.5$
                        chlorine $\underline{EN = 3.0}$
                                         $\Delta EN = 0.5$

Increasing ionic character: $CCl_4 \sim ClO_2 < TiCl_3 < BaCl_2$

5.6 H is positively polarized (blue). O is negatively polarized (red). This is consistent with the electronegativity values for O (3.5) and H (2.1). The more negatively polarized atom should be the one with the larger electronegativity.

5.7 (a) NCl_3, nitrogen trichloride (b) P_4O_6, tetraphosphorus hexoxide
(c) S_2F_2, disulfur difluoride (d) SeO_2, selenium dioxide

5.8 (a) disulfur dichloride, S_2Cl_2 (b) iodine monochloride, ICl
(c) nitrogen triiodide, NI_3

5.9 (a) PCl_5, phosphorus pentachloride (b) N_2O, dinitrogen monoxide

5.10 (a)  (b)

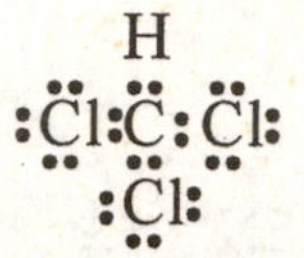

5.11
hydronium ion 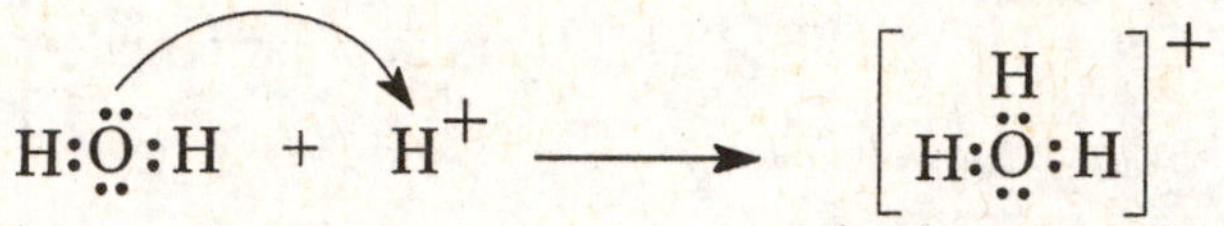

5.12 (a) 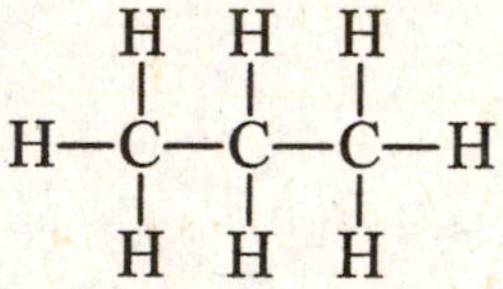 (b) 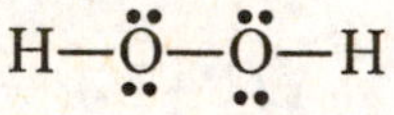 (c)

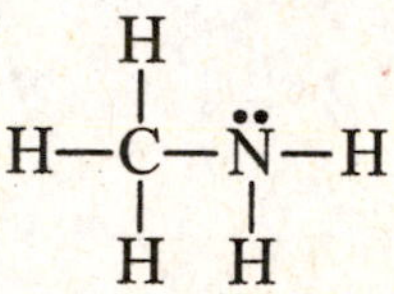

(d) 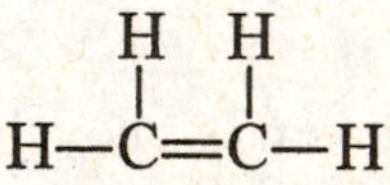 (e)  (f)

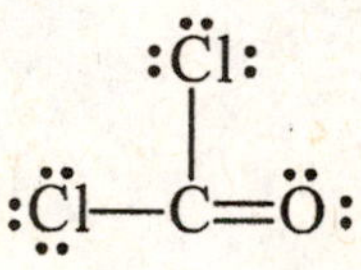

5.13

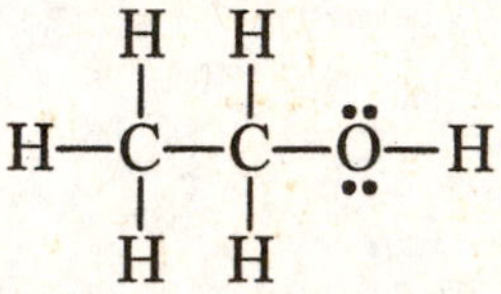

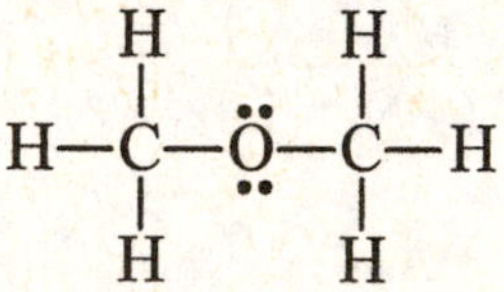

5.14 Molecular formula: $C_4H_5N_3O$;

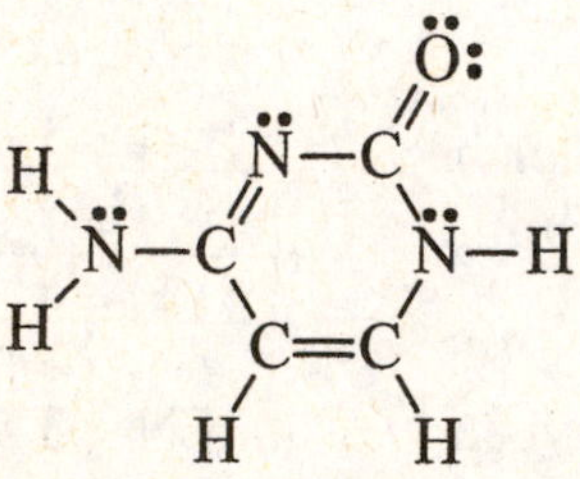

5.15

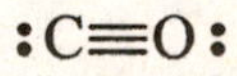

5.16 (a) 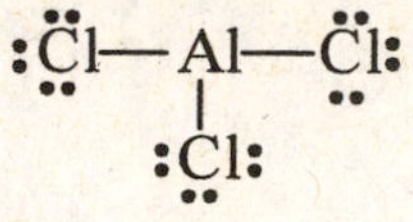 :Cl—Al—Cl: (b) 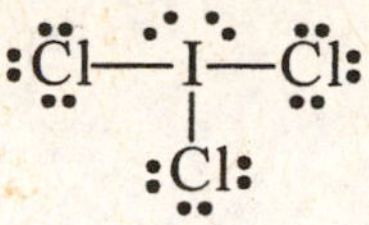 :Cl—I—Cl:
             | |
           :Cl: :Cl:

(c) 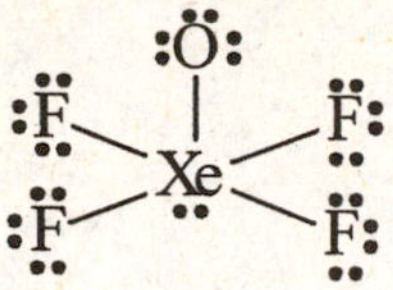   :O: (d) 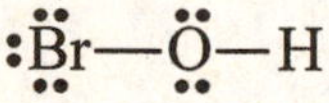 :Br—O—H
        :F   F:
          Xe
        :F   F:

5.17 (a) 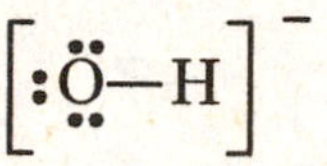 [:O—H]⁻   (b) 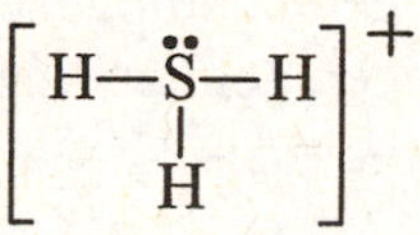 [H—S—H]⁺   (c) 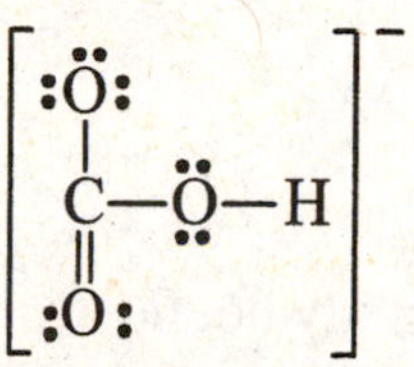 [:O:]⁻
                              | C—O—H
                              H   :O:

(d) 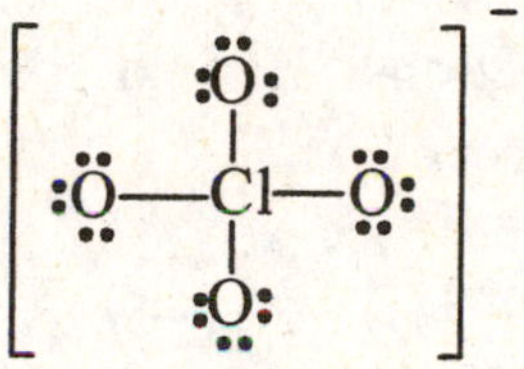 [:O:]⁻
      :O—Cl—O:
           :O:

5.18   :N=N=O: ⟷ :N≡N—O:

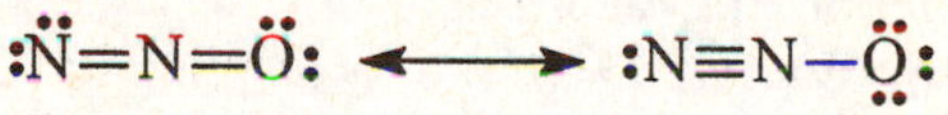

5.19 (a) 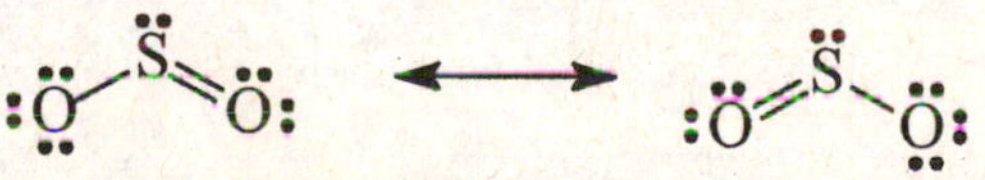   :O—S=O: ⟷ :O=S—O:

(b) 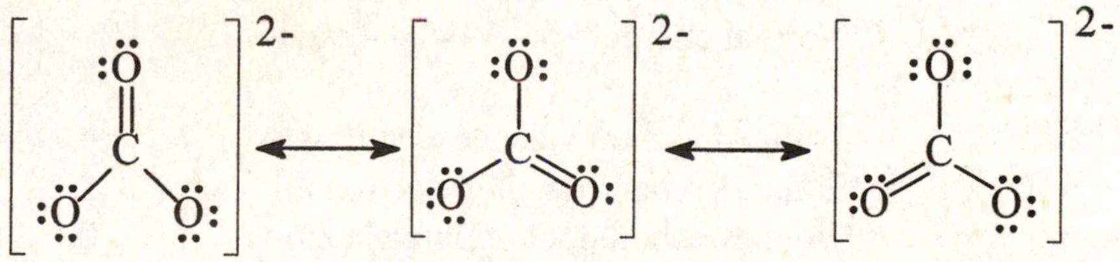 [:O:]²⁻   [:O:]²⁻   [:O:]²⁻
         C   ⟷   C   ⟷   C
      :O: O: :O: O: :O: O:

(c) [H—C=O:]⁻ ⟷ [H—C—O:]⁻
          O: O:

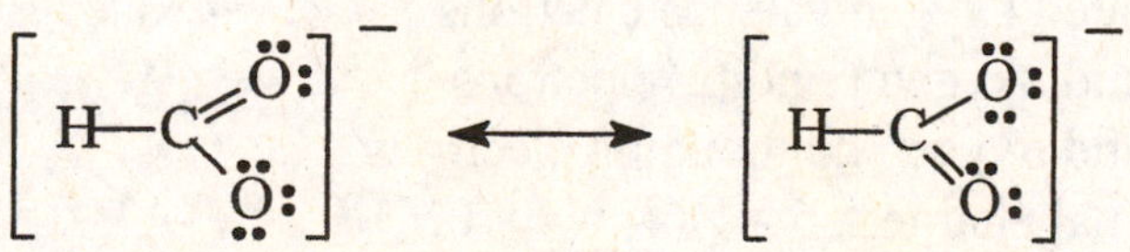

(d) 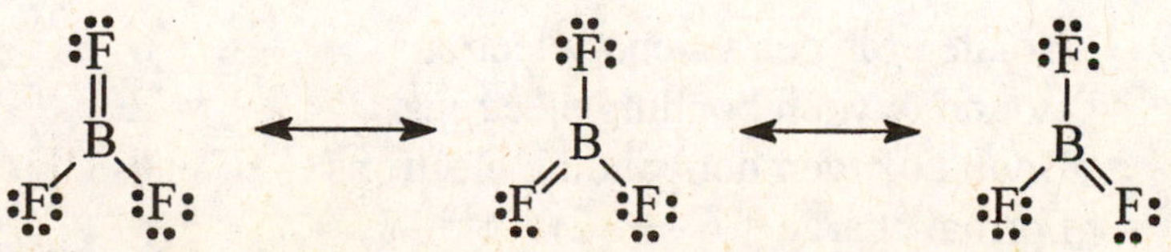   :F: :F: :F:
       B   ⟷   B   ⟷   B
    :F: :F: :F: :F: :F: :F:

5.20

5.21   For nitrogen:

	Isolated nitrogen valence electrons	5
	Bound nitrogen bonding electrons	8
	Bound nitrogen nonbonding electrons	0

Formal charge = $5 - \frac{1}{2}(8) - 0 = +1$

For singly bound
oxygen:

	Isolated oxygen valence electrons	6
	Bound oxygen bonding electrons	2
	Bound oxygen nonbonding electrons	6

Formal charge = $6 - \frac{1}{2}(2) - 6 = -1$

For doubly bound
oxygen:

	Isolated oxygen valence electrons	6
	Bound oxygen bonding electrons	4
	Bound oxygen nonbonding electrons	4

Formal charge = $6 - \frac{1}{2}(4) - 4 = 0$

5.22   (a) $\left[:\overset{..}{N}=C=\overset{..}{O}: \right]^{-}$

For nitrogen:

	Isolated nitrogen valence electrons	5
	Bound nitrogen bonding electrons	4
	Bound nitrogen nonbonding electrons	4

Formal charge = $5 - \frac{1}{2}(4) - 4 = -1$

For carbon:

	Isolated carbon valence electrons	4
	Bound carbon bonding electrons	8
	Bound carbon nonbonding electrons	0

Formal charge = $4 - \frac{1}{2}(8) - 0 = 0$

For oxygen:

	Isolated oxygen valence electrons	6
	Bound oxygen bonding electrons	4
	Bound oxygen nonbonding electrons	4

Formal charge = $6 - \frac{1}{2}(4) - 4 = 0$

(b) $:\overset{..}{\underset{..}{O}}-\overset{..}{O}=\overset{..}{O}:$

For left oxygen:

	Isolated oxygen valence electrons	6
	Bound oxygen bonding electrons	2
	Bound oxygen nonbonding electrons	6

Formal charge = $6 - \frac{1}{2}(2) - 6 = -1$

For central oxygen:	Isolated oxygen valence electrons	6
	Bound oxygen bonding electrons	6
	Bound oxygen nonbonding electrons	2

Formal charge = $6 - \frac{1}{2}(6) - 2 = +1$

For right oxygen:	Isolated oxygen valence electrons	6
	Bound oxygen bonding electrons	4
	Bound oxygen nonbonding electrons	4

Formal charge = $6 - \frac{1}{2}(4) - 4 = 0$

5.23

	Number of Bonded Atoms	Number of Lone Pairs	Shape
(a) O_3	2	1	bent
(b) H_3O^+	3	1	trigonal pyramidal
(c) XeF_2	2	3	linear
(d) PF_6^-	6	0	octahedral
(e) $XeOF_4$	5	1	square pyramidal
(f) AlH_4^-	4	0	tetrahedral
(g) BF_4^-	4	0	tetrahedral
(h) $SiCl_4$	4	0	tetrahedral
(i) ICl_4^-	4	2	square planar
(j) $AlCl_3$	3	0	trigonal planar

5.24

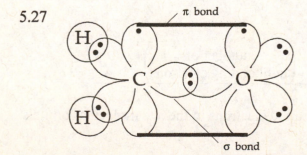

5.25　(a) tetrahedral　　(b) seesaw

5.26

Each C is sp^3 hybridized. The C–C bond is formed by the overlap of one singly occupied sp^3 hybrid orbital from each C. The C–H bonds are formed by the overlap of one singly occupied sp^3 orbital on C with a singly occupied H 1s orbital.

5.27

π bond

σ bond

The carbon in formaldehyde is sp^2 hybridized.

5.28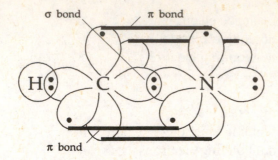

In HCN the carbon is sp hybridized.

5.29

In CO_2 the carbon is sp hybridized.

5.30

In Cl_2CO the carbon is sp2 hybridized.

5.31   (a) sp^2 (b) sp   (c) sp^3

5.32   For He_2^+ σ^*_{1s} $\underline{\uparrow\quad}$

                σ_{1s} $\underline{\uparrow\downarrow}$

He_2^+ Bond order = $\dfrac{\left(\begin{array}{c}\text{number of}\\\text{bonding electrons}\end{array}\right) - \left(\begin{array}{c}\text{number of}\\\text{antibonding electrons}\end{array}\right)}{2} = \dfrac{2-1}{2} = 1/2$

He_2^+ should be stable with a bond order of 1/2.

5.33   For B_2

σ^*_{2p} $\underline{\quad}$

π^*_{2p} $\underline{\quad}\ \underline{\quad}$

σ_{2p} $\underline{\quad}$

π_{2p} $\underline{\uparrow}\ \underline{\uparrow}$

σ^*_{2s} $\underline{\uparrow\downarrow}$

σ_{2s} $\underline{\uparrow\downarrow}$

B_2 Bond order = $\dfrac{\left(\begin{array}{c}\text{number of}\\\text{bonding electrons}\end{array}\right) - \left(\begin{array}{c}\text{number of}\\\text{antibonding electrons}\end{array}\right)}{2} = \dfrac{4-2}{2} = 1$

B_2 is paramagnetic because it has two unpaired electrons in the π_{2p} molecular orbitals.

For C_2

σ^*_{2p} ——

π^*_{2p} —— ——

σ_{2p} ——

π_{2p} ↑↓ ↑↓

σ^*_{2s} ↑↓

σ_{2s} ↑↓

C_2 Bond order = $\dfrac{6-2}{2}$ = 2; C_2 is diamagnetic because all electrons are paired.

5.34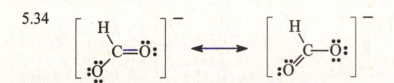

5.35 The mirror image of molecule (a) has the same shape as (a) and is identical to it in all respects, so there is no handedness associated with it. The mirror image of molecule (b) is different than (b) so there is a handedness to this molecule.

Key Concept Problems

5.36 (a) alanine, $C_3H_7NO_2$ (b) ethylene glycol, $C_2H_6O_2$ (c) acetic acid, $C_2H_4O_2$

5.37 As the electrostatic potential maps are drawn, the Li and Cl are at the tops of each map. The red area is for a negatively polarized region (associated with Cl). The blue area is for a positively polarized region (associated with Li). Map (a) is for CH_3Cl and Map (b) is for CH_3Li.

5.38 (a) square pyramidal   (b) trigonal pyramidal
     (c) square planar   (d) trigonal planar

5.39 (a) trigonal bipyramidal   (b) tetrahedral
     (c) square pyramidal (4 ligands in the horizontal plane, including one hidden)

5.40 Molecular model (c) does not have a tetrahedral central atom. It is square planar.

5.41 The expected hybridizations of C and N in urea are sp^2 and sp^3, respectively. The expected bond angles are (i) N-C-O and N-C-N, ~120°, and (ii) C-N-H and H-N-H, ~109°. Based on the molecular model, the C and N are both sp^2 hybridized and all bond angles are ~120°.

5.42 (a) $C_8H_9NO_2$
(b), (c) & (d)

5.43 (a) $C_{13}H_{10}N_2O_4$
(b), (c) & (d)

All carbons that have only single bonds are sp³ hybridized and have tetrahedral geometries. All carbons that have double bonds are sp² hybridized and have trigonal planar geometries.

Section Problems
Electronegativity and Polar Covalent Bonds (Sections 5.4–5.5)

5.44 Electronegativity increases from left to right across a period and decreases down a group.

5.45 Z = 119 would be below francium and have a very low electronegativity.

5.46 K < Li < Mg < Pb < C < Br

5.47 Cl > C > Cu > Ca > Cs

5.48 (a) HF   fluorine   EN = 4.0
               hydrogen $\underline{EN = 2.1}$
                              ΔEN = 1.9   HF is polar covalent.

      (b) HI   iodine   EN = 2.5
               hydrogen $\underline{EN = 2.1}$
                              ΔEN = 0.4   HI is polar covalent.

      (c) $PdCl_2$ chlorine   EN = 3.0
               palladium $\underline{EN = 2.2}$
                              ΔEN = 0.8 $PdCl_2$ is polar covalent.

      (d) BBr_3 bromine   EN = 2.8
               boron $\underline{EN = 2.0}$
                              ΔEN = 0.8 BBr_3 is polar covalent.

(e) NaOH   Na$^+$ – OH$^-$ is ionic
    OH$^-$ oxygen   EN = 3.5
              hydrogen   EN = 2.1
                          ΔEN = 1.4   OH$^-$ is polar covalent.

(f) CH$_3$Li   lithium   EN = 1.0
                 carbon   EN = 2.5
                             ΔEN = 1.5   CH$_3$Li is polar covalent.

5.49   The electronegativity for each element is shown in parentheses.
        (a) C (2.5), H (2.1), Cl (3.0): The C–Cl bond is more polar than the C–H bond because of the larger electronegativity difference between the bonded atoms.
        (b) Si (1.8), Li (1.0), Cl (3.0): The Si–Cl bond is more polar than the Si–Li bond because of the larger electronegativity difference between the bonded atoms.
        (c) N (3.0), Cl (3.0), Mg (1.2): The N–Mg bond is more polar than the N–Cl bond because of the larger electronegativity difference between the bonded atoms.

5.50   (a) $\overset{\delta-}{C} - \overset{\delta+}{H}$ $\overset{\delta+}{C} - \overset{\delta-}{Cl}$ (b) $\overset{\delta-}{Si} - \overset{\delta+}{Li}$ $\overset{\delta+}{Si} - \overset{\delta-}{Cl}$

        (c) N – Cl $\overset{\delta-}{N} - \overset{\delta+}{Mg}$

5.51   (a) $\overset{\delta-}{F} - \overset{\delta+}{H}$ (b) $\overset{\delta-}{I} - \overset{\delta+}{H}$ (c) $\overset{\delta-}{Cl} - \overset{\delta+}{Pd}$

        (d) $\overset{\delta-}{Br} - \overset{\delta+}{B}$ (e) $\overset{\delta-}{O} - \overset{\delta+}{H}$

5.52   (a) Phosphorus trichloride   (b) Dinitrogen trioxide   (c) Tetraphosphorus heptoxide
        (d) Bromine trifluoride   (e) Nitrogen trichloride   (f) Tetraphosphorus hexoxide;
        (g) Disulfur difluoride   (h) Selenium dioxide

5.53   (a) S_2Cl_2 (b) ICl   (c) NI_3 (d) Cl_2O (e) ClO_3 (f) S_4N_4

Electron-Dot Structures and Resonance (Sections 5.6–5.8)

5.54   The octet rule states that main-group elements tend to react so that they attain a noble gas electron configuration with filled s and p sublevels (8 electrons) in their valence electron shells. The transition metals are characterized by partially filled d orbitals that can be used to expand their valence shell beyond the normal octet of electrons.

5.55   (a) AlCl$_3$ Al has only 6 electrons around it.   (b) PCl$_5$ P has 10 electrons around it.

5.56 (a)

$$\ddot{B}r - \underset{\underset{:\ddot{B}r:}{|}}{\overset{:\ddot{B}r:}{\underset{}{C}}} - \ddot{B}r:$$

(b) $:\ddot{C}l - \underset{\underset{:\ddot{C}l:}{|}}{\ddot{N}} - \ddot{C}l:$

(c)

$$H - \underset{\underset{H}{|}}{\overset{H}{\underset{}{C}}} - \underset{\underset{H}{|}}{\overset{H}{\underset{}{C}}} - \ddot{C}l:$$

(d) $\left[\underset{\underset{:\ddot{F}:}{|}}{:\ddot{F} - B - \ddot{F}:} \right]^{-}$

(e) $\left[:\ddot{O} - \ddot{O}: \right]^{2-}$

(f) $\left[:N \equiv O: \right]^{+}$

5.57 (a) $:\ddot{C}l - \underset{\underset{:\ddot{C}l:}{|}}{\ddot{S}b} - \ddot{C}l:$

(b) $:\ddot{F} - \ddot{K}r - \ddot{F}:$

(c)

$$\overset{:\ddot{O}}{\underset{}{\cdot \ddot{C}l - \ddot{O}:}}$$

(d) $\overset{:\ddot{F}:}{\underset{:\ddot{F}:}{:\ddot{F} - \underset{|}{\overset{|}{P}} - \ddot{F}:}}$ with F above and below

(e)

$$H - \ddot{O} - \underset{\underset{O}{\overset{:\ddot{O}:}{||}}}{P} - \ddot{O} - H$$
$$\underset{H}{|}$$

(f) $\overset{:\ddot{O}:}{\underset{:\ddot{C}l:}{:\ddot{S}e - \ddot{C}l:}}$

5.58 (a) $H - N \equiv N - \ddot{N}: \longleftrightarrow H - \ddot{N} = N = \ddot{N}: \longleftrightarrow H - \ddot{N} - N \equiv N:$

(b)

$$\overset{:\ddot{O}}{\underset{:\ddot{O}:}{:\ddot{O} - S}} \longleftrightarrow \overset{:\ddot{O}:}{\underset{:\ddot{O}:}{:\ddot{O} - S - \ddot{O}:}} \longleftrightarrow \overset{:\ddot{O}:}{\underset{:\ddot{O}:}{:\ddot{O} - S = \ddot{O}:}}$$

(c) $\left[:N \equiv C - \ddot{S}: \right]^{-} \longleftrightarrow \left[:\ddot{N} = C = \ddot{S}: \right]^{-} \longleftrightarrow \left[:\ddot{N} - C \equiv S: \right]^{-}$

5.59 (a) $:N \equiv N - \ddot{O}: \longleftrightarrow :\ddot{N} = N = \ddot{O}: \longleftrightarrow :\ddot{N} - N \equiv O:$

(b) $\cdot \ddot{N} = \ddot{O}: \longleftrightarrow :\ddot{N} = \ddot{O} \cdot$

(c)

$$\cdot N \overset{\ddot{O}:}{\underset{\ddot{O}:}{}} \longleftrightarrow \cdot N \overset{\ddot{O}:}{\underset{\ddot{O}:}{}}$$

(d)

$$:\ddot{O} = \ddot{N} - \underset{\underset{\ddot{O}:}{}}{\overset{:\ddot{O}:}{N}} \longleftrightarrow :\ddot{O} = \ddot{N} - \underset{\underset{\ddot{O}:}{}}{\overset{\ddot{O}:}{N}} \longleftrightarrow :\ddot{O} = N = \underset{\underset{\ddot{O}:}{}}{\overset{\ddot{O}:}{N}}$$

5.60

$$:\overset{\displaystyle .\,.}{O}\quad\overset{\displaystyle .\,.}{O}:$$
$$\parallel\qquad\parallel$$
$$H-\overset{..}{\underset{..}{O}}-C-C-\overset{..}{\underset{..}{O}}-H$$

5.61 $:\overset{..}{S}=C=\overset{..}{S}:$; CS_2 has two double bonds.

5.62 (a) yes (b) yes (c) yes (d) yes

5.63 (a) yes (b) no (c) yes

5.64 (a) The anion has 32 valence electrons. Each Cl has seven valence electrons (28 total). The minus one charge on the anion accounts for one valence electron. This leaves three valence electrons for X. X is Al.
(b) The cation has eight valence electrons. Each H has one valence electron (4 total). X is left with four valence electrons. Since this is a cation, one valence electron was removed from X. X has five valence electrons. X is P.

5.65 (a) This fourth-row element has six valence electrons. It is Se.
(b) This fourth-row element has eight valence electrons. It is Kr.

5.66 (a)

$$\overset{\displaystyle\overset{..}{O}:}{\parallel}$$
$$:\overset{..}{\underset{..}{Cl}}-C-\overset{..}{\underset{..}{O}}-\overset{\overset{\textstyle H}{|}}{\underset{\underset{\textstyle H}{|}}{C}}-H$$

(b)

$$H-\overset{\overset{\textstyle H}{|}}{\underset{\underset{\textstyle H}{|}}{C}}-C\equiv C-H$$

5.67 (a)

$$\overset{\displaystyle :\overset{..}{O}\quad H}{}$$
$$\overset{\parallel\quad |}{}$$
$$H-C-\overset{|}{\underset{..}{N}}-H$$

(b)

$$H-\overset{\overset{\textstyle H}{|}}{\underset{\underset{\textstyle H}{|}}{C}}-C\equiv N-\overset{..}{\underset{..}{O}}:$$

Formal Charges (Section 5.9)

5.68 $:C\equiv O:$

For carbon:	Isolated carbon valence electrons	4
	Bound carbon bonding electrons	6
	Bound carbon nonbonding electrons	2
	Formal charge = $4 - \frac{1}{2}(6) - 2 = -1$	

For oxygen:	Isolated oxygen valence electrons	6
	Bound oxygen bonding electrons	6
	Bound oxygen nonbonding electrons	2
	Formal charge = $6 - \frac{1}{2}(6) - 2 = +1$	

5.69   (a)

$$
\begin{array}{c}
\text{H} \\
| \\
\text{H}-\ddot{\text{N}}-\ddot{\text{O}}-\text{H}
\end{array}
$$

For hydrogen:	Isolated hydrogen valence electrons	1
	Bound hydrogen bonding electrons	2
	Bound hydrogen nonbonding electrons	0
	Formal charge = $1 - \frac{1}{2}(2) - 0 = 0$	

For nitrogen:	Isolated nitrogen valence electrons	5
	Bound nitrogen bonding electrons	6
	Bound nitrogen nonbonding electrons	2
	Formal charge = $5 - \frac{1}{2}(6) - 2 = 0$	

For oxygen:	Isolated oxygen valence electrons	6
	Bound oxygen bonding electrons	4
	Bound oxygen nonbonding electrons	4
	Formal charge = $6 - \frac{1}{2}(4) - 4 = 0$	

(b)

$$
\left[
\begin{array}{c}
\text{H} \\
| \\
\text{H}-\ddot{\text{N}}-\overset{|}{\underset{|}{\text{C}}}-\text{H} \\
| \\
\text{H}
\end{array}
\right]^{-}
$$

For hydrogen:	Isolated hydrogen valence electrons	1
	Bound hydrogen bonding electrons	2
	Bound hydrogen nonbonding electrons	0
	Formal charge = $1 - \frac{1}{2}(2) - 0 = 0$	

For nitrogen:	Isolated nitrogen valence electrons	5
	Bound nitrogen bonding electrons	4
	Bound nitrogen nonbonding electrons	4
	Formal charge = $5 - \frac{1}{2}(4) - 4 = -1$	

For carbon:	Isolated carbon valence electrons	4
	Bound carbon bonding electrons	8
	Bound carbon nonbonding electrons	0
	Formal charge = $4 - \frac{1}{2}(8) - 0 = 0$	

(c)

$$
\begin{array}{c}
:\ddot{\text{O}}: \\
\| \\
:\ddot{\text{Cl}}-\text{P}-\ddot{\text{Cl}}: \\
| \\
:\ddot{\text{Cl}}:
\end{array}
$$

For chlorine:	Isolated chlorine valence electrons	7
	Bound chlorine bonding electrons	2
	Bound chlorine nonbonding electrons	6
	Formal charge = $7 - \frac{1}{2}(2) - 6 = 0$	

For oxygen:	Isolated oxygen valence electrons	6
Bound oxygen bonding electrons	2
Bound oxygen nonbonding electrons	6
Formal charge = $6 - \frac{1}{2}(2) - 6 = -1$

For phosphorus:	Isolated phosphorus valence electrons	5
Bound phosphorus bonding electrons	8
Bound phosphorus nonbonding electrons	0
Formal charge = $5 - \frac{1}{2}(8) - 0 = +1$

5.70

$$\left[\ddot{:O}-\overset{..}{\underset{..}{Cl}}-\ddot{O:}\right]^{-}$$

For both oxygens:	Isolated oxygen valence electrons	6
Bound oxygen bonding electrons	2
Bound oxygen nonbonding electrons	6
Formal charge = $6 - \frac{1}{2}(2) - 6 = -1$

For chlorine:	Isolated chlorine valence electrons	7
Bound chlorine bonding electrons	4
Bound chlorine nonbonding electrons	4
Formal charge = $7 - \frac{1}{2}(4) - 4 = +1$

$$\left[\ddot{:O}-\overset{..}{\underset{..}{Cl}}=\ddot{O}\right]^{-}$$

For left oxygen:	Isolated oxygen valence electrons	6
Bound oxygen bonding electrons	2
Bound oxygen nonbonding electrons	6
Formal charge = $6 - \frac{1}{2}(2) - 6 = -1$

For right oxygen:	Isolated oxygen valence electrons	6
Bound oxygen bonding electrons	4
Bound oxygen nonbonding electrons	4
Formal charge = $6 - \frac{1}{2}(4) - 4 = 0$

For chlorine:	Isolated chlorine valence electrons	7
Bound chlorine bonding electrons	6
Bound chlorine nonbonding electrons	4
Formal charge = $7 - \frac{1}{2}(6) - 4 = 0$

5.71

$$H\ddot{O}-\overset{\displaystyle :O:}{\underset{..}{\overset{\|}{S}}}-\ddot{O}H$$

For sulfur:	Isolated sulfur valence electrons	6
Bound sulfur bonding electrons	8
Bound sulfur nonbonding electrons	2
Formal charge = $6 - \frac{1}{2}(8) - 2 = 0$

For doubly bound oxygen:	Isolated oxygen valence electrons	6
	Bound oxygen bonding electrons	4
	Bound oxygen nonbonding electrons	4
	Formal charge = $6 - \frac{1}{2}(4) - 4 = 0$	

For oxygen bound to hydrogen:	Isolated oxygen valence electrons	6
	Bound oxygen bonding electrons	4
	Bound oxygen nonbonding electrons	4
	Formal charge = $6 - \frac{1}{2}(4) - 4 = 0$	

For hydrogen:	Isolated hydrogen valence electrons	1
	Bound hydrogen bonding electrons	2
	Bound hydrogen nonbonding electrons	0
	Formal charge = $1 - \frac{1}{2}(2) - 0 = 0$	

$$\ddot{O}:$$
$$H\ddot{O}—S—\ddot{O}H$$

For sulfur:	Isolated sulfur valence electrons	6
	Bound sulfur bonding electrons	6
	Bound sulfur nonbonding electrons	2
	Formal charge = $6 - \frac{1}{2}(6) - 2 = +1$	

For oxygen not bound to hydrogen:	Isolated oxygen valence electrons	6
	Bound oxygen bonding electrons	2
	Bound oxygen nonbonding electrons	6
	Formal charge = $6 - \frac{1}{2}(2) - 6 = -1$	

For oxygen bound to hydrogen:	Isolated oxygen valence electrons	6
	Bound oxygen bonding electrons	4
	Bound oxygen nonbonding electrons	4
	Formal charge = $6 - \frac{1}{2}(4) - 4 = 0$	

For hydrogen:	Isolated hydrogen valence electrons	1
	Bound hydrogen bonding electrons	2
	Bound hydrogen nonbonding electrons	0
	Formal charge = $1 - \frac{1}{2}(2) - 0 = 0$	

5.72 (a)

$$\begin{array}{c} H \\ \diagdown \\ C=N=\ddot{N}: \\ \diagup \\ H \end{array}$$

For hydrogen:	Isolated hydrogen valence electrons	1
	Bound hydrogen bonding electrons	2
	Bound hydrogen nonbonding electrons	0
	Formal charge = $1 - \frac{1}{2}(2) - 0 = 0$	

For nitrogen: (central)	Isolated nitrogen valence electrons	5
	Bound nitrogen bonding electrons	8
	Bound nitrogen nonbonding electrons	0
	Formal charge = $5 - \frac{1}{2}(8) - 0 = +1$	

For nitrogen: (terminal)	Isolated nitrogen valence electrons	5
	Bound nitrogen bonding electrons	4
	Bound nitrogen nonbonding electrons	4
	Formal charge = $5 - \frac{1}{2}(4) - 4 = -1$	

For carbon:	Isolated carbon valence electrons	4
	Bound carbon bonding electrons	8
	Bound carbon nonbonding electrons	0
	Formal charge = $4 - \frac{1}{2}(8) - 0 = 0$	

(b)

$$\begin{array}{c} H \\ \diagdown \\ \diagup\ C - \ddot{N} = \ddot{N} \colon \\ H \end{array}$$

For hydrogen:	Isolated hydrogen valence electrons	1
	Bound hydrogen bonding electrons	2
	Bound hydrogen nonbonding electrons	0
	Formal charge = $1 - \frac{1}{2}(2) - 0 = 0$	

For nitrogen: (central)	Isolated nitrogen valence electrons	5
	Bound nitrogen bonding electrons	6
	Bound nitrogen nonbonding electrons	2
	Formal charge = $5 - \frac{1}{2}(6) - 2 = 0$	

For nitrogen: (terminal)	Isolated nitrogen valence electrons	5
	Bound nitrogen bonding electrons	4
	Bound nitrogen nonbonding electrons	4
	Formal charge = $5 - \frac{1}{2}(4) - 4 = -1$	

For carbon:	Isolated carbon valence electrons	4
	Bound carbon bonding electrons	6
	Bound carbon nonbonding electrons	0
	Formal charge = $4 - \frac{1}{2}(6) - 0 = +1$	

Structure (a) is more important because of the octet of electrons around carbon.

5.73

For oxygen:	Isolated oxygen valence electrons	6
	Bound oxygen bonding electrons	4
	Bound oxygen nonbonding electrons	4
	Formal charge = $6 - \frac{1}{2}(4) - 4 = 0$

For left carbon:	Isolated carbon valence electrons	4
	Bound carbon bonding electrons	8
	Bound carbon nonbonding electrons	0
	Formal charge = $4 - \frac{1}{2}(8) - 0 = 0$

For right carbon:	Isolated carbon valence electrons	4
	Bound carbon bonding electrons	6
	Bound carbon nonbonding electrons	2
	Formal charge = $4 - \frac{1}{2}(6) - 2 = -1$

For oxygen:	Isolated oxygen valence electrons	6
	Bound oxygen bonding electrons	2
	Bound oxygen nonbonding electrons	6
	Formal charge = $6 - \frac{1}{2}(2) - 6 = -1$

For left carbon:	Isolated carbon valence electrons	4
	Bound carbon bonding electrons	8
	Bound carbon nonbonding electrons	0
	Formal charge = $4 - \frac{1}{2}(8) - 0 = 0$

For right carbon:	Isolated carbon valence electrons	4
	Bound carbon bonding electrons	8
	Bound carbon nonbonding electrons	0
	Formal charge = $4 - \frac{1}{2}(8) - 0 = 0$

The second structure is more important because of the –1 formal charge on the more electronegative oxygen.

The VSEPR Model (Section 5.10)

5.74   From data in Table 5.5:
        (a) trigonal planar   (b) trigonal bipyramidal   (c) linear   (d) octahedral

5.75   From data in Table 5.5:
        (a) T shaped   (b) bent   (c) square planar

5.76   From data in Table 5.5:
        (a) tetrahedral, 4   (b) octahedral, 6   (c) bent, 3 or 4
        (d) linear, 2 or 5   (e) square pyramidal, 6   (f) trigonal pyramidal, 4

5.77   From data in Table 5.5:
        (a) seesaw, 5   (b) square planar, 6   (c) trigonal bipyramidal, 5
        (d) T shaped, 5   (e) trigonal planar, 3   (f) linear, 2 or 5

5.78

	Number of Bonded Atoms	Number of Lone Pairs	Shape
(a) H_2Se	2	2	bent
(b) $TiCl_4$	4	0	tetrahedral
(c) O_3	2	1	bent
(d) GaH_3	3	0	trigonal planar

5.79

	Number of Bonded Atoms	Number of Lone Pairs	Shape
(a) XeO_4	4	0	tetrahedral
(b) SO_2Cl_2	4	0	tetrahedral
(c) OsO_4	4	0	tetrahedral
(d) SeO_2	2	1	bent

5.80

	Number of Bonded Atoms	Number of Lone Pairs	Shape
(a) SbF_5	5	0	trigonal bipyramidal
(b) IF_4^+	4	1	see saw
(c) SeO_3^{2-}	3	1	trigonal pyramidal
(d) CrO_4^{2-}	4	0	tetrahedral

5.81

	Number of Bonded Atoms	Number of Lone Pairs	Shape
(a) NO_3^-	3	0	trigonal planar
(b) NO_2^+	2	0	linear
(c) NO_2^-	2	1	bent

5.82

		Number of Bonded Atoms	Number of Lone Pairs	Shape
(a)	PO_4^{3-}	4	0	tetrahedral
(b)	MnO_4^-	4	0	tetrahedral
(c)	SO_4^{2-}	4	0	tetrahedral
(d)	SO_3^{2-}	3	1	trigonal pyramidal
(e)	ClO_4^-	4	0	tetrahedral
(f)	SCN^-	2	0	linear

(C is the central atom)

5.83

		Number of Bonded Atoms	Number of Lone Pairs	Shape
(a)	XeF_3^+	3	2	T shaped
(b)	SF_3^+	3	1	trigonal pyramidal
(c)	ClF_2^+	2	2	bent
(d)	CH_3^+	3	0	trigonal planar

5.84 (a) In SF_2 the sulfur is bound to two fluorines and contains two lone pairs of electrons. SF_2 is bent and the F–S–F bond angle is approximately $109°$.
(b) In N_2H_2 each nitrogen is bound to the other nitrogen and one hydrogen. Each nitrogen has one lone pair of electrons. The H–N–N bond angle is approximately $120°$.
(c) In KrF_4 the krypton is bound to four fluorines and contains two lone pairs of electrons. KrF_4 is square planar, and the F–Kr–F bond angle is $90°$.
(d) In $NOCl$ the nitrogen is bound to one oxygen and one chlorine and contains one lone pair of electrons. $NOCl$ is bent, and the Cl–N–O bond angle is approximately $120°$.

5.85 (a) In PCl_6^- the phosphorus is bound to six chlorines. There are no lone pairs of electrons on the phosphorus. PCl_6^- is octahedral, and the Cl–P–Cl bond angle is $90°$.
(b) In ICl_2^- the iodine is bound to two chlorines and contains three lone pairs of electrons. ICl_2^- is linear, and the Cl–I–Cl bond angle is $180°$.
(c) In SO_4^{2-} the sulfur is bound to four oxygens. There are no lone pairs of electrons on the sulfur. SO_4^{2-} is tetrahedral, and the O–S–O bond angle is $109.5°$.
(d) In BO_3^{3-} the boron is bound to three oxygens. There are no lone pairs of electrons on the boron. BO_3^{3-} is trigonal planar, and the O–B–O bond angle is $120°$.

5.86

$H–C_a–H$	$\sim 120°$	$C_b–C_c–N$	$180°$
$H–C_a–C_b$	$\sim 120°$	$C_a–C_b–H$	$\sim 120°$
$C_a–C_b–C_c$	$\sim 120°$	$H–C_b–C_c$	$\sim 120°$

5.87

5.88   All six carbons in cyclohexane are bonded to two other carbons and two hydrogens (i.e. four charge clouds). The geometry about each carbon is tetrahedral with a C–C–C bond angle of approximately 109°. Because the geometry about each carbon is tetrahedral, the cyclohexane ring cannot be flat.

5.89   All six carbon atoms are sp^2 hybridized and the bond angles are ~120°. The geometry about each carbon is trigonal planar.

Hybrid Orbitals and Molecular Orbital Theory (Sections 5.12–5.16)

5.90   In a π bond, the shared electrons occupy a region above and below a line connecting the two nuclei. A σ bond has its shared electrons located along the axis between the two nuclei.

5.91   Electrons in a bonding molecular orbital spend most of their time in the region between the two nuclei, helping to bond the atoms together. Electrons in an antibonding molecular orbital cannot occupy the central region between the nuclei and cannot contribute to bonding.

5.92   See Table 5.6.
(a) sp   (b) sp^2 (c) sp^3

5.93   See Table 5.6.
(a) tetrahedral   (b) trigonal planar   (c) linear

5.94   (a) sp^2 (b) sp^2 (c) sp^3 (d) sp^2

5.95   (a) sp^3 (b) sp^2 (c) sp^2 (d) sp^3

5.96

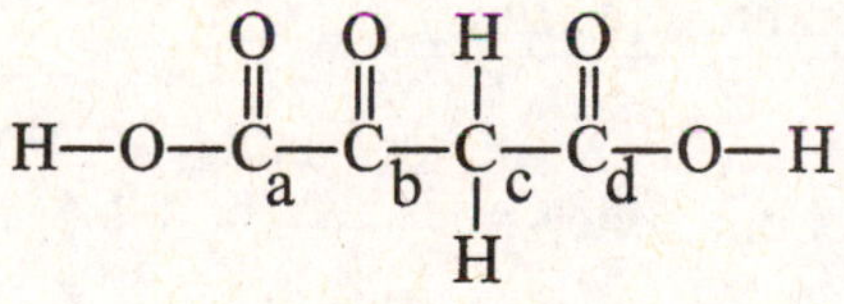

Carbons a, b, and d are sp^2 hybridized and carbon c is sp^3 hybridized.

The bond angles around carbons a, b, and d are ~120°. The bond angles around carbon c are ~109°. The terminal H-O-C bond angles are ~109°.

5.97   (a)

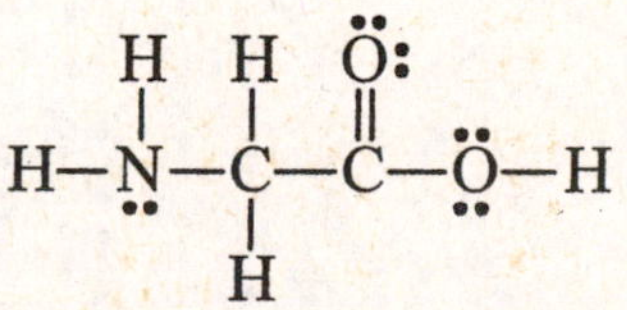

(b) H–C–H, ~109°;   O–C–O, ~120°;   H–N–H, ~107°
(c) N, sp^3; left C, sp^3; right C, sp^2

5.98

	O_2^+	O_2	O_2^-
σ^*_{2p}	—	—	—
π^*_{2p}	↑ / —	↑ / ↑	↑↓ / ↑
π_{2p}	↑↓ / ↑↓	↑↓ / ↑↓	↑↓ / ↑↓
σ_{2p}	↑↓	↑↓	↑↓
σ^*_{2s}	↑↓	↑↓	↑↓
σ_{2s}	↑↓	↑↓	↑↓

$$\text{Bond order} = \frac{\begin{pmatrix}\text{number of}\\ \text{bonding electrons}\end{pmatrix} - \begin{pmatrix}\text{number of}\\ \text{antibonding electrons}\end{pmatrix}}{2}$$

O_2^+ bond order $= \dfrac{8 - 3}{2} = 2.5$ $\qquad$ O_2 bond order $= \dfrac{8 - 4}{2} = 2$

O_2^- bond order $= \dfrac{8 - 5}{2} = 1.5$

All are stable with bond orders between 1.5 and 2.5. All have unpaired electrons.

5.99

	N_2^+	N_2	N_2^-
σ^*_{2p}	—	—	—
π^*_{2p}	— / —	— / —	↑ / —
σ_{2p}	↑	↑↓	↑↓
π_{2p}	↑↓ / ↑↓	↑↓ / ↑↓	↑↓ / ↑↓
σ^*_{2s}	↑↓	↑↓	↑↓
σ_{2s}	↑↓	↑↓	↑↓

$$\text{Bond order} = \frac{\begin{pmatrix}\text{number of}\\ \text{bonding electrons}\end{pmatrix} - \begin{pmatrix}\text{number of}\\ \text{antibonding electrons}\end{pmatrix}}{2}$$

N_2^+ bond order $= \dfrac{7 - 2}{2} = 2.5$ $\qquad$ N_2 bond order $= \dfrac{8 - 2}{2} = 3$

N_2^- bond order $= \dfrac{8 - 3}{2} = 2.5$

All are stable with bond orders of either 3 or 2.5. N_2^+ and N_2^- contain unpaired electrons.

5.100 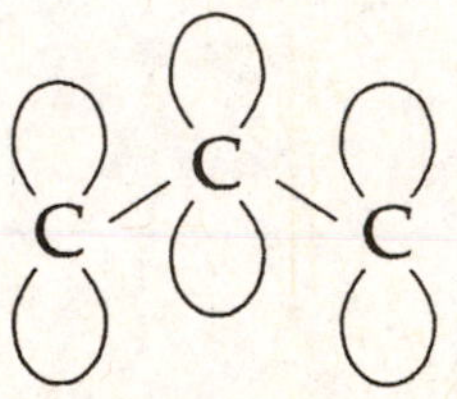

p orbitals in allyl cation

allyl cation showing only the σ bonds (each C is sp² hybridized)

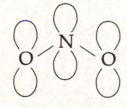

delocalized MO model for π bonding in the allyl cation

5.101

p orbitals in NO_2^-

NO_2^- showing only the σ bonds (N is sp² hybridized)

delocalized MO model for π bonding in NO_2^-

Chapter Problems

5.102

5.103 In ascorbic acid (Problem 5.102) all carbons that have only single bonds are sp³ hybridized. The three carbons that have double bonds are sp² hybridized.

5.104 Every carbon is sp² hybridized. There are 18 σ bonds and 5 π bonds.

5.105

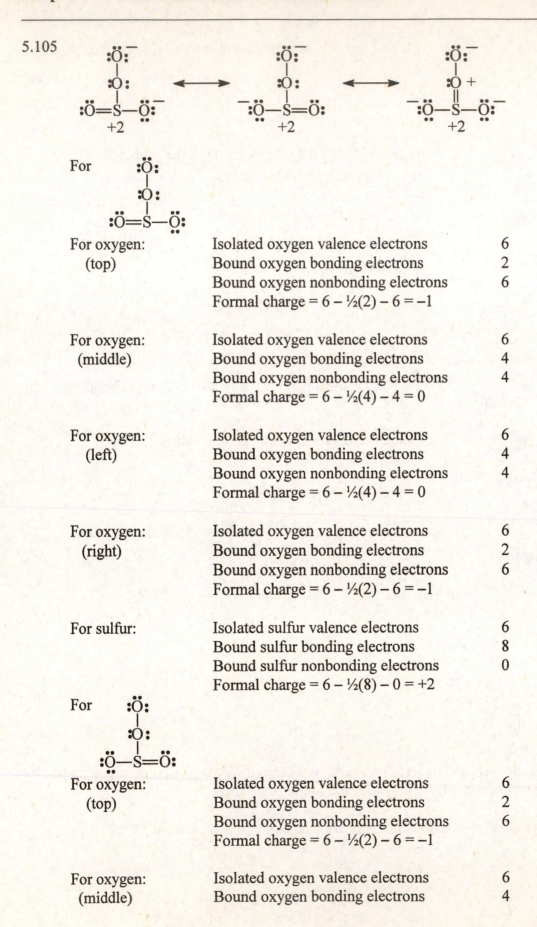

For oxygen: (top)	Isolated oxygen valence electrons	6
	Bound oxygen bonding electrons	2
	Bound oxygen nonbonding electrons	6
	Formal charge = $6 - \frac{1}{2}(2) - 6 = -1$	

For oxygen: (middle)	Isolated oxygen valence electrons	6
	Bound oxygen bonding electrons	4
	Bound oxygen nonbonding electrons	4
	Formal charge = $6 - \frac{1}{2}(4) - 4 = 0$	

For oxygen: (left)	Isolated oxygen valence electrons	6
	Bound oxygen bonding electrons	4
	Bound oxygen nonbonding electrons	4
	Formal charge = $6 - \frac{1}{2}(4) - 4 = 0$	

For oxygen: (right)	Isolated oxygen valence electrons	6
	Bound oxygen bonding electrons	2
	Bound oxygen nonbonding electrons	6
	Formal charge = $6 - \frac{1}{2}(2) - 6 = -1$	

For sulfur:	Isolated sulfur valence electrons	6
	Bound sulfur bonding electrons	8
	Bound sulfur nonbonding electrons	0
	Formal charge = $6 - \frac{1}{2}(8) - 0 = +2$	

For

For oxygen: (top)	Isolated oxygen valence electrons	6
	Bound oxygen bonding electrons	2
	Bound oxygen nonbonding electrons	6
	Formal charge = $6 - \frac{1}{2}(2) - 6 = -1$	

For oxygen: (middle)	Isolated oxygen valence electrons	6
	Bound oxygen bonding electrons	4

	Bound oxygen nonbonding electrons	4
	Formal charge = $6 - \frac{1}{2}(4) - 4 = 0$	

For oxygen: (left)

Isolated oxygen valence electrons	6
Bound oxygen bonding electrons	2
Bound oxygen nonbonding electrons	6
Formal charge = $6 - \frac{1}{2}(2) - 6 = -1$	

For oxygen: (right)

Isolated oxygen valence electrons	6
Bound oxygen bonding electrons	4
Bound oxygen nonbonding electrons	4
Formal charge = $6 - \frac{1}{2}(4) - 4 = 0$	

For sulfur:

Isolated sulfur valence electrons	6
Bound sulfur bonding electrons	8
Bound sulfur nonbonding electrons	0
Formal charge = $6 - \frac{1}{2}(8) - 0 = +2$	

For

$$:\ddot{O}:$$
$$|$$
$$:O$$
$$\|$$
$$:\ddot{O}-S-\ddot{O}:$$

For oxygen: (top)

Isolated oxygen valence electrons	6
Bound oxygen bonding electrons	2
Bound oxygen nonbonding electrons	6
Formal charge = $6 - \frac{1}{2}(2) - 6 = -1$	

For oxygen: (middle)

Isolated oxygen valence electrons	6
Bound oxygen bonding electrons	6
Bound oxygen nonbonding electrons	2
Formal charge = $6 - \frac{1}{2}(6) - 2 = +1$	

For oxygen: (left)

Isolated oxygen valence electrons	6
Bound oxygen bonding electrons	2
Bound oxygen nonbonding electrons	6
Formal charge = $6 - \frac{1}{2}(2) - 6 = -1$	

For oxygen: (right)

Isolated oxygen valence electrons	6
Bound oxygen bonding electrons	2
Bound oxygen nonbonding electrons	6
Formal charge = $6 - \frac{1}{2}(2) - 6 = -1$	

For sulfur:

Isolated sulfur valence electrons	6
Bound sulfur bonding electrons	8
Bound sulfur nonbonding electrons	0
Formal charge = $6 - \frac{1}{2}(8) - 0 = +2$	

5.106

$$H-\underset{\underset{H}{|}}{\overset{\overset{H}{|}}{C}}-\ddot{N}=C=\ddot{O}: \longleftrightarrow H-\underset{\underset{H}{|}}{\overset{\overset{H}{|}}{C}}-\overset{+}{N}\equiv C-\overset{-}{\underset{\cdot\cdot}{O}}:$$

For $H-\underset{\underset{H}{|}}{\overset{\overset{H}{|}}{C}}-\ddot{N}=C=\ddot{O}:$

For hydrogen:	Isolated hydrogen valence electrons	1
	Bound hydrogen bonding electrons	2
	Bound hydrogen nonbonding electrons	0
	Formal charge = $1 - \frac{1}{2}(2) - 0 = 0$	

For carbon: (left)	Isolated carbon valence electrons	4
	Bound carbon bonding electrons	8
	Bound carbon nonbonding electrons	0
	Formal charge = $4 - \frac{1}{2}(8) - 0 = 0$	

For nitrogen:	Isolated nitrogen valence electrons	5
	Bound nitrogen bonding electrons	6
	Bound nitrogen nonbonding electrons	2
	Formal charge = $5 - \frac{1}{2}(6) - 2 = 0$	

For carbon: (right)	Isolated carbon valence electrons	4
	Bound carbon bonding electrons	8
	Bound carbon nonbonding electrons	0
	Formal charge = $4 - \frac{1}{2}(8) - 0 = 0$	

For oxygen:	Isolated oxygen valence electrons	6
	Bound oxygen bonding electrons	4
	Bound oxygen nonbonding electrons	4
	Formal charge = $6 - \frac{1}{2}(4) - 4 = 0$	

For $H-\underset{\underset{H}{|}}{\overset{\overset{H}{|}}{C}}-\overset{+}{N}\equiv C-\overset{-}{\underset{\cdot\cdot}{O}}:$

For hydrogen:	Isolated hydrogen valence electrons	1
	Bound hydrogen bonding electrons	2
	Bound hydrogen nonbonding electrons	0
	Formal charge = $1 - \frac{1}{2}(2) - 0 = 0$	

For carbon: (left)	Isolated carbon valence electrons	4
	Bound carbon bonding electrons	8
	Bound carbon nonbonding electrons	0
	Formal charge = $4 - \frac{1}{2}(8) - 0 = 0$	

For nitrogen:	Isolated nitrogen valence electrons	5
	Bound nitrogen bonding electrons	8
	Bound nitrogen nonbonding electrons	0
	Formal charge = $5 - \frac{1}{2}(8) - 0 = +1$	

For carbon: (right)	Isolated carbon valence electrons	4
	Bound carbon bonding electrons	8
	Bound carbon nonbonding electrons	0
	Formal charge = $4 - \frac{1}{2}(8) - 0 = 0$	

For oxygen:	Isolated oxygen valence electrons	6
	Bound oxygen bonding electrons	2
	Bound oxygen nonbonding electrons	6
	Formal charge = $6 - \frac{1}{2}(2) - 6 = -1$	

5.107

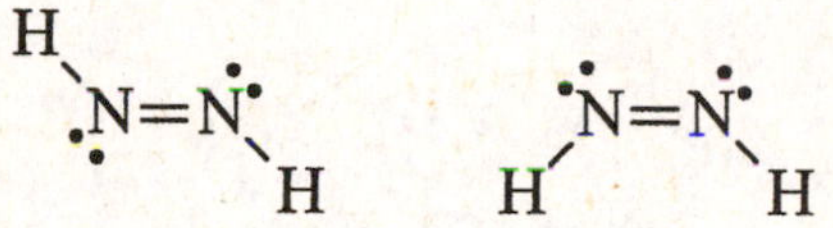

They are geometric isomers not resonance forms. In resonance forms the atoms have the same geometrical arrangement.

5.108 (a) reactants

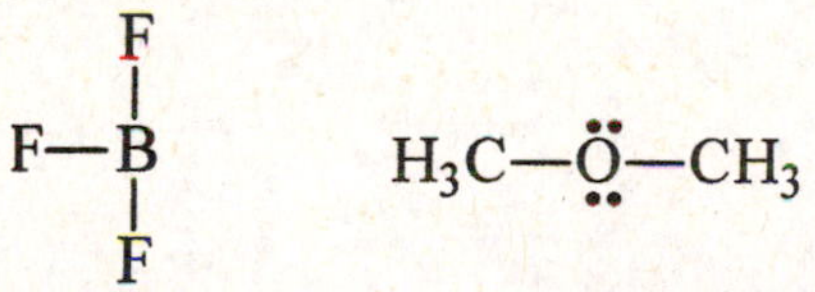

For boron:	Isolated boron valence electrons	3
	Bound boron bonding electrons	6
	Bound boron nonbonding electrons	0
	Formal charge = $3 - \frac{1}{2}(6) - 0 = 0$	

For oxygen:	Isolated oxygen valence electrons	6
	Bound oxygen bonding electrons	4
	Bound oxygen nonbonding electrons	4
	Formal charge = $6 - \frac{1}{2}(4) - 4 = 0$	

product

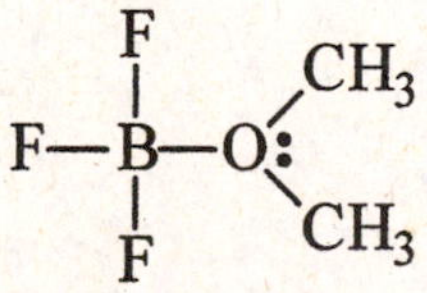

For boron:	Isolated boron valence electrons	3
	Bound boron bonding electrons	8
	Bound boron nonbonding electrons	0
	Formal charge = $3 - \frac{1}{2}(8) - 0 = -1$	

For oxygen:	Isolated oxygen valence electrons	6
	Bound oxygen bonding electrons	6
	Bound oxygen nonbonding electrons	2
	Formal charge = $6 - \frac{1}{2}(6) - 2 = +1$	

(b) In BF_3 the B has three bonding pairs of electrons and no lone pairs. The B is sp^2 hybridized and BF_3 is trigonal planar.

$H_3C—\overset{\cdot\cdot}{\underset{\cdot\cdot}{O}}—CH_3$ is bent about the oxygen because of two bonding pairs and two lone pairs of electrons. The O is sp^3 hybridized.

In the product, B is sp^3 hybridized (with four bonding pairs of electrons), and the geometry about it is tetrahedral. The O is also sp^3 hybridized (with three bonding pairs and one lone pair of electrons), and the geometry about it is trigonal pyramidal.

5.109 Both the B and N are sp^2 hybridized. All bond angles are ~120°. The overall geometry of the molecule is planar.

5.110 The triply bonded carbon atoms are sp hybridized. The theoretical bond angle for C–C≡C is 180°. Benzyne is so reactive because the C–C≡C bond angle is closer to 120° and is very strained.

5.111 (a) $H—C≡C—H$ (b) $H—\overset{\cdot\cdot}{N}=\overset{\cdot\cdot}{N}—H$ (c)

$$:\overset{\cdot\cdot}{\underset{\cdot\cdot}{O}}:$$
$$:\overset{\cdot\cdot}{\underset{\cdot\cdot}{Cl}}—\overset{|}{\underset{|}{S}}—\overset{\cdot\cdot}{\underset{\cdot\cdot}{Cl}}:$$

5.112

$$:\overset{\cdot\cdot}{\underset{\cdot\cdot}{Cl}}\quad :\overset{\cdot\cdot}{O}—H$$
$$:\overset{\cdot\cdot}{\underset{\cdot\cdot}{Cl}}—\overset{|}{\underset{|}{C}}—\overset{|}{\underset{|}{C}}—\overset{\cdot\cdot}{O}—H$$
$$:\overset{\cdot\cdot}{\underset{\cdot\cdot}{Cl}}\quad H$$

5.113 Li_2

$$
\begin{array}{ll}
\sigma^*_{2s} & \underline{\quad\quad} \\
\sigma_{2s} & \underline{\uparrow\downarrow} \\
\sigma^*_{1s} & \underline{\uparrow\downarrow} \\
\sigma_{1s} & \underline{\uparrow\downarrow}
\end{array}
$$

Li_2 Bond order $= \dfrac{\left(\begin{array}{c}\text{number of} \\ \text{bonding electrons}\end{array}\right) - \left(\begin{array}{c}\text{number of} \\ \text{antibonding electrons}\end{array}\right)}{2} = \dfrac{4-2}{2} = 1$

The bond order for Li_2 is 1, and the molecule is likely to be stable.

5.114 C_2^{2-}

$$
\begin{array}{ll}
\sigma^*_{2p} & \underline{\quad\quad} \\
\pi^*_{2p} & \underline{\quad}\ \ \underline{\quad} \\
\sigma_{2p} & \underline{\uparrow\downarrow} \\
\pi_{2p} & \underline{\uparrow\downarrow}\ \ \underline{\uparrow\downarrow} \\
\sigma^*_{2s} & \underline{\uparrow\downarrow} \\
\sigma_{2s} & \underline{\uparrow\downarrow}
\end{array}
$$

Bond order $= \dfrac{\left(\begin{array}{c}\text{number of} \\ \text{bonding electrons}\end{array}\right) - \left(\begin{array}{c}\text{number of} \\ \text{antibonding electrons}\end{array}\right)}{2}$

C_2^{2-} bond order $= \dfrac{8-2}{2} = 3$; there is a triple bond between the two carbons.

5.115 (a)

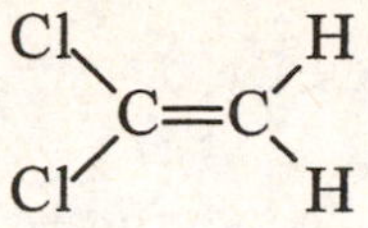

(b)

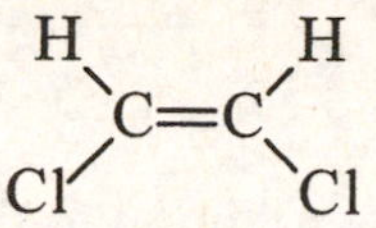

(c)

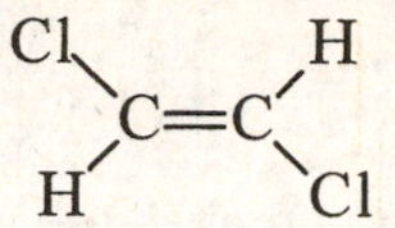

Structure (a) is different from structures (b) and (c) because both chlorines are on the same carbon. Structures (b) and (c) are different because in (b) both chlorines are on the same side of the molecule ("cis") and in (c) they are on opposite sides of the molecule ("trans"). There is no rotation around the carbon-carbon double bond.

5.116 $CH_4(g) + Cl_2(g) \rightarrow CH_3Cl(g) + HCl(g)$

Energy change = D (Reactant bonds) – D (Product bonds)

Energy change = $[4\, D_{C-H} + D_{Cl-Cl}] – [3\, D_{C-H} + D_{C-Cl} + D_{H-Cl}]$

Energy change = $[(4\text{ mol})(410\text{ kJ/mol}) + (1\text{ mol})(243\text{ kJ/mol}]$
        $– [(3\text{ mol})(410\text{ kJ/mol}) + (1\text{ mol})(330\text{ kJ/mol}) + (1\text{ mol})(432\text{ kJ/mol})] = -109\text{ kJ}$

5.117

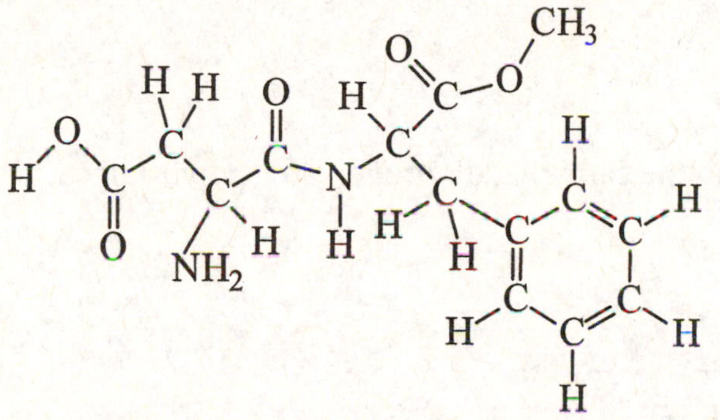

5.118 (a)

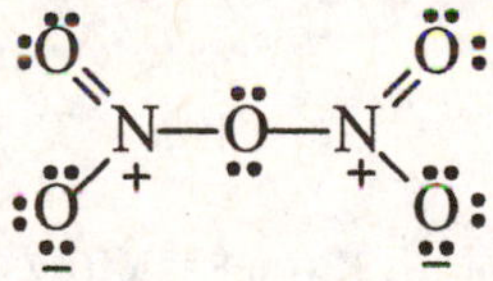

(b)

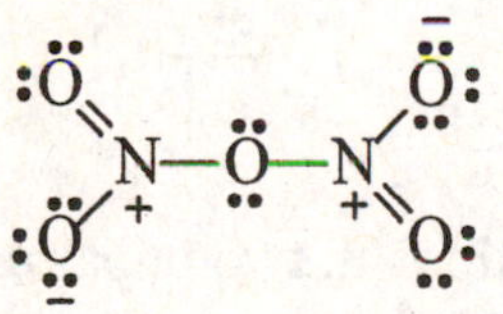

(c)

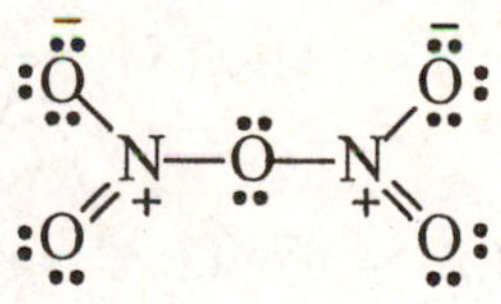

(d)

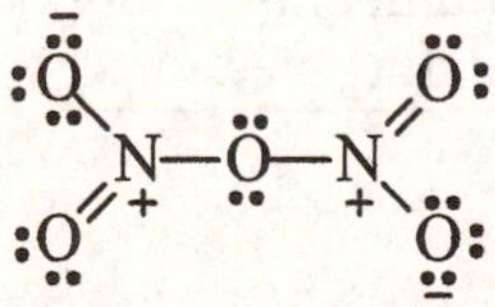

(e)

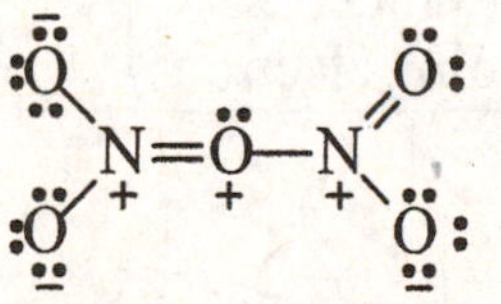

(f)

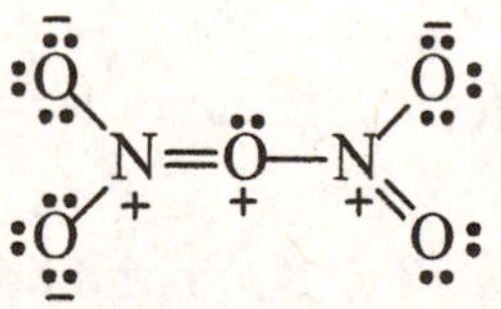

(g)

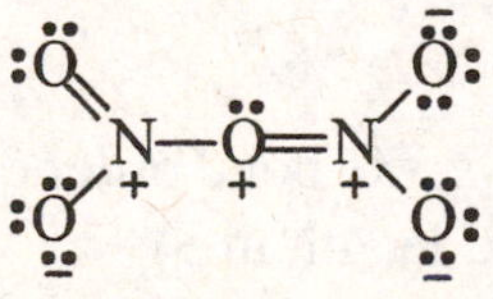

(h)

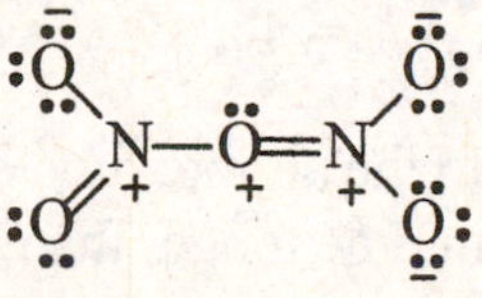

Structures (a) – (d) make more important contributions to the resonance hybrid because of only –1 and 0 formal charges on the oxygens. A +1 formal charge is unlikely.

5.119 (a) (1) 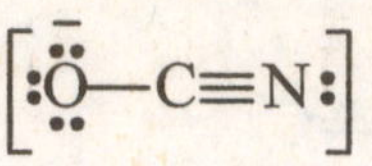 $\left[:\overset{..}{\overset{-}{O}}\!-\!C\!\equiv\!N: \right]^{-}$ (2) 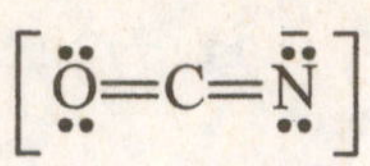 $\left[\overset{..}{O}\!=\!C\!=\!\overset{..}{\overset{-}{N}} \right]^{-}$ (3) 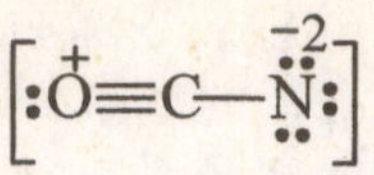 $\left[:\overset{+}{\overset{..}{O}}\!\equiv\!C\!-\!\overset{..}{\overset{-2}{N}}: \right]^{-}$

(b) Structure (1) makes the greatest contribution to the resonance hybrid because of the –1 formal charge on the oxygen. Structure (3) makes the least contribution to the resonance hybrid because of the +1 formal charge on the oxygen.

(c) and (d) OCN⁻ is linear because the C has 2 charge clouds. It is sp hybridized in all three resonance structures. It forms two π bonds.

5.120

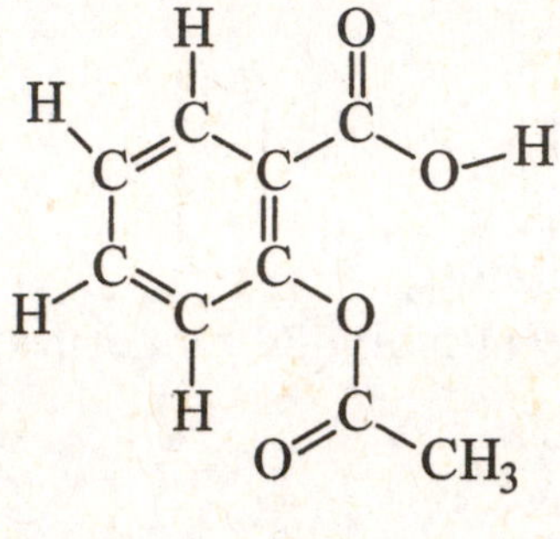

21 σ bonds
5 π bonds
Each C with a double bond is sp² hybridized.
The –CH₃ carbon is sp³ hybridized.

5.121 $\left[H\!-\!C\!\equiv\!N\!-\!\overset{..}{Xe}\!-\!\overset{..}{\overset{..}{F}}: \right]^{+}$ Both the carbon and nitrogen are sp hybridized.

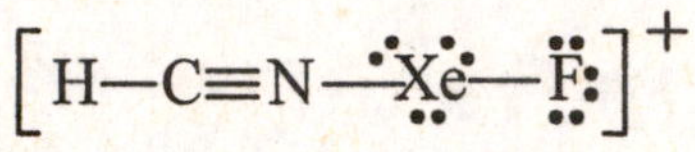

5.122 (a)

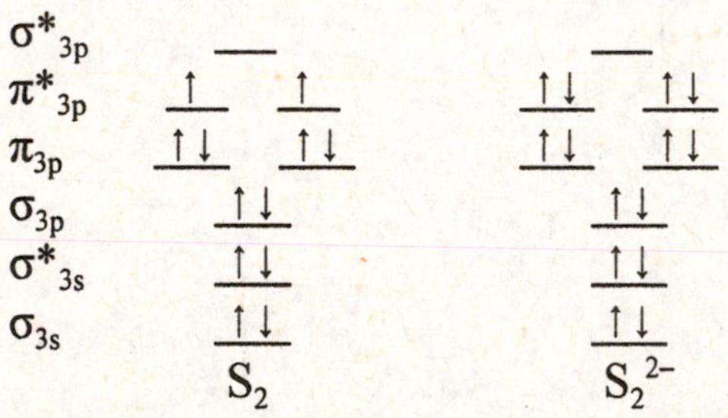

(b) S₂ would be paramagnetic with two unpaired electrons in the π*₃ₚ MOs.

(c) Bond order $= \dfrac{\left(\begin{array}{c}\text{number of} \\ \text{bonding electrons}\end{array}\right) - \left(\begin{array}{c}\text{number of} \\ \text{antibonding electrons}\end{array}\right)}{2}$

S₂ bond order $= \dfrac{8-4}{2} = 2$

(d) S₂²⁻ bond order $= \dfrac{8-6}{2} = 1$

The two added electrons go into the antibonding π*₃ₚ MOs, the bond order drops from 2 to 1, and the bond length in S₂²⁻ should be longer than the bond length in S₂.

5.123 (a) CO

σ^*_{2p} —

π^*_{2p} —   —

σ_{2p} $\uparrow\downarrow$

π_{2p} $\uparrow\downarrow$ $\uparrow\downarrow$

σ^*_{2s} $\uparrow\downarrow$

σ_{2s} $\uparrow\downarrow$

(b) All electrons are paired, CO is diamagnetic.

(c)

$$\text{Bond order} = \frac{\left(\begin{array}{c}\text{number of}\\\text{bonding electrons}\end{array}\right) - \left(\begin{array}{c}\text{number of}\\\text{antibonding electrons}\end{array}\right)}{2}$$

$$\text{CO bond order} = \frac{8-2}{2} = 3$$

The bond order here matches that predicted by the electron-dot structure (:C≡O:).

(d)

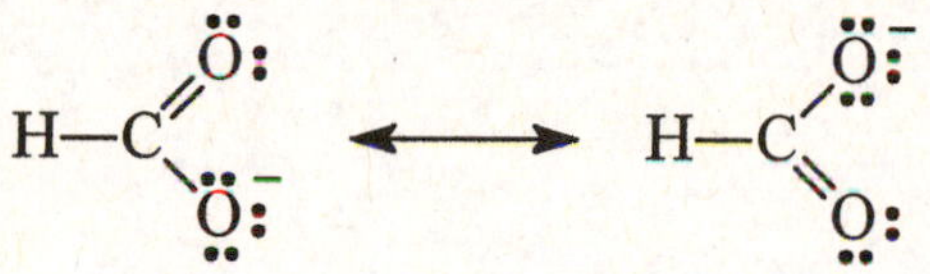

5.124 (a)

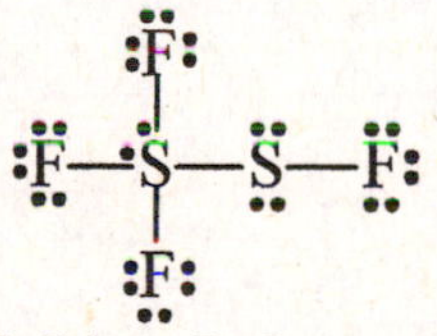

The left S has 5 electron clouds (4 bonding, 1 lone pair). The geometry about this S is seesaw. The right S has 4 electron clouds (2 bonding, 2 lone pairs). The geometry about this S is bent.

(b)

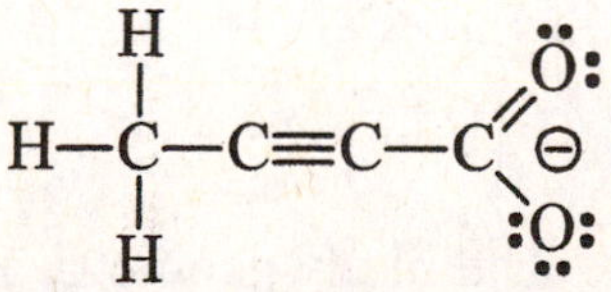

The left C has 4 electron clouds (4 bonding, 0 lone pairs). The geometry about this C is tetrahedral. The right C has 3 electron clouds (3 bonding, 0 lone pairs). The geometry about this C is trigonal planar. The central two C's have 2 electron clouds (2 bonding, 0 lone pairs). The geometry about these two C's is linear.

5.125

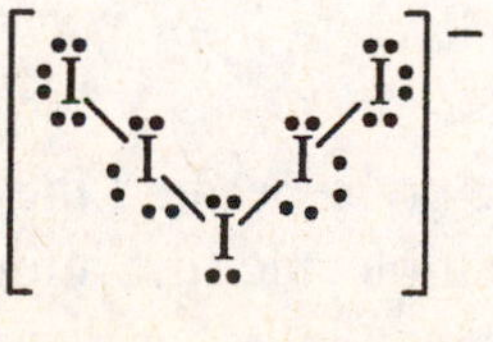

Multiconcept Problems

5.126 (a) $\left[\ddot{\text{:}\overset{..}{\text{O}}}\text{—H}\right]^{-}$ $\text{:}\overset{..}{\overset{.}{\text{O}}}\text{—H}$

(b) The oxygen in OH has a half-filled 2p orbital that can accept the additional electron. For a 2p orbital, $n = 2$ and $l = 1$.

(c) The electron affinity for OH is slightly more negative than for an O atom because when OH gains an additional electron, it achieves an octet configuration.

5.127 (a) (4 orbitals)(3 electrons) = 12 outer-shell electrons

(b) 3 electrons

(c) $1s^3\ 2s^3\ 2p^6$; $\text{:}\overset{..}{\underset{..}{\text{X}}}\text{:}$

(d) $\text{:}\overset{..}{\underset{..}{\text{X}}}\text{::}\overset{..}{\underset{..}{\text{X}}}\text{:}$

(e)

$$
\begin{array}{ll}
\sigma^*_{2p} & \underline{} \\[4pt]
\pi^*_{2p} & \underline{\uparrow\downarrow}\quad \underline{\uparrow} \\[4pt]
\pi_{2p} & \underline{\uparrow\downarrow\uparrow}\quad \underline{\uparrow\downarrow\uparrow} \\[4pt]
\sigma_{2p} & \underline{\uparrow\downarrow\uparrow} \\[4pt]
\sigma^*_{2s} & \underline{\uparrow\downarrow\uparrow} \\[4pt]
\sigma_{2s} & \underline{\uparrow\downarrow\uparrow}
\end{array}
$$

$$
\text{Bond order} = \dfrac{\left(\begin{array}{c}\text{number of}\\ \text{bonding electrons}\end{array}\right) - \left(\begin{array}{c}\text{number of}\\ \text{antibonding electrons}\end{array}\right)}{3}
$$

$$
X_2 \ \text{ bond order} = \dfrac{12-6}{3} = 2
$$

5.128 (a)

$$
\left[\ \begin{array}{c}
\quad\ \ \text{:}\overset{..}{\text{O}}\qquad\quad \text{:}\overset{..}{\text{O}} \\
\quad\ \ \|\qquad\qquad \| \\
\text{:O}=\overset{}{\text{Cr}}\text{—}\overset{..}{\underset{..}{\text{O}}}\text{—}\overset{}{\text{Cr}}=\text{O:} \\
\quad\ \ \overset{}{\underset{..}{\text{:O:}}}\qquad\quad \overset{}{\underset{..}{\text{:O:}}}
\end{array}\ \right]^{2-}
$$

(b) Each Cr atom has 6 pairs of electrons around it. The likely geometry about each Cr atom is tetrahedral because each Cr has 4 charge clouds.

5.129 (a)

polar   polar   nonpolar

(b) All three molecules are planar. The first two structures are polar because they both have an unsymmetrical distribution of atoms about the center of the molecule (the middle of the double bond), and bond polarities do not cancel. Structure 3 is nonpolar because the H's and

Cl's, respectively, are symmetrically distributed about the center of the molecule, both being opposite each other. In this arrangement, bond polarities cancel.

(c) 200 nm = 200 x 10^{-9} m

$$E = \frac{hc}{\lambda} = \frac{(6.626 \times 10^{-34} \text{ J·s})(3.00 \times 10^8 \text{ m/s})}{200 \times 10^{-9} \text{ m}} (6.022 \times 10^{23} \text{ /mol})$$

E = 5.99 x 10^5 J/mol = 599 kJ/mol

(d)

The π bond must be broken before rotation can occur.

5.130 (a) Each carbon is sp^2 hybridized.
(b) & (c)

antibonding	—	
antibonding	— —	
nonbonding	↑↓ ↑↓	
bonding	↑↓ ↑↓	
bonding	↑↓	

(d) The cyclooctatetraene dianion has only paired electrons and is diamagnetic.

6

Mass Relationships in Chemical Reactions

6.1 $2 NaClO_3 \rightarrow 2 NaCl + 3 O_2$

6.2   (a) $C_6H_{12}O_6 \rightarrow 2 C_2H_6O + 2 CO_2$
      (b) $6 CO_2 + 6 H_2O \rightarrow C_6H_{12}O_6 + 6 O_2$
      (c) $4 NH_3 + Cl_2 \rightarrow N_2H_4 + 2 NH_4Cl$

6.3 $3 A_2 + 2 B \rightarrow 2 BA_3$

6.4   (a) Fe_2O_3: $2(55.85) + 3(16.00) = 159.7$ amu
      (b) H_2SO_4: $2(1.01) + 1(32.07) + 4(16.00) = 98.1$ amu
      (c) $C_6H_8O_7$: $6(12.01) + 8(1.01) + 7(16.00) = 192.1$ amu
      (d) $C_{16}H_{18}N_2O_4S$: $16(12.01) + 18(1.01) + 2(14.01) + 4(16.00) + 1(32.07) = 334.4$ amu

6.5 $C_9H_8O_4$, 180.2 amu;   500 mg $= 500 \times 10^{-3}$ g $= 0.500$ g

$$0.500 \text{ g} \times \frac{1 \text{ mol}}{180.2 \text{ g}} = 2.77 \times 10^{-3} \text{ mol aspirin}$$

$$2.77 \times 10^{-3} \text{ mol} \times \frac{6.022 \times 10^{23} \text{ molecules}}{1 \text{ mol}} = 1.67 \times 10^{21} \text{ aspirin molecules}$$

6.6   salicylic acid, $C_7H_6O_3$, 138.1 amu;   acetic anhydride, $C_4H_6O_3$, 102.1 amu
      aspirin, $C_9H_8O_4$, 180.2 amu;   acetic acid, $C_2H_4O_2$, 60.1 amu

$$4.50 \text{ g } C_7H_6O_3 \times \frac{1 \text{ mol } C_7H_6O_3}{138.1 \text{ g } C_7H_6O_3} \times \frac{1 \text{ mol } C_4H_6O_3}{1 \text{ mol } C_7H_6O_3} \times \frac{102.1 \text{ g } C_4H_6O_3}{1 \text{ mol } C_4H_6O_3} = 3.33 \text{ g } C_4H_6O_3$$

$$4.50 \text{ g } C_7H_6O_3 \times \frac{1 \text{ mol } C_7H_6O_3}{138.1 \text{ g } C_7H_6O_3} \times \frac{1 \text{ mol } C_9H_8O_4}{1 \text{ mol } C_7H_6O_3} \times \frac{180.2 \text{ g } C_9H_8O_4}{1 \text{ mol } C_9H_8O_4} = 5.87 \text{ g } C_9H_8O_4$$

$$4.50 \text{ g } C_7H_6O_3 \times \frac{1 \text{ mol } C_7H_6O_3}{138.1 \text{ g } C_7H_6O_3} \times \frac{1 \text{ mol } C_2H_4O_2}{1 \text{ mol } C_7H_6O_3} \times \frac{60.1 \text{ g } C_2H_4O_2}{1 \text{ mol } C_2H_4O_2} = 1.96 \text{ g } C_2H_4O_2$$

6.7 C_2H_4, 28.1 amu; C_2H_6O, 46.1 amu

$$4.6 \text{ g } C_2H_4 \times \frac{1 \text{ mol } C_2H_4}{28.1 \text{ g } C_2H_4} \times \frac{1 \text{ mol } C_2H_6O}{1 \text{ mol } C_2H_4} \times \frac{46.1 \text{ g } C_2H_6O}{1 \text{ mol } C_2H_6O} = 7.5 \text{ g } C_2H_6O$$

(theoretical yield)

$$\text{Percent yield} = \frac{\text{Actual yield}}{\text{Theoretical yield}} \times 100\% = \frac{4.7 \text{ g}}{7.5 \text{ g}} \times 100\% = 63\%$$

6.8   CH$_4$, 16.04 amu;   CH$_2$Cl$_2$, 84.93 amu;   1.85 kg = 1850 g

1850 g CH$_4$ x $\dfrac{1 \text{ mol CH}_4}{16.04 \text{ g CH}_4}$ x $\dfrac{1 \text{ mol CH}_2\text{Cl}_2}{1 \text{ mol CH}_4}$ x $\dfrac{84.93 \text{ g CH}_2\text{Cl}_2}{1 \text{ mol CH}_2\text{Cl}_2}$ = 9800 g CH$_2$Cl$_2$
(theoretical yield)

Actual yield = (9800 g)(0.431) = 4220 g CH$_2$Cl$_2$

6.9   Li$_2$O, 29.9 amu: 65 kg = 65,000 g;   H$_2$O, 18.0 amu: 80.0 kg = 80,000 g

65,000 g Li$_2$O x $\dfrac{1 \text{ mol Li}_2\text{O}}{29.9 \text{ g Li}_2\text{O}}$ = 2.17 x 10^3 mol Li$_2$O

80,000 g H$_2$O x $\dfrac{1 \text{ mol H}_2\text{O}}{18.0 \text{ g H}_2\text{O}}$ = 4.44 x 10^3 mol H$_2$O

The reaction stoichiometry between Li$_2$O and H$_2$O is one to one. There are twice as many moles of H$_2$O as there are moles of Li$_2$O. Therefore, Li$_2$O is the limiting reactant.
(4.44 x 10^3 mol – 2.17 x 10^3 mol) = 2.27 x 10^3 mol H$_2$O remaining

2.27 x 10^3 mol H$_2$O x $\dfrac{18.0 \text{ g H}_2\text{O}}{1 \text{ mol H}_2\text{O}}$ = 40,860 g H$_2$O = 40.9 kg = 41 kg H$_2$O

6.10   LiOH, 23.9 amu; CO$_2$, 44.0 amu

500.0 g LiOH x $\dfrac{1 \text{ mol LiOH}}{23.9 \text{ g LiOH}}$ x $\dfrac{1 \text{ mol CO}_2}{1 \text{ mol LiOH}}$ x $\dfrac{44.0 \text{ g CO}_2}{1 \text{ mol CO}_2}$ = 921 g CO$_2$

6.11   (a) A + B$_2$ → AB$_2$
There is a 1:1 stoichiometry between the two reactants. A is the limiting reactant because there are fewer reactant A's than there are reactant B$_2$'s.
(b) 1.0 mol of AB$_2$ can be made from 1.0 mol of A and 1.0 mol of B$_2$.

6.12   (a) 125 mL = 0.125 L; (0.20 mol/L)(0.125 L) = 0.025 mol NaHCO$_3$
(b) 650.0 mL = 0.6500 L; (2.50 mol/L)(0.6500 L) = 1.62 mol H$_2$SO$_4$

6.13   (a) NaOH, 40.0 amu; 500.0 mL = 0.5000 L

1.25 $\dfrac{\text{mol NaOH}}{\text{L}}$ x 0.500 L x $\dfrac{40.0 \text{ g NaOH}}{1 \text{ mol NaOH}}$ = 25.0 g NaOH

(b) C$_6$H$_{12}$O$_6$, 180.2 amu

0.250 $\dfrac{\text{mol C}_6\text{H}_{12}\text{O}_6}{\text{L}}$ x 1.50 L x $\dfrac{180.2 \text{ g C}_6\text{H}_{12}\text{O}_6}{1 \text{ mol C}_6\text{H}_{12}\text{O}_6}$ = 67.6 g C$_6$H$_{12}$O$_6$

6.14   C$_6$H$_{12}$O$_6$, 180.2 amu;

25.0 g C$_6$H$_{12}$O$_6$ x $\dfrac{1 \text{ mol C}_6\text{H}_{12}\text{O}_6}{180.2 \text{ g C}_6\text{H}_{12}\text{O}_6}$ = 0.1387 mol C$_6$H$_{12}$O$_6$

0.1387 mol x $\dfrac{1 \text{ L}}{0.20 \text{ mol}}$ = 0.69 L; 0.69 L = 690 mL

6.15 $C_{27}H_{46}O$, 386.7 amu;   750 mL = 0.750 L

$$0.005 \ \frac{mol \ C_{27}H_{46}O}{L} \ \times \ 0.750 \ L \ \times \ \frac{386.7 \ g \ C_{27}H_{46}O}{1 \ mol \ C_{27}H_{46}O} = 1 \ g \ C_{27}H_{46}O$$

6.16 $M_i \times V_i = M_f \times V_f$; $M_f = \dfrac{M_i \times V_i}{V_f} = \dfrac{3.50 \ M \times 75.0 \ mL}{400.0 \ mL} = 0.656 \ M$

6.17 $M_i \times V_i = M_f \times V_f$; $V_i = \dfrac{M_f \times V_f}{M_i} = \dfrac{0.500 \ M \times 250.0 \ mL}{18.0 \ M} = 6.94 \ mL$

Dilute 6.94 mL of 18.0 M H_2SO_4 with enough water to make 250.0 mL of solution. The resulting solution will be 0.500 M H_2SO_4.

6.18   50.0 mL = 0.0500 L;   (0.100 mol/L)(0.0500 L) = 5.00 x 10^{-3} mol NaOH

$$5.00 \times 10^{-3} \ mol \ NaOH \times \frac{1 \ mol \ H_2SO_4}{2 \ mol \ NaOH} = 2.50 \times 10^{-3} \ mol \ H_2SO_4$$

$$volume = 2.50 \times 10^{-3} \ mol \times \frac{1 \ L}{0.250 \ mol} = 0.0100 \ L; \ 0.0100 \ L = 10.0 \ mL \ H_2SO_4$$

6.19 $HNO_3(aq) + KOH(aq) \rightarrow KNO_3(aq) + H_2O(l)$
       25.0 mL = 0.0250 L and 68.5 mL = 0.0685 L

$$0.150 \ \frac{mol \ KOH}{L} \ \times \ 0.0250 \ L \ \times \ \frac{1 \ mol \ HNO_3}{1 \ mol \ KOH} = 3.75 \times 10^{-3} \ mol \ HNO_3$$

$$HNO_3 \ molarity = \frac{3.75 \times 10^{-3} \ mol}{0.0685 \ L} = 5.47 \times 10^{-2} \ M$$

6.20   From the reaction stoichiometry, moles NaOH = moles CH_3CO_2H
       (0.200 mol/L)(0.0947 L) = 0.018 94 mol NaOH = 0.018 94 mol CH_3CO_2H

$$molarity = \frac{0.018 \ 94 \ mol}{0.0250 \ L} = 0.758 \ M$$

6.21   Because the two volumes are equal (let the volume = y L), the concentrations are proportional to the number of solute ions.

$$OH^- \ concentration = 1.00 \ M \times \frac{y \ L}{12 \ H^+} \times \frac{8 \ OH^-}{y \ L} = 0.67 \ M$$

6.22   For dimethylhydrazine, $C_2H_8N_2$, divide each subscript by 2 to obtain the empirical formula. The empirical formula is CH_4N. $C_2H_8N_2$, 60.1 amu or 60.1 g/mol

$$\% \ C = \frac{2 \times 12.0 \ g}{60.1 \ g} \times 100\% = 39.9\%$$

$$\% \ H = \frac{8 \times 1.01 \ g}{60.1 \ g} \times 100\% = 13.4\%$$

$$\% \ N = \frac{2 \times 14.0 \ g}{60.1 \ g} \times 100\% = 46.6\%$$

6.23   Assume a 100.0 g sample. From the percent composition data, a 100.0 g sample contains 14.25 g C, 56.93 g O, and 28.83 g Mg.

$$14.25 \text{ g C} \times \frac{1 \text{ mol C}}{12.0 \text{ g C}} = 1.19 \text{ mol C}$$

$$56.93 \text{ g O} \times \frac{1 \text{ mol O}}{16.0 \text{ g O}} = 3.56 \text{ mol O}$$

$$28.83 \text{ g Mg} \times \frac{1 \text{ mol Mg}}{24.3 \text{ g Mg}} = 1.19 \text{ mol Mg}$$

$Mg_{1.19}C_{1.19}O_{3.56}$; divide each subscript by the smallest, 1.19.
$Mg_{1.19/1.19}C_{1.19/1.19}O_{3.56/1.19}$
The empirical formula is $MgCO_3$.

6.24 $C_6H_8O_7$, 192.1 amu or 192.1 g/mol

$$\% \text{C} = \frac{6 \times 12.0 \text{ g}}{192.1 \text{ g}} \times 100\% = 37.5\%$$

$$\% \text{H} = \frac{8 \times 1.01 \text{ g}}{192.1 \text{ g}} \times 100\% = 4.21\%$$

$$\% \text{O} = \frac{7 \times 16.0 \text{ g}}{192.1 \text{ g}} \times 100\% = 58.3\%$$

6.25 $$1.161 \text{ g H}_2\text{O} \times \frac{1 \text{ mol H}_2\text{O}}{18.0 \text{ g H}_2\text{O}} \times \frac{2 \text{ mol H}}{1 \text{ mol H}_2\text{O}} = 0.129 \text{ mol H}$$

$$2.818 \text{ g CO}_2 \times \frac{1 \text{ mol CO}_2}{44.0 \text{ g CO}_2} \times \frac{1 \text{ mol C}}{1 \text{ mol CO}_2} = 0.0640 \text{ mol C}$$

$$0.129 \text{ mol H} \times \frac{1.01 \text{ g H}}{1 \text{ mol H}} = 0.130 \text{ g H}$$

$$0.0640 \text{ mol C} \times \frac{12.0 \text{ g C}}{1 \text{ mol C}} = 0.768 \text{ g C}$$

$$1.00 \text{ g total} - (0.130 \text{ g H} + 0.768 \text{ g C}) = 0.102 \text{ g O}$$

$$0.102 \text{ g O} \times \frac{1 \text{ mol O}}{16.0 \text{ g O}} = 0.006\ 38 \text{ mol O}$$

$C_{0.0640}H_{0.129}O_{0.006\ 38}$; divide each subscript by the smallest, 0.006 38.
$C_{0.0640/0.006\ 38}H_{0.129/0.006\ 38}O_{0.006\ 38/0.006\ 38}$
$C_{10.03}H_{20.22}O_1$
The empirical formula is $C_{10}H_{20}O$.

6.26   The empirical formula is CH_2O, 30 amu: molecular mass = 150 amu.

$$\frac{\text{molecular mass}}{\text{empirical formula mass}} = \frac{150 \text{ amu}}{30 \text{ amu}} = 5; \text{ therefore}$$

molecular formula = 5 x empirical formula = $C_{(5\times1)}H_{(5\times2)}O_{(5\times1)} = C_5H_{10}O_5$

6.27   (a) Assume a 100.0 g sample. From the percent composition data, a 100.0 g sample contains 21.86 g H and 78.14 g B.

$$21.86 \text{ g H } \times \frac{1 \text{ mol H}}{1.01 \text{ g H}} = 21.6 \text{ mol H}$$

$$78.14 \text{ g B } \times \frac{1 \text{ mol B}}{10.8 \text{ g B}} = 7.24 \text{ mol B}$$

$B_{7.24} H_{21.6}$; divide each subscript by the smaller, 7.24.

$B_{7.24 / 7.24} H_{21.6 / 7.24}$

The empirical formula is BH_3, 13.8 amu.

27.7 amu / 13.8 amu = 2; molecular formula = $B_{(2 \times 1)} H_{(2 \times 3)} = B_2 H_6$

(b) Assume a 100.0 g sample. From the percent composition data, a 100.0 g sample contains 6.71 g H, 40.00 g C, and 53.28 g O.

$$6.71 \text{ g H } \times \frac{1 \text{ mol H}}{1.01 \text{ g H}} = 6.64 \text{ mol H}$$

$$40.00 \text{ g C } \times \frac{1 \text{ mol C}}{12.0 \text{ g C}} = 3.33 \text{ mol C}$$

$$53.28 \text{ g O } \times \frac{1 \text{ mol O}}{16.0 \text{ g O}} = 3.33 \text{ mol O}$$

$C_{3.33} H_{6.64} O_{3.33}$; divide each subscript by the smallest, 3.33.

$C_{3.33 / 3.33} H_{6.64 / 3.33} O_{3.33 / 3.33}$

The empirical formula is CH_2O, 30.0 amu.

90.08 amu / 30.0 amu = 3; molecular formula = $C_{(3 \times 1)} H_{(3 \times 2)} O_{(3 \times 1)} = C_3 H_6 O_3$

6.28   Main sources of error in calculating Avogadro's number by spreading oil on a pond are:
(i)   the assumption that the oil molecules are tiny cubes
(ii) the assumption that the oil layer is one molecule thick
(iii) the assumption of a molecular mass of 900 amu for the oil

6.29   area of oil = $2.0 \times 10^7 \text{ cm}^2$

volume of oil = 4.9 cm^3 = area x 4 l = $(2.0 \times 10^7 \text{ cm}^2)$ x 4 l

$$l = \frac{4.9 \text{ cm}^3}{(2.0 \times 10^7 \text{ cm}^2)(4)} = 6.125 \times 10^{-8} \text{ cm}$$

area of oil = $2.0 \times 10^7 \text{ cm}^2 = l^2$ x N = $(6.125 \times 10^{-8} \text{ cm})^2$ x N

$$N = \frac{2.0 \times 10^7 \text{ cm}^2}{(6.125 \times 10^{-8} \text{ cm})^2} = 5.33 \times 10^{21} \text{ oil molecules}$$

moles of oil = (4.9 cm^3) x (0.95 g/cm^3) x $\dfrac{1 \text{ mol oil}}{900 \text{ g oil}}$ = 0.0052 mol oil

Avogadro's number = $\dfrac{5.33 \times 10^{21} \text{ molecules}}{0.0052 \text{ mol}}$ = 1.0×10^{24} molecules/mole

Key Concept Problems

6.30   The concentration of a solution is cut in half when the volume is doubled. This is best represented by box (b).

6.31   (c) $2 A + B_2 \rightarrow A_2B_2$

6.32   The molecular formula for cytosine is $C_4H_5N_3O$.

$$\text{mol } CO_2 = 0.001 \text{ mol cyt } \times \frac{4 C}{\text{cyt}} \times \frac{1 CO_2}{C} = 0.004 \text{ mol } CO_2$$

$$\text{mol } H_2O = 0.001 \text{ mol cyt } \times \frac{5 H}{\text{cyt}} \times \frac{1 H_2O}{2 H} = 0.0025 \text{ mol } H_2O$$

6.33   reactants, box (d), and products, box (c)

6.34 $C_{17}H_{18}F_3NO$ $17(12.01) + 18(1.01) + 3(19.00) + 1(14.01) + 1(16.00) = 309.36$ amu

6.35 $C_3H_7NO_2S$, 121.2 amu or 121.2 g/mol

$$\% C = \frac{3 \times 12.0 \text{ g}}{121.2 \text{ g}} \times 100\% = 29.7\%$$

$$\% H = \frac{7 \times 1.01 \text{ g}}{121.2 \text{ g}} \times 100\% = 5.83\%$$

$$\% N = \frac{1 \times 14.0 \text{ g}}{121.2 \text{ g}} \times 100\% = 11.6\%$$

$$\% O = \frac{2 \times 16.0 \text{ g}}{121.2 \text{ g}} \times 100\% = 26.4\%$$

$$\% S = \frac{1 \times 32.1 \text{ g}}{121.2 \text{ g}} \times 100\% = 26.5\%$$

6.36   (a) $A_2 + 3 B_2 \rightarrow 2 AB_3$; B_2 is the limiting reactant because it is completely consumed.
(b) For 1.0 mol of A_2, 3.0 mol of B_2 are required. Because only 1.0 mol of B_2 is available, B_2 is the limiting reactant.

$$1 \text{ mol } B_2 \times \frac{2 \text{ mol } AB_3}{3 \text{ mol } B_2} = 2/3 \text{ mol } AB_3$$

6.37 $C_xH_y \overset{O_2}{\rightarrow} 3 CO_2 + 4 H_2O$; x is equal to the coefficient for CO_2 and y is equal to 2 times the coefficient for H_2O. The empirical formula for the hydrocarbon is C_3H_8.

Section Problems
Balancing Equations (Section 6.1)

6.38   Equation (b) is balanced, (a) is not balanced.

6.39   (a) and (c) are not balanced, (b) is balanced.
(a) $2\,Al + Fe_2O_3 \rightarrow Al_2O_3 + 2\,Fe$ (balanced)
(c) $4\,Au + 8\,NaCN + O_2 + 2\,H_2O \rightarrow 4\,NaAu(CN)_2 + 4\,NaOH$ (balanced)

6.40   (a) $Mg + 2\,HNO_3 \rightarrow H_2 + Mg(NO_3)_2$
(b) $CaC_2 + 2\,H_2O \rightarrow Ca(OH)_2 + C_2H_2$
(c) $2\,S + 3\,O_2 \rightarrow 2\,SO_3$
(d) $UO_2 + 4\,HF \rightarrow UF_4 + 2\,H_2O$

6.41   (a) $2\,NH_4NO_3 \rightarrow 2\,N_2 + O_2 + 4\,H_2O$
(b) $C_2H_6O + O_2 \rightarrow C_2H_4O_2 + H_2O$
(c) $C_2H_8N_2 + 2\,N_2O_4 \rightarrow 3\,N_2 + 2\,CO_2 + 4\,H_2O$

Molecular Masses and Stoichiometry (Section 6.3)

6.42 Hg_2Cl_2: $2(200.59) + 2(35.45) = 472.1$ amu
$C_4H_8O_2$: $4(12.01) + 8(1.01) + 2(16.00) = 88.1$ amu
CF_2Cl_2: $1(12.01) + 2(19.00) + 2(35.45) = 120.9$ amu

6.43   (a) $(1 \times 30.97$ amu$) + (Y \times 35.45$ amu$) = 137.3$ amu; Solve for Y; $Y = 3$.
The formula is PCl_3.
(b) $(10 \times 12.01$ amu$) + (14 \times 1.008$ amu$) + (Z \times 14.01$ amu$) = 162.2$ amu
Solve for Z; $Z = 2$. The formula is $C_{10}H_{14}N_2$.

6.44   One mole equals the atomic mass or molecular mass in grams.
(a) Ti, 47.87 g   (b) Br_2, 159.81 g   (c) Hg, 200.59 g   (d) H_2O, 18.02 g

6.45   (a) 1.00 g Cr $\times \dfrac{1\text{ mol Cr}}{52.0\text{ g Cr}} = 0.0192$ mol Cr

(b) 1.00 g $Cl_2 \times \dfrac{1\text{ mol }Cl_2}{70.9\text{ g }Cl_2} = 0.0141$ mol Cl_2

(c) 1.00 g Au $\times \dfrac{1\text{ mol Au}}{197.0\text{ g Au}} = 0.005\ 08$ mol Au

(d) 1.00 g $NH_3 \times \dfrac{1\text{ mol }NH_3}{17.0\text{ g }NH_3} = 0.0588$ mol NH_3

6.46   There are 2 ions per formula unit of NaCl. (2.5 mol)(2 mol ions/mol) = 5.0 mol ions

6.47   There are 2 K^+ ions per formula unit of K_2SO_4.

$$1.45 \text{ mol } K_2SO_4 \ \times \ \frac{2 \text{ mol } K^+}{1 \text{ mol } K_2SO_4} \ = 2.90 \text{ mol } K^+$$

6.48   There are 3 ions (one Mg^{2+} and 2 Cl^-) per formula unit of $MgCl_2$.
$MgCl_2$, 95.2 amu

$$27.5 \text{ g } MgCl_2 \ \times \ \frac{1 \text{ mol } MgCl_2}{95.2 \text{ g } MgCl_2} \ \times \ \frac{3 \text{ mol ions}}{1 \text{ mol } MgCl_2} \ = 0.867 \text{ mol ions}$$

6.49   There are 3 F^- anions per formula unit of AlF_3.
AlF_3, 84.0 amu

$$35.6 \text{ g } AlF_3 \ \times \ \frac{1 \text{ mol } AlF_3}{84.0 \text{ g } AlF_3} \ \times \ \frac{3 \text{ mol anions}}{1 \text{ mol } AlF_3} \ = 1.27 \text{ mol } F^-$$

6.50   Molar mass = $\dfrac{3.28 \text{ g}}{0.0275 \text{ mol}}$ = 119 g/mol; molecular mass = 119 amu.

6.51   Molar mass = $\dfrac{221.6 \text{ g}}{0.5731 \text{ mol}}$ = 386.7 g/mol; molecular mass = 386.7 amu.

6.52 $FeSO_4$, 151.9 amu; 300 mg = 0.300 g

$$0.300 \text{ g } FeSO_4 \ \times \ \frac{1 \text{ mol } FeSO_4}{151.9 \text{ g } FeSO_4} \ = 1.97 \ \times \ 10^{-3} \text{ mol } FeSO_4$$

$$1.97 \ \times \ 10^{-3} \text{ mol } FeSO_4 \ \times \ \frac{6.022 \ \times \ 10^{23} \text{ Fe(II) atoms}}{1 \text{ mol } FeSO_4} \ = 1.19 \ \times \ 10^{21} \text{ Fe(II) atoms}$$

6.53 $0.0001 \text{ g C} \ \times \ \dfrac{1 \text{ mol C}}{12.0 \text{ g C}} \ \times \ \dfrac{6.02 \times 10^{23} \text{ C atoms}}{1 \text{ mol C}} \ = 5 \times 10^{18} \text{ C atoms}$

6.54 $C_8H_{10}N_4O_2$, 194.2 amu; 125 mg = 0.125 g

$$0.125 \text{ g caffeine} \ \times \ \frac{1 \text{ mol caffeine}}{194.2 \text{ g caffeine}} \ = 6.44 \times 10^{-4} \text{ mol caffeine}$$

$$0.125 \text{ g caffeine} \ \times \ \frac{1 \text{ mol caffeine}}{194.2 \text{ g caffeine}} \ \times \ \frac{6.022 \times 10^{23} \text{ molecules}}{1 \text{ mol}} \ = 3.88 \times 10^{20} \text{ caffeine molecules}$$

6.55 $45 \ \dfrac{\text{g}}{\text{egg}} \ \times \ \dfrac{6.02 \times 10^{23} \text{ eggs}}{1 \text{ mol eggs}} \ = 2.7 \times 10^{25} \text{ g/mol of eggs}$

6.56   (a) $1.0 \text{ g Li} \times \dfrac{1 \text{ mol Li}}{6.94 \text{ g Li}} = 0.14 \text{ mol Li}$

(b) $1.0 \text{ g Au} \times \dfrac{1 \text{ mol Au}}{197.0 \text{ g Au}} = 0.0051 \text{ mol Au}$

(c) penicillin G: $C_{16}H_{17}N_2O_4SK$, 372.5 amu

$1.0 \text{ g} \times \dfrac{1 \text{ mol penicillin G}}{372.5 \text{ g penicillin G}} = 2.7 \times 10^{-3} \text{ mol penicillin G}$

6.57   (a) $0.0015 \text{ mol Na} \times \dfrac{23.0 \text{ g Na}}{1 \text{ mol Na}} = 0.034 \text{ g Na}$

(b) $0.0015 \text{ mol Pb} \times \dfrac{207.2 \text{ g Pb}}{1 \text{ mol Pb}} = 0.31 \text{ g Pb}$

(c) $C_{16}H_{13}ClN_2O$, 284.7 amu

$0.0015 \text{ mol diazepam} \times \dfrac{284.7 \text{ g diazepam}}{1 \text{ mol diazepam}} = 0.43 \text{ g diazepam}$

6.58 TiO_2, 79.87 amu; $100.0 \text{ kg Ti} \times \dfrac{79.87 \text{ kg TiO}_2}{47.87 \text{ kg Ti}} = 166.8 \text{ kg TiO}_2$

6.59 Fe_2O_3, 159.7 amu; $\% \text{ Fe} = \dfrac{2(55.85 \text{ g}) \text{ Fe}}{159.7 \text{ g Fe}_2O_3} \times 100\% = 69.94\%$

mass Fe = (0.6994)(105 kg) = 73.4 kg

6.60   (a) $2 \text{ Fe}_2O_3 + 3 \text{ C} \rightarrow 4 \text{ Fe} + 3 \text{ CO}_2$

(b) Fe_2O_3, 159.7 amu; $525 \text{ g Fe}_2O_3 \times \dfrac{1 \text{ mol Fe}_2O_3}{159.7 \text{ g Fe}_2O_3} \times \dfrac{3 \text{ mol C}}{2 \text{ mol Fe}_2O_3} = 4.93 \text{ mol C}$

(c) $4.93 \text{ mol C} \times \dfrac{12.01 \text{ g C}}{1 \text{ mol C}} = 59.2 \text{ g C}$

6.61   (a) $Fe_2O_3 + 3 \text{ CO} \rightarrow 2 \text{ Fe} + 3 \text{ CO}_2$
(b) Fe_2O_3, 159.7 amu; CO, 28.01 amu

$3.02 \text{ g Fe}_2O_3 \times \dfrac{1 \text{ mol Fe}_2O_3}{159.7 \text{ g Fe}_2O_3} \times \dfrac{3 \text{ mol CO}}{1 \text{ mol Fe}_2O_3} \times \dfrac{28.01 \text{ g CO}}{1 \text{ mol CO}} = 1.59 \text{ g CO}$

(c) $1.68 \text{ mol Fe}_2O_3 \times \dfrac{3 \text{ mol CO}}{1 \text{ mol Fe}_2O_3} \times \dfrac{28.01 \text{ g CO}}{1 \text{ mol CO}} = 141 \text{ g CO}$

6.62 (a) $2 Mg + O_2 \rightarrow 2 MgO$

(b) Mg, 24.30 amu; O_2, 32.00 amu; MgO, 40.30 amu

$$25.0 \text{ g Mg} \times \frac{1 \text{ mol Mg}}{24.30 \text{ g Mg}} \times \frac{1 \text{ mol O}_2}{2 \text{ mol Mg}} \times \frac{32.00 \text{ g O}_2}{1 \text{ mol O}_2} = 16.5 \text{ g O}_2$$

$$25.0 \text{ g Mg} \times \frac{1 \text{ mol Mg}}{24.30 \text{ g Mg}} \times \frac{2 \text{ mol MgO}}{2 \text{ mol Mg}} \times \frac{40.30 \text{ g MgO}}{1 \text{ mol MgO}} = 41.5 \text{ g MgO}$$

(c) $$25.0 \text{ g O}_2 \times \frac{1 \text{ mol O}_2}{32.00 \text{ g O}_2} \times \frac{2 \text{ mol Mg}}{1 \text{ mol O}_2} \times \frac{24.30 \text{ g Mg}}{1 \text{ mol Mg}} = 38.0 \text{ g Mg}$$

$$25.0 \text{ g O}_2 \times \frac{1 \text{ mol O}_2}{32.00 \text{ g O}_2} \times \frac{2 \text{ mol MgO}}{1 \text{ mol O}_2} \times \frac{40.30 \text{ g MgO}}{1 \text{ mol MgO}} = 63.0 \text{ g MgO}$$

6.63 $C_2H_4 + H_2O \rightarrow C_2H_6O$; C_2H_4, 28.05 amu; H_2O, 18.02 amu; C_2H_6O, 46.07 amu

(a) $$0.133 \text{ mol H}_2O \times \frac{1 \text{ mol C}_2H_4}{1 \text{ mol H}_2O} \times \frac{28.05 \text{ g C}_2H_4}{1 \text{ mol C}_2H_4} = 3.73 \text{ g C}_2H_4$$

$$0.133 \text{ mol H}_2O \times \frac{1 \text{ mol C}_2H_6O}{1 \text{ mol H}_2O} \times \frac{46.07 \text{ g C}_2H_6O}{1 \text{ mol C}_2H_6O} = 6.13 \text{ g C}_2H_6O$$

(b) $$0.371 \text{ mol C}_2H_4 \times \frac{1 \text{ mol H}_2O}{1 \text{ mol C}_2H_4} \times \frac{18.02 \text{ g H}_2O}{1 \text{ mol H}_2O} = 6.69 \text{ g H}_2O$$

$$0.371 \text{ mol C}_2H_4 \times \frac{1 \text{ mol C}_2H_6O}{1 \text{ mol C}_2H_4} \times \frac{46.07 \text{ g C}_2H_6O}{1 \text{ mol C}_2H_6O} = 17.1 \text{ g C}_2H_6O$$

6.64 (a) $2 HgO \rightarrow 2 Hg + O_2$

(b) HgO, 216.6 amu; Hg, 200.6 amu; O_2, 32.0 amu

$$45.5 \text{ g HgO} \times \frac{1 \text{ mol HgO}}{216.6 \text{ g HgO}} \times \frac{2 \text{ mol Hg}}{2 \text{ mol HgO}} \times \frac{200.6 \text{ g Hg}}{1 \text{ mol Hg}} = 42.1 \text{ g Hg}$$

$$45.5 \text{ g HgO} \times \frac{1 \text{ mol HgO}}{216.6 \text{ g HgO}} \times \frac{1 \text{ mol O}_2}{2 \text{ mol HgO}} \times \frac{32.00 \text{ g O}_2}{1 \text{ mol O}_2} = 3.36 \text{ g O}_2$$

(c) $$33.3 \text{ g O}_2 \times \frac{1 \text{ mol O}_2}{32.00 \text{ g O}_2} \times \frac{2 \text{ mol HgO}}{1 \text{ mol O}_2} \times \frac{216.6 \text{ g HgO}}{1 \text{ mol HgO}} = 451 \text{ g HgO}$$

6.65 5.60 kg = 5600 g; $TiCl_4$, 189.7 amu; TiO_2, 79.87 amu

$$5600 \text{ g TiCl}_4 \times \frac{1 \text{ mol TiCl}_4}{189.7 \text{ g TiCl}_4} \times \frac{1 \text{ mol TiO}_2}{1 \text{ mol TiCl}_4} \times \frac{79.87 \text{ g TiO}_2}{1 \text{ mol TiO}_2} = 2358 \text{ g TiO}_2 = 2.36 \text{ kg TiO}_2$$

6.66　$2.00 \text{ g Ag } \times \dfrac{1 \text{ mol Ag}}{107.9 \text{ g Ag}} = 0.0185 \text{ mol Ag}; \quad 0.657 \text{ g Cl } \times \dfrac{1 \text{ mol Cl}}{35.45 \text{ g Cl}} = 0.0185 \text{ mol Cl}$

$Ag_{0.0185}Cl_{0.0185}$; divide both subscripts by 0.0185.
The empirical formula is AgCl.

6.67　$5.0 \text{ g Al } \times \dfrac{1 \text{ mol Al}}{27.0 \text{ g Al}} = 0.19 \text{ mol Al}; \quad 4.45 \text{ g O } \times \dfrac{1 \text{ mol O}}{16.0 \text{ g O}} = 0.28 \text{ mol O}$

$Al_{0.19}O_{0.28}$; divide both subscripts by the smaller, 0.19.
$Al_{0.19/0.19}O_{0.28/0.19}$
$Al_1O_{1.5}$; multiply both subscripts by 2 to obtain integers.
The empirical formula is Al_2O_3.

Limiting Reactants and Reaction Yield (Sections 6.4–6.5)

6.68　$3.44 \text{ mol N}_2 \times \dfrac{3 \text{ mol H}_2}{1 \text{ mol N}_2} = 10.3 \text{ mol H}_2 \text{ required.}$

Because there is only 1.39 mol H_2, H_2 is the limiting reactant.

$1.39 \text{ mol H}_2 \times \dfrac{2 \text{ mol NH}_3}{3 \text{ mol H}_2} \times \dfrac{17.03 \text{ g NH}_3}{1 \text{ mol NH}_3} = 15.8 \text{ g NH}_3$

$1.39 \text{ mol H}_2 \times \dfrac{1 \text{ mol N}_2}{3 \text{ mol H}_2} \times \dfrac{28.01 \text{ g N}_2}{1 \text{ mol N}_2} = 13.0 \text{ g N}_2 \text{ reacted}$

$3.44 \text{ mol N}_2 \times \dfrac{28.01 \text{ g N}_2}{1 \text{ mol N}_2} = 96.3 \text{ g N}_2 \text{ initially}$

$(96.3 \text{ g} - 13.0 \text{ g}) = 83.3 \text{ g N}_2 \text{ left over}$

6.69　H_2, 2.016 amu; Cl_2, 70.91 amu; HCl 36.46 amu

$3.56 \text{ g H}_2 \times \dfrac{1 \text{ mol H}_2}{2.016 \text{ g H}_2} = 1.77 \text{ mol H}_2$

$8.94 \text{ g Cl}_2 \times \dfrac{1 \text{ mol Cl}_2}{70.91 \text{ g Cl}_2} = 0.126 \text{ mol Cl}_2$

Because the reaction stoichiometry between H_2 and Cl_2 is one to one, Cl_2 is the limiting reactant.

$0.126 \text{ mol Cl}_2 \times \dfrac{2 \text{ mol HCl}}{1 \text{ mol Cl}_2} \times \dfrac{36.46 \text{ g HCl}}{1 \text{ mol HCl}} = 9.19 \text{ g HCl}$

6.70 C_2H_4, 28.05 amu; Cl_2, 70.91 amu; $C_2H_4Cl_2$, 98.96 amu

$$15.4 \text{ g } C_2H_4 \text{ x } \frac{1 \text{ mol } C_2H_4}{28.05 \text{ g } C_2H_4} = 0.549 \text{ mol } C_2H_4$$

$$3.74 \text{ g } Cl_2 \text{ x } \frac{1 \text{ mol } Cl_2}{70.91 \text{ g } Cl_2} = 0.0527 \text{ mol } Cl_2$$

Because the reaction stoichiometry between C_2H_4 and Cl_2 is one to one, Cl_2 is the limiting reactant.

$$0.0527 \text{ mol } Cl_2 \text{ x } \frac{1 \text{ mol } C_2H_4Cl_2}{1 \text{ mol } Cl_2} \text{ x } \frac{98.96 \text{ g } C_2H_4Cl_2}{1 \text{ mol } C_2H_4Cl_2} = 5.22 \text{ g } C_2H_4Cl_2$$

6.71 (a) NaCl, 58.44 amu; $AgNO_3$, 169.9 amu; AgCl, 143.3 amu; $NaNO_3$, 85.00 amu

$$NaCl + AgNO_3 \rightarrow AgCl + NaNO_3$$

$$1.3 \text{ g NaCl } \text{ x } \frac{1 \text{ mol NaCl}}{58.44 \text{ g NaCl}} = 0.022 \text{ mol NaCl}$$

$$3.5 \text{ g } AgNO_3 \text{ x } \frac{1 \text{ mol } AgNO_3}{169.9 \text{ g } AgNO_3} = 0.021 \text{ mol } AgNO_3$$

Because the reaction stoichiometry between NaCl and $AgNO_3$ is one to one, $AgNO_3$ is the limiting reactant.

$$0.021 \text{ mol } AgNO_3 \text{ x } \frac{1 \text{ mol AgCl}}{1 \text{ mol } AgNO_3} \text{ x } \frac{143.3 \text{ g AgCl}}{1 \text{ mol AgCl}} = 3.0 \text{ g AgCl}$$

$$0.021 \text{ mol } AgNO_3 \text{ x } \frac{1 \text{ mol } NaNO_3}{1 \text{ mol } AgNO_3} \text{ x } \frac{85.00 \text{ g } NaNO_3}{1 \text{ mol } NaNO_3} = 1.8 \text{ g } NaNO_3$$

$$0.021 \text{ mol } AgNO_3 \text{ x } \frac{1 \text{ mol NaCl}}{1 \text{ mol } AgNO_3} \text{ x } \frac{58.44 \text{ g NaCl}}{1 \text{ mol NaCl}} = 1.2 \text{ g NaCl reacted}$$

(1.3 g – 1.2 g) = 0.1 g NaCl left over

(b) $BaCl_2$, 208.2 amu; H_2SO_4, 98.08 amu; $BaSO_4$, 233.4 amu; HCl, 36.46 amu

$$BaCl_2 + H_2SO_4 \rightarrow BaSO_4 + 2 \text{ HCl}$$

$$2.65 \text{ g } BaCl_2 \text{ x } \frac{1 \text{ mol } BaCl_2}{208.2 \text{ g } BaCl_2} = 0.0127 \text{ mol } BaCl_2$$

$$6.78 \text{ g } H_2SO_4 \text{ x } \frac{1 \text{ mol } H_2SO_4}{98.08 \text{ g } H_2SO_4} = 0.0691 \text{ mol } H_2SO_4$$

Because the reaction stoichiometry between $BaCl_2$ and H_2SO_4 is one to one, $BaCl_2$ is the limiting reactant.

$$0.0127 \text{ mol BaCl}_2 \text{ x } \frac{1 \text{ mol BaSO}_4}{1 \text{ mol BaCl}_2} \text{ x } \frac{233.4 \text{ g BaSO}_4}{1 \text{ mol BaSO}_4} = 2.96 \text{ g BaSO}_4$$

$$0.0127 \text{ mol BaCl}_2 \text{ x } \frac{2 \text{ mol HCl}}{1 \text{ mol BaCl}_2} \text{ x } \frac{36.46 \text{ g HCl}}{1 \text{ mol HCl}} = 0.926 \text{ g HCl}$$

$$0.0127 \text{ mol BaCl}_2 \text{ x } \frac{1 \text{ mol H}_2\text{SO}_4}{1 \text{ mol BaCl}_2} \text{ x } \frac{98.1 \text{ g H}_2\text{SO}_4}{1 \text{ mol H}_2\text{SO}_4} = 1.25 \text{ g H}_2\text{SO}_4 \text{ reacted}$$

$(6.78 \text{ g} - 1.25 \text{ g}) = 5.53 \text{ g H}_2\text{SO}_4 \text{ left over}$

6.72 $CaCO_3$, 100.1 amu; HCl, 36.46 amu

$$CaCO_3 + 2 HCl \rightarrow CaCl_2 + H_2O + CO_2$$

$$2.35 \text{ g CaCO}_3 \text{ x } \frac{1 \text{ mol CaCO}_3}{100.1 \text{ g CaCO}_3} = 0.0235 \text{ mol CaCO}_3$$

$$2.35 \text{ g HCl} \text{ x } \frac{1 \text{ mol HCl}}{36.46 \text{ g HCl}} = 0.0645 \text{ mol HCl}$$

The reaction stoichiometry is 1 mole of $CaCO_3$ for every 2 moles of HCl. For 0.0235 mol $CaCO_3$, we only need 2(0.0235 mol) = 0.0470 mol HCl. We have 0.0645 mol HCl, therefore $CaCO_3$ is the limiting reactant.

$$0.0235 \text{ mol CaCO}_3 \text{ x } \frac{1 \text{ mol CO}_2}{1 \text{ mol CaCO}_3} \text{ x } \frac{22.4 \text{ L}}{1 \text{ mol CO}_2} = 0.526 \text{ L CO}_2$$

6.73 $2 NaN_3 \rightarrow 3 N_2 + 2 Na$; NaN_3, 65.01 amu; N_2, 28.01 amu

$$38.5 \text{ g NaN}_3 \text{ x } \frac{1 \text{ mol NaN}_3}{65.01 \text{ g NaN}_3} \text{ x } \frac{3 \text{ mol N}_2}{2 \text{ mol NaN}_3} \text{ x } \frac{47.0 \text{ L}}{1.00 \text{ mol N}_2} = 41.8 \text{ L}$$

6.74 $CH_3CO_2H + C_5H_{12}O \rightarrow C_7H_{14}O_2 + H_2O$

CH_3CO_2H, 60.05 amu; $C_5H_{12}O$, 88.15 amu; $C_7H_{14}O_2$, 130.19 amu

$$3.58 \text{ g CH}_3\text{CO}_2\text{H} \text{ x } \frac{1 \text{ mol CH}_3\text{CO}_2\text{H}}{60.05 \text{ g CH}_3\text{CO}_2\text{H}} = 0.0596 \text{ mol CH}_3\text{CO}_2\text{H}$$

$$4.75 \text{ g C}_5\text{H}_{12}\text{O} \text{ x } \frac{1 \text{ mol C}_5\text{H}_{12}\text{O}}{88.15 \text{ g C}_5\text{H}_{12}\text{O}} = 0.0539 \text{ mol C}_5\text{H}_{12}\text{O}$$

Because the reaction stoichiometry between CH_3CO_2H and $C_5H_{12}O$ is one to one, isopentyl alcohol ($C_5H_{12}O$) is the limiting reactant.

$$0.0539 \text{ mol C}_5\text{H}_{12}\text{O} \text{ x } \frac{1 \text{ mol C}_7\text{H}_{14}\text{O}_2}{1 \text{ mol C}_5\text{H}_{12}\text{O}} \text{ x } \frac{130.19 \text{ g C}_7\text{H}_{14}\text{O}_2}{1 \text{ mol C}_7\text{H}_{14}\text{O}_2} = 7.02 \text{ g C}_7\text{H}_{14}\text{O}_2$$

7.02 g $C_7H_{14}O_2$ is the theoretical yield. Actual yield = (7.02 g)(0.45) = 3.2 g.

6.75 $K_2PtCl_4 + 2 NH_3 \rightarrow 2 KCl + Pt(NH_3)_2Cl_2$

K_2PtCl_4, 415.1 amu; NH_3, 17.03 amu; $Pt(NH_3)_2Cl_2$, 300.0 amu

$$55.8 \text{ g } K_2PtCl_4 \times \frac{1 \text{ mol } K_2PtCl_4}{415.1 \text{ g } K_2PtCl_4} = 0.134 \text{ mol } K_2PtCl_4$$

$$35.6 \text{ g } NH_3 \times \frac{1 \text{ mol } NH_3}{17.03 \text{ g } NH_3} = 2.09 \text{ mol } NH_3$$

Only 2(0.134) = 0.268 mol NH_3 are needed to react with 0.134 mol K_2PtCl_4. Therefore, the NH_3 is in excess and K_2PtCl_4 is the limiting reactant.

$$0.134 \text{ mol } K_2PtCl_4 \times \frac{1 \text{ mol } Pt(NH_3)_2Cl_2}{1 \text{ mol } K_2PtCl_4} \times \frac{300.0 \text{ g } Pt(NH_3)_2Cl_2}{1 \text{ mol } Pt(NH_3)_2Cl_2} = 40.2 \text{ g } Pt(NH_3)_2Cl_2$$

40.2 g $Pt(NH_3)_2Cl_2$ is the theoretical yield.
Actual yield = (40.2 g)(0.95) = 38 g $Pt(NH_3)_2Cl_2$.

6.76 $CH_3CO_2H + C_5H_{12}O \rightarrow C_7H_{14}O_2 + H_2O$

CH_3CO_2H, 60.05 amu; $C_5H_{12}O$, 88.15 amu; $C_7H_{14}O_2$, 130.19 amu

$$1.87 \text{ g } CH_3CO_2H \times \frac{1 \text{ mol } CH_3CO_2H}{60.05 \text{ g } CH_3CO_2H} = 0.0311 \text{ mol } CH_3CO_2H$$

$$2.31 \text{ g } C_5H_{12}O \times \frac{1 \text{ mol } C_5H_{12}O}{88.15 \text{ g } C_5H_{12}O} = 0.0262 \text{ mol } C_5H_{12}O$$

Because the reaction stoichiometry between CH_3CO_2H and $C_5H_{12}O$ is one to one, isopentyl alcohol ($C_5H_{12}O$) is the limiting reactant.

$$0.0262 \text{ mol } C_5H_{12}O \times \frac{1 \text{ mol } C_7H_{14}O_2}{1 \text{ mol } C_5H_{12}O} \times \frac{130.19 \text{ g } C_7H_{14}O_2}{1 \text{ mol } C_7H_{14}O_2} = 3.41 \text{ g } C_7H_{14}O_2$$

3.41 g $C_7H_{14}O_2$ is the theoretical yield.

$$\% \text{ Yield} = \frac{\text{Actual yield}}{\text{Theoretical yield}} \times 100\% = \frac{2.96 \text{ g}}{3.41 \text{ g}} \times 100\% = 86.8\%$$

6.77 $K_2PtCl_4 + 2 NH_3 \rightarrow 2 KCl + Pt(NH_3)_2Cl_2$

K_2PtCl_4, 415.1 amu; NH_3, 17.03 amu; $Pt(NH_3)_2Cl_2$, 300.0 amu

$$3.42 \text{ g } K_2PtCl_4 \times \frac{1 \text{ mol } K_2PtCl_4}{415.1 \text{ g } K_2PtCl_4} = 0.008 \ 24 \text{ mol } K_2PtCl_4$$

$$1.61 \text{ g } NH_3 \times \frac{1 \text{ mol } NH_3}{17.03 \text{ g } NH_3} = 0.0945 \text{ mol } NH_3$$

Only 2 x (0.008 24) = 0.0165 mol of NH_3 are needed to react with 0.008 24 mol K_2PtCl_4. Therefore, the NH_3 is in excess and K_2PtCl_4 is the limiting reactant.

$$0.008\ 24\ \text{mol}\ K_2PtCl_4 \times \frac{1\ \text{mol}\ Pt(NH_3)_2Cl_2}{1\ \text{mol}\ K_2PtCl_4} \times \frac{300.0\ \text{g}\ Pt(NH_3)_2Cl_2}{1\ \text{mol}\ Pt(NH_3)_2Cl_2} = 2.47\ \text{g}\ Pt(NH_3)_2Cl_2$$

2.47 g $Pt(NH_3)_2Cl_2$ is the theoretical yield. 2.08 g $Pt(NH_3)_2Cl_2$ is the actual yield.

$$\%\ \text{Yield} = \frac{\text{Actual yield}}{\text{Theoretical yield}} \times 100\% = \frac{2.08\ \text{g}}{2.47\ \text{g}} \times 100\% = 84.2\%$$

Molarity, Solution Stoichiometry, Dilution, and Titration (Sections 6.6–6.9)

6.78 (a) 35.0 mL = 0.0350 L; $\dfrac{1.200\ \text{mol}\ HNO_3}{L} \times 0.0350\ \text{L} = 0.0420\ \text{mol}\ HNO_3$

(b) 175 mL = 0.175 L; $\dfrac{0.67\ \text{mol}\ C_6H_{12}O_6}{L} \times 0.175\ \text{L} = 0.12\ \text{mol}\ C_6H_{12}O_6$

6.79 (a) C_2H_6O, 46.07 amu; 250.0 mL = 0.2500 L

$\dfrac{0.600\ \text{mol}\ C_2H_6O}{L} \times 0.2500\ \text{L} = 0.150\ \text{mol}\ C_2H_6O$

(0.150 mol)(46.07 g/mol) = 6.91 g C_2H_6O

(b) H_3BO_3, 61.83 amu; 167 mL = 0.167 L

$\dfrac{0.200\ \text{mol}\ H_3BO_3}{L} \times 0.167\ \text{L} = 0.0334\ \text{mol}\ H_3BO_3$

(0.0334 mol)(61.83 g/mol) = 2.07 g H_3BO_3

6.80 $BaCl_2$, 208.2 amu

$15.0\ \text{g}\ BaCl_2 \times \dfrac{1\ \text{mol}\ BaCl_2}{208.2\ \text{g}\ BaCl_2} = 0.0720\ \text{mol}\ BaCl_2$

$0.0720\ \text{mol} \times \dfrac{1.0\ \text{L}}{0.45\ \text{mol}} = 0.16\ \text{L};\quad 0.16\ \text{L} = 160\ \text{mL}$

6.81 $0.0171\ \text{mol}\ KOH \times \dfrac{1.00\ \text{L}}{0.350\ \text{mol}\ KOH} = 0.0489\ \text{L};\quad 0.0489\ \text{L} = 48.9\ \text{mL}$

6.82 NaCl, 58.4 amu; 400 mg = 0.400 g; 100 mL = 0.100 L

$0.400\ \text{g}\ NaCl \times \dfrac{1\ \text{mol}\ NaCl}{58.4\ \text{g}\ NaCl} = 0.006\ 85\ \text{mol}\ NaCl$

$\text{molarity} = \dfrac{0.006\ 85\ \text{mol}}{0.100\ \text{L}} = 0.0685\ M$

6.83 $C_6H_{12}O_6$, 180.2 amu; 90 mg = 0.090 g; 100 mL = 0.100 L

$$0.090 \text{ g } C_6H_{12}O_6 \times \frac{1 \text{ mol } C_6H_{12}O_6}{180.2 \text{ g } C_6H_{12}O_6} = 0.000\ 50 \text{ mol } C_6H_{12}O_6$$

$$\text{molarity} = \frac{0.000\ 50 \text{ mol}}{0.100 \text{ L}} = 0.0050 \text{ M} = 5.0 \times 10^{-3} \text{ M}$$

6.84   NaCl, 58.4 amu; KCl, 74.6 amu; $CaCl_2$, 111.0 amu; 500 mL = 0.500 L

$$4.30 \text{ g NaCl} \times \frac{1 \text{ mol NaCl}}{58.4 \text{ g NaCl}} = 0.0736 \text{ mol NaCl}$$

$$0.150 \text{ g KCl} \times \frac{1 \text{ mol KCl}}{74.6 \text{ g KCl}} = 0.002\ 01 \text{ mol KCl}$$

$$0.165 \text{ g } CaCl_2 \times \frac{1 \text{ mol } CaCl_2}{111.0 \text{ g } CaCl_2} = 0.001\ 49 \text{ mol } CaCl_2$$

0.0736 mol + 0.002 01 mol + 2(0.001 49 mol) = 0.0786 mol Cl^-

$$Na^+ \text{ molarity} = \frac{0.0736 \text{ mol}}{0.500 \text{ L}} = 0.147 \text{ M}$$

$$Ca^{2+} \text{ molarity} = \frac{0.001\ 49 \text{ mol}}{0.500 \text{ L}} = 0.002\ 98 \text{ M}$$

$$K^+ \text{ molarity} = \frac{0.002\ 01 \text{ mol}}{0.500 \text{ L}} = 0.004\ 02 \text{ M}$$

$$Cl^- \text{ molarity} = \frac{0.0786 \text{ mol}}{0.500 \text{ L}} = 0.157 \text{ M}$$

6.85 $3.045 \text{ g Cu} \times \dfrac{1 \text{ mol Cu}}{63.546 \text{ g Cu}} = 0.047\ 92 \text{ mol Cu}$; 50.0 mL = 0.0500 L

$$Cu(NO_3)_2 \text{ molarity} = \frac{0.047\ 92 \text{ mol}}{0.0500 \text{ L}} = 0.958 \text{ M}$$

6.86 $M_f \times V_f = M_i \times V_i$; $M_f = \dfrac{M_i \times V_i}{V_f} = \dfrac{12.0 \text{ M} \times 35.7 \text{ mL}}{250.0 \text{ mL}} = 1.71 \text{ M HCl}$

6.87 $M_f \times V_f = M_i \times V_i$; $V_f = \dfrac{M_i \times V_i}{M_f} = \dfrac{0.0913 \text{ M} \times 70.00 \text{ mL}}{0.0150 \text{ M}} = 426 \text{ mL}$

6.88 $2\ HBr(aq)\ +\ K_2CO_3(aq)\ \rightarrow\ 2\ KBr(aq)\ +\ CO_2(g)\ +\ H_2O(l)$

K_2CO_3, 138.2 amu; 450 mL = 0.450 L

$$\frac{0.500\ \text{mol HBr}}{L}\ \text{x}\ 0.450\ L = 0.225\ \text{mol HBr}$$

$$0.225\ \text{mol HBr x}\ \frac{1\ \text{mol}\ K_2CO_3}{2\ \text{mol HBr}}\ \text{x}\ \frac{138.2\ \text{g}\ K_2CO_3}{1\ \text{mol}\ K_2CO_3} = 15.5\ \text{g}\ K_2CO_3$$

6.89 $2\ C_4H_{10}S\ +\ NaOCl\ \rightarrow\ C_8H_{18}S_2\ +\ NaCl\ +\ H_2O$

$C_4H_{10}S$, 90.19 amu; 5.00 mL = 0.005 00 L

$$\frac{0.0985\ \text{mol NaOCl}}{L}\ \text{x}\ 0.005\ 00\ L = 4.925\ \text{x}\ 10^{-4}\ \text{mol NaOCl}$$

$$4.925\ \text{x}\ 10^{-4}\ \text{mol NaOCl x}\ \frac{2\ \text{mol}\ C_4H_{10}S}{1\ \text{mol NaOCl}}\ \text{x}\ \frac{90.19\ \text{g}\ C_4H_{10}S}{1\ \text{mol}\ C_4H_{10}S} = 0.0888\ \text{g}\ C_4H_{10}S$$

6.90 $H_2C_2O_4$, 90.04 amu

$$3.225\ \text{g}\ H_2C_2O_4\ \text{x}\ \frac{1\ \text{mol}\ H_2C_2O_4}{90.04\ \text{g}\ H_2C_2O_4}\ \text{x}\ \frac{2\ \text{mol}\ KMnO_4}{5\ \text{mol}\ H_2C_2O_4} = 0.0143\ \text{mol}\ KMnO_4$$

$$0.0143\ \text{mol}\ \text{x}\ \frac{1\ L}{0.250\ \text{mol}} = 0.0572\ L = 57.2\ \text{mL}$$

6.91 $H_2C_2O_4$, 90.04 amu; 400.0 mL = 0.4000 L; 25.0 mL = 0.0250 L

$$12.0\ \text{g}\ H_2C_2O_4\ \text{x}\ \frac{1\ \text{mol}\ H_2C_2O_4}{90.04\ \text{g}\ H_2C_2O_4} = 0.133\ \text{mol}\ H_2C_2O_4$$

$$\text{molarity} = \frac{0.133\ \text{mol}}{0.4000\ L} = 0.333\ \text{M}\ H_2C_2O_4$$

$$H_2C_2O_4(aq)\ +\ 2\ KOH(aq)\ \rightarrow\ K_2C_2O_4(aq)\ +\ 2\ H_2O(l)$$

$$\frac{0.333\ \text{mol}\ C_2H_2O_4}{L}\ \text{x}\ 0.0250\ L = 0.008\ 32\ \text{mol}\ H_2C_2O_4$$

$$0.008\ 32\ \text{mol}\ H_2C_2O_4\ \text{x}\ \frac{2\ \text{mol KOH}}{1\ \text{mol}\ H_2C_2O_4} = 0.0166\ \text{mol KOH}$$

$$0.0166\ \text{mol}\ \text{x}\ \frac{1\ L}{0.100\ \text{mol}} = 0.166\ L;\ 0.166\ L = 166\ \text{mL}$$

Formulas and Elemental Analysis (Sections 6.10–6.11)

6.92 CH_4N_2O, 60.1 amu

$$\% \, C = \frac{12.0 \, g \, C}{60.1 \, g} \times 100\% = 20.0\%$$

$$\% \, H = \frac{4 \times 1.01 \, g \, H}{60.1 \, g} \times 100\% = 6.72\%$$

$$\% \, N = \frac{2 \times 14.0 \, g \, N}{60.1 \, g} \times 100\% = 46.6\%$$

$$\% \, O = \frac{16.0 \, g \, O}{60.1 \, g} \times 100\% = 26.6\%$$

6.93   (a) $Cu_2(OH)_2CO_3$, 221.1 amu

$$\% \, Cu = \frac{2 \times 63.5 \, g \, Cu}{221.1 \, g} \times 100\% = 57.4\%$$

$$\% \, O = \frac{5 \times 16.0 \, g \, O}{221.1 \, g} \times 100\% = 36.2\%$$

$$\% \, C = \frac{12.0 \, g \, C}{221.1 \, g} \times 100\% = 5.43\%$$

$$\% \, H = \frac{2 \times 1.01 \, g \, H}{221.1 \, g} \times 100\% = 0.914\%$$

(b) $C_8H_9NO_2$, 151.2 amu

$$\% \, C = \frac{8 \times 12.0 \, g \, C}{151.2 \, g} \times 100\% = 63.5\%$$

$$\% \, H = \frac{9 \times 1.01 \, g \, H}{151.2 \, g} \times 100\% = 6.01\%$$

$$\% \, N = \frac{14.0 \, g \, N}{151.2 \, g} \times 100\% = 9.26\%$$

$$\% \, O = \frac{2 \times 16.0 \, g \, O}{151.2 \, g} \times 100\% = 21.2\%$$

(c) $Fe_4[Fe(CN)_6]_3$, 859.2 amu

$$\% \, Fe = \frac{7 \times 55.85 \, g \, Fe}{859.2 \, g} \times 100\% = 45.50\%$$

$$\% \, C = \frac{18 \times 12.01 \, g \, C}{859.2 \, g} \times 100\% = 25.16\%$$

$$\% \, N = \frac{18 \times 14.01 \, g \, N}{859.2 \, g} \times 100\% = 29.35\%$$

6.94   Assume a 100.0 g sample. From the percent composition data, a 100.0 g sample contains 24.25 g F and 75.75 g Sn.

$$24.25 \text{ g F } \times \frac{1 \text{ mol F}}{19.00 \text{ g F}} = 1.276 \text{ mol F}$$

$$75.75 \text{ g Sn } \times \frac{1 \text{ mol Sn}}{118.7 \text{ g Sn}} = 0.6382 \text{ mol Sn}$$

$Sn_{0.6382}F_{1.276}$; divide each subscript by the smaller, 0.6382.
$Sn_{0.6382 / 0.6382}F_{1.276 / 0.6382}$
The empirical formula is SnF_2.

6.95   (a) Assume a 100.0 g sample of ibuprofen. From the percent composition data, a 100.0 g sample contains 75.69 g C, 15.51 g O, and 8.80 g H.

$$75.69 \text{ g C } \times \frac{1 \text{ mol C}}{12.01 \text{ g C}} = 6.302 \text{ mol C}$$

$$15.51 \text{ g O } \times \frac{1 \text{ mol O}}{16.00 \text{ g O}} = 0.9694 \text{ mol O}$$

$$8.80 \text{ g H } \times \frac{1 \text{ mol H}}{1.01 \text{ g H}} = 8.71 \text{ mol H}$$

$C_{6.302}H_{8.71}O_{0.9694}$; divide each subscript by the smallest, 0.9694.
$C_{6.302 / 0.9694}H_{8.71 / 0.9694}O_{0.9694 / 0.9694}$
$C_{6.5}H_9O$; multiply each subscript by 2 to obtain integers.
The empirical formula is $C_{13}H_{18}O_2$.

(b) Assume a 100.0 g sample of tetraethyllead. From the percent composition data, a 100.0 g sample contains 29.71 g C, 6.23 g H, and 64.06 g Pb.

$$29.71 \text{ g C } \times \frac{1 \text{ mol C}}{12.01 \text{ g C}} = 2.474 \text{ mol C}$$

$$6.23 \text{ g H } \times \frac{1 \text{ mol H}}{1.01 \text{ g H}} = 6.17 \text{ mol H}$$

$$64.06 \text{ g Pb } \times \frac{1 \text{ mol Pb}}{207.2 \text{ g Pb}} = 0.3092 \text{ mol Pb}$$

$Pb_{0.3092}C_{2.474}H_{6.17}$; divide each subscript by the smallest, 0.3092.
$Pb_{0.3092 / 0.3092}C_{2.474 / 0.3092}H_{6.17 / 0.3092}$
The empirical formula is PbC_8H_{20}.

(c) Assume a 100.0 g sample of zircon. From the percent composition data, a 100.0 g sample contains 34.91 g O, 15.32 g Si, and 49.76 g Zr.

$$34.91 \text{ g O } \times \frac{1 \text{ mol O}}{16.00 \text{ g O}} = 2.182 \text{ mol O}$$

$$15.32 \text{ g Si } \times \frac{1 \text{ mol Si}}{28.09 \text{ g Si}} = 0.5454 \text{ mol Si}$$

$$49.76 \text{ g Zr } \times \frac{1 \text{ mol Zr}}{91.22 \text{ g Zr}} = 0.5455 \text{ mol Zr}$$

$Zr_{0.5455}Si_{0.5454}O_{2.182}$; divide each subscript by the smallest, 0.5454.
$Zr_{0.5455/0.5454}Si_{0.5454/0.5454}O_{2.182/0.5454}$
The empirical formula is $ZrSiO_4$.

6.96   Mass of toluene sample = 45.62 mg = 0.045 62 g; mass of CO_2 = 152.5 mg = 0.1525 g;
mass of H_2O = 35.67 mg = 0.035 67 g

$$0.1525 \text{ g CO}_2 \times \frac{1 \text{ mol CO}_2}{44.01 \text{ g CO}_2} \times \frac{1 \text{ mol C}}{1 \text{ mol CO}_2} = 0.003 \ 465 \text{ mol C}$$

$$\text{mass C} = 0.003 \ 465 \text{ mol C} \times \frac{12.011 \text{ g C}}{1 \text{ mol C}} = 0.041 \ 62 \text{ g C}$$

$$0.035 \ 67 \text{ g H}_2O \times \frac{1 \text{ mol H}_2O}{18.02 \text{ g H}_2O} \times \frac{2 \text{ mol H}}{1 \text{ mol H}_2O} = 0.003 \ 959 \text{ mol H}$$

$$\text{mass H} = 0.003 \ 959 \text{ mol H} \times \frac{1.008 \text{ g H}}{1 \text{ mol H}} = 0.003 \ 991 \text{ g H}$$

The (mass C + mass H) = 0.041 62 g + 0.003 991 g = 0.045 61 g. The calculated mass of (C + H) essentially equals the mass of the toluene sample, this means that toluene contains only C and H and no other elements.
$C_{0.003\ 465}H_{0.003\ 959}$; divide each subscript by the smaller, 0.003 465.
$C_{0.003\ 465/0.003\ 465}H_{0.003\ 959/0.003\ 465}$
$CH_{1.14}$; multiply each subscript by 7 to obtain integers.
The empirical formula is C_7H_8.

6.97   5.024 mg = 0.005 024 g; 13.90 mg = 0.013 90 g; 6.048 mg = 0.006 048 g

$$0.013 \ 90 \text{ g CO}_2 \times \frac{1 \text{ mol CO}_2}{44.01 \text{ g CO}_2} \times \frac{1 \text{ mol C}}{1 \text{ mol CO}_2} = 3.158 \times 10^{-4} \text{ mol C}$$

$$0.006 \ 048 \text{ g H}_2O \times \frac{1 \text{ mol H}_2O}{18.02 \text{ g H}_2O} \times \frac{2 \text{ mol H}}{1 \text{ mol H}_2O} = 6.713 \times 10^{-4} \text{ mol H}$$

$$3.158 \times 10^{-4} \text{ mol C} \times \frac{12.01 \text{ g C}}{1 \text{ mol C}} = 0.003 \ 793 \text{ g C}$$

$$6.713 \times 10^{-4} \text{ mol H} \times \frac{1.008 \text{ g H}}{1 \text{ mol H}} = 0.000 \ 676 \ 7 \text{ g H}$$

mass N = 0.005 024 g − (0.003 793 g + 0.000 676 7 g) = 0.000 554 g N

$$0.000\ 554\ g\ N \times \frac{1\ mol\ N}{14.01\ g\ N} = 3.95 \times 10^{-5}\ mol\ N$$

Scale each mol quantity to eliminate exponents.

$C_{3.158}H_{6.713}N_{0.395}$; divide each subscript by the smallest, 0.395.

$C_{3.158\,/\,0.395}H_{6.713\,/\,0.395}N_{0.395\,/\,0.395}$

The empirical formula is $C_8H_{17}N$.

6.98 Let X equal the molecular mass of cytochrome c.

$$0.0043 = \frac{55.847\ amu}{X}; \quad X = \frac{55.847\ amu}{0.0043} = 13,000\ amu$$

6.99 Let X equal the molecular mass of nitrogenase.

$$0.000\ 872 = \frac{2 \times 95.94\ amu}{X}; \quad X = \frac{2 \times 95.94\ amu}{0.000\ 872} = 220,000\ amu$$

6.100 Let X equal the molecular mass of disilane.

$$0.9028 = \frac{2 \times 28.09\ amu}{X}; \quad X = \frac{2 \times 28.09\ amu}{0.9028} = 62.23\ amu$$

62.23 amu − 2(Si atomic mass) = 62.23 amu − 2(28.09 amu) = 6.05 amu

6.05 amu is the total mass of H atoms.

$$6.05\ amu \times \frac{1\ H\ atom}{1.01\ amu} = 6\ H\ atoms;\ \text{Disilane is } Si_2H_6.$$

6.101 Let X equal the molecular mass of MS_2.

$$0.4006 = \frac{2 \times 32.07\ amu}{X}; \quad X = \frac{2 \times 32.07\ amu}{0.4006} = 160.1\ amu$$

Atomic mass of M = 160.1 amu − 2(S atomic mass)

= 160.1 amu − 2(32.07 amu) = 95.96 amu

M is Mo.

Chapter Problems

6.102 (a) $C_6H_{12}O_6$, 180.2 amu

$$\%\,C = \frac{6 \times 12.01\ g\ C}{180.2\ g} \times 100\% = 39.99\%$$

$$\%\,H = \frac{12 \times 1.008\ g\ H}{180.2\ g} \times 100\% = 6.713\%$$

$$\%\,O = \frac{6 \times 16.00\ g\ O}{180.2\ g} \times 100\% = 53.27\%$$

(b) H_2SO_4, 98.08 amu

$$\%\,H = \frac{2 \times 1.008\ g\ H}{98.08\ g} \times 100\% = 2.055\%$$

$$\%\,S = \frac{32.07\ g\ S}{98.08\ g} \times 100\% = 32.70\%$$

$$\%\,O = \frac{4 \times 16.00\ g\ O}{98.08\ g} \times 100\% = 65.25\%$$

(c) $KMnO_4$, 158.0 amu

$$\%\,K = \frac{39.10\ g\ K}{158.0\ g} \times 100\% = 24.75\%$$

$$\%\,Mn = \frac{54.94\ g\ Mn}{158.0\ g} \times 100\% = 34.77\%$$

$$\%\,O = \frac{4 \times 16.00\ g\ O}{158.0\ g} \times 100\% = 40.51\%$$

(d) $C_7H_5NO_3S$, 183.2 amu

$$\%\,C = \frac{7 \times 12.01\ g\ C}{183.2\ g} \times 100\% = 45.89\%$$

$$\%\,H = \frac{5 \times 1.008\ g\ H}{183.2\ g} \times 100\% = 2.751\%$$

$$\%\,N = \frac{14.01\ g\ N}{183.2\ g} \times 100\% = 7.647\%$$

$$\%\,O = \frac{3 \times 16.00\ g\ O}{183.2\ g} \times 100\% = 26.20\%$$

$$\%\,S = \frac{32.07\ g\ S}{183.2\ g} \times 100\% = 17.51\%$$

6.103 (a) Assume a 100.0 g sample of aspirin. From the percent composition data, a 100.0 g sample contains 60.00 g C, 35.52 g O, and 4.48 g H.

$$60.00\ g\ C \times \frac{1\ mol\ C}{12.01\ g\ C} = 4.996\ mol\ C$$

$$35.52\ g\ O \times \frac{1\ mol\ O}{16.00\ g\ O} = 2.220\ mol\ O$$

$$4.48\ g\ H \times \frac{1\ mol\ H}{1.01\ g\ H} = 4.44\ mol\ H$$

$C_{4.996}H_{4.44}O_{2.220}$; divide each subscript by the smallest, 2.220.

$C_{4.996\,/\,2.220}H_{4.44\,/\,2.220}O_{2.220\,/\,2.220}$

$C_{2.25}H_2O_1$; multiply each subscript by 4 to obtain integers.
The empirical formula is $C_9H_8O_4$.

(b) Assume a 100.0 g sample of ilmenite. From the percent composition data, a 100.0 g sample contains 31.63 g O, 31.56 g Ti, and 36.81 g Fe.

$$31.63 \text{ g O} \ \times \ \frac{1 \text{ mol O}}{16.00 \text{ g O}} = 1.977 \text{ mol O}$$

$$31.56 \text{ g Ti} \ \times \ \frac{1 \text{ mol Ti}}{47.87 \text{ g Ti}} = 0.6593 \text{ mol Ti}$$

$$36.81 \text{ g Fe} \ \times \ \frac{1 \text{ mol Fe}}{55.85 \text{ g Fe}} = 0.6591 \text{ mol Fe}$$

$Fe_{0.6591}Ti_{0.6593}O_{1.977}$; divide each subscript by the smallest, 0.6591.
$Fe_{0.6591/0.6591}Ti_{0.6593/0.6591}O_{1.977/0.6591}$
The empirical formula is $FeTiO_3$.

(c) Assume a 100.0 g sample of sodium thiosulfate. From the percent composition data, a 100.0 g sample contains 30.36 g O, 29.08 g Na, and 40.56 g S.

$$30.36 \text{ g O} \ \times \ \frac{1 \text{ mol O}}{16.00 \text{ g O}} = 1.897 \text{ mol O}$$

$$29.08 \text{ g Na} \ \times \ \frac{1 \text{ mol Na}}{22.99 \text{ g Na}} = 1.265 \text{ mol Na}$$

$$40.56 \text{ g S} \ \times \ \frac{1 \text{ mol S}}{32.07 \text{ g S}} = 1.265 \text{ mol S}$$

$Na_{1.265}S_{1.265}O_{1.897}$; divide each subscript by the smallest, 1.265.
$Na_{1.265/1.265}S_{1.265/1.265}O_{1.897/1.265}$
$NaSO_{1.5}$; multiply each subscript by 2 to obtain integers.
The empirical formula is $Na_2S_2O_3$.

6.104 (a) $SiCl_4 + 2 H_2O \ \rightarrow \ SiO_2 + 4 HCl$
(b) $P_4O_{10} + 6 H_2O \ \rightarrow \ 4 H_3PO_4$
(c) $CaCN_2 + 3 H_2O \ \rightarrow \ CaCO_3 + 2 NH_3$
(d) $3 NO_2 + H_2O \ \rightarrow \ 2 HNO_3 + NO$

6.105 NaH, 24.00 amu; B_2H_6, 27.67 amu; $NaBH_4$, 37.83 amu
$2 NaH + B_2H_6 \rightarrow 2 NaBH_4$

$$8.55 \text{ g NaH} \times \frac{1 \text{ mol NaH}}{24.00 \text{ g NaH}} = 0.356 \text{ mol NaH}$$

$$6.75 \text{ g B}_2\text{H}_6 \times \frac{1 \text{ mol B}_2\text{H}_6}{27.67 \text{ g B}_2\text{H}_6} = 0.244 \text{ mol B}_2\text{H}_6$$

For 0.244 mol B_2H_6, 2 x (0.244) = 0.488 mol NaH are needed. Because only 0.356 mol of NaH is available, NaH is the limiting reactant.

$$0.356 \text{ mol NaH} \times \frac{2 \text{ mol NaBH}_4}{2 \text{ mol NaH}} \times \frac{37.83 \text{ g NaBH}_4}{1 \text{ mol NaBH}_4} = 13.5 \text{ g NaBH}_4 \text{ produced}$$

$$0.356 \text{ mol NaH} \times \frac{1 \text{ mol B}_2H_6}{2 \text{ mol NaH}} \times \frac{27.67 \text{ g B}_2H_6}{1 \text{ mol B}_2H_6} = 4.93 \text{ g B}_2H_6 \text{ reacted}$$

B_2H_6 left over = 6.75 g – 4.93 g = 1.82 g B_2H_6

6.106   Assume a 100.0 g sample of ferrocene. From the percent composition data, a 100.0 g sample contains 5.42 g H, 64.56 g C, and 30.02 g Fe.

$$5.42 \text{ g H} \times \frac{1 \text{ mol H}}{1.01 \text{ g H}} = 5.37 \text{ mol H}$$

$$64.56 \text{ g C} \times \frac{1 \text{ mol C}}{12.01 \text{ g C}} = 5.376 \text{ mol C}$$

$$30.02 \text{ g Fe} \times \frac{1 \text{ mol Fe}}{55.85 \text{ g Fe}} = 0.5375 \text{ mol Fe}$$

$C_{5.376}H_{5.37}Fe_{0.5375}$; divide each subscript by the smallest, 0.5375.
$C_{5.376/0.5375}H_{5.37/0.5375}Fe_{0.5375/0.5375}$
The empirical formula is $C_{10}H_{10}Fe$.

6.107   Mass of 1 HCl molecule = $(36.5 \frac{\text{amu}}{\text{molecule}})(1.6605 \times 10^{-24} \frac{\text{g}}{\text{amu}}) = 6.06 \times 10^{-23}$ g/molecule

$$\text{Avogadro's number} = \left(\frac{36.5 \text{ g/mol}}{6.06 \times 10^{-23} \text{ g/molecule}} \right) = 6.02 \times 10^{23} \text{ molecules/mol}$$

6.108 Na_2SO_4, 142.04 amu; Na_3PO_4, 163.94 amu; Li_2SO_4, 109.95 amu; 100.00 mL = 0.10000 L

$$0.550 \text{ g Na}_2SO_4 \times \frac{1 \text{ mol Na}_2SO_4}{142.04 \text{ g Na}_2SO_4} = 0.003 \ 872 \text{ mol Na}_2SO_4$$

$$1.188 \text{ g Na}_3PO_4 \times \frac{1 \text{ mol Na}_3PO_4}{163.94 \text{ g Na}_3PO_4} = 0.007 \ 247 \text{ mol Na}_3PO_4$$

$$0.223 \text{ g Li}_2SO_4 \times \frac{1 \text{ mol Li}_2SO_4}{109.95 \text{ g Li}_2SO_4} = 0.002 \ 028 \text{ mol Li}_2SO_4$$

$$Na^+ \text{ molarity} = \frac{(2 \times 0.003 \ 872 \text{ mol}) + (3 \times 0.007 \ 247 \text{ mol})}{0.100 \ 00 \text{ L}} = 0.295 \text{ M}$$

$$Li^+ \text{ molarity} = \frac{2 \times 0.002 \ 028 \text{ mol}}{0.100 \ 00 \text{ L}} = 0.0406 \text{ M}$$

$$SO_4^{2-} \text{ molarity} = \frac{(1 \times 0.003\,872 \text{ mol}) + (1 \times 0.002\,028 \text{ mol})}{0.100\,00 \text{ L}} = 0.0590 \text{ M}$$

$$PO_4^{3-} \text{ molarity} = \frac{1 \times 0.007\,247 \text{ mol}}{0.100\,00 \text{ L}} = 0.0725 \text{ M}$$

6.109 23.46 mg = 0.023 46 g; 20.42 mg = 0.02042 g; 33.27 mg = 0.033 27 g

$$0.033\,27 \text{ g CO}_2 \times \frac{1 \text{ mol CO}_2}{44.01 \text{ g CO}_2} \times \frac{1 \text{ mol C}}{1 \text{ mol CO}_2} = 7.560 \times 10^{-4} \text{ mol C}$$

$$0.020\,42 \text{ g H}_2\text{O} \times \frac{1 \text{ mol H}_2\text{O}}{18.02 \text{ g H}_2\text{O}} \times \frac{2 \text{ mol H}}{1 \text{ mol H}_2\text{O}} = 2.266 \times 10^{-3} \text{ mol H}$$

$$7.560 \times 10^{-4} \text{ mol C} \times \frac{12.01 \text{ g C}}{1 \text{ mol C}} = 0.009\,080 \text{ g C}$$

$$2.266 \times 10^{-3} \text{ mol H} \times \frac{1.008 \text{ g H}}{1 \text{ mol H}} = 0.002\,284 \text{ g H}$$

mass O = 0.023 46 g – (0.009 080 g + 0.002 284 g) = 0.012 10 g O

$$0.012\,10 \text{ g O} \times \frac{1 \text{ mol O}}{16.00 \text{ g O}} = 7.563 \times 10^{-4} \text{ mol O}$$

Scale each mol quantity to eliminate exponents.
$C_{0.7560}H_{2.266}O_{0.7563}$; divide each subscript by the smallest, 0.7560.
$C_{0.7560\,/\,0.7560}H_{2.266\,/\,0.7560}O_{0.7563\,/\,0.7560}$
The empirical formula is CH_3O, 31.0 amu.
62.0 amu / 31.0 amu = 2; molecular formula = $C_{(2\,\times\,1)}H_{(2\,\times\,3)}O_{(2\,\times\,1)} = C_2H_6O_2$

6.110 High resolution mass spectrometry is capable of measuring the mass of molecules with a particular isotopic composition.

6.111 (a) $CO(NH_2)_2(aq) + 6 \text{ HOCl}(aq) \rightarrow 2 \text{ NCl}_3(aq) + CO_2(aq) + 5 \text{ H}_2\text{O}(l)$
       (b) $2 \text{ Ca}_3(PO_4)_2(s) + 6 \text{ SiO}_2(s) + 10 \text{ C}(s) \rightarrow P_4(g) + 6 \text{ CaSiO}_3(l) + 10 \text{ CO}(g)$

6.112 The combustion reaction is: $2 \text{ C}_8H_{18} + 25 \text{ O}_2 \rightarrow 16 \text{ CO}_2 + 18 \text{ H}_2\text{O}$
       C_8H_{18}, 114.23 amu; CO_2, 44.01 amu

$$\text{pounds CO}_2 = 1.00 \text{ gal} \times \frac{3.7854 \text{ L}}{1 \text{ gal}} \times \frac{1000 \text{ mL}}{1 \text{ L}} \times \frac{0.703 \text{ g C}_8H_{18}}{1 \text{ mL}} \times \frac{1 \text{ mol C}_8H_{18}}{114.23 \text{ g C}_8H_{18}} \times$$

$$\frac{16 \text{ mol CO}_2}{2 \text{ mol C}_8H_{18}} \times \frac{44.01 \text{ g CO}_2}{1 \text{ mol CO}_2} \times \frac{1 \text{ lb}}{453.59 \text{ g}} = 18.1 \text{ pounds CO}_2$$

6.113   The reaction is: $CaCO_3 + 2 HCl \rightarrow CaCl_2 + CO_2 + H_2O$

$CaCO_3$, 100.09 amu; CO_2, 44.01 amu

$$\text{mol } CaCO_3 = 6.35 \text{ g } CaCO_3 \times \frac{1 \text{ mol } CaCO_3}{100.09 \text{ g } CaCO_3} = 0.0634 \text{ mol } CaCO_3$$

$$\text{mol } HCl = 500.0 \text{ mL } HCl \times \frac{1 \text{ L}}{1000 \text{ mL}} \times \frac{0.31 \text{ mol } HCl}{1 \text{ L}} = 0.155 \text{ mol } HCl$$

Determine the limiting reactant.

$$\text{mol } HCl \text{ needed} = 0.0634 \text{ mol } CaCO_3 \times \frac{2 \text{ mol } HCl}{1 \text{ mol } CaCO_3} = 0.127 \text{ mol } HCl \text{ needed}$$

Because we have excess HCl, $CaCO_3$ is the limiting reactant.

$$\text{mass } CO_2 = 0.0634 \text{ mol } CaCO_3 \times \frac{1 \text{ mol } CO_2}{1 \text{ mol } CaCO_3} \times \frac{44.01 \text{ g } CO_2}{1 \text{ mol } CO_2} = 2.79 \text{ g } CO_2$$

6.114   AgCl, 143.32 amu; CO_2, 44.01 amu; H_2O, 18.02 amu

$$\text{mol Cl in 1.00 g of X} = 1.95 \text{ g AgCl} \times \frac{1 \text{ mol AgCl}}{143.32 \text{ g AgCl}} \times \frac{1 \text{ mol Cl}}{1 \text{ mol AgCl}} = 0.0136 \text{ mol Cl}$$

$$\text{mass Cl} = 0.0136 \text{ mol Cl} \times \frac{35.453 \text{ g Cl}}{1 \text{ mol Cl}} = 0.482 \text{ g Cl}$$

$$\text{mol C in 1.00 g of X} = 0.900 \text{ g } CO_2 \times \frac{1 \text{ mol } CO_2}{44.01 \text{ g } CO_2} \times \frac{1 \text{ mol C}}{1 \text{ mol } CO_2} = 0.0204 \text{ mol C}$$

$$\text{mass C} = 0.0204 \text{ mol C} \times \frac{12.011 \text{ g C}}{1 \text{ mol C}} = 0.245 \text{ g C}$$

$$\text{mol H in 1.00 g of X} = 0.735 \text{ g } H_2O \times \frac{1 \text{ mol } H_2O}{18.02 \text{ g } H_2O} \times \frac{2 \text{ mol H}}{1 \text{ mol } H_2O} = 0.0816 \text{ mol H}$$

$$\text{mass H} = 0.0816 \text{ mol H} \times \frac{1.008 \text{ g H}}{1 \text{ mol H}} = 0.0823 \text{ g H}$$

mass N = 1.00 g – mass Cl – mass C – mass H = 1.00 – 0.482 g – 0.245 g – 0.0823 g = 0.19 g N

$$\text{mol N in 1.00 g of X} = 0.19 \text{ g N} \times \frac{1 \text{ mol N}}{14.01 \text{ g N}} = 0.014 \text{ mol N}$$

Determine empirical formula.

$C_{0.0204}H_{0.0816}N_{0.014}Cl_{0.0136}$; divide each subscript by the smallest, 0.0136.

$C_{0.0204 / 0.0136}H_{0.0816 / 0.0136}N_{0.014 / 0.0136}Cl_{0.0136 / 0.0136}$

$C_{1.5}H_6NCl$, multiply each subscript by 2 to get integers.

The empirical formula is $C_3H_{12}N_2Cl_2$.

6.115 $CaCO_3$, 100.09 amu

$$\% \text{ Ca} = \frac{40.08 \text{ g Ca}}{100.09 \text{ g}} \times 100\% = 40.04\%$$

$$\% \text{ C} = \frac{12.01 \text{ g C}}{100.09 \text{ g}} \times 100\% = 12.00\%$$

$$\% \text{ O} = \frac{3 \times 16.00 \text{ g O}}{100.09 \text{ g}} \times 100\% = 47.96\%$$

Because the mass %'s for the pulverized rock are different from the mass %'s for pure $CaCO_3$ calculated here, the pulverized rock cannot be pure $CaCO_3$.

6.116 Let SA stand for salicylic acid.

$$\text{mol C in 1.00 g of SA} = 2.23 \text{ g CO}_2 \times \frac{1 \text{ mol CO}_2}{44.01 \text{ g CO}_2} \times \frac{1 \text{ mol C}}{1 \text{ mol CO}_2} = 0.0507 \text{ mol C}$$

$$\text{mass C} = 0.0507 \text{ mol C} \times \frac{12.011 \text{ g C}}{1 \text{ mol C}} = 0.609 \text{ g C}$$

$$\text{mol H in 1.00 g of SA} = 0.39 \text{ g H}_2\text{O} \times \frac{1 \text{ mol H}_2\text{O}}{18.02 \text{ g H}_2\text{O}} \times \frac{2 \text{ mol H}}{1 \text{ mol H}_2\text{O}} = 0.043 \text{ mol H}$$

$$\text{mass H} = 0.043 \text{ mol H} \times \frac{1.008 \text{ g H}}{1 \text{ mol H}} = 0.043 \text{ g H}$$

$$\text{mass O} = 1.00 \text{ g} - \text{mass C} - \text{mass H} = 1.00 - 0.609 \text{ g} - 0.043 \text{ g} = 0.35 \text{ g O}$$

$$\text{mol O in 1.00 g of} = 0.35 \text{ g N} \times \frac{1 \text{ mol O}}{16.00 \text{ g O}} = 0.022 \text{ mol O}$$

Determine empirical formula.
$C_{0.0507}H_{0.043}O_{0.022}$; divide each subscript by the smallest, 0.022.
$C_{0.0507 / 0.022}H_{0.043 / 0.022}O_{0.022 / 0.022}$
$C_{2.3}H_2O$, multiply each subscript by 3 to get integers.
The empirical formula is $C_7H_6O_3$. The empirical formula mass = 138.12 g/mol.

Because salicylic acid has only one acidic hydrogen, there is a 1 to 1 mol ratio between salicylic acid and NaOH in the acid-base titration.

$$\text{mol SA in 1.00 g SA} = 72.4 \text{ mL} \times \frac{1 \text{ L}}{1000 \text{ mL}} \times \frac{0.100 \text{ mol NaOH}}{1 \text{ L}} \times \frac{1 \text{ mol SA}}{1 \text{ mol NaOH}} =$$

$$0.00724 \text{ mol SA}$$

$$\text{SA molar mass} = \frac{1.00 \text{ g}}{0.00724 \text{ mol}} = 138 \text{ g/mol}$$

Because the empirical formula mass and the molar mass are the same, the empirical formula is the molecular formula for salicylic acid.

6.117 (a) $\text{mol C} = 4.83 \text{ g CO}_2 \times \dfrac{1 \text{ mol CO}_2}{44.01 \text{ g CO}_2} \times \dfrac{1 \text{ mol C}}{1 \text{ mol CO}_2} = 0.110 \text{ mol C}$

$\text{mass C} = 0.110 \text{ mol C} \times \dfrac{12.011 \text{ g C}}{1 \text{ mol C}} = 1.32 \text{ g C}$

$\text{mol H} = 1.48 \text{ g H}_2\text{O} \times \dfrac{1 \text{ mol H}_2\text{O}}{18.02 \text{ g H}_2\text{O}} \times \dfrac{2 \text{ mol H}}{1 \text{ mol H}_2\text{O}} = 0.164 \text{ mol H}$

$\text{mass H} = 0.164 \text{ mol H} \times \dfrac{1.008 \text{ g H}}{1 \text{ mol H}} = 0.165 \text{ g H}$

$109.8 \text{ mL} = 0.1098 \text{ L}$

$\text{mol NaOH} = (0.1098 \text{ L})(1.00 \text{ mol/L}) = 0.110 \text{ mol NaOH}$

$\text{H}_2\text{SO}_4(aq) + 2 \text{ NaOH}(aq) \rightarrow \text{Na}_2\text{SO}_4(aq) + 2 \text{ H}_2\text{O}(l)$

$\text{mol H}_2\text{SO}_4 = 0.110 \text{ mol NaOH} \times \dfrac{1 \text{ mol H}_2\text{SO}_4}{2 \text{ mol NaOH}} = 0.0550 \text{ mol H}_2\text{SO}_4$

$\text{mol S} = 0.0550 \text{ mol H}_2\text{SO}_4 \times \dfrac{1 \text{ mol S}}{1 \text{ mol H}_2\text{SO}_4} = 0.0550 \text{ mol S}$

$\text{mass S} = 0.0550 \text{ mol S} \times \dfrac{32.06 \text{ g S}}{1 \text{ mol S}} = 1.76 \text{ g S}$

$\text{mass O} = 5.00 \text{ g} - \text{mass C} - \text{mass H} - \text{mass S} = 5.00 \text{ g} - 1.32 \text{ g} - 0.165 \text{ g} - 1.76 \text{ g} = 1.75 \text{ g O}$

$\text{mol O} = 1.75 \text{ g O} \times \dfrac{1 \text{ mol O}}{16.00 \text{ g O}} = 0.109 \text{ mol O}$

$\text{C}_{0.110}\text{H}_{0.164}\text{O}_{0.109}\text{S}_{0.0550}$; divide each subscript by the smallest, 0.0550.

$\text{C}_{0.110/0.0550}\text{H}_{0.164/0.0550}\text{O}_{0.109/0.0550}\text{S}_{0.0550/0.0550}$

The empirical formula is $\text{C}_2\text{H}_3\text{O}_2\text{S}$.

The empirical formula mass = 91.1 g/mol.

(b) $54.9 \text{ mL} = 0.0549 \text{ L}$

$\text{mol NaOH} = (0.0549 \text{ L})(1.00 \text{ mol/L}) = 0.0549 \text{ mol NaOH}$

Because X has two acidic hydrogens, two mol of NaOH are required to titrate 1 mol of X.

$\text{mol X} = 0.0549 \text{ mol NaOH} \times \dfrac{1 \text{ mol X}}{2 \text{ mol NaOH}} = 0.0274 \text{ mol X}$

$\text{X molar mass} = \dfrac{5.00 \text{ g}}{0.0274 \text{ mol}} = 182 \text{ g/mol}$

Because the molar mass is twice the empirical formula mass, the molecular formula is twice the empirical formula.

The molecular formula is $\text{C}_{(2 \times 2)}\text{H}_{(2 \times 3)}\text{O}_{(2 \times 2)}\text{S}_{(2 \times 1)} = \text{C}_4\text{H}_6\text{O}_4\text{S}_2$.

6.118 Let X equal the mass of benzoic acid and Y the mass of gallic acid in the 1.00 g mixture. Therefore, $X + Y = 1.00$ g.

Because both acids contain only one acidic hydrogen, there is a 1 to 1 mol ratio between each acid and NaOH in the acid-base titration.

In the titration, mol benzoic acid + mol gallic acid = mol NaOH.

Therefore, $X \times \dfrac{1 \text{ mol BA}}{122 \text{ g BA}} + Y \times \dfrac{1 \text{ mol GA}}{170 \text{ g GA}} = \text{mol NaOH}$

$\text{mol NaOH} = 14.7 \text{ mL} \times \dfrac{1 \text{ L}}{1000 \text{ mL}} \times \dfrac{0.500 \text{ mol NaOH}}{1 \text{ L}} = 0.00735 \text{ mol NaOH}$

We have two unknowns, X and Y, and two equations.

$X + Y = 1.00 \text{ g}$

$X \times \dfrac{1 \text{ mol BA}}{122 \text{ g BA}} + Y \times \dfrac{1 \text{ mol GA}}{170 \text{ g GA}} = 0.00735 \text{ mol NaOH}$

Rearrange to get $X = 1.00 \text{ g} - Y$ and then substitute it into the equation above to solve for Y.

$(1.00 \text{ g} - Y) \times \dfrac{1 \text{ mol BA}}{122 \text{ g BA}} + Y \times \dfrac{1 \text{ mol GA}}{170 \text{ g GA}} = 0.00735 \text{ mol NaOH}$

$\dfrac{1 \text{ mol}}{122} - \dfrac{Y \text{ mol}}{122 \text{ g}} + \dfrac{Y \text{ mol}}{170 \text{ g}} = 0.00735 \text{ mol}$

$-\dfrac{Y \text{ mol}}{122 \text{ g}} + \dfrac{Y \text{ mol}}{170 \text{ g}} = 0.00735 \text{ mol} - \dfrac{1 \text{ mol}}{122} = -8.47 \times 10^{-4} \text{ mol}$

$\dfrac{(-Y \text{ mol})(170 \text{ g}) + (Y \text{ mol})(122 \text{ g})}{(170 \text{ g})(122 \text{ g})} = -8.47 \times 10^{-4} \text{ mol}$

$\dfrac{-48 \ Y \text{ mol}}{20740 \text{ g}} = -8.47 \times 10^{-4} \text{ mol}; \quad \dfrac{48 \ Y}{20740 \text{ g}} = 8.47 \times 10^{-4}$

$Y = \dfrac{(20740 \text{ g})(8.47 \times 10^{-4})}{48} = 0.366 \text{ g}$

$X = 1.00 \text{ g} - 0.366 \text{ g} = 0.634 \text{ g}$

In the 1.00 g mixture there is 0.63 g of benzoic acid and 0.37 g of gallic acid.

6.119 C_2H_6O, 46.07 amu; H_2O, 18.02 amu

Let X = mass of H_2O in the 10.00 g sample.

Let Y = mass of ethanol (C_2H_6O) in the 10.00 g sample.

$X + Y = 10.00 \text{ g}$ and $Y = 10.00 \text{ g} - X$

mass of collected $H_2O = 11.27 \text{ g}$

$\text{mass of collected } H_2O = X + \left(Y \times \dfrac{1 \text{ mol } C_2H_6O}{46.07 \text{ g } C_2H_6O} \times \dfrac{3 \text{ mol } H_2O}{1 \text{ mol } C_2H_6O} \times \dfrac{18.02 \text{ g } H_2O}{1 \text{ mol } H_2O} \right)$

Substitute for Y.

$11.27 \text{ g} = X + \left((10.00 \text{ g} - X) \times \dfrac{1 \text{ mol } C_2H_6O}{46.07 \text{ g } C_2H_6O} \times \dfrac{3 \text{ mol } H_2O}{1 \text{ mol } C_2H_6O} \times \dfrac{18.02 \text{ g } H_2O}{1 \text{ mol } H_2O} \right)$

$11.27 \text{ g} = X + (10.00 \text{ g} - X)(1.173)$

$11.27 \text{ g} = X + 11.73 \text{ g} - 1.173 \text{ X}$

$0.173 \text{ X} = 11.73 \text{ g} - 11.27 \text{ g} = 0.46 \text{ g}$

$X = \dfrac{0.46 \text{ g}}{0.173} = 2.7 \text{ g } H_2O$

$Y = 10.00 \text{ g} - X = 10.00 \text{ g} - 2.7 \text{ g} = 7.3 \text{ g } C_2H_6O$

6.120   FeO, 71.85 amu; Fe_2O_3, 159.7 amu

Let X equal the mass of FeO and Y the mass of Fe_2O_3 in the 10.0 g mixture. Therefore, $X + Y = 10.0 \text{ g}$.

$\text{mol Fe} = 7.43 \text{ g} \times \dfrac{1 \text{ mol Fe}}{55.85 \text{ g Fe}} = 0.133 \text{ mol Fe}$

$\text{mol FeO} + 2 \times \text{mol } Fe_2O_3 = 0.133 \text{ mol Fe}$

$X \times \dfrac{1 \text{ mol FeO}}{71.85 \text{ g FeO}} + 2 \times \left(Y \times \dfrac{1 \text{ mol } Fe_2O_3}{159.7 \text{ g } Fe_2O_3} \right) = 0.133 \text{ mol Fe}$

Rearrange to get $X = 10.0 \text{ g} - Y$ and then substitute it into the equation above to solve for Y.

$(10.0 \text{ g} - Y) \times \dfrac{1 \text{ mol FeO}}{71.85 \text{ g FeO}} + 2 \times \left(Y \times \dfrac{1 \text{ mol } Fe_2O_3}{159.7 \text{ g } Fe_2O_3} \right) = 0.133 \text{ mol Fe}$

$\dfrac{10.0 \text{ mol}}{71.85} - \dfrac{Y \text{ mol}}{71.85 \text{ g}} + \dfrac{2 \text{ Y mol}}{159.7 \text{ g}} = 0.133 \text{ mol}$

$-\dfrac{Y \text{ mol}}{71.85 \text{ g}} + \dfrac{2 \text{ Y mol}}{159.7 \text{ g}} = 0.133 \text{ mol} - \dfrac{10.0 \text{ mol}}{71.85} = -0.0062 \text{ mol}$

$\dfrac{(-Y \text{ mol})(159.7 \text{ g}) + (2 \text{ Y mol})(71.85 \text{ g})}{(71.85 \text{ g})(159.7 \text{ g})} = -0.0062 \text{ mol}$

$\dfrac{-16.0 \text{ Y mol}}{11474 \text{ g}} = -0.0062 \text{ mol}; \quad \dfrac{16.0 \text{ Y}}{11474 \text{ g}} = 0.0062$

$Y = (0.0062)(11474 \text{ g})/16.0 = 4.44 \text{ g} = 4.4 \text{ g } Fe_2O_3$

$X = 10.0 \text{ g} - Y = 10.0 \text{ g} - 4.4 \text{ g} = 5.6 \text{ g FeO}$

6.121   AgCl, 143.32 amu

Find the mass of Cl in 1.68 g of AgCl.

$\text{mol Cl in 1.68 g of AgCl} = 1.68 \text{ g AgCl} \times \dfrac{1 \text{ mol AgCl}}{143.32 \text{ g AgCl}} \times \dfrac{1 \text{ mol Cl}}{1 \text{ mol AgCl}} = 0.0117 \text{ mol Cl}$

$\text{mass Cl} = 0.0117 \text{ mol Cl} \times \dfrac{35.453 \text{ g Cl}}{1 \text{ mol Cl}} = 0.415 \text{ g Cl}$

All of the Cl in AgCl came from XCl_3.

Find the mass of X in 0.634 g of XCl_3.

Mass of X = 0.634 g - 0.415 g = 0.219 g X

$$0.0117 \text{ mol Cl} \times \frac{1 \text{ mol X}}{3 \text{ mol Cl}} = 0.00390 \text{ mol X}$$

$$\text{molar mass of X} = \frac{0.219 \text{ g}}{0.00390 \text{ mol}} = 56.2 \text{ g/mol}; \quad X = Fe$$

6.122 $C_6H_{12}O_6 + 6 O_2 \rightarrow 6 CO_2 + 6 H_2O$; $C_6H_{12}O_6$, 180.16 amu; CO_2, 44.01 amu

$$66.3 \text{ g } C_6H_{12}O_6 \times \frac{1 \text{ mol } C_6H_{12}O_6}{180.16 \text{ g } C_6H_{12}O_6} \times \frac{6 \text{ mol } CO_2}{1 \text{ mol } C_6H_{12}O_6} \times \frac{44.01 \text{ g } CO_2}{1 \text{ mol } CO_2} = 97.2 \text{ g } CO_2$$

$$66.3 \text{ g } C_6H_{12}O_6 \times \frac{1 \text{ mol } C_6H_{12}O_6}{180.16 \text{ g } C_6H_{12}O_6} \times \frac{6 \text{ mol } CO_2}{1 \text{ mol } C_6H_{12}O_6} \times \frac{25.4 \text{ L } CO_2}{1 \text{ mol } CO_2} = 56.1 \text{ L } CO_2$$

6.123 $H_2C_2O_4$, 90.04 amu; 22.35 mL = 0.02235 L

$$0.5170 \text{ g } H_2C_2O_4 \times \frac{1 \text{ mol } H_2C_2O_4}{90.04 \text{ g } H_2C_2O_4} \times \frac{2 \text{ mol } KMnO_4}{5 \text{ mol } H_2C_2O_4} = 0.002\ 297 \text{ mol } KMnO_4$$

$$KMnO_4 \text{ molarity} = \frac{0.002\ 297 \text{ mol } KMnO_4}{0.022\ 35 \text{ L}} = 0.1028 \text{ M}$$

6.124 Mass of Cu = 2.196 g; mass of S = 2.748 g – 2.196 g = 0.552 g S

(a) $\%Cu = \dfrac{2.196 \text{ g}}{2.748 \text{ g}} \times 100\% = 79.91\%$

$\%S = \dfrac{0.552 \text{ g}}{2.748 \text{ g}} \times 100\% = 20.1\%$

(b) $2.196 \text{ g Cu} \times \dfrac{1 \text{ mol Cu}}{63.55 \text{ g Cu}} = 0.034\ 55 \text{ mol Cu}$

$0.552 \text{ g S} \times \dfrac{1 \text{ mol S}}{32.07 \text{ g S}} = 0.0172 \text{ mol S}$

$Cu_{0.03455}S_{0.0172}$; divide each subscript by the smaller, 0.0172.
$Cu_{0.03455\,/\,0.0172}S_{0.0172\,/\,0.0172}$
The empirical formula is Cu_2S.

(c) Cu_2S, 159.16 amu

$$\frac{5.6 \text{ g } Cu_2S}{1 \text{ cm}^3} \times \frac{1 \text{ mol } Cu_2S}{159.16 \text{ g } Cu_2S} \times \frac{2 \text{ mol } Cu^+ \text{ ions}}{1 \text{ mol } Cu_2S} \times \frac{6.022 \times 10^{23} \text{ } Cu^+ \text{ ions}}{1 \text{ mol } Cu^+ \text{ ions}}$$

$$= 4.2 \times 10^{22} \text{ } Cu^+ \text{ ions/cm}^3$$

6.125 Mass of added Cl = mass of XCl_5 – mass of XCl_3 = 13.233 g – 8.729 g = 4.504 g

mass of Cl in XCl_5 = 5 Cl's x $\dfrac{4.504 \text{ g}}{2 \text{ } Cl's}$ = 11.26 g Cl

mass of X in XCl_5 = 13.233 g – 11.26 g = 1.973 g X

11.26 g Cl x $\dfrac{1 \text{ mol } Cl}{35.45 \text{ g } Cl}$ = 0.3176 mol Cl

0.3176 mol Cl x $\dfrac{1 \text{ mol } X}{5 \text{ mol } Cl}$ = 0.063 52 mol X

molar mass of X = $\dfrac{1.973 \text{ g } X}{0.063 \text{ 52 mol } X}$ = 31.1 g/mol; atomic mass =31.1 amu, X = P

6.126 PCl_3, 137.33 amu; PCl_5, 208.24 amu

Let Y = mass of PCl_3 in the mixture, and $(10.00 - Y)$ = mass of PCl_5 in the mixture.

fraction Cl in PCl_3 = $\dfrac{(3)(35.453 \text{ g/mol})}{137.33 \text{ g/mol}}$ = 0.774 48

fraction Cl in PCl_5 = $\dfrac{(5)(35.453 \text{ g/mol})}{208.24 \text{ g/mol}}$ = 0.851 25

(mass of Cl in PCl_3) + (mass of Cl in PCl_5) = mass of Cl in the mixture

0.774 48Y + 0.851 25(10.00 g – Y) = (0.8104)(10.00 g)

Y = 5.32 g PCl_3 and 10.00 – Y = 4.68 g PCl_5

6.127 100.00 mL = 0.100 00 L; 71.02 mL = 0.071 02 L

mol H_2SO_4 = $\dfrac{0.1083 \text{ mol } H_2SO_4}{L}$ x 0.100 00 L = 0.010 83 mol H_2SO_4

mol $NaOH$ = $\dfrac{0.1241 \text{ mol } NaOH}{L}$ x 0.071 02 L = 0.008 814 mol $NaOH$

H_2SO_4 + 2 $NaOH$ $\rightarrow$ Na_2SO_4 + 2 H_2O

mol H_2SO_4 reacted with $NaOH$ = 0.008 814 mol $NaOH$ x $\dfrac{1 \text{ mol } H_2SO_4}{2 \text{ mol } NaOH}$ = 0.004 407 mol H_2SO_4

mol H_2SO_4 reacted with MCO_3 = 0.010 83 mol – 0.004 407 mol = 0.006 423 mol H_2SO_4

mol H_2SO_4 reacted with MCO_3 = mol CO_3^{2-} in MCO_3 = mol CO_2 produced = 0.006 423 mol CO_2

(a) CO_3^{2-}, 60.01 amu; 0.006 423 mol CO_3^{2-} x $\dfrac{60.01 \text{ g } CO_3^{2-}}{1 \text{ mol } CO_3^{2-}}$ = 0.3854 g CO_3^{2-}

mass of M = 1.268 g – 0.3854 g = 0.8826 g M

molar mass of M = $\dfrac{0.8826 \text{ g}}{0.006 \text{ 423 mol}}$ = 137.4 g/mol; M is Ba

(b) 0.006 423 mol CO_2 x $\dfrac{44.01 \text{ g } CO_2}{1 \text{ mol } CO_2}$ x $\dfrac{1 \text{ L}}{1.799 \text{ g}}$ = 0.1571 L CO_2

6.128 NH_4NO_3, 80.04 amu; $(NH_4)_2HPO_4$, 132.06 amu

Assume you have a 100.0 g sample of the mixture.

Let X = grams of NH_4NO_3 and $(100.0 - X)$ = grams of $(NH_4)_2HPO_4$.

Both compounds contain 2 nitrogen atoms per formula unit.

Because the mass % N in the sample is 30.43%, the 100.0 g sample contains 30.43 g N.

$$\text{mol } NH_4NO_3 = (X) \times \frac{1 \text{ mol } NH_4NO_3}{80.04 \text{ g}}$$

$$\text{mol } (NH_4)_2HPO_4 = (100.0 - X) \times \frac{1 \text{ mol } (NH_4)_2HPO_4}{132.06 \text{ g}}$$

$$\text{mass N} = \left(\left((X) \times \frac{1 \text{ mol } NH_4NO_3}{80.04 \text{ g}}\right) + \left((100.0 - X) \times \frac{1 \text{ mol } (NH_4)_2HPO_4}{132.06 \text{ g}}\right)\right) \times$$

$$\left(\frac{2 \text{ mol N}}{1 \text{ mol ammonium cmpds}}\right) \times \left(\frac{14.0067 \text{ g N}}{1 \text{ mol N}}\right) = 30.43 \text{ g}$$

Solve for X.

$$\left(\frac{X}{80.04} + \frac{100.0 - X}{132.06}\right)(2)(14.0067) = 30.43$$

$$\left(\frac{X}{80.04} + \frac{100.0 - X}{132.06}\right) = 1.08627$$

$$\frac{(132.06)(X) + (100.0 - X)(80.04)}{(80.04)(132.06)} = 1.08627$$

$(132.06)(X) + (100.0 - X)(80.04) = (1.08627)(80.04)(132.06)$

$132.06X + 8004 - 80.04X = 11481.96$

$132.06X - 80.04X = 11481.96 - 8004$

$52.02X = 3477.96$

$$X = \frac{3477.96}{52.02} = 66.86 \text{ g } NH_4NO_3$$

$(100.0 - X) = (100.0 - 66.86) = 33.14 \text{ g } (NH_4)_2HPO_4$

$$\frac{\text{mass}_{NH_4NO_3}}{\text{mass}_{(NH_4)_2HPO_4}} = \frac{66.86 \text{ g}}{33.14 \text{ g}} = 2.018$$

The mass ratio of NH_4NO_3 to $(NH_4)_2HPO_4$ in the mixture is 2 to 1.

6.129 $Na_2CO_3 \rightarrow Na_2O + CO_2$; Na_2CO_3, 106 amu; Na_2O, 62 amu

$CaCO_3 \rightarrow CaO + CO_2$, $CaCO_3$, 100 amu; CaO, 56 amu

In a 0.35 kg sample of glass there would be:

    0.12 x 0.35 kg = 0.042 kg = 42 g of Na_2O

    0.13 x 0.35 kg = 0.045 kg = 45 g of CaO

    350 g – 42 g – 45 g = 263 g of SiO_2

$$\text{mass } Na_2CO_3 = 42 \text{ g } Na_2O \times \frac{1 \text{ mol } Na_2O}{62 \text{ g } Na_2O} \times \frac{1 \text{ mol } Na_2CO_3}{1 \text{ mol } Na_2O} \times \frac{106 \text{ g } Na_2CO_3}{1 \text{ mol } Na_2CO_3} = 72 \text{ g } Na_2CO_3$$

$$\text{mass } CaCO_3 = 45 \text{ g CaO} \times \frac{1 \text{ mol CaO}}{56 \text{ g CaO}} \times \frac{1 \text{ mol } CaCO_3}{1 \text{ mol CaO}} \times \frac{100 \text{ g } CaCO_3}{1 \text{ mol } CaCO_3} = 80 \text{ g } CaCO_3$$

To make 0.35 kg of glass, start with 72 g Na_2CO_3, 80 g $CaCO_3$, and 263 g SiO_2.

6.130 (a) 56.0 mL = 0.0560 L

$$\text{mol } X_2 = (0.0560 \text{ L } X_2)\left(\frac{1 \text{ mol}}{22.41 \text{ L}}\right) = 0.00250 \text{ mol } X_2$$

$$\text{mass } X_2 = 1.12 \text{ g } MX_2 - 0.720 \text{ g } MX = 0.40 \text{ g } X_2$$

$$\text{molar mass } X_2 = \frac{0.40 \text{ g}}{0.00250 \text{ mol}} = 160 \text{ g/mol}$$

atomic mass of X = 160/2 = 80 amu;   X is Br.

(b) $\text{mol } MX = 0.00250 \text{ mol } X_2 \times \dfrac{2 \text{ mol } MX}{1 \text{ mol } X_2} = 0.00500 \text{ mol } MX$

$$\text{mass of X in } MX = 0.00500 \text{ mol } MX \times \frac{1 \text{ mol X}}{1 \text{ mol } MX} \times \frac{80 \text{ g X}}{1 \text{ mol X}} = 0.40 \text{ g X}$$

$$\text{mass of M in } MX = 0.720 \text{ g } MX - 0.40 \text{ g X} = 0.32 \text{ g M}$$

$$\text{molar mass M} = \frac{0.32 \text{ g}}{0.00500 \text{ mol}} = 64 \text{ g/mol}$$

atomic mass of X = 64 amu;   M is Cu.

Multiconcept Problems

6.131 (a) Cl_2, 70.91 amu

$$M + Cl_2 \;\rightarrow\; MCl_2$$

$$\text{mol } Cl_2 = 0.8092 \text{ g } Cl_2 \times \frac{1 \text{ mol } Cl_2}{70.91 \text{ g } Cl_2} = 0.01141 \text{ mol } Cl_2$$

$$\text{mol M} = 0.01141 \text{ mol } Cl_2 \times \frac{1 \text{ mol M}}{1 \text{ mol } Cl_2} = 0.01141 \text{ mol M}$$

$$\text{molar mass of M} = \frac{1.000 \text{ g}}{0.01141 \text{ mol}} = 87.64 \text{ g/mol}$$

atomic mass of M = 87.64 amu;   M = Sr

(b) $q = \dfrac{9.46 \text{ kJ}}{0.01141 \text{ mol}} = 829 \text{ kJ/mol}$

6.132   AgCl, 143.32 amu

(a) mass Cl in AgCl = 1.126 g AgCl x $\dfrac{35.453 \text{ g Cl}}{143.32 \text{ g AgCl}}$ = 0.279 g Cl

%Cl in alkaline earth chloride = $\dfrac{0.279 \text{ g Cl}}{0.436 \text{ g}}$ x 100% = 64.0% Cl

(b) Because M is an alkaline earth metal, M is a 2+ cation.

For MCl_2, mass of M = 0.436 g – 0.279 g = 0.157 g M

mol M = 0.279 g Cl x $\dfrac{1 \text{ mol Cl}}{35.453 \text{ g Cl}}$ x $\dfrac{1 \text{ mol M}}{2 \text{ mol Cl}}$ = 0.003 93 mol M

molar mass for M = $\dfrac{0.157 \text{ g}}{0.003 \text{ 93 mol}}$ = 39.9 g/mol;   M = Ca

(c) $Ca(s) + Cl_2(g) \rightarrow CaCl_2(s)$

$CaCl_2(aq) + 2\ AgNO_3(aq) \rightarrow 2\ AgCl(s) + Ca(NO_3)_2(aq)$

(d) 1.005 g Ca x $\dfrac{1 \text{ mol Ca}}{40.078 \text{ g Ca}}$ = 0.0251 mol Ca

1.91×10^{22} Cl_2 molecules x $\dfrac{1 \text{ mol } Cl_2}{6.022 \times 10^{23} \text{ } Cl_2 \text{ molecules}}$ = 0.0317 mol Cl_2

Because the stoichiometry between Ca and Cl_2 is one to one, the Cl_2 is in excess.

Mass Cl_2 unreacted = (0.0317 – 0.0251) mol Cl_2 x $\dfrac{70.91 \text{ g } Cl_2}{1 \text{ mol } Cl_2}$ = 0.47 g Cl_2 unreacted

6.133   (a) (i) $M_2O_3(s) + 3\ C(s) + 3\ Cl_2(g) \rightarrow 2\ MCl_3(l) + 3\ CO(g)$

(ii) $2\ MCl_3(l) + 3\ H_2(g) \rightarrow 2\ M(s) + 6\ HCl(g)$

(b) $HCl(aq) + NaOH(aq) \rightarrow H_2O(l) + NaCl(aq)$

144.2 mL = 0.1442 L

mol NaOH = (0.511 mol/L)(0.1442 L) = 0.07369 mol NaOH

mol HCl = 0.07369 mol NaOH x $\dfrac{1 \text{ mol HCl}}{1 \text{ mol NaOH}}$ = 0.07369 mol HCl

mol M = 0.07369 mol HCl x $\dfrac{2 \text{ mol M}}{6 \text{ mol HCl}}$ = 0.02456 mol M

mol M_2O_3 = 0.02456 mol M x $\dfrac{2 \text{ mol } MCl_3}{2 \text{ mol M}}$ x $\dfrac{1 \text{ mol } M_2O_3}{2 \text{ mol } MCl_3}$ = 0.01228 mol M_2O_3

molar mass M_2O_3 = $\dfrac{0.855 \text{ g}}{0.01228 \text{ mol}}$ = 69.6 g/mol; molecular mass M_2O_3 = 69.6 amu

atomic mass of M = $\dfrac{69.6 \text{ amu} - (3 \times 16.0 \text{ amu})}{2}$ = 10.8 amu;   M = B

(c) mass of M = 0.02456 mol M x $\dfrac{10.81 \text{ g M}}{1 \text{ mol M}}$ = 0.265 g M

6.134 (a)

$Sr(s) \rightarrow Sr(g)$	+164.44	kJ/mol
$Sr(g) \rightarrow Sr^+(g) + e^-$	+549.5	kJ/mol
$Sr^+(g) \rightarrow Sr^{2+}(g) + e^-$	+1064.2	kJ/mol
$Cl_2(g) \rightarrow 2\,Cl(g)$	+243	kJ/mol
$2[Cl(g) + e^- \rightarrow Cl^-(g)]$	2(−348.6)	kJ/mol
$Sr^{2+}(g) + 2\,Cl^-(g) \rightarrow SrCl_2(s)$	−2156	kJ/mol

$$\text{Sum} = -832 \quad \text{kJ/mol for } Sr(s) + Cl_2(g) \rightarrow SrCl_2(s)$$

(b) Sr, 87.62 amu; Cl_2, 70.91 amu; $SrCl_2$, 158.53 amu

$$20.0 \text{ g Sr} \times \frac{1 \text{ mol Sr}}{87.62 \text{ g Sr}} = 0.228 \text{ mol Sr} \quad \text{and} \quad 25.0 \text{ g Cl}_2 \times \frac{1 \text{ mol Cl}_2}{70.91 \text{ g Cl}_2} = 0.353 \text{ mol Cl}_2$$

Because there is a 1:1 stoichiometry between the reactants, the one with the smaller mole amount is the limiting reactant. Sr is the limiting reactant.

$$0.228 \text{ mol Sr} \times \frac{1 \text{ mol SrCl}_2}{1 \text{ mol Sr}} \times \frac{158.53 \text{ g SrCl}_2}{1 \text{ mol SrCl}_2} = 36.1 \text{ g SrCl}_2$$

(c) $0.228 \text{ mol SrCl}_2 \times \dfrac{-832 \text{ kJ}}{1 \text{ mol SrCl}_2} = -190 \text{ kJ}$

190 kJ is released during the reaction of 20.0 g of Sr with 25.0 g Cl_2.

6.135 (a) $XOCl_2 + 2\,H_2O \rightarrow 2\,HCl + H_2XO_3$

(b) 96.1 mL = 0.0961 L

mol NaOH = (0.1225 mol/L)(0.0961 L) = 0.01177 mol NaOH

$$\text{mol H}^+ = 0.01177 \text{ mol NaOH} \times \frac{1 \text{ mol H}^+}{1 \text{ mol NaOH}} = 0.01177 \text{ mol H}^+$$

Of the total H^+ concentration, half comes from HCl and half comes from H_2XO_3.

$$\text{mol H}_2XO_3 = \frac{0.01177 \text{ mol H}^+}{2} \times \frac{1 \text{ mol H}_2XO_3}{2 \text{ mol H}^+} = 2.943 \times 10^{-3} \text{ mol H}_2XO_3$$

$$\text{mol XOCl}_2 = 2.943 \times 10^{-3} \text{ mol H}_2XO_3 \times \frac{1 \text{ mol XOCl}_2}{1 \text{ mol H}_2XO_3} = 2.943 \times 10^{-3} \text{ mol XOCl}_2$$

$$\text{molar mass XOCl}_2 = \frac{0.350 \text{ g XOCl}_2}{2.943 \times 10^{-3} \text{ mol XOCl}_2} = 118.9 \text{ g/mol}$$

molecular mass of $XOCl_2$ = 118.9 amu

atomic mass of X = 118.9 amu − 16.0 amu − 2(35.45 amu) = 32.0 amu: X = S

(c)

$$:\ddot{O}:$$
$$|$$
$$:\ddot{C}l - S - \ddot{C}l:$$

(d) trigonal pyramidal

130

7 Reactions in Aqueous Solution

7.1   (a) precipitation   (b) redox   (c) acid-base neutralization

7.2 $FeBr_3$ contains 3 Br^- ions. The molar concentration of Br^- ions = 3 x 0.225 M = 0.675 M.

7.3 A_2Y is the strongest electrolyte because it is completely dissociated into ions.
       A_2X is the weakest electrolyte because it is the least dissociated of the three substances.

7.4   (a) Ionic equation:
       $2\,Ag^+(aq) + 2\,NO_3^-(aq) + 2\,Na^+(aq) + CrO_4^{2-}(aq) \rightarrow Ag_2CrO_4(s) + 2\,Na^+(aq) + 2\,NO_3^-(aq)$
       Delete spectator ions from the ionic equation to get the net ionic equation.
       Net ionic equation: $2\,Ag^+(aq) + CrO_4^{2-}(aq) \rightarrow Ag_2CrO_4(s)$
       (b) Ionic equation:
       $2\,H^+(aq) + SO_4^{2-}(aq) + MgCO_3(s) \rightarrow H_2O(l) + CO_2(g) + Mg^{2+}(aq) + SO_4^{2-}(aq)$
       Delete spectator ions from the ionic equation to get the net ionic equation.
       Net ionic equation: $2\,H^+(aq) + MgCO_3(s) \rightarrow H_2O(l) + CO_2(g) + Mg^{2+}(aq)$
       (c) Ionic equation:
       $Hg^{2+}(aq) + 2\,NO_3^-(aq) + 2\,NH_4^+(aq) + 2\,I^-(aq) \rightarrow HgI_2(s) + 2\,NH_4^+(aq) + 2\,NO_3^-(aq)$
       Delete spectator ions from the ionic equation to get the net ionic equation.
       Net ionic equation: $Hg^{2+}(aq) + 2\,I^-(aq) \rightarrow HgI_2(s)$

7.5   (a) $CdCO_3$, insoluble   (b) MgO, insoluble   (c) Na_2S, soluble
       (d) $PbSO_4$, insoluble   (e) $(NH_4)_3PO_4$, soluble   (f) $HgCl_2$, soluble

7.6   (a) Ionic equation:
       $Ni^{2+}(aq) + 2\,Cl^-(aq) + 2\,NH_4^+(aq) + S^{2-}(aq) \rightarrow NiS(s) + 2\,NH_4^+(aq) + 2\,Cl^-(aq)$
       Delete spectator ions from the ionic equation to get the net ionic equation.
       Net ionic equation: $Ni^{2+}(aq) + S^{2-}(aq) \rightarrow NiS(s)$
       (b) Ionic equation:
       $2\,Na^+(aq) + CrO_4^{2-}(aq) + Pb^{2+}(aq) + 2\,NO_3^-(aq) \rightarrow PbCrO_4(s) + 2\,Na^+(aq) + 2\,NO_3^-(aq)$
       Delete spectator ions from the ionic equation to get the net ionic equation.
       Net ionic equation: $Pb^{2+}(aq) + CrO_4^{2-}(aq) \rightarrow PbCrO_4(s)$
       (c) Ionic equation:
       $2\,Ag^+(aq) + 2\,ClO_4^-(aq) + Ca^{2+}(aq) + 2\,Br^-(aq) \rightarrow 2\,AgBr(s) + Ca^{2+}(aq) + 2\,ClO_4^-(aq)$
       Delete spectator ions from the ionic equation and reduce coefficients to get the net ionic
       equation.
       Net ionic equation: $Ag^+(aq) + Br^-(aq) \rightarrow AgBr(s)$
       (d) Ionic equation:
       $Zn^{2+}(aq) + 2\,Cl^-(aq) + 2\,K^+(aq) + CO_3^{2-}(aq) \rightarrow ZnCO_3(s) + 2\,K^+(aq) + 2\,Cl^-(aq)$
       Delete spectator ions from the ionic equation to get the net ionic equation.
       Net ionic equation: $Zn^{2+}(aq) + CO_3^{2-}(aq) \rightarrow ZnCO_3(s)$

7.7 $3\,CaCl_2(aq) + 2\,Na_3PO_4(aq) \rightarrow Ca_3(PO_4)_2(s) + 6\,NaCl(aq)$
Ionic equation:
$3\,Ca^{2+}(aq) + 6\,Cl^-(aq) + 6\,Na^+(aq) + 2\,PO_4^{3-}(aq) \rightarrow Ca_3(PO_4)_2(s) + 6\,Na^+(aq) + 6\,Cl^-(aq)$
Delete spectator ions from the ionic equation to get the net ionic equation.
Net ionic equation: $3\,Ca^{2+}(aq) + 2\,PO_4^{3-}(aq) \rightarrow Ca_3(PO_4)_2(s)$

7.8   A precipitate results from the reaction. The precipitate contains cations and anions in a 3:2 ratio. The precipitate is either $Mg_3(PO_4)_2$ or $Zn_3(PO_4)_2$.

7.9   (a) HIO_4, periodic acid   (b) $HBrO_2$, bromous acid   (c) H_2CrO_4, chromic acid

7.10   (a) H_3PO_3 (b) H_2Se

7.11   (a) Ionic equation:
$2\,Cs^+(aq) + 2\,OH^-(aq) + 2\,H^+(aq) + SO_4^{2-}(aq) \rightarrow 2\,Cs^+(aq) + SO_4^{2-}(aq) + 2\,H_2O(l)$
Delete spectator ions from the ionic equation and reduce coefficients to get the net ionic equation.
Net ionic equation: $H^+(aq) + OH^-(aq) \rightarrow H_2O(l)$

(b) Ionic equation:
$Ca^{2+}(aq) + 2\,OH^-(aq) + 2\,CH_3CO_2H(aq) \rightarrow Ca^{2+}(aq) + 2\,CH_3CO_2^-(aq) + 2\,H_2O(l)$
Delete spectator ions from the ionic equation and reduce coefficients to get the net ionic equation.
Net ionic equation: $CH_3CO_2H(aq) + OH^-(aq) \rightarrow CH_3CO_2^-(aq) + H_2O(l)$

7.12   HY is the strongest acid because it is completely dissociated.
HX is the weakest acid because it is the least dissociated.

7.13   (a) $SnCl_4$:   Cl –1, Sn +4   (b) CrO_3:   O –2, Cr +6
         (c) $VOCl_3$:   O –2, Cl –1, V +5   (d) V_2O_3:   O –2, V +3
         (e) HNO_3:   O –2, H +1, N +5   (f) $FeSO_4$:   O –2, S +6, Fe +2

7.14 $2\,Cu^{2+}(aq) + 4\,I^-(aq) \rightarrow 2\,CuI(s) + I_2(aq)$
oxidation numbers: Cu^{2+} +2; I^- –1;   CuI: Cu +1, I –1; I_2: 0
oxidizing agent (oxidation number decreases), Cu^{2+}
reducing agent (oxidation number increases), I^-

7.15   (a) $SnO_2(s) + 2\,C(s) \rightarrow Sn(s) + 2\,CO(g)$
C is oxidized (its oxidation number increases from 0 to +2). C is the reducing agent.
The Sn in SnO_2 is reduced (its oxidation number decreases from +4 to 0). SnO_2 is the oxidizing agent.
(b) $Sn^{2+}(aq) + 2\,Fe^{3+}(aq) \rightarrow Sn^{4+}(aq) + 2\,Fe^{2+}(aq)$
Sn^{2+} is oxidized (its oxidation number increases from +2 to +4). Sn^{2+} is the reducing agent.
Fe^{3+} is reduced (its oxidation number decreases from +3 to +2). Fe^{3+} is the oxidizing agent.

(c) $4 NH_3(g) + 5 O_2(g) \rightarrow 4 NO(g) + 6 H_2O(l)$
The N in NH_3 is oxidized (its oxidation number increases from -3 to $+2$). NH_3 is the reducing agent.
Each O in O_2 is reduced (its oxidation number decreases from 0 to -2). O_2 is the oxidizing agent.

7.16　(a) Pt is below H in the activity series; therefore NO REACTION.
　　　(b) Mg is below Ca in the activity series; therefore NO REACTION.

7.17　Because B will reduce A^+, B is above A in the activity series. Because B will not reduce C^+, C is above B in the activity series. Therefore C must be above A in the activity series and C will reduce A^+.

7.18　"Any element higher in the activity series will react with the ion of any element lower in the activity series."
$A + D^+ \rightarrow A^+ + D$; therefore A is higher than D.
$B^+ + D \rightarrow B + D^+$; therefore D is higher than B.
$C^+ + D \rightarrow C + D^+$; therefore D is higher than C.
$B + C^+ \rightarrow B^+ + C$; therefore B is higher than C.
The net result is A > D > B > C.

7.19　(a) $MnO_4^-(aq) \rightarrow MnO_2(s)$ 　　　　(reduction)
　　　　　$IO_3^-(aq) \rightarrow IO_4^-(aq)$ 　　　　(oxidation)

　　　(b) $NO_3^-(aq) \rightarrow NO_2(g)$ 　　　　(reduction)
　　　　　$SO_2(aq) \rightarrow SO_4^{2-}(aq)$ 　　　　(oxidation)

7.20　$NO_3^-(aq) + Cu(s) \rightarrow NO(g) + Cu^{2+}(aq)$
　　　$[Cu(s) \rightarrow Cu^{2+}(aq) + 2 e^-] \times 3$ 　　　　(oxidation half reaction)

　　　$NO_3^-(aq) \rightarrow NO(g)$
　　　$NO_3^-(aq) \rightarrow NO(g) + 2 H_2O(l)$
　　　$4 H^+(aq) + NO_3^-(aq) \rightarrow NO(g) + 2 H_2O(l)$
　　　$[3 e^- + 4 H^+(aq) + NO_3^-(aq) \rightarrow NO(g) + 2 H_2O(l)] \times 2$ 　　　(reduction half reaction)

　　　Combine the two half reactions.
　　　$2 NO_3^-(aq) + 8 H^+(aq) + 3 Cu(s) \rightarrow 3 Cu^{2+}(aq) + 2 NO(g) + 4 H_2O(l)$

7.21　$Fe(OH)_2(s) + O_2(g) \rightarrow Fe(OH)_3(s)$
　　　$[Fe(OH)_2(s) + OH^-(aq) \rightarrow Fe(OH)_3(s) + e^-] \times 4$ 　(oxidation half reaction)

　　　$O_2(g) \rightarrow 2 H_2O(l)$
　　　$4 H^+(aq) + O_2(g) \rightarrow 2 H_2O(l)$
　　　$4 e^- + 4 H^+(aq) + O_2(g) \rightarrow 2 H_2O(l)$
　　　$4 e^- + 4 H^+(aq) + 4 OH^-(aq) + O_2(g) \rightarrow 2 H_2O(l) + 4 OH^-(aq)$
　　　$4 e^- + 4 H_2O(l) + O_2(g) \rightarrow 2 H_2O(l) + 4 OH^-(aq)$
　　　$4 e^- + 2 H_2O(l) + O_2(g) \rightarrow 4 OH^-(aq)$ 　　　　(reduction half reaction)

Combine the two half reactions.

$4 \text{ Fe(OH)}_2(s) + 4 \text{ OH}^-(aq) + 2 \text{ H}_2\text{O}(l) + \text{O}_2(g) \rightarrow 4 \text{ Fe(OH)}_3(s) + 4 \text{ OH}^-(aq)$

$4 \text{ Fe(OH)}_2(s) + 2 \text{ H}_2\text{O}(l) + \text{O}_2(g) \rightarrow 4 \text{ Fe(OH)}_3(s)$

7.22 31.50 mL = 0.031 50 L; 10.00 mL = 0.010 00 L

$$0.031\ 50 \text{ L} \times \frac{0.105 \text{ mol BrO}_3^-}{1 \text{ L}} \times \frac{6 \text{ mol Fe}^{2+}}{1 \text{ mol BrO}_3^-} = 1.98 \times 10^{-2} \text{ mol Fe}^{2+}$$

$$\text{molarity} = \frac{1.98 \times 10^{-2} \text{ mol Fe}^{2+}}{0.010\ 00 \text{ L}} = 1.98 \text{ M Fe}^{2+} \text{ solution}$$

7.23 $\text{Pb}(s) + \text{HSO}_4^-(aq) \rightarrow \text{PbSO}_4(s)$

$\text{Pb}(s) + \text{HSO}_4^-(aq) \rightarrow \text{PbSO}_4(s) + \text{H}^+(aq)$

$\text{Pb}(s) + \text{HSO}_4^-(aq) \rightarrow \text{PbSO}_4(s) + \text{H}^+(aq) + 2 \text{ e}^-$ (oxidation half reaction)

$\text{PbO}_2(s) + \text{HSO}_4^-(aq) \rightarrow \text{PbSO}_4(s)$

$\text{PbO}_2(s) + \text{HSO}_4^-(aq) \rightarrow \text{PbSO}_4(s) + 2 \text{ H}_2\text{O}(l)$

$\text{PbO}_2(s) + \text{HSO}_4^-(aq) + 3 \text{ H}^+(aq) \rightarrow \text{PbSO}_4(s) + 2 \text{ H}_2\text{O}(l)$

$\text{PbO}_2(s) + \text{HSO}_4^-(aq) + 3 \text{ H}^+(aq) + 2 \text{ e}^- \rightarrow \text{PbSO}_4(s) + 2 \text{ H}_2\text{O}(l)$

(reduction half reaction)

Combine the two half reactions.

$\text{Pb}(s) + \text{PbO}_2(s) + 2 \text{ HSO}_4^-(aq) + 3 \text{ H}^+(aq) \rightarrow 2 \text{ PbSO}_4(s) + 2 \text{ H}_2\text{O}(l) + \text{H}^+(aq)$

$\text{Pb}(s) + \text{PbO}_2(s) + 2 \text{ HSO}_4^-(aq) + 2 \text{ H}^+(aq) \rightarrow 2 \text{ PbSO}_4(s) + 2 \text{ H}_2\text{O}(l)$

7.24 For a green process look for a solvent that is safe, non-toxic, non-polluting, and renewable. H_2O would be an excellent green solvent.

Key Concept Problems

7.25 (a) $2 \text{ Na}^+(aq) + \text{CO}_3^{2-}(aq)$ does not form a precipitate. This is represented by box (1).
(b) $\text{Ba}^{2+}(aq) + \text{CrO}_4^{2-}(aq) \rightarrow \text{BaCrO}_4(s)$. This is represented by box (2).
(c) $2 \text{ Ag}^+(aq) + \text{SO}_4^{2-}(aq) \rightarrow \text{Ag}_2\text{SO}_4(s)$. This is represented by box (3).

7.26 In the precipitate there are two cations (blue) for each anion (green). Looking at the ions in the list, the anion must have a -2 charge and the cation a $+1$ charge for charge neutrality of the precipitate. The cation must be Ag^+ because all Na^+ salts are soluble. Ag_2CrO_4 and Ag_2CO_3 are insoluble and consistent with the observed result.

7.27 One OH^- will react with each available H^+ on the acid forming H_2O. The acid is identified by how many of the 12 OH^- react with three molecules of each acid.
(a) Three HF's react with three OH^-, leaving nine OH^- unreacted (box 2).
(b) Three H_2SO_3's react with six OH^-, leaving six OH^- unreacted (box 3).
(c) Three H_3PO_4's react with nine OH^-, leaving three OH^- unreacted (box 1).

7.28   The concentration in the buret is three times that in the flask. The NaOCl concentration is 0.040 M. Because the I^- concentration in the buret is three times the OCl^- concentration in the flask and the reaction requires 2 I^- ions per OCl^- ion, 2/3 or 67% of the I^- solution from the buret must be added to the flask to react with all of the OCl^-.

7.29   (a) Ionic equation:
$K^+(aq) + Cl^-(aq) + Ag^+(aq) + NO_3^-(aq) \rightarrow AgCl(s) + K^+(aq) + NO_3^-(aq)$
(b) Ionic equation:
$HF(aq) + K^+(aq) + OH^-(aq) \rightarrow K^+(aq) + F^-(aq) + H_2O(l)$
(c) Ionic equation:
$Ba^{2+}(aq) + 2\ Cl^-(aq) + 2\ Na^+(aq) + SO_4^{2-}(aq) \rightarrow BaSO_4(s) + 2\ Na^+(aq) + 2\ Cl^-(aq)$

Reaction (c) would have the highest initial conductivity because of the 3 net ions for each $BaCl_2$ (a strong electrolyte).
Reaction (b) would have have the lowest (almost zero) initial conductivity because HF is a very weak acid/electrolyte.
Reaction (a) would have an intermediate initial conductivity between that for reactions (b) and (c).
Figure (1) is for reaction (a); figure (2) is for reaction (b); and figure (3) is for reaction (c).

7.30   (a) $Sr^+ + At \rightarrow Sr + At^+$ No reaction.
(b) $Si + At^+ \rightarrow Si^+ + At$ Reaction would occur.
(c) $Sr + Si^+ \rightarrow Sr^+ + Si$ Reaction would occur.

7.31   "Any element higher in the activity series will react with the ion of any element lower in the activity series."
$BLUE + RED^+ \rightarrow BLUE^+ + RED$; therefore BLUE is higher than RED.
$RED + GREEN^+ \rightarrow RED^+ + GREEN$; therefore RED is higher than GREEN.
The net result is BLUE > RED > GREEN. Because BLUE is above GREEN, the following reaction will occur: $BLUE + GREEN^+ \rightarrow BLUE^+ + GREEN$

Section Problems
Aqueous Reactions, Net Ionic Equations, and Electrolytes (Sections 7.2–7.3)

7.32   (a) precipitation   (b) redox   (c) acid-base neutralization

7.33   (a) redox   (b) precipitation   (c) acid-base neutralization

7.34   (a) Ionic equation:
$Hg^{2+}(aq) + 2\ NO_3^-(aq) + 2\ Na^+(aq) + 2\ I^-(aq) \rightarrow 2\ Na^+(aq) + 2\ NO_3^-(aq) + HgI_2(s)$
Delete spectator ions from the ionic equation to get the net ionic equation.
Net ionic equation: $Hg^{2+}(aq) + 2\ I^-(aq) \rightarrow HgI_2(s)$

(b) $2\ HgO(s) \xrightarrow{\text{Heat}} 2\ Hg(l) + O_2(g)$
(c) Ionic equation:
$H_3PO_4(aq) + 3\ K^+(aq) + 3\ OH^-(aq) \rightarrow 3\ K^+(aq) + PO_4^{3-}(aq) + 3\ H_2O(l)$
Delete spectator ions from the ionic equation to get the net ionic equation.
Net ionic equation: $H_3PO_4(aq) + 3\ OH^-(aq) \rightarrow PO_4^{3-}(aq) + 3\ H_2O(l)$

7.35   (a) $S_8(s) + 8 O_2(g) \rightarrow 8 SO_2(g)$
        (b) Ionic equation:
        $Ni^{2+}(aq) + 2 Cl^-(aq) + 2 Na^+(aq) + S^{2-}(aq) \rightarrow NiS(s) + 2 Na^+(aq) + 2 Cl^-(aq)$
        Delete spectator ions from the ionic equation to get the net ionic equation.
        Net ionic equation: $Ni^{2+}(aq) + S^{2-}(aq) \rightarrow NiS(s)$
        (c) Ionic equation:
        $2 CH_3CO_2H(aq) + Ba^{2+}(aq) + 2 OH^-(aq) \rightarrow 2 CH_3CO_2^-(aq) + Ba^{2+}(aq) + 2 H_2O(l)$
        Delete spectator ions from the ionic equation to get the net ionic equation.
        Net ionic equation: $CH_3CO_2H(aq) + OH^-(aq) \rightarrow CH_3CO_2^-(aq) + H_2O(l)$

7.36 $Ba(OH)_2$ is soluble in aqueous solution, dissociates into $Ba^{2+}(aq)$ and $2 OH^-(aq)$, and
        conducts electricity. In aqueous solution, H_2SO_4 dissociates into $H^+(aq)$ and $HSO_4^-(aq)$.
        H_2SO_4 solutions conduct electricity. When equal molar solutions of $Ba(OH)_2$ and H_2SO_4
        are mixed, the insoluble $BaSO_4$ is formed along with two H_2O. In water, $BaSO_4$ does not
        produce any appreciable amount of ions and the mixture does not conduct electricity.

7.37 H_2O is polar and a good H^+ acceptor. It allows the polar HCl to dissociate into ions in
        aqueous solution: $HCl + H_2O \rightarrow H_3O^+ + Cl^-$.
        $CHCl_3$ is not very polar and not a H^+ acceptor and so does not allow the polar HCl to
        dissociate into ions.

7.38   (a) HBr, strong electrolyte   (b) HF, weak electrolyte
        (c) $NaClO_4$, strong electrolyte   (d) $(NH_4)_2CO_3$, strong electrolyte
        (e) NH_3, weak electrolyte   (f) C_2H_5OH, nonelectrolyte

7.39   It is possible for a molecular compound to be a strong electrolyte. For example, HCl is a
        molecular compound when pure, but dissociates completely to give H^+ and Cl^- ions when
        it dissolves in water.

7.40   (a) K_2CO_3 contains 3 ions ($2 K^+$ and $1 CO_3^{2-}$).
        The molar concentration of ions = 3 x 0.750 M = 2.25 M.
        (b) $AlCl_3$ contains 4 ions ($1 Al^{3+}$ and $3 Cl^-$).
        The molar concentration of ions = 4 x 0.355 M = 1.42 M.

7.41   (a) CH_3OH is a nonelectrolyte. The ion concentration from CH_3OH is zero.
        (b) $HClO_4$ is a strong acid.
        $HClO_4(aq) \rightarrow H^+(aq) + ClO_4^-(aq)$
        In solution, there are 2 moles of ions per mole of $HClO_4$.
        The molar concentration of ions = 2×0.225 M = 0.450 M.

Precipitation Reactions and Solubility Guidelines (Section 7.4)

7.42   (a) Ag_2O, insoluble   (b) $Ba(NO_3)_2$, soluble
        (c) $SnCO_3$, insoluble   (d) Fe_2O_3, insoluble

7.43   (a) ZnS, insoluble   (b) $Au_2(CO_3)_3$, insoluble
        (c) $PbCl_2$, insoluble (soluble in hot water)   (d) MnO_2, insoluble

7.44   (a) No precipitate will form.   (b) $FeCl_2(aq) + 2\,KOH(aq) \rightarrow Fe(OH)_2(s) + 2\,KCl(aq)$
      (c) No precipitate will form.   (d) No precipitate will form.

7.45   (a) $MnCl_2(aq) + Na_2S(aq) \rightarrow MnS(s) + 2\,NaCl(aq)$
      (b) No precipitate will form.
      (c) $3\,Hg(NO_3)_2(aq) + 2\,Na_3PO_4(aq) \rightarrow Hg_3(PO_4)_2(s) + 6\,NaNO_3(aq)$
      (d) $Ba(NO_3)_2(aq) + 2\,KOH(aq) \rightarrow Ba(OH)_2(s) + 2\,KNO_3(aq)$

7.46   (a) $Pb(NO_3)_2(aq) + Na_2SO_4(aq) \rightarrow PbSO_4(s) + 2\,NaNO_3(aq)$
      (b) $3\,MgCl_2(aq) + 2\,K_3PO_4(aq) \rightarrow Mg_3(PO_4)_2(s) + 6\,KCl(aq)$
      (c) $ZnSO_4(aq) + Na_2CrO_4(aq) \rightarrow ZnCrO_4(s) + Na_2SO_4(aq)$

7.47   (a) $AlCl_3(aq) + 3\,NaOH(aq) \rightarrow Al(OH)_3(s) + 3\,NaCl(aq)$
      (b) $Fe(NO_3)_2(aq) + Na_2S(aq) \rightarrow FeS(s) + 2\,NaNO_3(aq)$
      (c) $CoSO_4(aq) + K_2CO_3(aq) \rightarrow CoCO_3(s) + K_2SO_4(aq)$

7.48   Add $HCl(aq)$; it will selectively precipitate $AgCl(s)$.

7.49   Add $Na_2SO_4(aq)$; it will selectively precipitate $BaSO_4(s)$.

7.50 Ag^+ is eliminated because it would have precipitated as $AgCl(s)$; Ba^{2+} is eliminated because it would have precipitated as $BaSO_4(s)$. The solution might contain Cs^+ and/or NH_4^+. Neither of these will precipitate with OH^-, SO_4^{2-}, or Cl^-.

7.51 Cl^- is eliminated because it would have precipitated as $AgCl(s)$. OH^- is eliminated because it would have precipitated as either $AgOH(s)$ or $Cu(OH)_2(s)$. SO_4^{2-} is eliminated because it would have precipitated as $BaSO_4(s)$. The solution might contain NO_3^- because all nitrates are soluble.

Acids, Bases, and Neutralization Reactions (Section 7.5)

7.52   Add the solution to an active metal, such as magnesium. Bubbles of H_2 gas indicate the presence of an acid.

7.53   We use a double arrow to show the dissociation of a weak acid or weak base in aqueous solution to indicate the equilibrium between reactants and products.

7.54   (a) $2\,H^+(aq) + 2\,ClO_4^-(aq) + Ca^{2+}(aq) + 2\,OH^-(aq) \rightarrow Ca^{2+}(aq) + 2\,ClO_4^-(aq) + 2\,H_2O(l)$
      (b) $CH_3CO_2H(aq) + Na^+(aq) + OH^-(aq) \rightarrow CH_3CO_2^-(aq) + Na^+(aq) + H_2O(l)$

7.55   (a) $2\,HF(aq) + Ca^{2+}(aq) + 2\,OH^-(aq) \rightarrow Ca^{2+}(aq) + 2\,F^-(aq) + 2\,H_2O(l)$
      (b) $Mg(OH)_2(s) + 2\,H^+(aq) + 2\,NO_3^-(aq) \rightarrow Mg^{2+}(aq) + 2\,NO_3^-(aq) + 2\,H_2O(l)$

7.56   (a) $LiOH(aq) + HI(aq) \rightarrow LiI(aq) + H_2O(l)$
      Ionic equation: $Li^+(aq) + OH^-(aq) + H^+(aq) + I^-(aq) \rightarrow Li^+(aq) + I^-(aq) + H_2O(l)$
      Delete spectator ions from the ionic equation to get the net ionic equation.
      Net ionic equation: $H^+(aq) + OH^-(aq) \rightarrow H_2O(l)$

(b) $2\ HBr(aq) + Ca(OH)_2(aq) \rightarrow CaBr_2(aq) + 2\ H_2O(l)$
Ionic equation:
$2\ H^+(aq) + 2\ Br^-(aq) + Ca^{2+}(aq) + 2\ OH^-(aq) \rightarrow Ca^{2+}(aq) + 2\ Br^-(aq) + 2\ H_2O(l)$
Delete spectator ions from the ionic equation to get the net ionic equation.
Net ionic equation: $H^+(aq) + OH^-(aq) \rightarrow H_2O(l)$

7.57   (a) $2\ Fe(OH)_3(s) + 3\ H_2SO_4(aq) \rightarrow Fe_2(SO_4)_3(aq) + 6\ H_2O(l)$
Ionic equation and net ionic equation are the same.
$2\ Fe(OH)_3(s) + 3\ H^+(aq) + 3\ HSO_4^-(aq) \rightarrow 2\ Fe^{3+}(aq) + 3\ SO_4^{2-}(aq) + 6\ H_2O(l)$

(b) $HClO_3(aq) + NaOH(aq) \rightarrow NaClO_3(aq) + H_2O(l)$
Ionic equation $H^+(aq) + ClO_3^-(aq) + Na^+(aq) + OH^-(aq) \rightarrow Na^+(aq) + ClO_3^-(aq) + H_2O(l)$
Delete spectator ions from the ionic equation to get the net ionic equation.
Net ionic equation: $H^+(aq) + OH^-(aq) \rightarrow H_2O(l)$

Redox Reactions and Oxidation Numbers (Sections 7.6–7.8)

7.58   The best reducing agents are at the bottom left of the periodic table. The best oxidizing agents are at the top right of the periodic table (excluding the noble gases).

7.59   The most easily reduced elements in the periodic table are in the top-right corner, excluding group 8A.
The most easily oxidized elements in the periodic table are in the bottom-left corner.

7.60   (a) An oxidizing agent gains electrons.
(b) A reducing agent loses electrons.
(c) A substance undergoing oxidation loses electrons.
(d) A substance undergoing reduction gains electrons.

7.61   (a) In a redox reaction, the oxidation number decreases for an oxidizing agent.
(b) In a redox reaction, the oxidation number increases for a reducing agent.
(c) In a redox reaction, the oxidation number increases for a substance undergoing oxidation.
(d) In a redox reaction, the oxidation number decreases for a substance undergoing reduction.

7.62   (a) NO_2 O -2, N $+4$ (b) SO_3 O -2, S $+6$
(c) $COCl_2$ O -2, Cl -1, C $+4$ (d) CH_2Cl_2 Cl -1, H $+1$, C 0
(e) $KClO_3$ O -2, K $+1$, Cl $+5$ (f) HNO_3 O -2, H $+1$, N $+5$

7.63   (a) $VOCl_3$ O -2,   Cl -1,   V $+5$
(b) $CuSO_4$ O -2,   S $+6$,   Cu $+2$
(c) CH_2O O -2,   H $+1$,   C 0
(d) Mn_2O_7 O -2,   Mn $+7$
(e) OsO_4 O -2,   Os $+8$
(f) H_2PtCl_6 Cl -1,   H $+1$,   Pt $+4$

7.64 (a) ClO_3^- O –2, Cl +5   (b) SO_3^{2-} O –2, S +4
      (c) $C_2O_4^{2-}$ O –2, C +3   (d) NO_2^- O –2, N +3
      (e) BrO^- O –2, Br +1   (f) AsO_4^{3-} O –2, As +5

7.65 (a) $Cr(OH)_4^-$ O –2,   H +1,   Cr +3
      (b) $S_2O_3^{2-}$ O –2,   S +2
      (c) NO_3^- O –2,   N +5
      (d) MnO_4^{2-} O –2,   Mn +6
      (e) HPO_4^{2-} O –2,   H +1,   P +5
      (f) $V_2O_7^{4-}$ O –2,   V +5

7.66 (a) $Ca(s) + Sn^{2+}(aq) \rightarrow Ca^{2+}(aq) + Sn(s)$
      Ca(s) is oxidized (oxidation number increases from 0 to +2).
      $Sn^{2+}(aq)$ is reduced (oxidation number decreases from +2 to 0).
      (b) $ICl(s) + H_2O(l) \rightarrow HCl(aq) + HOI(aq)$
      No oxidation numbers change. The reaction is not a redox reaction.

7.67 (a) $Si(s) + 2\,Cl_2(g) \rightarrow SiCl_4(l)$
      Si(s) is oxidized (oxidation number increases from 0 to +4).
      $Cl_2(g)$ is reduced (oxidation number decreases from 0 to –1).
      (b) $Cl_2(g) + 2\,NaBr(aq) \rightarrow Br_2(aq) + 2\,NaCl(aq)$
      $Br^-(aq)$ is oxidized (oxidation number increases from –1 to 0).
      $Cl_2(g)$ is reduced (oxidation number decreases from 0 to –1).

7.68 (a) Zn is below Na^+; therefore no reaction.
      (b) Pt is below H^+; therefore no reaction.
      (c) Au is below Ag^+; therefore no reaction.
      (d) Ag is above Au^{3+}; the reaction is $Au^{3+}(aq) + 3\,Ag(s) \rightarrow 3\,Ag^+(aq) + Au(s)$.

7.69 Sr is more metallic than Sb because it is in the same period and to the left of Sb on the periodic table. Sr is the better reducing agent.
      $2\,Sb^{3+}(aq) + 3\,Sr(s) \rightarrow 2\,Sb(s) + 3\,Sr^{2+}(aq)$ will occur, the reverse will not.

7.70 (a) "Any element higher in the activity series will react with the ion of any element lower in the activity series."
      $A + B^+ \rightarrow A^+ + B$; therefore A is higher than B.
      $C^+ + D \rightarrow$ no reaction; therefore C is higher than D.
      $B + D^+ \rightarrow B^+ + D$; therefore B is higher than D.
      $B + C^+ \rightarrow B^+ + C$; therefore B is higher than C.
      The net result is A > B > C > D.
      (b) (1) C is below A^+; therefore no reaction.
          (2) D is below A^+; therefore no reaction.

7.71 (a) "Any element higher in the activity series will react with the ion of any element lower in the activity series."
      $2\,A + B^{2+} \rightarrow 2\,A^+ + B$; therefore A is higher than B.

B + D^{2+} → B^{2+} + D; therefore B is higher than D.

A$^+$ + C → no reaction; therefore A is higher than C.

2 C + B^{2+} → 2 C$^+$ + B; therefore C is higher than B.

The net result is A > C > B > D.

(b) (1) D is below A$^+$; therefore no reaction.

(2) C is above D^{2+}; therefore the reaction will occur.

Balancing Redox Reactions (Section 7.9)

7.72 (a) N oxidation number decreases from +5 to +2; reduction.

(b) Zn oxidation number increases from 0 to +2; oxidation.

(c) Ti oxidation number increases from +3 to +4; oxidation.

(d) Sn oxidation number decreases from +4 to +2; reduction.

7.73 (a) O oxidation number decreases from 0 to –2; reduction.

(b) O oxidation number increases from –1 to 0; oxidation.

(c) Mn oxidation number decreases from +7 to +6; reduction.

(d) C oxidation number increases from –2 to 0; oxidation.

7.74 (a) $NO_3^-(aq) \rightarrow NO(g)$

$NO_3^-(aq) \rightarrow NO(g) + 2 H_2O(l)$

$4 H^+(aq) + NO_3^-(aq) \rightarrow NO(g) + 2 H_2O(l)$

$3 e^- + 4 H^+(aq) + NO_3^-(aq) \rightarrow NO(g) + 2 H_2O(l)$

(b) $Zn(s) \rightarrow Zn^{2+}(aq) + 2 e^-$

(c) $Ti^{3+}(aq) \rightarrow TiO_2(s)$

$Ti^{3+}(aq) + 2 H_2O(l) \rightarrow TiO_2(s)$

$Ti^{3+}(aq) + 2 H_2O(l) \rightarrow TiO_2(s) + 4 H^+(aq)$

$Ti^{3+}(aq) + 2 H_2O(l) \rightarrow TiO_2(s) + 4 H^+(aq) + e^-$

(d) $Sn^{4+}(aq) + 2 e^- \rightarrow Sn^{2+}(aq)$

7.75 (a) $O_2(g) \rightarrow OH^-(aq)$

$O_2(g) \rightarrow OH^-(aq) + H_2O(l)$

$3 H^+(aq) + O_2(g) \rightarrow OH^-(aq) + H_2O(l)$

$3 H^+(aq) + 3 OH^-(aq) + O_2(g) \rightarrow 4 OH^-(aq) + H_2O(l)$

$3 H_2O(l) + O_2(g) \rightarrow 4 OH^-(aq) + H_2O(l)$

$4 e^- + 2 H_2O(l) + O_2(g) \rightarrow 4 OH^-(aq)$

(b) $H_2O_2(aq) \rightarrow O_2(g)$

$H_2O_2(aq) \rightarrow O_2(g) + 2 H^+(aq)$

$2 OH^-(aq) + H_2O_2(aq) \rightarrow O_2(g) + 2 H^+(aq) + 2 OH^-(aq)$

$2 OH^-(aq) + H_2O_2(aq) \rightarrow O_2(g) + 2 H_2O(l) + 2 e^-$

(c) $MnO_4^-(aq) \rightarrow MnO_4^{2-}(aq)$

$MnO_4^-(aq) + e^- \rightarrow MnO_4^{2-}(aq)$

(d) $CH_3OH(aq) \rightarrow CH_2O(aq)$

$CH_3OH(aq) \rightarrow CH_2O(aq) + 2 H^+(aq)$

$CH_3OH(aq) + 2 OH^-(aq) \rightarrow CH_2O(aq) + 2 H^+(aq) + 2 OH^-(aq)$

$CH_3OH(aq) + 2 OH^-(aq) \rightarrow CH_2O(aq) + 2 H_2O(l)$

$CH_3OH(aq) + 2 OH^-(aq) \rightarrow CH_2O(aq) + 2 H_2O(l) + 2 e^-$

7.76 (a) $Te(s) + NO_3^-(aq) \rightarrow TeO_2(s) + NO(g)$
oxidation: $Te(s) \rightarrow TeO_2(s)$
reduction: $NO_3^-(aq) \rightarrow NO(g)$

(b) $H_2O_2(aq) + Fe^{2+}(aq) \rightarrow Fe^{3+}(aq) + H_2O(l)$
oxidation: $Fe^{2+}(aq) \rightarrow Fe^{3+}(aq)$
reduction: $H_2O_2(aq) \rightarrow H_2O(l)$

7.77 (a) $Mn(s) + NO_3^-(aq) \rightarrow Mn^{2+}(aq) + NO_2(g)$
oxidation: $Mn(s) \rightarrow Mn^{2+}(aq)$
reduction: $NO_3^-(aq) \rightarrow NO_2(g)$

(b) $Mn^{3+}(aq) \rightarrow MnO_2(s) + Mn^{2+}(aq)$
oxidation: $Mn^{3+}(aq) \rightarrow MnO_2(s)$
reduction: $Mn^{3+}(aq) \rightarrow Mn^{2+}(aq)$

7.78 (a) $Cr_2O_7^{2-}(aq) \rightarrow Cr^{3+}(aq)$
$Cr_2O_7^{2-}(aq) \rightarrow 2\ Cr^{3+}(aq)$
$Cr_2O_7^{2-}(aq) \rightarrow 2\ Cr^{3+}(aq) + 7\ H_2O(l)$
$14\ H^+(aq) + Cr_2O_7^{2-}(aq) \rightarrow 2\ Cr^{3+}(aq) + 7\ H_2O(l)$
$14\ H^+(aq) + Cr_2O_7^{2-}(aq) + 6\ e^- \rightarrow 2\ Cr^{3+}(aq) + 7\ H_2O(l)$

(b) $CrO_4^{2-}(aq) \rightarrow Cr(OH)_4^-(aq)$
$4\ H^+(aq) + CrO_4^{2-}(aq) \rightarrow Cr(OH)_4^-(aq)$
$4\ H^+(aq) + 4\ OH^-(aq) + CrO_4^{2-}(aq) \rightarrow Cr(OH)_4^-(aq) + 4\ OH^-(aq)$
$4\ H_2O(l) + CrO_4^{2-}(aq) \rightarrow Cr(OH)_4^-(aq) + 4\ OH^-(aq)$
$4\ H_2O(l) + CrO_4^{2-}(aq) + 3\ e^- \rightarrow Cr(OH)_4^-(aq) + 4\ OH^-(aq)$

(c) $Bi^{3+}(aq) \rightarrow BiO_3^-(aq)$
$Bi^{3+}(aq) + 3\ H_2O(l) \rightarrow BiO_3^-(aq)$
$Bi^{3+}(aq) + 3\ H_2O(l) \rightarrow BiO_3^-(aq) + 6\ H^+(aq)$
$Bi^{3+}(aq) + 3\ H_2O(l) + 6\ OH^-(aq) \rightarrow BiO_3^-(aq) + 6\ H^+(aq) + 6\ OH^-(aq)$
$Bi^{3+}(aq) + 3\ H_2O(l) + 6\ OH^-(aq) \rightarrow BiO_3^-(aq) + 6\ H_2O(l)$
$Bi^{3+}(aq) + 6\ OH^-(aq) \rightarrow BiO_3^-(aq) + 3\ H_2O(l)$
$Bi^{3+}(aq) + 6\ OH^-(aq) \rightarrow BiO_3^-(aq) + 3\ H_2O(l) + 2\ e^-$

(d) $ClO^-(aq) \rightarrow Cl^-(aq)$
$ClO^-(aq) \rightarrow Cl^-(aq) + H_2O(l)$
$2\ H^+(aq) + ClO^-(aq) \rightarrow Cl^-(aq) + H_2O(l)$
$2\ H^+(aq) + 2\ OH^-(aq) + ClO^-(aq) \rightarrow Cl^-(aq) + H_2O(l) + 2\ OH^-(aq)$
$2\ H_2O(l) + ClO^-(aq) \rightarrow Cl^-(aq) + H_2O(l) + 2\ OH^-(aq)$
$H_2O(l) + ClO^-(aq) \rightarrow Cl^-(aq) + 2\ OH^-(aq)$
$H_2O(l) + ClO^-(aq) + 2\ e^- \rightarrow Cl^-(aq) + 2\ OH^-(aq)$

7.79 (a) $VO^{2+}(aq) \rightarrow V^{3+}(aq)$
$VO^{2+}(aq) \rightarrow V^{3+}(aq) + H_2O(l)$
$2\ H^+(aq) + VO^{2+}(aq) \rightarrow V^{3+}(aq) + H_2O(l)$
$2\ H^+(aq) + VO^{2+}(aq) + e^- \rightarrow V^{3+}(aq) + H_2O(l)$

(b) $Ni(OH)_2(s) \rightarrow Ni_2O_3(s)$

$2\,Ni(OH)_2(s) \rightarrow Ni_2O_3(s) + H_2O(l)$

$2\,Ni(OH)_2(s) \rightarrow Ni_2O_3(s) + H_2O(l) + 2\,H^+(aq)$

$2\,Ni(OH)_2(s) + 2\,OH^-(aq) \rightarrow Ni_2O_3(s) + H_2O(l) + 2\,H^+(aq) + 2\,OH^-(aq)$

$2\,Ni(OH)_2(s) + 2\,OH^-(aq) \rightarrow Ni_2O_3(s) + 3\,H_2O(l) + 2\,e^-$

(c) $NO_3^-(aq) \rightarrow NO_2(g)$

$NO_3^-(aq) \rightarrow NO_2(g) + H_2O(l)$

$2\,H^+(aq) + NO_3^-(aq) \rightarrow NO_2(g) + H_2O(l)$

$2\,H^+(aq) + NO_3^-(aq) + e^- \rightarrow NO_2(g) + H_2O(l)$

(d) $Br_2(aq) \rightarrow BrO_3^-(aq)$

$Br_2(aq) \rightarrow 2\,BrO_3^-(aq)$

$Br_2(aq) + 6\,H_2O(l) \rightarrow 2\,BrO_3^-(aq)$

$Br_2(aq) + 6\,H_2O(l) \rightarrow 2\,BrO_3^-(aq) + 12\,H^+(aq)$

$Br_2(aq) + 6\,H_2O(l) + 12\,OH^-(aq) \rightarrow 2\,BrO_3^-(aq) + 12\,H^+(aq) + 12\,OH^-(aq)$

$Br_2(aq) + 6\,H_2O(l) + 12\,OH^-(aq) \rightarrow 2\,BrO_3^-(aq) + 12\,H_2O(l)$

$Br_2(aq) + 12\,OH^-(aq) \rightarrow 2\,BrO_3^-(aq) + 6\,H_2O(l) + 10\,e^-$

7.80   (a) $MnO_4^-(aq) \rightarrow MnO_2(s)$

$MnO_4^-(aq) \rightarrow MnO_2(s) + 2\,H_2O(l)$

$4\,H^+(aq) + MnO_4^-(aq) \rightarrow MnO_2(s) + 2\,H_2O(l)$

$[4\,H^+(aq) + MnO_4^-(aq) + 3\,e^- \rightarrow MnO_2(s) + 2\,H_2O(l)] \times 2$ (reduction half reaction)

$IO_3^-(aq) \rightarrow IO_4^-(aq)$

$H_2O(l) + IO_3^-(aq) \rightarrow IO_4^-(aq)$

$H_2O(l) + IO_3^-(aq) \rightarrow IO_4^-(aq) + 2\,H^+(aq)$

$[H_2O(l) + IO_3^-(aq) \rightarrow IO_4^-(aq) + 2\,H^+(aq) + 2\,e^-] \times 3$ (oxidation half reaction)

Combine the two half reactions.

$8\,H^+(aq) + 3\,H_2O(l) + 2\,MnO_4^-(aq) + 3\,IO_3^-(aq) \rightarrow$
$\qquad\qquad\qquad 6\,H^+(aq) + 4\,H_2O(l) + 2\,MnO_2(s) + 3\,IO_4^-(aq)$

$2\,H^+(aq) + 2\,MnO_4^-(aq) + 3\,IO_3^-(aq) \rightarrow 2\,MnO_2(s) + 3\,IO_4^-(aq) + H_2O(l)$

$2\,H^+(aq) + 2\,OH^-(aq) + 2\,MnO_4^-(aq) + 3\,IO_3^-(aq) \rightarrow$
$\qquad\qquad\qquad 2\,MnO_2(s) + 3\,IO_4^-(aq) + H_2O(l) + 2\,OH^-(aq)$

$2\,H_2O(l) + 2\,MnO_4^-(aq) + 3\,IO_3^-(aq) \rightarrow$
$\qquad\qquad\qquad 2\,MnO_2(s) + 3\,IO_4^-(aq) + H_2O(l) + 2\,OH^-(aq)$

$H_2O(l) + 2\,MnO_4^-(aq) + 3\,IO_3^-(aq) \rightarrow 2\,MnO_2(s) + 3\,IO_4^-(aq) + 2\,OH^-(aq)$

(b) $Cu(OH)_2(s) \rightarrow Cu(s)$

$Cu(OH)_2(s) \rightarrow Cu(s) + 2\,H_2O(l)$

$2\,H^+(aq) + Cu(OH)_2(s) \rightarrow Cu(s) + 2\,H_2O(l)$

$[2\,H^+(aq) + Cu(OH)_2(s) + 2\,e^- \rightarrow Cu(s) + 2\,H_2O(l)] \times 2$ (reduction half reaction)

$N_2H_4(aq) \rightarrow N_2(g)$

$N_2H_4(aq) \rightarrow N_2(g) + 4\,H^+(aq)$

$N_2H_4(aq) \rightarrow N_2(g) + 4\,H^+(aq) + 4\,e^-$ (oxidation half reaction)

Combine the two half reactions.

$4 H^+(aq) + 2 Cu(OH)_2(s) + N_2H_4(aq) \rightarrow 2 Cu(s) + 4 H_2O(l) + N_2(g) + 4 H^+(aq)$

$2 Cu(OH)_2(s) + N_2H_4(aq) \rightarrow 2 Cu(s) + 4 H_2O(l) + N_2(g)$

(c) $Fe(OH)_2(s) \rightarrow Fe(OH)_3(s)$

$Fe(OH)_2(s) + H_2O(l) \rightarrow Fe(OH)_3(s)$

$Fe(OH)_2(s) + H_2O(l) \rightarrow Fe(OH)_3(s) + H^+(aq)$

$[Fe(OH)_2(s) + H_2O(l) \rightarrow Fe(OH)_3(s) + H^+(aq) + e^-]$ x 3   (oxidation half reaction)

$CrO_4^{2-}(aq) \rightarrow Cr(OH)_4^-(aq)$

$4 H^+(aq) + CrO_4^{2-}(aq) \rightarrow Cr(OH)_4^-(aq)$

$4 H^+(aq) + CrO_4^{2-}(aq) + 3 e^- \rightarrow Cr(OH)_4^-(aq)$ (reduction half reaction)

Combine the two half reactions.

$3 Fe(OH)_2(s) + 3 H_2O(l) + 4 H^+(aq) + CrO_4^{2-}(aq) \rightarrow$
$\qquad 3 Fe(OH)_3(s) + 3 H^+(aq) + Cr(OH)_4^-(aq)$

$3 Fe(OH)_2(s) + 3 H_2O(l) + H^+(aq) + CrO_4^{2-}(aq) \rightarrow 3 Fe(OH)_3(s) + Cr(OH)_4^-(aq)$

$3 Fe(OH)_2(s) + 3 H_2O(l) + H^+(aq) + OH^-(aq) + CrO_4^{2-}(aq) \rightarrow$
$\qquad 3 Fe(OH)_3(s) + Cr(OH)_4^-(aq) + OH^-(aq)$

$3 Fe(OH)_2(s) + 4 H_2O(l) + CrO_4^{2-}(aq) \rightarrow 3 Fe(OH)_3(s) + Cr(OH)_4^-(aq) + OH^-(aq)$

(d) $ClO_4^-(aq) \rightarrow ClO_2^-(aq)$

$ClO_4^-(aq) \rightarrow ClO_2^-(aq) + 2 H_2O(l)$

$4 H^+(aq) + ClO_4^-(aq) \rightarrow ClO_2^-(aq) + 2 H_2O(l)$

$4 H^+(aq) + ClO_4^-(aq) + 4 e^- \rightarrow ClO_2^-(aq) + 2 H_2O(l)$ (reduction half reaction)

$H_2O_2(aq) \rightarrow O_2(g)$

$H_2O_2(aq) \rightarrow O_2(g) + 2 H^+(aq)$

$[H_2O_2(aq) \rightarrow O_2(g) + 2 H^+(aq) + 2 e^-]$ x 2   (oxidation half reaction)

Combine the two half reactions.

$4 H^+(aq) + ClO_4^-(aq) + 2 H_2O_2(aq) \rightarrow ClO_2^-(aq) + 2 H_2O(l) + 2 O_2(g) + 4 H^+(aq)$

$ClO_4^-(aq) + 2 H_2O_2(aq) \rightarrow ClO_2^-(aq) + 2 H_2O(l) + 2 O_2(g)$

7.81   (a) $S_2O_3^{2-}(aq) \rightarrow S_4O_6^{2-}(aq)$

$2 S_2O_3^{2-}(aq) \rightarrow S_4O_6^{2-}(aq)$

$2 S_2O_3^{2-}(aq) \rightarrow S_4O_6^{2-}(aq) + 2 e^-$ (oxidation half reaction)

$I_2(aq) \rightarrow I^-(aq)$

$I_2(aq) \rightarrow 2 I^-(aq)$

$I_2(aq) + 2 e^- \rightarrow 2 I^-(aq)$ (reduction half reaction)

Combine the two half reactions.

$2 S_2O_3^{2-}(aq) + I_2(aq) \rightarrow S_4O_6^{2-}(aq) + 2 I^-(aq)$

(b) $\quad Mn^{2+}(aq) \rightarrow MnO_2(s)$

$Mn^{2+}(aq) + 2\,H_2O(l) \rightarrow MnO_2(s)$

$Mn^{2+}(aq) + 2\,H_2O(l) \rightarrow MnO_2(s) + 4\,H^+(aq)$

$Mn^{2+}(aq) + 2\,H_2O(l) \rightarrow MnO_2(s) + 4\,H^+(aq) + 2\,e^-$ (oxidation half reaction)

$H_2O_2(aq) \rightarrow 2\,H_2O(l)$

$2\,H^+(aq) + H_2O_2(aq) \rightarrow 2\,H_2O(l)$

$2\,H^+(aq) + H_2O_2(aq) + 2\,e^- \rightarrow 2\,H_2O(l)$ (reduction half reaction)

Combine the two half reactions.

$Mn^{2+}(aq) + 2\,H_2O(l) + 2\,H^+(aq) + H_2O_2(aq) \rightarrow MnO_2(s) + 4\,H^+(aq) + 2\,H_2O(l)$

$Mn^{2+}(aq) + H_2O_2(aq) \rightarrow MnO_2(s) + 2\,H^+(aq)$

$Mn^{2+}(aq) + H_2O_2(aq) + 2\,OH^-(aq) \rightarrow MnO_2(s) + 2\,H^+(aq) + 2\,OH^-(aq)$

$Mn^{2+}(aq) + H_2O_2(aq) + 2\,OH^-(aq) \rightarrow MnO_2(s) + 2\,H_2O(l)$

(c) $\quad Zn(s) \rightarrow Zn(OH)_4^{2-}(aq)$

$4\,H_2O(l) + Zn(s) \rightarrow Zn(OH)_4^{2-}(aq)$

$4\,H_2O(l) + Zn(s) \rightarrow Zn(OH)_4^{2-}(aq) + 4\,H^+(aq)$

$[4\,H_2O(l) + Zn(s) \rightarrow Zn(OH)_4^{2-}(aq) + 4\,H^+(aq) + 2\,e^-] \times 4$ (oxidation half reaction)

$NO_3^-(aq) \rightarrow NH_3(aq)$

$NO_3^-(aq) \rightarrow NH_3(aq) + 3\,H_2O(l)$

$9\,H^+(aq) + NO_3^-(aq) \rightarrow NH_3(aq) + 3\,H_2O(l)$

$9\,H^+(aq) + NO_3^-(aq) + 8\,e^- \rightarrow NH_3(aq) + 3\,H_2O(l)$ (reduction half reaction)

Combine the two half reactions.

$16\,H_2O(l) + 4\,Zn(s) + 9\,H^+(aq) + NO_3^-(aq) \rightarrow$
$\qquad\qquad 4\,Zn(OH)_4^{2-}(aq) + 16\,H^+(aq) + NH_3(aq) + 3\,H_2O(l)$

$13\,H_2O(l) + 4\,Zn(s) + NO_3^-(aq) \rightarrow 4\,Zn(OH)_4^{2-}(aq) + 7\,H^+(aq) + NH_3(aq)$

$13\,H_2O(l) + 4\,Zn(s) + NO_3^-(aq) + 7\,OH^-(aq) \rightarrow$
$\qquad\qquad 4\,Zn(OH)_4^{2-}(aq) + 7\,H^+(aq) + 7\,OH^-(aq) + NH_3(aq)$

$13\,H_2O(l) + 4\,Zn(s) + NO_3^-(aq) + 7\,OH^-(aq) \rightarrow$
$\qquad\qquad 4\,Zn(OH)_4^{2-}(aq) + 7\,H_2O(l) + NH_3(aq)$

$6\,H_2O(l) + 4\,Zn(s) + NO_3^-(aq) + 7\,OH^-(aq) \rightarrow 4\,Zn(OH)_4^{2-}(aq) + NH_3(aq)$

(d) $\quad Bi(OH)_3(s) \rightarrow Bi(s)$

$Bi(OH)_3(s) \rightarrow Bi(s) + 3\,H_2O(l)$

$3\,H^+(aq) + Bi(OH)_3(s) \rightarrow Bi(s) + 3\,H_2O(l)$

$[3\,H^+(aq) + Bi(OH)_3(s) + 3\,e^- \rightarrow Bi(s) + 3\,H_2O(l)] \times 2$ (reduction half reaction)

$Sn(OH)_3^-(aq) \rightarrow Sn(OH)_6^{2-}(aq)$

$Sn(OH)_3^-(aq) + 3\,H_2O(l) \rightarrow Sn(OH)_6^{2-}(aq)$

$Sn(OH)_3^-(aq) + 3\,H_2O(l) \rightarrow Sn(OH)_6^{2-}(aq) + 3\,H^+(aq)$

$[Sn(OH)_3^-(aq) + 3\,H_2O(l) \rightarrow Sn(OH)_6^{2-}(aq) + 3\,H^+(aq) + 2\,e^-] \times 3$

$\qquad\qquad\qquad\qquad\qquad\qquad\qquad\qquad$ (oxidation half reaction)

Combine the two half reactions.

$6 H^+(aq) + 2 Bi(OH)_3(s) + 3 Sn(OH)_3^-(aq) + 9 H_2O(l) \rightarrow$
$\qquad\qquad 2 Bi(s) + 6 H_2O(l) + 3 Sn(OH)_6^{2-}(aq) + 9 H^+(aq)$

$2 Bi(OH)_3(s) + 3 Sn(OH)_3^-(aq) + 3 H_2O(l) \rightarrow 2 Bi(s) + 3 Sn(OH)_6^{2-}(aq) + 3 H^+(aq)$

$2 Bi(OH)_3(s) + 3 Sn(OH)_3^-(aq) + 3 H_2O(l) + 3 OH^-(aq) \rightarrow$
$\qquad\qquad 2 Bi(s) + 3 Sn(OH)_6^{2-}(aq) + 3 H^+(aq) + 3 OH^-(aq)$

$2 Bi(OH)_3(s) + 3 Sn(OH)_3^-(aq) + 3 H_2O(l) + 3 OH^-(aq) \rightarrow$
$\qquad\qquad 2 Bi(s) + 3 Sn(OH)_6^{2-}(aq) + 3 H_2O(l)$

$2 Bi(OH)_3(s) + 3 Sn(OH)_3^-(aq) + 3 OH^-(aq) \rightarrow 2 Bi(s) + 3 Sn(OH)_6^{2-}(aq)$

7.82 (a)

$Zn(s) \rightarrow Zn^{2+}(aq)$
$Zn(s) \rightarrow Zn^{2+}(aq) + 2 e^-$ \qquad (oxidation half reaction)

$VO^{2+}(aq) \rightarrow V^{3+}(aq)$
$VO^{2+}(aq) \rightarrow V^{3+}(aq) + H_2O(l)$
$2 H^+(aq) + VO^{2+}(aq) \rightarrow V^{3+}(aq) + H_2O(l)$
$[2 H^+(aq) + VO^{2+}(aq) + e^- \rightarrow V^{3+}(aq) + H_2O(l)] \times 2$ \qquad (reduction half reaction)

Combine the two half reactions.
$Zn(s) + 2 VO^{2+}(aq) + 4 H^+(aq) \rightarrow Zn^{2+}(aq) + 2 V^{3+}(aq) + 2 H_2O(l)$

(b)

$Ag(s) \rightarrow Ag^+(aq)$
$Ag(s) \rightarrow Ag^+(aq) + e^-$ \qquad (oxidation half reaction)

$NO_3^-(aq) \rightarrow NO_2(g)$
$NO_3^-(aq) \rightarrow NO_2(g) + H_2O(l)$
$2 H^+(aq) + NO_3^-(aq) \rightarrow NO_2(g) + H_2O(l)$
$2 H^+(aq) + NO_3^-(aq) + e^- \rightarrow NO_2(g) + H_2O(l)$ \quad (reduction half reaction)

Combine the two half reactions.
$2 H^+(aq) + Ag(s) + NO_3^-(aq) \rightarrow Ag^+(aq) + NO_2(g) + H_2O(l)$

(c)

$Mg(s) \rightarrow Mg^{2+}(aq)$
$[Mg(s) \rightarrow Mg^{2+}(aq) + 2 e^-] \times 3$ \qquad (oxidation half reaction)

$VO_4^{3-}(aq) \rightarrow V^{2+}(aq)$
$VO_4^{3-}(aq) \rightarrow V^{2+}(aq) + 4 H_2O(l)$
$8 H^+(aq) + VO_4^{3-}(aq) \rightarrow V^{2+}(aq) + 4 H_2O(l)$
$[8 H^+(aq) + VO_4^{3-}(aq) + 3 e^- \rightarrow V^{2+}(aq) + 4 H_2O(l)] \times 2$ \quad (reduction half reaction)

Combine the two half reactions.
$3 Mg(s) + 16 H^+(aq) + 2 VO_4^{3-}(aq) \rightarrow 3 Mg^{2+}(aq) + 2 V^{2+}(aq) + 8 H_2O(l)$

(d)

$I^-(aq) \rightarrow I_3^-(aq)$
$3 I^-(aq) \rightarrow I_3^-(aq)$
$[3 I^-(aq) \rightarrow I_3^-(aq) + 2 e^-] \times 8$ \qquad (oxidation half reaction)

$IO_3^-(aq) \rightarrow I_3^-(aq)$

$3 IO_3^-(aq) \rightarrow I_3^-(aq)$

$3 IO_3^-(aq) \rightarrow I_3^-(aq) + 9 H_2O(l)$

$18 H^+(aq) + 3 IO_3^-(aq) \rightarrow I_3^-(aq) + 9 H_2O(l)$

$18 H^+(aq) + 3 IO_3^-(aq) + 16 e^- \rightarrow I_3^-(aq) + 9 H_2O(l)$ (reduction half reaction)

Combine the two half reactions.

$18 H^+(aq) + 3 IO_3^-(aq) + 24 I^-(aq) \rightarrow 9 I_3^-(aq) + 9 H_2O(l)$

Divide each coefficient by 3.

$6 H^+(aq) + IO_3^-(aq) + 8 I^-(aq) \rightarrow 3 I_3^-(aq) + 3 H_2O(l)$

7.83   (a) $MnO_4^-(aq) \rightarrow Mn^{2+}(aq)$

$MnO_4^-(aq) \rightarrow Mn^{2+}(aq) + 4 H_2O(l)$

$8 H^+(aq) + MnO_4^-(aq) \rightarrow Mn^{2+}(aq) + 4 H_2O(l)$

$[8 H^+(aq) + MnO_4^-(aq) + 5 e^- \rightarrow Mn^{2+}(aq) + 4 H_2O(l)] \times 4$

(reduction half reaction)

$C_2H_5OH(aq) \rightarrow CH_3CO_2H(aq)$

$C_2H_5OH(aq) + H_2O(l) \rightarrow CH_3CO_2H(aq)$

$C_2H_5OH(aq) + H_2O(l) \rightarrow CH_3CO_2H(aq) + 4 H^+(aq)$

$[C_2H_5OH(aq) + H_2O(l) \rightarrow CH_3CO_2H(aq) + 4 H^+(aq) + 4 e^-] \times 5$

(oxidation half reaction)

Combine the two half reactions.

$32 H^+(aq) + 4 MnO_4^-(aq) + 5 C_2H_5OH(aq) + 5 H_2O(l) \rightarrow$

$4 Mn^{2+}(aq) + 16 H_2O(l) + 5 CH_3CO_2H(aq) + 20 H^+(aq)$

$12 H^+(aq) + 4 MnO_4^-(aq) + 5 C_2H_5OH(aq) \rightarrow$

$4 Mn^{2+}(aq) + 11 H_2O(l) + 5 CH_3CO_2H(aq)$

(b) $Cr_2O_7^{2-}(aq) \rightarrow Cr^{3+}(aq)$

$Cr_2O_7^{2-}(aq) \rightarrow 2 Cr^{3+}(aq)$

$Cr_2O_7^{2-}(aq) \rightarrow 2 Cr^{3+}(aq) + 7 H_2O(l)$

$14 H^+(aq) + Cr_2O_7^{2-}(aq) \rightarrow 2 Cr^{3+}(aq) + 7 H_2O(l)$

$14 H^+(aq) + Cr_2O_7^{2-}(aq) + 6 e^- \rightarrow 2 Cr^{3+}(aq) + 7 H_2O(l)$ (reduction half reaction)

$H_2O_2(aq) \rightarrow O_2(g)$

$H_2O_2(aq) \rightarrow O_2(g) + 2 H^+(aq)$

$[H_2O_2(aq) \rightarrow O_2(g) + 2 H^+(aq) + 2 e^-] \times 3$ (oxidation half reaction)

Combine the two half reactions.

$14 H^+(aq) + Cr_2O_7^{2-}(aq) + 3 H_2O_2(aq) \rightarrow$

$2 Cr^{3+}(aq) + 7 H_2O(l) + 3 O_2(g) + 6 H^+(aq)$

$8 H^+(aq) + Cr_2O_7^{2-}(aq) + 3 H_2O_2(aq) \rightarrow 2 Cr^{3+}(aq) + 7 H_2O(l) + 3 O_2(g)$

(c) $Sn^{2+}(aq) \rightarrow Sn^{4+}(aq)$

$[Sn^{2+}(aq) \rightarrow Sn^{4+}(aq) + 2 e^-] \times 4$ (oxidation half reaction)

$$IO_4^-(aq) \rightarrow I^-(aq)$$
$$IO_4^-(aq) \rightarrow I^-(aq) + 4 H_2O(l)$$
$$8 H^+(aq) + IO_4^-(aq) \rightarrow I^-(aq) + 4 H_2O(l)$$
$$8 H^+(aq) + IO_4^-(aq) + 8 e^- \rightarrow I^-(aq) + 4 H_2O(l) \quad \text{(reduction half reaction)}$$

Combine the two half reactions.
$$4 Sn^{2+}(aq) + 8 H^+(aq) + IO_4^-(aq) \rightarrow 4 Sn^{4+}(aq) + I^-(aq) + 4 H_2O(l)$$

(d) $\quad PbO_2(s) + Cl^-(aq) \rightarrow PbCl_2(s)$
$$PbO_2(s) + 2 Cl^-(aq) \rightarrow PbCl_2(s)$$
$$PbO_2(s) + 2 Cl^-(aq) \rightarrow PbCl_2(s) + 2 H_2O(l)$$
$$PbO_2(s) + 4 H^+(aq) + 2 Cl^-(aq) \rightarrow PbCl_2(s) + 2 H_2O(l)$$
$$[PbO_2(s) + 4 H^+(aq) + 2 Cl^-(aq) + 2 e^- \rightarrow PbCl_2(s) + 2 H_2O(l)] \times 2$$
$$\text{(reduction half reaction)}$$

$$H_2O(l) \rightarrow O_2(g)$$
$$2 H_2O(l) \rightarrow O_2(g)$$
$$2 H_2O(l) \rightarrow O_2(g) + 4 H^+(aq)$$
$$2 H_2O(l) \rightarrow O_2(g) + 4 H^+(aq) + 4 e^- \quad \text{(oxidation half reaction)}$$

Combine the two half reactions.
$$2 PbO_2(s) + 8 H^+(aq) + 4 Cl^-(aq) + 2 H_2O(l) \rightarrow$$
$$2 PbCl_2(s) + 4 H_2O(l) + O_2(g) + 4 H^+(aq)$$
$$2 PbO_2(s) + 4 H^+(aq) + 4 Cl^-(aq) \rightarrow 2 PbCl_2(s) + 2 H_2O(l) + O_2(g)$$

Redox Stoichiometry (Section 7.10)

7.84 $\quad I_2(aq) + 2 S_2O_3^{2-}(aq) \rightarrow S_4O_6^{2-}(aq) + 2 I^-(aq); \qquad 35.20 \text{ mL} = 0.032\ 50 \text{ L}$

$$0.035\ 20 \text{ L} \times \frac{0.150 \text{ mol } S_2O_3^{2-}}{L} \times \frac{1 \text{ mol } I_2}{2 \text{ mol } S_2O_3^{2-}} \times \frac{253.8 \text{ g } I_2}{1 \text{ mol } I_2} = 0.670 \text{ g } I_2$$

7.85 $\quad 2.486 \text{ g } I_2 \times \dfrac{1 \text{ mol } I_2}{253.8 \text{ g } I_2} \times \dfrac{2 \text{ mol } S_2O_3^{2-}}{1 \text{ mol } I_2} = 1.959 \times 10^{-2} \text{ mol } S_2O_3^{2-}$

$$1.959 \times 10^{-2} \text{ mol} \times \frac{1 \text{ L}}{0.250 \text{ mol}} = 0.0784 \text{ L}; \quad 0.0784 \text{ L} = 78.4 \text{ mL}$$

7.86 $\quad 3 H_3AsO_3(aq) + BrO_3^-(aq) \rightarrow Br^-(aq) + 3 H_3AsO_4(aq)$
$22.35 \text{ mL} = 0.022\ 35 \text{ L} \text{ and } 50.00 \text{ mL} = 0.050\ 00 \text{ L}$

$$0.022\ 35 \text{ L} \times \frac{0.100 \text{ mol } BrO_3^-}{L} \times \frac{3 \text{ mol } H_3AsO_3}{1 \text{ mol } BrO_3^-} = 6.70 \times 10^{-3} \text{ mol } H_3AsO_3$$

$$\text{molarity} = \frac{6.70 \times 10^{-3} \text{ mol}}{0.050\ 00 \text{ L}} = 0.134 \text{ M As(III)}$$

7.87 As_2O_3, 197.84 amu; 28.55 mL = 0.028 55 L

$$1.550 \text{ g } As_2O_3 \text{ x } \frac{1 \text{ mol } As_2O_3}{197.84 \text{ g } As_2O_3} \text{ x } \frac{2 \text{ mol } H_3AsO_3}{1 \text{ mol } As_2O_3} \text{ x } \frac{1 \text{ mol } BrO_3^-}{3 \text{ mol } H_3AsO_3}$$

$$= 5.223 \text{ x } 10^{-3} \text{ mol } BrO_3^-; \quad KBrO_3 \text{ molarity} = \frac{5.223 \text{ x } 10^{-3} \text{ mol}}{0.028 \text{ 55 L}} = 0.1829 \text{ M}$$

7.88 $2 \text{ Fe}^{3+}(aq) + Sn^{2+}(aq) \rightarrow 2 \text{ Fe}^{2+}(aq) + Sn^{4+}(aq);$ 13.28 mL = 0.013 28 L

$$0.013 \text{ 28 L x } \frac{0.1015 \text{ mol } Sn^{2+}}{L} \text{ x } \frac{2 \text{ mol } Fe^{3+}}{1 \text{ mol } Sn^{2+}} \text{ x } \frac{55.845 \text{ g } Fe^{3+}}{1 \text{ mol } Fe^{3+}} = 0.1506 \text{ g } Fe^{3+}$$

$$\text{mass \% Fe} = \frac{0.1506 \text{ g}}{0.1875 \text{ g}} \text{ x } 100\% = 80.32\%$$

7.89 Fe_2O_3, 159.69 amu; 23.84 mL = 0.023 84 L

$$1.4855 \text{ g } Fe_2O_3 \text{ x } \frac{1 \text{ mol } Fe_2O_3}{159.69 \text{ g } Fe_2O_3} \text{ x } \frac{2 \text{ mol } Fe^{3+}}{1 \text{ mol } Fe_2O_3} \text{ x } \frac{1 \text{ mol } Sn^{2+}}{2 \text{ mol } Fe^{3+}} = 0.009 \text{ 302 mol } Sn^{2+}$$

$$Sn^{2+} \text{ molarity} = \frac{0.009 \text{ 302 mol}}{0.023 \text{ 84 L}} = 0.3902 \text{ M}$$

7.90 $C_2H_5OH(aq) + 2 \text{ Cr}_2O_7^{2-}(aq) + 16 \text{ H}^+(aq) \rightarrow 2 \text{ CO}_2(g) + 4 \text{ Cr}^{3+}(aq) + 11 \text{ H}_2O(l)$
     C_2H_5OH, 46.07 amu; 8.76 mL = 0.008 76 L

$$0.008 \text{ 76 L x } \frac{0.049 \text{ 88 mol } Cr_2O_7^{2-}}{L} \text{ x } \frac{1 \text{ mol } C_2H_5OH}{2 \text{ mol } Cr_2O_7^{2-}} \text{ x } \frac{46.07 \text{ g } C_2H_5OH}{1 \text{ mol } C_2H_5OH}$$

$$= 0.010 \text{ 07 g } C_2H_5OH$$

$$\text{mass \% } C_2H_5OH = \frac{0.010 \text{ 07 g}}{10.002 \text{ g}} \text{ x } 100\% = 0.101\%$$

7.91   21.08 mL = 0.021 08 L

$$0.021 \text{ 08 L x } \frac{9.88 \text{ x } 10^{-4} \text{ mol } MnO_4^-}{L} \text{ x } \frac{5 \text{ mol } H_2C_2O_4}{2 \text{ mol } MnO_4^-} \text{ x } \frac{1 \text{ mol } Ca^{2+}}{1 \text{ mol } H_2C_2O_4} \text{ x } \frac{40.08 \text{ g } Ca^{2+}}{1 \text{ mol } Ca^{2+}}$$

$$= 0.002 \text{ 09 g} = 2.09 \text{ mg}$$

Chapter Problems

7.92   (a) $[Fe(CN)_6]^{3-}(aq) \rightarrow Fe(CN)_6^{4-}(aq)$
               $([Fe(CN)_6]^{3-}(aq) + e^- \rightarrow [Fe(CN)_6]^{4-}(aq)) \text{ x } 4$ (reduction half reaction)

$N_2H_4(aq) \rightarrow N_2(g)$

$N_2H_4(aq) \rightarrow N_2(g) + 4 H^+(aq)$

$N_2H_4(aq) \rightarrow N_2(g) + 4 H^+(aq) + 4 e^-$

$N_2H_4(aq) + 4 OH^-(aq) \rightarrow N_2(g) + 4 H^+(aq) + 4 OH^-(aq) + 4 e^-$

$N_2H_4(aq) + 4 OH^-(aq) \rightarrow N_2(g) + 4 H_2O(l) + 4 e^-$ (oxidation half reaction)

Combine the two half reactions.

$4 [Fe(CN)_6]^{3-}(aq) + N_2H_4(aq) + 4 OH^-(aq) \rightarrow$
$$4 [Fe(CN)_6]^{4-}(aq) + N_2(g) + 4 H_2O(l)$$

(b) $Cl_2(g) \rightarrow Cl^-(aq)$

$Cl_2(g) \rightarrow 2 Cl^-(aq)$

$Cl_2(g) + 2 e^- \rightarrow 2 Cl^-(aq)$ (reduction half reaction)

$SeO_3^{2-}(aq) \rightarrow SeO_4^{2-}(aq)$

$SeO_3^{2-}(aq) + H_2O(l) \rightarrow SeO_4^{2-}(aq)$

$SeO_3^{2-}(aq) + H_2O(l) \rightarrow SeO_4^{2-}(aq) + 2 H^+(aq)$

$SeO_3^{2-}(aq) + H_2O(l) \rightarrow SeO_4^{2-}(aq) + 2 H^+(aq) + 2 e^-$

$SeO_3^{2-}(aq) + H_2O(l) + 2 OH^-(aq) \rightarrow SeO_4^{2-}(aq) + 2 H^+(aq) + 2 OH^-(aq) + 2 e^-$

$SeO_3^{2-}(aq) + H_2O(l) + 2 OH^-(aq) \rightarrow SeO_4^{2-}(aq) + 2 H_2O(l) + 2 e^-$

$SeO_3^{2-}(aq) + 2 OH^-(aq) \rightarrow SeO_4^{2-}(aq) + H_2O(l) + 2 e^-$ (oxidation half reaction)

Combine the two half reactions.

$SeO_3^{2-}(aq) + Cl_2(g) + 2 OH^-(aq) \rightarrow SeO_4^{2-}(aq) + 2 Cl^-(aq) + H_2O(l)$

(c) $Co^{2+}(aq) \rightarrow Co(OH)_3(s)$

$Co^{2+}(aq) + 3 H_2O(l) \rightarrow Co(OH)_3(s)$

$Co^{2+}(aq) + 3 H_2O(l) \rightarrow Co(OH)_3(s) + 3 H^+(aq)$

$[Co^{2+}(aq) + 3 H_2O(l) \rightarrow Co(OH)_3(s) + 3 H^+(aq) + e^-] \times 2$
$$\text{(oxidation half reaction)}$$

$HO_2^-(aq) \rightarrow H_2O(l)$

$HO_2^-(aq) \rightarrow 2 H_2O(l)$

$3 H^+(aq) + HO_2^-(aq) \rightarrow 2 H_2O(l)$

$3 H^+(aq) + HO_2^-(aq) + 2 e^- \rightarrow 2 H_2O(l)$ (reduction half reaction)

Combine the two half reactions.

$2 Co^{2+}(aq) + 6 H_2O(l) + 3 H^+(aq) + HO_2^-(aq) \rightarrow$
$$2 Co(OH)_3(s) + 6 H^+(aq) + 2 H_2O(l)$$

$2 Co^{2+}(aq) + 4 H_2O(l) + HO_2^-(aq) \rightarrow 2 Co(OH)_3(s) + 3 H^+(aq)$

$2 Co^{2+}(aq) + 4 H_2O(l) + HO_2^-(aq) + 3 OH^-(aq) \rightarrow$
$$2 Co(OH)_3(s) + 3 H^+(aq) + 3 OH^-(aq)$$

$2 Co^{2+}(aq) + 4 H_2O(l) + HO_2^-(aq) + 3 OH^-(aq) \rightarrow$
$$2 Co(OH)_3(s) + 3 H_2O(l)$$

$2 Co^{2+}(aq) + H_2O(l) + HO_2^-(aq) + 3 OH^-(aq) \rightarrow 2 Co(OH)_3(s)$

7.93   57.91 mL = 0.057 91 L

$$0.057\ 91\ L \times \frac{0.1018\ mol\ Ce^{4+}}{L} \times \frac{1\ mol\ Fe^{2+}}{1\ mol\ Ce^{4+}} \times \frac{55.85\ g\ Fe^{2+}}{1\ mol\ Fe^{2+}} = 0.3292\ g\ Fe^{2+}$$

$$mass\ \%\ Fe = \frac{0.3292\ g}{1.2284\ g} \times 100\% = 26.80\%$$

7.94   (a) C_2H_6 H +1,   C –3
     (b) $Na_2B_4O_7$ O –2,   Na +1,   B +3
     (c) Mg_2SiO_4 O –2,   Mg +2,   Si +4

7.95   (a) $PbO_2(s) \rightarrow Pb^{2+}(aq)$
        $PbO_2(s) \rightarrow Pb^{2+}(aq) + 2\ H_2O(l)$
        $4\ H^+(aq) + PbO_2(s) \rightarrow Pb^{2+}(aq) + 2\ H_2O(l)$
        $[4\ H^+(aq) + PbO_2(s) + 2\ e^- \rightarrow Pb^{2+}(aq) + 2\ H_2O(l)] \times 5$ (reduction half reaction)

        $Mn^{2+}(aq) \rightarrow MnO_4^-(aq)$
        $4\ H_2O(l) + Mn^{2+}(aq) \rightarrow MnO_4^-(aq)$
        $4\ H_2O(l) + Mn^{2+}(aq) \rightarrow MnO_4^-(aq) + 8\ H^+(aq)$
        $[4\ H_2O(l) + Mn^{2+}(aq) \rightarrow MnO_4^-(aq) + 8\ H^+(aq) + 5\ e^-] \times 2$
                            (oxidation half reaction)
        Combine the two half reactions.
        $20\ H^+(aq) + 5\ PbO_2(s) + 8\ H_2O(l) + 2\ Mn^{2+}(aq) \rightarrow$
                    $5\ Pb^{2+}(aq) + 10\ H_2O(l) + 2\ MnO_4^-(aq) + 16\ H^+(aq)$
        $4\ H^+(aq) + 5\ PbO_2(s) + 2\ Mn^{2+}(aq) \rightarrow 5\ Pb^{2+}(aq) + 2\ H_2O(l) + 2\ MnO_4^-(aq)$

    (b) $As_2O_3(s) \rightarrow H_3AsO_4(aq)$
        $As_2O_3(s) \rightarrow 2\ H_3AsO_4(aq)$
        $5\ H_2O(l) + As_2O_3(s) \rightarrow 2\ H_3AsO_4(aq)$
        $5\ H_2O(l) + As_2O_3(s) \rightarrow 2\ H_3AsO_4(aq) + 4\ H^+(aq)$
        $5\ H_2O(l) + As_2O_3(s) \rightarrow 2\ H_3AsO_4(aq) + 4\ H^+(aq) + 4\ e^-$ (oxidation half reaction)

        $NO_3^-(aq) \rightarrow HNO_2(aq)$
        $NO_3^-(aq) \rightarrow HNO_2(aq) + H_2O(l)$
        $3\ H^+(aq) + NO_3^-(aq) \rightarrow HNO_2(aq) + H_2O(l)$

        $[3\ H^+(aq) + NO_3^-(aq) + 2\ e^- \rightarrow HNO_2(aq) + H_2O(l)] \times 2$
                          (reduction half reaction)
        Combine the two half reactions.
        $5\ H_2O(l) + As_2O_3(s) + 6\ H^+(aq) + 2\ NO_3^-(aq) \rightarrow$
                    $2\ H_3AsO_4(aq) + 4\ H^+(aq) + 2\ HNO_2(aq) + 2\ H_2O(l)$
        $3\ H_2O(l) + As_2O_3(s) + 2\ H^+(aq) + 2\ NO_3^-(aq) \rightarrow 2\ H_3AsO_4(aq) + 2\ HNO_2(aq)$

    (c) $Br_2(aq) \rightarrow Br^-(aq)$
        $Br_2(aq) \rightarrow 2\ Br^-(aq)$
        $Br_2(aq) + 2\ e^- \rightarrow 2\ Br^-(aq)$ (reduction half reaction)

$$SO_2(g) \rightarrow HSO_4^-(aq)$$
$$2\,H_2O(l) + SO_2(g) \rightarrow HSO_4^-(aq)$$
$$2\,H_2O(l) + SO_2(g) \rightarrow HSO_4^-(aq) + 3\,H^+(aq)$$
$$2\,H_2O(l) + SO_2(g) \rightarrow HSO_4^-(aq) + 3\,H^+(aq) + 2\,e^- \quad \text{(oxidation half reaction)}$$

Combine the two half reactions.
$$2\,H_2O(l) + Br_2(aq) + SO_2(g) \rightarrow 2\,Br^-(aq) + HSO_4^-(aq) + 3\,H^+(aq)$$

(d) $I^-(aq) \rightarrow I_2(s)$
$$2\,I^-(aq) \rightarrow I_2(s)$$
$$2\,I^-(aq) \rightarrow I_2(s) + 2\,e^- \quad \text{(oxidation half reaction)}$$

$$NO_2^-(aq) \rightarrow NO(g)$$
$$NO_2^-(aq) \rightarrow NO(g) + H_2O(l)$$
$$2\,H^+(aq) + NO_2^-(aq) \rightarrow NO(g) + H_2O(l)$$
$$[2\,H^+(aq) + NO_2^-(aq) + e^- \rightarrow NO(g) + H_2O(l)] \times 2 \quad \text{(reduction half reaction)}$$

Combine the two half reactions.
$$4\,H^+(aq) + 2\,NO_2^-(aq) + 2\,I^-(aq) \rightarrow 2\,NO(g) + I_2(s) + 2\,H_2O(l)$$

7.96 (a) "Any element higher in the activity series will react with the ion of any element lower in the activity series."
$C + B^+ \rightarrow C^+ + B$; therefore C is higher than B.
$A^+ + D \rightarrow$ no reaction; therefore A is higher than D.
$C^+ + A \rightarrow$ no reaction; therefore C is higher than A.
$D + B^+ \rightarrow D^+ + B$; therefore D is higher than B.
The net result is $C > A > D > B$.
(b) (1) The reaction, $A^+ + C \rightarrow A + C^+$, will occur because C is above A in the activity series.
(2) The reaction, $A^+ + B \rightarrow A + B^+$, will not occur because B is below A in the activity series.

7.97 (a) $K_{sp} = [Ag^+]^2[CrO_4^{2-}]$

(b) $Ag_2CrO_4(s) \rightleftarrows 2\,Ag^+(aq) + CrO_4^{2-}(aq)$
$$\qquad\qquad\qquad 2x \qquad\qquad x$$
In a saturated solution $2x = [Ag^+]$ and $x = [CrO_4^{2-}]$.
$K_{sp} = [Ag^+]^2[CrO_4^{2-}] = 1.1 \times 10^{-12} = (2x)^2(x) = 4x^3$; Solve for x; $x = 6.5 \times 10^{-5}$ M
$[Ag^+] = 2x = 2(6.5 \times 10^{-5}\,M) = 1.3 \times 10^{-4}\,M$; $[CrO_4^{2-}] = x = 6.5 \times 10^{-5}\,M$

7.98 $MgF_2(s) \rightleftarrows Mg^{2+}(aq) + 2\,F^-(aq)$
$$\qquad\qquad\qquad x \qquad\qquad 2x$$
$[Mg^{2+}] = x = 2.6 \times 10^{-4}\,M$ and $[F^-] = 2x = 2(2.6 \times 10^{-4}\,M) = 5.2 \times 10^{-4}\,M$ in a saturated solution.
$K_{sp} = [Mg^{2+}][F^-]^2 = (2.6 \times 10^{-4}\,M)(5.2 \times 10^{-4}\,M)^2 = 7.0 \times 10^{-11}$

7.99   65.20 mL = 0.065 20 L

1.926 g succinic acid x $\dfrac{1 \text{ mol succinic acid}}{118.1 \text{ g succinic acid}}$ = 0.016 31 mol succinic acid

$0.5000 \dfrac{\text{mol NaOH}}{1 \text{ L}}$ x 0.065 20 L = 0.032 60 mol NaOH

$\dfrac{0.032\,60 \text{ mol NaOH}}{0.016\,31 \text{ mol succinic acid}}$ = 2; therefore succinic acid has two acidic hydrogens.

7.100   (a) Add HCl to precipitate Hg_2Cl_2. $Hg_2^{2+}(aq) + 2Cl^-(aq) \rightarrow Hg_2Cl_2(s)$
        (b) Add H_2SO_4 to precipitate $PbSO_4$. $Pb^{2+}(aq) + SO_4^{2-}(aq) \rightarrow PbSO_4(s)$
        (c) Add Na_2CO_3 to precipitate $CaCO_3$. $Ca^{2+}(aq) + CO_3^{2-}(aq) \rightarrow CaCO_3(s)$
        (d) Add Na_2SO_4 to precipitate $BaSO_4$. $Ba^{2+}(aq) + SO_4^{2-}(aq) \rightarrow BaSO_4(s)$

7.101   (a) Add $AgNO_3$ to precipitate AgCl. $Ag^+(aq) + Cl^-(aq) \rightarrow AgCl(s)$
        (b) Add $NiCl_2$ to precipitate NiS. $Ni^{2+}(aq) + S^{2-}(aq) \rightarrow NiS(s)$
        (c) Add $CaCl_2$ to precipitate $CaCO_3$. $Ca^{2+}(aq) + CO_3^{2-}(aq) \rightarrow CaCO_3(s)$
        (d) Add $MgCl_2$ to precipitate $Mg(OH)_2$. $Mg^{2+}(aq) + 2 OH^-(aq) \rightarrow Mg(OH)_2(s)$

7.102   All four reactions are redox reactions.
        (a) $Mn(OH)_2(s) \rightarrow Mn(OH)_3(s)$
                $Mn(OH)_2(s) + OH^-(aq) \rightarrow Mn(OH)_3(s)$
                $[Mn(OH)_2(s) + OH^-(aq) \rightarrow Mn(OH)_3(s) + e^-]$ x 2   (oxidation half reaction)

                $H_2O_2(aq) \rightarrow 2 H_2O(l)$
                $2 H^+(aq) + H_2O_2(aq) \rightarrow 2 H_2O(l)$
                $2 e^- + 2 H^+(aq) + H_2O_2(aq) \rightarrow 2 H_2O(l)$
                $2 e^- + 2 OH^-(aq) + 2 H^+(aq) + H_2O_2(aq) \rightarrow 2 H_2O(l) + 2 OH^-(aq)$
                $2 e^- + 2 H_2O(l) + H_2O_2(aq) \rightarrow 2 H_2O(l) + 2 OH^-(aq)$
                $2 e^- + H_2O_2(aq) \rightarrow 2 OH^-(aq)$ (reduction half reaction)

                Combine the two half reactions.
                $2 Mn(OH)_2(s) + 2 OH^-(aq) + H_2O_2(aq) \rightarrow 2 Mn(OH)_3(s) + 2 OH^-(aq)$
                $2 Mn(OH)_2(s) + H_2O_2(aq) \rightarrow 2 Mn(OH)_3(s)$

        (b) $[MnO_4^{2-}(aq) \rightarrow MnO_4^-(aq) + e^-]$ x 2   (oxidation half reaction)

                $MnO_4^{2-}(aq) \rightarrow MnO_2(s)$
                $MnO_4^{2-}(aq) \rightarrow MnO_2(s) + 2 H_2O(l)$
                $4 H^+(aq) + MnO_4^{2-}(aq) \rightarrow MnO_2(s) + 2 H_2O(l)$
                $2 e^- + 4 H^+(aq) + MnO_4^{2-}(aq) \rightarrow MnO_2(s) + 2 H_2O(l)$ (reduction half reaction)

                Combine the two half reactions.
                $4 H^+(aq) + 3 MnO_4^{2-}(aq) \rightarrow MnO_2(s) + 2 MnO_4^-(aq) + 2 H_2O(l)$

(c) $I^-(aq) \rightarrow I_3^-(aq)$

        $3\ I^-(aq) \rightarrow I_3^-(aq)$

        $[3\ I^-(aq) \rightarrow I_3^-(aq) + 2\ e^-]\ x\ 8$ (oxidation half reaction)

        $IO_3^-(aq) \rightarrow I_3^-(aq)$

        $3\ IO_3^-(aq) \rightarrow I_3^-(aq)$

        $3\ IO_3^-(aq) \rightarrow I_3^-(aq) + 9\ H_2O(l)$

        $18\ H^+(aq) + 3\ IO_3^-(aq) \rightarrow I_3^-(aq) + 9\ H_2O(l)$

        $16\ e^- + 18\ H^+(aq) + 3\ IO_3^-(aq) \rightarrow I_3^-(aq) + 9\ H_2O(l)$ (reduction half reaction)

        Combine the two half reactions.

        $24\ I^-(aq) + 3\ IO_3^-(aq) + 18\ H^+(aq) \rightarrow 9\ I_3^-(aq) + 9\ H_2O(l)$

        Divide all coefficients by 3.

        $8\ I^-(aq) + IO_3^-(aq) + 6\ H^+(aq) \rightarrow 3\ I_3^-(aq) + 3\ H_2O(l)$

(d) $P(s) \rightarrow HPO_3^{2-}(aq)$

        $3\ H_2O(l) + P(s) \rightarrow HPO_3^{2-}(aq)$

        $3\ H_2O(l) + P(s) \rightarrow HPO_3^{2-}(aq) + 5\ H^+(aq)$

        $[3\ H_2O(l) + P(s) \rightarrow HPO_3^{2-}(aq) + 5\ H^+(aq) + 3\ e^-]\ x\ 2$

                                           (oxidation half reaction)

        $PO_4^{3-}(aq) \rightarrow HPO_3^{2-}(aq)$

        $PO_4^{3-}(aq) \rightarrow HPO_3^{2-}(aq) + H_2O(l)$

        $3\ H^+(aq) + PO_4^{3-}(aq) \rightarrow HPO_3^{2-}(aq) + H_2O(l)$

        $[2\ e^- + 3\ H^+(aq) + PO_4^{3-}(aq) \rightarrow HPO_3^{2-}(aq) + H_2O(l)]\ x\ 3$

                                           (reduction half reaction)

        Combine the two half reactions and add OH⁻.

        $6\ H_2O(l) + 2\ P(s) + 9\ H^+(aq) + 3\ PO_4^{3-}(aq) \rightarrow$

                                   $5\ HPO_3^{2-}(aq) + 10\ H^+(aq) + 3\ H_2O(l)$

        $3\ H_2O(l) + 2\ P(s) + 3\ PO_4^{3-}(aq) \rightarrow 5\ HPO_3^{2-}(aq) + H^+(aq)$

        $3\ H_2O(l) + 2\ P(s) + 3\ PO_4^{3-}(aq) + OH^-(aq) \rightarrow$

                                   $5\ HPO_3^{2-}(aq) + H^+(aq) + OH^-(aq)$

        $3\ H_2O(l) + 2\ P(s) + 3\ PO_4^{3-}(aq) + OH^-(aq) \rightarrow 5\ HPO_3^{2-}(aq) + H_2O(l)$

        $2\ H_2O(l) + 2\ P(s) + 3\ PO_4^{3-}(aq) + OH^-(aq) \rightarrow 5\ HPO_3^{2-}(aq)$

7.103   100.0 mL = 0.1000 L;   47.14 mL = 0.047 14 L

      mol HCl and HBr = mol H^+ = $0.1235\ \dfrac{\text{mol NaOH}}{1\ L}\ x\ 0.047\ 14\ L = 5.8218\ x\ 10^{-3}$ mol

      mass of AgCl and AgBr = 0.9974 g;   mol Ag = mol H^+ = $5.8218\ x\ 10^{-3}$ mol

      mass of Ag = $5.8218\ x\ 10^{-3}$ mol Ag $x\ \dfrac{107.87\ \text{g Ag}}{1\ \text{mol Ag}} = 0.6280$ g Ag

      mass of Cl and Br = 0.9974 g − 0.6280 g = 0.3694 g of Cl and Br

      Let Y = moles Cl and Z = moles Br in 0.3694 g of Cl and Br.

      Let (Y + Z) = moles Ag in 0.6280 g Ag.

      For Ag: 0.6280 g = (Y + Z) x 107.87 g

      For Cl and Br: 0.3694 g = (Y x 35.453 g) + (Z x 79.904 g)

Solve the simultaneous equations for Y and Z.

Rearrange the Ag equation: $\left(\dfrac{0.6280 \text{ g}}{107.87 \text{ g}} - Z \right) = Y$

Substitute for Y in the Cl and Br equation above and solve for Z.

$0.3694 \text{ g} = \left[\left(\dfrac{0.6280 \text{ g}}{107.87 \text{ g}} - Z \right) \times 35.453 \text{ g} \right] + (Z \times 79.904 \text{ g}); \quad Z = \dfrac{0.1630}{44.451} = 3.667 \times 10^{-3}$

$Y = \left(\dfrac{0.6280 \text{ g}}{107.87 \text{ g}} - Z \right) = \left(\dfrac{0.6280 \text{ g}}{107.87 \text{ g}} - 3.667 \times 10^{-3} \right) = 2.155 \times 10^{-3}$

$\text{HCl molarity} = \dfrac{2.155 \times 10^{-3} \text{ mol}}{0.1000 \text{ L}} = 0.021\ 55 \text{ M}$

$\text{HBr molarity} = \dfrac{3.667 \times 10^{-3} \text{ mol}}{0.1000 \text{ L}} = 0.036\ 67 \text{ M}$

7.104 (a) $S_4O_6^{2-}(aq) \rightarrow H_2S(aq)$
$S_4O_6^{2-}(aq) \rightarrow 4 H_2S(aq)$
$S_4O_6^{2-}(aq) \rightarrow 4 H_2S(aq) + 6 H_2O(l)$
$20 H^+(aq) + S_4O_6^{2-}(aq) \rightarrow 4 H_2S(aq) + 6 H_2O(l)$
$18 e^- + 20 H^+(aq) + S_4O_6^{2-}(aq) \rightarrow 4 H_2S(aq) + 6 H_2O(l)$ (reduction half reaction)

$Al(s) \rightarrow Al^{3+}(aq)$
$[Al(s) \rightarrow Al^{3+}(aq) + 3 e^-] \times 6$ (oxidation half reaction)

Combine the two half reactions.
$20 H^+(aq) + S_4O_6^{2-}(aq) + 6 Al(s) \rightarrow 4 H_2S(aq) + 6 Al^{3+}(aq) + 6 H_2O(l)$

(b) $S_2O_3^{2-}(aq) \rightarrow S_4O_6^{2-}(aq)$
$2 S_2O_3^{2-}(aq) \rightarrow S_4O_6^{2-}(aq)$
$[2 S_2O_3^{2-}(aq) \rightarrow S_4O_6^{2-}(aq) + 2 e^-] \times 3$ (oxidation half reaction)

$Cr_2O_7^{2-}(aq) \rightarrow Cr^{3+}(aq)$
$Cr_2O_7^{2-}(aq) \rightarrow 2 Cr^{3+}(aq)$
$Cr_2O_7^{2-}(aq) \rightarrow 2 Cr^{3+}(aq) + 7 H_2O(l)$
$14 H^+(aq) + Cr_2O_7^{2-}(aq) \rightarrow 2 Cr^{3+}(aq) + 7 H_2O(l)$
$6 e^- + 14 H^+(aq) + Cr_2O_7^{2-}(aq) \rightarrow 2 Cr^{3+}(aq) + 7 H_2O(l)$ (reduction half reaction)

Combine the two half reactions.
$14 H^+(aq) + 6 S_2O_3^{2-}(aq) + Cr_2O_7^{2-}(aq) \rightarrow 3 S_4O_6^{2-}(aq) + 2 Cr^{3+}(aq) + 7 H_2O(l)$

(c) $ClO_3^-(aq) \rightarrow Cl^-(aq)$
$ClO_3^-(aq) \rightarrow Cl^-(aq) + 3 H_2O(l)$
$6 H^+(aq) + ClO_3^-(aq) \rightarrow Cl^-(aq) + 3 H_2O(l)$
$[6 e^- + 6 H^+(aq) + ClO_3^-(aq) \rightarrow Cl^-(aq) + 3 H_2O(l)] \times 14$ (reduction half reaction)

$As_2S_3(s) \rightarrow H_2AsO_4^-(aq) + HSO_4^-(aq)$

$As_2S_3(s) \rightarrow 2\,H_2AsO_4^-(aq) + 3\,HSO_4^-(aq)$

$20\,H_2O(l) + As_2S_3(s) \rightarrow 2\,H_2AsO_4^-(aq) + 3\,HSO_4^-(aq)$

$20\,H_2O(l) + As_2S_3(s) \rightarrow 2\,H_2AsO_4^-(aq) + 3\,HSO_4^-(aq) + 33\,H^+(aq)$

$[20\,H_2O(l) + As_2S_3(s) \rightarrow 2\,H_2AsO_4^-(aq) + 3\,HSO_4^-(aq) + 33\,H^+(aq) + 28\,e^-]\times 3$

(oxidation half reaction)

Combine the two half reactions.

$84\,H^+(aq) + 60\,H_2O(l) + 14\,ClO_3^-(aq) + 3\,As_2S_3(s) \rightarrow$
$\qquad 14\,Cl^-(aq) + 6\,H_2AsO_4^-(aq) + 9\,HSO_4^-(aq) + 42\,H_2O(l) + 99\,H^+(aq)$

$18\,H_2O(l) + 14\,ClO_3^-(aq) + 3\,As_2S_3(s) \rightarrow$
$\qquad 14\,Cl^-(aq) + 6\,H_2AsO_4^-(aq) + 9\,HSO_4^-(aq) + 15\,H^+(aq)$

(d) $IO_3^-(aq) \rightarrow I^-(aq)$

$IO_3^-(aq) \rightarrow I^-(aq) + 3\,H_2O(l)$

$6\,H^+(aq) + IO_3^-(aq) \rightarrow I^-(aq) + 3\,H_2O(l)$

$[6\,e^- + 6\,H^+(aq) + IO_3^-(aq) \rightarrow I^-(aq) + 3\,H_2O(l)]\times 7$ (reduction half reaction)

$Re(s) \rightarrow ReO_4^-(aq)$

$4\,H_2O(l) + Re(s) \rightarrow ReO_4^-(aq)$

$4\,H_2O(l) + Re(s) \rightarrow ReO_4^-(aq) + 8\,H^+(aq)$

$[4\,H_2O(l) + Re(s) \rightarrow ReO_4^-(aq) + 8\,H^+(aq) + 7\,e^-]\times 6$ (oxidation half reaction)

Combine the two half reactions.

$42\,H^+(aq) + 24\,H_2O(l) + 7\,IO_3^-(aq) + 6\,Re(s) \rightarrow$
$\qquad 7\,I^-(aq) + 6\,ReO_4^-(aq) + 21\,H_2O(l) + 48\,H^+(aq)$

$3\,H_2O(l) + 7\,IO_3^-(aq) + 6\,Re(s) \rightarrow 7\,I^-(aq) + 6\,ReO_4^-(aq) + 6\,H^+(aq)$

(e) $HSO_4^-(aq) + Pb_3O_4(s) \rightarrow PbSO_4(s)$

$3\,HSO_4^-(aq) + Pb_3O_4(s) \rightarrow 3\,PbSO_4(s)$

$3\,HSO_4^-(aq) + Pb_3O_4(s) \rightarrow 3\,PbSO_4(s) + 4\,H_2O(l)$

$5\,H^+(aq) + 3\,HSO_4^-(aq) + Pb_3O_4(s) \rightarrow 3\,PbSO_4(s) + 4\,H_2O(l)$

$[2\,e^- + 5\,H^+(aq) + 3\,HSO_4^-(aq) + Pb_3O_4(s) \rightarrow 3\,PbSO_4(s) + 4\,H_2O(l)]\times 10$

(reduction half reaction)

$As_4(s) \rightarrow H_2AsO_4^-(aq)$

$As_4(s) \rightarrow 4\,H_2AsO_4^-(aq)$

$16\,H_2O(l) + As_4(s) \rightarrow 4\,H_2AsO_4^-(aq)$

$16\,H_2O(l) + As_4(s) \rightarrow 4\,H_2AsO_4^-(aq) + 24\,H^+(aq)$

$16\,H_2O(l) + As_4(s) \rightarrow 4\,H_2AsO_4^-(aq) + 24\,H^+(aq) + 20\,e^-$ (oxidation half reaction)

Combine the two half reactions.

$26\,H^+(aq) + 30\,HSO_4^-(aq) + As_4(s) + 10\,Pb_3O_4(s) \rightarrow$
$\qquad 4\,H_2AsO_4^-(aq) + 30\,PbSO_4(s) + 24\,H_2O(l)$

(f) $HNO_2(aq) \rightarrow NO_3^-(aq)$

$H_2O(l) + HNO_2(aq) \rightarrow NO_3^-(aq)$

$H_2O(l) + HNO_2(aq) \rightarrow NO_3^-(aq) + 3\,H^+(aq)$

$H_2O(l) + HNO_2(aq) \rightarrow NO_3^-(aq) + 3\,H^+(aq) + 2\,e^-$ (oxidation half reaction)

$HNO_2(aq) \rightarrow NO(g)$

$HNO_2(aq) \rightarrow NO(g) + H_2O(l)$

$H^+(aq) + HNO_2(aq) \rightarrow NO(g) + H_2O(l)$

$[1\,e^- + H^+(aq) + HNO_2(aq) \rightarrow NO(g) + H_2O(l)] \times 2$ (reduction half reaction)

Combine the two half reactions.

$3\,HNO_2(aq) \rightarrow NO_3^-(aq) + 2\,NO(g) + H_2O(l) + H^+(aq)$

7.105 (a) $C_4H_4O_6^{2-}(aq) \rightarrow CO_3^{2-}(aq)$

$C_4H_4O_6^{2-}(aq) \rightarrow 4\,CO_3^{2-}(aq)$

$C_4H_4O_6^{2-}(aq) + 6\,H_2O(l) \rightarrow 4\,CO_3^{2-}(aq)$

$C_4H_4O_6^{2-}(aq) + 6\,H_2O(l) \rightarrow 4\,CO_3^{2-}(aq) + 16\,H^+(aq)$

$[C_4H_4O_6^{2-}(aq) + 6\,H_2O(l) \rightarrow 4\,CO_3^{2-}(aq) + 16\,H^+(aq) + 10\,e^-] \times 3$

(oxidation half reaction)

$ClO_3^-(aq) \rightarrow Cl^-(aq)$

$ClO_3^-(aq) \rightarrow Cl^-(aq) + 3\,H_2O(l)$

$ClO_3^-(aq) + 6\,H^+(aq) \rightarrow Cl^-(aq) + 3\,H_2O(l)$

$[6\,e^- + ClO_3^-(aq) + 6\,H^+(aq) \rightarrow Cl^-(aq) + 3\,H_2O(l)] \times 5$ (reduction half reaction)

Combine the two half reactions.

$3\,C_4H_4O_6^{2-}(aq) + 18\,H_2O(l) + 5\,ClO_3^-(aq) + 30\,H^+(aq) \rightarrow$
$\qquad 12\,CO_3^{2-}(aq) + 48\,H^+(aq) + 5\,Cl^-(aq) + 15\,H_2O(l)$

$3\,C_4H_4O_6^{2-}(aq) + 3\,H_2O(l) + 5\,ClO_3^-(aq) \rightarrow 12\,CO_3^{2-}(aq) + 18\,H^+(aq) + 5\,Cl^-(aq)$

$3\,C_4H_4O_6^{2-}(aq) + 3\,H_2O(l) + 5\,ClO_3^-(aq) + 18\,OH^-(aq) \rightarrow$
$\qquad 12\,CO_3^{2-}(aq) + 18\,H^+(aq) + 18\,OH^-(aq) + 5\,Cl^-(aq)$

$3\,C_4H_4O_6^{2-}(aq) + 3\,H_2O(l) + 5\,ClO_3^-(aq) + 18\,OH^-(aq) \rightarrow$
$\qquad 12\,CO_3^{2-}(aq) + 18\,H_2O(aq) + 5\,Cl^-(aq)$

$3\,C_4H_4O_6^{2-}(aq) + 5\,ClO_3^-(aq) + 18\,OH^-(aq) \rightarrow 12\,CO_3^{2-}(aq) + 15\,H_2O(l) + 5\,Cl^-(aq)$

(b) $Al(s) \rightarrow Al(OH)_4^-(aq)$

$Al(s) + 4\,OH^-(aq) \rightarrow Al(OH)_4^-(aq)$

$[Al(s) + 4\,OH^-(aq) \rightarrow Al(OH)_4^-(aq) + 3\,e^-] \times 11$ (oxidation half reaction)

$BiONO_3(s) \rightarrow Bi(s) + NH_3(aq)$

$BiONO_3(s) \rightarrow Bi(s) + NH_3(aq) + 4\,H_2O(l)$

$BiONO_3(s) + 11\,H^+(aq) \rightarrow Bi(s) + NH_3(aq) + 4\,H_2O(l)$

$[BiONO_3(s) + 11\,H^+(aq) + 11\,e^- \rightarrow Bi(s) + NH_3(aq) + 4\,H_2O(l)] \times 3$

(reduction half reaction)

Combine the two half reactions.

$11\,Al(s) + 44\,OH^-(aq) + 3\,BiONO_3(s) + 33\,H^+(aq) \rightarrow$
$\qquad 11\,Al(OH)_4^-(aq) + 3\,Bi(s) + 3\,NH_3(aq) + 12\,H_2O(l)$

$11\,Al(s) + 11\,OH^-(aq) + 3\,BiONO_3(s) + 33\,H_2O(l) \rightarrow$
$\qquad 11\,Al(OH)_4^-(aq) + 3\,Bi(s) + 3\,NH_3(aq) + 12\,H_2O(l)$

$11\,Al(s) + 11\,OH^-(aq) + 3\,BiONO_3(s) + 21\,H_2O(l) \rightarrow$
$\qquad 11\,Al(OH)_4^-(aq) + 3\,Bi(s) + 3\,NH_3(aq)$

(c) $H_2O_2(aq) \rightarrow O_2(g)$

$H_2O_2(aq) \rightarrow O_2(g) + 2\,H^+(aq)$

$[H_2O_2(aq) \rightarrow O_2(g) + 2\,H^+(aq) + 2\,e^-] \times 4$ (oxidation half reaction)

$Cl_2O_7(aq) \rightarrow ClO_2^-(aq)$

$Cl_2O_7(aq) \rightarrow 2\,ClO_2^-(aq)$

$Cl_2O_7(aq) \rightarrow 2\,ClO_2^-(aq) + 3\,H_2O(l)$

$Cl_2O_7(aq) + 6\,H^+(aq) \rightarrow 2\,ClO_2^-(aq) + 3\,H_2O(l)$

$Cl_2O_7(aq) + 6\,H^+(aq) + 8\,e^- \rightarrow 2\,ClO_2^-(aq) + 3\,H_2O(l)$ (reduction half reaction)

Combine the two half reactions.

$4\,H_2O_2(aq) + Cl_2O_7(aq) + 6\,H^+(aq) \rightarrow 4\,O_2(g) + 8\,H^+(aq) + 2\,ClO_2^-(aq) + 3\,H_2O(l)$

$4\,H_2O_2(aq) + Cl_2O_7(aq) \rightarrow 4\,O_2(g) + 2\,H^+(aq) + 2\,ClO_2^-(aq) + 3\,H_2O(l)$

$4\,H_2O_2(aq) + Cl_2O_7(aq) + 2\,OH^-(aq) \rightarrow$
$$4\,O_2(g) + 2\,H^+(aq) + 2\,OH^-(aq) + 2\,ClO_2^-(aq) + 3\,H_2O(l)$$

$4\,H_2O_2(aq) + Cl_2O_7(aq) + 2\,OH^-(aq) \rightarrow 4\,O_2(g) + 2\,ClO_2^-(aq) + 5\,H_2O(l)$

(d) $Tl_2O_3(s) \rightarrow TlOH(s)$

$Tl_2O_3(s) \rightarrow 2\,TlOH(s)$

$Tl_2O_3(s) \rightarrow 2\,TlOH(s) + H_2O(l)$

$Tl_2O_3(s) + 4\,H^+(aq) \rightarrow 2\,TlOH(s) + H_2O(l)$

$Tl_2O_3(s) + 4\,H^+(aq) + 4\,e^- \rightarrow 2\,TlOH(s) + H_2O(l)$ (reduction half reaction)

$NH_2OH(aq) \rightarrow N_2(g)$

$2\,NH_2OH(aq) \rightarrow N_2(g)$

$2\,NH_2OH(aq) \rightarrow N_2(g) + 2\,H_2O(l)$

$2\,NH_2OH(aq) \rightarrow N_2(g) + 2\,H_2O(l) + 2\,H^+(aq)$

$[2\,NH_2OH(aq) \rightarrow N_2(g) + 2\,H_2O(l) + 2\,H^+(aq) + 2\,e^-] \times 2$ (oxidation half reaction)

Combine the two half reactions.

$Tl_2O_3(s) + 4\,H^+(aq) + 4\,NH_2OH(aq) \rightarrow 2\,TlOH(s) + 2\,N_2(g) + 5\,H_2O(l) + 4\,H^+(aq)$

$Tl_2O_3(s) + 4\,NH_2OH(aq) \rightarrow 2\,TlOH(s) + 2\,N_2(g) + 5\,H_2O(l)$

(e) $Cu(NH_3)_4^{2+}(aq) \rightarrow Cu(s) + 4\,NH_3(aq)$

$Cu(NH_3)_4^{2+}(aq) + 2\,e^- \rightarrow Cu(s) + 4\,NH_3(aq)$ (reduction half reaction)

$S_2O_4^{2-}(aq) \rightarrow SO_3^{2-}(aq)$

$S_2O_4^{2-}(aq) \rightarrow 2\,SO_3^{2-}(aq)$

$S_2O_4^{2-}(aq) + 2\,H_2O(l) \rightarrow 2\,SO_3^{2-}(aq)$

$S_2O_4^{2-}(aq) + 2\,H_2O(l) \rightarrow 2\,SO_3^{2-}(aq) + 4\,H^+(aq)$

$S_2O_4^{2-}(aq) + 2\,H_2O(l) \rightarrow 2\,SO_3^{2-}(aq) + 4\,H^+(aq) + 2\,e^-$ (oxidation half reaction)

Combine the two half reactions.

$Cu(NH_3)_4^{2+}(aq) + S_2O_4^{2-}(aq) + 2\,H_2O(l) \rightarrow Cu(s) + 4\,NH_3(aq) + 2\,SO_3^{2-}(aq) + 4\,H^+(aq)$

$Cu(NH_3)_4^{2+}(aq) + S_2O_4^{2-}(aq) + 2\,H_2O(l) + 4\,OH^-(aq) \rightarrow$
$$Cu(s) + 4\,NH_3(aq) + 2\,SO_3^{2-}(aq) + 4\,H^+(aq) + 4\,OH^-(aq)$$

$$Cu(NH_3)_4^{2+}(aq) + S_2O_4^{2-}(aq) + 2\ H_2O(l) + 4\ OH^-(aq) \rightarrow$$
$$Cu(s) + 4\ NH_3(aq) + 2\ SO_3^{2-}(aq) + 4\ H_2O(l)$$
$$Cu(NH_3)_4^{2+}(aq) + S_2O_4^{2-}(aq) + 4\ OH^-(aq) \rightarrow$$
$$Cu(s) + 4\ NH_3(aq) + 2\ SO_3^{2-}(aq) + 2\ H_2O(l)$$

(f) $Mn(OH)_2(s) \rightarrow MnO_2(s)$

$$Mn(OH)_2(s) \rightarrow MnO_2(s) + 2\ H^+(aq)$$
$$[Mn(OH)_2(s) \rightarrow MnO_2(s) + 2\ H^+(aq) + 2\ e^-]\ x\ 3 \qquad \text{(oxidation half reaction)}$$

$$MnO_4^-(aq) \rightarrow MnO_2(s)$$
$$MnO_4^-(aq) \rightarrow MnO_2(s) + 2\ H_2O(l)$$
$$MnO_4^-(aq) + 4\ H^+(aq) \rightarrow MnO_2(s) + 2\ H_2O(l)$$
$$[MnO_4^-(aq) + 4\ H^+(aq) + 3\ e^- \rightarrow MnO_2(s) + 2\ H_2O(l)]\ x\ 2 \quad \text{(reduction half reaction)}$$

Combine the two half reactions.
$$3\ Mn(OH)_2(s) + 2\ MnO_4^-(aq) + 8\ H^+(aq) \rightarrow 5\ MnO_2(s) + 6\ H^+(aq) + 4\ H_2O(l)$$
$$3\ Mn(OH)_2(s) + 2\ MnO_4^-(aq) + 2\ H^+(aq) \rightarrow 5\ MnO_2(s) + 4\ H_2O(l)$$
$$3\ Mn(OH)_2(s) + 2\ MnO_4^-(aq) + 2\ H^+(aq) + 2\ OH^-(aq) \rightarrow$$
$$5\ MnO_2(s) + 4\ H_2O(l) + 2\ OH^-(aq)$$
$$3\ Mn(OH)_2(s) + 2\ MnO_4^-(aq) + 2\ H_2O(l) \rightarrow 5\ MnO_2(s) + 4\ H_2O(l) + 2\ OH^-(aq)$$
$$3\ Mn(OH)_2(s) + 2\ MnO_4^-(aq) \rightarrow 5\ MnO_2(s) + 2\ H_2O(l) + 2\ OH^-(aq)$$

7.106 CuO, 79.55 amu; Cu_2O, 143.09 amu

Let X equal the mass of CuO and Y the mass of Cu_2O in the 10.50 g mixture. Therefore, X + Y = 10.50 g.

$$\text{mol Cu} = 8.66\ g\ x\ \frac{1\ mol\ Cu}{63.546\ g\ Cu} = 0.1363\ mol\ Cu$$

$$\text{mol CuO} + 2\ x\ \text{mol Cu}_2O = 0.1363\ mol\ Cu$$

$$X\ x\ \frac{1\ mol\ CuO}{79.55\ g\ CuO} + 2\ x\ \left(Y\ x\ \frac{1\ mol\ Cu_2O}{143.09\ g\ Cu_2O} \right) = 0.1363\ mol\ Cu$$

Rearrange to get X = 10.50 g – Y and then substitute it into the equation above to solve for Y.

$$(10.50\ g - Y)\ x\ \frac{1\ mol\ CuO}{79.55\ g\ CuO} + 2\ x\ \left(Y\ x\ \frac{1\ mol\ Cu_2O}{143.09\ g\ Cu_2O} \right) = 0.1363\ mol\ Cu$$

$$\frac{10.50\ mol}{79.55} - \frac{Y\ mol}{79.55\ g} + \frac{2\ Y\ mol}{143.09\ g} = 0.1363\ mol$$

$$-\frac{Y\ mol}{79.55\ g} + \frac{2\ Y\ mol}{143.09\ g} = 0.1363\ mol - \frac{10.50\ mol}{79.55} = 0.0043\ mol$$

$$\frac{(-\ Y\ mol)(143.09\ g) + (2\ Y\ mol)(79.55\ g)}{(79.55\ g)(143.09\ g)} = 0.0043\ mol$$

$$\frac{16.01\ Y\ mol}{11383\ g} = 0.0043\ mol; \quad \frac{16.01\ Y}{11383\ g} = 0.0043$$

Y = (0.0043)(11383 g)/16.01 = 3.06 g Cu_2O

X = 10.50 g – Y = 10.50 g – 3.06 g = 7.44 g CuO

7.107 (a) PbI_2, 461.01 amu

$Pb(NO_3)_2(aq) + 2 KI(aq) \rightarrow PbI_2(s) + 2 KNO_3(aq)$

75.0 mL = 0.0750 L and 100.0 mL = 0.1000 L

mol $Pb(NO_3)_2$ = (0.0750 L)(0.100 mol/L) = 7.50×10^{-3} mol $Pb(NO_3)_2$

mol KI = (0.1000 L)(0.190 mol/L) = 1.90×10^{-2} mol KI

mols KI needed = 7.50×10^{-3} mol $Pb(NO_3)_2 \times \dfrac{2 \text{ mol KI}}{1 \text{ mol } Pb(NO_3)_2}$ = 1.50×10^{-2} mol KI

There is an excess of KI, so $Pb(NO_3)_2$ is the limiting reactant.

mass PbI_2 = 7.50×10^{-3} mol $Pb(NO_3)_2 \times \dfrac{1 \text{ mol } PbI_2}{1 \text{ mol } Pb(NO_3)_2} \times \dfrac{461.01 \text{ g } PbI_2}{1 \text{ mol } PbI_2}$ = 3.46 g PbI_2

(b) Because $Pb(NO_3)_2$ is the limiting reactant, Pb^{2+} is totally consumed and $[Pb^{2+}] = 0$.

mol K^+ = mol KI = 1.90×10^{-2} mol

mol NO_3^- = 7.50×10^{-3} mol $Pb(NO_3)_2 \times \dfrac{2 \text{ mol } NO_3^-}{1 \text{ mol } Pb(NO_3)_2}$ = 0.0150 mol NO_3^-

mol I^- = (initial mol KI) − (mol KI needed) = 0.0190 mol − 0.0150 mol = 0.0040 mol I^-

total volume = 0.0750 L + 0.1000 L = 0.1750 L

$[K^+] = \dfrac{0.0190 \text{ mol}}{0.1750 \text{ L}}$ = 0.109 M

$[NO_3^-] = \dfrac{0.0150 \text{ mol}}{0.1750 \text{ L}}$ = 0.0857 M

$[I^-] = \dfrac{0.0040 \text{ mol}}{0.1750 \text{ L}}$ = 0.023 M

Multiconcept Problems

7.108 NaOH, 40.00 amu; $Ba(OH)_2$, 171.34 amu

Let X equal the mass of NaOH and Y the mass of $Ba(OH)_2$ in the 10.0 g mixture.
Therefore, X + Y = 10.0 g.

mol HCl = 108.9 mL $\times \dfrac{1 \text{ L}}{1000 \text{ mL}} \times \dfrac{1.50 \text{ mol HCl}}{1 \text{ L}}$ = 0.163 mol HCl

mol NaOH + 2 × mol $Ba(OH)_2$ = 0.163 mol HCl

$X \times \dfrac{1 \text{ mol NaOH}}{40.00 \text{ g NaOH}} + 2 \times \left(Y \times \dfrac{1 \text{ mol } Ba(OH)_2}{171.34 \text{ g } Ba(OH)_2} \right)$ = 0.163 mol HCl

Rearrange to get X = 10.0 g − Y and then substitute it into the equation above to solve for Y.

$(10.0 \text{ g} - Y) \times \dfrac{1 \text{ mol NaOH}}{40.00 \text{ g NaOH}} + 2 \times \left(Y \times \dfrac{1 \text{ mol } Ba(OH)_2}{171.34 \text{ g } Ba(OH)_2} \right)$ = 0.163 mol HCl

$\dfrac{10.00 \text{ mol}}{40.00} - \dfrac{Y \text{ mol}}{40.00 \text{ g}} + \dfrac{2 \text{ Y mol}}{171.34 \text{ g}}$ = 0.163 mol

$-\dfrac{Y \text{ mol}}{40.00 \text{ g}} + \dfrac{2 \text{ Y mol}}{171.34 \text{ g}}$ = 0.163 mol $-\dfrac{10.00 \text{ mol}}{40.00}$ = −0.087 mol

$$\frac{(-Y \text{ mol})(171.34 \text{ g}) + (2 Y \text{ mol})(40.00 \text{ g})}{(40.00 \text{ g})(171.34 \text{ g})} = -0.087 \text{ mol}$$

$$\frac{-91.34 \text{ Y mol}}{6853.6 \text{ g}} = -0.087 \text{ mol}; \quad \frac{91.34 \text{ Y}}{6853.6 \text{ g}} = 0.087$$

$Y = (0.087)(6853.6 \text{ g})/91.34 = 6.5 \text{ g Ba(OH)}_2$

$X = 10.0 \text{ g} - Y = 10.0 \text{ g} - 6.5 \text{ g} = 3.5 \text{ g NaOH}$

7.109 $100.0 \text{ mL} = 0.1000 \text{ L}$ and $50.0 \text{ mL} = 0.0500 \text{ L}$

mol $Na_2SO_4 = (0.1000 \text{ L})(0.100 \text{ mol/L}) = 0.0100$ mol Na_2SO_4

mol SO_4^{2-} = mol $Na_2SO_4 = 0.0100$ mol SO_4^{2-}

mol $Na^+ = 0.0100$ mol $Na_2SO_4 \times \dfrac{2 \text{ mol Na}^+}{1 \text{ mol Na}_2SO_4} = 0.0200$ mol Na^+

mol $ZnCl_2 = (0.0500 \text{ L})(0.300 \text{ mol/L}) = 0.0150$ mol $ZnCl_2$

mol Zn^{2+} = mol $ZnCl_2 = 0.0150$ mol Zn^{2+}

mol $Cl^- = 0.0150$ mol $ZnCl_2 \times \dfrac{2 \text{ mol Cl}^-}{1 \text{ mol ZnCl}_2} = 0.0300$ mol Cl^-

mol $Ba(CN)_2 = (0.1000 \text{ L})(0.200 \text{ mol/L}) = 0.0200$ mol $Ba(CN)_2$

mol Ba^{2+} = mol $Ba(CN)_2 = 0.0200$ mol Ba^{2+}

mol $CN^- = 0.0200$ mol $Ba(CN)_2 \times \dfrac{2 \text{ mol CN}^-}{1 \text{ mol Ba(CN)}_2} = 0.0400$ mol CN^-

The following two reactions will take place to form precipitates.

$Zn^{2+}(aq) + 2 CN^-(aq) \rightarrow Zn(CN)_2(s)$

$Ba^{2+}(aq) + SO_4^{2-}(aq) \rightarrow BaSO_4(s)$

For Zn^{2+}, mol CN^- needed $= 0.0150$ mol $Zn^{2+} \times \dfrac{2 \text{ mol CN}^-}{1 \text{ mol Zn}^{2+}} = 0.0300$ mol CN^- needed

CN^- is in excess, so Zn^{2+} is the limiting reactant and is totally consumed.

mol CN^- remaining after reaction $= 0.0400$ mol $- 0.0300$ mol $= 0.0100$ mol CN^-

For Ba^{2+}, mol SO_4^{2-} needed = mol $Ba^{2+} = 0.0200$ mol SO_4^{2-} needed

Ba^{2+} is in excess, so SO_4^{2-} is the limiting reactant and is totally consumed.

mol Ba^{2+} remaining after reaction $= 0.0200$ mol $- 0.0100$ mol $= 0.0100$ mol Ba^{2+}

total volume $= 0.1000 \text{ L} + 0.0500 \text{ L} + 0.1000 \text{ L} = 0.2500 \text{ L}$

$[Zn^{2+}] = 0$

$[SO_4^{2-}] = 0$

$[Na^+] = \dfrac{0.0200 \text{ mol}}{0.2500 \text{ L}} = 0.0800 \text{ M}$

$[Cl^-] = \dfrac{0.0300 \text{ mol}}{0.2500 \text{ L}} = 0.120 \text{ M}$

$[CN^-] = \dfrac{0.0100 \text{ mol}}{0.2500 \text{ L}} = 0.0400 \text{ M}$

$[Ba^{2+}] = \dfrac{0.0100 \text{ mol}}{0.2500 \text{ L}} = 0.0400 \text{ M}$

7.110 KNO_3, 101.10 amu; $BaCl_2$, 208.24 amu; NaCl, 58.44 amu; $BaSO_4$, 233.40 amu;
$AgCl$, 143.32 amu

(a) The two precipitates are $BaSO_4(s)$ and $AgCl(s)$.

(b) H_2SO_4 only reacts with $BaCl_2$.

$H_2SO_4(aq) + BaCl_2(aq) \rightarrow BaSO_4(s) + 2 HCl(aq)$

Calculate the number of moles of $BaCl_2$ in 100.0 g of the mixture.

$$\text{mol } BaCl_2 = 67.3 \text{ g } BaSO_4 \times \frac{1 \text{ mol } BaSO_4}{233.40 \text{ g } BaSO_4} \times \frac{1 \text{ mol } BaCl_2}{1 \text{ mol } BaSO_4} = 0.288 \text{ mol } BaCl_2$$

Calculate mass and moles of $BaCl_2$ in 250.0 g sample.

$$\text{mass } BaCl_2 = 0.288 \text{ mol } BaCl_2 \times \frac{208.24 \text{ g } BaCl_2}{1 \text{ mol } BaCl_2} \times \frac{250.0 \text{ g}}{100.0 \text{ g}} = 150. \text{ g } BaCl_2$$

$$\text{mol } BaCl_2 = 150. \text{ g } BaCl_2 \times \frac{1 \text{ mol } BaCl_2}{208.24 \text{ g } BaCl_2} = 0.720 \text{ mol } BaCl_2$$

$AgNO_3$ reacts with both NaCl and $BaCl_2$ in the remaining 150.0 g of the mixture.

$3 AgNO_3(aq) + NaCl(aq) + BaCl_2(aq) \rightarrow 3 AgCl(s) + NaNO_3(aq) + Ba(NO_3)_2(aq)$

Calculate the moles of AgCl that would have been produced from the 250.0 g mixture.

$$\text{mol } AgCl = 197.6 \text{ g } AgCl \times \frac{1 \text{ mol } AgCl}{143.32 \text{ g } AgCl} \times \frac{250.0 \text{ g}}{150.0 \text{ g}} = 2.30 \text{ mol } AgCl$$

$\text{mol } AgCl = 2 \times (\text{mol } BaCl_2) + \text{mol } NaCl$

Calculate the moles and mass of NaCl in the 250.0 g mixture.

$2.30 \text{ mol } AgCl = 2 \times 0.720 \text{ mol } BaCl_2 + \text{mol } NaCl$

$\text{mol } NaCl = 2.30 \text{ mol} - 2(0.720 \text{ mol}) = 0.86 \text{ mol } NaCl$

$$\text{mass } NaCl = 0.86 \text{ mol } NaCl \times \frac{58.44 \text{ g } NaCl}{1 \text{ mol } NaCl} = 50. \text{ g } NaCl$$

Calculate the mass of KNO_3 in the 250.0 g mixture.

total mass = mass $BaCl_2$ + mass NaCl + mass KNO_3

$250.0 \text{ g} = 150. \text{ g } BaCl_2 + 50. \text{ g } NaCl + \text{mass } KNO_3$

mass KNO_3 = 250.0 g – 150. g $BaCl_2$ – 50. g NaCl = 50. g KNO_3

7.111 100.0 mL = 0.1000 L; 50.0 mL = 0.0500 L; 250.0 mL = 0.2500 L

After step (2):

$BaCl_2(aq) + 2 AgNO_3(aq) \rightarrow AgCl(s) + Ba(NO_3)_2(aq)$

mol $BaCl_2$ = (0.1000 L)(0.100 mol/L) = 0.0100 mol $BaCl_2$

mol Ba^{2+} = mol $BaCl_2$ = 0.0100 mol Ba^{2+}

$$\text{mol } Cl^- = 0.0100 \text{ mol } BaCl_2 \times \frac{2 \text{ mol } Cl^-}{1 \text{ mol } BaCl_2} = 0.0200 \text{ mol } Cl^-$$

mol $AgNO_3$ = (0.0500 L)(0.100 mol/L) = 0.00500 mol $AgNO_3$

mol Ag^+ = mol $AgNO_3$ = 0.00500 mol Ag^+

mol NO_3^- = mol $AgNO_3$ = 0.00500 mol NO_3^-

0.00500 mol Ag^+ requires only 0.00500 mol Cl^-, so Ag^+ is the limiting reactant and totally consumed.

mol Cl^- remaining after reaction = 0.0200 mol – 0.00500 mol = 0.0150 mol Cl^-

After step (3):

$Ba^{2+}(aq) + H_2SO_4(aq) \rightarrow BaSO_4(s) + 2 H^+(aq)$

mol H_2SO_4 = (0.0500 L)(0.100 mol/L) = 0.00500 mol H_2SO_4

mol SO_4^{2-} = mol H_2SO_4 = 0.00500 mol SO_4^{2-}

mol H^+ = 0.00500 mol H_2SO_4 x $\dfrac{2\ mol\ H^+}{1\ mol\ H_2SO_4}$ = 0.0100 mol H^+

0.0100 mol Ba^{2+} requires 0.0100 mol SO_4^{2-}, so SO_4^{2-} is the limiting reactant and is totally consumed.

mol Ba^{2+} remaining after reaction = 0.0100 mol – 0.00500 mol = 0.00500 mol Ba^{2+}

After step (4):

$NH_3(aq) + H^+(aq) \rightarrow NH_4^+(aq)$

mol NH_3 = (0.2500 L)(0.100 mol/L) = 0.0250 mol NH_3

0.0250 mol NH_3 requires 0.0250 mol H^+, so H^+ is the limiting reactant and is totally consumed.

mol NH_3 remaining after reaction = 0.0250 mol – 0.0100 mol = 0.0150 mol NH_3

mol NH_4^+ = mol H^+ before reaction = 0.0100 mol NH_4^+

total volume = 0.1000 L + 0.0500 L + 0.0500 L + 0.2500 L = 0.4500 L

$[Ba^{2+}]$ = $\dfrac{0.00500\ mol}{0.4500\ L}$ = 0.0111 M

$[Cl^-]$ = $\dfrac{0.0150\ mol}{0.4500\ L}$ = 0.0333 M

$[NO_3^-]$ = $\dfrac{0.00500\ mol}{0.4500\ L}$ = 0.0111 M

$[NH_3]$ = $\dfrac{0.0150\ mol}{0.4500\ L}$ = 0.0333 M

$[NH_4^+]$ = $\dfrac{0.0100\ mol}{0.4500\ L}$ = 0.0222 M

7.112 (a) $Cr^{2+}(aq) + Cr_2O_7^{2-}(aq) \rightarrow Cr^{3+}(aq)$

   $[Cr^{2+}(aq) \rightarrow Cr^{3+}(aq) + e^-]$ x 6   (oxidation half reaction)

   $Cr_2O_7^{2-}(aq) \rightarrow Cr^{3+}(aq)$
   $Cr_2O_7^{2-}(aq) \rightarrow 2\ Cr^{3+}(aq)$
   $Cr_2O_7^{2-}(aq) \rightarrow 2\ Cr^{3+}(aq) + 7\ H_2O(l)$
   $14\ H^+(aq) + Cr_2O_7^{2-}(aq) \rightarrow 2\ Cr^{3+}(aq) + 7\ H_2O(l)$
   $6\ e^- + 14\ H^+(aq) + Cr_2O_7^{2-}(aq) \rightarrow 2\ Cr^{3+}(aq) + 7\ H_2O(l)$ (reduction half reaction)

   Combine the two half reactions.
   $14\ H^+(aq) + Cr_2O_7^{2-}(aq) + 6\ Cr^{2+}(aq) \rightarrow 8\ Cr^{3+}(aq) + 7\ H_2O(l)$

(b) total volume = 100.0 ml + 20.0 mL = 120.0 mL = 0.1200 L

Initial moles:

$$0.120 \; \frac{\text{mol } Cr(NO_3)_2}{1 \text{ L}} \times 0.1000 \text{ L} = 0.0120 \text{ mol } Cr(NO_3)_2$$

$$0.500 \; \frac{\text{mol } HNO_3}{1 \text{ L}} \times 0.1000 \text{ L} = 0.0500 \text{ mol } HNO_3$$

$$0.250 \; \frac{\text{mol } K_2Cr_2O_7}{1 \text{ L}} \times 0.0200 \text{ L} = 0.005 \, 00 \text{ mol } K_2Cr_2O_7$$

Check for the limiting reactant. 0.0120 mol of Cr^{2+} requires $(0.0120)/6 = 0.00200$ mol $Cr_2O_7^{2-}$ and $(14/6)(0.0120) = 0.0280$ mol H^+. Both are in excess of the required amounts, so Cr^{2+} is the limiting reactant.

$$14 \; H^+(aq) \; + \; Cr_2O_7^{2-}(aq) \; + \; 6 \; Cr^{2+}(aq) \; \rightarrow \; 8 \; Cr^{3+}(aq) \; + \; 7 \; H_2O(l)$$

	14 H⁺	Cr₂O₇²⁻	6 Cr²⁺	8 Cr³⁺
Initial moles	0.0500	0.00500	0.0120	0
Change	−14x	−x	−6x	+8x

Because Cr^{2+} is the limiting reactant, $6x = 0.0120$ and $x = 0.00200$

Final moles	0.0220	0.00300	0	0.0160

$$\text{mol } K^+ = 0.00500 \text{ mol } K_2Cr_2O_7 \times \frac{2 \text{ mol } K^+}{1 \text{ mol } K_2Cr_2O_7} = 0.0100 \text{ mol } K^+$$

$$\text{mol } NO_3^- = 0.0120 \text{ mol } Cr(NO_3)_2 \times \frac{2 \text{ mol } NO_3^-}{1 \text{ mol } Cr(NO_3)_2}$$

$$+ \; 0.0500 \text{ mol } HNO_3 \times \frac{1 \text{ mol } NO_3^-}{1 \text{ mol } HNO_3} = 0.0740 \text{ mol } NO_3^-$$

$\text{mol } H^+ = 0.0220 \text{ mol};$ $\text{mol } Cr_2O_7^{2-} = 0.00300 \text{ mol};$ $\text{mol } Cr^{3+} = 0.01600 \text{ mol}$

Check for charge neutrality.

Total moles of +charge = $0.0100 + 0.0220 + 3 \times (0.01600) = 0.0800$ mol +charge

Total moles of −charge = $0.0740 + 2 \times (0.00300) = 0.0800$ mol −charge

The charges balance and there is electrical neutrality in the solution after the reaction.

$$K^+ \text{ molarity} = \frac{0.0100 \text{ mol } K^+}{0.1200 \text{ L}} = 0.0833 \text{ M}$$

$$NO_3^- \text{ molarity} = \frac{0.0740 \text{ mol } NO_3^-}{0.1200 \text{ L}} = 0.617 \text{ M}$$

$$H^+ \text{ molarity} = \frac{0.0220 \text{ mol } H^+}{0.1200 \text{ L}} = 0.183 \text{ M}$$

$$Cr_2O_7^{2-} \text{ molarity} = \frac{0.00300 \text{ mol } Cr_2O_7^{2-}}{0.1200 \text{ L}} = 0.0250 \text{ M}$$

$$Cr^{3+} \text{ molarity} = \frac{0.0160 \text{ mol } Cr^{3+}}{0.1200 \text{ L}} = 0.133 \text{ M}$$

7.113 (a) (1) $I^-(aq) \rightarrow I_3^-(aq)$

$3\ I^-(aq) \rightarrow I_3^-(aq)$

$3\ I^-(aq) \rightarrow I_3^-(aq) + 2\ e^-$ (oxidation half reaction)

$HNO_2(aq) \rightarrow NO(g)$

$HNO_2(aq) \rightarrow NO(g) + H_2O(l)$

$H^+(aq) + HNO_2(aq) \rightarrow NO(g) + H_2O(l)$

$[e^- + H^+(aq) + HNO_2(aq) \rightarrow NO(g) + H_2O(l)] \times 2$ (reduction half reaction)

Combine the two half reactions.

$3\ I^-(aq) + 2\ H^+(aq) + 2\ HNO_2(aq) \rightarrow I_3^-(aq) + 2\ NO(g) + 2\ H_2O(l)$

(2) $S_2O_3^{2-}(aq) \rightarrow S_4O_6^{2-}(aq)$

$2\ S_2O_3^{2-}(aq) \rightarrow S_4O_6^{2-}(aq)$

$2\ S_2O_3^{2-}(aq) \rightarrow S_4O_6^{2-}(aq) + 2\ e^-$ (oxidation half reaction)

$I_3^-(aq) \rightarrow I^-(aq)$

$I_3^-(aq) \rightarrow 3\ I^-(aq)$

$2\ e^- + I_3^-(aq) \rightarrow 3\ I^-(aq)$ (reduction half reaction)

Combine the two half reactions.

$2\ S_2O_3^{2-}(aq) + I_3^-(aq) \rightarrow S_4O_6^{2-}(aq) + 3\ I^-(aq)$

(b) $18.77\ mL = 0.018\ 77\ L;\ NO_2^-,\ 46.01\ amu$

$0.1500\ \dfrac{mol\ S_2O_3^{2-}}{1\ L} \times 0.018\ 77\ L = 0.002\ 815\ 5\ mol\ S_2O_3^{2-}$

$mass\ NO_2^- = 0.002\ 815\ 5\ mol\ S_2O_3^{2-} \times \dfrac{1\ mol\ I_3^-}{2\ mol\ S_2O_3^{2-}} \times \dfrac{2\ mol\ NO_2^-}{1\ mol\ I_3^-} \times$

$\dfrac{46.01\ g\ NO_2^-}{1\ mol\ NO_2^-} = 0.1295\ g\ NO_2^-$

$mass\ \%\ NO_2^- = \dfrac{0.1295\ g}{2.935\ g} \times 100\% = 4.412\%$

7.114 (a) (1) $Cu(s) \rightarrow Cu^{2+}(aq)$

$[Cu(s) \rightarrow Cu^{2+}(aq) + 2\ e^-] \times 3$ (oxidation half reaction)

$NO_3^-(aq) \rightarrow NO(g)$

$NO_3^-(aq) \rightarrow NO(g) + 2\ H_2O(l)$

$4\ H^+(aq) + NO_3^-(aq) \rightarrow NO(g) + 2\ H_2O(l)$

$[3\ e^- + 4\ H^+(aq) + NO_3^-(aq) \rightarrow NO(g) + 2\ H_2O(l)] \times 2$ (reduction half reaction)

Combine the two half reactions.

$3\ Cu(s) + 8\ H^+(aq) + 2\ NO_3^-(aq) \rightarrow 3\ Cu^{2+}(aq) + 2\ NO(g) + 4\ H_2O(l)$

(2) $Cu^{2+}(aq) + SCN^-(aq) \rightarrow CuSCN(s)$
        $[e^- + Cu^{2+}(aq) + SCN^-(aq) \rightarrow CuSCN(s)] \times 2$ (reduction half reaction)

        $HSO_3^-(aq) \rightarrow HSO_4^-(aq)$
        $H_2O(l) + HSO_3^-(aq) \rightarrow HSO_4^-(aq)$
        $H_2O(l) + HSO_3^-(aq) \rightarrow HSO_4^-(aq) + 2\ H^+(aq)$
        $H_2O(l) + HSO_3^-(aq) \rightarrow HSO_4^-(aq) + 2\ H^+(aq) + 2\ e^-$
                                        (oxidation half reaction)

Combine the two half reactions.
$2\ Cu^{2+}(aq) + 2\ SCN^-(aq) + H_2O(l) + HSO_3^-(aq) \rightarrow$
$\qquad\qquad\qquad\qquad 2\ CuSCN(s) + HSO_4^-(aq) + 2\ H^+(aq)$

(3) $Cu^+(aq) \rightarrow Cu^{2+}(aq)$
        $[Cu^+(aq) \rightarrow Cu^{2+}(aq) + e^-] \times 10$ (oxidation half reaction)

        $IO_3^-(aq) \rightarrow I_2(aq)$
        $2\ IO_3^-(aq) \rightarrow I_2(aq)$
        $2\ IO_3^-(aq) \rightarrow I_2(aq) + 6\ H_2O(l)$
        $12\ H^+(aq) + 2\ IO_3^-(aq) \rightarrow I_2(aq) + 6\ H_2O(l)$
        $10\ e^- + 12\ H^+(aq) + 2\ IO_3^-(aq) \rightarrow I_2(aq) + 6\ H_2O(l)$
                                        (reduction half reaction)

Combine the two half reactions.
$10\ Cu^+(aq) + 12\ H^+(aq) + 2\ IO_3^-(aq) \rightarrow 10\ Cu^{2+}(aq) + I_2(aq) + 6\ H_2O(l)$

(4) $I_2(aq) \rightarrow I^-(aq)$
        $I_2(aq) \rightarrow 2\ I^-(aq)$
        $2\ e^- + I_2(aq) \rightarrow 2\ I^-(aq)$ (reduction half reaction)

        $S_2O_3^{2-}(aq) \rightarrow S_4O_6^{2-}(aq)$
        $2\ S_2O_3^{2-}(aq) \rightarrow S_4O_6^{2-}(aq)$
        $2\ S_2O_3^{2-}(aq) \rightarrow S_4O_6^{2-}(aq) + 2\ e^-$ (oxidation half reaction)

Combine the two half reactions.
$I_2(aq) + 2\ S_2O_3^{2-}(aq) \rightarrow 2\ I^-(aq) + S_4O_6^{2-}(aq)$

(5) $2\ ZnNH_4PO_4 \rightarrow Zn_2P_2O_7 + H_2O + 2\ NH_3$

(b)   10.82 mL = 0.01082 L
        mol $S_2O_3^{2-}$ = (0.1220 mol/L)(0.01082 L) = 0.00132 mol $S_2O_3^{2-}$

        mol I_2 = 0.00132 mol $S_2O_3^{2-}$ x $\dfrac{1\text{ mol }I_2}{2\text{ mol }S_2O_3^{2-}}$ = 6.60 x 10^{-4} mol I_2

        mol Cu^+ = 6.60 x 10^{-4} mol I_2 x $\dfrac{10\text{ mol }Cu^+}{1\text{ mol }I_2}$ = 6.60 x 10^{-3} mol Cu^+ (Cu)

        g Cu = (6.60 x 10^{-3} mol)(63.546 g/mol) = 0.419 g Cu

$$\text{mass \% Cu in brass} = \frac{0.419 \text{ g Cu}}{0.544 \text{ g brass}} \times 100\% = 77.1\% \text{ Cu}$$

(c) $Zn_2P_2O_7$, 304.72 amu

$$\text{mass \% Zn in } Zn_2P_2O_7 = \frac{2 \times 65.39 \text{ g}}{304.72 \text{ g}} \times 100\% = 42.92\%$$

mass of Zn in $Zn_2P_2O_7$ = (0.4292)(0.246 g) = 0.106 g Zn

$$\text{mass \% Zn in brass} = \frac{0.106 \text{ g Zn}}{0.544 \text{ g brass}} \times 100\% = 19.5\% \text{ Zn}$$

7.115 (a) $BaSO_4$, 233.38 amu

$$\text{mol S} = 7.19 \text{ g } BaSO_4 \times \frac{1 \text{ mol } BaSO_4}{233.38 \text{ g } BaSO_4} \times \frac{1 \text{ mol S}}{1 \text{ mol } BaSO_4} = 0.0308 \text{ mol S}$$

$$\text{theoretical mol S} = \frac{0.0308 \text{ mol S}}{0.913} = 0.0337 \text{ mol S}$$

(b) Assume n = 1:

$$\text{mol Cl in } MCl_5 = 0.0337 \text{ mol S} \times \frac{5 \text{ mol Cl}}{1 \text{ mol S}} = 0.168 \text{ mol Cl}$$

$$\text{mass Cl} = 0.168 \text{ mol Cl} \times \frac{35.453 \text{ g Cl}}{1 \text{ mol Cl}} = 5.97 \text{ g Cl}$$

This is impossible because the initial mass of MCl_5 was only 4.61 g.

Assume n = 2:

$$\text{mol Cl in } MCl_5 = 0.0337 \text{ mol S} \times \frac{5 \text{ mol Cl}}{2 \text{ mol S}} = 0.0842 \text{ mol Cl}$$

$$\text{mass Cl} = 0.0842 \text{ mol Cl} \times \frac{35.453 \text{ g Cl}}{1 \text{ mol Cl}} = 2.99 \text{ g Cl}$$

mass M = 4.61g – 2.99 g = 1.62 g M

$$\text{mol M} = 0.0337 \text{ mol S} \times \frac{1 \text{ mol M}}{2 \text{ mol S}} = 0.0168 \text{ mol}$$

$$\text{M molar mass} = \frac{1.62 \text{ g}}{0.0168 \text{ mol}} = 96.4 \text{ g/mol}; \quad \text{M atomic mass} = 96.4 \text{ amu}$$

96.4 is reasonable and suggests that M is Mo.

Assume n = 3:

$$\text{mol Cl in } MCl_5 = 0.0337 \text{ mol S} \times \frac{5 \text{ mol Cl}}{3 \text{ mol S}} = 0.0562 \text{ mol Cl}$$

$$\text{mass Cl} = 0.0562 \text{ mol Cl} \times \frac{35.453 \text{ g Cl}}{1 \text{ mol Cl}} = 1.99 \text{ g Cl}$$

mass M = 4.61g – 1.99 g = 2.62 g M

$$\text{mol M} = 0.0337 \text{ mol S} \times \frac{1 \text{ mol M}}{3 \text{ mol S}} = 0.0112 \text{ mol}$$

$$M \text{ molar mass} = \frac{2.62 \text{ g}}{0.0112 \text{ mol}} = 234 \text{ g/mol}; \quad M \text{ atomic mass} = 234 \text{ amu}$$

234 is between Pa and U, which is highly unlikely for a lubricant.

Assume n = 4:

$$\text{mol Cl in MCl}_5 = 0.0337 \text{ mol S} \times \frac{5 \text{ mol Cl}}{4 \text{ mol S}} = 0.0421 \text{ mol Cl}$$

$$\text{mass Cl} = 0.0421 \text{ mol Cl} \times \frac{35.453 \text{ g Cl}}{1 \text{ mol Cl}} = 1.49 \text{ g Cl}$$

mass M = 4.61 g − 1.49 g = 3.12 g M

$$\text{mol M} = 0.0337 \text{ mol S} \times \frac{1 \text{ mol M}}{4 \text{ mol S}} = 0.00842 \text{ mol}$$

$$M \text{ molar mass} = \frac{3.12 \text{ g}}{0.008\,42 \text{ mol}} = 371 \text{ g/mol}; \quad M \text{ atomic mass} = 371 \text{ amu}$$

No known elements have a mass as great as 371 amu.
(c) M is most likely Mo and the metal sulfide is MoS_2.
(d) (1) $2 \text{ MoCl}_5(s) + 5 \text{ Na}_2S(s) \rightarrow 2 \text{ MoS}_2(s) + S(l) + 10 \text{ NaCl}(s)$
    (2) $2 \text{ MoS}_2(s) + 7 \text{ O}_2(g) \rightarrow 2 \text{ MoO}_3(s) + 4 \text{ SO}_2(g)$
    (3) $SO_2(g) + 2 \text{ Fe}^{3+}(aq) + 2 \text{ H}_2O(l) \rightarrow 2 \text{ Fe}^{2+}(aq) + SO_4^{2-}(aq) + 4 \text{ H}^+(aq)$
    (4) $SO_4^{2-}(aq) + Ba^{2+}(aq) \rightarrow BaSO_4(s)$

7.116 (a) $H_3MO_3(aq) \rightarrow H_3MO_4(aq)$
$H_3MO_3(aq) + H_2O(l) \rightarrow H_3MO_4(aq)$
$H_3MO_3(aq) + H_2O(l) \rightarrow H_3MO_4(aq) + 2 \text{ H}^+(aq)$
$[H_3MO_3(aq) + H_2O(l) \rightarrow H_3MO_4(aq) + 2 \text{ H}^+(aq) + 2 \text{ e}^-] \times 5$ (oxidation half reaction)

$MnO_4^-(aq) \rightarrow Mn^{2+}(aq)$
$MnO_4^-(aq) \rightarrow Mn^{2+}(aq) + 4 \text{ H}_2O(l)$
$MnO_4^-(aq) + 8 \text{ H}^+(aq) \rightarrow Mn^{2+}(aq) + 4 \text{ H}_2O(l)$
$[MnO_4^-(aq) + 8 \text{ H}^+(aq) + 5 \text{ e}^- \rightarrow Mn^{2+}(aq) + 4 \text{ H}_2O(l)] \times 2$ (reduction half reaction)

Combine the two half reactions.
$5 \text{ H}_3MO_3(aq) + 5 \text{ H}_2O(l) + 2 \text{ MnO}_4^-(aq) + 16 \text{ H}^+(aq) \rightarrow$
$$5 \text{ H}_3MO_4(aq) + 10 \text{ H}^+(aq) + 2 \text{ Mn}^{2+}(aq) + 8 \text{ H}_2O(l)$$
$5 \text{ H}_3MO_3(aq) + 2 \text{ MnO}_4^-(aq) + 6 \text{ H}^+(aq) \rightarrow 5 \text{ H}_3MO_4(aq) + 2 \text{ Mn}^{2+}(aq) + 3 \text{ H}_2O(l)$

(b) 10.7 mL = 0.0107 L
$\text{mol MnO}_4^- = (0.0107 \text{ L})(0.100 \text{ mol/L}) = 1.07 \times 10^{-3} \text{ mol MnO}_4^-$

$$\text{mol H}_3MO_3 = 1.07 \times 10^{-3} \text{ mol MnO}_4^- \times \frac{5 \text{ mol H}_3MO_3}{2 \text{ mol MnO}_4^-} = 2.67 \times 10^{-3} \text{ mol H}_3MO_3$$

$$\text{mol M}_2O_3 = 2.67 \times 10^{-3} \text{ mol H}_3MO_3 \times \frac{1 \text{ mol M}_2O_3}{2 \text{ mol H}_3MO_3} = 1.34 \times 10^{-3} \text{ mol M}_2O_3$$

(c) mol M in M_2O_3 = 1.34 x 10^{-3} mol M_2O_3 x $\dfrac{2 \text{ mol M}}{1 \text{ mol } M_2O_3}$ = 2.68 x 10^{-3} mol M

M molar mass = $\dfrac{0.200 \text{ g}}{2.68 \times 10^{-3} \text{ mol}}$ = 74.6 g/mol;   M atomic mass = 74.6 amu

M is As.

(d) E = hv = (6.626 x 10^{-34} J·s)(9.07 x 10^{14} s^{-1}) = 6.01 x 10^{-19} J/photon
E = (6.01 x 10^{-19} J/photon)(6.022 x 10^{23} photons/mol)(1 kJ/1000 J) = 362 kJ/mol

8 Thermochemistry: Chemical Energy

8.1 (a) and (b) are state functions; (c) is not.

8.2 $\Delta V = (4.3 \text{ L} - 8.6 \text{ L}) = -4.3 \text{ L}$

$w = -P\Delta V = -(44 \text{ atm})(-4.3 \text{ L}) = +189.2 \text{ L} \cdot \text{atm}$

$w = (189.2 \text{ L} \cdot \text{atm})(101 \frac{\text{J}}{\text{L} \cdot \text{atm}}) = +1.9 \times 10^4 \text{ J}$

The positive sign for the work indicates that the surroundings do work on the system. Energy flows into the system.

8.3 $w = -P\Delta V = -(2.5 \text{ atm})(3 \text{ L} - 2 \text{ L}) = -2.5 \text{ L} \cdot \text{atm}$

$w = (-2.5 \text{ L} \cdot \text{atm})\left(101 \frac{\text{J}}{\text{L} \cdot \text{atm}}\right) = -252.5 \text{ J} = -250 \text{ J} = -0.25 \text{ kJ}$

The negative sign indicates that the expanding system loses work energy and does work on the surroundings.

8.4 (a) $w = -P\Delta V$ is positive and $P\Delta V$ is negative for this reaction because the system volume is decreased at constant pressure.

(b) $P\Delta V$ is small compared to ΔE.

$\Delta H = \Delta E + P\Delta V$; ΔH is negative. Its value is slightly more negative than ΔE.

8.5 $\Delta H^\circ = -484 \frac{\text{kJ}}{2 \text{ mol H}_2}$

$P\Delta V = (1.00 \text{ atm})(-5.6 \text{ L}) = -5.6 \text{ L} \cdot \text{atm}$

$P\Delta V = (-5.6 \text{ L} \cdot \text{atm})(101 \frac{\text{J}}{\text{L} \cdot \text{atm}}) = -565.6 \text{ J} = -570 \text{ J} = -0.57 \text{ kJ}$

$w = -P\Delta V = 570 \text{ J} = 0.57 \text{ kJ}$

$\Delta H = \frac{-121 \text{ kJ}}{0.50 \text{ mol H}_2}$

$\Delta E = \Delta H - P\Delta V = -121 \text{ kJ} - (-0.57 \text{ kJ}) = -120.43 \text{ kJ} = -120 \text{ kJ}$

8.6 $\Delta V = 448 \text{ L}$ and assume $P = 1.00 \text{ atm}$

$w = -P\Delta V = -(1.00 \text{ atm})(448 \text{ L}) = -448 \text{ L} \cdot \text{atm}$

$w = -(448 \text{ L} \cdot \text{atm})(101 \frac{\text{J}}{\text{L} \cdot \text{atm}}) = -4.52 \times 10^4 \text{ J}$

$w = -4.52 \times 10^4 \text{ J} \times \frac{1 \text{ kJ}}{1000 \text{ J}} = -45.2 \text{ kJ}$

8.7   (a) C_3H_8, 44.10 amu; $\Delta H° = -2220$ kJ/mol C_3H_8

$$15.5 \text{ g} \times \frac{1 \text{ mol } C_3H_8}{44.10 \text{ g } C_3H_8} \times \frac{-2220 \text{ kJ}}{1 \text{ mol } C_3H_8} = -780. \text{ kJ}$$

780. kJ of heat is evolved.

(b) $Ba(OH)_2 \cdot 8\,H_2O$, 315.5 amu; $\Delta H° = +80.3$ kJ/mol $Ba(OH)_2 \cdot 8\,H_2O$

$$4.88 \text{ g} \times \frac{1 \text{ mol } Ba(OH)_2 \cdot 8\,H_2O}{315.5 \text{ g } Ba(OH)_2 \cdot 8\,H_2O} \times \frac{80.3 \text{ kJ}}{1 \text{ mol } Ba(OH)_2 \cdot 8\,H_2O} = +1.24 \text{ kJ}$$

1.24 kJ of heat is absorbed.

8.8 CH_3NO_2, 61.04 amu

$$q = 100.0 \text{ g } CH_3NO_2 \times \frac{1 \text{ mol } CH_3NO_2}{61.04 \text{ g } CH_3NO_2} \times \frac{2441.6 \text{ kJ}}{4 \text{ mol } CH_3NO_2} = 1.000 \times 10^3 \text{ kJ}$$

8.9 $q = (\text{specific heat}) \times m \times \Delta T = (4.18 \frac{J}{g \cdot °C})(350 \text{ g})(3 \text{ °C} - 25 \text{ °C}) = -3.2 \times 10^4 \text{ J}$

$$q = -3.2 \times 10^4 \text{ J} \times \frac{1 \text{ kJ}}{1000 \text{ J}} = -32 \text{ kJ}$$

8.10 $q = (\text{specific heat}) \times m \times \Delta T$

$$\text{specific heat} = \frac{q}{m \times \Delta T} = \frac{97.2 \text{ J}}{(75.0 \text{ g})(10.0 °C)} = 0.130 \text{ J/(g} \cdot °C)$$

8.11   25.0 mL = 0.0250 L and 50.0 mL = 0.0500 L

mol H_2SO_4 = (1.00 mol/L)(0.0250 L) = 0.0250 mol H_2SO_4

mol NaOH = (1.00 mol/L)(0.0500 L) = 0.0500 mol NaOH

NaOH and H_2SO_4 are present in a 2:1 mol ratio. This matches the stoichiometric ratio in the balanced equation.

$q = (\text{specific heat}) \times m \times \Delta T$

m = (25.0 mL + 50.0 mL)(1.00 g/mL) = 75.0 g

$$q = (4.18 \frac{J}{g \cdot °C})(75.0 \text{ g})(33.9°C - 25.0°C) = 2790 \text{ J}$$

mol H_2SO_4 = 0.0250 L × 1.00 $\frac{mol}{L}$ H_2SO_4 = 0.0250 mol H_2SO_4

$$\text{Heat evolved per mole of } H_2SO_4 = \frac{2.79 \times 10^3 \text{ J}}{0.0250 \text{ mol } H_2SO_4} = 1.1 \times 10^5 \text{ J/mol } H_2SO_4$$

Because the reaction evolves heat, the sign for ΔH is negative.

$$\Delta H = -1.1 \times 10^5 \text{ J} \times \frac{1 \text{ kJ}}{1000 \text{ J}} = -1.1 \times 10^2 \text{ kJ}$$

8.12 $\qquad CH_4(g) + Cl_2(g) \rightarrow CH_3Cl(g) + HCl(g) \qquad \Delta H^\circ_1 = -98.3$ kJ

$\qquad\qquad \underline{CH_3Cl(g) + Cl_2(g) \rightarrow CH_2Cl_2(g) + HCl(g)} \qquad \Delta H^\circ_2 = -104$ kJ

Sum $\quad CH_4(g) + 2\,Cl_2(g) \rightarrow CH_2Cl_2(g) + 2\,HCl(g)$

$\qquad \Delta H^\circ = \Delta H^\circ_1 + \Delta H^\circ_2 = -202$ kJ

8.13 (a) $A + 2\,B \rightarrow D$; $\Delta H^\circ = -100$ kJ $+ (-50$ kJ$) = -150$ kJ

(b) The red arrow corresponds to step 1: $\quad A + B \rightarrow C$

The green arrow corresponds to step 2: $\quad C + B \rightarrow D$

The blue arrow corresponds to the overall reaction.

(c) The top energy level represents $A + 2\,B$.

The middle energy level represents $C + B$.

The bottom energy level represents D.

8.14 Reactants $CH_4 + 2\,Cl_2$

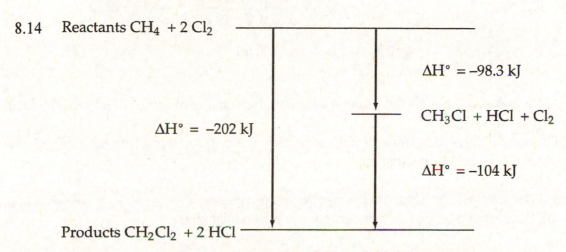

$\Delta H^\circ = -98.3$ kJ

$\Delta H^\circ = -202$ kJ

$CH_3Cl + HCl + Cl_2$

$\Delta H^\circ = -104$ kJ

Products $CH_2Cl_2 + 2\,HCl$

8.15 $4\,NH_3(g) + 5\,O_2(g) \rightarrow 4\,NO(g) + 6\,H_2O(g)$

$\Delta H^\circ_{rxn} = [4\,\Delta H^\circ_f\,(NO) + 6\,\Delta H^\circ_f\,(H_2O)] - [4\,\Delta H^\circ_f\,(NH_3)]$

$\Delta H^\circ_{rxn} = [(4\text{ mol})(91.3\text{ kJ/mol}) + (6\text{ mol})(-241.8\text{ kJ/mol})] - [(4\text{ mol})(-46.1\text{ kJ/mol})]$

$\Delta H^\circ_{rxn} = -901.2$ kJ

8.16 $6\,CO_2(g) + 6\,H_2O(l) \rightarrow C_6H_{12}O_6(s) + 6\,O_2(g)$

$\Delta H^\circ_{rxn} = \Delta H^\circ_f(C_6H_{12}O_6) - [6\,\Delta H^\circ_f(CO_2) + 6\,\Delta H^\circ_f(H_2O(l))]$

$\Delta H^\circ_{rxn} = [(1\text{ mol})(-1273.3\text{ kJ/mol})] - [(6\text{ mol})(-393.5\text{ kJ/mol}) + (6\text{ mol})(-285.8\text{ kJ/mol})]$

$\Delta H^\circ_{rxn} = +2802.5$ kJ $= +2803$ kJ

8.17 $H_2C{=}CH_2(g) + H_2O(g) \rightarrow C_2H_5OH(g)$

$\Delta H^\circ_{rxn} = D$ (Reactant bonds) $- D$ (Product bonds)

$\Delta H^\circ_{rxn} = (D_{C=C} + 4\,D_{C-H} + 2\,D_{O-H}) - (D_{C-C} + D_{C-O} + 5\,D_{C-H} + D_{O-H})$

$\Delta H^\circ_{rxn} = [(1\text{ mol})(611\text{ kJ/mol}) + (4\text{ mol})(410\text{ kJ/mol}) + (2\text{ mol})(460\text{ kJ/mol})]$

$- [(1\text{ mol})(350\text{ kJ/mol}) + (1\text{ mol})(350\text{ kJ/mol}) + (5\text{ mol})(410\text{ kJ/mol}) + (1\text{ mol})(460\text{ kJ/mol})]$

$\Delta H^\circ_{rxn} = -39$ kJ

8.18 $2 NH_3(g) + Cl_2(g) \rightarrow N_2H_4(g) + 2 HCl(g)$

$\Delta H^\circ_{rxn} = D$ (Reactant bonds) $- D$ (Product bonds)

$\Delta H^\circ_{rxn} = (6 D_{N-H} + D_{Cl-Cl}) - (D_{N-N} + 4 D_{N-H} + 2 D_{H-Cl})$

$\Delta H^\circ_{rxn} = [(6 \text{ mol})(390 \text{ kJ/mol}) + (1 \text{ mol})(243 \text{ kJ/mol})]$

$\qquad\qquad - [(1 \text{ mol})(240 \text{ kJ/mol}) + (4 \text{ mol})(390 \text{ kJ/mol}) + (2 \text{ mol})(432 \text{ kJ/mol})] = -81 \text{ kJ}$

8.19 $C_4H_{10}(l) + \dfrac{13}{2} O_2(g) \rightarrow 4 CO_2(g) + 5 H_2O(g)$

$\Delta H^\circ_{rxn} = [4 \Delta H^\circ_f (CO_2) + 5 \Delta H^\circ_f (H_2O)] - \Delta H^\circ_f (C_4H_{10})$

$\Delta H^\circ_{rxn} = [(4 \text{ mol})(-393.5 \text{ kJ/mol}) + (5 \text{ mol})(-241.8 \text{ kJ/mol})] - [(1 \text{ mol})(-147.5 \text{ kJ/mol})]$

$\Delta H^\circ_{rxn} = -2635.5 \text{ kJ}; \quad \Delta H^\circ_C = -2635.5 \text{ kJ/mol}$

$C_4H_{10}, 58.12 \text{ amu}; \qquad \Delta H^\circ_C = \left(-2635.5 \dfrac{\text{kJ}}{\text{mol}}\right)\left(\dfrac{1 \text{ mol}}{58.12 \text{ g}}\right) = -45.35 \text{ kJ/g}$

$\Delta H^\circ_C = \left(-45.35 \dfrac{\text{kJ}}{\text{g}}\right)\left(0.579 \dfrac{\text{g}}{\text{mL}}\right) = -26.3 \text{ kJ/mL}$

8.20 ΔS° is negative because the reaction decreases the number of moles of gaseous molecules.

8.21   The reaction proceeds from a solid and a gas (reactants) to all gas (product). Randomness increases, so ΔS° is positive.

8.22   (a) Because ΔG° is negative, the reaction is spontaneous.
       (b) Because ΔG° is positive, the reaction is nonspontaneous.

8.23 $\Delta G^\circ = \Delta H^\circ - T\Delta S^\circ = (-92.2 \text{ kJ}) - (298 \text{ K})(-0.199 \text{ kJ/K}) = -32.9 \text{ kJ}$
Because ΔG° is negative, the reaction is spontaneous.
Set $\Delta G^\circ = 0$ and solve for T.

$\Delta G^\circ = 0 = \Delta H^\circ - T\Delta S^\circ; \quad T = \dfrac{\Delta H^\circ}{\Delta S^\circ} = \dfrac{-92.2 \text{ kJ}}{-0.199 \text{ kJ/K}} = 463 \text{ K} = 190 \text{ °C}$

8.24   (a) $2 A_2 + B_2 \rightarrow 2 A_2B$
       (b) Because the reaction is exothermic, ΔH is negative. There are more reactant molecules than product molecules. The randomness of the system decreases on going from reactant to product, therefore ΔS is negative.
       (c) Because $\Delta G = \Delta H - T\Delta S$, a reaction with both ΔH and ΔS negative is favored at low temperatures where the negative ΔH term is larger than the positive $- T\Delta S$, and ΔG is negative.

8.25 $C_2H_6O + 3 O_2 \rightarrow 2 CO_2 + 3 H_2O$

$2 C_{19}H_{38}O_2 + 55 O_2 \rightarrow 38 CO_2 + 38 H_2O$

8.26   Because the standard heat of formation of $CO_2(g)$ (–393.5 kJ/mol) is more negative than that of $H_2O(g)$ (–241.8 kJ/mol), formation of CO_2 releases more heat than formation of H_2O. According to the balanced equations (Problem 8.25), combustion of ethanol yields a 2:3 ratio of CO_2 to H_2O, whereas combustion of biodiesel yields a 1:1 ratio. Thus biodiesel has a more favorable (more negative) combustion enthalpy per gram than ethanol.

Key Concept Problems

8.27 (a) $2\,AB_2 \rightarrow A_2 + 2\,B_2$
(b) Because the reaction is exothermic, ΔH is negative. Because the number of molecules increases going from reactants to products, ΔS is positive.
(c) When ΔH is negative and ΔS is positive, the reaction is likely to be spontaneous at all temperatures.

8.28. (a) $w = -P\Delta V$, $\Delta V > 0$; therefore $w < 0$ and the system is doing work on the surroundings.
(b) Since the temperature has increased there has been an enthalpy change. The system evolved heat, the reaction is exothermic, and $\Delta H < 0$.

8.29 (a)

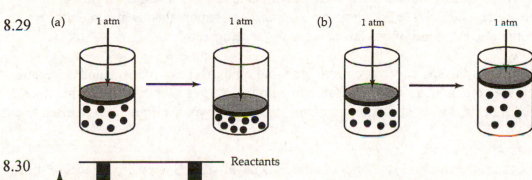

8.30

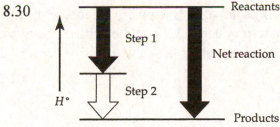

8.31   Reactants $CH_3CH_2OH + O_2$

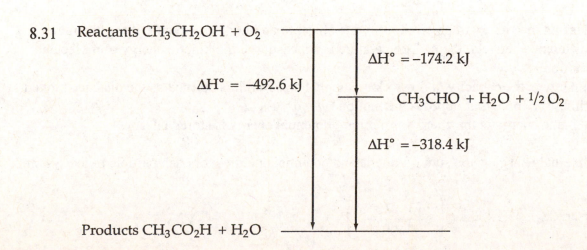

Products $CH_3CO_2H + H_2O$

8.32 $\Delta H = \Delta E + P\Delta V$

$\Delta H - \Delta E = P\Delta V$

$\Delta V = \dfrac{\Delta H - \Delta E}{P} = \dfrac{[-35.0 \text{ kJ} - (-34.8 \text{ kJ})]}{1 \text{ atm}} \times \dfrac{1 \text{ L} \cdot \text{atm}}{101 \times 10^{-3} \text{ kJ}} = -2 \text{ L}$

$\Delta V = -2 \text{ L} = V_{\text{final}} - V_{\text{initial}} = V_{\text{final}} - 5 \text{ L};$ $V_{\text{final}} = -2 \text{ L} - (-5 \text{ L}) = 3 \text{ L}$

The volume decreases from 5 L to 3 L.

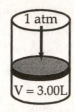

8.33 $\Delta H^{\circ} = +55 \text{ kJ}$

ΔS° is positive because randomness increases when a solid is converted to a gas.
$\Delta G^{\circ} = \Delta H^{\circ} - T\Delta S^{\circ}$; For the reaction to be spontaneous, ΔG° must be negative.
Because ΔH° and ΔS° are both positive, the reaction is spontaneous at some higher temperatures, but nonspontaneous at some lower temperatures.

8.34   The change is the spontaneous conversion of a liquid to a gas. ΔG is negative because the change is spontaneous. The conversion of a liquid to a gas is endothermic, therefore ΔH is positive. ΔS is positive because randomness increases when a liquid is converted to a gas.

8.35   (a) $2 \text{ A}_3 \rightarrow 3 \text{ A}_2$

(b) Because the reaction is spontaneous, ΔG is negative. ΔS is positive because the number of molecules increases in going from reactant to products. ΔH could be either positive or negative and the reaction would still be spontaneous. ΔH is probably positive because there is more bond breaking than bond making.

Section Problems
Heat, Work, and Energy (Sections 8.1–8.3)

8.36   Heat is the energy transferred from one object to another as the result of a temperature difference between them. Temperature is a measure of the kinetic energy of molecular motion.
Energy is the capacity to do work or supply heat. Work is defined as the distance moved times the force that opposes the motion ($w = d \times F$).
Kinetic energy is the energy of motion. Potential energy is stored energy.

8.37   Internal energy is the sum of kinetic and potential energies for each particle in the system.

8.38   Car: $E_K = \frac{1}{2}(1400 \text{ kg})\left(\dfrac{115 \times 10^3 \text{ m}}{3600 \text{ s}}\right)^2 = 7.1 \times 10^5 \text{ J}$

Truck: $E_K = \frac{1}{2}(12{,}000 \text{ kg})\left(\dfrac{38 \times 10^3 \text{ m}}{3600 \text{ s}}\right)^2 = 6.7 \times 10^5 \text{ J}$

The car has more kinetic energy.

8.39 Heat = $q = 7.1 \times 10^5$ J (from Problem 8.38)

$q = $ (specific heat) x m x ΔT

$m = \dfrac{q}{(\text{specific heat}) \times \Delta T} = \dfrac{7.1 \times 10^5 \text{ J}}{\left(4.18 \dfrac{\text{J}}{\text{g} \cdot {}^\circ\text{C}}\right)(50\,^\circ\text{C} - 20\,^\circ\text{C})} = 5.7 \times 10^3$ g of water

8.40 $w = -P\Delta V = -(3.6 \text{ atm})(3.4 \text{ L} - 3.2 \text{ L}) = -0.72 \text{ L} \cdot \text{atm}$

$w = (-0.72 \text{ L} \cdot \text{atm})\left(\dfrac{101 \text{ J}}{1 \text{ L} \cdot \text{atm}}\right) = -72.7 \text{ J} = -70 \text{ J};$ The energy change is negative.

8.41 $V_{initial} = 50.0 \text{ mL} + 50 \text{ mL} = 100.0 \text{ mL} = 0.1000 \text{ L}$
$V_{final} = 50.0 \text{ mL} = 0.0500 \text{ L}$
$\Delta V = V_{final} - V_{initial} = (0.0500 \text{ L} - 0.1000 \text{ L}) = -0.0500 \text{ L}$
$w = -P\Delta V = -(1.5 \text{ atm})(-0.0500 \text{ L}) = +0.075 \text{ L} \cdot \text{atm}$

$w = (+0.075 \text{ L} \cdot \text{atm})\left(101 \dfrac{\text{J}}{1 \text{ L} \cdot \text{atm}}\right) = +7.6 \text{ J}$

The positive sign for the work indicates that the surroundings do work on the system. Energy flows into the system.

Energy and Enthalpy (Sections 8.4–8.6)

8.42 $\Delta E = q_v$ is the heat change associated with a reaction at constant volume. Since $\Delta V = 0$, no PV work is done.
$\Delta H = q_p$ is the heat change associated with a reaction at constant pressure. Since $\Delta V \neq 0$, PV work can also be done.

8.43 ΔH is negative for an exothermic reaction. ΔH is positive for an endothermic reaction.

8.44 $\Delta H = \Delta E + P\Delta V$; ΔH and ΔE are nearly equal when there are no gases involved in a chemical reaction, or, if gases are involved, $\Delta V = 0$ (that is, there are the same number of reactant and product gas molecules).

8.45 Heat is lost on going from $H_2O(g) \rightarrow H_2O(l) \rightarrow H_2O(s)$.
$H_2O(g)$ has the highest enthalpy content. $H_2O(s)$ has the lowest enthalpy content.

8.46 $P\Delta V = -7.6$ J   (from Problem 8.41)
$\Delta H = \Delta E + P\Delta V$
$\Delta E = \Delta H - P\Delta V = -0.31 \text{ kJ} - (-7.6 \times 10^{-3} \text{ kJ}) = -0.30 \text{ kJ}$

47 $\Delta H = -244$ kJ and $w = -P\Delta V = 35$ kJ; therefore $P\Delta V = -35$ kJ
$\Delta E = \Delta H - P\Delta V = -244$ kJ $- (-35$ kJ$) = -209$ kJ
For the system: $\Delta H = -244$ kJ and $\Delta E = -209$ kJ
ΔH and ΔE for the surroundings are just the opposite of what they are for the system.
For the surroundings: $\Delta H = 244$ kJ and $\Delta E = 209$ kJ

48 $\Delta H = -1256.2$ kJ/mol C_2H_2; C_2H_2, 26.04 amu
$w = -P\Delta V = -(1.00$ atm$)(-2.80$ L$) = 2.80$ L·atm
$w = (2.80$ L·atm$)\left(\dfrac{101\text{ J}}{1\text{ L·atm}}\right) = 283$ J $= 0.283$ kJ
6.50 g x $\dfrac{1\text{ mol } C_2H_2}{26.04\text{ g } C_2H_2} = 0.250$ mol C_2H_2
$q = (-1256.2$ kJ/mol$)(0.250$ mol$) = -314$ kJ
$\Delta E = \Delta H - P\Delta V = -314$ kJ $- (-0.283$ kJ$) = -314$ kJ

49 C_2H_4, 28.05 amu; HCl, 36.46 amu
$w = -P\Delta V = -(1.00$ atm$)(-71.5$ L$) = 71.5$ L·atm
$w = (71.5$ L·atm$)\left(\dfrac{101\text{ J}}{1\text{ L·atm}}\right) = 7222$ J $= 7.22$ kJ
89.5 g C_2H_4 x $\dfrac{1\text{ mol } C_2H_4}{28.05\text{ g } C_2H_4} = 3.19$ mol C_2H_4; 125 g HCl x $\dfrac{1\text{ mol HCl}}{36.46\text{ g HCl}} = 3.43$ mol HCl
Because the reaction stoichiometry between C_2H_4 and HCl is 1:1, C_2H_4 is the limiting reactant.
$\Delta H° = -72.3$ kJ/mol C_2H_4
$q = (-72.3$ kJ/mol$)(3.19$ mol$) = -231$ kJ
$\Delta E = \Delta H - P\Delta V = -231$ kJ $- (-7.22$ kJ$) = -224$ kJ

50 $C_4H_{10}O$, 74.12 amu; mass of $C_4H_{10}O = (0.7138$ g/mL$)(100$ mL$) = 71.38$ g
mol $C_4H_{10}O = 71.38$ g x $\dfrac{1\text{ mol}}{74.12\text{ g}} = 0.9630$ mol
$q = n$ x $\Delta H_{vap} = 0.9630$ mol x 26.5 kJ/mol $= 25.5$ kJ

51 Assume 100 mL of $H_2O = 100$ g; H_2O, 18.02 amu
100 g x $\dfrac{1\text{ mol } H_2O}{18.02\text{ g } H_2O}$ x $\dfrac{40.7\text{ kJ}}{1\text{ mol } H_2O} = 226$ kJ
The heat to vaporize 100 mL of H_2O is much greater than that to vaporize 100 mL of diethyl ether.

52 Al, 26.98 amu
mol Al $= 5.00$ g x $\dfrac{1\text{ mol}}{26.98\text{ g}} = 0.1853$ mol
$q = n$ x $\Delta H° = 0.1853$ mol Al x $\dfrac{-1408.4\text{ kJ}}{2\text{ mol Al}} = -131$ kJ; 131 kJ is released.

8.53   Na, 22.99 amu; $\Delta H° = -368.4$ kJ/2 mol Na $= -184.2$ kJ/mol Na

$$1.00 \text{ g Na } \times \frac{1 \text{ mol Na}}{22.99 \text{ g Na}} \times \frac{-184.2 \text{ kJ}}{1 \text{ mol Na}} = -8.01 \text{ kJ}$$

8.01 kJ of heat is evolved. The reaction is exothermic.

8.54 Fe_2O_3, 159.7 amu

$$\text{mol } Fe_2O_3 = 2.50 \text{ g} \times \frac{1 \text{ mol}}{159.7 \text{ g}} = 0.015 \ 65 \text{ mol}$$

$$q = n \times \Delta H° = 0.015 \ 65 \text{ mol } Fe_2O_3 \times \frac{-24.8 \text{ kJ}}{1 \text{ mol } Fe_2O_3} = -0.388 \text{ kJ}; \ 0.388 \text{ kJ is evolved.}$$

Because ΔH is negative, the reaction is exothermic.

8.55   CaO, 56.08 amu

$$\text{mol CaO} = 233.0 \text{ g} \times \frac{1 \text{ mol}}{56.08 \text{ g}} = 4.155 \text{ mol}$$

$$q = n \times \Delta H° = 4.155 \text{ mol CaO} \times \frac{464.6 \text{ kJ}}{1 \text{ mol CaO}} = 1930 \text{ kJ}; \ 1930 \text{ kJ is absorbed.}$$

Because ΔH is positive, the reaction is endothermic.

Calorimetry and Heat Capacity (Section 8.7)

8.56   Heat capacity is the amount of heat required to raise the temperature of a substance a given amount. Specific heat is the amount of heat necessary to raise the temperature of exactly 1 g of a substance by exactly 1 °C.

8.57   A measurement carried out in a bomb calorimeter is done at constant volume and therefore ΔE is obtained.

8.58   Na, 22.99 amu

$$\text{specific heat} = 28.2 \ \frac{J}{\text{mol} \cdot °C} \times \frac{1 \text{ mol}}{22.99 \text{ g}} = 1.23 \text{ J/(g} \cdot °C)$$

8.59 $q = $ (specific heat) x m x ΔT

$$\text{specific heat} = \frac{q}{m \times \Delta T} = \frac{89.7 \text{ J}}{(33.0 \text{ g})(5.20 \ °C)} = 0.523 \text{ J/(g} \cdot °C)$$

$C_m = [0.523 \text{ J/(g} \cdot °C)](47.88 \text{ g/mol}) = 25.0 \text{ J/(mol} \cdot °C)$

8.60   Mass of solution = 50.0 g + 1.045 g = 51.0 g

$q = $ (specific heat) x m x ΔT

$$q = \left(4.18 \ \frac{J}{\text{g} \cdot °C}\right)(51.0 \text{ g})(32.3 \ °C - 25.0 \ °C) = 1.56 \times 10^3 \text{ J} = 1.56 \text{ kJ}$$

CaO, 56.08 amu; $\text{mol CaO} = 1.045 \text{ g} \times \dfrac{1 \text{ mol}}{56.08 \text{ g}} = 0.018 \ 63 \text{ mol}$

Heat evolved per mole of CaO $= \dfrac{1.56 \text{ kJ}}{0.018 \ 63 \text{ mol}} = 83.7 \text{ kJ/mol CaO}$

Because the reaction evolves heat, the sign for ΔH is negative. $\Delta H = -83.7 \text{ kJ}$

8.61 C_6H_6, 78.11 amu; $2 \ C_6H_6(l) + 15 \ O_2(g) \rightarrow 12 \ CO_2(g) + 6 \ H_2O(g)$

$\Delta E = q_v = -q_{H_2O} = -\left(4.18 \ \dfrac{J}{g \cdot {}^\circ C}\right)(250.0 \text{ g})(7.48 \ {}^\circ C) = -7817 \text{ J} = -7.82 \text{ kJ}$

$0.187 \text{ g } C_6H_6 \times \dfrac{1 \text{ mol } C_6H_6}{78.11 \text{ g } C_6H_6} = 0.002 \ 39 \text{ mol } C_6H_6$

$\Delta E(\text{per mole}) = (-7.82 \text{ kJ})/(0.002 \ 39 \text{ mol}) = -3.27 \times 10^3 \text{ kJ/mol}$
$\Delta E(\text{per gram } C_6H_6) = (-3.27 \times 10^3 \text{ kJ/mol})/(78.11 \text{ g/mol}) = -41.9 \text{ kJ/g}$

8.62 NaOH, 40.00 amu;   HCl, 36.46 amu
$8.00 \text{ g NaOH} \times \dfrac{1 \text{ mol NaOH}}{40.00 \text{ g NaOH}} = 0.200 \text{ mol NaOH}$

$8.00 \text{ g HCl} \times \dfrac{1 \text{ mol HCl}}{36.46 \text{ g HCl}} = 0.219 \text{ mol HCl}$

Because the reaction stoichiometry between NaOH and HCl is 1:1, the NaOH is the limiting reactant.

$q_P = -q_{soln} = -(\text{specific heat}) \times m \times \Delta T = -\left(4.18 \ \dfrac{J}{g \cdot {}^\circ C}\right)(316 \text{ g})(33.5 \ {}^\circ C - 25.0 \ {}^\circ C) = -11.2 \text{ kJ}$

$\Delta H = q_p/n = (-11.2 \text{ kJ})/(0.200 \text{ mol}) = -56 \text{ kJ/mol}$
When 10.00 g of HCl in 248.0 g of water is added, the same temperature increase is observed because the mass of NaOH is the same and it is still the limiting reactant. The mass of the solution is also the same.

8.63 NH_4NO_3, 80.04 amu;   assume 125 mL = 125 g H_2O
$50.0 \text{ g } NH_4NO_3 \times \dfrac{1 \text{ mol } NH_4NO_3}{80.04 \text{ g } NH_4NO_3} = 0.625 \text{ mol } NH_4NO_3$

$q_p = \Delta H \times n = (+25.7 \text{ kJ/mol})(0.625 \text{ mol}) = 16.1 \text{ kJ} = 16,100 \text{ J}$
$q_{soln} = -q_p = -16,100 \text{ J}$
$q_{soln} = (\text{specific heat}) \times m \times \Delta T$

$\Delta T = \dfrac{q_{soln}}{(\text{specific heat}) \times m} = \dfrac{-16,100 \text{ J}}{\left(4.18 \ \dfrac{J}{g \cdot {}^\circ C}\right)(50 \text{ g} + 125 \text{ g})} = -22.0 \ {}^\circ C$

$\Delta T = -22.0 \ {}^\circ C = T_{final} - T_{initial} = T_{final} - 25.0 \ {}^\circ C$
$T_{final} = -22.0 \ {}^\circ C + 25.0 \ {}^\circ C = 3.0 \ {}^\circ C$

Hess's Law and Heats of Formation (Sections 8.8–8.9)

8.64   The standard state of an element is its most stable form at 1 atm and the specified temperature, usually 25 °C.

8.65   A compound's standard heat of formation is the amount of heat associated with the formation of 1 mole of a compound from its elements (in their standard states).

8.66   Hess's Law – the overall enthalpy change for a reaction is equal to the sum of the enthalpy changes for the individual steps in the reaction.
Hess's Law works because of the law of conservation of energy.

8.67   Elements always have $\Delta H^{\circ}_f = 0$ because the standard state of elements is the reference point from which all enthalpy changes are measured.

8.68
$$S(s) + O_2(g) \rightarrow SO_2(g) \qquad \Delta H^{\circ}_1 = -296.8 \text{ kJ}$$
$$\underline{SO_2 + \tfrac{1}{2} O_2(g) \rightarrow SO_3(g)} \qquad \Delta H^{\circ}_2 = -98.9 \text{ kJ}$$

Sum $S(s) + 3/2\, O_2(g) \rightarrow SO_3(g) \qquad \Delta H^{\circ}_3 = \Delta H^{\circ}_1 + \Delta H^{\circ}_2$

$\Delta H^{\circ}_f = \Delta H^{\circ}_3 = -296.8 \text{ kJ} + (-98.9 \text{ kJ}) = -395.7 \text{ kJ/mol}$

8.69 $\Delta H^{\circ}_{rxn} = [12\, \Delta H^{\circ}_f(CO_2) + 6\, \Delta H^{\circ}_f(H_2O)] - [2\, \Delta H^{\circ}_f(C_6H_6)]$
$-6534 \text{ kJ} = [(12 \text{ mol})(-393.5 \text{ kJ/mol}) + (6 \text{ mol})(-285.8 \text{ kJ/mol})] - [(2 \text{ mol})(\Delta H^{\circ}_f(C_6H_6))]$
Solve for $\Delta H^{\circ}_f(C_6H_6)$.
$-6534 \text{ kJ} = -6436.8 \text{ kJ} - [(2 \text{ mol})(\Delta H^{\circ}_f(C_6H_6))]; \qquad 97.2 \text{ kJ} = (2 \text{ mol})(\Delta H^{\circ}_f(C_6H_6))$
$\Delta H^{\circ}_f(C_6H_6) = +48.6 \text{ kJ/mol}$

8.70
$$SO_3(g) + H_2O(l) \rightarrow H_2SO_4(aq) \qquad \Delta H^{\circ}_1 = -227.8 \text{ kJ}$$
$$H_2(g) + \tfrac{1}{2} O_2(g) \rightarrow H_2O(l) \qquad \Delta H^{\circ}_2 = \Delta H^{\circ}_f = -285.8 \text{ kJ}$$
$$\underline{S(s) + 3/2\, O_2(g) \rightarrow SO_3(g)} \qquad \Delta H^{\circ}_3 = \Delta H^{\circ}_f = -395.7$$

Sum $S(s) + H_2(g) + 2\, O_2(g) \rightarrow H_2SO_4(aq) \qquad \Delta H^{\circ}_f(H_2SO_4) = ?$

$\Delta H^{\circ}_f(H_2SO_4) = \Delta H^{\circ}_1 + \Delta H^{\circ}_2 + \Delta H^{\circ}_3 = -909.3 \text{ kJ}$

8.71 $\Delta H^{\circ}_{rxn} = [\Delta H^{\circ}_f(CH_3CO_2H) + \Delta H^{\circ}_f(H_2O)] - \Delta H^{\circ}_f(CH_3CH_2OH)$
$\Delta H^{\circ}_{rxn} = [(1 \text{ mol})(-484.5 \text{ kJ/mol}) + (1 \text{mol})(-285.8 \text{ kJ/mol})] - [(1 \text{ mol})(-277.7 \text{ kJ/mol})]$
$\Delta H^{\circ}_{rxn} = -492.6 \text{ kJ}$

8.72 $C_8H_8(l) + 10\, O_2(g) \rightarrow 8\, CO_2(g) + 4\, H_2O(l)$
$\Delta H^{\circ}_{rxn} = \Delta H^{\circ}_c = -4395 \text{ kJ}$
$\Delta H^{\circ}_{rxn} = [8\, \Delta H^{\circ}_f(CO_2) + 4\, \Delta H^{\circ}_f(H_2O)] - \Delta H^{\circ}_f(C_8H_8)$
$-4395 \text{ kJ} = [(8 \text{ mol})(-393.5 \text{ kJ/mol}) + (4 \text{ mol})(-285.8 \text{ kJ/mol})] - [(1 \text{ mol})(\Delta H^{\circ}_f(C_8H_8))]$
Solve for $\Delta H^{\circ}_f(C_8H_8)$
$-4395 \text{ kJ} = -4291.2 \text{ kJ} - (1 \text{ mol})(\Delta H^{\circ}_f(C_8H_8)); \qquad -103.8 \text{ kJ} = -(1 \text{ mol})(\Delta H^{\circ}_f(C_8H_8))$
$\Delta H^{\circ}_f(C_8H_8) = \dfrac{-103.8 \text{ kJ}}{-1 \text{ mol}} = +103.8 \text{ kJ/mol}$

8.73 $C_5H_{12}O(l) + 15/2\ O_2(g) \rightarrow 5\ CO_2(g) + 6\ H_2O(l)$

$\Delta H^\circ_{rxn} = [5\ \Delta H^\circ_f(CO_2) + 6\ \Delta H^\circ_f(H_2O)] - \Delta H^\circ_f(C_5H_{12}O)$

$\Delta H^\circ_{rxn} = [(5\ mol)(-393.5\ kJ/mol) + (6\ mol)(-285.8\ kJ/mol)] - [(1\ mol)(-313.6\ kJ/mol)]$

$\Delta H^\circ_{rxn} = -3369\ kJ$

8.74 $\Delta H^\circ_{rxn} = \Delta H^\circ_f(MTBE) - [\Delta H^\circ_f(\text{2-Methylpropene}) + \Delta H^\circ_f(CH_3OH)]$

$-57.5\ kJ = -313.6\ kJ - [(1\ mol)(\Delta H^\circ_f(\text{2-Methylpropene})) + (-239.2\ kJ)]$

Solve for ΔH°_f(2-Methylpropene).

$-16.9\ kJ = (1\ mol)(\Delta H^\circ_f(\text{2-Methylpropene}))$

$\Delta H^\circ_f(\text{2-Methylpropene}) = -16.9\ kJ/mol$

8.75 $C_{51}H_{88}O_6(l) + 70\ O_2(g) \rightarrow 51\ CO_2(g) + 44\ H_2O(l)$

$\Delta H^\circ_{rxn} = [51\ \Delta H^\circ_f(CO_2) + 44\ \Delta H^\circ_f(H_2O)] - \Delta H^\circ_f(C_{51}H_{88}O_6)$

$\Delta H^\circ_{rxn} = [(51\ mol)(-393.5\ kJ/mol) + (44\ mol)(-285.8\ kJ/mol)] - [(1\ mol)(-1310\ kJ/mol)]$

$\Delta H^\circ_{rxn} = -3.133 \times 10^4\ kJ/mol\ C_{51}H_{88}O_6$

$C_{51}H_{88}O_6$, 797.25 amu

$$q = -3.133 \times 10^4\ \frac{kJ}{mol} \times \frac{1\ mol}{797.25\ g} \times 0.94\ \frac{g}{mL} = -37\ kJ/mL\ ;\ 37\ kJ\ \text{released per mL}$$

Bond Dissociation Energies (Section 8.10)

8.76 $H_2C=CH_2(g) + H_2(g) \rightarrow CH_3CH_3(g)$

$\Delta H^\circ_{rxn} = D\ (\text{Reactant bonds}) - D\ (\text{Product bonds})$

$\Delta H^\circ_{rxn} = (D_{C=C} + 4\ D_{C-H} + D_{H-H}) - (6\ D_{C-H} + D_{C-C})$

$\Delta H^\circ_{rxn} = [(1\ mol)(611\ kJ/mol) + (4\ mol)(410\ kJ/mol) + (1\ mol)(436\ kJ/mol)]$

$\quad - [(6\ mol)(410\ kJ/mol) + (1\ mol)(350\ kJ/mol)] = -123\ kJ$

8.77 $CH_3CH=CH_2 + H_2O \rightarrow CH_3CH(OH)CH_3$

$\Delta H^\circ_{rxn} = D\ (\text{Reactant bonds}) - D\ (\text{Product bonds})$

$\Delta H^\circ_{rxn} = (D_{C=C} + D_{C-C} + 6\ D_{C-H} + 2\ D_{O-H}) - (2\ D_{C-C} + 7\ D_{C-H} + D_{C-O} + D_{O-H})$

$\Delta H^\circ_{rxn} = [(1\ mol)(611\ kJ/mol) + (1\ mol)(350\ kJ/mol) + (6\ mol)(410\ kJ/mol)$

$+ (2\ mol)(460\ kJ/mol)] - [(2\ mol)(350\ kJ/mol) + (7\ mol)(410\ kJ/mol)$

$+ (1\ mol)(350\ kJ/mol) + (1\ mol)(460\ kJ/mol)] = -39\ kJ$

8.78 $C_4H_{10} + 13/2\ O_2 \rightarrow 4\ CO_2 + 5\ H_2O$

$\Delta H^\circ_{rxn} = D\ (\text{Reactant bonds}) - D\ (\text{Product bonds})$

$\Delta H^\circ_{rxn} = (3\ D_{C-C} + 10\ D_{C-H} + 13/2\ D_{O=O}) - (8\ D_{C=O} + 10\ D_{O-H})$

$\Delta H^\circ_{rxn} = [(3\ mol)(350\ kJ/mol) + (10\ mol)(410\ kJ/mol) + (13/2\ mol)(498\ kJ/mol)]$

$\quad - [(8\ mol)(804\ kJ/mol) + (10\ mol)(460\ kJ/mol)] = -2645\ kJ$

8.79 $CH_3CO_2H + CH_3CH_2OH \rightarrow CH_3CO_2CH_2CH_3 + H_2O$

$\Delta H^\circ_{rxn} = D\ (\text{Reactant bonds}) - D\ (\text{Product bonds})$

$\Delta H^\circ_{rxn} = (D_{C=O} + 2\ D_{C-O} + 8\ D_{C-H} + 2\ D_{O-H}) - (D_{C=O} + 2\ D_{C-O} + 8\ D_{C-H} + 2\ D_{O-H}) = 0\ kJ$

Free Energy and Entropy (Sections 8.12–8.13)

8.80   Entropy is a measure of molecular randomness.

8.81 $\Delta G = \Delta H - T\Delta S$; ΔH is usually more important because it is usually much larger than $T\Delta S$.

8.82   A reaction can be spontaneous yet endothermic if ΔS is positive (more randomness) and the $T\Delta S$ term is larger than ΔH.

8.83   A reaction can be nonspontaneous yet exothermic if ΔS is negative (less randomness) and the temperature is high enough so that the $T\Delta S$ term is more negative than ΔH.

8.84   (a) positive   (more randomness)   (b) negative   (less randomness)

8.85   (a) positive   (more randomness)   (b) negative   (less randomness)
        (c) positive   (more randomness)

8.86   (a) zero   (equilibrium)   (b) zero   (equilibrium)
        (c) negative   (spontaneous)

8.87   Because the mixing of gas molecules is spontaneous, ΔG is negative. The mixture of gas molecules is more random so ΔS is positive. For the diffusion of gases, ΔH is approximately zero.

8.88 ΔS is positive. The reaction increases the total number of molecules.

8.89 $\Delta S < 0$. The reaction decreases the number of gas molecules.

8.90 $\Delta G = \Delta H - T\Delta S$
        (a) $\Delta G = -48 \text{ kJ} - (400 \text{ K})(135 \times 10^{-3} \text{ kJ/K}) = -102 \text{ kJ}$
        $\Delta G < 0$, spontaneous; $\Delta H < 0$, exothermic.
        (b) $\Delta G = -48 \text{ kJ} - (400 \text{ K})(-135 \times 10^{-3} \text{ kJ/K}) = +6 \text{ kJ}$
        $\Delta G > 0$, nonspontaneous; $\Delta H < 0$, exothermic.
        (c) $\Delta G = +48 \text{ kJ} - (400 \text{ K})(135 \times 10^{-3} \text{ kJ/K}) = -6 \text{ kJ}$
        $\Delta G < 0$, spontaneous; $\Delta H > 0$, endothermic.
        (d) $\Delta G = +48 \text{ kJ} - (400 \text{ K})(-135 \times 10^{-3} \text{ kJ/K}) = +102 \text{ kJ}$
        $\Delta G > 0$, nonspontaneous; $\Delta H > 0$, endothermic.

8.91 $\Delta G = \Delta H - T\Delta S$
        (a) $\Delta G = -128 \text{ kJ} - (500 \text{ K})(35 \times 10^{-3} \text{ kJ/K}) = -146 \text{ kJ}$
        $\Delta G < 0$, spontaneous; $\Delta H < 0$, exothermic
        (b) $\Delta G = +67 \text{ kJ} - (250 \text{ K})(-140 \times 10^{-3} \text{ kJ/K}) = +102 \text{ kJ}$
        $\Delta G > 0$, nonspontaneous; $\Delta H > 0$, endothermic
        (c) $\Delta G = +75 \text{ kJ} - (800 \text{ K})(95 \times 10^{-3} \text{ kJ/K}) = -1 \text{ kJ}$
        $\Delta G < 0$, spontaneous; $\Delta H > 0$, endothermic

8.92 $\Delta G = \Delta H - T\Delta S$;   Set $\Delta G = 0$ and solve for T (the crossover temperature).

$$T = \frac{\Delta H}{\Delta S} = \frac{-33 \text{ kJ}}{-0.058 \text{ kJ/K}} = 570 \text{ K}$$

8.93   Because $\Delta H > 0$ and $\Delta S < 0$, the reaction is nonspontaneous at all temperatures. There is no crossover temperature.

8.94   (a) $\Delta H < 0$ and $\Delta S > 0$; reaction is spontaneous at all temperatures.
(b) $\Delta H < 0$ and $\Delta S < 0$; reaction has a crossover temperature.
(c) $\Delta H > 0$ and $\Delta S > 0$; reaction has a crossover temperature.
(d) $\Delta H > 0$ and $\Delta S < 0$; reaction is nonspontaneous at all temperatures.

8.95   (a) $\Delta H < 0$ and $\Delta S < 0$. The reaction is favored by enthalpy but not by entropy.
$\Delta G° = \Delta H° - T\Delta S° = -217.5 \text{ kJ/mol} - (298 \text{ K})[-233.9 \times 10^{-3} \text{ kJ/(K} \cdot \text{mol)}] = -147.8 \text{ kJ}$
(b) The reaction has a crossover temperature. Set $\Delta G = 0$ and solve for T (the crossover temperature).
$\Delta G° = 0 = \Delta H° - T\Delta S°$

$$T = \frac{\Delta H°}{\Delta S°} = \frac{217.5 \text{ kJ/mol}}{233.9 \times 10^{-3} \text{ kJ/(K} \cdot \text{mol)}} = 929.9 \text{ K}$$

8.96 $T = -114.1 \text{ °C} = 273.15 + (-114.1) = 159.0 \text{ K}$
$\Delta G_{fus} = \Delta H_{fus} - T\Delta S_{fus}$; $\Delta G = 0$ at the melting point temperature.
Set $\Delta G = 0$ and solve for ΔS_{fus}.
$\Delta G = 0 = \Delta H_{fus} - T\Delta S_{fus}$

$$\Delta S_{fus} = \frac{\Delta H_{fus}}{T} = \frac{5.02 \text{ kJ/mol}}{159.0 \text{ K}} = 0.0316 \text{ kJ/(K·mol)} = 31.6 \text{ J/(K·mol)}$$

8.97 $T = 61.2 \text{ °C} = 273.15 + (61.2) = 334.4 \text{ K}$
$\Delta G_{vap} = \Delta H_{vap} - T\Delta S_{vap}$; $\Delta G = 0$ at the boiling point temperature.
Set $\Delta G = 0$ and solve for ΔS_{vap}.
$\Delta G = 0 = \Delta H_{vap} - T\Delta S_{vap}$

$$\Delta S_{vap} = \frac{\Delta H_{vap}}{T} = \frac{29.2 \text{ kJ/mol}}{334.4 \text{ K}} = 0.0873 \text{ kJ/(K·mol)} = 87.3 \text{ J/(K·mol)}$$

Chapter Problems

8.98 $Mg(s) + 2 HCl(aq) \rightarrow MgCl_2(aq) + H_2(g)$

$$\text{mol Mg} = 1.50 \text{ g} \times \frac{1 \text{ mol}}{24.3 \text{ g}} = 0.0617 \text{ mol Mg}$$

$$\text{mol HCl} = 0.200 \text{ L} \times 6.00 \frac{\text{mol}}{\text{L}} = 1.20 \text{ mol HCl}$$

There is an excess of HCl. Mg is the limiting reactant.

$$q = \left(4.18 \ \frac{J}{g \cdot {}^{\circ}C}\right)(200 \ g)(42.9 \ {}^{\circ}C - 25.0 \ {}^{\circ}C) + \left(776 \ \frac{J}{{}^{\circ}C}\right)(42.9 \ {}^{\circ}C - 25.0 \ {}^{\circ}C) = 2.89 \times 10^4 \ J$$

$$q = 2.89 \times 10^4 \ J \times \frac{1 \ kJ}{1000 \ J} = 28.9 \ kJ$$

Heat evolved per mole of Mg $= \dfrac{28.9 \ kJ}{0.0617 \ mol} = 468 \ kJ/mol$

Because the reaction evolves heat, the sign for ΔH is negative. $\Delta H = -468 \ kJ$

8.99 (a) $C(s) + CO_2(g) \rightarrow 2 \ CO(g)$
$\Delta H^{\circ}_{rxn} = [2 \ \Delta H^{\circ}_f(CO)] - \Delta H^{\circ}_f(CO_2)$
$\Delta H^{\circ}_{rxn} = [(2 \ mol)(-110.5 \ kJ/mol)] - [(1 \ mol)(-393.5 \ kJ/mol)] = +172.5 \ kJ$

(b) $2 \ H_2O_2(aq) \rightarrow 2 \ H_2O(l) + O_2(g)$
$\Delta H^{\circ}_{rxn} = [2 \ \Delta H^{\circ}_f(H_2O)] - [2 \ \Delta H^{\circ}_f(H_2O_2)]$
$\Delta H^{\circ}_{rxn} = [(2 \ mol)(-285.8 \ kJ/mol)] - [(2 \ mol)(-191.2 \ kJ/mol)] = -189.2 \ kJ$

(c) $Fe_2O_3(s) + 3 \ CO(g) \rightarrow 2 \ Fe(s) + 3 \ CO_2(g)$
$\Delta H^{\circ}_{rxn} = [3 \ \Delta H^{\circ}_f(CO_2)] - [\Delta H^{\circ}_f(Fe_2O_3) + 3 \ \Delta H^{\circ}_f(CO)]$
$\Delta H^{\circ}_{rxn} = [(3 \ mol)(-393.5 \ kJ/mol)]$
$\qquad - [(1 \ mol)(-824.2 \ kJ/mol) + (3 \ mol)(-110.5 \ kJ/mol)] = -24.8 \ kJ$

8.100

$2 \ NO(g) + O_2(g) \rightarrow 2 \ NO_2(g)$	$\Delta H^{\circ}_1 = 2(-58.1 \ kJ)$
$\underline{2 \ NO_2(g) \rightarrow N_2O_4(g)}$	$\Delta H^{\circ}_2 = -55.3 \ kJ$

Sum $\quad 2 \ NO(g) + O_2(g) \rightarrow N_2O_4(g)$
$\Delta H^{\circ} = \Delta H^{\circ}_1 + \Delta H^{\circ}_2 = -171.5 \ kJ$

8.101 $\Delta G = \Delta H - T\Delta S$; at equilibrium $\Delta G = 0$. Set $\Delta G = 0$ and solve for T.

$$\Delta G = 0 = \Delta H - T\Delta S; \quad T = \frac{\Delta H}{\Delta S} = \frac{30.91 \ kJ/mol}{93.2 \times 10^{-3} \ kJ/(K \cdot mol)} = 332 \ K = 59 \ {}^{\circ}C$$

8.102 $\Delta G_{fus} = \Delta H_{fus} - T\Delta S_{fus}$; at the melting point $\Delta G = 0$. Set $\Delta G = 0$ and solve for T (the melting point).

$$\Delta G = 0 = \Delta H_{fus} - T\Delta S_{fus}; \quad T = \frac{\Delta H_{fus}}{\Delta S_{fus}} = \frac{9.95 \ kJ}{0.0357 \ kJ/K} = 279 \ K$$

8.103 $HgS(s) + O_2(g) \rightarrow Hg(l) + SO_2(g)$
(a) $\Delta H^{\circ}_{rxn} = \Delta H^{\circ}_f(SO_2) - \Delta H^{\circ}_f(HgS)$
$\Delta H^{\circ}_{rxn} = [(1 \ mol)(-296.8 \ kJ/mol)] - [(1 \ mol)(-58.2 \ kJ/mol)] = -238.6 \ kJ$
(b) and (c) Because $\Delta H < 0$ and $\Delta S > 0$, the reaction is spontaneous at <u>all</u> temperatures.

8.104 $\Delta H^{\circ}_{rxn} = D$ (Reactant bonds) $- D$ (Product bonds)
(a) $2 \ CH_4(g) \rightarrow C_2H_6(g) + H_2(g)$
$\Delta H^{\circ}_{rxn} = (8 \ D_{C-H}) - (D_{C-C} + 6 \ D_{C-H} + D_{H-H})$
$\Delta H^{\circ}_{rxn} = [(8 \ mol)(410 \ kJ/mol)] - [(1 \ mol)(350 \ kJ/mol) + (6 \ mol)(410 \ kJ/mol)$
$+ (1 \ mol)(436 \ kJ/mol)] = +34 \ kJ$

(b) $C_2H_6(g) + F_2(g) \rightarrow C_2H_5F(g) + HF(g)$

$\Delta H°_{rxn} = (6\,D_{C-H} + D_{C-C} + D_{F-F}) - (5\,D_{C-H} + D_{C-C} + D_{C-F} + D_{H-F})$

$\Delta H°_{rxn} = [(6\text{ mol})(410\text{ kJ/mol}) + (1\text{ mol})(350\text{ kJ/mol}) + (1\text{ mol})(159\text{ kJ/mol})]$

$\qquad - [(5\text{ mol})(410\text{ kJ/mol}) + (1\text{ mol})(350\text{ kJ/mol}) + (1\text{ mol})(450\text{ kJ/mol})$

$\qquad\qquad + (1\text{ mol})(570\text{ kJ/mol})] = -451\text{ kJ}$

(c) $N_2(g) + 3\,H_2(g) \rightarrow 2\,NH_3(g)$

The bond dissociation energy for N_2 is 945 kJ/mol.

$\Delta H°_{rxn} = (D_{N_2} + 3\,D_{H-H}) - (6\,D_{N-H})$

$\Delta H°_{rxn} = [(1\text{ mol})(945\text{ kJ/mol}) + (3\text{ mol})(436\text{ kJ/mol})] - [(6\text{ mol})(390\text{ kJ/mol})] = -87\text{ kJ}$

8.105 (a) $\Delta H°_{rxn} = \Delta H°_f(CH_3OH) - \Delta H°_f(CO)$

$\Delta H°_{rxn} = [(1\text{ mol})(-238.7\text{ kJ/mol})] - [(1\text{ mol})(-110.5\text{ kJ/mol})] = -128.2\text{ kJ}$

(b) $\Delta G° = \Delta H° - T\Delta S° = -128.2\text{ kJ} - (298\text{ K})(-332 \times 10^{-3}\text{ kJ/K}) = -29.3\text{ kJ}$

(c) Step 1 is spontaneous since $\Delta G° < 0$.

(d) $\Delta H°$, because it is larger than $T\Delta S°$.

(e) Set $\Delta G = 0$ and solve for T.

$\Delta G = 0 = \Delta H - T\Delta S$

$T = \dfrac{\Delta H}{\Delta S} = \dfrac{128.2\text{ kJ}}{332 \times 10^{-3}\text{ kJ/K}} = 386\text{ K};$ The reaction is spontaneous below 386 K.

(f) $\Delta H°_{rxn} = \Delta H°_f(CH_4) - \Delta H°_f(CH_3OH)$

$\Delta H°_{rxn} = [(1\text{ mol})(-74.8\text{ kJ/mol})] - [(1\text{ mol})(-238.7\text{ kJ/mol})] = +163.9\text{ kJ}$

(g) $\Delta G° = \Delta H° - T\Delta S° = +163.9\text{ kJ} - (298\text{ K})(162 \times 10^{-3}\text{ kJ/K}) = +115.6\text{ kJ}$

(h) Step 2 is nonspontaneous since $\Delta G° > 0$.

(i) $\Delta H°$, because it is larger than $T\Delta S°$.

(j) Set $\Delta G = 0$ and solve for T.

$\Delta G = 0 = \Delta H - T\Delta S$

$T = \dfrac{\Delta H}{\Delta S} = \dfrac{163.9\text{ kJ}}{162 \times 10^{-3}\text{ kJ/K}} = 1012\text{ K};$ The reaction is spontaneous above 1012 K.

(k) $\Delta G°_{overall} = \Delta G°_1 + \Delta G°_2 = -29.3\text{ kJ} + 115.6\text{ kJ} = +86.3\text{ kJ}$

$\quad \Delta H°_{overall} = \Delta H°_1 + \Delta H°_2 = -128.2\text{ kJ} + 163.9\text{ kJ} = +35.7\text{ kJ}$

$\quad \Delta S°_{overall} = \Delta S°_1 + \Delta S°_2 = -332\text{ J/K} + 162\text{ J/K} = -170\text{ J/K}$

(l) The overall reaction is nonspontaneous since $\Delta G°_{overall} > 0$ at 298 K.

(m) The two reactions should be run separately. Run step 1 below 386 K and run step 2 above 1012 K.

8.106 (a) $2\,C_8H_{18}(l) + 25\,O_2(g) \rightarrow 16\,CO_2(g) + 18\,H_2O(l)$

(b) $C_8H_{18}(l) + 25/2\,O_2(g) \rightarrow 8\,CO_2(g) + 9\,H_2O(l)$

$\Delta H°_{rxn} = \Delta H°_c = -5461\text{ kJ}$

$\Delta H°_{rxn} = [8\,\Delta H°_f(CO_2) + 9\,\Delta H°_f(H_2O)] - \Delta H°_f(C_8H_{18})$

$-5461\text{ kJ} = [(8\text{ mol})(-393.5\text{ kJ/mol}) + (9\text{ mol})(-285.8\text{ kJ/mol})] - [(1\text{ mol})(\Delta H°_f(C_8H_{18}))]$

Solve for $\Delta H°_f(C_8H_{18})$.

$-5461\text{ kJ} = -5720.2\text{ kJ} - [(1\text{ mol})(\Delta H°_f(C_8H_{18}))]$

$259\text{ kJ} = -(1\text{ mol})(\Delta H°_f(C_8H_{18}))$

$\Delta H°_f(C_8H_{18}) = -259\text{ kJ/mol}$

8.107 Assume 1.00 kg of H_2O; $1 \text{ kg}\cdot\text{m}^2/\text{s}^2 = 1$ J

$E_p = (1.00 \text{ kg})(9.81 \text{ m/s}^2)(739 \text{ m}) = 7250 \text{ kg}\cdot\text{m}^2/\text{s}^2 = 7250$ J

q = specific heat x m x ΔT

$$\Delta T = \frac{q}{m \text{ x specific heat}} = \frac{7250 \text{ J}}{(1000 \text{ g})(4.18 \text{ J/(g}\cdot\text{°C)})} = 1.73 \text{ °C (temperature rise)}$$

8.108 (a) $\Delta S_{total} = \Delta S_{system} + \Delta S_{surr}$ and $\Delta S_{surr} = -\Delta H/T$

$\Delta S_{total} = \Delta S_{system} + (-\Delta H/T) = \Delta S_{system} - \Delta H/T$

$\Delta S_{system} = \Delta S_{total} + \Delta H/T$

$\Delta G = \Delta H - T\Delta S$ (substitute ΔS_{system} for ΔS in this equation)

$\Delta G = \Delta H - T(\Delta S_{total} + \Delta H/T) = -T\Delta S_{total}$

$\Delta G = -T\Delta S_{total}$ For a spontaneous reaction, if $\Delta S_{total} > 0$ then $\Delta G < 0$.

(b) $\Delta G° = \Delta H° - T\Delta S°$

$\Delta H° = \Delta G° + T\Delta S°$

$$\Delta S_{surr} = -\frac{\Delta H°}{T} = -\frac{[\Delta G° + T\Delta S°]}{T} = -\frac{[2879 \times 10^3 \text{ J/mol} + (298 \text{ K})(-210 \text{ J/(K}\cdot\text{mol))}]}{298 \text{ K}}$$

$\Delta S_{surr} = -9451 \text{ J/(K}\cdot\text{mol)}$

8.109

$3/2 \text{ NO}_2(g) + 1/2 \text{ H}_2O(l) \rightarrow \text{HNO}_3(aq) + 1/2 \text{ NO}(g)$	$\Delta H°_1 = \dfrac{-137.3 \text{ kJ}}{2}$
$3/2 \text{ NO}(g) + 3/4 \text{ O}_2(g) \rightarrow 3/2 \text{ NO}_2(g)$	$\Delta H°_2 = \dfrac{(3)(-116.2 \text{ kJ})}{4}$
$1/2 \text{ N}_2(g) + 3/2 \text{ H}_2(g) \rightarrow \text{NH}_3(g)$	$\Delta H°_3 = -46.1 \text{ kJ}$
$\text{NH}_3(g) + 5/4 \text{ O}_2(g) \rightarrow \text{NO}(g) + 6/4 \text{ H}_2O(l)$	$\Delta H°_4 = \dfrac{-1165.2 \text{ kJ}}{4}$
$\underline{\text{H}_2O(l) \rightarrow 1/2 \text{ O}_2(g) + \text{H}_2(g)}$	$\underline{\Delta H°_5 = +285.8 \text{ kJ}}$
Sum $\;1/2 \text{ H}_2(g) + 1/2 \text{ N}_2(g) + 3/2 \text{ O}_2(g) \rightarrow \text{HNO}_3(aq)$	$\Delta H° = -207.4 \text{ kJ}$

8.110

$2 \text{ CH}_4(g) + 4 \text{ O}_2(g) \rightarrow 2 \text{ CO}_2(g) + 4 \text{ H}_2O(l)$	$\Delta H°_1 = 2(-890.3 \text{ kJ})$
$\text{C}_2\text{H}_6(g) \rightarrow \text{C}_2\text{H}_4(g) + \text{H}_2(g)$	$\Delta H°_2 = +136.3 \text{ kJ}$
$2 \text{ CO}_2(g) + 3 \text{ H}_2O(l) \rightarrow \text{C}_2\text{H}_6(g) + 7/2 \text{ O}_2(g)$	$\Delta H°_3 = \dfrac{3120.8 \text{ kJ}}{2}$
$\underline{\text{H}_2O(l) \rightarrow \text{H}_2(g) + 1/2 \text{ O}_2(g)}$	$\underline{\Delta H°_4 = +285.8 \text{ kJ}}$
Sum $\;2 \text{ CH}_4(g) \rightarrow \text{C}_2\text{H}_4(g) + 2 \text{ H}_2(g)$	$\Delta H° = +201.9 \text{ kJ}$

8.111 $q_{Mo} = (110.0 \text{ g})(\text{specific heat Mo})(28.0 \text{ °C} - 100.0 \text{ °C})$

$q_{H_2O} = (150.0 \text{ g})[4.184 \text{ J/(g}\cdot\text{°C)}](28.0 \text{ °C} - 24.6 \text{ °C})$

$q_{Mo} = -q_{H_2O}$

$(110.0 \text{ g})(\text{specific heat Mo})(28.0 \text{ °C} - 100.0 \text{ °C}) = -(150.0 \text{ g})[4.184 \text{ J/(g}\cdot\text{°C)}](28.0 \text{ °C} - 24.6 \text{ °C})$

$$\text{specific heat Mo} = \frac{-(150.0 \text{ g})[4.184 \text{ J/(g}\cdot\text{°C)}](28.0 \text{ °C} - 24.6 \text{ °C})}{(110.0 \text{ g})(28.0 \text{ °C} - 100.0 \text{ °C})} = 0.27 \text{ J/(g}\cdot\text{°C)}$$

8.112 $q_{ice\ tea} = -q_{ice}$

$q_{ice\ tea} = (4.18\ \dfrac{J}{g \cdot {}^{\circ}C})(400.0\ g)(10.0\ {}^{\circ}C - 80.0\ {}^{\circ}C) = -1.17 \times 10^5\ J$

H_2O, 18.02 amu

$q_{ice} = 1.17 \times 10^5\ J = (6.01\ kJ/mol)\left(\dfrac{1000\ J}{1\ kJ}\right)\left(m_{ice} \times \dfrac{1\ mol\ H_2O}{18.02\ g\ H_2O}\right)$

$+\left(4.18\ \dfrac{J}{(g \cdot {}^{\circ}C)}\right)(m_{ice})(10.0\ {}^{\circ}C - 0.0\ {}^{\circ}C)$

Solve for the mass of ice, m_{ice}.

$1.17 \times 10^5\ J = (3.34 \times 10^2\ J/g)(m_{ice}) + (41.8\ J/g)(m_{ice}) = (3.76 \times 10^2\ J/g)(m_{ice})$

$m_{ice} = \dfrac{1.17 \times 10^5\ J}{3.76 \times 10^2\ J/g} = 311\ g$ of ice

8.113 There is a large excess of NaOH.   5.00 mL = 0.005 00 L
mol citric acid = (0.005 00 L)(0.64 mol/L) = 0.0032 mol citric acid

$q_{H_2O} = (51.6\ g)[4.0\ J/(g \cdot {}^{\circ}C)](27.9\ {}^{\circ}C - 26.0\ {}^{\circ}C) = 392\ J$

$q_{rxn} = -q_{H_2O} = -392\ J$

$\Delta H = -\dfrac{392\ J}{0.0032\ mol} \times \dfrac{1\ kJ}{1000\ J} = -123\ kJ/mol = -120\ kJ/mol$ citric acid

8.114 $CsOH(aq) + HCl(aq) \rightarrow CsCl(aq) + H_2O(l)$

mol CsOH = 0.100 L $\times \dfrac{0.200\ mol\ CsOH}{1.00\ L} = 0.0200\ mol\ CsOH$

mol HCl = 0.050 L $\times \dfrac{0.400\ mol\ HCl}{1.00\ L} = 0.0200\ mol\ HCl$

The reactants were mixed in equal mole amounts.
Total volume = 150 mL and has a mass of 150 g.

$q_{solution} = \left(4.2\ \dfrac{J}{(g \cdot {}^{\circ}C)}\right)(150\ g)(24.28\ {}^{\circ}C - 22.50\ {}^{\circ}C) = 1121\ J$

$q_{reaction} = -q_{solution} = -1121\ J$

$\Delta H = \dfrac{q_{reaction}}{mol\ CsOH} = \dfrac{-1121\ J}{0.0200\ mol\ CsOH} \times \dfrac{1\ kJ}{1000\ J} = -56\ kJ/mol\ CsOH$

8.115 $NaNO_3$, 84.99 amu; KF, 58.10 amu

For $NaNO_3(s) \rightarrow NaNO_3(aq)$, q = 20.4 kJ/mol $\times \dfrac{1\ mol\ NaNO_3}{84.99\ g\ NaNO_3} = 0.240\ kJ/g$

For $KF(s) \rightarrow KF(aq)$, q = −17.7 kJ/mol $\times \dfrac{1\ mol\ KF}{58.10\ g\ KF} = -0.305\ kJ/g$

$q_{soln} = (110.0\ g)[4.18\ J/(g \cdot {}^{\circ}C)](2.22\ {}^{\circ}C) = 1021\ J = 1.02\ kJ$

$q_{rxn} = -q_{soln} = -1.02$ kJ

Let X = mass of $NaNO_3$ and Y = mass of KF

X + Y = 10.0 g, so Y = 10.0 g − X

$q_{rxn} = -1.02$ kJ = X(0.240 kJ/g) + Y(− 0.305 kJ/g)   (substitute for Y and solve for X)

−1.02 kJ = X(0.240 kJ/g) + (10.0 g − X)(− 0.305 kJ/g)

−1.02 kJ = (0.240 kJ)X − 3.05 kJ + (0.305 kJ)X

2.03 kJ = (0.545 kJ)X

$X = \dfrac{2.03 \text{ kJ}}{0.545 \text{ kJ}} = 3.72$ g $NaNO_3$

Y = 10.0 g − X = 10.0 g − 3.72 g = 6.28 g KF = 6.3 g KF

8.116

$$\begin{array}{ll} & \Delta H \\ 4\ CO(g) + 2\ O_2(g) \rightarrow 4\ CO_2(g) & 2(-566.0 \text{ kJ}) \\ 2\ NO_2(g) \rightarrow 2\ NO(g) + O_2(g) & +116.2 \text{ kJ} \\ \underline{2\ NO(g) \rightarrow O_2(g) + N_2(g)} & \underline{2(-91.3 \text{ kJ})} \\ 4\ CO(g) + 2\ NO_2(g) \rightarrow 4\ CO_2(g) + N_2(g) & -1198.4 \text{ kJ} \end{array}$$

Multiconcept Problems

8.117 (a) Each S has 2 bonding pairs and 2 lone pairs of electrons. Each S is sp^3 hybridized and the geometry around each S is bent.

(b) $\Delta H = D$ (reactant bonds) − D (product bonds) = $(8\ D_{S-S}) - (4\ D_{S=S}) = +237$ kJ

$\Delta H = [(8 \text{ mol})(225 \text{ kJ/mol}) - [(4 \text{ mol})(D_{S=S})] = +237$ kJ

− (4 mol)($D_{S=S}$) = 237 kJ − 1800 kJ = −1563 kJ

$D_{S=S}$ = (1563 kJ)/(4 mol) = 391 kJ/mol

(c)

S_2 should be paramagnetic with two unpaired electrons in the π^*_{3p} MOs.

8.118 (a)

(b) C(g) + ½ O_2(g) + Cl_2(g) → $COCl_2$(g)

$\Delta H^{\circ}_f = \Delta H^{\circ}_f(C(g)) + (½\ D_{O=O} + D_{Cl-Cl}) - (D_{C=O} + 2\ D_{C-Cl})$

ΔH°_f = (716.7 kJ) + [(½ mol)(498 kJ/mol) + (1 mol)(243 kJ/mol)]

− [(1 mol)(732 kJ/mol) + (2 mol)(330 kJ/mol)]

ΔH°_{rxn} = − 183 kJ per mol $COCl_2$;   From Appendix B, $\Delta H^{\circ}_f(COCl_2)$ = −219.1 kJ/mol

The calculation of ΔH°_f from bond energies is only an estimate because the bond energies are average values derived from many different compounds.

8.119 (a) (1) $2 CH_3CO_2H(l) + Na_2CO_3(s) \rightarrow 2 CH_3CO_2Na(aq) + CO_2(g) + H_2O(l)$

(2) $CH_3CO_2H(l) + NaHCO_3(s) \rightarrow CH_3CO_2Na(aq) + CO_2(g) + H_2O(l)$

(b) CH_3CO_2H, 60.05 amu; Na_2CO_3, 105.99 amu; $NaHCO_3$, 84.01 amu

$$1 \text{ gal} \times \frac{3.7854 \text{ L}}{1 \text{ gal}} \times \frac{1000 \text{ mL}}{1 \text{ L}} \times \frac{1.049 \text{ g } CH_3CO_2H}{1 \text{ mL}} = 3971 \text{ g } CH_3CO_2H$$

$$3971 \text{ g } CH_3CO_2H \times \frac{1 \text{ mol } CH_3CO_2H}{60.05 \text{ g } CH_3CO_2H} = 66.13 \text{ mol } CH_3CO_2H$$

For reaction (1)

$$66.13 \text{ mol } CH_3CO_2H \times \frac{1 \text{ mol } Na_2CO_3}{2 \text{ mol } CH_3CO_2H} \times \frac{105.99 \text{ g } Na_2CO_3}{1 \text{ mol } Na_2CO_3} \times \frac{1 \text{ kg}}{1000 \text{ g}} = 3.505 \text{ kg } Na_2CO_3$$

For reaction (2)

$$66.13 \text{ mol } CH_3CO_2H \times \frac{1 \text{ mol } NaHCO_3}{1 \text{ mol } CH_3CO_2H} \times \frac{84.01 \text{ g } NaHCO_3}{1 \text{ mol } NaHCO_3} \times \frac{1 \text{ kg}}{1000 \text{ g}} = 5.556 \text{ kg } NaHCO_3$$

(c) $2 CH_3CO_2H(l) + Na_2CO_3(s) \rightarrow 2 CH_3CO_2Na(aq) + CO_2(g) + H_2O(l)$

$\Delta H^\circ_{rxn} = [2 \Delta H^\circ_f(CH_3CO_2Na) + \Delta H^\circ_f(CO_2) + \Delta H^\circ_f(H_2O)]$
$\qquad\qquad - [2 \Delta H^\circ_f(CH_3CO_2H) + \Delta H^\circ_f(Na_2CO_3)]$

$\Delta H^\circ_{rxn} = [(2 \text{ mol})(-726.1 \text{ kJ/mol}) + (1 \text{ mol})(-393.5 \text{ kJ/mol}) + (1 \text{ mol})(-285.8 \text{ kJ/mol})]$
$\qquad\qquad - [(2 \text{ mol})(-484.5 \text{ kJ/mol}) + (1 \text{ mol})(-1130.7 \text{ kJ/mol})]$

$\Delta H^\circ_{rxn} = -31.8 \text{ kJ for 2 mol } CH_3CO_2H$

$$\text{Heat} = -\frac{31.8 \text{ kJ}}{2 \text{ mol } CH_3CO_2H} \times 66.13 \text{ mol } CH_3CO_2H = -1050 \text{ kJ (liberated)}$$

$CH_3CO_2H(l) + NaHCO_3(s) \rightarrow CH_3CO_2Na(aq) + CO_2(g) + H_2O(l)$

$\Delta H^\circ_{rxn} = [\Delta H^\circ_f(CH_3CO_2Na) + \Delta H^\circ_f(CO_2) + \Delta H^\circ_f(H_2O)]$
$\qquad\qquad - [\Delta H^\circ_f(CH_3CO_2H) + \Delta H^\circ_f(NaHCO_3)]$

$\Delta H^\circ_{rxn} = [(1 \text{ mol})(-726.1 \text{ kJ/mol}) + (1 \text{ mol})(-393.5 \text{ kJ/mol}) + (1 \text{ mol})(-285.8 \text{ kJ/mol})]$
$\qquad\qquad - [(1 \text{ mol})(-484.5 \text{ kJ/mol}) + (1 \text{ mol})(-950.8 \text{ kJ/mol})]$

$\Delta H^\circ_{rxn} = +29.9 \text{ kJ for 1 mol } CH_3CO_2H$

$$q = \frac{29.9 \text{ kJ}}{1 \text{ mol } CH_3CO_2H} \times 66.13 \text{ mol } CH_3CO_2H = +1980 \text{ kJ (absorbed)}$$

8.120 (a) $2 K(s) + 2 H_2O(l) \rightarrow 2 KOH(aq) + H_2(g)$

(b) $\Delta H^\circ_{rxn} = [2 \Delta H^\circ_f(KOH)] - [2 \Delta H^\circ_f(H_2O)]$

$\quad \Delta H^\circ_{rxn} = [(2 \text{ mol})(-482.4 \text{ kJ/mol})] - [(2 \text{ mol})(-285.8 \text{ kJ/mol})] = -393.2 \text{ kJ}$

(c) The reaction produces 393.2 kJ/ 2 mol K = 196.6 kJ/ mol K. Assume that the mass of the water does not change and that the specific heat = 4.18 J/(g·°C) for the solution that is produced.

$$q = 7.55 \text{ g K} \times \frac{1 \text{ mol K}}{39.10 \text{ g K}} \times \frac{196.6 \text{ kJ}}{1 \text{ mol K}} \times \frac{1000 \text{ J}}{1 \text{ kJ}} = 3.80 \times 10^4 \text{ J}$$

$q = $ (specific heat) x m x ΔT

$$\Delta T = \frac{q}{(\text{specific heat}) \times m} = \frac{3.80 \times 10^4 \text{ J}}{[4.18 \text{ J/(g} \cdot {}^\circ\text{C)}](400.0 \text{ g})} = 22.7 \,{}^\circ\text{C}$$

$$\Delta T = T_{final} - T_{initial}$$

$$T_{final} = \Delta T + T_{initial} = 22.7 \,{}^\circ\text{C} + 25.0 \,{}^\circ\text{C} = 47.7 \,{}^\circ\text{C}$$

(d) $7.55 \text{ g K} \times \dfrac{1 \text{ mol K}}{39.10 \text{ g K}} \times \dfrac{2 \text{ mol KOH}}{2 \text{ mol K}} = 0.193 \text{ mol KOH}$

Assume that the mass of the solution does not change during the reaction and that the solution has a density of 1.00 g/mL.

$$\text{solution volume} = 400.0 \text{ g} \times \frac{1.00 \text{ mL}}{1 \text{ g}} \times \frac{1 \text{ L}}{1000 \text{ mL}} = 0.400 \text{ L}$$

$$\text{molarity} = \frac{0.193 \text{ mol KOH}}{0.400 \text{ L}} = 0.483 \text{ M}$$

$$2 \text{ KOH(aq)} + H_2SO_4\text{(aq)} \rightarrow K_2SO_4\text{(aq)} + 2 H_2O\text{(l)}$$

$$0.193 \text{ mol KOH} \times \frac{1 \text{ mol } H_2SO_4}{2 \text{ mol KOH}} \times \frac{1000 \text{ mL}}{0.554 \text{ mol } H_2SO_4} = 174 \text{ mL of } 0.554 \text{ M } H_2SO_4$$

8.121 (a)

$$\begin{array}{cc} H & H \\ | & | \\ H-\underset{\cdot\cdot}{N}-\underset{\cdot\cdot}{N}-H \end{array}$$

Each N is sp^3 hybridized and the geometry about each N is trigonal pyramidal.

(b) $2/4 \text{ NH}_3\text{(g)} + 3/4 \text{ N}_2\text{O(g)} \rightarrow \text{N}_2\text{(g)} + 3/4 \text{ H}_2\text{O(l)}$ $\qquad \Delta H^\circ_1 = \dfrac{-1011.2 \text{ kJ}}{4}$

$1/4 \text{ H}_2\text{O(l)} + 1/4 \text{ N}_2\text{H}_4\text{(l)} \rightarrow 1/8 \text{ O}_2\text{(g)} + 1/2 \text{ NH}_3\text{(g)}$ $\qquad \Delta H^\circ_2 = \dfrac{+286 \text{ kJ}}{8}$

$3/4 \text{ N}_2\text{H}_4\text{(l)} + 3/4 \text{ H}_2\text{O(l)} \rightarrow 3/4 \text{ N}_2\text{O(g)} + 9/4 \text{ H}_2\text{(g)}$ $\qquad \Delta H^\circ_3 = \dfrac{(3)(+317 \text{ kJ})}{4}$

$\underline{9/4 \text{ H}_2\text{(g)} + 9/8 \text{ O}_2\text{(g)} \rightarrow 9/4 \text{ H}_2\text{O(l)}} \qquad\qquad$ $\Delta H^\circ_4 = \dfrac{(9)(-285.8 \text{ kJ})}{4}$

Sum $\quad \text{N}_2\text{H}_4\text{(l)} + \text{O}_2\text{(g)} \rightarrow \text{N}_2\text{(g)} + 2 \text{ H}_2\text{O(l)}$ $\qquad \Delta H^\circ = -622 \text{ kJ}$

(c) N_2H_4, 32.045 amu

$$\text{mol N}_2\text{H}_4 = 100.0 \text{ g N}_2\text{H}_4 \times \frac{1 \text{ mol N}_2\text{H}_4}{32.045 \text{ g N}_2\text{H}_4} = 3.12 \text{ mol N}_2\text{H}_4$$

$q = (3.12 \text{ mol N}_2\text{H}_4)(622 \text{ kJ/mol}) = 1940 \text{ kJ}$

8.122 Assume 100.0 g of Y.

$$\text{mol F} = 61.7 \text{ g F} \times \frac{1 \text{ mol F}}{19.00 \text{ g F}} = 3.25 \text{ mol F}$$

$$\text{mol Cl} = 38.3 \text{ g Cl} \times \frac{1 \text{ mol Cl}}{35.45 \text{ g Cl}} = 1.08 \text{ mol Cl}$$

$Cl_{1.08}F_{3.25}$; divide each subscript by the smaller of the two, 1.08.

$Cl_{1.08/1.08}F_{3.25/1.08}$

ClF_3

(a) Y is ClF_3 and X is ClF

(b) 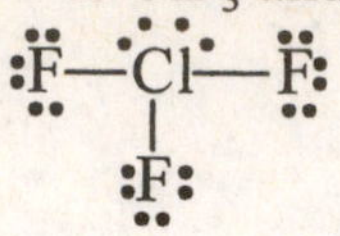 $:\ddot{F}—\overset{..}{Cl}—\ddot{F}:$
$\quad\quad\quad |$
$\quad\quad :\ddot{F}:$

There are five electron clouds around the Cl (3 bonding and 2 lone pairs). The geometry is T-shaped.

(c)

	ΔH
$Cl_2O(g) + 3\ OF_2(g) \rightarrow 2\ O_2(g) + 2\ ClF_3(g)$	−532.8 kJ
$2\ ClF(g) + O_2(g) \rightarrow Cl_2O(g) + OF_2(g)$	+205.4 kJ
$O_2(g) + 2\ F_2(g) \rightarrow 2\ OF_2(g)$	2(+24.5 kJ)
$2\ ClF(g) + 2\ F_2(g) \rightarrow 2\ ClF_3(g)$	−278.4 kJ

Divide reaction coefficients and ΔH by 2.

$$ClF(g) + F_2(g) \rightarrow ClF_3(g) \quad\quad \Delta H = -278.4\ kJ/2 = -139.2\ kJ/mol\ ClF_3$$

(d) ClF, 54.45 amu

$$q = 25.0\ g\ ClF \times \frac{1\ mol\ ClF}{54.45\ g\ ClF} \times \frac{-139.2\ kJ}{1\ mol\ ClF} \times 0.875 = -55.9\ kJ$$

55.9 kJ is released in this reaction.

9   Gases: Their Properties and Behavior

9.1   1.00 atm = 14.7 psi

$$1.00 \text{ mm Hg} \times \frac{1 \text{ atm}}{760 \text{ mm Hg}} \times \frac{14.7 \text{ psi}}{1 \text{ atm}} = 1.93 \times 10^{-2} \text{ psi}$$

9.2   1.00 atmosphere pressure can support a column of Hg 0.760 m high. Because the density of H_2O is 1.00 g/mL and that of Hg is 13.6 g/mL, 1.00 atmosphere pressure can support a column of H_2O 13.6 times higher than that of Hg. The column of H_2O supported by 1.00 atmosphere will be (0.760 m)(13.6) = 10.3 m.

9.3   The pressure in the flask is less than 0.975 atm because the liquid level is higher on the side connected to the flask. The 24.7 cm of Hg is the difference between the two pressures.

$$\text{Pressure difference} = 24.7 \text{ cm Hg} \times \frac{1.00 \text{ atm}}{76.0 \text{ cm Hg}} = 0.325 \text{ atm}$$

Pressure in flask = 0.975 atm – 0.325 atm = 0.650 atm

9.4   The pressure in the flask is greater than 750 mm Hg because the liquid level is lower on the side connected to the flask.

$$\text{Pressure difference} = 25 \text{ cm Hg} \times \frac{10 \text{ mm Hg}}{1 \text{ cm Hg}} = 250 \text{ mm Hg}$$

Pressure in flask = 750 mm Hg + 250 mm Hg = 1000 mm Hg

9.5   (a) Assume an initial volume of 1.00 L.
First consider the volume change resulting from a change in the number of moles with the pressure and temperature constant.

$$\frac{V_i}{n_i} = \frac{V_f}{n_f}; \quad V_f = \frac{V_i n_f}{n_i} = \frac{(1.00 \text{ L})(0.225 \text{ mol})}{0.3 \text{ mol}} = 0.75 \text{ L}$$

Now consider the volume change from 0.75 L as a result of a change in temperature with the number of moles and the pressure constant.

$$\frac{V_i}{T_i} = \frac{V_f}{T_f}; \quad V_f = \frac{V_i T_f}{T_i} = \frac{(0.75 \text{ L})(400 \text{ K})}{300 \text{ K}} = 1.0 \text{ L}$$

There is no net change in the volume as a result of the decrease in the number of moles of gas and a temperature increase.

(b) Assume an initial volume of 1.00 L.

First consider the volume change resulting from a change in the number of moles with the pressure and temperature constant.

$$\frac{V_i}{n_i} = \frac{V_f}{n_f}; \quad V_f = \frac{V_i n_f}{n_i} = \frac{(1.00 \text{ L})(0.225 \text{ mol})}{0.3 \text{ mol}} = 0.75 \text{ L}$$

Now consider the volume change from 0.75 L as a result of a change in temperature with the number of moles and the pressure constant.

$$\frac{V_i}{T_i} = \frac{V_f}{T_f}; \quad V_f = \frac{V_i T_f}{T_i} = \frac{(0.75 \text{ L})(200 \text{ K})}{300 \text{ K}} = 0.5 \text{ L}$$

The volume would be cut in half as a result of the decrease in the number of moles of gas and a temperature decrease.

9.6 $n = \dfrac{PV}{RT} = \dfrac{(1.000 \text{ atm})(1.000 \times 10^5 \text{ L})}{\left(0.082\ 06 \dfrac{\text{L} \cdot \text{atm}}{\text{K} \cdot \text{mol}}\right)(273.15 \text{ K})} = 4.461 \times 10^3 \text{ mol CH}_4$

CH_4, 16.04 amu; mass $CH_4 = (4.461 \times 10^3 \text{ mol})\left(\dfrac{16.04 \text{ g}}{1 \text{ mol}}\right) = 7.155 \times 10^4 \text{ g CH}_4$

9.7 C_3H_8, 44.10 amu; V = 350 mL = 0.350 L; T = 20 °C = 293 K

$n = 3.2 \text{ g} \times \dfrac{1 \text{ mol C}_3\text{H}_8}{44.10 \text{ g C}_3\text{H}_8} = 0.073 \text{ mol C}_3\text{H}_8$

$P = \dfrac{nRT}{V} = \dfrac{(0.073 \text{ mol})\left(0.082\ 06 \dfrac{\text{L} \cdot \text{atm}}{\text{K} \cdot \text{mol}}\right)(293 \text{ K})}{0.350 \text{ L}} = 5.0 \text{ atm}$

9.8 $P = 1.51 \times 10^4 \text{ kPa} \times \dfrac{1 \text{ atm}}{101.325 \text{ kPa}} = 149 \text{ atm}; \quad T = 25.0 \text{ °C} = 298 \text{ K}$

$n = \dfrac{PV}{RT} = \dfrac{(149 \text{ atm})(43.8 \text{ L})}{\left(0.082\ 06 \dfrac{\text{L} \cdot \text{atm}}{\text{K} \cdot \text{mol}}\right)(298 \text{ K})} = 267 \text{ mol He}$

9.9 The volume and number of moles of gas remain constant.

$\dfrac{nR}{V} = \dfrac{P_i}{T_i} = \dfrac{P_f}{T_f}; \quad T_f = \dfrac{P_f T_i}{P_i} = \dfrac{(2.37 \text{ atm})(273 \text{ K})}{2.15 \text{ atm}} = 301 \text{ K} = 28 \text{ °C}$

9.10 (a) The temperature has increased by about 10% (from 300 K to 325 K) while the amount and the pressure are unchanged. Thus, the volume should increase by about 10%.

(b) The temperature has increased by a factor of 1.5 (from 300 K to 450 K) and the pressure has increased by a factor of 3 (from 0.9 atm to 2.7 atm) while the amount is unchanged. Thus, the volume should decrease by half (1.5/3 = 0.5).

(c) Both the amount and the pressure have increased by a factor of 3 (from 0.075 mol to 0.22 mol and from 0.9 atm to 2.7 atm) while the temperature is unchanged. Thus, the volume is unchanged.

9.11 $CaCO_3(s) + 2 HCl(aq) \rightarrow CaCl_2(aq) + CO_2(g) + H_2O(l)$
$CaCO_3$, 100.1 amu; CO_2, 44.01 amu

$$\text{mole } CO_2 = 33.7 \text{ g } CaCO_3 \times \frac{1 \text{ mol } CaCO_3}{100.1 \text{ g } CaCO_3} \times \frac{1 \text{ mol } CO_2}{1 \text{ mol } CaCO_3} = 0.337 \text{ mol } CO_2$$

$$\text{mass } CO_2 = 0.337 \text{ mol } CO_2 \times \frac{44.01 \text{ g } CO_2}{1 \text{ mol } CO_2} = 14.8 \text{ g } CO_2$$

$$V = \frac{nRT}{P} = \frac{(0.337 \text{ mol})\left(0.082\ 06 \dfrac{L \cdot atm}{K \cdot mol}\right)(273 \text{ K})}{1.00 \text{ atm}} = 7.55 \text{ L}$$

9.12 $C_3H_8(g) + 5 O_2(g) \rightarrow 3 CO_2(g) + 4 H_2O(l)$

$$n_{propane} = \frac{PV}{RT} = \frac{(4.5 \text{ atm})(15.0 \text{ L})}{\left(0.082\ 06 \dfrac{L \cdot atm}{K \cdot mol}\right)(298 \text{ K})} = 2.76 \text{ mol } C_3H_8$$

$$2.76 \text{ mol } C_3H_8 \times \frac{3 \text{ mol } CO_2}{1 \text{ mol } C_3H_8} = 8.28 \text{ mol } CO_2$$

$$V = \frac{nRT}{P} = \frac{(8.28 \text{ mol})\left(0.082\ 06 \dfrac{L \cdot atm}{K \cdot mol}\right)(273 \text{ K})}{1.00 \text{ atm}} = 186 \text{ L} = 190 \text{ L}$$

9.13 $\quad n = \dfrac{PV}{RT} = \dfrac{(1.00 \text{ atm})(1.00 \text{ L})}{\left(0.082\ 06\ \dfrac{\text{L} \cdot \text{atm}}{\text{K} \cdot \text{mol}}\right)(273 \text{ K})} = 0.0446 \text{ mol}$

molar mass $= \dfrac{1.52 \text{ g}}{0.0446 \text{ mol}} = 34.1 \text{ g/mol}$; molecular mass $= 34.1$ amu

$Na_2S(aq) + 2 \text{ HCl}(aq) \rightarrow H_2S(g) + 2 \text{ NaCl}(aq)$

The foul-smelling gas is H_2S, hydrogen sulfide.

9.14 $\quad 12.45 \text{ g } H_2 \ \times \ \dfrac{1 \text{ mol } H_2}{2.016 \text{ g } H_2} = 6.176 \text{ mol } H_2$

$60.67 \text{ g } N_2 \ \times \ \dfrac{1 \text{ mol } N_2}{28.01 \text{ g } N_2} = 2.166 \text{ mol } N_2$

$2.38 \text{ g } NH_3 \ \times \ \dfrac{1 \text{ mol } NH_3}{17.03 \text{ g } NH_3} = 0.140 \text{ mol } NH_3$

$n_{total} = n_{H_2} + n_{N_2} + n_{NH_3} = 6.176 \text{ mol} + 2.166 \text{ mol} + 0.140 \text{ mol} = 8.482 \text{ mol}$

$X_{H_2} = \dfrac{6.176 \text{ mol}}{8.482 \text{ mol}} = 0.7281$; $X_{N_2} = \dfrac{2.166 \text{ mol}}{8.482 \text{ mol}} = 0.2554$; $X_{NH_3} = \dfrac{0.140 \text{ mol}}{8.482 \text{ mol}} = 0.0165$

9.15 $\quad n_{total} = 8.482 \text{ mol (from Problem 9.14)} \qquad T = 90 \text{ °C} = 363 \text{ K}$

$P_{total} = \dfrac{n_{total}RT}{V} = \dfrac{(8.482 \text{ mol})\left(0.082\ 06\ \dfrac{\text{L} \cdot \text{atm}}{\text{K} \cdot \text{mol}}\right)(363 \text{ K})}{10.00 \text{ L}} = 25.27 \text{ atm}$

$P_{H_2} = X_{H_2} \cdot P_{total} = (0.7281)(25.27 \text{ atm}) = 18.4 \text{ atm}$

$P_{N_2} = X_{N_2} \cdot P_{total} = (0.2554)(25.27 \text{ atm}) = 6.45 \text{ atm}$

$P_{NH_3} = X_{NH_3} \cdot P_{total} = (0.0165)(25.27 \text{ atm}) = 0.417 \text{ atm}$

9.16 $\quad P_{H_2O} = X_{H_2O} \cdot P_{Total} = (0.0287)(0.977 \text{ atm}) = 0.0280 \text{ atm}$

9.17 $\quad$ The number of moles of each gas is proportional to the number of each of the different gas molecules in the container.

$n_{total} = n_{red} + n_{yellow} + n_{green} = 6 + 2 + 4 = 12$

$X_{red} = \dfrac{n_{red}}{n_{total}} = \dfrac{6}{12} = 0.500$; $X_{yellow} = \dfrac{n_{yellow}}{n_{total}} = \dfrac{2}{12} = 0.167$; $X_{green} = \dfrac{n_{green}}{n_{total}} = \dfrac{4}{12} = 0.333$

$P_{red} = X_{red} \cdot P_{total} = (0.500)(600 \text{ mm Hg}) = 300 \text{ mm Hg}$

$P_{yellow} = X_{yellow} \cdot P_{total} = (0.167)(600 \text{ mm Hg}) = 100 \text{ mm Hg}$

$P_{green} = X_{green} \cdot P_{total} = (0.333)(600 \text{ mm Hg}) = 200 \text{ mm Hg}$

9.18 $u = \sqrt{\dfrac{3RT}{M}}$, M = molar mass, R = 8.314 J/(K · mol), 1 J = 1 kg · m^2/s^2

at 37 °C = 310 K, $u = \sqrt{\dfrac{3 \times 8.314 \text{ kg m}^2/(\text{s}^2 \text{ K mol}) \times 310 \text{ K}}{28.01 \times 10^{-3} \text{ kg/mol}}} = 525$ m/s

at –25 °C = 248 K, $u = \sqrt{\dfrac{3 \times 8.314 \text{ kg m}^2/(\text{s}^2 \text{ K mol}) \times 248 \text{ K}}{28.01 \times 10^{-3} \text{ kg/mol}}} = 470$ m/s

9.19 $u = \sqrt{\dfrac{3RT}{M}}$, M = molar mass, R = 8.314 J/(K · mol), 1 J = 1 kg · m^2/s^2

O_2, 32.00 amu, 32.00 x 10^{-3} kg/mol

$u = 580 \text{ mi/h} \times \dfrac{1.6093 \text{ km}}{1 \text{ mi}} \times \dfrac{1000 \text{ m}}{1 \text{ km}} \times \dfrac{1 \text{ hr}}{60 \text{ min}} \times \dfrac{1 \text{ min}}{60 \text{ s}} = 259$ m/s

$u = \sqrt{\dfrac{3RT}{M}}$; $u^2 = \dfrac{3RT}{M}$

$T = \dfrac{u^2 M}{3R} = \dfrac{(259 \text{ m/s})^2 (32.00 \times 10^{-3} \text{ kg/mol})}{(3)(8.314 \text{ kg·m}^2/\text{s}^2 \cdot \text{K·mol})} = 86.1$ K

$T = 86.1 - 273.15 = -187.0$ °C

9.20   (a) $\dfrac{\text{rate } O_2}{\text{rate Kr}} = \sqrt{\dfrac{M_{Kr}}{M_{O_2}}} = \sqrt{\dfrac{83.8}{32.0}}$; $\dfrac{\text{rate } O_2}{\text{rate Kr}} = 1.62$

O_2 diffuses 1.62 times faster than Kr.

(b) $\dfrac{\text{rate } C_2H_2}{\text{rate } N_2} = \sqrt{\dfrac{M_{N_2}}{M_{C_2H_2}}} = \sqrt{\dfrac{28.0}{26.0}}$; $\dfrac{\text{rate } C_2H_2}{\text{rate } N_2} = 1.04$

C_2H_2 diffuses 1.04 times faster than N_2.

9.21 $\dfrac{\text{rate } ^{20}\text{Ne}}{\text{rate} ^{22}\text{Ne}} = \sqrt{\dfrac{M\ ^{22}\text{Ne}}{M\ ^{20}\text{Ne}}} = \sqrt{\dfrac{22}{20}} = 1.05$; $\dfrac{\text{rate } ^{21}\text{Ne}}{\text{rate} ^{22}\text{Ne}} = \sqrt{\dfrac{M\ ^{22}\text{Ne}}{M\ ^{21}\text{Ne}}} = \sqrt{\dfrac{22}{21}} = 1.02$

Thus, the relative rates of diffusion are $^{20}\text{Ne}(1.05) > {}^{21}\text{Ne}(1.02) > {}^{22}\text{Ne}(1.00)$.

9.22 $P = \dfrac{nRT}{V} = \dfrac{(0.500 \text{ mol}) \left(0.082 \ 06 \ \dfrac{\text{L· atm}}{\text{K· mol}} \right) (300 \text{ K})}{(0.600 \text{ L})} = 20.5$ atm

$P = \dfrac{nRT}{V - nb} - \dfrac{an^2}{V^2}$

$$P = \frac{(0.500 \text{ mol})\left(0.082\ 06\ \dfrac{L \cdot atm}{K \cdot mol}\right)(300 \text{ K})}{[(0.600 \text{ L}) - (0.500 \text{ mol})(0.0387 \text{ L/mol})]} - \frac{\left(1.35\ \dfrac{L^2 \cdot atm}{mol^2}\right)(0.500 \text{ mol})^2}{(0.600 \text{ L})^2} = 20.3 \text{ atm}$$

9.23 The amount of ozone is assumed to be constant.

Therefore $nR = \dfrac{P_i V_i}{T_i} = \dfrac{P_f V_f}{T_f}$

Because $V \propto h$, then $\dfrac{P_i h_i}{T_i} = \dfrac{P_f h_f}{T_f}$ where h is the thickness of the O_3 layer.

$$h_f = \frac{P_i}{P_f} \times \frac{T_f}{T_i} \times h_i = \left(\frac{1.6 \times 10^{-9} \text{ atm}}{1 \text{ atm}}\right)\left(\frac{273 \text{ K}}{230 \text{ K}}\right)(20 \times 10^3 \text{ m}) = 3.8 \times 10^{-5} \text{ m}$$

(Actually, $V = 4\pi r^2 h$, where r = the radius of Earth. When you go out ~30 km to get to the ozone layer, the change in r^2 is less than 1%. Therefore you can neglect the change in r^2 and assume that V is proportional to h.)

9.24 For ether, the MAC = $\dfrac{15 \text{ mm Hg}}{760 \text{ mm Hg}} \times 100\% = 2.0\%$

9.25 (a) Let X = partial pressure of chloroform.

MAC = $\dfrac{X}{760 \text{ mm Hg}} \times 100\% = 0.77\%$

Solve for X. $X = 760 \text{ mm Hg} \times \dfrac{0.77\%}{100\%} = 5.9 \text{ mm Hg}$

(b) $CHCl_3$, 119.4 amu

$$PV = nRT; \quad n = \frac{PV}{RT} = \frac{\left(5.9 \text{ mm Hg} \times \dfrac{1.00 \text{ atm}}{760 \text{ mm Hg}}\right)(10.0 \text{ L})}{\left(0.082\ 06\ \dfrac{L \cdot atm}{K \cdot mol}\right)(273 \text{ K})} = 0.00347 \text{ mol } CHCl_3$$

mass $CHCl_3$ = 0.00347 mol $CHCl_3 \times \dfrac{119.4 \text{ g } CHCl_3}{1 \text{ mol } CHCl_3} = 0.41 \text{ g } CHCl_3$

Key Concept Problems

9.26 The picture on the right will be the same as that on the left, apart from random scrambling of the He and Ar atoms.

9.27 n_{Total} = (1 black) + (3 blue) + (4 red) + (6 green) = 14
P_{Total} = 420 mm Hg

$$P_x = P_{Total} \times \frac{n_x}{n_{Total}}$$

$$P_{black} = (420 \text{ mm Hg}) \times \left(\frac{1}{14}\right) = 30.0 \text{ mm Hg}; \quad P_{blue} = (420 \text{ mm Hg}) \times \left(\frac{3}{14}\right) = 90.0 \text{ mm Hg}$$

$$P_{red} = (420 \text{ mm Hg}) \times \left(\frac{4}{14}\right) = 120 \text{ mm Hg}; \quad P_{green} = (420 \text{ mm Hg}) \times \left(\frac{6}{14}\right) = 180 \text{ mm Hg}$$

9.28 (a) The volume of a gas is proportional to the kelvin temperature at constant pressure. As the temperature increases from 300 K to 450 K, the volume will increase by a factor of 1.5.

(b) The volume of a gas is inversely proportional to pressure at constant temperature. As the pressure increases from 1 atm to 2 atm, the volume will decrease by a factor of 2.

(c) $PV = nRT$;   The amount of gas (n) is constant.

Therefore $nR = \dfrac{P_i V_i}{T_i} = \dfrac{P_f V_f}{T_f}$.

Assume $V_i = 1$ L and solve for V_f.

$$\frac{P_i V_i T_f}{T_i P_f} = \frac{(3 \text{ atm})(1 \text{ L})(200 \text{ K})}{(300 \text{ K})(2 \text{ atm})} = V_f = 1 \text{ L}$$

There is no change in volume.

9.29 If the sample remains a gas at 150 K, then drawing (c) represents the gas at this temperature. The gas molecules still fill the container.

9.30 The two gases should mix randomly and homogeneously and this is best represented by drawing (c).

9.31 The two gases will be equally distributed among the three flasks.

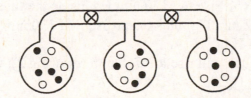

9.32 The gas pressure in the bulb in mm Hg is equal to the difference in the height of the Hg in the two arms of the manometer.

9.33 A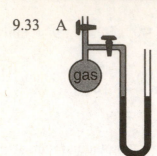

When stopcock A is opened, the pressure in the flask will equal the external pressure, and the level of mercury will be the same in both arms of the manometer.

9.34 (a) Because there are more yellow gas molecules than there are blue, the yellow gas molecules have the higher average speed.
(b) Each rate is proportional to the number of effused gas molecules of each type.
$M_{yellow} = 25$ amu

$$\frac{rate_{blue}}{rate_{yellow}} = \sqrt{\frac{M_{yellow}}{M_{blue}}}; \quad \frac{5}{6} = \sqrt{\frac{25 \text{ amu}}{M_{blue}}}; \quad \left(\frac{5}{6}\right)^2 = \frac{25 \text{ amu}}{M_{blue}}; \quad M_{blue} = \frac{25 \text{ amu}}{\left(\frac{5}{6}\right)^2} = 36 \text{ amu}$$

9.35

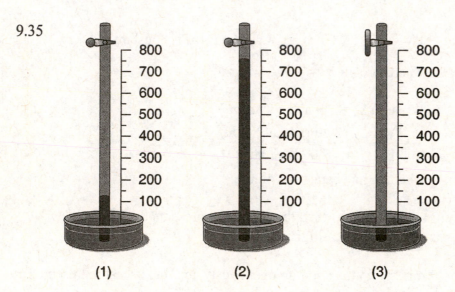

(1)   (2)   (3)

(a) $\dfrac{P_2}{T_2} = \dfrac{P_1}{T_1}$; $P_2 = \dfrac{(T_2)(P_1)}{T_1} = \dfrac{(248 \text{ K})(760 \text{ mm Hg})}{298 \text{ K}} = 632$ mm Hg

The column of Hg will rise to ~130 mm Hg inside the tube (drawing 1). The pressure inside the tube (632 mm Hg) plus the pressure of 130 mm Hg equals the external pressure, ~760 mm Hg.
(b) The column of Hg will rise to ~760 mm (drawing 2), which is equal to the external pressure, ~760 mm Hg.
(c) The pressure inside the tube is equal to the external pressure and the Hg level inside the tube will be the same as in the dish (drawing 3).

Section Problems
Gases and Gas Pressure (Section 9.1)

9.36   Temperature is a measure of the average kinetic energy of gas particles.

9.37   Gases are much more compressible than solids or liquids because there is a large amount
of empty space between individual gas molecules.

9.38 $P = 480 \text{ mm Hg} \times \dfrac{1.00 \text{ atm}}{760 \text{ mm Hg}} = 0.632 \text{ atm};$ $P = 480 \text{ mm Hg} \times \dfrac{101{,}325 \text{ Pa}}{760 \text{ mm Hg}} = 6.40 \times 10^4 \text{ Pa}$

9.39 $P = 352 \text{ torr} \times \dfrac{101{,}325 \text{ Pa}}{760 \text{ torr}} \times \dfrac{1 \text{ kPa}}{1000 \text{ Pa}} = 46.9 \text{ kPa}$

$P = 0.255 \text{ atm} \times \dfrac{760 \text{ mm Hg}}{1.00 \text{ atm}} = 194 \text{ mm Hg}$

$P = 0.0382 \text{ mm Hg} \times \dfrac{101{,}325 \text{ Pa}}{760 \text{ mm Hg}} = 5.09 \text{ Pa}$

9.40 $P_{flask} > 754.3 \text{ mm Hg};$ $P_{flask} = 754.3 \text{ mm Hg} + 176 \text{ mm Hg} = 930 \text{ mm Hg}$

9.41 $P_{flask} < 1.021 \text{ atm}$ (see Figure 9.4)

$P_{difference} = 28.3 \text{ cm Hg} \times \dfrac{1.00 \text{ atm}}{76.0 \text{ cm Hg}} = 0.372 \text{ atm}$

$P_{flask} = 1.021 \text{ atm} - P_{difference} = 1.021 \text{ atm} - 0.372 \text{ atm} = 0.649 \text{ atm}$

9.42 $P_{flask} > 752.3 \text{ mm Hg}$ (see Figure 9.4);   If the pressure in the flask can support a column
of ethyl alcohol (d = 0.7893 g/mL) 55.1 cm high, then it can only support a column of Hg
that is much shorter because of the higher density of Hg.

$55.1 \text{ cm} \times \dfrac{0.7893 \text{ g/mL}}{13.546 \text{ g/mL}} = 3.21 \text{ cm Hg} = 32.1 \text{ mm Hg}$

$P_{flask} = 752.3 \text{ mm Hg} + 32.1 \text{ mm Hg} = 784.4 \text{ mm Hg}$

$P_{flask} = 784.4 \text{ mm Hg} \times \dfrac{101{,}325 \text{ Pa}}{760 \text{ mm Hg}} = 1.046 \times 10^5 \text{ Pa}$

9.43   Compute the height of a column of $CHCl_3$ that 1.00 atm can support.

$760 \text{ mm Hg} \times \dfrac{13.546 \text{ g/mL}}{1.4832 \text{ g/mL}} = 6941 \text{ mm } CHCl_3;$ therefore $1.00 \text{ atm} = 6941 \text{ mm } CHCl_3$

The pressure in the flask is less than atmospheric pressure.

$P_{atm} - P_{flask} = 0.849 \text{ atm} - 0.788 \text{ atm} = 0.061 \text{ atm}$

$0.061 \text{ atm} \times \dfrac{6941 \text{ mm } CHCl_3}{1.00 \text{ atm}} = 423 \text{ mm } CHCl_3$

The chloroform will be 423 mm higher in the manometer arm connected to the flask.

9.44   % Volume

N_2	78.08
O_2	20.95
Ar	0.93
CO_2	0.037

The % volume for a particular gas is proportional to the number of molecules of that gas in a mixture of gases.

Average molecular mass of air

$= (0.7808)(\text{mol. mass } N_2) + (0.2095)(\text{mol. mass } O_2)$
$\qquad + (0.0093)(\text{at. mass Ar}) + (0.000\ 37)(\text{mol. mass } CO_2)$
$= (0.7808)(28.01\ \text{amu}) + (0.2095)(32.00\ \text{amu})$
$\qquad + (0.0093)(39.95\ \text{amu}) + (0.000\ 37)(44.01\ \text{amu}) = 28.96\ \text{amu}$

9.45   The % volume for a particular gas is proportional to the number of molecules of that gas in a mixture of gases.

Average molecular mass of a diving-gas $= (0.020)(\text{mol. mass } O_2) + (0.980)(\text{at. mass He})$
$= (0.020)(32.00\ \text{amu}) + (0.980)(4.00\ \text{amu}) = 4.56\ \text{amu}$

The Gas Laws (Sections 9.2–9.3)

9.46   (a) $\dfrac{nR}{V} = \dfrac{P_i}{T_i} = \dfrac{P_f}{T_f}$; $\dfrac{P_i T_f}{T_i} = P_f$

Let $P_i = 1$ atm, $T_i = 100$ K, $T_f = 300$ K

$P_f = \dfrac{P_i T_f}{T_i} = \dfrac{(1\ \text{atm})(300\ \text{K})}{(100\ \text{K})} = 3\ \text{atm}$

The pressure would triple.

(b) $\dfrac{RT}{V} = \dfrac{P_i}{n_i} = \dfrac{P_f}{n_f}$; $\dfrac{P_i n_f}{n_i} = P_f$

Let $P_i = 1$ atm, $n_i = 3$ mol, $n_f = 1$ mol

$P_f = \dfrac{P_i n_f}{n_i} = \dfrac{(1\ \text{atm})(1\ \text{mol})}{(3\ \text{mol})} = \dfrac{1}{3}\ \text{atm}$

The pressure would be $\dfrac{1}{3}$ the initial pressure.

(c) $nRT = P_i V_i = P_f V_f$; $\dfrac{P_i V_i}{V_f} = P_f$

Let $P_i = 1$ atm, $V_i = 1$ L, $V_f = 1 - 0.45$ L $= 0.55$ L

$P_f = \dfrac{P_i V_i}{V_f} = \dfrac{(1\ \text{atm})(1\ \text{L})}{(0.55\ \text{L})} = 1.8\ \text{atm}$

The pressure would increase by 1.8 times.

(d) $nR = \dfrac{P_i V_i}{T_i} = \dfrac{P_f V_f}{T_f}$; $\qquad \dfrac{P_i V_i T_f}{T_i V_f} = P_f$

Let $P_i = 1$ atm, $V_i = 1$ L, $T_i = 200$ K, $V_f = 3$ L, $T_i = 100$ K

$P_f = \dfrac{P_i V_i T_f}{T_i V_f} = \dfrac{(1 \text{ atm})(1 \text{ L})(100 \text{ K})}{(200 \text{ K})(3 \text{ L})} = 0.17$ atm

The pressure would be 0.17 times the initial pressure.

9.47 (a) $\quad \dfrac{nR}{P} = \dfrac{V_i}{T_i} = \dfrac{V_f}{T_f}$; $\qquad \dfrac{V_i T_f}{T_i} = V_f$

Let $V_i = 1$ L, $T_i = 400$ K, $T_f = 200$ K

$V_f = \dfrac{V_i T_f}{T_i} = \dfrac{(1 \text{ L})(200 \text{ K})}{(400 \text{ K})} = 0.5$ L

The volume would be halved.

(b) $\quad \dfrac{RT}{P} = \dfrac{V_i}{n_i} = \dfrac{V_f}{n_f}$; $\qquad \dfrac{V_i n_f}{n_i} = V_f$

Let $V_i = 1$ L, $n_i = 4$ mol, $n_f = 5$ mol

$V_f = \dfrac{V_i n_f}{n_i} = \dfrac{(1 \text{ L})(5 \text{ mol})}{(4 \text{ mol})} = 1.25$ L

The volume would increase by 1/4.

(c) $\quad nRT = P_i V_i = P_f V_f$; $\qquad \dfrac{P_i V_i}{P_f} = V_f$

Let $V_i = 1$ L, $P_i = 4$ atm, $P_f = 1$ atm

$V_f = \dfrac{P_i V_i}{P_f} = \dfrac{(4 \text{ atm})(1 \text{ L})}{(1 \text{ atm})} = 4$ L

The volume would increase by a factor of 4.

(d) $\quad nR = \dfrac{P_i V_i}{T_i} = \dfrac{P_f V_f}{T_f}$; $\qquad \dfrac{P_i V_i T_f}{P_f T_i} = V_f$

Let $V_i = 1$ L, $T_i = 200$ K, $T_f = 400$ K, $P_i = 1$ atm, $P_f = 2$ atm

$V_f = \dfrac{P_i V_i T_f}{P_f T_i} = \dfrac{(1 \text{ atm})(1 \text{ L})(400 \text{ K})}{(2 \text{ atm})(200 \text{ K})} = 1$ L

There is no volume change.

9.48 They all contain the same number of gas molecules.

9.49   For air, T = 50 °C = 323 K.

$$n = \frac{PV}{RT} = \frac{\left(750 \text{ mm Hg} \times \frac{1.00 \text{ atm}}{760 \text{ mm Hg}}\right)(2.50 \text{ L})}{\left(0.082\ 06\ \frac{L \cdot atm}{K \cdot mol}\right)(323 \text{ K})} = 0.0931 \text{ mol air}$$

For CO_2, T = –10 °C = 263 K

$$n = \frac{PV}{RT} = \frac{\left(765 \text{ mm Hg} \times \frac{1.00 \text{ atm}}{760 \text{ mm Hg}}\right)(2.16 \text{ L})}{\left(0.082\ 06\ \frac{L \cdot atm}{K \cdot mol}\right)(263 \text{ K})} = 0.101 \text{ mol } CO_2$$

Because the number of moles of CO_2 is larger than the number of moles of air, the CO_2 sample contains more molecules.

9.50   n and T are constant; therefore $nRT = P_i V_i = P_f V_f$

$$V_f = \frac{P_i V_i}{P_f} = \frac{(150 \text{ atm})(49.0 \text{ L})}{(1.02 \text{ atm})} = 7210 \text{ L}$$

n and P are constant; therefore $\dfrac{nR}{P} = \dfrac{V_i}{T_i} = \dfrac{V_f}{T_f}$

$$V_f = \frac{V_i T_f}{T_i} = \frac{(49.0 \text{ L})(308 \text{ K})}{(293 \text{ K})} = 51.5 \text{ L}$$

9.51 $T_i = 20$ °C = 293 K; $nR = \dfrac{P_i V_i}{T_i} = \dfrac{P_f V_f}{T_f}$

$$V_f = \frac{P_i V_i T_f}{T_i P_f} = \frac{(140 \text{ atm})(8.0 \text{ L})(273 \text{ K})}{(293 \text{ K})(1.00 \text{ atm})} = 1.0 \times 10^3 \text{ L}$$

9.52 $15.0 \text{ g } CO_2 \times \dfrac{1 \text{ mol } CO_2}{44.0 \text{ g } CO_2} = 0.341 \text{ mol } CO_2$

$$P = \frac{nRT}{V} = \frac{(0.341 \text{ mol})\left(0.082\ 06\ \frac{L \cdot atm}{K \cdot mol}\right)(300 \text{ K})}{(0.30 \text{ L})} = 27.98 \text{ atm}$$

$27.98 \text{ atm} \times \dfrac{760 \text{ mm Hg}}{1 \text{ atm}} = 2.1 \times 10^4 \text{ mm Hg}$

9.53 $2.0 \text{ g } N_2 \times \dfrac{1 \text{ mol } N_2}{28.0 \text{ g } N_2} = 0.0714 \text{ mol } N_2$

$$T = \frac{PV}{nR} = \frac{(6.0 \text{ atm})(0.40 \text{ L})}{(0.0714 \text{ mol})\left(0.082\ 06\ \frac{L \cdot atm}{K \cdot mol}\right)} = 410 \text{ K}$$

9.54
$$\frac{1 \text{ H atom}}{cm^3} \times \frac{1 \text{ mol H}}{6.02 \times 10^{23} \text{ atoms}} \times \frac{1000 \text{ cm}^3}{1 \text{ L}} = 1.7 \times 10^{-21} \text{ mol H/L}$$

$$P = \frac{nRT}{V} = \frac{(1.7 \times 10^{-21} \text{ mol})\left(0.082 \ 06 \ \frac{L \cdot atm}{K \cdot mol}\right)(100 \text{ K})}{(1 \text{ L})} = 1.4 \times 10^{-20} \text{ atm}$$

$$P = 1.4 \times 10^{-20} \text{ atm} \times \frac{760 \text{ mm Hg}}{1.0 \text{ atm}} = 1 \times 10^{-17} \text{ mm Hg}$$

9.55 CH_4, 16.04 amu; 5.54 kg = 5.54×10^3 g; T = 20 °C = 293 K

$$P = \frac{nRT}{V} = \frac{\left(5.54 \times 10^3 \text{ g} \times \frac{1 \text{ mol}}{16.04 \text{ g}}\right)\left(0.082 \ 06 \ \frac{L \cdot atm}{K \cdot mol}\right)(293 \text{ K})}{(43.8 \text{ L})} = 189.6 \text{ atm}$$

$$P = 189.6 \text{ atm} \times \frac{101,325 \text{ Pa}}{1 \text{ atm}} \times \frac{1 \text{ k Pa}}{1000 \text{ Pa}} = 1.92 \times 10^4 \text{ kPa}$$

9.56 $n = \frac{PV}{RT} = \dfrac{\left(17,180 \text{ kPa} \times \frac{1000 \text{ Pa}}{1 \text{ k Pa}} \times \frac{1 \text{ atm}}{101,325 \text{ Pa}}\right)(43.8 \text{ L})}{\left(0.082 \ 06 \ \frac{L \cdot atm}{K \cdot mol}\right)(293 \text{K})} = 308.9 \text{ mol}$

$$\text{mass Ar} = 308.9 \text{ mol} \times \frac{39.948 \text{ g}}{1 \text{ mol}} = 12340 \text{ g} = 1.23 \times 10^4 \text{ g}$$

9.57 $P = 13,800 \text{ kPa} \times \frac{1000 \text{ Pa}}{1 \text{ kPa}} \times \frac{1 \text{ atm}}{101,325 \text{ Pa}} = 136.2 \text{ atm}$

n and T are constant; therefore $nRT = P_iV_i = P_fV_f$

$$V_f = \frac{P_iV_i}{P_f} = \frac{(136.2 \text{ atm})(2.30 \text{ L})}{(1.25 \text{ atm})} = 250.6 \text{ L}$$

$$250.6 \text{ L} \times \frac{1 \text{ balloon}}{1.5 \text{ L}} = 167 \text{ balloons}$$

Gas Stoichiometry (Section 9.4)

9.58 For steam, T = 123.0 °C = 396 K
$$n = \frac{PV}{RT} = \frac{(0.93 \text{ atm})(15.0 \text{ L})}{\left(0.082 \ 06 \ \frac{L \cdot atm}{K \cdot mol}\right)(396 \text{ K})} = 0.43 \text{ mol steam}$$

For ice, H_2O, 18.02 amu; $n = 10.5 \text{ g} \times \frac{1 \text{ mol}}{18.02 \text{ g}} = 0.583 \text{ mol ice}$

Because the number of moles of ice is larger than the number of moles of steam, the ice contains more H_2O molecules.

9.59 $T = 85.0\ ^\circ C = 358\ K$

$$n_{Ar} = \frac{PV}{RT} = \frac{\left(1111\ mm\ Hg\ \times\ \dfrac{1.00\ atm}{760\ mm\ Hg}\right)(3.14\ L)}{\left(0.082\ 06\ \dfrac{L\cdot atm}{K\cdot mol}\right)(358\ K)} = 0.156\ mol\ Ar$$

Cl_2, 70.91 amu; $n_{Cl_2} = 11.07\ g\ Cl_2 \times \dfrac{1\ mol\ Cl_2}{70.91\ g\ Cl_2} = 0.156\ mol\ Cl_2$

There are equal numbers of Ar atoms and Cl_2 molecules in their respective samples.

9.60   The containers are identical. Both containers contain the same number of gas molecules. Weigh the containers. Because the molecular mass for O_2 is greater than the molecular mass for H_2, the heavier container contains O_2.

9.61   Assuming that you can see through the flask, Cl_2 gas is greenish and Ar is colorless.

9.62   room volume = 4.0 m x 5.0 m x 2.5 m x $\dfrac{1\ L}{10^{-3}\ m^3}$ = $5.0 \times 10^4\ L$

$$n_{total} = \frac{PV}{RT} = \frac{(1.0\ atm)(5.0\ \times\ 10^4\ L)}{\left(0.082\ 06\ \dfrac{L\cdot atm}{K\cdot mol}\right)(273\ K)} = 2.23 \times 10^3\ mol$$

$n_{O_2} = (0.2095)n_{total} = (0.2095)(2.23 \times 10^3\ mol) = 467\ mol\ O_2$

mass O_2 = 467 mol x $\dfrac{32.0\ g}{1\ mol}$ = $1.5 \times 10^4\ g\ O_2$

9.63   0.25 g O_2 x $\dfrac{1\ mol\ O_2}{32.0\ g\ O_2}$ = $7.8 \times 10^{-3}\ mol\ O_2$

$$V = \frac{nRT}{P} = \frac{(7.8\ \times\ 10^{-3}\ mol)\left(0.082\ 06\ \dfrac{L\cdot atm}{K\cdot mol}\right)(310\ K)}{1.0\ atm} = 0.198\ L = 0.200\ L = 200\ mL\ O_2$$

9.64   (a) CH_4, 16.04 amu; $d = \dfrac{16.04\ g}{22.4\ L} = 0.716\ g/L$

(b) CO_2, 44.01 amu; $d = \dfrac{44.01\ g}{22.4\ L} = 1.96\ g/L$

(c) O_2, 32.00 amu; $d = \dfrac{32.00\ g}{22.4\ L} = 1.43\ g/L$

(d) UF_6, 352.0 amu; $d = \dfrac{352.0\ g}{22.4\ L} = 15.7\ g/L$

9.65 Average molar mass = (0.270)(molar mass F_2) + (0.730)(molar mass He)

= (0.270)(38.00 g/mol) + (0.730)(4.003 g/mol) = 13.18 g/mol

Assume 1.00 mole of the gas mixture. T = 27.5 ºC = 300.6 K

$$V = \frac{nRT}{P} = \frac{(1.00 \text{ mol})\left(0.082\ 06\ \frac{\text{L} \cdot \text{atm}}{\text{K} \cdot \text{mol}}\right)(300.6 \text{ K})}{\left(714 \text{ mm Hg} \ \text{x} \ \frac{1.00 \text{ atm}}{760 \text{ mm Hg}}\right)} = 26.3 \text{ L}$$

$$d = \frac{13.18 \text{ g}}{26.3 \text{ L}} = 0.501 \text{ g/L}$$

9.66 $$n = \frac{PV}{RT} = \frac{\left(356 \text{ mm Hg} \ \text{x} \ \frac{1.00 \text{ atm}}{760 \text{ mm Hg}}\right)(1.500 \text{ L})}{\left(0.082\ 06\ \frac{\text{L} \cdot \text{atm}}{\text{K} \cdot \text{mol}}\right)(295.5 \text{ K})} = 0.0290 \text{ mol}$$

molar mass = $\dfrac{0.9847 \text{ g}}{0.0290 \text{ mol}}$ = 34.0 g/mol; molecular mass = 34.0 amu

9.67 (a) Assume 1.000 L gas sample

$$n = \frac{PV}{RT} = \frac{(1.00 \text{ atm})(1.000 \text{ L})}{\left(0.082\ 06\ \frac{\text{L} \cdot \text{atm}}{\text{K} \cdot \text{mol}}\right)(273 \text{ K})} = 0.0446 \text{ mol}$$

molar mass = $\dfrac{1.342 \text{ g}}{0.0446 \text{ mol}}$ = 30.1 g/mol; molecular mass = 30.1 amu

(b) Assume 1.000 L gas sample

$$n = \frac{PV}{RT} = \frac{\left(752 \text{ mm Hg} \ \text{x} \ \frac{1.00 \text{ atm}}{760 \text{ mm Hg}}\right)(1.000 \text{ L})}{\left(0.082\ 06\ \frac{\text{L} \cdot \text{atm}}{\text{K} \cdot \text{mol}}\right)(298 \text{ K})} = 0.0405 \text{ mol}$$

molar mass = $\dfrac{1.053 \text{ g}}{0.0405 \text{ mol}}$ = 26.0 g/mol; molecular mass = 26.0 amu

9.68 2 HgO(s) → 2 Hg(l) + O_2(g);   HgO, 216.59 amu

10.57 g HgO x $\dfrac{1 \text{ mol HgO}}{216.59 \text{ g HgO}}$ x $\dfrac{1 \text{ mol O}_2}{2 \text{ mol HgO}}$ = 0.024 40 mol O_2

$$V = \frac{nRT}{P} = \frac{(0.024\ 40 \text{ mol})\left(0.082\ 06\ \frac{\text{L} \cdot \text{atm}}{\text{K} \cdot \text{mol}}\right)(273.15 \text{ K})}{1.000 \text{ atm}} = 0.5469 \text{ L}$$

9.69 2 HgO(s) → 2 Hg(l) + O_2(g);   HgO, 216.59 amu

mass HgO = 0.0155 mol O_2 x $\dfrac{2 \text{ mol HgO}}{1 \text{ mol O}_2}$ x $\dfrac{216.59 \text{ g HgO}}{1 \text{ mol HgO}}$ = 6.71 g HgO

9.70 $Zn(s) + 2 HCl(aq) \rightarrow ZnCl_2(aq) + H_2(g)$

(a) $25.5 \text{ g Zn} \times \dfrac{1 \text{ mol Zn}}{65.39 \text{ g Zn}} \times \dfrac{1 \text{ mol H}_2}{1 \text{ mol Zn}} = 0.390 \text{ mol H}_2$

$$V = \frac{nRT}{P} = \frac{(0.390 \text{ mol})\left(0.082\ 06\ \dfrac{\text{L} \cdot \text{atm}}{\text{K} \cdot \text{mol}}\right)(288 \text{ K})}{\left(742 \text{ mm Hg} \times \dfrac{1.00 \text{ atm}}{760 \text{ mm Hg}}\right)} = 9.44 \text{ L}$$

(b) $n = \dfrac{PV}{RT} = \dfrac{\left(350 \text{ mm Hg} \times \dfrac{1.00 \text{ atm}}{760 \text{ mm Hg}}\right)(5.00 \text{ L})}{\left(0.082\ 06\ \dfrac{\text{L} \cdot \text{atm}}{\text{K} \cdot \text{mol}}\right)(303.15 \text{ K})} = 0.092\ 56 \text{ mol H}_2$

$0.092\ 56 \text{ mol H}_2 \times \dfrac{1 \text{ mol Zn}}{1 \text{ mol H}_2} \times \dfrac{65.39 \text{ g Zn}}{1 \text{ mol Zn}} = 6.05 \text{ g Zn}$

9.71 $2 NH_4NO_3(s) \rightarrow 2 N_2(g) + 4 H_2O(g) + O_2(g); \qquad NH_4NO_3, 80.04 \text{ amu}$

Total moles of gas = $450 \text{ g NH}_4\text{NO}_3 \times \dfrac{1 \text{ mol NH}_4\text{NO}_3}{80.04 \text{ g NH}_4\text{NO}_3} \times \dfrac{7 \text{ mol gas}}{2 \text{ mol NH}_4\text{NO}_3} = 19.68 \text{ mol}$

$T = 450\ ^\circ\text{C} = 723 \text{ K}$

$$V = \frac{nRT}{P} = \frac{(19.68 \text{ mol})\left(0.082\ 06\ \dfrac{\text{L} \cdot \text{atm}}{\text{K} \cdot \text{mol}}\right)(723 \text{ K})}{(1.00 \text{ atm})} = 1.17 \times 10^3 \text{ L}$$

9.72   (a) $V_{24h} = (4.50 \text{ L/min})(60 \text{ min/h})(24 \text{ h/day}) = 6480 \text{ L}$

$V_{CO_2} = (0.034)V_{24h} = (0.034)(6480 \text{ L}) = 220 \text{ L}$

$n = \dfrac{PV}{RT} = \dfrac{\left(735 \text{ mm Hg} \times \dfrac{1.00 \text{ atm}}{760 \text{ mm Hg}}\right)(220 \text{ L})}{\left(0.082\ 06\ \dfrac{\text{L} \cdot \text{atm}}{\text{K} \cdot \text{mol}}\right)(298 \text{ K})} = 8.70 \text{ mol CO}_2$

$8.70 \text{ mol CO}_2 \times \dfrac{44.01 \text{ g CO}_2}{1 \text{ mol CO}_2} = 383 \text{ g} = 380 \text{ g CO}_2$

(b) $2 Na_2O_2(s) + 2 CO_2(g) \rightarrow 2 Na_2CO_3(s) + O_2(g); \qquad Na_2O_2, 77.98 \text{ amu}$

$3.65 \text{ kg} = 3650 \text{ g}$

$3650 \text{ g Na}_2\text{O}_2 \times \dfrac{1 \text{ mol Na}_2\text{O}_2}{77.98 \text{ g Na}_2\text{O}_2} \times \dfrac{2 \text{ mol CO}_2}{2 \text{ mol Na}_2\text{O}_2} \times \dfrac{1 \text{ day}}{8.70 \text{ mol CO}_2} = 5.4 \text{ days}$

9.73 $2\ TiCl_4(g)\ +\ H_2(g)\ \rightarrow\ 2\ TiCl_3(s)\ +\ 2\ HCl(g);\quad TiCl_4,\ 189.69\ amu$

(a) $T = 435\ °C = 708\ K$

$$n_{H_2} = \frac{PV}{RT} = \frac{\left(795\ mm\ Hg\ \times\ \dfrac{1.00\ atm}{760\ mm\ Hg}\right)(155\ L)}{\left(0.082\ 06\ \dfrac{L \cdot atm}{K \cdot mol}\right)(708\ K)} = 2.79\ mol\ H_2$$

$$2.79\ mol\ H_2 \times \frac{2\ mol\ TiCl_4}{1\ mol\ H_2} \times \frac{189.69\ g\ TiCl_4}{1\ mol\ TiCl_4} = 1058\ g = 1060\ g\ TiCl_4$$

(b) $n_{HCl} = 2.79\ mol\ H_2 \times \dfrac{2\ mol\ HCl}{1\ mol\ H_2} = 5.58\ mol\ HCl$

$$V = \frac{n_{HCl}RT}{P} = \frac{(5.58\ mol)\left(0.082\ 06\ \dfrac{L \cdot atm}{K \cdot mol}\right)(273\ K)}{(1.00\ atm)} = 125\ L\ HCl$$

Dalton's Law and Mole Fraction (Section 9.5)

9.74 Because of Avogadro's Law ($V \propto n$), the % volumes are also % moles.

	% mole
N_2	78.08
O_2	20.95
Ar	0.93
CO_2	0.037

In decimal form, % mole = mole fraction.

$P_{N_2} = X_{N_2} \cdot P_{total} = (0.7808)(1.000\ atm) = 0.7808\ atm$

$P_{O_2} = X_{O_2} \cdot P_{total} = (0.2095)(1.000\ atm) = 0.2095\ atm$

$P_{Ar} = X_{Ar} \cdot P_{total} = (0.0093)(1.000\ atm) = 0.0093\ atm$

$P_{CO_2} = X_{CO_2} \cdot P_{total} = (0.000\ 37)(1.000\ atm) = 0.000\ 37\ atm$

Pressures of the rest are negligible.

9.75 $X_{CH_4} = \dfrac{94\ mol}{100\ mol} = 0.94;\quad P_{CH_4} = X_{CH_4} \cdot P_{total} = (0.94)(1.48\ atm) = 1.4\ atm$

$X_{C_2H_6} = \dfrac{4\ mol}{100\ mol} = 0.040;\quad P_{C_2H_6} = X_{C_2H_6} \cdot P_{total} = (0.040)(1.48\ atm) = 0.059\ atm$

$X_{C_3H_8} = \dfrac{1.5\ mol}{100\ mol} = 0.015;\quad P_{C_3H_8} = X_{C_3H_8} \cdot P_{total} = (0.015)(1.48\ atm) = 0.022\ atm$

$X_{C_4H_{10}} = \dfrac{0.5\ mol}{100\ mol} = 0.0050;\quad P_{C_4H_{10}} = X_{C_4H_{10}} \cdot P_{total} = (0.0050)(1.48\ atm) = 0.0074\ atm$

9.76   Assume a 100.0 g sample. $g\ CO_2 = 1.00\ g$ and $g\ O_2 = 99.0\ g$

$mol\ CO_2 = 1.00\ g\ CO_2 \times \dfrac{1\ mol\ CO_2}{44.01\ g\ CO_2} = 0.0227\ mol\ CO_2$

$mol\ O_2 = 99.0\ g\ O_2 \times \dfrac{1\ mol\ O_2}{32.00\ g\ O_2} = 3.094\ mol\ O_2$

$n_{total} = 3.094\ mol + 0.0227\ mol = 3.117\ mol$

$X_{O_2} = \dfrac{3.094\ mol}{3.117\ mol} = 0.993;\quad X_{CO_2} = \dfrac{0.0227\ mol}{3.117\ mol} = 0.007\ 28$

$P_{O_2} = X_{O_2} \cdot P_{total} = (0.993)(0.977\ atm) = 0.970\ atm$

$P_{CO_2} = X_{CO_2} \cdot P_{total} = (0.007\ 28)(0.977\ atm) = 0.007\ 11\ atm$

9.77   From Problem 9.78: $X_{HCl} = 0.026,\ X_{H_2} = 0.094,\ X_{Ne} = 0.88$

$P_{HCl} = X_{HCl} \cdot P_{total} = (0.026)(13,800\ kPa) = 3.6 \times 10^2\ kPa$

$P_{H_2} = X_{H_2} \cdot P_{total} = (0.094)(13,800\ kPa) = 1.3 \times 10^3\ kPa$

$P_{Ne} = X_{Ne} \cdot P_{total} = (0.88)(13,800\ kPa) = 1.2 \times 10^4\ kPa$

9.78   Assume a 100.0 g sample.

$g\ HCl = (0.0500)(100.0\ g) = 5.00\ g;\quad 5.00\ g\ HCl \times \dfrac{1\ mol\ HCl}{36.5\ g\ HCl} = 0.137\ mol\ HCl$

$g\ H_2 = (0.0100)(100.0\ g) = 1.00\ g;\quad 1.00\ g\ H_2 \times \dfrac{1\ mol\ H_2}{2.016\ g\ H_2} = 0.496\ mol\ H_2$

$g\ Ne = (0.94)(100.0\ g) = 94\ g;\quad 94\ g\ Ne \times \dfrac{1\ mol\ Ne}{20.18\ g\ Ne} = 4.66\ mol\ Ne$

$n_{total} = 0.137 + 0.496 + 4.66 = 5.3\ mol$

$X_{HCl} = \dfrac{0.137\ mol}{5.3\ mol} = 0.026;\quad X_{H_2} = \dfrac{0.496\ mol}{5.3\ mol} = 0.094;\quad X_{Ne} = \dfrac{4.66\ mol}{5.3\ mol} = 0.88$

9.79   Assume a 1.000 L gas sample.

$n = \dfrac{PV}{RT} = \dfrac{(1.000\ atm)(1.000\ L)}{\left(0.082\ 06\ \dfrac{L \cdot atm}{K \cdot mol}\right)(273.15\ K)} = 0.044\ 61\ mol$

$average\ molar\ mass = \dfrac{1.413\ g}{0.044\ 61\ mol} = 31.67\ g/mol$

$31.67 = x \cdot M_{Ar} + (1 - x) \cdot M_{N_2}$
$31.67 = (x)(39.948) + (1 - x)(28.013)$
Solve for x: $x = 0.3064,\quad 1 - x = 0.6936$
The mixture contains 30.64% Ar and 69.36% N_2.
Assume 100 moles of gas.

$X_{Ar} = \dfrac{30.64\ mol}{100\ mol} = 0.3064;\quad X_{N_2} = \dfrac{69.36\ mol}{100\ mol} = 0.6936$

9.80 $P_{total} = P_{H_2} + P_{H_2O}$; $P_{H_2} = P_{total} - P_{H_2O} = 747$ mm Hg $- 23.8$ mm Hg $= 723$ mm Hg

$$n = \frac{PV}{RT} = \frac{\left(723 \text{ mm Hg} \times \dfrac{1.00 \text{ atm}}{760 \text{ mm Hg}}\right)(3.557 \text{ L})}{\left(0.082\ 06 \dfrac{\text{L} \cdot \text{atm}}{\text{K} \cdot \text{mol}}\right)(298 \text{ K})} = 0.1384 \text{ mol } H_2$$

$0.1384 \text{ mol } H_2 \times \dfrac{1 \text{ mol Mg}}{1 \text{ mol } H_2} \times \dfrac{24.3 \text{ g Mg}}{1 \text{ mol Mg}} = 3.36 \text{ g Mg}$

9.81 $P_{total} = P_{Cl_2} + P_{H_2O} = 755$ mm Hg

$P_{Cl_2} = P_{total} - P_{H_2O} = 755$ mm Hg $- 28.7$ mm Hg $= 726.3$ mm Hg

(a) $X_{Cl_2} = \dfrac{P_{Cl_2}}{P_{total}} = \dfrac{726.3 \text{ mm Hg}}{755 \text{ mm Hg}} = 0.962$

(b)   NaCl, 58.44 amu

$$n_{Cl_2} = \frac{PV}{RT} = \frac{\left(726.3 \text{ mm Hg} \times \dfrac{1.00 \text{ atm}}{760 \text{ mm Hg}}\right)(0.597 \text{ L})}{\left(0.082\ 06 \dfrac{\text{L} \cdot \text{atm}}{\text{K} \cdot \text{mol}}\right)(300 \text{ K})} = 0.0232 \text{ mol } Cl_2$$

$0.0232 \text{ mol } Cl_2 \times \dfrac{2 \text{ mol NaCl}}{1 \text{ mol } Cl_2} \times \dfrac{58.44 \text{ g NaCl}}{1 \text{ mol NaCl}} = 2.71 \text{ g NaCl}$

Kinetic-Molecular Theory and Graham's Law (Sections 9.6–9.7)

9.82   The kinetic-molecular theory is based on the following assumptions:
1. A gas consists of tiny particles, either atoms or molecules, moving about at random.
2. The volume of the particles themselves is negligible compared with the total volume of the gas; most of the volume of a gas is empty space.
3. The gas particles act independently; there are no attractive or repulsive forces between particles.
4. Collisions of the gas particles, either with other particles or with the walls of the container, are elastic; that is, the total kinetic energy of the gas particles is constant at constant T.
5. The average kinetic energy of the gas particles is proportional to the Kelvin temperature of the sample.

9.83   Diffusion – The mixing of different gases by random molecular motion and with frequent collisions.
Effusion – The process in which gas molecules escape through a tiny hole in a membrane without collisions.

9.84   Heat is the energy transferred from one object to another as the result of a temperature difference between them. Temperature is a measure of the kinetic energy of molecular motion.

9.85 The atomic mass of He is much less than the molecular mass of the major components of air (N_2 and O_2). The rate of effusion of He through the balloon skin is much faster.

9.86 $u = \sqrt{\dfrac{3\,RT}{M}} = \sqrt{\dfrac{3 \times 8.314\ kg\,m^2/(s^2\,K\,mol) \times 220\ K}{28.0 \times 10^{-3}\ kg/mol}} = 443\ m/s$

9.87 For Br_2: $u = \sqrt{\dfrac{3RT}{M}} = \sqrt{\dfrac{3 \times 8.314\ kg\,m^2/(s^2\,K\,mol) \times 293\ K}{159.8 \times 10^{-3}\ kg/mol}} = 214\ m/s$

For Xe: $u = 214\ m/s = \sqrt{\dfrac{3 \times 8.314\ kg\,m^2/(s^2\,K\,mol) \times T}{131.3 \times 10^{-3}\ kg/mol}}$

Square both sides of the equation and solve for T.

$45796\ m^2/s^2 = \dfrac{3 \times 8.314\ kg\,m^2/(s^2\,K\,mol) \times T}{131.3 \times 10^{-3}\ kg/mol}$

$T = 241\ K = -32\ ^\circ C$

9.88 For H_2, $u = \sqrt{\dfrac{3\,RT}{M}} = \sqrt{\dfrac{3 \times 8.314\ kg\,m^2/(s^2\,K\,mol) \times 150\ K}{2.02 \times 10^{-3}\ kg/mol}} = 1360\ m/s$

For He, $u = \sqrt{\dfrac{3 \times 8.314\ kg\,m^2/(s^2\,K\,mol) \times 648\ K}{4.00 \times 10^{-3}\ kg/mol}} = 2010\ m/s$

He at 375 ºC has the higher average speed.

9.89 UF_6, 352.02 amu; $\quad$ T = 25 ºC = 298 K

$u = \sqrt{\dfrac{3\,RT}{M}} = \sqrt{\dfrac{3 \times 8.314\ kg\,m^2/(s^2\,K\,mol) \times 298\ K}{352.02 \times 10^{-3}\ kg/mol}} = 145\ m/s$

Ferrari

$145\ \dfrac{mi}{hr} \times \dfrac{1.6093\ km}{1\ mi} \times \dfrac{1000\ m}{1\ km} \times \dfrac{1\ hr}{3600\ s} = 64.8\ m/s$

The UF_6 molecule has the higher average speed.

9.90 $\dfrac{rate_{H_2}}{rate_X} = \sqrt{\dfrac{M_X}{M_{H_2}}}; \qquad \dfrac{2.92}{1} = \dfrac{\sqrt{M_X}}{\sqrt{2.02}}; \qquad 2.92\sqrt{2.02} = \sqrt{M_X}$

$M_X = (2.92\sqrt{2.02})^2 = 17.2\ g/mol;$ molecular mass = 17.2 amu

9.91 $\dfrac{rate\ Xe}{rate\ Z} = \sqrt{\dfrac{M_Z}{M_{Xe}}}; \qquad \dfrac{1}{1.86} = \dfrac{\sqrt{M_Z}}{\sqrt{131.29}}; \qquad \dfrac{\sqrt{131.29}}{1.86} = \sqrt{M_Z}; \qquad \dfrac{131.29}{(1.86)^2} = M_Z$

$M_Z = 37.9\ g/mol;$ molecular mass = 37.9 amu; The gas could be F_2.

9.92   HCl, 36.5 amu; F_2, 38.0 amu; Ar, 39.9 amu

$$\frac{\text{rate HCl}}{\text{rate Ar}} = \sqrt{\frac{M_{Ar}}{M_{HCl}}} = \sqrt{\frac{39.9}{36.5}} = 1.05; \qquad \frac{\text{rate } F_2}{\text{rate Ar}} = \sqrt{\frac{M_{Ar}}{M_{F_2}}} = \sqrt{\frac{39.9}{38.0}} = 1.02$$

The relative rates of diffusion are HCl(1.05) > F_2(1.02) > Ar(1.00).

9.93   Because CO and N_2 have the same mass, they will have the same diffusion rates.

9.94 $$u = 45 \text{ m/s} = \sqrt{\frac{3 \times 8.314 \text{ kg m}^2/(s^2 \text{ K mol}) \times T}{4.00 \times 10^{-3} \text{ kg/mol}}}$$

Square both sides of the equation and solve for T.

$$2025 \text{ m}^2/s^2 = \frac{3 \times 8.314 \text{ kg m}^2/(s^2 \text{ K mol}) \times T}{4.00 \times 10^{-3} \text{ kg/mol}}$$

$T = 0.325 \text{ K} = -272.83 \text{ °C}$ (near absolute zero)

9.95 $$230 \text{ km/h} \times \frac{1000 \text{ m}}{1 \text{ km}} \times \frac{1 \text{ h}}{3600 \text{ s}} = 63.9 \text{ m/s}$$

$$u = 63.9 \text{ m/s} = \sqrt{\frac{3 \times 8.314 \text{ kg m}^2/(s^2 \text{ K mol}) \times T}{32.0 \times 10^{-3} \text{ kg/mol}}}$$

Square both sides of the equation and solve for T.

$$4083 \text{ m}^2/s^2 = \frac{3 \times 8.314 \text{ kg m}^2/(s^2 \text{ K mol}) \times T}{32.0 \times 10^{-3} \text{ kg/mol}}$$

$T = 5.24 \text{ K} = -268 \text{ °C}$

Chapter Problems

9.96 $$\frac{\text{rate } ^{35}Cl_2}{\text{rate } ^{37}Cl_2} = \sqrt{\frac{M \ ^{37}Cl_2}{M \ ^{35}Cl_2}} = \sqrt{\frac{74.0}{70.0}} = 1.03$$

$$\frac{\text{rate } ^{35}Cl\ ^{37}Cl}{\text{rate } ^{37}Cl_2} = \sqrt{\frac{M \ ^{37}Cl_2}{M \ ^{35}Cl\ ^{37}Cl}} = \sqrt{\frac{74.0}{72.0}} = 1.01$$

The relative rates of diffusion are $^{35}Cl_2$(1.03) > $^{35}Cl^{37}Cl$(1.01) > $^{37}Cl_2$(1.00).

9.97   Average molecular mass of air = 28.96 amu; CO_2, 44.01 amu

$$P = 760 \text{ mm Hg} \times \frac{44.01 \text{ g/mol}}{28.96 \text{ g/mol}} = 1155 \text{ mm Hg}$$

9.98 $$V = \frac{nRT}{P} = \frac{(1.00 \text{ mol})\left(0.082 \ 06 \ \dfrac{\text{L} \cdot \text{atm}}{\text{K} \cdot \text{mol}}\right)(1050 \text{ K})}{(75 \text{ atm})} = 1.1 \text{ L}$$

9.99 $\quad 1 \text{ atm} = \left(\dfrac{14.70 \text{ lb}}{\text{in.}^2}\right)\left(\dfrac{1.00 \text{ in.}}{2.54 \text{ cm}}\right)^2\left(\dfrac{453.59 \text{ g}}{1 \text{ lb}}\right) = 1033.5 \text{ g/cm}^2$

column height $= (1033.5 \text{ g/cm}^2)(1 \text{ cm}^3/0.89\text{g}) = 1161 \text{ cm} = 1200 \text{ cm} = 12 \text{ m}$

9.100 $\quad n = \dfrac{PV}{RT} = \dfrac{(2.15 \text{ atm})(7.35 \text{ L})}{\left(0.082\ 06 \dfrac{\text{L} \cdot \text{atm}}{\text{K} \cdot \text{mol}}\right)(293 \text{ K})} = 0.657 \text{ mol Ar}$

$0.657 \text{ mol Ar} \times \dfrac{39.948 \text{ g Ar}}{1 \text{ mol Ar}} = 26.2 \text{ g Ar}$

$m_{total} = 478.1 \text{ g} + 26.2 \text{ g} = 504.3 \text{ g}$

9.101 This is initially a Boyle's Law problem, because only P and V are changing while n and T remain fixed. The initial volume for each gas is the volume of their individual bulbs. The final volume for each gas is the total volume of the three bulbs.

$nRT = P_iV_i = P_fV_f; \quad V_f = 1.50 + 1.00 + 2.00 = 4.50 \text{ L}$

For CO_2: $\quad P_f = \dfrac{P_iV_i}{V_f} = \dfrac{(2.13 \text{ atm})(1.50 \text{ L})}{(4.50 \text{ L})} = 0.710 \text{ atm}$

For H_2: $\quad P_f = \dfrac{P_iV_i}{V_f} = \dfrac{(0.861 \text{ atm})(1.00 \text{ L})}{(4.50 \text{ L})} = 0.191 \text{ atm}$

For Ar: $\quad P_f = \dfrac{P_iV_i}{V_f} = \dfrac{(1.15 \text{ atm})(2.00 \text{ L})}{(4.50 \text{ L})} = 0.511 \text{ atm}$

From Dalton's Law, $P_{total} = P_{CO_2} + P_{H_2} + P_{Ar}$

$P_{total} = 0.710 \text{ atm} + 0.191 \text{ atm} + 0.511 \text{ atm} = 1.412 \text{ atm}$

9.102 (a) Bulb A contains $CO_2(g)$ and $N_2(g)$; $\quad$ Bulb B contains $CO_2(g)$, $N_2(g)$, and $H_2O(s)$.

(b) Initial moles of gas $= n = \dfrac{PV}{RT} = \dfrac{\left(564 \text{ mm Hg} \times \dfrac{1.00 \text{ atm}}{760 \text{ mm Hg}}\right)(1.000 \text{ L})}{\left(0.082\ 06 \dfrac{\text{L} \cdot \text{atm}}{\text{K} \cdot \text{mol}}\right)(298 \text{ K})}$

Initial moles of gas $= 0.030\ 35 \text{ mol}$

mol gas in Bulb A $= n = \dfrac{PV}{RT} = \dfrac{\left(219 \text{ mm Hg} \times \dfrac{1.00 \text{ atm}}{760 \text{ mm Hg}}\right)(1.000 \text{ L})}{\left(0.082\ 06 \dfrac{\text{L} \cdot \text{atm}}{\text{K} \cdot \text{mol}}\right)(298 \text{ K})} = 0.011\ 78 \text{ mol}$

mol gas in Bulb B $= n = \dfrac{PV}{RT} = \dfrac{\left(219 \text{ mm Hg} \times \dfrac{1.00 \text{ atm}}{760 \text{ mm Hg}}\right)(1.000 \text{ L})}{\left(0.082\ 06 \dfrac{\text{L} \cdot \text{atm}}{\text{K} \cdot \text{mol}}\right)(203 \text{ K})} = 0.017\ 30 \text{ mol}$

$n_{H_2O} = n_{initial} - n_A - n_B = 0.030\ 35 - 0.011\ 78 - 0.017\ 30 = 0.001\ 27\ mol = 0.0013\ mol\ H_2O$

(c) Bulb A contains $N_2(g)$.
Bulb B contains $N_2(g)$ and $H_2O(s)$.
Bulb C contains $N_2(g)$ and $CO_2(s)$.

(d) $n_A = \dfrac{PV}{RT} = \dfrac{\left(33.5\ mm\ Hg\ \times\ \dfrac{1.00\ atm}{760\ mm\ Hg}\right)(1.000\ L)}{\left(0.082\ 06\ \dfrac{L \cdot atm}{K \cdot mol}\right)(298\ K)} = 0.001\ 803\ mol$

$n_B = \dfrac{PV}{RT} = \dfrac{\left(33.5\ mm\ Hg\ \times\ \dfrac{1.00\ atm}{760\ mm\ Hg}\right)(1.000\ L)}{\left(0.082\ 06\ \dfrac{L \cdot atm}{K \cdot mol}\right)(203\ K)} = 0.002\ 646\ mol$

$n_C = \dfrac{PV}{RT} = \dfrac{\left(33.5\ mm\ Hg\ \times\ \dfrac{1.00\ atm}{760\ mm\ Hg}\right)(1.000\ L)}{\left(0.082\ 06\ \dfrac{L \cdot atm}{K \cdot mol}\right)(83\ K)} = 0.006\ 472\ mol$

$n_{N_2} = n_A + n_B + n_C = 0.001\ 803 + 0.002\ 646 + 0.006\ 472 = 0.010\ 92\ mol\ N_2$

(e) $n_{CO_2} = n_{initial} - n_{H_2O} - n_{N_2} = 0.030\ 35 - 0.0013 - 0.010\ 92 = 0.0181\ mol\ CO_2$

9.103 $C_3H_5N_3O_9$, 227.1 amu

(a) moles $C_3H_5N_3O_9 = 1.00\ g \times \dfrac{1\ mol}{227.1\ g} = 0.004\ 40\ mol$

$n_{air} = \dfrac{PV}{RT} = \dfrac{(1.00\ atm)(0.500\ L)}{\left(0.082\ 06\ \dfrac{L \cdot atm}{K \cdot mol}\right)(293\ K)} = 0.0208\ mol\ air$

(b) moles gas from $C_3H_5N_3O_9 = 0.004\ 40\ mol \times \dfrac{29\ mol\ gas}{4\ mol\ nitro}$

moles gas from $C_3H_5N_3O_9 = 0.0319\ mol\ gas\ from\ C_3H_5N_3O_9$
$n_{total} = 0.0319\ mol + 0.0208\ mol = 0.0527\ mol$

(c) $P = \dfrac{nRT}{V} = \dfrac{(0.0527\ mol)\left(0.082\ 06\ \dfrac{L \cdot atm}{K \cdot mol}\right)(698\ K)}{(0.500\ L)} = 6.04\ atm$

9.104 NH_3, 17.03 amu;   mol $NH_3 = 45.0\ g \times \dfrac{1\ mol}{17.03\ g} = 2.64\ mol$

$P = \dfrac{nRT}{V}$ or $P = \dfrac{nRT}{(V - nb)} - \dfrac{an^2}{V^2}$

(a) At T = 0 °C = 273 K

$$P = \frac{(2.64 \text{ mol})\left(0.082\ 06\ \frac{L \cdot atm}{K \cdot mol}\right)(273 \text{ K})}{(1.000 \text{ L})} = 59.1 \text{ atm}$$

$$P = \frac{(2.64 \text{ mol})\left(0.082\ 06\ \frac{L \cdot atm}{K \cdot mol}\right)(273 \text{ K})}{[(1.000 \text{ L}) - (2.64 \text{ mol})(0.0371 \text{ L/mol})]} - \frac{\left(4.17\ \frac{L^2 \cdot atm}{mol^2}\right)(2.64 \text{ mol})^2}{(1.000 \text{ L})^2}$$

P = 65.6 atm – 29.1 atm = 36.5 atm

(b) At T = 50 °C = 323 K

$$P = \frac{(2.64 \text{ mol})\left(0.082\ 06\ \frac{L \cdot atm}{K \cdot mol}\right)(323 \text{ K})}{(1.000 \text{ L})} = 70.0 \text{ atm}$$

$$P = \frac{(2.64 \text{ mol})\left(0.082\ 06\ \frac{L \cdot atm}{K \cdot mol}\right)(323 \text{ K})}{[(1.000 \text{ L}) - (2.64 \text{ mol})(0.0371 \text{ L/mol})]} - \frac{\left(4.17\ \frac{L^2 \cdot atm}{mol^2}\right)(2.64 \text{ mol})^2}{(1.000 \text{ L})^2}$$

P = 77.6 atm – 29.1 atm = 48.5 atm

(c) At T = 100 °C = 373 K

$$P = \frac{(2.64 \text{ mol})\left(0.082\ 06\ \frac{L \cdot atm}{K \cdot mol}\right)(373 \text{ K})}{(1.000 \text{ L})} = 80.8 \text{ atm}$$

$$P = \frac{(2.64 \text{ mol})\left(0.082\ 06\ \frac{L \cdot atm}{K \cdot mol}\right)(373 \text{ K})}{[(1.000 \text{ L}) - (2.64 \text{ mol})(0.0371 \text{ L/mol})]} - \frac{\left(4.17\ \frac{L^2 \cdot atm}{mol^2}\right)(2.64 \text{ mol})^2}{(1.000 \text{ L})^2}$$

P = 89.6 atm – 29.1 atm = 60.5 atm

At the three temperatures, the van der Waals equation predicts a much lower pressure than does the ideal gas law. This is likely due to the fact that NH_3 can hydrogen bond leading to strong intermolecular forces.

9.105 (a) $n_{total} = \frac{PV}{RT} = \frac{\left(258 \text{ mm Hg} \times \frac{1.00 \text{ atm}}{760 \text{ mm Hg}}\right)(0.500 \text{ L})}{\left(0.082\ 06\ \frac{L \cdot atm}{K \cdot mol}\right)(293 \text{ K})} = 0.007\ 06 \text{ mol}$

(b) $n_B = \frac{PV}{RT} = \frac{\left(344 \text{ mm Hg} \times \frac{1 \text{ atm}}{760 \text{ mm Hg}}\right)(0.250 \text{ L})}{\left(0.082\ 06\ \frac{L \cdot atm}{K \cdot mol}\right)(293 \text{ K})} = 0.004\ 71 \text{ moles}$

(c) $d = \frac{0.218 \text{ g}}{0.250 \text{ L}} = 0.872 \text{ g/L}$

214

(d) molar mass $= \dfrac{0.218 \text{ g}}{0.004\ 71 \text{ mol}} = 46.3 \text{ g/mol},\ NO_2;$ mol. mass $= 46.3$ amu

(e) $Hg_2CO_3(s) + 6\ HNO_3(aq) \rightarrow 2\ Hg(NO_3)_2(aq) + 3\ H_2O(l) + CO_2(g) + 2\ NO_2(g)$

9.106 CO_2, 44.01 amu

mol $CO_2 = 500.0$ g $CO_2 \times \dfrac{1 \text{ mol } CO_2}{44.01 \text{ g } CO_2} = 11.36$ mol CO_2

$PV = nRT$

$P = \dfrac{nRT}{V} = \dfrac{(11.36 \text{ mol})\left(0.082\ 06\ \dfrac{L \cdot atm}{K \cdot mol}\right)(700 \text{ K})}{(0.800 \text{ L})} = 816$ atm

9.107 (a) Let $x = $ mol C_nH_{2n+2} in the reaction mixture.

Combustion of $C_nH_{2n+2} \rightarrow nCO_2 + (n+1)H_2O$ needs $n + \left(\dfrac{n+1}{2}\right) = \dfrac{3n+1}{2}$ mol O_2

Balanced equation is: $C_nH_{2n+2}(g) + \left(\dfrac{3n+1}{2}\right)O_2(g) \rightarrow nCO_2(g) + (n+1)H_2O(g)$

In going from reactants to products, the increase in the number of moles is

$[n + (n+1)] - \left[1 + \dfrac{3n+1}{2}\right] = \dfrac{n-1}{2}$ per mol of C_nH_{2n+2} reacted.

Before reaction: total mol $= \dfrac{PV}{RT} = \dfrac{(2.000 \text{ atm})(0.4000 \text{ L})}{\left(0.082\ 06\ \dfrac{L \cdot atm}{K \cdot mol}\right)(298.15 \text{ K})} = 0.032\ 70$ mol

After reaction: total mol $= \dfrac{PV}{RT} = \dfrac{(2.983 \text{ atm})(0.4000 \text{ L})}{\left(0.082\ 06\ \dfrac{L \cdot atm}{K \cdot mol}\right)(398.15 \text{ K})} = 0.036\ 52$ mol

Difference $= 0.032\ 70$ mol $- 0.036\ 52$ mol $= 0.003\ 82$ mol

Increase in number of mol $= \left(\dfrac{n-1}{2}\right)x = 0.003\ 82$ mol; $x = \dfrac{2(0.003\ 82)}{n-1}$

Also $x = \dfrac{\text{g } C_nH_{2n+2}}{\text{molar mass}} = \dfrac{0.148 \text{ g}}{[12.01n + 1.008(2n+2)] \text{ g/mol}}$

So $\dfrac{2(0.003\ 82)}{n-1} = \dfrac{0.148}{14.026n + 2.016}$; $0.148\ n - 0.148 = 0.107\ n + 0.0154$

$0.041\ n = 0.163;\ n = \dfrac{0.163}{0.041} = 4.0$

C_nH_{2n+2} is C_4H_{10} (butane); molar mass $= (4)(12.01) + (10)(1.008) = 58.12$ g/mol

(b) $0.148 \text{ g } C_4H_{10} \times \dfrac{1 \text{ mol } C_4H_{10}}{58.12 \text{ g } C_4H_{10}} = 0.002\ 55 \text{ mol } C_4H_{10}$

mol O_2 initially = total mol − mol C_4H_{10} = 0.032 70 mol − 0.002 55 mol = 0.030 15 mol O_2

$P_{C_4H_{10}} = \left(\dfrac{n_{C_4H_{10}}}{n_{total}}\right) P_{initial} = \left(\dfrac{0.002\ 55 \text{ mol}}{0.032\ 70 \text{ mol}}\right)(2.000 \text{ atm}) = 0.156 \text{ atm}$

$P_{O_2} = \left(\dfrac{n_{O_2}}{n_{total}}\right) P_{initial} = \left(\dfrac{0.030\ 15 \text{ mol}}{0.032\ 70 \text{ mol}}\right)(2.000 \text{ atm}) = 1.844 \text{ atm}$

(c) $C_4H_{10}(g) + \dfrac{13}{2} O_2 \rightarrow 4 CO_2(g) + 5 H_2O(g)$

$0.002\ 55 \text{ mol } C_4H_{10} \times \dfrac{4 \text{ mol } CO_2}{1 \text{ mol } C_4H_{10}} = 0.0102 \text{ mol } CO_2$

$0.002\ 55 \text{ mol } C_4H_{10} \times \dfrac{5 \text{ mol } H_2O}{1 \text{ mol } C_4H_{10}} = 0.012\ 75 \text{ mol } H_2O$

mol O_2 unreacted = total mol after reaction − mol CO_2 − mol H_2O
$= 0.03652 \text{ mol} − 0.0102 \text{ mol} − 0.01275 = 0.01357 \text{ mol } O_2$

$P_{CO_2} = \left(\dfrac{n_{CO_2}}{n_{total}}\right) P_{final} = \left(\dfrac{0.0102 \text{ mol}}{0.036\ 52 \text{ mol}}\right)(2.983 \text{ atm}) = 0.833 \text{ atm}$

$P_{H_2O} = \left(\dfrac{n_{H_2O}}{n_{total}}\right) P_{final} = \left(\dfrac{0.012\ 75 \text{ mol}}{0.036\ 52 \text{ mol}}\right)(2.983 \text{ atm}) = 1.041 \text{ atm}$

$P_{O_2} = \left(\dfrac{n_{O_2}}{n_{total}}\right) P_{final} = \left(\dfrac{0.013\ 57 \text{ mol}}{0.036\ 52 \text{ mol}}\right)(2.983 \text{ atm}) = 1.108 \text{ atm}$

9.108 (a) average molecular mass for natural gas
$= (0.915)(16.04 \text{ amu}) + (0.085)(30.07 \text{ amu}) = 17.2 \text{ amu}$

total moles of gas $= 15.50 \text{ g} \times \dfrac{1 \text{ mol gas}}{17.2 \text{ g gas}} = 0.901 \text{ mol gas}$

(b) $P = \dfrac{(0.901 \text{ mol})\left(0.082\ 06 \dfrac{L \cdot atm}{K \cdot mol}\right)(293 \text{ K})}{(15.00 \text{ L})} = 1.44 \text{ atm}$

(c) $P_{CH_4} = X_{CH_4} \cdot P_{total} = (1.44 \text{ atm})(0.915) = 1.32 \text{ atm}$

$P_{C_2H_6} = X_{C_2H_6} \cdot P_{total} = (1.44 \text{ atm})(0.085) = 0.12 \text{ atm}$

(d) $\Delta H_{combustion}(CH_4) = -890.3 \text{ kJ/mol}$ and $\Delta H_{combustion}(C_2H_6) = -1427.7 \text{ kJ/mol}$

Heat liberated $= (0.915)(0.901 \text{ mol})(-890.3 \text{ kJ/mol})$
$$+ (0.085)(0.901)(-1427.7 \text{ kJ/mol}) = -843 \text{ kJ}$$

9.109 $PV = nRT$

$$n_{total\,(initial)} = \frac{PV}{RT} = \frac{(3.00 \text{ atm})(10.0 \text{ L})}{\left(0.082\ 06 \dfrac{\text{L} \cdot \text{atm}}{\text{K} \cdot \text{mol}}\right)(373.1 \text{ K})} = 0.980 \text{ mol}$$

$$n_{total\,(final)} = \frac{PV}{RT} = \frac{(2.40 \text{ atm})(10.0 \text{ L})}{\left(0.082\ 06 \dfrac{\text{L} \cdot \text{atm}}{\text{K} \cdot \text{mol}}\right)(373.1 \text{ K})} = 0.784 \text{ mol}$$

	$CS_2(g)$	$+$	$3 O_2(g)$	$\rightarrow$	$CO_2(g)$	$+$	$2 SO_2(g)$
before reaction (mol)	y		$0.980 - y$		0		0
change (mol)	$-x$		$-3x$		$+x$		$+2x$
after reaction (mol)	$y - x = 0$		$0.980 - y - 3x$		x		2x

$n_{total\,(final)} = (y - x) + (0.980 - y - 3x) + x + 2x = 0.784 \text{ mol}$

$0.980 \text{ mol} - 4x + 3x = 0.784 \text{ mol}$

$x = 0.980 \text{ mol} - 0.784 \text{ mol} = 0.196 \text{ mol}$

$\text{mol } CO_2 = x = 0.196 \text{ mol}$

$$P_{CO_2} = \frac{nRT}{V} = \frac{(0.196 \text{ mol})\left(0.082\ 06 \dfrac{\text{L} \cdot \text{atm}}{\text{K} \cdot \text{mol}}\right)(373.1 \text{ K})}{(10.0 \text{ L})} = 0.600 \text{ atm}$$

$\text{mol } SO_2 = 2x = 2(0.196 \text{ mol}) = 0.392 \text{ mol}$

$$P_{SO_2} = \frac{nRT}{V} = \frac{(0.392 \text{ mol})\left(0.082\ 06 \dfrac{\text{L} \cdot \text{atm}}{\text{K} \cdot \text{mol}}\right)(373.1 \text{ K})}{(10.0 \text{ L})} = 1.20 \text{ atm}$$

$\text{mol } O_2 = 0.980 \text{ mol} - y - 3x = 0.980 \text{ mol} - x - 3x = 0.980 - 4(0.196 \text{ mol}) = 0.196 \text{ mol}$

$P_{O_2} = P_{CO_2} = 0.600 \text{ atm}$

9.110 (a) $T = 0 \text{ °C} = 273 \text{ K};\ PV = nRT$

$$n_Q = \frac{PV}{RT} = \frac{(0.229 \text{ atm})(0.0500 \text{ L})}{\left(0.082\ 06 \dfrac{\text{L} \cdot \text{atm}}{\text{K} \cdot \text{mol}}\right)(273 \text{ K})} = 5.11 \times 10^{-4} \text{ mol Q}$$

$$Q \text{ molar mass} = \frac{0.100 \text{ g Q}}{5.11 \times 10^{-4} \text{ mol Q}} = 196 \text{ g/mol}$$

Xe molar mass = 131.3 g/mol

O_n molar mass = 196 g/mol – 131.3 g/mol = 65 g/mol; 65/16 ≈ 4

So, n = 4 and XeO_4 is the likely formula for Q.

(b) The decomposition reaction is $XeO_4(g) \rightarrow Xe(g) + 2 O_2(g)$.

After decomposition $n_{Xe} = n_{XeO_4} = 5.11 \times 10^{-4} \text{ mol}$ and $n_{O_2} = 2 \times n_{XeO_4} = 1.02 \times 10^{-3} \text{ mol}$

217

$T = 100 \ °C = 373 \ K$

$$P_{Xe} = \frac{nRT}{V} = \frac{(5.11 \times 10^{-4} \ mol)\left(0.082 \ 06 \ \frac{L \cdot atm}{K \cdot mol}\right)(373 \ K)}{(0.0500 \ L)} = 0.313 \ atm$$

$P_{O_2} = 2 \times P_{Xe} = 2 \times 0.313 \ atm = 0.626 \ atm$

$P_{total} = P_{Xe} + P_{O_2} = 0.313 \ atm + 0.626 \ atm = 0.939 \ atm$

9.111 $Ca(ClO_3)_2$, 206.98 amu; $Ca(ClO)_2$, 142.98 amu
(a) $Ca(ClO_3)_2(s) \rightarrow CaCl_2(s) + 3 \ O_2(g)$
$Ca(ClO)_2(s) \rightarrow CaCl_2(s) + O_2(g)$
(b) $T = 700 \ °C = 700 + 273 = 973 \ K$
$PV = nRT$

$$n_{O_2} = \frac{PV}{RT} = \frac{(1.00 \ atm)(10.0 \ L)}{\left(0.082 \ 06 \ \frac{L \cdot atm}{K \cdot mol}\right)(973 \ K)} = 0.125 \ mol \ O_2$$

Let X = mol $Ca(ClO_3)_2$ and let Y = mol $Ca(ClO)_2$
X(206.98 g/mol) + Y(142.98 g/mol) = 10.0 g
3X + Y = 0.125 mol, so Y = 0.125 mol – 3X (substitute for Y and solve for X)
X(206.98 g/mol) + (0.125 mol – 3X)(142.98 g/mol) = 10.0 g
X(206.98 g/mol) + 17.9 g – X(428.94 g/mol) = 10.0 g
X(206.98 g/mol) – X(428.94 g/mol) = 10.0 g – 17.9 g = –7.9 g
X(–221.96 g/mol) = –7.9 g
X = (–7.9 g)/(–221.96 g/mol) = 0.0356 mol $Ca(ClO_3)_2$
Y = 0.125 mol – 3X; Y = 0.125 mol – 3(0.0356 mol) = 0.0182 mol $Ca(ClO)_2$

$$\text{mass } Ca(ClO_3)_2 = 0.0356 \ mol \ Ca(ClO_3)_2 \times \frac{206.98 \ g \ Ca(ClO_3)_2}{1 \ mol \ Ca(ClO_3)_2} = 7.4 \ g \ Ca(ClO_3)_2$$

mass $Ca(ClO)_2$ = 10.0 g – 7.4 g = 2.6 g $Ca(ClO)_2$

9.112 PCl_3, 137.3 amu; O_2, 32.00 amu; $POCl_3$, 153.3 amu
$2 \ PCl_3(g) + O_2(g) \rightarrow 2 \ POCl_3(g)$

$$\text{mol } PCl_3 = 25.0 \ g \times \frac{1 \ mol \ PCl_3}{137.3 \ g \ PCl_3} = 0.182 \ mol \ PCl_3$$

$$\text{mol } O_2 = 3.00 \ g \times \frac{1 \ mol \ O_2}{32.00 \ g \ O_2} = 0.0937 \ mol \ O_2$$

Check for limiting reactant.

$$\text{mol } O_2 \ needed = 0.182 \ mol \ PCl_3 \times \frac{1 \ mol \ O_2}{2 \ mol \ PCl_3} = 0.0910 \ mol \ O_2 \ needed$$

There is a slight excess of O_2. PCl_3 is the limiting reactant.

$$\text{mol } POCl_3 = 0.182 \ mol \ PCl_3 \times \frac{2 \ mol \ POCl_3}{2 \ mol \ PCl_3} = 0.182 \ mol \ POCl_3$$

mol O_2 left over $= 0.0937$ mol $- 0.0910$ mol $= 0.0027$ mol O_2 left over

$T = 200.0$ °C $= 200.0 + 273.15 = 473.1$ K; $PV = nRT$

$$P = \frac{nRT}{V} = \frac{(0.182 \text{ mol} + 0.0027 \text{ mol})\left(0.082\ 06\ \dfrac{L \cdot atm}{K \cdot mol}\right)(473.1 \text{ K})}{(5.00 \text{ L})} = 1.43 \text{ atm}$$

9.113 (a) $T = 225$ °C $= 225 + 273 = 498$ K

$$PV = nRT \qquad P^o_{NOCl} = \frac{nRT}{V} = \frac{(2.00 \text{ mol})\left(0.082\ 06\ \dfrac{L \cdot atm}{K \cdot mol}\right)(498 \text{ K})}{(400.0 \text{ L})} = 0.204 \text{ atm}$$

	2 NOCl(g)	$\rightarrow$	2 NO(g)	$+$	$Cl_2(g)$
initial (atm)	0.204		0		0
change (atm)	$-2x$		$+2x$		$+x$
equil (atm)	$0.204 - 2x$		$2x$		x

P_{total} (after rxn) $= (0.204 \text{ atm} - 2x) + 2x + x = 0.246$ atm

$x = 0.246$ atm $- 0.204$ atm $= 0.042$ atm

$P_{NO} = 2x = 2(0.042) = 0.084$ atm; $P_{Cl_2} = x = 0.042$ atm

$P_{NOCl} = 0.204 - 2x = 0.204 - 2(0.042) = 0.120$ atm

(b) % NOCl decomposed $= \dfrac{2x}{P^o_{NOCl}} \times 100\% = \dfrac{0.084 \text{ atm}}{0.204 \text{ atm}} \times 100\% = 41\%$

9.114 O_2, 32.00 amu; O_3, 48.00 amu

	$3 \text{ O}_2\text{(g)}$	$\rightarrow$	$2 \text{ O}_3\text{(g)}$
initial (atm)	32.00		0
change (atm)	$-3x$		$+2x$
after rxn (atm)	$32.00 - 3x$		$2x$

$P_{Total} = P_{O_2} + P_{O_3} = 30.64$ atm $= 32.00$ atm $- 3x + 2x = 32.00$ atm $- x$

$x = 32.00$ atm $- 30.64$ atm $= 1.36$ atm

$P_{O_2} = 32.00 - 3x = 32.00 - 3(1.36 \text{ atm}) = 27.92$ atm

$P_{O_3} = 2x = 2(1.36 \text{ atm}) = 2.72$ atm

$T = 25$ °C $= 25 + 273 = 298$ K; $PV = nRT$

$$n_{O_2} = \frac{PV}{RT} = \frac{(27.92 \text{ atm})(10.00 \text{ L})}{\left(0.082\ 06\ \dfrac{L \cdot atm}{K \cdot mol}\right)(298 \text{ K})} = 11.42 \text{ mol } O_2$$

$$n_{O_3} = \frac{PV}{RT} = \frac{(2.72 \text{ atm})(10.00 \text{ L})}{\left(0.082\ 06\ \dfrac{L \cdot atm}{K \cdot mol}\right)(298 \text{ K})} = 1.11 \text{ mol } O_3$$

$$\text{mass } O_2 = 11.42 \text{ mol } O_2 \text{ x } \frac{32.00 \text{ g } O_2}{1 \text{ mol } O_2} = 365.4 \text{ g } O_2$$

$$\text{mass } O_3 = 1.11 \text{ mol } O_3 \text{ x } \frac{48.00 \text{ g } O_3}{1 \text{ mol } O_3} = 53.3 \text{ g } O_3$$

total mass = 365.4 g + 53.3 g = 418.7 g

$$\text{mass \% } O_3 = \frac{\text{mass } O_3}{\text{total mass}} = \frac{53.3 \text{ g}}{418.7 \text{ g}} \text{ x } 100\% = 12.7\%$$

9.115 $CaCO_3$, 100.09 amu; CaO, 56.08 amu

$$\text{mol CaO (or } CO_2) = 25.0 \text{ g } CaCO_3 \text{ x } \frac{1 \text{ mol } CaCO_3}{100.09 \text{ g } CaCO_3} \text{ x } \frac{1 \text{ mol CaO or } CO_2}{1 \text{ mol } CaCO_3} = 0.250 \text{ mol}$$

$$\text{mass CaO} = 0.250 \text{ mol CaO x } \frac{56.08 \text{ g CaO}}{1 \text{ mol CaO}} = 14.02 \text{ g CaO}$$

(a) 500.0 mL = 0.5000 L

$PV = nRT$; $n_{CO_2} = 0.250$ mol

$$P_{CO_2} = \frac{nRT}{V} = \frac{(0.250 \text{ mol})\left(0.082\ 06 \frac{\text{L} \cdot \text{atm}}{\text{K} \cdot \text{mol}}\right)(1500 \text{ K})}{(0.5000 \text{ L})} = 61.5 \text{ atm}$$

(b) $V_{CaO} = (14.02 \text{ g})/(3.34 \text{ g/mL}) = 4.20 \text{ mL}$

V = 500.0 mL – 4.20 mL = 495.8 mL = 0.4958 L

$$P = \frac{nRT}{V - nb} - \frac{an^2}{V^2}$$

$$P = \frac{(0.250 \text{ mol})\left(0.082\ 06 \frac{\text{L} \cdot \text{atm}}{\text{K} \cdot \text{mol}}\right)(1500 \text{ K})}{[(0.4958 \text{ L}) - (0.250 \text{ mol})(0.0427 \text{ L/mol})]} - \frac{\left(3.59 \frac{\text{L}^2 \cdot \text{atm}}{\text{mol}^2}\right)(0.250 \text{ mol})^2}{(0.4958 \text{ L})^2} = 62.5 \text{ atm}$$

Multiconcept Problems

9.116 CO_2, 44.01 amu

$CH_4(g) + 2\ O_2(g) \rightarrow CO_2(g) + 2\ H_2O(g)$ $\qquad \Delta H° = -802$ kJ

(a) 1.00 atm of CH_4 only requires 2.00 atm O_2, therefore O_2 is in excess.

T = 300 °C = 300 + 273 = 573 K; PV = nRT

$$n_{CH_4} = \frac{PV}{RT} = \frac{(1.00 \text{ atm})(4.00 \text{ L})}{\left(0.082\ 06 \frac{\text{L} \cdot \text{atm}}{\text{K} \cdot \text{mol}}\right)(573 \text{ K})} = 0.0851 \text{ mol } CH_4$$

$$n_{O_2} = \frac{PV}{RT} = \frac{(4.00 \text{ atm})(4.00 \text{ L})}{\left(0.082\ 06 \frac{\text{L} \cdot \text{atm}}{\text{K} \cdot \text{mol}}\right)(573 \text{ K})} = 0.340 \text{ mol } O_2$$

$$\text{mass } CO_2 = 0.0851 \text{ mol } CH_4 \times \frac{1 \text{ mol } CO_2}{1 \text{ mol } CH_4} \times \frac{44.01 \text{ g } CO_2}{1 \text{ mol } CO_2} = 3.75 \text{ g } CO_2$$

(b) $q_{rxn} = 0.0851 \text{ mol } CH_4 \times \dfrac{-802 \text{ kJ}}{1 \text{ mol } CH_4} = -68.3 \text{ kJ}$

	$CH_4(g)$	+	$2 O_2(g)$	→	$CO_2(g)$	+	$2 H_2O(g)$
initial (mol)	0.0851		0.340		0		0
change (mol)	–0.0851		–2(0.0851)		+0.0851		+2(0.0851)
after rxn (mol)	0		0.340 – 2(0.0851)		0.0851		0.170

total moles of gas = 0.340 ml – 2(0.0851) mol + 0.0851 mol + 0.170 mol = 0.425 mol gas

$$q_{rxn} = -68.3 \text{ kJ} \times \frac{1000 \text{ J}}{1 \text{ kJ}} = -68,300 \text{ J}$$

$$q_{vessel} = -q_{rxn} = 68,300 \text{ J} = (0.425 \text{ mol})(21 \text{ J/(mol} \cdot \text{°C)})(t_f - 300 \text{ °C}) +$$

$$(14.500 \text{ kg})\left(\frac{1000 \text{ g}}{1 \text{ kg}}\right)(0.449 \text{ J/(g} \cdot \text{°C)})(t_f - 300 \text{ °C})$$

Solve for t_f.

68,300 J = (8.925 J/°C + 6510 J/°C)(t_f – 300 °C) = (6519 J/°C)(t_f – 300 °C)

$$\frac{68,300 \text{ J}}{6519 \text{ J/°C}} = 10.5 \text{ °C} = (t_f - 300 \text{ °C})$$

300 °C + 10.5 °C = t_f

t_f = 310 °C

(c) T = 310 °C = 310 + 273 = 583 K

$$P_{CO_2} = \frac{nRT}{V} = \frac{(0.0851 \text{ mol})\left(0.082 \ 06 \ \dfrac{L \cdot atm}{K \cdot mol}\right)(583 \text{ K})}{(4.00 \text{ L})} = 1.02 \text{ atm}$$

9.117 $X + 3 O_2 \rightarrow 2 CO_2 + 3 H_2O$

(a) X = C_2H_6O

$C_2H_6O + 3 O_2 \rightarrow 2 CO_2 + 3 H_2O$

(b) It is an empirical formula because it is the smallest whole number ratio of atoms. It is also a molecular formula because any higher multiple such as $C_4H_{12}O_2$ does not correspond to a stable electron-dot structure.

(c)

(d) C_2H_6O, 46.07 amu

$$\text{mol } C_2H_6O = 5.000 \text{ g } C_2H_6O \times \frac{1 \text{ mol } C_2H_6O}{46.07 \text{ g } C_2H_6O} = 0.1085 \text{ mol } C_2H_6O$$

$$\Delta H_{combustion} = \frac{-144.2 \text{ kJ}}{0.1085 \text{ mol}} = -1328.6 \text{ kJ/mol}$$

$$\Delta H_{combustion} = [2 \, \Delta H^\circ_f(CO_2) + 3 \, \Delta H^\circ_f(H_2O)] - \Delta H^\circ_f(C_2H_6O)$$

$$\Delta H^\circ_f(C_2H_6O) = [2 \, \Delta H^\circ_f(CO_2) + 3 \, \Delta H^\circ_f(H_2O)] - \Delta H_{combustion}$$

$$= [(2 \, mol)(-393.5 \, kJ/mol) + (3 \, mol)(-241.8 \, kJ/mol)] - (-1328.6 \, kJ)$$

$$= -183.8 \, kJ/mol$$

9.118 (a) $2 \, C_8H_{18}(l) + 25 \, O_2(g) \rightarrow 16 \, CO_2(g) + 18 \, H_2O(g)$

(b) $4.6 \times 10^{10} \, L \, C_8H_{18} \times \dfrac{1000 \, mL}{1 \, L} \times \dfrac{0.792 \, g}{1 \, mL} = 3.64 \times 10^{13} \, g \, C_8H_{18}$

$3.64 \times 10^{13} \, g \, C_8H_{18} \times \dfrac{1 \, mol \, C_8H_{18}}{114.2 \, g \, C_8H_{18}} \times \dfrac{16 \, mol \, CO_2}{2 \, mol \, C_8H_{18}} = 2.55 \times 10^{12} \, mol \, CO_2$

$2.55 \times 10^{12} \, mol \, CO_2 \times \dfrac{44.0 \, g \, CO_2}{1 \, mol \, CO_2} \times \dfrac{1 \, kg}{1000 \, g} = 1.1 \times 10^{11} \, kg \, CO_2$

(c) $V = \dfrac{nRT}{P} = \dfrac{(2.55 \times 10^{12} \, mol)\left(0.082 \, 06 \, \dfrac{L \cdot atm}{K \cdot mol}\right)(273 \, K)}{(1.00 \, atm)} = 5.7 \times 10^{13} \, L \text{ of } CO_2$

(d) 12.5 moles of O_2 are needed for each mole of isooctane (from part a).

$12.5 \, mol \, O_2 = (0.210)(n_{air}); \qquad n_{air} = \dfrac{12.5 \, mol}{0.210} = 59.5 \, mol \, air$

$V = \dfrac{nRT}{P} = \dfrac{(59.5 \, mol)\left(0.082 \, 06 \, \dfrac{L \cdot atm}{K \cdot mol}\right)(273 \, K)}{(1.00 \, atm)} = 1.33 \times 10^3 \, L$

9.119 (a) Freezing point of H_2O on the Rankine scale is $(9/5)(273.15) = 492 \, °R$.

(b) $R = \dfrac{PV}{nT} = \dfrac{(1.00 \, atm)(22.414 \, L)}{(1.00 \, mol)(492 \, °R)} = 0.0456 \, \dfrac{L \cdot atm}{°R \cdot mol}$

(c) $P = \dfrac{(2.50 \, mol)\left(0.0456 \, \dfrac{L \cdot atm}{°R \cdot mol}\right)(525 \, °R)}{[(0.4000 \, L) - (2.50 \, mol)(0.04278 \, L/mol)]} - \dfrac{\left(2.253 \, \dfrac{L^2 \cdot atm}{mol^2}\right)(2.50 \, mol)^2}{(0.4000 \, L)^2}$

$P = 204.2 \, atm - 88.0 \, atm = 116 \, atm$

9.120 $n = \dfrac{PV}{RT} = \dfrac{(1 \, atm)(1323 \, L)}{\left(0.082 \, 06 \, \dfrac{L \cdot atm}{K \cdot mol}\right)(2223 \, K)} = 7.25 \, mol \text{ of all gases}$

(a) $0.004 \, 00 \, mol \text{ "nitro"} \times \dfrac{7.25 \, mol \, gases}{1 \, mol \text{ "nitro"}} = 0.0290 \, mol \text{ hot gases}$

(b) $n = \dfrac{PV}{RT} = \dfrac{\left(623 \, mm \, Hg \times \dfrac{1.00 \, atm}{760 \, mm \, Hg}\right)(0.500 \, L)}{\left(0.082 \, 06 \, \dfrac{L \cdot atm}{K \cdot mol}\right)(263 \, K)} = 0.0190 \, mol \, B + C + D$

$n_A = n_{total} - n_{(B+C+D)} = 0.0290 - 0.0190 = 0.0100$ mol A; A = H_2O

(c) $n = \dfrac{PV}{RT} = \dfrac{\left(260 \text{ mm Hg} \times \dfrac{1.00 \text{ atm}}{760 \text{ mm Hg}}\right)(0.500 \text{ L})}{\left(0.082\ 06 \dfrac{\text{L} \cdot \text{atm}}{\text{K} \cdot \text{mol}}\right)(298 \text{ K})} = 0.007\ 00$ mol C + D

$n_B = n_{(B+C+D)} - n_{(C+D)} = 0.0190 - 0.007\ 00 = 0.0120$ mol B; B = CO_2

(d) $n = \dfrac{PV}{RT} = \dfrac{\left(223 \text{ mm Hg} \times \dfrac{1.00 \text{ atm}}{760 \text{ mm Hg}}\right)(0.500 \text{ L})}{\left(0.082\ 06 \dfrac{\text{L} \cdot \text{atm}}{\text{K} \cdot \text{mol}}\right)(298 \text{ K})} = 0.006\ 00$ mol D

$n_C = n_{(C+D)} - n_D = 0.007\ 00 - 0.006\ 00 = 0.001\ 00$ mol C; C = O_2

molar mass D = $\dfrac{0.168 \text{ g}}{0.006\ 00 \text{ mol}} = 28.0$ g/mol; D = N_2

(e) $0.004\ C_3H_5N_3O_9(l) \rightarrow 0.0100\ H_2O(g) + 0.012\ CO_2(g) + 0.001\ O_2(g) + 0.006\ N_2(g)$
Multiply each coefficient by 1000 to obtain integers.
$4\ C_3H_5N_3O_9(l) \rightarrow 10\ H_2O(g) + 12\ CO_2(g) + O_2(g) + 6\ N_2(g)$

9.121 CO_2, 44.01 amu; H_2O, 18.02 amu

(a) mol C = 0.3744 g CO_2 x $\dfrac{1 \text{ mol } CO_2}{44.01 \text{ g } CO_2}$ x $\dfrac{1 \text{ mol C}}{1 \text{ mol } CO_2}$ = 0.008 507 mol C

mass C = 0.008 507 mol C x $\dfrac{12.011 \text{ g C}}{1 \text{ mol C}}$ = 0.1022 g C

mol H = 0.1838 g H_2O x $\dfrac{1 \text{ mol } H_2O}{18.02 \text{ g } H_2O}$ x $\dfrac{2 \text{ mol H}}{1 \text{ mol } H_2O}$ = 0.020 400 mol H

mass H = 0.020 400 mol H x $\dfrac{1.008 \text{ g H}}{1 \text{ mol H}}$ = 0.02056 g H

mass O = 0.1500 g – 0.1022 g – 0.02056 g = 0.0272 g O

mol O = 0.0272 g O x $\dfrac{1 \text{ mol O}}{16.00 \text{ g O}}$ = 0.001 70 mol O

$C_{0.008\ 507}H_{0.020\ 400}O_{0.001\ 70}$; divide each subscript by the smallest, 0.001 70.
$C_{0.008\ 507\,/\,0.001\ 70}H_{0.020\ 400\,/\,0.001\ 70}O_{0.001\ 70\,/\,0.001\ 70}$
The empirical formula is $C_5H_{12}O$.
The empirical formula mass is 88 g/mol.

(b) 1 atm = 101,325 Pa; T = 54.8 °C = 54.8 + 273.15 = 327.9 K

$PV = nRT$

$$n = \frac{PV}{RT} = \frac{\left(100.0 \text{ kPa} \times \dfrac{1.00 \text{ atm}}{101.325 \text{ kPa}}\right)(1.00 \text{ L})}{\left(0.082\ 06 \dfrac{\text{L} \cdot \text{atm}}{\text{K} \cdot \text{mol}}\right)(327.9 \text{ K})} = 0.0367 \text{ mol methyl } \textit{tert}\text{-butyl ether}$$

methyl *tert*-butyl ether molar mass $= \dfrac{3.233 \text{ g}}{0.0367 \text{ mol}} = 88.1 \text{ g/mol}$

The empirical formula mass and the molar mass are the same, so the molecular formula and empirical formula are the same. $C_5H_{12}O$ is the molecular formula and 88.15 amu is the molecular mass for methyl *tert*-butyl ether.

(c) $C_5H_{12}O(l) + 15/2\ O_2(g) \rightarrow 5\ CO_2(g) + 6\ H_2O(l)$

(d) $\Delta H°_{combustion} = [5\ \Delta H°_f\ (CO_2) + 6\ \Delta H°_f\ (H_2O(l))] - \Delta H°_f\ (C_5H_{12}O) = -3368.7 \text{ kJ}$

$-3368.7 \text{ kJ} = [(5 \text{ mol})(-393.5 \text{ kJ/mol}) + (6 \text{ mol})(-285.8 \text{ kJ/mol})] - (1 \text{ mol})\Delta H°_f\ (C_5H_{12}O)$

$(1 \text{ mol})\Delta H°_f\ (C_5H_{12}O) = [(5 \text{ mol})(-393.5 \text{ kJ/mol}) + (6 \text{ mol})(-285.8 \text{ kJ/mol})] + 3368.7 \text{ kJ}$

$\Delta H°_f\ (C_5H_{12}O) = -313.6 \text{ kJ/mol}$

10 Liquids, Solids, and Phase Changes

10.1 $\mu = Q \times r = (1.60 \times 10^{-19}\ C)(92 \times 10^{-12}\ m)\left(\dfrac{1\ D}{3.336 \times 10^{-30}\ C\cdot m}\right) = 4.41\ D$

% ionic character for HF $= \dfrac{1.83\ D}{4.41\ D} \times 100\% = 41\%$

HF has more ionic character than HCl. HCl has only 18% ionic character.

10.2 (a) SF_6 has polar covalent bonds but the molecule is symmetrical (octahedral). The individual bond polarities cancel, and the molecule has no dipole moment.
(b) $H_2C=CH_2$ can be assumed to have nonpolar C–H bonds. In addition, the molecule is symmetrical. The molecule has no dipole moment.

(c) The C–Cl bonds in $CHCl_3$ are polar covalent bonds, and the molecule is polar.

(d) The C–Cl bonds in CH_2Cl_2 are polar covalent bonds, and the molecule is polar.

10.3 net

10.4 The N atom is electron rich (red) because of its high electronegativity. The C and H atoms are electron poor (blue) because they are less electronegative.

10.5 (a) Of the four substances, only HNO_3 has a net dipole moment.
(b) Only HNO_3 can hydrogen bond.
(c) Ar has fewer electrons than Cl_2 and CCl_4, and has the smallest dispersion forces.

10.6 H_2S dipole-dipole, dispersion
CH_3OH hydrogen bonding, dipole-dipole, dispersion
C_2H_6 dispersion
Ar   dispersion
$Ar < C_2H_6 < H_2S < CH_3OH$

10.7 (a) $CO_2(s) \rightarrow CO_2(g)$, ΔS is positive (b) $H_2O(g) \rightarrow H_2O(l)$, ΔS is negative
(c) ΔS is positive (more randomness)

10.8 $\Delta G = \Delta H - T\Delta S$; at the boiling point (phase change), $\Delta G = 0$.

$$\Delta H = T\Delta S; \quad T = \frac{\Delta H_{vap}}{\Delta S_{vap}} = \frac{29.2 \text{ kJ/mol}}{87.5 \times 10^{-3} \text{ kJ/(K} \cdot \text{mol)}} = 334 \text{ K}$$

10.9 The boiling point is the temperature where the vapor pressure of a liquid equals the external pressure.
$P_1 = 760$ mm Hg; $P_2 = 260$ mm Hg; $T_1 = 80.1$ °C
$\Delta H_{vap} = 30.7$ kJ/mol

$$\ln P_2 = \ln P_1 + \frac{\Delta H_{vap}}{R}\left(\frac{1}{T_1} - \frac{1}{T_2}\right)$$

$$(\ln P_2 - \ln P_1)\left(\frac{R}{\Delta H_{vap}}\right) = \frac{1}{T_1} - \frac{1}{T_2}$$

Solve for T_2 (the boiling point for benzene at 260 mm Hg).

$$\frac{1}{T_1} - (\ln P_2 - \ln P_1)\left(\frac{R}{\Delta H_{vap}}\right) = \frac{1}{T_2}$$

$$\frac{1}{353.2 \text{ K}} - [\ln(260) - \ln(760)]\left(\frac{8.3145 \dfrac{\text{J}}{\text{K} \cdot \text{mol}}}{30,700 \text{ J/mol}}\right) = \frac{1}{T_2}$$

$$\frac{1}{T_2} = 0.003\ 122 \text{ K}^{-1}; \ T_2 = 320 \text{ K} = 47\ ^\circ\text{C} \quad \text{(boiling point is lower at lower pressure)}$$

10.10 $$\Delta H_{vap} = \frac{(\ln P_2 - \ln P_1)(R)}{\left(\dfrac{1}{T_1} - \dfrac{1}{T_2}\right)}$$

$P_1 = 400$ mm Hg; $T_1 = 41.0$ °C $= 314.2$ K
$P_2 = 760$ mm Hg; $T_2 = 331.9$ K

$$\Delta H_{vap} = \frac{[\ln(760) - \ln(400)]\left(8.3145 \dfrac{\text{J}}{\text{K} \cdot \text{mol}}\right)}{\left(\dfrac{1}{314.2 \text{ K}} - \dfrac{1}{331.9 \text{ K}}\right)} = 31,442 \text{ J/mol} = 31.4 \text{ kJ/mol}$$

10.11 (a) 1/8 atom at 8 corners and 1 atom at body center = 2 atoms
(b) 1/8 atom at 8 corners and 1/2 atom at 6 faces = 4 atoms

10.12 For a simple cube, $d = 2r$; $\quad r = \dfrac{d}{2} = \dfrac{334 \text{ pm}}{2} = 167$ pm

10.13   For a simple cube, there is one atom per unit cell.

mass of one Po atom = 209 g/mol x $\dfrac{1\ mol}{6.022 \times 10^{23}\ atoms}$ = 3.4706 x 10^{-22} g/atom

unit cell edge = d = 334 pm = 334 x 10^{-12} m = 3.34 x 10^{-8} cm
unit cell volume = d^3 = (3.34 x 10^{-8} cm)^3 = 3.7260 x 10^{-23} cm^3

density = $\dfrac{mass}{volume}$ = $\dfrac{3.4706 \times 10^{-22}\ g}{3.7260 \times 10^{-23}\ cm^3}$ = 9.31 g/cm^3

10.14   There are several possibilities. Here's one:

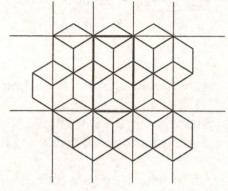

10.15   For CuCl:
1/8 Cl$^-$ at 8 corners and 1/2 Cl$^-$ at 6 faces = 4 Cl$^-$ (4 minuses)
4 Cu$^+$ inside (4 pluses)
For BaCl$_2$:
1/8 Ba^{2+} at 8 corners and 1/2 Ba^{2+} at 6 faces = 4 Ba^{2+} (8 pluses)
8 Cl$^-$ inside (8 minuses)

10.16   (a) In the unit cell there is a rhenium atom at each corner of the cube. The number of
rhenium atoms in the unit cell = 1/8 Re at 8 corners = 1 Re atom.
In the unit cell there is an oxygen atom in the center of each edge of the cube. The
number of oxygen atoms in the unit cell = 1/4 O on 12 edges = 3 O atoms.
(b) ReO$_3$
(c) Each oxide has a –2 charge and there are three of them for a total charge of –6. The
charge (oxidation state) of rhenium must be +6 to balance the negative charge of the oxides.
(d) Each oxygen atom is surrounded by two rhenium atoms. The geometry is linear.
(e) Each rhenium atom is surrounded by six oxygen atoms. The geometry is octahedral.

10.17   The minimum pressure at which liquid CO$_2$ can exist is its triple point pressure of 5.11 atm.

10.18   (a) CO$_2$(s) → CO$_2$(g)
(b) CO$_2$(l) → CO$_2$(g)
(c) CO$_2$(g) → CO$_2$(l) → supercritical CO$_2$

10.19 (a)

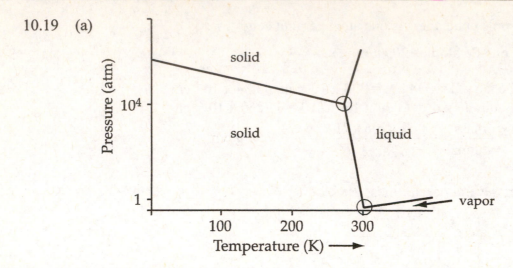

(b) Gallium has two triple points. The one below 1 atm is a solid, liquid, vapor triple point. The one at 10^4 atm is a solid(1), solid(2), liquid triple point.
(c) Increasing the pressure favors the liquid phase, giving the solid/liquid boundary a negative slope. At 1 atm pressure the liquid phase is more dense than the solid phase.

10.20 As the name implies, the constituent particles in an ionic liquid are cations and anions rather than molecules.

10.21 In ionic liquids the cation has an irregular shape and one or both of the ions are large and bulky to disperse charges over a large volume. Both factors minimize the crystal lattice energy, making the solid less stable and favoring the liquid.

Key Concept Problems

10.22 The electronegative O atoms are electron rich (red), while the rest of the molecule is electron poor (blue).

10.23 (a) cubic closest-packed   (b) simple cubic
     (c) hexagonal closest-packed   (d) body-centered cubic

10.24 (a) cubic closest-packed
(b) 1/8 S^{2-} at 8 corners and 1/2 S^{2-} at 6 faces = 4 S^{2-};   4 Zn^{2+} inside

10.25 (a) 1/8 Ca^{2+} at 8 corners = 1 Ca^{2+}; 1/2 O^{2-} at 6 faces = 3 O^{2-}; 1 Ti^{4+} inside
The formula for perovskite is $CaTiO_3$.
(b) The oxidation number of Ti is +4 to maintain charge neutrality in the unit cell.

10.26 (a) normal boiling point ≈ 300 K; normal melting point ≈ 180 K
(b) (i) solid   (ii) gas   (iii) supercritical fluid

10.27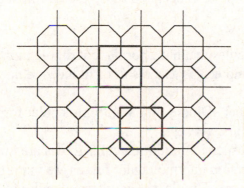

10.28 Here are two possibilities:

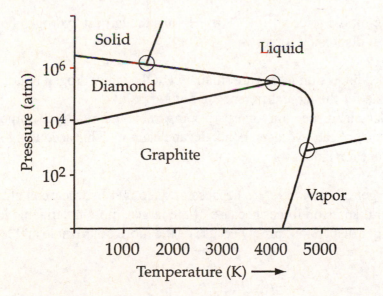

10.29 (a), (c), (d)

(b) There are three triple points.

(e) The solid phase that is stable at the higher pressure is more dense. The more dense phase is diamond.

10.30

10.31   After the volume is decreased, equilibrium is reestablished and the equilibrium vapor
        pressure will still be 28.0 mm Hg. Vapor pressure depends only on temperature, not on
        volume.

Section Problems
Dipole Moments and Intermolecular Forces (Sections 10.1–10.2)

10.32   If a molecule has polar covalent bonds, the molecular shape (and location of lone pairs of
        electrons) determines whether the bond dipoles cancel and thus whether the molecule
        has a dipole moment.

10.33   Dipole-dipole forces arise between molecules that have permanent dipole moments.
        London dispersion forces arise between molecules as a result of induced temporary dipoles.

10.34   (a) $CHCl_3$ has a permanent dipole moment. Dipole-dipole intermolecular forces are
        important. London dispersion forces are also present.
        (b) O_2 has no dipole moment. London dispersion intermolecular forces are important.
        (c) Polyethylene, C_nH_{2n+2}. London dispersion intermolecular forces are important.
        (d) CH_3OH has a permanent dipole moment. Dipole-dipole intermolecular forces and
        hydrogen bonding are important. London dispersion forces are also present.

10.35   (a) Xe has no dipole-dipole forces   (b) HF has the largest hydrogen bond forces
        (c) Xe has the largest dispersion forces

10.36   For CH_3OH and CH_4, dispersion forces are small. CH_3OH can hydrogen bond; CH_4
        cannot. This accounts for the large difference in boiling points.
        For 1-decanol and decane, dispersion forces are comparable and relatively large along
        the C–H chain. 1-decanol can hydrogen bond; decane cannot. This accounts for the 57
        °C higher boiling point for 1-decanol.

10.37   (a) C_8H_{18} has the larger dispersion forces because of its longer hydrocarbon chain.
        (b) HI has the larger dispersion forces because of the larger, more polarizable iodine.
        (c) H_2Se has the larger dispersion forces because of the more polarizable and less
        electronegative Se.

10.38   (a)   (b)

230

(c) 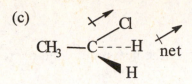 (d)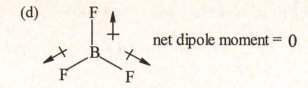

net dipole moment = 0

10.39 (a) 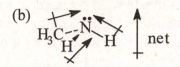 net (b) 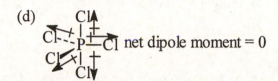 net

(c)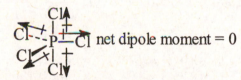

F—Xe—F
net dipole moment = 0

(d)

Cl---P≡Cl net dipole moment = 0

10.40

O=S=O ↓ net   O=C=O
net dipole moment = 0

SO$_2$ is bent and the individual bond dipole moments add to give the molecule a net dipole moment.
CO$_2$ is linear and the individual bond dipole moments point in opposite directions to cancel each other out. CO$_2$ has no net dipole moment.

10.41

Cl---P Cl ↓ net   Cl---P≡Cl net dipole moment = 0

In both PCl$_3$ and PCl$_5$ the P–Cl bond is polar covalent. PCl$_3$ is trigonal pyramidal and the bond dipoles add to give the molecule a net dipole moment. PCl$_5$ is trigonal bipyramidal and the bond dipoles cancel. PCl$_5$ has no dipole moment.

10.42

:N—H---:N—H
hydrogen bond

10.43

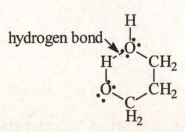

hydrogen bond

Vapor Pressure and Phase Changes (Sections 10.4–10.5)

10.44 ΔH_{vap} is usually larger than ΔH_{fusion} because ΔH_{vap} is the heat required to overcome all intermolecular forces.

10.45 Sublimation is the direct conversion of a solid to a gas. A solid can also be converted to a gas in two steps; melting followed by vaporization. The energy to convert a solid to a gas must be the same regardless of the path. Therefore $\Delta H_{subl} = \Delta H_{fusion} + \Delta H_{vap}$.

10.46 (a) $Hg(l) \rightarrow Hg(g)$ (b) no change of state, Hg remains a liquid
(c) $Hg(g) \rightarrow Hg(l) \rightarrow Hg(s)$

10.47 (a) solid I_2 melts to form liquid I_2 (b) no change of state, I_2 remains a liquid

10.48 As the pressure over the liquid H_2O is lowered, H_2O vapor is removed by the pump. As H_2O vapor is removed, more of the liquid H_2O is converted to H_2O vapor. This conversion is an endothermic process and the temperature decreases. The combination of both a decrease in pressure and temperature takes the system across the liquid/solid boundary in the phase diagram so the H_2O that remains turns to ice.

10.49 The normal boiling point for ether is relatively low (34.6 °C). As the pressure is reduced by the pump, the relatively high vapor pressure of the ether equals the external pressure produced by the pump, and the liquid boils.

10.50 H_2O, 18.02 amu; $5.00 \text{ g } H_2O \times \dfrac{1 \text{ mol } H_2O}{18.02 \text{ g } H_2O} = 0.2775 \text{ mol } H_2O$

$q_1 = (0.2775 \text{ mol})[36.6 \times 10^{-3} \text{ kJ/(K} \cdot \text{mol)}](273 \text{ K} - 263 \text{ K}) = 0.1016 \text{ kJ}$
$q_2 = (0.2775 \text{ mol})(6.01 \text{ kJ/mol}) = 1.668 \text{ kJ}$
$q_3 = (0.2775 \text{ mol})(75.3 \times 10^{-3} \text{ kJ/(K} \cdot \text{mol)}](303 \text{ K} - 273 \text{ K}) = 0.6269 \text{ kJ}$
$q_{total} = q_1 + q_2 + q_3 = 2.40 \text{ kJ}$;   2.40 kJ of heat is required.

10.51 H_2O, 18.02 amu; $15.3 \text{ g } H_2O \times \dfrac{1 \text{ mol } H_2O}{18.02 \text{ g } H_2O} = 0.8491 \text{ mol } H_2O$

$q_1 = (0.8491 \text{ mol})[33.6 \times 10^{-3} \text{ kJ/(K} \cdot \text{mol)}](373 \text{ K} - 388 \text{ K}) = -0.4279 \text{ kJ}$
$q_2 = -(0.8491 \text{ mol})(40.67 \text{ kJ/mol}) = -34.53 \text{ kJ}$
$q_3 = (0.8491 \text{ mol})[75.3 \times 10^{-3} \text{ kJ/(K} \cdot \text{mol)}](348 \text{ K} - 373 \text{ K}) = -1.598 \text{ kJ}$
$q_{total} = q_1 + q_2 + q_3 = -36.6 \text{ kJ}$;   36.6 kJ of heat is released.

10.52 H_2O, 18.02 amu; $7.55 \text{ g } H_2O \times \dfrac{1 \text{ mol } H_2O}{18.02 \text{ g } H_2O} = 0.4190 \text{ mol } H_2O$

$q_1 = (0.4190 \text{ mol})[75.3 \times 10^{-3} \text{ kJ/(K} \cdot \text{mol)}](273.15 \text{ K} - 306.65 \text{ K}) = -1.057 \text{ kJ}$
$q_2 = -(0.4190 \text{ mol})(6.01 \text{ kJ/mol}) = -2.518 \text{ kJ}$
$q_3 = (0.4190 \text{ mol})[36.6 \times 10^{-3} \text{ kJ/(K} \cdot \text{mol)}](263.15 \text{ K} - 273.15 \text{ K}) = -0.1534 \text{ kJ}$
$q_{total} = q_1 + q_2 + q_3 = -3.73 \text{ kJ}$;   3.73 kJ of heat is released.

10.53 C_2H_5OH, 46.07 amu;   25.0 g C_2H_5OH x $\dfrac{\text{1 mol } C_2H_5OH}{\text{46.07 g } C_2H_5OH}$ = 0.543 mol C_2H_5OH

$q_1 = (0.543 \text{ mol})[65.6 \times 10^{-3} \text{ kJ/(K} \cdot \text{mol)}](351.45 \text{ K} - 366.15 \text{ K}) = -0.524 \text{ kJ}$

$q_2 = -(0.543 \text{ mol})(38.56 \text{ kJ/mol}) = -20.94 \text{ kJ}$

$q_3 = (0.543 \text{ mol})[112.3 \times 10^{-3} \text{ kJ/(K} \cdot \text{mol)}](263.15 \text{ K} - 351.45 \text{ K}) = -5.38 \text{ kJ}$

$q_{total} = q_1 + q_2 + q_3 = -26.8 \text{ kJ}$;   26.8 kJ of heat is released.

10.54

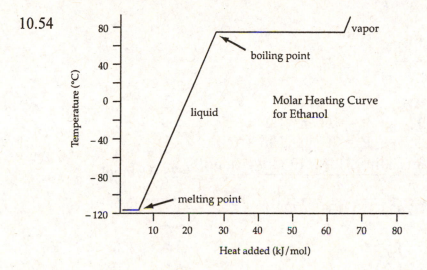

10.55

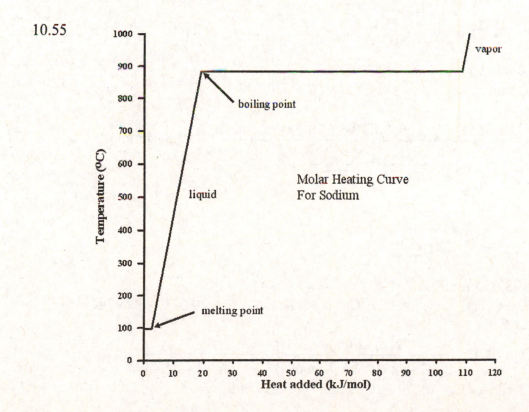

10.56   boiling point = 218 °C = 491 K

$\Delta G = \Delta H_{vap} - T\Delta S_{vap}$; At the boiling point (phase change), $\Delta G = 0$

$\Delta H_{vap} = T\Delta S_{vap}$; $\Delta S_{vap} = \dfrac{\Delta H_{vap}}{T} = \dfrac{43.3 \text{ kJ/mol}}{491 \text{ K}} = 0.0882 \text{ kJ/(K}\cdot\text{mol)} = 88.2 \text{ J/(K}\cdot\text{mol)}$

10.57 $\Delta S_{fus} = \dfrac{\Delta H_{fus}}{T} = \dfrac{2.64 \text{ kJ/mol}}{371 \text{ K}} = 0.007\,12 \text{ kJ/(K}\cdot\text{mol)} = 7.12 \text{ J/(K}\cdot\text{mol)}$

10.58 $\Delta H_{vap} = \dfrac{(\ln P_2 - \ln P_1)(R)}{\left(\dfrac{1}{T_1} - \dfrac{1}{T_2}\right)}$

$T_1 = -5.1 \text{ °C} = 268.0 \text{ K};$ $P_1 = 100 \text{ mm Hg}$
$T_2 = 46.5 \text{ °C} = 319.6 \text{ K};$ $P_2 = 760 \text{ mm Hg}$

$\Delta H_{vap} = \dfrac{[\ln(760) - \ln(100)][8.3145 \times 10^{-3} \text{ kJ/(K}\cdot\text{mol)}]}{\left(\dfrac{1}{268.0 \text{ K}} - \dfrac{1}{319.6 \text{ K}}\right)} = 28.0 \text{ kJ/mol}$

10.59 $\Delta H_{vap} = \dfrac{(\ln P_2 - \ln P_1)(R)}{\left(\dfrac{1}{T_1} - \dfrac{1}{T_2}\right)}$

$P_1 = 100 \text{ mm Hg};$ $T_1 = 5.4 \text{ °C} = 278.6 \text{ K}$
$P_2 = 760 \text{ mm Hg};$ $T_2 = 57.7 \text{ °C} = 330.8 \text{ K}$

$\Delta H_{vap} = \dfrac{[\ln(760) - \ln(100)][8.3145 \times 10^{-3} \text{ kJ/(K}\cdot\text{mol)}]}{\left(\dfrac{1}{278.6 \text{ K}} - \dfrac{1}{330.8 \text{ K}}\right)} = 29.8 \text{ kJ/mol}$

10.60 $\ln P_2 = \ln P_1 + \dfrac{\Delta H_{vap}}{R}\left(\dfrac{1}{T_1} - \dfrac{1}{T_2}\right)$

$\Delta H_{vap} = 28.0 \text{ kJ/mol}$
$P_1 = 100 \text{ mm Hg};$ $T_1 = -5.1 \text{ °C} = 268.0 \text{ K};$ $T_2 = 20.0 \text{ °C} = 293.2 \text{ K}$
Solve for P_2.

$\ln P_2 = \ln(100) + \dfrac{28.0 \text{ kJ/mol}}{[8.3145 \times 10^{-3} \text{ kJ/(K}\cdot\text{mol)}]}\left(\dfrac{1}{268.0 \text{ K}} - \dfrac{1}{293.2 \text{ K}}\right)$

$\ln P_2 = 5.6852$; $P_2 = e^{5.6852} = 294.5 \text{ mm Hg} = 294 \text{ mm Hg}$

10.61 $\ln P_2 = \ln P_1 + \dfrac{\Delta H_{vap}}{R}\left(\dfrac{1}{T_1} - \dfrac{1}{T_2}\right)$

$\Delta H_{vap} = 29.8$ kJ/mol

$P_1 = 100$ mm Hg; $T_1 = 5.4\ °C = 278.6$ K; $T_2 = 30.0\ °C = 303.2$ K

Solve for P_2.

$\ln P_2 = \ln(100) + \dfrac{29.8 \text{ kJ/mol}}{[8.3145 \times 10^{-3} \text{ kJ/(K·mol)}]}\left(\dfrac{1}{278.6 \text{ K}} - \dfrac{1}{303.2 \text{ K}}\right)$

$\ln P_2 = 5.6489;\quad P_2 = e^{5.6489} = 284$ mm Hg

10.62

T(K)	P_{vap}(mm Hg)	$\ln P_{vap}$	1/T
263	80.1	4.383	0.003 802
273	133.6	4.8949	0.003 663
283	213.3	5.3627	0.003 534
293	329.6	5.7979	0.003 413
303	495.4	6.2054	0.003 300
313	724.4	6.5853	0.003 195

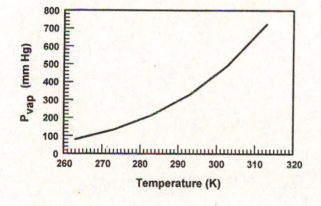
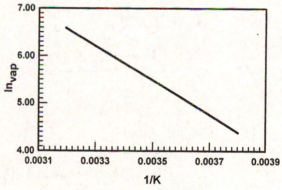

$\ln P_{vap} = \left(-\dfrac{\Delta H_{vap}}{R}\right)\dfrac{1}{T} + C;\quad C = 18.2$

$\text{slope} = -3628 \text{ K} = -\dfrac{\Delta H_{vap}}{R}$

$\Delta H_{vap} = (3628 \text{ K})(R) = (3628 \text{ K})[8.3145 \times 10^{-3} \text{ kJ/(K·mol)}] = 30.1 \text{ kJ/mol}$

10.63

T(K)	P_{vap}(mm Hg)	$\ln P_{vap}$	1/T
500	39.3	3.671	0.002 000
520	68.5	4.227	0.001 923
540	114.4	4.7397	0.001 852
560	191.6	5.2554	0.001 786
580	286.4	5.6574	0.001 724
600	432.3	6.0691	0.001 667

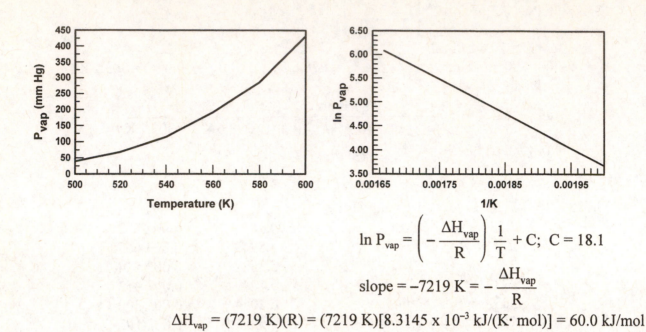

$$\ln P_{vap} = \left(-\frac{\Delta H_{vap}}{R}\right)\frac{1}{T} + C; \quad C = 18.1$$

$$\text{slope} = -7219 \text{ K} = -\frac{\Delta H_{vap}}{R}$$

$$\Delta H_{vap} = (7219 \text{ K})(R) = (7219 \text{ K})[8.3145 \times 10^{-3} \text{ kJ/(K} \cdot \text{mol})] = 60.0 \text{ kJ/mol}$$

10.64 $\Delta H_{vap} = 30.1$ kJ/mol

10.65 $\Delta H_{vap} = 60.0$ kJ/mol

10.66 $\Delta H_{vap} = \dfrac{(\ln P_2 - \ln P_1)(R)}{\left(\dfrac{1}{T_1} - \dfrac{1}{T_2}\right)}$

$P_1 = 80.1$ mm Hg; $\qquad T_1 = 263$ K
$P_2 = 724.4$ mm Hg; $\qquad T_2 = 313$ K

$$\Delta H_{vap} = \frac{[\ln(724.4) - \ln(80.1)][8.3145 \times 10^{-3} \text{ kJ/(K} \cdot \text{mol})]}{\left(\dfrac{1}{263 \text{ K}} - \dfrac{1}{313 \text{ K}}\right)} = 30.1 \text{ kJ/mol}$$

The calculated ΔH_{vap} and that obtained from the plot in Problem 10.62 are the same.

10.67 $\Delta H_{vap} = \dfrac{(\ln P_2 - \ln P_1)(R)}{\left(\dfrac{1}{T_1} - \dfrac{1}{T_2}\right)}$

$P_1 = 39.3$ mm Hg; $\qquad T_1 = 500$ K
$P_2 = 432.3$ mm Hg; $\qquad T_2 = 600$ K

$$\Delta H_{vap} = \frac{[\ln(432.3) - \ln(39.3)][8.3145 \times 10^{-3} \text{ kJ/(K} \cdot \text{mol})]}{\left(\dfrac{1}{500 \text{ K}} - \dfrac{1}{600 \text{ K}}\right)} = 59.8 \text{ kJ/mol}$$

The calculated ΔH_{vap} and that obtained from the plot in Problem 10.63 are consistent with each other. The value from the slope is 60.0 kJ/mol

Structures of Solids (Sections 10.8–10.9)

10.68 molecular solid, CO_2, I_2; metallic solid, any metallic element;
covalent network solid, diamond; ionic solid, NaCl

10.69 molecular solid, covalent molecules; metallic solid, metal atoms;
covalent network solid, nonmetal atoms; ionic solid, cations and anions

10.70 The unit cell is the smallest repeating unit in a crystal.

10.71 From Table 10.10.
Hexagonal and cubic closest packing are the most efficient because 74% of the available
space is used.
Simple cubic packing is the least efficient because only 52% of the available space is used.

10.72 Cu is face-centered cubic. $d = 362$ pm; $r = \sqrt{\dfrac{d^2}{8}} = \sqrt{\dfrac{(362 \text{ pm})^2}{8}} = 128$ pm

362 pm = 362 x 10^{-12} m = 3.62 x 10^{-8} cm
unit cell volume = (3.62 x 10^{-8} cm)^3 = 4.74 x 10^{-23} cm^3

mass of one Cu atom = 63.55 g/mol x $\dfrac{1 \text{ mol}}{6.022 \times 10^{23} \text{ atom}}$ = 1.055 x 10^{-22} g/atom

Cu is face-centered cubic; there are, therefore four Cu atoms in the unit cell.
unit cell mass = (4 atoms)(1.055 x 10^{-22} g/atom) = 4.22 x 10^{-22} g

density = $\dfrac{\text{mass}}{\text{volume}}$ = $\dfrac{4.22 \times 10^{-22} \text{g}}{4.74 \times 10^{-23} \text{ cm}^3}$ = 8.90 g/cm^3

10.73 Pb is face-centered cubic. $d = 495$ pm = 4.95 x 10^{-8} cm

$r = \sqrt{\dfrac{d^2}{8}} = \sqrt{\dfrac{(495 \text{ pm})^2}{8}} = 175$ pm

unit cell volume = (4.95 x 10^{-8} cm)^3 = 1.2129 x 10^{-22} cm^3

mass of one Pb atom = 207.2 g/mol x $\dfrac{1 \text{ mol}}{6.022 \times 10^{23} \text{ atoms}}$ = 3.4407 x 10^{-22} g/atom

Pb is face-centered cubic; there are, therefore four Pb atoms in the unit cell.

density = $\dfrac{\text{mass}}{\text{volume}}$ = $\dfrac{4(3.4407 \times 10^{-22} \text{ g})}{1.2129 \times 10^{-22} \text{ cm}^3}$ = 11.3 g/cm^3

10.74 mass of one Al atom = 26.98 g/mol x $\dfrac{1 \text{ mol}}{6.022 \times 10^{23} \text{ atom}}$ = 4.480 x 10^{-23} g/atom

Al is face-centered cubic; there are, therefore four Al atoms in the unit cell.
unit cell mass = (4 atoms)(4.480 x 10^{-23} g/atom) = 1.792 x 10^{-22} g

density = $\dfrac{\text{mass}}{\text{volume}}$

$$\text{unit cell volume} = \frac{\text{unit cell mass}}{\text{density}} = \frac{1.792 \times 10^{-22} \text{ g}}{2.699 \text{ g/cm}^3} = 6.640 \times 10^{-23} \text{ cm}^3$$

$$\text{unit cell edge} = d = \sqrt[3]{6.640 \times 10^{-23} \text{ cm}^3} = 4.049 \times 10^{-8} \text{ cm}$$

$$d = 4.049 \times 10^{-8} \text{ cm} \times \frac{1 \text{ m}}{100 \text{ cm}} = 4.049 \times 10^{-10} \text{ m} = 404.9 \times 10^{-12} \text{ m} = 404.9 \text{ pm}$$

10.75　W is body-centered cubic. d = 317 pm
a = edge = d; b = face diagonal; c = body diagonal
$b^2 = 2a^2$
$c^2 = a^2 + b^2$
$c^2 = a^2 + 2a^2 = 3a^2$
$c = \sqrt{3}\, a$

unit cell body diagonal = $\sqrt{3}\, d = \sqrt{3}$ (317 pm) = 549 pm

10.76　unit cell body diagonal = 4r = 549 pm

For W, $r = \frac{549 \text{ pm}}{4} = 137$ pm

10.77　mass of one Na atom = 23.0 g/mol x $\frac{1 \text{ mol}}{6.022 \times 10^{23} \text{ atoms}} = 3.82 \times 10^{-23}$ g/atom

Because Na is body-centered cubic; there are two Na atoms in the unit cell.
unit cell mass = 2(3.82 x 10⁻²³ g) = 7.64 x 10⁻²³ g

$$\text{unit cell volume} = \frac{\text{unit cell mass}}{\text{density}} = \frac{7.64 \times 10^{-23} \text{ g}}{0.971 \text{ g/cm}^3} = 7.87 \times 10^{-23} \text{ cm}^3$$

$$\text{unit cell edge} = d = \sqrt[3]{7.87 \times 10^{-23} \text{ cm}^3} = 4.29 \times 10^{-8} \text{ cm} = 429 \text{ pm}$$

$$4r = \sqrt{3}\, d; \quad r = \frac{\sqrt{3}\, d}{4} = \frac{\sqrt{3}\,(429 \text{ pm})}{4} = 186 \text{ pm}$$

10.78　mass of one Ti atom = 47.88 g/mol x $\frac{1 \text{ mol}}{6.022 \times 10^{23} \text{ atoms}} = 7.951 \times 10^{-23}$ g/atom

r = 144.8 pm = 144.8 x 10⁻¹² m

r = 144.8 x 10⁻¹² m x $\frac{100 \text{ cm}}{1 \text{ m}}$ = 1.448 x 10⁻⁸ cm

Calculate the volume and then the density for Ti assuming it is primitive cubic, body-centered cubic, and face-centered cubic. Compare the calculated density with the actual density to identify the unit cell.

For primitive cubic:
　　d = 2r; volume = d³ = [2(1.448 x 10⁻⁸ cm)]³ = 2.429 x 10⁻²³ cm³

$$\text{density} = \frac{\text{unit cell mass}}{\text{volume}} = \frac{7.951 \times 10^{-23} \text{ g}}{2.429 \times 10^{-23} \text{ cm}^3} = 3.273 \text{ g/cm}^3$$

For face-centered cubic:

$$d = 2\sqrt{2}\,r;\ \text{volume} = d^3 = [2\sqrt{2}\,(1.448 \times 10^{-8}\text{ cm})]^3 = 6.870 \times 10^{-23}\text{ cm}^3$$

$$\text{density} = \frac{4(7.951 \times 10^{-23}\text{ g})}{6.870 \times 10^{-23}\text{ cm}^3} = 4.630\text{ g/cm}^3$$

For body-centered cubic:

From Problems 10.75 and 10.76,

$$d = \frac{4r}{\sqrt{3}};\ \text{volume} = d^3 = \left[\frac{4(1.448 \times 10^{-8}\text{ cm})}{\sqrt{3}}\right]^3 = 3.739 \times 10^{-23}\text{ cm}^3$$

$$\text{density} = \frac{2(7.951 \times 10^{-23}\text{ g})}{3.739 \times 10^{-23}\text{ cm}^3} = 4.253\text{ g/cm}^3$$

The calculated density for a face-centered cube (4.630 g/cm^3) is closest to the actual density of 4.54 g/cm^3. Ti crystallizes in the face-centered cubic unit cell.

10.79   mass of one Ca = 40.08 g/mol x $\dfrac{1\text{ mol}}{6.022 \times 10^{23}\text{ atom}}$ = 6.656×10^{-23} g/atom

unit cell edge = d = 558.2 pm = 5.582×10^{-8} cm
unit cell volume = d^3 = $(5.582 \times 10^{-8}\text{ cm})^3 = 1.739 \times 10^{-22}\text{ cm}^3$
unit cell mass = $(1.739 \times 10^{-22}\text{ cm}^3)(1.55\text{ g/cm}^3) = 2.695 \times 10^{-22}$ g

(a) number of Ca atoms in unit cell = $\dfrac{\text{unit cell mass}}{\text{mass of one Ca atom}}$

$$= \frac{2.695 \times 10^{-22}\text{ g}}{6.656 \times 10^{-23}\text{ g/atom}} = 4.05 = 4\text{ Ca atoms}$$

(b) Because the unit cell contains 4 Ca atoms, the unit cell is face-centered cubic.

10.80   Six Na^+ ions touch each H^- ion and six H^- ions touch each Na^+ ion.

10.81   For CsCl: (1/8 x 8 corners), so 1 Cl^- and 1 minus per unit cell
1 Cs^+ inside, so 1 plus per unit cell

10.82 Na^+ H^- Na^+
$\leftarrow$ 488 pm $\rightarrow$ unit cell edge = d = 488 pm; Na–H bond = d/2 = 244 pm

10.83   See Problem 10.75.
body diagonal = $\sqrt{3}\,d$ = $\sqrt{3}$ (412.3 pm) = 714.12 pm
Cs–Cl bond = body diagonal/2 = (714.12 pm)/2 = 357.1 pm
Cs–Cl bond length = $r_{Cs^+} + r_{Cl^-}$
357.1 pm = $r_{Cs^+} + r_{Cl^-}$
357.1 pm = r_{Cs^+} + 181 pm
r_{Cs^+} = 357.1 pm − 181 pm = 176 pm

Phase Diagrams (Section 10.11)

10.84   (a) gas   (b) liquid   (c) solid

10.85   (a) $H_2O(l) \rightarrow H_2O(s)$
      (b) 380 °C is above the critical temperature; therefore, the water cannot be liquefied. At the higher pressure, it will behave as a supercritical fluid.

10.86

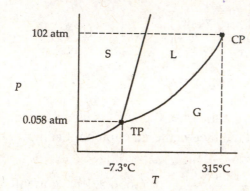

10.87

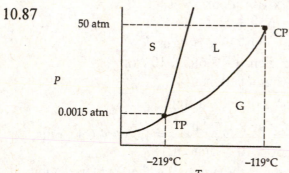

10.88   (a) $Br_2(s)$ (b) $Br_2(l)$

10.89   (a) $O_2(l)$ (b) O_2 - supercritical fluid

10.90   Solid O_2 does not melt when pressure is applied because the solid is denser than the liquid, and the solid/liquid boundary in the phase diagram slopes to the right.

10.91   Ammonia can be liquefied at 25 °C because this temperature is below T_c (132.5 °C). Methane cannot be liquefied at 25 °C because this temperature is above T_c (−82.1 °C). Sulfur dioxide can be liquefied at 25 °C because this temperature is below T_c (157.8 °C).

10.92

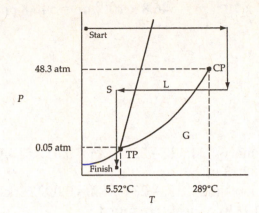

The starting phase is benzene as a solid, and the final phase is benzene as a gas.

10.93

The starting phase is a gas, and the final phase is a liquid.

10.94   solid → liquid → supercritical fluid → liquid → solid → gas

10.95   gas → solid → liquid → gas → liquid

Chapter Problems

10.96   Because chlorine is larger than fluorine, the charge separation is larger in CH_3Cl compared to CH_3F, resulting in CH_3Cl having a slightly larger dipole moment.

10.97   Because Ar crystallizes in a face-centered cubic unit cell, there are four Ar atoms in the unit cell.

mass of one Ar atom $= 39.95$ g/mol x $\dfrac{1 \text{ mol}}{6.022 \times 10^{23} \text{ atom}} = 6.634 \times 10^{-23}$ g/atom

unit cell mass $= 4$ atoms x mass of one Ar atom
$= 4$ atoms x 6.634×10^{-23} g/atom $= 2.654 \times 10^{-22}$ g

density $= \dfrac{\text{mass}}{\text{volume}}$

unit cell volume $= \dfrac{\text{unit cell mass}}{\text{density}} = \dfrac{2.654 \times 10^{-22} \text{ g}}{1.623 \text{ g/cm}^3} = 1.635 \times 10^{-22}$ cm^3

unit cell edge $= d = \sqrt[3]{1.635 \times 10^{-22} \text{ cm}^3} = 5.468 \times 10^{-8}$ cm

$$d = 5.468 \times 10^{-8} \text{ cm} \times \frac{1\text{m}}{100 \text{ cm}} = 5.468 \times 10^{-10} \text{ m} = 546.8 \times 10^{-12} \text{ m} = 546.8 \text{ pm}$$

$$r = \sqrt{\frac{d^2}{8}} = \sqrt{\frac{(546.8 \text{ pm})^2}{8}} = 193.3 \text{ pm}$$

10.98 $7.50 \text{ g} \times \dfrac{1 \text{ mol}}{200.6 \text{ g}} = 0.037\ 39 \text{ mol Hg}$

$q_1 = (0.037\ 39 \text{ mol})[28.2 \times 10^{-3} \text{ kJ/(K} \cdot \text{mol)}](234.3 \text{ K} - 223.1 \text{ K}) = 0.011\ 81 \text{ kJ}$
$q_2 = (0.037\ 39 \text{ mol})(2.33 \text{ kJ/mol}) = 0.087\ 12 \text{ kJ}$
$q_3 = (0.037\ 39 \text{ mol})[27.9 \times 10^{-3} \text{ kJ/(K} \cdot \text{mol)}](323.1 \text{ K} - 234.3 \text{ K}) = 0.092\ 63 \text{ kJ}$
$q_{total} = q_1 + q_2 + q_3 = 0.192 \text{ kJ};$ \qquad 0.192 kJ of heat is required.

10.99

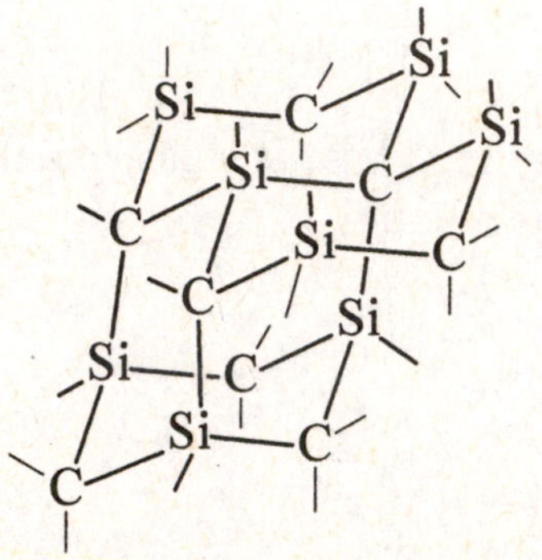

10.100 $\ln P_2 = \ln P_1 + \dfrac{\Delta H_{vap}}{R}\left(\dfrac{1}{T_1} - \dfrac{1}{T_2}\right)$

$\Delta H_{vap} = 40.67$ kJ/mol
At 1 atm, H_2O boils at 100 ºC; therefore set
$T_1 = 100$ ºC $= 373$ K, and $P_1 = 1.00$ atm.
Let $T_2 = 95$ ºC $= 368$ K, and solve for P_2.   (P_2 is the atmospheric pressure in Denver.)

$$\ln P_2 = \ln(1) + \frac{40.67 \text{ kJ/mol}}{[8.3145 \times 10^{-3} \text{ kJ/(K} \cdot \text{mol)}]}\left(\frac{1}{373 \text{ K}} - \frac{1}{368 \text{ K}}\right)$$

$\ln P_2 = -0.1782;$ $P_2 = e^{-0.1782} = 0.837$ atm

10.101

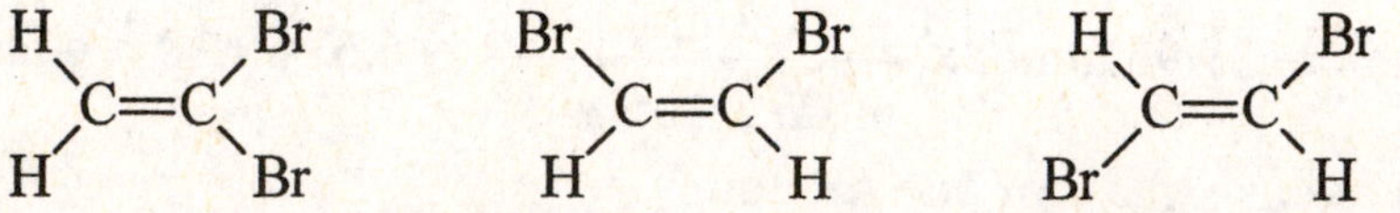

No dipole moment.

10.102 $\Delta G = \Delta H - T\Delta S$; at the melting point (phase change), $\Delta G = 0$.

$$\Delta H = T\Delta S; \quad T = \frac{\Delta H_{fus}}{\Delta S_{fus}} = \frac{9.037 \text{ kJ/mol}}{9.79 \times 10^{-3} \text{ kJ/(K} \cdot \text{mol)}} = 923 \text{ K} = 650 \text{ °C}$$

10.103 melting point = $-23.2\ °C = 250.0\ K$

$\Delta G = \Delta H_{fusion} - T\Delta S_{fusion}$

At the melting point (phase change), $\Delta G = 0$

$\Delta H_{fusion} = T\Delta S_{fusion}$

$$\Delta S_{fusion} = \frac{\Delta H_{fusion}}{T} = \frac{9.37\ kJ/mol}{250.0\ K} = 0.0375\ kJ/(K \cdot mol) = 37.5\ J/(K \cdot mol)$$

10.104 $\Delta H_{vap} = \dfrac{(\ln P_2 - \ln P_1)(R)}{\left(\dfrac{1}{T_1} - \dfrac{1}{T_2}\right)}$

$P_1 = 40.0\ mm\ Hg;$ $\qquad$ $T_1 = -81.6\ °C = 191.6\ K$
$P_2 = 400\ mm\ Hg;$ $\qquad$ $T_2 = -43.9\ °C = 229.2\ K$

$$\Delta H_{vap} = \frac{[\ln(400) - \ln(40.0)]\left(8.3145 \times 10^{-3}\dfrac{kJ}{K \cdot mol}\right)}{\left(\dfrac{1}{191.6\ K} - \dfrac{1}{229.2\ K}\right)} = 22.36\ kJ/mol$$

Using $\Delta H_{vap} = 22.36\ kJ/mol$

$$\ln P_2 = \ln P_1 + \frac{\Delta H_{vap}}{R}\left(\frac{1}{T_1} - \frac{1}{T_2}\right)$$

$$(\ln P_2 - \ln P_1)\left(\frac{R}{\Delta H_{vap}}\right) = \frac{1}{T_1} - \frac{1}{T_2}$$

$$\frac{1}{T_1} - (\ln P_2 - \ln P_1)\left(\frac{R}{\Delta H_{vap}}\right) = \frac{1}{T_2}$$

$P_1 = 40.0\ mm\ Hg;$ $\qquad$ $T_1 = 191.6\ K$
$P_2 = 760\ mm\ Hg$

Solve for T_2, the normal boiling point.

$$\frac{1}{191.6\ K} - [\ln(760) - \ln(40.0)]\left(\frac{8.3145 \times 10^{-3}\dfrac{kJ}{K \cdot mol}}{22.36\ kJ/mol}\right) = \frac{1}{T_2}$$

$$\frac{1}{T_2} = 0.004\ 124\ 33;\ \ T_2 = 242.46\ K = -30.7\ °C$$

10.105 (a) $\ln P_2 = \ln P_1 + \dfrac{\Delta H_{vap}}{R}\left(\dfrac{1}{T_1} - \dfrac{1}{T_2}\right)$

$$(\ln P_2 - \ln P_1)\left(\frac{R}{\Delta H_{vap}}\right) = \frac{1}{T_1} - \frac{1}{T_2}$$

$$\frac{1}{T_1} - (\ln P_2 - \ln P_1)\left(\frac{R}{\Delta H_{vap}}\right) = \frac{1}{T_2}$$

$P_1 = 100.0$ mm Hg; $T_1 = -23$ °C = 250 K
$P_2 = 760.0$ mm Hg
Solve for T_2, the normal boiling point for CCl_3F.

$$\frac{1}{250 \text{ K}} - [\ln(760.0) - \ln(100.0)]\left(\frac{8.3145 \times 10^{-3}\dfrac{\text{kJ}}{\text{K} \cdot \text{mol}}}{24.77 \text{ kJ/mol}}\right) = \frac{1}{T_2}$$

$\dfrac{1}{T_2} = 0.003\ 319;$ $\qquad\qquad$ $T_2 = 301.3$ K = 28.1 °C

(b) $\Delta S_{vap} = \dfrac{\Delta H_{vap}}{T} = \dfrac{24.77 \text{ kJ/mol}}{301.3 \text{ K}} = 0.082\ 21 \text{ kJ/(K} \cdot \text{mol)} = 82.2 \text{ J/(K} \cdot \text{mol)}$

10.106 $\Delta H_{vap} = \dfrac{(\ln P_2 - \ln P_1)(R)}{\left(\dfrac{1}{T_1} - \dfrac{1}{T_2}\right)}$

$P_1 = 100$ mm Hg; $\qquad\qquad$ $T_1 = -110.3$ °C = 162.85 K
$P_2 = 760$ mm Hg; $\qquad\qquad$ $T_2 = -88.5$ °C = 184.65 K

$$\Delta H_{vap} = \frac{[\ln(760) - \ln(100)]\left(8.3145 \times 10^{-3}\dfrac{\text{kJ}}{\text{K} \cdot \text{mol}}\right)}{\left(\dfrac{1}{162.85 \text{ K}} - \dfrac{1}{184.65 \text{ K}}\right)} = 23.3 \text{ kJ/mol}$$

10.107 $\ln P_2 = \ln P_1 + \dfrac{\Delta H_{vap}}{R}\left(\dfrac{1}{T_1} - \dfrac{1}{T_2}\right)$

$(\ln P_2 - \ln P_1)\left(\dfrac{R}{\Delta H_{vap}}\right) = \dfrac{1}{T_1} - \dfrac{1}{T_2}$

$\dfrac{1}{T_1} - (\ln P_2 - \ln P_1)\left(\dfrac{R}{\Delta H_{vap}}\right) = \dfrac{1}{T_2}$

$P_1 = 760$ mm Hg; $\qquad\qquad$ $T_1 = 56.1$ °C = 329.2 K
$P_2 = 105$ mm Hg $\qquad\qquad$ Solve for T_2.

$$\frac{1}{329.2 \text{ K}} - [\ln(105) - \ln(760)]\left(\frac{8.3145 \times 10^{-3}\dfrac{\text{kJ}}{\text{K} \cdot \text{mol}}}{29.1 \text{ kJ/mol}}\right) = \frac{1}{T_2}$$

$\dfrac{1}{T_2} = 0.003\ 603;$ $\qquad\qquad$ $T_2 = 277.5$ K = 4.4 °C

10.108

Kr cannot be liquified at room temperature because room temperature is above T_c (−63 °C).

10.109 (a) Kr(l)   (b) supercritical Kr

10.110 For a body-centered cube:

$$4r = \sqrt{3} \text{ edge}; \qquad \text{edge} = \frac{4r}{\sqrt{3}}$$

$$\text{volume of sphere} = \frac{4}{3}\pi r^3$$

$$\text{volume of unit cell} = \left(\frac{4r}{\sqrt{3}}\right)^3 = \frac{64\,r^3}{3\sqrt{3}}$$

$$\text{volume of 2 spheres} = 2\left(\frac{4}{3}\pi r^3\right) = \frac{8}{3}\pi r^3$$

$$\% \text{ volume occupied} = \frac{\left(\frac{8}{3}\pi r^3\right)}{\left(\frac{64r^3}{3\sqrt{3}}\right)} \times 100\% = 68\%$$

10.111 From Problem 10.75, $4r = \sqrt{3}\,d$; $r = \dfrac{\sqrt{3}\,d}{4} = \dfrac{\sqrt{3}(287 \text{ pm})}{4} = 124$ pm

10.112 unit cell edge = d = 287 pm = 287 x 10^{-12} m = 2.87 x 10^{-8} cm
unit cell volume = d^3 = (2.87 x 10^{-8} cm)^3 = 2.364 x 10^{-23} cm^3
unit cell mass = (2.364 x 10^{-23} cm^3)(7.86 g/cm^3) = 1.858 x 10^{-22} g
Fe is body-centered cubic; therefore there are two Fe atoms per unit cell.

$$\text{mass of one Fe atom} = \frac{1.858 \times 10^{-22} \text{ g}}{2 \text{ Fe atoms}} = 9.290 \times 10^{-23} \text{ g/atom}$$

$$\text{Avogadro's number} = 55.85 \text{ g/mol} \times \frac{1 \text{ atom}}{9.290 \times 10^{-23} \text{ g}} = 6.01 \times 10^{23} \text{ atoms/mol}$$

10.113   unit cell edge = d = 408 pm = 408 x 10^{-12} m = 4.08 x 10^{-8} cm
unit cell volume = (4.08 x 10^{-8} cm)^3 = 6.792 x 10^{-23} cm^3
unit cell mass = (10.50 g/cm^3)(6.792 x 10^{-23} cm^3) = 7.132 x 10^{-22} g
Ag is face-centered cubic; therefore there are four Ag atoms in the unit cell.

mass of one Ag atom = $\dfrac{7.132 \times 10^{-22} \text{g}}{4 \text{ Ag atoms}}$ = 1.783 x 10^{-22} g/atom

Avogadro's number = 107.9 g/mol x $\dfrac{1 \text{ atom}}{1.783 \times 10^{-22} \text{ g}}$ = 6.05 x 10^{23} atoms/mol

10.114   (a) unit cell edge = $2r_{Cl^-} + 2r_{Na^+}$ = 2(181 pm) + 2(97 pm) = 556 pm

(b) unit cell edge = d = 556 pm = 556 x 10^{-12} m = 5.56 x 10^{-8} cm
unit cell volume = (5.56 x 10^{-8} cm)^3 = 1.719 x 10^{-22} cm^3
The unit cell contains 4 Na^+ ions and 4 Cl^- ions.

mass of one Na^+ ion = 22.99 g/mol x $\dfrac{1 \text{ mol}}{6.022 \times 10^{23} \text{ ions}}$ = 3.818 x 10^{-23} g/Na^+

mass of one Cl^- ion = 35.45 g/mol x $\dfrac{1 \text{ mol}}{6.022 \times 10^{23} \text{ ions}}$ = 5.887 x 10^{-23} g/Cl^-

unit cell mass = 4(3.818 x 10^{-23} g) + 4(5.887 x 10^{-23} g) = 3.882 x 10^{-22} g

density = $\dfrac{\text{unit cell mass}}{\text{unit cell volume}}$ = $\dfrac{3.882 \times 10^{-22} \text{ g}}{1.719 \times 10^{-22} \text{ cm}^3}$ = 2.26 g/cm^3

10.115   (a) (1/2 Nb/face)(6 faces) = 3 Nb; (1/4 O/edge)(12 edges) = 3 O
(b) NbO
(c) The oxidation state of Nb is +2.

10.116 Al_2O_3, ionic (greater lattice energy than NaCl because of higher ion charges);
F_2, dispersion; H_2O, H–bonding, dipole-dipole; Br_2, dispersion (larger and more
polarizable than F_2), ICl, dipole-dipole, NaCl, ionic

rank according to normal boiling points: $F_2 < Br_2 < ICl < H_2O < NaCl < Al_2O_3$

10.117 Ag_2Te, 343. 33 amu; 529 pm = 529 x 10^{-12} m = 529 x 10^{-10} cm
unit cell volume = (529 x 10^{-10} cm)^3 = 1.48 x 10^{-22} cm^3
unit cell mass = (1.48 x 10^{-22} cm^3)(7.70 g/cm^3) = 1.14 x 10^{-21} g

mass of one Ag_2Te = $\dfrac{343.33 \text{ g } Ag_2Te/\text{mol}}{6.022 \times 10^{23} \ Ag_2Te \text{ formula units}/\text{mol}}$ = 5.70 x 10^{-22} g Ag_2Te/formula unit

Ag_2Te formula units/unit cell = $\dfrac{1.14 \times 10^{-21} \text{ g/unit cell}}{5.70 \times 10^{-22} \text{ g/}Ag_2Te}$ = 2 Ag_2Te/unit cell

Ag/unit cell = $\dfrac{2 \ Ag_2Te}{\text{unit cell}}$ x $\dfrac{2 \text{ Ag}}{Ag_2Te}$ = 4 Ag/unit cell

10.118 (a)

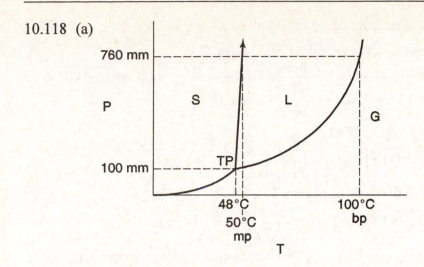

(b) (i) solid (ii) gas (iii) liquid (iv) liquid (v) solid

Multiconcept Problems

10.119 $C_2H_5OH(l) \rightarrow C_2H_5OH(g)$

Calculate ΔH and ΔS for this process and assume they do not change as a function of temperature.

$\Delta H^o = \Delta H^o_f(C_2H_5OH(g)) - \Delta H^o_f(C_2H_5OH(l))$

$\Delta H^o = [(1 \text{ mol})(-235.1 \text{ kJ/mol})] - [(1 \text{ mol})(-277.7 \text{ kJ/mol})] = 42.6 \text{ kJ}$

$\Delta S^o = S^o(C_2H_5OH(g)) - S^o(C_2H_5OH(l))$

$\Delta S^o = [(1 \text{ mol})(282.6 \text{ J/(K} \cdot \text{mol)})] - [(1 \text{ mol})(161 \text{ J/(K} \cdot \text{mol)})] = 122 \text{ J/K}$

$\Delta S^o = 122 \times 10^{-3} \text{ kJ/K})$

$\Delta G^o = \Delta H^o - T\Delta S^o$ and at the boiling point, $\Delta G = 0$

$0 = \Delta H^o - T_{bp}\Delta S^o$

$T_{bp}\Delta S^o = \Delta H^o$

$T_{bp} = \dfrac{\Delta H^o}{\Delta S^o} = \dfrac{42.6 \text{ kJ}}{122 \times 10^{-3} \text{ kJ/K}} = 349 \text{ K}$

$T_{bp} = 349 - 273 = 76 \text{ }^oC$

$\ln P_2 = \ln P_1 + \dfrac{\Delta H_{vap}}{R}\left(\dfrac{1}{T_1} - \dfrac{1}{T_2}\right)$

$\Delta H_{vap} = 42.6 \text{ kJ/mol}$

At 1 atm, C_2H_5OH boils at 349 K; therefore set

$T_1 = 349$ K, and $P_1 = 1.00$ atm.

Let $T_2 = 25 \text{ }^oC = 298$ K, and solve for P_2

P_2 is the vapor pressure of C_2H_5OH at 25 ^oC.

$\ln P_2 = \ln(1.00) + \dfrac{42.6 \text{ kJ/mol}}{[8.3145 \times 10^{-3} \text{ kJ/(K} \cdot \text{mol)}]}\left(\dfrac{1}{349 \text{ K}} - \dfrac{1}{298 \text{ K}}\right)$

$\ln P_2 = -2.512; \quad P_2 = e^{-2.512} = 0.0811 \text{ atm}$

$P_2 = 0.0811 \text{ atm} \times \dfrac{760 \text{ mm Hg}}{1.00 \text{ atm}} = 61.6 \text{ mm Hg}$

10.120 (a) Let the formula of magnetite be Fe_xO_y, then $Fe_xO_y + y\ CO \rightarrow x\ Fe + y\ CO_2$

$$n_{CO_2} = y = \frac{PV}{RT} = \frac{\left(751\ mm\ Hg \times \dfrac{1.00\ atm}{760\ mm\ Hg}\right)(1.136\ L)}{\left(0.082\ 06\ \dfrac{L \cdot atm}{K \cdot mol}\right)(298\ K)} = 0.04590\ mol\ CO_2$$

0.04590 mol CO_2 = mol of O in Fe_xO_y

mass of O in Fe_xO_y = 0.04590 mol O $\times \dfrac{16.0\ g\ O}{1\ mol\ O}$ = 0.7345 g O

mass of Fe in Fe_xO_y = 2.660 g – 0.7345 g = 1.926 g Fe

(b) mol Fe in magnetite = 1.926 g Fe $\times \dfrac{1\ mol\ Fe}{55.85\ g\ Fe}$ = 0.0345 mol Fe

formula of magnetite: $Fe_{0.0345}O_{0.0459}$ (divide each subscript by the smaller)

$Fe_{0.0345/0.0345}O_{0.0459/0.0345}$

$FeO_{1.33}$ (multiply both subscripts by 3)

$Fe_{(1 \times 3)}O_{(1.33 \times 3)};$ Fe_3O_4

(c) unit cell edge = d = 839 pm = 839 x 10^{-12} m

d = 839 x 10^{-12} m $\times \dfrac{100\ cm}{1\ m}$ = 8.39 x 10^{-8} cm

unit cell volume = d^3 = (8.39 x 10^{-8} cm)^3 = 5.91 x 10^{-22} cm^3

unit cell mass = (5.91 x 10^{-22} cm^3)(5.20 g/cm^3) = 3.07 x 10^{-21} g

mass of Fe in unit cell = $\left(\dfrac{1.926\ g\ Fe}{2.660\ g}\right)$(3.07 x 10^{-21} g) = 2.22 x 10^{-21} g Fe

mass of O in unit cell = $\left(\dfrac{0.7345\ g\ O}{2.660\ g}\right)$(3.07 x 10^{-21} g) = 8.47 x 10^{-22} g O

Fe atoms in unit cell = 2.22 x 10^{-21} g $\times \dfrac{6.022\ x\ 10^{23}\ atoms/mol}{55.847\ g/mol}$ = 24 Fe atoms

O atoms in unit cell = 8.47 x 10^{-22} g $\times \dfrac{6.022\ x\ 10^{23}\ atoms/mol}{16.00\ g/mol}$ = 32 O atoms

10.121 (a) $n_{H_2} = \dfrac{PV}{RT} = \dfrac{\left(740\ mm\ Hg \times \dfrac{1.00\ atm}{760\ mm\ Hg}\right)(4.00\ L)}{\left(0.08206\ \dfrac{L \cdot atm}{K \cdot mol}\right)(296\ K)}$ = 0.160 mol H_2

M = Group 3A metal; $2\ M(s) + 6\ H^+(aq) \rightarrow 2\ M^{3+}(aq) + 3\ H_2(g)$

n_M = 0.160 mol H_2 $\times \dfrac{2\ mol\ M}{3\ mol\ H_2}$ = 0.107 mol M

mass M = 1.07 cm^3 x 2.70 g/cm^3 = 2.89 g M

molar mass M = $\dfrac{2.89\ g\ M}{0.107\ mol\ M}$ = 27.0 g/mol; The Group 3A metal is Al

(b) mass of one Al atom = 26.98 g/mol x $\dfrac{1 \text{ mol}}{6.022 \text{ x } 10^{23} \text{ atoms}}$ = 4.48 x 10^{-23} g/atom

unit cell edge = d = 404 pm = 404 x 10^{-12} m

d = 404 x 10^{-12} m x $\dfrac{100 \text{ cm}}{1 \text{ m}}$ = 4.04 x 10^{-8} cm

unit cell volume = d^3 = (4.04 x 10^{-8} cm)^3 = 6.59 x 10^{-23} cm^3

Calculate the density of Al assuming it is primitive cubic, body-centered cubic, and face-centered cubic. Compare the calculated density with the actual density to identify the unit cell.

For primitive cubic:

density = $\dfrac{\text{unit cell mass}}{\text{unit cell volume}}$ = $\dfrac{(1 \text{ Al})(4.48 \text{ x } 10^{-23} \text{ g/Al atom})}{6.59 \text{ x } 10^{-23} \text{ cm}^3}$ = 0.680 g/cm^3

For body-centered cubic:

density = $\dfrac{\text{unit cell mass}}{\text{unit cell volume}}$ = $\dfrac{(2 \text{ Al})(4.48 \text{ x } 10^{-23} \text{ g/Al atom})}{6.59 \text{ x } 10^{-23} \text{ cm}^3}$ = 1.36 g/cm^3

For face-centered cubic:

density = $\dfrac{\text{unit cell mass}}{\text{unit cell volume}}$ = $\dfrac{(4 \text{ Al})(4.48 \text{ x } 10^{-23} \text{ g/Al atom})}{6.59 \text{ x } 10^{-23} \text{ cm}^3}$ = 2.72 g/cm^3

The calculated density for a face-centered cube (2.72 g/cm^3) is closest to the actual density of 2.70 g/cm^3. Al crystallizes in the face-centered cubic unit cell.

(c) $r = \sqrt{\dfrac{d^2}{8}} = \sqrt{\dfrac{(404 \text{ pm})^2}{8}}$ = 143 pm

10.122 (a) M = alkali metal; 500.0 mL = 0.5000 L; 802 °C = 1075 K

$n_M = \dfrac{PV}{RT} = \dfrac{\left(12.5 \text{ mm Hg x } \dfrac{1.00 \text{ atm}}{760 \text{ mm Hg}}\right)(0.5000 \text{ L})}{\left(0.082\,06 \dfrac{\text{L} \cdot \text{atm}}{\text{K} \cdot \text{mol}}\right)(1075 \text{ K})}$ = 9.32 x 10^{-5} mol M

1.62 mm = 1.62 x 10^{-3} m; crystal volume = (1.62 x 10^{-3} m)^3 = 4.25 x 10^{-9} m^3

M atoms in crystal = (9.32 x 10^{-5} mol)(6.022 x 10^{23} atoms/mol) = 5.61 x 10^{19} M atoms

Because M is body-centered cubic, only 68% (Table 10.10) of the total volume is occupied by M atoms.

volume of M atom = $\dfrac{(0.68)(4.25 \text{ x } 10^{-9} \text{ m})}{5.61 \text{ x } 10^{19} \text{ M atoms}}$ = 5.15 x 10^{-29} m^3/M atom

volume of a sphere = $\dfrac{4}{3}\pi r^3$

$r_M = \sqrt[3]{\dfrac{3(\text{volume})}{4\pi}} = \sqrt[3]{\dfrac{3(5.15 \text{ x } 10^{-29} \text{ m}^3)}{4\pi}}$ = 2.31 x 10^{-10} m = 231 x 10^{-12} m = 231 pm

(b) The radius of 231 pm is closest to that of K.

(c) 1.62 mm = 0.162 cm

$$\text{density of solid} = \frac{(9.32 \times 10^{-5}\ \text{mol})(39.1\ \text{g/mol})}{(0.162\ \text{cm})^3} = 0.857\ \text{g/cm}^3$$

$$\text{density of vapor} = \frac{(9.32 \times 10^{-5}\ \text{mol})(39.1\ \text{g/mol})}{500.0\ \text{cm}^3} = 7.29 \times 10^{-6}\ \text{g/cm}^3$$

10.123 (a)

$$n_{X_2} = \frac{PV}{RT} = \frac{\left(755\ \text{mm Hg} \times \dfrac{1.00\ \text{atm}}{760\ \text{mm Hg}}\right)(0.500\ \text{L})}{\left(0.082\ 06\ \dfrac{\text{L} \cdot \text{atm}}{\text{K} \cdot \text{mol}}\right)(298\ \text{K})} = 0.0203\ \text{mol}\ X_2$$

$$M(s)\ +\ 1/2\ X_2(g)\ \rightarrow\ MX(s)$$

$$\text{mol M} = 0.0203\ \text{mol}\ X_2\ \times\ \frac{1\ \text{mol M}}{1/2\ \text{mol}\ X_2} = 0.0406\ \text{mol M}$$

$$\text{molar mass M} = \frac{1.588\ \text{g M}}{0.0406\ \text{mol M}} = 39.1\ \text{g/mol};\ \ \text{atomic mass} = 39.1\ \text{amu}\ ;\ \ M = K$$

(b) From Figure 4.4, the radius for K^+ is ~140 pm.
unit cell edge = 535 pm = $2r_{K^+} + 2r_{X^-}$

$$r_{X^-} = \frac{535\ \text{pm} - 2r_{K^+}}{2} = \frac{535\ \text{pm} - 2(140\ \text{pm})}{2} = 128\ \text{pm}$$

From Figure 4.5, $X^- = F^-$

(c) Because the cation and anion are of comparable size, the anions are not in contact with each other.

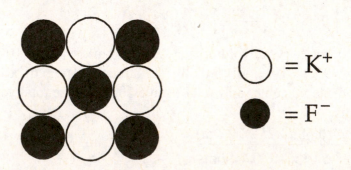

(d) unit cell contents: 1/8 F^- at 8 corners and 1/2 F^- at 6 faces = 4 F^-
                 1/4 K^+ at 12 edges and 1 K^+ inside = 4 K^+

$$\text{mass of one } K^+ = \frac{39.098\ \text{g/mol}}{6.022 \times 10^{23}\ K^+/\text{mol}} = 6.493 \times 10^{-23}\ \text{g}/K^+$$

$$\text{mass of one } F^- = \frac{18.998 \text{ g/mol}}{6.022 \times 10^{23} \text{ F}^-/\text{mol}} = 3.155 \times 10^{-23} \text{ g/F}^-$$

unit cell mass $= (4 \text{ K}^+)(6.493 \times 10^{-23} \text{ g/K}^+) + (4 \text{ F}^-)(3.155 \times 10^{-23} \text{ g/F}^-) = 3.859 \times 10^{-22} \text{ g}$

unit cell volume $= [(535 \times 10^{-12} \text{ m})(100 \text{ cm/m})]^3 = 1.531 \times 10^{-22} \text{ cm}^3$

$$\text{density} = \frac{\text{mass of unit cell}}{\text{volume of unit cell}} = \frac{3.859 \times 10^{-22} \text{ g}}{1.531 \times 10^{-22} \text{ cm}^3} = 2.52 \text{ g/cm}^3$$

(e) $K(s) + 1/2 \, F_2(g) \rightarrow KF(s)$ is a formation reaction.

$$\Delta H^o_f(KF) = \frac{-22.83 \text{ kJ}}{0.0406 \text{ mol}} = -562 \text{ kJ/mol}$$

11 Solutions and Their Properties

11.1 Toluene is nonpolar and is insoluble in water.
Br_2 is nonpolar but because of its size, is polarizable and is soluble in water.
KBr is an ionic compound and is very soluble in water.
toluene $<$ Br_2 $<$ KBr   (solubility in H_2O)

11.2 (a) Na^+ has the larger (more negative) hydration energy because the Na^+ ion is smaller than the Cs^+ ion and water molecules can approach more closely and bind more tightly to the Na^+ ion.
(b) Ba^{2+} has the larger (more negative) hydration energy because of its higher charge.

11.3 NaCl, 58.44 amu; 1.00 mol NaCl = 58.44 g
1.00 L H_2O = 1000 mL = 1000 g (assuming a density of 1.00 g/mL)

$$\text{mass \% NaCl} = \frac{58.44 \text{ g}}{1000 \text{ g} + 58.44 \text{ g}} \times 100\% = 5.52 \text{ mass \%}$$

11.4 $$\text{ppm} = \frac{\text{mass of } CO_2}{\text{total mass of solution}} \times 10^6 \text{ ppm}$$

total mass of solution = density x volume = (1.3 g/L)(1.0 L) = 1.3 g

$$35 \text{ ppm} = \frac{\text{mass of } CO_2}{1.3 \text{ g}} \times 10^6 \text{ ppm}$$

$$\text{mass of } CO_2 = \frac{(35 \text{ ppm})(1.3 \text{ g})}{10^6 \text{ ppm}} = 4.6 \times 10^{-5} \text{ g } CO_2$$

11.5 Assume 1.00 L of sea water.
mass of 1.00 L = (1000 mL)(1.025 g/mL) = 1025 g

$$\frac{\text{mass NaCl}}{1025 \text{ g}} \times 100\% = 3.50 \text{ mass \%}; \qquad \text{mass NaCl} = \frac{1025 \text{ g} \times 3.50}{100} = 35.88 \text{ g}$$

There are 35.88 g NaCl per 1.00 L of solution.

$$M = \frac{\left(35.88 \text{ g NaCl} \times \dfrac{1 \text{ mol NaCl}}{58.44 \text{ g NaCl}} \right)}{1.00 \text{ L}} = 0.614 \text{ M}$$

11.6 $C_{27}H_{46}O$, 386.7 amu; $CHCl_3$, 119.4 amu; $40.0 \text{ g} \times \dfrac{1 \text{ kg}}{1000 \text{ g}} = 0.0400 \text{ kg}$

$$\text{molality} = \frac{\text{mol } C_{27}H_{46}O}{\text{kg } CHCl_3} = \frac{\left(0.385 \text{ g} \times \dfrac{1 \text{ mol}}{386.7 \text{ g}} \right)}{0.0400 \text{ kg}} = 0.0249 \text{ mol/kg} = 0.0249 \text{ } m$$

$$X_{C_{27}H_{46}O} = \frac{\text{mol } C_{27}H_{46}O}{\text{mol } C_{27}H_{46}O + \text{mol } CHCl_3}$$

$$X_{C_{27}H_{46}O} = \frac{\left(0.385 \text{ g} \times \dfrac{1 \text{ mol}}{386.7 \text{ g}} \right)}{\left[\left(0.385 \text{ g} \times \dfrac{1 \text{ mol}}{386.7 \text{ g}} \right) + \left(40.0 \text{ g} \times \dfrac{1 \text{ mol}}{119.4 \text{ g}} \right) \right]} = 2.96 \times 10^{-3}$$

11.7 CH_3CO_2Na, 82.03 amu

$$\text{kg } H_2O = (0.150 \text{ mol } CH_3CO_2Na)\left(\frac{1 \text{ kg } H_2O}{0.500 \text{ mol } CH_3CO_2Na} \right) = 0.300 \text{ kg } H_2O$$

$$\text{mass } CH_3CO_2Na = 0.150 \text{ mol } CH_3CO_2Na \times \frac{82.03 \text{ g } CH_3CO_2Na}{1 \text{ mol } CH_3CO_2Na} = 12.3 \text{ g } CH_3CO_2Na$$

mass of solution needed = 300 g + 12.3 g = 312 g

11.8   Assume you have a solution with 1.000 kg (1000 g) of H_2O. If this solution is 0.258 m, then it must also contain 0.258 mol glucose.

$$\text{mass of glucose} = 0.258 \text{ mol} \times \frac{180.2 \text{ g}}{1 \text{ mol}} = 46.5 \text{ g glucose}$$

mass of solution = 1000 g + 46.5 g = 1046.5 g
density = 1.0173 g/mL

$$\text{volume of solution} = 1046.5 \text{ g} \times \frac{1 \text{ mL}}{1.0173 \text{ g}} = 1028.7 \text{ mL}$$

$$\text{volume} = 1028.7 \text{ mL} \times \frac{1 \text{ L}}{1000 \text{ mL}} = 1.029 \text{ L}; \qquad \text{molarity} = \frac{0.258 \text{ mol}}{1.029 \text{ L}} = 0.251 \text{ M}$$

11.9   Assume 1.00 L of solution.
mass of 1.00 L = (1.0042 g/mL)(1000 mL) = 1004.2 g of solution

$$0.500 \text{ mol } CH_3CO_2H \times \frac{60.05 \text{ g } CH_3CO_2H}{1 \text{ mol } CH_3CO_2H} = 30.02 \text{ g } CH_3CO_2H$$

$$1004.2 \text{ g} - 30.02 \text{ g} = 974.2 \text{ g} = 0.9742 \text{ kg of } H_2O; \qquad \text{molality} = \frac{0.500 \text{ mol}}{0.9742 \text{ kg}} = 0.513 \text{ } m$$

11.10   Assume you have 100.0 g of seawater.
mass NaCl = (0.0350)(100.0 g) = 3.50 g NaCl
mass H_2O = 100.0 g – 3.50 g = 96.5 g H_2O

NaCl, 58.44 amu; $\text{mol NaCl} = 3.50 \text{ g} \times \dfrac{1 \text{ mol}}{58.44 \text{ g}} = 0.0599 \text{ mol NaCl}$

$$\text{mass } H_2O = 96.5 \text{ g} \times \frac{1 \text{ kg}}{1000 \text{ g}} = 0.0965 \text{ kg } H_2O; \qquad \text{molality} = \frac{0.0599 \text{ mol}}{0.0965 \text{ kg}} = 0.621 \text{ } m$$

11.11 $M = k \cdot P$; $k = \dfrac{M}{P} = \dfrac{3.2 \times 10^{-2} \, M}{1.0 \, atm} = 3.2 \times 10^{-2} \, mol/(L \cdot atm)$

11.12 (a) $M = k \cdot P = [3.2 \times 10^{-2} \, mol/(L \cdot atm)](2.5 \, atm) = 0.080 \, M$

(b) $M = k \cdot P = [3.2 \times 10^{-2} \, mol/(L \cdot atm)](4.0 \times 10^{-4} \, atm) = 1.3 \times 10^{-5} \, M$

11.13 $C_7H_6O_2$, 122.1 amu; C_2H_6O, 46.07 amu

$$X_{solv} = \frac{mol \ C_2H_6O}{mol \ C_2H_6O + mol \ C_7H_6O_2} = \frac{\left(100 \ g \times \dfrac{1 \ mol}{46.07 \ g}\right)}{\left(100 \ g \times \dfrac{1 \ mol}{46.07 \ g}\right) + \left(5.00 \ g \times \dfrac{1 \ mol}{122.1 \ g}\right)} = 0.981$$

$P_{soln} = P_{solv} \cdot X_{solv} = (100.5 \ mm \ Hg)(0.981) = 98.6 \ mm \ Hg$

11.14 $P_{soln} = P_{solv} \cdot X_{solv}$; $\quad X_{solv} = \dfrac{P_{soln}}{P_{solv}} = \dfrac{(55.3 - 1.30) \ mm \ Hg}{55.3 \ mm \ Hg} = 0.976$

NaBr dissociates into two ions in aqueous solution.

$$X_{solv} = \frac{mol \ H_2O}{mol \ H_2O + mol \ Na^+ + mol \ Br^-}$$

$$X_{solv} = 0.976 = \frac{\left(250 \ g \times \dfrac{1 \ mol}{18.02 \ g}\right)}{\left(250 \ g \times \dfrac{1 \ mol}{18.02 \ g}\right) + x \ mol \ Na^+ + x \ mol \ Br^-}$$

$0.976 = \dfrac{13.9 \ mol}{13.9 \ mol + 2x \ mol}$; $\quad$ solve for x.

$0.976(13.9 \ mol + 2x \ mol) = 13.9 \ mol$

$13.566 \ mol + 1.952 \ x \ mol = 13.9 \ mol$

$1.952 \ x \ mol = 13.9 \ mol - 13.566 \ mol$

$x \ mol = \dfrac{13.9 \ mol - 13.566 \ mol}{1.952} = 0.171 \ mol$

$x = 0.171 \ mol \ Na^+ = 0.171 \ mol \ Br^- = 0.171 \ mol \ NaBr$

NaBr, 102.9 amu; $\quad$ mass NaBr $= 0.171 \ mol \times \dfrac{102.9 \ g}{1 \ mol} = 17.6 \ g \ NaBr$

11.15 At any given temperature, the vapor pressure of a solution is lower than the vapor pressure of the pure solvent. The upper curve represents the vapor pressure of the pure solvent. The lower curve represents the vapor pressure of the solution.

11.16 C_2H_5OH, 46.07 amu; H_2O, 18.02 amu

(a) $25.0 \ g \ C_2H_5OH \times \dfrac{1 \ mol \ C_2H_5OH}{46.07 \ g \ C_2H_5OH} = 0.5426 \ mol \ C_2H_5OH$

$$100.0 \text{ g H}_2\text{O} \times \frac{1 \text{ mol H}_2\text{O}}{18.02 \text{ g H}_2\text{O}} = 5.549 \text{ mol H}_2\text{O}$$

$$X_{\text{C}_2\text{H}_5\text{OH}} = \frac{0.5426 \text{ mol}}{0.5426 \text{ mol} + 5.549 \text{ mol}} = 0.08907$$

$$X_{\text{H}_2\text{O}} = \frac{5.549 \text{ mol}}{0.5426 \text{ mol} + 5.549 \text{ mol}} = 0.9109$$

$$P_{\text{soln}} = X_{\text{C}_2\text{H}_5\text{OH}} P^{\text{o}}_{\text{C}_2\text{H}_5\text{OH}} + X_{\text{H}_2\text{O}} P^{\text{o}}_{\text{H}_2\text{O}}$$

$$P_{\text{soln}} = (0.08907)(61.2 \text{ mm Hg}) + (0.9109)(23.8 \text{ mm Hg}) = 27.1 \text{ mm Hg}$$

(b) $100 \text{ g C}_2\text{H}_5\text{OH} \times \dfrac{1 \text{ mol C}_2\text{H}_5\text{OH}}{46.07 \text{ g C}_2\text{H}_5\text{OH}} = 2.171 \text{ mol C}_2\text{H}_6\text{O}$

$$25.0 \text{ g H}_2\text{O} \times \frac{1 \text{ mol H}_2\text{O}}{18.02 \text{ g H}_2\text{O}} = 1.387 \text{ mol H}_2\text{O}$$

$$X_{\text{C}_2\text{H}_5\text{OH}} = \frac{2.171 \text{ mol}}{2.171 \text{ mol} + 1.387 \text{ mol}} = 0.6102$$

$$X_{\text{H}_2\text{O}} = \frac{1.387 \text{ mol}}{2.171 \text{ mol} + 1.387 \text{ mol}} = 0.3898$$

$$P_{\text{soln}} = X_{\text{C}_2\text{H}_5\text{OH}} P^{\text{o}}_{\text{C}_2\text{H}_5\text{OH}} + X_{\text{H}_2\text{O}} P^{\text{o}}_{\text{H}_2\text{O}}$$

$$P_{\text{soln}} = (0.6102)(61.2 \text{ mm Hg}) + (0.3898)(23.8 \text{ mm Hg}) = 46.6 \text{ mm Hg}$$

11.17 (a) Because the vapor pressure of the solution (red curve) is higher than that of the first liquid (green curve), the vapor pressure of the second liquid must be higher than that of the solution (red curve). Because the second liquid has a higher vapor pressure than the first liquid, the second liquid has a lower boiling point.

(b)

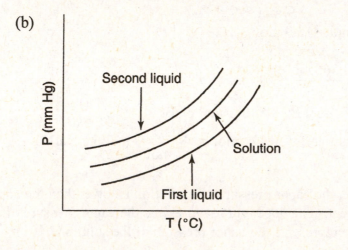

11.18 $\text{C}_9\text{H}_8\text{O}_4$, 180.2 amu; CHCl_3 is the solvent. For CHCl_3, $K_b = 3.63 \dfrac{^\circ\text{C} \cdot \text{kg}}{\text{mol}}$

$$75.00 \text{ g} \times \frac{1 \text{ kg}}{1000 \text{ g}} = 0.075\,00 \text{ kg}$$

$$\Delta T_b = K_b \cdot m = \left(3.63 \ \frac{°C \cdot kg}{mol}\right)\left(\frac{\left(1.50 \ g \ x \ \frac{1 \ mol}{180.2 \ g}\right)}{0.075 \ 00 \ kg}\right) = 0.40 \ °C$$

Solution boiling point = 61.7 °C + ΔT_b = 61.7 °C + 0.40 °C = 62.1 °C

11.19 $MgCl_2$, 95.21 amu

$$110 \ g \ x \ \frac{1 \ kg}{1000 \ g} = 0.110 \ kg$$

$$\Delta T_f = K_f \cdot m \cdot i = \left(1.86 \ \frac{°C \cdot kg}{mol}\right)\left(\frac{\left(7.40 \ g \ x \ \frac{1 \ mol}{95.21 \ g}\right)}{0.110 \ kg}\right)(2.7) = 3.55 \ °C$$

Solution freezing point = 0.00 °C – ΔT_f = 0.00 °C – 3.55 °C = –3.55 °C

11.20 $\Delta T_f = K_f \cdot m \cdot i$;   For KBr, i = 2.
Solution freezing point = –2.95 °C = 0.00 °C – ΔT_f; ΔT_f = 2.95 °C

$$m = \frac{\Delta T_f}{K_f \cdot i} = \frac{2.95 \ °C}{\left(1.86 \ \frac{°C \cdot kg}{mol}\right)(2)} = 0.793 \ mol/kg = 0.793 \ m$$

11.21   HCl, 36.46 amu; $\Delta T_f = K_f \cdot m \cdot i$

$$190 \ g \ x \ \frac{1 \ kg}{1000 \ g} = 0.190 \ kg$$

Solution freezing point = – 4.65 °C = 0.00 °C – ΔT_f; ΔT_f = 4.65 °C

$$i = \frac{\Delta T_f}{K_f \cdot m} = \frac{4.65 \ °C}{\left(1.86 \ \frac{°C \cdot kg}{mol}\right)\left(\frac{9.12 \ g \ x \ \frac{1 \ mol}{36.46 \ g}}{0.190 \ kg}\right)} = 1.9$$

11.22   The red curve represents the vapor pressure of pure chloroform.
(a) The normal boiling point for a liquid is the temperature where the vapor pressure of the liquid equals 1 atm (760 mm Hg). The approximate boiling point of pure chloroform is 62 °C.

(b) The approximate boiling point of the solution is 69 °C.
ΔT_b = 69 °C – 62 °C = 7 °C
$\Delta T_b = K_b \cdot m$

$$m = \frac{\Delta T_b}{K_b} = \frac{7 \ °C}{3.63 \ \frac{°C \cdot kg}{mol}} = 2 \ mol/kg = 2 \ m$$

11.23 For $CaCl_2$ there are 3 ions (solute particles)/$CaCl_2$
$\Pi = MRT$; For $CaCl_2$, $\Pi = 3MRT$

$$\Pi = (3)(0.125 \text{ mol/L})\left(0.082\ 06\ \frac{L \cdot atm}{K \cdot mol}\right)(310 \text{ K}) = 9.54 \text{ atm}$$

11.24 $\Pi = MRT$; $M = \dfrac{\Pi}{RT} = \dfrac{(3.85 \text{ atm})}{\left(0.082\ 06\ \dfrac{L \cdot atm}{K \cdot mol}\right)(300 \text{ K})} = 0.156 \text{ M}$

11.25 $\Delta T_f = K_f \cdot m$; $m = \dfrac{\Delta T_f}{K_f} = \dfrac{2.10\ ^\circ C}{37.7\ \dfrac{^\circ C \cdot kg}{mol}} = 0.0557 \text{ mol/kg} = 0.0557\ m$

$$35.00 \text{ g} \times \frac{1 \text{ kg}}{1000 \text{ g}} = 0.03500 \text{ kg}$$

$$mol = 0.0557\ \frac{mol}{kg} \times 0.03500 \text{ kg} = 0.001\ 95 \text{ mol naphthalene}$$

$$\text{molar mass of naphthalene} = \frac{0.250 \text{ g naphthalene}}{0.001\ 95 \text{ mol naphthalene}} = 128 \text{ g/mol}$$

11.26 $\Pi = MRT$; $M = \dfrac{\Pi}{RT} = \dfrac{\left(149 \text{ mm Hg} \times \dfrac{1 \text{ atm}}{760 \text{ mm Hg}}\right)}{\left(0.08206\ \dfrac{L \cdot atm}{K \cdot mol}\right)(298 \text{ K})} = 8.02 \times 10^{-3} \text{ M}$

$300.0 \text{ mL} = 0.3000 \text{ L}$
$(8.02 \times 10^{-3} \text{ mol/L})(0.3000 \text{ L}) = 0.002\ 406 \text{ mol sucrose}$

$$\text{molar mass of sucrose} = \frac{0.822 \text{ g sucrose}}{0.002\ 406 \text{ mol sucrose}} = 342 \text{ g/mol}$$

11.27 (a) and (c)

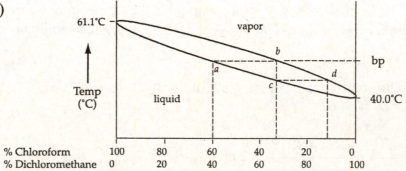

(b) The mixture will begin to boil at ~50 °C.
(d) After two cycles of boiling and condensing, the approximate composition of the liquid would be 90% dichloromethane and 10% chloroform.

11.28 Both solvent molecules and small solute particles can pass through a semipermeable dialysis membrane. Only large colloidal particles such as proteins can't pass through. Only solvent molecules can pass through a semipermeable membrane used for osmosis.

Key Concept Problems

11.29 The upper curve is pure ether.
(a) The normal boiling point for ether is the temperature where the upper curve intersects the 760 mm Hg line, ~ 37 °C.
(b) $\Delta T_b \approx 3$ °C

$$\Delta T_b = K_b \cdot m; \qquad m = \frac{\Delta T_b}{K_b} = \frac{3 \text{ °C}}{2.02 \dfrac{\text{°C} \cdot \text{kg}}{\text{mol}}} \approx 1.5 \text{ mol/kg} \approx 1.5 \ m$$

11.30 (a) < (b) < (c)

11.31 At any given temperature, the vapor pressure of a mixture of two pure liquids falls between the individual vapor pressures of the two pure liquids themselves. Because the vapor pressure of the mixture is greater than the vapor pressure of the solvent, the second liquid is more volatile (has a higher vapor pressure) than the solvent.

11.32 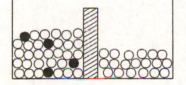 Assume that only the blue (open) spheres (solvent) can pass through the semipermeable membrane. There will be a net transfer of solvent from the right compartment (pure solvent) to the left compartment (solution) to achieve equilibrium.

11.33 At point 1, the temperature should be near the boiling point of the lower boiling solvent, $CHCl_3$, approximately 62 °C.
At point 3, the temperature should be about halfway between the two boiling points at approximately 70 °C.
At point 2, the temperature should be about halfway between the temperatures at points 1 and 3, approximately 66 °C.

11.34 The vapor pressure of the NaCl solution is lower than that of pure H_2O. More H_2O molecules will go into the vapor from the pure H_2O than from the NaCl solution. More H_2O vapor molecules will go into the NaCl solution than into pure H_2O. The result is represented by (b).

11.35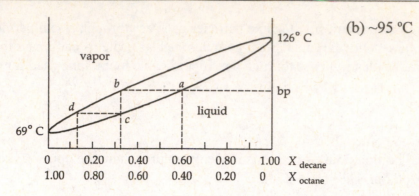

(b) ~95 °C

11.36 (a) Solution B (i = 3) is represented by the red line because its freezing point depressed the most. Solution A (i = 1) is represented by the blue line.
(b) There is a difference of 3.0 °C between the red and blue line. The 3.0 °C equates to 2 i units ($i_{red} - i_{blue} = 3 - 1 = 2$) so 1 i unit is worth 1.5 °C. The blue line (i = 1) is at 14.0 °C, so the freezing point of the pure liquid is 14.0 °C + 1.5 °C = 15.5 °C.
(c) Both the solutions are the same concentration.

$$\Delta T = K_f \cdot m; \quad m = \frac{\Delta T}{K_f} = \frac{1.5 \, ^\circ C}{3.0 \, ^\circ C/m} = 0.50 \, m$$

11.37 (a) The red curve represents the solution of a volatile solute and the green curve represents the solution of a nonvolatile solute.
(b) & (d)

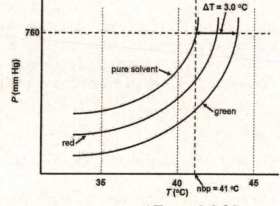

(c) $\Delta T = K_f \cdot m; \quad m = \dfrac{\Delta T}{K_f} = \dfrac{3.0 \, ^\circ C}{2.0 \, ^\circ C/m} = 1.5 \, m$

Section Problems
Solutions and Energy Changes (Sections 11.1–11.2)

11.38 The surface area of a solid plays an important role in determining how rapidly a solid dissolves. The larger the surface area, the more solid-solvent interactions, and the more rapidly the solid will dissolve. Powdered NaCl has a much larger surface area than a large block of NaCl, and it will dissolve more rapidly.

11.39 (a) a gas in a liquid – carbonated soft drink
       (b) a solid in a solid – metal alloys (14-karat gold)
       (c) a liquid in a solid – dental amalgam (Hg in Ag)

11.40 Substances tend to dissolve when the solute and solvent have the same type and
       magnitude of intermolecular forces; thus the rule of thumb "like dissolves like."

11.41 Both Br_2 and CCl_4 are nonpolar, and intermolecular forces for both are dispersion forces.
       H_2O is a polar molecule with dipole-dipole forces and hydrogen bonding. Therefore, Br_2
       is more soluble in CCl_4.

11.42 Energy is required to overcome intermolecular forces holding solute particles together in
       the crystal. For an ionic solid, this is the lattice energy. Substances with higher lattice
       energies tend to be less soluble than substances with lower lattice energies.

11.43 SO_4^{2-} has the larger hydration energy because of its higher charge. Both SO_4^{2-} and ClO_4^-
       are comparable in size, so size is not a factor.

11.44 Ethyl alcohol and water are both polar with small dispersion forces. They both can
       hydrogen bond, and are miscible.
       Pentyl alcohol is slightly polar and can hydrogen bond. It has, however, a relatively
       large dispersion force because of its size, which limits its water solubility.

11.45 The intermolecular forces associated with octane are dispersion forces. Both pentyl
       alcohol and methyl alcohol can hydrogen bond. Pentyl alcohol has relatively large
       dispersion forces because of its size. Methyl alcohol does not. Pentyl alcohol is soluble
       in octane; methyl alcohol is not.

11.46 $CaCl_2$, 110.98 amu
       For a 1.00 m solution:
              heat released = 81,300 J
              mass of solution = 1000 g H_2O + 110.98 g $CaCl_2$ = 1110.98 g

$$\Delta T = \frac{q}{(\text{specific heat})(\text{mass of solution})} = \frac{81,300 \text{ J}}{[4.18 \text{ J/(K} \cdot \text{g)}](1110.98 \text{ g})} = 17.5 \text{ K} = 17.5 \text{ °C}$$

       Final temperature = 25.0 °C + 17.5 °C = 42.5 °C

11.47 NH_4ClO_4, 117.48 amu
       For a 1.00 m solution:
              heat absorbed = 33,500 J
              mass of solution = 1000 g H_2O + 117.48 g NH_4ClO_4 = 1117.48 g

$$\Delta T = \frac{q}{(\text{specific heat})(\text{mass of solution})} = \frac{-33,500 \text{ J}}{[4.18 \text{ J/(K} \cdot \text{g)}](1117.48 \text{ g})} = -7.2 \text{ K} = -7.2 \text{ °C}$$

       Final temperature = 25.0 °C − 7.2 °C = 17.8 °C

Units of Concentration (Section 11.3)

11.48 $\text{molarity} = \dfrac{\text{moles of solute}}{\text{liters of solution}}$; $\text{molality} = \dfrac{\text{moles of solute}}{\text{kg of solvent}}$

11.49   A saturated solution contains enough solute so that there is an equilibrium between dissolved solute and undissolved solid.
A supersaturated solution contains a greater-than-equilibrium amount of solute.

11.50   (a) Dissolve 0.150 mol of glucose in water; dilute to 1.00 L.
(b) Dissolve 1.135 mol of KBr in 1.00 kg of H_2O.
(c) Mix together 0.15 mol of CH_3OH with 0.85 mol of H_2O.

11.51   (a) Dissolve 15.5 mg urea in 100 mL water
(b) Choose a K^+ salt, say KCl, and dissolve 0.0075 mol (0.559 g) in water; dilute to 100 mL.

11.52 $C_7H_6O_2$, 122.12 amu, 165 mL = 0.165 L
mol $C_7H_6O_2$ = (0.0268 mol/L)(0.165 L) = 0.004 42 mol

mass $C_7H_6O_2$ = 0.004 42 mol x $\dfrac{122.12 \text{ g}}{1 \text{ mol}}$ = 0.540 g

Dissolve 4.42×10^{-3} mol (0.540 g) of $C_7H_6O_2$ in enough $CHCl_3$ to make 165 mL of solution.

11.53 $C_7H_6O_2$, 122.12 amu

0.0268 mol $C_7H_6O_2$ x $\dfrac{122.12 \text{ g } C_7H_6O_2}{1 \text{ mol } C_7H_6O_2}$ = 3.27 g $C_7H_6O_2$

Dissolve 3.27 g of $C_7H_6O_2$ in 1.000 kg of $CHCl_3$, and take 165 mL of the solution.

11.54   (a) KCl, 74.6 amu
A 0.500 M KCl solution contains 37.3 g of KCl per 1.00 L of solution.
A 0.500 mass % KCl solution contains 5.00 g of KCl per 995 g of water.
The 0.500 M KCl solution is more concentrated (that is, it contains more solute per amount of solvent).
(b) Both solutions contain the same amount of solute. The 1.75 M solution contains less solvent than the 1.75 m solution. The 1.75 M solution is more concentrated.

11.55   (a) KI, 166.00 amu; KBr, 119.00 amu; assume 1.000 L = 1000 mL = 1000 g solution

10 ppm = $\dfrac{\text{mass KI}}{1000 \text{ g}}$ x 10^6 ;   mass KI = 0.010 g

10,000 ppb = $\dfrac{\text{mass KBr}}{1000 \text{ g}}$ x 10^9 ;   mass KBr = 0.010 g

Both solutions contain the same mass of solute in the same amount of solvent. Because the molar mass of KBr is less than that of KI, the number of moles of KBr is larger than the number of moles of KI. The KBr solution has a higher molarity than the KI solution.

(b) Because the mass % of the two solutions is the same, they both contain the same mass of solute and solution. Because the molar mass of KCl is less than that of citric acid, the number of moles of KCl is larger than the number of moles of citric acid. The KCl solution has a higher molarity than the citric acid solution.

11.56 (a) $C_6H_8O_7$, 192.12 amu

$$0.655 \text{ mol } C_6H_8O_7 \times \frac{192.12 \text{ g } C_6H_8O_7}{1 \text{ mol } C_6H_8O_7} = 126 \text{ g } C_6H_8O_7$$

$$\text{mass \% } C_6H_8O_7 = \frac{126 \text{ g}}{126 \text{ g} + 1000 \text{ g}} \times 100\% = 11.2 \text{ mass \%}$$

(b) $0.135 \text{ mg} = 0.135 \times 10^{-3}$ g

$(5.00 \text{ mL } H_2O)(1.00 \text{ g/mL}) = 5.00 \text{ g } H_2O$

$$\text{mass \% KBr} = \frac{0.135 \times 10^{-3} \text{ g}}{(0.135 \times 10^{-3} \text{ g}) + 5.00 \text{ g}} \times 100\% = 0.002 \, 70 \text{ mass \% KBr}$$

(c) $\text{mass \% aspirin} = \frac{5.50 \text{ g}}{5.50 \text{ g} + 145 \text{ g}} \times 100\% = 3.65 \text{ mass \% aspirin}$

11.57 (a) $\text{molality} = \frac{0.655 \text{ mol}}{1.00 \text{ kg}} = 0.655 \, m$

(b) KBr, 119.00 amu; $5.00 \text{ g} = 0.005 \, 00$ kg

$$\text{molality} = \frac{\left(0.135 \times 10^{-3} \text{ g} \times \dfrac{1 \text{ mol}}{119.00 \text{ g}}\right)}{0.005 \, 00 \text{ kg}} = 2.27 \times 10^{-4} \text{ mol/kg} = 2.27 \times 10^{-4} \, m$$

(c) $C_9H_8O_4$, 180.16 amu; $145 \text{ g} = 0.145$ kg

$$\text{molality} = \frac{\left(5.50 \text{ g} \times \dfrac{1 \text{ mol}}{180.16 \text{ g}}\right)}{0.145 \text{ kg}} = 0.211 \text{ mol/kg} = 0.211 \, m$$

11.58 $P_{O_3} = P_{\text{total}} \cdot X_{O_3}$

$$X_{O_3} = \frac{P_{O_3}}{P_{\text{total}}} = \frac{1.6 \times 10^{-9} \text{ atm}}{1.3 \times 10^{-2} \text{ atm}} = 1.2 \times 10^{-7}$$

Assume one mole of air (29 g/mol)

$\text{mol } O_3 = n_{\text{air}} \cdot X_{O_3} = (1 \text{ mol})(1.2 \times 10^{-7}) = 1.2 \times 10^{-7} \text{ mol } O_3$

O_3, 48.00 amu; $\quad \text{mass } O_3 = 1.2 \times 10^{-7} \text{ mol} \times \dfrac{48.0 \text{ g}}{1 \text{ mol}} = 5.8 \times 10^{-6} \text{ g } O_3$

$$\text{ppm } O_3 = \frac{5.8 \times 10^{-6} \text{ g}}{29 \text{ g}} \times 10^6 = 0.20 \text{ ppm}$$

11.59 Assume 1 mL of blood weighs 1 g. 1 dL = 0.1 L = 100 mL = 100 g

$$ppb = \frac{10 \times 10^{-6}\,g}{100\,g} \times 10^9 = 100\ ppb$$

11.60 (a) H_2SO_4, 98.08 amu; $\quad$ molality $= \dfrac{\left(25.0\,g \times \dfrac{1\ mol}{98.08\,g}\right)}{1.30\ kg} = 0.196\ mol/kg = 0.196\ m$

(b) $C_{10}H_{14}N_2$, 162.23 amu; $\ CH_2Cl_2$, 84.93 amu

$$2.25\,g\ C_{10}H_{14}N_2 \ \times\ \frac{1\ mol\ C_{10}H_{14}N_2}{162.23\ g\ C_{10}H_{14}N_2} = 0.0139\ mol\ C_{10}H_{14}N_2$$

$$80.0\,g\ CH_2Cl_2\ \times\ \frac{1\ mol\ CH_2Cl_2}{84.93\ g\ CH_2Cl_2} = 0.942\ mol\ CH_2Cl_2$$

$$X_{C_{10}H_{14}N_2} = \frac{0.0139\ mol}{0.942\ mol + 0.0139\ mol} = 0.0145$$

$$X_{CH_2Cl_2} = \frac{0.942\ mol}{0.942\ mol + 0.0139\ mol} = 0.985$$

11.61 NaOCl, 74.44 amu
A 5.0 mass % aqueous solution of NaOCl contains 5.0 g NaOCl and 95 g H_2O.

$$molality = \frac{\left(5.0\,g \times \dfrac{1\ mol}{74.44\,g}\right)}{0.095\ kg} = 0.71\ mol/kg = 0.71\ m$$

$$5.0\,g\ NaOCl\ \times\ \frac{1\ mol\ NaOCl}{74.44\ g\ NaOCl} = 0.0672\ mol\ NaOCl$$

$$95\,g\ H_2O\ \times\ \frac{1\ mol\ H_2O}{18.02\ g\ H_2O} = 5.27\ mol\ H_2O$$

$$X_{NaOCl} = \frac{0.0672\ mol}{5.27\ mol + 0.0672\ mol} = 0.013$$

11.62 16.0 mass % $= \dfrac{16.0\ g\ H_2SO_4}{16.0\ g\ H_2SO_4\ +\ 84.0\ g\ H_2O}$

H_2SO_4, 98.08 amu; $\qquad$ density = 1.1094 g/mL

volume of solution $= 100.0\,g\ \times\ \dfrac{1\ mL}{1.1094\,g} = 90.14\ mL = 0.090\ 14\ L$

$$molality = \frac{\left(16.0\,g \times \dfrac{1\ mol}{98.08\,g}\right)}{0.090\ 14\ L} = 1.81\ M$$

11.63 $C_2H_6O_2$, 62.07 amu
A 40.0 mass % aqueous solution of $C_2H_6O_2$ contains 40.0 g $C_2H_6O_2$ and 60.0 g H_2O.
density = 1.0514 g/mL

$$\text{volume of solution} = 100.0 \text{ g } \times \frac{1 \text{ mL}}{1.0514 \text{ g}} = 95.1 \text{ mL} = 0.0951 \text{ L}$$

$$\text{molarity} = \frac{\left(40.0 \text{ g} \times \dfrac{1 \text{ mol}}{62.07 \text{ g}}\right)}{0.0951 \text{ L}} = 6.78 \text{ M}$$

11.64 $$\text{molality} = \frac{\left(40.0 \text{ g} \times \dfrac{1 \text{ mol}}{62.07 \text{ g}}\right)}{0.0600 \text{ kg}} = 10.7 \text{ mol/kg} = 10.7 \ m$$

11.65 $$\text{molality} = \frac{\left(16.0 \text{ g} \times \dfrac{1 \text{ mol}}{98.08 \text{ g}}\right)}{0.0840 \text{ kg}} = 1.94 \text{ mol/kg} = 1.94 \ m$$

11.66 $C_{19}H_{21}NO_3$, 311.38 amu; 1.5 mg = 1.5×10^{-3} g

$$1.3 \times 10^{-3} \text{ mol/kg} = \frac{\left(1.5 \times 10^{-3} \text{ g} \times \dfrac{1 \text{ mol}}{311.38 \text{ g}}\right)}{\text{kg of solvent}}; \quad \text{solve for kg of solvent.}$$

$$\text{kg of solvent} = \frac{\left(1.5 \times 10^{-3} \text{ g} \times \dfrac{1 \text{ mol}}{311.38 \text{ g}}\right)}{1.3 \times 10^{-3} \text{ mol/kg}} = 0.0037 \text{ kg}$$

Because the solution is very dilute, kg of solvent ≈ kg of solution.

$$\text{g of solution} = (0.0037 \text{ kg})\left(\frac{1000 \text{ g}}{1 \text{ kg}}\right) = 3.7 \text{ g}$$

11.67 $C_{12}H_{22}O_{11}$, 342.30 amu

$$32.5 \text{ g } C_{12}H_{22}O_{11} \times \frac{1 \text{ mol } C_{12}H_{22}O_{11}}{342.30 \text{ g } C_{12}H_{22}O_{11}} = 0.0949 \text{ mol } C_{12}H_{22}O_{11}$$

$$0.850 \ m = 0.850 \text{ mol/kg} = \frac{0.0949 \text{ mol}}{\text{kg of } H_2O}; \quad \text{kg of } H_2O = \frac{0.0949 \text{ mol}}{0.850 \text{ mol/kg}} = 0.112 \text{ kg}$$

$$\text{mass of } H_2O = 0.112 \text{ kg} \times \frac{1000 \text{ g}}{1 \text{ kg}} = 112 \text{ g } H_2O$$

11.68 $C_6H_{12}O_6$, 180.16 amu; H_2O, 18.02 amu;   Assume 1.00 L of solution.
mass of solution = (1000 mL)(1.0624 g/mL) = 1062.4 g

$$\text{mass of solute} = 0.944 \text{ mol} \times \frac{180.16 \text{ g}}{1 \text{ mol}} = 170.1 \text{ g } C_6H_{12}O_6$$

mass of H_2O = 1062.4 g − 170.1 g = 892.3 g H_2O

mol $C_6H_{12}O_6$ = 0.944 mol;   mol H_2O = 892.3 g x $\dfrac{1 \text{ mol}}{18.02 \text{ g}}$ = 49.5 mol

(a) $X_{C_6H_{12}O_6}$ = $\dfrac{\text{mol } C_6H_{12}O_6}{\text{mol } C_6H_{12}O_6 + \text{mol } H_2O}$ = $\dfrac{0.944 \text{ mol}}{0.944 \text{ mol} + 49.5 \text{ mol}}$ = 0.0187

(b) mass % = $\dfrac{\text{mass } C_6H_{12}O_6}{\text{total mass of solution}}$ x 100% = $\dfrac{170.1 \text{ g}}{1062.4 \text{ g}}$ x 100% = 16.0%

(c) molality = $\dfrac{\text{mol } C_6H_{12}O_6}{\text{kg } H_2O}$ = $\dfrac{0.944 \text{ mol}}{0.8923 \text{ kg}}$ = 1.06 mol/kg = 1.06 m

11.69 $C_{12}H_{22}O_{11}$, 342.30 amu;   Assume 1.00 L of solution.
mass of solution = (1000 mL)(1.0432 g/mL) = 1043.2 g

mass of solute = 0.335 mol $C_{12}H_{22}O_{11}$ x $\dfrac{342.30 \text{ g } C_{12}H_{22}O_{11}}{1 \text{ mol } C_{12}H_{22}O_{11}}$ = 114.7 g $C_{12}H_{22}O_{11}$

mass of H_2O = 1043.2 g – 114.7 g = 928.5 g H_2O

mol $C_{12}H_{22}O_{11}$ = 0.335 mol;   928.5 g H_2O x $\dfrac{1 \text{ mol } H_2O}{18.02 \text{ g } H_2O}$ = 51.53 mol H_2O

(a) $X_{C_{12}H_{22}O_{11}}$ = $\dfrac{0.335 \text{ mol}}{51.53 \text{ mol} + 0.335 \text{ mol}}$ = 0.006 46

(b) mass % $C_{12}H_{22}O_{11}$ = $\dfrac{114.7 \text{ g}}{1043.2 \text{ g}}$ x 100% = 11.0 mass % $C_{12}H_{22}O_{11}$

(c) molality = $\dfrac{0.335 \text{ mol}}{0.9285 \text{ kg}}$ = 0.361 mol/kg = 0.361 m

Solubility and Henry's Law (Section 11.4)

11.70   M = k · P = (0.091 $\dfrac{\text{mol}}{\text{L} \cdot \text{atm}}$)(0.75 atm) = 0.068 M

11.71   M = k · P;   k = $\dfrac{\text{M}}{\text{P}}$ = $\dfrac{0.195 \text{ M}}{1.00 \text{ atm}}$ = 0.195 mol/(L· atm)

P = 25.5 mm Hg x $\dfrac{1.00 \text{ atm}}{760 \text{ mm Hg}}$ = 0.0336 atm

M = k · P = (0.195 $\dfrac{\text{mol}}{\text{L} \cdot \text{atm}}$)(0.0336 atm) = 6.55 x 10^{-3} M

11.72   M = k · P

Calculate k: k = $\dfrac{\text{M}}{\text{P}}$ = $\dfrac{2.21 \text{ x } 10^{-3} \text{ mol/L}}{1.00 \text{ atm}}$ = 2.21 x 10^{-3} $\dfrac{\text{mol}}{\text{L} \cdot \text{atm}}$

Convert 4 mg/L to mol/L:
4 mg = 4 x 10^{-3} g

$$O_2 \text{ molarity} = \dfrac{\left(4 \times 10^{-3} \text{ g} \times \dfrac{1 \text{ mol}}{32.00 \text{ g}}\right)}{1.00 \text{ L}} = 1.25 \times 10^{-4} \text{ M}$$

$$P_{O_2} = \dfrac{M}{k} = \dfrac{1.25 \times 10^{-4} \dfrac{\text{mol}}{\text{L}}}{2.21 \times 10^{-3} \dfrac{\text{mol}}{\text{L} \cdot \text{atm}}} = 0.06 \text{ atm}$$

11.73 $k = 1.93 \times 10^{-3} \text{ mol/(L} \cdot \text{atm)}$

$$M = k \cdot P = [1.93 \times 10^{-3} \text{ mol/(L} \cdot \text{atm)}]\left(68 \text{ mm Hg} \times \dfrac{1.00 \text{ atm}}{760 \text{ mm Hg}}\right) = 1.73 \times 10^{-4} \text{ mol/L}$$

$$1.73 \times 10^{-4} \text{ mol/L} \times \dfrac{32.00 \text{ g O}_2}{1 \text{ mol O}_2} \times \dfrac{1 \text{ mg}}{1 \times 10^{-3} \text{ g}} = 5.5 \text{ mg/L}$$

11.74 $[Xe] = 10 \text{ mmol/L} = 0.010 \text{ M at STP}$

$$M = k \cdot P; \ k = \dfrac{M}{P} = \dfrac{0.010 \text{ M}}{1.0 \text{ atm}} = 0.010 \text{ mol/(L} \cdot \text{atm)}$$

11.75   Assuming H_2O as the solvent, NH_3 does not obey Henry's law because NH_3 can both hydrogen bond and react with H_2O.

Colligative Properties (Sections 11.5–11.8)

11.76   The difference in entropy between the solvent in a solution and a pure solvent is responsible for colligative properties.

11.77   Osmotic pressure is the amount of pressure that needs to be applied to cause osmosis to stop.

11.78   NaCl is a nonvolatile solute. Methyl alcohol is a volatile solute. When NaCl is added to water, the vapor pressure of the solution is decreased, which means that the boiling point of the solution will increase. When methyl alcohol is added to water, the vapor pressure of the solution is increased which means that the boiling point of the solution will decrease.

11.79   When 100 mL of 9 M H_2SO_4 at 0 °C is added to 100 mL of liquid water at 0 °C, the temperature rises because ΔH_{soln} for H_2SO_4 is exothermic.
When 100 mL of 9 M H_2SO_4 at 0 °C is added to 100 g of solid ice at 0 °C, some of the ice will melt (an endothermic process) and the temperature will fall because the H_2SO_4 (solute) lowers the freezing point of the ice/water mixture.

11.80

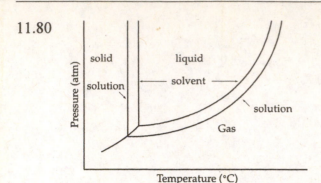

11.81   Molality is a temperature independent concentration unit. For freezing point depression and boiling point elevation, molality is used so that the solute concentration is independent of temperature changes. Molarity is temperature dependent. Molarity can be used for osmotic pressure because osmotic pressure is measured at a fixed temperature.

11.82   (a) CH_4N_2O, 60.06 amu; H_2O, 18.02 amu

$$10.0 \text{ g } CH_4N_2O \times \frac{1 \text{ mol } CH_4N_2O}{60.06 \text{ g } CH_4N_2O} = 0.167 \text{ mol } CH_4N_2O$$

$$150.0 \text{ g } H_2O \times \frac{1 \text{ mol } H_2O}{18.02 \text{ g } H_2O} = 8.32 \text{ mol } H_2O$$

$$X_{H_2O} = \frac{8.32 \text{ mol}}{8.32 \text{ mol } + 0.167 \text{ mol}} = 0.980$$

$$P_{soln} = P^o_{H_2O} \cdot X_{H_2O} = (71.93 \text{ mm Hg})(0.980) = 70.5 \text{ mm Hg}$$

(b) LiCl, 42.39 amu; $10.0 \text{ g LiCl} \times \dfrac{1 \text{ mol LiCl}}{42.39 \text{ g LiCl}} = 0.236 \text{ mol LiCl}$

LiCl dissociates into Li^+(aq) and Cl^-(aq) in H_2O.
mol Li^+ = mol Cl^- = mol LiCl = 0.236 mol

$$150.0 \text{ g } H_2O \times \frac{1 \text{ mol } H_2O}{18.02 \text{ g } H_2O} = 8.32 \text{ mol } H_2O$$

$$X_{H_2O} = \frac{8.32 \text{ mol}}{8.32 \text{ mol} + 0.236 \text{ mol} + 0.236 \text{ mol}} = 0.946$$

$$P_{soln} = P^o_{H_2O} \cdot X_{H_2O} = (71.93 \text{ mm Hg})(0.946) = 68.0 \text{ mm Hg}$$

11.83 $C_6H_{12}O_6$, 180.16 amu; CH_3OH, 32.04 amu

$$16.0 \text{ g } C_6H_{12}O_6 \times \frac{1 \text{ mol } C_6H_{12}O_6}{180.16 \text{ g } C_6H_{12}O_6} = 0.0888 \text{ mol } C_6H_{12}O_6$$

$$80.0 \text{ g } CH_3OH \times \frac{1 \text{ mol } CH_3OH}{32.04 \text{ g } CH_3OH} = 2.50 \text{ mol } CH_3OH$$

$$X_{CH_3OH} = \frac{2.50 \text{ mol}}{2.50 \text{ mol} + 0.0888 \text{ mol}} = 0.966$$

$$P_{soln} = P^o_{CH_3OH} \cdot X_{CH_3OH} = (140 \text{ mm Hg})(0.966) = 135 \text{ mm Hg}$$

11.84 For H_2O, $K_b = 0.51 \dfrac{^oC \cdot kg}{mol}$; 150.0 g = 0.1500 kg

(a) $\Delta T_b = K_b \cdot m = \left(0.51 \dfrac{^oC \cdot kg}{mol}\right)\left(\dfrac{0.167 \text{ mol}}{0.1500 \text{ kg}}\right) = 0.57 \,^oC$

Solution boiling point = 100.00 ^oC + ΔT_b = 100.00 ^oC + 0.57 ^oC = 100.57 ^oC

(b) $\Delta T_b = K_b \cdot m = \left(0.51 \dfrac{^oC \cdot kg}{mol}\right)\left(\dfrac{2(0.236 \text{ mol})}{0.1500 \text{ kg}}\right) = 1.6 \,^oC$

Solution boiling point = 100.00 ^oC + ΔT_b = 100.00 ^oC + 1.6 ^oC = 101.6 ^oC

11.85 For H_2O, $K_f = 1.86 \dfrac{^oC \cdot kg}{mol}$; 150.0 g = 0.1500 kg

(a) $\Delta T_f = K_f \cdot m = \left(1.86 \dfrac{^oC \cdot kg}{mol}\right)\left(\dfrac{0.167 \text{ mol}}{0.1500 \text{ kg}}\right) = 2.07 \,^oC$

Solution freezing point = 0.00 ^oC − ΔT_f = 0.00 ^oC − 2.07 ^oC = −2.07 ^oC

(b) $\Delta T_f = K_f \cdot m = \left(1.86 \dfrac{^oC \cdot kg}{mol}\right)\left(\dfrac{2(0.236 \text{ mol})}{0.1500 \text{ kg}}\right) = 5.85 \,^oC$

Solution freezing point = 0.00 ^oC − ΔT_f = 0.00 ^oC − 5.85 ^oC = −5.85 ^oC

11.86 $\Delta T_f = K_f \cdot m \cdot i$

Solution freezing point = − 4.3 ^oC = 0.00 ^oC − ΔT_f; ΔT_f = 4.3 ^oC

$$i = \frac{\Delta T_f}{K_f \cdot m} = \frac{4.3 \,^oC}{\left(1.86 \dfrac{^oC \cdot kg}{mol}\right)(1.0 \text{ mol/kg})} = 2.3$$

11.87 $\Delta T_b = K_b \cdot m \cdot i = \left(0.51 \dfrac{^oC \cdot kg}{mol}\right)(0.75 \text{ mol/kg})(1.85) = 0.71 \,^oC$

Solution boiling point = 100.00 ^oC + ΔT_b = 100.00 ^oC + 0.71 ^oC = 100.71 ^oC

11.88 Acetone, C_3H_6O, 58.08 amu, $P^o_{C_3H_6O}$ = 285 mm Hg

Ethyl acetate, $C_4H_8O_2$, 88.11 amu, $P^o_{C_4H_8O_2}$ = 118 mm Hg

$$25.0 \text{ g } C_3H_6O \times \frac{1 \text{ mol } C_3H_6O}{58.08 \text{ g } C_3H_6O} = 0.430 \text{ mol } C_3H_6O$$

$$25.0 \text{ g C}_4\text{H}_8\text{O}_2 \times \frac{1 \text{ mol C}_4\text{H}_8\text{O}_2}{88.11 \text{ g C}_4\text{H}_8\text{O}_2} = 0.284 \text{ mol C}_4\text{H}_8\text{O}_2$$

$$X_{C_3H_6O} = \frac{0.430 \text{ mol}}{0.430 \text{ mol} + 0.284 \text{ mol}} = 0.602; \quad X_{C_4H_8O_2} = \frac{0.284 \text{ mol}}{0.430 \text{ mol} + 0.284 \text{ mol}} = 0.398$$

$$P_{soln} = P^o_{C_3H_6O} \cdot X_{C_3H_6O} + P^o_{C_4H_8O_2} \cdot X_{C_4H_8O_2}$$

$$P_{soln} = (285 \text{ mm Hg})(0.602) + (118 \text{ mm Hg})(0.398) = 219 \text{ mm Hg}$$

11.89 $CHCl_3$, 119.38 amu, $P^o_{CHCl_3} = 205$ mm Hg; CH_2Cl_2, 84.93 amu, $P^o_{CH_2Cl_2} = 415$ mm Hg

$$15.0 \text{ g CHCl}_3 \times \frac{1 \text{ mol CHCl}_3}{119.38 \text{ g CHCl}_3} = 0.126 \text{ mol CHCl}_3$$

$$37.5 \text{ g CH}_2\text{Cl}_2 \times \frac{1 \text{ mol CH}_2\text{Cl}_2}{84.93 \text{ g CH}_2\text{Cl}_2} = 0.442 \text{ mol CH}_2\text{Cl}_2$$

$$X_{CHCl_3} = \frac{0.126 \text{ mol}}{0.126 \text{ mol} + 0.442 \text{ mol}} = 0.222; \quad X_{CH_2Cl_2} = \frac{0.442 \text{ mol}}{0.126 \text{ mol} + 0.442 \text{ mol}} = 0.778$$

$$P_{soln} = P^o_{CHCl_3} \cdot X_{CHCl_3} + P^o_{CH_2Cl_2} \cdot X_{CH_2Cl_2}$$

$$P_{soln} = (205 \text{ mm Hg})(0.222) + (415 \text{ mm Hg})(0.778) = 368 \text{ mm Hg}$$

11.90 In the liquid, $X_{acetone} = 0.602$ and $X_{ethyl\ acetate} = 0.398$
In the vapor, $P_{Total} = 219$ mm Hg

$$P_{acetone} = P^o_{acetone} \cdot X_{acetone} = (285 \text{ mm Hg})(0.602) = 172 \text{ mm Hg}$$

$$P_{ethyl\ acetate} = P^o_{ethyl\ acetate} \cdot X_{ethyl\ acetate} = (118 \text{ mm Hg})(0.398) = 47 \text{ mm Hg}$$

$$X_{acetone} = \frac{P_{acetone}}{P_{total}} = \frac{172 \text{ mm Hg}}{219 \text{ mm Hg}} = 0.785; \quad X_{ethyl\ acetate} = \frac{P_{ethyl\ acetate}}{P_{total}} = \frac{47 \text{ mm Hg}}{219 \text{ mm Hg}} = 0.215$$

11.91 In the liquid, $X_{CHCl_3} = 0.222$ and $X_{CH_2Cl_2} = 0.778$
In the vapor, $P_{total} = 368$ mm Hg

$$P_{CHCl_3} = P^o_{CHCl_3} \cdot X_{CHCl_3} = (205 \text{ mm Hg})(0.222) = 45.5 \text{ mm Hg}$$

$$P_{CH_2Cl_2} = P^o_{CH_2Cl_2} \cdot X_{CH_2Cl_2} = (415 \text{ mm Hg})(0.778) = 323 \text{ mm Hg}$$

$$X_{CHCl_3} = \frac{P_{CHCl_3}}{P_{total}} = \frac{45.5 \text{ mm Hg}}{368 \text{ mm Hg}} = 0.124; \quad X_{CH_2Cl_2} = \frac{P_{CH_2Cl_2}}{P_{total}} = \frac{323 \text{ mm Hg}}{368 \text{ mm Hg}} = 0.876$$

11.92 $C_9H_8O_4$, 180.16 amu; 215 g = 0.215 kg

$$\Delta T_b = K_b \cdot m = 0.47 \text{ °C}; \quad K_b = \frac{\Delta T_b}{m} = \frac{0.47 \text{ °C}}{\left(\dfrac{5.00 \text{ g} \times \dfrac{1 \text{ mol}}{180.16 \text{ g}}}{0.215 \text{ kg}}\right)} = 3.6 \frac{\text{°C} \cdot \text{kg}}{\text{mol}}$$

11.93 $C_6H_8O_6$, 176.13 amu; 50.0 g = 0.0500 kg

$$\Delta T_f = K_f \cdot m = 1.33 \ ^\circ C; \qquad K_f = \frac{\Delta T_f}{m} = \frac{1.33 \ ^\circ C}{\left(\dfrac{3.00 \ g \times \dfrac{1 \ mol}{176.13 \ g}}{0.0500 \ kg}\right)} = 3.90 \ \frac{^\circ C \cdot kg}{mol}$$

11.94 $\Delta T_b = K_b \cdot m = 1.76 \ ^\circ C; \qquad m = \dfrac{\Delta T_b}{K_b} = \dfrac{1.76 \ ^\circ C}{3.07 \ \dfrac{^\circ C \cdot kg}{mol}} = 0.573 \ m$

11.95 $C_6H_{12}O_6$, 180.16 amu

For ethyl alcohol, $K_b = 1.22 \ \dfrac{^\circ C \cdot kg}{mol}$; 285 g = 0.285 kg

$$\Delta T_b = K_b \cdot m = \left(1.22 \ \frac{^\circ C \cdot kg}{mol}\right)\left(\frac{26.0 \ g \times \dfrac{1 \ mol}{180.16 \ g}}{0.285 \ kg}\right) = 0.618 \ ^\circ C$$

Solution boiling point = normal boiling point + ΔT_b = 79.1 °C
Normal boiling point = 79.1 °C – ΔT_b = 79.1 °C – 0.618 °C = 78.5 °C

11.96 $\Pi = MRT$
(a) NaCl 58.44 amu; 350.0 mL = 0.3500 L
There are 2 moles of ions/mole of NaCl

$$\Pi = (2)\left(\frac{5.00 \ g \times \dfrac{1 \ mol}{58.44 \ g}}{0.3500 \ L}\right)\left(0.082 \ 06 \ \frac{L \cdot atm}{K \cdot mol}\right)(323 \ K) = 13.0 \ atm$$

(b) CH_3CO_2Na, 82.03 amu; 55.0 mL = 0.0550 L
There are 2 moles of ions/mole of CH_3CO_2Na

$$\Pi = (2)\left(\frac{6.33 \ g \times \dfrac{1 \ mol}{82.03 \ g}}{0.0550 \ L}\right)\left(0.082 \ 06 \ \frac{L \cdot atm}{K \cdot mol}\right)(283K) = 65.2 \ atm$$

11.97 $\Pi = MRT = \left(\dfrac{11.5 \times 10^{-3} \ g \times \dfrac{1 \ mol}{5990 \ g}}{0.006 \ 60 \ L}\right)\left(0.082 \ 06 \ \dfrac{L \cdot atm}{K \cdot mol}\right)(298K) = 0.007 \ 11 \ atm$

$$\Pi = 0.007 \ 11 \ atm \times \frac{760 \ mm \ Hg}{1 \ atm} = 5.41 \ mm \ Hg$$

height of H_2O column = 5.41 mm Hg x $\dfrac{13.534 \ mm \ H_2O}{1.00 \ mm \ Hg}$ = 73.2 mm

height of H_2O column = 73.2 mm x $\dfrac{1 \ m}{1000 \ mm}$ = 0.0732 m

11.98 $\Pi = MRT$; $\quad M = \dfrac{\Pi}{RT} = \dfrac{4.85 \text{ atm}}{\left(0.082\ 06 \dfrac{L \cdot atm}{K \cdot mol}\right)(300 \text{ K})} = 0.197 \text{ M}$

11.99 $\Pi = MRT$; $\quad M = \dfrac{\Pi}{RT} = \dfrac{7.7 \text{ atm}}{\left(0.082\ 06 \dfrac{L \cdot atm}{K \cdot mol}\right)(310 \text{ K})} = 0.30 \text{ M}$

Uses of Colligative Properties (Section 11.9)

11.100 Osmotic pressure is most often used for the determination of molecular mass because, of the four colligative properties, osmotic pressure gives the largest colligative property change per mole of solute.

11.101 $C_6H_{12}O_6$ does not dissociate in aqueous solution. LiCl and NaCl both dissociate into two solute particles per formula unit in aqueous solution. $CaCl_2$ dissociates into three solute particles per formula unit in aqueous solution. Assume that you have 1.00 g of each substance. Calculate the number of moles of solute particles in 1.00 g of each substance.

$C_6H_{12}O_6$, 180.2 amu; moles solute particles = $1.00 \text{ g} \times \dfrac{1 \text{ mol}}{180.2 \text{ g}} = 0.005\ 55$ moles

LiCl, 42.4 amu; moles solute particles = $2\left(1.00 \text{ g} \times \dfrac{1 \text{ mol}}{42.4 \text{ g}}\right) = 0.0472$ moles

NaCl, 58.4 amu; moles solute particles = $2\left(1.00 \text{ g} \times \dfrac{1 \text{ mol}}{58.4 \text{ g}}\right) = 0.0342$ moles

$CaCl_2$, 111.0 amu; moles solute particles = $3\left(1.00 \text{ g} \times \dfrac{1 \text{ mol}}{111.0 \text{ g}}\right) = 0.0270$ moles

LiCl produces more solute particles/gram than any of the other three substances. LiCl would be the most efficient per unit mass.

11.102 $\Pi = 407.2 \text{ mm Hg} \times \dfrac{1 \text{ atm}}{760 \text{ mm Hg}} = 0.5358 \text{ atm}$

$\Pi = MRT$; $M = \dfrac{\Pi}{RT} = \dfrac{0.5358 \text{ atm}}{\left(0.082\ 06 \dfrac{L \cdot atm}{K \cdot mol}\right)(298.15 \text{ K})} = 0.021\ 90 \text{ M}$

$200.0 \text{ mL} \times \dfrac{1 \text{ L}}{1000 \text{ mL}} = 0.2000 \text{ L}$

mol cellobiose = $(0.2000 \text{ L})(0.021\ 90 \text{ mol/L}) = 4.380 \times 10^{-3} \text{ mol}$

molar mass of cellobiose = $\dfrac{1.500 \text{ g cellobiose}}{4.380 \times 10^{-3} \text{ mol cellobiose}} = 342.5 \text{ g/mol}$

molecular mass = 342.5 amu

11.103 height of Hg column = 32.9 cm H_2O x $\dfrac{1.00 \text{ cm Hg}}{13.534 \text{ cm } H_2O}$ = 2.43 cm Hg

Π = 2.43 cm Hg x $\dfrac{1.00 \text{ atm}}{76.0 \text{ cm Hg}}$ = 0.0320 atm

Π = MRT;   M = $\dfrac{\Pi}{RT}$ = $\dfrac{0.0320 \text{ atm}}{\left(0.082\ 06 \dfrac{L \cdot atm}{K \cdot mol}\right)(298 \text{ K})}$ = 0.001 31 M

20.0 mL x $\dfrac{1 \text{ L}}{1000 \text{ mL}}$ = 0.0200 L

15.0 mg x $\dfrac{1.00 \text{ g}}{1000 \text{ mg}}$ = 0.0150 g

mol met-enkephalin = (0.0200 L)(0.001 31 mol/L) = 2.62 x 10^{-5} mol

molar mass of met-enkephalin = $\dfrac{0.0150 \text{ g met-enkephalin}}{2.62 \times 10^{-5} \text{ mol met-enkephalin}}$ = 573 g/mol

molecular mass = 573 amu

11.104 HCl is a strong electrolyte in H_2O and completely dissociates into two solute particles per each HCl.
HF is a weak electrolyte in H_2O. Only a few percent of the HF molecules dissociates into ions.

11.105 Na_2SO_4, 142.0 amu; $m = \dfrac{\left(71 \text{ g} \times \dfrac{1 \text{ mol}}{142.0 \text{ g}}\right)}{1.00 \text{ kg}}$ = 0.50 mol/kg = 0.50 m

$\Delta T_b = K_b \cdot m = \left(0.51 \dfrac{^{\circ}C \cdot kg}{mol}\right)(0.50\ m)$ = 0.26 $^{\circ}$C

The experimental ΔT is approximately 3 times that predicted by the equation above because Na_2SO_4 dissociates into three solute particles (2 Na^+ and SO_4^{2-}) in aqueous solution.

11.106 First, determine the empirical formula:
Assume 100.0 g of β-carotene.

10.51% H   10.51 g H x $\dfrac{1 \text{ mol H}}{1.008 \text{ g H}}$ = 10.43 mol H

89.49% C   89.49 g C x $\dfrac{1 \text{ mol C}}{12.01 \text{ g C}}$ = 7.45 mol C

$C_{7.45}H_{10.43}$; divide each subscript by the smaller, 7.45.
$C_{7.45/7.45}H_{10.43/7.45}$
$CH_{1.4}$
Multiply each subscript by 5 to obtain integers.
Empirical formula is C_5H_7, 67.1 amu.

Second, calculate the molecular mass:

$$\Delta T_f = K_f \cdot m; \quad m = \frac{\Delta T_f}{K_f} = \frac{1.17\ ^\circ C}{37.7\ \dfrac{^\circ C \cdot kg}{mol}} = 0.0310\ mol/kg = 0.0310\ m$$

$$1.50\ g \times \frac{1 kg}{1000\ g} = 1.50 \times 10^{-3}\ kg$$

mol β-carotene $= (1.50 \times 10^{-3}\ kg)(0.0310\ mol/kg) = 4.65 \times 10^{-5}\ mol$

molar mass of β-carotene $= \dfrac{0.0250\ g\ \beta\text{-carotene}}{4.65 \times 10^{-5}\ mol\ \beta\text{-carotene}} = 538\ g/mol$

molecular mass = 538 amu

Finally, determine the molecular formula:
Divide the molecular mass by the empirical formula mass.

$$\frac{538\ amu}{67.1\ amu} = 8; \quad \text{molecular formula is } C_{(8 \times 5)}H_{(8 \times 7)}, \text{ or } C_{40}H_{56}$$

11.107 First, determine the empirical formula:
Assume a 100.0 g sample of lysine.

49.29% C $\qquad$ $49.29\ g\ C \times \dfrac{1\ mol\ C}{12.011\ g\ C} = 4.10\ mol\ C$

9.65% H $\qquad$ $9.65\ g\ H \times \dfrac{1\ mol\ H}{1.008\ g\ H} = 9.57\ mol\ H$

19.16% N $\qquad$ $19.16\ g\ N \times \dfrac{1\ mol\ N}{14.007\ g\ N} = 1.37\ mol\ N$

21.89% O $\qquad$ $21.89\ g\ O \times \dfrac{1\ mol\ O}{15.999\ g\ O} = 1.37\ mol\ O$

$C_{4.10}H_{9.57}N_{1.37}O_{1.37}$; divide each subscript by the smallest, 1.37.
$C_{4.10/1.37}H_{9.57/1.37}N_{1.37/1.37}O_{1.37/1.37}$
Empirical formula is C_3H_7NO, 73.09 amu
Second, calculate the molecular mass:

$$\Delta T_f = K_f \cdot m = 1.37\ ^\circ C; \quad m = \frac{\Delta T_f}{K_f} = \frac{1.37\ ^\circ C}{8.00\ \dfrac{^\circ C \cdot kg}{mol}} = 0.171\ mol/kg = 0.171\ m$$

$$1.200\ g \times \frac{1 kg}{1000\ g} = 1.200 \times 10^{-3}\ kg; \quad 30.0\ mg \times \frac{1.00\ g}{1000\ mg} = 0.0300\ g$$

mol lysine $= (1.200 \times 10^{-3}\ kg)(0.171\ mol/kg) = 2.05 \times 10^{-4}\ mol$

molar mass of lysine $= \dfrac{0.0300\ g\ lysine}{2.05 \times 10^{-4}\ mol\ lysine} = 146\ g/mol$

molecular mass = 146 amu

Finally, determine the molecular formula:

Divide the molecular mass by the empirical formula mass.

$$\frac{146 \text{ amu}}{73.09 \text{ amu}} = 2; \quad \text{molecular formula is } C_{(2 \times 3)}H_{(2 \times 7)}N_{(2 \times 1)}O_{(2 \times 1)}, \text{ or } C_6H_{14}N_2O_2$$

Chapter Problems

11.108 K_f for snow (H_2O) is 1.86 $\dfrac{°C \cdot kg}{mol}$. Reasonable amounts of salt are capable of lowering the freezing point (ΔT_f) of the snow below an air temperature of −2 °C. Reasonable amounts of salt, however, are not capable of causing a ΔT_f of more than 30 °C which would be required if it is to melt snow when the air temperature is −30 °C.

11.109 KBr, 119.00 amu; for KBr, i = 2

$$125 \text{ g} \times \frac{1 \text{kg}}{1000 \text{ g}} = 0.125 \text{ kg}$$

$$\Delta T_b = 103.2 \text{ °C} - 100.0 \text{ °C} = 3.2 \text{ °C}$$

$$\Delta T_b = K_b \cdot m \cdot i; \quad m = \frac{\Delta T_b}{K_b \cdot 2} = \frac{3.2 \text{ °C}}{\left(0.51 \dfrac{°C \cdot kg}{mol}\right)(2)} = 3.137 \text{ mol/kg} = 3.137 \, m$$

mol KBr = (0.125 kg)(3.137 mol/kg) = 0.392 mol KBr

$$\text{mass of KBr} = 0.392 \text{ mol KBr} \times \frac{119.00 \text{ g KBr}}{1 \text{ mol KBr}} = 47 \text{ g KBr}$$

11.110 $C_2H_6O_2$, 62.07 amu; $\quad \Delta T_f = 22.0$ °C

$$\Delta T_f = K_f \cdot m; \quad m = \frac{\Delta T_f}{K_f} = \frac{22.0 \text{ °C}}{1.86 \dfrac{°C \cdot kg}{mol}} = 11.8 \text{ mol/kg} = 11.8 \, m$$

mol $C_2H_6O_2$ = (3.55 kg)(11.8 mol/kg) = 41.9 mol $C_2H_6O_2$

$$\text{mass } C_2H_6O_2 = 41.9 \text{ mol } C_2H_6O_2 \times \frac{62.07 \text{ g } C_2H_6O_2}{1 \text{ mol } C_2H_6O_2} = 2.60 \times 10^3 \text{ g } C_2H_6O_2$$

11.111 The vapor pressure of toluene is lower than the vapor pressure of benzene at the same temperature. When 1 mL of toluene is added to 100 mL of benzene, the vapor pressure of the solution decreases, which means that the boiling point of the solution will increase. When 1 mL of benzene is added to 100 mL of toluene, the vapor pressure of the solution increases, which means that the boiling point of the solution will decrease.

11.112 When solid $CaCl_2$ is added to liquid water, the temperature rises because ΔH_{soln} for $CaCl_2$ is exothermic.

When solid $CaCl_2$ is added to ice at 0 °C, some of the ice will melt (an endothermic process) and the temperature will fall because the $CaCl_2$ lowers the freezing point of an ice/water mixture.

11.113 AgCl, 143.32 amu; there are 2 ions/AgCl; $\Pi = 2MRT$

$$\Pi = 2\left(\frac{0.007 \times 10^{-3} \text{ g} \times \dfrac{1 \text{ mol}}{143.32 \text{ g}}}{0.001 \text{ L}}\right)\left(0.082\ 06\ \frac{\text{L} \cdot \text{atm}}{\text{K} \cdot \text{mol}}\right)(278 \text{ K}) = 0.002 \text{ atm}$$

11.114 $C_{10}H_8$, 128.17 amu; $\Delta T_f = 0.35$ °C

$$\Delta T_f = K_f \cdot m; \qquad m = \frac{\Delta T_f}{K_f} = \frac{0.35\ ^{\circ}\text{C}}{5.12\ \dfrac{^{\circ}\text{C} \cdot \text{kg}}{\text{mol}}} = 0.0684 \text{ mol/kg} = 0.0684\ m$$

$$150.0 \text{ g} \times \frac{1 \text{kg}}{1000 \text{ g}} = 0.1500 \text{ kg}$$

$$\text{mol } C_{10}H_8 = (0.1500 \text{ kg})(0.0684 \text{ mol/kg}) = 0.0103 \text{ mol } C_{10}H_8$$

$$\text{mass } C_{10}H_8 = 0.0103 \text{ mol } C_{10}H_8 \times \frac{128.17 \text{ g } C_{10}H_8}{1 \text{ mol } C_{10}H_8} = 1.3 \text{ g } C_{10}H_8$$

11.115 Br_2, 159.81 amu; CCl_4, 153.82 amu

$$P^{o}_{Br_2} = 30.5 \text{ kPa} \times \frac{760 \text{ mm Hg}}{101.325 \text{ kPa}} = 228.8 \text{ mm Hg}$$

$$P^{o}_{CCl_4} = 16.5 \text{ kPa} \times \frac{760 \text{ mm Hg}}{101.325 \text{ kPa}} = 123.8 \text{ mm Hg}$$

$$1.50 \text{ g } Br_2 \times \frac{1 \text{ mol } Br_2}{159.81 \text{ g } Br_2} = 9.39 \times 10^{-3} \text{ mol } Br_2$$

$$145.0 \text{ g } CCl_4 \times \frac{1 \text{ mol } CCl_4}{153.82 \text{ g } CCl_4} = 0.943 \text{ mol } CCl_4$$

$$X_{Br_2} = \frac{9.39 \times 10^{-3} \text{ mol}}{(0.943 \text{ mol}) + (9.39 \times 10^{-3} \text{ mol})} = 0.009\ 86$$

$$X_{CCl_4} = \frac{0.943 \text{ mol}}{(0.943 \text{ mol}) + (9.39 \times 10^{-3} \text{ mol})} = 0.990$$

$$P_{soln} = P^{o}_{Br_2} \cdot X_{Br_2} + P^{o}_{CCl_4} \cdot X_{CCl_4}$$

$$P_{soln} = (228.8 \text{ mm Hg})(0.009\ 86) + (123.8 \text{ mm Hg})(0.990) = 125 \text{ mm Hg}$$

11.116 NaCl, 58.44 amu; there are 2 ions/NaCl

A 3.5 mass % aqueous solution of NaCl contains 3.5 g NaCl and 96.5 g H_2O.

$$\text{molality} = \frac{\left(3.5 \text{ g} \times \dfrac{1 \text{ mol}}{58.44 \text{ g}}\right)}{0.0965 \text{ kg}} = 0.62 \text{ mol/kg} = 0.62\ m$$

$$\Delta T_f = K_f \cdot 2 \cdot m = \left(1.86\ \frac{^{\circ}\text{C} \cdot \text{kg}}{\text{mol}}\right)(2)(0.62 \text{ mol/kg}) = 2.3\ ^{\circ}\text{C}$$

Solution freezing point = 0.0 °C $- \Delta T_f$ = 0.0 °C $-$ 2.3 °C = $-$2.3 °C

$$\Delta T_b = K_b \cdot 2 \cdot m = \left(0.51 \, \frac{°C \cdot kg}{mol} \right)(2)(0.62 \, mol/kg) = 0.63 \, °C$$

Solution boiling point = $100.00 \, °C + \Delta T_b = 100.00 \, °C + 0.63 \, °C = 100.63 \, °C$

11.117 (a) Assume a total mass of solution of 1000.0 g.

$$ppm = \frac{\text{mass of solute ion}}{\text{total mass of solution}} \times 10^6$$

For each ion: mass of solute ion = $\dfrac{(ppm)(1000.0 \, g)}{10^6}$

Ion	Mass	Moles
Cl^-	19.0 g	0.536 mol
Na^+	10.5 g	0.457 mol
SO_4^{2-}	2.65 g	0.0276 mol
Mg^{2+}	1.35 g	0.0555 mol
Ca^{2+}	0.400 g	0.009 98 mol
K^+	0.380 g	0.009 72 mol
HCO_3^-	0.140 g	0.002 29 mol
Br^-	0.065 g	0.000 81 mol
Total	34.5 g	1.099 mol

Mass of H_2O = 1000.0 g − 34.5 g = 965.5 g H_2O = 0.9655 kg H_2O

$$molality = \frac{1.099 \, mol}{0.9655 \, kg} = 1.138 \, mol/kg = 1.138 \, m$$

(b) Assume M = m for a dilute solution.

$$\Pi = MRT = (1.138 \, mol/L)\left(0.082 \, 06 \, \frac{L \cdot atm}{K \cdot mol} \right)(300 \, K) = 28.0 \, atm$$

11.118 (a) 90 mass % isopropyl alcohol = $\dfrac{10.5 \, g}{10.5 \, g + \text{mass of } H_2O} \times 100\%$

Solve for the mass of H_2O.

$$\text{mass of } H_2O = \left(10.5 \, g \times \frac{100}{90} \right) - 10.5 \, g = 1.2 \, g$$

mass of solution = 10.5 g + 1.2 g = 11.7 g
11.7 g of rubbing alcohol contains 10.5 g of isopropyl alcohol.
(b) C_3H_8O, 60.10 amu
mass C_3H_8O = (0.90)(50.0 g) = 45 g

$$45 \, g \, C_3H_8O \times \frac{1 \, mol \, C_3H_8O}{60.10 \, g \, C_3H_8O} = 0.75 \, mol \, C_3H_8O$$

11.119 $C_6H_{12}O_6$, 180.16 amu; 50.0 mL $= 0.0500$ L; 17.5 mg $= 17.5 \times 10^{-3}$ g

$\Pi = MRT$

$$T = \frac{\Pi}{MR} = \frac{\left(37.8 \text{ mm Hg} \times \dfrac{1 \text{ atm}}{760 \text{ mm Hg}}\right)}{\left(\dfrac{\left(17.5 \times 10^{-3}\text{g} \times \dfrac{1 \text{ mol}}{180.16 \text{ g}}\right)}{0.0500 \text{ L}}\right)\left(0.082\ 06 \dfrac{\text{L} \cdot \text{atm}}{\text{K} \cdot \text{mol}}\right)} = 312 \text{ K}$$

11.120 First, determine the empirical formula.

3.47 mg $= 3.47 \times 10^{-3}$ g sample

10.10 mg $= 10.10 \times 10^{-3}$ g CO_2

2.76 mg $= 2.76 \times 10^{-3}$ g H_2O

mass C $= 10.10 \times 10^{-3}$ g CO_2 x $\dfrac{12.01 \text{ g C}}{44.01 \text{ g } CO_2}$ $= 2.76 \times 10^{-3}$ g C

mass H $= 2.76 \times 10^{-3}$ g H_2O x $\dfrac{2 \times 1.008 \text{ g H}}{18.02 \text{ g } H_2O}$ $= 3.09 \times 10^{-4}$ g H

mass O $= 3.47 \times 10^{-3}$ g $- 2.76 \times 10^{-3}$ g C $- 3.09 \times 10^{-4}$ g H $= 4.01 \times 10^{-4}$ g O

2.76 $\times 10^{-3}$ g C x $\dfrac{1 \text{ mol C}}{12.01 \text{ g C}}$ $= 2.30 \times 10^{-4}$ mol C

3.09 $\times 10^{-4}$ g H x $\dfrac{1 \text{ mol H}}{1.008 \text{ g H}}$ $= 3.07 \times 10^{-4}$ mol H

4.01 $\times 10^{-4}$ g O x $\dfrac{1 \text{ mol O}}{16.00 \text{ g O}}$ $= 2.51 \times 10^{-5}$ mol O $= 0.251 \times 10^{-4}$ mol O

To simplify the empirical formula, divide each mol quantity by 10^{-4}.

$C_{2.30}H_{3.07}O_{0.251}$; divide all subscripts by the smallest, 0.251.

$C_{2.30 / 0.251}H_{3.07 / 0.251}O_{0.251 / 0.251}$

$C_{9.16}H_{12.23}O$

Empirical formula is $C_9H_{12}O$, 136 amu.

Second, determine the molecular mass.

7.55 mg $= 7.55 \times 10^{-3}$ g estradiol; 0.500 g x $\dfrac{1 \text{ kg}}{1000 \text{ g}}$ $= 5.00 \times 10^{-4}$ kg camphor

$$\Delta T_f = K_f \cdot m; \quad m = \frac{\Delta T_f}{K_f} = \frac{2.10 \text{ °C}}{37.7 \dfrac{\text{°C} \cdot \text{kg}}{\text{mol}}} = 0.0557 \text{ mol/kg} = 0.0557 \, m$$

$$m = \frac{\text{mol estradiol}}{\text{kg solvent}}$$

mol estradiol $= m$ x (kg solvent) $= (0.0557 \text{ mol/kg})(5.00 \times 10^{-4} \text{ kg}) = 2.79 \times 10^{-5}$ mol

molar mass $= \dfrac{7.55 \times 10^{-3} \text{ g estradiol}}{2.79 \times 10^{-5} \text{ mol estradiol}}$ $= 271$ g/mol; molecular mass $= 271$ amu

Finally, determine the molecular formula:

Divide the molecular mass by the empirical formula mass.

$$\frac{271 \text{ amu}}{136 \text{ amu}} = 2$$

molecular formula is $C_{(2 \times 9)}H_{(2 \times 12)}O_{(2 \times 1)}$, or $C_{18}H_{24}O_2$

11.121 $CCl_3CO_2H(aq) \rightleftarrows H^+(aq) + CCl_3CO_2^-(aq)$

$\quad\quad 1.00 - x \quad\quad\quad\quad x \quad\quad\quad\quad x$

$$\Delta T_f = K_f \cdot m; \quad m = \frac{\Delta T_f}{K_f} = \frac{2.53 \text{ °C}}{1.86 \dfrac{\text{°C} \cdot \text{kg}}{\text{mol}}} = 1.36 \ m$$

$1.36 = 1.00 - x + x + x = 1 + x; \quad x = 0.36$

36% of the acid molecules are dissociated.

11.122 (a) H_2SO_4, 98.08 amu; $2.238 \text{ mol } H_2SO_4 \times \dfrac{98.08 \text{ g } H_2SO_4}{1 \text{ mol } H_2SO_4} = 219.50 \text{ g } H_2SO_4$

mass of 2.238 m solution = $219.50 \text{ g } H_2SO_4 + 1000 \text{ g } H_2O = 1219.50 \text{ g}$

volume of 2.238 m solution = $1219.50 \text{ g} \times \dfrac{1.0000 \text{ mL}}{1.1243 \text{ g}} = 1084.68 \text{ mL} = 1.0847 \text{ L}$

molarity of 2.238 m solution = $\dfrac{2.238 \text{ mol}}{1.0847 \text{ L}} = 2.063 \text{ M}$

The molarity of the H_2SO_4 solution is less than the molarity of the $BaCl_2$ solution. Because equal volumes of the two solutions are mixed, H_2SO_4 is the limiting reactant and the number of moles of H_2SO_4 determines the number of moles of $BaSO_4$ produced as the white precipitate.

$(0.05000 \text{ L}) \times (2.063 \text{ mol } H_2SO_4/L) \times \dfrac{1 \text{ mol } BaSO_4}{1 \text{ mol } H_2SO_4} \times \dfrac{233.39 \text{ g } BaSO_4}{1 \text{ mol } BaSO_4} = 24.07 \text{ g } BaSO_4$

(b) More precipitate will form because of the excess $BaCl_2$ in the solution.

11.123 KCl, 74.55amu; KNO_3, 101.10 amu; $Ba(NO_3)_2$, 261.34 amu

$$\Pi = MRT; \quad M = \frac{\Pi}{RT} = \frac{\left(744.7 \text{ mm Hg} \times \dfrac{1.00 \text{ atm}}{760 \text{ mm Hg}}\right)}{\left(0.082 \ 06 \dfrac{\text{L} \cdot \text{atm}}{\text{K} \cdot \text{mol}}\right)(298 \text{ K})} = 0.040 \ 07 \text{ M}$$

$0.040 \ 07 \text{ M} = \dfrac{\text{mol ions}}{0.500 \text{ L}};$ mol ions = $(0.040 \ 07 \text{ mol/L})(0.500 \text{ L}) = 0.020 \ 035$ mol ions

mass Cl = $1.000 \text{ g} \times 0.2092 = 0.2092 \text{ g Cl}$

mass KCl = $0.2092 \text{ g Cl} \times \dfrac{1 \text{ mol Cl}}{35.453 \text{ g Cl}} \times \dfrac{1 \text{ mol KCl}}{1 \text{ mol Cl}} \times \dfrac{74.55 \text{ g KCl}}{1 \text{ mol KCl}} = 0.440 \text{ g KCl}$

mol ions from KCl = $0.440 \text{ g KCl} \times \dfrac{1 \text{ mol KCl}}{74.55 \text{ g KCl}} \times \dfrac{2 \text{ mol ions}}{1 \text{ mol KCl}} = 0.0118$ mol ions mol ions from KNO_3 and $Ba(NO_3)_2 = 0.020\ 035 - 0.0118 = 0.008\ 235$ mol ions

Let x = mass KNO_3 and y = mass $Ba(NO_3)_2$

$x + y = 1.000\ g - 0.440\ g = 0.560\ g$

$\left(\dfrac{x}{101.10}\right)2 + \left(\dfrac{y}{261.34}\right)3 = 0.008\ 235$ mol ions

$x = 0.560 - y$

$0.0198x + 0.0115y = 0.008\ 235$

$0.0198(0.560 - y) + 0.0115y = 0.008\ 235$

$0.011\ 09 - 0.0198y + 0.0115y = 0.008\ 235$

$0.002\ 855 = 0.0083y;\ y = \dfrac{0.002\ 855}{0.0083} = 0.3440;\ x = 0.560 - 0.3440 = 0.216$

mass % KCl $= \dfrac{0.440\ g}{1.000\ g} \times 100\% = 44.0\%$

mass % $KNO_3 = \dfrac{0.216\ g}{1.000\ g} \times 100\% = 21.6\%$

mass % $Ba(NO_3)_2 = \dfrac{0.344\ g}{1.000\ g} \times 100\% = 34.4\%$

11.124 Let $x = X_{H_2O}$ and $y = X_{CH_3OH}$ and assume $n_{total} = 1.00$ mol

$(14.5\ mm\ Hg)x + (82.5\ mm\ Hg)y = 39.4\ mm\ Hg$

$(26.8\ mm\ Hg)x + (140.3\ mm\ Hg)y = 68.2\ mm\ Hg$

$x = \dfrac{68.2 - 140.3y}{26.8}$

$\dfrac{14.5(68.2 - 140.3y)}{26.8} + 82.5y = 39.4$

$\dfrac{(988.9 - 2034.35y)}{26.8} + 82.5y = 39.4$

$36.90 - 75.91y + 82.5y = 39.4;\ 6.59y = 2.5;\ y = \dfrac{2.5}{6.59} = 0.3794$

$x = \dfrac{[68.2 - 140.3(0.3794)]}{26.8} = 0.5586$

$X_{LiCl} = 1 - X_{H_2O} - X_{CH_3OH} = 1 - 0.5586 - 0.3794 = 0.0620$

The mole fraction equals the number of moles of each component because $n_{total} = 1.00$ mol.

mass LiCl $= 0.0620$ mol LiCl $\times \dfrac{42.39\ g\ LiCl}{1\ mol\ LiCl} = 2.6\ g\ LiCl$

mass $H_2O = 0.5588$ mol $H_2O \times \dfrac{18.02\ g\ H_2O}{1\ mol\ H_2O} = 10.1\ g\ H_2O$

mass $CH_3OH = 0.3794$ mol $CH_3OH \times \dfrac{32.04\ g\ CH_3OH}{1\ mol\ CH_3OH} = 12.2\ g\ CH_3OH$

total mass $= 2.6\ g + 10.1\ g + 12.2\ g = 24.9\ g$

$$\text{mass \% LiCl} = \frac{2.6\ \text{g}}{24.9\ \text{g}} \times 100\% = 10\%$$

$$\text{mass \% H}_2\text{O} = \frac{10.1\ \text{g}}{24.9\ \text{g}} \times 100\% = 41\%$$

$$\text{mass \% CH}_3\text{OH} = \frac{12.2\ \text{g}}{24.9\ \text{g}} \times 100\% = 49\%$$

11.125 KI, 166.00 amu

$$\Delta T_f = K_f \cdot m \cdot i; \quad m = \frac{\Delta T_f}{K_f \cdot i} = \frac{1.95\ ^\circ\text{C}}{\left(1.86\ \dfrac{^\circ\text{C} \cdot \text{kg}}{\text{mol}}\right)(2)} = 0.524\ \text{mol/kg} = 0.524\ m$$

$$\Pi = i \cdot MRT; \quad M = \frac{\Pi}{i \cdot RT} = \frac{25.0\ \text{atm}}{(2)\left(0.082\,06\ \dfrac{\text{L} \cdot \text{atm}}{\text{K} \cdot \text{mol}}\right)(298\ \text{K})} = 0.511\ \text{M} = 0.511\ \text{mol/L}$$

1.000 L of solution contains 0.511 mol KI and is 0.524 m.

$$\text{mass KI} = 0.511\ \text{mol KI} \times \frac{166.00\ \text{g KI}}{1\ \text{mol KI}} = 84.83\ \text{g KI}$$

Calculate the mass of solvent in this solution.

$$0.524\ m = 0.524\ \text{mol/kg} = \frac{0.511\ \text{mol}}{\text{mass of solvent}}$$

$$\text{mass of solvent} = \frac{0.511\ \text{mol}}{0.524\ \text{mol/kg}} = 0.9752\ \text{kg} = 975.2\ \text{g}$$

$$\text{mass of solution} = \text{mass KI} + \text{mass of solvent} = 84.83\ \text{g} + 975.2\ \text{g} = 1060\ \text{g}$$

$$\text{density} = \frac{1060\ \text{g}}{1000\ \text{mL}} = 1.06\ \text{g/mL}$$

11.126 Solution freezing point = $-1.03\ ^\circ\text{C} = 0.00\ ^\circ\text{C} - \Delta T_f; \quad \Delta T_f = 1.03\ ^\circ\text{C}$

$$\Delta T_f = K_f \cdot m; \quad m = \frac{\Delta T_f}{K_f} = \frac{1.03\ ^\circ\text{C}}{1.86\ \dfrac{^\circ\text{C} \cdot \text{kg}}{\text{mol}}} = 0.554\ \text{mol/kg} = 0.554\ m$$

$$\Pi = MRT; \quad M = \frac{\Pi}{RT} = \frac{(12.16\ \text{atm})}{\left(0.082\,06\ \dfrac{\text{L} \cdot \text{atm}}{\text{K} \cdot \text{mol}}\right)(298\ \text{K})} = 0.497\ \frac{\text{mol}}{\text{L}}$$

Assume 1.000 L = 1000 mL of solution.

mass of solution = (1000 mL)(1.063 g/mL) = 1063 g

$$\text{mass of H}_2\text{O in 1000 mL of solution} = \frac{1000\ \text{g H}_2\text{O}}{0.554\ \text{mol of solute}} \times 0.497\ \text{mol} = 897\ \text{g H}_2\text{O}$$

mass of solute = total mass − mass of H_2O = 1063 g − 897 g = 166 g solute

$$\text{molar mass} = \frac{166\ \text{g}}{0.497\ \text{mol}} = 334\ \text{g/mol}$$

11.127 C_6H_6, 78.11 amu

$$299 \text{ mm Hg} = P^o_{C_6H_6} \cdot X_{C_6H_6} + P^o_X \cdot X_X$$

$$299 \text{ mm Hg} = (395 \text{ mm Hg}) \cdot X_{C_6H_6} + (96 \text{ mm Hg}) \cdot X_X$$

$$X_{C_6H_6} + X_X = 1; \quad X_X = 1 - X_{C_6H_6}$$

$$299 \text{ mm Hg} = (395 \text{ mm Hg}) \cdot X_{C_6H_6} + (96 \text{ mm Hg})(1 - X_{C_6H_6})$$

$$299 \text{ mm Hg} = (395 \text{ mm Hg}) \cdot X_{C_6H_6} + 96 \text{ mm Hg} - (96 \text{ mm Hg}) \cdot X_{C_6H_6}$$

$$299 \text{ mm Hg} - 96 \text{ mm Hg} = (395 \text{ mm Hg}) \cdot X_{C_6H_6} - (96 \text{ mm Hg}) \cdot X_{C_6H_6}$$

$$203 \text{ mm Hg} = (299 \text{ mm Hg}) \cdot X_{C_6H_6}$$

$$X_{C_6H_6} = 203 \text{ mm Hg}/299 \text{ mm Hg} = 0.679$$

Assume the mixture contains 1.00 mol (78.11 g) of C_6H_6. Then a 50/50 mixture will also contain 78.11 g of X.

$$X_{C_6H_6} = \frac{1 \text{ mol } C_6H_6}{1 \text{ mol } C_6H_6 + \text{mol X}} = 0.679$$

$$1 \text{ mol } C_6H_6 = (0.679)(1 \text{ mol } C_6H_6 + \text{mol X})$$

$$\frac{1 \text{ mol } C_6H_6}{0.679} = 1 \text{ mol } C_6H_6 + \text{mol X}$$

$$\frac{1 \text{ mol } C_6H_6}{0.679} - 1 \text{ mol } C_6H_6 = \text{mol X}$$

mol X = 0.473 mol

molar mass X = 78.11 g/0.473 mol = 165 g/mol

11.128 (a) NaCl, 58.44 amu; $CaCl_2$, 110.98 amu; H_2O, 18.02 amu

$$\text{mol NaCl} = 100.0 \text{ g NaCl} \times \frac{1 \text{ mol NaCl}}{58.44 \text{ g NaCl}} = 1.711 \text{ mol NaCl}$$

$$\text{mol CaCl}_2 = 100.0 \text{ g CaCl}_2 \times \frac{1 \text{ mol CaCl}_2}{110.98 \text{ g CaCl}_2} = 0.9011 \text{ mol CaCl}_2$$

mass of solution = (1000 mL)(1.15 g/mL) = 1150 g

mass of H_2O in solution = mass of solution – mass NaCl – mass $CaCl_2$

$$= 1150 \text{ g} - 100.0 \text{ g} - 100.0 \text{ g} = 950 \text{ g}$$

$$= 950 \text{ g} \times \frac{1 \text{ kg}}{1000 \text{ g}} = 0.950 \text{ kg}$$

$$\Delta T_b = K_b \cdot (m_{NaCl} \cdot i + m_{CaCl_2} \cdot i)$$

$$\Delta T_b = \left(0.51 \frac{^oC \cdot kg}{mol} \right) \left(\frac{(1.711 \text{ mol NaCl} \cdot 2) + (0.9011 \text{ mol CaCl}_2 \cdot 3)}{0.950 \text{ kg}} \right) = 3.3 \ ^oC$$

solution boiling point = 100.0 °C + ΔT_b = 100.0 °C + 3.3 °C = 103.3 °C

(b) mol H_2O = 950 g H_2O x $\dfrac{1\ mol\ H_2O}{18.02\ g\ H_2O}$ = 52.7 mol H_2O

$P_{Solution} = P^o \cdot X_{H_2O}$

$P_{Solution} = P^o \cdot \left(\dfrac{52.7\ mol\ H_2O}{(52.7\ mol\ H_2O) + (1.711\ mol\ NaCl \cdot 2) + (0.9011\ mol\ CaCl_2 \cdot 3)} \right)$

$P_{Solution}$ = (23.8 mm Hg)(0.896) = 21.3 mm Hg

11.129 HIO_3, 175.91 amu

mass of 1.00 L solution = $(1.00 \times 10^3\ mL)(1.07\ g/mL) = 1.07 \times 10^3$ g

1.00 L of solution contains 1 mol (175.91 g) HIO_3.

mass of H_2O = 1070 g – 175.91 g = 894 g = 0.894 kg

m = 1.00 mol/0.894 kg = 1.12 m

$\Delta T_f = K_f \cdot m \cdot i$

$i = \dfrac{\Delta T_f}{K_f \cdot m} = \dfrac{2.78\ ^oC}{\left(1.86\ \dfrac{^oC \cdot kg}{mol} \right)(1.12\ mol/kg)}$ = 1.33;   The acid is 33% dissociated.

1.130 (a) KI, 166.00 amu

Assume you have 1.000 L of 1.24 M solution.

mass of solution = (1000 mL)(1.15 g/mL) = 1150 g

mass of KI in solution = 1.24 mol KI x $\dfrac{166.00\ g\ KI}{1\ mol\ KI}$ = 206 g KI

mass of H_2O in solution = mass of solution – mass KI = 1150 g – 206 g = 944 g

$= 944\ g \times \dfrac{1\ kg}{1000\ g}$ = 0.944 kg

molality = $\dfrac{1.24\ mol\ KI}{0.944\ kg\ H_2O}$ = 1.31 m

(b) For KI, i = 2 assuming complete dissociation.

$\Delta T_f = K_f \cdot m \cdot i = \left(1.86\ \dfrac{^oC \cdot kg}{mol} \right)(1.31\ m)(2)$ = 4.87 ^oC

Solution freezing point = 0.00 ^oC – ΔT_f = 0.00 ^oC – 4.87 ^oC = – 4.87 ^oC

(c) $i = \dfrac{\Delta T_f}{K_f \cdot m} = \dfrac{4.46\ ^oC}{\left(1.86\ \dfrac{^oC \cdot kg}{mol} \right)(1.31\ mol/kg)}$ = 1.83

Because the calculated i is only 1.83 and not 2, the percent dissociation for KI is 83%.

1.131 (a) For NaCl, i = 2 and for $MgCl_2$, i = 3;   T = 25 ^oC = 25 + 273 = 298 K

$\Pi = i \cdot MRT = [(2)(0.470\ mol/L) + (3)(0.068\ mol/L)]\left(0.082\ 06\ \dfrac{L \cdot atm}{K \cdot mol} \right)(298\ K)$ = 28.0 atm

(b) Calculate the molarity for an osmotic pressure = 100.0 atm.

$$\Pi = MRT; \quad M = \frac{\Pi}{RT} = \frac{(100.0 \text{ atm})}{\left(0.082\ 06\ \dfrac{L \cdot atm}{K \cdot mol}\right)(298 \text{ K})} = 4.09 \text{ mol/L}$$

$$M_{conc} \times V_{conc} = M_{dil} \times V_{dil}$$

$$V_{conc} = \frac{M_{dil} \times V_{dil}}{M_{conc}} = \frac{[(2)(0.470 \text{ mol/L}) + (3)(0.068 \text{ mol/L})](1.00 \text{ L})}{4.09 \text{ mol/L}} = 0.28 \text{ L}$$

A volume of 1.00 L of seawater can be reduced to 0.28 L by an osmotic pressure of 100.0 atm. The volume of fresh water that can be obtained is (1.00 L − 0.28 L) = 0.72 L.

11.132 NaCl, 58.44 amu; $C_{12}H_{22}O_{11}$, 342.3 amu

Let X = mass NaCl and Y = mass $C_{12}H_{22}O_{11}$, then X + Y = 100.0 g.

$$500.0 \text{ g} \times \frac{1 \text{ kg}}{1000 \text{ g}} = 0.5000 \text{ kg}$$

Solution freezing point = −2.25 °C = 0.00 °C − ΔT_f, = ΔT_f = 0.00 °C + 2.25 °C = 2.25 °C

$$\Delta T_f = K_f \cdot (m_{NaCl} \cdot i + m_{C_{12}H_{22}O_{11}})$$

$$\Delta T_b = \left(1.86 \frac{°C \cdot kg}{mol}\right)\left(\frac{(\text{mol NaCl} \cdot 2) + (\text{mol } C_{12}H_{22}O_{11})}{0.5000 \text{ kg}}\right) = 2.25 \text{ °C}$$

$$\text{mol NaCl} = X \text{ g NaCl} \times \frac{1 \text{ mol NaCl}}{58.44 \text{ g NaCl}} = X/58.44 \text{ mol}$$

$$\text{mol } C_{12}H_{22}O_{11} = Y \text{ g } C_{12}H_{22}O_{11} \times \frac{1 \text{ mol } C_{12}H_{22}O_{11}}{342.3 \text{ g } C_{12}H_{22}O_{11}} = Y/342.3 \text{ mol}$$

$$\Delta T_b = \left(1.86 \frac{°C \cdot kg}{mol}\right)\left(\frac{((X/58.44) \cdot 2 \text{ mol}) + ((Y/342.3) \text{ mol})}{0.5000 \text{ kg}}\right) = 2.25 \text{ °C}$$

X = 100 − Y

$$\left(1.86 \frac{°C \cdot kg}{mol}\right)\left(\frac{\{[(100 − Y)/58.44] \cdot 2 \text{ mol}]\} + [(Y/342.3) \text{ mol}]}{0.5000 \text{ kg}}\right) = 2.25 \text{ °C}$$

$$\left(\frac{[(200/58.44) − (2Y/58.44) + (Y/342.3)] \text{ mol}}{0.5000 \text{ kg}}\right) = \frac{2.25 \text{ °C}}{\left(1.86 \dfrac{°C \cdot kg}{mol}\right)} = 1.21 \text{ mol/kg}$$

$$\left(\frac{[(3.42) − (0.0313Y)] \text{ mol}}{0.5000 \text{ kg}}\right) = 1.21 \text{ mol/kg}$$

[(3.42) − (0.0313Y)] = (0.5000 kg)(1.21) = 0.605
− 0.0313 Y = 0.605 − 3.42 = − 2.81
Y = (− 2.81)/(− 0.0313) = 89.8 g of $C_{12}H_{22}O_{11}$
X = 100.0 g − Y = 100.0 g − 89.8 g = 10.2 g of NaCl

Multiconcept Problems

11.133   (a) 382.6 mL = 0.3826 L; 20.0 °C = 293.2 K
   PV = nRT

$$n_{H_2} = \frac{PV}{RT} = \frac{\left(755 \text{ mm Hg} \times \dfrac{1.0 \text{ atm}}{760 \text{ mm Hg}}\right)(0.3826 \text{ L})}{\left(0.082\ 06 \dfrac{L \cdot atm}{K \cdot mol}\right)(293.2 \text{ K})} = 0.0158 \text{ mol } H_2$$

   (b) $M + x \text{ HCl} \rightarrow x/2\ H_2 + MCl_x$

   moles HCl reacted = $0.0158 \text{ mol } H_2 \times \dfrac{x \text{ mol HCl}}{x/2 \text{ mol } H_2} = 0.0316$ mol HCl

   moles Cl reacted = moles HCl reacted = 0.0316 mol Cl

   mass Cl = $0.0316 \text{ mol Cl} \times \dfrac{35.453 \text{ g Cl}}{1 \text{ mol Cl}} = 1.120$ g Cl

   mass MCl_x = mass M + mass Cl = 1.385 g + 1.120 g = 2.505 g MCl_x

   (c) $\Delta T_f = K_f \cdot m$; $m = \dfrac{\Delta T_f}{K_f} = \dfrac{3.53 \text{ °C}}{1.86 \dfrac{°C \cdot kg}{mol}} = 1.90 \text{ mol/kg} = 1.90\ m$

   (d) 25.0 g = 0.0250 kg

   $1.90\ m = 1.90 \dfrac{mol}{kg} = \dfrac{x \text{ mol ions}}{0.0250 \text{ kg}}$

   mol ions = (1.90 mol/kg)(0.0250 kg) = 0.0475 mol ions
   (e) mol M = mol ions – mol Cl = 0.0475 mol – 0.0316 mol = 0.0159 mol M

   $\dfrac{Cl}{M} = \dfrac{0.0316 \text{ mol}}{0.0159 \text{ mol}} = 2$, the formula is MCl_2.

   molar mass = $\dfrac{2.505 \text{ g}}{0.0159 \text{ mol}} = 157.5$ g/mol; molecular mass = 157.5 amu

   (f) atomic mass of M = 157.5 amu – 2(35.453 amu) = 86.6 amu; M = Sr

11.134   (a) 20.00 mL = 0.02000 L
   mol NaOH = (0.02000 L)(2.00 mol/L) = 0.0400 mol NaOH

   mol CO_2 = $0.0400 \text{ mol NaOH} \times \dfrac{1 \text{ mol } CO_2}{2 \text{ mol NaOH}} = 0.0200$ mol CO_2

   mol C = $0.0200 \text{ mol } CO_2 \times \dfrac{1 \text{ mol C}}{1 \text{ mol } CO_2} = 0.0200$ mol C

   mass C = $0.0200 \text{ mol C} \times \dfrac{12.011 \text{ g C}}{1 \text{ mol C}} = 0.240$ g C

   mass H = mass of compound – mass of C = 0.270 g – 0.240 g = 0.030 g H

   mol H = $0.030 \text{ g H} \times \dfrac{1 \text{ mol H}}{1.008 \text{ g H}} = 0.030$ mol H

   The mole ratio of C and H in the molecule is $C_{0.0200} H_{0.030}$.
   $C_{0.0200} H_{0.030}$; divide both subscripts by the smaller of the two, 0.0200.

$C_{0.0200\,/\,0.0200}\,H_{0.030\,/\,0.0200}$

$C_1H_{1.5}$, multiply both subscripts by 2.

$C_{(2\,x\,1)}\,H_{(2\,x\,1.5)}$

C_2H_3 (27.05 amu) is the empirical formula.

(b) $\Delta T_f = K_f \cdot m$; $m = \dfrac{\Delta T_f}{K_f} = \dfrac{(179.8\ °C - 177.9\ °C)}{37.7\,\dfrac{°C \cdot kg}{mol}} = 0.050$ mol/kg $= 0.050\ m$

$50.0\ g \times \dfrac{1\ kg}{1000\ g} = 0.0500\ kg$

mol solute $= (0.050$ mol/kg$)(0.0500$ kg$) = 0.0025$ mol

molar mass $= \dfrac{0.270\ g}{0.0025\ mol} = 108$ g/mol; molecular mass $= 108$ amu

(c) To find the molecular formula, first divide the molecular mass by the mass of the empirical formula unit.

$\dfrac{108}{27} = 4$

Multiply the subscripts in the empirical formula by the result of this division, 4.

$C_{(4\,x\,2)}\,H_{(4\,x\,3)}$

C_8H_{12} is the molecular formula of the compound.

11.135 CO_2, 44.01 amu; H_2O, 18.02 amu

mol C $= 106.43$ mg $CO_2 \times \dfrac{1\ g}{1000\ mg} \times \dfrac{1\ mol\ CO_2}{44.01\ g\ CO_2} \times \dfrac{1\ mol\ C}{1\ mol\ CO_2} = 0.002\ 418$ mol C

mass C $= 0.002\ 418$ mol C $\times \dfrac{12.011\ g\ C}{1\ mol\ C} = 0.029\ 04$ g C

mol H $= 32.100$ mg $H_2O \times \dfrac{1\ g}{1000\ mg} \times \dfrac{1\ mol\ H_2O}{18.02\ g\ H_2O} \times \dfrac{2\ mol\ H}{1\ mol\ H_2O} = 0.003\ 563$ mol H

mass H $= 0.003\ 563$ mol H $\times \dfrac{1.008\ g\ H}{1\ mol\ H} = 0.003\ 592$ g H

mass O $= \left(36.72\ mg \times \dfrac{1\ g}{1000\ mg} \right) - 0.029\ 04$ g C $- 0.003\ 592$ g H $= 0.004\ 088$ g O

mol O $= 0.004\ 088$ g O $\times \dfrac{1\ mol\ O}{16.00\ g\ O} = 0.000\ 255\ 5$ mol O

$C_{0.002\ 418}H_{0.003\ 563}O_{0.000\ 255\ 5}$; divide all subscripts by the smallest, 0.000 255 5.

$C_{0.002\ 418\,/\,0.000\ 255\ 5}H_{0.003\ 563\,/\,0.000\ 255\ 5}O_{0.000\ 255\ 5\,/\,0.000\ 255\ 5}$

$C_{9.5}H_{14}O$; multiply all subscripts by 2.

$C_{(9.5\,x\,2)}H_{(14\,x\,2)}O_{(1\,x\,2)}$

Empirical formula is $C_{19}H_{28}O_2$, 288 amu

$T = 25\ °C = 25 + 273 = 298$ K

$$\Pi = MRT; \quad M = \frac{\Pi}{RT} = \frac{\left(21.5 \text{ mm Hg} \times \frac{1 \text{ atm}}{760 \text{ mm Hg}}\right)}{\left(0.082\,06 \frac{L \cdot atm}{K \cdot mol}\right)(298 \text{ K})} = 0.001\,16 \text{ mol/L}$$

15.0 mL = 0.0150 L

mol solute = (0.001 16 mol/L)(0.0150 L) = 1.74 x 10^{-5} mol

$$\text{molar mass} = \frac{\left(5.00 \text{ mg} \times \frac{1 \text{ g}}{1000 \text{ mg}}\right)}{1.74 \times 10^{-5} \text{ mol}} = 287 \text{ g/mol}$$

The molar mass and the empirical formula mass are essentially identical, so the molecular formula and the empirical formula are the same. The molecular formula is $C_{19}H_{28}O_2$.

11.136   AgCl, 143.32 amu

Solution freezing point = – 4.42 °C = 0.00 °C – ΔT_f; ΔT_f = 0.00 °C + 4.42 °C = 4.42 °C

$\Delta T_f = K_f \cdot m$

$$\text{total ion } m = \frac{\Delta T_f}{K_f} = \frac{4.42 \text{ °C}}{1.86 \frac{\text{°C} \cdot kg}{mol}} = 2.376 \text{ mol/kg} = 2.376 \text{ } m$$

$$150.0 \text{ g} \times \frac{1 \text{ kg}}{1000 \text{ g}} = 0.1500 \text{ kg}$$

total mol of ions = (2.376 mol/kg)(0.1500 kg) = 0.3564 mol of ions

An excess of $AgNO_3$ reacts with all Cl$^-$ to produce 27.575 g AgCl.

$$\text{total mol Cl}^- = 27.575 \text{ g AgCl} \times \frac{1 \text{ mol AgCl}}{143.32 \text{ g AgCl}} \times \frac{1 \text{ mol Cl}^-}{1 \text{ mol AgCl}} = 0.1924 \text{ mol Cl}^-$$

Let P = mol XCl and Q = mol YCl_2.

0.3564 mol ions = 2 x mol XCl + 3 x mol YCl_2 = (2 x P) + (3 x Q)

0.1924 mol Cl$^-$ = mol XCl + 2 x mol YCl_2 = P + (2 x Q)

P = 0.1924 – (2 x Q)

0.3564 = 2 x [0.1924 – (2 x Q)] + (3 x Q) = 0.3848 – (4 x Q) + (3 x Q)

Q = 0.3848 – 0.3564 = 0.0284 mol YCl_2

P = 0.1924 – (2 x Q) = 0.1924 – (2 x 0.0284) = 0.1356 mol XCl

$$\text{mass Cl in XCl} = 0.1356 \text{ mol XCl} \times \frac{1 \text{ mol Cl}}{1 \text{ mol XCl}} \times \frac{35.453 \text{ g Cl}}{1 \text{ mol Cl}} = 4.81 \text{ g Cl}$$

$$\text{mass Cl in YCl}_2 = 0.0284 \text{ mol YCl}_2 \times \frac{2 \text{ mol Cl}}{1 \text{ mol YCl}_2} \times \frac{35.453 \text{ g Cl}}{1 \text{ mol Cl}} = 2.01 \text{ g Cl}$$

total mass of XCl and YCl_2 = 8.900 g

mass of X + Y = total mass – mass Cl = 8.900 g – 4.81 g – 2.01 g = 2.08 g

X is an alkali metal and there are 0.1356 mol of X in XCl.

If X = Li, then mass of X = (0.1356 mol)(6.941 g/mol) = 0.941 g

If X = Na, then mass of X = (0.1356 mol)(22.99 g/mol) = 3.12 g but this is not possible because 3.12 g is greater than the total mass of X + Y. Therefore, X is Li.

mass of Y = 2.08 – mass of X = 2.08 g – 0.941 g = 1.14 g

Y is an alkaline earth metal and there are 0.0284 mol of Y in YCl_2.

molar mass of Y = 1.14 g/0.0284 mol = 40.1 g/mol. Therefore, Y is Ca.

$$\text{mass LiCl} = 0.1356 \text{ mol LiCl} \times \frac{42.39 \text{ g LiCl}}{1 \text{ mol LiCl}} = 5.75 \text{ g LiCl}$$

$$\text{mass CaCl}_2 = 0.0284 \text{ mol CaCl}_2 \times \frac{110.98 \text{ g CaCl}_2}{1 \text{ mol CaCl}_2} = 3.15 \text{ g CaCl}_2$$

12 Chemical Kinetics

12.1 $3\ I^-(aq) + H_3AsO_4(aq) + 2\ H^+(aq) \rightarrow I_3^-(aq) + H_3AsO_3(aq) + H_2O(l)$

(a) $-\dfrac{\Delta[I^-]}{\Delta t} = 4.8 \times 10^{-4}\ M/s$

$$\dfrac{\Delta[I_3^-]}{\Delta t} = \dfrac{1}{3}\left(-\dfrac{\Delta[I^-]}{\Delta t}\right) = \left(\dfrac{1}{3}\right)(4.8 \times 10^{-4}\ M/s) = 1.6 \times 10^{-4}\ M/s$$

(b) $-\dfrac{\Delta[H^+]}{\Delta t} = 2\left(\dfrac{\Delta[I_3^-]}{\Delta t}\right) = (2)(1.6 \times 10^{-4}\ M/s) = 3.2 \times 10^{-4}\ M/s$

12.2 $2\ N_2O_5(g) \rightarrow 4\ NO_2(g) + O_2(g)$

time	$[N_2O_5]$	$[O_2]$
200 s	0.0142 M	0.0029 M
300 s	0.0120 M	0.0040 M

Rate of decomposition of $N_2O_5 = -\dfrac{\Delta[N_2O_5]}{\Delta t} = -\dfrac{0.0120\ M - 0.0142\ M}{300\ s - 200\ s} = 2.2 \times 10^{-5}\ M/s$

Rate of formation of $O_2 = \dfrac{\Delta[O_2]}{\Delta t} = \dfrac{0.0040\ M - 0.0029\ M}{300\ s - 200\ s} = 1.1 \times 10^{-5}\ M/s$

12.3 $Rate = k[BrO_3^-][Br^-][H^+]^2$,
1st order in BrO_3^-, 1st order in Br^-, 2nd order in H^+, 4th order overall
$Rate = k[H_2][I_2]$, 1st order in H_2, 1st order in I_2, 2nd order overall

12.4 $H_2O_2(aq) + 3\ I^-(aq) + 2\ H^+(aq) \rightarrow I_3^-(aq) + 2\ H_2O(l)$

$$Rate = \dfrac{\Delta[I_3^-]}{\Delta t} = k[H_2O_2]^m[I^-]^n$$

(a) $\dfrac{Rate_3}{Rate_1} = \dfrac{2.30 \times 10^{-4}\ M/s}{1.15 \times 10^{-4}\ M/s} = 2 \qquad \dfrac{[H_2O_2]_3}{[H_2O_2]_1} = \dfrac{0.200\ M}{0.100\ M} = 2$

Because both ratios are the same, m = 1.

$\dfrac{Rate_2}{Rate_1} = \dfrac{2.30 \times 10^{-4}\ M/s}{1.15 \times 10^{-4}\ M/s} = 2 \qquad \dfrac{[I^-]_2}{[I^-]_1} = \dfrac{0.200\ M}{0.100\ M} = 2$

Because both ratios are the same, n = 1.
The rate law is: $Rate = k[H_2O_2][I^-]$

(b) $k = \dfrac{\text{Rate}}{[H_2O_2][I^-]}$

Using data from Experiment 1: $\quad k = \dfrac{1.15 \times 10^{-4}\ M/s}{(0.100\ M)(0.100\ M)} = 1.15 \times 10^{-2}\ /(M \cdot s)$

(c) Rate $= k[H_2O_2][I^-] = [1.15 \times 10^{-2}/(M \cdot s)](0.300\ M)(0.400\ M) = 1.38 \times 10^{-3}\ M/s$

12.5

Rate Law	Units of k
Rate $= k[(CH_3)_3CBr]$	$1/s$
Rate $= k[Br_2]$	$1/s$
Rate $= k[BrO_3^-][Br^-][H^+]^2$	$1/(M^3 \cdot s)$
Rate $= k[H_2][I_2]$	$1/(M \cdot s)$

12.6 (a) The reactions in vessels (a) and (b) have the same rate, the same number of B molecules, but different numbers of A molecules. Therefore, the rate does not depend on A and its reaction order is zero. The same conclusion can be drawn from the reactions in vessels (c) and (d).

The rate for the reaction in vessel (c) is four times the rate for the reaction in vessel (a). Vessel (c) has twice as many B molecules than does vessel (a). Because the rate quadruples when the concentration of B doubles, the reaction order for B is two.

(b) rate $= k[B]^2$

12.7 (a) $\ln \dfrac{[Co(NH_3)_5Br^{2+}]_t}{[Co(NH_3)_5Br^{2+}]_o} = -kt$

$k = 6.3 \times 10^{-6}/s; \qquad t = 10.0\ h \times \dfrac{3600\ s}{1\ h} = 36{,}000\ s$

$\ln[Co(NH_3)_5Br^{2+}]_t = -kt + \ln[Co(NH_3)_5Br^{2+}]_o$

$\ln[Co(NH_3)_5Br^{2+}]_t = -(6.3 \times 10^{-6}/s)(36{,}000\ s) + \ln(0.100)$

$\ln[Co(NH_3)_5Br^{2+}]_t = -2.5294; \qquad$ After 10.0 h, $[Co(NH_3)_5Br^{2+}] = e^{-2.5294} = 0.080\ M$

(b) $[Co(NH_3)_5Br^{2+}]_o = 0.100\ M$

If 75% of the $Co(NH_3)_5Br^{2+}$ reacts then 25% remains.

$[Co(NH_3)_5Br^{2+}]_t = (0.25)(0.100\ M) = 0.025\ M$

$\ln \dfrac{[Co(NH_3)_5Br^{2+}]_t}{[Co(NH_3)_5Br^{2+}]_o} = -kt; \quad t = \dfrac{\ln \dfrac{[Co(NH_3)_5Br^{2+}]_t}{[Co(NH_3)_5Br^{2+}]_o}}{-k}$

$t = \dfrac{\ln\left(\dfrac{0.025}{0.100}\right)}{-(6.3 \times 10^{-6}/s)} = 2.2 \times 10^5\ s; \quad t = 2.2 \times 10^5\ s \times \dfrac{1\ h}{3600\ s} = 61\ h$

12.8

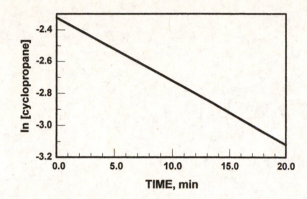

Slope = –0.03989/min = –6.6 x 10^{-4}/s and k = –slope
A plot of ln[cyclopropane] versus time is linear, indicating that the data fit the equation
for a first-order reaction. k = 6.6 x 10^{-4}/s (0.040/min)

12.9   (a) k = 1.8 x 10^{-5}/s

$$t_{1/2} = \frac{0.693}{k} = \frac{0.693}{1.8 \times 10^{-5}/s} = 38,500 \text{ s}; \qquad t_{1/2} = 38,500 \text{ s} \times \frac{1 \text{ h}}{3600 \text{ s}} = 11 \text{ h}$$

(b) 0.30 M $\xrightarrow{t_{1/2}}$ 0.15 M $\xrightarrow{t_{1/2}}$ 0.075 M $\xrightarrow{t_{1/2}}$ 0.0375 M $\xrightarrow{t_{1/2}}$ 0.019 M

(c) Because 25% of the initial concentration corresponds to 1/4 or $(1/2)^2$ of the initial
concentration, the time required is two half-lives: t = $2t_{1/2}$ = 2(11 h) = 22 h

12.10   After one half-life, there would be four A molecules remaining. After two half-lives,
there would be two A molecules remaining. This is represented by the drawing at t = 10
min. 10 min is equal to two half-lives, therefore, $t_{1/2}$ = 5 min for this reaction. After 15
min (three half-lives) only one A molecule would remain.

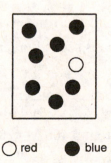

○ red   ● blue

12.11 $t_{1/2} = \dfrac{0.693}{k} = \dfrac{0.693}{1.08 \times 10^{-2}\,h^{-1}} = 64.2\,h$

12.12 $k = \dfrac{0.693}{t_{1/2}} = \dfrac{0.693}{5715\,y} = 1.21 \times 10^{-4}\,y^{-1}$

12.13 $\ln\left(\dfrac{N}{N_0}\right) = -0.693\left(\dfrac{t}{t_{1/2}}\right) = -0.693\left(\dfrac{16{,}230\,y}{5715\,y}\right) = -1.968$

$\dfrac{N}{N_0} = e^{-1.968} = 0.140; \qquad \dfrac{N}{100\%} = 0.140; \qquad N = 14.0\%$

12.14 $\ln\left(\dfrac{N}{N_0}\right) = (-0.693)\left(\dfrac{t}{t_{1/2}}\right);$ $\qquad \dfrac{N}{N_0} = \dfrac{\text{Decay rate at time t}}{\text{Decay rate at t} = 0}$

$\ln\left(\dfrac{10{,}860}{16{,}800}\right) = (-0.693)\left(\dfrac{28.0\,d}{t_{1/2}}\right);$ $\qquad -0.436 = (-0.693)\left(\dfrac{28.0\,d}{t_{1/2}}\right)$

$t_{1/2} = \dfrac{(-0.693)(28.0\,d)}{(-0.436)} = 44.5\,d$

12.15

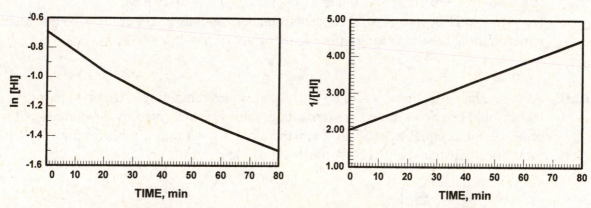

(a) A plot of 1/[HI] versus time is linear. The reaction is second-order.

(b) $k = \text{slope} = 0.0308/(M \cdot min)$

(c) $t = \dfrac{1}{k}\left[\dfrac{1}{[HI]_t} - \dfrac{1}{[HI]_o}\right] = \dfrac{1}{0.0308/(M \cdot min)}\left[\dfrac{1}{0.100\,M} - \dfrac{1}{0.500\,M}\right] = 260\,min$

(d) It requires one half-life ($t_{1/2}$) for the [HI] to drop from 0.400 M to 0.200 M.

$t_{1/2} = \dfrac{1}{k[HI]_o} = \dfrac{1}{[0.0308/(M \cdot min)](0.400\,M)} = 81.2\,min$

12.16 (a)

$$NO_2(g) + F_2(g) \rightarrow NO_2F(g) + F(g)$$
$$\underline{F(g) + NO_2(g) \rightarrow NO_2F(g)}$$

Overall reaction $\quad 2 NO_2(g) + F_2(g) \rightarrow 2 NO_2F(g)$

Because F(g) is produced in the first reaction and consumed in the second, it is a reaction intermediate.

(b) In each reaction there are two reactants, so each elementary reaction is bimolecular.

12.17 (a) Rate = $k[O_3][O]$ (b) Rate = $k[Br]^2[Ar]$ (c) Rate = $k[Co(CN)_5(H_2O)^{2-}]$

12.18

$$Co(CN)_5(H_2O)^{2-}(aq) \rightarrow Co(CN)_5^{2-}(aq) + H_2O(l) \qquad \text{(slow)}$$
$$\underline{Co(CN)_5^{2-}(aq) + I^-(aq) \rightarrow Co(CN)_5I^{3-}(aq)} \qquad \text{(fast)}$$

Overall reaction $\quad Co(CN)_5(H_2O)^{2-}(aq) + I^-(aq) \rightarrow Co(CN)_5I^{3-}(aq) + H_2O(l)$

The predicted rate law for the overall reaction is the rate law for the first (slow) elementary reaction: Rate = $k[Co(CN)_5(H_2O)^{2-}]$
The predicted rate law is in accord with the observed rate law.

12.19 (a) $2 NO(g) + O_2(g) \rightarrow 2 NO_2(g)$
(b) $\text{Rate}_{forward} = k_1[NO][O_2]$ and $\text{Rate}_{reverse} = k_{-1}[NO_3]$
Because of the equilibrium, $\text{Rate}_{forward} = \text{Rate}_{reverse}$, and $k_1[NO][O_2] = k_{-1}[NO_3]$.

$$[NO_3] = \frac{k_1}{k_{-1}}[NO][O_2]$$

The rate law for the rate determining step is Rate = $k_2[NO_3][NO]$. Because 2 NO disappear in the overall reaction for every NO that reacts in the second step, the rate law for the overall reaction is Rate = $-\Delta[NO]/\Delta t = 2k_2[NO_3][NO]$. In this rate law substitute for $[NO_3]$.

Rate = $2k_2 \dfrac{k_1}{k_{-1}}[NO]^2[O_2]$, which is consistent with the experimental rate law.

(c) $k = \dfrac{2k_2 k_1}{k_{-1}}$

12.20 (a) $E_a = 100 \text{ kJ/mol} - 20 \text{ kJ/mol} = 80 \text{ kJ/mol}$
(b) The reaction is endothermic because the energy of the products is higher than the energy of the reactants.
(c)
$$\begin{array}{cc} A & \cdots C \\ \vdots & \vdots \\ B & \cdots D \end{array}$$

12.21 (a) $\ln\left(\dfrac{k_2}{k_1}\right) = \left(\dfrac{-E_a}{R}\right)\left(\dfrac{1}{T_2} - \dfrac{1}{T_1}\right)$

$k_1 = 3.7 \times 10^{-5}/s$, $T_1 = 25 \,°C = 298 \text{ K}$
$k_2 = 1.7 \times 10^{-3}/s$, $T_2 = 55 \,°C = 328 \text{ K}$

$$E_a = -\frac{[\ln k_2 - \ln k_1]R}{\left(\dfrac{1}{T_2} - \dfrac{1}{T_1}\right)}$$

$$E_a = -\frac{[\ln(1.7 \times 10^{-3}) - \ln(3.7 \times 10^{-5})][8.314 \times 10^{-3}\ \text{kJ/(K·mol)}]}{\left(\dfrac{1}{328\ \text{K}} - \dfrac{1}{298\ \text{K}}\right)} = 104\ \text{kJ/mol}$$

(b) $k_1 = 3.7 \times 10^{-5}/\text{s}$, $T_1 = 25\ °\text{C} = 298\ \text{K}$

solve for k_2, $T_2 = 35\ °\text{C} = 308\ \text{K}$

$$\ln k_2 = \left(\frac{-E_a}{R}\right)\left(\frac{1}{T_2} - \frac{1}{T_1}\right) + \ln k_1$$

$$\ln k_2 = \left(\frac{-104\ \text{kJ/mol}}{8.314 \times 10^{-3}\ \text{kJ/(K·mol)}}\right)\left(\frac{1}{308\ \text{K}} - \frac{1}{298\ \text{K}}\right) + \ln(3.7 \times 10^{-5})$$

$\ln k_2 = -8.84$; $k_2 = e^{-8.84} = 1.4 \times 10^{-4}/\text{s}$

12.22 Assume that concentration is proportional to the number of each molecule in a box.
(a) Comparing boxes (1) and (2), the concentration of A doubles, B and C_2 remain the same and the rate does not change. This means the reaction is zeroth-order in A.
Comparing boxes (1) and (3), the concentration of C_2 doubles, A and B remain the same and the rate doubles. This means the reaction is first-order in C_2.
Comparing boxes (1) and (4), the concentration of B triples, A and C_2 remain the same and the rate triples. This means the reaction is first-order in B.
(b) Rate $= k\,[B][C_2]$
(c)

$$
\begin{array}{lll}
\text{B} + \text{C}_2 & \rightarrow \ \text{BC}_2 & \text{(slow)} \\
\text{A} + \text{BC}_2 & \rightarrow \ \text{AC} + \text{BC} & \\
\underline{\text{A} + \text{BC} \ \ \rightarrow \ \text{AC} + \text{B}} & & \\
2\,\text{A} + \text{C}_2 & \rightarrow \ 2\,\text{AC} & \text{(overall)}
\end{array}
$$

(d) B doesn't appear in the overall reaction because it is consumed in the first step and regenerated in the third step. B is therefore a catalyst. BC_2 and BC are intermediates because they are formed in one step and then consumed in a subsequent step in the reaction.

12.23

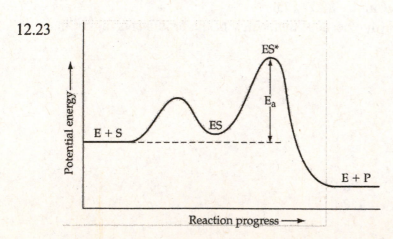

Key Concept Problems

12.24 (a) Because Rate = k[A][B], the rate is proportional to the product of the number of A molecules and the number of B molecules. The relative rates of the reaction in vessels (a) – (d) are 2 : 1 : 4 : 2.
(b) Because the same reaction takes place in each vessel, the k's are all the same.

12.25 (a) Because Rate = k[A], the rate is proportional to the number of A molecules in each reaction vessel. The relative rates of the reaction are 2 : 4 : 3.
(b) For a first-order reaction, half-lives are independent of concentration. The half-lives are the same.
(c) Concentrations will double, rates will double, and half-lives will be unaffected.

12.26 (a) For the first-order reaction, half of the A molecules are converted to B molecules each minute.

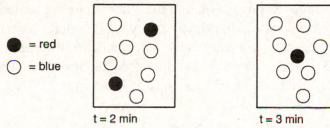

(b) Because half of the A molecules are converted to B molecules in 1 min, the half-life is 1 minute.

12.27 (a) Two molecules of A are converted to two molecules of B every minute. This means the rate is constant throughout the course of the reaction. The reaction is zeroth-order.
(b)

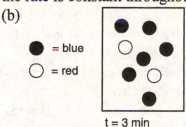

(c) Rate = k

$$k = \frac{(2)\left(\dfrac{6.0 \times 10^{21}\ \text{molecules}}{1.0\ \text{L}}\right)\left(\dfrac{1\ \text{mol}}{6.022 \times 10^{23}\ \text{molecules}}\right)}{\text{min}} \times \frac{1\ \text{min}}{60\ \text{s}} = 3.3 \times 10^{-4}\ \text{M/s}$$

12.28 (a) Because the half-life is inversely proportional to the concentration of A molecules, the reaction is second-order in A.
(b) Rate = k[A]^2
(c) The second box represents the passing of one half-life, and the third box represents the passing of a second half-life for a second-order reaction. A relative value of k can be calculated.

$$k = \frac{1}{t_{1/2}[A]} = \frac{1}{(1)(16)} = 0.0625$$

$t_{1/2}$ in going from box 3 to box 4 is: $t_{1/2} = \dfrac{1}{k[A]} = \dfrac{1}{(0.0625)(4)} = 4$ min

(For fourth box, t = 7 min)

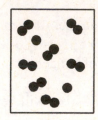

t = 3 min + 4 min = 7 min

12.29 (a) bimolecular   (b) unimolecular   (c) termolecular

12.30 Assume that concentration is proportional to the number of each molecule in a box.
(a) Comparing boxes (1) and (2), the concentration of B doubles, A remains the same and the rate does not change. This means the reaction is zeroth-order in B.
Comparing boxes (3) and (2), the concentration of A doubles, B remains the same and the rate quadruples. This means the reaction is second-order in A.
(b) Rate = k [A]^2.
(c)   2 A → A$_2$ (slow)
$\underline{A_2 + B \;\rightarrow\; AB + A}$
A + B → AB   (overall)
(d) A$_2$ is an intermediate because it is formed in one step and then consumed in a subsequent step in the reaction.

12.31 Assume that concentration is proportional to the number of each molecule in a box.
(a) Comparing boxes (2) and (1), the concentration of C doubles, B and AB remain the same and the rate doubles. This means the reaction is first-order in C.
Comparing boxes (2) and (3), the concentration of AB doubles, B and C remain the same and the rate remains the same. This means the reaction is zeroth-order in AB.
Comparing boxes (1) and (4), the concentration of B doubles, C and AB remain the same and the rate doubles. This means the reaction is first-order in B.
Rate = k [B][C]
(b)   B + C → BC   (slow)
$\underline{BC + AB \;\rightarrow\; A + B_2 + C}$
AB + B → A + B$_2$ (overall)
(c) C doesn't appear in the overall reaction because it is consumed in the first step and regenerated in the second step. Therefore, C is a catalyst. BC is an intermediate because it is formed in one step and then consumed in a subsequent step in the reaction.

12.32 (a) BC + D → B + CD
(b) 1. B–C + D (reactants), A (catalyst); 2. B---C---A (transition state), D (reactant); 3. A–C (intermediate), B (product), D (reactant); 4. A---C---D (transition state), B (product); 5. A (catalyst), C–D + B (products)

(c) The first step is rate determining because the first maximum in the potential energy curve is greater than the second (relative) maximum; Rate = k[A][BC]

(d) Endothermic

12.33

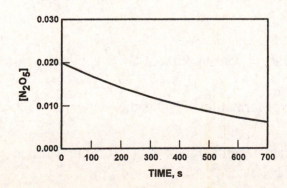

Section Problems
Reaction Rates (Section 12.1)

12.34 M/s or $\dfrac{mol}{L \cdot s}$

12.35 molecules/$(cm^3 \cdot s)$

12.36 (a) Rate = $\dfrac{-\Delta[cyclopropane]}{\Delta t}$ = $-\dfrac{0.080\ M - 0.098\ M}{5.0\ min - 0.0\ min}$ = 3.6×10^{-3} M/min

Rate = $3.6 \times 10^{-3}\ \dfrac{M}{min}$ x $\dfrac{1\ min}{60\ s}$ = 6.0×10^{-5} M/s

(b) Rate = $\dfrac{-\Delta[cyclopropane]}{\Delta t}$ = $-\dfrac{0.044\ M - 0.054\ M}{20.0\ min - 15.0\ min}$ = 2.0×10^{-3} M/min

Rate = $2.0 \times 10^{-3}\ \dfrac{M}{min}$ x $\dfrac{1\ min}{60\ s}$ = 3.3×10^{-5} M/s

Ordinarily the reaction rate decreases as a reaction proceeds, so the average rate is different for different time periods during the reaction.

12.37 (a) Rate = $-\dfrac{\Delta[NO_2]}{\Delta t}$ = $-\dfrac{(5.59 \times 10^{-3}\ M) - (6.58 \times 10^{-3}\ M)}{100\ s - 50\ s}$ = 2.0×10^{-5} M/s

(b) Rate = $-\dfrac{\Delta[NO_2]}{\Delta t}$ = $-\dfrac{(4.85 \times 10^{-3}\ M) - (5.59 \times 10^{-3}\ M)}{150\ s - 100\ s}$ = 1.5×10^{-5} M/s

Ordinarily the reaction rate decreases as a reaction proceeds, so the average rate is different for different time periods during the reaction.

12.38

(a) The instantaneous rate of decomposition of N_2O_5 at t = 200 s is determined from the slope of the curve at t = 200 s.

$$\text{Rate} = -\frac{\Delta[N_2O_5]}{\Delta t} = -\text{slope} = -\frac{(1.20 \times 10^{-2}\text{ M}) - (1.69 \times 10^{-2}\text{ M})}{300\text{ s} - 100\text{ s}} = 2.4 \times 10^{-5}\text{ M/s}$$

(b) The initial rate of decomposition of N_2O_5 is determined from the slope of the curve at t = 0 s. This is equivalent to the slope of the curve from 0 s to 100 s because in this time interval the curve is almost linear.

$$\text{Initial rate} = -\frac{\Delta[N_2O_5]}{\Delta t} = -\text{slope} = -\frac{(1.69 \times 10^{-2}\text{ M}) - (2.00 \times 10^{-2}\text{ M})}{100\text{ s} - 0\text{ s}} = 3.1 \times 10^{-5}\text{ M/s}$$

12.39

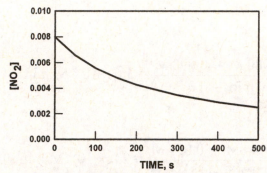

(a) The instantaneous rate of decomposition of NO_2 at t = 100 s is determined from the slope of the curve at t = 100 s.

$$\text{Rate} = -\frac{\Delta[NO_2]}{\Delta t} = -\text{slope} = -\frac{(4.00 \times 10^{-3}\text{ M}) - (7.00 \times 10^{-3}\text{ M})}{190\text{ s} - 20\text{ s}} = 1.8 \times 10^{-5}\text{ M/s}$$

(b) The initial rate of decomposition of NO_2 is determined from the slope of the curve at t = 0 s. This is equivalent to the slope of the curve from 0 s to 50 s because in this time interval the curve is almost linear.

$$\text{Initial rate} = -\frac{\Delta[NO_2]}{\Delta t} = -\text{slope} = -\frac{(6.58 \times 10^{-3}\text{ M}) - (8.00 \times 10^{-3}\text{ M})}{50\text{ s} - 0\text{ s}} = 2.8 \times 10^{-5}\text{ M/s}$$

12.40 (a) $-\dfrac{\Delta[H_2]}{\Delta t} = -3\dfrac{\Delta[N_2]}{\Delta t}$; The rate of consumption of H_2 is 3 times faster.

(b) $\dfrac{\Delta[NH_3]}{\Delta t} = -2\dfrac{\Delta[N_2]}{\Delta t}$; The rate of formation of NH_3 is 2 times faster.

12.41 (a) $-\dfrac{\Delta[O_2]}{\Delta t} = -\dfrac{5}{4}\dfrac{\Delta[NH_3]}{\Delta t}$; The rate of consumption of O_2 is 1.25 times faster.

(b) $\dfrac{\Delta[NO]}{\Delta t} = -\dfrac{\Delta[NH_3]}{\Delta t}$; The rate of formation of NO is the same.

$\dfrac{\Delta[H_2O]}{\Delta t} = -\dfrac{6}{4}\dfrac{\Delta[NH_3]}{\Delta t}$; The rate of formation of H_2O is 1.5 times faster.

12.42 $N_2(g) + 3 H_2(g) \rightarrow 2 NH_3(g);$ $-\dfrac{\Delta[N_2]}{\Delta t} = -\dfrac{1}{3}\dfrac{\Delta[H_2]}{\Delta t} = \dfrac{1}{2}\dfrac{\Delta[NH_3]}{\Delta t}$

12.43 (a) $-\dfrac{\Delta[I^-]}{\Delta t} = -3\dfrac{\Delta[S_2O_8{}^{2-}]}{\Delta t} = 3(1.5 \times 10^{-3}\ M/s) = 4.5 \times 10^{-3}\ M/s$

(b) $\dfrac{\Delta[SO_4{}^{2-}]}{\Delta t} = -2\dfrac{\Delta[S_2O_8{}^{2-}]}{\Delta t} = 2(1.5 \times 10^{-3}\ M/s) = 3.0 \times 10^{-3}\ M/s$

Rate Laws (Sections 12.2–12.3)

12.44 Rate = $k[NO]^2[Br_2]$; 2nd order in NO; 1st order in Br_2; 3rd order overall

12.45 Rate = $k[CHCl_3][Cl_2]^{1/2}$; 1st order in $CHCl_3$; 1/2 order in Cl_2; 3/2 order overall

12.46 Rate = $k[H_2][ICl]$; units for k are $\dfrac{L}{mol \cdot s}$ or $1/(M \cdot s)$

12.47 Rate = $k[NO]^2[H_2]$, units for k are $1/(M^2 \cdot s)$

12.48 (a) Rate = $k[CH_3Br][OH^-]$
(b) Because the reaction is first-order in OH^-, if the $[OH^-]$ is decreased by a factor of 5, the rate will also decrease by a factor of 5.
(c) Because the reaction is first-order in each reactant, if both reactant concentrations are doubled, the rate will increase by a factor of $2 \times 2 = 4$.

12.49 (a) Rate = $k[Br^-][BrO_3^-][H^+]^2$
(b) The overall reaction order is $1 + 1 + 2 = 4$.
(c) Because the reaction is second-order in H^+, if the $[H^+]$ is tripled, the rate will increase by a factor of $3^2 = 9$.
(d) Because the reaction is first-order in both Br^- and BrO_3^-, if both reactant concentrations are halved, the rate will decrease by a factor of 4 ($1/2 \times 1/2 = 1/4$).

12.50 (a) Rate = $k[CH_3COCH_3]^m$

$$m = \dfrac{\ln\left(\dfrac{Rate_2}{Rate_1}\right)}{\ln\left(\dfrac{[CH_3COCH_3]_2}{[CH_3COCH_3]_1}\right)} = \dfrac{\ln\left(\dfrac{7.8 \times 10^{-5}}{5.2 \times 10^{-5}}\right)}{\ln\left(\dfrac{9.0 \times 10^{-3}}{6.0 \times 10^{-3}}\right)} = 1;\quad Rate = k[CH_3COCH_3]$$

(b) From Experiment 1: $k = \dfrac{Rate}{[CH_3COCH_3]} = \dfrac{5.2 \times 10^{-5}\ M/s}{6.0 \times 10^{-3}\ M} = 8.7 \times 10^{-3}/s$

(c) Rate = $k[CH_3COCH_3] = (8.7 \times 10^{-3}/s)(1.8 \times 10^{-3}M) = 1.6 \times 10^{-5}\ M/s$

12.51 (a) Rate $= k[CH_3NNCH_3]^m$

$$m = \frac{\ln\left(\dfrac{Rate_2}{Rate_1}\right)}{\ln\left(\dfrac{[CH_3NNCH_3]_2}{[CH_3NNCH_3]_1}\right)} = \frac{\ln\left(\dfrac{2.0 \times 10^{-6}}{6.0 \times 10^{-6}}\right)}{\ln\left(\dfrac{8.0 \times 10^{-3}}{2.4 \times 10^{-2}}\right)} = 1; \quad Rate = k[CH_3NNCH_3]$$

(b) From Experiment 1: $k = \dfrac{Rate}{[CH_3NNCH_3]} = \dfrac{6.0 \times 10^{-6} \text{ M/s}}{2.4 \times 10^{-2} \text{ M}} = 2.5 \times 10^{-4}\text{/s}$

(c) Rate $= k[CH_3NNCH_3] = (2.5 \times 10^{-4}\text{/s})(0.020 \text{ M}) = 5.0 \times 10^{-6}$ M/s

12.52 (a) Rate $= k[NH_4^+]^m[NO_2^-]^n$

$$m = \frac{\ln\left(\dfrac{Rate_2}{Rate_1}\right)}{\ln\left(\dfrac{[NH_4^+]_2}{[NH_4^+]_1}\right)} = \frac{\ln\left(\dfrac{3.6 \times 10^{-6}}{7.2 \times 10^{-6}}\right)}{\ln\left(\dfrac{0.12}{0.24}\right)} = 1; \quad n = \frac{\ln\left(\dfrac{Rate_3}{Rate_2}\right)}{\ln\left(\dfrac{[NO_2^-]_3}{[NO_2^-]_2}\right)} = \frac{\ln\left(\dfrac{5.4 \times 10^{-6}}{3.6 \times 10^{-6}}\right)}{\ln\left(\dfrac{0.15}{0.10}\right)} = 1$$

Rate $= k[NH_4^+][NO_2^-]$

(b) From Experiment 1: $k = \dfrac{Rate}{[NH_4^+][NO_2^-]} = \dfrac{7.2 \times 10^{-6} \text{ M/s}}{(0.24 \text{ M})(0.10 \text{ M})} = 3.0 \times 10^{-4}/(\text{M} \cdot \text{s})$

(c) Rate $= k[NH_4^+][NO_2^-] = [3.0 \times 10^{-4}/(\text{M} \cdot \text{s})](0.39 \text{ M})(0.052 \text{ M}) = 6.1 \times 10^{-6}$ M/s

12.53 (a) Rate $= k[NO]^m[Cl_2]^n$

$$m = \frac{\ln\left(\dfrac{Rate_2}{Rate_1}\right)}{\ln\left(\dfrac{[NO]_2}{[NO]_1}\right)} = \frac{\ln\left(\dfrac{4.0 \times 10^{-2}}{1.0 \times 10^{-2}}\right)}{\ln\left(\dfrac{0.26}{0.13}\right)} = 2; \quad n = \frac{\ln\left(\dfrac{Rate_3}{Rate_1}\right)}{\ln\left(\dfrac{[Cl_2]_3}{[Cl_2]_1}\right)} = \frac{\ln\left(\dfrac{5.0 \times 10^{-3}}{1.0 \times 10^{-2}}\right)}{\ln\left(\dfrac{0.10}{0.20}\right)} = 1$$

Rate $= k[NO]^2[Cl_2]$

(b) From Experiment 1: $k = \dfrac{Rate}{[NO]^2[Cl_2]} = \dfrac{1.0 \times 10^{-2} \text{ M/s}}{(0.13 \text{ M})^2(0.20 \text{ M})} = 3.0/(\text{M}^2 \cdot \text{s})$

(c) Rate $= k[NO]^2[Cl_2] = [3.0/(\text{M}^2 \cdot \text{s})](0.12 \text{ M})^2(0.12 \text{ M}) = 5.2 \times 10^{-3}$ M/s

Integrated Rate Law; Half-Life (Sections 12.4–12.5, 12.7–12.8)

12.54 $\ln \dfrac{[C_3H_6]_t}{[C_3H_6]_0} = -kt$, $k = 6.7 \times 10^{-4}/s$

(a) $t = 30$ min $\times \dfrac{60 \text{ s}}{1 \text{ min}} = 1800$ s

$\ln[C_3H_6]_t = -kt + \ln[C_3H_6]_0 = -(6.7 \times 10^{-4}/s)(1800 \text{ s}) + \ln(0.0500) = -4.202$

$[C_3H_6]_t = e^{-4.202} = 0.015$ M

(b) $t = \dfrac{\ln \dfrac{[C_3H_6]_t}{[C_3H_6]_0}}{-k} = \dfrac{\ln\left(\dfrac{0.0100}{0.0500}\right)}{-(6.7 \times 10^{-4}/s)} = 2402$ s; $t = 2402$ s $\times \dfrac{1 \text{ min}}{60 \text{ s}} = 40$ min

(c) $[C_3H_6]_0 = 0.0500$ M; If 25% of the C_3H_6 reacts then 75% remains.

$[C_3H_6]_t = (0.75)(0.0500 \text{ M}) = 0.0375$ M

$t = \dfrac{\ln \dfrac{[C_3H_6]_t}{[C_3H_6]_0}}{-k} = \dfrac{\ln\left(\dfrac{0.0375}{0.0500}\right)}{-(6.7 \times 10^{-4}/s)} = 429$ s; $t = 429$ s $\times \dfrac{1 \text{ min}}{60 \text{ s}} = 7.2$ min

12.55 $\ln \dfrac{[CH_3NC]_t}{[CH_3NC]_0} = -kt$, $k = 5.11 \times 10^{-5}/s$

(a) $t = 2.00$ hr $\times \dfrac{60 \text{ min}}{1 \text{ hr}} \times \dfrac{60 \text{ s}}{1 \text{ min}} = 7200$ s

$\ln[CH_3NC]_t = -kt + \ln[CH_3NC]_0 = -(5.11 \times 10^{-5}/s)(7200 \text{ s}) + \ln(0.0340) = -3.749$

$[CH_3NC]_t = e^{-3.749} = 0.0235$ M

(b) $t = \dfrac{\ln \dfrac{[CH_3NC]_t}{[CH_3NC]_0}}{-k} = \dfrac{\ln\left(\dfrac{0.0300}{0.0340}\right)}{-5.11 \times 10^{-5}/s)} = 2449$ s; $t = 2449$ s $\times \dfrac{1 \text{ min}}{60 \text{ s}} = 40.8$ min

(c) $[CH_3NC]_0 = 0.0340$ M; If 20% of the CH_3NC reacts then 80% remains.

$[CH_3NC]_t = (0.80)(0.0340 \text{ M}) = 0.0272$ M

$t = \dfrac{\ln \dfrac{[CH_3NC]_t}{[CH_3NC]_0}}{-k} = \dfrac{\ln\left(\dfrac{0.0272}{0.0340}\right)}{-(5.11 \times 10^{-5}/s)} = 4367$ s; $t = 4367$ s $\times \dfrac{1 \text{ min}}{60 \text{ s}} = 72.8$ min

12.56 $t_{1/2} = \dfrac{0.693}{k} = \dfrac{0.693}{6.7 \times 10^{-4}/s} = 1034\ s = 17\ min$

$t = \dfrac{\ln \dfrac{[C_3H_6]_t}{[C_3H_6]_o}}{-k} = \dfrac{\ln \dfrac{(0.0625)(0.0500)}{(0.0500)}}{-6.7 \times 10^{-4}/s} = 4140\ s$

$t = 4140\ s \times \dfrac{1\ min}{60\ s} = 69\ min$

This is also 4 half-lives. $100 \overset{t_{1/2}}{\to} 50 \overset{t_{1/2}}{\to} 25 \overset{t_{1/2}}{\to} 12.5 \overset{t_{1/2}}{\to} 6.25$

12.57 $t_{1/2} = \dfrac{0.693}{k} = \dfrac{0.693}{5.11 \times 10^{-5}/s} = 13,562\ s$

$t_{1/2} = 13,562\ s \times \dfrac{1\ min}{60\ s} \times \dfrac{1\ hr}{60\ min} = 3.77\ hr$

$t = \dfrac{\ln \dfrac{[CH_3NC]_t}{[CH_3NC]_o}}{-k} = \dfrac{\ln \dfrac{(0.125)(0.0340)}{(0.0340)}}{-5.11 \times 10^{-5}/s} = 40,694\ s$

$t = 40,694\ s \times \dfrac{1\ min}{60\ s} \times \dfrac{1\ hr}{60\ min} = 11.3\ hr$

This is also 3 half-lives. $100 \overset{t_{1/2}}{\to} 50 \overset{t_{1/2}}{\to} 25 \overset{t_{1/2}}{\to} 12.5$

12.58 $t_{1/2} = 8.0\ h$

$0.60\ M \overset{t_{1/2}}{\to} 0.30\ M \overset{t_{1/2}}{\to} 0.15\ M$ requires 2 half-lives so it will take 16.0 h.

12.59 $t_{1/2} = 3.33\ h$

$0.800\ M \overset{t_{1/2}}{\to} 0.400\ M \overset{t_{1/2}}{\to} 0.200\ M \overset{t_{1/2}}{\to} 0.100\ M \overset{t_{1/2}}{\to} 0.0500\ M$
requires 4 half-lives so it will take 13.3 h.

12.60 $kt = \dfrac{1}{[C_4H_6]_t} - \dfrac{1}{[C_4H_6]_0}, \qquad k = 4.0 \times 10^{-2}/(M \cdot s)$

(a) $t = 1.00\ h \times \dfrac{60\ min}{1\ hr} \times \dfrac{60\ s}{1\ min} = 3600\ s$

$\dfrac{1}{[C_4H_6]_t} = kt + \dfrac{1}{[C_4H_6]_0} = (4.0 \times 10^{-2}/(M \cdot s))(3600\ s) + \dfrac{1}{0.0200\ M}$

$\dfrac{1}{[C_4H_6]_t} = 194/M$ and $[C_4H_6] = 5.2 \times 10^{-3}\ M$

(b) $\quad t = \dfrac{1}{k}\left[\dfrac{1}{[C_4H_6]_t} - \dfrac{1}{[C_4H_6]_0}\right]$

$t = \dfrac{1}{4.0 \times 10^{-2}/(M \cdot s)}\left[\dfrac{1}{(0.0020\ M)} - \dfrac{1}{(0.0200\ M)}\right] = 11{,}250\ s$

$t = 11{,}250\ s \times \dfrac{1\ min}{60\ s} \times \dfrac{1\ hr}{60\ min} = 3.1\ h$

12.61 $\quad kt = \dfrac{1}{[HI]_t} - \dfrac{1}{[HI]_0},\quad k = 9.7 \times 10^{-6}/(M \cdot s)$

(a) $\quad t = 6.00\ day \times \dfrac{24\ hr}{1\ day} \times \dfrac{60\ min}{1\ hr} \times \dfrac{60\ s}{1\ min} = 518{,}400\ s$

$\dfrac{1}{[HI]_t} = kt + \dfrac{1}{[HI]_0} = (9.7 \times 10^{-6}/(M \cdot s))(518{,}400\ s) + \dfrac{1}{0.100\ M}$

$\dfrac{1}{[HI]_t} = 15.03/M \quad \text{and} \quad [HI] = 0.067\ M$

(b) $\quad t = \dfrac{1}{k}\left[\dfrac{1}{[HI]_t} - \dfrac{1}{[HI]_0}\right]$

$t = \dfrac{1}{9.7 \times 10^{-6}/(M \cdot s)}\left[\dfrac{1}{(0.085\ M)} - \dfrac{1}{(0.100\ M)}\right] = 181{,}928\ s$

$t = 181{,}928\ s \times \dfrac{1\ min}{60\ s} \times \dfrac{1\ hr}{60\ min} \times \dfrac{1\ day}{24\ hr} = 2.1\ days$

12.62 $\quad t_{1/2} = \dfrac{1}{k[C_4H_6]_0} = \dfrac{1}{[4.0 \times 10^{-2}/(M \cdot s)](0.0200\ M)} = 1250\ s = 21\ min$

$t = t_{1/2} = \dfrac{1}{k[C_4H_6]_0} = \dfrac{1}{[4.0 \times 10^{-2}/(M \cdot s)](0.0100\ M)} = 2500\ s = 42\ min$

12.63 $\quad t_{1/2} = \dfrac{1}{k[HI]_0} = \dfrac{1}{[9.7 \times 10^{-6}/(M \cdot s)](0.100\ M)} = 1{,}030{,}928\ s$

$1{,}030{,}928\ s \times \dfrac{1\ min}{60\ s} \times \dfrac{1\ hr}{60\ min} \times \dfrac{1\ day}{24\ hr} = 12\ days$

$t = t_{1/2} = \dfrac{1}{k[HI]_0} = \dfrac{1}{[9.7 \times 10^{-6}/(M \cdot s)](0.200\ M)} = 515{,}464\ s$

$515{,}464\ s \times \dfrac{1\ min}{60\ s} \times \dfrac{1\ hr}{60\ min} \times \dfrac{1\ day}{24\ hr} = 6.0\ days$

12.64

time (min)	[N₂O]	ln[N₂O]	1/[N₂O]
0	0.250	−1.386	4.00
60	0.218	−1.523	4.59
90	0.204	−1.590	4.90
120	0.190	−1.661	5.26
180	0.166	−1.796	6.02

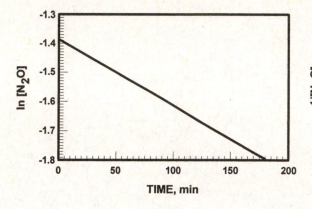

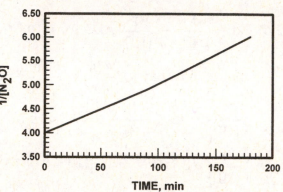

A plot of ln [N₂O] versus time is linear. The reaction is first-order in N₂O.

$k = -\text{slope} = -(-2.28 \times 10^{-3}/\text{min}) = 2.28 \times 10^{-3}/\text{min}$

$k = 2.28 \times 10^{-3}/\text{min} \times \dfrac{1\ \text{min}}{60\ \text{s}} = 3.79 \times 10^{-5}/\text{s}$

12.65

time (s)	[NOBr]	ln[NOBr]	1/[NOBr]
0	0.0400	−3.219	25.0
10	0.0303	−3.497	33.0
20	0.0244	−3.713	41.0
30	0.0204	−3.892	49.0
40	0.0175	−4.046	57.1

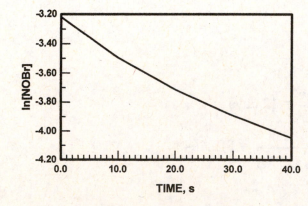

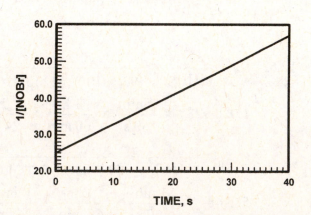

A plot of 1/[NOBr] versus time is linear. The reaction is second-order in NOBr.

$k = \text{slope} = 0.80/(\text{M} \cdot \text{s})$

12.66 $k = \dfrac{0.693}{t_{1/2}} = \dfrac{0.693}{248\ s} = 2.79 \times 10^{-3}/s$

12.67 $t_{1/2} = \dfrac{1}{k[A]_o}$; $t_{1/2} = 25\ min \times \dfrac{60\ s}{1\ min} = 1500\ s$

$k = \dfrac{1}{t_{1/2}[A]_0} = \dfrac{1}{(1500\ s)(0.036\ M)} = 1.8 \times 10^{-2}\ M^{-1}\,s^{-1}$

12.68 (a) The units for the rate constant, k, indicate the reaction is zeroth-order.
(b) For a zeroth-order reaction, $[A]_t - [A]_o = -kt$

$t = 30\ min \times \dfrac{60\ s}{1\ min} = 1800\ s$

$[A]_t = -kt + [A]_o = -(3.6 \times 10^{-5}\ M/s)(1800\ s) + 0.096\ M = 0.031\ M$
(c) Let $[A]_t = [A]_o/2$

$t_{1/2} = \dfrac{[A]_o/2 - [A]_o}{-k} = \dfrac{0.096/2\ M - 0.096\ M}{-3.6 \times 10^{-5}\ M/s} = 1333\ s$

$t_{1/2} = 1333\ s \times \dfrac{1\ min}{60\ s} = 22\ min$

12.69 (a)

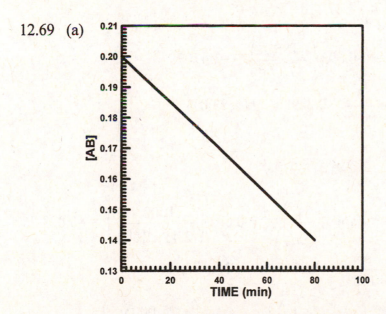

A plot of [AB] versus time is linear. The reaction is zeroth-order and k = – slope.

$k = -\dfrac{0.140\ M - 0.200\ M}{80.0\ min - 0\ min} = 7.50 \times 10^{-4}\ M/min$

$k = 7.50 \times 10^{-4}\ M/min \times \dfrac{1\ min}{60\ s} = 1.25 \times 10^{-5}\ M/s$

(b) $[A]_t - [A]_o = -kt$
$[A]_t = -kt + [A]_o = -(7.50 \times 10^{-4}\ M/min)(126\ min) + 0.200\ M = 0.105\ M$

(c) $[A]_t - [A]_o = -kt$

$$t = \frac{[A]_t - [A]_o}{-k} = \frac{0.100\ M - 0.200\ M}{-7.50 \times 10^{-4}\ M/min} = 133\ min$$

Radioactive Decay Rates (Section 12.6)

12.70 $k = \dfrac{0.693}{t_{1/2}} = \dfrac{0.693}{2.805\ d} = 0.247\ d^{-1}$

12.71 $k = \dfrac{0.693}{t_{1/2}} = \dfrac{0.693}{78.25\ h} = 8.86 \times 10^{-3}\ h^{-1}$

12.72 $t_{1/2} = \dfrac{0.693}{k} = \dfrac{0.693}{0.228\ d^{-1}} = 3.04\ d$

12.73 $t_{1/2} = \dfrac{0.693}{k} = \dfrac{0.693}{2.88 \times 10^{-5}\ y^{-1}} = 2.41 \times 10^4\ y$

12.74 $t_{1/2} = \dfrac{0.693}{k} = \dfrac{0.693}{7.95 \times 10^{-3}\ d^{-1}} = 87.17\ d$

$$\ln\left(\frac{N}{N_0}\right) = (-0.693)\left(\frac{t}{t_{1/2}}\right) = (-0.693)\left(\frac{185\ d}{87.17\ d}\right) = -1.4707$$

$\dfrac{N}{N_0} = e^{-1.4707} = 0.2298;$ $\qquad \dfrac{N}{100\%} = 0.2298;$ $\quad N = 23.0\%$

12.75 $t_{1/2} = \dfrac{0.693}{k} = \dfrac{0.693}{2.88 \times 10^{-5}\ y^{-1}} = 2.41 \times 10^4\ y$

After 1000 y: $\ln\left(\dfrac{N}{N_0}\right) = (-0.693)\left(\dfrac{t}{t_{1/2}}\right) = (-0.693)\left(\dfrac{1000\ y}{2.41 \times 10^4\ y}\right) = -0.028\,76$

$\dfrac{N}{N_0} = e^{-0.02876} = 0.9717;$ $\qquad \dfrac{N}{100\%} = 0.9717;$ $\qquad N = 97.17\%$

After 25,000 y: $\ln\left(\dfrac{N}{N_0}\right) = (-0.693)\left(\dfrac{t}{t_{1/2}}\right) = (-0.693)\left(\dfrac{25,000\ y}{2.41 \times 10^4\ y}\right) = -0.7189$

$\dfrac{N}{N_0} = e^{-0.7189} = 0.4873;$ $\qquad \dfrac{N}{100\%} = 0.4873;$ $\qquad N = 48.73\%$

After 100,000 y: $\ln\left(\dfrac{N}{N_0}\right) = (-0.693)\left(\dfrac{t}{t_{1/2}}\right) = (-0.693)\left(\dfrac{100,000\ y}{2.41 \times 10^4\ y}\right) = -2.876$

$\dfrac{N}{N_0} = e^{-2.876} = 0.0564;$ $\qquad \dfrac{N}{100\%} = 0.0564;$ $\qquad N = 5.64\%$

12.76 $t_{1/2} = (102\ y)(365\ d/y)(24\ h/d)(3600\ s/h) = 3.2167 \times 10^9\ s$

$$k = \frac{0.693}{t_{1/2}} = \frac{0.693}{3.2167 \times 10^9\ s} = 2.1544 \times 10^{-10}\ s^{-1}$$

$$N = (1.0 \times 10^{-9}\ g)\left(\frac{1\ mol\ Po}{209\ g\ Po}\right)(6.022 \times 10^{23}\ atoms/mol) = 2.881 \times 10^{12}\ atoms$$

Decay rate = kN = $(2.1544 \times 10^{-10}\ s^{-1})(2.881 \times 10^{12}\ atoms) = 6.21 \times 10^2\ s^{-1}$
621 α particles are emitted in 1.0 s.

12.77 $t_{1/2} = (3.0 \times 10^5\ y)(365\ d/y)(24\ h/d)(60\ min/h) = 1.6 \times 10^{11}\ min$

$$k = \frac{0.693}{t_{1/2}} = \frac{0.693}{1.6 \times 10^{11}\ min} = 4.3 \times 10^{-12}\ min^{-1}$$

$$N = (5.0 \times 10^{-3}\ g)\left(\frac{1\ mol\ {}^{36}Cl}{36\ g}\right)(6.022 \times 10^{23}\ atoms/mol) = 8.4 \times 10^{19}\ atoms$$

Decay rate = kN = $(4.3 \times 10^{-12}\ min^{-1})(8.4 \times 10^{19}\ atoms) = 3.6 \times 10^8\ min^{-1}$

12.78 Decay rate = kN

$$N = (1.0 \times 10^{-3}\ g)\left(\frac{1\ mol\ {}^{79}Se}{79\ g}\right)(6.022 \times 10^{23}\ atoms/mol) = 7.6 \times 10^{18}\ atoms$$

$$k = \frac{Decay\ rate}{N} = \frac{1.5 \times 10^5/s}{7.6 \times 10^{18}} = 2.0 \times 10^{-14}\ s^{-1}$$

$$t_{1/2} = \frac{0.693}{k} = \frac{0.693}{2.0 \times 10^{-14}\ s^{-1}} = 3.5 \times 10^{13}\ s$$

$$t_{1/2} = (3.5 \times 10^{13}\ s)\left(\frac{1\ h}{3600\ s}\right)\left(\frac{1\ d}{24\ h}\right)\left(\frac{1\ y}{365\ d}\right) = 1.1 \times 10^6\ y$$

12.79 Decay rate = kN

$$N = (1.0 \times 10^{-9}\ g)\left(\frac{1\ mol\ Ti}{44\ g\ Ti}\right)(6.022 \times 10^{23}\ atoms/mol) = 1.37 \times 10^{13}\ atoms$$

$$k = \frac{Decay\ rate}{N} = \frac{4.8 \times 10^3\ s^{-1}}{1.37 \times 10^{13}} = 3.50 \times 10^{-10}\ s^{-1}$$

$k = (3.50 \times 10^{-10}\ s^{-1})(3600\ s/h)(24\ h/d)(365\ d/y) = 1.10 \times 10^{-2}\ y^{-1}$

$$t_{1/2} = \frac{0.693}{k} = \frac{0.693}{1.10 \times 10^{-2}\ y^{-1}} = 63\ y$$

12.80 $\ln\left(\dfrac{N}{N_0}\right) = (-0.693)\left(\dfrac{t}{t_{1/2}}\right)$; $\dfrac{N}{N_0} = \dfrac{Decay\ rate\ at\ time\ t}{Decay\ rate\ at\ time\ t\ =\ 0}$

$\ln\left(\dfrac{6990}{8540}\right) = (-0.693)\left(\dfrac{10.0\ d}{t_{1/2}}\right)$; $t_{1/2} = 34.6\ d$

12.81 $\ln\left(\dfrac{N}{N_0}\right) = (-0.693)\left(\dfrac{t}{t_{1/2}}\right);$ $\qquad \dfrac{N}{N_0} = \dfrac{\text{Decay rate at time } t}{\text{Decay rate at time } t = 0}$

$\ln\left(\dfrac{10,980}{53,500}\right) = (-0.693)\left(\dfrac{48.0\text{ h}}{t_{1/2}}\right);$ $\quad t_{1/2} = 21.0\text{ h}$

Reaction Mechanisms (Sections 12.9–12.11)

12.82 An elementary reaction is a description of an individual molecular event that involves the breaking and/or making of chemical bonds. By contrast, the overall reaction describes the stoichiometry of the overall process but provides no information about how the reaction occurs.

12.83 Molecularity is the number of reactant molecules or atoms for an elementary reaction. Reaction order is the sum of the exponents of the concentration terms in the rate law.

12.84 There is no relationship between the coefficients in a balanced chemical equation for an overall reaction and the exponents in the rate law unless the overall reaction occurs in a single elementary step, in which case the coefficients in the balanced equation are the exponents in the rate law.

12.85 The rate-determining step is the slowest step in a multistep reaction. The coefficients in the balanced equation for the rate-determining step are the exponents in the rate law.

12.86 (a) $\qquad\qquad H_2(g) + ICl(g) \rightarrow HI(g) + HCl(g)$
$\underline{\qquad\qquad HI(g) + ICl(g) \rightarrow I_2(g) + HCl(g)\qquad}$
Overall reaction $\qquad H_2(g) + 2\,ICl(g) \rightarrow I_2(g) + 2\,HCl(g)$
(b) Because HI(g) is produced in the first step and consumed in the second step, it is a reaction intermediate.
(c) In each reaction there are two reactant molecules, so each elementary reaction is bimolecular.

12.87 (a) $\qquad\qquad NO(g) + Cl_2(g) \rightarrow NOCl_2(g)$
$\underline{\qquad\qquad NOCl_2(g) + NO(g) \rightarrow 2\,NOCl(g)\qquad}$
Overall reaction $\qquad 2\,NO(g) + Cl_2(g) \rightarrow 2\,NOCl(g)$
(b) Because $NOCl_2$ is produced in the first step and consumed in the second step, $NOCl_2$ is a reaction intermediate.
(c) Each elementary step is bimolecular.

12.88 (a) bimolecular, Rate $= k[O_3][Cl]$ $\qquad\qquad$ (b) unimolecular, Rate $= k[NO_2]$
(c) bimolecular, Rate $= k[ClO][O]$ $\qquad\qquad$ (d) termolecular, Rate $= k[Cl]^2[N_2]$

12.89 (a) unimolecular, Rate $= k[I_2]$ $\qquad\qquad$ (b) termolecular, Rate $= k[NO]^2[Br_2]$
(c) bimolecular, Rate $= k[CH_3Br][OH^-]$

12.90 (a) $\qquad\qquad\qquad$ $NO_2Cl(g) \;\rightarrow\; NO_2(g) + Cl(g)$

$\qquad\qquad\qquad\qquad$ $\underline{Cl(g) + NO_2Cl(g) \;\rightarrow\; NO_2(g) + Cl_2(g)}$

$\qquad$ Overall reaction $\qquad$ $2\,NO_2Cl(g) \;\rightarrow\; 2\,NO_2(g) + Cl_2(g)$

$\qquad$ (b) 1. unimolecular; 2. bimolecular

$\qquad$ (c) Rate $= k[NO_2Cl]$

12.91 (a) $\qquad\qquad\qquad$ $Mo(CO)_6 \;\rightarrow\; Mo(CO)_5 + CO$

$\qquad\qquad\qquad\qquad$ $\underline{Mo(CO)_5 + L \;\rightarrow\; Mo(CO)_5L}$

$\qquad$ Overall reaction $\qquad$ $Mo(CO)_6 + L \;\rightarrow\; Mo(CO)_5L + CO$

$\qquad$ (b) 1. unimolecular; 2. bimolecular

$\qquad$ (c) Rate $= k[Mo(CO)_6]$

12.92 $NO_2(g) + F_2(g) \rightarrow NO_2F(g) + F(g)$ $\qquad$ (slow)

$\qquad$ $F(g) + NO_2(g) \rightarrow NO_2F(g)$ $\qquad$ (fast)

12.93 $O_3(g) + NO(g) \rightarrow O_2(g) + NO_2(g)$ $\qquad$ (slow)

$\qquad$ $NO_2(g) + O(g) \rightarrow O_2(g) + NO(g)$ $\qquad$ (fast)

12.94 (a) $2\,NO(g) + O_2(g) \rightarrow 2\,NO_2(g)$

(b) $Rate_{forward} = k_1[NO]^2$ and $Rate_{reverse} = k_{-1}[N_2O_2]$

Because of the equilibrium, $Rate_{forward} = Rate_{reverse}$, and $k_1[NO]^2 = k_{-1}[N_2O_2]$.

$$[N_2O_2] = \frac{k_1}{k_{-1}}[NO]^2$$

The rate law for the rate determining step is Rate $= -\Delta[NO]/\Delta t = 2k_2[N_2O_2][O_2]$ because two NO molecules are consumed in the overall reaction for every N_2O_2 that reacts in the second step. In this rate law substitute for $[N_2O_2]$. $\quad$ Rate $= 2k_2\,\dfrac{k_1}{k_{-1}}[NO]^2[O_2]$

(c) $k = \dfrac{2k_2k_1}{k_{-1}}$

12.95 (a) $2\,N_2O_5(g) + 4\,NO_2(g) + O_2(g)$

(b) An intermediate is formed in one step and then consumed in a subsequent step in the reaction. NO_2 is an intermediate as well as a product. NO_3 and NO are intermediates.

(c) $Rate_{forward} = k_1[N_2O_5]$ and $Rate_{reverse} = k_{-1}[NO_2][NO_3]$

Because of the equilibrium, $Rate_{forward} = Rate_{reverse}$, and $k_1[N_2O_5] = k_{-1}[NO_2][NO_3]$.

$$\frac{k_1}{k_{-1}}[N_2O_5] = [NO_2][NO_3]$$

The rate law for the rate determining step is Rate $= k_2[NO_2][NO_3]$. In this rate law substitute for $[NO_2][NO_3]$.

$$Rate = k_2\,\frac{k_1}{k_{-1}}[N_2O_5]$$

(d) $k = \dfrac{k_2k_1}{k_{-1}}$

The Arrhenius Equation (Sections 12.12–12.13)

12.96 Very few collisions involve a collision energy greater than or equal to the activation energy, and only a fraction of those have the proper orientation for reaction.

12.97 The two reactions have frequency factors that differ by a factor of 10.

12.98 Plot ln k versus 1/T to determine the activation energy, E_a.

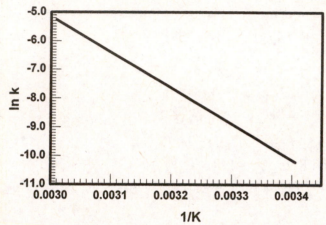

Slope $= -1.25 \times 10^4$ K

$E_a = -R(\text{slope}) = -[8.314 \times 10^{-3} \text{ kJ/(K} \cdot \text{mol)}](-1.25 \times 10^4 \text{ K}) = 104$ kJ/mol

12.99 Plot ln k versus 1/T to determine the activation energy, E_a.

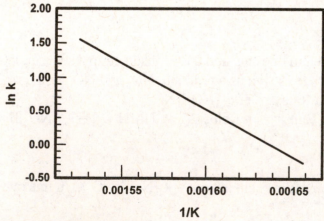

Slope $= -1.359 \times 10^4$ K

$E_a = -R(\text{slope}) = -[8.314 \times 10^{-3} \text{ kJ/(K} \cdot \text{mol)}](-1.359 \times 10^4 \text{ K}) = 113$ kJ/mol

12.100 (a) $\ln\left(\dfrac{k_2}{k_1}\right) = \left(\dfrac{-E_a}{R}\right)\left(\dfrac{1}{T_2} - \dfrac{1}{T_1}\right)$

$k_1 = 1.3/(M \cdot s)$, $T_1 = 700$ K
$k_2 = 23.0/(M \cdot s)$, $T_2 = 800$ K

$E_a = - \dfrac{[\ln k_2 - \ln k_1](R)}{\left(\dfrac{1}{T_2} - \dfrac{1}{T_1}\right)}$

$E_a = - \dfrac{[\ln(23.0) - \ln(1.3)][8.314 \times 10^{-3}\ kJ/(K \cdot mol)]}{\left(\dfrac{1}{800\ K} - \dfrac{1}{700\ K}\right)} = 134\ kJ/mol$

(b) $k_1 = 1.3/(M \cdot s)$, $T_1 = 700$ K
solve for k_2, $T_2 = 750$ K

$\ln k_2 = \left(\dfrac{-E_a}{R}\right)\left(\dfrac{1}{T_2} - \dfrac{1}{T_1}\right) + \ln k_1$

$\ln k_2 = \left(\dfrac{-133.8\ kJ/mol}{8.314 \times 10^{-3}\ kJ/(K \cdot mol)}\right)\left(\dfrac{1}{750\ K} - \dfrac{1}{700\ K}\right) + \ln(1.3) = 1.795$

$k_2 = e^{1.795} = 6.0/(M \cdot s)$

12.101 $\ln\left(\dfrac{k_2}{k_1}\right) = \left(\dfrac{-E_a}{R}\right)\left[\dfrac{1}{T_2} - \dfrac{1}{T_1}\right]$

(a) Because the rate doubles, $k_2 = 2k_1$
$k_1 = 1.0 \times 10^{-3}/s$, $T_1 = 25\ °C = 298$ K
$k_2 = 2.0 \times 10^{-3}/s$, $T_2 = 35\ °C = 308$ K

$E_a = - \dfrac{[\ln k_2 - \ln k_1](R)}{\left(\dfrac{1}{T_2} - \dfrac{1}{T_1}\right)}$

$E_a = - \dfrac{[\ln(2.0 \times 10^{-3}) - \ln(1.0 \times 10^{-3})][8.314 \times 10^{-3}\ kJ/(K \cdot mol)]}{\left(\dfrac{1}{308\ K} - \dfrac{1}{298\ K}\right)} = 53\ kJ/mol$

(b) Because the rate triples, $k_2 = 3k_1$
$k_1 = 1.0 \times 10^{-3}/s$, $T_1 = 25\ °C = 298$ K
$k_2 = 3.0 \times 10^{-3}/s$, $T_2 = 35\ °C = 308$ K

$E_a = - \dfrac{[\ln(3.0 \times 10^{-3}) - \ln(1.0 \times 10^{-3})][8.314 \times 10^{-3}\ kJ/(K \cdot mol)]}{\left(\dfrac{1}{308\ K} - \dfrac{1}{298\ K}\right)} = 84\ kJ/mol$

12.102 $\ln\left(\dfrac{k_2}{k_1}\right) = \left(\dfrac{-E_a}{R}\right)\left(\dfrac{1}{T_2} - \dfrac{1}{T_1}\right)$

assume $k_1 = 1.0/(M \cdot s)$ at $T_1 = 25 \text{ °C} = 298 \text{ K}$
assume $k_2 = 15/(M \cdot s)$ at $T_2 = 50 \text{ °C} = 323 \text{ K}$

$E_a = -\dfrac{[\ln k_2 - \ln k_1](R)}{\left(\dfrac{1}{T_2} - \dfrac{1}{T_1}\right)}$

$E_a = -\dfrac{[\ln(15) - \ln(1.0)][8.314 \times 10^{-3} \text{ kJ}/(K \cdot \text{mol})]}{\left(\dfrac{1}{323 \text{ K}} - \dfrac{1}{298 \text{ K}}\right)} = 87 \text{ kJ/mol}$

12.103 $\ln\left(\dfrac{k_2}{k_1}\right) = \left(\dfrac{-E_a}{R}\right)\left(\dfrac{1}{T_2} - \dfrac{1}{T_1}\right)$

assume $k_1 = 1.0/(M \cdot s)$ at $T_1 = 15 \text{ °C} = 288 \text{ K}$
assume $k_2 = 6.37/(M \cdot s)$ at $T_2 = 45 \text{ °C} = 318 \text{ K}$

$E_a = -\dfrac{[\ln k_2 - \ln k_1](R)}{\left(\dfrac{1}{T_2} - \dfrac{1}{T_1}\right)}$

$E_a = -\dfrac{[\ln(6.37) - \ln(1.0)][8.314 \times 10^{-3} \text{ kJ}/(K \cdot \text{mol})]}{\left(\dfrac{1}{318 \text{ K}} - \dfrac{1}{288 \text{ K}}\right)} = 47.0 \text{ kJ/mol}$

12.104

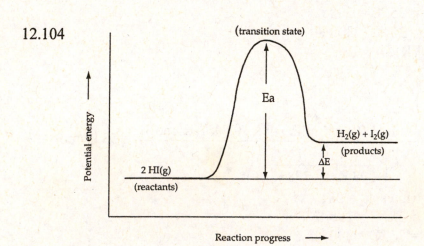

312

12.105   (a)

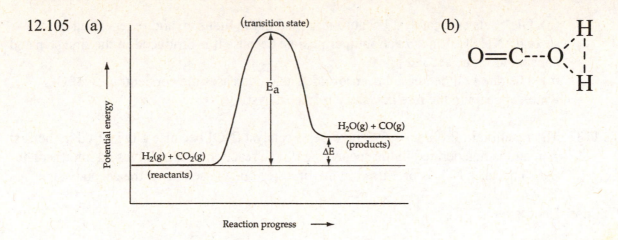

(b)

Catalysis (Sections 12.14–12.15)

12.106   A catalyst does participate in the reaction, but it is not consumed because it reacts in one step of the reaction and is regenerated in a subsequent step.

12.107   A catalyst doesn't appear in the chemical equation for a reaction because a catalyst reacts in one step of the reaction but is regenerated in a subsequent step.

12.108   A catalyst increases the rate of a reaction by changing the reaction mechanism and lowering the activation energy.

12.109   A homogeneous catalyst is one that exists in the same phase as the reactants.
Example: $NO(g)$ acts as a homogeneous catalyst for the conversion of $O_2(g)$ to $O_3(g)$.
A heterogeneous catalyst is one that exists in a different phase from the reactants.
Example: solid Ni, Pd, or Pt for catalytic hydrogenation, $C_2H_4(g) + H_2(g) \rightarrow C_2H_6(g)$.

12.110   (a) $O_3(g) + O(g) \rightarrow 2 O_2(g)$
(b) Cl acts as a catalyst.
(c) ClO is a reaction intermediate.
(d) A catalyst reacts in one step and is regenerated in a subsequent step. A reaction intermediate is produced in one step and consumed in another.

12.111   (a)

$$2 SO_2(g) + 2 NO_2(g) \rightarrow 2 SO_3(g) + 2 NO(g)$$
$$\underline{2 NO(g) + O_2(g) \rightarrow 2 NO_2(g)}$$

Overall reaction $2 SO_2(g) + O_2(g) \rightarrow 2 SO_3(g)$

(b) $NO_2(g)$ acts as a catalyst because it is used in the first step and regenerated in the second. $NO(g)$ is a reaction intermediate because it is produced in the first step and consumed in the second.

12.112   (a)

$$NH_2NO_2(aq) + OH^-(aq) \rightarrow NHNO_2^-(aq) + H_2O(l)$$
$$\underline{NHNO_2^-(aq) \rightarrow N_2O(g) + OH^-(aq)}$$

Overall reaction $NH_2NO_2(aq) \rightarrow N_2O(g) + H_2O(l)$

(b) OH^- acts as a catalyst because it is used in the first step and regenerated in the second. $NHNO_2^-$ is a reaction intermediate because it is produced in the first step and consumed in the second.

(c) The rate will decrease because added acid decreases the concentration of OH^-, which appears in the rate law since it is a catalyst.

12.113   The reaction in Problem 12.93 involves a catalyst (NO) because NO is used in the first step and is regenerated in the second step. The reaction also involves an intermediate (NO_2) because NO_2 is produced in the first step and is used up in the second step.

Chapter Problems

12.114 $2\,AB_2 \rightarrow A_2 + 2\,B_2$
(a) Measure the change in the concentration of AB_2 as a function of time.
(b) and (c) If a plot of $[AB_2]$ versus time is linear, the reaction is zeroth-order and $k = -$slope. If a plot of $\ln[AB_2]$ versus time is linear, the reaction is first-order and $k = -$slope. If a plot of $1/[AB_2]$ versus time is linear, the reaction is second-order and $k =$ slope.

12.115 $A \rightarrow B + C$
(a) Measure the change in the concentration of A as a function of time at several different temperatures.
(b) Plot $\ln[A]$ versus time, for each temperature. Straight line graphs will result and k at each temperature equals $-$slope. Graph $\ln k$ versus $1/K$, where K is the kelvin temperature. Determine the slope of the line. $E_a = -R$(slope) where $R = 8.314 \times 10^{-3}$ kJ/(K · mol).

12.116   (a) Rate $= k[B_2][C]$
(b) $B_2 + C \rightarrow CB + B$ (slow)
      $CB + A \rightarrow AB + C$ (fast)
(c) C is a catalyst. C does not appear in the chemical equation because it is consumed in the first step and regenerated in the second step.

12.117   (a)

(b) Reaction 2 is the fastest (smallest E_a), and reaction 3 is the slowest (largest E_a).
(c) Reaction 3 is the most endothermic (positive ΔE), and reaction 1 is the most exothermic (largest negative ΔE).

12.118   The first maximum represents the potential energy of the transition state for the first step. The second maximum represents the potential energy of the transition state for the second step. The saddle point between the two maxima represents the potential energy of the intermediate products.

12.119   Because 0.060 M is half of 0.120 M, 5.2 h is the half-life.
For a first-order reaction, the half-life is independent of initial concentration. Because 0.015 M is half of 0.030 M, it will take one half-life, 5.2 h.

$$k = \frac{0.693}{t_{1/2}} = \frac{0.693}{5.2 \text{ h}} = 0.133/\text{h}$$

$$\ln\frac{[N_2O_5]_t}{[N_2O_5]_0} = -kt$$

$$t = \frac{\ln\dfrac{[N_2O_5]_t}{[N_2O_5]_0}}{-k} = \frac{\ln\left(\dfrac{0.015}{0.480}\right)}{-(0.133/\text{h})} = 26 \text{ h} \quad \text{(Note that t is five half-lives.)}$$

12.120   (a) The reaction rate will increase with an increase in temperature at constant volume.
(b) The reaction rate will decrease with an increase in volume at constant temperature because reactant concentrations will decrease.
(c) The reaction rate will increase with the addition of a catalyst.
(d) Addition of an inert gas at constant volume will not affect the reaction rate.

12.121   As the temperature of a gas is raised by 10 °C, even though the collision frequency increases by only ~2%, the reaction rate increases by 100% or more because there is an exponential increase in the fraction of the collisions that leads to products.

12.122   (a) Rate $= k[C_2H_4Br_2]^m[I^-]^n$

$$m = \frac{\ln\left(\dfrac{\text{Rate}_2}{\text{Rate}_1}\right)}{\ln\left(\dfrac{[C_2H_4Br_2]_2}{[C_2H_4Br_2]_1}\right)} = \frac{\ln\left(\dfrac{1.74 \times 10^{-4}}{6.45 \times 10^{-5}}\right)}{\ln\left(\dfrac{0.343}{0.127}\right)} = 1$$

$$n = \frac{\ln\left(\dfrac{\text{Rate}_3 \cdot [C_2H_4Br_2]_2}{\text{Rate}_2 \cdot [C_2H_4Br_2]_3}\right)}{\ln\left(\dfrac{[I^-]_3}{[I^-]_2}\right)} = \frac{\ln\left(\dfrac{(1.26 \times 10^{-4})(0.343)}{(1.74 \times 10^{-4})(0.203)}\right)}{\ln\left(\dfrac{0.125}{0.102}\right)} = 1$$

Rate $= k[C_2H_4Br_2][I^-]$

(b) From Experiment 1:

$$k = \frac{\text{Rate}}{[C_2H_4Br_2][I^-]} = \frac{6.45 \times 10^{-5} \text{ M/s}}{(0.127 \text{ M})(0.102 \text{ M})} = 4.98 \times 10^{-3}/(\text{M} \cdot \text{s})$$

(c) Rate = $k[C_2H_4Br_2][I^-] = [4.98 \times 10^{-3}(\text{M} \cdot \text{s})](0.150 \text{ M})(0.150 \text{ M}) = 1.12 \times 10^{-4}$ M/s

12.123 (a) From the data in the table for Experiment 1, we see that 0.20 mol of A reacts with 0.10 mol of B to produce 0.10 mol of D. The balanced equation for the reaction is:
2 A + B → D

(b) From the data in the table, initial Rates = $-\dfrac{\Delta A}{\Delta t}$ have been calculated.

For example, from Experiment 1:

$$\text{Initial rate} = -\frac{\Delta A}{\Delta t} = -\frac{(4.80 \text{ M} - 5.00 \text{ M})}{60 \text{ s}} = 3.33 \times 10^{-3} \text{ M/s}$$

Initial concentrations and initial rate data have been collected in the table below.

EXPT	$[A]_o$ (M)	$[B]_o$ (M)	$[C]_o$ (M)	Initial Rate (M/s)
1	5.00	2.00	1.00	3.33×10^{-3}
2	10.00	2.00	1.00	6.66×10^{-3}
3	5.00	4.00	1.00	3.33×10^{-3}
4	5.00	2.00	2.00	6.66×10^{-3}

Rate = $k[A]^m[B]^n[C]^p$
From Expts 1 and 2, [A] doubles and the initial rate doubles; therefore m = 1.
From Expts 1 and 3, [B] doubles but the initial rate does not change; therefore n = 0.
From Expts 1 and 4, [C] doubles and the initial rate doubles; therefore p = 1.
The reaction is: first-order in A; zeroth-order in B; first-order in C; second-order overall.
(c) Rate = $k[A][C]$
(d) C is a catalyst. C appears in the rate law, but it is not consumed in the reaction.
(e) A + C → AC   (slow)
    AC + B → AB + C (fast)
    A + AB → D   (fast)
(f) From data in Expt 1:

$$k = \frac{\text{Rate}}{[A][C]} = \frac{\Delta D/\Delta t}{[A][C]} = \frac{0.10 \text{ M}/60 \text{ s}}{(5.00 \text{ M})(1.00 \text{ M})} = 3.4 \times 10^{-4}/(\text{M} \cdot \text{s})$$

12.124 For E_a = 50 kJ/mol

$$f = e^{-E_a/RT} = \exp\left\{\frac{-50 \text{ kJ/mol}}{[8.314 \times 10^{-3} \text{ kJ/(K} \cdot \text{mol})](300 \text{ K})}\right\} = 2.0 \times 10^{-9}$$

For E_a = 100 kJ/mol

$$f = e^{-E_a/RT} = \exp\left\{\frac{-100 \text{ kJ/mol}}{[8.314 \times 10^{-3} \text{ kJ/(K} \cdot \text{mol})](300 \text{ K})}\right\} = 3.9 \times 10^{-18}$$

12.125 $\ln\left(\dfrac{k_2}{k_1}\right) = \left(\dfrac{-E_a}{R}\right)\left(\dfrac{1}{T_2} - \dfrac{1}{T_1}\right)$

$k_2 = 2.5k_1$

$k_1 = 1.0, \quad T_1 = 20\ °C = 293\ K$

$k_2 = 2.5, \quad T_2 = 30\ °C = 303\ K$

$E_a = -\dfrac{[\ln k_2 - \ln k_1](R)}{\left(\dfrac{1}{T_2} - \dfrac{1}{T_1}\right)}$

$E_a = -\dfrac{[\ln(2.5) - \ln(1.0)][8.314 \times 10^{-3}\ kJ/(K \cdot mol)]}{\left(\dfrac{1}{303\ K} - \dfrac{1}{293\ K}\right)} = 68\ kJ/mol$

$k_1 = 1.0, \quad T_1 = 120\ °C = 393\ K$

$k_2 = ?, \quad T_2 = 130\ °C = 403\ K$

Solve for k_2.

$\ln k_2 = \dfrac{-E_a}{R}\left(\dfrac{1}{T_2} - \dfrac{1}{T_1}\right) + \ln k_1$

$\ln k_2 = \dfrac{-68\ kJ/mol}{[8.314 \times 10^{-3}\ kJ/(K \cdot mol)]}\left(\dfrac{1}{403\ K} - \dfrac{1}{393\ K}\right) + \ln(1.0) = 0.516$

$k_2 = e^{0.516} = 1.7;$ The rate increases by a factor of 1.7.

12.126 (a) $2\ NO(g) + Br_2(g) \rightarrow 2\ NOBr(g)$

(b) Since $NOBr_2$ is generated in the first step and consumed in the second step, $NOBr_2$ is a reaction intermediate.

(c) Rate = $k[NO][Br_2]$

(d) It can't be the first step. It must be the second step.

12.127 $\text{Rate}_{forward} = k_1[NO][Br_2]$ and $\text{Rate}_{reverse} = k_{-1}[NOBr_2]$

Because of the equilibrium, $\text{Rate}_{forward} = \text{Rate}_{reverse}$, and $k_1[NO][Br_2] = k_{-1}[NOBr_2]$.

$\dfrac{k_1}{k_{-1}}[NO][Br_2] = [NOBr_2]$

The rate law for the rate determining step is Rate = $k_2[NOBr_2][NO]$. In this rate law substitute for $[NOBr_2]$.

Rate = $k_2\ \dfrac{k_1}{k_{-1}}[NO]^2[Br_2]$ and $k = \dfrac{k_2 \cdot k_1}{k_{-1}}$

12.128 $[A] = -kt + [A]_o$

$[A]_o/2 = -kt_{1/2} + [A]_o$

$[A]_o/2 - [A]_o = -kt_{1/2}$

$-[A]_o/2 = -kt_{1/2}$

$[A]_o/2 = kt_{1/2}$

For a zeroth-order reaction, $t_{1/2} = \dfrac{[A]_o}{2\,k}$

For a zeroth-order reaction, each half-life is half of the previous one.
For a first-order reaction, each half-life is the same as the previous one.
For a second-order reaction, each half-life is twice the previous one.

12.129 (a) $I^-(aq) + OCl^-(aq) \rightarrow Cl^-(aq) + OI^-(aq)$

(b) From the data in the table, initial rates $= -\dfrac{\Delta[I^-]}{\Delta t}$ have been calculated.

For example, from Experiment 1:

$$\text{Initial rate} = -\frac{\Delta[I^-]}{\Delta t} = -\frac{(2.17 \times 10^{-4}\ M - 2.40 \times 10^{-4}\ M)}{10\ s} = 2.30 \times 10^{-6}\ M/s$$

Initial concentrations and initial rate data have been collected in the table below.

EXPT	$[I^-]_o$ (M)	$[OCl^-]_o$ (M)	$[OH^-]_o$ (M)	Initial Rate (M/s)
1	2.40×10^{-4}	1.60×10^{-4}	1.00	2.30×10^{-6}
2	1.20×10^{-4}	1.60×10^{-4}	1.00	1.20×10^{-6}
3	2.40×10^{-4}	4.00×10^{-5}	1.00	6.00×10^{-7}
4	1.20×10^{-4}	1.60×10^{-4}	2.00	6.00×10^{-7}

$\text{Rate} = k[I^-]^m[OCl^-]^n[OH^-]^p$

From Expts 1 and 2, $[I^-]$ is cut in half and the initial rate is cut in half; therefore $m = 1$.
From Expts 1 and 3, $[OCl^-]$ is reduced by a factor of four and the initial rate is reduced by a factor of four; therefore $n = 1$.
From Expts 2 and 4, $[OH^-]$ is doubled and the initial rate is cut in half; therefore $p = -1$.

$\text{Rate} = k\dfrac{[I^-][OCl^-]}{[OH^-]}$

From data in Expt 1:

$$k = \frac{\text{Rate}\,[OH^-]}{[I^-][OCl^-]} = \frac{(2.30 \times 10^{-6}\ M/s)(1.00\ M)}{(2.40 \times 10^{-4}\ M)(1.60 \times 10^{-4}\ M)} = 60/s$$

(c) The reaction does not occur by a single-step mechanism because OH^- appears in the rate law but not in the overall reaction.

(d)

$\qquad\qquad OCl^-(aq) + H_2O(l) \rightleftharpoons HOCl(aq) + OH^-(aq)$ \qquad (fast)
$\qquad\qquad HOCl(aq) + I^-(aq) \rightarrow HOI(aq) + Cl^-(aq)$ \qquad (slow)
$\qquad\qquad \underline{HOI(aq) + OH^-(aq) \rightarrow H_2O(l) + OI^-(aq)}$ \qquad (fast)

Overall reaction $\quad I^-(aq) + OCl^-(aq) \rightarrow Cl^-(aq) + OI^-(aq)$

Because the forward and reverse rates in step 1 are equal, $k_1[OCl^-][H_2O] = k_{-1}[HOCl][OH^-]$. Solving for $[HOCl]$ and substituting into the rate law for the second step gives

$$\text{Rate} = k_2[HOCl][I^-] = \frac{k_1 k_2}{k_{-1}} \frac{[OCl^-][H_2O][I^-]}{[OH^-]}$$

$[H_2O]$ is constant and can be combined into k.

Because the rate law for the overall reaction is equal to the rate law for the rate-determining step, the rate law for the overall reaction is

$$\text{Rate} = k \frac{[OCl^-][I^-]}{[OH^-]} \quad \text{where } k = \frac{k_1 k_2 [H_2O]}{k_{-1}}$$

12.130 (a) $\text{Rate}_f = k_f[A]$ and $\text{Rate}_r = k_r[B]$

(b)

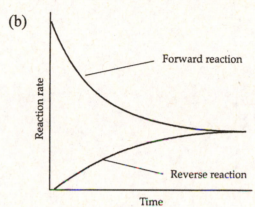

(c) When $\text{Rate}_f = \text{Rate}_r$, $k_f[A] = k_r[B]$, and $\dfrac{[B]}{[A]} = \dfrac{k_f}{k_r} = \dfrac{(3.0 \times 10^{-3})}{(1.0 \times 10^{-3})} = 3$

12.131 (a) $1 \rightarrow 1/2 \rightarrow 1/4 \rightarrow 1/8$

After three half-lives, 1/8 of the strontium-90 will remain.

(b) $k = \dfrac{0.693}{t_{1/2}} = \dfrac{0.693}{29 \text{ y}} = 0.0239/\text{y} = 0.024/\text{y}$

(c) $t = \dfrac{\ln\dfrac{(Sr-90)_t}{(Sr-90)_o}}{-k} = \dfrac{\ln\dfrac{(0.01)}{(1)}}{-0.0239/\text{y}} = 193 \text{ y}$

12.132 $k = \dfrac{0.693}{t_{1/2}} = \dfrac{0.693}{5715 \text{ y}} = 1.21 \times 10^{-4}/\text{y}$

$$t = \dfrac{\ln\dfrac{(^{14}C)_t}{(^{14}C)_o}}{-k} = \dfrac{\ln\dfrac{(2.3)}{(15.3)}}{-1.21 \times 10^{-4}/\text{y}} = 1.6 \times 10^4 \text{ y}$$

12.133 Decay rate $= kN$

$$k = \dfrac{0.693}{t_{1/2}} = \dfrac{0.693}{1.25 \times 10^9 \text{ y}} = 5.54 \times 10^{-10} \text{ y}^{-1}$$

KCl, 74.55 amu

N = number of $^{40}K^+$ ions in a 1.00 g sample of KCl

$$N = (0.000\ 117)(1.00\ \text{g})\left(\frac{1\ \text{mol KCl}}{74.55\ \text{g}}\right)\left(\frac{1\ \text{mol K}^+}{1\ \text{mol KCl}}\right)(6.022 \times 10^{23}\ \text{mol}^{-1})$$

$N = 9.45 \times 10^{17}\ ^{40}K^+$ ions

Decay rate = $kN = (5.54 \times 10^{-10}\ \text{y}^{-1})(9.45 \times 10^{17}) = 5.24 \times 10^8/\text{y}$

Disintegration/s $= (5.24 \times 10^8/\text{y})\left(\dfrac{1\ \text{y}}{365\ \text{d}}\right)\left(\dfrac{1\ \text{d}}{24\ \text{h}}\right)\left(\dfrac{1\ \text{h}}{3600\ \text{s}}\right) = 16.6/\text{s}$

12.134 $\ln\left(\dfrac{N}{N_0}\right) = (-0.693)\left(\dfrac{t}{t_{1/2}}\right)$

$\dfrac{N}{N_0} = \dfrac{\text{Decay rate at time } t}{\text{Decay rate at time } t = 0}$

$\ln\left(\dfrac{100 - 99.99}{100}\right) = (-0.693)\left(\dfrac{t}{1.53\ \text{s}}\right)$; $\quad t = 20.3\ \text{s}$

12.135 $t_{1/2} = \dfrac{0.693}{k} = \dfrac{0.693}{0.063\ \text{s}^{-1}} = 11\ \text{s}$

$\ln\left(\dfrac{N}{N_0}\right) = (-0.693)\left(\dfrac{t}{t_{1/2}}\right)$; $\qquad \dfrac{N}{N_0} = \dfrac{\text{Decay rate at time } t}{\text{Decay rate at time } t = 0}$

$\ln\left(\dfrac{100 - 99.99}{100}\right) = (-0.693)\left(\dfrac{t}{11\ \text{s}}\right)$; $\qquad t = 150\ \text{s}$

12.136 $t_{1/2} = 1.1 \times 10^{20}\ \text{y} = (1.1 \times 10^{20}\ \text{y})(365\ \text{d/y}) = 4.0 \times 10^{22}\ \text{d}$

$k = \dfrac{0.693}{t_{1/2}} = \dfrac{0.693}{4.0 \times 10^{22}\ \text{d}} = 1.7 \times 10^{-23}\ \text{d}^{-1}$ and $N = 6.02 \times 10^{23}$ atoms

Decay rate = $kN = (1.7 \times 10^{-23}\ \text{d}^{-1})(6.02 \times 10^{23}\ \text{atoms}) = 10/\text{d}$

There are 10 disintegrations per day.

12.137 Rate = $k\ [N_2O_4]$; $k = \dfrac{\text{Rate}}{[N_2O_4]}$

At 25 °C, $k_1 = \dfrac{5.0 \times 10^3\ \text{M/s}}{0.10\ \text{M}} = 5.0 \times 10^4\ \text{s}^{-1}$

At 40 °C, $k_2 = \dfrac{2.3 \times 10^4\ \text{M/s}}{0.15\ \text{M}} = 1.5 \times 10^5\ \text{s}^{-1}$

25 °C = 25 + 273 = 298 K and 40 °C = 40 + 273 = 313 K

$$\ln\left(\frac{k_2}{k_1}\right) = \left(\frac{-E_a}{R}\right)\left(\frac{1}{T_2} - \frac{1}{T_1}\right)$$

$$E_a = -\frac{[\ln k_2 - \ln k_1](R)}{\left(\frac{1}{T_2} - \frac{1}{T_1}\right)}$$

$$E_a = -\frac{[\ln(1.5 \times 10^5) - \ln(5.0 \times 10^4)][8.314 \times 10^{-3}\ kJ/(K \cdot mol)]}{\left(\frac{1}{313\ K} - \frac{1}{298\ K}\right)} = 56.8\ kJ/mol$$

12.138 $X \rightarrow$ products is a first-order reaction

$$t = 60\ min \times \frac{60\ s}{1\ min} = 3600\ s$$

$$\ln\frac{[X]_t}{[X]_o} = -kt; \qquad k = \frac{\ln\dfrac{[X]_t}{[X]_o}}{-t}$$

At 25 °C, calculate k_1: $\quad k_1 = \dfrac{\ln\left(\dfrac{0.600\ M}{1.000\ M}\right)}{-3600\ s} = 1.42 \times 10^{-4}\ s^{-1}$

At 35 °C, calculate k_2: $\quad k_2 = \dfrac{\ln\left(\dfrac{0.200\ M}{0.600\ M}\right)}{-3600\ s} = 3.05 \times 10^{-4}\ s^{-1}$

At an unknown temperature calculate k_3: $\quad k_3 = \dfrac{\ln\left(\dfrac{0.010\ M}{0.200\ M}\right)}{-3600\ s} = 8.32 \times 10^{-4}\ s^{-1}$

$T_1 = 25\ °C = 25 + 273 = 298\ K$
$T_2 = 35\ °C = 35 + 273 = 308\ K$

Calculate E_a using k_1 and k_2.

$$\ln\left(\frac{k_2}{k_1}\right) = \left(\frac{-E_a}{R}\right)\left(\frac{1}{T_2} - \frac{1}{T_1}\right)$$

$$E_a = -\frac{[\ln k_2 - \ln k_1](R)}{\left(\frac{1}{T_2} - \frac{1}{T_1}\right)}$$

$$E_a = -\frac{[\ln(3.05 \times 10^{-4}) - \ln(1.42 \times 10^{-4})][8.314 \times 10^{-3}\ kJ/(K \cdot mol)]}{\left(\frac{1}{308\ K} - \frac{1}{298\ K}\right)} = 58.3\ kJ/mol$$

Use E_a, k_1, and k_3 to calculate T_3.

$$\frac{1}{T_3} = \frac{\ln\left(\dfrac{k_3}{k_1}\right)}{\left(\dfrac{-E_a}{R}\right)} + \frac{1}{T_1} = \frac{\ln\left(\dfrac{8.32 \times 10^{-4}}{1.42 \times 10^{-4}}\right)}{\left(\dfrac{-58.3 \text{ kJ/mol}}{8.314 \times 10^{-3} \text{ kJ/(K·mol)}}\right)} + \frac{1}{298 \text{ K}} = 0.003104/K$$

$$T_3 = \frac{1}{0.003104/K} = 322 \text{ K} = 322 - 273 = 49 \text{ °C}$$

At 3:00 p.m. raise the temperature to 49 °C to finish the reaction by 4:00 p.m.

12.139

$$N_2O_4(g) \rightarrow 2\,NO_2(g)$$

	$N_2O_4(g)$	$2\,NO_2(g)$
before (mm Hg)	17.0	0
change (mm Hg)	$-x$	$+2x$
after (mm Hg)	$17.0 - x$	$2x$

$2x = 1.3$ mm Hg

$x = 1.3$ mm Hg/2 = 0.65 mm Hg

$P_t(N_2O_4) = 17.0 - x = 17.0 - 0.65 = 16.35$ mm Hg

$$k = \frac{0.693}{t_{1/2}} = \frac{0.693}{1.3 \times 10^{-5} \text{ s}} = 5.3 \times 10^4 \text{ s}^{-1}$$

$$\ln\frac{P_t}{P_o} = -kt; \quad t = \frac{\ln\dfrac{P_t}{P_o}}{-k} = \frac{\ln\dfrac{16.35}{17}}{-5.3 \times 10^4 \text{ s}^{-1}} = 7.4 \times 10^{-7} \text{ s}$$

12.140 (a) When equal volumes of two solutions are mixed, both concentrations are cut in half.

$[H_3O^+]_o = [OH^-]_o = 1.0$ M

When 99.999% of the acid is neutralized, $[H_3O^+] = [OH^-] = 1.0$ M $- (1.0$ M $\times 0.99999)$

$= 1.0 \times 10^{-5}$ M

Using the 2nd order integrated rate law:

$$kt = \frac{1}{[H_3O^+]_t} - \frac{1}{[H_3O^+]_o}; \qquad t = \frac{1}{k}\left[\frac{1}{[H_3O^+]_t} - \frac{1}{[H_3O^+]_o}\right]$$

$$t = \frac{1}{(1.3 \times 10^{11} \text{ M}^{-1}\text{s}^{-1})}\left[\frac{1}{(1.0 \times 10^{-5} \text{ M})} - \frac{1}{(1.0 \text{ M})}\right] = 7.7 \times 10^{-7} \text{ s}$$

(b) The rate of an acid-base neutralization reaction would be limited by the speed of mixing, which is much slower than the intrinsic rate of the reaction itself.

12.141 $$\frac{1}{([O_2])^2} = 8\,kt + \frac{1}{([O_2]_o)^2}$$

$$\frac{1}{([O_2])^2} = 8(25 \text{ M}^{-2}\text{s}^{-1})(100.0 \text{ s}) + \frac{1}{(0.0100 \text{ M})^2}$$

$$\frac{1}{([O_2])^2} = 30{,}000 \text{ M}^{-2}$$

$$[O_2] = \sqrt{\frac{1}{30{,}000 \text{ M}^{-2}}} = 0.005\,77 \text{ M}$$

$$
\begin{array}{lccc}
 & 2\,NO(g) & +\quad O_2(g) & \rightarrow \quad 2\,NO_2(g) \\
\text{before (M)} & 0.0200 & 0.0100 & 0 \\
\text{change (M)} & -2x & -x & +2x \\
\text{after (M)} & 0.0200 - 2x & 0.0100 - x & 2x
\end{array}
$$

$[O_2] = 0.005\,77 \text{ M} = 0.0100 \text{ M} - x$

$x = 0.0100 \text{ M} - 0.005\,77 \text{ M} = 0.004\,23 \text{ M}$

$[NO] = 0.0200 \text{ M} - 2x = 0.0200 \text{ M} - 2(0.004\,23 \text{ M}) = 0.0115 \text{ M}$

$[O_2] = 0.005\,77 \text{ M}$

$[NO_2] = 2x = 2(0.004\,23 \text{ M}) = 0.008\,46 \text{ M}$

12.142 Looking at the two experiments at 600 K, when the NO_2 concentration is doubled, the rate increased by a factor of 4. Therefore, the reaction is 2nd order.

Rate = k $[NO_2]^2$

Calculate k_1 at 600 K: $k_1 = $ Rate/$[NO_2]^2 = 5.4 \times 10^{-7}$ M s^{-1}/(0.0010 M)$^2 = 0.54$ M^{-1} s^{-1}

Calculate k_2 at 700 K: $k_2 = $ Rate/$[NO_2]^2 = 5.2 \times 10^{-6}$ M s^{-1}/(0.0020 M)$^2 = 13$ M^{-1} s^{-1}

Calculate E_a using k_1 and k_2.

$$\ln\left(\frac{k_2}{k_1}\right) = \left(\frac{-E_a}{R}\right)\left(\frac{1}{T_2} - \frac{1}{T_1}\right)$$

$$E_a = -\frac{[\ln k_2 - \ln k_1](R)}{\left(\dfrac{1}{T_2} - \dfrac{1}{T_1}\right)}$$

$$E_a = -\frac{[\ln(13) - \ln(0.54)][8.314 \times 10^{-3} \text{ kJ/(K·mol)}]}{\left(\dfrac{1}{700 \text{ K}} - \dfrac{1}{600 \text{ K}}\right)} = 111 \text{ kJ/mol}$$

Calculate k_3 at 650 K using E_a and k_1.

Solve for k_3.

$$\ln k_3 = \frac{-E_a}{R}\left(\frac{1}{T_3} - \frac{1}{T_1}\right) + \ln k_1$$

$$\ln k_3 = \frac{-111 \text{ kJ/mol}}{[8.314 \times 10^{-3} \text{ kJ/(K·mol)}]}\left(\frac{1}{650 \text{ K}} - \frac{1}{600 \text{ K}}\right) + \ln(0.54) = 1.0955$$

$k_3 = e^{1.0955} = 3.0$ M^{-1} s^{-1}

$$k_3 t = \frac{1}{[NO_2]_t} - \frac{1}{[NO_2]_o} ; \qquad t = \frac{1}{k_3}\left[\frac{1}{[NO_2]_t} - \frac{1}{[NO_2]_o}\right]$$

$$t = \frac{1}{(3.0\ M^{-1}s^{-1})}\left[\frac{1}{(0.0010\ M)} - \frac{1}{(0.0050\ M)}\right] = 2.7\ x\ 10^2\ s$$

12.143 Rate = k [A]^x[B]^y

Comparing Experiments 1 and 2, the concentration of B does not change, the concentration of A doubles, and the rate doubles. This means the reaction is first-order in A (x = 1).

Comparing Experiments 1 and 3, the rate would drop to 0.9 x 10^{-5} M/s as a result of the concentration of A being cut in half. Then with the the concentration of B doubling, the rate increases by a factor of 4, to 3.6 x 10^{-5} M/s. This means the reaction is second-order in B (y = 2).

$$\text{At 600 K, } k_1 = \frac{Rate}{[A][B]^2} = \frac{4.3\ x\ 10^{-5}\ M/s}{(0.50\ M)(0.50\ M)^2} = 3.4\ x\ 10^{-4}\ M^{-2}\ s^{-1}$$

$$\text{At 700 K, } k_2 = \frac{Rate}{[A][B]^2} = \frac{1.8\ x\ 10^{-5}\ M/s}{(0.20\ M)(0.10\ M)^2} = 9.0\ x\ 10^{-3}\ M^{-2}\ s^{-1}$$

$$\ln\left(\frac{k_2}{k_1}\right) = \left(\frac{-E_a}{R}\right)\left(\frac{1}{T_2} - \frac{1}{T_1}\right)$$

$$E_a = -\frac{[\ln k_2 - \ln k_1](R)}{\left(\dfrac{1}{T_2} - \dfrac{1}{T_1}\right)}$$

$$E_a = -\frac{[\ln(9.0\ x\ 10^{-3}) - \ln(3.4\ x\ 10^{-4})][8.314\ x\ 10^{-3}\ kJ/(K\cdot mol)]}{\left(\dfrac{1}{700\ K} - \dfrac{1}{600\ K}\right)} = 114\ kJ/mol$$

12.144 A → C is a first-order reaction.

The reaction is complete at 200 s when the absorbance of C reaches 1.200.

Because there is a one to one stoichiometry between A and C, the concentration of A must be proportional to 1.200 – absorbance of C. Any two data points can be used to find k. Let [A]$_o$ ∝ 1.200 and at 100 s, [A]$_t$ ∝ 1.200 – 1.188 = 0.012

$$\ln\frac{[A]_t}{[A]_o} = -kt; \qquad k = \frac{\ln\dfrac{[A]_t}{[A]_o}}{-t}; \qquad k = \frac{\ln\left(\dfrac{0.012\ M}{1.200\ M}\right)}{-100\ s} = 0.0461\ s^{-1}$$

$$t_{1/2} = \frac{0.693}{k} = \frac{0.693}{0.0461\ s^{-1}} = 15\ s$$

12.145 Rate = k [HI]^x

$$\frac{Rate_2}{Rate_1} = \frac{k[0.30]^x}{k[0.10]^x}$$

$$x = \frac{\log \dfrac{Rate_2}{Rate_1}}{\log \dfrac{(0.30)}{(0.10)}} = \frac{\log \dfrac{1.6 \times 10^{-4}}{1.8 \times 10^{-5}}}{\log \dfrac{(0.30)}{(0.10)}} = \frac{0.949}{0.477} = 2$$

Rate = k [HI]^2

At 700 K, $k_1 = \dfrac{Rate}{[HI]^2} = \dfrac{1.8 \times 10^{-5} \text{ M/s}}{(0.10 \text{ M})^2} = 1.8 \times 10^{-3} \text{ M}^{-1} \text{ s}^{-1}$

At 800 K, $k_2 = \dfrac{Rate}{[HI]^2} = \dfrac{3.9 \times 10^{-3} \text{ M/s}}{(0.20 \text{ M})^2} = 9.7 \times 10^{-2} \text{ M}^{-1} \text{ s}^{-1}$

$$\ln\left(\frac{k_2}{k_1}\right) = \left(\frac{-E_a}{R}\right)\left(\frac{1}{T_2} - \frac{1}{T_1}\right)$$

$$E_a = - \frac{[\ln k_2 - \ln k_1](R)}{\left(\dfrac{1}{T_2} - \dfrac{1}{T_1}\right)}$$

$$E_a = - \frac{[\ln(9.7 \times 10^{-2}) - \ln(1.8 \times 10^{-3})][8.314 \times 10^{-3} \text{ kJ/(K} \cdot \text{mol)}]}{\left(\dfrac{1}{800 \text{ K}} - \dfrac{1}{700 \text{ K}}\right)} = 186 \text{ kJ/mol}$$

Calculate k_4 at 650 K using E_a and k_1.
Solve for k_4.

$$\ln k_4 = \frac{-E_a}{R}\left(\frac{1}{T_4} - \frac{1}{T_1}\right) + \ln k_1$$

$$\ln k_4 = \frac{-186 \text{ kJ/mol}}{[8.314 \times 10^{-3} \text{ kJ/(K} \cdot \text{mol)}]}\left(\frac{1}{650 \text{ K}} - \frac{1}{700 \text{ K}}\right) + \ln(1.8 \times 10^{-3}) = -8.778$$

$k_4 = e^{-8.778} = 1.5 \times 10^{-4} \text{ M}^{-1} \text{ s}^{-1}$

$$[HI] = \sqrt{\frac{Rate}{k_4}} = \sqrt{\frac{1.0 \times 10^{-5} \text{ M/s}}{1.5 \times 10^{-4} \text{ M}^{-1}\text{s}^{-1}}} = 0.26 \text{ M}$$

12.146 For radioactive decay, $\ln \dfrac{N}{N_o} = -kt$

For ^{235}U, $k_1 = \dfrac{0.693}{t_{1/2}} = \dfrac{0.693}{7.04 \times 10^8 \text{ y}} = 9.84 \times 10^{-10} \text{ y}^{-1}$

For ^{238}U, $k_2 = \dfrac{0.693}{t_{1/2}} = \dfrac{0.693}{4.47 \times 10^9 \text{ y}} = 1.55 \times 10^{-10} \text{ y}^{-1}$

For ^{235}U, $\ln \dfrac{N_1}{N_{o1}} = -k_1 t$ and $\ln \dfrac{N_1}{N_{o1}} + k_1 t = 0$

For ^{238}U, $\ln \dfrac{N_2}{N_{o2}} = -k_2 t$ and $\ln \dfrac{N_2}{N_{o2}} + k_2 t = 0$

Set the two equations that are equal to zero equal to each other and solve for t.

$$\ln \dfrac{N_1}{N_{o1}} + k_1 t = \ln \dfrac{N_2}{N_{o2}} + k_2 t$$

$$\ln \dfrac{N_1}{N_{o1}} - \ln \dfrac{N_2}{N_{o2}} = k_2 t - k_1 t = (k_2 - k_1)t$$

$$\ln \dfrac{\left(\dfrac{N_1}{N_{o1}}\right)}{\left(\dfrac{N_2}{N_{o2}}\right)} = (k_2 - k_1)t, \text{ now } N_{o1} = N_{o2}, \text{ so } \ln \dfrac{N_1}{N_2} = (k_2 - k_1)t$$

$$\dfrac{N_1}{N_2} = 7.25 \times 10^{-3}, \text{ so } \ln(7.25 \times 10^{-3}) = (1.55 \times 10^{-10} \text{ y}^{-1} - 9.84 \times 10^{-10} \text{ y}^{-1})t$$

$$t = \dfrac{-4.93}{-8.29 \times 10^{-10} \text{ y}^{-1}} = 5.9 \times 10^9 \text{ y}$$

The age of the elements is 5.9×10^9 y (6 billion years).

Multiconcept Problems

12.147 (a) $k = A e^{-\frac{E_a}{RT}} = (6.0 \times 10^8/(M \cdot s)) \, e^{-\frac{6.3 \text{ kJ/mol}}{[8.314 \times 10^{-3} \text{ kJ/(K} \cdot \text{mol)}](298 \text{ K})}} = 4.7 \times 10^7/(M \cdot s)$

(b) $:\!\ddot{O}\!=\!\ddot{N}\!-\!\ddot{\underset{\cdot\cdot}{F}}\!:$ N has 3 electron clouds, is sp^2 hybridized, and the molecule is bent.

(c) O—N---F---F

(d) The reaction has such a low activation energy because the F–F bond is very weak and the N–F bond is relatively strong.

12.148 $2 \, HI(g) \; \rightarrow \; H_2(g) \; + \; I_2(g)$

(a) mass HI $= 1.50 \text{ L} \times \dfrac{1000 \text{ mL}}{1 \text{ L}} \times \dfrac{0.0101 \text{ g}}{1 \text{ mL}} = 15.15 \text{ g HI}$

$15.15 \text{ g HI} \times \dfrac{1 \text{ mol HI}}{127.91 \text{ g HI}} = 0.118 \text{ mol HI}$

$[HI] = \dfrac{0.118 \text{ mol}}{1.50 \text{ L}} = 0.0787 \text{ mol/L}$

$-\dfrac{\Delta [HI]}{\Delta t} = k[HI]^2 = (0.031/(M \cdot \text{min}))(0.0787 \text{ M})^2 = 1.92 \times 10^{-4} \text{ M/min}$

$2 \, HI(g) \; \rightarrow \; H_2(g) \; + \; I_2(g)$

326

$$\frac{\Delta [I_2]}{\Delta t} = \frac{1}{2}\left(-\frac{\Delta [HI]}{\Delta t}\right) = \frac{1.92 \times 10^{-4} \text{ M/min}}{2} = 9.60 \times 10^{-5} \text{ M/min}$$

$(9.60 \times 10^{-5} \text{ M/min})(1.50 \text{ L})(6.022 \times 10^{23} \text{ molecules/mol}) = 8.7 \times 10^{19} \text{ molecules/min}$

(b) Rate $= k[HI]^2$

$$\frac{1}{[HI]_t} = kt + \frac{1}{[HI]_o} = (0.031/(M \cdot min))\left(8.00 \text{ h} \times \frac{60.0 \text{ min}}{1 \text{ h}}\right) + \frac{1}{0.0787 \text{ M}} = 27.59/M$$

$$[HI]_t = \frac{1}{27.59/M} = 0.0362 \text{ M}$$

From stoichiometry, $[H_2]_t = 1/2 ([HI]_o - [HI]_t) = 1/2 (0.0787 \text{ M} - 0.0362 \text{ M}) = 0.0212 \text{ M}$
410 °C = 683 K
PV = nRT

$$P_{H_2} = \left(\frac{n}{V}\right)RT = (0.0212 \text{ mol/L})\left(0.082\ 06 \frac{L \cdot atm}{K \cdot mol}\right)(683 \text{ K}) = 1.2 \text{ atm}$$

12.149 $2 NO_2(g) \rightarrow 2 NO(g) + O_2(g)$
$k = 4.7/(M \cdot s)$
(a) The units for k indicate a second-order reaction.
(b) 383 °C = 656 K
PV = nRT

$$[NO_2]_o = \frac{n}{V} = \frac{P}{RT} = \frac{\left(746 \text{ mm Hg} \times \dfrac{1.000 \text{ atm}}{760 \text{ mm Hg}}\right)}{\left(0.082\ 06 \dfrac{L \cdot atm}{K \cdot mol}\right)(656 \text{ K})} = 0.01823 \text{ mol/L}$$

initial rate $= k[NO_2]_o^2 = [4.7/(M \cdot s)](0.01823 \text{ mol/L})^2 = 1.56 \times 10^{-3} \text{ mol/(L} \cdot \text{s)}$

initial rate for $O_2 = \dfrac{\text{initial rate for NO}_2}{2} = \dfrac{1.56 \times 10^{-3} \text{ mol/(L} \cdot \text{s)}}{2} = 7.80 \times 10^{-4} \text{ mol/(L} \cdot \text{s)}$

initial rate for $O_2 = [7.80 \times 10^{-4} \text{ mol/(L} \cdot \text{s)}](32.00 \text{ g/mol}) = 0.025 \text{ g/(L} \cdot \text{s)}$

(c) $\dfrac{1}{[NO_2]_t} = kt + \dfrac{1}{[NO_2]_0} = [4.7/(M \cdot s)](60 \text{ s}) + \dfrac{1}{0.01823 \text{ M}}$

$\dfrac{1}{[NO_2]_t} = 336.9/M$ and $[NO_2] = 0.00297 \text{ M}$

	$2 NO_2(g)$	$\rightarrow$	$2 NO(g)$	+	$O_2(g)$
before reaction (M)	0.01823		0		0
change (M)	−2x		+2x		+x
after 1.00 min (M)	0.01823 − 2x		2x		x

after 1.00 min $[NO_2] = 0.00297 \text{ M} = 0.01823 - 2x$
$x = 0.00763 \text{ M} = [O_2]$
mass $O_2 = (0.00763 \text{ mol/L})(5.00 \text{ L})(32.00 \text{ g/mol}) = 1.22 \text{ g } O_2$

12.150 (a) N_2O_5, 108.01 amu

$$[N_2O_5]_o = \frac{\left(2.70 \text{ g } N_2O_5 \times \dfrac{1 \text{ mol } N_2O_5}{108.01 \text{ g } N_2O_5}\right)}{2.00 \text{ L}} = 0.0125 \text{ mol/L}$$

$\ln [N_2O_5]_t = -kt + \ln [N_2O_5]_o = -(1.7 \times 10^{-3} \text{ s}^{-1})\left(13.0 \text{ min} \times \dfrac{60.0 \text{ s}}{1 \text{ min}}\right) + \ln (0.0125) = -5.71$

$[N_2O_5]_t = e^{-5.71} = 3.31 \times 10^{-3} \text{ mol/L}$

After 13.0 min, mol $N_2O_5 = (3.31 \times 10^{-3} \text{ mol/L})(2.00 \text{ L}) = 6.62 \times 10^{-3}$ mol N_2O_5

	$N_2O_5(g)$	$\rightarrow$	$2 NO_2(g)$	$+$	$1/2 O_2(g)$
before reaction (mol)	0.0250		0		0
change (mol)	$-x$		$+2x$		$+1/2x$
after reaction (mol)	$0.0250 - x$		$2x$		$1/2x$

After 13.0 min, mol $N_2O_5 = 6.62 \times 10^{-3} = 0.0250 - x$
$x = 0.0184$ mol

After 13.0 min, $n_{total} = n_{N_2O_5} + n_{NO_2} + n_{O_2} = (6.62 \times 10^{-3}) + 2(0.0184) + 1/2(0.0184)$

$n_{total} = 0.0526$ mol
$55\ ^\circ C = 328$ K

$$PV = nRT; \quad P_{total} = \frac{nRT}{V} = \frac{(0.0526 \text{ mol})\left(0.082\ 06 \dfrac{\text{L} \cdot \text{atm}}{\text{K} \cdot \text{mol}}\right)(328 \text{ K})}{2.00 \text{ L}} = 0.71 \text{ atm}$$

(b) $N_2O_5(g) \rightarrow 2 NO_2(g) + 1/2 O_2(g)$
$\Delta H^\circ_{rxn} = 2 \Delta H^\circ_f(NO_2) - \Delta H^\circ_f(N_2O_5)$
$\Delta H^\circ_{rxn} = (2 \text{ mol})(33.2 \text{ kJ/mol}) - (1 \text{ mol})(11 \text{ kJ/mol}) = 55.4 \text{ kJ} = 5.54 \times 10^4 \text{ J}$
initial rate $= k[N_2O_5]_o = (1.7 \times 10^{-3} \text{ s}^{-1})(0.0125 \text{ mol/L}) = 2.125 \times 10^{-5} \text{ mol/(L} \cdot \text{s)}$
initial rate absorbing heat $= [2.125 \times 10^{-5} \text{ mol/(L} \cdot \text{s)}](2.00 \text{ L})(5.54 \times 10^4 \text{ J/mol}) = 2.4 \text{ J/s}$

(c)

$\ln [N_2O_5]_t = -kt + \ln [N_2O_5]_o = -(1.7 \times 10^{-3} \text{ s}^{-1})\left(10.0 \text{ min} \times \dfrac{60.0 \text{ s}}{1 \text{ min}}\right) + \ln (0.0125) = -5.40$

$[N_2O_5]_t = e^{-5.40} = 4.52 \times 10^{-3} \text{ mol/L}$
After 10.0 min, mol $N_2O_5 = (4.52 \times 10^{-3} \text{ mol/L})(2.00 \text{ L}) = 9.03 \times 10^{-3}$ mol N_2O_5

	$N_2O_5(g)$	$\rightarrow$	$2 NO_2(g)$	$+$	$1/2 O_2(g)$
before reaction (mol)	0.0250		0		0
change (mol)	$-x$		$+2x$		$+1/2x$
after reaction (mol)	$0.0250 - x$		$2x$		$1/2x$

After 10.0 min, mol $N_2O_5 = 9.03 \times 10^{-3} = 0.0250 - x$
$x = 0.0160$ mol
heat absorbed $= (0.0160 \text{ mol})(55.4 \text{ kJ/mol}) = 0.89 \text{ kJ}$

12.151 $2 N_2O(g) \rightarrow 2 N_2(g) + O_2(g)$

P_{O_2} (in exit gas) = 1.0 mm Hg; P_{total} = 1.50 atm = 1140 mm Hg

From the reaction stoichiometry:

P_{N_2} (in exit gas) = 2 P_{O_2} = 2.0 mm Hg

P_{N_2O} (in exit gas) = $P_{total} - P_{N_2} - P_{O_2}$ = 1140 − 2.0 − 1.0 = 1137 mm Hg

Assume P_{N_2O} (initial) = P_{total} = 1140 mm Hg   (In assuming a constant total pressure in

the   tube, we are neglecting the slight change in pressure due to the reaction.)

Volume of tube = $\pi r^2 l$ = $\pi (1.25 \text{ cm})^2 (20 \text{ cm})$ = 98.2 cm³ = 0.0982 L

Time, t, gases are in the tube = $\dfrac{\text{volume of tube}}{\text{flow rate}}$ x $\dfrac{0.0982 \text{ L}}{0.75 \text{ L/min}}$ x $\dfrac{60 \text{ s}}{1 \text{ min}}$ = 7.86 s

At time t, $\dfrac{[N_2O]_t}{[N_2O]_0} = \dfrac{P_{N_2O} \text{ (in exit gas)}}{P_{N_2O} \text{ (initial)}} = \dfrac{1137 \text{ mm Hg}}{1140 \text{ mm Hg}}$ = 0.997 37

Because $k = A e^{-\frac{E_a}{RT}}$ and A = 4.2 x 10⁹ s⁻¹, k has units of s⁻¹. Therefore, this is a first-

order reaction and the appropriate integrated rate law is $\ln \dfrac{[N_2O]_t}{[N_2O]_0} = -kt$.

$$k = \frac{-\ln \dfrac{[N_2O]_t}{[N_2O]_0}}{t} = \frac{-\ln(0.99737)}{7.86 \text{ s}} = 3.35 \times 10^{-4} \text{ s}^{-1}$$

From the Arrhenius equation, $\ln k = \ln A - \dfrac{E_a}{RT}$

$$T = \frac{E_a}{(R)[\ln A - \ln k]} = \frac{222 \text{ kJ/mol}}{(8.314 \times 10^{-3} \text{ kJ/(K} \cdot \text{mol)})[(22.16) - (-8.00)]} = 885 \text{ K}$$

12.152 H_2O_2, 34.01 amu

mass H_2O_2 = (0.500 L)(1000 mL/1 L)(1.00 g/ 1 mL)(0.0300) = 15.0 g H_2O_2

mol H_2O_2 = 15.0 g H_2O_2 x $\dfrac{1 \text{ mol } H_2O_2}{34.01 \text{ g } H_2O_2}$ = 0.441 H_2O_2

$[H_2O_2]_o$ = $\dfrac{0.441 \text{ mol}}{0.500 \text{ L}}$ = 0.882 mol /L

$k = \dfrac{0.693}{t_{1/2}} = \dfrac{0.693}{10.7 \text{ h}}$ = 6.48 x 10⁻²/h

$\ln [H_2O_2]_t = -kt + \ln [H_2O_2]_o$

$\ln [H_2O_2]_t = -(6.48 \times 10^{-2}/\text{h})(4.02 \text{ h}) + \ln (0.882)$

$\ln [H_2O_2]_t = -0.386$; $[H_2O_2]_t = e^{-0.386}$ = 0.680 mol/L

mol H_2O_2 = (0.680 mol/L)(0.500 L) = 0.340 mol

$$2 H_2O_2(aq) \rightarrow 2 H_2O(l) + O_2(g)$$

before reaction (mol)	0.441	0	0
change (mol)	$-2x$	$+2x$	$+x$
after reaction (mol)	$0.441 - 2x$	$2x$	x

After 4.02 h, mol H_2O_2 = 0.340 mol = 0.441 − 2x; solve for x.

$2x = 0.101$

$x = 0.0505$ mol = mol O_2

$P = 738$ mm Hg $\times \dfrac{1.00 \text{ atm}}{760 \text{ mm Hg}} = 0.971$ atm

$PV = nRT$

$$V = \frac{nRT}{P} = \frac{(0.0505 \text{ mol})\left(0.082\ 06\ \dfrac{L \cdot atm}{K \cdot mol}\right)(293 \text{ K})}{0.971 \text{ atm}} = 1.25 \text{ L}$$

$P\Delta V = (0.971 \text{ atm})(1.25 \text{ L}) = 1.21 \text{ L} \cdot \text{atm}$

$w = -P\Delta V = -1.21 \text{ L·atm} = (-1.21 \text{ L} \cdot \text{atm})\left(101 \dfrac{J}{L \cdot atm}\right) = -122 \text{ J}$

12.153 (a)

$$CH_3CHO(g) \rightarrow CH_4(g) + CO(g)$$

before (atm)	0.500	0	0
change (atm)	$-x$	$+x$	$+x$
after (atm)	$0.500 - x$	x	x

At 605 s, $P_{total} = P_{CH_3CHO} + P_{CH_4} + P_{CO} = (0.500 \text{ atm} - x) + x + x = 0.808$ atm

$x = 0.808$ atm $- 0.500$ atm $= 0.308$ atm

The integrated rate law for a second-order reaction in terms of molar concentrations is

$\dfrac{1}{[A]_t} = kt + \dfrac{1}{[A]_o}$. The ideal gas law, PV = nRT, can be rearranged to show how P is

proportional to the molar concentration of a gas.

$P = \dfrac{n}{V} RT$ (R and T are constant), so $P \propto \dfrac{n}{V}$ = molar concentration

Because of this relationship, the second-order integrated rate law can be rewritten in terms of partial pressures.

$$\frac{1}{P_t} = kt + \frac{1}{P_o}; \qquad \frac{1}{P_t} - \frac{1}{P_o} = kt; \qquad \frac{\left(\dfrac{1}{P_t} - \dfrac{1}{P_o}\right)}{t} = k$$

P is the partial pressure of CH_3CHO.

At t = 0, P_o = 0.500 and at t = 605 s, P_t = 0.500 atm − 0.308 atm = 0.192 atm

$$k = \frac{\left(\dfrac{1}{0.192 \text{ atm}} - \dfrac{1}{0.500 \text{ atm}}\right)}{605 \text{ s}} = 5.30 \times 10^{-3} \text{ atm}^{-1} \text{ s}^{-1}$$

(b) Use the ideal gas law to convert atm^{-1} to M^{-1}.

$$P = \frac{n}{V}RT; \quad \frac{P}{RT} = \frac{n}{V}; \quad \frac{1}{P}RT = \frac{V}{n} = M^{-1}$$

So, multiply k by RT to convert atm^{-1} s^{-1} to M^{-1} s^{-1}.

$k = (5.30 \times 10^{-3} \text{ atm}^{-1} \text{ s}^{-1})RT$

$$k = (5.30 \times 10^{-3} \text{ atm}^{-1} \text{ s}^{-1})\left(0.082\ 06\ \frac{\text{L} \cdot \text{atm}}{\text{K} \cdot \text{mol}}\right)(791 \text{ K}) = 0.344\ \frac{\text{L}}{\text{mol} \cdot \text{s}} = 0.344 \text{ M}^{-1} \text{ s}^{-1}$$

(c) $CH_3CHO(g) \rightarrow CH_4(g) + CO(g)$

$\Delta H^{\circ}_{rxn} = [\Delta H^{\circ}_f(CH_4) + \Delta H^{\circ}_f(CO)] - \Delta H^{\circ}_f(CH_3CHO)]$

$\Delta H^{\circ}_{rxn} = [(1 \text{ mol})(-74.8 \text{ kJ/mol}) + (1 \text{ mol})(-110.5 \text{ kJ/mol})] - (1 \text{ mol})(-166.2 \text{ kJ/mol})$

$\Delta H^{\circ}_{rxn} = -19.1$ kJ per mole of CH_3CHO that decomposes

$PV = nRT$

$$\text{mol } CH_3CHO \text{ reacted} = \frac{PV}{RT} = \frac{(0.308 \text{ atm})(1.00 \text{ L})}{\left(0.082\ 06\ \dfrac{\text{L} \cdot \text{atm}}{\text{K} \cdot \text{mol}}\right)(791 \text{ K})} = 0.004\ 74 \text{ mol}$$

$q = (0.004\ 74 \text{ mol})(19.1 \text{ kJ/mol})(1000 \text{ J/kJ}) = 90.6$ J liberated after a reaction time of 605 s.

13

Chemical Equilibrium

13.1 (a) $K_c = \dfrac{[SO_3]^2}{[SO_2]^2[O_2]}$ (b) $K_c = \dfrac{[SO_2]^2[O_2]}{[SO_3]^2}$

13.2 (a) $K_c = \dfrac{[SO_3]^2}{[SO_2]^2[O_2]} = \dfrac{(5.0 \times 10^{-2})^2}{(3.0 \times 10^{-3})^2(3.5 \times 10^{-3})} = 7.9 \times 10^4$

(b) $K_c = \dfrac{[SO_2]^2[O_2]}{[SO_3]^2} = \dfrac{(3.0 \times 10^{-3})^2(3.5 \times 10^{-3})}{(5.0 \times 10^{-2})^2} = 1.3 \times 10^{-5}$

13.3 (a) $K_c = \dfrac{[H^+][C_3H_5O_3^-]}{[C_3H_6O_3]}$

(b) $K_c = \dfrac{[(0.100)(0.0365)]^2}{[0.100 - (0.100)(0.0365)]} = 1.38 \times 10^{-4}$

13.4 From (1), $K_c = \dfrac{[AB][B]}{[A][B_2]} = \dfrac{(1)(2)}{(1)(2)} = 1$

For a mixture to be at equilibrium, $\dfrac{[AB][B]}{[A][B_2]}$ must be equal to 1.

For (2), $\dfrac{[AB][B]}{[A][B_2]} = \dfrac{(2)(1)}{(2)(1)} = 1$. This mixture is at equilibrium.

For (3), $\dfrac{[AB][B]}{[A][B_2]} = \dfrac{(1)(1)}{(4)(2)} = 0.125$. This mixture is not at equilibrium.

For (4), $\dfrac{[AB][B]}{[A][B_2]} = \dfrac{(2)(2)}{(4)(1)} = 1$. This mixture is at equilibrium.

13.5 $K_p = \dfrac{(P_{CO_2})(P_{H_2})}{(P_{CO})(P_{H_2O})} = \dfrac{(6.12)(20.3)}{(1.31)(10.0)} = 9.48$

13.6 $2\,NO(g) + O_2 \rightleftharpoons 2\,NO_2(g);\ \Delta n = 2 - 3 = -1$
$K_p = K_c(RT)^{\Delta n}, \qquad K_c = K_p(1/RT)^{\Delta n}$
at 500 K: $\quad K_p = (6.9 \times 10^5)[(0.082\ 06)(500)]^{-1} = 1.7 \times 10^4$

at 1000 K: $\quad K_c = (1.3 \times 10^{-2})\left(\dfrac{1}{(0.082\ 06)(1000)}\right)^{-1} = 1.1$

13.7 (a) $K_c = \dfrac{[H_2]^3}{[H_2O]^3}$, $K_p = \dfrac{(P_{H_2})^3}{(P_{H_2O})^3}$, $\Delta n = (3) - (3) = 0$ and $K_p = K_c$

(b) $K_c = [H_2]^2[O_2]$, $K_p = (P_{H_2})^2(P_{O_2})$, $\Delta n = (3) - (0) = 3$ and $K_p = K_c(RT)^3$

(c) $K_c = \dfrac{[HCl]^4}{[SiCl_4][H_2]^2}$, $K_p = \dfrac{(P_{HCl})^4}{(P_{SiCl_4})(P_{H_2})^2}$, $\Delta n = (4) - (3) = 1$ and $K_p = K_c(RT)$

(d) $K_c = \dfrac{1}{[Hg_2^{2+}][Cl^-]^2}$

13.8 $K_c = 1.2 \times 10^{-42}$. Because K_c is very small, the equilibrium mixture contains mostly H_2 molecules. H is in periodic group 1A. A very small value of K_c is consistent with strong bonding between 2 H atoms, each with one valence electron.

13.9 The container volume of 5.0 L must be included to calculate molar concentrations.

(a) $Q_c = \dfrac{[NO_2]_t^2}{[NO]_t^2[O_2]_t} = \dfrac{(0.80 \text{ mol}/5.0 \text{ L})^2}{(0.060 \text{ mol}/5.0 \text{ L})^2(1.0 \text{ mol}/5.0 \text{ L})} = 890$

Because $Q_c < K_c$, the reaction is not at equilibrium. The reaction will proceed to the right to reach equilibrium.

(b) $Q_c = \dfrac{[NO_2]_t^2}{[NO]_t^2[O_2]_t} = \dfrac{(4.0 \text{ mol}/5.0 \text{ L})^2}{(5.0 \times 10^{-3} \text{ mol}/5.0 \text{ L})^2(0.20 \text{ mol}/5.0 \text{ L})} = 1.6 \times 10^7$

Because $Q_c > K_c$, the reaction is not at equilibrium. The reaction will proceed to the left to reach equilibrium.

13.10 $K_c = \dfrac{[AB]^2}{[A_2][B_2]} = 4$; For a mixture to be at equilibrium, $\dfrac{[AB]^2}{[A_2][B_2]}$ must be equal to 4.

For (1), $Q_c = \dfrac{[AB]^2}{[A_2][B_2]} = \dfrac{(6)^2}{(1)(1)} = 36$, $Q_c > K_c$

For (2), $Q_c = \dfrac{[AB]^2}{[A_2][B_2]} = \dfrac{(4)^2}{(2)(2)} = 4$, $Q_c = K_c$

For (3), $Q_c = \dfrac{[AB]^2}{[A_2][B_2]} = \dfrac{(2)^2}{(3)(3)} = 0.44$, $Q_c < K_c$

(a) (2)   (b) (1), reverse; (3), forward

13.11 $K_c = \dfrac{[H]^2}{[H_2]} = 1.2 \times 10^{-42}$

(a) $[H] = \sqrt{K_c[H_2]} = \sqrt{(1.2 \times 10^{-42})(0.10)} = 3.5 \times 10^{-22}$ M

(b) H atoms $= (3.5 \times 10^{-22} \text{ mol/L})(1.0 \text{ L})(6.022 \times 10^{23} \text{ atoms/mol}) = 210$ H atoms

H_2 molecules $= (0.10 \text{ mol/L})(1.0 \text{ L})(6.022 \times 10^{23} \text{ molecules/mol}) = 6.0 \times 10^{22}$ H_2 molecules

13.12

	CO(g) +	H$_2$O(g) $\rightleftharpoons$	CO$_2$(g) +	H$_2$(g)
initial (M)	0.150	0.150	0	0
change (M)	$-x$	$-x$	$+x$	$+x$
equil (M)	0.150 $-$ x	0.150 $-$ x	x	x

$$K_c = 4.24 = \frac{[CO_2][H_2]}{[CO][H_2O]} = \frac{x^2}{(0.150 - x)^2}$$

Take the square root of both sides and solve for x.

$$\sqrt{4.24} = \sqrt{\frac{x^2}{(0.150 - x)^2}}; \quad 2.06 = \frac{x}{0.150 - x}; \quad x = 0.101$$

At equilibrium, $[CO_2] = [H_2] = x = 0.101$ M
$[CO] = [H_2O] = 0.150 - x = 0.150 - 0.101 = 0.049$ M

13.13

	N$_2$O$_4$(g) $\rightleftharpoons$	2 NO$_2$(g)
initial (M)	0.0500	0
change (M)	$-x$	$+2x$
equil (M)	0.0500 $-$ x	2x

$$K_c = 4.64 \times 10^{-3} = \frac{[NO_2]^2}{[N_2O_4]} = \frac{(2x)^2}{(0.0500 - x)}$$

$4x^2 + (4.64 \times 10^{-3})x - (2.32 \times 10^{-4}) = 0$

Use the quadratic formula to solve for x.

$$x = \frac{-(4.64 \times 10^{-3}) \pm \sqrt{(4.64 \times 10^{-3})^2 - 4(4)(-2.32 \times 10^{-4})}}{2(4)} = \frac{-0.00464 \pm 0.06110}{8}$$

x = −0.008 22 and 0.007 06

Discard the negative solution (−0.008 22) because it leads to a negative concentration of NO$_2$ and that is impossible.
$[N_2O_4] = 0.0500 - x = 0.0500 - 0.007\,06 = 0.0429$ M
$[NO_2] = 2x = 2(0.007\,06) = 0.0141$ M

13.14 N$_2$O$_4$(g) $\rightleftharpoons$ 2 NO$_2$(g)

$$Q_c = \frac{[NO_2]_t^2}{[N_2O_4]_t} = \frac{(0.0300 \text{ mol/L})^2}{(0.0200 \text{ mol/L})} = 0.0450; \quad Q_c > K_c$$

The reaction will approach equilibrium by going from right to left.

	N$_2$O$_4$(g) $\rightleftharpoons$	2 NO$_2$(g)
initial (M)	0.0200	0.0300
change (M)	$+x$	$-2x$
equil (M)	0.0200 $+$ x	0.0300 $-$ 2x

$$K_c = 4.64 \times 10^{-3} = \frac{[NO_2]^2}{[N_2O_4]} = \frac{(0.0300 - 2x)^2}{(0.0200 + x)}$$

$4x^2 - 0.1246x + (8.072 \times 10^{-4}) = 0$
Use the quadratic formula to solve for x.

$$x = \frac{-(-0.1246) \pm \sqrt{(-0.1246)^2 - 4(4)(8.072 \times 10^{-4})}}{2(4)} = \frac{0.1246 \pm 0.05109}{8}$$

x = 0.0220 and 0.009 19

Discard the larger solution (0.0220) because it leads to a negative concentration of NO_2, and that is impossible.

$[N_2O_4]$ = 0.0200 + x = 0.0200 + 0.009 19 = 0.0292 M

$[NO_2]$ = 0.0300 – 2x = 0.0300 – 2(0.009 19) = 0.0116 M

13.15 $K_p = \dfrac{(P_{CO})(P_{H_2})}{(P_{H_2O})} = 2.44$, $Q_p = \dfrac{(1.00)(1.40)}{(1.20)} = 1.17$, $Q_p < K_p$ and the reaction goes to

the right to reach equilibrium.

	C(s)	+	H_2O(g)	⇌	CO(g)	+	H_2(g)
initial (atm)			1.20		1.00		1.40
change (atm)			–x		+x		+x
equil (atm)			1.20 – x		1.00 + x		1.40 + x

$$K_p = \frac{(P_{CO})(P_{H_2})}{(P_{H_2O})} = 2.44 = \frac{(1.00 + x)(1.40 + x)}{(1.20 - x)}$$

$x^2 + 4.84x - 1.53 = 0$

Use the quadratic formula to solve for x.

$$x = \frac{-(4.84) \pm \sqrt{(4.84)^2 - 4(1)(-1.53)}}{2(1)} = \frac{-4.84 \pm 5.44}{2}$$

x = –5.14 and 0.300

Discard the negative solution (–5.14) because it leads to negative partial pressures and that is impossible.

P_{H_2O} = 1.20 – x = 1.20 – 0.300 = 0.90 atm

P_{CO} = 1.00 + x = 1.00 + 0.300 = 1.30 atm

P_{H_2} = 1.40 + x = 1.40 + 0.300 = 1.70 atm

13.16   (a)   CO (reactant) added, H_2 concentration increases.

        (b) CO_2 (product) added, H_2 concentration decreases.

        (c) H_2O (reactant) removed, H_2 concentration decreases.

        (d) CO_2 (product) removed, H_2 concentration increases.

        At equilibrium, $Q_c = K_c = \dfrac{[CO_2][H_2]}{[CO][H_2O]}$.   If some CO_2 is removed from the

equilibrium mixture, the numerator in Q_c is decreased, which means that $Q_c < K_c$ and the reaction will shift to the right, increasing the H_2 concentration.

13.17   (a)   Because there are 2 mol of gas on both sides of the balanced equation, the composition of the equilibrium mixture is unaffected by a change in pressure. The number of moles of reaction products remains the same.

(b) Because there are 2 mol of gas on the left side and 1 mol of gas on the right side of the balanced equation, the stress of an increase in pressure is relieved by a shift in the reaction to the side with fewer moles of gas (in this case, to products). The number of moles of reaction products increases.

(c) Because there is 1 mol of gas on the left side and 2 mol of gas on the right side of the balanced equation, the stress of an increase in pressure is relieved by a shift in the reaction to the side with fewer moles of gas (in this case, to reactants). The number of moles of reaction product decreases.

13.18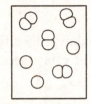

13.19   Le Châtelier's principle predicts that a stress of added heat will be relieved by net reaction in the direction that absorbs the heat. Since the reaction is endothermic, the equilibrium will shift from left to right (K_c will increase) with an increase in temperature. Therefore, the equilibrium mixture will contain more of the offending NO, the higher the temperature.

13.20   The reaction is exothermic. As the temperature is increased the reaction shifts from right to left. The amount of ethyl acetate decreases.

$$K_c = \frac{[CH_3CO_2C_2H_5][H_2O]}{[CH_3CO_2H][C_2H_5OH]}$$

As the temperature is decreased, the reaction shifts from left to right. The product concentrations increase, and the reactant concentrations decrease. This corresponds to an increase in K_c.

13.21   There are more AB(g) molecules at the higher temperature. The equilibrium shifted to the right at the higher temperature, which means the reaction is endothermic.

13.22   (a) A catalyst does not affect the equilibrium composition. The amount of CO remains the same.
(b) The reaction is exothermic. An increase in temperature shifts the reaction toward reactants. The amount of CO increases.
(c) Because there are 3 mol of gas on the left side and 2 mol of gas on the right side of the balanced equation, the stress of an increase in pressure is relieved by a shift in the reaction to the side with fewer moles of gas (in this case, to products). The amount of CO decreases.
(d) An increase in pressure as a result of the addition of an inert gas (with no volume change) does not affect the equilibrium composition. The amount of CO remains the same.
(e) Adding O_2 increases the O_2 concentration and shifts the reaction toward products. The amount of CO decreases.

13.23 (a) Because K_c is so large, k_f is larger than k_r.

(b) $K_c = \dfrac{k_f}{k_r}$; $\quad k_r = \dfrac{k_f}{K_c} = \dfrac{8.5 \times 10^6 \ M^{-1} s^{-1}}{3.4 \times 10^{34}} = 2.5 \times 10^{-28} \ M^{-1} \ s^{-1}$

(c) Because the reaction is exothermic, E_a (forward) is less than E_a (reverse). Consequently, as the temperature decreases, k_r decreases more than k_f decreases, and therefore $K_c = \dfrac{k_f}{k_r}$ increases.

13.24 $Hb + O_2 \rightleftharpoons Hb(O_2)$

If CO binds to Hb, Hb is removed from the reaction and the reaction will shift to the left, resulting in O_2 being released from $Hb(O_2)$. This will decrease the effectiveness of Hb for carrying O_2.

13.25 The equilibrium shifts to the left because at the higher altitude the concentration of O_2 is decreased.

13.26 There are 26 π electrons.

13.27 The partial pressure of O_2 in the atmosphere is 0.2095 atm.

$PV = nRT$

$n = \dfrac{PV}{RT} = \dfrac{(0.2095 \ \text{atm})(0.500 \ L)}{\left(0.082 \ 06 \ \dfrac{L \cdot \text{atm}}{K \cdot \text{mol}} \right)(298 \ K)} = 4.28 \times 10^{-3} \ \text{mol} \ O_2$

$4.28 \times 10^{-3} \ \text{mol} \ O_2 \times \dfrac{6.022 \times 10^{23} \ O_2 \ \text{molecules}}{1 \ \text{mol} \ O_2} = 2.58 \times 10^{21} \ O_2 \ \text{molecules}$

Key Concept Problems

13.28 (a) (1) and (3) because the number of A's and B's are the same in the third and fourth box.

(b) $K_c = \dfrac{[B]}{[A]} = \dfrac{6}{4} = 1.5$

(c) Because the same number of molecules appear on both sides of the equation, the volume terms in K_c all cancel. Therefore, we can calculate K_c without including the volume.

13.29 (a) $A_2 + C_2 \rightleftharpoons 2 \ AC$ (most product molecules)

(b) $A_2 + B_2 \rightleftharpoons 2 \ AB$ (fewest product molecules)

13.30 (a) Only reaction (3), $K_c = \dfrac{[A][AB]}{[A_2][B]} = \dfrac{(2)(4)}{(2)(2)} = 2$, is at equilibrium.

(b) $Q_c = \dfrac{[A][AB]}{[A_2][B]} = \dfrac{(3)(5)}{(1)(1)} = 15$ for reaction (1). Because $Q_c > K_c$, the reaction will go in the reverse direction to reach equilibrium.

$$Q_c = \frac{[A][AB]}{[A_2][B]} = \frac{(1)(3)}{(3)(3)} = 1/3 \text{ for reaction (2). Because } Q_c < K_c, \text{ the reaction will go in}$$

the forward direction to reach equilibrium.

13.31   (a) $A_2 + 2\,B \rightleftharpoons 2\,AB$
(b) The number of AB molecules will increase, because as the volume is decreased at constant temperature, the pressure will increase and the reaction will shift to the side of fewer molecules to reduce the pressure.

13.32   When the stopcock is opened, the reaction will go in the reverse direction because there will be initially an excess of AB molecules.

13.33   As the temperature is raised, the reaction proceeds in the reverse direction. This is consistent with an exothermic reaction where "heat" can be considered as a product.

13.34   (a) $AB \rightarrow A + B$
(b) The reaction is endothermic because a stress of added heat (higher temperature) shifts the $AB \rightleftharpoons A + B$ equilibrium to the right.
(c) If the volume is increased, the pressure is decreased. The stress of decreased pressure will be relieved by a shift in the equilibrium from left to right, thus increasing the number of A atoms.

13.35   Heat $+ BaCO_3(s) \rightleftharpoons BaO(s) + CO_2(g)$

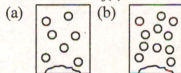

13.36   (a)

13.37   This equilibrium mixture has a $K_c \propto \dfrac{(2)(2)}{(3)^2}$ and is less than 1. This means that $k_f < k_r$.

13.38   (a) $A \rightarrow 2\,B$
(b) (1) The reaction is exothermic. As the temperature is increased, the reaction shifts to the left. A increases, B decreases, and K_c decreases.
(2) When the volume decreases, the reaction shifts to the side with fewer gas molecules, which is towards the reactant. The amount of A increases.
(3) If there is no volume change, there is no change in the equilibrium composition, and the amount of A remains the same.
(4) A catalyst does not change the equilibrium composition, and the amount of A remains the same.

13.39 (a) $2\,A + B_2 \;\rightarrow\; 2\,AB$

(b) (1) Increasing the partial pressure of B_2 (a reactant) shifts the reaction towards products and the amount of AB increases.

(2) Adding solid A does not change the equilibrium composition, and the amount of AB remains the same.

(3) When the volume increases, the reaction shifts to the side with more gas molecules, which is towards the product. The amount of AB increases.

(4) The reaction is exothermic. As the temperature is increased, the reaction shifts to the left. B_2 increases, AB decreases, and K_c decreases.

Section Problems
Equilibrium Expressions and Equilibrium Constants (Sections 13.1–13.4)

13.40 (a) $K_c = \dfrac{[CO][H_2]^3}{[CH_4][H_2O]}$
  (b) $K_c = \dfrac{[ClF_3]^2}{[F_2]^3[Cl_2]}$
  (c) $K_c = \dfrac{[HF]^2}{[H_2][F_2]}$

13.41 (a) $K_c = \dfrac{[CH_3CHO]^2}{[C_2H_4]^2[O_2]}$
  (b) $K_c = \dfrac{[N_2][O_2]}{[NO]^2}$
  (c) $K_c = \dfrac{[NO]^4[H_2O]^6}{[NH_3]^4[O_2]^5}$

13.42 (a) $K_p = \dfrac{(P_{CO})(P_{H_2})^3}{(P_{CH_4})(P_{H_2O})}$, $\Delta n = 2$ and $K_p = K_c(RT)^2$

(b) $K_p = \dfrac{(P_{ClF_3})^2}{(P_{F_2})^3(P_{Cl_2})}$, $\Delta n = -2$ and $K_p = K_c(RT)^{-2}$

(c) $K_p = \dfrac{(P_{HF})^2}{(P_{H_2})(P_{F_2})}$, $\Delta n = 0$ and $K_p = K_c$

13.43 (a) $K_p = \dfrac{(P_{CH_3CHO})^2}{(P_{C_2H_4})^2(P_{O_2})}$, $\Delta n = -1$ and $K_p = K_c(RT)^{-1}$

(b) $K_p = \dfrac{(P_{N_2})(P_{O_2})}{(P_{NO})^2}$, $\Delta n = 0$ and $K_p = K_c$

(c) $K_p = \dfrac{(P_{NO})^4(P_{H_2O})^6}{(P_{NH_3})^4(P_{O_2})^5}$, $\Delta n = 1$ and $K_p = K_c(RT)$

13.44 $K_c = \dfrac{[C_2H_5OC_2H_5][H_2O]}{[C_2H_5OH]^2}$

13.45 $\quad K_c = \dfrac{[HOCH_2CH_2OH]}{[C_2H_4O][H_2O]}$

13.46 $\quad K_c = \dfrac{[Isocitrate]}{[Citrate]}$

13.47 $\quad K_c = \dfrac{[citric\ acid]}{[acetic\ acid][oxaloacetic\ acid]}$

13.48 $\quad$ The two reactions are the reverse of each other.

$$K_c(reverse) = \frac{1}{K_c(forward)} = \frac{1}{7.5 \times 10^{-9}} = 1.3 \times 10^8$$

13.49 $\quad$ The two reactions are the reverse of each other.

$$K_p(reverse) = \frac{1}{K_p(forward)} = \frac{1}{50.2} = 1.99 \times 10^{-2}$$

13.50 $\quad K_c = \dfrac{[PCl_3][Cl_2]}{[PCl_5]} = \dfrac{(1.5 \times 10^{-2})(3.2 \times 10^{-2})}{(8.3 \times 10^{-3})} = 0.058$

13.51 $\quad K_p = \dfrac{(P_{ClNO})^2}{(P_{NO})^2(P_{Cl_2})} = \dfrac{(1.35)^2}{(0.240)^2(0.608)} = 52.0$

13.52 $\quad$ The container volume of 2.00 L must be included to calculate molar concentrations.
Initial [HI] = 9.30 x 10^{-3} mol/2.00 L = 4.65 x 10^{-3} M = 0.004 65 M

	$H_2(g)$	+	$I_2(g)$	$\rightleftharpoons$	$2\ HI(g)$
initial (M)	0		0		0.004 65
change (M)	+x		+x		−2x
equil (M)	x		x		0.004 65 − 2x

x = [H$_2$] = [I$_2$] = 6.29 x 10^{-4} M = 0.000 629 M
[HI] = 0.004 65 − 2x = 0.004 65 − 2(0.000 629) = 0.003 39 M

$$K_c = \frac{[HI]^2}{[H_2][I_2]} = \frac{(0.003\ 39)^2}{(0.000\ 629)^2} = 29.0$$

13.53 $\quad$ (a) $\quad K_c = \dfrac{[H^+][CH_3CO_2^-]}{[CH_3CO_2H]}$

(b)

	$CH_3CO_2H(aq)$	$\rightleftharpoons$	$H^+(aq)$	+	$CH_3CO_2^-(aq)$
initial (M)	1.0		0		0
change (M)	−0.0042		+0.0042		+0.0042
equil (M)	1.0 − 0.0042		0.0042		0.0042

$$K_c = \frac{(0.0042)(0.0042)}{(1.0 - 0.0042)} = 1.8 \times 10^{-5}$$

13.54 (a) $K_c = \dfrac{[CH_3CO_2C_2H_5][H_2O]}{[CH_3CO_2H][C_2H_5OH]}$

(b) $\qquad\qquad CH_3CO_2H(soln) + C_2H_5OH(soln) \rightleftharpoons CH_3CO_2C_2H_5(soln) + H_2O(soln)$

initial (mol)	1.00	1.00	0	0
change (mol)	$-x$	$-x$	$+x$	$+x$
equil (mol)	$1.00 - x$	$1.00 - x$	x	x

$x = 0.65$ mol; $\quad 1.00 - x = 0.35$ mol; $\qquad K_c = \dfrac{(0.65)^2}{(0.35)^2} = 3.4$

Because there are the same number of molecules on both sides of the equation, the volume terms in K_c cancel. Therefore, we can calculate K_c without including the volume.

13.55 $CH_3CO_2C_2H_5(soln) + H_2O(soln) \rightleftharpoons CH_3CO_2H(soln) + C_2H_5OH(soln)$

$$K_c(\text{hydrolysis}) = \frac{1}{K_c(\text{forward})} = \frac{1}{3.4} = 0.29$$

13.56 $\Delta n = 1$ and $K_p = K_c(RT) = (0.575)(0.082\ 06)(500) = 23.6$

13.57 $2\ SO_2(g) + O_2(g) \rightleftharpoons 2\ SO_3(g); \qquad \Delta n = 2 - (2 + 1) = -1$ and $K_p = 3.30$

$$K_c = K_p\left(\frac{1}{RT}\right)^{\Delta n} = (3.30)\left(\frac{1}{(0.082\ 06)(1000)}\right)^{-1} = 271$$

13.58 $K_p = P_{H_2O} = 0.0313$ atm; $\Delta n = 1$

$$K_c = K_p\left(\frac{1}{RT}\right) = (0.0313)\left(\frac{1}{(0.082\ 06)(298)}\right) = 1.28 \times 10^{-3}$$

13.59 $P_{C_{10}H_8} = 0.10$ mm Hg $\times \dfrac{1\ \text{atm}}{760\ \text{mm Hg}} = 1.3 \times 10^{-4}$ atm

$K_p = P_{C_{10}H_8} = 1.3 \times 10^{-4}; \qquad \Delta n = 1 - 0 = 1, \ T = 27\ ^\circ C = 300\ K$

$$K_c = K_p\left(\frac{1}{RT}\right)^{\Delta n} = (1.3 \times 10^{-4})\left(\frac{1}{(0.082\ 06)(300)}\right) = 5.3 \times 10^{-6}$$

13.60 (a) $K_c = \dfrac{[CO_2]^3}{[CO]^3}, \quad K_p = \dfrac{(P_{CO_2})^3}{(P_{CO})^3}$ $\qquad$ (b) $K_c = \dfrac{1}{[O_2]^3}, \quad K_p = \dfrac{1}{(P_{O_2})^3}$

(c) $K_c = [SO_3], \quad K_p = P_{SO_3}$ $\qquad\qquad$ (d) $K_c = [Ba^{2+}][SO_4^{2-}]$

13.61 (a) $K_c = \dfrac{[H_2O]^3}{[H_2]^3}$, $\quad K_p = \dfrac{(P_{H_2O})^3}{(P_{H_2})^3}$ $\qquad$ (b) $K_c = \dfrac{1}{[Ag^+][Cl^-]}$

$\quad$ (c) $K_c = \dfrac{[HCl]^6}{[H_2O]^3}$, $\quad K_p = \dfrac{(P_{HCl})^6}{(P_{H_2O})^3}$ $\qquad$ (d) $K_c = [CO_2]$, $\qquad K_p = P_{CO_2}$

Using the Equilibrium Constant (Section 13.5)

13.62 (a) Because K_c is very large, the equilibrium mixture contains mostly product.
(b) Because K_c is very small, the equilibrium mixture contains mostly reactants.

13.63 (a) proceeds hardly at all toward completion
(b) goes almost all the way to completion

13.64 (a) Because K_c is very small, the equilibrium mixture contains mostly reactant.
(b) Because K_c is very large, the equilibrium mixture contains mostly product.
(c) Because $K_c = 1.8$, the equilibrium mixture contains an appreciable concentration of both reactants and products.

13.65 (a) Because K_c is very large, the equilibrium mixture contains mostly product.
(b) Because $K_c = 7.5 \times 10^{-3}$, the equilibrium mixture contains an appreciable concentration of both reactants and products.
(c) Because K_c is very small, the equilibrium mixture contains mostly reactant.

13.66 $K_c = 1.2 \times 10^{82}$ is very large. When equilibrium is reached, very little if any ethanol will remain because the reaction goes to completion.

13.67 Because K_c is very small, pure air will contain very little O_3 (ozone) at equilibrium.

$$3\,O_2(g) \rightleftharpoons 2\,O_3(g); \quad K_c = \dfrac{[O_3]^2}{[O_2]^3} = 1.7 \times 10^{-56}; \quad [O_2] = 8 \times 10^{-3}\,M$$

$$[O_3] = \sqrt{[O_2]^3 \times K_c} = \sqrt{(8 \times 10^{-3})^3 (1.7 \times 10^{-56})} = 9 \times 10^{-32}\,M$$

13.68 The container volume of 10.0 L must be included to calculate molar concentrations.

$$Q_c = \dfrac{[CS_2]_t[H_2]_t^4}{[CH_4]_t[H_2S]_t^2} = \dfrac{(3.0\text{ mol}/10.0\text{ L})(3.0\text{ mol}/10.0\text{ L})^4}{(2.0\text{ mol}/10.0\text{ L})(4.0\text{ mol}/10.0\text{ L})^2} = 7.6 \times 10^{-2}; \quad K_c = 2.5 \times 10^{-3}$$

The reaction is not at equilibrium because $Q_c > K_c$. The reaction will proceed from right to left to reach equilibrium.

13.69 $Q_c = \dfrac{[CO]_t[H_2]_t^3}{[H_2O]_t[CH_4]_t} = \dfrac{(0.15)(0.20)^3}{(0.035)(0.050)} = 0.69; \quad K_c = 4.7$

The reaction is not at equilibrium because $Q_c < K_c$. The reaction will proceed from left to right to reach equilibrium.

13.70 $\quad K_c = \dfrac{[NH_3]^2}{[N_2][H_2]^3} = 0.29;$ $\quad$ At equilibrium, $[N_2] = 0.036$ M and $[H_2] = 0.15$ M

$\quad [NH_3] = \sqrt{[N_2] \times [H_2]^3 \times K_c} = \sqrt{(0.036)(0.15)^3(0.29)} = 5.9 \times 10^{-3}$ M

13.71 $\quad K_c = 2.7 \times 10^2 = \dfrac{[SO_3]^2}{[SO_2]^2[O_2]};$ $\quad$ Because $[SO_3] = [SO_2]$, then $2.7 \times 10^2 = \dfrac{1}{[O_2]}$

$\quad [O_2] = 3.7 \times 10^{-3}$ M

13.72

$$N_2(g) \quad + \quad O_2(g) \quad \rightleftharpoons \quad 2\,NO(g)$$

initial (M)	1.40	1.40	0
change (M)	$-x$	$-x$	$+2x$
equil (M)	$1.40 - x$	$1.40 - x$	$2x$

$K_c = 1.7 \times 10^{-3} = \dfrac{[NO]^2}{[N_2][O_2]} = \dfrac{(2x)^2}{(1.40-x)^2}$

Take the square root of both sides and solve for x.

$\sqrt{1.7 \times 10^{-3}} = \sqrt{\dfrac{(2x)^2}{(1.40-x)^2}};$ $\quad 4.1 \times 10^{-2} = \dfrac{2x}{1.40-x};$ $\quad x = 2.8 \times 10^{-2}$

At equilibrium, $[NO] = 2x = 2(2.8 \times 10^{-2}) = 0.056$ M

$[N_2] = [O_2] = 1.40 - x = 1.40 - (2.8 \times 10^{-2}) = 1.37$ M

13.73

$$N_2(g) \quad + \quad O_2(g) \quad \rightleftharpoons \quad 2\,NO(g)$$

initial (M)	2.24	0.56	0
change (M)	$-x$	$-x$	$+2x$
equil (M)	$2.24 - x$	$0.56 - x$	$2x$

$K_c = \dfrac{[NO]^2}{[N_2][O_2]} = 1.7 \times 10^{-3} = \dfrac{(2x)^2}{(2.24-x)(0.56-x)}$

$4x^2 + (4.8 \times 10^{-3})x - (2.1 \times 10^{-3}) = 0$

Use the quadratic formula to solve for x.

$x = \dfrac{-(4.8 \times 10^{-3}) \pm \sqrt{(4.8 \times 10^{-3})^2 - 4(4)(-2.1 \times 10^{-3})}}{2(4)} = \dfrac{-0.0048 \pm 0.1834}{8}$

$x = -0.0235$ and 0.0223

Discard the negative solution (-0.0235) because it gives a negative NO concentration and that is impossible.

$[N_2] = 2.24 - x = 2.24 - 0.0223 = 2.22$ M

$[O_2] = 0.56 - x = 0.56 - 0.0223 = 0.54$ M; $\quad [NO] = 2x = 2(0.0223) = 0.045$ M

13.74

$$PCl_5(g) \quad \rightleftharpoons \quad PCl_3(g) \quad + \quad Cl_2(g)$$

initial (M)	0.160	0	0
change (M)	$-x$	$+x$	$+x$
equil (M)	$0.160 - x$	x	x

$$K_c = \frac{[PCl_3][Cl_2]}{[PCl_5]} = 5.8 \times 10^{-2} = \frac{x^2}{0.160 - x}$$

$x^2 + (5.8 \times 10^{-2})x - 0.00928 = 0$

Use the quadratic formula to solve for x.

$$x = \frac{(-5.8 \times 10^{-2}) \pm \sqrt{(5.8 \times 10^{-2})^2 - 4(1)(-0.00928)}}{2(1)} = \frac{(-5.8 \times 10^{-2}) \pm 0.20}{2}$$

$x = 0.071$ and -0.129

Discard the negative solution (-0.129) because it gives negative concentrations of PCl_3 and Cl_2 and that is impossible.

$[PCl_3] = [Cl_2] = x = 0.071$ M; $\qquad [PCl_5] = 0.160 - x = 0.160 - 0.071 = 0.089$ M

13.75 $\quad Q_c = \dfrac{(0.100)(0.040)}{(0.200)} = 0.020$, $Q_c < K_c$ therefore the reaction proceeds from reactants to products to reach equilibrium.

	$PCl_5(g)$	$\rightleftarrows$	$PCl_3(g)$	+	$Cl_2(g)$
initial (M)	0.200		0.100		0.040
change (M)	$-x$		$+x$		$+x$
equil (M)	$0.200 - x$		$0.100 + x$		$0.040 + x$

$$K_c = \frac{[PCl_3][Cl_2]}{[PCl_5]} = 5.8 \times 10^{-2} = \frac{(0.100 + x)(0.040 + x)}{0.200 - x}$$

$x^2 + 0.198x - (7.60 \times 10^{-3}) = 0$

Use the quadratic formula to solve for x.

$$x = \frac{(-0.198) \pm \sqrt{(0.198)^2 - 4(1)(-7.60 \times 10^{-3})}}{2(1)} = \frac{(-0.198) \pm 0.264}{2}$$

$x = 0.033$ and -0.231

Discard the negative solution (-0.231) because it gives negative concentrations of PCl_3 and Cl_2 and that is impossible.

$[PCl_3] = 0.100 + x = 0.100 + 0.033 = 0.133$ M
$[Cl_2] = 0.040 + x = 0.040 + 0.033 = 0.073$ M
$[PCl_5] = 0.200 - x = 0.200 - 0.033 = 0.167$ M

13.76 (a) $\quad K_c = \dfrac{[CH_3CO_2C_2H_5][H_2O]}{[CH_3CO_2H][C_2H_5OH]} = 3.4 = \dfrac{(x)(12.0)}{(4.0)(6.0)}; \qquad x = 6.8$ moles $CH_3CO_2C_2H_5$

Note that the volume cancels because the same number of molecules appear on both sides of the chemical equation.

(b)	$CH_3CO_2H(soln)$	+	$C_2H_5OH(soln)$	$\rightleftarrows$	$CH_3CO_2C_2H_5(soln)$	+	$H_2O(soln)$
initial (mol)	1.00		10.00		0		0
change (mol)	$-x$		$-x$		$+x$		$+x$
equil (mol)	$1.00 - x$		$10.00 - x$		x		x

$$K_c = 3.4 = \frac{x^2}{(1.00 - x)(10.00 - x)}$$

$2.4x^2 - 37.4x + 34 = 0$

Use the quadratic formula to solve for x.

$$x = \frac{-(-37.4) \pm \sqrt{(-37.4)^2 - 4(2.4)(34)}}{2(2.4)} = \frac{37.4 \pm 32.75}{4.8}$$

x = 0.969 and 14.6

Discard the larger solution (14.6) because it leads to negative concentrations and that is impossible.

mol CH_3CO_2H = 1.00 – x = 1.00 – 0.969 = 0.03 mol

mol C_2H_5OH = 10.00 – x = 10.00 – 0.969 = 9.03 mol

mol $CH_3CO_2C_2H_5$ = mol H_2O = x = 0.97 mol

13.77 When equal volumes of two solutions are mixed together, their concentrations are cut in half.

	$CH_3Cl(aq)$ +	$OH^-(aq)$ ⇌	$CH_3OH(aq)$ +	$Cl^-(aq)$
initial (M)	0.05	0.1	0	0
assume complete reaction (M)	0	0.05	0.05	0.05
assume small back reaction (M)	+x	+x	–x	–x
equil (M)	x	0.05 + x	0.05 – x	0.05 – x

$$K_c = \frac{[CH_3OH][Cl^-]}{[CH_3Cl][OH^-]} = 10^{16} = \frac{(0.05-x)^2}{x(0.05+x)} ; \qquad \text{Because } K_c \text{ is very large, } x \ll 0.05.$$

$$10^{16} \approx \frac{(0.05)^2}{x(0.05)} ; \qquad x = 5 \times 10^{-18}$$

$[CH_3Cl] = x = 5 \times 10^{-18}$ M; $[OH^-] = [CH_3OH] = [Cl^-] \approx 0.05$ M

13.78

	$ClF_3(g)$ ⇌	$ClF(g)$ +	$F_2(g)$
initial (atm)	1.47	0	0
change (atm)	–x	+x	+x
equil (atm)	1.47 – x	x	x

$$K_p = \frac{(P_{ClF})(P_{F_2})}{(P_{ClF_3})} = 0.140 = \frac{(x)(x)}{1.47-x} ; \text{ solve for x.}$$

$$x^2 + 0.140x - 0.2058 = 0$$

Use the quadratic formula to solve for x.

$$x = \frac{-(0.140) \pm \sqrt{(0.140)^2 - (4)(1)(-0.2058)}}{2(1)}$$

$$x = \frac{-0.140 \pm 0.918}{2}$$

x = 0.389 and –0.529

Discard the negative solution (–0.529) because it gives negative partial pressures and that is impossible.

$P_{ClF} = P_{F_2} = x = 0.389$ atm

$P_{ClF_3} = 1.47 - x = 1.47 - 0.389 = 1.08$ atm

13.79 $$Fe_2O_3(s) + 3\,CO(g) \rightleftharpoons 2\,Fe(s) + 3\,CO_2(g)$$

initial (atm)	0.978	0
change (atm)	−3x	+3x
equil (atm)	0.978 − 3x	3x

$$K_p = \frac{(P_{CO_2})^3}{(P_{CO})^3} = 19.9 = \frac{(3x)^3}{(0.978 - 3x)^3};\ \text{take the cube root of both sides and solve for x.}$$

$$\sqrt[3]{19.9} = 2.71 = \frac{3x}{(0.978 - 3x)}$$

$2.65 − 8.13x = 3x$

$2.65 = 11.13x$

$x = 2.65/11.13 = 0.238$ atm

$P_{CO} = 0.978 − 3x = 0.978 − 3(0.238) = 0.264$ atm

$P_{CO_2} = 3x = 3(0.238) = 0.714$ atm

Le Châtelier's Principle (Sections 13.6–13.10)

13.80 (a) Cl^- (reactant) added, $AgCl(s)$ increases
 (b) Ag^+ (reactant) added, $AgCl(s)$ increases
 (c) Ag^+ (reactant) removed, $AgCl(s)$ decreases
 (d) Cl^- (reactant) removed, $AgCl(s)$ decreases

Disturbing the equilibrium by decreasing $[Cl^-]$ increases Q_c $\left(Q_c = \dfrac{1}{[Ag^+]_t[Cl^-]_t}\right)$ to a

value greater than K_c. To reach a new state of equilibrium, Q_c must decrease, which means that the denominator must increase; that is, the reaction must go from right to left, thus decreasing the amount of solid AgCl.

13.81 (a) $ClNO$ (product) added, NO_2 concentration decreases
 (b) NO (reactant) added, NO_2 concentration increases
 (c) NO (reactant) removed, NO_2 concentration decreases
 (d) $ClNO_2$ (reactant) added, NO_2 concentration increases

Adding $ClNO_2$ decreases the value of Q_c $\left(Q_c = \dfrac{[ClNO][NO_2]}{[ClNO_2][NO]}\right)$. To reach a new state

of equilibrium, the reaction must go from left to right, thus increasing the concentration of NO_2.

13.82 (a) Because there are 2 mol of gas on the left side and 3 mol of gas on the right side of the balanced equation, the stress of an increase in pressure is relieved by a shift in the reaction to the side with fewer moles of gas (in this case, to reactants). The number of moles of reaction products decreases.
 (b) Because there are 2 mol of gas on both sides of the balanced equation, the composition of the equilibrium mixture is unaffected by a change in pressure. The number of moles of reaction product remains the same.

(c) Because there are 2 mol of gas on the left side and 1 mol of gas on the right side of the balanced equation, the stress of an increase in pressure is relieved by a shift in the reaction to the side with fewer moles of gas (in this case, to products). The number of moles of reaction products increases.

13.83 As the volume increases, the pressure decreases at constant temperature.
(a) Because there is 1 mol of gas on the left side and 2 mol of gas on the right side of the balanced equation, the stress of an increase in volume (decrease in pressure) is relieved by a shift in the reaction to the side with the larger number of moles of gas (in this case, to products).
(b) Because there are 3 mol of gas on the left side and 2 mol of gas on the right side of the balanced equation, the stress of an increase in volume (decrease in pressure) is relieved by a shift in the reaction to the side with the larger number of moles of gas (in this case, to reactants).
(c) Because there are 3 mol of gas on both sides of the balanced equation, the composition of the equilibrium mixture is unaffected by an increase in volume (decrease in pressure). There is no net reaction in either direction.

13.84 $CO(g) + H_2O(g) \rightleftharpoons CO_2(g) + H_2(g)$ $\Delta H° = -41.2 \text{ kJ}$
The reaction is exothermic. $[H_2]$ decreases when the temperature is increased.
As the temperature is decreased, the reaction shifts to the right. $[CO_2]$ and $[H_2]$ increase, $[CO]$ and $[H_2O]$ decrease, and K_c increases.

13.85 Because $\Delta H°$ is positive, the reaction is endothermic.
heat $+ 3 O_2(g) \rightleftharpoons 2 O_3(g)$

$$K_c = \frac{[O_3]^2}{[O_2]^3}$$

As the temperature increases, heat is added to the reaction, causing a shift to the right. The $[O_3]$ increases, and the $[O_2]$ decreases. This results in an increase in K_c.

13.86 (a) HCl is a source of Cl^- (product), the reaction shifts left, the equilibrium $[CoCl_4^{2-}]$ increases.
(b) $Co(NO_3)_2$ is a source of $Co(H_2O)_6^{2+}$ (product), the reaction shifts left, the equilibrium $[CoCl_4^{2-}]$ increases.
(c) All concentrations will initially decrease and the reaction will shift to the right; the equilibrium $[CoCl_4^{2-}]$ decreases.
(d) For an exothermic reaction, the reaction shifts to the left when the temperature is increased; the equilibrium $[CoCl_4^{2-}]$ increases.

13.87 (a) $Fe(NO_3)_3$ is a source of Fe^{3+}. Fe^{3+} (reactant) added; the $FeCl^{2+}$ concentration increases.
(b) Cl^- (reactant) removed; the $FeCl^{2+}$ concentration decreases.
(c) An endothermic reaction shifts to the right as the temperature increases; the $FeCl^{2+}$ concentration increases.
(d) A catalyst does not affect the composition of the equilibrium mixture; no change in $FeCl^{2+}$ concentration.

13.88   (a) The reaction is exothermic. The amount of CH_3OH (product) decreases as the temperature increases.
(b) When the volume decreases, the reaction shifts to the side with fewer gas molecules. The amount of CH_3OH increases.
(c) Addition of an inert gas (He) does not affect the equilibrium composition. There is no change.
(d) Addition of CO (reactant) shifts the reaction toward product. The amount of CH_3OH increases.
(e) Addition or removal of a catalyst does not affect the equilibrium composition. There is no change.

13.89   (a) An endothermic reaction shifts to the right as the temperature increases. The amount of acetone increases.
(b) Because there is 1 mol of gas on the left side and 2 mol of gas on the right side of the balanced equation, the stress of an increase in volume (decrease in pressure) is relieved by a shift in the reaction to the side with the larger number of moles of gas (in this case, to products). The amount of acetone increases.
(c) The addition of Ar (an inert gas) with no volume change does not affect the composition of the equilibrium mixture. The amount of acetone does not change.
(d) H_2 (product) added; the amount of acetone decreases.
(e) A catalyst does not affect the composition of the equilibrium mixture. The amount of acetone does not change.

Chemical Equilibrium and Chemical Kinetics (Section 13.11)

13.90 $A + B \rightleftarrows C$
$rate_f = k_f[A][B]$ and $rate_r = k_r[C]$;   at equilibrium, $rate_f = rate_r$
$$k_f[A][B] = k_r[C]; \qquad \frac{k_f}{k_r} = \frac{[C]}{[A][B]} = K_c$$

13.91   An equilibrium mixture that contains large amounts of reactants and small amounts of products has a small K_c. A small K_c has $k_f < k_r$ (c).

13.92 $K_c = \dfrac{k_f}{k_r} = \dfrac{0.13}{6.2 \times 10^{-4}} = 210$

13.93 $K_c = \dfrac{k_f}{k_r}; \qquad k_r = \dfrac{k_f}{K_c} = \dfrac{6 \times 10^{-6}\ M^{-1}s^{-1}}{10^{16}} = 6 \times 10^{-22}\ M^{-1}s^{-1}$

13.94 k_r increases more than k_f, this means that E_a (reverse) is greater than E_a (forward). The reaction is exothermic when E_a (reverse) $> E_a$ (forward).

13.95 k_f increases more than k_r, this means that E_a (forward) is greater than E_a (reverse). The reaction is endothermic when E_a (forward) $> E_a$ (reverse).

Chapter Problems

13.96 $K_c = \dfrac{[NH_3]^2}{[N_2][H_2]^3} = 0.291$

At equilibrium, $[N_2] = 1.0 \times 10^{-3}$ M and $[H_2] = 2.0 \times 10^{-3}$ M

$[NH_3] = \sqrt{[N_2] \times [H_2]^3 \times K_c} = \sqrt{(1.0 \times 10^{-3})(2.0 \times 10^{-3})^3(0.291)} = 1.5 \times 10^{-6}$ M

13.97 K_c is very large. The reaction goes essentially to completion.

$K_c = \dfrac{1}{[CO_2]}$; $[CO_2] = \dfrac{1}{K_c} = \dfrac{1}{1.6 \times 10^{24}} = 6.2 \times 10^{-25}$ M

13.98 $2\ HI(g) \rightleftharpoons H_2(g) + I_2(g)$

Calculate K_c. $K_c = \dfrac{[H_2][I_2]}{[HI]^2} = \dfrac{(0.13)(0.70)}{(2.1)^2} = 0.0206$

$[HI] = \dfrac{0.20\ mol}{0.5000\ L} = 0.40$ M

	$2\ HI(g)$	$\rightleftharpoons$	$H_2(g)$	$+$	$I_2(g)$
initial (M)	0.40		0		0
change (M)	$-2x$		$+x$		$+x$
equil (M)	$0.40 - 2x$		x		x

$K_c = 0.0206 = \dfrac{[H_2][I_2]}{[HI]^2} = \dfrac{x^2}{(0.40 - 2x)^2}$

Take the square root of both sides, and solve for x.

$\sqrt{0.0206} = \sqrt{\dfrac{x^2}{(0.40 - 2x)^2}}$; $0.144 = \dfrac{x}{0.40 - 2x}$; $x = 0.045$

At equilibrium, $[H_2] = [I_2] = x = 0.045$ M; $[HI] = 0.40 - 2x = 0.40 - 2(0.045) = 0.31$ M

13.99 Note the container volume is 5.00 L
$[H_2] = [I_2] = 1.00\ mol/5.00\ L = 0.200$ M; $[HI] = 2.50\ mol/5.00\ L = 0.500$ M

	$H_2(g)$	$+$	$I_2(g)$	$\rightleftharpoons$	$2\ HI(g)$
initial (M)	0.200		0.200		0.500
change (M)	$-x$		$-x$		$+2x$
equil (M)	$0.200 - x$		$0.200 - x$		$0.500 + 2x$

$K_c = \dfrac{[HI]^2}{[H_2][I_2]} = 129 = \dfrac{(0.500 + 2x)^2}{(0.200 - x)^2}$

Take the square root of both sides, and solve for x.

$\sqrt{129} = \sqrt{\dfrac{(0.500 + 2x)^2}{(0.200 - x)^2}}$; $11.4 = \dfrac{0.500 + 2x}{0.200 - x}$; $x = 0.133$

$[H_2] = [I_2] = 0.200 - x = 0.200 - 0.133 = 0.067$ M
$[HI] = 0.500 + 2x = 0.500 + 2(0.133) = 0.766$ M

13.100 $[H_2O] = \dfrac{6.00 \text{ mol}}{5.00 \text{ L}} = 1.20$ M

$$
\begin{array}{ccccccc}
 & C(s) & + & H_2O(g) & \rightleftharpoons & CO(g) & + & H_2(g) \\
\text{initial (M)} & & & 1.20 & & 0 & & 0 \\
\text{change (M)} & & & -x & & +x & & +x \\
\text{equil (M)} & & & 1.20-x & & x & & x
\end{array}
$$

$K_c = \dfrac{[CO][H_2]}{[H_2O]} = 3.0 \times 10^{-2} = \dfrac{x^2}{1.20-x}$

$x^2 + (3.0 \times 10^{-2})x - 0.036 = 0$

Use the quadratic formula to solve for x.

$x = \dfrac{-(0.030) \pm \sqrt{(0.030)^2 - 4(-0.036)}}{2(1)} = \dfrac{-0.030 \pm 0.381}{2}$

$x = 0.176$ and -0.206

Discard the negative solution (–0.206) because it leads to negative concentrations and that is impossible.

$[CO] = [H_2] = x = 0.18$ M; $[H_2O] = 1.20 - x = 1.20 - 0.18 = 1.02$ M

13.101 (a) Because K_p is larger at the higher temperature, the reaction has shifted toward products at the higher temperature, which means the reaction is endothermic.
(b) (i) Increasing the volume causes the reaction to shift toward the side with more mol of gas (product side). The equilibrium amounts of PCl_3 and Cl_2 increase while that of PCl_5 decreases.
(ii) If there is no volume change, there is no change in equilibrium concentrations.
(iii) Addition of a catalyst does not affect the equilibrium concentrations.

13.102 A decrease in volume (a) and the addition of reactants (c) will affect the composition of the equilibrium mixture, but leave the value of K_c unchanged.
A change in temperature (b) affects the value of K_c.
Addition of a catalyst (d) or an inert gas (e) affects neither the composition of the equilibrium mixture nor the value of K_c.

13.103 (a) Addition of a solid does not affect the equilibrium composition. There is no change in the number of moles of CO_2.
(b) Adding a product causes the reaction to shift toward reactants. The number of moles of CO_2 decreases.
(c) Decreasing the volume causes the reaction to shift toward the side with fewer mol of gas (reactant side). The number of moles of CO_2 decreases.
(d) The reaction is endothermic. An increase in temperature shifts the reaction toward products. The number of moles of CO_2 increases.

13.104 (a) $[PCl_5] = 1.000$ mol/5.000 L $= 0.2000$ M

$$
\begin{array}{cccc}
 & PCl_5(g) & \rightleftharpoons \quad PCl_3(g) & + \quad Cl_2(g) \\
\text{initial (M)} & 0.2000 & 0 & 0 \\
\text{change (M)} & -(0.2000)(0.7850) & +(0.2000)(0.7850) & +(0.2000)(0.7850) \\
\text{equil (M)} & 0.0430 & 0.1570 & 0.1570
\end{array}
$$

$$K_c = \frac{[PCl_3][Cl_2]}{[PCl_5]} = \frac{(0.1570)(0.1570)}{(0.0430)} = 0.573$$

$\Delta n = 1$ and $K_p = K_c(RT) = (0.573)(0.082\ 06)(500) = 23.5$

(b) $Q_c = \dfrac{[PCl_3][Cl_2]}{[PCl_5]} = \dfrac{(0.150)(0.600)}{(0.500)} = 0.18$

Because $Q_c < K_c$, the reaction proceeds to the right to reach equilibrium.

	$PCl_5(g)$	$\rightleftharpoons$	$PCl_3(g)$	+	$Cl_2(g)$
initial (M)	0.500		0.150		0.600
change (M)	$-x$		$+x$		$+x$
equil (M)	$0.500 - x$		$0.150 + x$		$0.600 + x$

$$K_c = \frac{[PCl_3][Cl_2]}{[PCl_5]} = 0.573 = \frac{(0.150 + x)(0.600 + x)}{(0.500 - x)};\ \text{solve for x.}$$

$x^2 + 1.323x - 0.1965 = 0$

$$x = \frac{-(1.323) \pm \sqrt{(1.323)^2 - (4)(1)(-0.1965)}}{2(1)} = \frac{-1.323 \pm 1.593}{2}$$

$x = -1.458$ and 0.135

Discard the negative solution (-1.458) because it will lead to negative concentrations and that is impossible.

$[PCl_5] = 0.500 - x = 0.500 - 0.135 = 0.365$ M

$[PCl_3] = 0.150 + x = 0.150 + 0.135 = 0.285$ M

$[Cl_2] = 0.600 + x = 0.600 + 0.135 = 0.735$ M

13.105 $Q_p = \dfrac{(2.00)(1.50)}{(3.00)} = 1.00$, $Q_p < K_p$ therefore the reaction proceeds from reactants to products to reach equilibrium.

	$PCl_5(g)$	$\rightleftharpoons$	$PCl_3(g)$	+	$Cl_2(g)$
initial (atm)	3.00		2.00		1.50
change (atm)	$-x$		$+x$		$+x$
equil (atm)	$3.00 - x$		$2.00 + x$		$1.50 + x$

$$K_p = \frac{(P_{PCl_3})(P_{Cl_2})}{(P_{PCl_5})} = 1.42 = \frac{(2.00 + x)(1.50 + x)}{3.00 - x}$$

$x^2 + 4.92x - 1.26 = 0$

Use the quadratic formula to solve for x.

$$x = \frac{(-4.92) \pm \sqrt{(4.92)^2 - 4(1)(-1.26)}}{2(1)} = \frac{(-4.92) \pm 5.41}{2}$$

$x = 0.245$ and -5.165

Discard the negative solution (-5.165) because it gives negative partial pressures and that is impossible.

$P_{PCl_5} = 3.00 - x = 3.00 - 0.245 = 2.76$ atm

$P_{PCl_3} = 2.00 + x = 2.00 + 0.245 = 2.24$ atm

$P_{Cl_2} = 1.50 + x = 1.50 + 0.245 = 1.74$ atm

$P_{total} = P_{PCl_5} + P_{PCl_3} + P_{Cl_2} = 2.76 + 2.24 + 1.74 = 6.74$ atm

13.106 (a) $K_c = \dfrac{[C_2H_6][C_2H_4]}{[C_4H_{10}]}$ $\qquad K_p = \dfrac{(P_{C_2H_6})(P_{C_2H_4})}{P_{C_4H_{10}}}$

(b) $K_p = 12$; $\Delta n = 1$; $K_c = K_p\left(\dfrac{1}{RT}\right) = (12)\left(\dfrac{1}{(0.082\ 06)(773)}\right) = 0.19$

(c)

	$C_4H_{10}(g)$	$\rightleftharpoons$	$C_2H_6(g)$	$+$	$C_2H_4(g)$
initial (atm)	50		0		0
change (atm)	$-x$		$+x$		$+x$
equil (atm)	$50 - x$		x		x

$K_p = 12 = \dfrac{x^2}{50 - x};\qquad x^2 + 12x - 600 = 0$

Use the quadratic formula to solve for x.

$x = \dfrac{(-12) \pm \sqrt{(12)^2 - 4(1)(-600)}}{2(1)} = \dfrac{-12 \pm 50.44}{2}$

$x = -31.22$ and 19.22

Discard the negative solution (−31.22) because it leads to negative concentrations and that is impossible.

% C_4H_{10} converted $= \dfrac{19.22}{50} \times 100\% = 38\%$

$P_{total} = P_{C_4H_{10}} + P_{C_2H_6} + P_{C_2H_4} = (50 - x) + x + x = (50 - 19) + 19 + 19 = 69$ atm

(d) A decrease in volume would decrease the % conversion of C_4H_{10}.

13.107

	$C(s)$	$+$	$CO_2(g)$	$\rightleftharpoons$	$2\ CO(g)$
initial (M)	excess		1.50 mol/20.0 L		0
change (M)			$-x$		$+2x$
equil (M)			$0.0750 - x$		$2x$

$[CO] = 2x = 7.00 \times 10^{-2}$ M; $x = 0.0350$ M

(a) $[CO_2] = 0.0750 - x = 0.0750 - 0.0350 = 0.0400$ M

(b) $K_c = \dfrac{[CO]^2}{[CO_2]} = \dfrac{(7.00 \times 10^{-2})^2}{(0.0400)} = 0.122$

13.108 (a) $K_p = 3.45$; $\Delta n = 1$; $K_c = K_p\left(\dfrac{1}{RT}\right) = (3.45)\left(\dfrac{1}{(0.082\ 06)(500)}\right) = 0.0840$

(b) $[(CH_3)_3CCl] = 1.00$ mol/5.00 L $= 0.200$ M

	$(CH_3)_3CCl(g)$	$\rightleftharpoons$	$(CH_3)_2C{=}CH_2(g)$	$+$	$HCl(g)$
initial (M)	0.200		0		0
change (M)	$-x$		$+x$		$+x$
equil (M)	$0.200 - x$		x		x

$$K_c = 0.0840 = \frac{x^2}{0.200 - x}; \quad x^2 + 0.0840x - 0.0168 = 0$$

Use the quadratic formula to solve for x.

$$x = \frac{(-0.0840) \pm \sqrt{(0.0840)^2 - 4(1)(-0.0168)}}{2(1)} = \frac{-0.0840 \pm 0.272}{2}$$

$x = -0.178$ and 0.094

Discard the negative solution (-0.178) because it leads to negative concentrations and that is impossible.

$[(CH_3)_2C=CCH_2] = [HCl] = x = 0.094$ M

$[(CH_3)_3CCl] = 0.200 - x = 0.200 - 0.094 = 0.106$ M

(c) $K_p = 3.45$

	$(CH_3)_3CCl(g)$	$\rightleftharpoons$	$(CH_3)_2C=CH_2(g)$	+	$HCl(g)$
initial (atm)	0		0.400		0.600
change (atm)	+x		−x		−x
equil (atm)	x		0.400 − x		0.600 − x

$$K_p = 3.45 = \frac{(0.400 - x)(0.600 - x)}{x}$$

$$x^2 - 4.45x + 0.240 = 0$$

Use the quadratic formula to solve for x.

$$x = \frac{-(-4.45) \pm \sqrt{(-4.45)^2 - 4(1)(0.240)}}{2(1)} = \frac{4.45 \pm 4.34}{2}$$

$x = 0.055$ and 4.40

Discard the larger solution (4.40) because it leads to negative partial pressures and that is impossible.

$P_{\text{t-butyl chloride}} = x = 0.055$ atm; $\quad P_{\text{isobutylene}} = 0.400 - x = 0.400 - 0.055 = 0.345$ atm

$P_{HCl} = 0.600 - x = 0.600 - 0.055 = 0.545$ atm

13.109 (a) The Arrhenius equation gives for the forward and reverse reactions:

$$k_f = A_f e^{-E_{a,f}/RT} \quad \text{and} \quad k_r = A_r e^{-E_{a,r}/RT}$$

Addition of a catalyst decreases the activation energies by ΔE_a, so

$$k_f = A_f e^{-(E_{a,f} - \Delta E_a)/RT} = A_f e^{-E_{a,f}/RT} \times e^{\Delta E_a/RT}$$

and

$$k_r = A_r e^{-(E_{a,r} - \Delta E_a)/RT} = A_r e^{-E_{a,r}/RT} \times e^{\Delta E_a/RT}$$

Therefore, the rate constants for both the forward and reverse reactions increase by the same factor, $e^{\Delta E_a/RT}$.

(b) The equilibrium constant is given by $K_c = \dfrac{k_f}{k_r} = \dfrac{A_f e^{-E_{a,f}/RT}}{A_r e^{-E_{a,r}/RT}} = \dfrac{A_f}{A_r} e^{-\Delta E/RT}$

where $\Delta E = E_{a,f} - E_{a,r}$. Addition of a catalyst decreases the activation energies by ΔE_a.

So, $K_c = \dfrac{k_f}{k_r} = \dfrac{A_f e^{-E_{a,f}/RT} \times e^{\Delta E_a/RT}}{A_r e^{-E_{a,r}/RT} \times e^{\Delta E_a/RT}} = \dfrac{A_f}{A_r} e^{-\Delta E/RT}$

K_c is unchanged because of cancellation of $e^{\Delta E_a/RT}$, the factor by which the two rate constants increase.

13.110 The activation energy (E_a) is positive, and for an exothermic reaction, $E_{a,r} > E_{a,f}$.

$$k_f = A_f\, e^{-E_{a,f}/RT}, \quad k_r = A_r\, e^{-E_{a,r}/RT}$$

$$K_c = \frac{k_f}{k_r} = \frac{A_f e^{-E_{a,f}/RT}}{A_r e^{-E_{a,r}/RT}} = \frac{A_f}{A_r} e^{(E_{a,r} - E_{a,f})/RT}$$

$(E_{a,r} - E_{a,f})$ is positive, so the exponent is always positive. As the temperature increases, the exponent, $(E_{a,r} - E_{a,f})/RT$, decreases and the value for K_c decreases as well.

13.111 (a) $PV = nRT$

$$P_{Br_2}^o = \frac{nRT}{V} = \frac{(0.974\ \text{mol})\left(0.082\ 06\ \dfrac{\text{L}\cdot\text{atm}}{\text{K}\cdot\text{mol}}\right)(1000\ \text{K})}{1.00\ \text{L}} = 80\ \text{atm}$$

$$P_{H_2}^o = \frac{nRT}{V} = \frac{(1.22\ \text{mol})\left(0.082\ 06\ \dfrac{\text{L}\cdot\text{atm}}{\text{K}\cdot\text{mol}}\right)(1000\ \text{K})}{1.00\ \text{L}} = 100\ \text{atm}$$

Because K_p is very large, assume first that the reaction goes to completion and then is followed by a small back reaction.

	$H_2(g)$	$+$	$Br_2(g)$	$\rightleftarrows$	$2\ HBr(g)$
before (atm)	100		80		0
100% reaction (atm)	-80		-80		$+2(80)$
after (atm)	20		0		160
back reaction (atm)	$+x$		$+x$		$-2x$
after (atm)	$20 + x$		x		$160 - 2x$

$P_{H_2} = 20 + x \approx 20\ \text{atm}$

$P_{HBr} = 160 - 2x \approx 160\ \text{atm}$

$$K_p = \frac{(P_{HBr})^2}{(P_{H_2})(P_{Br_2})} = 2.1 \times 10^6 = \frac{(160)^2}{(20)(x)}$$

$$P_{Br_2} = x = \frac{(160)^2}{(20)(2.1 \times 10^6)} = 6.1 \times 10^{-4}\ \text{atm}$$

(b) (ii) Adding Br_2 will cause the greatest increase in the pressure of HBr. The very large value of K_p means that the reaction goes essentially to completion. Therefore, the reaction stops when the limiting reactant, Br_2, is essentially consumed. No matter how much H_2 is added or how far the the equilibrium is shifted (by lowering the temperature) to favor the formation of HBr, the amount of HBr will ultimately be limited by the amount of Br_2 present. Therefore, more Br_2 must be added in order to produce more HBr.

13.112 (a) $PV = nRT$, $n_{total} = \dfrac{PV}{RT} = \dfrac{(0.588 \text{ atm})(1.00 \text{ L})}{\left(0.082\ 06\ \dfrac{\text{L} \cdot \text{atm}}{\text{K} \cdot \text{mol}}\right)(300 \text{ K})} = 0.0239 \text{ mol}$

	2 NOBr(g)	$\rightleftharpoons$	2 NO(g)	$+$	$Br_2\text{(g)}$
initial (mol)	0.0200		0		0
change (mol)	$-2x$		$+2x$		$+x$
equil (mol)	$0.0200 - 2x$		$2x$		x

$n_{total} = 0.0239 \text{ mol} = (0.0200 - 2x) + 2x + x = 0.0200 + x$

$x = 0.0239 - 0.0200 = 0.0039 \text{ mol}$

Because the volume is 1.00 L, the molarity equals the number of moles.

$[NOBr] = 0.0200 - 2x = 0.0200 - 2(0.0039) = 0.0122 \text{ M}$

$[NO] = 2x = 2(0.0039) = 0.0078 \text{ M}$

$[Br_2] = x = 0.0039 \text{ M}$

$K_c = \dfrac{[NO]^2[Br_2]}{[NOBr]^2} = \dfrac{(0.0078)^2(0.0039)}{(0.0122)^2} = 1.6 \times 10^{-3}$

(b) $\Delta n = (3) - (2) = 1$, $K_p = K_c(RT) = (1.6 \times 10^{-3})(0.082\ 06)(300) = 0.039$

13.113 NO_2, 46.01 amu

$\text{mol } NO_2 = 4.60 \text{ g } NO_2 \times \dfrac{1 \text{ mol } NO_2}{46.01 \text{ g } NO_2} = 0.100 \text{ mol } NO_2$

$[NO_2] = 0.100 \text{ mol}/10.0 \text{ L} = 0.0100 \text{ M}$

	$2 \text{ NO}_2\text{(g)}$	$\rightleftharpoons$	$N_2O_4\text{(g)}$
initial (M)	0.0100		0
change (M)	$-2x$		$+x$
equil (M)	$0.0100 - 2x$		x

$K_c = \dfrac{[N_2O_4]}{[NO_2]^2} = 4.72 = \dfrac{x}{(0.0100 - 2x)^2}$

$18.88x^2 - 1.189x + 0.000\ 472 = 0$

Use the quadratic formula to solve for x.

$x = \dfrac{-(-1.189) \pm \sqrt{(-1.189)^2 - 4(18.88)(0.000\ 472)}}{2(18.88)} = \dfrac{1.189 \pm 1.1739}{37.76}$

$x = 0.0626$ and 4.00×10^{-4}

Discard the larger solution (0.0626) because it leads to a negative concentration of NO_2 and that is impossible.

$n_{total} = (0.0100 - 2x + x)(10.0 \text{ L}) = (0.0100 - x)(10.0 \text{ L})$

$= [0.0100 \text{ mol/L} - (4.00 \times 10^{-4} \text{ mol/L})](10.0 \text{ L}) = 0.0960 \text{ mol}$

$100 \text{ °C} = 100 + 273 = 373 \text{ K}$

$$P_{total} = \frac{n_{total}RT}{V} = \frac{(0.0960 \text{ mol})\left(0.082\ 06\ \dfrac{\text{L} \cdot \text{atm}}{\text{K} \cdot \text{mol}}\right)(373 \text{ K})}{10.0 \text{ L}} = 0.294 \text{ atm}$$

13.114 (a) $W(s) + 4 Br(g) \rightleftharpoons WBr_4(g)$

$$K_p = \frac{P_{WBr_4}}{(P_{Br})^4} = 100, \quad P_{WBr_4} = (P_{Br})^4(100) = (0.010 \text{ atm})^4(100) = 1.0 \times 10^{-6} \text{ atm}$$

(b) Because K_p is smaller at the higher temperature, the reaction has shifted toward reactants at the higher temperature, which means the reaction is exothermic.

(c) At 2800 K, $Q_p = \dfrac{(1.0 \times 10^{-6})}{(0.010)^4} = 100$, $Q_p > K_p$ so the reaction will go from products to reactants, depositing tungsten back onto the filament.

13.115 (a)

$(NH_4)(NH_2CO_2)(s)$	$\rightleftharpoons$	$2 NH_3(g)$	$+$	$CO_2(g)$
initial (atm)		0		0
change (atm)		+2x		+x
equil (atm)		2x		x

$P_{total} = 2x + x = 3x = 0.116 \text{ atm}$

$x = 0.116 \text{ atm}/3 = 0.0387 \text{ atm}$

$P_{NH_3} = 2x = 2(0.0387 \text{ atm}) = 0.0774 \text{ atm}$

$P_{CO_2} = x = 0.0387 \text{ atm}$

$K_p = (P_{NH_3})^2(P_{CO_2}) = (0.0774)^2(0.0387) = 2.32 \times 10^{-4}$

(b) (i) The total quantity of NH_3 would decrease. When product, CO_2, is added, the equilibrium will shift to the left.

(ii) The total quantity of NH_3 would remain unchanged. Adding a pure solid, $(NH_4)(NH_2CO_2)$, to a heterogeneous equilibrium will not affect the position of the equilibrium.

(iii) The total quantity of NH_3 would increase. When product, CO_2, is removed, the equilibrium will shift to the right.

(iv) The total quantity of NH_3 would increase. When the total volume is increased, the reaction will shift to the side with more total moles of gas, which in this case is the product side.

(v) The total quantity of NH_3 would remain unchanged. Neon is an inert gas which will have no effect on the reaction or on the position of equilibrium.

(vi) The total quantity of NH_3 would increase. Because the reaction is endothermic, an increase in temperature will shift the equilibrium to to the right.

13.116 $2 NO_2(g) \rightleftharpoons N_2O_4(g)$

$\Delta n = (1) - (2) = -1$ and $K_p = K_c(RT)^{-1} = (216)[(0.082\ 06)(298)]^{-1} = 8.83$

$$K_p = \frac{P_{N_2O_4}}{(P_{NO_2})^2} = 8.83$$

Let $X = P_{N_2O_4}$ and $Y = P_{NO_2}$

$P_{total} = 1.50$ atm $= X + Y$ and $\dfrac{X}{Y^2} = 8.83$. Use these two equations to solve for X and Y.

$X = 1.50 - Y$

$$\frac{1.50 - Y}{Y^2} = 8.83$$

$8.83 Y^2 + Y - 1.50 = 0$

Use the quadratic formula to solve for Y.

$$Y = \frac{-(1) \pm \sqrt{(1)^2 - 4(8.83)(-1.50)}}{2(8.83)} = \frac{-1 \pm 7.35}{17.7}$$

$Y = -0.472$ and 0.359

Discard the negative solution (−0.472) because it leads to a negative partial pressure of NO_2 and that is impossible.

$Y = P_{NO_2} = 0.359$ atm

$X = P_{N_2O_4} = 1.50$ atm $- Y = 1.50$ atm $- 0.359$ atm $= 1.14$ atm

13.117 $500\ ^{\circ}C = 500 + 273 = 773$ K and $840\ ^{\circ}C = 840 + 273 = 1113$ K

Calculate the undissociated pressure of F_2 at 1113 K.

$$\frac{P_2}{T_2} = \frac{P_1}{T_1}; \quad P_2 = \frac{P_1 T_2}{T_1} = \frac{(0.600\ \text{atm})(1113\ \text{K})}{773\ \text{K}} = 0.864\ \text{atm}$$

	$F_2(g)$	$\rightleftharpoons$	$2 F(g)$
initial (atm)	0.864		0
change (atm)	−x		+2x
equil (atm)	0.864 − x		2x

$P_{total} = (0.864\ \text{atm} - x) + 2x = 0.864\ \text{atm} + x = 0.984\ \text{atm}$

$x = 0.984\ \text{atm} - 0.864\ \text{atm} = 0.120\ \text{atm}$

$P_{F_2} = (0.864\ \text{atm} - x) = (0.864\ \text{atm} - 0.120\ \text{atm}) = 0.744\ \text{atm}$

$P_F = 2x = 2(0.120\ \text{atm}) = 0.240\ \text{atm}$

$$K_p = \frac{(P_F)^2}{P_{F_2}} = \frac{(0.240)^2}{0.744} = 0.0774$$

13.118

	$N_2(g)$	$+$	$3 H_2(g)$	$\rightleftharpoons$	$2 NH_3(g)$
initial (mol)	0		0		X
change (mol)	+y		+3y		−2y
equil (mol)	y		3y		X − 2y

y = 0.200 mol
Because the volume is 1.00 L, the molarity equals the number of moles.
$[N_2] = y = 0.200$ M; $[H_2] = 3y = 3(0.200) = 0.600$ M

$$K_c = \frac{[NH_3]^2}{[N_2][H_2]^3} = \frac{[NH_3]^2}{(0.200)(0.600)^3} = 4.20 \text{, solve for } [NH_3]_{eq}$$

$$[NH_3]_{eq}^2 = [N_2][H_2]^3(4.20) = (0.200)(0.600)^3(4.20)$$

$$[NH_3]_{eq} = \sqrt{[N_2][H_2]^3(4.20)} = \sqrt{(0.200)(0.600)^3(4.20)} = 0.426 \text{ M}$$

$[NH_3]_{eq} = 0.426$ M $= X - 2(0.200) = [NH_3]_o - 2(0.200)$
$[NH_3]_o = 0.426 + 2(0.200) = 0.826$ M
0.826 mol of NH_3 were placed in the 1.00 L reaction vessel.

13.119 N_2O_4, 92.01 amu

$$\text{mol } N_2O_4 = 46.0 \text{ g} \times \frac{1 \text{ mol } N_2O_4}{92.01 \text{ g}} = 0.500 \text{ mol } N_2O_4$$

45 °C = 318 K
$K_p = K_c(RT)^{\Delta n}$; $\Delta n = 2 - 1 = 1$; $K_p = K_c(RT) = (0.619)(0.08\ 206)(318) = 16.2$

$$P_{N_2O_4} = \frac{nRT}{V} = \frac{(0.500 \text{ mol})\left(0.082\ 06 \dfrac{L \cdot atm}{K \cdot mol}\right)(318 \text{ K})}{2.00 \text{ L}} = 6.52 \text{ atm}$$

$$N_2O_4(g) \rightleftharpoons 2\ NO_2(g)$$

	$N_2O_4(g)$	$\rightleftharpoons$ $2\ NO_2(g)$
initial (atm)	6.52	0
change (atm)	–x	+2x
equil (atm)	6.52 – x	2x

$$K_p = \frac{(P_{NO_2})^2}{P_{N_2O_4}} = 16.2 = \frac{(2x)^2}{6.52 - x}$$

$(16.2)(6.52 - x) = 4x^2$
$105.6 - 16.2x = 4x^2$
$4x^2 + 16.2x - 105.6 = 0$
Use the quadratic formula to solve for x.

$$x = \frac{-(16.2) \pm \sqrt{(16.2)^2 - 4(4)(-105.6)}}{2(4)} = \frac{-16.2 \pm 44.2}{8}$$

x = –7.55 and 3.50
Discard the negative solution (–7.55) because it leads to a negative partial pressure and that is impossible.
$P_{N_2O_4} = 6.52 - x = 6.52 - 3.50 = 3.02$ atm

$P_{NO_2} = 2x = 2(3.50) = 7.00$ atm

13.120 ClF_3, 92.45 amu

(a) $mol\ ClF_3 = 9.25\ g \times \dfrac{1\ mol\ ClF_3}{92.45\ g} = 0.100\ mol\ ClF_3$

$[ClF_3] = \dfrac{0.100\ mol\ ClF_3}{2.00\ L} = 0.0500\ M$

	$ClF_3(g)$	$\rightleftharpoons$	$ClF(g)$	+	$F_2(g)$
initial (M)	0.0500		0		0
change (M)	–x		+x		+x
equil (M)	0.0500 – x		x		x
	0.0401		0.009 90		0.009 90

where x = 0.0500 x 0.198 = 0.009 90

$K_c = \dfrac{[ClF][F_2]}{[ClF_3]} = \dfrac{(0.009\ 90)^2}{0.0401} = 0.002\ 44$

(b) $K_p = K_c(RT)^{\Delta n}$; $\Delta n = 2 – 1 = 1$; $K_p = K_c(RT) = (0.002\ 44)(0.082\ 06)(700) = 0.140$

(c) $mol\ ClF_3 = 39.4\ g \times \dfrac{1\ mol\ ClF_3}{92.45\ g} = 0.426\ mol\ ClF_3$

$[ClF_3] = \dfrac{0.426\ mol\ ClF_3}{2.00\ L} = 0.213\ M$

	$ClF_3(g)$	$\rightleftharpoons$	$ClF(g)$	+	$F_2(g)$
initial (M)	0.213		0		0
change (M)	–x		+x		+x
equil (M)	0.213 – x		x		x

$K_c = \dfrac{[ClF][F_2]}{[ClF_3]} = 0.00\ 244 = \dfrac{x^2}{0.213 – x}$

$(0.002\ 44)(0.213 – x) = x^2$

$5.20 \times 10^{-4} – 0.002\ 44x = x^2$

$x^2 + 0.002\ 44x – (5.20 \times 10^{-4}) = 0$

Use the quadratic formula to solve for x.

$x = \dfrac{-(0.002\ 44) \pm \sqrt{(0.002\ 44)^2 – 4(1)(-5.20 \times 10^{-4})}}{2(1)} = \dfrac{-0.002\ 44 \pm 0.0457}{2}$

x = –0.0241 and 0.0216

Discard the negative solution (–0.0241) because it leads to a negative partial pressure and that is impossible.

$[ClF_3] = 0.213 – x = 0.213 – 0.0216 = 0.191\ M$

$[ClF] = [F_2] = x = 0.0216\ M$

Multiconcept Problems

13.121 (a) $K_p = \dfrac{(P_F)^2}{P_{F_2}} = 7.83$ at 1500 K

$P_F = \sqrt{K_p P_{F_2}} = \sqrt{(7.83)(0.200)} = 1.25$ atm

(b)
	$F_2(g)$	$\rightleftharpoons$	$2\,F(g)$
initial (atm)	x		0
change (atm)	−y		+2y
equil (atm)	x − y		2y

$2y = 1.25$ so $y = 0.625$

$x - y = 0.200$; $x = 0.200 + y = 0.200 + 0.625 = 0.825$

$f = \dfrac{0.625}{0.825} = 0.758$

(c) The shorter bond in F_2 is expected to be stronger. However, because of the small size of F, repulsion between the lone pairs of the two halogen atoms are much greater in F_2 than in Cl_2.

13.122 (a) $[N_2O_4] = \dfrac{0.500 \text{ mol}}{4.00 \text{ L}} = 0.125$ M

	$N_2O_4(g)$	$\rightleftharpoons$	$2\,NO_2(g)$
initial (M)	0.125		0
change (M)	−(0.793)(0.125)		+(2)(0.793)(0.125)
equil (M)	0.125 − (0.793)(0.125)		(2)(0.793)(0.125)

At equilibrium, $[N_2O_4] = 0.125 - (0.793)(0.125) = 0.0259$ M

$[NO_2] = (2)(0.793)(0.125) = 0.198$ M

$K_c = \dfrac{[NO_2]^2}{[N_2O_4]} = \dfrac{(0.198)^2}{(0.0259)} = 1.51$

$\Delta n = 2 - 1 = 1$ and $K_p = K_c(RT)^{\Delta n}$; $K_p = K_c(RT) = (1.51)(0.082\,06)(400) = 49.6$

(b)

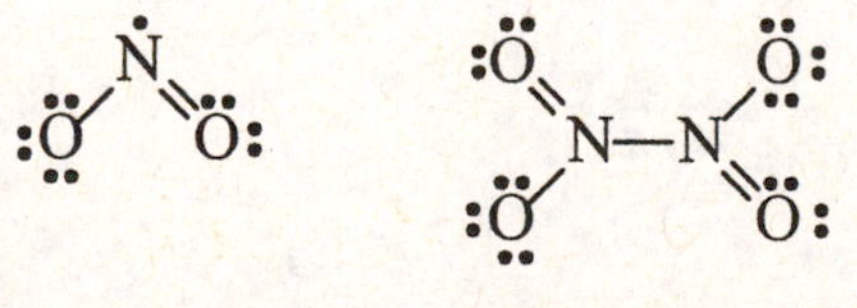

13.123 (a) Because $\Delta n = 0$, $K_p = K_c = 1.0 \times 10^5$

(b) $K_p = 1.0 \times 10^5 = \dfrac{P_{propene}}{P_{cyclopropane}}$; $P_{cyclopropane} = \dfrac{P_{propene}}{1.0 \times 10^5} = \dfrac{5.0 \text{ atm}}{1.0 \times 10^5} = 5.0 \times 10^{-5}$ atm

(c) The ratio of the two concentrations is equal to K_c. The ratio (K_c) cannot be changed by adding cyclopropane. The individual concentrations can change, but the ratio of the concentrations can't.

Because there is one mole of gas on each side of the balanced equation, the composition of the equilibrium mixture is unaffected by a decrease in volume. The ratio of the two concentrations will not change.

(d) Because K_c is large, $k_f > k_r$.

(e) Because the C–C–C bond angles are 60° and the angles between sp^3 hybrid orbitals are 109.5°, the hybrid orbitals are not oriented along the bond directions. Their overlap is therefore poor, and the C–C bonds are correspondingly weak.

13.124 2 monomer $\rightleftharpoons$ dimer

(a) In benzene, $K_c = 1.51 \times 10^2$

	2 monomer	$\rightleftharpoons$	dimer
initial (M)	0.100		0
change (M)	−2x		+x
equil (M)	0.100 − 2x		x

$$K_c = \frac{[\text{dimer}]}{[\text{monomer}]^2} = 1.51 \times 10^2 = \frac{x}{(0.100 - 2x)^2}$$

$604x^2 - 61.4x + 1.51 = 0$

Use the quadratic formula to solve for x.

$$x = \frac{-(-61.4) \pm \sqrt{(-61.4)^2 - (4)(604)(1.51)}}{2(604)} = \frac{61.4 \pm 11.04}{1208}$$

$x = 0.0600$ and 0.0417

Discard the larger solution (0.0600) because it gives a negative concentration of the monomer and that is impossible.

$[\text{monomer}] = 0.100 - 2x = 0.100 - 2(0.0417) = 0.017$ M; $[\text{dimer}] = x = 0.0417$ M

$$\frac{[\text{dimer}]}{[\text{monomer}]} = \frac{0.0417 \text{ M}}{0.017 \text{ M}} = 2.5$$

(b) In H_2O, $K_c = 3.7 \times 10^{-2}$

	2 monomer	$\rightleftharpoons$	dimer
initial (M)	0.100		0
change (M)	−2x		+x
equil (M)	0.100 − 2x		x

$$K_c = \frac{[\text{dimer}]}{[\text{monomer}]^2} = 3.7 \times 10^{-2} = \frac{x}{(0.100 - 2x)^2}$$

$0.148x^2 - 1.0148x + 0.000\,37 = 0$

Use the quadratic formula to solve for x.

$$x = \frac{-(-1.0148) \pm \sqrt{(-1.0148)^2 - (4)(0.148)(0.00037)}}{2(0.148)} = \frac{1.0148 \pm 1.0147}{0.296}$$

$x = 6.86$ and 3.4×10^{-4}

Discard the larger solution (6.86) because it gives a negative concentration of the monomer and that is impossible.

$[\text{monomer}] = 0.100 - 2x = 0.100 - 2(3.4 \times 10^{-4}) = 0.099$ M; $[\text{dimer}] = x = 3.4 \times 10^{-4}$ M

$$\frac{[\text{dimer}]}{[\text{monomer}]} = \frac{3.4 \times 10^{-4}\,M}{0.099\,M} = 0.0034$$

(c) K_c for the water solution is so much smaller than K_c for the benzene solution because H_2O can hydrogen bond with acetic acid, thus preventing acetic acid dimer formation. Benzene cannot hydrogen bond with acetic acid.

13.125 (a) H_2O, 18.015 amu; $125.4\text{ g }H_2O \times \dfrac{1\text{ mol }H_2O}{18.015\text{ g }H_2O} = 6.96\text{ mol }H_2O$

Given that mol CO = mol H_2O = 6.96 mol

$$P_{CO} = P_{H_2O} = \frac{nRT}{V} = \frac{(6.96\text{ mol})\left(0.082\,06\,\dfrac{\text{L}\cdot\text{atm}}{\text{K}\cdot\text{mol}}\right)(700\text{ K})}{10.0\text{ L}} = 40.0\text{ atm}$$

	$CO(g)$	+	$H_2O(g)$	⇌	$CO_2(g)$	+	$H_2(g)$
initial (atm)	40.0		40.0		0		0
equil (atm)	9.80		9.80		40.0 − 9.80		40.0 − 9.80
					= 30.2		= 30.2

$$K_p = \frac{(P_{CO_2})(P_{H_2})}{(P_{CO})(P_{H_2O})} = \frac{(30.2)(30.2)}{(9.80)(9.80)} = 9.50$$

(b) $31.4\text{ g }H_2O \times \dfrac{1\text{ mol }H_2O}{18.015\text{ g }H_2O} = 1.743\text{ mol }H_2O$

$$P_{H_2O} = \frac{nRT}{V} = \frac{(1.743\text{ mol})\left(0.082\,06\,\dfrac{\text{L}\cdot\text{atm}}{\text{K}\cdot\text{mol}}\right)(700\text{ K})}{10.0\text{ L}} = 10.0\text{ atm}$$

P_{H_2O} has been increased by 10.0 atm; a new equilibrium will be established.

	$CO(g)$	+	$H_2O(g)$	⇌	$CO_2(g)$	+	$H_2(g)$
initial (atm)	9.80		9.80 +10.0		30.2		30.2
change (atm)	−x		−x		+x		+x
equil (atm)	9.80 − x		19.8 − x		30.2 + x		30.2 + x

$$K_p = \frac{(P_{CO_2})(P_{H_2})}{(P_{CO})(P_{H_2O})} = 9.50 = \frac{(30.2 + x)(30.2 + x)}{(9.80 - x)(19.80 - x)}$$

$8.50x^2 - 341.6x + 931.34 = 0$

Use the quadratic formula to solve for x.

$$x = \frac{-(-341.6) \pm \sqrt{(-341.6)^2 - 4(8.50)(931.34)}}{2(8.50)} = \frac{341.6 \pm 291.6}{17.0}$$

x = 37.25 and 2.94

Discard the larger solution (37.25) because it leads to negative partial pressures and that is impossible.

$P_{CO} = 9.80 - x = 9.80 - 2.94 = 6.86$ atm

$P_{H_2O} = 19.8 - x = 19.8 - 2.94 = 16.9$ atm

$P_{CO_2} = P_{H_2} = 30.2 + x = 30.2 + 2.94 = 33.1$ atm

$$n_{H_2} = \frac{PV}{RT} = \frac{(33.1 \text{ atm})(10.0 \text{ L})}{\left(0.082\,06 \, \dfrac{\text{L} \cdot \text{atm}}{\text{K} \cdot \text{mol}}\right)(700 \text{ K})} = 5.76 \text{ mol } H_2$$

$$\frac{5.76 \text{ mol } H_2}{10.0 \text{ L}} \times \frac{1 \text{ L}}{1000 \text{ mL}} \times \frac{1 \text{ mL}}{1 \text{ cm}^3} \times \frac{6.022 \times 10^{23} \, H_2 \text{ molecules}}{1 \text{ mol } H_2} = 3.47 \times 10^{20} \, H_2 \text{ molecules/cm}^3$$

13.126 (a) CO_2, 44.01 amu; CO, 28.01 amu

$$79.2 \text{ g } CO_2 \times \frac{1 \text{ mol } CO_2}{44.01 \text{ g } CO_2} = 1.80 \text{ mol } CO_2$$

$$CO_2(g) \quad + \quad C(s) \quad \rightleftarrows \quad 2 \, CO(g)$$

initial (mol)	1.80		0
change (mol)	−x		+2x
equil (mol)	1.80 − x		2x

total mass of gas in flask = (16.3 g/L)(5.00 L) = 81.5 g

81.5 = (1.80 − x)(44.01) + (2x)(28.01)

81.5 = 79.22 − 44.01x + 56.02x; 2.28 = 12.01x; x = 2.28/12.01 = 0.19

$n_{CO_2} = 1.80 - x = 1.80 - 0.19 = 1.61$ mol CO_2; $n_{CO} = 2x = 2(0.19) = 0.38$ mol CO

$$P_{CO_2} = \frac{nRT}{V} = \frac{(1.61 \text{ mol})\left(0.082\,06 \, \dfrac{\text{L} \cdot \text{atm}}{\text{K} \cdot \text{mol}}\right)(1000 \text{ K})}{5.0 \text{ L}} = 26.4 \text{ atm}$$

$$P_{CO} = \frac{nRT}{V} = \frac{(0.38 \text{ mol})\left(0.082\,06 \, \dfrac{\text{L} \cdot \text{atm}}{\text{K} \cdot \text{mol}}\right)(1000 \text{ K})}{5.0 \text{ L}} = 6.24 \text{ atm}$$

$$K_p = \frac{(P_{CO})^2}{(P_{CO_2})} = \frac{(6.24)^2}{(26.4)} = 1.47$$

(b) At 1100K, the total mass of gas in flask = (16.9 g/L)(5.00 L) = 84.5 g

84.5 = (1.80 − x)(44.01) + (2x)(28.01)

84.5 = 79.22 − 44.01x + 56.02x; 5.28 = 12.01x; x = 5.28/12.01 = 0.44

$n_{CO_2} = 1.80 - x = 1.80 - 0.44 = 1.36$ mol CO_2; $n_{CO} = 2x = 2(0.44) = 0.88$ mol CO

$$P_{CO_2} = \frac{nRT}{V} = \frac{(1.36 \text{ mol})\left(0.082\,06 \, \dfrac{\text{L} \cdot \text{atm}}{\text{K} \cdot \text{mol}}\right)(1100 \text{ K})}{5.0 \text{ L}} = 24.6 \text{ atm}$$

$$P_{CO} = \frac{nRT}{V} = \frac{(0.88 \text{ mol})\left(0.082\,06 \, \dfrac{\text{L} \cdot \text{atm}}{\text{K} \cdot \text{mol}}\right)(1100 \text{ K})}{5.0 \text{ L}} = 15.9 \text{ atm}$$

$$K_p = \frac{(P_{CO})^2}{(P_{CO_2})} = \frac{(15.9)^2}{(24.6)} = 10.3$$

(c) In agreement with Le Châtelier's principle, the reaction is endothermic because K_p increases with increasing temperature.

13.127 CO_2, 44.01 amu; CO, 28.01 amu; $BaCO_3$, 197.34 amu

$$1.77 \text{ g CO}_2 \times \frac{1 \text{ mol CO}_2}{44.01 \text{ g CO}_2} = 0.0402 \text{ mol CO}_2$$

	$CO_2(g)$	+	C(s)	$\rightleftharpoons$	2 CO(g)
initial (mol)	0.0402				0
change (mol)	−x				+2x
equil (mol)	0.0402 − x				2x

$$3.41 \text{ g BaCO}_3 \times \frac{1 \text{ mol BaCO}_3}{197.34 \text{ g BaCO}_3} \times \frac{1 \text{ mol CO}_2}{1 \text{ mol BaCO}_3} = 0.0173 \text{ mol CO}_2$$

mol CO_2 = 0.0173 = 0.0402 − x;   x = 0.0229
mol CO = 2x = 2(0.0229) = 0.0458 mol CO

$$P_{CO_2} = \frac{nRT}{V} = \frac{(0.0173 \text{ mol})\left(0.082\,06\,\frac{\text{L} \cdot \text{atm}}{\text{K} \cdot \text{mol}}\right)(1100 \text{ K})}{1.000 \text{ L}} = 1.562 \text{ atm}$$

$$P_{CO} = \frac{nRT}{V} = \frac{(0.0458 \text{ mol})\left(0.082\,06\,\frac{\text{L} \cdot \text{atm}}{\text{K} \cdot \text{mol}}\right)(1100 \text{ K})}{1.000 \text{ L}} = 4.134 \text{ atm}$$

$$K_p = \frac{(P_{CO})^2}{(P_{CO_2})} = \frac{(4.134)^2}{(1.562)} = 11.0$$

13.128 (a) N_2O_4, 92.01 amu

$$14.58 \text{ g N}_2O_4 \times \frac{1 \text{ mol N}_2O_4}{92.01 \text{ g N}_2O_4} = 0.1585 \text{ mol N}_2O_4$$

$$PV = nRT \qquad P_{N_2O_4} = \frac{nRT}{V} = \frac{(0.1585 \text{ mol})\left(0.082\,06\,\frac{\text{L} \cdot \text{atm}}{\text{K} \cdot \text{mol}}\right)(400 \text{ K})}{1.000 \text{ L}} = 5.20 \text{ atm}$$

	$N_2O_4(g)$	$\rightleftharpoons$	2 NO$_2$(g)
initial (atm)	5.20		0
change (atm)	−x		+2x
equil (atm)	5.20 − x		2x

$P_{total} = P_{N_2O_4} + P_{NO_2} = (5.20 - x) + (2x) = 9.15 \text{ atm}$

5.20 + x = 9.15 atm
x = 3.95 atm

$P_{N_2O_4} = 5.20 - x = 5.20 - 3.95 = 1.25$ atm

$P_{NO_2} = 2x = 2(3.95) = 7.90$ atm

$K_p = \dfrac{(P_{NO_2})^2}{(P_{N_2O_4})} = \dfrac{(7.90)^2}{(1.25)} = 49.9$

$\Delta n = 1$ and $K_c = K_p\left(\dfrac{1}{RT}\right) = \dfrac{(49.9)}{(0.082\ 06)(400)} = 1.52$

(b) $\Delta H^\circ_{rxn} = [2\ \Delta H^\circ_f(NO_2)] - \Delta H^\circ_f(N_2O_4)$

$\Delta H^\circ_{rxn} = [(2\ mol)(33.2\ kJ/mol)] - [(1mol)(11.1\ kJ/mol)] = 55.3$ kJ

moles N_2O_4 reacted $= n = \dfrac{PV}{RT} = \dfrac{(3.95\ atm)(1.000\ L)}{\left(0.082\ 06\ \dfrac{L\cdot atm}{K\cdot mol}\right)(400\ K)} = 0.1203$ mol N_2O_4

$q = (55.3\ kJ/mol\ N_2O_4)(0.1203\ mol\ N_2O_4) = 6.65$ kJ

13.129 $C_{10}H_8(s) \rightleftharpoons C_{10}H_8(g)$

(a) $K_c = [C_{10}H_8] = 5.40 \times 10^{-6}$

room volume $= 8.0\ ft \times 10.0\ ft \times 8.0\ ft = 640\ ft^3$

room volume $= 640\ ft^3 \times \left(\dfrac{12\ in.}{1\ ft}\right)^3 \times \left(\dfrac{2.54\ cm}{1\ in.}\right)^3 \times \left(\dfrac{1.0\ L}{1000\ cm^3}\right) = 18{,}122.8$ L

$C_{10}H_8$ molecules $= (5.40 \times 10^{-6}\ mol/L)(18{,}122.8\ L)(6.022 \times 10^{23}\ molecules/mol)$

$= 5.89 \times 10^{22}\ C_{10}H_8$ molecules

(b) $C_{10}H_8$, 128.17 amu

mol $C_{10}H_8 = (5.40 \times 10^{-6}\ mol/L)(18{,}122.8\ L) = 0.0979$ mol $C_{10}H_8$

mass $C_{10}H_8 = 0.0979\ mol\ C_{10}H_8 \times \dfrac{128.17\ g\ C_{10}H_8}{1\ mol\ C_{10}H_8} = 12.55\ g\ C_{10}H_8$

moth ball: $r = 12.0\ mm/2 = 6.0\ mm = 0.60$ cm

volume of moth ball $= (4/3)\pi r^3 = (4/3)\pi(0.60cm)^3 = 0.905\ cm^3$

mass of moth ball $= (0.905\ cm^3/moth\ ball)(1.16\ g/cm^3) = 1.05$ g/moth ball

number of moth balls $= \dfrac{12.55\ g\ C_{10}H_8}{1.05\ g\ C_{10}H_8/moth\ ball} = 12$ moth balls

13.130 The atmosphere is 21% (0.21) O_2; $P_{O_2} = (0.21)\left(720\ mm\ Hg \times \dfrac{1\ atm}{760\ mm\ Hg}\right) = 0.199$ atm

$2\ O_3(g) \rightleftharpoons 3\ O_2(g)$

$K_p = \dfrac{(P_{O_2})^3}{(P_{O_3})^2};\qquad P_{O_3} = \sqrt{\dfrac{(P_{O_2})^3}{K_p}} = \sqrt{\dfrac{(0.199)^3}{1.3 \times 10^{57}}} = 2.46 \times 10^{-30}$ atm

$vol = 10 \times 10^6\ m^3 \times \left(\dfrac{100\ cm}{1\ m}\right)^3 \times \dfrac{1\ L}{1000\ cm^3} = 1.0 \times 10^{10}$ L

$$n_{O_3} = \frac{PV}{RT} = \frac{(2.46 \times 10^{-30} \text{ atm})(1.0 \times 10^{10} \text{ L})}{\left(0.082\ 06\ \frac{\text{L} \cdot \text{atm}}{\text{K} \cdot \text{mol}}\right)(298 \text{ K})} = 1.0 \times 10^{-21} \text{ mol O}_3$$

$$\text{O}_3 \text{ molecules} = 1.0 \times 10^{-21} \text{ mol O}_3 \times \frac{6.022 \times 10^{23} \text{ O}_3 \text{ molecules}}{1 \text{ mol O}_3} = 6.0 \times 10^2 \text{ O}_3 \text{ molecules}$$

13.131 250.0 mL = 0.2500 L

$[CH_3CO_2H] = 0.0300 \text{ mol}/0.2500 \text{ L} = 0.120 \text{ M}$

$$2 \text{ CH}_3\text{CO}_2\text{H} \rightleftharpoons (\text{CH}_3\text{CO}_2\text{H})_2$$

	2 CH₃CO₂H	(CH₃CO₂H)₂
initial (M)	0.120	0
change (M)	−2x	+x
equil (M)	0.120 − 2x	x

$$K_c = \frac{[\text{dimer}]}{[\text{monomer}]^2} = 1.51 \times 10^2 = \frac{x}{(0.120 - 2x)^2}$$

$604x^2 - 73.48x + 2.1744 = 0$

Use the quadratic formula to solve for x.

$$x = \frac{-(-73.48) \pm \sqrt{(-73.48)^2 - 4(604)(2.1744)}}{2(604)} = \frac{73.48 \pm 12.08}{1208}$$

x = 0.0708 and 0.0508

Discard the larger solution (0.0708) because it leads to a negative concentration and that is impossible.

[monomer] = 0.120 − 2x = 0.120 − 2(0.0508) = 0.0184 M

[dimer] = x = 0.0508 M

(b) 25 °C = 25 + 273 = 298 K

$$\Pi = MRT = (0.0184 \text{ M} + 0.0508 \text{ M})\left(0.082\ 06\ \frac{\text{L} \cdot \text{atm}}{\text{K} \cdot \text{mol}}\right)(298\text{K}) = 1.69 \text{ atm}$$

13.132 $\text{PCl}_5(g) \rightleftharpoons \text{PCl}_3(g) + \text{Cl}_2(g)$

$\Delta n = (2) - (1) = 1$ and at 700 K, $K_p = K_c(RT) = (46.9)(0.082\ 06)(700) = 2694$

(a) Because K_p is larger at the higher temperature, the reaction has shifted toward products at the higher temperature, which means the reaction is endothermic. Because the reaction involves breaking two P–Cl bonds and forming just one Cl–Cl bond, it should be endothermic.

(b) PCl₅, 208.24 amu

$$\text{mol PCl}_5 = 1.25 \text{ g PCl}_5 \times \frac{1 \text{ mol PCl}_5}{208.24 \text{ g PCl}_5} = 6.00 \times 10^{-3} \text{ mol}$$

$$PV = nRT, \quad P_{PCl_5} = \frac{nRT}{V} = \frac{(6.00 \times 10^{-3} \text{ mol})\left(0.082\ 06\ \frac{\text{L} \cdot \text{atm}}{\text{K} \cdot \text{mol}}\right)(700 \text{ K})}{0.500 \text{ L}} = 0.689 \text{ atm}$$

Because K_p is so large, first assume the reaction goes to completion and then allow for a small back reaction.

$$\begin{array}{lccc}
 & PCl_5(g) & \rightleftharpoons \quad PCl_3(g) & + \quad Cl_2(g) \\
\text{before rxn (atm)} & 0.689 & 0 & 0 \\
\text{100\% rxn (atm)} & -0.689 & +0.689 & +0.689 \\
\text{after rxn (atm)} & 0 & 0.689 & 0.689 \\
\text{back rxn (atm)} & +x & -x & -x \\
\text{equil (atm)} & x & 0.689 - x & 0.689 - x
\end{array}$$

$$K_p = \frac{(P_{PCl_3})(P_{Cl_2})}{P_{PCl_5}} = 2694 = \frac{(0.689-x)^2}{x} \approx \frac{(0.689)^2}{x}$$

$$x = P_{PCl_5} = \frac{(0.689)^2}{2694} = 1.76 \times 10^{-4} \text{ atm}$$

$$P_{total} = P_{PCl_5} + P_{PCl_3} + P_{Cl_2}$$

$$P_{total} = x + (0.689 - x) + (0.689 - x) = 0.689 + 0.689 - 1.76 \times 10^{-4} = 1.38 \text{ atm}$$

$$\% \text{ dissociation} = \frac{(P_{PCl_5})_o - (P_{PCl_5})}{(P_{PCl_5})_o} \times 100\% = \frac{0.689 - (1.76 \times 10^{-4})}{0.689} \times 100\% = 99.97\%$$

(c)

The molecular geometry is trigonal bipyramidal. There is no dipole moment because of a symmetrical distribution of Cl's around the central P.

The molecular geometry is trigonal pyramidal. There is a dipole moment because of the lone pair of electrons on the P and an unsymmetrical distribution of Cl's around the central P.

14   Aqueous Equilibria: Acids and Bases

14.1   (a) $H_2SO_4(aq) + H_2O(l) \rightleftharpoons H_3O^+(aq) + HSO_4^-(aq)$

conjugate base

(b) $HSO_4^-(aq) + H_2O(l) \rightleftharpoons H_3O^+(aq) + SO_4^{2-}(aq)$

conjugate base

(c) $H_3O^+(aq) + H_2O(l) \rightleftharpoons H_3O^+(aq) + H_2O(l)$

conjugate base

(d) $NH_4^+(aq) + H_2O(l) \rightleftharpoons H_3O^+(aq) + NH_3(aq)$

conjugate base

14.2   (a) $HCO_3^-(aq) + H_2O(l) \rightleftharpoons H_2CO_3(aq) + OH^-(aq)$

conjugate acid

(b) $CO_3^{2-}(aq) + H_2O(l) \rightleftharpoons HCO_3^-(aq) + OH^-(aq)$

conjugate acid

(c) $OH^-(aq) + H_2O(l) \rightleftharpoons H_2O(l)(aq) + OH^-(aq)$

conjugate acid

(d) $H_2PO_4^-(aq) + H_2O(l) \rightleftharpoons H_3PO_4(aq) + OH^-(aq)$

conjugate acid

14.3 $HCl(aq) + NH_3(aq) \rightleftharpoons NH_4^+(aq) + Cl^-(aq)$

acid   base   acid   base

conjugate acid-base pairs

14.4   (a) $HF(aq) + NO_3^-(aq) \rightleftharpoons HNO_3(aq) + F^-(aq)$

HNO_3 is a stronger acid than HF, and F^- is a stronger base than NO_3^- (see Table 14.1). Because proton transfer occurs from the stronger acid to the stronger base, the reaction proceeds from right to left.

(b) $NH_4^+(aq) + CO_3^{2-}(aq) \rightleftharpoons HCO_3^-(aq) + NH_3(aq)$

NH_4^+ is a stronger acid than HCO_3^-, and CO_3^{2-} is a stronger base than NH_3 (see Table 14.1). Because proton transfer occurs from the stronger acid to the stronger base, the reaction proceeds from left to right.

14.5   (a) Both HX and HY have the same initial concentration. HY is more dissociated than HX. Therefore, HY is the stronger acid.

(b) The conjugate base (X^-) of the weaker acid (HX) is the stronger base.

(c) $HX + Y^- \rightleftharpoons HY + X^-$; Proton transfer occurs from the stronger acid to the stronger base. The reaction proceeds to the left.

369

14.6 $[H_3O^+] = \dfrac{K_w}{[OH^-]} = \dfrac{1.0 \times 10^{-14}}{5.0 \times 10^{-6}} = 2.0 \times 10^{-9}$ M

Because $[OH^-] > [H_3O^+]$, the solution is basic.

14.7 $K_w = [H_3O^+][OH^-];$ In a neutral solution, $[H_3O^+] = [OH^-]$

At 50 °C, $[H_3O^+] = [OH^-] = \sqrt{K_w} = \sqrt{5.5 \times 10^{-14}} = 2.3 \times 10^{-7}$ M

14.8   (a) $[H_3O^+] = \dfrac{K_w}{[OH^-]} = \dfrac{1.0 \times 10^{-14}}{1.58 \times 10^{-6}} = 6.3 \times 10^{-9}$ M

pH $= -\log[H_3O^+] = -\log(6.3 \times 10^{-9}) = 8.20$

(b)   pH $= -\log[H_3O^+] = -\log(6.0 \times 10^{-5}) = 4.22$

14.9   (a) $[H_3O^+] = 10^{-pH} = 10^{-7.40} = 4.0 \times 10^{-8}$ M

$[OH^-] = \dfrac{K_w}{[H_3O^+]} = \dfrac{1.0 \times 10^{-14}}{4.0 \times 10^{-8}} = 2.5 \times 10^{-7}$ M

(b) $[H_3O^+] = 10^{-pH} = 10^{-2.8} = 2 \times 10^{-3}$ M

$[OH^-] = \dfrac{K_w}{[H_3O^+]} = \dfrac{1.0 \times 10^{-14}}{2 \times 10^{-3}} = 5 \times 10^{-12}$ M

14.10   (a)   Because $HClO_4$ is a strong acid, $[H_3O^+] = 0.050$ M.

pH $= -\log[H_3O^+] = -\log(0.050) = 1.30$

(b)   Because HCl is a strong acid, $[H_3O^+] = 6.0$ M.

pH $= -\log[H_3O^+] = -\log(6.0) = -0.78$

(c)   Because KOH is a strong base, $[OH^-] = 4.0$ M.

$[H_3O^+] = \dfrac{K_w}{[OH^-]} = \dfrac{1.0 \times 10^{-14}}{4.0} = 2.5 \times 10^{-15}$ M

pH $= -\log[H_3O^+] = -\log(2.5 \times 10^{-15}) = 14.60$

(d)   Because $Ba(OH)_2$ is a strong base, $[OH^-] = 2(0.010$ M$) = 0.020$ M.

$[H_3O^+] = \dfrac{K_w}{[OH^-]} = \dfrac{1.0 \times 10^{-14}}{0.020} = 5.0 \times 10^{-13}$ M

pH $= -\log[H_3O^+] = -\log(5.0 \times 10^{-13}) = 12.30$

14.11 $BaO(s) + H_2O(l) \rightarrow Ba(OH)_2(aq);$ BaO, 153.33 amu

0.25 g BaO $\times \dfrac{1 \text{ mol BaO}}{153.33 \text{ g BaO}} \times \dfrac{1 \text{ mol } Ba(OH)_2}{1 \text{ mol BaO}} \times \dfrac{2 \text{ mol } OH^-}{1 \text{ mol } Ba(OH)_2} = 3.26 \times 10^{-3}$ mol OH^-

$[OH^-] = \dfrac{3.26 \times 10^{-3} \text{ mol } OH^-}{0.500 \text{ L}} = 6.52 \times 10^{-3}$ M

$[H_3O^+] = \dfrac{K_w}{[OH^-]} = \dfrac{1.0 \times 10^{-14}}{6.52 \times 10^{-3}} = 1.53 \times 10^{-12}$ M

pH $= -\log[H_3O^+] = -\log(1.53 \times 10^{-12}) = 11.81$

14.12 $$HOCl(aq) + H_2O(l) \rightleftharpoons H_3O^+(aq) + OCl^-(aq)$$

initial (M)	0.10	~0	0
change (M)	$-x$	$+x$	$+x$
equil (M)	$0.10 - x$	x	x

$x = [H_3O^+] = 10^{-pH} = 10^{-4.23} = 5.9 \times 10^{-5}$ M

$[OCl^-] = x = 5.9 \times 10^{-5}$ M; $[HOCl] = 0.10 - x = (0.10 - 5.9 \times 10^{-5})$ M

$$K_a = \frac{[H_3O^+][OCl^-]}{[HOCl]} = \frac{(5.9 \times 10^{-5})(5.9 \times 10^{-5})}{(0.10 - 5.9 \times 10^{-5})} = 3.5 \times 10^{-8}$$

This value of K_a agrees with the value in Table 14.2.

14.13 (a) HZ is completely dissociated. HX and HY are at the same concentration and HX is more dissociated than HY. The strongest acid is HZ, the weakest is HY.
K_a (HY) $<$ K_a (HX) $<$ K_a (HZ)
(b) HZ
(c) HY has the highest pH; HX has the lowest pH (highest $[H_3O^+]$).

14.14 (a) $$CH_3CO_2H(aq) + H_2O(l) \rightleftharpoons H_3O^+(aq) + CH_3CO_2^-(aq)$$

initial (M)	1.00	~0	0
change (M)	$-x$	$+x$	$+x$
equil (M)	$1.00 - x$	x	x

$$K_a = \frac{[H_3O^+][CH_3CO_2^-]}{[CH_3CO_2H]} = 1.8 \times 10^{-5} = \frac{x^2}{1.00 - x} \approx \frac{x^2}{1.00}$$

Solve for x. $x = [H_3O^+] = 4.2 \times 10^{-3}$ M

$pH = -\log[H_3O^+] = -\log(4.2 \times 10^{-3}) = 2.38$

$[CH_3CO_2^-] = x = 4.2 \times 10^{-3}$ M; $[CH_3CO_2H] = 1.00 - x = 1.00$ M

$$[OH^-] = \frac{K_w}{[H_3O^+]} = \frac{1.0 \times 10^{-14}}{4.2 \times 10^{-3}} = 2.4 \times 10^{-12}$$ M

(b) $$CH_3CO_2H(aq) + H_2O(l) \rightleftharpoons H_3O^+(aq) + CH_3CO_2^-(aq)$$

initial (M)	0.0100	~0	0
change (M)	$-x$	$+x$	$+x$
equil (M)	$0.0100 - x$	x	x

$$K_a = \frac{[H_3O^+][CH_3CO_2^-]}{[CH_3CO_2H]} = 1.8 \times 10^{-5} = \frac{x^2}{0.0100 - x}$$

$x^2 + (1.8 \times 10^{-5})x - (1.8 \times 10^{-7}) = 0$

Use the quadratic formula to solve for x.

$$x = \frac{-(1.8 \times 10^{-5}) \pm \sqrt{(1.8 \times 10^{-5})^2 - 4(-1.8 \times 10^{-7})}}{2(1)} = \frac{(-1.8 \times 10^{-5}) \pm (8.5 \times 10^{-4})}{2}$$

$x = 4.2 \times 10^{-4}$ and -4.3×10^{-4}

Of the two solutions for x, only the positive value of x has physical meaning because x is the $[H_3O^+]$.

$x = [H_3O^+] = 4.2 \times 10^{-4}$ M

$pH = -\log[H_3O^+] = -\log(4.2 \times 10^{-4}) = 3.38$

$[CH_3CO_2^-] = x = 4.2 \times 10^{-4} \ M$

$[CH_3CO_2H] = 0.0100 - x = 0.0100 - (4.2 \times 10^{-4}) = 0.0096 \ M$

$[OH^-] = \dfrac{K_w}{[H_3O^+]} = \dfrac{1.0 \times 10^{-14}}{4.2 \times 10^{-4}} = 2.4 \times 10^{-11} \ M$

14.15 $C_6H_8O_6$, 176.13 amu; 250 mg = 0.250 g; 250 mL = 0.250 L

$$[C_6H_8O_6] = \dfrac{\left(0.250 \ g \times \dfrac{1 \ mol}{176.13 \ g}\right)}{0.250 \ L} = 5.68 \times 10^{-3} \ M$$

$$C_6H_8O_6(aq) + H_2O(l) \rightleftharpoons H_3O^+(aq) + C_6H_7O_6^-(aq)$$

initial (M)	5.68×10^{-3}	~0	0
change (M)	$-x$	$+x$	$+x$
equil (M)	$(5.68 \times 10^{-3}) - x$	x	x

$K_a = \dfrac{[H_3O^+][C_6H_7O_6^-]}{[C_6H_8O_6]} = 8.0 \times 10^{-5} = \dfrac{x^2}{(5.68 \times 10^{-3}) - x}$

$x^2 + (8.0 \times 10^{-5})x - (4.54 \times 10^{-7}) = 0$

Use the quadratic formula to solve for x.

$$x = \dfrac{-(8.0 \times 10^{-5}) \pm \sqrt{(8.0 \times 10^{-5})^2 - (4)(-4.54 \times 10^{-7})}}{2(1)} = \dfrac{(-8.0 \times 10^{-5}) \pm 0.001 \ 35}{2}$$

$x = 6.35 \times 10^{-4}$ and -7.15×10^{-4}

Of the two solutions for x, only the positive value of x has physical meaning because x is the $[H_3O^+]$.

$x = [H_3O^+] = 6.35 \times 10^{-4} \ M$

$pH = -\log[H_3O^+] = -\log(6.35 \times 10^{-4}) = 3.20$

14.16 (a) From Example 14.10 in the text:

$[H_3O^+] = [HF]_{diss} = 4.0 \times 10^{-3} \ M$

% dissociation $= \dfrac{[HF]_{diss}}{[HF]_{initial}} \times 100\% = \dfrac{4.0 \times 10^{-3} \ M}{0.050 \ M} \times 100\% = 8.0\%$ dissociation

(b) $$HF(aq) + H_2O(l) \rightleftharpoons H_3O^+(aq) + F^-(aq)$$

initial (M)	0.50	~0	0
change (M)	$-x$	$+x$	$+x$
equil (M)	$0.50 - x$	x	x

$K_a = \dfrac{[H_3O^+][F^-]}{[HF]} = 3.5 \times 10^{-4} = \dfrac{x^2}{0.50 - x}$

$x^2 + (3.5 \times 10^{-4})x - (1.75 \times 10^{-4}) = 0$

Use the quadratic formula to solve for x.

$$x = \dfrac{-(3.5 \times 10^{-4}) \pm \sqrt{(3.5 \times 10^{-4})^2 - 4(1)(-1.75 \times 10^{-4})}}{2(1)} = \dfrac{(-3.5 \times 10^{-4}) \pm 0.0265}{2}$$

x = 0.0131 and −0.0134

Of the two solutions for x, only the positive value of x has physical meaning, because x is the $[H_3O^+]$.

$[H_3O^+] = [HF]_{diss} = 0.013$ M

% dissociation $= \dfrac{[HF]_{diss}}{[HF]_{initial}} \times 100\% = \dfrac{0.013\ M}{0.50\ M} \times 100\% = 2.6\%$ dissociation

14.17

$$H_2SO_3(aq) + H_2O(l) \rightleftharpoons H_3O^+(aq) + HSO_3^-(aq)$$

initial (M)	0.10	~0	0
change (M)	−x	+x	+x
equil (M)	0.10 − x	x	x

$K_{a1} = \dfrac{[H_3O^+][HSO_3^-]}{[H_2SO_3]} = 1.5 \times 10^{-2} = \dfrac{x^2}{0.10 - x}$

$x^2 + 0.015x − 0.0015 = 0$

Use the quadratic formula to solve for x.

$x = \dfrac{-(0.015) \pm \sqrt{(0.015)^2 - (4)(-0.0015)}}{2(1)} = \dfrac{-0.015 \pm 0.079}{2}$

x = 0.032 and −0.047

Of the two solutions for x, only the positive value of x has physical meaning since x is the $[H_3O^+]$.

x = $[H_3O^+]$ = $[HSO_3^-]$ = 0.032 M; $[H_2SO_3]$ = 0.10 − x = 0.10 − 0.032 = 0.07 M

The second dissociation of H_2SO_3 produces a negligible amount of H_3O^+ compared with that from the first dissociation.

$HSO_3^-(aq) + H_2O(l) \rightleftharpoons H_3O^+(aq) + SO_3^{2-}(aq)$

$K_{a2} = \dfrac{[H_3O^+][SO_3^{2-}]}{[HSO_3^-]} = 6.3 \times 10^{-8} = \dfrac{(0.032)[SO_3^{2-}]}{(0.032)}$

$[SO_3^{2-}] = K_{a2} = 6.3 \times 10^{-8}$ M

$[OH^-] = \dfrac{K_w}{[H_3O^+]} = \dfrac{1.0 \times 10^{-14}}{0.032} = 3.1 \times 10^{-13}$ M

pH = −log$[H_3O^+]$ = −log(0.032) = 1.49

14.18 From the complete dissociation of the first proton, $[H_3O^+] = [HSeO_4^-] = 0.50$ M.

For the dissociation of the second proton, the following equilibrium must be considered:

$$HSeO_4^-(aq) + H_2O(l) \rightleftharpoons H_3O^+(aq) + SeO_4^{2-}(aq)$$

initial (M)	0.50	0.50	0
change (M)	−x	+x	+x
equil (M)	0.50 − x	0.50 + x	x

$K_{a2} = \dfrac{[H_3O^+][SeO_4^{2-}]}{[HSeO_4^-]} = 1.2 \times 10^{-2} = \dfrac{(0.50 + x)(x)}{0.50 - x}$

$x^2 + 0.512x − 0.0060 = 0$

Use the quadratic formula to solve for x.

$$x = \frac{-(0.512) \pm \sqrt{(0.512)^2 - 4(-0.0060)}}{2(1)} = \frac{-0.512 \pm 0.535}{2}$$

x = 0.011 and −0.524

Of the two solutions for x, only the positive value of x has physical meaning, since x is the $[SeO_4^{2-}]$.

$[H_2SeO_4] = 0$ M; $[HSeO_4^-] = 0.50 - x = 0.49$ M; $[SeO_4^{2-}] = x = 0.011$ M

$[H_3O^+] = 0.50 + x = 0.51$ M

pH = $-\log[H_3O^+] = -\log(0.51) = 0.29$

$$[OH^-] = \frac{K_w}{[H_3O^+]} = \frac{1.0 \times 10^{-14}}{0.51} = 2.0 \times 10^{-14}\text{ M}$$

14.19
$$NH_3(aq) + H_2O(l) \rightleftarrows NH_4^+(aq) + OH^-(aq)$$

initial (M)	0.40	0	~0
change (M)	−x	+x	+x
equil (M)	0.40 − x	x	x

$$K_b = \frac{[NH_4^+][OH^-]}{[NH_3]} = 1.8 \times 10^{-5} = \frac{x^2}{0.40 - x} \approx \frac{x^2}{0.40}$$

Solve for x. x = $[OH^-] = 2.7 \times 10^{-3}$ M

$[NH_4^+] = x = 2.7 \times 10^{-3}$ M; $[NH_3] = 0.40 - x = 0.40$ M

$$[H_3O^+] = \frac{K_w}{[OH^-]} = \frac{1.0 \times 10^{-14}}{2.7 \times 10^{-3}} = 3.7 \times 10^{-12}\text{ M}$$

pH = $-\log[H_3O^+] = -\log(3.7 \times 10^{-12}) = 11.43$

14.20 $C_{21}H_{22}N_2O_2$, 334.42 amu; 16 mg = 0.016 g

$$\text{molarity} = \frac{\left(0.016\text{ g} \times \dfrac{1\text{ mol}}{334.42\text{ g}}\right)}{0.100\text{ L}} = 4.8 \times 10^{-4}\text{ M}$$

$$C_{21}H_{22}N_2O_2(aq) + H_2O(l) \rightleftarrows C_{21}H_{23}N_2O_2^+(aq) + OH^-(aq)$$

initial (M)	4.8×10^{-4}	0	~0
change (M)	−x	+x	+x
equil (M)	$(4.8 \times 10^{-4}) - x$	x	x

$$K_b = \frac{[C_{21}H_{23}N_2O_2^+][OH^-]}{[C_{21}H_{22}N_2O_2]} = 1.8 \times 10^{-6} = \frac{x^2}{(4.8 \times 10^{-4}) - x}$$

$x^2 + (1.8 \times 10^{-6})x - (8.6 \times 10^{-10}) = 0$

Use the quadratic formula to solve for x.

$$x = \frac{-(1.8 \times 10^{-6}) \pm \sqrt{(1.8 \times 10^{-6})^2 - (4)(-8.6 \times 10^{-10})}}{2(1)} = \frac{(-1.8 \times 10^{-6}) \pm (5.87 \times 10^{-5})}{2}$$

x = 2.84×10^{-5} and -3.02×10^{-5}

Of the two solutions for x, only the positive value of x has physical meaning, because x is the $[OH^-]$.

$[OH^-] = 2.84 \times 10^{-5}$ M

$[H_3O^+] = \dfrac{K_w}{[OH^-]} = \dfrac{1.0 \times 10^{-14}}{2.84 \times 10^{-5}} = 3.52 \times 10^{-10}$ M

pH $= -\log[H_3O^+] = -\log(3.52 \times 10^{-10}) = 9.45$

14.21 (a) $K_a = \dfrac{K_w}{K_b \text{ for } C_5H_{11}N} = \dfrac{1.0 \times 10^{-14}}{1.3 \times 10^{-3}} = 7.7 \times 10^{-12}$

(b) $K_b = \dfrac{K_w}{K_a \text{ for HOCl}} = \dfrac{1.0 \times 10^{-14}}{3.5 \times 10^{-8}} = 2.9 \times 10^{-7}$

(c) pK_b for HCO_2^- = 14.00 − pK_a = 14.00 − 3.74 = 10.26

14.22 (a) 0.25 M NH_4Br

NH_4^+ is an acidic cation. Br^- is a neutral anion. The salt solution is acidic.

For NH_4^+, $K_a = \dfrac{K_w}{K_b \text{ for } NH_3} = \dfrac{1.0 \times 10^{-14}}{1.8 \times 10^{-5}} = 5.6 \times 10^{-10}$

	$NH_4^+(aq)$ + $H_2O(l)$ ⇌	$H_3O^+(aq)$ +	$NH_3(aq)$
initial (M)	0.25	~0	0
change (M)	−x	+x	+x
equil (M)	0.25 − x	x	x

$K_a = \dfrac{[H_3O^+][NH_3]}{[NH_4^+]} = 5.6 \times 10^{-10} = \dfrac{x^2}{0.25-x} \approx \dfrac{x^2}{0.25}$

Solve for x. $x = [H_3O^+] = 1.2 \times 10^{-5}$ M

pH $= -\log[H_3O^+] = -\log(1.2 \times 10^{-5}) = 4.92$

(b) 0.40 M $ZnCl_2$

Zn^{2+} is an acidic cation. Cl^- is a neutral anion. The salt solution is acidic.

	$Zn(H_2O)_6^{2+}(aq)$ + $H_2O(l)$ ⇌	$H_3O^+(aq)$ +	$Zn(H_2O)_5(OH)^+(aq)$
initial (M)	0.40	~0	0
change (M)	−x	+x	+x
equil(M)	0.40 − x	x	x

$K_a = \dfrac{[H_3O^+][Zn(H_2O)_5(OH)^+]}{[Zn(H_2O)_6^{2+}]} = 2.5 \times 10^{-10} = \dfrac{x^2}{0.40-x} \approx \dfrac{x^2}{0.40}$

Solve for x. $x = [H_3O^+] = 1.0 \times 10^{-5}$ M

pH $= -\log[H_3O^+] = -\log(1.0 \times 10^{-5}) = 5.00$

14.23 For NO_2^-, $K_b = \dfrac{K_w}{K_a \text{ for } HNO_2} = \dfrac{1.0 \times 10^{-14}}{4.6 \times 10^{-4}} = 2.2 \times 10^{-11}$

	$NO_2^-(aq)$ + $H_2O(l)$ ⇌	$HNO_2(aq)$ +	$OH^-(aq)$
initial (M)	0.20	0	~0
change (M)	−x	+x	+x
equil (M)	0.20 − x	x	x

$$K_b = \frac{[HNO_2][OH^-]}{[NO_2^-]} = 2.2 \times 10^{-11} = \frac{x^2}{0.20 - x} \approx \frac{x^2}{0.20}$$

Solve for x. $x = [OH^-] = 2.1 \times 10^{-6}$ M

$$[H_3O^+] = \frac{K_w}{[OH^-]} = \frac{1.0 \times 10^{-14}}{2.1 \times 10^{-6}} = 4.8 \times 10^{-9}$$ M

$$pH = -\log[H_3O^+] = -\log(4.8 \times 10^{-9}) = 8.32$$

14.24 For NH_4^+, $K_a = \dfrac{K_w}{K_b \text{ for } NH_3} = \dfrac{1.0 \times 10^{-14}}{1.8 \times 10^{-5}} = 5.6 \times 10^{-10}$

For CN^-, $K_b = \dfrac{K_w}{K_a \text{ for } HCN} = \dfrac{1.0 \times 10^{-14}}{4.9 \times 10^{-10}} = 2.0 \times 10^{-5}$

Because $K_b > K_a$, the solution is basic.

14.25 (a) KBr: K^+, neutral cation; Br^-, neutral anion; solution is neutral
(b) $NaNO_2$: Na^+, neutral cation; NO_2^-, basic anion; solution is basic
(c) NH_4Br: NH_4^+, acidic cation; Br^-, neutral anion; solution is acidic
(d) $ZnCl_2$: Zn^{2+}, acidic cation; Cl^-, neutral anion; solution is acidic
(e) NH_4F

For NH_4^+, $K_a = \dfrac{K_w}{K_b \text{ for } NH_3} = \dfrac{1.0 \times 10^{-14}}{1.8 \times 10^{-5}} = 5.6 \times 10^{-10}$

For F^-, $K_b = \dfrac{K_w}{K_a \text{ for } HF} = \dfrac{1.0 \times 10^{-14}}{3.5 \times 10^{-4}} = 2.9 \times 10^{-11}$

Because $K_a > K_b$, the solution is acidic.

14.26 (a) H_2Se is a stronger acid than H_2S because Se is below S in the 6A group and the H–Se bond is weaker than the H–S bond.
(b) HI is a stronger acid than H_2Te because I is to the right of Te in the same row of the periodic table, I is more electronegative than Te, and the H–I bond is more polar.
(c) HNO_3 is a stronger acid than HNO_2 because acid strength increases with increasing oxidation number of N. The oxidation number for N is +5 in HNO_3 and +3 in HNO_2.
(d) H_2SO_3 is a stronger acid than H_2SeO_3 because acid strength increases with increasing electronegativity of the central atom. S is more electronegative than Se.

14.27 (a) Lewis acid, $AlCl_3$; Lewis base, Cl^- (b) Lewis acid, Ag^+; Lewis base, NH_3
(c) Lewis acid, SO_2; Lewis base, OH^- (d) Lewis acid, Cr^{3+}; Lewis base, H_2O

14.28

376

14.29 Lewis acids include not only H^+ but also other cations and neutral molecules having vacant valence orbitals that can accept a share in a pair of electrons donated by a Lewis base. The O^{2-} from CaO is the Lewis base and SO_2 is the Lewis acid.

$$:\overset{..}{\underset{..}{O}}-\overset{..}{\underset{..}{S}}-\overset{..}{\underset{..}{O}}: \ + \ :\overset{..}{\underset{..}{O}}:^{2-} \longrightarrow \ :SO_3^{2-}$$

14.30 NO_2, 46.01 amu

$$\text{mol } NO_2 = 5.47 \text{ mg} \times \frac{1.00 \times 10^{-3} \text{ g}}{1 \text{ mg}} \times \frac{1 \text{ mol } NO_2}{46.01 \text{ g } NO_2} = 1.19 \times 10^{-4} \text{ mol } NO_2$$

$$3 NO_2(g) + H_2O(l) \rightarrow 2 HNO_3(aq) + NO(g)$$

$$\text{mol } HNO_3 = 1.19 \times 10^{-4} \text{ mol } NO_2 \times \frac{2 \text{ mol } HNO_3}{3 \text{ mol } NO_2} = 7.93 \times 10^{-5} \text{ mol } HNO_3$$

$$[HNO_3] = \frac{7.93 \times 10^{-5} \text{ mol } HNO_3}{1.00 \text{ L}} = 7.93 \times 10^{-5} \text{ M}$$

HNO_3 is a strong acid and is completely dissociated therefore $[H_3O^+] = 7.93 \times 10^{-5} \text{ M}$
$pH = -\log [H_3O^+] = -\log (7.93 \times 10^{-5} \text{ M}) = 4.101$

Key Concept Problems

14.31 (a) acids, HCO_3^- and H_3O^+; bases, H_2O and CO_3^{2-}
(b) acids, HF and H_2CO_3; bases HCO_3^- and F^-

14.32 (a) X^-, Y^-, Z^- (b) HX < HZ < HY (c) HY (d) HX (e) (2/10) x 100% = 20%

14.33 (c) represents a solution of a weak diprotic acid, H_2A. Because K_{a2} is always less than K_{a1}, (a) and (d) represent impossible situations. (b) contains no H_2A.

14.34 For H_2SO_4, there is complete dissociation of only the first H^+. At equilibrium, there should be H_3O^+, HSO_4^-, and a small amount of SO_4^{2-}. This is best represented by (b).

14.35 (a) The weakest acid has the strongest conjugate base. HY is the weakest acid because it is the least dissociated. Therefore, Y^- has the largest K_b.
(b) The strongest acid has the weakest conjugate base. HX is the strongest acid because it is the most dissociated. Therefore, X^- is the weakest base.

14.36 (a) $Y^- < Z^- < X^-$
(b) The weakest base, Y^-, has the strongest conjugate acid.
(c) X^- is the strongest conjugate base and has the smallest pK_b.
(d) The numbers of HA molecules and OH^- ions are equal because the reaction of A^- with water has a 1:1 stoichiometry: $A^- + H_2O \rightleftharpoons HA + OH^-$

14.37 (a) $M(H_2O)_6^{n+}(aq) + H_2O(l) \rightleftharpoons H_3O^+(aq) + M(H_2O)_5(OH)^{(n-1)+}(aq)$

$$K_a = \frac{[H_3O^+][M(H_2O)_5(OH)^{(n-1)+}]}{[M(H_2O)_6^{n+}]}$$

(b) As the charge of the metal cation increases, the equilibrium constant increases because the OH bonds in coordinated water molecules are more polar, facilitating the dissociation of a H^+.

(c) $M(H_2O)_6^{3+}$ is the strongest acid. $M(H_2O)_6^+$, the weakest acid, has the strongest conjugate base.

14.38

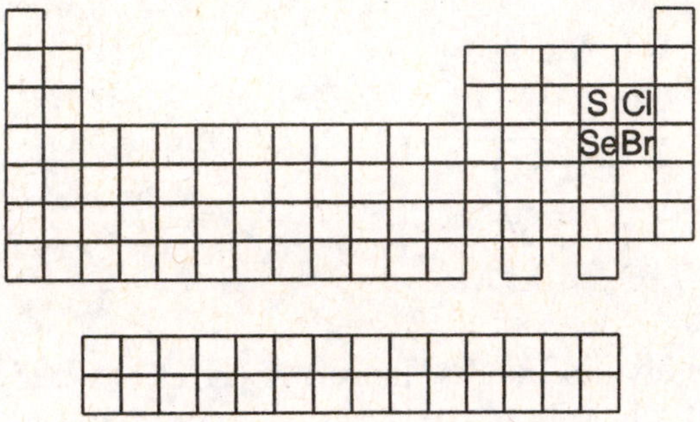

(a) H_2S, weakest; HBr, strongest. Acid strength for H_nX increases with increasing polarity of the H–X bond and with increasing size of X.

(b) H_2SeO_3, weakest; $HClO_3$, strongest. Acid strength for H_nYO_3 increases with increasing electronegativity of Y.

14.39 (a) Brønsted-Lowry acids: NH_4^+, $H_2PO_4^-$; Brønsted-Lowry bases: SO_3^{2-}, OCl^-, $H_2PO_4^-$

(b) Lewis acids: Fe^{3+}, BCl_3; Lewis bases: SO_3^{2-}, OCl^-, $H_2PO_4^-$

14.40 (a) $H_3BO_3(aq) + H_2O(l) \rightleftharpoons H_3O^+(aq) + H_2BO_3^-(aq)$

(b) $H_3BO_3(aq) + 2 H_2O(l) \rightleftharpoons H_3O^+(aq) + B(OH)_4^-(aq)$

14.41

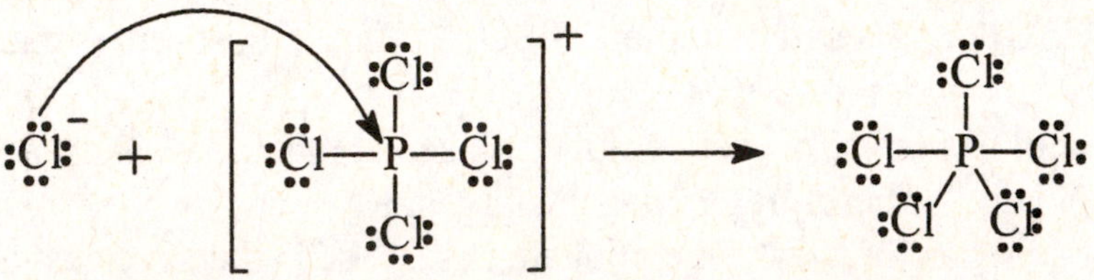

Section Problems
Acid–Base Concepts (Sections 14.1–14.2)

14.42 NH_3, CN^-, and NO_2^-

14.43 HCO_3^-

$HCO_3^-(aq) + OH^-(aq) \rightleftharpoons CO_3^{2-}(aq) + H_2O(l)$

$HCO_3^-(aq) + H_3O^+(aq) \rightleftharpoons H_2CO_3(aq) + H_2O(l)$

14.44 (a) SO_4^{2-} (b) HSO_3^- (c) HPO_4^{2-} (d) NH_3 (e) OH^- (f) NH_2^-

14.45 (a) HSO_3^- (b) H_3O^+ (c) $CH_3NH_3^+$ (d) H_2O (e) H_2CO_3 (f) H_2

14.46 (a) $CH_3CO_2H(aq) + NH_3(aq) \rightleftarrows NH_4^+(aq) + CH_3CO_2^-(aq)$
  acid   base ——— acid   base

(b) $CO_3^{2-}(aq) + H_3O^+(aq) \rightleftarrows H_2O(l) + HCO_3^-(aq)$
  base   acid ——— base   acid

(c) $HSO_3^-(aq) + H_2O(l) \rightleftarrows H_3O^+(aq) + SO_3^{2-}(aq)$
  acid   base ——— acid   base

(d) $HSO_3^-(aq) + H_2O(l) \rightleftarrows H_2SO_3(aq) + OH^-(aq)$
  base   acid   acid   base

14.47 (a) $CN^-(aq) + H_2O(l) \rightleftarrows HCN(aq) + OH^-(aq)$
  base   acid   acid   base

(b) $H_2PO_4^-(aq) + H_2O(l) \rightleftarrows H_3O^+(aq) + HPO_4^{2-}(aq)$
  acid   base ——— acid   base

(c) $HPO_4^{2-}(aq) + H_2O(l) \rightleftarrows H_2PO_4^-(aq) + OH^-(aq)$
  base   acid   acid   base

(d) $NH_4^+(aq) + NO_2^-(aq) \rightleftarrows HNO_2(aq) + NH_3(aq)$
  acid   base ——— acid   base

14.48 From data in Table 14.1: Strong acids: HNO_3 and H_2SO_4; Strong bases: H^- and O^{2-}

14.49 The weaker acid of the pair has the stronger conjugate base.
(a) H_2CO_3 is a weak acid. H_2SO_4 is a strong acid. H_2CO_3 has the stronger conjugate base.
(b) HCl is a strong acid. HF is a weak acid. HF has the stronger conjugate base.
(c) NH_4^+ is a weaker acid than HF. NH_4^+ has the stronger conjugate base.
(d) HCN is a weaker acid than HSO_4^-. HCN has the stronger conjugate base.

14.50 (a) left, HCO_3^- is the stronger base   (b) left, F^- is the stronger base
(c) right, NH_3 is the stronger base   (d) right, CN^- is the stronger base

379

14.51 (a) right, $CH_3CO_2^-$ is the stronger base (b) left, NO_2^- is the stronger base
(c) left, CO_3^{2-} is the stronger base (d) right, CN^- is the stronger base

Dissociation of Water; pH (Sections 14.4–14.5)

14.52 If $[H_3O^+] > 1.0 \times 10^{-7}$ M, solution is acidic.
If $[H_3O^+] < 1.0 \times 10^{-7}$ M, solution is basic.
If $[H_3O^+] = [OH^-] = 1.0 \times 10^{-7}$ M, solution is neutral.
If $[OH^-] > 1.0 \times 10^{-7}$ M, solution is basic
If $[OH^-] < 1.0 \times 10^{-7}$ M, solution is acidic.

(a) $[OH^-] = \dfrac{K_w}{[H_3O^+]} = \dfrac{1.0 \times 10^{-14}}{3.4 \times 10^{-9}} = 2.9 \times 10^{-6}$ M, basic

(b) $[H_3O^+] = \dfrac{K_w}{[OH^-]} = \dfrac{1.0 \times 10^{-14}}{0.010} = 1.0 \times 10^{-12}$ M, basic

(c) $[H_3O^+] = \dfrac{K_w}{[OH^-]} = \dfrac{1.0 \times 10^{-14}}{1.0 \times 10^{-10}} = 1.0 \times 10^{-4}$ M, acidic

(d) $[OH^-] = \dfrac{K_w}{[H_3O^+]} = \dfrac{1.0 \times 10^{-14}}{1.0 \times 10^{-7}} = 1.0 \times 10^{-7}$ M, neutral

(e) $[OH^-] = \dfrac{K_w}{[H_3O^+]} = \dfrac{1.0 \times 10^{-14}}{8.6 \times 10^{-5}} = 1.2 \times 10^{-10}$ M, acidic

14.53 If $[H_3O^+] > 1.0 \times 10^{-7}$ M, solution is acidic.
If $[H_3O^+] < 1.0 \times 10^{-7}$ M, solution is basic.
If $[H_3O^+] = [OH^-] = 1.0 \times 10^{-7}$ M, solution is neutral.
If $[OH^-] > 1.0 \times 10^{-7}$ M, solution is basic
If $[OH^-] < 1.0 \times 10^{-7}$ M, solution is acidic.

(a) $[OH^-] = \dfrac{K_w}{[H_3O^+]} = \dfrac{1.0 \times 10^{-14}}{2.5 \times 10^{-4}} = 4.0 \times 10^{-11}$ M, acidic

(b) $[OH^-] = \dfrac{K_w}{[H_3O^+]} = \dfrac{1.0 \times 10^{-14}}{2.0} = 5.0 \times 10^{-15}$ M, acidic

(c) $[H_3O^+] = \dfrac{K_w}{[OH^-]} = \dfrac{1.0 \times 10^{-14}}{5.6 \times 10^{-9}} = 1.8 \times 10^{-6}$ M, acidic

(d) $[H_3O^+] = \dfrac{K_w}{[OH^-]} = \dfrac{1.0 \times 10^{-14}}{1.5 \times 10^{-3}} = 6.7 \times 10^{-12}$ M, basic

(e) $[H_3O^+] = \dfrac{K_w}{[OH^-]} = \dfrac{1.0 \times 10^{-14}}{1.0 \times 10^{-7}} = 1.0 \times 10^{-7}$ M, neutral

14.54 (a) $pH = -\log[H_3O^+] = -\log(2.0 \times 10^{-5}) = 4.70$

(b) $[H_3O^+] = \dfrac{K_w}{[OH^-]} = \dfrac{1.0 \times 10^{-14}}{4 \times 10^{-3}} = 2.5 \times 10^{-12}$ M

$pH = -\log[H_3O^+] = -\log(2.5 \times 10^{-12}) = 11.6$

(c) $pH = -\log[H_3O^+] = -\log(3.56 \times 10^{-9}) = 8.449$

(d) $pH = -\log[H_3O^+] = -\log(10^{-3}) = 3$

(e) $[H_3O^+] = \dfrac{K_w}{[OH^-]} = \dfrac{1.0 \times 10^{-14}}{12} = 8.3 \times 10^{-16}$ M

$pH = -\log[H_3O^+] = -\log(8.3 \times 10^{-16}) = 15.08$

14.55 (a) $[H_3O^+] = \dfrac{K_w}{[OH^-]} = \dfrac{1.0 \times 10^{-14}}{7.6 \times 10^{-3}} = 1.3 \times 10^{-12}$ M

$pH = -\log[H_3O^+] = -\log(1.3 \times 10^{-12}) = 11.88$

(b) $pH = -\log[H_3O^+] = -\log(10^{-8}) = 8$

(c) $pH = -\log[H_3O^+] = -\log(5.0) = -0.70$

(d) $[H_3O^+] = \dfrac{K_w}{[OH^-]} = \dfrac{1.0 \times 10^{-14}}{1.0 \times 10^{-7}} = 1.0 \times 10^{-7}$ M

$pH = -\log[H_3O^+] = -\log(1.0 \times 10^{-7}) = 7.00$

(e) $pH = -\log[H_3O^+] = -\log(2.18 \times 10^{-10}) = 9.662$

14.56 $[H_3O^+] = 10^{-pH}$;

(a) 8×10^{-5} M   (b) 1.5×10^{-11} M   (c) 1.0 M

(d) 5.6×10^{-15} M   (e) 10 M   (f) 5.78×10^{-6} M

14.57 $[H_3O^+] = 10^{-pH}$;

(a) 1×10^{-9} M   (b) 1.0×10^{-7} M   (c) 2 M

(d) 6.6×10^{-16} M   (e) 2.3×10^{-3} M   (f) 1.75×10^{-11} M

14.58 $\Delta pH = \log(\Delta[H_3O^+])$;   (a) $\Delta pH = \log(1000) = 3$

(b) $\Delta pH = \log(1.0 \times 10^5) = 5.00$ (c) $\Delta pH = \log(2.0) = 0.30$

14.59 (a) $10^1 = 10$ (b) 1×10^{10} (c) $10^{0.10} = 1.3$

14.60 (a) $pH = -\log[H_3O^+] = -\log(10^{-2}) = 2$

(b) $pH = -\log[H_3O^+] = -\log(4 \times 10^{-8}) = 7.4$

(c) $[H_3O^+] = \dfrac{K_w}{[OH^-]} = \dfrac{1.0 \times 10^{-14}}{8 \times 10^{-8}} = 1.25 \times 10^{-7}$ M

$pH = -\log[H_3O^+] = -\log(1.25 \times 10^{-7}) = 6.9$

(d) $[H_3O^+] = \dfrac{K_w}{[OH^-]} = \dfrac{1.0 \times 10^{-14}}{6 \times 10^{-10}} = 1.7 \times 10^{-5}$ M

$pH = -\log[H_3O^+] = -\log(1.7 \times 10^{-5}) = 4.8$

$[H_3O^+] = \dfrac{K_w}{[OH^-]} = \dfrac{1.0 \times 10^{-14}}{2 \times 10^{-6}} = 5 \times 10^{-9}$ M

$pH = -\log[H_3O^+] = -\log(5 \times 10^{-9}) = 8.3$;   pH = 4.8 to 8.3

(e) $[H_3O^+] = \dfrac{K_w}{[OH^-]} = \dfrac{1.0 \times 10^{-14}}{2 \times 10^{-7}} = 5 \times 10^{-8}$ M

$pH = -\log[H_3O^+] = -\log(5 \times 10^{-8}) = 7.3$

14.61 (a) $pH = -\log[H_3O^+] = -\log(3 \times 10^{-4}) = 3.5$

(b) $[H_3O^+] = \dfrac{K_w}{[OH^-]} = \dfrac{1.0 \times 10^{-14}}{6 \times 10^{-7}} = 1.7 \times 10^{-8}$ M

$pH = -\log[H_3O^+] = -\log(1.7 \times 10^{-8}) = 7.8$

(c) $[H_3O^+] = \dfrac{K_w}{[OH^-]} = \dfrac{1.0 \times 10^{-14}}{10^{-11}} = 0.001$ M

$pH = -\log[H_3O^+] = -\log(0.001) = 3$

(d) $pH = -\log[H_3O^+] = -\log(1.3 \times 10^{-2}) = 1.89$

(e) $[H_3O^+] = \dfrac{K_w}{[OH^-]} = \dfrac{1.0 \times 10^{-14}}{6.3 \times 10^{-9}} = 1.6 \times 10^{-6}$ M

$pH = -\log[H_3O^+] = -\log(1.6 \times 10^{-6}) = 5.80$

Strong Acids and Strong Bases (Section 14.7)

14.62 (a) $[H_3O^+] = 0.40$ M; $pH = -\log[H_3O^+] = -\log(0.40) = 0.40$
(b) $[OH^-] = 3.7 \times 10^{-4}$ M

$[H_3O^+] = \dfrac{K_w}{[OH^-]} = \dfrac{1.0 \times 10^{-14}}{3.7 \times 10^{-4}} = 2.7 \times 10^{-11}$ M

$pH = -\log[H_3O^+] = -\log(2.7 \times 10^{-11}) = 10.57$
(c) $[OH^-] = 2(5.0 \times 10^{-5}$ M$) = 1.0 \times 10^{-4}$ M

$[H_3O^+] = \dfrac{K_w}{[OH^-]} = \dfrac{1.0 \times 10^{-14}}{1.0 \times 10^{-4}} = 1.0 \times 10^{-10}$ M

$pH = -\log[H_3O^+] = -\log(1.0 \times 10^{-10}) = 10.00$

14.63 (a) $[H_3O^+] = 1.8$ M; $pH = -\log[H_3O^+] = -\log(1.8) = -0.26$
(b) $[OH^-] = 1.2$ M

$[H_3O^+] = \dfrac{K_w}{[OH^-]} = \dfrac{1.0 \times 10^{-14}}{1.2} = 8.3 \times 10^{-15}$ M

$pH = -\log[H_3O^+] = -\log(8.3 \times 10^{-15}) = 14.08$
(c) $[OH^-] = 2(5.3 \times 10^{-3}$ M$) = 1.06 \times 10^{-2}$ M

$[H_3O^+] = \dfrac{K_w}{[OH^-]} = \dfrac{1.0 \times 10^{-14}}{1.06 \times 10^{-2}} = 9.4 \times 10^{-13}$ M

$pH = -\log[H_3O^+] = -\log(9.4 \times 10^{-13}) = 12.03$

14.64 (a) LiOH, 23.95 amu; 250 mL = 0.250 L

$$\text{molarity of LiOH(aq)} = \frac{\left(4.8 \text{ g} \times \dfrac{1 \text{ mol}}{23.95 \text{ g}}\right)}{0.250 \text{ L}} = 0.80 \text{ M}$$

LiOH is a strong base; therefore $[OH^-] = 0.80$ M

$$[H_3O^+] = \frac{K_w}{[OH^-]} = \frac{1.0 \times 10^{-14}}{0.80} = 1.25 \times 10^{-14} \text{ M}$$

$$pH = -\log[H_3O^+] = -\log(1.25 \times 10^{-14}) = 13.90$$

(b) HCl, 36.46 amu

$$\text{molarity of HCl(aq)} = \frac{\left(0.93 \text{ g} \times \dfrac{1 \text{ mol}}{36.46 \text{ g}}\right)}{0.40 \text{ L}} = 0.064 \text{ M}$$

HCl is a strong acid; therefore $[H_3O^+] = 0.064$ M

$$pH = -\log[H_3O^+] = -\log(0.064) = 1.19$$

(c) $M_f \cdot V_f = M_i \cdot V_i$

$$M_f = \frac{M_i \cdot V_i}{V_f} = \frac{(0.10 \text{ M})(50 \text{ mL})}{(1000 \text{ mL})} = 5.0 \times 10^{-3} \text{ M}$$

$$pH = -\log[H_3O^+] = -\log(5.0 \times 10^{-3}) = 2.30$$

(d) For HCl, $M_f = \dfrac{M_i \cdot V_i}{V_f} = \dfrac{(2.0 \times 10^{-3} \text{ M})(100 \text{ mL})}{(500 \text{ mL})} = 4.0 \times 10^{-4} \text{ M}$

For HClO$_4$, $M_f = \dfrac{M_i \cdot V_i}{V_f} = \dfrac{(1.0 \times 10^{-3} \text{ M})(400 \text{ mL})}{(500 \text{ mL})} = 8.0 \times 10^{-4} \text{ M}$

$$[H_3O^+] = (4.0 \times 10^{-4} \text{ M}) + (8.0 \times 10^{-4} \text{ M}) = 1.2 \times 10^{-3} \text{ M}$$

$$pH = -\log[H_3O^+] = -\log(1.2 \times 10^{-3}) = 2.92$$

14.65 (a) Na$_2$O, 61.98 amu; 100 mL = 0.100 L

$$\text{moles of Na}_2\text{O} = 0.20 \text{ g} \times \frac{1 \text{ mol}}{61.98 \text{ g}} = 0.0032 \text{ mol}$$

$$O^{2-}(aq) + H_2O(l) \xrightarrow{100\%} 2 \, OH^-(aq)$$

moles of $OH^- = (2)(0.0032 \text{ mol}) = 0.0064$ mol

$$[OH^-] = \frac{0.0064 \text{ mol}}{0.100 \text{ L}} = 0.064 \text{ M}$$

$$[H_3O^+] = \frac{K_w}{[OH^-]} = \frac{1.0 \times 10^{-14}}{0.064} = 1.6 \times 10^{-13} \text{ M}$$

$$pH = -\log[H_3O^+] = -\log(1.6 \times 10^{-13}) = 12.80$$

(b) HNO_3, 63.01

$$\text{molarity of } HNO_3(aq) = \frac{\left(1.26 \text{ g} \times \dfrac{1 \text{ mol}}{63.01 \text{ g}}\right)}{0.500 \text{ L}} = 0.0400 \text{ M}$$

$pH = -\log[H_3O^+] = -\log(0.0400) = 1.398$

(c) $[OH^-] = 2(0.075 \text{ M}) = 0.15 \text{ M}$

$M_f \cdot V_f = M_i \cdot V_i$

$$M_f = \frac{M_i \cdot V_i}{V_f} = \frac{(0.15 \text{ M})(40.0 \text{ mL})}{(300.0 \text{ mL})} = 0.020 \text{ M}$$

$$[H_3O^+] = \frac{K_w}{[OH^-]} = \frac{1.0 \times 10^{-14}}{0.020} = 5.0 \times 10^{-13} \text{ M}$$

$pH = -\log[H_3O^+] = -\log(5.0 \times 10^{-13}) = 12.30$

(d) On mixing equal volumes of the two strong acids, both acid concentrations are cut in half.

$[H_3O^+] = 0.10 \text{ M} + 0.25 \text{ M} = 0.35 \text{ M}$

$pH = -\log[H_3O^+] = -\log(0.35) = 0.46$

Weak Acids (Sections 14.8–14.10)

14.66 (a) $HClO_2(aq) + H_2O(l) \rightleftharpoons H_3O^+(aq) + ClO_2^-(aq)$; $K_a = \dfrac{[H_3O^+][ClO_2^-]}{[HClO_2]}$

(b) $HOBr(aq) + H_2O(l) \rightleftharpoons H_3O^+(aq) + OBr^-(aq)$; $K_a = \dfrac{[H_3O^+][OBr^-]}{[HOBr]}$

(c) $HCO_2H(aq) + H_2O(l) \rightleftharpoons H_3O^+(aq) + HCO_2^-(aq)$; $K_a = \dfrac{[H_3O^+][HCO_2^-]}{[HCO_2H]}$

14.67 (a) $HN_3(aq) + H_2O(l) \rightleftharpoons H_3O^+(aq) + N_3^-(aq)$; $K_a = \dfrac{[H_3O^+][N_3^-]}{[HN_3]}$

(b) $C_6H_5CO_2H(aq) + H_2O(l) \rightleftharpoons H_3O^+(aq) + C_6H_5CO_2^-(aq)$; $K_a = \dfrac{[H_3O^+][C_6H_5CO_2^-]}{[C_6H_5CO_2H]}$

(c) $H_2O_2(aq) + H_2O(l) \rightleftharpoons H_3O^+(aq) + HO_2^-(aq)$; $K_a = \dfrac{[H_3O^+][HO_2^-]}{[H_2O_2]}$

14.68 (a) The larger the K_a, the stronger the acid.
$C_6H_5OH < HOCl < CH_3CO_2H < HNO_3$
(b) The larger the K_a, the larger the percent dissociation for the same concentration.
$HNO_3 > CH_3CO_2H > HOCl > C_6H_5OH$
1 M HNO_3, $[H_3O^+] = 1$ M
1 M CH_3CO_2H, $[H_3O^+] = \sqrt{[HA] \times K_a} = \sqrt{(1 \text{ M})(1.8 \times 10^{-5})} = 4 \times 10^{-3}$ M

1 M HOCl, $[H_3O^+] = \sqrt{[HA] \times K_a} = \sqrt{(1 \text{ M})(3.5 \times 10^{-8})} = 2 \times 10^{-4}$ M

1 M C_6H_5OH, $[H_3O^+] = \sqrt{[HA] \times K_a} = \sqrt{(1 \text{ M})(1.3 \times 10^{-10})} = 1 \times 10^{-5}$ M

14.69   (a) The larger the K_a, the stronger the acid.
$HCN < HOBr < HCO_2H < HClO_4$
(b) The larger the K_a, the larger the percent dissociation for the same concentration.
$HClO_4 > HCO_2H > HOBr > HCN$
1 M $HClO_4$, $[H_3O^+] = 1$ M

1 M HCO_2H, $[H_3O^+] = \sqrt{[HA] \times K_a} = \sqrt{(1 \text{ M})(1.8 \times 10^{-4})} = 1 \times 10^{-2}$ M

1 M HOBr, $[H_3O^+] = \sqrt{[HA] \times K_a} = \sqrt{(1 \text{ M})(2.0 \times 10^{-9})} = 4 \times 10^{-5}$ M

1 M HCN, $[H_3O^+] = \sqrt{[HA] \times K_a} = \sqrt{(1 \text{ M})(4.9 \times 10^{-10})} = 2 \times 10^{-5}$ M

14.70

	$HOBr(aq)$	$+$	$H_2O(l)$	$\rightleftharpoons$	$H_3O^+(aq)$	$+$	$OBr^-(aq)$
initial (M)	0.040				~0		0
change (M)	−x				+x		+x
equil (M)	0.040 − x				x		x

$x = [H_3O^+] = 10^{-pH} = 10^{-5.05} = 8.9 \times 10^{-6}$ M

$$K_a = \frac{[H_3O^+][OBr^-]}{[HOBr]} = \frac{x^2}{0.040 - x} = \frac{(8.9 \times 10^{-6})^2}{0.040 - (8.9 \times 10^{-6})} = 2.0 \times 10^{-9}$$

14.71

	$C_3H_6O_3(aq)$	$+$	$H_2O(l)$	$\rightleftharpoons$	$H_3O^+(aq)$	$+$	$C_3H_5O_3^-(aq)$
initial (M)	0.10				~0		0
change (M)	−x				+x		+x
equil (M)	0.10 − x				x		x

$x = [H_3O^+] = 10^{-pH} = 10^{-2.43} = 3.7 \times 10^{-3}$ M

$[C_3H_5O_3^-] = x = 3.7 \times 10^{-3}$ M; $[C_3H_6O_3] = 0.10 - x = 0.10 - (3.7 \times 10^{-3})$ M $= 0.10$ M

$$K_a = \frac{[H_3O^+][C_3H_5O_3^-]}{[C_3H_6O_3]} = \frac{(3.7 \times 10^{-3})(3.7 \times 10^{-3})}{0.10} = 1.4 \times 10^{-4}$$

$pK_a = -\log K_a = -\log(1.4 \times 10^{-4}) = 3.85$

14.72

	$C_6H_5OH(aq)$	$+$	$H_2O(l)$	$\rightleftharpoons$	$H_3O^+(aq)$	$+$	$C_6H_5O^-(aq)$
initial (M)	0.10				~0		0
change (M)	−x				+x		+x
equil (M)	0.10 − x				x		x

$$K_a = \frac{[H_3O^+][C_6H_5O^-]}{[C_6H_5OH]} = 1.3 \times 10^{-10} = \frac{x^2}{0.10 - x} \approx \frac{x^2}{0.10}$$

Solve for x. $x = 3.6 \times 10^{-6}$ M $= [H_3O^+] = [C_6H_5O^-]$
$[C_6H_5OH] = 0.10 - x = 0.10$ M
$pH = -\log[H_3O^+] = -\log(3.6 \times 10^{-6}) = 5.44$

$$[OH^-] = \frac{K_w}{[H_3O^+]} = \frac{1.0 \times 10^{-14}}{3.6 \times 10^{-6}} = 2.8 \times 10^{-9} \text{ M}$$

$$\% \text{ dissociation} = \frac{[C_6H_5OH]_{diss}}{[C_6H_5OH]_{initial}} \times 100\% = \frac{3.6 \times 10^{-6} \text{ M}}{0.10 \text{ M}} \times 100\% = 0.0036\%$$

14.73 $\quad\quad\quad\quad$ $HCO_2H(aq) + H_2O(l) \rightleftharpoons H_3O^+(aq) + HCO_2^-(aq)$

initial (M)	0.20	~0	0
change (M)	−x	+x	+x
equil (M)	0.20 − x	x	x

$$K_a = \frac{[H_3O^+][HCO_2^-]}{[HCO_2H]} = 1.8 \times 10^{-4} = \frac{x^2}{0.20 - x}$$

$x^2 + (1.8 \times 10^{-4})x - (3.6 \times 10^{-5}) = 0$

Use the quadratic formula to solve for x.

$$x = \frac{-(1.8 \times 10^{-4}) \pm \sqrt{(1.8 \times 10^{-4})^2 - (4)(-3.6 \times 10^{-5})}}{2(1)} = \frac{(-1.8 \times 10^{-4}) \pm 0.012}{2}$$

$x = 0.0059$ and -0.0061

Of the two solutions for x, only the positive value has physical meaning because x is the $[H_3O^+]$.

$[H_3O^+] = [HCO_2^-] = x = 0.0059$ M

$[HCO_2H] = 0.20 - x = 0.20 - 0.0059 = 0.19$ M

$$[OH^-] = \frac{K_w}{[H_3O^+]} = \frac{1.0 \times 10^{-14}}{0.0059} = 1.7 \times 10^{-12} \text{ M}$$

$pH = -\log[H_3O^+] = -\log(0.0059) = 2.23$

$$\% \text{ dissociation} = \frac{[HCO_2H]_{diss}}{[HCO_2H]_{initial}} \times 100\% = \frac{0.0059 \text{ M}}{0.20 \text{ M}} \times 100\% = 3.0\%$$

14.74 $\quad\quad\quad\quad$ $HNO_2(aq) + H_2O(l) \rightleftharpoons H_3O^+(aq) + NO_2^-(aq)$

initial (M)	1.5	~0	0
change (M)	−x	+x	+x
equil (M)	1.5 − x	x	x

$$K_a = \frac{[H_3O^+][NO_2^-]}{[HNO_2]} = 4.5 \times 10^{-4} = \frac{x^2}{1.5 - x} \approx \frac{x^2}{1.5}$$

Solve for x. $x = 0.026$ M $= [H_3O^+]$

$pH = -\log[H_3O^+] = -\log(0.026) = 1.59$

$$\% \text{ dissociation} = \frac{[HNO_2]_{diss}}{[HNO_2]_{initial}} \times 100\% = \frac{0.026 \text{ M}}{1.5 \text{ M}} \times 100\% = 1.7\%$$

14.75 $C_9H_8O_4$, 180.16 amu; 300 mL = 0.300 L;   324 mg = 0.324 g

$$[C_9H_8O_4] = \frac{\left((2)(0.324 \text{ g}) \times \dfrac{1 \text{ mol}}{180.16 \text{ g}} \right)}{0.300 \text{ L}} = 0.0120 \text{ M}$$

	$C_9H_8O_4(aq)$ + $H_2O(l)$ ⇌	$H_3O^+(aq)$ +	$C_9H_7O_4^-(aq)$
initial (M)	0.0120	~0	0
change (M)	−x	+x	+x
equil (M)	0.0120 − x	x	x

$$K_a = \frac{[H_3O^+][C_9H_7O_4^-]}{[C_9H_8O_4]} = 3.0 \times 10^{-4} = \frac{x^2}{0.0120 - x}$$

$x^2 + (3.0 \times 10^{-4})x - (3.6 \times 10^{-6}) = 0$

Use the quadratic formula to solve for x.

$$x = \frac{-(3.0 \times 10^{-4}) \pm \sqrt{(3.0 \times 10^{-4})^2 - (4)(-3.6 \times 10^{-6})}}{2(1)} = \frac{(-3.0 \times 10^{-4}) \pm (3.8 \times 10^{-3})}{2}$$

$x = 1.8 \times 10^{-3}$ and -2.1×10^{-3}

Of the two solutions for x, only the positive value of x has physical meaning because x is the $[H_3O^+]$.

$[H_3O^+] = x = 1.8 \times 10^{-3}$ M

$pH = -\log[H_3O^+] = -\log(1.8 \times 10^{-3}) = 2.74$

$$\% \text{ dissociation} = \frac{[C_9H_8O_4]_{diss}}{[C_9H_8O_4]_{initial}} \times 100\% = \frac{1.8 \times 10^{-3} \text{ M}}{0.0120 \text{ M}} \times 100\% = 15\%$$

Polyprotic Acids (Section 14.11)

14.76 $H_2SeO_4(aq)$ + $H_2O(l)$ ⇌ $H_3O^+(aq)$ + $HSeO_4^-(aq)$; $K_{a1} = \dfrac{[H_3O^+][HSeO_4^-]}{[H_2SeO_4]}$

$HSeO_4^-(aq)$ + $H_2O(l)$ ⇌ $H_3O^+(aq)$ + $SeO_4^{2-}(aq)$; $K_{a2} = \dfrac{[H_3O^+][SeO_4^{2-}]}{[HSeO_4^-]}$

14.77   (a) $H_3PO_4(aq)$ + $H_2O(l)$ ⇌ $H_3O^+(aq)$ + $H_2PO_4^-(aq)$; $K_{a1} = \dfrac{[H_3O^+][H_2PO_4^-]}{[H_3PO_4]}$

(b) $H_2PO_4^-(aq)$ + $H_2O(l)$ ⇌ $H_3O^+(aq)$ + $HPO_4^{2-}(aq)$; $K_{a2} = \dfrac{[H_3O^+][HPO_4^{2-}]}{[H_2PO_4^-]}$

(c) $HPO_4^{2-}(aq)$ + $H_2O(l)$ ⇌ $H_3O^+(aq)$ + $PO_4^{3-}(aq)$; $K_{a3} = \dfrac{[H_3O^+][PO_4^{3-}]}{[HPO_4^{2-}]}$

14.78

	$H_2CO_3(aq)$ + $H_2O(l)$ ⇌	$H_3O^+(aq)$ +	$HCO_3^-(aq)$
initial (M)	0.010	~0	0
change (M)	−x	+x	+x
equil (M)	0.010 − x	x	x

$$K_{a1} = \frac{[H_3O^+][HCO_3^-]}{[H_2CO_3]} = 4.3 \times 10^{-7} = \frac{x^2}{0.010 - x} \approx \frac{x^2}{0.010}$$

Solve for x. $x = 6.6 \times 10^{-5}$

$[H_3O^+] = [HCO_3^-] = x = 6.6 \times 10^{-5}$ M; $\qquad [H_2CO_3] = 0.010 - x = 0.010$ M

The second dissociation of H_2CO_3 produces a negligible amount of H_3O^+ compared with that from the first dissociation.

$$HCO_3^-(aq) + H_2O(l) \rightleftharpoons H_3O^+(aq) + CO_3^{2-}(aq)$$

$$K_{a2} = \frac{[H_3O^+][CO_3^{2-}]}{[HCO_3^-]} = 5.6 \times 10^{-11} = \frac{(6.6 \times 10^{-5})[CO_3^{2-}]}{(6.6 \times 10^{-5})}$$

$[CO_3^{2-}] = K_{a2} = 5.6 \times 10^{-11}$ M

$$[OH^-] = \frac{K_w}{[H_3O^+]} = \frac{1.0 \times 10^{-14}}{6.6 \times 10^{-5}} = 1.5 \times 10^{-10} \text{ M}$$

$pH = -\log[H_3O^+] = -\log(6.6 \times 10^{-5}) = 4.18$

14.79
$$H_2SO_3(aq) + H_2O(l) \rightleftharpoons H_3O^+(aq) + HSO_3^-(aq)$$

	H_2SO_3		H_3O^+	HSO_3^-
initial (M)	0.025		~0	0
change (M)	−x		+x	+x
equil (M)	0.025 − x		x	x

$$K_{a1} = \frac{[H_3O^+][HSO_3^-]}{[H_2SO_3]} = 1.5 \times 10^{-2} = \frac{x^2}{0.025 - x}$$

$x^2 + 0.015x - (3.75 \times 10^{-4}) = 0$

Use the quadratic formula to solve for x.

$$x = \frac{-(0.015) \pm \sqrt{(0.015)^2 - (4)(-3.75 \times 10^{-4})}}{2(1)} = \frac{-0.015 \pm 0.0415}{2}$$

$x = 0.013$ and -0.028

Of the two solutions for x, only the positive value of x has physical meaning because x is the $[H_3O^+]$.

$x = [H_3O^+] = [HSO_3^-] = 0.013$ M

$[H_2SO_3] = 0.025 - x = 0.025 - 0.013 = 0.012$ M

The second dissociation of H_2SO_3 produces a negligible amount of H_3O^+ compared with that from the first dissociation.

$$HSO_3^-(aq) + H_2O(l) \rightleftharpoons H_3O^+(aq) + SO_3^{2-}(aq)$$

$$K_{a2} = \frac{[H_3O^+][SO_3^{2-}]}{[HSO_3^-]} = 6.3 \times 10^{-8} = \frac{(0.013)[SO_3^{2-}]}{(0.013)}$$

$[SO_3^{2-}] = K_{a2} = 6.3 \times 10^{-8}$ M

$pH = -\log[H_3O^+] = -\log(0.013) = 1.89$

14.80   For the dissociation of the first proton, the following equilibrium must be considered:

$$H_2C_2O_4(aq) + H_2O(l) \rightleftharpoons H_3O^+(aq) + HC_2O_4^-(aq)$$

initial (M)	0.20	~0	0
change (M)	$-x$	$+x$	$+x$
equil (M)	$0.20 - x$	x	x

$$K_{a1} = \frac{[H_3O^+][HC_2O_4^-]}{[H_2C_2O_4]} = 5.9 \times 10^{-2} = \frac{x^2}{0.20 - x}$$

$x^2 + 0.059x - 0.0118 = 0$

Use the quadratic formula to solve for x.

$$x = \frac{-(0.059) \pm \sqrt{(0.059)^2 - 4(1)(-0.0118)}}{2(1)} = \frac{-0.059 \pm 0.225}{2}$$

$x = 0.083$ and -0.142

Of the two solutions for x, only the positive value of x has physical meaning, because x is the $[H_3O^+]$.

$[H_3O^+] = [HC_2O_4^-] = 0.083$ M

For the dissociation of the second proton, the following equilibrium must be considered:

$$HC_2O_4^-(aq) + H_2O(l) \rightleftharpoons H_3O^+(aq) + C_2O_4^{2-}(aq)$$

initial (M)	0.083	0.083	0
change (M)	$-x$	$+x$	$+x$
equil (M)	$0.083 - x$	$0.083 + x$	x

$$K_{a2} = \frac{[H_3O^+][C_2O_4^{2-}]}{[HC_2O_4^-]} = 6.4 \times 10^{-5} = \frac{(0.083 + x)(x)}{0.083 - x} \approx \frac{(0.083)(x)}{0.083} = x$$

$[H_3O^+] = 0.083 + x = 0.083$ M

$pH = -\log[H_3O^+] = -\log(0.083) = 1.08$

$[C_2O_4^{2-}] = x = 6.4 \times 10^{-5}$ M

14.81   Mixing equal volumes of the two strong acids cuts the initial acid concentrations in half.

$[H_3O^+] = 0.1$ M (from HCl) $+ 0.3$ M (from H_2SO_4) $= 0.4$ M

For the dissociation of the second proton from H_2SO_4, the following equilibrium must be considered:

$$HSO_4^-(aq) + H_2O(l) \rightleftharpoons H_3O^+(aq) + SO_4^{2-}(aq)$$

initial (M)	0.3	0.4	0
change (M)	$-x$	$+x$	$+x$
equil (M)	$0.3 - x$	$0.4 + x$	x

$$K_{a2} = \frac{[H_3O^+][SO_4^{2-}]}{[HSO_4^-]} = 1.2 \times 10^{-2} = \frac{(0.4 + x)(x)}{0.3 - x}$$

$x^2 + 0.412x - 0.0036 = 0$

Use the quadratic formula to solve for x.

$$x = \frac{-(0.412) \pm \sqrt{(0.412)^2 - 4(1)(-0.0036)}}{2(1)} = \frac{-0.412 \pm 0.429}{2}$$

$x = 0.0085$ and -0.420

Of the two solutions for x, only the positive value of x has physical meaning, because x is the $[SO_4^{2-}]$.

$[H_3O^+] = 0.4 + x = 0.4$ M

$[SO_4^{2-}] = x = 0.008$ M

Weak Bases; Relation Between K_a and K_b (Sections 14.12–14.13)

14.82 (a) $(CH_3)_2NH(aq) + H_2O(l) \rightleftharpoons (CH_3)_2NH_2^+(aq) + OH^-(aq);$ $K_b = \dfrac{[(CH_3)_2NH_2^+][OH^-]}{[(CH_3)_2NH]}$

(b) $C_6H_5NH_2(aq) + H_2O(l) \rightleftharpoons C_6H_5NH_3^+(aq) + OH^-(aq);$ $K_b = \dfrac{[C_6H_5NH_3^+][OH^-]}{[C_6H_5NH_2]}$

(c) $CN^-(aq) + H_2O(l) \rightleftharpoons HCN(aq) + OH^-(aq);$ $K_b = \dfrac{[HCN][OH^-]}{[CN^-]}$

14.83 (a) $C_5H_5N(aq) + H_2O(l) \rightleftharpoons C_5H_5NH^+(aq) + OH^-(aq);$ $K_b = \dfrac{[C_5H_5NH^+][OH^-]}{[C_5H_5N]}$

(b) $C_2H_5NH_2(aq) + H_2O(l) \rightleftharpoons C_2H_5NH_3^+(aq) + OH^-(aq);$ $K_b = \dfrac{[C_2H_5NH_3^+][OH^-]}{[C_2H_5NH_2]}$

(c) $CH_3CO_2^-(aq) + H_2O(l) \rightleftharpoons CH_3CO_2H(aq) + OH^-(aq);$ $K_b = \dfrac{[CH_3CO_2H][OH^-]}{[CH_3CO_2^-]}$

14.84 $[H_3O^+] = 10^{-pH} = 10^{-9.5} = 3.16 \times 10^{-10}$ M

$$[OH^-] = \frac{K_w}{[H_3O^+]} = \frac{1.0 \times 10^{-14}}{3.16 \times 10^{-10}} = 3.16 \times 10^{-5} \text{ M}$$

	$C_{17}H_{19}NO_3(aq)$	$+ H_2O(l) \rightleftharpoons$	$C_{17}H_{20}NO_3^+(aq)$	$+ OH^-(aq)$
initial (M)	7.0×10^{-4}		0	~0
change (M)	$-x$		$+x$	$+x$
equil (M)	$(7.0 \times 10^{-4}) - x$		x	x

$x = [OH^-] = 3.16 \times 10^{-5}$ M

$$K_b = \frac{[C_{17}H_{20}NO_3^+][OH^-]}{[C_{17}H_{19}NO_3]} = \frac{x^2}{(7.0 \times 10^{-4}) - x} = \frac{(3.16 \times 10^{-5})^2}{(7.0 \times 10^{-4}) - (3.16 \times 10^{-5})} = 1.49 \times 10^{-6}$$

$pK_b = -\log K_b = -\log(1.49 \times 10^{-6}) = 5.827 = 5.8$

14.85 $[H_3O^+] = 10^{-pH} = 10^{-9.75} = 1.78 \times 10^{-10}$ M

$$[OH^-] = \frac{K_w}{[H_3O^+]} = \frac{1.0 \times 10^{-14}}{1.78 \times 10^{-10}} = 5.62 \times 10^{-5} \text{ M}$$

	quinine(aq) + H$_2$O(l) $\rightleftharpoons$	quinineH$^+$(aq) +	OH$^-$(aq)
initial (M)	1.00×10^{-3}	0	~0
change (M)	$-x$	$+x$	$+x$
equil (M)	$(1.00 \times 10^{-3}) - x$	x	x

$x = [OH^-] = 5.62 \times 10^{-5}$ M

$$K_b = \frac{[\text{quinineH}^+][OH^-]}{[\text{quinine}]} = \frac{x^2}{(1.00 \times 10^{-3}) - x} = \frac{(5.62 \times 10^{-5})^2}{[(1.00 \times 10^{-3}) - (5.62 \times 10^{-5})]} = 3.3 \times 10^{-6}$$

$pK_b = -\log K_b = -\log(3.3 \times 10^{-6}) = 5.48$

14.86 (a)

	CH$_3$NH$_2$(aq) + H$_2$O(l) $\rightleftharpoons$	CH$_3$NH$_3^+$(aq) +	OH$^-$(aq)
initial (M)	0.24	0	~0
change (M)	$-x$	$+x$	$+x$
equil (M)	$0.24 - x$	x	x

$$K_b = \frac{[CH_3NH_3^+][OH^-]}{[CH_3NH_2]} = 3.7 \times 10^{-4} = \frac{x^2}{0.24 - x}$$

$x^2 + (3.7 \times 10^{-4})x - (8.9 \times 10^{-5}) = 0$

Use the quadratic formula to solve for x.

$$x = \frac{-(3.7 \times 10^{-4}) \pm \sqrt{(3.7 \times 10^{-4})^2 - (4)(-8.9 \times 10^{-5})}}{2(1)} = \frac{(-3.7 \times 10^{-4}) \pm 0.0189}{2}$$

$x = 0.0093$ and -0.0096

Of the two solutions for x, only the positive value of x has physical meaning because x is the $[OH^-]$.

$[OH^-] = x = 0.0093$ M

$$[H_3O^+] = \frac{K_w}{[OH^-]} = \frac{1.0 \times 10^{-14}}{0.0093} = 1.1 \times 10^{-12} \text{ M}$$

$pH = -\log[H_3O^+] = -\log(1.1 \times 10^{-12}) = 11.96$

(b)

	C$_5$H$_5$N(aq) + H$_2$O(l) $\rightleftharpoons$	C$_5$H$_5$NH$^+$(aq) +	OH$^-$(aq)
initial (M)	0.040	0	~0
change (M)	$-x$	$+x$	$+x$
equil (M)	$0.040 - x$	x	x

$$K_b = \frac{[C_5H_5NH^+][OH^-]}{[C_5H_5N]} = 1.8 \times 10^{-9} = \frac{x^2}{0.040 - x} \approx \frac{x^2}{0.040}$$

Solve for x. $x = [OH^-] = 8.5 \times 10^{-6}$ M

$$[H_3O^+] = \frac{K_w}{[OH^-]} = \frac{1.0 \times 10^{-14}}{8.5 \times 10^{-6}} = 1.2 \times 10^{-9} \text{ M}$$

$pH = -\log[H_3O^+] = -\log(1.2 \times 10^{-9}) = 8.92$

(c) $\qquad$ $NH_2OH(aq) + H_2O(l) \rightleftharpoons NH_3OH^+(aq) + OH^-(aq)$

initial (M) $\quad$ 0.075 $\qquad\qquad\qquad$ 0 $\qquad\qquad$ ~0

change (M) $\quad$ –x $\qquad\qquad\qquad\qquad$ +x $\qquad\qquad$ +x

equil (M) $\quad$ 0.075 – x $\qquad\qquad\qquad$ x $\qquad\qquad$ x

$$K_b = \frac{[NH_3OH^+][OH^-]}{[NH_2OH]} = 9.1 \times 10^{-9} = \frac{x^2}{0.075 - x} \approx \frac{x^2}{0.075}$$

Solve for x. $\quad x = [OH^-] = 2.6 \times 10^{-5}$ M

$$[H_3O^+] = \frac{K_w}{[OH^-]} = \frac{1.0 \times 10^{-14}}{2.6 \times 10^{-5}} = 3.8 \times 10^{-10}$$ M

$pH = -\log[H_3O^+] = -\log(3.8 \times 10^{-10}) = 9.42$

14.87 $\qquad$ $C_6H_5NH_2(aq) + H_2O(l) \rightleftharpoons C_6H_5NH_3^+(aq) + OH^-(aq)$

initial (M) $\quad$ 0.15 $\qquad\qquad\qquad\qquad$ 0 $\qquad\qquad$ ~0

change (M) $\quad$ –x $\qquad\qquad\qquad\qquad$ +x $\qquad\qquad$ +x

equil (M) $\quad$ 0.15 – x $\qquad\qquad\qquad\qquad$ x $\qquad\qquad$ x

$$K_b = \frac{[C_6H_5NH_3^+][OH^-]}{[C_6H_5NH_2]} = 4.3 \times 10^{-10} = \frac{x^2}{0.15 - x} \approx \frac{x^2}{0.15}$$

Solve for x. $\quad x = 8.0 \times 10^{-6}$ M $= [C_6H_5NH_3^+] = [OH^-]$

$[C_6H_5NH_2] = 0.15 - x = 0.15$ M

$$[H_3O^+] = \frac{K_w}{[OH^-]} = \frac{1.0 \times 10^{-14}}{8.0 \times 10^{-6}} = 1.25 \times 10^{-9}$$ M $= 1.2 \times 10^{-9}$ M

$pH = -\log[H_3O^+] = -\log(1.25 \times 10^{-9}) = 8.90$

14.88 (a) $K_a = \dfrac{K_w}{K_b \text{ for } C_3H_7NH_2} = \dfrac{1.0 \times 10^{-14}}{5.1 \times 10^{-4}} = 2.0 \times 10^{-11}$

(b) $K_a = \dfrac{K_w}{K_b \text{ for } NH_2OH} = \dfrac{1.0 \times 10^{-14}}{9.1 \times 10^{-9}} = 1.1 \times 10^{-6}$

(c) $K_a = \dfrac{K_w}{K_b \text{ for } C_6H_5NH_2} = \dfrac{1.0 \times 10^{-14}}{4.3 \times 10^{-10}} = 2.3 \times 10^{-5}$

(d) $K_a = \dfrac{K_w}{K_b \text{ for } C_5H_5N} = \dfrac{1.0 \times 10^{-14}}{1.8 \times 10^{-9}} = 5.6 \times 10^{-6}$

14.89 (a) $K_b = \dfrac{K_w}{K_a \text{ for } HF} = \dfrac{1.0 \times 10^{-14}}{3.5 \times 10^{-4}} = 2.9 \times 10^{-11}$

(b) $K_b = \dfrac{K_w}{K_a \text{ for } HOBr} = \dfrac{1.0 \times 10^{-14}}{2.0 \times 10^{-9}} = 5.0 \times 10^{-6}$

(c) $K_b = \dfrac{K_w}{K_a \text{ for } H_2S} = \dfrac{1.0 \times 10^{-14}}{1.0 \times 10^{-7}} = 1.0 \times 10^{-7}$

(d) $K_b = \dfrac{K_w}{K_a \text{ for } HS^-} = \dfrac{1.0 \times 10^{-14}}{1 \times 10^{-19}} = 1 \times 10^5$

Acid–Base Properties of Salts (Section 14.14)

14.90 (a) $CH_3NH_3^+(aq) + H_2O(l) \rightleftarrows H_3O^+(aq) + CH_3NH_2(aq)$

      acid   base ——— acid   base

(b) $Cr(H_2O)_6^{3+}(aq) + H_2O(l) \rightleftarrows H_3O^+(aq) + Cr(H_2O)_5(OH)^{2+}(aq)$

      acid   base —— acid   base

(c) $CH_3CO_2^-(aq) + H_2O(l) \rightleftarrows CH_3CO_2H(aq) + OH^-(aq)$

      base   acid   acid   base

(d) $PO_4^{3-}(aq) + H_2O(l) \rightleftarrows HPO_4^{2-}(aq) + OH^-(aq)$

      base   acid   acid   base

14.91 (a) Na_2CO_3; Na^+ is a neutral cation. CO_3^{2-} is a basic anion.
$CO_3^{2-}(aq) + H_2O(l) \rightleftarrows HCO_3^-(aq) + OH^-(aq)$

base   acid   acid   base

(b) NH_4NO_3; NH_4^+ is an acidic cation. NO_3^- is a neutral anion.
$NH_4^+(aq) + H_2O(l) \rightleftarrows H_3O^+(aq) + NH_3(aq)$

acid   base ——— acid   base

(c) $NaCl$; Na^+ is a neutral cation. Cl^- is a neutral anion.
$H_2O(l) + H_2O(l) \rightleftarrows H_3O^+(aq) + OH^-(aq)$

acid   base ——— acid   base

(d) $ZnCl_2$; Zn^{2+} is an acidic cation. Cl^- is a neutral anion.
$Zn(H_2O)_6^{2+}(aq) + H_2O(l) \rightleftarrows H_3O^+(aq) + Zn(H_2O)_5(OH)^+(aq)$

acid   base ——— acid   base

14.92 (a) F^- (conjugate base of a weak acid), basic solution
(b) Br^- (anion of a strong acid), neutral solution
(c) NH_4^+ (conjugate acid of a weak base), acidic solution

(d) $K(H_2O)_6^+$ (neutral cation), neutral solution

(e) SO_3^{2-} (conjugate base of a weak acid), basic solution

(f) $Cr(H_2O)_6^{3+}$ (acidic cation), acidic solution

14.93 (a) $Fe(NO_3)_3$: Fe^{3+}, acidic cation; NO_3^-, neutral anion; solution is acidic

(b) $Ba(NO_3)_2$: Ba^{2+}, neutral cation; NO_3^-, neutral anion; solution is neutral

(c) $NaOCl$: Na^+, neutral cation; OCl^-, basic anion; solution is basic

(d) NH_4I: NH_4^+, acidic cation; I^-, neutral anion; solution is acidic

(e) NH_4NO_2; For NH_4^+, $K_a = 5.6 \times 10^{-10}$; for NO_2^-, $K_b = 2.2 \times 10^{-11}$
Because $K_a > K_b$, the solution is acidic.

(f) $(CH_3NH_3)Cl$: $CH_3NH_3^+$, acidic cation; Cl^-, neutral anion; solution is acidic

14.94 (a) $(C_2H_5NH_3)NO_3$: $C_2H_5NH_3^+$, acidic cation; NO_3^-, neutral anion
$C_2H_5NH_2$, $K_b = 6.4 \times 10^{-4}$

$$C_2H_5NH_3^+, \; K_a = \frac{K_w}{K_b \text{ for } C_2H_5NH_2} = \frac{1.0 \times 10^{-14}}{6.4 \times 10^{-4}} = 1.56 \times 10^{-11}$$

$$C_2H_5NH_3^+(aq) + H_2O(l) \rightleftharpoons H_3O^+(aq) + C_2H_5NH_2(aq)$$

initial (M)	0.10	~0	0
change (M)	−x	+x	+x
equil (M)	0.10 − x	x	x

$$K_a = \frac{[H_3O^+][C_2H_5NH_2]}{[C_2H_5NH_3^+]} = 1.56 \times 10^{-11} = \frac{x^2}{0.10 - x} \approx \frac{x^2}{0.10}$$

Solve for x. $x = 1.25 \times 10^{-6} \text{ M} = 1.2 \times 10^{-6} \text{ M} = [H_3O^+] = [C_2H_5NH_2]$
$pH = -\log[H_3O^+] = -\log(1.25 \times 10^{-6}) = 5.90$
$[C_2H_5NH_3^+] = 0.10 - x = 0.10 \text{ M};$ $\quad [NO_3^-] = 0.10 \text{ M}$

$$[OH^-] = \frac{K_w}{[H_3O^+]} = \frac{1.0 \times 10^{-14}}{1.25 \times 10^{-6} \text{ M}} = 8.0 \times 10^{-9}$$

(b) $Na(CH_3CO_2)$: Na^+, neutral cation; $CH_3CO_2^-$, basic anion
CH_3CO_2H, $K_a = 1.8 \times 10^{-5}$

$$CH_3CO_2^-, \; K_b = \frac{K_w}{K_a \text{ for } CH_3CO_2H} = \frac{1.0 \times 10^{-14}}{1.8 \times 10^{-5}} = 5.6 \times 10^{-10}$$

$$CH_3CO_2^-(aq) + H_2O(aq) \rightleftharpoons CH_3CO_2H(aq) + OH^-(aq)$$

initial (M)	0.10	0	~0
change (M)	−x	+x	+x
equil (M)	0.10 − x	x	x

$$K_b = \frac{[CH_3CO_2H][OH^-]}{[CH_3CO_2^-]} = 5.6 \times 10^{-10} = \frac{x^2}{0.10 - x} \approx \frac{x^2}{0.10}$$

Solve for x. $x = 7.5 \times 10^{-6} \text{ M} = [CH_3CO_2H] = [OH^-]$
$[CH_3CO_2^-] = 0.10 - x = 0.10 \text{ M};$ $\quad [Na^+] = 0.10 \text{ M}$

$$[H_3O^+] = \frac{K_w}{[OH^-]} = \frac{1.0 \times 10^{-14}}{7.5 \times 10^{-6}} = 1.3 \times 10^{-9} \text{ M}$$

$pH = -\log[H_3O^+] = -\log(1.3 \times 10^{-9}) = 8.89$

(c) $NaNO_3$: Na^+, neutral cation; NO_3^-, neutral anion

$[Na^+] = [NO_3^-] = 0.10$ M

$[H_3O^+] = [OH^-] = 1.0 \times 10^{-7}$ M; $pH = 7.00$

14.95   (a) $\qquad Fe(H_2O)_6^{2+}(aq) + H_2O(l) \rightleftharpoons H_3O^+(aq) + Fe(H_2O)_5(OH)^+(aq)$

initial (M) $\quad$ 0.020 $\qquad\qquad\qquad\qquad\qquad$ ~0 $\qquad\qquad$ 0

change (M) $\quad$ $-x$ $\qquad\qquad\qquad\qquad\qquad\quad$ $+x$ $\qquad\qquad$ $+x$

equil (M) $\quad$ $0.020 - x$ $\qquad\qquad\qquad\qquad\quad$ x $\qquad\qquad$ x

$$K_a = \frac{[H_3O^+][Fe(H_2O)_5(OH)^+]}{[Fe(H_2O)_6^{2+}]} = 3.2 \times 10^{-10} = \frac{x^2}{0.020 - x} \approx \frac{x^2}{0.020}$$

Solve for x. $x = [H_3O^+] = 2.5 \times 10^{-6}$ M

$pH = -\log[H_3O^+] = -\log(2.5 \times 10^{-6}) = 5.60$

$$\% \text{ dissociation} = \frac{[Fe(H_2O)_6^{2+}]_{diss}}{[Fe(H_2O)_6^{2+}]_{initial}} \times 100\% = \frac{2.5 \times 10^{-6} \text{ M}}{0.020 \text{ M}} \times 100\% = 0.012\%$$

(b) $\qquad Fe(H_2O)_6^{3+}(aq) + H_2O(l) \rightleftharpoons H_3O^+(aq) + Fe(H_2O)_5(OH)^{2+}(aq)$

initial (M) $\quad$ 0.020 $\qquad\qquad\qquad\qquad\qquad$ ~0 $\qquad\qquad$ 0

change (M) $\quad$ $-x$ $\qquad\qquad\qquad\qquad\qquad\quad$ $+x$ $\qquad\qquad$ $+x$

equil (M) $\quad$ $0.020 - x$ $\qquad\qquad\qquad\qquad\quad$ x $\qquad\qquad$ x

$$K_a = \frac{[H_3O^+][Fe(H_2O)_5(OH)^{2+}]}{[Fe(H_2O)_6^{3+}]} = 6.3 \times 10^{-3} = \frac{x^2}{0.020 - x}$$

$x^2 + (6.3 \times 10^{-3})x - (1.26 \times 10^{-4}) = 0$

Use the quadratic formula to solve for x.

$$x = \frac{-(6.3 \times 10^{-3}) \pm \sqrt{(6.3 \times 10^{-3})^2 - (4)(-1.26 \times 10^{-4})}}{2(1)} = \frac{(-6.3 \times 10^{-3}) \pm 0.0233}{2}$$

$x = 0.0085$ and -0.0148

Of the two solutions for x, only the positive value of x has physical meaning because x is the $[H_3O^+]$.

$x = [H_3O^+] = 0.0085$ M

$pH = -\log[H_3O^+] = -\log(0.0085) = 2.07$

$$\% \text{ dissociation} = \frac{[Fe(H_2O)_6^{3+}]_{diss}}{[Fe(H_2O)_6^{3+}]_{initial}} \times 100\% = \frac{0.0085 \text{ M}}{0.020 \text{ M}} \times 100\% = 42\%$$

Factors That Affect Acid Strength (Section 14.15)

14.96   (a) $PH_3 < H_2S < HCl$; electronegativity increases from P to Cl

(b) $NH_3 < PH_3 < AsH_3$; X–H bond strength decreases from N to As (down a group)

(c) $HBrO < HBrO_2 < HBrO_3$; acid strength increases with the number of O atoms

14.97   (a) $H_2Se > H_2S > H_2O$; X–H bond strength decreases from O to Se (down a group)

(b) $HClO_3 > HBrO_3 > HIO_3$; electronegativity increases from I to Cl

(c) $HCl > H_2S > PH_3$; electronegativity increases from P to Cl

14.98   (a) HCl; The strength of a binary acid H_nA increases as A moves from left to right and from top to bottom in the periodic table.
(b) $HClO_3$; The strength of an oxoacid increases with increasing electronegativity and increasing oxidation state of the central atom.
(c) HBr; The strength of a binary acid H_nA increases as A moves from left to right and from top to bottom in the periodic table.

14.99   (a) H_2SO_3; $HClO_3$ is weaker than $HClO_4$ because of the lower oxidation number of the Cl, and H_2SO_3 is weaker than $HClO_3$ because S is less electronegative than Cl.
(b) NH_3; H_2O is weaker than H_2S because of the smaller size of O and the stronger O–H bond. NH_3 is weaker than H_2O because N is less electronegative than O, and the N–H bond is therefore less polar.
(c) $Ga(OH)_3$; The acid strength of an oxoacid decreases with decreasing electronegativity of the central atom.

14.100 (a) H_2Te, weaker X–H bond
(b) H_3PO_4, P has higher electronegativity
(c) $H_2PO_4^-$, lower negative charge
(d) NH_4^+, higher positive charge and N is more electronegative than C

14.101 (a) ClO_2^- (conjugate base of $HClO_2$) is a stronger base than ClO_3^- (conjugate base of $HClO_3$) because $HClO_2$ is the weaker acid.
(b) $HSeO_4^-$ (conjugate base of H_2SeO_4) is a stronger base than HSO_4^- (conjugate base of H_2SO_4) because H_2SeO_4 is the weaker acid.
(c) OH^- (conjugate base of H_2O) is a stronger base than HS^- (conjugate base of H_2S) because H_2O is the weaker acid.
(d) HS^- is the stronger base because Br^- has no basic properties. HBr is a strong acid.

Lewis Acids and Bases (Section 14.16)

14.102 (a) Lewis acid, SiF_4; Lewis base, F^- (b) Lewis acid, Zn^{2+}; Lewis base, NH_3
(c) Lewis acid, $HgCl_2$; Lewis base, Cl^- (d) Lewis acid, CO_2; Lewis base, H_2O

14.103 (a) Lewis acid, $BeCl_2$; Lewis base, Cl^- (b) Lewis acid, Mg^{2+}; Lewis base, H_2O
(c) Lewis acid, SO_3; Lewis base, OH^- (d) Lewis acid, BF_3; Lewis base, F^-

14.104 (a) $2\ :\!\ddot{\underset{..}{F}}\!:^- \ +\ SiF_4 \longrightarrow SiF_6^{2-}$
(b) $4\ \ddot{N}H_3 \ +\ Zn^{2+} \longrightarrow Zn(NH_3)_4^{2+}$
(c) $2\ :\!\ddot{\underset{..}{Cl}}\!:^- \ +\ HgCl_2 \longrightarrow HgCl_4^{2-}$
(d) $H_2\ddot{O}\!: \ +\ CO_2 \longrightarrow H_2CO_3$

14.105 (a) $2\ :\!\ddot{\underset{..}{Cl}}\!:^- \ +\ BeCl_2 \longrightarrow BeCl_4^{2-}$
(b) $6\ H_2\ddot{O}\!: \ +\ Mg^{2+} \longrightarrow Mg(H_2O)_6^{2+}$

(c)

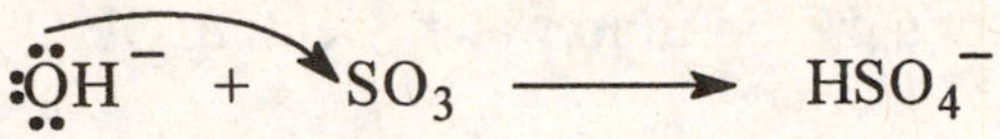

$:\overset{..}{\underset{..}{O}}H^- + \ SO_3 \longrightarrow \ HSO_4^-$

(d)

$:\overset{..}{\underset{..}{F}}:^- + \ BF_3 \longrightarrow \ BF_4^-$

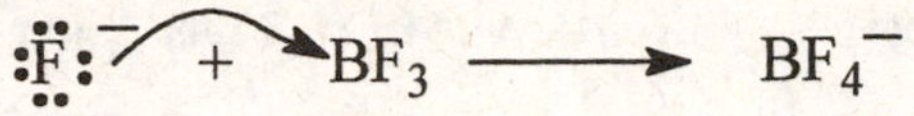

14.106 (a) CN^-, Lewis base   (b) H^+, Lewis acid   (c) H_2O, Lewis base
       (d) Fe^{3+}, Lewis acid   (e) OH^-, Lewis base   (f) CO_2, Lewis acid
       (g) $P(CH_3)_3$, Lewis base   (h) $B(CH_3)_3$, Lewis acid

14.107 (a) BF_3, F is more electronegative than H
       (b) SO_3, more O atoms draw electron density away from S
       (c) Sn^{4+}, higher charge
       (d) CH_3^+, electron deficient

Chapter Problems

14.108 In aqueous solution:
       H_2S acts as an acid only.
       HS^- can act as both an acid and a base.
       S^{2-} can act as a base only.
       H_2O can act as both an acid and a base.
       H_3O^+ acts as an acid only.
       OH^- acts as a base only.

14.109 $[H_3O^+] = 10^{-pH}$ and $[OH^-] = \dfrac{K_w}{[H_3O^+]}$

(a) $[H_3O^+] = 10^{-3.2} = 6 \times 10^{-4}$ M, $[OH^-] = \dfrac{1.0 \times 10^{-14}}{6 \times 10^{-4}} = 2 \times 10^{-11}$ M

(b) $[H_3O^+] = 10^{-7.8} = 2 \times 10^{-8}$ M, $[OH^-] = \dfrac{1.0 \times 10^{-14}}{2 \times 10^{-8}} = 5 \times 10^{-7}$ M

(c) $[H_3O^+] = 10^{-3.1} = 8 \times 10^{-4}$ M, $[OH^-] = \dfrac{1.0 \times 10^{-14}}{8 \times 10^{-4}} = 1 \times 10^{-11}$ M

(d) $[H_3O^+] = 10^{-6.4} = 4 \times 10^{-7}$ M, $[OH^-] = \dfrac{1.0 \times 10^{-14}}{4 \times 10^{-7}} = 2 \times 10^{-8}$ M

(e) $[H_3O^+] = 10^{-4.2} = 6 \times 10^{-5}$ M, $[OH^-] = \dfrac{1.0 \times 10^{-14}}{6 \times 10^{-5}} = 2 \times 10^{-10}$ M

(f) $[H_3O^+] = 10^{-1.9} = 1 \times 10^{-2}$ M, $[OH^-] = \dfrac{1.0 \times 10^{-14}}{1 \times 10^{-2}} = 1 \times 10^{-12}$ M

14.110

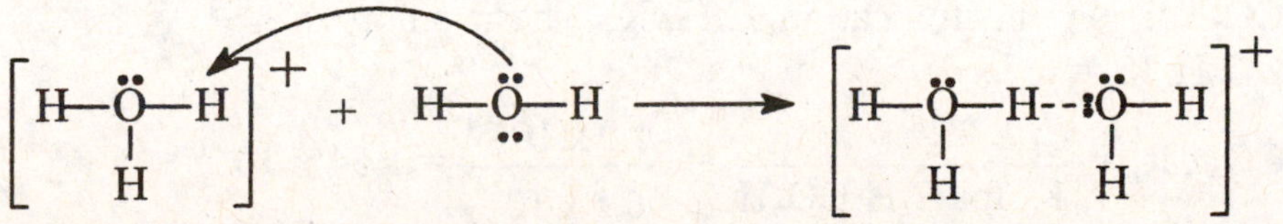

H_3O^+ can hydrogen bond with additional H_2O molecules.

14.111 OH^- because stronger bases accept a proton from water, yielding OH^-.

14.112 $HCO_3^-(aq) + Al(H_2O)_6^{3+}(aq) \rightarrow H_2O(l) + CO_2(g) + Al(H_2O)_5(OH)^{2+}(aq)$

14.113 (a) $Zn(NO_3)_2$, weakly acidic salt   (b) Na_2O, strong base
(c) $NaOCl$, weakly basic salt   (d) $NaClO_4$, neutral salt
(e) $HClO_4$, strong acid
$Na_2O < NaOCl < NaClO_4 < Zn(NO_3)_2 < HClO_4$

14.114 H_2O, 18.02 amu

at 0 °C, $[H_2O] = \dfrac{\left(0.9998 \text{ g} \times \dfrac{1 \text{ mol}}{18.02 \text{ g}} \right)}{0.001 \text{ L}} = 55.48 \text{ M}$

$K_w = [H_3O^+][OH^-]$, for a neutral solution $[H_3O^+] = [OH^-]$

$[H_3O^+] = \sqrt{K_w} = \sqrt{1.14 \times 10^{-15}} = 3.376 \times 10^{-8} \text{ M}$

$pH = -\log[H_3O^+] = -\log(3.376 \times 10^{-8}) = 7.472$

fraction dissociated $= \dfrac{[H_2O]_{diss}}{[H_2O]_{initial}} = \dfrac{3.376 \times 10^{-8} \text{ M}}{55.48 \text{ M}} = 6.09 \times 10^{-10}$

% dissociation $= \dfrac{[H_2O]_{diss}}{[H_2O]_{initial}} \times 100\% = \dfrac{3.376 \times 10^{-8} \text{ M}}{55.48 \text{ M}} \times 100\% = 6.09 \times 10^{-8}\%$

14.115 $HCN(aq) + H_2O(l) \rightleftharpoons H_3O^+(aq) + CN^-(aq)$; $K_a = \dfrac{[H_3O^+][CN^-]}{[HCN]}$

$CN^-(aq) + H_2O(l) \rightleftharpoons HCN(aq) + OH^-(aq)$; $K_b = \dfrac{[HCN][OH^-]}{[CN^-]}$

$K_a \times K_b = \left(\dfrac{[H_3O^+][CN^-]}{[HCN]} \right) \left(\dfrac{[HCN][OH^-]}{[CN^-]} \right) = [H_3O^+][OH^-] = K_w$

14.116 For $C_{10}H_{14}N_2H^+$, $K_{a1} = \dfrac{K_w}{K_{b1} \text{ for } C_{10}H_{14}N_2} = \dfrac{1.0 \times 10^{-14}}{1.0 \times 10^{-6}} = 1.0 \times 10^{-8}$

For $C_{10}H_{14}N_2H_2^{2+}$, $K_{a2} = \dfrac{K_w}{K_{b2} \text{ for } C_{10}H_{14}N_2H^+} = \dfrac{1.0 \times 10^{-14}}{1.3 \times 10^{-11}} = 7.7 \times 10^{-4}$

14.117 $C_6H_5CO_2Na$: Na^+, neutral cation, $C_6H_5CO_2^-$, basic anion
$C_6H_5CO_2H$, $K_a = 6.5 \times 10^{-5}$

$C_6H_5CO_2^-$, $K_b = \dfrac{K_w}{K_a \text{ for } C_6H_6CO_2H} = \dfrac{1.0 \times 10^{-14}}{6.5 \times 10^{-5}} = 1.54 \times 10^{-10}$

$$C_6H_5CO_2^-(aq) \; + \; H_2O(l) \; \rightleftharpoons \; C_6H_5CO_2H(aq) \; + \; OH^-(aq)$$

initial (M)	0.050	0	~0
change (M)	−x	+x	+x
equil (M)	0.050 − x	x	x

$$K_b = \frac{[C_6H_5CO_2H][OH^-]}{[C_6H_5CO_2^-]} = 1.54 \times 10^{-10} = \frac{x^2}{0.050 - x} \approx \frac{x^2}{0.050}$$

Solve for x. $x = 2.77 \times 10^{-6} = 2.8 \times 10^{-6} \, M = [C_6H_5CO_2H] = [OH^-]$

$[Na^+] = 0.050 \, M; \quad [C_6H_5CO_2^-] = 0.050 - x = 0.050 \, M$

$$[H_3O^+] = \frac{K_w}{[OH^-]} = \frac{1.0 \times 10^{-14}}{2.77 \times 10^{-6}} = 3.61 \times 10^{-9} \, M = 3.6 \times 10^{-9} \, M$$

$pH = -\log[H_3O^+] = -\log(3.61 \times 10^{-9}) = 8.44$

14.118 (a) $A^-(aq) \; + \; H_2O(l) \; \rightleftharpoons \; HA(aq) \; + \; OH^-(aq)$; basic

(b) $M(H_2O)_6^{3+}(aq) \; + \; H_2O(l) \; \rightleftharpoons \; H_3O^+(aq) \; + \; M(H_2O)_5(OH)^{2+}(aq)$; acidic

(c) $2 \, H_2O(l) \; \rightleftharpoons \; H_3O^+(aq) \; + \; OH^-(aq)$; neutral

(d) $M(H_2O)_6^{3+}(aq) \; + \; A^-(aq) \; \rightleftharpoons \; HA(aq) \; + \; M(H_2O)_5(OH)^{2+}(aq)$;
acidic because K_a for $M(H_2O)_6^{3+}$ (10^{-4}) is greater than K_b for A^- (10^{-9})

14.119 HCl is a strong acid. $[H_3O^+]_{initial} = 0.10 \, M$
The dissociation of HF produces a negligible amount of H_3O^+ compared with that produced from HCl.
$HF(aq) \; + \; H_2O(l) \; \rightleftharpoons \; H_3O^+(aq) \; + \; F^-(aq)$

$$K_a = \frac{[H_3O^+][F^-]}{[HF]} = 3.5 \times 10^{-4} = \frac{(0.10)[F^-]}{(0.10)}$$

$[F^-] = K_a = 3.5 \times 10^{-4} \, M; \quad [HF] = 0.10 \, M; \quad [H_3O^+] = 0.10 \, M$

$$[OH^-] = \frac{K_w}{[H_3O^+]} = \frac{1.0 \times 10^{-14}}{0.10} = 1.0 \times 10^{-13} \, M$$

$pH = -\log[H_3O^+] = -\log(0.10) = 1.00$

14.120

$$HIO_3(aq) \; + \; H_2O(l) \; \rightleftharpoons \; H_3O^+(aq) \; + \; IO_3^-(aq)$$

initial (M)	0.0500	~0	0
change (M)	−x	+x	+x
equil (M)	0.0500 − x	x	x

$$K_a = \frac{[H_3O^+][IO_3^-]}{[HIO_3]} = 1.7 \times 10^{-1} = \frac{x^2}{0.0500 - x}$$

$x^2 + 0.17x - 0.0085 = 0$
Use the quadratic formula to solve for x.

$$x = \frac{-(0.17) \pm \sqrt{(0.17)^2 - (4)(1)(-0.0085)}}{2(1)} = \frac{(-0.17) \pm 0.251}{2}$$

$x = -0.210$ and 0.0405

Of the two solutions for x, only the positive value of x has physical meaning because x is the $[H_3O^+]$.

$x = [H_3O^+] = 0.0405$ M $= 0.040$ M

pH $= -\log[H_3O^+] = -\log(0.040) = 1.39$

$[HIO_3] = 0.0500 - x = 0.0500 - 0.040 = 0.010$ M

$[IO_3^-] = x = 0.040$ M

$$[OH^-] = \frac{K_w}{[H_3O^+]} = \frac{1.0 \times 10^{-14}}{0.040} = 2.5 \times 10^{-13} \text{ M}$$

14.121 $[OCl^-] = 0.100$ mol/L $Ca(OCl)_2 \times \dfrac{2 \text{ mol } OCl^-}{1 \text{ mol } Ca(OCl)_2} = 0.200$ M

OCl^- is the conjugate base of HOCl.

$$K_b(OCl^-) = \frac{K_w}{K_a(HOCl)} = \frac{1.0 \times 10^{-14}}{3.5 \times 10^{-8}} = 2.9 \times 10^{-7}$$

	$OCl^-(aq)$	$+$	$H_2O(l)$	$\rightleftharpoons$	$HOCl(aq)$	$+$	$OH^-(aq)$
initial (M)	0.200				0		0
change (M)	$-x$				$+x$		$+x$
equil (M)	$0.200 - x$				x		x

$$K_b = \frac{[HOCl][OH^-]}{[OCl^-]} = 2.9 \times 10^{-7} = \frac{x^2}{0.200 - x} \approx \frac{x^2}{0.200}$$

$x = [OH^-] = \sqrt{(2.9 \times 10^{-7})(0.200)} = 2.4 \times 10^{-4}$

$$[H_3O^+] = \frac{K_w}{[OH^-]} = \frac{1.0 \times 10^{-14}}{2.4 \times 10^{-4}} = 4.2 \times 10^{-11} \text{ M}$$

pH $= -\log[H_3O^+] = -\log(4.2 \times 10^{-11}) = 10.38$

14.122 For $H_2C_2O_4$, $K_{a1} = 5.9 \times 10^{-2}$ and $K_{a2} = 6.4 \times 10^{-5}$.

For $C_2O_4^{2-}$, $K_{b1} = \dfrac{K_w}{K_{a2}} = \dfrac{1.0 \times 10^{-14}}{6.4 \times 10^{-5}} = 1.6 \times 10^{-10}$

For $HC_2O_4^-$, $K_{b2} = \dfrac{K_w}{K_{a1}} = \dfrac{1.0 \times 10^{-14}}{5.9 \times 10^{-2}} = 1.7 \times 10^{-13}$

	$C_2O_4^{2-}(aq)$	$+$	$H_2O(l)$	$\rightleftharpoons$	$HC_2O_4^-(aq)$	$+$	$OH^-(aq)$
initial (M)	0.100				0		~0
change (M)	$-x$				$+x$		$+x$
equil (M	$0.100 - x$				x		x

$$K_{b1} = \frac{[HC_2O_4^-][OH^-]}{[C_2O_4^{2-}]} = 1.6 \times 10^{-10} = \frac{x^2}{0.100 - x} \approx \frac{x^2}{0.100}$$

$[OH^-] = x = \sqrt{(1.6 \times 10^{-10})(0.100)} = 4.0 \times 10^{-6}$ M

$$[H_3O^+] = \frac{K_w}{[OH^-]} = \frac{1.0 \times 10^{-14}}{4.0 \times 10^{-6}} = 2.5 \times 10^{-9} \text{ M}$$

The second dissociation produces a negligible additional amount of OH^-.

$pH = -\log[H_3O^+] = -\log(2.5 \times 10^{-9}) = 8.60$

14.123 $K = 1.33 = \dfrac{[H_2SO_3]}{P_{SO_2}}$

$[H_2SO_3] = 1.33 \times P_{SO_2}$, and because P_{SO_2} is maintained at 1.00 atm, the $[H_2SO_3]$ remains constant at 1.33 M.

$$H_2SO_3(aq) + H_2O(l) \rightleftharpoons H_3O^+(aq) + HSO_3^-(aq)$$

equil (M)	1.33		x	x

$$K_{a1} = \frac{[H_3O^+][HSO_3^-]}{[H_2SO_3]} = 1.5 \times 10^{-2} = \frac{x^2}{1.33}$$

$x = [H_3O^+] = [HSO_3^-] = \sqrt{(1.5 \times 10^{-2})(1.33)} = 0.14 \text{ M}$

$$HSO_3^-(aq) + H_2O(l) \rightleftharpoons H_3O^+(aq) + SO_3^{2-}(aq)$$

initial (M)	0.14		0.14	0
change (M)	−y		+y	+y
equil (M)	0.14 − y		0.14 + y	y

$$K_{a2} = \frac{[H_3O^+][SO_3^{2-}]}{[HSO_3^-]} = 6.3 \times 10^{-8} = \frac{(0.14 + y)y}{(0.14 - y)} \approx \frac{(0.14)y}{0.14} = y$$

$y = [SO_3^{2-}] = 6.3 \times 10^{-8} \text{ M}$

$pH = -\log[H_3O^+] = -\log(0.14) = 0.85$

14.124   (a) NH_4F;   For NH_4^+, $K_a = 5.6 \times 10^{-10}$ and for F^-, $K_b = 2.9 \times 10^{-11}$
Because $K_a > K_b$, the salt solution is acidic.
(b) $(NH_4)_2SO_3$;   For NH_4^+, $K_a = 5.6 \times 10^{-10}$ and for SO_3^{2-}, $K_b = 1.6 \times 10^{-7}$
Because $K_b > K_a$, the salt solution is basic.

14.125   (a)

$$HOCl(aq) + H_2O(l) \rightleftharpoons H_3O^+(aq) + OCl^-(aq)$$

initial (M)	2.0	~0	0
change (M)	−x	+x	+x
equil (M)	2.0 − x	x	x

$$K_a = \frac{[H_3O^+][OCl^-]}{[HOCl]} = 3.5 \times 10^{-8} = \frac{x^2}{2.0 - x} \approx \frac{x^2}{2.0}$$

Solve for x. $x = 2.6 \times 10^{-4} \text{ M}$

$$\% \text{ dissociation} = \frac{[\text{HOCl}]_{\text{diss}}}{[\text{HOCl}]_{\text{initial}}} \times 100\% = \frac{2.6 \times 10^{-4} \text{ M}}{2.0 \text{ M}} \times 100\% = 0.013\%$$

(b) $\quad\quad\quad \text{HOCl(aq)} + \text{H}_2\text{O(l)} \rightleftharpoons \text{H}_3\text{O}^+(\text{aq}) + \text{OCl}^-(\text{aq})$

initial (M)	0.020	~0	0
change (M)	−x	+x	+x
equil (M)	0.020 − x	x	x

$$K_a = \frac{[\text{H}_3\text{O}^+][\text{OCl}^-]}{[\text{HOCl}]} = 3.5 \times 10^{-8} = \frac{x^2}{0.020 - x} \approx \frac{x^2}{0.020}$$

Solve for x. $\ x = 2.6 \times 10^{-5}$ M

$$\% \text{ dissociation} = \frac{[\text{HOCl}]_{\text{diss}}}{[\text{HOCl}]_{\text{initial}}} \times 100\% = \frac{2.6 \times 10^{-5} \text{ M}}{0.020 \text{ M}} \times 100\% = 0.13\%$$

(c) $\quad\quad\quad \text{HF(aq)} + \text{H}_2\text{O(l)} \rightleftharpoons \text{H}_3\text{O}^+(\text{aq}) + \text{F}^-(\text{aq})$

initial (M)	2.0	~0	0
change (M)	−x	+x	+x
equil (M)	2.0 − x	x	x

$$K_a = \frac{[\text{H}_3\text{O}^+][\text{F}^-]}{[\text{HF}]} = 3.5 \times 10^{-4} = \frac{x^2}{2.0 - x} \approx \frac{x^2}{2.0}$$

Solve for x. $\ x = 0.026$ M

$$\% \text{ dissociation} = \frac{[\text{HF}]_{\text{diss}}}{[\text{HF}]_{\text{initial}}} \times 100\% = \frac{0.026 \text{ M}}{2.0 \text{ M}} \times 100\% = 1.3\%$$

The percent dissociation increases by a factor of 10 when the acid concentration decreases by a factor of 100 for the same acid. $\% \text{ dissociation} \sim \dfrac{1}{\sqrt{\text{concentration}}}$.

The percent dissociation increases by a factor of 10^2 when K_a increases by a factor of 10^4 for the same concentration of acid. $\% \text{ dissociation} \sim \sqrt{K_a}$.

14.126　Fraction dissociated $= \dfrac{[\text{HA}]_{\text{diss}}}{[\text{HA}]_{\text{initial}}}$

For a weak acid, $[\text{HA}]_{\text{diss}} = [\text{H}_3\text{O}^+] = [\text{A}^-]$

$$K_a = \frac{[\text{H}_3\text{O}^+][\text{A}^-]}{[\text{HA}]} = \frac{[\text{H}_3\text{O}^+]^2}{[\text{HA}]}; \quad [\text{H}_3\text{O}^+] = \sqrt{K_a [\text{HA}]}$$

$$\text{Fraction dissociated} = \frac{[\text{HA}]_{\text{diss}}}{[\text{HA}]} = \frac{[\text{H}_3\text{O}^+]}{[\text{HA}]} = \frac{\sqrt{K_a [\text{HA}]}}{[\text{HA}]} = \sqrt{\frac{K_a}{[\text{HA}]}}$$

When the concentration of HA that dissociates is negligible compared with its initial concentration, the equilibrium concentration, [HA], equals the initial concentration, $[\text{HA}]_{\text{initial}}$.

$$\% \text{ dissociation} = \sqrt{\frac{K_a}{[\text{HA}]_{\text{initial}}}} \times 100\%$$

14.127

$$NH_4^+(aq) + H_2O(l) \rightleftharpoons H_3O^+(aq) + NH_3(aq) \qquad K_1 = K_a$$

$$F^-(aq) + H_2O(l) \rightleftharpoons HF(aq) + OH^-(aq) \qquad K_2 = K_b$$

$$\underline{H_3O^+(aq) + OH^-(aq) \rightleftharpoons 2\,H_2O(l)} \qquad K_3 = 1/K_w$$

Overall reaction $NH_4^+(aq) + F^-(aq) \rightleftharpoons HF(aq) + NH_3(aq) \qquad K = K_1K_2K_3$

$$K = K_1K_2K_3 = \frac{K_aK_b}{K_w}$$

The equilibrium constant for the transfer of a proton from the cation of a salt to the anion of the salt is equal to $(K_aK_b)/K_w$.

(a)
$$NH_4^+(aq) \;+\; F^-(aq) \;\rightleftharpoons\; HF(aq) \;+\; NH_3(aq)$$

	NH_4^+	F^-	HF	NH_3
initial (M)	0.25	0.25	0	0
change (M)	−x	−x	+x	+x
equil (M)	0.25 − x	0.25 − x	x	x

$$K = \frac{K_aK_b}{K_w} = \frac{(5.56 \times 10^{-10})(2.86 \times 10^{-11})}{(1.0 \times 10^{-14})} = \frac{[HF][NH_3]}{[NH_4^+][F^-]} = \frac{x^2}{(0.25-x)^2}$$

where K_a is K_a for NH_4^+ and K_b is K_b for F^-.
Take the square root of both sides and solve for x.
$x = 3.15 \times 10^{-4}$
$[NH_4^+] = [F^-] = 0.25 - x = 0.25$ M; $\qquad [NH_3] = [HF] = x = 3.15 \times 10^{-4}$ M

$$K_a(NH_4^+) = \frac{[H_3O^+][NH_3]}{[NH_4^+]} = \frac{[H_3O^+](3.15 \times 10^{-4})}{0.25} = 5.56 \times 10^{-10}$$

$[H_3O^+] = 4.4 \times 10^{-7}$ M

$$[OH^-] = \frac{K_w}{[H_3O^+]} = \frac{1.0 \times 10^{-14}}{4.4 \times 10^{-7}} = 2.3 \times 10^{-8} \text{ M}$$

$pH = -\log[H_3O^+] = -\log(4.4 \times 10^{-7}) = 6.36$

(b)
$$NH_4^+(aq) \;+\; SO_3^{2-}(aq) \;\rightleftharpoons\; HSO_3^-(aq) \;+\; NH_3(aq)$$

	NH_4^+	SO_3^{2-}	HSO_3^-	NH_3
initial (M)	0.50	0.25	0	0
change (M)	−x	−x	+x	+x
equil (M)	0.50 − x	0.25 − x	x	x

$$K = \frac{K_aK_b}{K_w} = \frac{(5.56 \times 10^{-10})(1.59 \times 10^{-7})}{(1.0 \times 10^{-14})} = \frac{[HSO_3^-][NH_3]}{[NH_4^+][SO_3^{2-}]} = \frac{x^2}{(0.50-x)(0.25-x)}$$

where K_a is K_a for NH_4^+ and K_b is K_b for SO_3^{2-}.
$0.9912x^2 + 0.006\,63x - 0.001\,11 = 0$
Use the quadratic formula to solve for x.

$$x = \frac{-(0.006\,63) \pm \sqrt{(0.006\,63)^2 - 4(0.9912)(-0.001\,11)}}{2(0.9912)} = \frac{-0.006\,63 \pm 0.066\,67}{2(0.9912)}$$

x = 0.0303 and −0.0370

Of the two solutions for x, only the positive value of x has physical meaning, because x is the $[NH_3]$ and $[HSO_3^-]$.

x = 0.030

$[NH_4^+] = 0.50 - x = 0.47$ M; $[SO_3^{2-}] = 0.25 - x = 0.22$ M

$[NH_3] = [HSO_3^-] = x = 0.030$ M

$$K_b(SO_3^{2-}) = \frac{[HSO_3^-][OH^-]}{[SO_3^{2-}]} = \frac{(0.030)[OH^-]}{0.22} = 1.59 \times 10^{-7}$$

$[OH^-] = 1.166 \times 10^{-6}$ M $= 1.2 \times 10^{-6}$ M

$$[H_3O^+] = \frac{K_w}{[OH^-]} = \frac{1.0 \times 10^{-14}}{1.166 \times 10^{-6}} = 8.6 \times 10^{-9} \text{ M}$$

pH = $-\log[H_3O^+] = -\log(8.6 \times 10^{-9}) = 8.07$

14.128 Both reactions occur together.

Let x = $[H_3O^+]$ from CH_3CO_2H and y = $[H_3O^+]$ from $C_6H_5CO_2H$

The following two equilibria must be considered:

$$CH_3CO_2H(aq) + H_2O(l) \rightleftharpoons H_3O^+(aq) + CH_3CO_2^-(aq)$$

initial (M)	0.10	y	0
change (M)	−x	+x	+x
equil (M)	0.10 − x	x + y	x

$$C_6H_5CO_2H(aq) + H_2O(l) \rightleftharpoons H_3O^+(aq) + C_6H_5CO_2^-(aq)$$

initial (M)	0.10	x	0
change (M)	−y	+y	+y
equil (M)	0.10 − y	x + y	y

$$K_a(\text{for } CH_3CO_2H) = \frac{[H_3O^+][CH_3CO_2^-]}{[CH_3CO_2H]} = 1.8 \times 10^{-5} = \frac{(x+y)(x)}{0.10-x} \approx \frac{(x+y)(x)}{0.10}$$

$1.8 \times 10^{-6} = (x+y)(x)$

$$K_a(\text{for } C_6H_5CO_2H) = \frac{[H_3O^+][C_6H_5CO_2^-]}{[C_6H_5CO_2H]} = 6.5 \times 10^{-5} = \frac{(x+y)(y)}{0.10-y} \approx \frac{(x+y)(y)}{0.10}$$

$6.5 \times 10^{-6} = (x+y)(y)$

$1.8 \times 10^{-6} = (x+y)(x)$

$6.5 \times 10^{-6} = (x+y)(y)$

These two equations must be solved simultaneously for x and y. Divide the first equation by the second.

$\dfrac{x}{y} = \dfrac{1.8 \times 10^{-6}}{6.5 \times 10^{-6}}$; x = 0.277y

$6.5 \times 10^{-6} = (x+y)(y)$; substitute x = 0.277y into this equation and solve for y.

$6.5 \times 10^{-6} = (0.277y + y)(y) = 1.277y^2$

$y = 0.002\ 256$

$x = 0.277y = (0.277)(0.002\ 256) = 0.000\ 624\ 9$

$[H_3O^+] = (x + y) = (0.000\ 624\ 9 + 0.002\ 256) = 0.002\ 881\ M$

$pH = -\log[H_3O^+] = -\log(0.002\ 881) = 2.54$

14.129 For $1.0 \times 10^{-10}\ M$ HCl, pH = 7.00, and the principal source of H_3O^+ is the dissociation of H_2O.

For $1.0 \times 10^{-7}\ M$ HCl,

$$2\ H_2O(l) \rightleftharpoons H_3O^+(aq) + OH^-(aq)$$

initial (M)	1.0×10^{-7}	~0
change (M)	+x	+x
equil (M)	$(1.0 \times 10^{-7}) + x$	x

$K_w = 1.0 \times 10^{-14} = [H_3O^+][OH^-] = [(1.0 \times 10^{-7}) + x](x)$

$x^2 + (1.0 \times 10^{-7})x - (1.0 \times 10^{-14}) = 0$

Solve for x using the quadratic formula.

$$x = \frac{-(1.0 \times 10^{-7}) \pm \sqrt{(1.0 \times 10^{-7})^2 - (4)(-1.0 \times 10^{-14})}}{2(1)} = \frac{(-1.0 \times 10^{-7}) \pm (2.236 \times 10^{-7})}{2}$$

$x = 6.18 \times 10^{-8}$ and -1.62×10^{-7}

Of the two solutions for x, only the positive value of x has physical meaning, because x is the $[OH^-]$.

$[H_3O^+] = (1.0 \times 10^{-7}) + x = (1.0 \times 10^{-7}) + (6.18 \times 10^{-8}) = 1.618 \times 10^{-7}\ M$

$pH = -\log[H_3O^+] = -\log(1.618 \times 10^{-7}) = 6.79$

14.130 $2\ NO_2(g) + H_2O(l) \rightarrow HNO_3(aq) + HNO_2(aq)$

$$\text{mol } HNO_3 = 0.0500 \text{ mol } NO_2 \times \frac{1 \text{ mol } HNO_3}{2 \text{ mol } NO_2} = 0.0250 \text{ mol } HNO_3$$

$$\text{mol } HNO_2 = 0.0500 \text{ mol } NO_2 \times \frac{1 \text{ mol } HNO_2}{2 \text{ mol } NO_2} = 0.0250 \text{ mol } HNO_2$$

Because the volume is 1.00 L, mol and molarity are the same.

HNO_3 is a strong acid and completely dissociated. From HNO_3, $[NO_3^-] = [H_3O^+] = 0.0250\ M$.

$$HNO_2(aq) + H_2O(l) \rightleftharpoons H_3O^+(aq) + NO_2^-(aq)$$

initial (M)	0.0250	0.0250	0
change (M)	−x	+x	+x
equil (M)	0.0250 − x	0.0250 + x	x

$K_a = \dfrac{[H_3O^+][NO_2^-]}{[HNO_2]} = 4.5 \times 10^{-4} = \dfrac{(0.0250 + x)x}{0.0250 - x}$

$x^2 + 0.02545x - 1.125 \times 10^{-5} = 0$

Solve for x using the quadratic formula.

$$x = \frac{-(0.025\ 45) \pm \sqrt{(0.025\ 45)^2 - (4)(-1.125 \times 10^{-5})}}{2(1)} = \frac{(-0.025\ 45) \pm (0.026\ 32)}{2}$$

$x = -0.0259$ and 4.35×10^{-4}

Of the two solutions for x, only the positive value of x has physical meaning, because x is the $[NO_2^-]$.

$[H_3O^+] = (0.0250) + x = (0.0250) + (4.35 \times 10^{-4}) = 0.0254$ M

$pH = -\log[H_3O^+] = -\log(0.02543) = 1.59$

$[OH^-] = \dfrac{K_w}{[H_3O^+]} = \dfrac{1.0 \times 10^{-14}}{0.0254} = 3.9 \times 10^{-13}$ M

$[NO_3^-] = 0.0250$ M

$[HNO_2] = 0.0250 - x = 0.0250 - 4.35 \times 10^{-4} = 0.0246$ M; $[NO_2^-] = x = 4.3 \times 10^{-4}$ M

14.131 (a) $2 \text{ HSol} \rightleftharpoons H_2Sol^+ + Sol^-$

$K_{HSol} = [H_2Sol^+][Sol^-] = 1 \times 10^{-35}$

(b) $K_b(CN^-) = \dfrac{K_{HSol}}{K_a(HCN)} = \dfrac{1 \times 10^{-35}}{1.3 \times 10^{-13}} = 7.7 \times 10^{-23}$

$$CN^- + HSol \rightleftharpoons HCN + Sol^-$$

	CN^-		HCN	Sol^-
initial (M)	0.010		0	0
change (M)	$-x$		$+x$	$+x$
equil (M)	$0.010 - x$		x	x

$K_b = \dfrac{[HCN][Sol^-]}{[CN^-]} = 7.7 \times 10^{-23} = \dfrac{x^2}{0.010 - x} \approx \dfrac{x^2}{0.010}$

$x = [Sol^-] = \sqrt{(7.7 \times 10^{-23})(0.010)} = 8.8 \times 10^{-13}$ M

$[H_2Sol^+] = \dfrac{K_{HSol}}{[Sol^-]} = \dfrac{1 \times 10^{-35}}{8.8 \times 10^{-13}} = 1 \times 10^{-23}$ M

Multiconcept Problems

14.132 H_3PO_4, 98.00 amu

Assume 1.000 L of solution.

Mass of solution $= 1.000$ L $\times \dfrac{1000 \text{ mL}}{1 \text{ L}} \times \dfrac{1.0353 \text{ g}}{1 \text{ mL}} = 1035.3$ g

Mass $H_3PO_4 = (0.070)(1035.3 \text{ g}) = 72.47$ g H_3PO_4

mol $H_3PO_4 = 72.47$ g $H_3PO_4 \times \dfrac{1 \text{ mol } H_3PO_4}{98.00 \text{ g } H_3PO_4} = 0.740$ mol H_3PO_4

$[H_3PO_4] = \dfrac{0.740 \text{ mol } H_3PO_4}{1.000 \text{ L}} = 0.740$ M

For the dissociation of the first proton, the following equilibrium must be considered:

$$H_3PO_4(aq) + H_2O(l) \rightleftharpoons H_3O^+(aq) + H_2PO_4^-(aq)$$

	H_3PO_4		H_3O^+	$H_2PO_4^-$
initial (M)	0.740		~0	0
change (M)	$-x$		$+x$	$+x$
equil (M)	$0.740 - x$		x	x

$K_{a1} = \dfrac{[H_3O^+][H_2PO_4^-]}{[H_3PO_4]} = 7.5 \times 10^{-3} = \dfrac{x^2}{0.740 - x}$

$x^2 + (7.5 \times 10^{-3})x - (5.55 \times 10^{-3}) = 0$

Solve for x using the quadratic formula.

$$x = \frac{-(7.5 \times 10^{-3}) \pm \sqrt{(7.5 \times 10^{-3})^2 - (4)(-5.55 \times 10^{-3})}}{2(1)} = \frac{(-7.5 \times 10^{-3}) \pm 0.149}{2}$$

$x = 0.0708$ and -0.0783

Of the two solutions for x, only the positive value of x has physical meaning, because x is the $[H_3O^+]$.

$x = 0.0708 \text{ M} = [H_2PO_4^-] = [H_3O^+]$

For the dissociation of the second proton, the following equilibrium must be considered:

$$H_2PO_4^-(aq) + H_2O(l) \rightleftharpoons H_3O^+(aq) + HPO_4^{2-}(aq)$$

initial (M)	0.0708	0.0708	0
change (M)	$-y$	$+y$	$+y$
equil (M)	$0.0708 - y$	$0.0708 + y$	y

$$K_{a2} = \frac{[H_3O^+][HPO_4^{2-}]}{[H_2PO_4^-]} = 6.2 \times 10^{-8} = \frac{(0.0708 + y)(y)}{0.0708 - y} \approx \frac{(0.0708)(y)}{0.0708} = y$$

$y = 6.2 \times 10^{-8} \text{ M} = [HPO_4^{2-}]$

For the dissociation of the third proton, the following equilibrium must be considered:

$$HPO_4^{2-}(aq) + H_2O(l) \rightleftharpoons H_3O^+(aq) + PO_4^{3-}(aq)$$

initial (M)	6.2×10^{-8}	0.0708	0
change (M)	$-z$	$+z$	$+z$
equil (M)	$(6.2 \times 10^{-8}) - z$	$0.0708 + z$	z

$$K_{a3} = \frac{[H_3O^+][PO_4^{3-}]}{[HPO_4^{2-}]} = 4.8 \times 10^{-13} = \frac{(0.0708 + z)(z)}{(6.2 \times 10^{-8}) - z} \approx \frac{(0.0708)(z)}{6.2 \times 10^{-8}}$$

$z = 4.2 \times 10^{-19} \text{ M} = [PO_4^{3-}]$

$[H_3PO_4] = 0.740 - x = 0.740 - 0.0708 = 0.67 \text{ M}$

$[H_2PO_4^-] = [H_3O^+] = 0.0708 \text{ M} = 0.071 \text{ M}$

$[HPO_4^{2-}] = 6.2 \times 10^{-8} \text{ M};\qquad [PO_4^{3-}] = 4.2 \times 10^{-19} \text{ M}$

$$[OH^-] = \frac{K_w}{[H_3O^+]} = \frac{1.0 \times 10^{-14}}{0.0708} = 1.4 \times 10^{-13} \text{ M}$$

$\text{pH} = -\log[H_3O^+] = -\log(0.0708) = 1.15$

14.133 $C_7H_5NO_3S$, 183.19 amu; 348 mg = 0.348 g

$$[C_7H_5NO_3S] = \frac{\left(0.348 \text{ g} \times \dfrac{1 \text{ mol}}{183.19 \text{ g}}\right)}{0.100 \text{ L}} = 0.0190 \text{ M}$$

Let $C_7H_5NO_3S = HSac$

Let $x = [H_3O^+]$ from HSac and $y = [H_3O^+]$ from H_2O

$$HSac(aq) + H_2O(l) \rightleftharpoons H_3O^+(aq) + Sac^-(aq)$$

initial (M)	0.0190	y	0
change (M)	–x	+x	+x
equil (M)	0.0190 – x	x + y	x

$$K_a = \frac{[H_3O^+][Sac^-]}{[HSac]} = 2.1 \times 10^{-12} = \frac{(x+y)(x)}{0.0190 - x} \approx \frac{(x+y)(x)}{0.0190}$$

$$4.0 \times 10^{-14} = (x+y)(x)$$

$$2 H_2O(l) \rightleftharpoons H_3O^+(aq) + OH^-(aq)$$

initial (M)		x	~0
change (M)		+y	+y
equil (M)		x + y	y

$$K_w = [H_3O^+][OH^-] = 1.0 \times 10^{-14} = (x+y)(y)$$

$$4.0 \times 10^{-14} = (x+y)(x)$$

$$1.0 \times 10^{-14} = (x+y)(y)$$

Solve these two equations simultaneously for x and y. Divide the first equation by the second.

$$\frac{x}{y} = \frac{4.0 \times 10^{-14}}{1.0 \times 10^{-14}} = 4.0; \qquad x = 4.0y$$

$$1.0 \times 10^{-14} = (x+y)(y) = (4.0y + y)y = 5.0y^2$$

$$y = \sqrt{\frac{1.0 \times 10^{-14}}{5.0}} = 4.5 \times 10^{-8}$$

$$x = 4.0y = (4.0)(4.5 \times 10^{-8}) = 1.8 \times 10^{-7}$$

$$[H_3O^+] = (x+y) = (1.8 \times 10^{-7}) + (4.5 \times 10^{-8}) = 2.25 \times 10^{-7} \text{ M} = 2.2 \times 10^{-7} \text{ M}$$

$$pH = -\log[H_3O^+] = -\log(2.25 \times 10^{-7}) = 6.65$$

14.134 $[H_3O^+] = 10^{-pH} = 10^{-9.07} = 8.51 \times 10^{-10}$ M

$$[H_3O^+][OH^-] = K_w$$

$$[OH^-] = \frac{K_w}{[H_3O^+]} = \frac{1.0 \times 10^{-14}}{8.51 \times 10^{-10}} = 1.18 \times 10^{-5} \text{ M} = x \text{ below.}$$

$$K_a = 1.8 \times 10^{-5} \text{ for } CH_3CO_2H \text{ and } K_b = \frac{K_w}{K_a} = \frac{1.0 \times 10^{-14}}{1.8 \times 10^{-5}} = 5.56 \times 10^{-10}$$

Use the equilibrium associated with a weak base to solve for $[CH_3CO_2^-] = y$ below.

$$CH_3CO_2^-(aq) + H_2O(l) \rightleftharpoons CH_3CO_2H(aq) + OH^-(aq)$$

initial (M)	y	~0	0
change (M)	–x	+x	+x
equil (M)	y – x	x	x

$$K_b = \frac{[CH_3CO_2H][OH^-]}{[CH_3CO_2^-]} = 5.56 \times 10^{-10} = \frac{(1.18 \times 10^{-5})^2}{[y - (1.18 \times 10^{-5})]}$$

Solve for y.

$(5.56 \times 10^{-10})[y - (1.18 \times 10^{-5})] = (1.18 \times 10^{-5})^2$

$(5.56 \times 10^{-10})y - 6.56 \times 10^{-15} = 1.39 \times 10^{-10}$

$(5.56 \times 10^{-10})y = 1.39 \times 10^{-10}$

$y = (1.39 \times 10^{-10})/(5.56 \times 10^{-10}) = [CH_3CO_2^-] = 0.25 \text{ M}$

In 1.00 L of solution, the mass of CH_3CO_2Na solute $= (0.25 \text{ mol/L})\left(\dfrac{82.035 \text{ g } CH_3CO_2Na}{1 \text{ mol } CH_3CO_2Na}\right) = 20.5 \text{ g}$

mass of solution $= (1000 \text{ mL})\left(\dfrac{1.0085 \text{ g}}{1 \text{ mL}}\right) = 1008.5 \text{ g}$

mass of solvent $= 1008.5 \text{ g} - 20.5 \text{ g} = 988 \text{ g} = 0.988 \text{ kg}$

$m = \dfrac{0.25 \text{ mol } CH_3CO_2Na}{0.988 \text{ kg}} = 0.25 \text{ } m$

Because CH_3CO_2Na is a strong electrolyte, the ionic compound is completely dissociated and $[CH_3CO_2^-] = [Na^+]$. The contribution of CH_3CO_2H and OH^- to the total molality of the solution is negligible.

$\Delta T_f = K_f \cdot (2 \cdot m) = (1.86 \text{ °C/}m)(2)(0.25 \text{ } m) = 0.93 \text{ °C}$

Solution freezing point $= 0.00 \text{ °C} - \Delta T_f = 0.00 \text{ °C} - 0.93 \text{ °C} = -0.93 \text{ °C}$

14.135 (a) SO_2, 64.06 amu; $CaCO_3$, 100.09 amu

mass SO_2 in g $= 4.0 \times 10^6$ tons x $\dfrac{2000 \text{ lb}}{1 \text{ ton}}$ x $\dfrac{453.59 \text{ g}}{1 \text{ lb}} = 3.63 \times 10^{12}$ g SO_2

mol $SO_2 = 3.63 \times 10^{12}$ g SO_2 x $\dfrac{1 \text{ mol } SO_2}{64.06 \text{ g } SO_2} = 5.67 \times 10^{10}$ mol SO_2

mol $CaCO_3$ reacted $= 5.67 \times 10^{10}$ mol SO_2 x $\dfrac{1 \text{ mol } SO_3}{1 \text{ mol } SO_2}$ x $\dfrac{1 \text{ mol } H_2SO_4}{1 \text{ mol } SO_3}$ x $\dfrac{1 \text{ mol } CaCO_3}{1 \text{ mol } H_2SO_4}$

$= 5.67 \times 10^{10}$ mol $CaCO_3$

mass $CaCO_3 = 5.67 \times 10^{10}$ mol $CaCO_3$ x $\dfrac{100.09 \text{ g } CaCO_3}{1 \text{ mol } CaCO_3}$ x $\dfrac{1 \text{ lb}}{453.59 \text{ g}} = 1.25 \times 10^{10}$ lbs

$(500 \text{ lbs})(0.03) = 15 \text{ lbs}$

statues damaged $= \dfrac{1.25 \times 10^{10} \text{ lbs}}{15 \text{ lbs/statue}} = 8.3 \times 10^8$ statues

(b) mol $CO_2 = 5.67 \times 10^{10}$ mol $CaCO_3$ x $\dfrac{1 \text{ mol } CO_2}{1 \text{ mol } CaCO_3} = 5.67 \times 10^{10}$ mol CO_2

$PV = nRT; \quad V = \dfrac{nRT}{P} = \dfrac{(5.67 \times 10^{10} \text{ mol})\left(0.082\ 06 \dfrac{L \cdot atm}{K \cdot mol}\right)(293 \text{ K})}{\left(735 \text{ mm Hg x } \dfrac{1.00 \text{ atm}}{760 \text{ mm Hg}}\right)} = 1.4 \times 10^{12}$ L CO_2

(c) The cation is H_3O^+. O has four charge clouds (one being a lone pair) and is sp^3 hybridized. The geometry of H_3O^+ is trigonal pyramidal.

14.136 Na_3PO_4, 163.94 amu

$$3.28 \text{ g } Na_3PO_4 \times \frac{1 \text{ mol } Na_3PO_4}{163.94 \text{ g } Na_3PO_4} = 0.0200 \text{ mol} = 20.0 \text{ mmol } Na_3PO_4$$

300.0 mL x 0.180 mmol/mL = 54.0 mmol HCl

	$H_3O^+(aq)$	+ $PO_4^{3-}(aq)$	$\rightleftharpoons$ $HPO_4^{2-}(aq)$	+ $H_2O(l)$
before (mmol)	54.0	20.0	0	
change (mmol)	−20.0	−20.0	+20.0	
after (mmol)	34.0	0	20.0	

	$H_3O^+(aq)$	+ $HPO_4^{2-}(aq)$	$\rightleftharpoons$ $H_2PO_4^-(aq)$	+ $H_2O(l)$
before (mmol)	34.0	20.0	0	
change (mmol)	−20.0	−20.0	+20.0	
after (mmol)	14.0	0	20.0	

	$H_3O^+(aq)$	+ $H_2PO_4^-(aq)$	$\rightleftharpoons$ $H_3PO_4(aq)$	+ $H_2O(l)$
before (mmol)	14.0	20.0	0	
change (mmol)	−14.0	−14.0	+14.0	
after (mmol)	0	6.0	14.0	

$$[H_3PO_4] = \frac{14.0 \text{ mmol}}{300.0 \text{ mL}} = 0.047 \text{ M}; \quad [H_2PO_4^-] = \frac{6.0 \text{ mmol}}{300.0 \text{ mL}} = 0.020 \text{ M}$$

	$H_3PO_4(aq)$	+ $H_2O(l)$	$\rightleftharpoons$ $H_3O^+(aq)$	+ $H_2PO_4^-(aq)$
initial (M)	0.047		~0	0.020
change (M)	−x		+x	+x
equil (M)	0.047 − x		x	0.020 + x

$$K_a = \frac{[H_3O^+][H_2PO_4^{2-}]}{[H_3PO_4]} = 7.5 \times 10^{-3} = \frac{x(0.020 + x)}{(0.047 - x)}$$

$x^2 + 0.0275x - (3.525 \times 10^{-4}) = 0$

Solve for x using the quadratic formula.

$$x = \frac{-(0.0275) \pm \sqrt{(0.0275)^2 - (4)(-3.525 \times 10^{-4})}}{2(1)} = \frac{-0.0275 \pm 0.0465}{2}$$

x = 0.009 52 and −0.0370

Of the two solutions for x, only the positive value of x has physical meaning, because x is the $[H_3O^+]$.

pH = −log$[H_3O^+]$ = −log(0.009 52) = 2.02

14.137 Let x = $[H_3O^+]$ from $Li(H_2O)_4^+$ and y = $[H_3O^+]$ from H_2O

	$Li(H_2O)_4^+(aq)$	+ $H_2O(l)$	$\rightleftharpoons$ $H_3O^+(aq)$	+ $Li(H_2O)_3(OH)(aq)$
initial (M)	0.10		y	0
change (M)	−x		+x	+x
equil (M)	0.10 − x		x	x

$$K_a = \frac{[H_3O^+][Li(H_2O)_3(OH)]}{[Li(H_2O)_4^+]} = 2.5 \times 10^{-14} = \frac{(x+y)(x)}{0.10-x} \approx \frac{(x+y)(x)}{0.10}$$

$$2.5 \times 10^{-15} = (x+y)(x)$$

$$2\,H_2O(l) \rightleftarrows H_3O^+(aq) + OH^-(aq)$$

initial (M)	x	~0
change (M)	+y	+y
equil (M)	x + y	y

$$K_w = [H_3O^+][OH^-] = 1.0 \times 10^{-14} = (x+y)(y)$$

$$2.5 \times 10^{-15} = (x+y)(x)$$
$$1.0 \times 10^{-14} = (x+y)(y)$$

Solve these two equations simultaneously for x and y. Divide the first equation by the second.

$$\frac{x}{y} = \frac{2.5 \times 10^{-15}}{1.0 \times 10^{-14}} = 0.25 \qquad x = 0.25y$$

$$1.0 \times 10^{-14} = (x+y)(y) = (0.25y + y)y = (1.25y)y = 1.25y^2$$

$$y = \sqrt{\frac{1.0 \times 10^{-14}}{1.25}} = 8.9 \times 10^{-8}$$

$$x = 0.25y = (0.25)(8.9 \times 10^{-8}) = 2.24 \times 10^{-8}$$

$$[H_3O^+] = (x+y) = (2.2 \times 10^{-8}) + (8.9 \times 10^{-8}) = 1.11 \times 10^{-7}\,M$$

$$pH = -\log[H_3O^+] = -\log(1.11 \times 10^{-7}) = 6.95$$

14.138 (a) $PV = nRT$; $\quad n = \dfrac{PV}{RT} = \dfrac{(0.601\text{ atm})(1.000\text{ L})}{\left(0.082\,06\,\dfrac{L \cdot atm}{K \cdot mol}\right)(293.1\text{ K})} = 0.0250$ mol HF

$$50.0\text{ mL} \times \frac{1.00\text{ L}}{1000\text{ mL}} = 0.0500\text{ L}$$

$$[HF] = \frac{0.0250\text{ mol HF}}{0.0500\text{ L}} = 0.500\text{ M}$$

$$HF(aq) + H_2O(l) \rightleftarrows H_3O^+(aq) + F^-(aq)$$

initial (M)	0.500	~0	0
change (M)	−x	+x	+x
equil (M)	0.500 − x	x	x

$$K_a = \frac{[H_3O^+][F^-]}{[HF]} = 3.5 \times 10^{-4} = \frac{x^2}{0.500-x}$$

$$x^2 + (3.5 \times 10^{-4})x - (1.75 \times 10^{-4}) = 0$$

Solve for x using the quadratic formula.

$$x = \frac{-(3.5 \times 10^{-4}) \pm \sqrt{(3.5 \times 10^{-4})^2 - (4)(-1.75 \times 10^{-4})}}{2(1)} = \frac{(-3.5 \times 10^{-4}) \pm 0.0265}{2}$$

$$x = -0.0134 \text{ and } 0.0131$$

Of the two solutions for x, only the positive value of x has physical meaning, because x is the $[H_3O^+]$.

pH = $-\log[H_3O^+] = -\log(0.0131) = 1.883 = 1.88$

(b) % dissociation = $\dfrac{0.0131\ M}{0.500} \times 100\% = 2.62\% = 2.6\%$

New % dissociation = $(3)(2.62\%) = 7.86\%$

Let X equal the concentration of HF dissociated and Y the new volume (in liters) that would triple the % dissociation.

$$K_a = \frac{X^2}{(0.0250/Y) - X} = 3.5 \times 10^{-4}$$

% dissociation = $\dfrac{X}{(0.0250/Y)} \times 100\% = 7.86\%$ and $\dfrac{X}{(0.0250/Y)} = 0.0786$

$X = 1.965 \times 10^{-3}/Y$

Substitute X into the K_a equation.

$$\frac{(1.965 \times 10^{-3}/Y)^2}{(0.0250/Y) - (1.965 \times 10^{-3}/Y)} = 3.5 \times 10^{-4}$$

$$\frac{3.861 \times 10^{-6}/Y^2}{0.0230/Y} = 3.5 \times 10^{-4}$$

$$\frac{3.861 \times 10^{-6}/Y}{0.0230} = 3.5 \times 10^{-4}$$

$$\frac{3.861 \times 10^{-6}}{8.05 \times 10^{-6}} = Y = 0.48\ L$$

The result in Problem 14.126 can't be used here because the concentration of HF that dissociates can't be neglected compared with the initial HF concentration.

14.139 PV = nRT; 25 °C = 25 + 273 = 298 K

$$n_{NH_3} = \frac{PV}{RT} = \frac{\left(650.8\ mm\ Hg \times \dfrac{1.00\ atm}{760\ mm\ Hg}\right)(2.000\ L)}{\left(0.082\ 06\ \dfrac{L \cdot atm}{K \cdot mol}\right)(298\ K)} = 0.0700\ mol$$

200.0 mL = 0.2000 L; mol CH_3CO_2H = (0.350 mol/L)(0.2000 L) = 0.0700 mol

There are equal moles of NH_3 and CH_3CO_2H.

$[NH_3]$ = 0.0700 mol/0.2000 L = 0.350 M

$CH_3CO_2H(aq) + H_2O(l) \rightleftharpoons H_3O^+(aq) + CH_3CO_2^-(aq)$	$K_a = 1.8 \times 10^{-5}$
$NH_3(aq) + H_2O(l) \rightleftharpoons NH_4^+(aq) + OH^-(aq)$	$K_b = 1.8 \times 10^{-5}$
$\underline{H_3O^+(aq) + OH^-(aq) \rightleftharpoons 2\ H_2O(l)}$	$1/K_w$
$CH_3CO_2H(aq) + NH_3(aq) \rightleftharpoons NH_4^+(aq) + CH_3CO_2^-(aq)$	$K = K_aK_b/K_w = 3.2 \times 10^4$

Because of the magnitude of K, assume 100% neutralization followed by a small amount of the back reaction.

$$CH_3CO_2H(aq) \; + \; NH_3(aq) \; \rightleftharpoons \; NH_4^+(aq) \; + \; CH_3CO_2^-(aq)$$

	CH_3CO_2H	NH_3	NH_4^+	$CH_3CO_2^-$
before reaction (M)	0.350	0.350	0	0
assume 100% reaction	−0.350	−0.350	+0.350	+0.350
after reaction (M)	0	0	0.350	0.350
assume small back rxn	+x	+x	−x	−x
equil (M)	x	x	0.350 − x	0.350 − x

$$K = K_aK_b/K_w = \frac{[NH_4^+][CH_3CO_2^-]}{[CH_3CO_2H][NH_3]} = 3.2 \times 10^4 = \frac{(0.350 - x)^2}{x^2}$$

$$\sqrt{3.2 \times 10^4} = 179 = \frac{0.350 - x}{x}$$

$179x = 0.350 - x$

$180x = 0.350$

$x = 0.350/180 = 1.9 \times 10^{-3}$

$[CH_3CO_2H] = [NH_3] = x = 1.9 \times 10^{-3}$ M

$[NH_4^+] = [CH_3CO_2^-] = 0.350 - x = 0.350 - (1.9 \times 10^{-3}) = 0.348$ M

Because K_a and K_b are equal, the solution is neutral and $[H_3O^+] = [OH^-] = 1.0 \times 10^{-7}$ M

$pH = -\log[H_3O^+] = -\log(1.0 \times 10^{-7}) = 7.00$

14.140 (a) Rate $= k[OCl^-]^x[NH_3]^y[OH^-]^z$

From experiments 1 & 2, the $[OCl^-]$ doubles and the rate doubles, therefore x = 1.

From experiments 2 & 3, the $[NH_3]$ triples and the rate triples, therefore y = 1.

From experiments 3 & 4, the $[OH^-]$ goes up by a factor of 10 and the rate goes down by a factor of 10, therefore z = −1.

$$Rate = k \frac{[OCl^-][NH_3]}{[OH^-]}$$

$$[H_3O^+] = 10^{-pH} = 10^{-12} = 1 \times 10^{-12} \text{ M}$$

$$[OH^-] = \frac{K_w}{[H_3O^+]} = \frac{1.0 \times 10^{-14}}{1 \times 10^{-12}} = 0.01 \text{ M}$$

Using experiment 1: $k = \dfrac{(Rate)[OH^-]}{[OCl^-][NH_3]} = \dfrac{(0.017 \text{ M/s})(0.01 \text{ M})}{(0.001 \text{ M})(0.01 \text{ M})} = 17 \text{ s}^{-1}$

(b) $K_1 = \dfrac{[HOCl][OH^-]}{[OCl^-]} = K_b(OCl^-) = \dfrac{K_w}{K_a(HOCl)} = \dfrac{1.0 \times 10^{-14}}{3.5 \times 10^{-8}} = 2.9 \times 10^{-7}$

For second step, Rate $= k_2 [HOCl][NH_3]$

Multiply the Rate by $\dfrac{K_1}{K_1} = \dfrac{K_1}{\left(\dfrac{[HOCl][OH^-]}{[OCl^-]} \right)}$

$$\text{Rate} = K_1 k_2 \frac{[HOCl][NH_3]}{\left(\dfrac{[HOCl][OH^-]}{[OCl^-]}\right)} = K_1 k_2 \frac{[HOCl][NH_3][OCl^-]}{[HOCl][OH^-]} = K_1 k_2 \frac{[OCl^-][NH_3]}{[OH^-]}$$

$$K_1 k_2 = k = 17 \text{ s}^{-1}$$

$$k_2 = \frac{17 \text{ s}^{-1}}{2.9 \times 10^{-7} \text{ M}} = 5.9 \times 10^7 \text{ M}^{-1}\text{s}^{-1}$$

15

Applications of Aqueous Equilibria

15.1 (a) $HNO_2(aq) + OH^-(aq) \rightleftharpoons NO_2^-(aq) + H_2O(l)$; NO_2^- (basic anion), pH > 7.00

(b) $H_3O^+(aq) + NH_3(aq) \rightleftharpoons NH_4^+(aq) + H_2O(l)$; NH_4^+ (acidic cation), pH < 7.00

(c) $OH^-(aq) + H_3O^+(aq) \rightleftharpoons 2 H_2O(l)$; pH = 7.00

15.2 (a) $HF(aq) + OH^-(aq) \rightleftharpoons H_2O(l) + F^-(aq)$

$$K_n = \frac{K_a}{K_w} = \frac{3.5 \times 10^{-4}}{1.0 \times 10^{-14}} = 3.5 \times 10^{10}$$

(b) $H_3O^+(aq) + OH^-(aq) \rightleftharpoons 2 H_2O(l)$

$$K_n = \frac{1}{K_w} = \frac{1}{1.0 \times 10^{-14}} = 1.0 \times 10^{14}$$

(c) $HF(aq) + NH_3(aq) \rightleftharpoons NH_4^+(aq) + F^-(aq)$

$$K_n = \frac{K_a K_b}{K_w} = \frac{(3.5 \times 10^{-4})(1.8 \times 10^{-5})}{1.0 \times 10^{-14}} = 6.3 \times 10^5$$

The tendency to proceed to completion is determined by the magnitude of K_n. The larger the value of K_n, the further does the reaction proceed to completion.

The tendency to proceed to completion is: reaction (c) < reaction (a) < reaction (b)

15.3

	$HCN(aq)$ + $H_2O(l)$ $\rightleftharpoons$	$H_3O^+(aq)$ +	$CN^-(aq)$
initial (M)	0.025	~0	0.010
change (M)	−x	+x	+x
equil (M)	0.025 − x	x	0.010 + x

$$K_a = \frac{[H_3O^+][CN^-]}{[HCN]} = 4.9 \times 10^{-10} = \frac{x(0.010 + x)}{0.025 - x} \approx \frac{x(0.010)}{0.025}$$

Solve for x. $x = 1.23 \times 10^{-9}$ M = 1.2×10^{-9} M = $[H_3O^+]$

$pH = -\log[H_3O^+] = -\log(1.23 \times 10^{-9}) = 8.91$

$$[OH^-] = \frac{K_w}{[H_3O^+]} = \frac{1.0 \times 10^{-14}}{1.23 \times 10^{-9}} = 8.2 \times 10^{-6} \text{ M}$$

$[Na^+] = [CN^-] = 0.010$ M; $[HCN] = 0.025$ M

$$\% \text{ dissociation} = \frac{[HCN]_{diss}}{[HCN]_{initial}} \times 100\% = \frac{1.23 \times 10^{-9} \text{ M}}{0.025 \text{ M}} \times 100\% = 4.9 \times 10^{-6} \%$$

15.4 From $NH_4Cl(s)$, $[NH_4^+]_{initial} = \dfrac{0.10 \text{ mol}}{0.500 \text{ L}} = 0.20$ M

$$NH_3(aq) + H_2O(l) \rightleftharpoons NH_4^+(aq) + OH^-(aq)$$

initial (M)	0.40	0.20	~0
change (M)	–x	+x	+x
equil (M)	0.40 – x	0.20 + x	x

$K_b = \dfrac{[NH_4^+][OH^-]}{[NH_3]} = 1.8 \times 10^{-5} = \dfrac{(0.20 + x)(x)}{(0.40 - x)} \approx \dfrac{(0.20)(x)}{(0.40)}$

Solve for x. $x = [OH^-] = 3.6 \times 10^{-5}$ M

$[H_3O^+] = \dfrac{K_w}{[OH^-]} = \dfrac{1.0 \times 10^{-14}}{3.6 \times 10^{-5}} = 2.8 \times 10^{-10}$ M

$pH = -\log[H_3O^+] = -\log(2.8 \times 10^{-10}) = 9.55$

15.5 Each solution contains the same number of B molecules. The presence of BH^+ from BHCl lowers the percent dissociation of B. Solution (2) contains no BH^+, therefore it has the largest percent dissociation. BH^+ is the conjugate acid of B. Solution (1) has the largest amount of BH^+, and it would be the most acidic solution and have the lowest pH.

15.6 (a) (1) and (3). Both pictures show equal concentrations of HA and A^-.
(b) (3). It contains a higher concentration of HA and A^-.

15.7
$$HF(aq) + H_2O(l) \rightleftharpoons H_3O^+(aq) + F^-(aq)$$

initial (M)	0.25	~0	0.50
change (M)	–x	+x	+x
equil (M)	0.25 – x	x	0.50 + x

$K_a = \dfrac{[H_3O^+][F^-]}{[HF]} = 3.5 \times 10^{-4} = \dfrac{x(0.50 + x)}{0.25 - x} \approx \dfrac{x(0.50)}{0.25}$

Solve for x. $x = 1.75 \times 10^{-4}$ M $= [H_3O^+]$
For the buffer, $pH = -\log[H_3O^+] = -\log(1.75 \times 10^{-4}) = 3.76$

(a) mol HF = 0.025 mol; mol F^- = 0.050 mol; vol = 0.100 L

$$\overset{100\%}{F^-(aq) + H_3O^+(aq) \rightarrow HF(aq) + H_2O(l)}$$

before (mol)	0.050	0.002	0.025
change (mol)	–0.002	–0.002	+0.002
after (mol)	0.048	0	0.027

$[H_3O^+] = K_a \dfrac{[HF]}{[F^-]} = (3.5 \times 10^{-4})\left(\dfrac{0.27}{0.48}\right) = 1.97 \times 10^{-4}$ M

$pH = -\log[H_3O^+] = -\log(1.97 \times 10^{-4}) = 3.71$

(b) mol HF = 0.025 mol; mol F$^-$ = 0.050 mol; vol = 0.100 L

$$\overset{100\%}{HF(aq) \; + \; OH^-(aq) \; \rightarrow \; F^-(aq) \; + \; H_2O(l)}$$

before (mol)	0.025	0.004	0.050
change (mol)	−0.004	−0.004	+0.004
after (mol)	0.021	0	0.054

$$[H_3O^+] = K_a \frac{[HF]}{[F^-]} = (3.5 \times 10^{-4})\left(\frac{0.21}{0.54}\right) = 1.36 \times 10^{-4} \; M$$

$$pH = -\log[H_3O^+] = -\log(1.36 \times 10^{-4}) = 3.87$$

15.8

$$HF(aq) \; + \; H_2O(l) \; \rightleftharpoons \; H_3O^+(aq) \; + \; F^-(aq)$$

initial (M)	0.050	~0	0.100
change (M)	−x	+x	+x
equil (M)	0.050 − x	x	0.100 + x

$$K_a = \frac{[H_3O^+][F^-]}{[HF]} = 3.5 \times 10^{-4} = \frac{x(0.100 + x)}{0.050 - x} \approx \frac{x(0.100)}{0.050}$$

Solve for x. x = [H$_3$O$^+$] = 1.75 × 10^{-4} M

pH = −log[H$_3$O$^+$] = −log(1.75 × 10^{-4}) = 3.76

mol HF = 0.050 mol/L × 0.100 L = 0.0050 mol HF

mol F$^-$ = 0.100 mol/L × 0.100 L = 0.0100 mol F$^-$

mol HNO$_3$ = mol H$_3$O$^+$ = 0.002 mol

$$\text{Neutralization reaction:} \quad \overset{100\%}{F^-(aq) \; + \; H_3O^+(aq) \; \rightarrow \; HF(aq) \; + \; H_2O(l)}$$

before reaction (mol)	0.0100	0.002	0.0050
change (mol)	−0.002	−0.002	+0.002
after reaction (mol)	0.008	0	0.007

$$[HF] = \frac{0.007 \; mol}{0.100 \; L} = 0.07 \; M; \qquad [F^-] = \frac{0.008 \; mol}{0.100 \; L} = 0.08 \; M$$

$$[H_3O^+] = K_a \frac{[HF]}{[F^-]} = (3.5 \times 10^{-4})\frac{(0.07)}{(0.08)} = 3 \times 10^{-4} \; M$$

pH = −log[H$_3$O$^+$] = −log(3 × 10^{-4}) = 3.5

This solution has less buffering capacity than the solution in Problem 15.7 because it contains less HF and F$^-$ per 100 mL. Note that the change in pH is greater than that in Problem 15.7.

15.9 When equal volumes of two solutions are mixed together, the concentration of each solution is cut in half.

$$pH = pK_a + \log \frac{[base]}{[acid]} = pK_a + \log \frac{[CO_3^{2-}]}{[HCO_3^-]}$$

For HCO$_3^-$, K$_a$ = 5.6 × 10^{-11}, pK$_a$ = −log K$_a$ = −log(5.6 × 10^{-11}) = 10.25

$$pH = 10.25 + \log\left(\frac{0.050}{0.10}\right) = 10.25 - 0.30 = 9.95$$

15.10 $\quad pH = pK_a + \log\dfrac{[\text{base}]}{[\text{acid}]} = pK_a + \log\dfrac{[CO_3^{2-}]}{[HCO_3^-]}$

For HCO_3^-, $K_a = 5.6 \times 10^{-11}$, $pK_a = -\log K_a = -\log(5.6 \times 10^{-11}) = 10.25$

$10.40 = 10.25 + \log\dfrac{[CO_3^{2-}]}{[HCO_3^-]}; \qquad \log\dfrac{[CO_3^{2-}]}{[HCO_3^-]} = 10.40 - 10.25 = 0.15$

$\dfrac{[CO_3^{2-}]}{[HCO_3^-]} = 10^{0.15} = 1.4$

To obtain a buffer solution with pH 10.40, make the Na_2CO_3 concentration 1.4 times the concentration of $NaHCO_3$.

15.11 $\quad$ Look for an acid with pK_a near the required pH of 7.50.
$K_a = 10^{-pH} = 10^{-7.50} = 3.2 \times 10^{-8}$
Suggested buffer system: $HOCl$ ($K_a = 3.5 \times 10^{-8}$) and $NaOCl$.

15.12 $\quad$ (a) serine is 66% dissociated at $pH = 9.15 + \log\left(\dfrac{66}{34}\right) = 9.44$

$\qquad$ (b) serine is 5% dissociated at $pH = 9.15 + \log\left(\dfrac{5}{95}\right) = 7.87$

15.13 $\quad$ (a) mol HCl = mol H_3O^+ = 0.100 mol/L x 0.0400 L = 0.004 00 mol
mol NaOH = mol OH^- = 0.100 mol/L x 0.0350 L = 0.003 50 mol

Neutralization reaction:	H_3O^+(aq) +	OH^-(aq)	→	2 H_2O(l)
before reaction (mol)	0.004 00	0.003 50		
change (mol)	−0.003 50	−0.003 50		
after reaction (mol)	0.000 50	0		

$[H_3O^+] = \dfrac{0.000\ 50\ \text{mol}}{(0.0400\ \text{L} + 0.0350\ \text{L})} = 6.7 \times 10^{-3}$ M

$pH = -\log[H_3O^+] = -\log(6.7 \times 10^{-3}) = 2.17$

(b) mol HCl = mol H_3O^+ = 0.100 mol/L x 0.0400 L = 0.004 00 mol
mol NaOH = mol OH^- = 0.100 mol/L x 0.0450 L = 0.004 50 mol

Neutralization reaction:	H_3O^+(aq) +	OH^-(aq)	→	2 H_2O(l)
before reaction (mol)	0.004 00	0.004 50		
change (mol)	−0.004 00	−0.004 00		
after reaction (mol)	0	0.000 50		

$[OH^-] = \dfrac{0.000\ 50\ \text{mol}}{(0.0400\ \text{L} + 0.0450\ \text{L})} = 5.9 \times 10^{-3}$ M

$[H_3O^+] = \dfrac{K_w}{[OH^-]} = \dfrac{1.0 \times 10^{-14}}{5.9 \times 10^{-3}} = 1.7 \times 10^{-12}$ M

$pH = -\log[H_3O^+] = -\log(1.7 \times 10^{-12}) = 11.77$
The results obtained here are consistent with the pH data in Table 15.1.

15.14 (a) mol NaOH = mol OH⁻ = 0.100 mol/L x 0.0400 L = 0.004 00 mol

mol HCl = mol H₃O⁺ = 0.0500 mol/L x 0.0600 L = 0.003 00 mol

Neutralization reaction: $H_3O^+(aq) + OH^-(aq) \rightarrow 2 H_2O(l)$

before reaction (mol)	0.003 00	0.004 00
change (mol)	−0.003 00	−0.003 00
after reaction (mol)	0	0.001 00

$$[OH^-] = \frac{0.001\ 00\ mol}{(0.0400\ L\ +\ 0.0600\ L)} = 1.0 \times 10^{-2}\ M$$

$$[H_3O^+] = \frac{K_w}{[OH^-]} = \frac{1.0 \times 10^{-14}}{1.0 \times 10^{-2}} = 1.0 \times 10^{-12}\ M$$

$$pH = -\log[H_3O^+] = -\log(1.0 \times 10^{-12}) = 12.00$$

(b) mol NaOH = mol OH⁻ = 0.100 mol/L x 0.0400 L = 0.004 00 mol

mol HCl = mol H₃O⁺ = 0.0500 mol/L x 0.0802 L = 0.004 01 mol

Neutralization reaction: $H_3O^+(aq) + OH^-(aq) \rightarrow 2 H_2O(l)$

before reaction (mol)	0.004 01	0.004 00
change (mol)	−0.004 00	−0.004 00
after reaction (mol)	0.000 01	0

$$[H_3O^+] = \frac{0.000\ 01\ mol}{(0.0400\ L\ +\ 0.0802\ L)} = 8.3 \times 10^{-5}\ M$$

$$pH = -\log[H_3O^+] = -\log(8.3 \times 10^{-5}) = 4.08$$

(c) mol NaOH = mol OH⁻ = 0.100 mol/L x 0.0400 L = 0.004 00 mol

mol HCl = mol H₃O⁺ = 0.0500 mol/L x 0.1000 L = 0.005 00 mol

Neutralization reaction: $H_3O^+(aq) + OH^-(aq) \rightarrow 2 H_2O(l)$

before reaction (mol)	0.005 00	0.004 00
change (mol)	−0.004 00	−0.004 00
after reaction (mol)	0.001 00	0

$$[H_3O^+] = \frac{0.001\ 00\ mol}{(0.0400\ L\ +\ 0.1000\ L)} = 7.1 \times 10^{-3}\ M$$

$$pH = -\log[H_3O^+] = -\log(7.1 \times 10^{-3}) = 2.15$$

15.15 (a) (3), only HA present   (b) (1), HA and A⁻ present

(c) (4), only A⁻ present   (d) (2), A⁻ and OH⁻ present

15.16 $$\text{mol NaOH required} = \left(\frac{0.016\ mol\ HOCl}{L}\right)(0.100\ L)\left(\frac{1\ mol\ NaOH}{1\ mol\ HOCl}\right) = 0.0016\ mol$$

$$\text{vol NaOH required} = (0.0016\ mol)\left(\frac{1\ L}{0.0400\ mol}\right) = 0.040\ L = 40\ mL$$

40 mL of 0.0400 M NaOH are required to reach the equivalence point.

(a) mmol HOCl = 0.016 mmol/mL x 100.0 mL = 1.6 mmol

mmol NaOH = mmol OH⁻ = 0.0400 mmol/mL x 10.0 mL = 0.400 mmol

Neutralization reaction: $\quad HOCl(aq) + OH^-(aq) \rightarrow OCl^-(aq) + H_2O(l)$

before reaction (mmol)	1.6	0.400	0
change (mmol)	-0.400	-0.400	$+0.400$
after reaction (mmol)	1.2	0	0.400

$$[HOCl] = \frac{1.2 \text{ mmol}}{(100.0 \text{ mL} + 10.0 \text{ mL})} = 1.09 \times 10^{-2} \text{ M}$$

$$[OCl^-] = \frac{0.400 \text{ mmol}}{(100.0 \text{ mL} + 10.0 \text{ mL})} = 3.64 \times 10^{-3} \text{ M}$$

$$HOCl(aq) + H_2O(l) \rightleftharpoons H_3O^+(aq) + OCl^-(aq)$$

initial (M)	0.0109	~0	0.003 64
change (M)	$-x$	$+x$	$+x$
equil (M)	$0.0109 - x$	x	$0.003\ 64 + x$

$$K_a = \frac{[H_3O^+][OCl^-]}{[HOCl]} = 3.5 \times 10^{-8} = \frac{x(0.003\ 64 + x)}{0.0109 - x} \approx \frac{x(0.003\ 64)}{0.0109}$$

Solve for x. $x = [H_3O^+] = 1.05 \times 10^{-7} \text{ M}$

$pH = -\log[H_3O^+] = -\log(1.05 \times 10^{-7}) = 6.98$

(b) Halfway to the equivalence point, $[OCl^-] = [HOCl]$

$pH = pK_a = -\log K_a = -\log(3.5 \times 10^{-8}) = 7.46$

(c) At the equivalence point the solution contains the salt, NaOCl.

mol NaOCl = initial mol HOCl = 0.0016 mol = 1.6 mmol

$$[OCl^-] = \frac{1.6 \text{ mmol}}{(100.0 \text{ mL} + 40.0 \text{ mL})} = 1.1 \times 10^{-2} \text{ M}$$

For OCl^-, $K_b = \dfrac{K_w}{K_a \text{ for HOCl}} = \dfrac{1.0 \times 10^{-14}}{3.5 \times 10^{-8}} = 2.9 \times 10^{-7}$

$$OCl^-(aq) + H_2O(l) \rightleftharpoons HOCl(aq) + OH^-(aq)$$

initial (M)	0.011	0	~0
change (M)	$-x$	$+x$	$+x$
equil (M)	$0.011 - x$	x	x

$$K_b = \frac{[HOCl][OH^-]}{[OCl^-]} = 2.9 \times 10^{-7} = \frac{x^2}{0.011 - x} \approx \frac{x^2}{0.011}$$

Solve for x. $x = [OH^-] = 5.65 \times 10^{-5} \text{ M}$

$$[H_3O^+] = \frac{K_w}{[OH^-]} = \frac{1.0 \times 10^{-14}}{5.65 \times 10^{-5}} = 1.77 \times 10^{-10} = 1.8 \times 10^{-10} \text{ M}$$

$pH = -\log[H_3O^+] = -\log(1.77 \times 10^{-10}) = 9.75$

15.17 From Problem 15.16, pH = 9.75 at the equivalence point.
Use thymolphthalein (pH 9.4 – 10.6). Bromthymol blue is unacceptable because it changes color halfway to the equivalence point.

15.18   (a) mol NaOH required to reach first equivalence point

$$= \left(\frac{0.0800 \text{ mol } H_2SO_3}{L}\right)(0.0400 \text{ L})\left(\frac{1 \text{ mol NaOH}}{1 \text{ mol } H_2SO_3}\right) = 0.003\ 20 \text{ mol}$$

vol NaOH required to reach first equivalence point

$$= (0.003\ 20 \text{ mol})\left(\frac{1 \text{ L}}{0.160 \text{ mol}}\right) = 0.020 \text{ L} = 20.0 \text{ mL}$$

20.0 mL is enough NaOH solution to reach the first equivalence point for the titration of the diprotic acid, H_2SO_3.
For H_2SO_3,

$K_{a1} = 1.5 \times 10^{-2}$, $pK_{a1} = -\log K_{a1} = -\log(1.5 \times 10^{-2}) = 1.82$

$K_{a2} = 6.3 \times 10^{-8}$, $pK_{a2} = -\log K_{a2} = -\log(6.3 \times 10^{-8}) = 7.20$

At the first equivalence point, pH $= \dfrac{pK_{a1} + pK_{a2}}{2} = \dfrac{1.82 + 7.20}{2} = 4.51$

(b) mol NaOH required to reach second equivalence point

$$= \left(\frac{0.0800 \text{ mol } H_2SO_3}{L}\right)(0.0400 \text{ L})\left(\frac{2 \text{ mol NaOH}}{1 \text{ mol } H_2SO_3}\right) = 0.006\ 40 \text{ mol}$$

vol NaOH required to reach second equivalence point

$$= (0.006\ 40 \text{ mol})\left(\frac{1 \text{ L}}{0.160 \text{ mol}}\right) = 0.040 \text{ L} = 40.0 \text{ mL}$$

30.0 mL is enough NaOH solution to reach halfway to the second equivalent point.
Halfway to the second equivalence point
pH $= pK_{a2} = -\log K_{a2} = -\log(6.3 \times 10^{-8}) = 7.20$

(c) mmol $HSO_3^- = 0.0800$ mmol/mL x 40.0 mL = 3.20 mmol
volume NaOH added after first equivalence point = 35.0 mL − 20.0 mL = 15.0 mL
mmol NaOH = mmol $OH^- = 0.160$ mmol/L x 15.0 mL = 2.40 mmol

Neutralization reaction:	$HSO_3^-(aq)$ +	$OH^-(aq)$	$\rightleftharpoons$	$SO_3^{2-}(aq)$ +	$H_2O(l)$
before reaction (mmol)	3.20	2.40		0	
change (mmol)	−2.40	−2.40		+2.40	
after reaction (mmol)	0.80	0		2.40	

$[HSO_3^-] = \dfrac{0.80 \text{ mmol}}{(40.0 \text{ mL } + 35.0 \text{ mL})} = 0.0107 \text{ M}$

$[SO_3^{2-}] = \dfrac{2.40 \text{ mmol}}{(40.0 \text{ mL } + 35.0 \text{ mL})} = 0.0320 \text{ M}$

	$HSO_3^-(aq)$ +	$H_2O(l)$	$\rightleftharpoons$	$H_3O^+(aq)$ +	$SO_3^{2-}(aq)$
initial (M)	0.0107			~0	0.0320
change (M)	−x			+x	+x
equil (M)	0.0107 − x			x	0.0320 + x

$$K_a = \frac{[H_3O^+][SO_3^{2-}]}{[HSO_3^-]} = 6.3 \times 10^{-8} = \frac{x(0.0320 + x)}{0.0107 - x} \approx \frac{x(0.0320)}{0.0107}$$

Solve for x. $x = [H_3O^+] = 2.1 \times 10^{-8}$ M
$pH = -\log[H_3O^+] = -\log(2.1 \times 10^{-8}) = 7.68$

15.19 Let H_2A^+ = valine cation
(a) mol NaOH required to reach first equivalence point

$$= \left(\frac{0.0250 \text{ mol } H_2A^+}{L}\right)(0.0400 \text{ L})\left(\frac{1 \text{ mol NaOH}}{1 \text{ mol } H_2A^+}\right) = 0.001\ 00 \text{ mol}$$

vol NaOH required to reach first equivalence point

$$= (0.001\ 00 \text{ mol})\left(\frac{1 \text{ L}}{0.100 \text{ mol}}\right) = 0.0100 \text{ L} = 10.0 \text{ mL}$$

10.0 mL is enough NaOH solution to reach the first equivalence point for the titration of the diprotic acid, H_2A^+.
For H_2A^+,
$K_{a1} = 4.8 \times 10^{-3}$, $pK_{a1} = -\log K_{a1} = -\log(4.8 \times 10^{-3}) = 2.32$
$K_{a2} = 2.4 \times 10^{-10}$, $pK_{a2} = -\log K_{a2} = -\log(2.4 \times 10^{-10}) = 9.62$

At the first equivalence point, $pH = \dfrac{pK_{a1} + pK_{a2}}{2} = \dfrac{2.32 + 9.62}{2} = 5.97$

(b) mol NaOH required to reach second equivalence point

$$= \left(\frac{0.0250 \text{ mol } H_2A^+}{L}\right)(0.0400 \text{ L})\left(\frac{2 \text{ mol NaOH}}{1 \text{ mol } H_2A^+}\right) = 0.002\ 00 \text{ mol}$$

vol NaOH required to reach second equivalence point

$$= (0.002\ 00 \text{ mol})\left(\frac{1 \text{ L}}{0.100 \text{ mol}}\right) = 0.0200 \text{ L} = 20.0 \text{ mL}$$

15.0 mL is enough NaOH solution to reach halfway to the second equivalent point.
Halfway to the second equivalence point
$pH = pK_{a2} = -\log K_{a2} = -\log(2.4 \times 10^{-10}) = 9.62$

(c) 20.0 mL is enough NaOH to reach the second equivalence point.
At the second equivalence point
mmol A^- = (0.0250 mmol/mL)(40.0 mL) = 1.00 mmol A^-
solution volume = 40.0 mL + 20.0 mL = 60.0 mL

$$[A^-] = \frac{1.00 \text{ mmol}}{60.0 \text{ mL}} = 0.0167 \text{ M}$$

	$A^-(aq)$ +	$H_2O(l)$ ⇌	$HA(aq)$ +	$OH^-(aq)$
initial (M)	0.0167		0	~0
change (M)	−x		+x	+x
equil (M)	0.0167 − x		x	x

$$K_b = \frac{K_w}{K_a \text{ for HA}} = \frac{K_w}{K_{a2}} = \frac{1.0 \times 10^{-14}}{2.4 \times 10^{-10}} = 4.17 \times 10^{-5}$$

$$K_b = \frac{[HA][OH^-]}{[A^-]} = 4.17 \times 10^{-5} = \frac{x^2}{0.0167 - x}$$

$$x^2 + (4.17 \times 10^{-5})x - (6.964 \times 10^{-7}) = 0$$

Use the quadratic formula to solve for x.

$$x = \frac{-(4.17 \times 10^{-5}) \pm \sqrt{(4.17 \times 10^{-5})^2 - (4)(1)(-6.964 \times 10^{-7})}}{2(1)} = \frac{(-4.17 \times 10^{-5}) \pm (1.67 \times 10^{-3})}{2}$$

$$x = 8.14 \times 10^{-4} \text{ and } -8.56 \times 10^{-4}$$

Of the two solutions for x, only the positive value has physical meaning because x is the $[OH^-]$.

$$x = [OH^-] = 8.14 \times 10^{-4} \text{ M}$$

$$[H_3O^+] = \frac{K_w}{[OH^-]} = \frac{1.0 \times 10^{-14}}{8.14 \times 10^{-4}} = 1.23 \times 10^{-11} \text{ M}$$

$$pH = -\log[H_3O^+] = -\log(1.23 \times 10^{-11}) = 10.91$$

15.20   (a) $K_{sp} = [Ag^+][Cl^-]$ (b) $K_{sp} = [Pb^{2+}][I^-]^2$
           (c) $K_{sp} = [Ca^{2+}]^3[PO_4^{3-}]^2$ (d) $K_{sp} = [Cr^{3+}][OH^-]^3$

15.21 $K_{sp} = [Ca^{2+}]^3[PO_4^{3-}]^2 = (2.01 \times 10^{-8})^3(1.6 \times 10^{-5})^2 = 2.1 \times 10^{-33}$

15.22 $[Ba^{2+}] = [SO_4^{2-}] = 1.05 \times 10^{-5}$ M; $K_{sp} = [Ba^{2+}][SO_4^{2-}] = (1.05 \times 10^{-5})^2 = 1.10 \times 10^{-10}$

15.23   (a) $AgCl(s) \rightleftharpoons Ag^+(aq) + Cl^-(aq)$

equil (M)   x   x

$K_{sp} = [Ag^+][Cl^-] = 1.8 \times 10^{-10} = (x)(x)$

molar solubility = $x = \sqrt{K_{sp}} = 1.3 \times 10^{-5}$ mol/L

AgCl, 143.32 amu,   solubility = $\dfrac{\left(1.3 \times 10^{-5} \text{ mol} \times \dfrac{143.32 \text{ g}}{1 \text{ mol}}\right)}{1 \text{ L}} = 0.0019$ g/L

(b) $Ag_2CrO_4(s) \rightleftharpoons 2 Ag^+(aq) + CrO_4^{2-}(aq)$

equil (M)   2x   x

$K_{sp} = [Ag^+]^2[CrO_4^{2-}] = 1.1 \times 10^{-12} = (2x)^2(x) = 4x^3$

molar solubility = $x = \sqrt[3]{\dfrac{1.1 \times 10^{-12}}{4}} = 6.5 \times 10^{-5}$ mol/L

Ag_2CrO_4, 331.73 amu,   solubility = $\dfrac{\left(6.5 \times 10^{-5} \text{ mol} \times \dfrac{331.73 \text{ g}}{1 \text{ mol}}\right)}{1 \text{ L}} = 0.022$ g/L

Ag_2CrO_4 has both the higher molar and gram solubility, despite its smaller value of K_{sp}.

15.24 Let the number of ions be proportional to its concentration.
For AgX, $K_{sp} = [Ag^+][X^-] \propto (4)(4) = 16$
For AgY, $K_{sp} = [Ag^+][Y^-] \propto (1)(9) = 9$
For AgZ, $K_{sp} = [Ag^+][Z^-] \propto (3)(6) = 18$
(a) AgZ   (b) AgY

15.25 $[Mg^{2+}]_0$ is from 0.10 M $MgCl_2$.

$$MgF_2(s) \rightleftharpoons Mg^{2+}(aq) + 2\,F^-(aq)$$

initial (M)	0.10	0
change (M)	+x	+2x
equil (M)	0.10 + x	2x

$K_{sp} = 7.4 \times 10^{-11} = [Mg^{2+}][F^-]^2 = (0.10 + x)(2x)^2 \approx (0.10)(4x^2)$
$x = 1.4 \times 10^{-5}$, molar solubility = $x = 1.4 \times 10^{-5}$ M

15.26 Compounds that contain basic anions are more soluble in acidic solution than in pure water. $AgCN$, $Al(OH)_3$, and ZnS all contain basic anions.

15.27 $[Cu^{2+}] = (5.0 \times 10^{-3}\,mol)/(0.500\,L) = 0.010$ M

$$Cu^{2+}(aq) + 4\,NH_3(aq) \rightleftharpoons Cu(NH_3)_4^{2+}(aq)$$

before reaction (M)	0.010	0.40	0
assume 100 % reaction (M)	−0.010	− 4(0.010)	+0.010
after reaction (M)	0	0.36	0.010
assume small back reaction (M)	+x	+4x	−x
equil (M)	x	0.36 + 4x	0.010 − x

$$K_f = \frac{[Cu(NH_3)_4^{2+}]}{[Cu^{2+}][NH_3]^4} = 5.6 \times 10^{11} = \frac{(0.010 - x)}{(x)(0.36 + 4x)^4} \approx \frac{0.010}{x(0.36)^4}$$

Solve for x. $x = [Cu^{2+}] = 1.1 \times 10^{-12}$ M

15.28

$AgBr(s) \rightleftharpoons Ag^+(aq) + Br^-(aq)$	$K_{sp} = 5.4 \times 10^{-13}$
$Ag^+(aq) + 2\,S_2O_3^{2-} \rightarrow Ag(S_2O_3)_2^{3-}(aq)$	$K_f = 4.7 \times 10^{13}$

dissolution $AgBr(s) + 2\,S_2O_3^{2-}(aq) \rightleftharpoons Ag(S_2O_3)_2^{3-}(aq) + Br^-(aq)$
reaction

$K = (K_{sp})(K_f) = (5.4 \times 10^{-13})(4.7 \times 10^{13}) = 25.4$

$$AgBr(s) + 2\,S_2O_3^{2-}(aq) \rightleftharpoons Ag(S_2O_3)_2^{3-}(aq) + Br^-(aq)$$

initial (M)	0.10	0	0
change (M)	−2x	x	x
equil (M)	0.10 − 2x	x	x

$$K = \frac{[Ag(S_2O_3)_2^{3-}][Br^-]}{[S_2O_3^{2-}]^2} = 25.4 = \frac{x^2}{(0.10 - 2x)^2}$$

Take the square root of both sides and solve for x.

$$\sqrt{25.4} = \sqrt{\frac{x^2}{(0.10 - 2x)^2}}; \quad 5.04 = \frac{x}{0.10 - 2x}; \quad x = \text{molar solubility} = 0.045\,\text{mol/L}$$

15.29   On mixing equal volumes of two solutions, the concentrations of both solutions are cut in half.
For $BaCO_3$, $K_{sp} = 2.6 \times 10^{-9}$
(a) $IP = [Ba^{2+}][CO_3^{2-}] = (1.5 \times 10^{-3})(1.0 \times 10^{-3}) = 1.5 \times 10^{-6}$
     $IP > K_{sp}$; a precipitate of $BaCO_3$ will form.
(b) $IP = [Ba^{2+}][CO_3^{2-}] = (5.0 \times 10^{-6})(2.0 \times 10^{-5}) = 1.0 \times 10^{-10}$
     $IP < K_{sp}$; no precipitate will form.

15.30 $pH = pK_a + \log \dfrac{[base]}{[acid]} = pK_a + \log \dfrac{[NH_3]}{[NH_4^+]}$

For NH_4^+, $K_a = 5.6 \times 10^{-10}$, $pK_a = -\log K_a = -\log(5.6 \times 10^{-10}) = 9.25$

$pH = 9.25 + \log \dfrac{(0.20)}{(0.20)} = 9.25$; $[H_3O^+] = 10^{-pH} = 10^{-9.25} = 5.6 \times 10^{-10}$ M

$[OH^-] = \dfrac{K_w}{[H_3O^+]} = \dfrac{1.0 \times 10^{-14}}{5.6 \times 10^{-10}} = 1.8 \times 10^{-5}$ M

$[Fe^{2+}] = [Mn^{2+}] = \dfrac{(25 \text{ mL})(1.0 \times 10^{-3} \text{ M})}{250 \text{ mL}} = 1.0 \times 10^{-4}$ M

For $Mn(OH)_2$, $K_{sp} = 2.1 \times 10^{-13}$
     $IP = [Mn^{2+}][OH^-]^2 = (1.0 \times 10^{-4})(1.8 \times 10^{-5})^2 = 3.2 \times 10^{-14}$
     $IP < K_{sp}$; no precipitate will form.
For $Fe(OH)_2$, $K_{sp} = 4.9 \times 10^{-17}$
     $IP = [Fe^{2+}][OH^-]^2 = (1.0 \times 10^{-4})(1.8 \times 10^{-5})^2 = 3.2 \times 10^{-14}$
     $IP > K_{sp}$; a precipitate of $Fe(OH)_2$ will form.

15.31 $MS(s) + 2 H_3O^+(aq) \rightleftharpoons M^{2+}(aq) + H_2S(aq) + 2 H_2O(l)$

$K_{spa} = \dfrac{[M^{2+}][H_2S]}{[H_3O^+]^2}$

For ZnS, $K_{spa} = 3 \times 10^{-2}$;   for CdS, $K_{spa} = 8 \times 10^{-7}$
$[Cd^{2+}] = [Zn^{2+}] = 0.005$ M
Because the two cation concentrations are equal, Q_c is the same for both.

$Q_c = \dfrac{[M^{2+}]_t[H_2S]_t}{[H_3O^+]_t^2} = \dfrac{(0.005)(0.10)}{(0.3)^2} = 6 \times 10^{-3}$

$Q_c > K_{spa}$ for CdS; CdS will precipitate. $Q_c < K_{spa}$ for ZnS; Zn^{2+} will remain in solution.

15.32   This protein has both acidic and basic sites. H_3PO_4–$H_2PO_4^-$ is an acidic buffer. It protonates the basic sites in the protein making them positive, and the protein migrates towards the negative electrode. H_3BO_3–$H_2BO_3^-$ is a basic buffer. At basic pH's, the acidic sites in the protein are dissociated making them negative, and the protein migrates towards the positive electrode.

15.33   To increase the rate at which the proteins migrate toward the negative electrode, increase the number of basic sites that are protonated by lowering the pH. Decrease the $[HPO_4^{2-}]/[H_2PO_4^-]$ ratio (less HPO_4^{2-}, more $H_2PO_4^-$) to lower the pH.

Key Concept Problems

15.34   A buffer solution contains a conjugate acid-base pair in about equal concentrations.
        (a) (1), (3), and (4)
        (b) (4) because it has the highest buffer concentration.

15.35   (a) (2) has the highest pH, $[A^-] > [HA]$
            (3) has the lowest pH, $[HA] > [A^-]$

     (b) 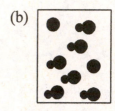   (c)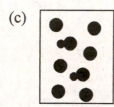

15.36   (4); only A^- and water should be present

15.37   (a) (1) corresponds to (iii); (2) to (i); (3) to (iv); and (4) to (ii)
        (b) Solution (3) has the highest pH; solution (2) has the lowest pH.

15.38   (a) (1) corresponds to (iii); (2) to (i); (3) to (ii); and (4) to (iv)
        (b)

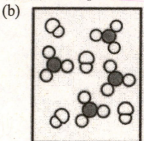

15.39   (a) (i) (1), only B present   (ii) (4), equal amounts of B and BH^+ present
        (iii) (3), only BH^+ present   (iv) (2), BH^+ and H_3O^+ present
        (b) The pH is less than 7 because BH^+ is an acidic cation.

15.40   (2) is supersaturated; (3) is unsaturated; (4) is unsaturated

15.41   Let the number of ions be proportional to its concentration.
        For Ag_2CrO_4, $K_{sp} = [Ag^+]^2[CrO_4^{2-}] \propto (4)^2(2) = 32$
        For (2), IP $= [Ag^+]^2[CrO_4^{2-}] \propto (2)^2(4) = 16$
        For (3), IP $= [Ag^+]^2[CrO_4^{2-}] \propto (6)^2(2) = 72$
        For (4), IP $= [Ag^+]^2[CrO_4^{2-}] \propto (2)^2(6) = 24$
        A precipitate will form when IP > K_{sp}. A precipitate will form only in (3).

15.42 (a) The lower curve represents the titration of a strong acid; the upper curve represents the titration of a weak acid.
(b) pH = 7 for titration of the strong acid; pH = 10 for titration of the weak acid.
(c) Halfway to the equivalence point, the pH = $pK_a \sim 6.3$.

15.43 (a) There are two equivalence points. It takes 40.0 mL of base to reach the first equivalence point and 80.0 mL to reach the second.
(b) At the first equivalence point the pH is approximately 7.5. At the second equivalence point the pH is approximately 11.
(c) pK_{a1} is equal to the pH halfway to the first equivalence point, $pK_{a1} = 5$. pK_{a2} is equal to the pH halfway between the first and second equivalence point, $pK_{a2} = 10$.

Section Problems
Neutralization Reactions (Section 15.1)

15.44 (a) $HI(aq) + NaOH(aq) \rightarrow H_2O(l) + NaI(aq)$
net ionic equation: $H_3O^+(aq) + OH^-(aq) \rightarrow 2\ H_2O(l)$
The solution at neutralization contains a neutral salt (NaI); pH = 7.00.
(b) $2\ HOCl(aq) + Ba(OH)_2(aq) \rightarrow 2\ H_2O(l) + Ba(OCl)_2(aq)$
net ionic equation: $HOCl(aq) + OH^-(aq) \rightarrow H_2O(l) + OCl^-(aq)$
The solution at neutralization contains a basic anion (OCl^-); pH > 7.00
(c) $HNO_3(aq) + C_6H_5NH_2(aq) \rightarrow C_6H_5NH_3NO_3(aq)$
net ionic equation: $H_3O^+(aq) + C_6H_5NH_2(aq) \rightarrow H_2O(l) + C_6H_5NH_3^+(aq)$
The solution at neutralization contains an acidic cation ($C_6H_5NH_3^+$); pH < 7.00.
(d) $C_6H_5CO_2H(aq) + KOH(aq) \rightarrow H_2O(l) + C_6H_5CO_2K(aq)$
net ionic equation: $C_6H_5CO_2H(aq) + OH^-(aq) \rightarrow H_2O(l) + C_6H_5CO_2^-(aq)$
The solution at neutralization contains a basic anion ($C_6H_5CO_2^-$); pH > 7.00.

15.45 (a) $HNO_2(aq) + CsOH(aq) \rightarrow H_2O(l) + CsNO_2(aq)$
net ionic equation: $HNO_2(aq) + OH^-(aq) \rightarrow H_2O(l) + NO_2^-(aq)$
The solution at neutralization contains a basic anion (NO_2^-); pH > 7.00
(b) $HBr(aq) + NH_3(aq) \rightarrow NH_4Br(aq)$
net ionic equation: $H_3O^+(aq) + NH_3(aq) \rightarrow H_2O(l) + NH_4^+(aq)$
The solution at neutralization contains an acidic cation (NH_4^+); pH < 7.00
(c) $HClO_4(aq) + KOH(aq) \rightarrow H_2O(l) + KClO_4(aq)$
net ionic equation: $H_3O^+(aq) + OH^-(aq) \rightarrow 2\ H_2O(l)$
The solution at neutralization contains a neutral salt ($KClO_4$); pH = 7.00
(d) $HOBr(aq) + NH_3(aq) \rightarrow NH_4OBr(aq)$
net ionic equation: $HOBr(aq) + NH_3(aq) \rightarrow NH_4^+(aq) + OBr^-(aq)$
The solution at neutralization contains the salt NH_4OBr.
$K_a(NH_4^+) = 5.6 \times 10^{-10}$ and $K_b(OBr^-) = 5.0 \times 10^{-5}$; $K_b(OBr^-) > K_a(NH_4^+)$; pH > 7.00

15.46 (a) Strong acid - strong base reaction
$$K_n = \frac{1}{K_w} = \frac{1}{1.0 \times 10^{-14}} = 1.0 \times 10^{14}$$

(b) Weak acid - strong base reaction
$$K_n = \frac{K_a}{K_w} = \frac{3.5 \times 10^{-8}}{1.0 \times 10^{-14}} = 3.5 \times 10^{6}$$

(c) Strong acid - weak base reaction
$$K_n = \frac{K_b}{K_w} = \frac{4.3 \times 10^{-10}}{1.0 \times 10^{-14}} = 4.3 \times 10^{4}$$

(d) Weak acid - strong base reaction
$$K_n = \frac{K_a}{K_w} = \frac{6.5 \times 10^{-5}}{1.0 \times 10^{-14}} = 6.5 \times 10^{9}$$

(c) < (b) < (d) < (a)

15.47 (a) Weak acid - strong base reaction $K_n = \dfrac{K_a}{K_w} = \dfrac{4.5 \times 10^{-4}}{1.0 \times 10^{-14}} = 4.5 \times 10^{10}$

(b) Strong acid - weak base reaction
$$K_n = \frac{K_b}{K_w} = \frac{1.8 \times 10^{-5}}{1.0 \times 10^{-14}} = 1.8 \times 10^{9}$$

(c) Strong acid - strong base reaction
$$K_n = \frac{1}{K_w} = \frac{1}{1.0 \times 10^{-14}} = 1.0 \times 10^{14}$$

(d) Weak acid - weak base reaction
$$K_n = \frac{K_a K_b}{K_w} = \frac{(2.0 \times 10^{-9})(1.8 \times 10^{-5})}{1.0 \times 10^{-14}} = 3.6$$

(d) < (b) < (a) < (c)

15.48 (a) After mixing, the solution contains the basic salt, NaF; pH > 7.00
(b) After mixing, the solution contains the neutral salt, NaCl; pH = 7.00
Solution (a) has the higher pH.

15.49 (a) After mixing, the solution contains the neutral salt, $NaClO_4$; pH = 7.00
(b) After mixing, the solution contains the acidic salt, NH_4ClO_4; pH < 7.00
Solution (b) has the lower pH.

15.50 Weak acid - weak base reaction $\quad K_n = \dfrac{K_a K_b}{K_w} = \dfrac{(1.3 \times 10^{-10})(1.8 \times 10^{-9})}{1.0 \times 10^{-14}} = 2.3 \times 10^{-5}$

K_n is small so the neutralization reaction does not proceed very far to completion.

15.51 Weak acid - weak base reaction $\quad K_n = \dfrac{K_a K_b}{K_w} = \dfrac{(8.0 \times 10^{-5})(4.3 \times 10^{-10})}{1.0 \times 10^{-14}} = 3.4$

Because K_n is close to 1, there will be an appreciable amount of aniline present at equilibrium.

The Common-Ion Effect (Section 15.2)

15.52 $HNO_2(aq) + H_2O(l) \rightleftharpoons H_3O^+(aq) + NO_2^-(aq)$
(a) $NaNO_2$ is a source of NO_2^- (reaction product). The equilibrium shifts towards reactants, and the percent dissociation of HNO_2 decreases.
(c) HCl is a source of H_3O^+ (reaction product). The equilibrium shifts towards reactants, and the percent dissociation of HNO_2 decreases.
(d) $Ba(NO_2)_2$ is a source of NO_2^- (reaction product). The equilibrium shifts towards reactants, and the percent dissociation of HNO_2 decreases.

15.53 $NH_3(aq) + H_2O(l) \rightleftharpoons NH_4^+(aq) + OH^-(aq)$
(a) KOH is a strong base, and it increases the $[OH^-]$. The pH increases.
(b) NH_4NO_3 is a source of NH_4^+ (reaction product). The equilibrium shifts towards reactants, and the $[OH^-]$ decreases. The pH decreases.
(c) NH_4Br is a source of NH_4^+ (reaction product). The equilibrium shifts towards reactants, and the $[OH^-]$ decreases. The pH decreases.
(d) KBr does not affect the pH of the solution.

15.54   (a) $HF(aq) + H_2O(l) \rightleftharpoons H_3O^+(aq) + F^-(aq)$
LiF is a source of F^- (reaction product). The equilibrium shifts toward reactants, and the $[H_3O^+]$ decreases. The pH increases.
(b) Because HI is a strong acid, addition of KI, a neutral salt, does not change the pH.
(c) $NH_3(aq) + H_2O(l) \rightleftharpoons NH_4^+(aq) + OH^-(aq)$
NH_4Cl is a source of NH_4^+ (reaction product). The equilibrium shifts toward reactants, and the $[OH^-]$ decreases. The pH decreases.

15.55   (a) $NH_3(aq) + H_2O(l) \rightleftharpoons NH_4^+(aq) + OH^-(aq)$
NH_4NO_3 is a source of NH_4^+ (reaction product). The equilibrium shifts toward reactants, and the $[OH^-]$ decreases. The pH decreases.
(b) $HCO_3^-(aq) + H_2O(l) \rightleftharpoons H_3O^+(aq) + CO_3^{2-}(aq)$
Na_2CO_3 is a source of CO_3^{2-} (reaction product). The equilibrium shifts toward reactants, and the $[H_3O^+]$ decreases. The pH increases.
(c) Because NaOH is a strong base, addition of $NaClO_4$, a neutral salt, does not change the pH.

15.56   For 0.25 M HF and 0.10 M NaF

	$HF(aq) + H_2O(l) \rightleftharpoons$	$H_3O^+(aq) +$	$F^-(aq)$
initial (M)	0.25	~0	0.10
change (M)	$-x$	$+x$	$+x$
equil (M)	$0.25 - x$	x	$0.10 + x$

$$K_a = \frac{[H_3O^+][F^-]}{[HF]} = 3.5 \times 10^{-4} = \frac{x(0.10+x)}{0.25-x} \approx \frac{x(0.10)}{0.25}$$

Solve for x. $x = [H_3O^+] = 8.8 \times 10^{-4}$ M

$pH = -\log[H_3O^+] = -\log(8.8 \times 10^{-4}) = 3.06$

15.57   On mixing equal volumes of two solutions, both concentrations are cut in half.
$[CH_3NH_2] = 0.10$ M; $[CH_3NH_3Cl] = 0.30$ M

$$CH_3NH_2(aq) \ + \ H_2O(l) \ \rightleftharpoons \ CH_3NH_3^+(aq) \ + \ OH^-(aq)$$

initial (M)	0.10	0.30	~0
change (M)	−x	+x	+x
equil (M)	0.10 − x	0.30 + x	x

$$K_b = \frac{[CH_3NH_3^+][OH^-]}{[CH_3NH_2]} = 3.7 \times 10^{-4} = \frac{(0.30+x)x}{0.10-x} \approx \frac{(0.30)x}{0.10}$$

Solve for x. $x = [OH^-] = 1.2 \times 10^{-4}$ M

$$[H_3O^+] = \frac{K_w}{[OH^-]} = \frac{1.0 \times 10^{-14}}{1.2 \times 10^{-4}} = 8.1 \times 10^{-11} \text{ M}$$

$pH = -\log[H_3O^+] = -\log(8.1 \times 10^{-11}) = 10.09$

15.58   For 0.10 M HN_3:

$$HN_3(aq) \ + \ H_2O(l) \ \rightleftharpoons \ H_3O^+(aq) \ + \ N_3^-(aq)$$

initial (M)	0.10	~0	0
change (M)	−x	+x	+x
equil (M)	0.10 − x	x	x

$$K_a = \frac{[H_3O^+][N_3^-]}{[HN_3]} = 1.9 \times 10^{-5} = \frac{x^2}{0.10-x} \approx \frac{x^2}{0.10}$$

Solve for x. $x = 1.4 \times 10^{-3}$ M

$$\% \text{ dissociation} = \frac{[HN_3]_{diss}}{[HN_3]_{initial}} \times 100\% = \frac{1.4 \times 10^{-3} \text{ M}}{0.10 \text{ M}} \times 100\% = 1.4\%$$

For 0.10 M HN_3 in 0.10 M HCl:

$$HN_3(aq) \ + \ H_2O(l) \ \rightleftharpoons \ H_3O^+(aq) \ + \ N_3^-(aq)$$

initial (M)	0.10	0.10	0
change (M)	−x	+x	+x
equil (M)	0.10 − x	0.10 + x	x

$$K_a = \frac{[H_3O^+][N_3^-]}{[HN_3]} = 1.9 \times 10^{-5} = \frac{(0.10+x)(x)}{0.10-x} \approx \frac{(0.10)(x)}{0.10} = x$$

Solve for x. $x = 1.9 \times 10^{-5}$ M

$$\% \text{ dissociation} = \frac{[HN_3]_{diss}}{[HN_3]_{initial}} \times 100\% = \frac{1.9 \times 10^{-5} \text{ M}}{0.10 \text{ M}} \times 100\% = 0.019\%$$

The % dissociation is less because of the common ion (H_3O^+) effect.

15.59
$$NH_3(aq) + H_2O(l) \rightleftharpoons NH_4^+(aq) + OH^-(aq)$$

initial (M)	0.30	0	~0
change (M)	−x	+x	+x
equil (M)	0.30 − x	x	x

$$K_b = \frac{[NH_4^+][OH^-]}{[NH_3]} = 1.8 \times 10^{-5} = \frac{x^2}{0.30 - x} \approx \frac{x^2}{0.30}$$

Solve for x. $x = [OH^-] = 2.3 \times 10^{-3}$ M

$$[H_3O^+] = \frac{K_w}{[OH^-]} = \frac{1.0 \times 10^{-14}}{2.3 \times 10^{-3}} = 4.3 \times 10^{-12} \text{ M}$$

$pH = -\log[H_3O^+] = -\log(4.3 \times 10^{-12}) = 11.37$

Add 4.0 g of NH_4NO_3.

NH_4NO_3, 80.04 amu; $[NH_4^+]$ = molarity of NH_4NO_3 = $\dfrac{\left(4.0 \text{ g} \times \dfrac{1 \text{ mol}}{80.04 \text{ g}}\right)}{0.100 \text{ L}} = 0.50$ M

$$NH_3(aq) + H_2O(l) \rightleftharpoons NH_4^+(aq) + OH^-(aq)$$

initial (M)	0.30	0.50	~0
change (M)	−x	+x	+x
equil (M)	0.30 − x	0.50 + x	x

$$K_b = \frac{[NH_4^+][OH^-]}{[NH_3]} = 1.8 \times 10^{-5} = \frac{(0.50 + x)x}{0.30 - x} \approx \frac{(0.50)x}{0.30}$$

Solve for x. $x = [OH^-] = 1.1 \times 10^{-5}$ M

$[H_3O^+] = \dfrac{K_w}{[OH^-]} = \dfrac{1.0 \times 10^{-14}}{1.1 \times 10^{-5}} = 9.1 \times 10^{-10}$ M; $pH = -\log[H_3O^+] = -\log(9.1 \times 10^{-10}) = 9.04$

The % dissociation decreases because of the common ion (NH_4^+) effect.

Buffer Solutions (Sections 15.3–15.4)

15.60 Solutions (a), (c), and (d) are buffer solutions. Neutralization reactions for (c) and (d) result in solutions with equal concentrations of HF and F^-.

15.61 Solutions (b), (c), and (d) are buffer solutions. Neutralization reactions for (b) and (d) result in solutions with equal concentrations of NH_3 and NH_4^+.

15.62 Both solutions buffer at the same pH because in both cases the $[NO_2^-]/[HNO_2] = 1$. Solution (a), however, has a higher concentration of both HNO_2 and NO_2^-, and therefore it has the greater buffer capacity.

15.63 Both solutions buffer at the same pH because in both cases the $[NH_3]/[NH_4^+] = 1.5$. Solution (b), however, has a higher concentration of both NH_3 and NH_4^+, therefore it has the greater buffer capacity.

15.64 When blood absorbs acid, the equilibrium shifts to the left, decreasing the pH, but not by much because the $[HCO_3^-]/[H_2CO_3]$ ratio remains nearly constant. When blood absorbs base, the equilibrium shifts to the right, increasing the pH, but not by much because the $[HCO_3^-]/[H_2CO_3]$ ratio remains nearly constant.

15.65 $H_2PO_4^-(aq) + H_2O(l) \rightleftharpoons H_3O^+(aq) + HPO_4^{2-}(aq)$

For $H_2PO_4^-$, $K_{a2} = 6.2 \times 10^{-8}$, $pK_{a2} = -\log K_{a2} = 7.21$

$$pH = 7.4 = pK_{a2} + \log\frac{[HPO_4^{2-}]}{[H_2PO_4^-]} = 7.21 + \log\frac{[HPO_4^{2-}]}{[H_2PO_4^-]}$$

To maintain pH near 7.4, you need $\log\dfrac{[HPO_4^{2-}]}{[H_2PO_4^-]} = 0.19$ and $\dfrac{[HPO_4^{2-}]}{[H_2PO_4^-]} = 10^{0.19} = 1.5$

The principal buffer reactions are:
$H_3O^+(aq) + HPO_4^{2-}(aq) \rightarrow H_2PO_4^-(aq) + H_2O(l)$
$OH^-(aq) + H_2PO_4^-(aq) \rightarrow HPO_4^{2-}(aq) + H_2O(l)$

15.66 $pH = pK_a + \log\dfrac{[base]}{[acid]} = pK_a + \log\dfrac{[CN^-]}{[HCN]}$
For HCN, $K_a = 4.9 \times 10^{-10}$, $pK_a = -\log K_a = -\log(4.9 \times 10^{-10}) = 9.31$
$pH = 9.31 + \log\left(\dfrac{0.12}{0.20}\right) = 9.09$

The pH of a buffer solution will not change on dilution because the acid and base concentrations will change by the same amount and their ratio will remain the same.

15.67 $NaHCO_3$, 84.01 amu; Na_2CO_3, 105.99 amu

$$[HCO_3^-] = \text{molarity of } NaHCO_3 = \frac{\left(4.2 \text{ g} \times \dfrac{1 \text{ mol}}{84.01 \text{ g}}\right)}{0.20 \text{ L}} = 0.25 \text{ M}$$

$$[CO_3^{2-}] = \text{molarity of } Na_2CO_3 = \frac{\left(5.3 \text{ g} \times \dfrac{1 \text{ mol}}{105.99 \text{ g}}\right)}{0.20 \text{ L}} = 0.25 \text{ M}$$

$$pH = pK_a + \log\frac{[base]}{[acid]} = pK_a + \log\frac{[CO_3^{2-}]}{[HCO_3^-]}$$

For HCO_3^-, $K_{a2} = 5.6 \times 10^{-11}$, $pK_{a2} = -\log K_{a2} = -\log(5.6 \times 10^{-11}) = 10.25$

$pH = 10.25 + \log\dfrac{[0.25]}{[0.25]} = 10.25$

The pH of a buffer solution will not change on dilution because the acid and base concentrations will change by the same amount and their ratio will remain the same.

15.68 $pH = pK_a + \log \dfrac{[base]}{[acid]} = pK_a + \log \dfrac{[NH_3]}{[NH_4^+]}$

For NH_4^+, $K_a = 5.6 \times 10^{-10}$, $pK_a = -\log K_a = -\log(5.6 \times 10^{-10}) = 9.25$

For the buffer: $pH = 9.25 + \log \dfrac{(0.200)}{(0.200)} = 9.25$

(a) add 0.0050 mol NaOH, $[OH^-] = 0.0050$ mol/0.500 L $= 0.010$ M

$$NH_4^+(aq) + OH^-(aq) \rightleftharpoons NH_3(aq) + H_2O(l)$$

	NH_4^+	OH^-	NH_3
before reaction (M)	0.200	0.010	0.200
change (M)	−0.010	−0.010	+0.010
after reaction (M)	0.200 − 0.010	0	0.200 + 0.010

$pH = 9.25 + \log \dfrac{[NH_3]}{[NH_4^+]} = 9.25 + \log \dfrac{(0.200 + 0.010)}{(0.200 - 0.010)} = 9.29$

(b) add 0.020 mol HCl, $[H_3O^+] = 0.020$ mol/0.500 L $= 0.040$ M

$$NH_3(aq) + H_3O^+(aq) \rightleftharpoons NH_4^+(aq) + H_2O(l)$$

	NH_3	H_3O^+	NH_4^+
before reaction (M)	0.200	0.040	0.200
change (M)	−0.040	−0.040	+0.040
after reaction (M)	0.200 − 0.040	0	0.200 + 0.040

$pH = 9.25 + \log \dfrac{[NH_3]}{[NH_4^+]} = 9.25 + \log \dfrac{(0.200 - 0.040)}{(0.200 + 0.040)} = 9.07$

15.69 $pH = pK_a + \log \dfrac{[base]}{[acid]} = pK_a + \log \dfrac{[SO_3^{2-}]}{[HSO_3^-]}$

For HSO_3^-, $K_a = 6.3 \times 10^{-8}$, $pK_a = -\log K_a = -\log(6.3 \times 10^{-8}) = 7.20$

For the buffer: $pH = 7.20 + \log \dfrac{(0.300)}{(0.500)} = 6.98$

(a) add $(0.0050 \text{ L})(0.20 \text{ mol/L}) = 0.0010$ mol HCl $= 0.0010$ mol H_3O^+

mol $HSO_3^- = (0.300 \text{ L})(0.500 \text{ mol/L}) = 0.150$ mol

mol $SO_3^{2-} = (0.300 \text{ L})(0.300 \text{ mol/L}) = 0.0900$ mol

$$SO_3^{2-}(aq) + H_3O^+(aq) \rightleftharpoons HSO_3^-(aq) + H_2O(l)$$

	SO_3^{2-}	H_3O^+	HSO_3^-
before reaction (mol)	0.0900	0.0010	0.150
change (mol)	−0.0010	−0.0010	+0.0010
after reaction (mol)	0.0900 − 0.0010	0	0.150 + 0.0010

$pH = 7.20 + \log \dfrac{[SO_3^{2-}]}{[HSO_3^-]} = 7.20 + \log \dfrac{(0.0900 - 0.0010)}{(0.150 + 0.0010)} = 6.97$

(b) add $(0.0050 \text{ L})(0.10 \text{ mol/L}) = 0.00050$ mol NaOH $= 0.00050$ mol OH^-

$$HSO_3^-(aq) + OH^-(aq) \rightleftharpoons SO_3^{2-}(aq) + H_2O(l)$$

	HSO_3^-	OH^-	SO_3^{2-}
before reaction (mol)	0.150	0.00050	0.0900
change (mol)	−0.00050	−0.00050	+0.00050
after reaction (mol)	0.150 − 0.00050	0	0.0900 + 0.00050

$$pH = 7.20 + \log \frac{[SO_3^{2-}]}{[HSO_3^-]} = 7.20 + \log \frac{(0.0900 + 0.00050)}{(0.150 - 0.00050)} = 6.98$$

15.70

	Acid	K_a	$pK_a = -\log K_a$
(a)	H_3BO_3	5.8×10^{-10}	9.24
(b)	HCO_2H	1.8×10^{-4}	3.74
(c)	$HOCl$	3.5×10^{-8}	7.46

The stronger the acid (the larger the K_a), the smaller is the pK_a.

15.71 (a) $K_a = 10^{-pK_a} = 10^{-5.00} = 1.0 \times 10^{-5}$ (b) $K_a = 10^{-pK_a} = 10^{-8.70} = 2.0 \times 10^{-9}$

(b) is the weaker acid

15.72 $pH = pK_a + \log \frac{[base]}{[acid]} = pK_a + \log \frac{[HCO_2^-]}{[HCO_2H]}$

For HCO_2H, $K_a = 1.8 \times 10^{-4}$; $pK_a = -\log K_a = -\log(1.8 \times 10^{-4}) = 3.74$

$pH = 3.74 + \log \frac{(0.50)}{(0.25)} = 4.04$

15.73 $pH = pK_a + \log \frac{[base]}{[acid]} = pK_a + \log \frac{[HCO_3^-]}{[H_2CO_3]}$

For H_2CO_3, $K_a = 4.3 \times 10^{-7}$; $pK_a = -\log K_a = -\log(4.3 \times 10^{-7}) = 6.37$

$7.40 = 6.37 + \log \frac{[HCO_3^-]}{[H_2CO_3]}$; $1.03 = \log \frac{[HCO_3^-]}{[H_2CO_3]}$

$\frac{[HCO_3^-]}{[H_2CO_3]} = 10^{1.03} = 10.7$; $\frac{[H_2CO_3]}{[HCO_3^-]} = 0.093$

15.74 $pH = pK_a + \log \frac{[base]}{[acid]} = pK_a + \log \frac{[NH_3]}{[NH_4^+]}$

For NH_4^+, $K_a = 5.6 \times 10^{-10}$; $pK_a = -\log K_a = -\log(5.6 \times 10^{-10}) = 9.25$

$9.80 = 9.25 + \log \frac{[NH_3]}{[NH_4^+]}$; $0.550 = \log \frac{[NH_3]}{[NH_4^+]}$; $\frac{[NH_3]}{[NH_4^+]} = 10^{0.55} = 3.5$

The volume of the 1.0 M NH_3 solution should be 3.5 times the volume of the 1.0 M NH_4Cl solution so that the mixture will buffer at pH 9.80.

15.75 $pH = pK_a + \log \frac{[base]}{[acid]} = pK_a + \log \frac{[CH_3CO_2^-]}{[CH_3CO_2H]}$

For CH_3CO_2H, $K_a = 1.8 \times 10^{-5}$; $pK_a = -\log K_a = -\log(1.8 \times 10^{-5}) = 4.74$

$$4.44 = 4.74 + \log\frac{[CH_3CO_2^-]}{[CH_3CO_2H]} \; ; \qquad -0.30 = \log\frac{[CH_3CO_2^-]}{[CH_3CO_2H]}$$

$$\frac{[CH_3CO_2^-]}{[CH_3CO_2H]} = 10^{-0.30} = 0.50$$

The solution should have 0.50 mol of $CH_3CO_2^-$ per mole of CH_3CO_2H. For example, you could dissolve 41g of CH_3CO_2Na in 1.00 L of 1.00 M CH_3CO_2H.

15.76 H_3PO_4, $K_{a1} = 7.5 \times 10^{-3}$; $pK_{a1} = -\log K_{a1} = 2.12$

$H_2PO_4^-$, $K_{a2} = 6.2 \times 10^{-8}$; $pK_{a2} = -\log K_{a2} = 7.21$

HPO_4^{2-}, $K_{a3} = 4.8 \times 10^{-13}$; $pK_{a3} = -\log K_{a3} = 12.32$

The buffer system of choice for pH 7.00 is (b) $H_2PO_4^- - HPO_4^{2-}$ because the pK_a for $H_2PO_4^-$ (7.21) is closest to 7.00.

15.77 HSO_4^-, $K_{a2} = 1.2 \times 10^{-2}$; $pK_{a2} = -\log K_{a2} = 1.92$
$HOCl$, $K_a = 3.5 \times 10^{-8}$; $pK_a = -\log K_a = 7.56$
$C_6H_5CO_2H$, $K_a = 6.5 \times 10^{-5}$; $pK_a = -\log K_a = 4.19$
The buffer system of choice for pH = 4.50 is (c) $C_6H_5CO_2H - C_6H_5CO_2^-$ because the pK_a for $C_6H_5CO_2H$ (4.19) is closest to 4.50.

pH Titration Curves (Sections 15.5–15.9)

15.78   (a) $(0.060 \text{ L})(0.150 \text{ mol/L})(1000 \text{ mmol/mol}) = 9.00 \text{ mmol } HNO_3$

(b) vol NaOH $= (9.00 \text{ mmol } HNO_3)\left(\dfrac{1 \text{ mmol NaOH}}{1 \text{ mmol } HNO_3}\right)\left(\dfrac{1 \text{ mL NaOH}}{0.450 \text{ mmol NaOH}}\right) = 20.0 \text{ mL NaOH}$

(c) At the equivalence point the solution contains the neutral salt $NaNO_3$. The pH is 7.00.

(d)

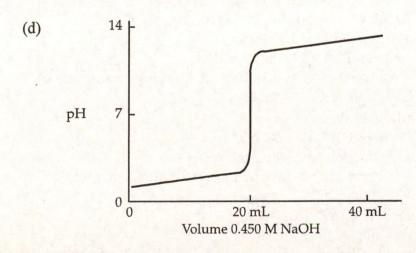

Volume 0.450 M NaOH

15.79

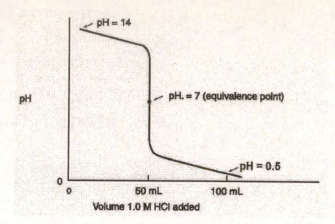

mmol NaOH = (50.0 mL)(1.0 mmol/mL) = 50 mmol

mmol HCl = mmol NaOH = 50 mmol

$$\text{vol HCl} = (50 \text{ mmol})\left(\frac{1.0 \text{ mL}}{1.0 \text{ mmol}}\right) = 50 \text{ mL}$$

50 mL of 1.0 M HCl is needed to reach the equivalence point.

15.80   mmol OH^- = (20.0 mL)(0.150 mmol/mL) = 3.00 mmol

mmol acid present = mmol OH^- added = 3.00 mmol

$$[\text{acid}] = \frac{3.00 \text{ mmol}}{60.0 \text{ mL}} = 0.0500 \text{ M}$$

15.81   mmol OH^- = (60.0 mL)(0.240 mmol/mL) = 14.4 mmol

$$\text{mmol acid present} = 14.4 \text{ mmol } OH^- \times \frac{1 \text{ mmol acid}}{2 \text{ mmol } OH^-} = 7.20 \text{ mmol acid}$$

$$[\text{acid}] = \frac{7.20 \text{ mmol}}{25.0 \text{ mL}} = 0.288 \text{ M}$$

15.82 $HBr(aq) + NaOH(aq) \rightarrow Na^+(aq) + Br^-(aq) + H_2O(l)$

(a) $[H_3O^+]$ = 0.120 M; pH = $-\log[H_3O^+]$ = $-\log$ (0.120) = 0.92

(b) (50.0 mL)(0.120 mmol/mL) = 6.00 mmol HBr

(20.0 mL)(0.240 mmol/mL) = 4.80 mmol NaOH

6.00 mmol HBr – 4.80 mmol NaOH = 1.20 mmol HBr after neutralization

$$[H_3O^+] = \frac{1.20 \text{ mmol}}{(50.0 \text{ mL} + 20.0 \text{ mL})} = 0.0171 \text{ M}$$

pH = $-\log[H_3O^+]$ = $-\log(0.0171)$ = 1.77

(c) (24.9 mL)(0.240 mmol/mL) = 5.98 mmol NaOH

6.00 mmol HBr – 5.98 mmol NaOH = 0.02 mmol HBr after neutralization

$$[H_3O^+] = \frac{0.02 \text{ mmol}}{(50.0 \text{ mL} + 24.9 \text{ mL})} = 3 \times 10^{-4} \text{ M}$$

pH = $-\log[H_3O^+]$ = $-\log(3 \times 10^{-4})$ = 3.5

(d) The titration reaches the equivalence point when 25.0 mL of 0.240 M NaOH is added. At the equivalence point the solution contains the neutral salt NaBr. The pH is 7.00.

(e) $(25.1 \text{ mL})(0.240 \text{ mmol/mL}) = 6.024 \text{ mmol NaOH}$

$6.024 \text{ mmol NaOH} - 6.00 \text{ mmol HBr} = 0.024 \text{ mmol NaOH after neutralization}$

$$[OH^-] = \frac{0.024 \text{ mmol}}{(50.0 \text{ mL} + 25.1 \text{ mL})} = 3.2 \times 10^{-4} \text{ M}$$

$$[H_3O^+] = \frac{K_w}{[OH^-]} = \frac{1.0 \times 10^{-14}}{3.2 \times 10^{-4}} = 3.1 \times 10^{-11} \text{ M}$$

$pH = -\log[H_3O^+] = -\log(3.1 \times 10^{-11}) = 10.5$

(f) $(40.0 \text{ mL})(0.240 \text{ mmol/mL}) = 9.60 \text{ mmol NaOH}$

$9.60 \text{ mmol NaOH} - 6.00 \text{ mmol HBr} = 3.60 \text{ mmol NaOH after neutralization}$

$$[OH^-] = \frac{3.60 \text{ mmol}}{(50.0 \text{ mL} + 40.0 \text{ mL})} = 0.040 \text{ M}$$

$$[H_3O^+] = \frac{K_w}{[OH^-]} = \frac{1.0 \times 10^{-14}}{0.040} = 2.5 \times 10^{-13} \text{ M}$$

$pH = -\log[H_3O^+] = -\log(2.5 \times 10^{-13}) = 12.60$

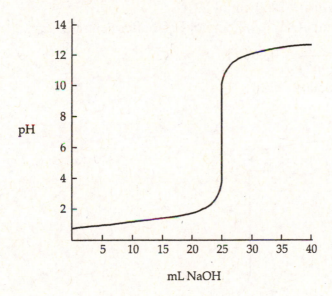

15.83 $Ba(OH)_2(aq) + 2 HNO_3(aq) \rightarrow Ba^{2+}(aq) + 2 NO_3^-(aq) + 2 H_2O(l)$

(a) $[OH^-] = 2(0.150 \text{ M}) = 0.300 \text{ M}$

$$[H_3O^+] = \frac{K_w}{[OH^-]} = \frac{1.0 \times 10^{-14}}{0.300} = 3.33 \times 10^{-14} \text{ M}$$

$pH = -\log[H_3O^+] = -\log(3.33 \times 10^{-14}) = 13.48$

(b) $(40.0 \text{ mL})(0.150 \text{ mmol/mL}) = 6.00 \text{ mmol Ba(OH)}_2$

$$6.00 \text{ mmol Ba(OH)}_2 \times \frac{2 \text{ mmol OH}^-}{1 \text{ mmol Ba(OH)}_2} = 12.0 \text{ mmol OH}^-$$

$(10.0 \text{ mL})(0.400 \text{ mmol/mL}) = 4.00 \text{ mmol HNO}_3$

$12.0 \text{ mmol OH}^- - 4.00 \text{ mmol HNO}_3 = 8.00 \text{ mmol OH}^-$ after neutralization

$$[OH^-] = \frac{8.00 \text{ mmol}}{(40.0 \text{ mL} + 10.0 \text{ mL})} = 0.160 \text{ M}$$

$$[H_3O^+] = \frac{K_w}{[OH^-]} = \frac{1.0 \times 10^{-14}}{0.160} = 6.25 \times 10^{-14} \text{ M}$$

$$pH = -\log[H_3O^+] = -\log(6.25 \times 10^{-14}) = 13.20$$

(c) $(20.0 \text{ mL})(0.400 \text{ mmol/mL}) = 8.00 \text{ mmol HNO}_3$

$12.0 \text{ mmol OH}^- - 8.00 \text{ mmol HNO}_3 = 4.00 \text{ mmol OH}^-$ after neutralization

$$[OH^-] = \frac{4.00 \text{ mmol}}{(40.0 \text{ mL} + 20.0 \text{ mL})} = 0.0667 \text{ M}$$

$$[H_3O^+] = \frac{K_w}{[OH^-]} = \frac{1.0 \times 10^{-14}}{0.0667} = 1.50 \times 10^{-13} \text{ M}$$

$$pH = -\log[H_3O^+] = -\log(1.50 \times 10^{-13}) = 12.82$$

(d) The titration reaches the equivalence point when 30.0 mL of 0.400 M HNO$_3$ is added. At the equivalence point, the solution contains the neutral salt Ba(NO$_3$)$_2$. The pH is 7.00.

(e) $(40.0 \text{ mL})(0.400 \text{ mmol/mL}) = 16.0 \text{ mmol HNO}_3$

$16.0 \text{ mmol HNO}_3 - 12.0 \text{ mmol OH}^- = 4.00 \text{ mmol H}_3\text{O}^+$ after neutralization

$$[H_3O^+] = \frac{4.00 \text{ mmol}}{(40.0 \text{ mL} + 40.0 \text{ mL})} = 0.0500 \text{ M}$$

$$pH = -\log[H_3O^+] = -\log(0.0500) = 1.30$$

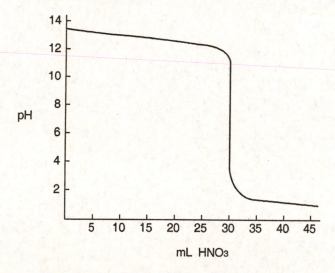

15.84   mmol HF = (40.0 mL)(0.250 mmol/mL) = 10.0 mmol

mmol NaOH required = mmol HF = 10.0 mmol

$$\text{mL NaOH required} = (10.0 \text{ mmol})\left(\frac{1.00 \text{ mL}}{0.200 \text{ mmol}}\right) = 50.0 \text{ mL}$$

50.0 mL of 0.200 M NaOH is required to reach the equivalence point.

For HF, $K_a = 3.5 \times 10^{-4}$; $pK_a = -\log K_a = -\log(3.5 \times 10^{-4}) = 3.46$

(a) mmol HF = 10.0 mmol

mmol NaOH = (0.200 mmol/mL)(10.0 mL) = 2.00 mmol

Neutralization reaction: $HF(aq) + OH^-(aq) \rightarrow F^-(aq) + H_2O(l)$

before reaction (mmol)	10.0	2.00	0
change (mmol)	−2.00	−2.00	+2.00
after reaction (mmol)	8.0	0	2.00

$$[HF] = \frac{8.0 \text{ mmol}}{(40.0 \text{ mL} + 10.0 \text{ mL})} = 0.16 \text{ M}; \quad [F^-] = \frac{2.00 \text{ mmol}}{(40.0 \text{ mL} + 10.0 \text{ mL})} = 0.0400 \text{ M}$$

$$HF(aq) + H_2O(l) \rightleftharpoons H_3O^+(aq) + F^-(aq)$$

initial (M)	0.16	~0	0.0400
change (M)	−x	+x	+x
equil (M)	0.16 − x	x	0.0400 + x

$$K_a = \frac{[H_3O^+][F^-]}{[HF]} = 3.5 \times 10^{-4} = \frac{x(0.0400 + x)}{0.16 - x} \approx \frac{x(0.0400)}{0.16}$$

Solve for x. $x = [H_3O^+] = 1.4 \times 10^{-3} \text{ M}$

$pH = -\log[H_3O^+] = -\log(1.4 \times 10^{-3}) = 2.85$

(b) Halfway to the equivalence point,

$pH = pK_a = -\log K_a = -\log(3.5 \times 10^{-4}) = 3.46$

(c) At the equivalence point only the salt NaF is in solution.

$$[F^-] = \frac{10.0 \text{ mmol}}{(40.0 \text{ mL} + 50.0 \text{ mL})} = 0.111 \text{ M}$$

$$F^-(aq) + H_2O(l) \rightleftharpoons HF(aq) + OH^-(aq)$$

initial (M)	0.111	0	~0
change (M)	−x	+x	+x
equil (M)	0.111 − x	x	x

For F^-, $K_b = \dfrac{K_w}{K_a \text{ for HF}} = \dfrac{1.0 \times 10^{-14}}{3.5 \times 10^{-4}} = 2.9 \times 10^{-11}$

$$K_b = \frac{[HF][OH^-]}{[F^-]} = 2.9 \times 10^{-11} = \frac{x^2}{0.111 - x} \approx \frac{x^2}{0.111}$$

Solve for x. $x = [OH^-] = 1.8 \times 10^{-6} \text{ M}$

$$[H_3O^+] = \frac{K_w}{[OH^-]} = \frac{1.0 \times 10^{-14}}{1.8 \times 10^{-6}} = 5.6 \times 10^{-9} \text{ M}$$

$pH = -\log[H_3O^+] = -\log(5.6 \times 10^{-9}) = 8.25$

(d) mmol HF = 10.0 mmol

mol NaOH = (0.200 mmol/mL)(80.0 mL) = 16.0 mmol

Neutralization reaction: $HF(aq) + OH^-(aq) \rightarrow F^-(aq) + H_2O(l)$

before reaction (mmol)	10.0	16.0	0
change (mmol)	−10.0	−10.0	+10.0
after reaction (mmol)	0	6.0	10.0

After the equivalence point, the pH of the solution is determined by the $[OH^-]$.

$$[OH^-] = \frac{6.0 \text{ mmol}}{(40.0 \text{ mL} + 80.0 \text{ mL})} = 5.0 \times 10^{-2} \text{ M}$$

$$[H_3O^+] = \dfrac{K_w}{[OH^-]} = \dfrac{1.0 \times 10^{-14}}{5.0 \times 10^{-2}} = 2.0 \times 10^{-13} \text{ M}$$

$$pH = -\log[H_3O^+] = -\log(2.0 \times 10^{-13}) = 12.70$$

15.85   mmol CH_3NH_2 = (100.0 mL)(0.100 mmol/mL) = 10.0 mmol

mmol HNO_3 required = mmol CH_3NH_2 = 10.0 mmol

$$\text{vol } HNO_3 \text{ required} = (10.0 \text{ mmol})\left(\dfrac{1.00 \text{ mL}}{0.250 \text{ mmol}}\right) = 40.0 \text{ mL}$$

40.0 mL of 0.250 M HNO_3 are required to reach the equivalence point.

(a) $CH_3NH_2(aq) + H_2O(l) \rightleftharpoons CH_3NH_3^+(aq) + OH^-(aq)$

initial (M)	0.100	0	~0
change (M)	−x	+x	+x
equil (M)	0.100 − x	x	x

$$K_b = \dfrac{[CH_3NH_3^+][OH^-]}{[CH_3NH_2]} = 3.7 \times 10^{-4} = \dfrac{x^2}{0.100 - x}$$

$x^2 + (3.7 \times 10^{-4})x - (3.7 \times 10^{-5}) = 0$

Use the quadratic formula to solve for x.

$$x = \dfrac{-(3.7 \times 10^{-4}) \pm \sqrt{(3.7 \times 10^{-4})^2 - (4)(-3.7 \times 10^{-5})}}{2(1)} = \dfrac{-3.7 \times 10^{-4} \pm 0.0122}{2}$$

x = 0.0059 and −0.0063

Of the two solutions for x, only the positive value of x has physical meaning because x is the $[OH^-]$.

$[OH^-]$ = x = 0.0059 M

$$[H_3O^+] = \dfrac{K_w}{[OH^-]} = \dfrac{1.0 \times 10^{-14}}{5.9 \times 10^{-3}} = 1.7 \times 10^{-12} \text{ M}$$

$pH = -\log[H_3O^+] = -\log(1.7 \times 10^{-12}) = 11.77$

(b) 20.0 mL of HNO_3 is halfway to the equivalence point.

$$\text{For } CH_3NH_3^+, \ K_a = \dfrac{K_w}{K_b \text{ for } CH_3NH_2} = \dfrac{1.0 \times 10^{-14}}{3.7 \times 10^{-4}} = 2.7 \times 10^{-11}$$

$pH = pK_a = -\log(2.7 \times 10^{-11}) = 10.57$

(c) At the equivalence point only the salt $CH_3NH_3NO_3$ is in solution.

mmol $CH_3NH_3NO_3$ = (0.100 mmol/mL)(100.0 mL) = 10.0 mmol

$$[CH_3NH_3^+] = \dfrac{10.0 \text{ mmol}}{(100.0 \text{ mL} + 40.0 \text{ mL})} = 0.0714 \text{ M}$$

$CH_3NH_3^+(aq) + H_2O(l) \rightleftharpoons H_3O^+(aq) + CH_3NH_2(aq)$

initial (M)	0.0714	~0	0
change (M)	−x	+x	+x
equil (M)	0.0714 − x	x	x

$$K_a = \dfrac{[H_3O^+][CH_3NH_2]}{[CH_3NH_3^+]} = 2.7 \times 10^{-11} = \dfrac{x^2}{0.0714 - x} \approx \dfrac{x^2}{0.0714}$$

Solve for x. $x = [H_3O^+] = 1.4 \times 10^{-6}$ M

$pH = -\log[H_3O^+] = -\log(1.4 \times 10^{-6}) = 5.85$

(d) mmol $CH_3NH_2 = (0.100 \text{ mmol/mL})(100.0 \text{ mL}) = 10.0$ mmol

mmol $HNO_3 = (0.250 \text{ mmol/mL})(60.0 \text{ mL}) = 15.0$ mmol

Neutralization reaction:	$CH_3NH_2(aq)$ +	$H_3O^+(aq)$ $\rightarrow$	$CH_3NH_3^+(aq)$ +	$H_2O(l)$
before reaction (mmol)	10.0	15.0	0	
change (mmol)	−10.0	−10.0	+10.0	
after reaction (mmol)	0	5.0	10.0	

After the equivalence point the pH of the solution is determined by the $[H_3O^+]$.

$$[H_3O^+] = \frac{5.0 \text{ mmol}}{(100.0 \text{ mL} + 60.0 \text{ mL})} = 3.1 \times 10^{-2} \text{ M}$$

$pH = -\log[H_3O^+] = -\log(3.1 \times 10^{-2}) = 1.51$

15.86 For H_2A^+, $K_{a1} = 4.6 \times 10^{-3}$ and $K_{a2} = 2.0 \times 10^{-10}$

(a) $(10.0 \text{ mL})(0.100 \text{ mmol/mL}) = 1.00$ mmol NaOH added = 1.00 mmol HA produced.

$(50.0 \text{ mL})(0.100 \text{ mmol/mL}) = 5.00$ mmol H_2A^+

5.00 mmol H_2A^+ − 1.00 mmol NaOH = 4.00 mmol H_2A^+ after neutralization

$$[H_2A^+] = \frac{4.00 \text{ mmol}}{(50.0 \text{ mL} + 10.0 \text{ mL})} = 6.67 \times 10^{-2} \text{ M}$$

$$[HA] = \frac{1.00 \text{ mmol}}{(50.0 \text{ mL} + 10.0 \text{ mL})} = 1.67 \times 10^{-2} \text{ M}$$

$$pH = pK_{a1} + \log\frac{[HA]}{[H_2A^+]} = -\log(4.6 \times 10^{-3}) + \log\left(\frac{1.67 \times 10^{-2}}{6.67 \times 10^{-2}}\right) = 1.74$$

(b) Halfway to the first equivalence point, $pH = pK_{a1} = 2.34$

(c) At the first equivalence point, $pH = \dfrac{pK_{a1} + pK_{a2}}{2} = 6.02$

(d) Halfway between the first and second equivalence points, $pH = pK_{a2} = 9.70$

(e) At the second equivalence point only the basic salt, NaA, is in solution.

$$K_b = \frac{K_w}{K_a \text{ for HA}} = \frac{K_w}{K_{a2}} = \frac{1.0 \times 10^{-14}}{2.0 \times 10^{-10}} = 5.0 \times 10^{-5}$$

mmol $A^- = (50.0 \text{ mL})(0.100 \text{ mmol/mL}) = 5.00$ mmol

$$[A^-] = \frac{5.0 \text{ mmol}}{(50.0 \text{ mL} + 100.0 \text{ mL})} = 3.3 \times 10^{-2} \text{ M}$$

	$A^-(aq)$ +	$H_2O(l)$ $\rightleftharpoons$	$HA(aq)$ +	$OH^-(aq)$
initial (M)	0.033		0	~0
change (M)	−x		+x	+x
equil (M)	0.033 − x		x	x

$$K_b = \frac{[HA][OH^-]}{[A^-]} = 5.0 \times 10^{-5} = \frac{(x)(x)}{0.033 - x} \approx \frac{x^2}{0.033}$$

Solve for x.

$$x = [OH^-] = \sqrt{(5.0 \times 10^{-5})(0.033)} = 1.3 \times 10^{-3} \text{ M}$$

$$[H_3O^+] = \frac{K_w}{[OH^-]} = \frac{1.0 \times 10^{-14}}{1.3 \times 10^{-3}} = 7.7 \times 10^{-12} \text{ M}$$

$$pH = -\log[H_3O^+] = -\log(7.7 \times 10^{-12}) = 11.11$$

15.87 For H_2CO_3, $K_{a1} = 4.3 \times 10^{-7}$ and $K_{a2} = 5.6 \times 10^{-11}$

(a) $(25.0 \text{ mL})(0.0200 \text{ mmol/mL}) = 0.500 \text{ mmol } H_2CO_3$

$(10.0 \text{ mL})(0.0250 \text{ mmol/mL}) = 0.250 \text{ mmol KOH added}$

$0.500 \text{ mmol } H_2CO_3 - 0.250 \text{ mmol KOH} = 0.250 \text{ mmol } HCO_3^- \text{ produced}$

This is halfway to the first equivalence point where $pH = pK_{a1} = -\log(4.3 \times 10^{-7}) = 6.37$

(b) At the first equivalence point, $pH = \dfrac{pK_{a1} + pK_{a2}}{2} = 8.31$

(c) Halfway between the first and second equivalence points, $pH = pK_{a2} = 10.25$

(d) At the second equivalence point only the basic salt, K_2CO_3, is in solution.

$$K_b = \frac{K_w}{K_a \text{ for } HCO_3^-} = \frac{K_w}{K_{a2}} = \frac{1.0 \times 10^{-14}}{5.6 \times 10^{-11}} = 1.8 \times 10^{-4}$$

$$\text{mmol } CO_3^{2-} = \text{mmol } H_2CO_3 = 0.500 \text{ mmol}$$

$$[CO_3^{2-}] = \frac{0.500 \text{ mmol}}{(25.0 \text{ mL} + 40.0 \text{ mL})} = 0.00769 \text{ M}$$

$$CO_3^{2-}(aq) + H_2O(l) \rightleftharpoons HCO_3^-(aq) + OH^-(aq)$$

	CO_3^{2-}	HCO_3^-	OH^-
initial (M)	0.00769	0	~0
change (M)	$-x$	$+x$	$+x$
equil (M)	$0.00769 - x$	x	x

$$K_b = \frac{[HCO_3^-][OH^-]}{[CO_3^{2-}]} = 1.8 \times 10^{-4} = \frac{(x)(x)}{0.00769 - x}$$

Use the quadratic formula to solve for x.

$$x = \frac{-(1.8 \times 10^{-4}) \pm \sqrt{(1.8 \times 10^{-4})^2 - (4)(1)(-1.4 \times 10^{-6})}}{2(1)} = \frac{(-1.8 \times 10^{-4}) \pm (2.37 \times 10^{-3})}{2}$$

$x = -1.27 \times 10^{-3}$ and 1.09×10^{-3}

Of the two solutions for x, only the positive value of x has physical meaning because x is the $[OH^-]$.

$[OH^-] = x = 1.09 \times 10^{-3} \text{ M}$

$$[H_3O^+] = \frac{K_w}{[OH^-]} = \frac{1.0 \times 10^{-14}}{1.09 \times 10^{-3}} = 9.2 \times 10^{-12} \text{ M}$$

$pH = -\log[H_3O^+] = -\log(9.2 \times 10^{-12}) = 11.04$

(e) excess KOH

$(50.0 \text{ mL} - 40.0 \text{ mL})(0.025 \text{ mmol/mL}) = 0.250 \text{ mmol KOH} = 0.250 \text{ mmol } OH^-$

$$[OH^-] = \frac{0.250 \text{ mmol}}{(25.0 \text{ mL} + 50.0 \text{ mL})} = 3.33 \times 10^{-3} \text{ M}$$

$$[H_3O^+] = \frac{K_w}{[OH^-]} = \frac{1.0 \times 10^{-14}}{3.33 \times 10^{-3}} = 3.0 \times 10^{-12} \text{ M}$$

$$pH = -\log[H_3O^+] = -\log(3.0 \times 10^{-12}) = 11.52$$

15.88   When equal volumes of acid and base react, all concentrations are cut in half.
(a) At the equivalence point, only the salt $NaNO_2$ is in solution.
$[NO_2^-] = 0.050$ M

$$\text{For } NO_2^-, K_b = \frac{K_w}{K_a \text{ for } HNO_2} = \frac{1.0 \times 10^{-14}}{4.5 \times 10^{-4}} = 2.2 \times 10^{-11}$$

$$NO_2^-(aq) + H_2O(l) \rightleftharpoons HNO_2(aq) + OH^-(aq)$$

	NO_2^-	HNO_2	OH^-
initial (M)	0.050	0	~0
change (M)	–x	+x	+x
equil (M)	0.050 – x	x	x

$$K_b = \frac{[HNO_2][OH^-]}{[NO_2^-]} = 2.2 \times 10^{-11} = \frac{(x)(x)}{0.050 - x} \approx \frac{x^2}{0.050}$$

Solve for x. $x = [OH^-] = 1.1 \times 10^{-6}$ M

$$[H_3O^+] = \frac{K_w}{[OH^-]} = \frac{1.0 \times 10^{-14}}{1.1 \times 10^{-6}} = 9.1 \times 10^{-9} \text{ M}$$

$$pH = -\log[H_3O^+] = -\log(9.1 \times 10^{-9}) = 8.04$$

Phenol red would be a suitable indicator. (see Figure 15.4)
(b) The pH is 7.00 at the equivalence point for the titration of a strong acid (HI) with a strong base (NaOH).
Bromthymol blue or phenol red would be suitable indicators. (Any indicator that changes color in the pH range 4 – 10 is satisfactory for a strong acid – strong base titration.)
(c) At the equivalence point only the salt CH_3NH_3Cl is in solution.
$[CH_3NH_3^+] = 0.050$ M

$$\text{For } CH_3NH_3^+, K_a = \frac{K_w}{K_b \text{ for } CH_3NH_2} = \frac{1.0 \times 10^{-14}}{3.7 \times 10^{-4}} = 2.7 \times 10^{-11}$$

$$CH_3NH_3^+(aq) + H_2O(l) \rightleftharpoons H_3O^+(aq) + CH_3NH_2(aq)$$

	$CH_3NH_3^+$	H_3O^+	CH_3NH_2
initial (M)	0.050	~0	0
change (M)	–x	+x	+x
equil (M)	0.050 – x	x	x

$$K_a = \frac{[H_3O^+][CH_3NH_2]}{[CH_3NH_3^+]} = 2.7 \times 10^{-11} = \frac{(x)(x)}{0.050 - x} \approx \frac{x^2}{0.050}$$

Solve for x. $x = [H_3O^+] = 1.2 \times 10^{-6}$ M; $pH = -\log[H_3O^+] = -\log(1.2 \times 10^{-6}) = 5.92$
Chlorphenol red would be a suitable indicator.

15.89   When equal volumes of acid and base react, all concentrations are cut in half.
(a) At the equivalence point only the salt $C_5H_{11}NHNO_3$ is in solution.
$[C_5H_{11}NH^+] = 0.10$ M

For $C_5H_{11}NH^+$, $K_a = \dfrac{K_w}{K_b \text{ for } C_5H_{11}N} = \dfrac{1.0 \times 10^{-14}}{1.3 \times 10^{-3}} = 7.7 \times 10^{-12}$

$$C_5H_{11}NH^+(aq) + H_2O(l) \rightleftharpoons H_3O^+(aq) + C_5H_{11}N(aq)$$

initial (M)	0.10	~0	0
change (M)	−x	+x	+x
equil (M)	0.10 − x	x	x

$K_a = \dfrac{[H_3O^+][C_5H_{11}N]}{[C_5H_{11}NH^+]} = 7.7 \times 10^{-12} = \dfrac{(x)(x)}{0.10-x} \approx \dfrac{x^2}{0.10}$

Solve for x. $x = [H_3O^+] = 8.8 \times 10^{-7}$ M

$pH = -\log[H_3O^+] = -\log(8.8 \times 10^{-7}) = 6.06$

Alizarin would be a suitable indicator (see Figure 15.4).

(b) At the equivalence point only the salt Na_2SO_3 is in solution.

$[SO_3^{2-}] = 0.10$ M

For SO_3^{2-}, $K_b = \dfrac{K_w}{K_a \text{ for } HSO_3^-} = \dfrac{1.0 \times 10^{-14}}{6.3 \times 10^{-8}} = 1.6 \times 10^{-7}$

$$SO_3^{2-}(aq) + H_2O(l) \rightleftharpoons HSO_3^-(aq) + OH^-(aq)$$

Initial (M)	0.10	0	~0
change (M)	−x	+x	+x
equil (M)	0.10 − x	x	x

$K_b = \dfrac{[HSO_3^-][OH^-]}{[SO_3^{2-}]} = 1.6 \times 10^{-7} = \dfrac{(x)(x)}{0.10-x} \approx \dfrac{x^2}{0.10}$

Solve for x. $x = [OH^-] = 1.26 \times 10^{-4}$ M

$[H_3O^+] = \dfrac{K_w}{[OH^-]} = \dfrac{1.0 \times 10^{-14}}{1.26 \times 10^{-4}} = 7.9 \times 10^{-11}$ M

$pH = -\log[H_3O^+] = -\log(7.9 \times 10^{-11}) = 10.10$

Thymolphthalein would be a suitable indicator.

(c) The pH is 7.00 at the equivalence point for the titration of a strong acid (HBr) with a strong base ($Ba(OH)_2$).

Alizarin, bromthymol blue, or phenol red would be suitable indicators. (Any indicator that changes color in the pH range 4–10 is satisfactory for a strong acid–strong base titration.)

Solubility Equilibria (Sections 15.10–15.11)

15.90 (a) $Ag_2CO_3(s) \rightleftharpoons 2 Ag^+(aq) + CO_3^{2-}(aq)$ $K_{sp} = [Ag^+]^2[CO_3^{2-}]$

(b) $PbCrO_4(s) \rightleftharpoons Pb^{2+}(aq) + CrO_4^{2-}(aq)$ $K_{sp} = [Pb^{2+}][CrO_4^{2-}]$

(c) $Al(OH)_3(s) \rightleftharpoons Al^{3+}(aq) + 3 OH^-(aq)$ $K_{sp} = [Al^{3+}][OH^-]^3$

(d) $Hg_2Cl_2(s) \rightleftharpoons Hg_2^{2+}(aq) + 2 Cl^-(aq)$ $K_{sp} = [Hg_2^{2+}][Cl^-]^2$

15.91 (a) $K_{sp} = [Ca^{2+}][OH^-]^2$ (b) $K_{sp} = [Ag^+]^3[PO_4^{3-}]$
       (c) $K_{sp} = [Ba^{2+}][CO_3^{2-}]$ (d) $K_{sp} = [Ca^{2+}]^5[PO_4^{3-}]^3[OH^-]$

15.92 (a) $K_{sp} = [Pb^{2+}][I^-]^2 = (5.0 \times 10^{-3})(1.3 \times 10^{-3})^2 = 8.4 \times 10^{-9}$

       (b) $[I^-] = \sqrt{\dfrac{K_{sp}}{[Pb^{2+}]}} = \sqrt{\dfrac{(8.4 \times 10^{-9})}{(2.5 \times 10^{-4})}} = 5.8 \times 10^{-3}$ M

       (c) $[Pb^{2+}] = \dfrac{K_{sp}}{[I^-]^2} = \dfrac{(8.4 \times 10^{-9})}{(2.5 \times 10^{-4})^2} = 0.13$ M

15.93 (a) $K_{sp} = [Ca^{2+}]^3[PO_4^{2-}]^2 = (2.9 \times 10^{-7})^3(2.9 \times 10^{-7})^2 = 2.1 \times 10^{-33}$

       (b) $[Ca^{2+}] = \sqrt[3]{\dfrac{K_{sp}}{[PO_4^{2-}]^2}} = \sqrt[3]{\dfrac{2.1 \times 10^{-33}}{(0.010)^2}} = 2.8 \times 10^{-10}$ M

       (c) $[PO_4^{2-}] = \sqrt{\dfrac{K_{sp}}{[Ca^{2+}]^3}} = \sqrt{\dfrac{2.1 \times 10^{-33}}{(0.010)^3}} = 4.6 \times 10^{-14}$ M

15.94 $Ag_2CO_3(s) \rightleftharpoons 2\,Ag^+(aq) + CO_3^{2-}(aq)$
       equil (M)   2x   x
       $[Ag^+] = 2x = 2.56 \times 10^{-4}$ M; $[CO_3^{2-}] = x = (2.56 \times 10^{-4} \text{ M})/2 = 1.28 \times 10^{-4}$ M
       $K_{sp} = [Ag^+]^2[CO_3^{2-}] = (2.56 \times 10^{-4})^2(1.28 \times 10^{-4}) = 8.39 \times 10^{-12}$

15.95 (a) $[Cd^{2+}] = [CO_3^{2-}] = 1.0 \times 10^{-6}$ M
       $K_{sp} = [Cd^{2+}][CO_3^{2-}] = (1.0 \times 10^{-6})^2 = 1.0 \times 10^{-12}$
       (b) $[Ca^{2+}] = 1.06 \times 10^{-2}$ M
       $[OH^-] = 2[Ca^{2+}] = 2(1.06 \times 10^{-2} \text{ M}) = 2.12 \times 10^{-2}$ M
       $K_{sp} = [Ca^{2+}][OH^-]^2 = (1.06 \times 10^{-2})(2.12 \times 10^{-2})^2 = 4.76 \times 10^{-6}$
       (c) $PbBr_2$, 367.01 amu

       $[Pb^{2+}] = \text{molarity of } PbBr_2 = \dfrac{\left(4.34 \text{ g} \times \dfrac{1 \text{ mol}}{367.01 \text{ g}}\right)}{1 \text{ L}} = 1.18 \times 10^{-2}$ M

       $[Br^-] = 2[Pb^{2+}] = 2(1.18 \times 10^{-2} \text{ M}) = 2.36 \times 10^{-2}$ M
       $K_{sp} = [Pb^{2+}][Br^-]^2 = (1.18 \times 10^{-2})(2.36 \times 10^{-2})^2 = 6.57 \times 10^{-6}$
       (d) $BaCrO_4$, 253.32 amu

       $[Ba^{2+}] = [CrO_4^{2-}] = \text{molarity of } BaCrO_4 = \dfrac{\left(2.8 \times 10^{-3} \text{ g} \times \dfrac{1 \text{ mol}}{253.32 \text{ g}}\right)}{1 \text{ L}} = 1.1 \times 10^{-5}$ M

       $K_{sp} = [Ba^{2+}][CrO_4^{2-}] = (1.1 \times 10^{-5})^2 = 1.2 \times 10^{-10}$

15.96 (a) $$BaCrO_4(s) \rightleftharpoons Ba^{2+}(aq) + CrO_4^{2-}(aq)$$
equil (M)   x   x

$$K_{sp} = [Ba^{2+}][CrO_4^{2-}] = 1.2 \times 10^{-10} = (x)(x)$$

molar solubility $= x = \sqrt{1.2 \times 10^{-10}} = 1.1 \times 10^{-5}$ M

(b) $$Mg(OH)_2(s) \rightleftharpoons Mg^{2+}(aq) + 2\,OH^-(aq)$$
equil (M)   x   2x

$$K_{sp} = [Mg^{2+}][OH^-]^2 = 5.6 \times 10^{-12} = x(2x)^2 = 4x^3$$

molar solubility $= x = \sqrt[3]{\dfrac{5.6 \times 10^{-12}}{4}} = 1.1 \times 10^{-4}$ M

(c) $$Ag_2SO_3(s) \rightleftharpoons 2\,Ag^+(aq) + SO_3^{2-}(aq)$$
equil (M)   2x   x

$$K_{sp} = [Ag^+]^2[SO_3^{2-}] = 1.5 \times 10^{-14} = (2x)^2 x = 4x^3$$

molar solubility $= x = \sqrt[3]{\dfrac{1.5 \times 10^{-14}}{4}} = 1.6 \times 10^{-5}$ M

15.97 (a) $$Ag_2CO_3(s) \rightleftharpoons 2\,Ag^+(aq) + CO_3^{2-}(aq)$$
equil (M)   2x   x

$$K_{sp} = [Ag^+]^2[CO_3^{2-}] = 8.4 \times 10^{-12} = (2x)^2(x) = 4x^3$$

molar solubility $= x = \sqrt[3]{\dfrac{8.4 \times 10^{-12}}{4}} = 1.3 \times 10^{-4}$ M

Ag_2CO_3, 275.75 amu

solubility $= (1.3 \times 10^{-4}$ mol/L$)(275.75$ g/mol$) = 0.036$ g/L

(b) $$CuBr(s) \rightleftharpoons Cu^+(aq) + Br^-(aq)$$
equil (M)   x   x

$$K_{sp} = [Cu^+][Br^-] = 6.3 \times 10^{-9} = (x)(x)$$

molar solubility $= x = \sqrt{6.3 \times 10^{-9}} = 7.9 \times 10^{-5}$ M

CuBr, 143.45 amu

solubility $= (7.9 \times 10^{-5}$ mol/L$)(143.45$ g/mol$) = 0.011$ g/L

(c) $$Cu_3(PO_4)_2(s) \rightleftharpoons 3\,Cu^{2+}(aq) + 2\,PO_4^{3-}(aq)$$
equil (M)   3x   2x

$$K_{sp} = [Cu^{2+}]^3[PO_4^{3-}]^2 = 1.4 \times 10^{-37} = (3x)^3(2x)^2 = 108x^5$$

molar solubility $= x = \sqrt[5]{\dfrac{1.4 \times 10^{-37}}{108}} = 1.7 \times 10^{-8}$

$Cu_3(PO_4)_2$, 380.58 amu

solubility $= (1.7 \times 10^{-8}$ mol/L$)(380.58$ g/mol$) = 6.5 \times 10^{-6}$ g/L

Factors That Affect Solubility (Section 15.12)

15.98 $Ag_2CO_3(s) \rightleftharpoons 2\,Ag^+(aq) + CO_3^{2-}(aq)$
(a) $AgNO_3$, source of Ag^+; equilibrium shifts left
(b) HNO_3, source of H_3O^+, removes CO_3^{2-}; equilibrium shifts right
(c) Na_2CO_3, source of CO_3^{2-}; equilibrium shifts left
(d) NH_3, forms $Ag(NH_3)_2^+$, removes Ag^+; equilibrium shifts right

15.99 $BaF_2(s) \rightleftharpoons Ba^{2+}(aq) + 2\,F^-(aq)$
(a) H^+ from HCl reacts with F^- forming the weak acid HF. The equilibrium shifts to the right increasing the solubility of BaF_2.
(b) KF, source of F^-; equilibrium shifts left, solubility of BaF_2 decreases.
(c) No change in solubility.
(d) $Ba(NO_3)_2$, source of Ba^{2+}; equilibrium shifts left, solubility of BaF_2 decreases.

15.100 (a)
$$PbCrO_4(s) \rightleftharpoons Pb^{2+}(aq) + CrO_4^{2-}(aq)$$
equil (M) $\qquad\qquad$ x $\qquad$ x
$K_{sp} = [Pb^{2+}][CrO_4^{2-}] = 2.8 \times 10^{-13} = (x)(x)$
molar solubility $= x = \sqrt{2.8 \times 10^{-13}} = 5.3 \times 10^{-7}$ M
(b)
$$PbCrO_4(s) \rightleftharpoons Pb^{2+}(aq) + CrO_4^{2-}(aq)$$
initial(M) $\qquad\qquad$ 0 $\qquad$ 1.0×10^{-3}
equil (M) $\qquad\qquad$ x $\qquad$ $1.0 \times 10^{-3} + x$
$K_{sp} = [Pb^{2+}][CrO_4^{2-}] = 2.8 \times 10^{-13} = (x)(1.0 \times 10^{-3} + x) \approx (x)(1.0 \times 10^{-3})$
molar solubility $= x = \dfrac{2.8 \times 10^{-13}}{1 \times 10^{-3}} = 2.8 \times 10^{-10}$ M

15.101 (a)
$$SrF_2(s) \rightleftharpoons Sr^{2+}(aq) + 2\,F^-(aq)$$
initial (M) $\qquad\qquad$ 0.010 $\qquad$ 0
equil (M) $\qquad\qquad$ $0.010 + x$ $\quad$ 2x
$K_{sp} = [Sr^{2+}][F^-]^2 = 4.3 \times 10^{-9} = (0.010 + x)(2x)^2 \approx (0.010)(2x)^2 = 0.040\,x^2$
molar solubility $= x = \sqrt{\dfrac{4.3 \times 10^{-9}}{0.040}} = 3.3 \times 10^{-4}$ M
(b)
$$SrF_2(s) \rightleftharpoons Sr^{2+}(aq) + 2\,F^-(aq)$$
initial (M) $\qquad\qquad$ 0 $\qquad$ 0.010
equil (M) $\qquad\qquad$ x $\qquad$ $0.010 + 2x$
$K_{sp} = [Sr^{2+}][F^-]^2 = 4.3 \times 10^{-9} = (x)(0.010 + 2x)^2 \approx (x)(0.010)^2 = x(0.00010)$
molar solubility $= x = \dfrac{4.3 \times 10^{-9}}{0.00010} = 4.3 \times 10^{-5}$ M

15.102 (b), (c), and (d) are more soluble in acidic solution.
(a) $AgBr(s) \rightleftharpoons Ag^+(aq) + Br^-(aq)$
(b) $CaCO_3(s) + H_3O^+(aq) \rightleftharpoons Ca^{2+}(aq) + HCO_3^-(aq) + H_2O(l)$
(c) $Ni(OH)_2(s) + 2\,H_3O^+(aq) \rightleftharpoons Ni^{2+}(aq) + 4\,H_2O(l)$
(d) $Ca_3(PO_4)_2(s) + 2\,H_3O^+(aq) \rightleftharpoons 3\,Ca^{2+}(aq) + 2\,HPO_4^{2-}(aq) + 2\,H_2O(l)$

15.103 (a), (b), and (d) are more soluble in acidic solution.

(a) $MnS(s) + 2 H_3O^+(aq) \rightleftharpoons Mn^{2+}(aq) + H_2S(aq) + 2 H_2O(l)$

(b) $Fe(OH)_3(s) + 3 H_3O^+(aq) \rightleftharpoons Fe^{3+}(aq) + 6 H_2O(l)$

(c) $AgCl(s) \rightleftharpoons Ag^+(aq) + Cl^-(aq)$

(d) $BaCO_3(s) + H_3O^+(aq) \rightleftharpoons Ba^{2+}(aq) + HCO_3^-(aq) + H_2O(l)$

15.104 On mixing equal volumes of two solutions, the concentrations of both solutions are cut in half.

	$Ag^+(aq)$	+	$2 CN^-(aq)$	$\rightleftharpoons$	$Ag(CN)_2^-(aq)$
before reaction (M)	0.0010		0.10		0
assume 100% reaction	−0.0010		−2(0.0010)		0.0010
after reaction (M)	0		0.098		0.0010
assume small back rxn	+x		+2x		−x
equil (M)	x		0.098 + 2x		0.0010 − x

$$K_f = 3.0 \times 10^{20} = \frac{[Ag(CN)_2^-]}{[Ag^+][CN^-]^2} = \frac{(0.0010 - x)}{x(0.098 + 2x)^2} \approx \frac{0.0010}{x(0.098)^2}$$

Solve for x. $x = [Ag^+] = 3.5 \times 10^{-22}$ M

15.105

	$Cr^{3+}(aq)$	+	$4 OH^-(aq)$	$\rightleftharpoons$	$Cr(OH)_4^-(aq)$
before reaction (M)	0.0050		1.0		0
assume 100% reaction	−0.0050		−(4)(0.0050)		+0.0050
after reaction(M)	0		0.98		0.0050
assume small back rxn	+x		+4x		−x
equil (M)	x		0.98 + 4x		0.0050 − x

$$K_f = \frac{[Cr(OH)_4^-]}{[Cr^{3+}][OH^-]^4} = 8 \times 10^{29} = \frac{(0.0050 - x)}{(x)(0.98 + 4x)^4} \approx \frac{(0.0050)}{(x)(0.98)^4}$$

Solve for x. $x = [Cr^{3+}] = 6.8 \times 10^{-33}$ M $= 7 \times 10^{-33}$ M

$$\text{fraction uncomplexed } Cr^{3+} = \frac{[Cr^{3+}]}{[Cr(OH)_4^-]} = \frac{7 \times 10^{-33} \text{ M}}{0.0050 \text{ M}} = 1.4 \times 10^{-30} = 1 \times 10^{-30}$$

15.106 (a)

$AgI(s) \rightleftharpoons Ag^+(aq) + I^-(aq)$ $K_{sp} = 8.5 \times 10^{-17}$

$\underline{Ag^+(aq) + 2 CN^-(aq) \rightarrow Ag(CN)_2^-(aq)}$ $K_f = 3.0 \times 10^{20}$

dissolution rxn $AgI(s) + 2 CN^-(aq) \rightleftharpoons Ag(CN)_2^-(aq) + I^-(aq)$

$K = (K_{sp})(K_f) = (8.5 \times 10^{-17})(3.0 \times 10^{20}) = 2.6 \times 10^4$

(b)

$Al(OH)_3(s) \rightleftharpoons Al^{3+}(aq) + 3 OH^-(aq)$ $K_{sp} = 1.9 \times 10^{-33}$

$\underline{Al^{3+}(aq) + 4 OH^-(aq) \rightarrow Al(OH)_4^-(aq)}$ $K_f = 3 \times 10^{33}$

dissolution rxn $Al(OH)_3(s) + OH^-(aq) \rightleftharpoons Al(OH)_4^-(aq)$

$K = (K_{sp})(K_f) = (1.9 \times 10^{-33})(3 \times 10^{33}) = 6$

(c)

$Zn(OH)_2(s) \rightleftharpoons Zn^{2+}(aq) + 2 OH^-(aq)$ $K_{sp} = 4.1 \times 10^{-17}$

$\underline{Zn^{2+}(aq) + 4 NH_3(aq) \rightarrow Zn(NH_3)_4^{2+}(aq)}$ $K_f = 7.8 \times 10^8$

dissolution rxn $Zn(OH)_2(s) + 4 NH_3(aq) \rightleftharpoons Zn(NH_3)_4^{2+} + 2 OH^-(aq)$

$K = (K_{sp})(K_f) = (4.1 \times 10^{-17})(7.8 \times 10^8) = 3.2 \times 10^{-8}$

15.107 (a) $Zn(OH)_2(s) \rightleftharpoons Zn^{2+}(aq) + 2\,OH^-(aq)$ $K_{sp} = 4.1 \times 10^{-17}$

$\underline{Zn^{2+}(aq) + 4\,OH^-(aq) \rightarrow Zn(OH)_4^{2-}(aq)}$ $K_f = 3 \times 10^{15}$

dissolution rxn $Zn(OH)_2(s) + 2\,OH^-(aq) \rightleftharpoons Zn(OH)_4^{2-}(aq)$

$K = (K_{sp})(K_f) = (4.1 \times 10^{-17})(3 \times 10^{15}) = 0.1$

(b) $Cu(OH)_2(s) \rightleftharpoons Cu^{2+}(aq) + 2\,OH^-(aq)$ $K_{sp} = 1.6 \times 10^{-19}$

$\underline{Cu^{2+}(aq) + 4\,NH_3(aq) \rightarrow Cu(NH_3)_4^{2+}(aq)}$ $K_f = 5.6 \times 10^{11}$

dissolution rxn $Cu(OH)_2(s) + 4\,NH_3(aq) \rightleftharpoons Cu(NH_3)_4^{2+} + 2\,OH^-(aq)$

$K = (K_{sp})(K_f) = (1.6 \times 10^{-19})(5.6 \times 10^{11}) = 9.0 \times 10^{-8}$

(c) $AgBr(s) \rightleftharpoons Ag^+(aq) + Br^-(aq)$ $K_{sp} = 5.4 \times 10^{-13}$

$\underline{Ag^+(aq) + 2\,NH_3(aq) \rightarrow Ag(NH_3)_2^+(aq)}$ $K_f = 1.7 \times 10^7$

dissolution rxn $AgBr(s) + 2\,NH_3(aq) \rightleftharpoons Ag(NH_3)_2^+(aq) + Br^-(aq)$

$K = (K_{sp})(K_f) = (5.4 \times 10^{-13})(1.7 \times 10^7) = 9.2 \times 10^{-6}$

15.108 (a) $AgI(s) \rightleftharpoons Ag^+(aq) + I^-(aq)$

equil (M)   x   x

$K_{sp} = [Ag^+][I^-] = 8.5 \times 10^{-17} = (x)(x)$

molar solubility $= x = \sqrt{8.5 \times 10^{-17}} = 9.2 \times 10^{-9}$ M

(b) $AgI(s) + 2\,CN^-(aq) \rightleftharpoons Ag(CN)_2^-(aq) + I^-(aq)$

initial (M)	0.10	0	0
change (M)	$-2x$	$+x$	$+x$
equil (M)	$0.10 - 2x$	x	x

$K = (K_{sp})(K_f) = (8.5 \times 10^{-17})(3.0 \times 10^{20}) = 2.6 \times 10^4$

$K = 2.6 \times 10^4 = \dfrac{[Ag(CN)_2^-][I^-]}{[CN^-]^2} = \dfrac{x^2}{(0.10 - 2x)^2}$

Take the square root of both sides and solve for x.

molar solubility $= x = 0.050$ M

15.109 $Cr(OH)_3(s) + OH^-(aq) \rightleftharpoons Cr(OH)_4^-(aq)$

initial (M)	0.50	0
change (M)	$-x$	$+x$
equil (M)	$0.50 - x$	x

$K = (K_{sp})(K_f) = (6.7 \times 10^{-31})(8 \times 10^{29}) = 0.54$

$K = 0.54 = \dfrac{[Cr(OH)_4^-]}{[OH^-]} = \dfrac{x}{0.50 - x}$

$0.27 - 0.54x = x$

$0.27 = 1.54x$

molar solubility $= x = \dfrac{0.27}{1.54} = 0.2$ M

Precipitation; Qualitative Analysis (Sections 15.13–15.15)

15.110 For $BaSO_4$, $K_{sp} = 1.1 \times 10^{-10}$

Total volume = 300 mL + 100 mL = 400 mL

$$[Ba^{2+}] = \frac{(4.0 \times 10^{-3} \text{ M})(100 \text{ mL})}{(400 \text{ mL})} = 1.0 \times 10^{-3} \text{ M}$$

$$[SO_4^{2-}] = \frac{(6.0 \times 10^{-4} \text{ M})(300 \text{ mL})}{(400 \text{ mL})} = 4.5 \times 10^{-4} \text{ M}$$

IP = $[Ba^{2+}]_t[SO_4^{2-}]_t = (1.0 \times 10^{-3})(4.5 \times 10^{-4}) = 4.5 \times 10^{-7}$

IP > K_{sp}; $BaSO_4$(s) will precipitate.

15.111 On mixing equal volumes of two solutions, the concentrations of both solutions are cut in half.
For $PbCl_2$, $K_{sp} = 1.2 \times 10^{-5} = [Pb^{2+}][Cl^-]^2$
IP = $(0.0050)(0.0050)^2 = 1.2 \times 10^{-7}$
IP < K_{sp}; no precipitate will form.

$$[Cl^-] = \sqrt{\frac{K_{sp}}{[Pb^{2+}]}} = \sqrt{\frac{1.2 \times 10^{-5}}{5.0 \times 10^{-3}}} = 0.049 \text{ M}$$

A $[Cl^-]$ just greater than 0.049 M will result in precipitation.

15.112 $BaSO_4$, $K_{sp} = 1.1 \times 10^{-10}$; $Fe(OH)_3$, $K_{sp} = 2.6 \times 10^{-39}$
Total volume = 80 mL + 20 mL = 100 mL

$$[Ba^{2+}] = \frac{(1.0 \times 10^{-5} \text{ M})(80 \text{ mL})}{(100 \text{ mL})} = 8.0 \times 10^{-6} \text{ M}$$

$$[OH^-] = 2[Ba^{2+}] = 2(8.0 \times 10^{-6}) = 1.6 \times 10^{-5} \text{ M}$$

$$[Fe^{3+}] = \frac{2(1.0 \times 10^{-5} \text{ M})(20 \text{ mL})}{(100 \text{ mL})} = 4.0 \times 10^{-6} \text{ M}$$

$$[SO_4^{2-}] = \frac{3(1.0 \times 10^{-5} \text{ M})(20 \text{ mL})}{(100 \text{ mL})} = 6.0 \times 10^{-6} \text{ M}$$

For $BaSO_4$, IP = $[Ba^{2+}]_t[SO_4^{2-}]_t = (8.0 \times 10^{-6})(6.0 \times 10^{-6}) = 4.8 \times 10^{-11}$
IP < K_{sp}; $BaSO_4$ will not precipitate.
For $Fe(OH)_3$, IP = $[Fe^{3+}]_t[OH^-]_t^3 = (4.0 \times 10^{-6})(1.6 \times 10^{-5})^3 = 1.6 \times 10^{-20}$
IP > K_{sp}; $Fe(OH)_3$(s) will precipitate.

15.113 (a) $$[CO_3^{2-}] = \frac{(2.0 \times 10^{-3} \text{ M})(0.10 \text{ mL})}{(250 \text{ mL})} = 8.0 \times 10^{-7} \text{ M}$$

$K_{sp} = 5.0 \times 10^{-9} = [Ca^{2+}][CO_3^{2-}]$
IP = $[Ca^{2+}][CO_3^{2-}] = (8.0 \times 10^{-4})(8.0 \times 10^{-7}) = 6.4 \times 10^{-10}$
IP < K_{sp}; no precipitate will form.

(b) Na_2CO_3, 106 amu; 10 mg = 0.010 g

$$[CO_3^{2-}] = \frac{\left(0.010 \text{ g} \times \dfrac{1 \text{ mol}}{106 \text{ g}}\right)}{0.250 \text{ L}} = 3.8 \times 10^{-4} \text{ M}$$

IP = $[Ca^{2+}][CO_3^{2-}] = (8.0 \times 10^{-4})(3.8 \times 10^{-4}) = 3.0 \times 10^{-7}$

IP > K_{sp}; $CaCO_3(s)$ will precipitate.

15.114 pH = 10.80; $[H_3O^+] = 10^{-pH} = 10^{-10.80} = 1.6 \times 10^{-11}$ M

$$[OH^-] = \frac{K_w}{[H_3O^+]} = \frac{1.0 \times 10^{-14}}{1.6 \times 10^{-11}} = 6.2 \times 10^{-4} \text{ M}$$

For $Mg(OH)_2$, $K_{sp} = 5.6 \times 10^{-12}$
IP = $[Mg^{2+}]_t[OH^-]_t^2 = (2.5 \times 10^{-4})(6.2 \times 10^{-4})^2 = 9.6 \times 10^{-11}$
IP > K_{sp}; $Mg(OH)_2(s)$ will precipitate

15.115 $Mg(OH)_2$, $K_{sp} = 5.6 \times 10^{-12}$; $Al(OH)_3$, $K_{sp} = 1.9 \times 10^{-33}$
pH = 8; $[H_3O^+] = 10^{-pH} = 10^{-8} = 1 \times 10^{-8}$ M

$$[OH^-] = \frac{K_w}{[H_3O^+]} = \frac{1.0 \times 10^{-14}}{1 \times 10^{-8}} = 1 \times 10^{-6} \text{ M}$$

For $Mg(OH)_2$, IP = $[Mg^{2+}][OH^-]^2 = (0.01)(1 \times 10^{-6})^2 = 1 \times 10^{-14}$
IP < K_{sp}; $Mg(OH)_2$ will not precipitate.
For $Al(OH)_3$, IP = $[Al^{3+}][OH^-]^3 = (0.01)(1 \times 10^{-6})^3 = 1 \times 10^{-20}$
IP > K_{sp}; $Al(OH)_3$ will precipitate.

15.116 $K_{spa} = \dfrac{[M^{2+}][H_2S]}{[H_3O^+]^2}$; FeS, $K_{spa} = 6 \times 10^2$; SnS, $K_{spa} = 1 \times 10^{-5}$

Fe^{2+} and Sn^{2+} can be separated by bubbling H_2S through an acidic solution containing the two cations because their K_{spa} values are so different.
For FeS and SnS, $Q_c = \dfrac{(0.01)(0.10)}{(0.3)^2} = 1.1 \times 10^{-2}$

For FeS, $Q_c < K_{spa}$, and FeS will not precipitate.
For SnS, $Q_c > K_{spa}$, and SnS will precipitate.

15.117 CoS, $K_{spa} = \dfrac{[Co^{2+}][H_2S]}{[H_3O^+]^2} = 3$

(i) In 0.5 M HCl, $[H_3O^+] = 0.5$ M

$$Q_c = \frac{[Co^{2+}]_t[H_2S]_t}{[H_3O^+]_t^2} = \frac{(0.10)(0.10)}{(0.5)^2} = 0.04; \quad Q_c < K_{spa}; \text{ CoS will not precipitate}$$

(ii) pH = 8; $[H_3O^+] = 10^{-pH} = 10^{-8} = 1 \times 10^{-8}$ M

$$Q_c = \frac{[Co^{2+}]_t[H_2S]_t}{[H_3O^+]_t^2} = \frac{(0.10)(0.10)}{(1 \times 10^{-8})^2} = 1 \times 10^{14}; \quad Q_c > K_{spa}; \text{ CoS(s) will precipitate}$$

15.118 (a) add Cl^- to precipitate $AgCl$

(b) add CO_3^{2-} to precipitate $CaCO_3$

(c) add H_2S to precipitate MnS

(d) add NH_3 and NH_4Cl to precipitate $Cr(OH)_3$

(Need buffer to control $[OH^-]$; excess OH^- produces the soluble $Cr(OH)_4^-$.)

15.119 (a) add Cl^- to precipitate Hg_2Cl_2

(b) add $(NH_4)_2HPO_4$ to precipitate $MgNH_4PO_4$

(c) add HCl and H_2S to precipitate HgS

(d) add Cl^- to precipitate $PbCl_2$

Chapter Problems

15.120 Prepare aqueous solutions of the three salts. Add a solution of $(NH_4)_2HPO_4$. If a white precipitate forms, the solution contains Mg^{2+}. Perform flame test on the other two solutions. A yellow flame test indicates Na^+. A violet flame test indicates K^+.

15.121 (a), solution contains HCN and CN^-

(c), solution can contain HCN and CN^-

(e), solution can contain HCN and CN^-

15.122 (a), solution contains H_2CO_3 and HCO_3^-

(b), solution contains HCO_3^- and CO_3^{2-}

(d), solution contains HCO_3^- and CO_3^{2-}

15.123

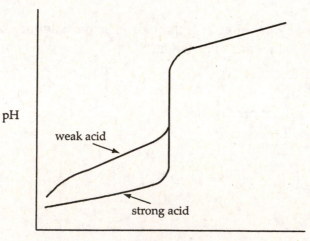

(a) The pH for the weak acid is higher.

(b) Initially, the pH rises more quickly for the weak acid, but then the curve becomes more level in the region halfway to the equivalence point.

(c) The pH is higher at the equivalence point for the weak acid.

(d) Both curves are identical beyond the equivalence point because the pH is determined by the excess $[OH^-]$.

(e) If the acid concentrations are the same, the volume of base needed to reach the equivalence point is the same.

15.124 (a)

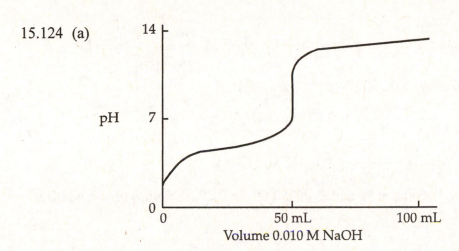

Volume 0.010 M NaOH

(b) mol NaOH required $= \left(\dfrac{0.010 \text{ mol HA}}{L} \right)(0.0500 \text{ L})\left(\dfrac{1 \text{ mol NaOH}}{1 \text{ mol HA}} \right) = 0.000\ 50 \text{ mol}$

vol NaOH required $= (0.000\ 50 \text{ mol})\left(\dfrac{1 \text{ L}}{0.010 \text{ mol}} \right) = 0.050 \text{ L} = 50 \text{ mL}$

(c) A basic salt is present at the equivalence point; pH > 7.00

(d) Halfway to the equivalence point, the pH $= pK_a = 4.00$

15.125 (a) $AgBr(s) \rightleftharpoons Ag^+(aq) + Br^-(aq)$

(i) HBr is a source of Br^- (reaction product). The solubility of AgBr is decreased.

(ii) unaffected

(iii) $AgNO_3$ is a source of Ag^+ (reaction product). The solubility of AgBr is decreased.

(iv) NH_3 forms a complex with Ag^+, removing it from solution. The solubility of AgBr is increased.

(b) $BaCO_3(s) \rightleftharpoons Ba^{2+}(aq) + CO_3^{2-}(aq)$

(i) HNO_3 reacts with CO_3^{2-}, removing it from the solution. The solubility of $BaCO_3$ is increased.

(ii) $Ba(NO_3)_2$ is a source of Ba^{2+} (reaction product). The solubility of $BaCO_3$ is decreased.

(iii) Na_2CO_3 is a source of CO_3^{2-} (reaction product). The solubility of $BaCO_3$ is decreased.

(iv) CH_3CO_2H reacts with CO_3^{2-}, removing it from the solution. The solubility of $BaCO_3$ is increased.

15.126 For NH_4^+, $K_a = \dfrac{K_w}{K_b \text{ for } NH_3} = \dfrac{1.0 \times 10^{-14}}{1.8 \times 10^{-5}} = 5.6 \times 10^{-10}$

$pK_a = -\log K_a = -\log(5.6 \times 10^{-10}) = 9.25$

$$pH = pK_a + \log \frac{[NH_3]}{[NH_4^+]}; \qquad 9.40 = 9.25 + \log \frac{[NH_3]}{[NH_4^+]}$$

$$\log \frac{[NH_3]}{[NH_4^+]} = 9.40 - 9.25 = 0.15; \qquad \frac{[NH_3]}{[NH_4^+]} = 10^{0.15} = 1.41$$

Because the volume is the same for both NH_3 and NH_4^+, $\dfrac{\text{mol } NH_3}{\text{mol } NH_4^+} = 1.41$.

mol NH_3 = (0.20 mol/L)(0.250 L) = 0.050 mol NH_3

$$\text{mol } NH_4^+ = \frac{\text{mol } NH_3}{1.41} = \frac{0.050}{1.41} = 0.035 \text{ mol } NH_4^+$$

$$\text{vol } NH_4^+ = (0.035 \text{ mol})\left(\frac{1 \text{ L}}{3.0 \text{ mol}}\right) = 0.012 \text{ L} = 12 \text{ mL}$$

12 mL of 3.0 M NH_4Cl must be added to 250 mL of 0.20 M NH_3 to obtain a buffer solution having a pH = 9.40.

15.127 $H_2PO_4^-(aq) + H_2O(l) \rightleftharpoons H_3O^+(aq) + HPO_4^{2-}(aq)$
   (a) Na_2HPO_4, source of HPO_4^{2-}, equilibrium shifts left, pH increases.
   (b) Addition of the strong acid, HBr, decreases the pH.
   (c) Addition of the strong base, KOH, increases the pH.
   (d) There is no change in the pH with the addition of the neutral salt KI.
   (e) H_3PO_4, source of $H_2PO_4^-$, equilibrium shifts right, pH decreases.
   (f) Na_3PO_4, source of PO_4^{3-}, decreases $[H_3O^+]$ by forming HPO_4^{2-}, pH increases.

15.128 pH = 10.35; $[H_3O^+] = 10^{-pH} = 10^{-10.35} = 4.5 \times 10^{-11}$ M

$$[OH^-] = \frac{K_w}{[H_3O^+]} = \frac{1.0 \times 10^{-14}}{4.5 \times 10^{-11}} = 2.2 \times 10^{-4} \text{ M}$$

$$[Mg^{2+}] = \frac{[OH^-]}{2} = \frac{2.2 \times 10^{-4}}{2} = 1.1 \times 10^{-4} \text{ M}$$

$$K_{sp} = [Mg^{2+}][OH^-]^2 = (1.1 \times 10^{-4})(2.2 \times 10^{-4})^2 = 5.3 \times 10^{-12}$$

15.129 mmol Hg_2^{2+} = (0.010 mmol/mL)(1.0 mL) = 0.010 mmol
   mmol Cl^- = (6 mmol/mL)(0.05 mL) = 0.3 mmol
   Assume complete reaction.

	$Hg_2^{2+}(aq)$ +	2 $Cl^-(aq)$ $\rightarrow$ $Hg_2Cl_2(s)$
before reaction (mmol)	0.010	0.3
change (mmol)	−0.010	−2(0.010)
after reaction (mmol)	0	0.28

$$[Cl^-] = \frac{0.28 \text{ mmol}}{1.05 \text{ mL}} = 0.27 \text{ M}; \quad \text{Allow } Hg_2Cl_2 \text{ to establish a new equilibrium.}$$

	$Hg_2Cl_2(s)$ $\rightleftharpoons$	$Hg_2^{2+}(aq)$ +	2 $Cl^-(aq)$
initial (M)		0	0.27
equil (M)		x	0.27 + 2x

$$K_{sp} = [Hg_2{}^{2+}][Cl^-]^2 = 1.4 \times 10^{-18} = x(0.27 + 2x)^2 \approx x(0.27)^2$$

$$x = [Hg_2{}^{2+}] = \frac{1.4 \times 10^{-18}}{(0.27)^2} = 2 \times 10^{-17} \text{ mol/L}$$

$Hg_2{}^{2+}$, 401.18 amu

$Hg_2{}^{2+}$ concentration = $(2 \times 10^{-17} \text{ mol/L})(401.18 \text{ g/mol}) = 8 \times 10^{-15}$ g/L

15.130 NaOH, 40.0 amu; $20 \text{ g} \times \dfrac{1 \text{ mol}}{40.0 \text{ g}} = 0.50$ mol NaOH

$(0.500 \text{ L})(1.5 \text{ mol/L}) = 0.75$ mol NH_4Cl

	$NH_4{}^+$(aq)	+	OH^-(aq)	$\rightleftharpoons$	NH_3(aq)	+	H_2O(l)
before reaction (mol)	0.75		0.50		0		
change (mol)	−0.50		−0.50		+0.50		
after reaction (mol)	0.25		0		0.50		

This reaction produces a buffer solution.

$[NH_4{}^+] = 0.25$ mol/0.500 L = 0.50 M; $[NH_3] = 0.50$ mol/0.500 L = 1.0 M

$$pH = pK_a + \log\frac{[base]}{[acid]} = pK_a + \log\frac{[NH_3]}{[NH_4{}^+]}$$

For $NH_4{}^+$, $K_a = \dfrac{K_w}{K_b \text{ for } NH_3} = \dfrac{1.0 \times 10^{-14}}{1.8 \times 10^{-5}} = 5.6 \times 10^{-10}$; $pK_a = -\log K_a = 9.25$

$$pH = 9.25 + \log\left(\frac{1.0}{0.5}\right) = 9.55$$

15.131 (a) AgCl, $K_{sp} = [Ag^+][Cl^-] = 1.8 \times 10^{-10}$

$$[Cl^-] = \frac{K_{sp}}{[Ag^+]} = \frac{1.8 \times 10^{-10}}{0.030} = 6.0 \times 10^{-9} \text{ M}$$

(b) Hg_2Cl_2, $K_{sp} = [Hg_2{}^{2+}][Cl^-]^2 = 1.4 \times 10^{-18}$

$$[Cl^-] = \sqrt{\frac{K_{sp}}{[Hg_2{}^{2+}]}} = \sqrt{\frac{1.4 \times 10^{-18}}{0.030}} = 6.8 \times 10^{-9} \text{ M}$$

(c) $PbCl_2$, $K_{sp} = [Pb^{2+}][Cl^-]^2 = 1.2 \times 10^{-5}$

$$[Cl^-] = \sqrt{\frac{K_{sp}}{[Pb^{2+}]}} = \sqrt{\frac{1.2 \times 10^{-5}}{0.030}} = 0.020 \text{ M}$$

AgCl(s) will begin to precipitate when the $[Cl^-]$ just exceeds 6.0×10^{-9} M. At this Cl^- concentration, IP < K_{sp} for $PbCl_2$ so all of the Pb^{2+} will remain in solution.

15.132 For $NH_4{}^+$, $K_a = \dfrac{K_w}{K_b \text{ for } NH_3} = \dfrac{1.0 \times 10^{-14}}{1.8 \times 10^{-5}} = 5.6 \times 10^{-10}$; $pK_a = -\log K_a = 9.25$

$$pH = pK_a + \log\frac{[NH_3]}{[NH_4{}^+]} = 9.25 + \log\frac{(0.50)}{(0.30)} = 9.47$$

$$[H_3O^+] = 10^{-pH} = 10^{-9.47} = 3.4 \times 10^{-10} \text{ M}$$

For MnS, $K_{spa} = \dfrac{[Mn^{2+}][H_2S]}{[H_3O^+]^2} = 3 \times 10^7$

$$\text{molar solubility} = [Mn^{2+}] = \dfrac{K_{spa}[H_3O^+]^2}{[H_2S]} = \dfrac{(3 \times 10^7)(3.4 \times 10^{-10})^2}{(0.10)} = 3.5 \times 10^{-11} \text{ M}$$

MnS, 87.00 amu; solubility $= (3.5 \times 10^{-11} \text{ mol/L})(87.00 \text{ g/mol}) = 3 \times 10^{-9} \text{ g/L}$

15.133   pH $= 9.00$; $[H_3O^+] = 10^{-pH} = 10^{-9.00} = 1.0 \times 10^{-9} \text{ M}$

$$[OH^-] = \dfrac{K_w}{[H_3O^+]} = \dfrac{1.0 \times 10^{-14}}{1.0 \times 10^{-9}} = 1.0 \times 10^{-5} \text{ M}$$

$$Mg(OH)_2(s) \rightleftharpoons Mg^{2+}(aq) + 2\,OH^-(aq)$$

equil (M) x 1.0×10^{-5} (fixed by buffer)

$$K_{sp} = [Mg^{2+}][OH^-]^2 = 5.6 \times 10^{-12} = x(1.0 \times 10^{-5})^2$$

$$\text{molar solubility} = x = \dfrac{5.6 \times 10^{-12}}{(1.0 \times 10^{-5})^2} = 0.056 \text{ M}$$

15.134   60.0 mL $= 0.0600$ L

$$\text{mol } H_3PO_4 = 0.0600 \text{ L} \times \dfrac{1.00 \text{ mol } H_3PO_4}{1.00 \text{ L}} = 0.0600 \text{ mol } H_3PO_4$$

$$\text{mol LiOH} = 1.00 \text{ L} \times \dfrac{0.100 \text{ mol LiOH}}{1.00 \text{ L}} = 0.100 \text{ mol LiOH}$$

	$H_3PO_4(aq)$ +	$OH^-(aq)$ →	$H_2PO_4^-(aq)$ +	$H_2O(l)$
before reaction (mol)	0.0600	0.100	0	
change (mol)	−0.0600	−0.0600	+0.0600	
after reaction (mol)	0	0.040	0.0600	

	$H_2PO_4^-(aq)$ +	$OH^-(aq)$ →	$HPO_4^{2-}(aq)$ +	$H_2O(l)$
before reaction (mol)	0.0600	0.040	0	
change (mol)	−0.040	−0.040	+0.040	
after reaction (mol)	0.020	0	0.040	

The resulting solution is a buffer because it contains the conjugate acid-base pair, $H_2PO_4^-$ and HPO_4^{2-}, at acceptable buffer concentrations.
For $H_2PO_4^-$, $K_{a2} = 6.2 \times 10^{-8}$ and $pK_{a2} = -\log K_{a2} = -\log (6.2 \times 10^{-8}) = 7.21$

$$pH = pK_{a2} + \log\dfrac{[HPO_4^{2-}]}{[H_2PO_4^-]} = 7.21 + \log\dfrac{(0.040 \text{ mol}/1.06 \text{ L})}{(0.020 \text{ mol}/1.06 \text{ L})}$$

$$pH = 7.21 + \log\dfrac{(0.040)}{(0.020)} = 7.21 + 0.30 = 7.51$$

15.135 (a) The mixture of 0.100 mol H_3PO_4 and 0.150 mol NaOH is a buffer and contains mainly $H_2PO_4^-$ and HPO_4^{2-} from the reactions:

$$H_3PO_4(aq) + OH^-(aq) \rightarrow H_2PO_4^-(aq) + H_2O(l)$$

before (mol)	0.100	0.150	0
change (mol)	–0.100	–0.100	+0.100
after (mol)	0	0.050	0.100

$$H_2PO_4^-(aq) + OH^-(aq) \rightarrow HPO_4^{2-}(aq) + H_2O(l)$$

before (mol)	0.100	0.050	0
change (mol)	–0.050	–0.050	+0.050
after (mol)	0.050	0	0.050

If water were used to dilute the solution instead of HCl, the pH would be equal to pK_{a2} because $[H_2PO_4^-] = [HPO_4^{2-}] = 0.050$ mol/1.00 L = 0.050 M

$$H_2PO_4^-(aq) + H_2O(l) \rightleftharpoons H_3O^+(aq) + HPO_4^{2-}(aq) \qquad K_{a2} = 6.2 \times 10^{-8}$$

$$pK_{a2} = -\log K_{2a} = -\log(6.2 \times 10^{-8}) = 7.21$$

$$pH = pK_{a2} + \log \frac{[HPO_4^{2-}]}{[H_2PO_4^-]} = pK_{a2} + \log(1) = pK_{a2} = 7.21$$

The pH is lower (6.73) because the added HCl converts some HPO_4^{2-} to $H_2PO_4^-$.

$$HPO_4^{2-}(aq) + H_3O^+(aq) \rightarrow H_2PO_4^-(aq) + H_2O(l)$$

before (M)	0.050	x	0.050
change (M)	–x	–x	+x
after (M)	0.050 – x	0	0.050 + x

$$[HPO_4^{2-}] + [H_2PO_4^-] = (0.050 - x) + (0.050 + x) = 0.100 \text{ M}$$

$$pH = pK_{a2} + \log \frac{[HPO_4^{2-}]}{[H_2PO_4^-]}$$

$$[HPO_4^{2-}] = 0.100 - [H_2PO_4^-]$$

$$6.73 = 7.21 + \log \frac{(0.100 - [H_2PO_4^-])}{[H_2PO_4^-]}$$

$$6.73 - 7.21 = -0.48 = \log \frac{(0.100 - [H_2PO_4^-])}{[H_2PO_4^-]}$$

$$10^{-0.48} = 0.331 = \frac{(0.100 - [H_2PO_4^-])}{[H_2PO_4^-]}$$

$$(0.331)[H_2PO_4^-] = 0.100 - [H_2PO_4^-]$$
$$(1.331)[H_2PO_4^-] = 0.100$$
$$[H_2PO_4^-] = 0.100/1.331 = 0.075 \text{ M}$$
$$[HPO_4^{2-}] = 0.100 - [H_2PO_4^-] = 0.100 - 0.075 = 0.025 \text{ M}$$

$$H_3PO_4(aq) + H_2O(l) \rightleftharpoons H_3O^+(aq) + H_2PO_4^-(aq) \qquad K_{a1} = 7.5 \times 10^{-3}$$

$$K_{a1} = \frac{[H_3O^+][H_2PO_4^-]}{[H_3PO_4]}$$

$$[H_3PO_4] = \frac{[H_3O^+][H_2PO_4^-]}{K_{a1}}$$

$$[H_3O^+] = 10^{-pH} = 10^{-6.73} = 1.86 \times 10^{-7} \text{ M}$$

$$[H_3PO_4] = \frac{(1.86 \times 10^{-7})(0.075)}{7.5 \times 10^{-3}} = 1.9 \times 10^{-6} \text{ M}$$

(b)   If distilled water were used and not HCl, the mole amounts of both $H_2PO_4^-$ and HPO_4^{2-} would be 0.050 mol. The HCl converted some HPO_4^{2-} to $H_2PO_4^-$.

$$HPO_4^{2-}(aq) + H_3O^+(aq) \rightarrow H_2PO_4^-(aq) + H_2O(l)$$

before (mol) 0.050   x   0.050
change (mol)   −x   −x   +x
after (mol)   0.050 − x   0   0.050 + x

From part (a), $[HPO_4^{2-}] = 0.025$ M

mol HPO_4^{2-} = (0.025 mol/L)(1.00 L) = 0.025 mol = 0.050 − x

x = mol H_3O^+ = mol HCl inadvertently added = 0.050 − 0.025 = 0.025 mol HCl

15.136 For CH_3CO_2H, $K_a = 1.8 \times 10^{-5}$ and $pK_a = -\log K_a = -\log(1.8 \times 10^{-5}) = 4.74$

The mixture will be a buffer solution containing the conjugate acid-base pair, CH_3CO_2H and $CH_3CO_2^-$, having a pH near the pK_a of CH_3CO_2H.

$$pH = pK_a + \log \frac{[CH_3CO_2^-]}{[CH_3CO_2H]}$$

$$4.85 = 4.74 + \log \frac{[CH_3CO_2^-]}{[CH_3CO_2H]}; \qquad 4.85 - 4.74 = \log \frac{[CH_3CO_2^-]}{[CH_3CO_2H]}$$

$$0.11 = \log \frac{[CH_3CO_2^-]}{[CH_3CO_2H]}; \qquad \frac{[CH_3CO_2^-]}{[CH_3CO_2H]} = 10^{0.11} = 1.3$$

In the Henderson-Hasselbalch equation, moles can be used in place of concentrations because both components are in the same volume so the volume terms cancel.
20.0 mL = 0.0200 L

Let X equal the volume of 0.10 M CH_3CO_2H and Y equal the volume of 0.15 M $CH_3CO_2^-$. Therefore, X + Y = 0.0200 L and

$$\frac{Y \times [CH_3CO_2^-]}{X \times [CH_3CO_2H]} = \frac{Y(0.15 \text{ mol/L})}{X(0.10 \text{ mol/L})} = 1.3$$

X = 0.0200 − Y

$$\frac{Y(0.15 \text{ mol/L})}{(0.020 - Y)(0.10 \text{ mol/L})} = 1.3$$

$$\frac{0.15Y}{0.0020 - 0.10Y} = 1.3$$

$0.15Y = 1.3(0.0020 - 0.10Y)$

$0.15Y = 0.0026 - 0.13Y$

$0.15Y + 0.13Y = 0.0026$

$0.28Y = 0.0026$

$Y = 0.0026/0.28 = 0.0093$ L

$X = 0.0200 - Y = 0.0200 - 0.0093 = 0.0107$ L

$X = 0.0107$ L $= 10.7$ mL and $Y = 0.0093$ L $= 9.3$ mL

You need to mix together 10.7 mL of 0.10 M CH_3CO_2H and 9.3 mL of 0.15 M $NaCH_3CO_2$ to prepare 20.0 mL of a solution with a pH of 4.85.

15.137 $[H_3O^+] = 10^{-pH} = 10^{-2.37} = 0.004\ 27$ M

$H_3Cit(aq) + H_2O(l) \rightleftharpoons H_3O^+(aq) + H_2Cit^-(aq)$

$K_{a1} = 7.1 \times 10^{-4} = \dfrac{[H_3O^+][H_2Cit^-]}{[H_3Cit]}$

$(7.1 \times 10^{-4})[H_3Cit] = (0.004\ 27)[H_2Cit^-]$

$[H_3Cit] = (0.004\ 27)[H_2Cit^-]/(7.1 \times 10^{-4}) = (6.01)[H_2Cit^-]$

$H_2Cit^-(aq) + H_2O(l) \rightleftharpoons H_3O^+(aq) + HCit^{2-}(aq)$

$K_{a2} = 1.7 \times 10^{-5} = \dfrac{[H_3O^+][HCit^{2-}]}{[H_2Cit^-]}$

$(1.7 \times 10^{-5})[H_2Cit^-] = (0.004\ 27)[HCit^{2-}]$

$[HCit^{2-}] = (1.7 \times 10^{-5})[H_2Cit^-]/(0.004\ 27) = (0.003\ 98)[H_2Cit^-]$

$[H_3Cit] + [H_2Cit^-] + [HCit^{2-}] + [Cit^{3-}] = 0.350$ M

Now assume $[Cit^{3-}] \approx 0$, so $[H_3Cit] + [H_2Cit^-] + [HCit^{2-}] = 0.350$ M and then by substitution:

$(6.01)[H_2Cit^-] + [H_2Cit^-] + (0.003\ 98)[H_2Cit^-] = 0.350$ M

$(7.01)[H_2Cit^-] = 0.350$ M

$[H_2Cit^-] = 0.350$ M$/7.01 = 0.050$ M

$[H_3Cit] = (6.01)[H_2Cit^-] = (6.01)(0.050$ M$) = 0.30$ M

$[HCit^{2-}] = (0.003\ 98)[H_2Cit^-] = (0.003\ 98)(0.050$ M$) = 2.0 \times 10^{-4}$ M

$HCit^{2-}(aq) + H_2O(l) \rightleftharpoons H_3O^+(aq) + Cit^{3-}(aq)$

$K_{a3} = 4.1 \times 10^{-7} = \dfrac{[H_3O^+][Cit^{3-}]}{[HCit^{2-}]}$

$[Cit^{3-}] = \dfrac{(K_{a3})[HCit^{2-}]}{[H_3O^+]} = \dfrac{(4.1 \times 10^{-7})(2.0 \times 10^{-4})}{(0.004\ 27)} = 1.9 \times 10^{-8}$ M

15.138 (a) HCl is a strong acid. HCN is a weak acid with $K_a = 4.9 \times 10^{-10}$. Before the titration, the $[H_3O^+] = 0.100$ M. The HCN contributes an insignificant amount of additional H_3O^+, so the pH $= -\log[H_3O^+] = -\log(0.100) = 1.00$

459

(b) 100.0 mL = 0.1000 L

$$\text{mol } H_3O^+ = 0.1000 \text{ L} \times \frac{0.100 \text{ mol HCl}}{1.00 \text{ L}} = 0.0100 \text{ mol } H_3O^+$$

add 75.0 mL of 0.100 M NaOH; 75.0 mL = 0.0750 L

$$\text{mol } OH^- = 0.0750 \text{ L} \times \frac{0.100 \text{ mol NaOH}}{1.00 \text{ L}} = 0.00750 \text{ mol } OH^-$$

	$H_3O^+(aq)$	+	$OH^-(aq)$	$\rightarrow$	$2 H_2O(l)$
before reaction (mol)	0.0100		0.0075		
change (mol)	−0.0075		−0.0075		
after reaction (mol)	0.0025		0		

$$[H_3O^+] = \frac{0.0025 \text{ mol } H_3O^+}{0.1000 \text{ L} + 0.0750 \text{ L}} = 0.0143 \text{ M}$$

$$pH = -\log[H_3O^+] = -\log(0.0143) = 1.84$$

(c) 100.0 mL of 0.100 M NaOH will completely neutralize all of the H_3O^+ from 100.0 mL of 0.100 M HCl. Only NaCl and HCN remain in the solution. NaCl is a neutral salt and does not affect the pH of the solution. [HCN] changes because of dilution. Because the solution volume is doubled, [HCN] is cut in half.

[HCN] = 0.100 M/2 = 0.0500 M

	$HCN(aq)$	+	$H_2O(l)$	$\rightleftharpoons$	$H_3O^+(aq)$	+	$CN^-(aq)$
initial (M)	0.0500				~0		0
change (M)	−x				+x		+x
equil (M)	0.0500 − x				x		x

$$K_a = \frac{[H_3O^+][CN^-]}{HCN} = 4.9 \times 10^{-10} = \frac{x^2}{0.0500 - x} \approx \frac{x^2}{0.0500}$$

$$[H_3O^+] = x = \sqrt{(0.0500)(4.9 \times 10^{-10})} = 4.95 \times 10^{-6} \text{ M}$$

$$pH = -\log[H_3O^+] = -\log(4.95 \times 10^{-6}) = 5.31$$

(d) Add an additional 25.0 mL of 0.100 M NaOH.

25.0 mL = 0.0250 L

$$\text{additional mol } OH^- = 0.0250 \text{ L} \times \frac{0.100 \text{ mol NaOH}}{1.00 \text{ L}} = 0.00250 \text{ mol } OH^-$$

$$\text{mol HCN} = 0.200 \text{ L} \times \frac{0.0500 \text{ mol HCN}}{1.00 \text{ L}} = 0.0100 \text{ mol HCN}$$

	$HCN(aq)$	+	$OH^-(aq)$	$\rightarrow$	$CN^-(aq)$	+	$H_2O(l)$
before reaction (mol)	0.0100		0.00250		0		
change (mol)	−0.00250		−0.00250		+0.00250		
after reaction (mol)	0.0075		0		0.00250		

The resulting solution is a buffer because it contains the conjugate acid-base pair, HCN and CN^-, at acceptable buffer concentrations.

For HCN, $K_a = 4.9 \times 10^{-10}$ and $pK_a = -\log K_a = -\log (4.9 \times 10^{-10}) = 9.31$

$$pH = pK_a + \log \frac{[CN^-]}{[HCN]} = 9.31 + \log \frac{(0.00250 \text{ mol}/0.2250 \text{ L})}{(0.0075 \text{ mol}/0.2250 \text{ L})}$$

$$pH = 9.31 + \log \frac{(0.00250)}{(0.0075)} = 9.31 - 0.48 = 8.83$$

15.139 (a) $\quad\quad Cd(OH)_2(s) \rightleftharpoons Cd^{2+}(aq) + 2 OH^-(aq)$

initial (M) $\quad\quad\quad\quad\quad\quad 0 \quad\quad\quad\quad \sim 0$

equil (M) $\quad\quad\quad\quad\quad\quad\quad x \quad\quad\quad\quad 2x$

$K_{sp} = [Cd^{2+}][OH^-]^2 = 5.3 \times 10^{-15} = (x)(2x)^2 = 4x^3$

$$\text{molar solubility} = x = \sqrt[3]{\frac{5.3 \times 10^{-15}}{4}} = 1.1 \times 10^{-5} \text{ M}$$

$[OH^-] = 2x = 2(1.1 \times 10^{-5} \text{ M}) = 2.2 \times 10^{-5} \text{ M}$

$$[H_3O^+] = \frac{1.0 \times 10^{-14}}{2.2 \times 10^{-5}} = 4.5 \times 10^{-10} \text{ M}$$

$pH = -\log[H_3O^+] = -\log(4.5 \times 10^{-10}) = 9.35$

(b) 90.0 mL = 0.0900 L

mol HNO_3 = (0.100 mol/L)(0.0900 L) = 0.009 00 mol HNO_3

The addition of HNO_3 dissolves some $Cd(OH)_2(s)$.

$\quad\quad\quad\quad\quad\quad Cd(OH)_2(s) + 2 HNO_3(aq) \rightarrow Cd^{2+}(aq) + 2 H_2O(l)$

before (mol)   0.100   0.009 00 1.1×10^{-5}

change (mol) -0.0045 $-2(0.0045)$ $1.1 \times 10^{-5} + 0.0045$

after (mol)   0.0955   0 ~ 0.0045

total volume = 100.0 mL + 90.0 mL = 190.0 mL = 0.1900 L

$[Cd^{2+}]$ = 0.0045 mol/0.1900 L = 0.024 M

$K_{sp} = 5.3 \times 10^{-15} = [Cd^{2+}][OH^-]^2 = (0.024)[OH^-]^2$

$$[OH^-] = \sqrt{\frac{5.3 \times 10^{-15}}{0.024}} = 4.7 \times 10^{-7} \text{ M}$$

$$[H_3O^+] = \frac{1.0 \times 10^{-14}}{4.7 \times 10^{-7}} = 2.1 \times 10^{-8} \text{ M}$$

$pH = -\log[H_3O^+] = -\log(2.1 \times 10^{-8}) = 7.68$

(c) volume HNO_3 = 0.0100 mol $Cd(OH)_2$ $\times$ $\dfrac{2 \text{ mol } HNO_3}{1 \text{ mol } Cd(OH)_2}$ $\times$ $\dfrac{1.00 \text{ L}}{0.100 \text{ mol}}$ $\times$ $\dfrac{1000 \text{ mL}}{1.00 \text{ L}}$ = 200 mL

15.140 (a) $\quad\quad Zn(OH)_2(s) \rightleftharpoons Zn^{2+}(aq) + 2 OH^-(aq)$

initial (M) $\quad\quad\quad\quad\quad\quad 0 \quad\quad\quad\quad \sim 0$

equil (M) $\quad\quad\quad\quad\quad\quad\quad x \quad\quad\quad\quad 2x$

$K_{sp} = [Zn^{2+}][OH^-]^2 = 4.1 \times 10^{-17} = (x)(2x)^2 = 4x^3$

$$\text{molar solubility} = x = \sqrt[3]{\frac{4.1 \times 10^{-17}}{4}} = 2.2 \times 10^{-6} \text{ M}$$

(b) $[OH^-] = 2x = 2(2.2 \times 10^{-6} \text{ M}) = 4.4 \times 10^{-6} \text{ M}$

$[H_3O^+] = \dfrac{1.0 \times 10^{-14}}{4.4 \times 10^{-6}} = 2.3 \times 10^{-9} \text{ M}$

$pH = -\log[H_3O^+] = -\log(2.3 \times 10^{-9}) = 8.64$

(c)

$Zn(OH)_2(s) \rightleftharpoons Zn^{2+}(aq) + 2\,OH^-(aq)$		$K_{sp} = 4.1 \times 10^{-17}$
$Zn^{2+}(aq) + 4\,OH^-(aq) \rightleftharpoons Zn(OH)_4^{2-}(aq)$		$K_f = 3 \times 10^{15}$
$Zn(OH)_2(s) + 2\,OH^-(aq) \rightleftharpoons Zn(OH)_4^{2-}(aq)$		$K = K_{sp} \cdot K_f = 0.123$

initial (M)	0.10	0
change (M)	$-2x$	$+x$
equil (M)	$0.10 - 2x$	x

$K = \dfrac{[Zn(OH)_4^{2-}]}{[OH^-]^2} = 0.123 = \dfrac{x}{(0.10 - 2x)^2}$

$0.492x^2 - 1.0492x + 0.00123 = 0$

Use the quadratic formula to solve for x.

$x = \dfrac{-(-1.0492) \pm \sqrt{(-1.0492)^2 - (4)(0.492)(0.00123)}}{2(0.492)} = \dfrac{1.0492 \pm 1.0480}{0.984}$

$x = 2.1$ and 1.2×10^{-3}

Of the two solutions for x, only 1.2×10^{-3} has physical meaning because the other solution leads to a negative $[OH^-]$.

molar solubility of $Zn(OH)_4^{2-}$ in 0.10 M NaOH = $x = 1.2 \times 10^{-3}$ M

15.141 (a)

$Fe(OH)_3(s) \rightleftharpoons Fe^{3+}(aq) + 3\,OH^-(aq)$		$K_{sp} = 2.6 \times 10^{-39}$
$H_3Cit(aq) + H_2O(l) \rightleftharpoons H_3O^+(aq) + H_2Cit^-(aq)$		$K_{a1} = 7.1 \times 10^{-4}$
$H_2Cit^-(aq) + H_2O(l) \rightleftharpoons H_3O^+(aq) + HCit^{2-}(aq)$		$K_{a2} = 1.7 \times 10^{-5}$
$HCit^{2-}(aq) + H_2O(l) \rightleftharpoons H_3O^+(aq) + Cit^{3-}(aq)$		$K_{a3} = 4.1 \times 10^{-7}$
$Fe^{3+}(aq) + Cit^{3-}(aq) \rightleftharpoons Fe(Cit)(aq)$		$K_f = 6.3 \times 10^{11}$
$3\,[H_3O^+(aq) + OH^-(aq) \rightleftharpoons 2\,H_2O(l)]$		$(1/K_w)^3 = 1.0 \times 10^{42}$
$Fe(OH)_3(s) + H_3Cit(aq) \rightleftharpoons Fe(Cit)(aq) + 3\,H_2O(l)$		

$K = K_{sp} K_{a1} K_{a2} K_{a3} K_f (1/K_w)^3 = 8.1$

(b)

$Fe(OH)_3(s) + H_3Cit(aq) \rightleftharpoons Fe(Cit)(aq) + 3\,H_2O(l)$

initial (M)	0.500	0
change (M)	$-x$	$+x$
equil (M)	$0.500 - x$	x

$K = \dfrac{[Fe(Cit)]}{[H_3Cit]} = 8.1 = \dfrac{x}{0.500 - x}$

$8.1(0.500 - x) = x$

$4.05 - 8.1x = x$

$4.05 = 9.1x$

$x = \text{molar solubility} = 4.05/9.1 = 0.45$ M

Multiconcept Problems

15.142 (a) $HA^-(aq) + H_2O(l) \rightleftharpoons H_3O^+(aq) + A^{2-}(aq)$ $\qquad$ $K_{a2} = 10^{-10}$

$\qquad$ $HA^-(aq) + H_2O(l) \rightleftharpoons H_2A(aq) + OH^-(aq)$ $\qquad$ $K_b = \dfrac{K_w}{K_{a1}} = 10^{-10}$

$\qquad$ $2\,HA^-(aq) \rightleftharpoons H_2A(aq) + A^{2-}(aq)$ $\qquad$ $K = \dfrac{K_{a2}}{K_{a1}} = 10^{-6}$

$\qquad$ $2\,H_2O(l) \rightleftharpoons H_3O^+(aq) + OH^-(aq)$ $\qquad$ $K_w = 1.0 \times 10^{-14}$

The principal reaction of the four is the one with the largest K, and that is the third reaction.

(b) $K_{a1} = \dfrac{[H_3O^+][HA^-]}{[H_2A]}$ and $K_{a2} = \dfrac{[H_3O^+][A^{2-}]}{[HA^-]}$

$[H_3O^+] = \dfrac{K_{a1}[H_2A]}{[HA^-]}$ and $[H_3O^+] = \dfrac{K_{a2}[HA^-]}{[A^{2-}]}$

$\dfrac{K_{a1}[H_2A]}{[HA^-]} \times \dfrac{K_{a2}[HA^-]}{[A^{2-}]} = [H_3O^+]^2$; $\qquad$ $\dfrac{K_{a1}K_{a2}[H_2A]}{[A^{2-}]} = [H_3O^+]^2$

Because the principal reaction is $2\,HA^-(aq) \rightleftharpoons H_2A(aq) + A^{2-}(aq)$, $[H_2A] = [A^{2-}]$.

$K_{a1}K_{a2} = [H_3O^+]^2$

$\log K_{a1} + \log K_{a2} = 2\log[H_3O^+]$

$\dfrac{\log K_{a1} + \log K_{a2}}{2} = \log[H_3O^+]$; $\qquad$ $\dfrac{-\log K_{a1} + (-\log K_{a2})}{2} = -\log[H_3O^+]$

$\dfrac{pK_{a1} + pK_{a2}}{2} = pH$

(c) $\qquad\qquad$ $2\,HA^-(aq) \rightleftharpoons H_2A(aq) + A^{2-}(aq)$

initial (M)	1.0	0	0
change (M)	$-2x$	$+x$	$+x$
equil (M)	$1.0 - 2x$	x	x

$K = \dfrac{[H_2A][A^{2-}]}{[HA^-]^2} = 1 \times 10^{-6} = \dfrac{x^2}{(1.0 - 2x)^2}$

Take the square root of both sides and solve for x.

$x = [A^{2-}] = 1 \times 10^{-3}$ M

mol $A^{2-} = (1 \times 10^{-3}\ \text{mol/L})(0.0500\ \text{L}) = 5 \times 10^{-5}$ mol A^{2-}

number of A^{2-} ions $= (5 \times 10^{-5}\ \text{mol}\ A^{2-})(6.022 \times 10^{23}\ \text{ions/mol}) = 3 \times 10^{19}\ A^{2-}$ ions

15.143 (a) (i)

$$\text{en(aq)} + \text{H}_2\text{O(l)} \rightleftharpoons \text{enH}^+\text{(aq)} + \text{OH}^-\text{(aq)}$$

	en(aq)		enH$^+$(aq)	OH$^-$(aq)
initial (M)	0.100		0	~0
change (M)	$-x$		$+x$	$+x$
equil (M)	$0.100 - x$		x	x

$$K_b = \frac{[\text{enH}^+][\text{OH}^-]}{[\text{en}]} = 5.2 \times 10^{-4} = \frac{(x)(x)}{0.100 - x}$$

$$x^2 + (5.2 \times 10^{-4})x - (5.2 \times 10^{-5}) = 0$$

Use the quadratic formula to solve for x.

$$x = \frac{-(5.2 \times 10^{-4}) \pm \sqrt{(5.2 \times 10^{-4})^2 - 4(1)(-5.2 \times 10^{-5})}}{2(1)} = \frac{-5.2 \times 10^{-4} \pm 0.01443}{2}$$

$x = -0.0075$ and 0.0070

Of the two solutions for x, only the positive value of x has physical meaning because x is the [OH$^-$].

$[\text{OH}^-] = x = 0.0070$ M

$$[\text{H}_3\text{O}^+] = \frac{K_w}{[\text{OH}^-]} = \frac{1.0 \times 10^{-14}}{0.0070} = 1.43 \times 10^{-12} \text{ M}$$

$\text{pH} = -\log[\text{H}_3\text{O}^+] = -\log(1.43 \times 10^{-12}) = 11.84$

(ii) (30.0 mL)(0.100 mmol/mL) = 3.00 mmol en

(15.0 mL)(0.100 mmol/mL) = 1.50 mmol HCl

Halfway to the first equivalence point, [OH$^-$] = K_{b1}

$$[\text{H}_3\text{O}^+] = \frac{K_w}{[\text{OH}^-]} = \frac{1.0 \times 10^{-14}}{5.2 \times 10^{-4}} = 1.92 \times 10^{-11} \text{ M}$$

$\text{pH} = -\log[\text{H}_3\text{O}^+] = -\log(1.92 \times 10^{-11}) = 10.72$

(iii) At the first equivalence point $\text{pH} = \dfrac{pK_{a1} + pK_{a2}}{2} = 9.14$

(iv) Halfway between the first and second equivalence points, [OH$^-$] = K_{b2} = 3.7×10^{-7} M

$$[\text{H}_3\text{O}^+] = \frac{K_w}{[\text{OH}^-]} = \frac{1.0 \times 10^{-14}}{3.7 \times 10^{-7}} = 2.70 \times 10^{-8} \text{ M}$$

$\text{pH} = -\log[\text{H}_3\text{O}^+] = -\log(2.70 \times 10^{-8}) = 7.57$

(v) At the second equivalence point only the acidic enH$_2$Cl$_2$ is in solution.

For enH$_2$$^{2+}$, $K_a = \dfrac{K_w}{K_b \text{ for enH}^+} = \dfrac{K_w}{K_{b2}} = \dfrac{1.0 \times 10^{-14}}{3.7 \times 10^{-7}} = 2.70 \times 10^{-8}$

$$[\text{enH}_2{}^{2+}] = \frac{3.00 \text{ mmol}}{(30.0 \text{ mL} + 60.0 \text{ mL})} = 0.0333 \text{ M}$$

$$\text{enH}_2{}^{2+}\text{(aq)} + \text{H}_2\text{O(l)} \rightleftharpoons \text{H}_3\text{O}^+\text{(aq)} + \text{enH}^+\text{(aq)}$$

	enH$_2$$^{2+}$(aq)		H$_3$O$^+$(aq)	enH$^+$(aq)
initial (M)	0.0333		~0	0
change (M)	$-x$		$+x$	$+x$
equil (M)	$0.0333 - x$		x	x

$$K_a = \frac{[H_3O^+][enH^+]}{[enH_2^{2+}]} = 2.70 \times 10^{-8} = \frac{(x)(x)}{0.0333 - x} \approx \frac{x^2}{0.0333}$$

Solve for x. $x = [H_3O^+] = \sqrt{(2.70 \times 10^{-8})(0.0333)} = 3.00 \times 10^{-5}$ M
$pH = -\log[H_3O^+] = -\log(3.00 \times 10^{-5}) = 4.52$

(vi) excess HCl
$(75.0 \text{ mL} - 60.0 \text{ mL})(0.100 \text{ mmol/mL}) = 1.50 \text{ mmol HCl} = 1.50 \text{ mmol } H_3O^+$

$$[H_3O^+] = \frac{1.50 \text{ mmol}}{(30.0 \text{ mL} + 75.0 \text{ mL})} = 0.0143 \text{ M}$$

$pH = -\log[H_3O^+] = -\log(0.0143) = 1.84$

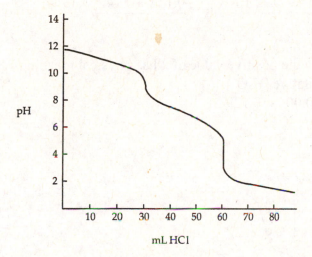

(b)

Each of the two nitrogens in ethylenediamine can accept a proton.

(c) Each nitrogen is sp^3 hybridized.

15.144 (a) The first equivalence point is reached when all the H_3O^+ from the HCl, and the H_3O^+ from the first ionization of H_3PO_4, is consumed.

At the first equivalence point $pH = \dfrac{pK_{a1} + pK_{a2}}{2} = 4.66$

$[H_3O^+] = 10^{-pH} = 10^{(-4.66)} = 2.2 \times 10^{-5}$ M
$(88.0 \text{ mL})(0.100 \text{ mmol/mL}) = 8.80 \text{ mmol NaOH}$ are used to get to the first equivalence point

(b) mmol $(HCl + H_3PO_4) = $ mmol NaOH $= 8.8$ mmol
mmol $H_3PO_4 = (126.4 \text{ mL} - 88.0 \text{ mL})(0.100 \text{ mmol/mL}) = 3.84$ mmol
mmol HCl $= (8.8 - 3.84) = 4.96$ mmol

$$[HCl] = \frac{4.96 \text{ mmol}}{40.0 \text{ mL}} = 0.124 \text{ M}; \qquad [H_3PO_4] = \frac{3.84 \text{ mmol}}{40.0 \text{ mL}} = 0.0960 \text{ M}$$

(c) 100% of the HCl is neutralized at the first equivalence point.

(d)

	$H_3PO_4(aq)$	$+ H_2O(l)$	$\rightleftharpoons$	$H_3O^+(aq)$	$+ H_2PO_4^-(aq)$
initial (M)	0.0960			0.124	0
change (M)	$-x$			$+x$	$+x$
equil (M)	$0.0960 - x$			$0.124 + x$	x

$$K_{a1} = \frac{[H_3O^+][H_2PO_4^-]}{[H_3PO_4]} = 7.5 \times 10^{-3} = \frac{(0.124 + x)(x)}{0.0960 - x}$$

$$x^2 + 0.132x - (7.2 \times 10^{-4}) = 0$$

Use the quadratic formula to solve for x.

$$x = \frac{-(0.132) \pm \sqrt{(0.132)^2 - 4(1)(-7.2 \times 10^{-4})}}{2(1)} = \frac{-0.132 \pm 0.142}{2}$$

$x = -0.137$ and 0.005

Of the two solutions for x, only the positive value of x has physical meaning because the other solution would give a negative $[H_3O^+]$.

$[H_3O^+] = 0.124 + x = 0.124 + 0.005 = 0.129 \text{ M}$

$pH = -\log[H_3O^+] = -\log(0.129) = 0.89$

(e)

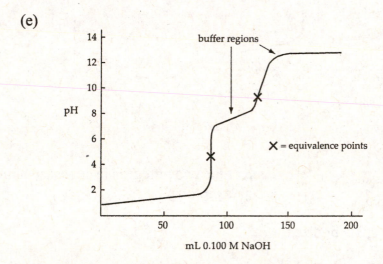

(f) Bromcresol green or methyl orange are suitable indicators for the first equivalence point. Thymolphthalein is a suitable indicator for the second equivalence point.

15.145 (a) $PV = nRT$; $25 \text{ °C} = 298 \text{ K}$

$$n_{HCl} = \frac{PV}{RT} = \frac{\left(732 \text{ mm Hg} \times \dfrac{1.00 \text{ atm}}{760 \text{ mm Hg}}\right)(1.000 \text{ L})}{\left(0.082\ 06 \dfrac{\text{L} \cdot \text{atm}}{\text{K} \cdot \text{mol}}\right)(298 \text{ K})} = 0.0394 \text{ mol HCl}$$

Na_2CO_3, 105.99 amu

$$mol\ Na_2CO_3 = 6.954\ g\ Na_2CO_3 \times \frac{1\ mol\ Na_2CO_3}{105.99\ g\ Na_2CO_3} = 0.0656\ mol\ Na_2CO_3$$

	$CO_3^{2-}(aq)$ +	$H_3O^+(aq)$	$\rightarrow$	$HCO_3^-(aq)$ +	$H_2O(l)$
before reaction (mol)	0.0656	0.0394		0	
change (mol)	−0.0394	−0.0394		+0.0394	
after reaction (mol)	0.0656 − 0.0394	0		0.0394	

mol CO_3^{2-} = 0.0656 − 0.0394 = 0.0262 mol and mol HCO_3^- = 0.0394 mol
Therefore, we have an HCO_3^-/CO_3^{2-} buffer solution.

$$pH = pK_{a2} + \log \frac{[CO_3^{2-}]}{[HCO_3^-]} = -\log(5.6 \times 10^{-11}) + \log \frac{0.0262\ mol/V}{0.0394\ mol/V}$$

$$pH = 10.25 - 0.177 = 10.08$$

(b) mol Na^+ = 2(0.0656 mol) = 0.1312 mol
mol CO_3^{2-} = 0.0262 mol
mol HCO_3^- = 0.0394 mol
<u>mol Cl^- = 0.0394 mol</u>
total ion moles = 0.2362 mol

$$\Delta T_f = K_f \cdot m, \qquad \Delta T_f = \left(1.86\ \frac{^\circ C \cdot kg}{mol}\right)\left(\frac{0.2362\ mol}{0.2500\ kg}\right) = 1.76\ ^\circ C$$

Solution freezing point = 0 °C − ΔT_f = −1.76 °C

(c) H_2O, 18.02 amu

$$mol\ H_2O = 250.0\ g \times \frac{1\ mol\ H_2O}{18.02\ g\ H_2O} = 13.87\ mol\ H_2O$$

$$X_{solv} = \frac{mol\ H_2O}{mol\ H_2O\ +\ mol\ ions} = \frac{13.87\ mol}{13.87\ mol\ +\ 0.2362\ mol} = 0.9833$$

$$P_{soln} = P_{solv} \cdot X_{solv} = (23.76\ mm\ Hg)(0.9833) = 23.36\ mm\ Hg$$

15.146 25 °C = 298 K

$$\Pi = 2MRT; \quad M = \frac{\Pi}{2RT} = \frac{\left(74.4\ mm\ Hg \times \dfrac{1.00\ atm}{760\ mm\ Hg}\right)}{(2)\left(0.082\ 06\ \dfrac{L \cdot atm}{K \cdot mol}\right)(298\ K)} = 0.00200\ M$$

$[M^+] = [X^-] = 0.00200\ M$
$K_{sp} = [M^+][X^-] = (0.00200)^2 = 4.00 \times 10^{-6}$

15.147 (a) $HCO_3^-(aq) + OH^-(aq) \rightarrow CO_3^{2-}(aq) + H_2O(l)$

(b) mol $HCO_3^- = (0.560 \text{ mol/L})(0.0500 \text{ L}) = 0.0280 \text{ mol } HCO_3^-$

mol $OH^- = (0.400 \text{ mol/L})(0.0500 \text{ L}) = 0.0200 \text{ mol } OH^-$

	$HCO_3^-(aq)$	$+ OH^-(aq)$	$\rightarrow CO_3^{2-}(aq)$	$+ H_2O(l)$
before reaction (mol)	0.0280	0.0200	0	
change (mol)	−0.0200	−0.0200	+0.0200	
after reaction (mol)	0.0280 − 0.0200	0	0.0200	

mol $HCO_3^- = 0.0280 - 0.0200 = 0.0080 \text{ mol}$

$$[HCO_3^-] = \frac{0.0080 \text{ mol}}{0.1000 \text{ L}} = 0.080 \text{ M} \qquad [CO_3^{2-}] = \frac{0.0200 \text{ mol}}{0.1000 \text{ L}} = 0.200 \text{ M}$$

	$HCO_3^-(aq)$	$+ H_2O(l)$	$\rightleftharpoons H_3O^+(aq)$	$+ CO_3^{2-}(aq)$
initial (M)	0.080		~0	0.200
change (M)	−x		+x	+x
equil (M)	0.080 − x		x	0.200 + x

$$K_a = \frac{[H_3O^+][CO_3^{2-}]}{[HCO_3^-]} = 5.6 \times 10^{-11} = \frac{x(0.200 + x)}{0.080 - x} \approx \frac{x(0.200)}{0.080}$$

Solve for x. $x = [H_3O^+] = 2.24 \times 10^{-11} \text{ M}$

$pH = -\log[H_3O^+] = -\log(2.24 \times 10^{-11}) = 10.65$

Because this solution contains both a weak acid (HCO_3^-) and its conjugate base, the solution is a buffer.

(c) $HCO_3^-(aq) + OH^-(aq) \rightarrow CO_3^{2-}(aq) + H_2O(l)$

$\Delta H^\circ_{rxn} = [\Delta H^\circ_f(CO_3^{2-}) + \Delta H^\circ_f(H_2O)] - [\Delta H^\circ_f(HCO_3^-) + \Delta H^\circ_f(OH^-)]$

$\Delta H^\circ_{rxn} = [(1 \text{ mol})(-677.1 \text{ kJ/mol}) + (1 \text{ mol})(-285.8 \text{ kJ/mol})]$
$\qquad - [(1 \text{ mol})(-692.0 \text{ kJ/mol}) + (1 \text{ mol})(-230 \text{ kJ/mol})]$

$\Delta H^\circ_{rxn} = -40.9 \text{ kJ}$

0.0200 moles each of HCO_3^- and OH^- reacted.

heat produced = q = (0.0200 mol)(40.9 kJ/mol) = 0.818 kJ = 818 J

(d) q = m x specific heat x ΔT

$$\Delta T = \frac{q}{m \text{ x specific heat}} = \frac{818 \text{ J}}{(100.0 \text{ g})[4.18 \text{ J/(g} \cdot {}^\circ\text{C)}]} = 2.0 \, {}^\circ\text{C}$$

Final temperature = 25 °C + 2.0 °C = 27 °C

15.148 (a) species present initially:

NH_4^+	CO_3^{2-}	H_2O
acid	base	acid or base

$2H_2O(l) \rightleftharpoons H_3O^+(aq) + OH^-(aq)$

$NH_4^+(aq) + H_2O(l) \rightleftharpoons NH_3(aq) + H_3O^+(aq)$

$CO_3^{2-}(aq) + H_2O(l) \rightleftharpoons HCO_3^-(aq) + OH^-(aq)$

NH_3, $K_b = 1.8 \times 10^{-5}$
NH_4^+, $K_a = 5.6 \times 10^{-10}$
CO_3^{2-}, $K_b = 1.8 \times 10^{-4}$
HCO_3^-, $K_a = 5.6 \times 10^{-11}$

In the mixture, proton transfer takes place from the stronger acid to the stronger base, so the principal reaction is $NH_4^+(aq) + CO_3^{2-}(aq) \rightleftharpoons HCO_3^-(aq) + NH_3(aq)$

(b)

$$NH_4^+(aq) + OH^-(aq) \rightleftharpoons NH_3(aq) + H_2O(l) \qquad K_1 = 1/K_b(NH_3)$$
$$\underline{CO_3^{2-}(aq) + H_2O(l) \rightleftharpoons HCO_3^-(aq) + OH^-(aq) \qquad K_2 = K_b(CO_3^{2-})}$$
$$NH_4^+(aq) + CO_3^{2-}(aq) \rightleftharpoons HCO_3^-(aq) + NH_3(aq) \qquad K = K_1 \cdot K_2$$

	NH_4^+	CO_3^{2-}	HCO_3^-	NH_3
initial (M)	0.16	0.080	0	0.16
change (M)	$-x$	$-x$	$+x$	$+x$
equil (M)	$0.16 - x$	$0.080 - x$	x	$0.16 + x$

$$K = \frac{[HCO_3^-][NH_3]}{[NH_4^+][CO_3^{2-}]} = \frac{1.8 \times 10^{-4}}{1.8 \times 10^{-5}} = 10 = \frac{x(0.16 + x)}{(0.16 - x)(0.080 - x)}$$

$9x^2 - 2.56x + 0.128 = 0$
Use the quadratic formula to solve for x.

$$x = \frac{-(-2.56) \pm \sqrt{(-2.56)^2 - (4)(9)(0.128)}}{2(9)} = \frac{2.56 \pm 1.395}{18}$$

$x = 0.220$ and 0.0647

Of the two solutions for x, only 0.00647 has physical meaning because 0.220 leads to negative concentrations.

$[NH_4^+] = 0.16 - x = 0.16 - 0.0647 = 0.0953 \text{ M} = 0.095 \text{ M}$
$[NH_3] = 0.16 + x = 0.16 + 0.0647 = 0.225 \text{ M} = 0.22 \text{ M}$
$[CO_3^{2-}] = 0.080 - x = 0.080 - 0.0647 = 0.0153 \text{ M} = 0.015 \text{ M}$
$[HCO_3^-] = x = 0.0647 \text{ M} = 0.065 \text{ M}$

The solution is a buffer containing two different sets of conjugate acid-base pairs. Either pair can be used to calculate the pH.

For NH_4^+, $K_a = 5.6 \times 10^{-10}$ and $pK_a = 9.25$

$$pH = pK_a + \log \frac{[NH_3]}{[NH_4^+]} = 9.25 + \log \frac{(0.225)}{(0.0953)} = 9.62$$

$[H_3O^+] = 10^{-pH} = 10^{-9.62} = 2.4 \times 10^{-10} \text{ M}$

$$[OH^-] = \frac{1.0 \times 10^{-14}}{2.4 \times 10^{-10}} = 4.2 \times 10^{-5} \text{ M}$$

$$[H_2CO_3] = \frac{[HCO_3^-][H_3O^+]}{K_a} = \frac{(0.647)(2.4 \times 10^{-10})}{(4.3 \times 10^{-7})} = 3.6 \times 10^{-4} \text{ M}$$

(c) For MCO_3, $IP = [M^{2+}][CO_3^{2-}] = (0.010)(0.0153) = 1.5 \times 10^{-4}$
$K_{sp}(CaCO_3) = 5.0 \times 10^{-9}$, $10^3 K_{sp} = 5.0 \times 10^{-6}$
$K_{sp}(BaCO_3) = 2.6 \times 10^{-9}$, $10^3 K_{sp} = 2.6 \times 10^{-6}$
$K_{sp}(MgCO_3) = 6.8 \times 10^{-6}$, $10^3 K_{sp} = 6.8 \times 10^{-3}$

IP > 10^3 K_{sp} for $CaCO_3$ and $BaCO_3$, but IP < 10^3 K_{sp} for $MgCO_3$ so the $[CO_3^{2-}]$ is large enough to give observable precipitation of $CaCO_3$ and $BaCO_3$, but not $MgCO_3$.

(d) For $M(OH)_2$, IP = $[M^{2+}][OH^-]^2$ = $(0.010)(4.17 \times 10^{-5})^2$ = 1.7×10^{-11}

$K_{sp}(Ca(OH)_2)$ = 4.7×10^{-6}, 10^3 K_{sp} = 4.7×10^{-3}

$K_{sp}(Ba(OH)_2)$ = 5.0×10^{-3}, 10^3 K_{sp} = 5.0

$K_{sp}(Mg(OH)_2)$ = 5.6×10^{-12}, 10^3 K_{sp} = 5.6×10^{-9}

IP < 10^3 K_{sp} for all three $M(OH)_2$. None precipitate.

(e) $\qquad CO_3^{2-}(aq) \ + \ H_2O(l) \ \rightleftharpoons \ HCO_3^-(aq) \ + \ OH^-(aq)$

initial (M)	0.08	0	~0
change (M)	–x	+x	+x
equil (M)	0.08 – x	x	x

$$K_b = \frac{[HCO_3^-][OH^-]}{[CO_3^{2-}]} = 1.8 \times 10^{-4} = \frac{x^2}{(0.08 - x)}$$

$x^2 + (1.8 \times 10^{-4})x - (1.44 \times 10^{-5}) = 0$

Use the quadratic formula to solve for x.

$$x = \frac{-(1.8 \times 10^{-4}) \pm \sqrt{(1.8 \times 10^{-4})^2 - (4)(1)(-1.44 \times 10^{-5})}}{2(1)} = \frac{-(1.8 \times 10^{-4}) \pm 7.59 \times 10^{-3}}{2}$$

x = 0.0037 and –0.0039

Of the two solutions for x, only 0.0037 has physical meaning because –0.0039 leads to negative concentrations.

$[OH^-]$ = x = 3.7×10^{-3} M

For MCO_3, IP = $[M^{2+}][CO_3^{2-}]$ = $(0.010)(0.08)$ = 8.0×10^{-4}

For $M(OH)_2$, IP = $[M^{2+}][OH^-]^2$ = $(0.010)(3.7 \times 10^{-3})^2$ = 1.4×10^{-7}

Comparing IP's here and 10^3 K_{sp}'s in (c) and (d) above, Ca^{2+} and Ba^{2+} cannot be separated from Mg^{2+} using 0.08 M Na_2CO_3. Na_2CO_3 is more basic than $(NH_4)_2CO_3$ and $Mg(OH)_2$ would precipitate along with $CaCO_3$ and $BaCO_3$.

15.149 (a) H_2SO_4, 98.09 amu

Assume 1.00 L = 1000 mL of solution.

mass of solution = (1000 mL)(1.836 g/mL) = 1836 g

mass H_2SO_4 = (0.980)(1836 g) = 1799 g H_2SO_4

$$\text{mol } H_2SO_4 = 1799 \text{ g } H_2SO_4 \times \frac{1 \text{ mol } H_2SO_4}{98.09 \text{ g } H_2SO_4} = 18.3 \text{ mol } H_2SO_4$$

$[H_2SO_4]$ = 18.3 mol/ 1.00 L = 18.3 M

(b) Na_2CO_3, 105.99 amu; 1 kg = 1000 g = 2.2046 lb

$H_2SO_4(aq) \ + \ Na_2CO_3(s) \ \rightarrow \ Na_2SO_4(aq) \ + \ H_2O(l) \ + \ CO_2(g)$

$$\text{mass } H_2SO_4 = (0.980)(36 \text{ tons}) \times \frac{2000 \text{ lb}}{1 \text{ ton}} \times \frac{1000 \text{ g}}{2.2046 \text{ lb}} = 3.20 \times 10^7 \text{ g } H_2SO_4$$

$$\text{mol } H_2SO_4 = 3.20 \times 10^7 \text{ g } H_2SO_4 \times \frac{1 \text{ mol } H_2SO_4}{98.09 \text{ g } H_2SO_4} = 3.26 \times 10^5 \text{ mol } H_2SO_4$$

$$\text{mass Na}_2\text{CO}_3 = 3.26 \times 10^5 \text{ mol H}_2\text{SO}_4 \times \frac{1 \text{ mol Na}_2\text{CO}_3}{1 \text{ mol H}_2\text{SO}_4} \times \frac{105.99 \text{ g Na}_2\text{CO}_3}{1 \text{ mol Na}_2\text{CO}_3} \times$$

$$\frac{1 \text{ kg}}{1000 \text{ g}} = 3.5 \times 10^4 \text{ kg Na}_2\text{CO}_3$$

(c) $\text{mol CO}_2 = 3.26 \times 10^5 \text{ mol H}_2\text{SO}_4 \times \dfrac{1 \text{ mol CO}_2}{1 \text{ mol H}_2\text{SO}_4} = 3.26 \times 10^5 \text{ mol CO}_2$

$18 \,^\circ\text{C} = 18 + 273 = 291 \text{ K}$

$PV = nRT$

$$V = \frac{nRT}{P} = \frac{(3.26 \times 10^5 \text{ mol})\left(0.082\ 06 \dfrac{\text{L} \cdot \text{atm}}{\text{K} \cdot \text{mol}}\right)(291 \text{ K})}{\left(745 \text{ mm Hg} \times \dfrac{1.00 \text{ atm}}{760 \text{ mm Hg}}\right)} = 7.9 \times 10^6 \text{ L}$$

15.150 $\text{Pb(CH}_3\text{CO}_2)_2$, 325.29 amu; PbS, 239.27 amu

(a) $\text{mass PbS} = (2 \text{ mL})(1 \text{ g/mL})(0.003) \times \dfrac{1 \text{ mol Pb(CH}_3\text{CO}_2)_2}{325.29 \text{ g Pb(CH}_3\text{CO}_2)_2} \times$

$\dfrac{1 \text{ mol PbS}}{1 \text{ mol Pb(CH}_3\text{CO}_2)_2} \times \dfrac{239.27 \text{ g PbS}}{1 \text{ mol PbS}} \times (30/100) = 0.0013 \text{ g}$

$= 1.3 \text{ mg PbS per dye application}$

(b) $[\text{H}_3\text{O}^+] = 10^{-\text{pH}} = 10^{-5.50} = 3.16 \times 10^{-6} \text{ M}$

	PbS(s) +	$2 \text{ H}_3\text{O}^+(\text{aq})$ ⇌	$\text{Pb}^{2+}(\text{aq})$ +	$\text{H}_2\text{S(aq)}$ +	$2 \text{ H}_2\text{O(l)}$
initial (M)		3.16×10^{-6}	0	0	
change (M)		$-2x$	$+x$	$+x$	
equil (M)		$3.16 \times 10^{-6} - 2x$	x	x	

$$K_{spa} = \frac{[\text{Pb}^{2+}][\text{H}_2\text{S}]}{[\text{H}_3\text{O}^+]^2} = \frac{x^2}{(3.16 \times 10^{-6} - 2x)^2} \approx \frac{x^2}{(3.16 \times 10^{-6})^2} = 3 \times 10^{-7}$$

$x^2 = (3.16 \times 10^{-6})^2(3 \times 10^{-7}) = 3.0 \times 10^{-18}$

$x = 1.7 \times 10^{-9} \text{ M} = [\text{Pb}^{2+}]$ for a saturated solution.

mass of PbS dissolved per washing =

$(3 \text{ gal})(3.7854 \text{ L/1 gal})(1.7 \times 10^{-9} \text{ mol/L}) \times \dfrac{239.27 \text{ g PbS}}{1 \text{ mol PbS}} = 4.6 \times 10^{-6} \text{ g PbS/washing}$

Number of washings required to remove 50% of the PbS from one application =

$$\frac{(0.0013 \text{ g PbS})(50/100)}{(4.6 \times 10^{-6} \text{ g PbS/washing})} = 1.4 \times 10^2 \text{ washings}$$

(c) The number of washings does not look reasonable. It seems too high considering that frequent dye application is recommended. If the PbS is located mainly on the surface of the hair, as is believed to be the case, solid particles of PbS can be lost by abrasion during shampooing.

16 Thermodynamics: Entropy, Free Energy, and Equilibrium

16.1 (a) spontaneous; (b), (c), and (d) nonspontaneous

16.2 (a) $H_2O(g) \rightarrow H_2O(l)$
A liquid has less randomness than a gas. Therefore, ΔS is negative.
(b) $I_2(g) \rightarrow 2\ I(g)$
ΔS is positive because the reaction increases the number of gaseous particles from 1 mol to 2 mol.
(c) $CaCO_3(s) \rightarrow CaO(s) + CO_2(g)$
ΔS is positive because the reaction increases the number of gaseous molecules.
(d) $Ag^+(aq) + Br^-(aq) \rightarrow AgBr(s)$
A solid has less randomness than +1 and −1 charged ions in an aqueous solution. Therefore, ΔS is negative.

16.3 (a) $A_2 + AB_3 \rightarrow 3\ AB$
(b) ΔS is positive because the reaction increases the number of gaseous molecules.

16.4 (a) disordered N_2O (more randomness)
(b) quartz glass (amorphous solid, more randomness)
(c) 1 mole N_2 at STP (larger volume, more randomness)
(d) 1 mole N_2 at 273 K and 0.25 atm (larger volume, more randomness)

16.5 $CaCO_3(s) \rightarrow CaO(s) + CO_2(g)$
$\Delta S^\circ = [S^\circ(CaO) + S^\circ(CO_2)] - S^\circ(CaCO_3)$
$\Delta S^\circ = [(1\ mol)(38.1\ J/(K \cdot mol)) + (1\ mol)(213.6\ J/(K \cdot mol))]$
$\qquad\qquad\qquad - (1\ mol)(91.7\ J/(K \cdot mol)) = +160.0\ J/K$

16.6 From Problem 16.5, $\Delta S_{sys} = \Delta S^\circ = 160.0\ J/K$
$CaCO_3(s) \rightarrow CaO(s) + CO_2(g)$
$\Delta H^\circ = [\Delta H^\circ_f(CaO) + \Delta H^\circ_f(CO_2)] - \Delta H^\circ_f(CaCO_3)$
$\Delta H^\circ = [(1\ mol)(-634.9\ kJ/mol) + (1\ mol)(-393.5\ kJ/mol)]$
$\qquad\qquad\qquad - (1\ mol)(-1207.6\ kJ/mol) = +179.2\ kJ$

$\Delta S_{surr} = \dfrac{-\Delta H^\circ}{T} = \dfrac{-179,200\ J}{298\ K} = -601\ J/K$

$\Delta S_{total} = \Delta S_{sys} + \Delta S_{surr} = 160.0\ J/K + (-601\ J/K) = -441\ J/K$
Because ΔS_{total} is negative, the reaction is not spontaneous under standard-state conditions at 25 °C.

16.7 (a) $\Delta G = \Delta H - T\Delta S = 55.3\ kJ - (298\ K)(0.1757\ kJ/K) = +2.9\ kJ$
Because $\Delta G > 0$, the reaction is nonspontaneous at 25 °C (298 K)

(b) Set $\Delta G = 0$ and solve for T.

$$0 = \Delta H - T\Delta S; \qquad T = \frac{\Delta H}{\Delta S} = \frac{55.3 \text{ kJ}}{0.1757 \text{ kJ/K}} = 315 \text{ K} = 42 \text{ °C}$$

16.8   (a) $\Delta G = \Delta H - T\Delta S = 59.11 \text{ kJ/mol} - (598 \text{ K})[0.0939 \text{ kJ/(K} \cdot \text{mol)}] = +3.0 \text{ kJ/mol}$
Because $\Delta G > 0$, Hg does not boil at 325 °C and 1 atm.
(b) The boiling point (phase change) is associated with an equilibrium. Set $\Delta G = 0$ and solve for T, the boiling point.

$$0 = \Delta H_{vap} - T\Delta S_{vap}; \qquad T_{bp} = \frac{\Delta H_{vap}}{\Delta S_{vap}} = \frac{59.11 \text{ kJ/mol}}{0.0939 \text{ kJ/(K} \cdot \text{mol)}} = 629 \text{ K} = 356 \text{ °C}$$

16.9 $\Delta H < 0$ (reaction involves bond making - exothermic)
$\Delta S < 0$ (the reaction has less randomness in going from reactants (2 atoms) to
          products (1 molecule)
$\Delta G < 0$ (the reaction is spontaneous)

16.10   From Problems 16.5 and 16.6: $\Delta H° = 179.2 \text{ kJ}$ and $\Delta S° = 160.0 \text{ J/K} = 0.1600 \text{ kJ/K}$
(a) $\Delta G° = \Delta H° - T\Delta S° = 179.2 \text{ kJ} - (298 \text{ K})(0.1600 \text{ kJ/K}) = +131.5 \text{ kJ}$
(b) Because $\Delta G > 0$, the reaction is nonspontaneous at 25 °C (298 K).
(c) Set $\Delta G = 0$ and solve for T, the temperature above which the reaction becomes spontaneous.

$$0 = \Delta H - T\Delta S; \qquad T = \frac{\Delta H}{\Delta S} = \frac{179.2 \text{ kJ}}{0.1600 \text{ kJ/K}} = 1120 \text{ K} = 847 \text{ °C}$$

16.11 $2 \text{ AB}_2 \rightarrow \text{ A}_2 + 2 \text{ B}_2$
(a) $\Delta S°$ is positive because the reaction increases the number of molecules.
(b) $\Delta H°$ is positive because the reaction is endothermic.
$\Delta G° = \Delta H° - T\Delta S°$
For the reaction to be spontaneous, $\Delta G°$ must be negative. This will only occur at high temperature where $T\Delta S°$ is greater than $\Delta H°$.

16.12   (a) $\text{CaC}_2(s) + 2 \text{ H}_2\text{O}(l) \rightarrow \text{ C}_2\text{H}_2(g) + \text{ Ca(OH)}_2(s)$
$\Delta G° = [\Delta G°_f(\text{C}_2\text{H}_2) + \Delta G°_f(\text{Ca(OH)}_2)] - [\Delta G°_f(\text{CaC}_2) + 2 \Delta G°_f(\text{H}_2\text{O})]$
$\Delta G° = [(1 \text{ mol})(209.9 \text{ kJ/mol}) + (1 \text{ mol})(-897.5 \text{ kJ/mol})]$
$\qquad - [(1 \text{ mol})(-64.8 \text{ kJ/mol}) + (2 \text{ mol})(-237.2 \text{ kJ/mol})] = -148.4 \text{ kJ}$
This reaction can be used for the synthesis of C_2H_2 because $\Delta G < 0$.
(b) It is not possible to synthesize acetylene from solid graphite and gaseous H_2 at 25 °C and 1 atm because $\Delta G°_f(\text{C}_2\text{H}_2) > 0$.

16.13 $\text{C}(s) + 2 \text{ H}_2(g) \rightarrow \text{ C}_2\text{H}_4(g)$

$$Q_p = \frac{P_{\text{C}_2\text{H}_4}}{(P_{\text{H}_2})^2} = \frac{(0.10)}{(100)^2} = 1.0 \times 10^{-5}$$

$\Delta G = \Delta G^{\circ} + RT \ln Q_p$

$\Delta G = 68.1 \text{ kJ/mol} + [8.314 \times 10^{-3} \text{ kJ/(K} \cdot \text{mol)}](298 \text{ K})\ln(1.0 \times 10^{-5}) = +39.6 \text{ kJ/mol}$

Because $\Delta G > 0$, the reaction is spontaneous in the reverse direction.

16.14 $\Delta G = \Delta G^{\circ} + RT \ln Q$ and $\Delta G^{\circ} = 15 \text{ kJ}$

For $A_2(g) + B_2(g) \rightleftharpoons 2 AB(g)$, $Q_p = \dfrac{(P_{AB})^2}{(P_{A_2})(P_{B_2})}$

Let the number of molecules be proportional to the partial pressure.
(1) $Q_p = 1.0$ (2) $Q_p = 0.0667$ (3) $Q_p = 18$
(a) Reaction (3) has the largest ΔG because Q_p is the largest. Reaction (2) has the smallest ΔG because Q_p is the smallest.
(b) $\Delta G = \Delta G^{\circ} = 15 \text{ kJ}$ because $Q_p = 1$ and $\ln(1) = 0$.

16.15   From Problem 16.10, $\Delta G^{\circ} = +131.5 \text{ kJ}$
$\Delta G^{\circ} = -RT \ln K_p$

$\ln K_p = \dfrac{-\Delta G^{\circ}}{RT} = \dfrac{-131.5 \text{ kJ/mol}}{[8.314 \times 10^{-3} \text{ kJ/(K} \cdot \text{mol)}](298 \text{ K})} = -53.1$

$K_p = e^{-53.1} = 9 \times 10^{-24}$

16.16 $H_2O(l) \rightleftharpoons H_2O(g)$
$K_p = P_{H_2O}$; K_p is equal to the vapor pressure for H_2O.

$\Delta G^{\circ} = \Delta G^{\circ}_f(H_2O(g)) - \Delta G^{\circ}_f(H_2O(l))$
$\Delta G^{\circ} = (1 \text{ mol})(-228.6 \text{ kJ/mol}) - (1 \text{ mol})(-237.2 \text{ kJ/mol}) = +8.6 \text{ kJ}$
$\Delta G^{\circ} = -RT \ln K_p$

$\ln K_p = \dfrac{-\Delta G^{\circ}}{RT} = \dfrac{-8.6 \text{ kJ/mol}}{[8.314 \times 10^{-3} \text{ kJ/(K} \cdot \text{mol)}](298 \text{ K})} = -3.5$

$K_p = P_{H_2O} = e^{-3.5} = 0.03 \text{ atm}$

16.17 $\Delta G^{\circ} = -RT \ln K = -[8.314 \times 10^{-3} \text{ kJ/(K} \cdot \text{mol)}](298 \text{ K}) \ln (1.0 \times 10^{-14}) = 80 \text{ kJ/mol}$

16.18   Photosynthetic cells in plants use the sun's energy to make glucose, which is then used by animals as their primary source of energy. The energy an animal obtains from glucose is then used to build and organize complex molecules, resulting in a decrease in entropy for the animal. At the same time, however, the entropy of the surroundings increases as the animal releases small, simple waste products such as CO_2 and H_2O. Furthermore, heat is released by the animal, further increasing the entropy of the surroundings. Thus, an organism pays for its decrease in entropy by increasing the entropy of the rest of the universe.

16.19   You would expect to see violations of the second law if you watched a movie run backwards. Consider an action-adventure movie with a lot of explosions. An explosion is a spontaneous process that increases the entropy of the universe. You would see an explosion go backwards if you run the the movie backwards but this is impossible because it would decrease the entropy of the universe.

Key Concept Problems

16.20 (a)

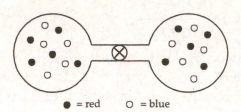

$\bullet$ = red $\qquad$ O = blue

(b) $\Delta H = 0$ (no heat is gained or lost in the mixing of ideal gases)

$\Delta S > 0$ (the mixture of the two gases has more randomness)

$\Delta G < 0$ (the mixing of the two gases is spontaneous)

(c) For an isolated system, $\Delta S_{surr} = 0$ and $\Delta S_{sys} = \Delta S_{Total} > 0$ for the spontaneous process.

(d) $\Delta G > 0$ and the process is nonspontaneous.

16.21 $\Delta H > 0$ (heat is absorbed during sublimation)

$\Delta S > 0$ (gas has more randomness than solid)

$\Delta G < 0$ (the reaction is spontaneous)

16.22 $\Delta H < 0$ (heat is lost during condensation)

$\Delta S < 0$ (liquid has less randomness than vapor)

$\Delta G < 0$ (the reaction is spontaneous)

16.23 $\Delta H = 0$ (system is an ideal gas at constant temperature)

$\Delta S < 0$ (there is less randomness in the smaller volume)

$\Delta G > 0$ (compression of a gas is not spontaneous)

16.24 (a) $2 A_2 + B_2 \rightarrow 2 A_2B$

(b) $\Delta H < 0$ (because ΔS is negative, ΔH must also be negative in order for ΔG to be negative)

$\Delta S < 0$ (the mixture becomes less random going from reactants (3 molecules) to products (2 molecules))

$\Delta G < 0$ (the reaction is spontaneous)

16.25 (a) For initial state 1, $Q_p < K_p$

(more reactant (A_2) than product (A) compared to the equilibrium state)

For initial state 2, $Q_p > K_p$

(more product (A) than reactant (A_2) compared to the equilibrium state)

(b) $\Delta H > 0$ (reaction involves bond breaking - endothermic)

$\Delta S > 0$ (equilibrium state has more randomness than initial state 1)

$\Delta G < 0$ (reaction spontaneously proceeds toward equilibrium)

(c) $\Delta H < 0$ (reaction involves bond making - exothermic)

$\Delta S < 0$ (equilibrium state has less randomness than initial state 2)

$\Delta G < 0$ (reaction spontaneously proceeds toward equilibrium)

(d) State 1 lies to the left of the minimum in Figure 16.10. State 2 lies to the right of the minimum.

16.26   (a) $\Delta H^\circ > 0$ (reaction involves bond breaking - endothermic)

$\Delta S^\circ > 0$ (2 A's have more randomness than A_2)

(b) ΔS° is for the complete conversion of 1 mole of A_2 in its standard state to 2 moles of A in its standard state.

(c) There is not enough information to say anything about the sign of ΔG°. ΔG° decreases (becomes less positive or more negative) as the temperature increases.

(d) K_p increases as the temperature increases. As the temperature increases there will be more A and less A_2.

(e) $\Delta G = 0$ at equilibrium.

16.27   (a) Because the free energy decreases as pure reactants form products and also decreases as pure products form reactants, the free energy curve must go through a minimum somewhere between pure reactants and pure products. At the minimum point, $\Delta G = 0$ and the system is at equilibrium.

(b) The minimum in the plot is on the left side of the graph because $\Delta G^\circ > 0$ and the equilibrium composition is rich in reactants.

16.28 $\Delta G^\circ = -RT \ln K$ where $K = \dfrac{[X]}{[A]}$ or $\dfrac{[Y]}{[A]}$ or $\dfrac{[Z]}{[A]}$

Let the number of molecules be proportional to the concentration.

(1) $K = 1$, $\ln K = 0$, and $\Delta G^\circ = 0$.

(2) $K > 1$, $\ln K$ is positive, and ΔG° is negative.

(3) $K < 1$, $\ln K$ is negative, and ΔG° is positive.

16.29   The equilibrium mixture is richer in reactant A at the higher temperature. This means the reaction is exothermic ($\Delta H < 0$). At 25 °C, $\Delta G^\circ < 0$ because $K > 1$ and at 45 °C, $\Delta G^\circ > 0$ because $K < 1$. Using the relationship. $\Delta G^\circ = \Delta H^\circ - T\Delta S^\circ$, with $\Delta H^\circ < 0$, ΔG° will become positive at the higher temperature only if ΔS° is negative.

Section Problems
Spontaneous Processes (Section 16.1)

16.30   A spontaneous process is one that proceeds on its own without any external influence.

For example: $H_2O(s) \rightarrow H_2O(l)$ at 25 °C

A nonspontaneous process takes place only in the presence of some continuous external influence.

For example: $2 NaCl(s) \rightarrow 2 Na(s) + Cl_2(g)$

16.31 Spontaneous does not mean instantaneous. Even though the decomposition can occur (is spontaneous), the rate of decomposition is determined by the kinetics of the reaction.

16.32 (a) and (d) nonspontaneous; (b) and (c) spontaneous

16.33 (a) and (c) spontaneous; (b) and (d) nonspontaneous

16.34 (b) and (d) spontaneous (because of the large positive K_p's)

16.35 (a) and (d) nonspontaneous (because of the small K's)

Entropy (Sections 16.2–16.4)

16.36 Molecular randomness is called entropy. For the following reaction, the entropy increases: $H_2O(s) \rightarrow H_2O(l)$ at 25 °C.

16.37 Exothermic reactions can become nonspontaneous at high temperatures if ΔS is negative. Endothermic reactions can become spontaneous at high temperatures if ΔS is positive.

16.38 (a) + (solid → gas)   (b) – (liquid → solid)
       (c) – (aqueous ions → solid)   (d) + ($CO_2(aq) \rightarrow CO_2(g)$)

16.39 (a) + (increase in moles of gas)
       (b) – (decrease in moles of gas and formation of liquid)
       (c) + (aqueous ions to gas)
       (d) – (decrease in moles of gas)

16.40 (a) – (liquid → solid)
       (b) – (decrease in number of O_2 molecules)
       (c) + (gas has more randomness in larger volume)
       (d) – (aqueous ions → solid)

16.41 (a) + (solid dissolved in water)   (b) + (increase in moles of gas)
       (c) + (mixed gases have more randomness) (d) + (liquid to gas)

16.42 $S = k \ln W$, $k = 1.38 \times 10^{-23}$ J/K
       (a) $S = (1.38 \times 10^{-23}$ J/K$) \ln (4^{12}) = 2.30 \times 10^{-22}$ J/K
       (b) $S = (1.38 \times 10^{-23}$ J/K$) \ln (4^{120}) = 2.30 \times 10^{-21}$ J/K
       (c) $S = (1.38 \times 10^{-23}$ J/K$) \ln (4^{6.02 \times 10^{23}}) = (1.38 \times 10^{-23}$ J/K$)(6.022 \times 10^{23})\ln 4 = 11.5$ J/K
       If all C–D bonds point in the same direction, $S = 0$.

16.43 $S = k \ln W$, $k = 1.38 \times 10^{-23}$ J/K
       (a) $W = 1$; $S = k \ln (1) = 0$
       (b) $W = 3^2$, $= 9$; $S = k \ln (3^2) = 3.03 \times 10^{-23}$ J/K
       (c) $W = 1$; $S = k \ln (1) = 0$

(d) $W = 3^3 = 27$; $S = k \ln (3^3) = 4.55 \times 10^{-23}$ J/K

(e) $W = 1$; $S = k \ln (1) = 0$

(f) $W = 3^{6.02 \times 10^{23}}$; $S = k \ln (3^{6.02 \times 10^{23}}) = 9.13$ J/K

$$\Delta S = R \ln \left(\frac{V_f}{V_i} \right) = (8.314 \text{ J/K})\ln 3 = 9.13 \text{ J/K} \qquad \text{The results are the same.}$$

16.44 (a) H_2 at 25 °C in 50 L (larger volume)

(b) O_2 at 25 °C, 1 atm (larger volume)

(c) H_2 at 100 °C, 1 atm (larger volume and higher T)

(d) CO_2 at 100 °C, 0.1 atm (larger volume and higher T)

16.45 (a) Ice at 0 °C, because of the higher temperature.

(b) N_2 at STP, because it has the larger volume.

(c) N_2 at 0 °C and 50 L, because it has the larger volume.

(d) Water vapor at 150 °C and 1 atm, because it has a larger volume and higher temperature.

16.46 $\Delta S = nR \ln \left(\frac{V_f}{V_i} \right) = (0.050 \text{ mol})(8.314 \text{ J/K} \cdot \text{mol})\ln \left(\frac{3.5 \text{ L}}{2.5 \text{ L}} \right) = 0.14$ J/K

16.47 $\Delta S = nR \ln \left(\frac{V_f}{V_i} \right) = (0.15 \text{ mol})(8.314 \text{ J/K} \cdot \text{mol})\ln \left(\frac{20.0 \text{ L}}{30.0 \text{ L}} \right) = -0.51$ J/K

Standard Molar Entropies and Standard Entropies of Reaction (Section 16.5)

16.48 The standard molar entropy of a substance is the entropy of 1 mol of the pure substance at 1 atm pressure and 25 °C. $\Delta S° = S°(\text{products}) - S°(\text{reactants})$

16.49 (a) Units of $S° = \dfrac{J}{K \cdot \text{mol}}$ \qquad (b) Units of $\Delta S° = $ J/K

Standard molar entropies are called absolute entropies because they are measured with respect to an absolute reference point, the entropy of the substance at 0 K.

$S° = 0 \dfrac{J}{K \cdot \text{mol}}$ at T = 0 K.

16.50 (a) $C_2H_6(g)$; more atoms/molecule

(b) $CO_2(g)$; more atoms/molecule

(c) $I_2(g)$; gas has more randomness than the solid

(d) $CH_3OH(g)$; gas has more randomness than the liquid

16.51 (a) $NO_2(g)$; more atoms/molecule

(b) $CH_3CO_2H(l)$; more atoms/molecule

(c) $Br_2(l)$; liquid has more randomness than the solid

(d) $SO_3(g)$; gas has more randomness than the solid

16.52 (a) $2 H_2O_2(l) \rightarrow 2 H_2O(l) + O_2(g)$

$\Delta S^\circ = [2 \, S^\circ(H_2O(l)) + S^\circ(O_2)] - 2 \, S^\circ(H_2O_2)$

$\Delta S^\circ = [(2 \text{ mol})(69.9 \text{ J/(K} \cdot \text{mol)}) + (1 \text{ mol})(205.0 \text{ J/(K} \cdot \text{mol)})]$

$\qquad - (2 \text{ mol})(110 \text{ J/(K} \cdot \text{mol)}) = +125 \text{ J/K}$ (+, because moles of gas increase)

(b) $2 Na(s) + Cl_2(g) \rightarrow 2 NaCl(s)$

$\Delta S^\circ = 2 \, S^\circ(NaCl) - [2 \, S^\circ(Na) + S^\circ(Cl_2)]$

$\Delta S^\circ = (2 \text{ mol})(72.1 \text{ J/(K} \cdot \text{mol)}) - [(2 \text{ mol})(51.2 \text{ J/(K} \cdot \text{mol)}) + (1 \text{ mol})(223.0 \text{ J/(K} \cdot \text{mol)})]$

$\Delta S^\circ = -181.2 \text{ J/K}$ (–, because moles of gas decrease)

(c) $2 O_3(g) \rightarrow 3 O_2(g)$

$\Delta S^\circ = 3 \, S^\circ(O_2) - 2 \, S^\circ(O_3)$

$\Delta S^\circ = (3 \text{ mol})(205.0 \text{ J/(K} \cdot \text{mol)}) - (2 \text{ mol})(238.8 \text{ J/(K} \cdot \text{mol)})$

$\Delta S^\circ = +137.4 \text{ J/K}$ (+, because moles of gas increase)

(d) $4 Al(s) + 3 O_2(g) \rightarrow 2 Al_2O_3(s)$

$\Delta S^\circ = 2 \, S^\circ(Al_2O_3) - [4 \, S^\circ(Al) + 3 \, S^\circ(O_2)]$

$\Delta S^\circ = (2 \text{ mol})(50.9 \text{ J/(K} \cdot \text{mol)}) - [(4 \text{ mol})(28.3 \text{ J/(K} \cdot \text{mol)}) + (3 \text{ mol})(205.0 \text{ J/(K} \cdot \text{mol)})]$

$\Delta S^\circ = -626.4 \text{ J/K}$ (–, because moles of gas decrease)

16.53 (a) $2 S(s) + 3 O_2(g) \rightarrow 2 SO_3(g)$

$\Delta S^\circ = 2 \, S^\circ(SO_3) - [2 \, S^\circ(S) + 3 \, S^\circ(O_2)]$

$\Delta S^\circ = (2 \text{ mol})(256.6 \text{ J/(K} \cdot \text{mol)}) - [(2 \text{ mol})(31.8 \text{ J/(K} \cdot \text{mol)}) + (3 \text{ mol})(205.0 \text{ J/(K} \cdot \text{mol)})]$

$\Delta S^\circ = -165.4 \text{ J/K}$ (–, because moles of gas decrease)

(b) $SO_3(g) + H_2O(l) \rightarrow H_2SO_4(aq)$

$\Delta S^\circ = S^\circ(H_2SO_4) - [S^\circ(SO_3) + S^\circ(H_2O)]$

$\Delta S^\circ = (1 \text{ mol})(20 \text{ J/(K} \cdot \text{mol)}) - [(1 \text{ mol})(256.6 \text{ J/(K} \cdot \text{mol)}) + (1 \text{ mol})(69.9 \text{ J/(K} \cdot \text{mol)})]$

$\Delta S^\circ = -306 \text{ J/K}$ (–, because of the conversion of a gas and water to an aqueous solution)

(c) $AgCl(s) \rightarrow Ag^+(aq) + Cl^-(aq)$

$\Delta S^\circ = [S^\circ(Ag^+) + S^\circ(Cl^-)] - S^\circ(AgCl)$

$\Delta S^\circ = [(1 \text{ mol})(72.7 \text{ J/(K} \cdot \text{mol)}) + (1 \text{ mol})(56.5 \text{ J/(K} \cdot \text{mol)})] - (1 \text{ mol})(96.2 \text{ J/(K} \cdot \text{mol)})$

$\Delta S^\circ = +33.0 \text{ J/K}$ (+, because a solid is converted to ions in aqueous solution)

(d) $NH_4NO_3(s) \rightarrow N_2O(g) + 2 H_2O(g)$

$\Delta S^\circ = [S^\circ(N_2O) + 2 \, S^\circ(H_2O)] - S^\circ(NH_4NO_3)$

$\Delta S^\circ = [(1 \text{ mol})(219.7 \text{ J/(K} \cdot \text{mol)}) + (2 \text{ mol})(188.7 \text{ J/(K} \cdot \text{mol)})] - (1 \text{ mol})(151.1 \text{ J/(K} \cdot \text{mol)})$

$\Delta S^\circ = +446.0 \text{ J/K}$ (+, because moles of gas increase)

Entropy and the Second Law of Thermodynamics (Section 16.6)

16.54 In any spontaneous process, the total entropy of a system and its surroundings always increases.

16.55 For a spontaneous process, $\Delta S_{total} = \Delta S_{sys} + \Delta S_{surr} > 0$. For an isolated system, $\Delta S_{surr} = 0$, and so $\Delta S_{sys} > 0$ is the criterion for spontaneous change. An example of a spontaneous process in an isolated system is the mixing of two gases.

16.56 $\Delta S_{surr} = \dfrac{-\Delta H}{T}$; the temperature (T) is always positive.

(a) For an exothermic reaction, ΔH is negative and ΔS_{surr} is positive.
(b) For an endothermic reaction, ΔH is positive and ΔS_{surr} is negative.

16.57 $\Delta S_{surr} \propto \dfrac{1}{T}$

Consider the surroundings as an infinitely large constant-temperature bath to which heat can be added without changing its temperature. If the surroundings has a low temperature, it has only a small amount of randomness, in which case addition of a given quantity of heat results in a substantial increase in the amount of randomness (a relatively large value of ΔS_{surr}). If the surroundings has a high temperature, it already has a large amount of randomness, and addition of the same quantity of heat produces only a marginal increase in the randomness (a relatively small value of ΔS_{surr}). Thus, we expect ΔS_{surr} to vary inversely with temperature.

16.58 $N_2(g) + 2 O_2(g) \rightarrow N_2O_4(g)$
$\Delta H° = \Delta H°_f(N_2O_4) = 11.1$ kJ
$\Delta S_{sys} = \Delta S° = S°(N_2O_4) - [S°(N_2) + 2 S°(O_2)]$
$\Delta S_{sys} = (1 \text{ mol})(304.3 \text{ J/(K} \cdot \text{mol)})$
$\qquad\qquad - [(1 \text{ mol})(191.5 \text{ J/(K} \cdot \text{mol)}) + (2 \text{ mol})(205.0 \text{ J/(K} \cdot \text{mol)})] = -297.2 \text{ J/K}$

$\Delta S_{surr} = \dfrac{-\Delta H°}{T} = \dfrac{-11.1 \text{ kJ}}{298 \text{ K}} = -0.0372 \text{ kJ/K} = -37.2 \text{ J/K}$

$\Delta S_{total} = \Delta S_{sys} + \Delta S_{surr} = -297.2 \text{ J/K} + (-37.2 \text{ J/K}) = -334.4 \text{ J/K}$
Because $\Delta S_{total} < 0$, the reaction is nonspontaneous.

16.59 $Cu_2S(s) + O_2(g) \rightarrow 2 Cu(s) + SO_2(g)$
$\Delta H° = \Delta H°_f(SO_2) - \Delta H°_f(Cu_2S)$
$\Delta H° = (1 \text{ mol})(-296.8 \text{ kJ/mol}) - (1 \text{ mol})(-79.5 \text{ kJ/mol}) = -217.3 \text{ kJ}$
$\Delta S_{sys} = \Delta S° = [2 S°(Cu) + S°(SO_2)] - [S°(Cu_2S) + S°(O_2)]$
$\Delta S_{sys} = [(2 \text{ mol})(33.1 \text{ J/(K} \cdot \text{mol)}) + (1 \text{ mol})(248.1 \text{ J/(K} \cdot \text{mol)})]$
$\qquad\qquad - [(1 \text{ mol})(120.9 \text{ J/(K} \cdot \text{mol)}) + (1 \text{ mol})(205.0 \text{ J/(K} \cdot \text{mol)})] = -11.6 \text{ J/K}$

$\Delta S_{surr} = \dfrac{-\Delta H°}{T} = \dfrac{-(-217,300 \text{ J})}{298.15 \text{ K}} = +728.8 \text{ J/K}$

$\Delta S_{total} = \Delta S_{sys} + \Delta S_{surr} = -11.6 \text{ J/K} + 728.8 \text{ J/K} = +717.2 \text{ J/K}$
Because ΔS_{total} is positive, the reaction is spontaneous under standard-state conditions at 25 °C.

16.60 (a) $\Delta S_{surr} = \dfrac{-\Delta H_{vap}}{T} = \dfrac{-30,700 \text{ J/mol}}{343 \text{ K}} = -89.5 \text{ J/(K} \cdot \text{mol)}$

$\Delta S_{total} = \Delta S_{vap} + \Delta S_{surr} = 87.0 \text{ J/(K} \cdot \text{mol)} + (-89.5 \text{ J/(K} \cdot \text{mol)}) = -2.5 \text{ J/(K} \cdot \text{mol)}$

(b) $\Delta S_{surr} = \dfrac{-\Delta H_{vap}}{T} = \dfrac{-30,700 \text{ J/mol}}{353 \text{ K}} = -87.0 \text{ J/(K} \cdot \text{mol)}$

$\Delta S_{total} = \Delta S_{vap} + \Delta S_{surr} = 87.0 \text{ J/(K} \cdot \text{mol)} + (-87.0 \text{ J/(K} \cdot \text{mol)}) = 0$

(c) $\Delta S_{surr} = \dfrac{-\Delta H_{vap}}{T} = \dfrac{-30{,}700 \text{ J/mol}}{363 \text{ K}} = -84.6 \text{ J/(K} \cdot \text{mol)}$

$\Delta S_{total} = \Delta S_{vap} + \Delta S_{surr} = 87.0 \text{ J/(K} \cdot \text{mol)} + (-84.6 \text{ J/(K} \cdot \text{mol)}) = +2.4 \text{ J/(K} \cdot \text{mol)}$
Benzene does not boil at 70 °C (343 K) because ΔS_{total} is negative.
The normal boiling point for benzene is 80 °C (353 K), where $\Delta S_{total} = 0$.

16.61 (a) $\Delta S_{surr} = \dfrac{-\Delta H_{fusion}}{T} = \dfrac{-28{,}160 \text{ J/mol}}{1050 \text{ K}} = -26.8 \text{ J/(K} \cdot \text{mol)}$

$\Delta S_{total} = \Delta S_{sys} + \Delta S_{surr} = 26.22 \text{ J/(K} \cdot \text{mol)} + (-26.8 \text{ J/(K} \cdot \text{mol)}) = -0.6 \text{ J/(K} \cdot \text{mol)}$

(b) $\Delta S_{surr} = \dfrac{-\Delta H_{fusion}}{T} = \dfrac{-28{,}160 \text{ J/mol}}{1074 \text{ K}} = -26.22 \text{ J/(K} \cdot \text{mol)}$

$\Delta S_{total} = \Delta S_{sys} + \Delta S_{surr} = 26.22 \text{ J/(K} \cdot \text{mol)} + (-26.22 \text{ J/(K} \cdot \text{mol)}) = 0$

(c) $\Delta S_{surr} = \dfrac{-\Delta H_{fusion}}{T} = \dfrac{-28{,}160 \text{ J/mol}}{1100 \text{ K}} = -25.6 \text{ J/(K} \cdot \text{mol)}$

$\Delta S_{total} = \Delta S_{sys} + \Delta S_{surr} = 26.22 \text{ J/(K} \cdot \text{mol)} + (-25.6 \text{ J/(K} \cdot \text{mol)}) = +0.6 \text{ J/(K} \cdot \text{mol)}$
NaCl melts at 1100 K because $\Delta S_{total} > 0$.
The melting point of NaCl is 1074 K, where $\Delta S_{total} = 0$.

Free Energy (Section 16.7)

16.62 $\underline{\Delta H}$	$\underline{\Delta S}$	$\underline{\Delta G = \Delta H - T\Delta S}$	$\underline{\text{Reaction Spontaneity}}$
–	+	–	Spontaneous at all temperatures
–	–	– or +	Spontaneous at low temperatures where $\lvert\Delta H\rvert > \lvert T\Delta S\rvert$ Nonspontaneous at high temperatures where $\lvert\Delta H\rvert < \lvert T\Delta S\rvert$
+	–	+	Nonspontaneous at all temperatures
+	+	– or +	Spontaneous at high temperatures where $T\Delta S > \Delta H$ Nonspontaneous at low temperature where $T\Delta S < \Delta H$

16.63 When ΔH and ΔS are both positive or both negative, the temperature determines the direction of spontaneous reaction. See Problem 16.62 for an explanation.

16.64 (a) 0 °C (temperature is below mp); $\Delta H > 0$, $\Delta S > 0$, $\Delta G > 0$
(b) 15 °C (temperature is above mp); $\Delta H > 0$, $\Delta S > 0$, $\Delta G < 0$

16.65 (a) $\Delta H = 0$

$\Delta S = R \ln \dfrac{V_{final}}{V_{initial}} = (8.314 \text{ J/K}) \ln 2 = 5.76 \text{ J/K}$

$\Delta G = \Delta H - T\Delta S$

Because $\Delta H = 0$, $\Delta G = -T\Delta S = -(298\ K)(5.76\ J/K) = -1717\ J = -1.72\ kJ$

(b) For a process in an isolated system, $\Delta S_{surr} = 0$. Therefore, $\Delta S_{total} = \Delta S_{sys} > 0$, and the process is spontaneous.

16.66 $\Delta H_{vap} = 30.7\ kJ/mol$

$\Delta S_{vap} = 87.0\ J/(K \cdot mol) = 87.0\ x\ 10^{-3}\ kJ/(K \cdot mol)$

$\Delta G_{vap} = \Delta H_{vap} - T\Delta S_{vap}$

(a) $\Delta G_{vap} = 30.7\ kJ/mol - (343\ K)(87.0\ x\ 10^{-3}\ kJ/(K \cdot mol)) = +0.9\ kJ/mol$

At 70 ºC (343 K), benzene does not boil because ΔG_{vap} is positive.

(b) $\Delta G_{vap} = 30.7\ kJ/mol - (353\ K)(87.0\ x\ 10^{-3}\ kJ/(K \cdot mol)) = 0$

80 ºC (353 K) is the boiling point for benzene because $\Delta G_{vap} = 0$

(c) $\Delta G_{vap} = 30.7\ kJ/mol - (363\ K)(87.0\ x\ 10^{-3}\ kJ/(K \cdot mol)) = -0.9\ kJ/mol$

At 90 ºC (363 K), benzene boils because ΔG_{vap} is negative.

16.67 $\Delta H_{fusion} = 28.16\ kJ/mol;$ $\Delta S_{fusion} = 26.22\ x\ 10^{-3}\ kJ/(K \cdot mol)$

$\Delta G_{fusion} = \Delta H_{fusion} - T\Delta S_{fusion}$

(a) $\Delta G_{fusion} = 28.16\ kJ/mol - (1050\ K)(26.22\ x\ 10^{-3}\ kJ/(K \cdot mol)) = +0.63\ kJ/mol$

At 1050 K, NaCl does not melt because ΔG_{fusion} is positive.

(b) $\Delta G_{fusion} = 28.16\ kJ/mol - (1074\ K)(26.22\ x\ 10^{-3}\ kJ/(K \cdot mol)) = 0.0$

1074 K is the melting point for NaCl because $\Delta G_{fusion} = 0$.

(c) $\Delta G_{fusion} = 28.16\ kJ/mol - (1100\ K)(26.22\ x\ 10^{-3}\ kJ/(K \cdot mol)) = -0.68\ kJ/mol$

At 1100 K, NaCl does melt because ΔG_{fusion} is negative.

16.68   At the melting point (phase change), $\Delta G_{fusion} = 0$.

$\Delta G_{fusion} = \Delta H_{fusion} - T\Delta S_{fusion}$

$0 = \Delta H_{fusion} - T\Delta S_{fusion};$ $T = \dfrac{\Delta H_{fusion}}{\Delta S_{fusion}} = \dfrac{18.02\ kJ/mol}{45.56\ x\ 10^{-3}\ kJ/(K \cdot mol)} = 395.5\ K = 122.4\ ºC$

16.69   128 ºC = 401 K

At the melting point (phase change), $\Delta G_{fusion} = 0$.

$\Delta G_{fusion} = \Delta H_{fusion} - T\Delta S_{fusion}$

$0 = \Delta H_{fusion} - T\Delta S_{fusion}$

$\Delta H_{fusion} = T\Delta S_{fusion} = (401\ K)[47.7\ x\ 10^{-3}\ kJ/(K \cdot mol)] = 19.1\ kJ/mol$

Standard Free-Energy Changes and Standard Free Energies of Formation (Sections 16.8–16.9)

16.70   (a) $\Delta G°$ is the change in free energy that occurs when reactants in their standard states are converted to products in their standard states.

(b) $\Delta G°_f$ is the free-energy change for formation of one mole of a substance in its standard state from the most stable form of the constituent elements in their standard states.

16.71   The standard state of a substance (solid, liquid, or gas) is the most stable form of a pure substance at 25 ºC and 1 atm pressure. For solutes, the condition is 1 M at 25 ºC.

16.72 (a) $N_2(g) + 2 O_2(g) \rightarrow 2 NO_2(g)$

$\Delta H° = 2 \Delta H°_f(NO_2) = (2 \text{ mol})(33.2 \text{ kJ/mol}) = 66.4 \text{ kJ}$

$\Delta S° = 2 S°(NO_2) - [S°(N_2) + 2 S°(O_2)]$

$\Delta S° = (2 \text{ mol})(240.0 \text{ J/(K} \cdot \text{mol)}) - [(1 \text{ mol})(191.5 \text{ J/(K} \cdot \text{mol)}) + (2 \text{ mol})(205.0 \text{ J/(K} \cdot \text{mol)})]$

$\Delta S° = -121.5 \text{ J/K} = -121.5 \times 10^{-3} \text{ kJ/K}$

$\Delta G° = \Delta H° - T\Delta S° = 66.4 \text{ kJ} - (298 \text{ K})(-121.5 \times 10^{-3} \text{ kJ/K}) = +102.6 \text{ kJ}$

Because $\Delta G°$ is positive, the reaction is nonspontaneous under standard-state conditions at 25 °C.

(b) $2 KClO_3(s) \rightarrow 2 KCl(s) + 3 O_2(g)$

$\Delta H° = 2 \Delta H°_f(KCl) - 2 \Delta H°_f(KClO_3)$

$\Delta H° = (2 \text{ mol})(-436.5 \text{ kJ/mol}) - (2 \text{ mol})(-397.7 \text{ kJ/mol}) = -77.6 \text{ kJ}$

$\Delta S° = [2 S°(KCl) + 3 S°(O_2)] - 2 S°(KClO_3)$

$\Delta S° = [(2 \text{ mol})(82.6 \text{ J/(K} \cdot \text{mol)}) + (3 \text{ mol})(205.0 \text{ J/(K} \cdot \text{mol)})] - (2 \text{ mol})(143.1 \text{ J/(K} \cdot \text{mol)})$

$\Delta S° = 494.0 \text{ J/(K} \cdot \text{mol)} = 494.0 \times 10^{-3} \text{ kJ/(K} \cdot \text{mol)}$

$\Delta G° = \Delta H° - T\Delta S° = -77.6 \text{ kJ} - (298 \text{ K})(494.0 \times 10^{-3} \text{ kJ/(K} \cdot \text{mol)}) = -224.8 \text{ kJ}$

Because $\Delta G°$ is negative, the reaction is spontaneous under standard-state conditions at 25 °C.

(c) $CH_3CH_2OH(l) + O_2(g) \rightarrow CH_3CO_2H(l) + H_2O(l)$

$\Delta H° = [\Delta H°_f(CH_3CO_2H) + \Delta H°_f(H_2O)] - \Delta H°_f(CH_3CH_2OH)$

$\Delta H° = [(1 \text{ mol})(-484.5 \text{ kJ/mol}) + (1 \text{ mol})(-285.8 \text{ kJ/mol})] - (1 \text{ mol})(-277.7 \text{ kJ/mol}) = -492.6 \text{ kJ}$

$\Delta S° = [S°(CH_3CO_2H) + S°(H_2O)] - [S°(CH_3CH_2OH) + S°(O_2)]$

$\Delta S° = [(1 \text{ mol})(160 \text{ J/(K} \cdot \text{mol)}) + (1 \text{ mol})(69.9 \text{ J/(K} \cdot \text{mol)})]$

$\qquad\qquad - [(1 \text{ mol})(161 \text{ J/(K} \cdot \text{mol)}) + (1 \text{ mol})(205.0 \text{ J/(K} \cdot \text{mol)})]$

$\Delta S° = -136.1 \text{ J/(K} \cdot \text{mol)} = -136.1 \times 10^{-3} \text{ kJ/(K} \cdot \text{mol)}$

$\Delta G° = \Delta H° - T\Delta S° = -492.6 \text{ kJ} - (298 \text{ K})(-136.1 \times 10^{-3} \text{ kJ/(K} \cdot \text{mol)}) = -452.0 \text{ kJ}$

Because $\Delta G°$ is negative, the reaction is spontaneous under standard-state conditions at 25 °C.

16.73 (a) $2 SO_2(g) + O_2(g) \rightarrow 2 SO_3(g)$

$\Delta H° = 2 \Delta H°_f(SO_3) - 2 \Delta H°_f(SO_2)$

$\Delta H° = (2 \text{ mol})(-395.7 \text{ kJ/mol}) - (2 \text{ mol})(-296.8 \text{ kJ/mol}) = -197.8 \text{ kJ}$

$\Delta S° = 2 S°(SO_3) - [2 S°(SO_2) + S°(O_2)]$

$\Delta S° = (2 \text{ mol})(256.6 \text{ J/(K} \cdot \text{mol)}) - [(2 \text{ mol})(248.1 \text{ J/(K} \cdot \text{mol)}) + (1 \text{ mol})(205.0 \text{ J/(K} \cdot \text{mol)})]$

$\Delta S° = -188.0 \text{ J/K} = -188.0 \times 10^{-3} \text{ kJ/K}$

$\Delta G° = \Delta H° - T\Delta S° = -197.8 \text{ kJ} - (298 \text{ K})(-188.0 \times 10^{-3} \text{ kJ/K}) = -141.8 \text{ kJ}$

Because $\Delta G°$ is negative, the reaction is spontaneous under standard-state conditions at 25 °C.

(b) $N_2(g) + 2 H_2(g) \rightarrow N_2H_4(l)$

$\Delta H° = \Delta H°_f(N_2H_4)$

$\Delta H° = (1 \text{ mol})(50.6 \text{ kJ/mol}) = 50.6 \text{ kJ}$

$\Delta S° = S°(N_2H_4) - [S°(N_2) + 2 S°(H_2)]$

$\Delta S° = (1 \text{ mol})(121.2 \text{ J/(K} \cdot \text{mol)}) - [(1 \text{ mol})(191.5 \text{ J/(K} \cdot \text{mol)}) + (2 \text{ mol})(130.6 \text{ J/(K} \cdot \text{mol)})]$

$\Delta S° = -331.5 \text{ J/K} = -331.5 \times 10^{-3} \text{ kJ/K}$

$\Delta G° = \Delta H° - T\Delta S° = 50.6 \text{ kJ} - (298 \text{ K})(-331.5 \times 10^{-3} \text{ kJ/K}) = +149.4 \text{ kJ}$

Because $\Delta G°$ is positive, the reaction is nonspontaneous under standard-state conditions at 25 °C.

(c) $CH_3OH(l) + O_2(g) \rightarrow HCO_2H(l) + H_2O(l)$

$\Delta H° = [\Delta H°_f(HCO_2H) + \Delta H°_f(H_2O)] - \Delta H°_f(CH_3OH)$

$\Delta H° = [(1 \text{ mol})(-424.7 \text{ kJ/mol}) + (1 \text{ mol})(-285.8 \text{ kJ/mol})] - (1 \text{ mol})(-239.2 \text{ kJ/mol}) = -471.3 \text{ kJ}$

$\Delta S° = [S°(HCO_2H) + S°(H_2O)] - [S°(CH_3OH) + S°(O_2)]$

$\Delta S° = [(1 \text{ mol})(129.0 \text{ J/(K} \cdot \text{mol})) + (1 \text{ mol})(69.9 \text{ J/(K} \cdot \text{mol}))]$

$\qquad\qquad\qquad - [(1 \text{ mol})(127 \text{ J/(K} \cdot \text{mol})) + (1 \text{ mol})(205.0 \text{ J/(K} \cdot \text{mol}))]$

$\Delta S° = -133.1 \text{ J/K} = -133.1 \times 10^{-3} \text{ kJ/K}$

$\Delta G° = \Delta H° - T\Delta S° = -471.3 \text{ kJ} - (298 \text{ K})(-133.1 \times 10^{-3} \text{ kJ/K}) = -431.6 \text{ kJ}$

Because $\Delta G°$ is negative, the reaction is spontaneous under standard-state conditions at 25 °C.

16.74 (a) $N_2(g) + 2 O_2(g) \rightarrow 2 NO_2(g)$

$\Delta G° = 2 \Delta G°_f(NO_2) = (2 \text{ mol})(51.3 \text{ kJ/mol}) = +102.6 \text{ kJ}$

(b) $2 KClO_3(s) \rightarrow 2 KCl(s) + 3 O_2(g)$

$\Delta G° = 2 \Delta G°_f(KCl) - 2 \Delta G°_f(KClO_3)$

$\Delta G° = (2 \text{ mol})(-408.5 \text{ kJ/mol}) - (2 \text{ mol})(-296.3 \text{ kJ/mol}) = -224.4 \text{ kJ}$

(c) $CH_3CH_2OH(l) + O_2(g) \rightarrow CH_3CO_2H(l) + H_2O(l)$

$\Delta G° = [\Delta G°_f(CH_3CO_2H) + \Delta G°_f(H_2O)] - \Delta G°_f(CH_3CH_2OH)$

$\Delta G° = [(1 \text{ mol})(-390 \text{ kJ/mol}) + (1 \text{ mol})(-237.2 \text{ kJ/mol})] - (1 \text{ mol})(-174.9 \text{ kJ/mol}) = -452 \text{ kJ}$

16.75 (a) $2 SO_2(g) + O_2(g) \rightarrow 2 SO_3(g)$

$\Delta G° = 2 \Delta G°_f(SO_3) - 2 \Delta G°_f(SO_2)$

$\Delta G° = (2 \text{ mol})(-371.1 \text{ kJ/mol}) - (2 \text{ mol})(-300.2 \text{ kJ/mol}) = -141.8 \text{ kJ}$

(b) $N_2(g) + 2 H_2(g) \rightarrow N_2H_4(l)$

$\Delta G° = \Delta G°_f(N_2H_4) = (1 \text{ mol})(149.2 \text{ kJ/mol}) = 149.2 \text{ kJ}$

(c) $CH_3OH(l) + O_2(g) \rightarrow HCO_2H(l) + H_2O(l)$

$\Delta G° = [\Delta G°_f(HCO_2H) + \Delta G°_f(H_2O)] - \Delta G°_f(CH_3OH)$

$\Delta G° = [(1 \text{ mol})(-361.4 \text{ kJ/mol}) + (1 \text{ mol})(-237.2 \text{ kJ/mol})] - (1 \text{ mol})(-166.6 \text{ kJ/mol})$

$\Delta G° = -432.0 \text{ kJ}$

16.76 A compound is thermodynamically stable with respect to its constituent elements at 25 °C if $\Delta G°_f$ is negative.

	$\Delta G°_f$ (kJ/mol)	Stable
(a) $BaCO_3(s)$	−1134.4	yes
(b) $HBr(g)$	−53.4	yes
(c) $N_2O(g)$	+104.2	no
(d) $C_2H_4(g)$	+68.1	no

16.77 A compound is thermodynamically stable with respect to its constituent elements at 25 °C if $\Delta G°_f$ is negative.

	$\Delta G°_f$ (kJ/mol)	Stable
(a) $C_6H_6(l)$	+124.5	no
(b) $NO(g)$	+87.6	no
(c) $PH_3(g)$	+13.5	no
(d) $FeO(s)$	−255	yes

16.78 $CH_2=CH_2(g) + H_2O(l) \rightarrow CH_3CH_2OH(l)$

$\Delta H° = \Delta H°_f(CH_3CH_2OH) - [\Delta H°_f(CH_2=CH_2) + \Delta H°_f(H_2O)]$

$\Delta H° = (1 \text{ mol})(-277.7 \text{ kJ/mol}) - [(1 \text{ mol})(52.3 \text{ kJ/mol}) + (1 \text{ mol})(-285.8 \text{ kJ/mol})]$

$\Delta H° = -44.2 \text{ kJ}$

$\Delta S° = S°(CH_3CH_2OH) - [S°(CH_2=CH_2) + S°(H_2O)]$

ΔS° = (1 mol)(161 J/(K · mol)) – [(1 mol)(219.5 J/(K · mol)) + (1 mol)(69.9 J/(K · mol))]

ΔS° = –128 J/(K · mol) = –128 x 10^{-3} kJ/(K · mol)

$\Delta G^\circ = \Delta H^\circ - T\Delta S^\circ$ = – 44.2 kJ – (298 K)(–128 x 10^{-3} kJ/K) = –6.1 kJ

Because ΔG° is negative, the reaction is spontaneous under standard-state conditions at 25 °C.

The reaction becomes nonspontaneous at high temperatures because ΔS° is negative.

To find the crossover temperature, set ΔG = 0 and solve for T.

$$T = \frac{\Delta H^\circ}{\Delta S^\circ} = \frac{-44{,}200 \text{ J}}{-128 \text{ J/K}} = 345 \text{ K} = 72 \text{ °C}$$

The reaction becomes nonspontaneous at 72 °C.

5.79 $2 H_2S(g) + SO_2(g) \rightarrow 3 S(s) + 2 H_2O(g)$

$\Delta H^\circ = 2 \Delta H^\circ_f(H_2O) - [2 \Delta H^\circ_f(H_2S) + \Delta H^\circ_f(SO_2)]$

ΔH° = (2 mol)(–241.8 kJ/mol) – [(2 mol)(–20.6 kJ/mol) + (1 mol)(–296.8 kJ/mol)] = –145.6 kJ

$\Delta S^\circ = [3 S^\circ(S) + 2 S^\circ(H_2O)] - [2 S^\circ(H_2S) + S^\circ(SO_2)]$

ΔS° = [(3 mol)(31.8 J/(K · mol)) + (2 mol)(188.7 J/(K · mol))]

$\qquad\qquad$ – [(2 mol)(205.7 J/(K · mol)) + (1 mol)(248.1 J/(K · mol))]

ΔS° = –186.7 J/K = –186.7 x 10^{-3} kJ/K

$\Delta G^\circ = \Delta H^\circ - T\Delta S^\circ$ = –145.6 kJ – (298 K)(–186.7 x 10^{-3} kJ/K) = –90.0 kJ

Because ΔG° is negative, the reaction is spontaneous under standard-state conditions at 25 °C.

The reaction becomes nonspontaneous at high temperatures because ΔS° is negative.

To find the crossover temperature set ΔG = 0 and solve for T.

$$T = \frac{\Delta H^\circ}{\Delta S^\circ} = \frac{-145{,}600 \text{ J}}{-186.7 \text{ J/K}} = 780 \text{ K} = 507 \text{ °C.}$$ The reaction becomes nonspontaneous at 507 °C.

5.80 $3 C_2H_2(g) \rightarrow C_6H_6(l)$

$\Delta G^\circ = \Delta G^\circ_f(C_6H_6) - 3 \Delta G^\circ_f(C_2H_2)$

ΔG° = (1 mol)(124.5 kJ/mol) – (3 mol)(209.9 kJ/mol) = –505.2 kJ

Because ΔG° is negative, the reaction is possible. Look for a catalyst.

Because ΔG°_f for benzene is positive (+124.5 kJ/mol), the synthesis of benzene from graphite and gaseous H_2 at 25 °C and 1 atm pressure is not possible.

5.81 $CH_2ClCH_2Cl(l) \rightarrow CH_2{=}CHCl(g) + HCl(g)$

$\Delta G^\circ = [\Delta G^\circ_f(CH_2{=}CHCl) + \Delta G^\circ_f(HCl)] - \Delta G^\circ_f(CH_2ClCH_2Cl)$

ΔG° = [(1 mol)(51.9 kJ/mol) + (1 mol)(–95.3 kJ/mol)] – (1 mol)(–79.6 kJ/mol) = +36.2 kJ

Because ΔG° is positive, the reaction is nonspontaneous under standard-state conditions at 25 °C.

$CH_2ClCH_2Cl(l) \rightarrow CH_2{=}CHCl(g) + HCl(g)$

um: $\underline{NaOH(aq) + HCl(g) \rightarrow Na^+(aq) + Cl^-(aq) + H_2O(l)}$

$CH_2ClCH_2Cl(l) + NaOH(aq) \rightarrow CH_2{=}CHCl(g) + Na^+(aq) + Cl^-(aq) + H_2O(l)$

$\Delta G^\circ = [\Delta G^\circ_f(CH_2{=}CHCl) + \Delta G^\circ_f(Na^+) + \Delta G^\circ_f(Cl^-) + \Delta G^\circ_f(H_2O)]$

$\qquad\qquad - [\Delta G^\circ_f(CH_2ClCH_2Cl) + \Delta G^\circ_f(NaOH)]$

ΔG° = [(1 mol)(51.9 kJ/mol) + (1 mol)(–261.9 kJ/mol)

$\qquad$ + (1 mol)(–131.3 kJ/mol) + (1 mol)(–237.2 kJ/mol)]

$\qquad\qquad$ – [(1 mol)(–79.6 kJ/mol) + (1 mol)(–419.2 kJ/mol)] = –79.7 kJ

486

Using NaOH(aq), $\Delta G^\circ = -79.7$ kJ and the reaction is spontaneous. (More generally, base removes HCl, driving the reaction to the right.)

The synthesis of a compound from its constituent elements is thermodynamically feasible at 25 °C and 1 atm pressure if ΔG°_f is negative.

Because $\Delta G^\circ_f(CH_2=CHCl) = +51.9$ kJ, the synthesis of vinyl chloride from its elements is not possible at 25 °C and 1 atm pressure.

Free Energy, Composition, and Chemical Equilibrium (Sections 16.10–16.11)

16.82 $\Delta G = \Delta G^\circ + RT \ln Q$

16.83 $\Delta G = \Delta G^\circ + RT \ln Q$
(a) If $Q < 1$, then $RT \ln Q$ is negative and $\Delta G < \Delta G^\circ$.
(b) If $Q = 1$, then $RT \ln Q = 0$ and $\Delta G = \Delta G^\circ$.
(c) If $Q > 1$, then $RT \ln Q$ is positive and $\Delta G > \Delta G^\circ$.
As Q increases the thermodynamic tendency for the reaction to occur decreases.

16.84 $\Delta G = \Delta G^\circ + RT \ln \left[\dfrac{(P_{SO_3})^2}{(P_{SO_2})^2(P_{O_2})} \right]$

(a) $\Delta G = (-141.8 \text{ kJ/mol}) + [8.314 \times 10^{-3} \text{ kJ/(K} \cdot \text{mol)}](298 \text{ K}) \ln \left[\dfrac{(1.0)^2}{(100)^2(100)} \right] = -176.0$ kJ/mol

(b) $\Delta G = (-141.8 \text{ kJ/mol}) + [8.314 \times 10^{-3} \text{ kJ/(K} \cdot \text{mol)}](298 \text{ K}) \ln \left[\dfrac{(10)^2}{(2.0)^2(1.0)} \right] = -133.8$ kJ/mol

(c) $Q = 1$, $\ln Q = 0$, $\Delta G = \Delta G^\circ = -141.8$ kJ/mol

16.85 $\Delta G = \Delta G^\circ + RT \ln \left[\dfrac{[NH_2CONH_2]}{(P_{NH_3})^2(P_{CO_2})} \right]$

(a) $\Delta G = -13.6 \text{ kJ/mol} + [8.314 \times 10^{-3} \text{ kJ/(K} \cdot \text{mol)}](298 \text{ K}) \ln \left[\dfrac{1.0}{(10)^2(10)} \right] = -30.7$ kJ/mol

Because ΔG is negative, the reaction is spontaneous.

(b) $\Delta G = -13.6 \text{ kJ/mol} + [8.314 \times 10^{-3} \text{ kJ/(K} \cdot \text{mol)}](298 \text{ K}) \ln \left[\dfrac{1.0}{(0.10)^2(0.10)} \right] = +3.5$ kJ/mol

Because ΔG is positive, the reaction is nonspontaneous.

16.86 $\Delta G^\circ = -RT \ln K$
(a) If $K > 1$, ΔG° is negative.   (b) If $K = 1$, $\Delta G^\circ = 0$.
(c) If $K < 1$, ΔG° is positive.

16.87 $K = e^{\frac{-\Delta G^\circ}{RT}}$ (a) If ΔG° is positive, K is small.   (b) If ΔG° is negative, K is large.

16.88 $\Delta G^{\circ} = -RT \ln K_p = -141.8$ kJ

$\ln K_p = \dfrac{-\Delta G^{\circ}}{RT} = \dfrac{-(-141.8 \text{ kJ/mol})}{[8.314 \times 10^{-3} \text{ kJ/(K} \cdot \text{mol)}](298 \text{ K})} = 57.23$

$K_p = e^{57.23} = 7.2 \times 10^{24}$

16.89 $\Delta G^{\circ} = -RT \ln K = -13.6$ kJ

$\ln K = \dfrac{-\Delta G^{\circ}}{RT} = \dfrac{-(-13.6 \text{ kJ/mol})}{[8.314 \times 10^{-3} \text{ kJ/(K} \cdot \text{mol)}](298 \text{ K})} = 5.49$

$K = e^{5.49} = 2.4 \times 10^{2}$

16.90 $C_2H_5OH(l) \rightleftharpoons C_2H_5OH(g)$

$\Delta G^{\circ} = \Delta G^{\circ}_f(C_2H_5OH(g)) - \Delta G^{\circ}_f(C_2H_5OH(l))$

$\Delta G^{\circ} = (1 \text{ mol})(-167.9 \text{ kJ/mol}) - (1 \text{ mol})(-174.9 \text{ kJ/mol}) = +7.0$ kJ

$\Delta G^{\circ} = -RT \ln K$

$\ln K = \dfrac{-\Delta G^{\circ}}{RT} = \dfrac{-(7.0 \text{ kJ/mol})}{[8.314 \times 10^{-3} \text{ kJ/(K} \cdot \text{mol)}](298 \text{ K})} = -2.83$

$K = e^{-2.83} = 0.059;$ $K = K_p = P_{C_2H_5OH} = 0.059$ atm

16.91 $\Delta G^{\circ} = -RT \ln K_a$

$\Delta G^{\circ} = -[8.314 \times 10^{-3} \text{ kJ/(K} \cdot \text{mol)}](298 \text{ K}) \ln(3.0 \times 10^{-4}) = +20.1$ kJ/mol

16.92 $2 CH_2=CH_2(g) + O_2(g) \rightarrow 2 C_2H_4O(g)$

$\Delta G^{\circ} = 2 \Delta G^{\circ}_f(C_2H_4O) - 2 \Delta G^{\circ}_f(CH_2=CH_2)$

$\Delta G^{\circ} = (2 \text{ mol})(-13.1 \text{ kJ/mol}) - (2 \text{ mol})(68.1 \text{ kJ/mol}) = -162.4$ kJ

$\Delta G^{\circ} = -RT \ln K$

$\ln K = \dfrac{-\Delta G^{\circ}}{RT} = \dfrac{-(-162.4 \text{ kJ/mol})}{[8.314 \times 10^{-3} \text{ kJ/(K} \cdot \text{mol)}](298 \text{ K})} = 65.55$

$K = K_p = e^{65.55} = 2.9 \times 10^{28}$

16.93 $TiO_2(s) + 2 Cl_2(g) + 2 C(s) \rightarrow TiCl_4(l) + 2 CO(g)$

$\Delta G^{\circ} = [\Delta G^{\circ}_f(TiCl_4) + 2 \Delta G^{\circ}_f(CO)] - \Delta G^{\circ}_f(TiO_2)$

$\Delta G^{\circ} = [(1 \text{ mol})(-737.2 \text{ kJ/mol}) + (2 \text{ mol})(-137.2 \text{ kJ/mol})] - (1 \text{ mol})(-888.8 \text{ kJ/mol})$

$\Delta G^{\circ} = -122.8$ kJ

$\Delta G^{\circ} = -RT \ln K$

$\ln K = \dfrac{-\Delta G^{\circ}}{RT} = \dfrac{-(-122.8 \text{ kJ/mol})}{[8.314 \times 10^{-3} \text{ kJ/(K} \cdot \text{mol)}](298 \text{ K})} = 49.56$

$K = K_p = e^{49.56} = 3.3 \times 10^{21}$

Chapter Problems

16.94 C_3H_8, 44.10 amu; 20 °C = 293 K

$$\text{mol } C_3H_8 = 1.32 \text{ g x } \frac{1 \text{ mol } C_3H_8}{44.10 \text{ g}} = 0.0300 \text{ mol } C_3H_8$$

$$V = \frac{nRT}{P} = \frac{(0.0300 \text{ mol})\left(0.082\ 06 \dfrac{L \cdot atm}{K \cdot mol}\right)(293 \text{ K})}{0.100 \text{ atm}} = 7.21 \text{ L}$$

Compress 7.21 L by a factor of 5 (7.21/5) to 1.44 L

$$\Delta S = nR \ln\left(\frac{V_f}{V_i}\right) = (0.0300 \text{ mol})(8.314 \text{ J/K} \cdot \text{mol})\ln\left(\frac{1.44 \text{ L}}{7.21 \text{ L}}\right) = -0.401 \text{ J/K}$$

16.95 (a), (c), and (d) are nonspontaneous;   (b) is spontaneous.

16.96 (a) Spontaneous does not mean fast, just possible.
(b) For a spontaneous reaction $\Delta S_{total} > 0$. ΔS_{sys} can be positive or negative.
(c) An endothermic reaction can be spontaneous if $\Delta S_{sys} > 0$.
(d) This statement is true because the sign of ΔG changes when the direction of a reaction is reversed.

16.97
Point Total	Possible Ways	Number of Ways
2	(1+1)	1
3	(2+1)(1+2)	2
4	(1+3)(2+2)(3+1)	3
5	(1+4)(2+3)(3+2)(4+1)	4
6	(1+5)(2+4)(3+3)(4+2)(5+1)	5
7	(1+6)(2+5)(3+4)(4+3)(5+2)(6+1)	6
8	(2+6)(3+5)(4+4)(5+3)(6+2)	5
9	(3+6)(4+5)(5+4)(6+3)	4
10	(4+6)(5+5)(6+4)	3
11	(6+5)(5+6)	2
12	(6+6)	1

Because a point total of 7 can be rolled in the most ways, it is the most probable point total.

16.98

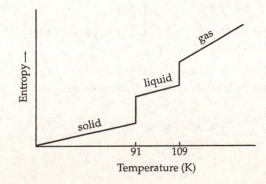

16.99 (a) $Q = 1$, $\ln Q = 0$, $\Delta G = \Delta G° = +79.9$ kJ
Because ΔG is positive, the reaction is spontaneous in the reverse direction.
(b) $\Delta G = \Delta G° + RT \ln Q$; $Q = [H_3O^+][OH^-] = (1.0 \times 10^{-7})^2 = 1.0 \times 10^{-14}$
$\Delta G = 79.9$ kJ/mol $+ [8.314 \times 10^{-3}$ kJ/(K $\cdot$ mol)$](298$ K$) \ln(1.0 \times 10^{-14}) = 0$
Because $\Delta G = 0$, the reaction is at equilibrium.
(c) $\Delta G = \Delta G° + RT \ln Q$
$Q = [H_3O^+][OH^-] = (1.0 \times 10^{-7})(1.0 \times 10^{-10}) = 1.0 \times 10^{-17}$
$\Delta G = 79.9$ kJ/mol $+ [8.314 \times 10^{-3}$ kJ/(K $\cdot$ mol)$](298$ K$) \ln(1.0 \times 10^{-17}) = -17.1$ kJ/mol
Because ΔG is negative, the reaction is spontaneous in the forward direction.
The results are consistent with Le Châtelier's principle. When the $[H_3O^+]$ and $[OH^-]$ are larger than the equilibrium concentrations (a), the reverse reaction takes place. When the product of $[H_3O^+]$ and the $[OH^-]$ is less than the equilibrium value, the forward reaction is spontaneous.
$\Delta G° = -RT \ln K$

$$\ln K = \frac{-\Delta G°}{RT} = \frac{-79.9 \text{ kJ/mol}}{[8.314 \times 10^{-3} \text{ kJ/(K} \cdot \text{mol)}](298 \text{ K})} = -32.25$$

$K = K_a = e^{-32.25} = 9.9 \times 10^{-15}$

16.100 At the normal boiling point, $\Delta G = 0$.

$$\Delta G_{vap} = \Delta H_{vap} - T\Delta S_{vap}; \qquad T = \frac{\Delta H_{vap}}{\Delta S_{vap}} = \frac{38,600 \text{ J}}{110 \text{ J/K}} = 351 \text{ K} = 78 \text{ °C}$$

16.101 At the normal boiling point, $\Delta G_{vap} = 0$.　　61 °C = 334 K

$$\Delta G_{vap} = \Delta H_{vap} - T\Delta S_{vap}; \qquad \Delta S_{vap} = \frac{\Delta H_{vap}}{T} = \frac{29,240 \text{ J}}{334 \text{ K}} = 87.5 \text{ J/K}$$

16.102 $\Delta G = \Delta H - T\Delta S$
(a) ΔH must be positive (endothermic) and greater than $T\Delta S$ in order for ΔG to be positive (nonspontaneous reaction).
(b) Set $\Delta G = 0$ and solve for ΔH.
$\Delta G = 0 = \Delta H - T\Delta S = \Delta H - (323$ K$)(104$ J/K$) = \Delta H - (33592$ J$) = \Delta H - (33.6$ kJ$)$
$\Delta H = 33.6$ kJ
ΔH must be greater than 33.6 kJ.

16.103 $NH_4NO_3(s) \rightarrow N_2O(g) + 2 H_2O(g)$
(a) $\Delta G° = [\Delta G°_f(N_2O) + 2 \Delta G°_f(H_2O)] - \Delta G°_f(NH_4NO_3)$
$\Delta G° = [(1$ mol$)(104.2$ kJ/mol$) + (2$ mol$)(-228.6$ kJ/mol$)] - (1$ mol$)(-184.0$ kJ/mol$)$
$\Delta G° = -169.0$ kJ
Because $\Delta G°$ is negative, the reaction is spontaneous.
(b) Because the reaction increases the number of moles of gas, $\Delta S°$ is positive.
$\Delta G° = \Delta H° - T\Delta S°$
As the temperature is raised, $\Delta G°$ becomes more negative.

(c) $\Delta G^\circ = -RT \ln K$

$$\ln K = \frac{-\Delta G^\circ}{RT} = \frac{-(-169.0 \text{ kJ/mol})}{[8.314 \times 10^{-3} \text{ kJ/(K} \cdot \text{mol})](298 \text{ K})} = 68.21$$

$K = K_p = e^{68.21} = 4.2 \times 10^{29}$

(d) $Q = (P_{N_2O})(P_{H_2O})^2 = (30)(30)^2 = (30)^3$

$\Delta G = \Delta G^\circ + RT \ln Q$

$\Delta G = -169.0 \text{ kJ/mol} + [8.314 \times 10^{-3}\text{kJ/(K} \cdot \text{mol})](298 \text{ K}) \ln[(30)^3] = -143.7 \text{ kJ/mol}$

16.104 (a) $2 \text{ Mg(s)} + O_2\text{(g)} \rightarrow 2 \text{ MgO(s)}$

$\Delta H^\circ = 2 \Delta H^\circ_f(\text{MgO}) = (2 \text{ mol})(-601.7 \text{ kJ/mol}) = -1203.4 \text{ kJ}$

$\Delta S^\circ = 2 S^\circ(\text{MgO}) - [2 S^\circ(\text{Mg}) + S^\circ(O_2)]$

$\Delta S^\circ = (2 \text{ mol})(26.9 \text{ J/(K} \cdot \text{mol})) - [(2 \text{ mol})(32.7 \text{ J/(K} \cdot \text{mol})) + (1 \text{ mol})(205.0 \text{ J/(K} \cdot \text{mol}))]$

$\Delta S^\circ = -216.6 \text{ J/K} = -216.6 \times 10^{-3} \text{ kJ/K}$

$\Delta G^\circ = \Delta H^\circ - T\Delta S^\circ = -1203.4 \text{ kJ} - (298 \text{ K})(-216.6 \times 10^{-3} \text{ kJ/K}) = -1138.8 \text{ kJ}$

Because ΔG° is negative, the reaction is spontaneous at 25 °C. ΔG° becomes less negative as the temperature is raised.

(b) $\text{MgCO}_3\text{(s)} \rightarrow \text{MgO(s)} + CO_2\text{(g)}$

$\Delta H^\circ = [\Delta H^\circ_f(\text{MgO}) + \Delta H^\circ_f(CO_2)] - \Delta H^\circ_f(\text{MgCO}_3)$

$\Delta H^\circ = [(1 \text{ mol})(-601.1 \text{ kJ/mol}) + (1 \text{ mol})(-393.5 \text{ kJ/mol})] - (1 \text{ mol})(-1096 \text{ kJ/mol}) = +101 \text{ kJ}$

$\Delta S^\circ = [S^\circ(\text{MgO}) + S^\circ(CO_2)] - S^\circ(\text{MgCO}_3)$

$\Delta S^\circ = [(1 \text{ mol})(26.9 \text{ J/(K} \cdot \text{mol})) + (1 \text{ mol})(213.6 \text{ J/(K} \cdot \text{mol}))] - (1 \text{ mol})(65.7 \text{ J/(K} \cdot \text{mol}))$

$\Delta S^\circ = 174.8 \text{ J/K} = 174.8 \times 10^{-3} \text{ kJ/K}$

$\Delta G^\circ = \Delta H^\circ - T\Delta S^\circ = 101 \text{ kJ} - (298 \text{ K})(174.8 \times 10^{-3} \text{ kJ/K}) = +49 \text{ kJ}$

Because ΔG° is positive, the reaction is not spontaneous at 25 °C. ΔG° becomes less positive as the temperature is raised.

(c) $\text{Fe}_2O_3\text{(s)} + 2 \text{ Al(s)} \rightarrow \text{Al}_2O_3\text{(s)} + 2 \text{ Fe(s)}$

$\Delta H^\circ = \Delta H^\circ_f(\text{Al}_2O_3) - \Delta H^\circ_f(\text{Fe}_2O_3)$

$\Delta H^\circ = (1 \text{ mol})(-1676 \text{ kJ/mol}) - (1 \text{ mol})(-824.2 \text{ kJ/mol}) = -852 \text{ kJ}$

$\Delta S^\circ = [S^\circ(\text{Al}_2O_3) + 2 S^\circ(\text{Fe})] - [S^\circ(\text{Fe}_2O_3) + 2 S^\circ(\text{Al})]$

$\Delta S^\circ = [(1 \text{ mol})(50.9 \text{ J/(K} \cdot \text{mol})) + (2 \text{ mol})(27.3 \text{ J/(K} \cdot \text{mol}))]$
$\qquad\qquad\qquad\qquad - [(1 \text{ mol})(87.4 \text{ J/(K} \cdot \text{mol})) + (2 \text{ mol})(28.3 \text{ J/(K} \cdot \text{mol}))]$

$\Delta S^\circ = -38.5 \text{ J/K} = -38.5 \times 10^{-3} \text{ kJ/K}$

$\Delta G^\circ = \Delta H^\circ - T\Delta S^\circ = -852 \text{ kJ} - (298 \text{ K})(-38.5 \times 10^{-3} \text{ kJ/K}) = -840 \text{ kJ}$

Because ΔG° is negative, the reaction is spontaneous at 25 °C. ΔG° becomes less negative as the temperature is raised.

(d) $2 \text{ NaHCO}_3\text{(s)} \rightarrow \text{Na}_2\text{CO}_3\text{(s)} + CO_2\text{(g)} + H_2O\text{(g)}$

$\Delta H^\circ = [\Delta H^\circ_f(\text{Na}_2\text{CO}_3) + \Delta H^\circ_f(CO_2) + \Delta H^\circ_f(H_2O)] - 2 \Delta H^\circ_f(\text{NaHCO}_3)$

$\Delta H^\circ = [(1 \text{ mol})(-1130.7 \text{ kJ/mol}) + (1 \text{ mol})(-393.5 \text{ kJ/mol})$
$\qquad\qquad\qquad + (1 \text{ mol})(-241.8 \text{ kJ/mol})] - (2 \text{ mol})(-950.8 \text{ kJ/mol}) = +135.6 \text{ kJ}$

$\Delta S^\circ = [S^\circ(\text{Na}_2\text{CO}_3) + S^\circ(CO_2) + S^\circ(H_2O)] - 2 S^\circ(\text{NaHCO}_3)$

$\Delta S^\circ = [(1 \text{ mol})(135.0 \text{ J/(K} \cdot \text{mol})) + (1 \text{ mol})(213.6 \text{ J/(K} \cdot \text{mol}))$
$\qquad\qquad\qquad + (1 \text{ mol})(188.7 \text{ J/(K} \cdot \text{mol}))] - (2 \text{ mol})(102 \text{ J/(K} \cdot \text{mol}))$

$\Delta S^\circ = +333 \text{ J/K} = +333 \times 10^{-3} \text{ kJ/K}$

$\Delta G^\circ = \Delta H^\circ - T\Delta S^\circ = +135.6 \text{ kJ} - (298 \text{ K})(+333 \times 10^{-3} \text{ kJ/K}) = +36.4 \text{ kJ}$

Because $\Delta G°$ is positive, the reaction is not spontaneous at 25 °C. $\Delta G°$ becomes less positive as the temperature is raised.

16.105 (a)

	$\Delta H_{vap}/T_{bp}$
ammonia	98 J/K
benzene	87 J/K
carbon tetrachloride	85 J/K
chloroform	87 J/K
mercury	94 J/K

(b) All processes are the conversion of a liquid to a gas at the boiling point. They should all have similar ΔS values. $\Delta H_{vap}/T_{bp}$ is equal to ΔS_{vap}.

(c) NH_3 deviates from Trouton's rule because of hydrogen bonding. NH_3(l) has less randomness and ΔS_{vap} is larger. Hg has metallic bonding which also leads to less randomness of the liquid.

16.106 (a) $6\ C(s)\ +\ 3\ H_2(g)\ \rightarrow\ C_6H_6(l)$

$\Delta S°_f = S°(C_6H_6)\ -\ [6\ S°(C)\ +\ 3\ S°(H_2)]$

$\Delta S°_f = (1\ mol)(173.4\ J/(K \cdot mol))\ -\ [(6\ mol)(5.7\ J/(K \cdot mol))\ +\ (3\ mol)(130.6\ J/(K \cdot mol))]$

$\Delta S°_f = -253\ J/K = -253\ J/(K \cdot mol)$

$\Delta G°_f = \Delta H°_f\ -\ T\Delta S°_f$

$\Delta S°_f = \dfrac{\Delta H°_f\ -\ \Delta G°_f}{T} = \dfrac{49.0\ kJ/mol\ -\ 124.5\ kJ/mol}{298\ K} = -0.2526\ kJ/(K \cdot mol)$

$\Delta S°_f = -252.6\ J/(K \cdot mol)$

Both calculations lead to the same value of $\Delta S°_f$.

(b) $Ca(s)\ +\ S(s)\ +\ 2\ O_2(g)\ \rightarrow\ CaSO_4(s)$

$\Delta S°_f = S°(CaSO_4)\ -\ [S°(Ca)\ +\ S°(S)\ +\ 2\ S°(O_2)]$

$\Delta S°_f = (1\ mol)(107\ J/(K \cdot mol))$

$\quad -\ [(1\ mol)(41.4\ J/(K \cdot mol))\ +\ (1\ mol)(31.8\ J/(K \cdot mol))\ +\ (2\ mol)(205.0\ J/(K \cdot mol))]$

$\Delta S°_f = -376\ J/K = -376\ J/(K \cdot mol)$

$\Delta G°_f = \Delta H°_f\ -\ T\Delta S°_f$

$\Delta S°_f = \dfrac{\Delta H°_f\ -\ \Delta G°_f}{T} = \dfrac{-1434.1\ kJ/mol\ -\ (-1321.9\ kJ/mol)}{298\ K} = -0.376\ kJ/(K \cdot mol)$

$\Delta S°_f = -376\ J/(K \cdot mol)$

Both calculations lead to the same value of $\Delta S°_f$.

(c) $2\ C(s)\ +\ 3\ H_2(g)\ +\ 1/2\ O_2(g)\ \rightarrow\ C_2H_5OH(l)$

$\Delta S°_f = S°(C_2H_5OH)\ -\ [S°(C)\ +\ S°(H_2)\ +\ 1/2\ S°(O_2)]$

$\Delta S°_f = (1\ mol)(161\ J/(K \cdot mol))$

$\quad -\ [(2\ mol)(5.7\ J/(K \cdot mol))\ +\ (3\ mol)(130.6\ J/(K \cdot mol))\ +\ (0.5\ mol)(205.0\ J/(K \cdot mol))]$

$\Delta S°_f = -345\ J/K = -345\ J/(K \cdot mol)$

$\Delta G°_f = \Delta H°_f\ -\ T\Delta S°_f$

$\Delta S°_f = \dfrac{\Delta H°_f\ -\ \Delta G°_f}{T} = \dfrac{-277.7\ kJ/mol\ -\ (-174.9\ kJ/mol)}{298\ K} = -0.345\ kJ/(K \cdot mol)$

$\Delta S°_f = -345\ J/(K \cdot mol)$

Both calculations lead to the same value of $\Delta S°_f$.

16.107 $MgCO_3(s) \rightarrow MgO(s) + CO_2(g)$
From Problem 16.104(b)
$\Delta H^\circ = +101 \text{ kJ}$; $\Delta S^\circ = 174.8 \text{ J/K} = 174.8 \times 10^{-3} \text{ kJ/K}$
The equilibrium pressure of CO_2 is equal to $K_p = P_{CO_2}$. K_p is not affected by the

quantities of $MgCO_3$ and MgO present. K_p can be calculated from ΔG°.
$\Delta G^\circ = \Delta H^\circ - T\Delta S^\circ$
$\Delta G^\circ = -RT \ln K_p$
(a) $\Delta G^\circ = 101 \text{ kJ} - (298 \text{ K})(174.8 \times 10^{-3} \text{ kJ/K}) = +49 \text{ kJ}$

$$\ln K_p = \frac{-\Delta G^\circ}{RT} = \frac{-49 \text{ kJ/mol}}{[8.314 \times 10^{-3} \text{ kJ/(K} \cdot \text{mol)}](298 \text{ K})} = -19.8$$

$K_p = P_{CO_2} = e^{-19.8} = 3 \times 10^{-9} \text{ atm}$
(b) $\Delta G^\circ = 101 \text{ kJ} - (553 \text{ K})(174.8 \times 10^{-3} \text{ kJ/K}) = 4.3 \text{ kJ}$

$$\ln K_p = \frac{-\Delta G^\circ}{RT} = \frac{-4.3 \text{ kJ/mol}}{[8.314 \times 10^{-3} \text{ kJ/(K} \cdot \text{mol)}](553 \text{ K})} = -0.94$$

$K_p = P_{CO_2} = e^{-0.94} = 0.39 \text{ atm}$

(c) $P_{CO_2} = 0.39 \text{ atm}$ because the temperature is the same as in (b).

16.108 $\Delta G^\circ = -RT \ln K_b$
At 20 °C: $\Delta G^\circ = -[8.314 \times 10^{-3} \text{ kJ/(K} \cdot \text{mol)}](293 \text{ K}) \ln(1.710 \times 10^{-5}) = +26.74 \text{ kJ/mol}$
At 50 °C: $\Delta G^\circ = -[8.314 \times 10^{-3} \text{ kJ/(K} \cdot \text{mol)}](323 \text{ K}) \ln(1.892 \times 10^{-5}) = +29.20 \text{ kJ/mol}$
$\Delta G^\circ = \Delta H^\circ - T\Delta S^\circ$

$26.74 = \Delta H^\circ - 293\Delta S^\circ$
$29.20 = \Delta H^\circ - 323\Delta S^\circ$ Solve these two equations simultaneously for ΔH° and ΔS°.

$26.74 + 293\Delta S^\circ = \Delta H^\circ$
$29.20 + 323\Delta S^\circ = \Delta H^\circ$ Set these two equations equal to each other.

$26.74 + 293\Delta S^\circ = 29.20 + 323\Delta S^\circ$
$26.74 - 29.20 = 323\Delta S^\circ - 293 \Delta S^\circ$
$-2.46 = 30\Delta S^\circ$
$\Delta S^\circ = -2.46/30 = -0.0820 = -0.0820 \text{ kJ/K} = -82.0 \text{ J/K}$
$26.74 + 293\Delta S^\circ = 26.74 + 293(-0.0820) = \Delta H^\circ = +2.71 \text{ kJ}$

16.109 (a) $\Delta H^\circ = 2 \Delta H^\circ_f(NH_3) = (2 \text{ mol})(-46.1 \text{ kJ/mol}) = -92.2 \text{ kJ}$
$\Delta G^\circ = 2 \Delta G^\circ_f(NH_3) = (2 \text{ mol})(-16.5 \text{ kJ/mol}) = -33.0 \text{ kJ}$
$\Delta G^\circ = \Delta H^\circ - T\Delta S^\circ$
$\Delta H^\circ - \Delta G^\circ = T\Delta S^\circ$

$$\Delta S^\circ = \frac{\Delta H^\circ - \Delta G^\circ}{T} = \frac{-92.2 \text{ kJ} - (-33.0 \text{ kJ})}{298 \text{ K}} = -0.199 \text{ kJ/K} = -199 \text{ J/K}$$

(b) ΔS° is negative because the number of mol of gas molecules decreases from 4 mol
to 2 mol on going from reactants to products.
(c) The reaction is spontaneous because ΔG° is negative.

(d) $\Delta G^\circ = \Delta H^\circ - T\Delta S^\circ = -92.2 \text{ kJ} - (350 \text{ K})(-0.199 \text{ kJ/K}) = -22.55 \text{ kJ}$

$\Delta G^\circ = -RT \ln K_p$

$\ln K_p = \dfrac{-\Delta G^\circ}{RT} = \dfrac{-(-22.55 \text{ kJ/mol})}{[8.314 \times 10^{-3} \text{ kJ/(K} \cdot \text{mol)}](350 \text{ K})} = 7.749$

$K_p = e^{7.749} = 2.3 \times 10^3$

$\Delta n = 2 - (1 + 3) = -2$

$K_c = K_p\left(\dfrac{1}{RT}\right)^{\Delta n} = (2.3 \times 10^3)\left(\dfrac{1}{RT}\right)^{-2} = (2.3 \times 10^3)(RT)^2$

$K_c = (2.3 \times 10^3)[(0.082\ 06)(350)]^2 = 1.9 \times 10^6$

16.110 (a) $\Delta H^\circ = [\Delta H^\circ_f(Ag^+(aq)) + \Delta H^\circ_f(Br^-(aq))] - \Delta H^\circ_f(AgBr(s))$

$\Delta H^\circ = [(1 \text{ mol})(105.6 \text{ kJ/mol}) + (1 \text{ mol})(-121.5 \text{ kJ/mol})] - (1 \text{ mol})(-100.4 \text{ kJ/mol}) = +84.5 \text{ kJ}$

$\Delta S^\circ = [S^\circ(Ag^+(aq)) + S^\circ(Br^-(aq))] - S^\circ(AgBr(s))$

$\Delta S^\circ = [(1 \text{ mol})(72.7 \text{ J/(K} \cdot \text{mol)}) + (1 \text{ mol})(82.4 \text{ J/(K} \cdot \text{mol)})]$
$\qquad\qquad\qquad - (1 \text{ mol})(107.1 \text{ J/(K} \cdot \text{mol)}) = +48.0 \text{ J}$

$\Delta G^\circ = \Delta H^\circ - T\Delta S^\circ = 84.5 \text{ kJ} - (298 \text{ K})(48.0 \times 10^{-3} \text{ kJ/K}) = +70.2 \text{ kJ}$

(b) $\Delta G^\circ = -RT \ln K_{sp}$

$\ln K_{sp} = \dfrac{-\Delta G^\circ}{RT} = \dfrac{-70.2 \text{ kJ/mol}}{[8.314 \times 10^{-3} \text{ kJ/(K} \cdot \text{mol)}](298 \text{ K})} = -28.3$

$K_{sp} = e^{-28.3} = 5 \times 10^{-13}$

(c) $Q = [Ag^+][Br^-] = (1.00 \times 10^{-5})(1.00 \times 10^{-5}) = 1.00 \times 10^{-10}$

$\Delta G = \Delta G^\circ + RT \ln Q$

$\Delta G = 70.2 \text{ kJ/mol} + [8.314 \times 10^{-3} \text{ kJ/(K} \cdot \text{mol)}](298 \text{ K}) \ln(1.00 \times 10^{-10}) = 13.2 \text{ kJ/mol}$

A positive value of ΔG means that the forward reaction is nonspontaneous under these conditions. The reverse reaction is therefore spontaneous, which is consistent with the fact that $Q > K_{sp}$.

16.111 (a) $\Delta G^\circ = \Delta H^\circ - T\Delta S^\circ$ and $\Delta G^\circ = -RT \ln K$

Set the two equations equal to each other.

$-RT \ln K = \Delta H^\circ - T\Delta S^\circ$

$\ln K = \dfrac{\Delta H^\circ - T\Delta S^\circ}{-RT}$

$\ln K = \dfrac{-\Delta H^\circ}{RT} + \dfrac{T\Delta S^\circ}{RT}$

$\ln K = \dfrac{-\Delta H^\circ}{RT} + \dfrac{\Delta S^\circ}{R}$

$\ln K = \dfrac{-\Delta H^\circ}{R}\left(\dfrac{1}{T}\right) + \dfrac{\Delta S^\circ}{R}$ This is the equation for a straight line ($y = mx + b$).

$y = \ln K;\qquad m = -\dfrac{\Delta H^\circ}{R} = \text{slope};\qquad x = \dfrac{1}{T};\qquad b = \dfrac{\Delta S^\circ}{R} = \text{intercept}$

(b) Plot ln K versus 1/T

$\Delta H° = -R(\text{slope})$ $\qquad\qquad$ $\Delta S° = R(\text{intercept})$

(c) For a reaction where K increases with increasing temperature, the following plot would be obtained:

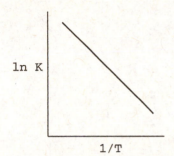

ln K

1/T

The slope is negative.
Because $\Delta H° = -R(\text{slope})$, $\Delta H°$ is positive, and the reaction is endothermic.

This prediction is in accord with LeChâtelier's principle because when you add heat (raise the temperature) for an endothermic reaction, the reaction in the forward direction takes place, the product concentrations increase and the reactant concentrations decrease. This results in an increase in K.

16.112 $\quad Br_2(l) \rightleftharpoons Br_2(g)$

$\Delta S° = S°(Br_2(g)) - S°(Br_2(l))$

$\Delta S° = (1 \text{ mol})(245.4 \text{ J/(K} \cdot \text{mol)}) - (1 \text{ mol})(152.2 \text{ J/(K} \cdot \text{mol)}) = 93.2 \text{ J/K} = 93.2 \times 10^{-3} \text{ kJ/K}$

$\Delta G = \Delta H° - T\Delta S°$

At the boiling point, $\Delta G = 0$.

$0 = \Delta H° - T_{bp}\Delta S°$

$T_{bp} = \dfrac{\Delta H°}{\Delta S°}$

$\Delta H° = T_{bp} \Delta S° = (332 \text{ K})(93.2 \times 10^{-3} \text{ kJ/K}) = 30.9 \text{ kJ}$

$K_p = P_{Br_2} = \left(227 \text{ mm Hg} \times \dfrac{1 \text{ atm}}{760 \text{ mm Hg}} \right) = 0.299 \text{ atm}$

$\Delta G° = -RT \ln K_p$ and $\Delta G° = \Delta H° - T\Delta S°$ $\quad$ (set equations equal to each other)

$\Delta H° - T\Delta S° = -RT \ln K_p$ $\quad$ (rearrange)

$\ln K_p = \dfrac{-\Delta H°}{R} \dfrac{1}{T} + \dfrac{\Delta S°}{R}$ $\quad$ (solve for T)

$$T = \dfrac{\left(\dfrac{-\Delta H°}{R} \right)}{\left(\ln K_p - \dfrac{\Delta S°}{R} \right)} = \dfrac{\left(\dfrac{-30.9 \text{ kJ/mol}}{8.314 \times 10^{-3} \text{ kJ/(K} \cdot \text{mol)}} \right)}{\left(\ln(0.299) - \dfrac{93.2 \times 10^{-3} \text{ kJ/(K} \cdot \text{mol)}}{8.314 \times 10^{-3} \text{ kJ/(K} \cdot \text{mol)}} \right)} = 299 \text{ K} = 26 \text{ °C}$$

$Br_2(l)$ has a vapor pressure of 227 mm Hg at 26 °C.

16.113   For PbI_2, $K_{sp} = [Pb^{2+}][I^-]^2$

$$PbI_2(s) \rightleftharpoons Pb^{2+}(aq) + 2\,I^-(aq)$$

initial (M)   0   0
equil (M)   x   2x

$K_{sp} = x(2x)^2 = 4x^3$, where x = molar solubility

At 20 °C = 20 + 273 = 293 K, $K_{sp} = 4(1.45 \times 10^{-3})^3 = 1.22 \times 10^{-8}$
At 80 °C = 80 + 273 = 353 K, $K_{sp} = 4(6.85 \times 10^{-3})^3 = 1.29 \times 10^{-6}$

From problem 16.111, $\ln K = \dfrac{-\Delta H^\circ}{RT} + \dfrac{\Delta S^\circ}{R}$

$$\ln K_1 - \ln K_2 = \frac{-\Delta H^\circ}{RT_1} + \frac{\Delta S^\circ}{R} - \left(\frac{-\Delta H^\circ}{RT_2} + \frac{\Delta S^\circ}{R} \right)$$

$$\ln \frac{K_1}{K_2} = \frac{-\Delta H^\circ}{RT_1} - \frac{-\Delta H^\circ}{RT_2} = \frac{-\Delta H^\circ}{R}\left(\frac{1}{T_1} - \frac{1}{T_2} \right) = \frac{\Delta H^\circ}{R}\left(\frac{1}{T_2} - \frac{1}{T_1} \right)$$

$$\Delta H^\circ = \frac{[\ln K_1 - \ln K_2]R}{\left(\dfrac{1}{T_2} - \dfrac{1}{T_1} \right)}$$

$$\Delta H^\circ = \frac{[\ln(1.22 \times 10^{-8}) - \ln(1.29 \times 10^{-6})][8.314 \times 10^{-3}\ kJ/(K \cdot mol)]}{\left(\dfrac{1}{353\ K} - \dfrac{1}{293\ K} \right)} = 66.8\ kJ/mol$$

$\Delta G^\circ = -RT \ln K_{sp} = -[8.314 \times 10^{-3}\ kJ/(K \cdot mol)](293\ K) \ln(1.22 \times 10^{-8}) = 44.4\ kJ/mol$

$\Delta G^\circ = \Delta H^\circ - T\Delta S^\circ$

$\Delta H^\circ - \Delta G^\circ = T\Delta S^\circ$

$\Delta S^\circ = \dfrac{\Delta H^\circ - \Delta G^\circ}{T}$

$\Delta S^\circ = \dfrac{66.8\ kJ/mol - 44.4\ kJ/mol}{293\ K} = 0.0765\ kJ/(K \cdot mol) = 76.5\ J/(K \cdot mol)$

16.114 $\Delta H^\circ = [2\ \Delta H^\circ_f(Cl^-(aq))] - [2\ \Delta H^\circ_f(Br^-(aq))]$
$\Delta H^\circ = [(2\ mol)(-167.2\ kJ/mol)] - [(2\ mol)(-121.5\ kJ/mol)] = -91.4\ kJ$
$\Delta S^\circ = [S^\circ(Br_2(l)) + 2\ S^\circ(Cl^-(aq))] - [2\ S^\circ(Br^-(aq)) + S^\circ(Cl_2(g))]$
$\Delta S^\circ = [(1\ mol)(152.2\ J/(K \cdot mol)) + (2\ mol)(56.5\ J/(K \cdot mol))]$
$\qquad\qquad - [(2\ mol)(82.4\ J/(K \cdot mol)) + (1\ mol)(223.0\ J/(K \cdot mol))] = -122.6\ J/K$
80 °C = 80 + 273 = 353 K
$\Delta G^\circ = \Delta H^\circ - T\Delta S^\circ = -91.4\ kJ - (353\ K)(-122.6 \times 10^{-3}\ kJ/K) = -48.1\ kJ$

$\Delta G^\circ = -RT \ln K$

$\ln K = \dfrac{-\Delta G^\circ}{RT} = \dfrac{-(-48.1\ kJmol)}{[8.314 \times 10^{-3}\ kJ/(K \cdot mol)](353\ K)} = 16.4$

$K = e^{16.4} = 1.3 \times 10^7$

16.115 $CS_2(l) \rightleftharpoons CS_2(g)$

$\Delta H^\circ = \Delta H^\circ_f(CS_2(g)) - \Delta H^\circ_f(CS_2(l))$

$\Delta H^\circ = [(1\ mol)(116.7\ kJ/mol)] - [(1\ mol)(89.0\ kJ/mol)] = 27.7\ kJ$

$\Delta S^\circ = S^\circ(CS_2(g)) - S^\circ(CS_2(l))$

$\Delta S^\circ = [(1\ mol)(237.7\ J/(K \cdot mol))] - [(1\ mol)(151.3\ J/(K \cdot mol))] = 86.4\ J/K$

$\Delta G = \Delta H^\circ - T\Delta S^\circ$

At the boiling point, $\Delta G = 0$.

$0 = \Delta H^\circ - T_{bp}\Delta S^\circ$

$$T_{bp} = \frac{\Delta H^\circ}{\Delta S^\circ} = \frac{27.7\ kJ}{86.4 \times 10^{-3}\ kJ/K} = 321\ K$$

$T_{bp} = 321\ K = 321 - 273 = 48\ ^\circ C$

16.116 $35\ ^\circ C = 35 + 273 = 308\ K$

$\Delta G^\circ = \Delta H^\circ - T\Delta S^\circ = -352\ kJ - (308\ K)(-899 \times 10^{-3}\ kJ/K) = -75.1\ kJ$

$\Delta G^\circ = -RT \ln K_p$

$$\ln K_p = \frac{-\Delta G^\circ}{RT} = \frac{-(-75.1\ kJ/mol)}{[8.314 \times 10^{-3}\ kJ/(K \cdot mol)](308\ K)} = 29.33$$

$K_p = e^{29.33} = 5.5 \times 10^{12}$

$$K_p = \frac{1}{(P_{H_2O})^6} = 5.5 \times 10^{12}$$

$$P_{H_2O} = \sqrt[6]{\frac{1}{5.5 \times 10^{12}}} = 0.0075\ atm$$

$$P_{H_2O} = 0.0075\ atm \times \frac{760\ mm\ Hg}{1\ atm} = 5.7\ mm\ Hg$$

16.117 $2\ KClO_3(s) \rightarrow 2\ KCl(s) + 3\ O_2(g)$

$\Delta H^\circ = 2\ \Delta H^\circ_f(KCl) - 2\ \Delta H^\circ_f(KClO_3)$

$\Delta H^\circ = (2\ mol)(-436.5\ kJ) - (2\ mol)(-397.7\ kJ) = -77.6\ kJ$

$25\ ^\circ C = 25 + 273 = 298\ K$

$\Delta G^\circ = \Delta H^\circ - T\Delta S^\circ$

$\Delta H^\circ - \Delta G^\circ = T\Delta S^\circ$

$$\Delta S^\circ = \frac{\Delta H^\circ - \Delta G^\circ}{T} = \frac{-77.6\ kJ - (-224.4\ kJ)}{298\ K} = 0.493\ kJ/K = 493\ J/K$$

$\Delta S^\circ = [2\ S^\circ(KCl) + 3\ S^\circ(O_2)] - 2\ S^\circ(KClO_3)$

$493\ J/K = [(2\ mol)(82.6\ J/(K \cdot mol)) + (3\ mol)S^\circ(O_2)] - (2\ mol)(143.1\ J/(K \cdot mol))$

$(3\ mol)S^\circ(O_2) = 493\ J/K - (2\ mol)(82.6\ J/(K \cdot mol)) + (2\ mol)(143.1\ J/(K \cdot mol))$

$(3\ mol)S^\circ(O_2) = 614\ J/K$

$S^\circ(O_2) = (614\ J/K)/(3\ mol) = 204.7\ J/(K \cdot mol) = 205\ J/(K \cdot mol)$

16.118 $N_2O_4(g) \rightleftharpoons 2\ NO_2(g)$

$\Delta H^\circ = 2\ \Delta H^\circ_f(NO_2) - \Delta H^\circ_f(N_2O_4) = (2\ mol)(33.2\ kJ) - (1\ mol)(11.1\ kJ) = 55.3\ kJ$

$\Delta S^\circ = 2\ S^\circ(NO_2) - S^\circ(N_2O_4) = (2\ mol)(240.0\ J/(K \cdot mol)) - (1\ mol)(304.3\ J/(K \cdot mol))$

$\Delta S^\circ = 175.7 \text{ J/K} = 175.7 \times 10^{-3} \text{ kJ/K}$

$\Delta G^\circ = \Delta H^\circ - T\Delta S^\circ$ and $\Delta G^\circ = -RT \ln K_p$; Set these two equations equal to each other and solve for T.

$\Delta H^\circ - T\Delta S^\circ = -RT \ln K_p$

$\Delta H^\circ = T\Delta S^\circ - RT \ln K_p = T(\Delta S^\circ - R \ln K_p)$

$$T = \frac{\Delta H^\circ}{\Delta S^\circ - R \ln K_p}$$

(a) $P_{N_2O_4} + P_{NO_2} = 1.00 \text{ atm}$ and $P_{NO_2} = 2 P_{N_2O_4}$

$P_{N_2O_4} + 2 P_{N_2O_4} = 3 P_{N_2O_4} = 1.00 \text{ atm}$

$P_{N_2O_4} = 1.00 \text{ atm}/3 = 0.333 \text{ atm}$

$P_{NO_2} = 1.00 \text{ atm} - P_{N_2O_4} = 1.00 - 0.333 = 0.667 \text{ atm}$

$$K_p = \frac{(P_{NO_2})^2}{P_{N_2O_4}} = \frac{(0.667)^2}{(0.333)} = 1.34$$

$$T = \frac{\Delta H^\circ}{\Delta S^\circ - R \ln K_p}$$

$$T = \frac{55.3 \text{ kJ/mol}}{[175.7 \times 10^{-3} \text{ kJ/(K} \cdot \text{mol)}] - [8.314 \times 10^{-3} \text{ kJ/(K} \cdot \text{mol)}] \ln(1.34)} = 319 \text{ K}$$

$T = 319 \text{ K} = 319 - 273 = 46 \text{ °C}$

(b) $P_{N_2O_4} + P_{NO_2} = 1.00 \text{ atm}$ and $P_{NO_2} = P_{N_2O_4}$ so $P_{NO_2} = P_{N_2O_4} = 0.50 \text{ atm}$

$$K_p = \frac{(P_{NO_2})^2}{P_{N_2O_4}} = \frac{(0.500)^2}{(0.500)} = 0.500$$

$$T = \frac{\Delta H^\circ}{\Delta S^\circ - R \ln K_p}$$

$$T = \frac{55.3 \text{ kJ/mol}}{[175.7 \times 10^{-3} \text{ kJ/(K} \cdot \text{mol)}] - [8.314 \times 10^{-3} \text{ kJ/(K} \cdot \text{mol)}] \ln(0.500)} = 305 \text{ K}$$

$T = 305 \text{ K} = 305 - 273 = 32 \text{ °C}$

Multiconcept Problems

16.119 The kinetic parameters [(a), (b), and (h)] are affected by a catalyst. The thermodynamic and equilibrium parameters [(c), (d), (e), (f), and (g)] are not affected by a catalyst.

16.120 $N_2(g) + 3 H_2(g) \rightleftharpoons 2 NH_3(g)$

$\Delta H^\circ = 2 \Delta H^\circ_f(NH_3) - [\Delta H^\circ_f(N_2) + 3 \Delta H^\circ_f(H_2)] = (2 \text{ mol})(-46.1 \text{ kJ}) - [0] = -92.2 \text{ kJ}$

$\Delta S^\circ = 2 S^\circ(NH_3) - [S^\circ(N_2) + 3 S^\circ(H_2)]$

$\Delta S^\circ = (2 \text{ mol})(192.3 \text{ J/(K} \cdot \text{mol)})$
$\qquad - [(1 \text{ mol})(191.5 \text{ J/(K} \cdot \text{mol)}) + (3 \text{ mol})(130.6 \text{ J/(K} \cdot \text{mol)})] = -198.7 \text{ J/K}$

$\Delta G^\circ = \Delta H^\circ - T\Delta S^\circ = -92.2 \text{ kJ} - (673 \text{ K})(-198.7 \times 10^{-3} \text{ kJ/K}) = 41.5 \text{ kJ}$

$\Delta G^\circ = -RT \ln K_p$

$\ln K_p = \dfrac{-\Delta G^\circ}{RT} = \dfrac{-41.5 \text{ kJ/mol}}{[8.314 \times 10^{-3} \text{ kJ/(K} \cdot \text{mol)}](673 \text{ K})} = -7.42$

$K_p = e^{-7.42} = 6.0 \times 10^{-4}$

Because $K_p = K_c(RT)^{\Delta n}$, $K_c = K_p(RT)^{-\Delta n}$

$K_c = K_p(RT)^2 = (6.0 \times 10^{-4})[(0.082\ 06)(673)]^2 = 1.83$

N_2, 28.01 amu; H_2, 2.016 amu

Initial concentrations:

$$[N_2] = \frac{(14.0 \text{ g})\left(\dfrac{1 \text{ mol}}{28.01 \text{ g}}\right)}{5.00 \text{ L}} = 0.100 \text{ M} \quad \text{and} \quad [H_2] = \frac{(3.024 \text{ g})\left(\dfrac{1 \text{ mol}}{2.016 \text{ g}}\right)}{5.00 \text{ L}} = 0.300 \text{ M}$$

$$
\begin{array}{lccc}
 & N_2(g) & + \ 3 H_2(g) & \rightleftharpoons \ 2 NH_3(g) \\
\text{initial (M)} & 0.100 & 0.300 & 0 \\
\text{change (M)} & -x & -3x & +2x \\
\text{equil (M)} & 0.100 - x & 0.300 - 3x & 2x \\
\end{array}
$$

$K_c = \dfrac{[NH_3]^2}{[N_2][H_2]^3} = \dfrac{(2x)^2}{(0.100 - x)(0.300 - 3x)^3} = \dfrac{4x^2}{27(0.100 - x)^4} = 1.83$

$\left(\dfrac{x}{(0.100 - x)^2}\right)^2 = \dfrac{(27)(1.83)}{4} = 12.35; \quad \dfrac{x}{(0.100 - x)^2} = \sqrt{12.35} = 3.514$

$3.514x^2 - 1.703x + 0.03514 = 0$

Use the quadratic formula to solve for x.

$x = \dfrac{-(-1.703) \pm \sqrt{(-1.703)^2 - (4)(3.514)(0.03514)}}{2(3.514)} = \dfrac{1.703 \pm 1.551}{7.028}$

$x = 0.463$ and 0.0216

Of the two solutions for x, only 0.0216 has physical meaning because 0.463 would lead to negative concentrations of N_2 and H_2.

$[N_2] = 0.100 - x = 0.100 - 0.0216 = 0.078 \text{ M}$

$[H_2] = 0.300 - 3x = 0.300 - 3(0.0216) = 0.235 \text{ M}$

$[NH_3] = 2x = 2(0.0216) = 0.043 \text{ M}$

16.121   (a) $2 SO_2(g) + O_2(g) \rightleftharpoons 2 SO_3(g)$

$\Delta H^\circ = 2 \Delta H^\circ_f(SO_3) - 2 \Delta H^\circ_f(SO_2)$

$\Delta H^\circ = (2 \text{ mol})(-395.7 \text{ kJ/mol}) - (2 \text{ mol})(-296.8 \text{ kJ/mol}) = -197.8 \text{ kJ}$

$\Delta S^\circ = 2 S^\circ(SO_3) - [2 S^\circ(SO_2) + S^\circ(O_2)]$

$\Delta S^\circ = (2 \text{ mol})(256.6 \text{ J/(K} \cdot \text{mol)}) - [(2 \text{ mol})(248.1 \text{ J/(K} \cdot \text{mol)}) + (1 \text{ mol})(205.0 \text{ J/(K} \cdot \text{mol)})]$

$\Delta S^\circ = -188.0 \text{ J/K} = -188.0 \times 10^{-3} \text{ kJ/K}$

$\Delta G^\circ = \Delta H^\circ - T\Delta S^\circ = -197.8 \text{ kJ} - (800 \text{ K})(-188.0 \times 10^{-3} \text{ kJ/K}) = -47.4 \text{ kJ}$

$\Delta G^\circ = -RT \ln K_p$

$\ln K_p = \dfrac{-\Delta G^\circ}{RT} = \dfrac{-(-47.4 \text{ kJ/mol})}{[8.314 \times 10^{-3} \text{ kJ/(K} \cdot \text{mol)}](800 \text{ K})} = 7.13$

$K_p = e^{7.13} = 1249$

SO_2, 64.06 amu; O_2, 32.00 amu

At 800 K:

$$P_{SO_2} = \frac{nRT}{V} = \frac{\left(192 \text{ g} \times \frac{1 \text{ mol}}{64.06 \text{ g}}\right)\left(0.082\,06\,\frac{L \cdot atm}{K \cdot mol}\right)(800 \text{ K})}{15.0 \text{ L}} = 13.1 \text{ atm}$$

$$P_{O_2} = \frac{nRT}{V} = \frac{\left(48.0 \text{ g} \times \frac{1 \text{ mol}}{32.00 \text{ g}}\right)\left(0.082\,06\,\frac{L \cdot atm}{K \cdot mol}\right)(800 \text{ K})}{15.0 \text{ L}} = 6.56 \text{ atm}$$

	2 SO_2(g)	+	O_2(g)	⇌	2 SO_3(g)
initial (atm)	13.1		6.56		0
assume complete rxn (atm)	0		0		13.1
assume a small back rxn	+2x		+x		–2x
equil (atm)	2x		x		13.1 – 2x

$$K_p = 1249 = \frac{[SO_3]^2}{[SO_2]^2[O_2]} = \frac{(13.1 - 2x)^2}{(2x)^2(x)} \approx \frac{(13.1)^2}{(2x)^2(x)}$$

Solve for x. $x^3 = 0.0343$; $x = 0.325$

Use successive approximations to solve for x because 2x is not negligible compared with 13.1.

Second approximation:

$$1249 = \frac{[13.1 - (2)(0.325)]^2}{(2x)^2(x)};$$ Solve for x. $x^3 = 0.0310$; $x = 0.314$

Third approximation:

$$1249 = \frac{[13.1 - (2)(0.314)]^2}{(2x)^2(x)};$$ Solve for x. $x^3 = 0.0311$; $x = 0.315$ (x has converged)

$P_{SO_2} = 2x = 2(0.315) = 0.63$ atm

$P_{O_2} = x = 0.32$ atm

$P_{SO_3} = 13.1 - 2x = 13.1 - 2(0.315) = 12.5$ atm

(b) The % yield of SO_3 decreases with increasing temperature because $\Delta S°$ is negative. $\Delta G°$ becomes less negative and K_p gets smaller as the temperature increases.

(c) At 1000 K:

$\Delta G° = \Delta H° - T\Delta S° = -197.8 \text{ kJ} - (1000 \text{ K})(-188.0 \times 10^{-3} \text{ kJ/K}) = -9.8 \text{ kJ}$

$\Delta G° = -RT \ln K_p$

$$\ln K_p = \frac{-\Delta G°}{RT} = \frac{-(-9.8 \text{ kJ/mol})}{[8.314 \times 10^{-3} \text{ kJ/(K} \cdot \text{mol)}](1000 \text{ K})} = 1.179$$

$K_p = e^{1.179} = 3.25$

$$P_{SO_2} = \frac{nRT}{V} = \frac{\left(192 \text{ g} \times \frac{1 \text{ mol}}{64.06 \text{ g}}\right)\left(0.082\,06\, \frac{L \cdot atm}{K \cdot mol}\right)(1000 \text{ K})}{15.0 \text{ L}} = 16.4 \text{ atm}$$

$$P_{O_2} = \frac{nRT}{V} = \frac{\left(48.0 \text{ g} \times \frac{1 \text{ mol}}{32.00 \text{ g}}\right)\left(0.082\,06\, \frac{L \cdot atm}{K \cdot mol}\right)(1000 \text{ K})}{15.0 \text{ L}} = 8.2 \text{ atm}$$

	$2 SO_2(g)$	$+$	$O_2(g)$	$\rightleftharpoons$	$2 SO_3(g)$
initial (atm)	16.4		8.2		0
assume complete rxn (atm)	0		0		16.4
assume a small back rxn	$+2x$		$+x$		$-2x$
equil (atm)	$2x$		x		$16.4 - 2x$

$$K_p = 3.25 = \frac{[SO_3]^2}{[SO_2]^2[O_2]} = \frac{(16.4 - 2x)^2}{(2x)^2(x)} \approx \frac{(16.4)^2}{(2x)^2(x)}$$

Solve for x. $x^3 = 20.7$; $x = 2.7$

Use successive approximations to solve for x because 2x is not negligible compared with 16.4.

Second approximation:

$$3.25 = \frac{[16.4 - (2)(2.7)]^2}{(2x)^2(x)}; \text{ Solve for x. } x^3 = 9.31; \ x = 2.1$$

Third approximation:

$$3.25 = \frac{[16.4 - (2)(2.1)]^2}{(2x)^2(x)}; \text{ Solve for x. } x^3 = 11.4; \ x = 2.3$$

Fourth approximation:

$$3.25 = \frac{[16.4 - (2)(2.3)]^2}{(2x)^2(x)}; \text{ Solve for x. } x^3 = 10.7; \ x = 2.2 \text{ (x has converged)}$$

$P_{SO_2} = 2x = 2(2.2) = 4.4 \text{ atm}$

$P_{O_2} = x = 2.2 \text{ atm}$

$P_{SO_3} = 16.4 - 2x = 16.4 - 2(2.2) = 12.0 \text{ atm}$

$P_{total} = P_{SO_2} + P_{O_2} + P_{SO_3} = 4.4 + 2.2 + 12.0 = 18.6 \text{ atm}$

On going from 800 K to 1000 K, P_{total} increases to 18.6 atm (because K_p decreases, but P increases with temperature at constant volume).

16.122 $Pb(s) + PbO_2(s) + 2 H^+(aq) + 2 HSO_4^-(aq) \rightarrow 2 PbSO_4(s) + 2 H_2O(l)$
(a) $\Delta G° = [2 \Delta G°_f(PbSO_4) + 2 \Delta G°_f(H_2O)] - [\Delta G°_f(PbO_2) + 2 \Delta G°_f(HSO_4^-)]$
$\Delta G° = (2 \text{ mol})(-813.2 \text{ kJ/mol}) + (2 \text{ mol})(-237.2 \text{ kJ/mol})]$
$\qquad - [(1 \text{ mol})(-217.4 \text{ kJ/mol}) + (2 \text{ mol})(-756.0 \text{ kJ/mol})] = -371.4 \text{ kJ}$

(b) $°C = 5/9(°F - 32) = 5/9(10 - 32) = -12.2 °C$; $-12.2 °C = 261 K$

$\Delta H° = [2 \Delta H°_f(PbSO_4) + 2 \Delta H°_f(H_2O)] - [\Delta H°_f(PbO_2) + 2 \Delta H°_f(HSO_4^-)]$

$\Delta H° = [(2 \text{ mol})(-919.9 \text{ kJ/mol}) + (2 \text{ mol})(-285.8 \text{ kJ/mol})]$

$$- [(1 \text{ mol})(-277 \text{ kJ/mol}) + (2 \text{ mol})(-887.3 \text{ kJ/mol})] = -359.8 \text{ kJ}$$

$\Delta S° = [2 S°(PbSO_4) + 2 S°(H_2O)] - [S°(Pb) + S°(PbO_2) + 2 S°(H^+) + 2 S°(HSO_4^-)]$

$\Delta S° = [(2 \text{ mol})(148.6 \text{ J/(K} \cdot \text{mol)}) + (2 \text{ mol})(69.9 \text{ J/(K} \cdot \text{mol)})]$

$$- [(1 \text{ mol})(64.8 \text{ J/(K} \cdot \text{mol)}) + (1 \text{ mol})(68.6 \text{ J/(K} \cdot \text{mol)})$$

$$+ (2 \text{ mol})(132 \text{ J/(K} \cdot \text{mol)})] = 39.6 \text{ J/K} = 39.6 \times 10^{-3} \text{ kJ/K}$$

$\Delta G° = \Delta H° - T\Delta S° = -359.8 \text{ kJ} - (261 \text{ K})(39.6 \times 10^{-3} \text{ kJ/K}) = -370.1 \text{ kJ at } 261 \text{ K}$

$$HSO_4^-(aq) + H_2O(l) \rightleftharpoons H_3O^+(aq) + SO_4^{2-}(aq)$$

initial (M)	0.100	0.100	0
change (M)	$-x$	$+x$	$+x$
equil (M)	$0.100 - x$	$0.100 + x$	x

$K_{a2} = \dfrac{[H_3O^+][SO_4^{2-}]}{[HSO_4^-]} = 1.2 \times 10^{-2} = \dfrac{(0.100 + x)x}{0.100 - x}$

$x^2 + 0.112x - (1.2 \times 10^{-3}) = 0$

Use the quadratic formula to solve for x.

$x = \dfrac{-(0.112) \pm \sqrt{(0.112)^2 - (4)(1)(-1.2 \times 10^{-3})}}{2(1)} = \dfrac{-0.112 \pm 0.132}{2}$

$x = -0.122$ and 0.010

Of the two solutions for x, only 0.010 has physical meaning because –0.122 would lead to negative concentrations of H_3O^+ and SO_4^{2-}.

$[H^+] = 0.100 + x = 0.100 + 0.010 = 0.110 \text{ M}$

$[HSO_4^-] = 0.100 - x = 0.100 - 0.010 = 0.090 \text{ M}$

$\Delta G = \Delta G° + RT \ln \dfrac{1}{[H^+]^2[HSO_4^-]^2}$

$\Delta G = (-370.1 \text{ kJ/mol}) + [8.314 \times 10^{-3} \text{ kJ/(K} \cdot \text{mol)}](261 \text{ K}) \ln \dfrac{1}{(0.110)^2(0.090)^2}$

$\Delta G = -350.1 \text{ kJ/mol}$

16.123 $CaCO_3(s) \rightleftharpoons Ca^{2+}(aq) + CO_3^{2-}(aq)$

$\Delta H° = [\Delta H°_f(Ca^{2+}) + \Delta H°_f(CO_3^{2-})] - \Delta H°_f(CaCO_3)$

$\Delta H° = [(1 \text{ mol})(-542.8 \text{ kJ/mol}) + (1 \text{ mol})(-677.1 \text{ kJ/mol})] - (1 \text{ mol})(-1207.6 \text{ kJ/mol})$

$\Delta H° = -12.3 \text{ kJ}$

$\Delta S° = [S°(Ca^{2+}) + S°(CO_3^{2-})] - S°(CaCO_3)$

$\Delta S° = [(1 \text{ mol})(-53.1 \text{ J/(K} \cdot \text{mol)}) + (1 \text{ mol})(-56.9 \text{ J/(K} \cdot \text{mol)})] - (1 \text{ mol})(91.7 \text{ J/(K} \cdot \text{mol)})$

$\Delta S° = -201.7 \text{ J/K} = -201.7 \times 10^{-3} \text{ kJ/K}$

$50 °C = 50 + 273 = 323 \text{ K}$

$\Delta G = \Delta H° - T\Delta S° = -12.3 \text{ kJ} - (323 \text{ K})(-201.7 \times 10^{-3} \text{ kJ/K}) = +52.85 \text{ kJ}$

$\Delta G = -RT \ln K_{sp}$

$$\ln K_{sp} = \frac{-\Delta G}{RT} = \frac{-52.85 \text{ J/mol}}{[8.314 \times 10^{-3} \text{ kJ/(K} \cdot \text{mol)}](323 \text{ K})} = -19.68$$

$$K_{sp} = e^{-19.68} = 2.8 \times 10^{-9}$$

$20 \text{ °C} = 20 + 273 = 293 \text{ K}$

$$n_{CO_2} = \frac{PV}{RT} = \frac{\left(731 \text{ mm Hg} \times \dfrac{1.00 \text{ atm}}{760 \text{ mm Hg}}\right)(1.000 \text{ L})}{\left(0.082 \ 06 \ \dfrac{\text{L} \cdot \text{atm}}{\text{K} \cdot \text{mol}}\right)(293 \text{ K})} = 0.0400 \text{ mol } CO_2$$

$Ca(OH)_2$, 74.09 amu

$$\text{mol } Ca(OH)_2 = 3.335 \text{ g } Ca(OH)_2 \ \times \ \frac{1 \text{ mol } Ca(OH)_2}{74.09 \text{ g } Ca(OH)_2} = 0.0450 \text{ mol } Ca(OH)_2$$

$$CO_2(g) \ + \ H_2O(l) \ \rightarrow \ H_2CO_3(aq)$$

$$Ca(OH)_2(aq) \ + \ H_2CO_3(aq) \ \rightarrow \ CaCO_3(s) \ + \ 2 \ H_2O(l)$$

	$Ca(OH)_2(aq)$	$H_2CO_3(aq)$	$CaCO_3(s)$	
before (mol)	0.0450	0.0400	0	
change (mol)	−0.0400	−0.0400	+0.0400	
after (mol)	0.0050	0	0.0400	

500.0 mL = 0.5000 L

$[Ca(OH)_2] = [Ca^{2+}] = 0.0050 \text{ mol}/0.5000 \text{ L} = 0.010 \text{ M}$

$$CaCO_3(s) \ \rightleftharpoons \ Ca^{2+}(aq) \ + \ CO_3^{2-}(aq)$$

	Ca^{2+}	CO_3^{2-}
initial (M)	0.010	0
change (M)	+x	+x
equil (M)	0.010 + x	x

$K_{sp} = [Ca^{2+}][CO_3^{2-}] = 2.8 \times 10^{-9} = (0.010 + x)x \approx 0.010x$

$x = \text{molar solubility} = 2.8 \times 10^{-9}/0.010 = 2.8 \times 10^{-7} \text{ M}$

Because $\Delta H°$ is negative (exothermic), the solubility of $CaCO_3$ is lower at 50 °C.

16.124 $PV = nRT$

$$n_{NH_3} = \frac{PV}{RT} = \frac{\left(744 \text{ mm Hg} \times \dfrac{1.00 \text{ atm}}{760 \text{ mm Hg}}\right)(1.00 \text{ L})}{\left(0.082 \ 06 \ \dfrac{\text{L} \cdot \text{atm}}{\text{K} \cdot \text{mol}}\right)(298.1 \text{ K})} = 0.0400 \text{ mol } NH_3$$

500.0 mL = 0.5000 L

$[NH_3] = 0.0400 \text{ mol}/0.5000 \text{ L} = 0.0800 \text{ M}$

$$NH_3(aq) \ + \ H_2O(l) \ \rightleftharpoons \ NH_4^+(aq) \ + \ OH^-(aq)$$

$\Delta H° = [\Delta H°_f(NH_4^+) \ + \ \Delta H°_f(OH^-)] \ - \ [\Delta H°_f(NH_3) \ + \ \Delta H°_f(H_2O)]$

$\Delta H° = [(1 \text{ mol})(-132.5 \text{ kJ/mol}) + (1 \text{ mol})(-230.0 \text{ kJ/mol})]$

$\qquad\qquad - [(1 \text{ mol})(-80.3 \text{ kJ/mol}) + (1 \text{ mol})(-285.8 \text{ kJ/mol})] = +3.6 \text{ kJ}$

$\Delta S° = [S°(NH_4^+) \ + \ S°(OH^-)] \ - \ [S°(NH_3) \ + \ S°(H_2O)]$

$\Delta S^\circ = [(1 \text{ mol})(113 \text{ J/(K} \cdot \text{mol)}) + (1 \text{ mol})(-10.8 \text{ J/(K} \cdot \text{mol)})]$

$\qquad - [(1 \text{ mol})(111 \text{ J/(K} \cdot \text{mol)}) + (1 \text{ mol})(69.9 \text{ J/(K} \cdot \text{mol)})] = -78.7 \text{ J/K}$

$T = 2.0 \text{ °C} = 2.0 + 273.1 = 275.1 \text{ K}$

$\Delta G^\circ = \Delta H^\circ - T\Delta S^\circ = 3.6 \text{ kJ} - (275.1 \text{ K})(-78.7 \times 10^{-3} \text{ kJ/K}) = 25.3 \text{ kJ}$

$\Delta G^\circ = -RT \ln K_b$

$\ln K_b = \dfrac{-\Delta G^\circ}{RT} = \dfrac{-25.3 \text{ kJ/mol}}{[8.314 \times 10^{-3} \text{ kJ/(K} \cdot \text{mol)}](275.1 \text{ K})} = -11.06$

$K_b = e^{-11.06} = 1.6 \times 10^{-5}$

$$\begin{array}{lcccc} & NH_3(aq) + H_2O(l) & \rightleftharpoons & NH_4^+(aq) + & OH^-(aq) \end{array}$$

initial (M)	0.0800	0	~0
change (M)	–x	+x	+x
equil (M)	0.0800 – x	x	x

at 2 °C, $K_b = \dfrac{[NH_4^+][OH^-]}{[NH_3]} = 1.6 \times 10^{-5} = \dfrac{x^2}{0.0800 - x} \approx \dfrac{x^2}{0.0800}$

$x^2 = (1.6 \times 10^{-5})(0.0800)$

$x = [OH^-] = \sqrt{(1.6 \times 10^{-5})(0.0800)} = 1.13 \times 10^{-3} \text{ M}$

$[H_3O^+] = \dfrac{1.0 \times 10^{-14}}{1.13 \times 10^{-3}} = 8.85 \times 10^{-12} \text{ M}$

$pH = -\log[H_3O^+] = -\log(8.85 \times 10^{-12}) = 11.05$

16.125 (a) $I_2(s) \rightarrow 2\,I^-(aq)$

$[I_2(s) + 2\,e^- \rightarrow 2\,I^-(aq)] \times 5$ reduction half reaction

$I_2(s) \rightarrow 2\,IO_3^-(aq)$

$I_2(s) + 6\,H_2O(l) \rightarrow 2\,IO_3^-(aq)$

$I_2(s) + 6\,H_2O(l) \rightarrow 2\,IO_3^-(aq) + 12\,H^+(aq)$

$I_2(s) + 6\,H_2O(l) \rightarrow 2\,IO_3^-(aq) + 12\,H^+(aq) + 10\,e^-$ oxidation half reaction

Combine the two half reactions.

$6\,I_2(s) + 6\,H_2O(l) \rightarrow 10\,I^-(aq) + 2\,IO_3^-(aq) + 12\,H^+(aq)$

Divide all coefficients by 2.

$3\,I_2(s) + 3\,H_2O(l) \rightarrow 5\,I^-(aq) + IO_3^-(aq) + 6\,H^+(aq)$

$3\,I_2(s) + 3\,H_2O(l) + 6\,OH^-(aq) \rightarrow 5\,I^-(aq) + IO_3^-(aq) + 6\,H^+(aq) + 6\,OH^-(aq)$

$3\,I_2(s) + 3\,H_2O(l) + 6\,OH^-(aq) \rightarrow 5\,I^-(aq) + IO_3^-(aq) + 6\,H_2O(l)$

$3\,I_2(s) + 6\,OH^-(aq) \rightarrow 5\,I^-(aq) + IO_3^-(aq) + 3\,H_2O(l)$

(b) $\Delta G^\circ = [5\,\Delta G^\circ_f(I^-) + \Delta G^\circ_f(IO_3^-) + 3\,\Delta G^\circ_f(H_2O(l))] - 6\,\Delta G^\circ_f(OH^-)$

$\Delta G^\circ = [(5 \text{ mol})(-51.6 \text{ kJ/mol}) + (1 \text{ mol})(-128.0 \text{ kJ/mol}) + (3 \text{ mol})(-237.2 \text{ kJ/mol})]$

$\qquad - (6 \text{ mol})(-157.3 \text{ kJ/mol}) = -153.8 \text{ kJ}$

(c) The reaction is spontaneous because ΔG° is negative.

(d) 25 °C = 25 + 273 = 298 K

$\Delta G° = -RT \ln K_c$

$\ln K_c = \dfrac{-\Delta G°}{RT} = \dfrac{-(-153.8 \text{ kJ/mol})}{[8.314 \times 10^{-3} \text{ kJ/(K} \cdot \text{mol)}](298 \text{ K})} = 62.077$

$K_c = e^{62.077} = 9.1 \times 10^{26}$

$K_c = \dfrac{[I^-]^5[IO_3^-]}{[OH^-]^6} = 9.1 \times 10^{26} = \dfrac{(0.10)^5(0.50)}{[OH^-]^6}$

$[OH^-] = \sqrt[6]{\dfrac{(0.10)^5(0.50)}{9.1 \times 10^{26}}} = 4.2 \times 10^{-6} \text{ M}$

$[H_3O^+] = \dfrac{1.0 \times 10^{-14}}{4.2 \times 10^{-6}} = 2.38 \times 10^{-9} \text{ M}$

$\text{pH} = -\log[H_3O^+] = -\log(2.38 \times 10^{-9}) = 8.62$

16.126 $\quad N_2O_4(g) \rightleftharpoons 2\,NO_2(g)$

$\Delta H° = 2\,\Delta H°_f(NO_2) - \Delta H°_f(N_2O_4) = (2 \text{ mol})(33.2 \text{ kJ/mol}) - (1 \text{ mol})(11.1 \text{ kJ/mol}) = 55.3 \text{ kJ}$

$\Delta S° = 2\,S°(NO_2) - S°(N_2O_4) = (2 \text{ mol})(240.0 \text{ J/(K} \cdot \text{mol)}) - (1 \text{ mol})(304.3 \text{ J/(K} \cdot \text{mol)})$

$\Delta S° = 175.7 \text{ J/K} = 175.7 \times 10^{-3} \text{ kJ/K}$

$\Delta G° = \Delta H° - T\Delta S° = 55.3 \text{ kJ} - (373 \text{ K})(175.7 \times 10^{-3} \text{ kJ/K}) = -10.2 \text{ kJ}$

$K_p = \dfrac{(P_{NO_2})^2}{P_{N_2O_4}}$

$\Delta G° = -RT \ln K_p$

$\ln K_p = \dfrac{-\Delta G°}{RT} = \dfrac{-(-10.2 \text{ kJ/mol})}{[8.314 \times 10^{-3} \text{ kJ/(K} \cdot \text{mol)}](373 \text{ K})} = 3.29$

$K_p = e^{3.29} = 27$

	$N_2O_4(g)$	$\rightleftharpoons$	$2\,NO_2(g)$
initial (atm)	1.00		1.00
change (atm)	$-x$		$+2x$
equil (atm)	$1.00 - x$		$1.00 + 2x$

$K_p = \dfrac{(P_{NO_2})^2}{P_{N_2O_4}} = 27 = \dfrac{(1.00 + 2x)^2}{(1.00 - x)}$

$4x^2 + 31x - 26 = 0$

Use the quadratic formula to solve for x.

$x = \dfrac{-(31) \pm \sqrt{(31)^2 - (4)(4)(-26)}}{2(4)} = \dfrac{-31 \pm 37.1}{8}$

$x = 0.76$ and -8.5

Of the two solutions for x, only 0.76 has physical meaning because -8.5 would lead to a negative partial pressure for NO_2.

$P_{N_2O_4} = 1.00 - x = 1.00 - 0.76 = 0.24 \text{ atm}; \quad P_{NO_2} = 1.00 + 2x = 1.00 + 2(0.76) = 2.52 \text{ atm}$

17 Electrochemistry

17.1 $2 Ag^+(aq) + Ni(s) \rightarrow 2 Ag(s) + Ni^{2+}(aq)$

There is a Ni anode in an aqueous solution of Ni^{2+}, and a Ag cathode in an aqueous solution of Ag^+. A salt bridge connects the anode and cathode compartment. The electrodes are connected through an external circuit.

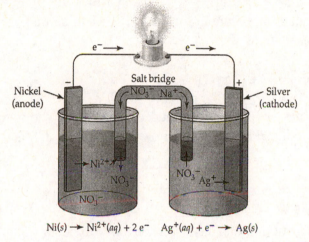

Ni(s) → Ni²⁺(aq) + 2 e⁻   Ag⁺(aq) + e⁻ → Ag(s)

17.2 $Pb(s) + Br_2(l) \rightarrow Pb^{2+}(aq) + 2 Br^-(aq)$

There is a Pb anode in an aqueous solution of Pb^{2+}. The cathode is a Pt wire that dips into a pool of liquid Br_2 and an aqueous solution that is saturated with Br_2. A salt bridge connects the anode and cathode compartment. The electrodes are connected through an external circuit.

17.3 $Fe(s)|Fe^{2+}(aq)\|Sn^{2+}(aq)|Sn(s)$

17.4 (a) and (b)

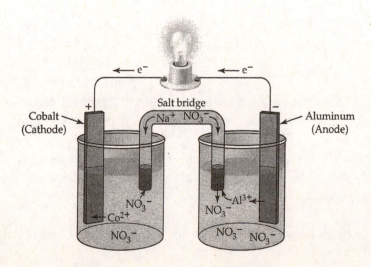

507

(c) $2 Al(s) + 3 Co^{2+}(aq) \rightarrow 2 Al^{3+}(aq) + 3 Co(s)$

(d) $Al(s) | Al^{3+}(aq) \| Co^{2+}(aq) | Co(s)$

17.5 $Al(s) + Cr^{3+}(aq) \rightarrow Al^{3+}(aq) + Cr(s)$

$$\Delta G° = -nFE° = -(3 \text{ mol e}^-)\left(\frac{96,500 \text{ C}}{1 \text{ mol e}^-}\right)(0.92 \text{ V})\left(\frac{1 \text{ J}}{1 \text{ C} \cdot \text{V}}\right) = -266,340 \text{ J} = -270 \text{ kJ}$$

17.6 oxidation: $Al(s) \rightarrow Al^{3+}(aq) + 3 e^-$ $E° = 1.66$ V

reduction: $\underline{Cr^{3+}(aq) + 3 e^- \rightarrow Cr(s)}$ $\underline{E° = ?}$

overall $Al(s) + Cr^{3+}(aq) \rightarrow Al^{3+}(aq) + Cr(s)$ $E° = 0.92$ V

The standard reduction potential for the Cr^{3+}/Cr half cell is:

$E° = 0.92 - 1.66 = -0.74$ V

17.7 (a) $Cl_2(g) + 2 e^- \rightarrow 2 Cl^-(aq)$ $E° = 1.36$ V

     $Ag^+(aq) + e^- \rightarrow Ag(s)$ $E° = 0.80$ V

Cl_2 has the greater tendency to be reduced (larger $E°$). The species that has the greater tendency to be reduced is the stronger oxidizing agent. Cl_2 is the stronger oxidizing agent.

(b) $Fe^{2+}(aq) + 2 e^- \rightarrow Fe(s)$ $E° = -0.45$ V

     $Mg^{2+}(aq) + 2 e^- \rightarrow Mg(s)$ $E° = -2.37$ V

The second half-reaction has the lesser tendency to occur in the forward direction (more negative $E°$) and the greater tendency to occur in the reverse direction. Therefore, Mg is the stronger reducing agent.

17.8 (a) $2 Fe^{3+}(aq) + 2 I^-(aq) \rightarrow 2 Fe^{2+}(aq) + I_2(s)$

reduction: $Fe^{3+}(aq) + e^- \rightarrow Fe^{2+}(aq)$ $E° = 0.77$ V

oxidation: $2 I^-(aq) \rightarrow I_2(s) + 2 e^-$ $\underline{E° = -0.54 \text{ V}}$

                 overall $E° = 0.23$ V

Because $E°$ for the overall reaction is positive, this reaction can occur under standard-state conditions.

(b) $3 Ni(s) + 2 Al^{3+}(aq) \rightarrow 3 Ni^{2+}(aq) + 2 Al(s)$

oxidation: $Ni(s) \rightarrow Ni^{2+}(aq) + 2 e^-$ $E° = 0.26$ V

reduction: $Al^{3+}(aq) + 3 e^- \rightarrow Al(s)$ $\underline{E° = -1.66 \text{ V}}$

                 overall $E° = -1.40$ V

Because $E°$ for the overall reaction is negative, this reaction cannot occur under standard-state conditions. This reaction can occur in the reverse direction.

17.9 (a) D is the strongest reducing agent. D^+ has the most negative standard reduction potential. A^{3+} is the strongest oxidizing agent. It has the most positive standard reduction potential.

(b) An oxidizing agent can oxidize any reducing agent that is below it in the table. B^{2+} can oxidize C and D.

A reducing agent can reduce any oxidizing agent that is above it in the table. C can reduce A^{3+} and B^{2+}.

(c) Use the two half-reactions that have the most positive and the most negative standard reduction potentials, respectively.

$$
\begin{array}{ll}
A^{3+} + 2\,e^- \rightarrow A^+ & 1.47\ \text{V} \\
\underline{2 \times (D \rightarrow D^+ + e^-)} & \underline{1.38\ \text{V}} \\
A^{3+} + 2\,D \rightarrow A^+ + 2\,D^+ & 2.85\ \text{V}
\end{array}
$$

17.10 $Cu(s) + 2\,Fe^{3+}(aq) \rightarrow Cu^{2+}(aq) + 2\,Fe^{2+}(aq)$

$$E^\circ = E^{\,o}_{Cu \rightarrow Cu^{2+}} + E^{\,o}_{Fe^{3+} \rightarrow Fe^{2+}} = -0.34\ \text{V} + 0.77\ \text{V} = 0.43\ \text{V}; \qquad n = 2\ \text{mol}\ e^-$$

$$E = E^\circ - \frac{0.0592\ \text{V}}{n} \log \frac{[Cu^{2+}][Fe^{2+}]^2}{[Fe^{3+}]^2} = 0.43\ \text{V} - \frac{(0.0592\ \text{V})}{2} \log \frac{(0.25)(0.20)^2}{(1.0 \times 10^{-4})^2} = 0.25\ \text{V}$$

17.11
$5\,[Cu(s) \rightarrow Cu^{2+}(aq) + 2\,e^-]$ (oxidation half reaction)
$2\,[5\,e^- + 8\,H^+(aq) + MnO_4^-(aq) \rightarrow Mn^{2+}(aq) + 4\,H_2O(l)]$ (reduction half reaction)

$5\,Cu(s) + 16\,H^+(aq) + 2\,MnO_4^-(aq) \rightarrow 5\,Cu^{2+}(aq) + 2\,Mn^{2+}(aq) + 8\,H_2O(l)$

$$\Delta E = -\frac{0.0592\ \text{V}}{n} \log \frac{[Cu^{2+}]^5[Mn^{2+}]^2}{[MnO_4^-]^2[H^+]^{16}}$$

(a) The anode compartment contains Cu^{2+}.

$$\Delta E = -\frac{0.0592\ \text{V}}{10} \log \frac{(0.01)^5(1)^2}{(1)^2(1)^{16}} = +0.059\ \text{V}$$

(b) The cathode compartment contains Mn^{2+}, MnO_4^-, and H^+.

$$\Delta E = -\frac{0.0592\ \text{V}}{10} \log \frac{(1)^5(0.01)^2}{(0.01)^2(0.01)^{16}} = -0.19\ \text{V}$$

17.12 $H_2(g) + Pb^{2+}(aq) \rightarrow 2\,H^+(aq) + Pb(s)$

$$E^\circ = E^{\,o}_{H_2 \rightarrow H^+} + E^{\,o}_{Pb^{2+} \rightarrow Pb} = 0\ \text{V} + (-0.13\ \text{V}) = -0.13\ \text{V}; \qquad n = 2\ \text{mol}\ e^-$$

$$E = E^\circ - \frac{0.0592\ \text{V}}{n} \log \frac{[H_3O^+]^2}{[Pb^{2+}](P_{H_2})}$$

$$0.28\ \text{V} = -0.13\ \text{V} - \frac{(0.0592\ \text{V})}{2} \log \frac{[H_3O^+]^2}{(1)(1)} = -0.13\ \text{V} - (0.0592\ \text{V}) \log [H_3O^+]$$

$pH = -\log[H_3O^+]$ therefore $0.28\ \text{V} = -0.13\ \text{V} + (0.0592\ \text{V})\ pH$

$$pH = \frac{(0.28\ \text{V} + 0.13\ \text{V})}{0.0592\ \text{V}} = 6.9$$

17.13 $4 Fe^{2+}(aq) + O_2(g) + 4 H^+(aq) \rightarrow 4 Fe^{3+}(aq) + 2 H_2O(l)$

$E^\circ = E^\circ_{Fe^{2+} \rightarrow Fe^{3+}} + E^\circ_{O_2 \rightarrow H_2O} = -0.77 \text{ V} + 1.23 \text{ V} = 0.46 \text{ V}; \quad n = 4 \text{ mol } e^-$

$E^\circ = \dfrac{0.0592 \text{ V}}{n} \log K; \quad \log K = \dfrac{nE^\circ}{0.0592 \text{ V}} = \dfrac{(4)(0.46 \text{ V})}{0.0592 \text{ V}} = 31; \quad K = 10^{31} \text{ at } 25 \text{ °C}$

17.14 $E^\circ = \dfrac{0.0592 \text{ V}}{n} \log K = \dfrac{0.0592 \text{ V}}{2} \log(1.8 \times 10^{-5}) = -0.140 \text{ V}$

17.15 (a) $Zn(s) + 2 MnO_2(s) + 2 NH_4^+(aq) \rightarrow Zn^{2+}(aq) + Mn_2O_3(s) + 2 NH_3(aq) + H_2O(l)$
(b) $Zn(s) + 2 MnO_2(s) \rightarrow ZnO(s) + Mn_2O_3(s)$
(c) $Cd(s) + 2 NiO(OH)(s) + 2 H_2O(l) \rightarrow Cd(OH)_2(s) + 2 Ni(OH)_2(s)$
(d) $x Li(s) + MnO_2(s) \rightarrow Li_xMnO_2(s)$
(e) $Li_xC_6(s) + Li_{1-x}CoO_2(s) \rightarrow 6 C(s) + LiCoO_2(s)$

17.16 A fuel cell and a battery are both galvanic cells that convert chemical energy into electrical energy utilizing a spontaneous redox reaction. A fuel cell differs from an ordinary battery in that the reactants are not contained within the cell but instead are continuously supplied from an external resevoir.

17.17 (a) anode reaction $2 H_2(g) + 4 OH^-(aq) \rightarrow 4 H_2O(l) + 4 e^-$ $E^\circ = 0.83 \text{ V}$
     cathode reaction $\underline{O_2(g) + 2 H_2O(l) + 4 e^- \rightarrow 4 OH^-(aq)}$ $E^\circ = 0.40 \text{ V}$
     overall reaction $2 H_2(g) + O_2(g) \rightarrow 2 H_2O(l)$ $E^\circ = 1.23 \text{ V}$

     (b) anode reaction $2 H_2(g) \rightarrow 4 H^+(aq) + 4 e^-$ $E^\circ = 0.00 \text{ V}$
     cathode reaction $\underline{O_2(g) + 4 H^+(aq) + 4 e^- \rightarrow 2 H_2O(l)}$ $E^\circ = 1.23 \text{ V}$
     overall reaction $2 H_2(g) + O_2(g) \rightarrow 2 H_2O(l)$ $E^\circ = 1.23 \text{ V}$

17.18 (a) $[Mg(s) \rightarrow Mg^{2+}(aq) + 2 e^-] \times 2$
     $\underline{O_2(g) + 4 H^+(aq) + 4 e^- \rightarrow 2 H_2O(l)}$
     $2 Mg(s) + O_2(g) + 4 H^+(aq) \rightarrow 2 Mg^{2+}(aq) + 2 H_2O(l)$

     (b) $[Fe(s) \rightarrow Fe^{2+}(aq) + 2 e^-] \times 4$
     $[O_2(g) + 4 H^+(aq) + 4 e^- \rightarrow 2 H_2O(l)] \times 2$
     $4 Fe^{2+}(aq) + O_2(g) + 4 H^+(aq) \rightarrow 4 Fe^{3+}(aq) + 2 H_2O(l)$
     $\underline{[2 Fe^{3+}(aq) + 4 H_2O(l) \rightarrow Fe_2O_3 \cdot H_2O(s) + 6 H^+(aq)] \times 2}$
     $4 Fe(s) + 3 O_2(g) + 2 H_2O(l) \rightarrow 2 Fe_2O_3 \cdot H_2O(s)$

17.19   (a)

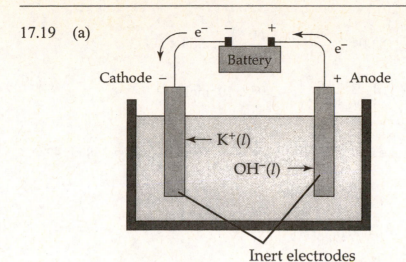

Inert electrodes

(b)   anode reaction $4\ OH^-(l) \rightarrow O_2(g) + 2\ H_2O(l) + 4\ e^-$
      cathode reaction $4\ K^+(l) + 4\ e^- \rightarrow 4\ K(l)$
      overall reaction $4\ K^+(l) + 4\ OH^-(l) \rightarrow 4\ K(l) + O_2(g) + 2\ H_2O(l)$

17.20   (a)   anode reaction $2\ Cl^-(aq) \rightarrow Cl_2(g) + 2\ e^-$
              cathode reaction $2\ H_2O(l) + 2\ e^- \rightarrow H_2(g) + 2\ OH^-(aq)$
              overall reaction $2\ Cl^-(aq) + 2\ H_2O(l) \rightarrow Cl_2(g) + H_2(g) + 2\ OH^-(aq)$

        (b)   anode reaction $2\ H_2O(l) \rightarrow O_2(g) + 4\ H^+(aq) + 4\ e^-$
              cathode reaction $2\ Cu^{2+}(aq) + 4\ e^- \rightarrow 2\ Cu(s)$
              overall reaction $2\ Cu^{2+}(aq) + 2\ H_2O(l) \rightarrow 2\ Cu(s) + O_2(g) + 4\ H^+(aq)$

        (c)   anode reaction $2\ H_2O(l) \rightarrow O_2(g) + 4\ H^+(aq) + 4\ e^-$
              cathode reaction $4\ H_2O(l) + 4\ e^- \rightarrow 2\ H_2(g) + 4\ OH^-(aq)$
              overall reaction $2\ H_2O(l) \rightarrow 2\ H_2(g) + O_2(g)$

17.21

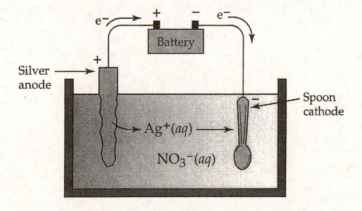

anode reaction $Ag(s) \rightarrow Ag^+(aq) + e^-$
cathode reaction $Ag^+(aq) + e^- \rightarrow Ag(s)$
The overall reaction is transfer of silver metal from the silver anode to the spoon.

17.22 Charge $= \left(1.00 \times 10^5 \dfrac{C}{s}\right)(8.00 \text{ h})\left(\dfrac{60 \text{ min}}{h}\right)\left(\dfrac{60 \text{ s}}{\text{min}}\right) = 2.88 \times 10^9 \text{ C}$

Moles of $e^- = (2.88 \times 10^9 \text{ C})\left(\dfrac{1 \text{ mol } e^-}{96{,}500 \text{ C}}\right) = 2.98 \times 10^4 \text{ mol } e^-$

cathode reaction: $Al^{3+} + 3 e^- \rightarrow Al$

mass Al $= (2.98 \times 10^4 \text{ mol } e^-) \times \dfrac{1 \text{ mol Al}}{3 \text{ mol } e^-} \times \dfrac{26.98 \text{ g Al}}{1 \text{ mol Al}} \times \dfrac{1 \text{ kg}}{1000 \text{ g}} = 268 \text{ kg Al}$

17.23 $3.00 \text{ g Ag} \times \dfrac{1 \text{ mol Ag}}{107.9 \text{ g Ag}} = 0.0278 \text{ mol Ag}$

cathode reaction: $Ag^+(aq) + e^- \rightarrow Ag(s)$

Charge $= (0.0278 \text{ mol Ag})\left(\dfrac{1 \text{ mol } e^-}{1 \text{ mol Ag}}\right)\left(\dfrac{96{,}500 \text{ C}}{1 \text{ mol } e^-}\right) = 2682.7 \text{ C}$

Time $= \dfrac{C}{A} = \left(\dfrac{2682.7 \text{ C}}{0.100 \text{ C/s}} \times \dfrac{1 \text{ h}}{3600 \text{ s}}\right) = 7.45 \text{ h}$

17.24 When a beam of white light strikes the anodized surface, part of the light is reflected from the outer TiO_2, while part penetrates through the semitransparent TiO_2 and is reflected from the inner metal. If the two reflections of a particular wavelength are out of phase, they interfere destructively and that wavelength is canceled from the reflected light. Because $n\lambda = 2d \times \sin\theta$, the canceled wavelength depends on the thickness of the TiO_2 layer.

17.25 volume $= \left(0.0100 \text{ mm} \times \dfrac{1 \text{ cm}}{10 \text{ mm}}\right)(10.0 \text{ cm})^2 = 0.100 \text{ cm}^3$

mol $Al_2O_3 = (0.100 \text{ cm}^3)(3.97 \text{ g/cm}^3)\dfrac{1 \text{ mol } Al_2O_3}{102.0 \text{ g } Al_2O_3} = 3.892 \times 10^{-3} \text{ mol } Al_2O_3$

mole $e^- = 3.892 \times 10^{-3} \text{ mol } Al_2O_3 \times \dfrac{6 \text{ mol } e^-}{1 \text{ mol } Al_2O_3} = 0.02335 \text{ mol } e^-$

coulombs $= 0.02335 \text{ mol } e^- \times \dfrac{96{,}500 \text{ C}}{1 \text{ mol } e^-} = 2253 \text{ C}$

time $= \dfrac{C}{A} = \dfrac{2253 \text{ C}}{0.600 \text{ C/s}} \times \dfrac{1 \text{ min}}{60 \text{ s}} = 62.6 \text{ min}$

Key Concept Problems

17.26 (a) - (d)

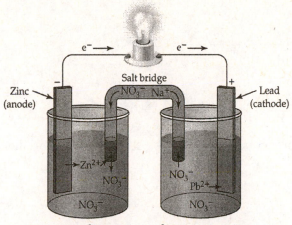

$$Zn(s) \rightarrow Zn^{2+}(aq) + 2\,e^- \quad Pb^{2+}(aq) + 2\,e^- \rightarrow Pb(s)$$

(e) anode reaction $Zn(s) \rightarrow Zn^{2+}(aq) + 2\,e^-$
    cathode reaction $\underline{Pb^{2+}(aq) + 2\,e^- \rightarrow Pb(s)}$
    overall reaction $Zn(s) + Pb^{2+}(aq) \rightarrow Zn^{2+}(aq) + Pb(s)$

17.27 (a) anode is Ni; cathode is Pt
    (b) anode reaction $3\,Ni(s) \rightarrow 3\,Ni^{2+}(aq) + 6\,e^-$
        cathode reaction $\underline{Cr_2O_7^{2-}(aq) + 14\,H^+(aq) + 6\,e^- \rightarrow 2\,Cr^{3+}(aq) + 7\,H_2O(l)}$
        overall reaction $Cr_2O_7^{2-}(aq) + 3\,Ni(s) + 14\,H^+(aq) \rightarrow$
$$2\,Cr^{3+}(aq) + 3\,Ni^{2+}(aq) + 7\,H_2O(l)$$

    (c) $Ni(s)\,|\,Ni^{2+}(aq)\,\|\,Cr_2O_7^{2-}(aq),\,Cr^{3+}\,|\,Pt(s)$

17.28 (a) The three cell reactions are the same except for cation concentrations.
    anode reaction $Cu(s) \rightarrow Cu^{2+}(aq) + 2\,e^-$ $E° = -0.34$ V
    cathode reaction $\underline{2\,Fe^{3+}(aq) + 2\,e^- \rightarrow 2\,Fe^{2+}(aq)}$ $\underline{E° = 0.77}$ V
    overall reaction $Cu(s) + 2\,Fe^{3+}(aq) \rightarrow Cu^{2+}(aq) + 2\,Fe^{2+}(aq)$ $E° = 0.43$ V

    (b)

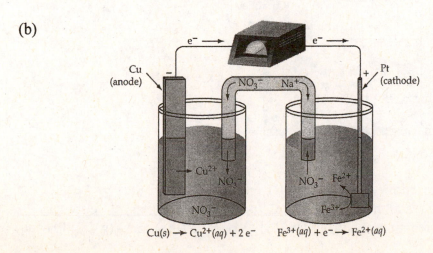

$$Cu(s) \rightarrow Cu^{2+}(aq) + 2\,e^- \quad Fe^{3+}(aq) + e^- \rightarrow Fe^{2+}(aq)$$

(c) $E = E^o - \dfrac{0.0592\,V}{n}\,\log\dfrac{[Cu^{2+}][Fe^{2+}]^2}{[Fe^{2+}]^2};\quad n = 2\ \text{mol}\ e^-$

(1) $E = E^o = 0.43\ V$ because all cation concentrations are 1 M.

(2) $E = E^o - \dfrac{0.0592\,V}{2}\,\log\dfrac{(1)(5)^2}{(1)^2} = 0.39\ V$

(3) $E = E^o - \dfrac{0.0592\,V}{2}\,\log\dfrac{(0.1)(0.1)^2}{(0.1)^2} = 0.46\ V$

Cell (3) has the largest potential, while cell (2) has the smallest as calculated from the Nernst equation.

17.29 (a) - (b)

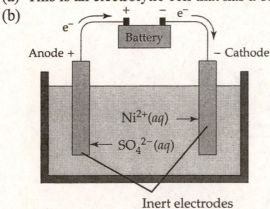

(c) anode reaction $2\,Br^-(aq) \rightarrow Br_2(aq) + 2\,e^-$
cathode reaction $\underline{Cu^{2+}(aq) + 2\,e^- \rightarrow Cu(s)}$
overall reaction $Cu^{2+}(aq) + 2\,Br^-(aq) \rightarrow Cu(s) + Br_2(aq)$

17.30 (a) This is an electrolytic cell that has a battery connected between two inert electrodes.
(b)

Inert electrodes

(c) anode reaction $2\,H_2O(l) \rightarrow O_2(g) + 4\,H^+(aq) + 4\,e^-$
cathode reaction $\underline{Ni^{2+}(aq) + 2\,e^- \rightarrow Ni(s)}$
overall reaction $2\,Ni^{2+}(aq) + 2\,H_2O(l) \rightarrow 2\,Ni(s) + O_2(g) + 4\,H^+(aq)$

17.31   (a) & (b)

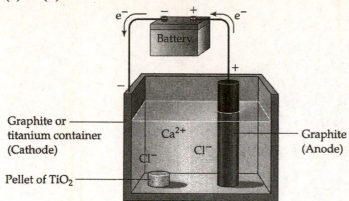

Graphite or titanium container (Cathode)

Ca^{2+}

Cl^- Cl^-

Graphite (Anode)

Pellet of TiO_2

(c)   anode reaction $2\,O^{2-} \rightarrow O_2(g) + 4\,e^-$

cathode reaction $\underline{TiO_2(s) + 4\,e^- \rightarrow Ti(s) + 2\,O^{2-}}$

overall reaction $TiO_2(s) \rightarrow Ti(s) + O_2(g)$

17.32 $Zn(s) + Cu^{2+}(aq) \rightarrow Zn^{2+}(aq) + Cu(s); \quad E = E^\circ - \dfrac{0.0592\ V}{2} \log \dfrac{[Zn^{2+}]}{[Cu^{2+}]}$

(a) E increases because increasing $[Cu^{2+}]$ decreases $\log \dfrac{[Zn^{2+}]}{[Cu^{2+}]}$.

(b) E will decrease because addition of H_2SO_4 increases the volume which decreases $[Cu^{2+}]$ and increases $\log \dfrac{[Zn^{2+}]}{[Cu^{2+}]}$.

(c) E decreases because increasing $[Zn^{2+}]$ increases $\log \dfrac{[Zn^{2+}]}{[Cu^{2+}]}$.

(d) Because there is no change in $[Zn^{2+}]$, there is no change in E.

17.33 $Cu(s) + 2\,Ag^+(aq) \rightarrow Cu^{2+}(aq) + 2\,Ag(s); \quad E = E^\circ - \dfrac{0.0592\ V}{2} \log \dfrac{[Cu^{2+}]}{[Ag^+]^2}$

(a) E decreases because addition of NaCl precipitates AgCl which decreases $[Ag^+]$ and increases $\log \dfrac{[Cu^{2+}]}{[Ag^+]^2}$.

(b) E increases because addition of NaCl increases the volume which decreases $[Cu^{2+}]$ and decreases $\log \dfrac{[Cu^{2+}]}{[Ag^+]^2}$.

(c) E decreases because addition of NH_3 complexes Ag^+, yielding $Ag(NH_3)_2^+$, which decreases $[Ag^+]$ and increases $\log \dfrac{[Cu^{2+}]}{[Ag^+]^2}$.

(d) E increases because addition of NH_3 complexes Cu^{2+}, yielding $Cu(NH_3)_4^{2+}$, which decreases $[Cu^{2+}]$ and decreases $\log \dfrac{[Cu^{2+}]}{[Ag^+]^2}$.

17.34 (a), (b) & (c)

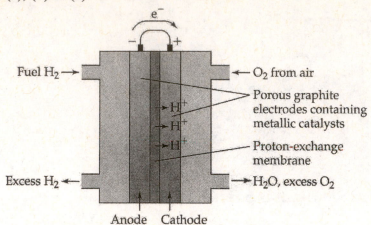

(d) anode reaction	$2\,H_2(g) \rightarrow 4\,H^+(aq) + 4\,e^-$	$E° = 0.00\ V$
cathode reaction	$O_2(g) + 4\,H^+(aq) + 4\,e^- \rightarrow 2\,H_2O(l)$	$E° = 1.23\ V$
overall reaction	$2\,H_2(g) + O_2(g) \rightarrow 2\,H_2O(l)$	$E° = 1.23\ V$

17.35 (a) A^+ is the strongest oxidizing agent. It has the most positive standard reduction potential. D is the strongest reducing agent. D^{3+} has the most negative standard reduction potential.

(b) An oxidizing agent can oxidize any reducing agent that is below it in the table. B^{2+} can oxidize C^- and D.

A reducing agent can reduce any oxidizing agent that is above it in the table. D can reduce A^+, B^{2+}, and C_2.

(c)

$3 \times (C_2 + 2\,e^- \rightarrow 2\,C^-)$	$0.17\ V$
$2 \times (D \rightarrow D^{3+} + 3\,e^-)$	$1.36\ V$
$3\,C_2 + 2\,D \rightarrow 6\,C^- + 2\,D^{3+}$	$1.53\ V$

Section Problems
Galvanic Cells (Sections 17.1–17.2)

17.36 The electrode where oxidation takes place is called the anode. For example, the lead electrode in the lead storage battery.
The electrode where reduction takes place is called the cathode. For example, the PbO_2 electrode in the lead storage battery.

17.37 The oxidizing agent gets reduced and reduction takes place at the cathode.

17.38 The cathode of a galvanic cell is considered to be the positive electrode because electrons flow through the external circuit toward the positive electrode (the cathode).

17.39 The salt bridge maintains charge neutrality in both the anode and cathode compartments of a galvanic cell.

17.40   (a) $Cd(s) + Sn^{2+}(aq) \rightarrow Cd^{2+}(aq) + Sn(s)$

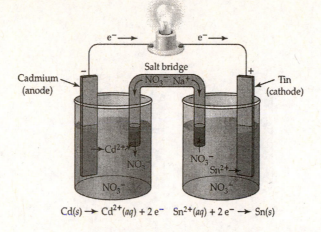

$Cd(s) \rightarrow Cd^{2+}(aq) + 2\,e^- \qquad Sn^{2+}(aq) + 2\,e^- \rightarrow Sn(s)$

(b) $2\,Al(s) + 3\,Cd^{2+}(aq) \rightarrow 2\,Al^{3+}(aq) + 3\,Cd(s)$

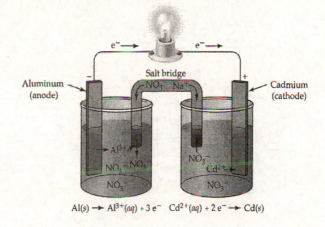

$Al(s) \rightarrow Al^{3+}(aq) + 3\,e^- \qquad Cd^{2+}(aq) + 2\,e^- \rightarrow Cd(s)$

(c) $6\,Fe^{2+}(aq) + Cr_2O_7^{2-}(aq) + 14\,H^+(aq) \rightarrow 6\,Fe^{3+}(aq) + 2\,Cr^{3+}(aq) + 7\,H_2O(l)$

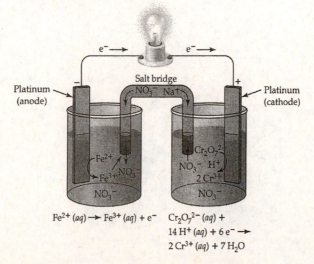

$Fe^{2+}(aq) \rightarrow Fe^{3+}(aq) + e^- \qquad Cr_2O_7^{2-}(aq) +$
$14\,H^+(aq) + 6\,e^- \rightarrow$
$2\,Cr^{3+}(aq) + 7\,H_2O$

17.41 (a) $3 Cu^{2+}(aq) + 2 Cr(s) \rightarrow 3 Cu(s) + 2 Cr^{3+}(aq)$

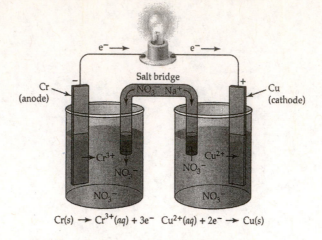

$$Cr(s) \rightarrow Cr^{3+}(aq) + 3e^- \quad Cu^{2+}(aq) + 2e^- \rightarrow Cu(s)$$

(b) $Pb(s) + 2 H^+(aq) \rightarrow Pb^{2+}(aq) + H_2(g)$

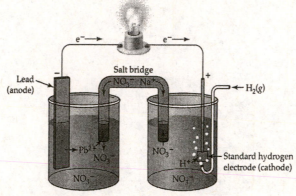

$$Pb(s) \rightarrow Pb^{2+}(aq) + 2e^- \quad 2 H^+(aq) + 2e^- \rightarrow H_2(g)$$

(c) $Cl_2(g) + Sn^{2+}(aq) \rightarrow Sn^{4+}(aq) + 2 Cl^-(aq)$

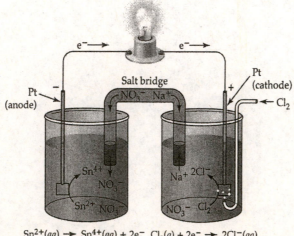

$$Sn^{2+}(aq) \rightarrow Sn^{4+}(aq) + 2e^- \quad Cl_2(q) + 2e^- \rightarrow 2Cl^-(aq)$$

17.42 (a) $Cd(s)|Cd^{2+}(aq)\|Sn^{2+}(aq)|Sn(s)$
(b) $Al(s)|Al^{3+}(aq)\|Cd^{2+}(aq)|Cd(s)$
(c) $Pt(s)|Fe^{2+}(aq), Fe^{3+}(aq)\|Cr_2O_7^{2-}(aq), Cr^{3+}(aq)|Pt(s)$

17.43 (a) $Cr(s)|Cr^{3+}(aq)\|Cu^{2+}(aq)|Cu(s)$
(b) $Pb(s)|Pb^{2+}(aq)\|H^+(aq)|H_2(g)|Pt(s)$
(c) $Pt(s)|Sn^{2+}(aq), Sn^{4+}(aq)\|Cl_2(g), Cl^-(aq)|Pt(s)$

17.44 (a)

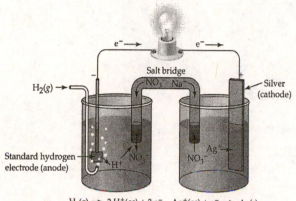

(b) anode reaction $H_2(g) \rightarrow 2H^+(aq) + 2e^-$
cathode reaction $\underline{2Ag^+(aq) + 2e^- \rightarrow 2Ag(s)}$
overall reaction $H_2(g) + 2Ag^+(aq) \rightarrow 2H^+(aq) + 2Ag(s)$

(c) $Pt(s)|H_2(g)|H^+(aq)\|Ag^+(aq)|Ag(s)$

17.45 (a)

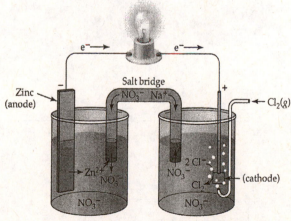

(b) anode reaction $Zn(s) \rightarrow Zn^{2+}(aq) + 2e^-$
cathode reaction $\underline{Cl_2(g) + 2e^- \rightarrow 2Cl^-(aq)}$
overall reaction $Zn(s) + Cl_2(g) \rightarrow Zn^{2+}(aq) + 2Cl^-(aq)$

(c) $Zn(s)|Zn^{2+}(aq)\|Cl_2(g)|Cl^-(aq)|C(s)$

17.46   (a)   anode reaction $Co(s) \rightarrow Co^{2+}(aq) + 2\ e^-$
            cathode reaction $\underline{Cu^{2+}(aq) + 2\ e^- \rightarrow Cu(s)}$
            overall reaction $Co(s) + Cu^{2+}(aq) \rightarrow Co^{2+}(aq) + Cu(s)$

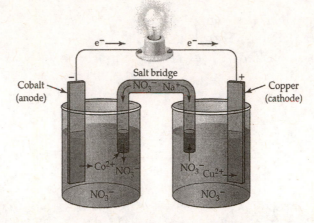

         (b)   anode reaction $2\ Fe(s) \rightarrow 2\ Fe^{2+}(aq) + 4\ e^-$
            cathode reaction $\underline{O_2(g) + 4\ H^+(aq) + 4\ e^- \rightarrow 2\ H_2O(l)}$
            overall reaction $2\ Fe(s) + O_2(g) + 4\ H^+(aq) \rightarrow 2\ Fe^{2+}(aq) + 2\ H_2O(l)$

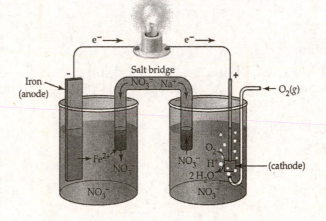

17.47   (a)   anode reaction $Mn(s) \rightarrow Mn^{2+}(aq) + 2\ e^-$
            cathode reaction $\underline{Pb^{2+}(aq) + 2\ e^- \rightarrow Pb(s)}$
            overall reaction $Mn(s) + Pb^{2+}(aq) \rightarrow Mn^{2+}(aq) + Pb(s)$

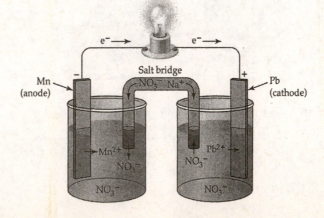

   (b) anode reaction $H_2(g) \rightarrow 2\,H^+(aq) + 2\,e^-$
        cathode reaction $\underline{2\,AgCl(s) + 2\,e^- \rightarrow 2\,Ag(s) + 2\,Cl^-(aq)}$
        overall reaction $H_2(g) + 2\,AgCl(s) \rightarrow 2\,Ag(s) + 2\,H^+(aq) + 2\,Cl^-(aq)$

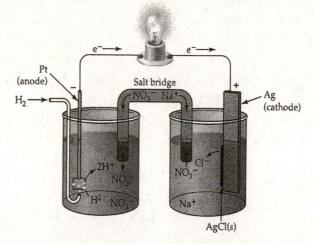

Cell Potentials and Free-Energy Changes; Standard Reduction Potentials (Sections 17.3–17.5)

17.48   The SI unit of electrical potential is the volt (V).
       The SI unit of charge is the coulomb (C).
       The SI unit of energy is the joule (J).
       $1\,J = 1\,C \cdot 1\,V$

17.49 $\Delta G = -nFE$; ΔG is the free energy change for the cell reaction
       n is the number of moles of e^-
       F is the Faraday (96,500 C/mol e^-)
       E is the galvanic cell potential

17.50   E is the standard cell potential ($E°$) when all reactants and products are in their standard states--solutes at 1 M concentrations, gases at a partial pressure of 1 atm, solids and liquids in pure form, all at 25 °C.

17.51   The standard reduction potential is the potential of the reduction half reaction in a galvanic cell where the other electrode is the standard hydrogen electrode.

17.52 $Zn(s) + Ag_2O(s) \rightarrow ZnO(s) + 2\,Ag(s)$; $n = 2$ mol e^-

$$\Delta G = -nFE = -(2\ \text{mol}\ e^-)\left(\frac{96,500\ \text{C}}{1\ \text{mol}\ e^-}\right)(1.60\ \text{V})\left(\frac{1\ \text{J}}{1\ \text{C}\cdot\text{V}}\right) = -308,800\ \text{J} = -309\ \text{kJ}$$

17.53 $Pb(s) + PbO_2(s) + 2\,H^+(aq) + 2\,HSO_4^-(aq) \rightarrow 2\,PbSO_4(s) + 2\,H_2O(l)$
       $n = 2$ mol e^-

$$\Delta G° = -nFE° = -(2\ \text{mol}\ e^-)\left(\frac{96,500\ \text{C}}{1\ \text{mol}\ e^-}\right)(1.924\ \text{V})\left(\frac{1\ \text{J}}{1\ \text{C}\cdot\text{V}}\right) = -371,300\ \text{J} = -371\ \text{kJ}$$

17.54 $2 H_2(g) + O_2(g) \rightarrow 2 H_2O(l);$ $n = 4$ mol e^- and 1 V $= 1$ J/C

$\Delta G^\circ = 2\ \Delta G^\circ_f(H_2O(l)) = (2\ \text{mol})(-237.2\ \text{kJ/mol}) = -474.4\ \text{kJ}$

$\Delta G^\circ = -nFE^\circ$ $E^\circ = \dfrac{-\Delta G^\circ}{nF} = \dfrac{-(-474{,}400\ \text{J})}{(4\ \text{mol }e^-)\left(\dfrac{96{,}500\ \text{C}}{1\ \text{mol }e^-}\right)} = +1.23\ \text{J/C} = +1.23\ \text{V}$

17.55 $CH_4(g) + 2 O_2(g) \rightarrow CO_2(g) + 2 H_2O(l);$ $n = 8$ mol e^- and 1 V $= 1$ J/C

$\Delta G^\circ = [\Delta G^\circ_f(CO_2) + 2\ \Delta G^\circ_f(H_2O(l))] - \Delta G^\circ_f(CH_4)$

$\Delta G^\circ = [(1\ \text{mol})(-394.4\ \text{kJ/mol}) + (2\ \text{mol})(-237.2\ \text{kJ/mol})]$
$\qquad\qquad - (1\ \text{mol})(-50.8\ \text{kJ/mol}) = -818.0\ \text{kJ}$

$\Delta G^\circ = -nFE^\circ$ $E^\circ = \dfrac{-\Delta G^\circ}{nF} = \dfrac{-(-818{,}000\ \text{J})}{(8\ \text{mol }e^-)\left(\dfrac{96{,}500\ \text{C}}{1\ \text{mol }e^-}\right)} = +1.06\ \text{J/C} = +1.06\ \text{V}$

17.56   oxidation: $Zn(s) \rightarrow Zn^{2+}(aq) + 2 e^-$ $E^\circ = 0.76$ V

        reduction: $\underline{Eu^{3+}(aq) + e^- \rightarrow Eu^{2+}(aq)}$ $\underline{E^\circ = ?}$

        overall $Zn(s) + 2 Eu^{3+}(aq) \rightarrow Zn^{2+}(aq) + 2 Eu^{2+}(aq)$ $E^\circ = 0.40$ V

        The standard reduction potential for the Eu^{3+}/Eu^{2+} half cell is:

        $E^\circ = 0.40 - 0.76 = -0.36$ V

17.57   oxidation: $2 Ag(s) + 2 Br^-(aq) \rightarrow 2 AgBr(s) + 2 e^-$ $E^\circ = ?$

        reduction: $\underline{Cu^{2+}(aq) + 2 e^- \rightarrow Cu(s)}$ $\underline{E^\circ = 0.34\ \text{V}}$

        overall $Cu^{2+}(aq) + 2 Ag(s) + 2 Br^-(aq) \rightarrow Cu(s) + 2 AgBr(s)$ $E^\circ = 0.27$ V

        E° for the oxidation half reaction $= 0.27 - 0.34 = -0.07$ V

        For $AgBr(s) + e^- \rightarrow Ag(s) + Br^-(aq),\ E^\circ = -(-0.07\ \text{V}) = +0.07$ V

17.58 $Sn^{4+}(aq) < Br_2(aq) < MnO_4^-$

17.59 $Pb(s) < Fe(s) < Al(s)$

17.60 $Cr_2O_7^{2-}(aq)$ is highest in the table of standard reduction potentials, therefore it is the strongest oxidizing agent.

        $Fe^{2+}(aq)$ is lowest in the table of standard reduction potentials, therefore it is the weakest oxidizing agent.

17.61   From Table 17.1:

        Sn^{2+} is the strongest reducing agent and Fe^{2+} is the weakest reducing agent.

17.62   (a) $Cd(s) + Sn^{2+}(aq) \rightarrow Cd^{2+}(aq) + Sn(s)$

        oxidation: $Cd(s) \rightarrow Cd^{2+}(aq) + 2 e^-$ $E^\circ = 0.40$ V

        reduction: $\underline{Sn^{2+}(aq) + 2 e^- \rightarrow Sn(s)}$ $\underline{E^\circ = -0.14\ \text{V}}$

                                        overall $E^\circ = 0.26$ V

        $n = 2$ mol e^-

$$\Delta G° = -nFE° = -(2 \text{ mol e}^-)\left(\frac{96,500 \text{ C}}{1 \text{ mol e}^-}\right)(0.26 \text{ V})\left(\frac{1 \text{ J}}{1 \text{ C} \cdot \text{V}}\right) = -50,180 \text{ J} = -50 \text{ kJ}$$

(b) $2 \text{ Al(s)} + 3 \text{ Cd}^{2+}(aq) \rightarrow 2 \text{ Al}^{3+}(aq) + 3 \text{ Cd(s)}$

oxidation:	$2 \text{ Al(s)} \rightarrow 2 \text{ Al}^{3+}(aq) + 6 \text{ e}^-$	$E° = 1.66 \text{ V}$
reduction:	$3 \text{ Cd}^{2+}(aq) + 6 \text{ e}^- \rightarrow 3 \text{ Cd(s)}$	$E° = -0.40 \text{ V}$
		overall $E° = 1.26 \text{ V}$

$n = 6 \text{ mol e}^-$

$$\Delta G° = -nFE° = -(6 \text{ mol e}^-)\left(\frac{96,500 \text{ C}}{1 \text{ mol e}^-}\right)(1.26 \text{ V})\left(\frac{1 \text{ J}}{1 \text{ C} \cdot \text{V}}\right) = -729,540 \text{ J} = -730 \text{ kJ}$$

(c) $6 \text{ Fe}^{2+}(aq) + \text{Cr}_2\text{O}_7^{2-}(aq) + 14 \text{ H}^+(aq) \rightarrow 6 \text{ Fe}^{3+}(aq) + 2 \text{ Cr}^{3+}(aq) + 7 \text{ H}_2\text{O(l)}$

oxidation:	$6 \text{ Fe}^{2+}(aq) \rightarrow 6 \text{ Fe}^{3+}(aq) + 6 \text{ e}^-$	$E° = -0.77 \text{ V}$
reduction:	$\text{Cr}_2\text{O}_7^{2-}(aq) + 14 \text{ H}^+(aq) + 6 \text{ e}^- + \rightarrow 2 \text{ Cr}^{3+}(aq) + 7 \text{ H}_2\text{O(l)}$	$E° = 1.36 \text{ V}$
		overall $E° = 0.59 \text{ V}$

$n = 6 \text{ mol e}^-$

$$\Delta G° = -nFE° = -(6 \text{ mol e}^-)\left(\frac{96,500 \text{ C}}{1 \text{ mol e}^-}\right)(0.59 \text{ V})\left(\frac{1 \text{ J}}{1 \text{ C} \cdot \text{V}}\right) = -341,610 \text{ J} = -342 \text{ kJ}$$

17.63   (a) $3 \text{ Cu}^{2+}(aq) + 2 \text{ Cr(s)} \rightarrow 3 \text{ Cu(s)} + 2 \text{ Cr}^{3+}(aq)$

oxidation	$2 \text{ Cr(s)} \rightarrow 2 \text{ Cr}^{3+}(aq) + 6 \text{ e}^-$	$E° = 0.74 \text{ V}$
reduction	$3 \text{ Cu}^{2+}(aq) + 6 \text{ e}^- \rightarrow 3 \text{ Cu(s)}$	$E° = 0.34 \text{ V}$
		overall $E° = 1.08 \text{ V}$

$n = 6 \text{ mol e}^-$

$$\Delta G° = -nFE° = -(6 \text{ mol e}^-)\left(\frac{96,500 \text{ C}}{1 \text{ mol e}^-}\right)(1.08 \text{ V})\left(\frac{1 \text{ J}}{1 \text{ C} \cdot \text{V}}\right) = -625,320 \text{ J} = -625 \text{ kJ}$$

(b) $\text{Pb(s)} + 2 \text{ H}^+(aq) \rightarrow \text{Pb}^{2+}(aq) + \text{H}_2(g)$

oxidation:	$\text{Pb(s)} \rightarrow \text{Pb}^{2+}(aq) + 2 \text{ e}^-$	$E° = 0.13 \text{ V}$
reduction:	$2 \text{ H}^+(aq) + 2 \text{ e}^- \rightarrow \text{H}_2(g)$	$E° = 0.00 \text{ V}$
		overall $E° = 0.13 \text{ V}$

$n = 2 \text{ mol e}^-$

$$\Delta G° = -nFE° = -(2 \text{ mol e}^-)\left(\frac{96,500 \text{ C}}{1 \text{ mol e}^-}\right)(0.13 \text{ V})\left(\frac{1 \text{ J}}{1 \text{ C} \cdot \text{V}}\right) = -25,090 \text{ J} = -25 \text{ kJ}$$

(c) $\text{Cl}_2(g) + \text{Sn}^{2+}(aq) \rightarrow \text{Sn}^{4+}(aq) + 2 \text{ Cl}^-(aq)$

oxidation	$\text{Sn}^{2+}(aq) \rightarrow \text{Sn}^{4+}(aq) + 2 \text{ e}^-$	$E° = -0.15 \text{ V}$
reduction	$\text{Cl}_2(g) + 2 \text{ e}^- \rightarrow 2 \text{ Cl}^-(aq)$	$E° = 1.36 \text{ V}$
		overall $E° = 1.21 \text{ V}$

$n = 2 \text{ mol e}^-$

$$\Delta G° = -nFE° = -(2 \text{ mol e}^-)\left(\frac{96,500 \text{ C}}{1 \text{ mol e}^-}\right)(1.21 \text{ V})\left(\frac{1 \text{ J}}{1 \text{ C} \cdot \text{V}}\right) = -233,530 \text{ J} = -234 \text{ kJ}$$

17.64 (a) $2 Fe^{2+}(aq) + Pb^{2+}(aq) \rightarrow 2 Fe^{3+}(aq) + Pb(s)$

oxidation: $2 Fe^{2+}(aq) \rightarrow 2 Fe^{3+}(aq) + 2 e^-$ $E^o = -0.77$ V
reduction: $Pb^{2+}(aq) + 2 e^- \rightarrow Pb(s)$ $\underline{E^o = -0.13 \text{ V}}$
overall $E^o = -0.90$ V

Because E^o is negative, this reaction is nonspontaneous.

(b) $Mg(s) + Ni^{2+}(aq) \rightarrow Mg^{2+}(aq) + Ni(s)$

oxidation: $Mg(s) \rightarrow Mg^{2+}(aq) + 2 e^-$ $E^o = 2.37$ V
reduction: $Ni^{2+}(aq) + 2 e^- \rightarrow Ni(s)$ $\underline{E^o = -0.26 \text{ V}}$
overall $E^o = 2.11$ V

Because E^o is positive, this reaction is spontaneous.

17.65 (a) $5 Ag^+(aq) + Mn^{2+}(aq) + 4 H_2O(l) \rightarrow 5 Ag(s) + MnO_4^-(aq) + 8 H^+(aq)$

oxidation: $Mn^{2+}(aq) + 4 H_2O(l) \rightarrow MnO_4^-(aq) + 8 H^+(aq) + 5 e^-$ $E^o = -1.51$ V
reduction: $5 Ag^+(aq) + 5 e^- \rightarrow 5 Ag(s)$ $\underline{E^o = 0.80 \text{ V}}$
overall $E^o = -0.71$ V

Because E^o is negative, this reaction is nonspontaneous.

(b) $2 H_2O_2(aq) \rightarrow O_2(g) + 2 H_2O(l)$

oxidation: $H_2O_2(aq) \rightarrow O_2(g) + 2 H^+(aq) + 2 e^-$ $E^o = -0.70$ V
reduction: $H_2O_2(aq) + 2 H^+(aq) + 2 e^- \rightarrow 2 H_2O(l)$ $\underline{E^o = 1.78 \text{ V}}$
overall $E^o = 1.08$ V

Because E^o is positive, this reaction is spontaneous.

17.66 (a) oxidation: $Sn^{2+}(aq) \rightarrow Sn^{4+}(aq) + 2 e^-$ $E^o = -0.15$ V
       reduction: $Br_2(aq) + 2 e^- \rightarrow 2 Br^-(aq)$ $\underline{E^o = 1.09 \text{ V}}$
overall $E^o = +0.94$ V

Because the overall E^o is positive, $Sn^{2+}(aq)$ can be oxidized by $Br_2(aq)$.

(b) oxidation: $Sn^{2+}(aq) \rightarrow Sn^{4+}(aq) + 2 e^-$ $E^o = -0.15$ V
     reduction: $Ni^{2+}(aq) + 2 e^- \rightarrow Ni(s)$ $\underline{E^o = -0.26 \text{ V}}$
overall $E^o = -0.41$ V

Because the overall E^o is negative, $Ni^{2+}(aq)$ cannot be reduced by $Sn^{2+}(aq)$.

(c) oxidation: $2 Ag(s) \rightarrow 2 Ag^+(aq) + 2 e^-$ $E^o = -0.80$ V
     reduction: $Pb^{2+}(aq) + 2 e^- \rightarrow Pb(s)$ $\underline{E^o = -0.13 \text{ V}}$
overall $E^o = -0.93$ V

Because the overall E^o is negative, $Ag(s)$ cannot be oxidized by $Pb^{2+}(aq)$.

(d) oxidation: $H_2SO_3(aq) + H_2O(l) \rightarrow SO_4^{2-}(aq) + 4 H^+(aq) + 2 e^-$ $E^o = -0.17$ V
     reduction: $I_2(s) + 2 e^- \rightarrow 2 I^-(aq)$ $\underline{E^o = 0.54 \text{ V}}$
overall $E^o = +0.37$ V

Because the overall E^o is positive, $I_2(s)$ can be reduced by H_2SO_3.

17.67 (a) oxidation: $\quad$ $Zn(s) \rightarrow Zn^{2+}(aq) + 2\,e^-$ $\quad$ $E^o = 0.76$ V
$\qquad$ reduction: $\quad$ $Pb^{2+}(aq) + 2\,e^- \rightarrow Pb(s)$ $\quad$ $\underline{E^o = -0.13\ V}$
$\qquad\qquad\qquad\qquad\qquad\qquad\qquad\qquad\qquad\qquad$ overall $E^o = 0.63$ V

$Zn(s) + Pb^{2+}(aq) \rightarrow Zn^{2+}(aq) + Pb(s)$
The reaction is spontaneous because E^o is positive.

$\quad$ (b) oxidation: $\quad$ $4\,Fe^{2+}(aq) \rightarrow 4\,Fe^{3+}(aq) + 4\,e^-$ $\quad$ $E^o = -0.77$ V
$\qquad$ reduction: $\quad$ $O_2(g) + 4\,H^+(aq) + 4\,e^- \rightarrow 2\,H_2O(l)$ $\quad$ $\underline{E^o = 1.23\ V}$
$\qquad\qquad\qquad\qquad\qquad\qquad\qquad\qquad\qquad\qquad$ overall $E^o = 0.46$ V

$4\,Fe^{2+}(aq) + O_2(g) + 4\,H^+(aq) \rightarrow 4\,Fe^{3+}(aq) + 2\,H_2O(l)$
The reaction is spontaneous because E^o is positive.

$\quad$ (c) oxidation: $\quad$ $2\,Ag(s) \rightarrow 2\,Ag^+(aq) + 2\,e^-$ $\quad$ $E^o = -0.80$ V
$\qquad$ reduction: $\quad$ $Ni^{2+}(aq) + 2\,e^- \rightarrow Ni(s)$ $\quad$ $\underline{E^o = -0.26\ V}$
$\qquad\qquad\qquad\qquad\qquad\qquad\qquad\qquad\qquad\qquad$ overall $E^o = -1.06$ V

There is no reaction because E^o is negative.

$\quad$ (d) oxidation: $\quad$ $H_2(g) \rightarrow 2\,H^+(aq) + 2\,e^-$ $\quad$ $E^o = 0.00$ V
$\qquad$ reduction: $\quad$ $Cd^{2+}(aq) + 2\,e^- \rightarrow Cd(s)$ $\quad$ $\underline{E^o = -0.40\ V}$
$\qquad\qquad\qquad\qquad\qquad\qquad\qquad\qquad\qquad\qquad$ overall $E^o = -0.40$ V

There is no reaction because E^o is negative.

The Nernst Equation (Sections 17.6–17.7)

17.68 $\quad$ $2\,Ag^+(aq) + Sn(s) \rightarrow 2\,Ag(s) + Sn^{2+}(aq)$
$\qquad$ oxidation: $\quad$ $Sn(s) \rightarrow Sn^{2+}(aq) + 2\,e^-$ $\quad$ $E^o = 0.14$ V
$\qquad$ reduction: $\quad$ $2\,Ag^+(aq) + 2\,e^- \rightarrow 2\,Ag(s)$ $\quad$ $\underline{E^o = 0.80\ V}$
$\qquad\qquad\qquad\qquad\qquad\qquad\qquad\qquad$ overall $E^o = 0.94$ V

$$E = E^o - \frac{0.0592\,V}{n} \log \frac{[Sn^{2+}]}{[Ag^+]^2} = 0.94\,V - \frac{(0.0592\,V)}{2} \log \frac{(0.020)}{(0.010)^2} = 0.87\,V$$

17.69 $\quad$ $2\,Fe^{2+}(aq) + Cl_2(g) \rightarrow 2\,Fe^{3+}(aq) + 2\,Cl^-(aq)$
$\qquad$ oxidation: $\quad$ $2\,Fe^{2+}(aq) \rightarrow 2\,Fe^{3+}(aq) + 2\,e^-$ $\quad$ $E^o = -0.77$ V
$\qquad$ reduction: $\quad$ $Cl_2(g) + 2\,e^- \rightarrow 2\,Cl^-(aq)$ $\quad$ $\underline{E^o = 1.36\ V}$
$\qquad\qquad\qquad\qquad\qquad\qquad\qquad\qquad$ overall $E^o = 0.59$ V

$$E = E^o - \frac{0.0592\,V}{n} \log \frac{[Fe^{3+}]^2[Cl^-]^2}{[Fe^{2+}]^2 P_{Cl_2}} = 0.59\,V - \frac{(0.0592\,V)}{2} \log \frac{(0.0010)^2(0.0030)^2}{(1.0)^2(0.50)} = 0.91\,V$$

17.70 $\quad$ $Pb(s) + Cu^{2+}(aq) \rightarrow Pb^{2+}(aq) + Cu(s)$
$\qquad$ oxidation: $\quad$ $Pb(s) \rightarrow Pb^{2+}(aq) + 2\,e^-$ $\quad$ $E^o = 0.13$ V
$\qquad$ reduction: $\quad$ $Cu^{2+}(aq) + 2\,e^- \rightarrow Cu(s)$ $\quad$ $\underline{E^o = 0.34\ V}$
$\qquad\qquad\qquad\qquad\qquad\qquad\qquad\qquad$ overall $E^o = 0.47$ V

$$E = E^o - \frac{0.0592\,V}{n} \log \frac{[Pb^{2+}]}{[Cu^{2+}]} = 0.47\,V - \frac{(0.0592\,V)}{2} \log \frac{1.0}{(1.0 \times 10^{-4})} = 0.35\,V$$

When $E = 0$, $0 = E^\circ - \dfrac{0.0592\,V}{n} \log \dfrac{[Pb^{2+}]}{[Cu^{2+}]} = 0.47\,V - \dfrac{(0.0592\,V)}{2} \log \dfrac{1.0}{[Cu^{2+}]}$

$0 = 0.47\,V + \dfrac{(0.0592\,V)}{2} \log [Cu^{2+}]$

$\log [Cu^{2+}] = (-0.47\,V)\left(\dfrac{2}{0.0592\,V}\right) = -15.88$

$[Cu^{2+}] = 10^{-15.88} = 1 \times 10^{-16}\,M$

17.71 $Fe(s) + Cu^{2+}(aq) \rightarrow Fe^{2+}(aq) + Cu(s)$
   oxidation: $Fe(s) \rightarrow Fe^{2+}(aq) + 2\,e^-$ $E^\circ = 0.45\,V$
   reduction: $Cu^{2+}(aq) + 2\,e^- \rightarrow Cu(s)$ $\underline{E^\circ = 0.34\,V}$
                                                                     overall $E^\circ = 0.79\,V$

$E = 0.67\,V = E^\circ - \dfrac{0.0592\,V}{n} \log \dfrac{[Fe^{2+}]}{[Cu^{2+}]} = 0.79\,V - \dfrac{(0.0592\,V)}{2} \log\left(\dfrac{0.10}{[Cu^{2+}]}\right)$

$0.67\,V = 0.79\,V - \dfrac{(0.0592\,V)}{2}(\log(0.10) - \log[Cu^{2+}])$

$\log[Cu^{2+}] = -5.05;$ $[Cu^{2+}] = 10^{-5.05} = 8.9 \times 10^{-6}\,M$

17.72 (a) $E = E^\circ - \dfrac{0.0592\,V}{n} \log [I^-]^2 = 0.54\,V - \dfrac{(0.0592\,V)}{2} \log (0.020)^2 = 0.64\,V$

(b) $E = E^\circ - \dfrac{0.0592\,V}{n} \log \dfrac{[Fe^{2+}]}{[Fe^{3+}]} = 0.77\,V - \dfrac{(0.0592\,V)}{1} \log\left(\dfrac{0.10}{0.10}\right) = 0.77\,V$

(c) $E = E^\circ - \dfrac{0.0592\,V}{n} \log \dfrac{[Sn^{4+}]}{[Sn^{2+}]} = -0.15\,V - \dfrac{(0.0592\,V)}{2} \log\left(\dfrac{0.40}{0.0010}\right) = -0.23\,V$

(d) $E = E^\circ - \dfrac{0.0592\,V}{n} \log \dfrac{[Cr_2O_7^{2-}][H^+]^{14}}{[Cr^{3+}]^2} = -1.36\,V - \dfrac{(0.0592\,V)}{6} \log\left(\dfrac{(1.0)(0.010)^{14}}{1.0}\right)$

$E = -1.36\,V - \dfrac{(0.0592\,V)}{6}(14)\log(0.010) = -1.08\,V$

17.73 $E = E^\circ - \dfrac{0.0592\,V}{n} \log \dfrac{P_{H_2}}{[H_3O^+]^2};$ $E^\circ = 0$, $n = 2$ mol e^-, and $P_{H_2} = 1$ atm

(a) $[H_3O^+] = 1.0\,M;$ $E = -\dfrac{0.0592\,V}{2} \log \dfrac{1}{(1.0)^2} = 0$

(b) $pH = 4.00$, $[H_3O^+] = 10^{-4.00} = 1.0 \times 10^{-4}\,M$
$E = -\dfrac{0.0592\,V}{2} \log \dfrac{1}{(1.0 \times 10^{-4})^2} = -0.24\,V$

(c) $[H_3O^+] = 1.0 \times 10^{-7}\,M;$ $E = -\dfrac{0.0592\,V}{2} \log \dfrac{1}{(1.0 \times 10^{-7})^2} = -0.41\,V$

(d) $[OH^-] = 1.0$ M; $[H_3O^+] = \dfrac{K_w}{[OH^-]} = \dfrac{1.0 \times 10^{-14}}{1.0} = 1.0 \times 10^{-14}$ M

$$E = -\frac{0.0592\,V}{2}\log\frac{1}{(1.0 \times 10^{-14})^2} = -0.83\ V$$

17.74 $H_2(g) + Ni^{2+}(aq) \rightarrow 2\,H^+(aq) + Ni(s)$

$$E^\circ = E^\circ_{H_2 \rightarrow H^+} + E^\circ_{Ni^{2+} \rightarrow Ni} = 0\ V + (-0.26\ V) = -0.26\ V$$

$$E = E^\circ - \frac{0.0592\,V}{n}\log\frac{[H_3O^+]^2}{[Ni^{2+}](P_{H_2})}$$

$$0.27\ V = -0.26\ V - \frac{(0.0592\ V)}{2}\log\frac{[H_3O^+]^2}{(1)(1)}$$

$$0.27\ V = -0.26\ V - (0.0592\ V)\log[H_3O^+]$$

$pH = -\log[H_3O^+]$ therefore $0.27\ V = -0.26\ V + (0.0592\ V)\,pH$

$$pH = \frac{(0.27\ V + 0.26\ V)}{0.0592\ V} = 9.0$$

17.75 $Zn(s) + 2\,H^+(aq) \rightarrow Zn^{2+}(aq) + H_2(g)$

$$E^\circ = E^\circ_{H^+ \rightarrow H_2} + E^\circ_{Zn \rightarrow Zn^{2+}} = 0\ V + 0.76\ V = 0.76\ V$$

$$E = E^\circ - \frac{0.0592\,V}{n}\log\frac{[Zn^{2+}](P_{H_2})}{[H_3O^+]^2}$$

$$0.58\ V = 0.76\ V - \frac{(0.0592\ V)}{2}\log\frac{(1)(1)}{[H_3O^+]^2}$$

$$0.58\ V = 0.76\ V + (0.0592\ V)\log[H_3O^+]$$

$pH = -\log[H_3O^+]$ therefore $0.58\ V = 0.76\ V - (0.0592\ V)\,pH$

$$pH = \frac{-(0.58\ V - 0.76\ V)}{0.0592\ V} = 3.0$$

Standard Cell Potentials and Equilibrium Constants (Section 17.8)

17.76 $\Delta G^\circ = -nFE^\circ$

Because n and F are always positive, ΔG° is negative when E° is positive because of the negative sign in the equation.

$$E^\circ = \frac{0.0592\,V}{n}\log K; \quad \log K = \frac{nE^\circ}{0.0592\,V}; \quad K = 10^{\frac{nE^\circ}{0.0592}}$$

If E° is positive, the exponent is positive (because n is positive), and K is greater than 1.

17.77 If $K < 1$, $E^\circ < 0$. When $E^\circ = 0$, $K = 1$.

17.78 $Ni(s) + 2 Ag^+(aq) \rightarrow Ni^{2+}(aq) + 2 Ag(s)$

oxidation: $Ni(s) \rightarrow Ni^{2+}(aq) + 2 e^-$ $E^\circ = 0.26$ V

reduction: $2 Ag^+(aq) + 2 e^- \rightarrow 2 Ag(s)$ $\underline{E^\circ = 0.80\ V}$

                                    overall $E^\circ = 1.06$ V

$$E^\circ = \frac{0.0592\ V}{n} \log K;\quad \log K = \frac{nE^\circ}{0.0592\ V} = \frac{(2)(1.06\ V)}{0.0592\ V} = 35.8;\quad K = 10^{35.8} = 6 \times 10^{35}$$

17.79 $2 MnO_4^-(aq) + 10 Cl^-(aq) + 16 H^+(aq) \rightarrow 2 Mn^{2+}(aq) + 5 Cl_2(g) + 8 H_2O(l)$

oxidation: $10 Cl^-(aq) \rightarrow 5 Cl_2(g) + 10 e^-$ $E^\circ = -1.36$ V

reduction: $2 MnO_4^-(aq) + 16 H^+(aq) + 10 e^- \rightarrow 2 Mn^{2+}(aq) + 8 H_2O(l)$ $\underline{E^\circ = 1.51\ V}$

                                   overall $E^\circ = 0.15$ V

$$E^\circ = \frac{0.0592\ V}{n} \log K;\quad \log K = \frac{nE^\circ}{0.0592\ V} = \frac{(10)(0.15\ V)}{0.0592\ V} = 25.3;\quad K = 10^{25.3} = 2 \times 10^{25}$$

17.80 E° and n are from Problem 17.62.

$$E^\circ = \frac{0.0592\ V}{n} \log K;\quad \log K = \frac{nE^\circ}{0.0592\ V}$$

(a) $Cd(s) + Sn^{2+}(aq) \rightarrow Cd^{2+}(aq) + Sn(s)$; $E^\circ = 0.26$ V and $n = 2$ mol e^-

$\log K = \dfrac{(2)(0.26\ V)}{0.0592\ V} = 8.8$; $K = 10^{8.8} = 6 \times 10^8$

(b) $2 Al(s) + 3 Cd^{2+}(aq) \rightarrow 2 Al^{3+}(aq) + 3 Cd(s)$; $E^\circ = 1.26$ V and $n = 6$ mol e^-

$\log K = \dfrac{(6)(1.26\ V)}{0.0592\ V} = 128$; $K = 10^{128}$

(c) $6 Fe^{2+}(aq) + Cr_2O_7^{2-}(aq) + 14 H^+(aq) \rightarrow 6 Fe^{3+}(aq) + 2 Cr^{3+}(aq) + 7 H_2O(l)$
$E^\circ = 0.59$ V and $n = 6$ mol e^-

$\log K = \dfrac{(6)(0.59\ V)}{0.0592\ V} = 60$; $K = 10^{60}$

17.81 E° and n are from Problem 17.63.

$$E^\circ = \frac{0.0592\ V}{n} \log K;\quad \log K = \frac{nE^\circ}{0.0592\ V}$$

(a) $3 Cu^{2+}(aq) + 2 Cr(s) \rightarrow 3 Cu(s) + 2 Cr^{3+}(aq)$; $E^\circ = 1.08$ V and $n = 6$ mol e^-

$\log K = \dfrac{(6)(1.08\ V)}{0.0592\ V} = 109$; $K = 10^{109}$

(b) $Pb(s) + 2 H^+(aq) \rightarrow Pb^{2+}(aq) + H_2(g)$; $E^\circ = 0.13$ V and $n = 2$ mol e^-

$\log K = \dfrac{(2)(0.13\ V)}{0.0592\ V} = 4.4$; $K = 10^{4.4} = 3 \times 10^4$

(c) $Cl_2(g) + Sn^{2+}(aq) \rightarrow Sn^{4+}(aq) + 2 Cl^-(aq)$; $E^\circ = 1.21$ V and $n = 2$ mol e^-

$\log K = \dfrac{(2)(1.21\ V)}{0.0592\ V} = 40.9$; $K = 10^{40.9} = 8 \times 10^{40}$

17.82 $Hg_2^{2+}(aq) \rightarrow Hg(l) + Hg^{2+}(aq)$

oxidation: $\frac{1}{2}[Hg_2^{2+}(aq) \rightarrow 2\ Hg^{2+}(aq) + 2\ e^-]$ $E° = -0.92$ V
reduction: $\frac{1}{2}[Hg_2^{2+}(aq) + 2\ e^- \rightarrow 2\ Hg(l)]$ $\underline{E° =\ \ 0.80\ V}$
overall $E° = -0.12$ V

$$E° = \frac{0.0592\ V}{n} \log K$$

$$\log K = \frac{nE°}{0.0592\ V} = \frac{(1)(-0.12\ V)}{0.0592\ V} = -2.027; \qquad K = 10^{-2.027} = 9 \times 10^{-3}$$

17.83 $2\ H_2O_2(aq) \rightarrow 2\ H_2O(l) + O_2(g)$

oxidation: $H_2O_2(aq) \rightarrow O_2(g) + 2\ H^+(aq) + 2e^-$ $E° = -0.70$ V
reduction: $H_2O_2(aq) + 2\ H^+(aq) + 2e^- \rightarrow 2\ H_2O(l)$ $\underline{E° =\ \ 1.78\ V}$
overall $E° =\ \ 1.08$ V

$$E° = \frac{0.0592\ V}{n} \log K; \quad \log K = \frac{nE°}{0.0592\ V} = \frac{(2)(1.08\ V)}{0.0592\ V} = 36.5; \quad K = 10^{36.5} = 3 \times 10^{36}$$

Batteries; Fuel Cells; Corrosion (Sections 17.9–17.11)

17.84   (a)

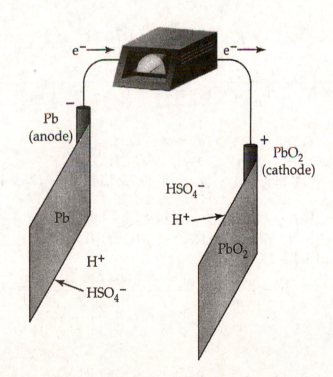

(b)   anode: $Pb(s) + HSO_4^-(aq) \rightarrow PbSO_4(s) + H^+(aq) + 2\ e^-$ $E° = 0.296$ V
cathode: $\underline{PbO_2(s) + 3\ H^+(aq) + HSO_4^-(aq) + 2\ e^- \rightarrow PbSO_4(s) + 2\ H_2O(l)}$ $\underline{E° = 1.628\ V}$
overall $Pb(s) + PbO_2(s) + 2\ H^+(aq) + 2\ HSO_4^-(aq) \rightarrow 2\ PbSO_4(s) + 2\ H_2O(l)$ $E° = 1.924$ V

(c) $E° = \dfrac{0.0592\ V}{n} \log K; \quad \log K = \dfrac{nE°}{0.0592\ V} = \dfrac{(2)(1.924\ V)}{0.0592\ V} = 65.0; \quad K = 1 \times 10^{65}$

(d) When the cell reaction reaches equilibrium the cell voltage = 0.

17.85 (a) anode: $Zn(s) + 2 OH^-(aq) \rightarrow ZnO(s) + H_2O(l) + 2 e^-$ $E° = 1.260$ V

cathode: $\underline{HgO(s) + H_2O(l) + 2 e^- \rightarrow Hg(l) + 2 OH^-(aq)}$ $\underline{E° = 0.098\ V}$

overall: $Zn(s) + HgO(s) \rightarrow ZnO(s) + Hg(l)$ $E° = 1.358$ V

(b) $n = 2$ mol e^- and 1 J $= 1$ C x 1 V

$$\Delta G° = -nFE° = -(2\ \text{mol } e^-)\left(\frac{96,500\ \text{C}}{1\ \text{mol } e^-}\right)(1.358\ \text{V})\left(\frac{1\ \text{J}}{1\ \text{C} \cdot \text{V}}\right) = -262,094\ \text{J} = -262\ \text{kJ}$$

$$E° = \frac{0.0592\ \text{V}}{n}\log K; \quad \log K = \frac{nE°}{0.0592\ \text{V}} = \frac{(2)(1.358\ \text{V})}{0.0592\ \text{V}} = 45.9; \quad K = 8 \times 10^{45}$$

(c) Because OH^- does not occur in the overall cell reaction, a change in KOH concentration does not change the cell voltage.

17.86 anode: $2 H_2(g) + 4 OH^-(aq) \rightarrow 4 H_2O(l) + 4 e^-$ $E° = 0.83$ V

cathode: $\underline{O_2(g) + 2 H_2O(l) + 4 e^- \rightarrow 4 OH^-(aq)}$ $\underline{E° = 0.40\ V}$

overall: $H_2(g) + O_2(g) \rightarrow 2 H_2O(l)$ $E° = 1.23$ V

$n = 4$ mol e^- and 1 J $= 1$ C x 1 V

$$\Delta G° = -nFE° = -(4\ \text{mol } e^-)\left(\frac{96,500\ \text{C}}{1\ \text{mol } e^-}\right)(1.23\ \text{V})\left(\frac{1\ \text{J}}{1\ \text{C} \cdot \text{V}}\right) = -474,780\ \text{J} = -475\ \text{kJ}$$

$$E° = \frac{0.0592\ \text{V}}{n}\log K; \quad \log K = \frac{nE°}{0.0592\ \text{V}} = \frac{(4)(1.23\ \text{V})}{0.0592\ \text{V}} = 83.1; \quad K = 10^{83.1} = 1 \times 10^{83}$$

$$E = E° - \frac{0.0592\ \text{V}}{n}\log\frac{1}{(P_{H_2})^2(P_{O_2})} = 1.23\ \text{V} - \frac{0.0592\ \text{V}}{4}\log\frac{1}{(25)^2(25)} = 1.29\ \text{V}$$

17.87 $2 CH_3OH(l) + 3 O_2(g) \rightarrow 2 CO_2(g) + 4 H_2O(l)$

$\Delta G° = [2\ \Delta G°_f(CO_2) + 4\ \Delta G°_f(H_2O)] - [2\ \Delta G°_f(CH_3OH)]$

$\Delta G° = [(2\ \text{mol})(-394.4\ \text{kJ/mol}) + (4\ \text{mol})(-237.2\ \text{kJ/mol})] - (2\ \text{mol})(-166.6\ \text{kJ/mol})$

$\Delta G° = -1404$ kJ

anode: $2 CH_3OH(l) + 2 H_2O(l) \rightarrow 2 CO_2(g) + 12 H^+(aq) + 12 e^-$

cathode: $3 O_2(g) + 12 H^+(aq) + 12 e^- \rightarrow 6 H_2O(l)$

$n = 12$ mol e^- and 1 J $= 1$ C x 1 V

$$\Delta G° = -nFE° \qquad E° = \frac{-\Delta G°}{nF} = \frac{-(-1,404,000\ \text{J})}{(12\ \text{mol } e^-)\left(\dfrac{96,500\ \text{C}}{1\ \text{mol } e^-}\right)} = +1.21\ \text{J/C} = +1.21\ \text{V}$$

$$E° = \frac{0.0592\ \text{V}}{n}\log K; \quad \log K = \frac{nE°}{0.0592\ \text{V}} = \frac{(12)(1.21\ \text{V})}{0.0592\ \text{V}} = 245; \quad K = 10^{245} = 1 \times 10^{245}$$

17.88 $Zn(s) + HgO(s) \rightarrow ZnO(s) + Hg(l);$ Zn, 65.39 amu; HgO, 216.59 amu

$$\text{mass HgO} = 2.00\ \text{g Zn} \times \frac{1\ \text{mol Zn}}{65.39\ \text{g Zn}} \times \frac{1\ \text{mol HgO}}{1\ \text{mol Zn}} \times \frac{216.59\ \text{g HgO}}{1\ \text{mol HgO}} = 6.62\ \text{g HgO}$$

17.89 $Cd(OH)_2(s) + 2 Ni(OH)_2(s) \rightarrow Cd(s) + 2 NiO(OH)(s) + 2 H_2O(l)$
$Ni(OH)_2$, 92.71 amu; Cd, 112.41 amu

mass Cd = 10.0 g $Ni(OH)_2$ x $\dfrac{1 \text{ mol } Ni(OH)_2}{92.71 \text{ g } Ni(OH)_2}$ x $\dfrac{1 \text{ mol Cd}}{2 \text{ mol } Ni(OH)_2}$ x $\dfrac{112.41 \text{ g Cd}}{1 \text{ mol Cd}}$ = 6.06 g Cd

17.90   Rust is a hydrated form of iron(III) oxide ($Fe_2O_3 \cdot H_2O$). Rust forms from the oxidation of Fe in the presence of O_2 and H_2O. Rust can be prevented by coating Fe with Zn (galvanizing).

17.91   Cr forms a protective oxide coating similar to Al.

17.92   Cathodic protection is the attachment of a more easily oxidized metal to the metal you want to protect. This forces the metal you want to protect to be the cathode, hence the name, cathodic protection. Zn and Al can offer cathodic protection to Fe (Ni and Sn cannot).

17.93   A sacrificial anode is a metal used for cathodic protection. It behaves as an anode and is more easily oxidized than the metal it is protecting. An example of a sacrificial anode is Zn for protecting Fe (galvanizing).

Electrolysis (Sections 17.12–17.14)

17.94   (a)

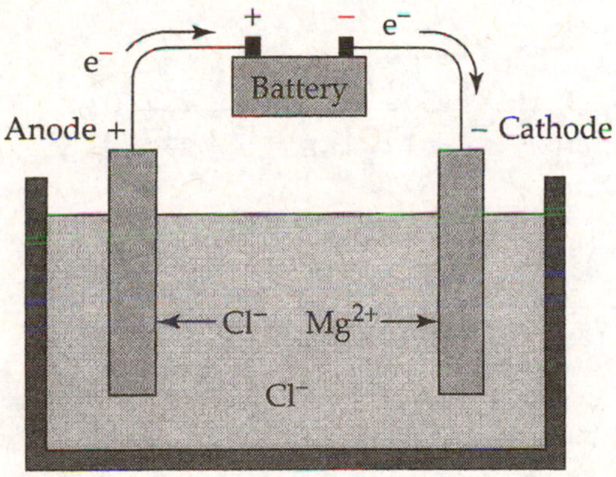

(b) anode: $2 Cl^-(l) \rightarrow Cl_2(g) + 2 e^-$
cathode: $\underline{Mg^{2+}(l) + 2 e^- \rightarrow Mg(l)}$
overall: $Mg^{2+}(l) + 2 Cl^-(l) \rightarrow Mg(l) + Cl_2(g)$

17.95   (a)

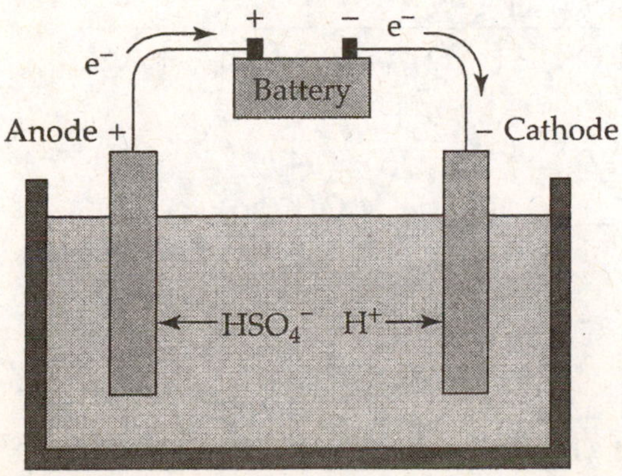

(b) anode: $2 H_2O(l) \rightarrow O_2(g) + 4 H^+(aq) + 4 e^-$

cathode: $\underline{4 H^+(aq) + 4 e^- \rightarrow 2 H_2(g)}$

overall: $2 H_2O(l) \rightarrow O_2(g) + 2 H_2(g)$

17.96 Possible anode reactions:

$$2 Cl^-(aq) \rightarrow Cl_2(g) + 2 e^-$$
$$2 H_2O(l) \rightarrow O_2(g) + 4 H^+(aq) + 4 e^-$$

Possible cathode reactions:

$$2 H_2O(l) + 2 e^- \rightarrow H_2(g) + 2 OH^-(aq)$$
$$Mg^{2+}(aq) + 2 e^- \rightarrow Mg(s)$$

Actual reactions:

anode: $2 Cl^-(aq) \rightarrow Cl_2(g) + 2 e^-$

cathode: $2 H_2O(l) + 2 e^- \rightarrow H_2(g) + 2 OH^-(aq)$

This anode reaction takes place instead of $2 H_2O(l) \rightarrow O_2(g) + 4 H^+(aq) + 4 e^-$ because of a high overvoltage for formation of gaseous O_2.

This cathode reaction takes place instead of $Mg^{2+}(aq) + 2 e^- \rightarrow Mg(s)$ because H_2O is easier to reduce than Mg^{2+}.

17.97 (a) $K(l)$ and $Cl_2(g)$ (b) $H_2(g)$ and $Cl_2(g)$. Solvent H_2O is reduced in preference to K^+.

17.98 (a) NaBr

anode: $2 Br^-(aq) \rightarrow Br_2(l) + 2 e^-$

cathode: $\underline{2 H_2O(l) + 2 e^- \rightarrow H_2(g) + 2 OH^-(aq)}$

overall: $2 H_2O(l) + 2 Br^-(aq) \rightarrow Br_2(l) + H_2(g) + 2 OH^-(aq)$

(b) $CuCl_2$

anode: $2 Cl^-(aq) \rightarrow Cl_2(g) + 2 e^-$

cathode: $\underline{Cu^{2+}(aq) + 2 e^- \rightarrow Cu(s)}$

overall: $Cu^{2+}(aq) + 2 Cl^-(aq) \rightarrow Cu(s) + Cl_2(g)$

(c) LiOH

anode: $4 OH^-(aq) \rightarrow O_2(g) + 2 H_2O(l) + 4 e^-$

cathode: $\underline{4 H_2O(l) + 4 e^- \rightarrow 2 H_2(g) + 4 OH^-(aq)}$

overall: $2 H_2O(l) \rightarrow O_2(g) + 2 H_2(g)$

17.99 (a) Ag_2SO_4

anode: $2 H_2O(l) \rightarrow O_2(g) + 4 H^+(aq) + 4 e^-$

cathode: $\underline{4 Ag^+(aq) + 4 e^- \rightarrow 4 Ag(s)}$

overall: $4 Ag^+(aq) + 2 H_2O(l) \rightarrow O_2(g) + 4 H^+(aq) + 4 Ag(s)$

(b) $Ca(OH)_2$

anode: $4 OH^-(aq) \rightarrow O_2(g) + 2 H_2O(l) + 4 e^-$

cathode: $\underline{4 H_2O(l) + 4 e^- \rightarrow 2 H_2(g) + 4 OH^-(aq)}$

overall: $2 H_2O(l) \rightarrow O_2(g) + 2 H_2(g)$

(c) KI

anode: $2 I^-(aq) \rightarrow I_2(s) + 2 e^-$

cathode: $\underline{2 H_2O(l) + 2 e^- \rightarrow H_2(g) + 2 OH^-(aq)}$

overall: $2 I^-(aq) + 2 H_2O(l) \rightarrow I_2(s) + H_2(g) + 2 OH^-(aq)$

17.100 $Ag^+(aq) + e^- \rightarrow Ag(s);$ $1\ A = 1\ C/s$

$$\text{mass Ag} = 2.40\ \frac{C}{s} \times 20.0\ \text{min} \times \frac{60\ s}{1\ \text{min}} \times \frac{1\ \text{mol}\ e^-}{96,500\ C} \times \frac{1\ \text{mol Ag}}{1\ \text{mol}\ e^-} \times \frac{107.87\ \text{g Ag}}{1\ \text{mol Ag}} = 3.22\ g$$

17.101 $Cu^{2+}(aq) + 2\ e^- \rightarrow Cu(s)$

$$\text{mol}\ e^- = 100.0\ \frac{C}{s} \times 24.0\ h \times \frac{60\ \text{min}}{h} \times \frac{60\ s}{\text{min}} \times \frac{1\ \text{mol}\ e^-}{96,500\ C} = 89.5\ \text{mol}\ e^-$$

$$\text{mass Cu} = 89.5\ \text{mol}\ e^- \times \frac{1\ \text{mol Cu}}{2\ \text{mol}\ e^-} \times \frac{63.54\ \text{g Cu}}{1\ \text{mol Cu}} \times \frac{1\ kg}{1000\ g} = 2.84\ \text{kg Cu}$$

17.102 $2\ Na^+(l) + 2\ Cl^-(l) \rightarrow 2\ Na(l) + Cl_2(g)$
$Na^+(l) + e^- \rightarrow Na(l);$ $1\ A = 1\ C/s;$ $1.00 \times 10^3\ kg = 1.00 \times 10^6\ g$

$$\text{Charge} = 1.00 \times 10^6\ \text{g Na} \times \frac{1\ \text{mol Na}}{22.99\ \text{g Na}} \times \frac{1\ \text{mol}\ e^-}{1\ \text{mol Na}} \times \frac{96,500\ C}{1\ \text{mol}\ e^-} = 4.20 \times 10^9\ C$$

$$\text{Time} = \frac{4.20 \times 10^9\ C}{30,000\ C/s} \times \frac{1\ h}{3600\ s} = 38.9\ h$$

$$1.00 \times 10^6\ \text{g Na} \times \frac{1\ \text{mol Na}}{22.99\ \text{g Na}} \times \frac{1\ \text{mol Cl}_2}{2\ \text{mol Na}} = 21,748.6\ \text{mol Cl}_2$$

$PV = nRT$

$$V = \frac{nRT}{P} = \frac{(21,748.6\ \text{mol})\left(0.082\,06\ \dfrac{L \cdot atm}{K \cdot mol}\right)(273.15\ K)}{1.00\ atm} = 4.87 \times 10^5\ \text{L Cl}_2$$

17.103 $Al^{3+} + 3\ e^- \rightarrow Al;$ $40.0\ kg = 40,000\ g;\ 1\ h = 3600\ s$

$$\text{Charge} = 40,000\ \text{g Al} \times \frac{1\ \text{mol Al}}{26.98\ \text{g Al}} \times \frac{3\ \text{mol}\ e^-}{1\ \text{mol Al}} \times \frac{96,500\ C}{1\ \text{mol}\ e^-} = 4.29 \times 10^8\ C$$

$$\text{Current} = \frac{4.29 \times 10^8\ C}{3600\ s} = 1.19 \times 10^5\ A$$

17.104 $PbSO_4(s) + H^+(aq) + 2\ e^- \rightarrow Pb(s) + HSO_4^-(aq)$

$$\text{mass PbSO}_4 = 10.0\ \frac{C}{s} \times 1.50\ h \times \frac{3600\ s}{1\ h} \times \frac{1\ \text{mol}\ e^-}{96,500\ C} \times \frac{1\ \text{mol PbSO}_4}{2\ \text{mol}\ e^-} \times \frac{303.3\ \text{g PbSO}_4}{1\ \text{mol PbSO}_4}$$

$\text{mass PbSO}_4 = 84.9\ \text{g PbSO}_4$

17.105 $Cr^{3+}(aq) + 3\ e^- \rightarrow Cr(s)$

$$\text{Charge} = 125\ \text{g Cr} \times \frac{1\ \text{mol Cr}}{52.00\ \text{g Cr}} \times \frac{3\ \text{mol}\ e^-}{1\ \text{mol Cr}} \times \frac{96,500\ C}{1\ \text{mol}\ e^-} = 6.96 \times 10^5\ C$$

$$\text{Time} = \frac{6.96 \times 10^5\ C}{200.0\ C/s} \times \frac{1\ \text{min}}{60\ s} = 58.0\ \text{min}$$

Chapter Problems

17.106 (a) $2 \text{ MnO}_4^-(aq) + 16 \text{ H}^+(aq) + 5 \text{ Sn}^{2+}(aq) \rightarrow 2 \text{ Mn}^{2+}(aq) + 5 \text{ Sn}^{4+}(aq) + 8 \text{ H}_2\text{O}(l)$
(b) MnO_4^- is the oxidizing agent; Sn^{2+} is the reducing agent.
(c) $E^\circ = 1.51 \text{ V} + (-0.15 \text{ V}) = 1.36 \text{ V}$

17.107 $2 \text{ Mn}^{3+}(aq) + 2 \text{ H}_2\text{O}(l) \rightarrow \text{Mn}^{2+}(aq) + \text{MnO}_2(s) + 4 \text{ H}^+(aq)$
$E^\circ = 1.51 \text{ V} + (-0.95 \text{ V}) = +0.56 \text{ V}$
Because E° is positive, the disproportionation is spontaneous under standard-state conditions.

17.108 (a) Ag^+ is the strongest oxidizing agent because Ag^+ has the most positive standard reduction potential.
Pb is the strongest reducing agent because Pb^{2+} has the most negative standard reduction potential.
(b)

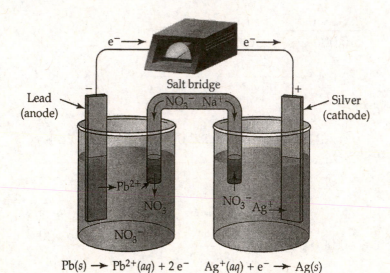

$$\text{Pb}(s) \rightarrow \text{Pb}^{2+}(aq) + 2 \text{ e}^- \qquad \text{Ag}^+(aq) + \text{e}^- \rightarrow \text{Ag}(s)$$

(c) $\text{Pb}(s) + 2 \text{ Ag}^+(aq) \rightarrow \text{Pb}^{2+}(aq) + 2 \text{ Ag}(s); \qquad n = 2 \text{ mol e}^-$
$E^\circ = E^\circ_{ox} + E^\circ_{red} = 0.13 \text{ V} + 0.80 \text{ V} = 0.93 \text{ V}$

$$\Delta G^\circ = -nFE^\circ = -(2 \text{ mol e}^-)\left(\frac{96{,}500 \text{ C}}{1 \text{ mol e}^-}\right)(0.93 \text{ V})\left(\frac{1 \text{ J}}{1 \text{ C} \cdot \text{V}}\right) = -179{,}490 \text{ J} = -180 \text{ kJ}$$

$$E^\circ = \frac{0.0592 \text{ V}}{n} \log K; \quad \log K = \frac{nE^\circ}{0.0592 \text{ V}} = \frac{(2)(0.93 \text{ V})}{0.0592 \text{ V}} = 31; \quad K = 10^{31}$$

(d) $E = E^\circ - \dfrac{0.0592 \text{ V}}{n} \log \dfrac{[\text{Pb}^{2+}]}{[\text{Ag}^+]^2} = 0.93 \text{ V} - \dfrac{0.0592 \text{ V}}{2} \log\left(\dfrac{0.01}{(0.01)^2}\right) = 0.87 \text{ V}$

17.109 For Pb^{2+}, $E = -0.13 - \dfrac{0.0592\,V}{2} \log \dfrac{1}{[Pb^{2+}]}$

For Cd^{2+}, $E = -0.40 - \dfrac{0.0592\,V}{2} \log \dfrac{1}{[Cd^{2+}]}$

Set these two equations for E equal to each other and solve for $[Cd^{2+}]/[Pb^{2+}]$.

$$-0.13 - \frac{0.0592\,V}{2} \log \frac{1}{[Pb^{2+}]} = -0.40 - \frac{0.0592\,V}{2} \log \frac{1}{[Cd^{2+}]}$$

$$0.27 = \frac{0.0592\,V}{2}(\log[Cd^{2+}] - \log[Pb^{2+}]) = \frac{0.0592\,V}{2} \log \frac{[Cd^{2+}]}{[Pb^{2+}]}$$

$$\log \frac{[Cd^{2+}]}{[Pb^{2+}]} = \frac{(0.27)(2)}{0.0592} = 9.1$$

$$\frac{[Cd^{2+}]}{[Pb^{2+}]} = 10^{9.1} = 1 \times 10^{9}$$

17.110 (a)

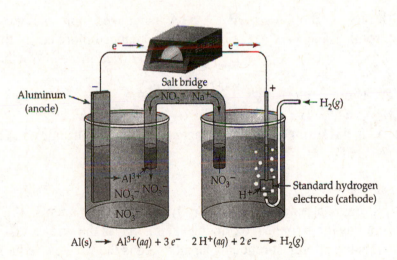

$Al(s) \longrightarrow Al^{3+}(aq) + 3\,e^{-}$ $2\,H^{+}(aq) + 2\,e^{-} \longrightarrow H_2(g)$

(b) $2\,Al(s) + 6\,H^{+}(aq) \rightarrow 2\,Al^{3+}(aq) + 3\,H_2(g)$
$E° = E°_{ox} + E°_{red} = 1.66\,V + 0.00\,V = 1.66\,V$

(c)

$$E = E° - \frac{0.0592\,V}{n} \log \frac{[Al^{3+}]^2(P_{H_2})^3}{[H^{+}]^6} = 1.66\,V - \frac{(0.0592\,V)}{6} \log\left(\frac{(0.10)^2(10.0)^3}{(0.10)^6}\right) = 1.59\,V$$

(d) $\Delta G° = -nFE° = -(6\text{ mol }e^{-})\left(\dfrac{96{,}500\,C}{1\text{ mol }e^{-}}\right)(1.66\,V)\left(\dfrac{1\,J}{1\,C\cdot V}\right) = -961{,}140\,J = -961\,kJ$

$E° = \dfrac{0.0592\,V}{n} \log K;$ $\log K = \dfrac{nE°}{0.0592\,V} = \dfrac{(6)(1.66\,V)}{0.0592\,V} = 168;$ $K = 10^{168}$

(e) mass $Al = 10.0\,\dfrac{C}{s} \times 25.0\text{ min} \times \dfrac{60\,s}{1\text{ min}} \times \dfrac{1\text{ mol }e^{-}}{96{,}500\,C} \times \dfrac{1\text{ mol Al}}{3\text{ mol }e^{-}} \times \dfrac{26.98\text{ g Al}}{1\text{ mol Al}} = 1.40\,g$

17.111 $Zn(s) \rightarrow Zn^{2+}(aq) + 2\,e^-$

$$\text{mass Zn} = 0.100\,\frac{C}{s} \times 200.0\,h \times \frac{60\,\min}{h} \times \frac{60\,s}{\min} \times \frac{1\,\text{mol}\,e^-}{96,500\,C} \times \frac{1\,\text{mol Zn}}{2\,\text{mol}\,e^-} \times \frac{65.39\,\text{g Zn}}{1\,\text{mol Zn}}$$

mass Zn = 24.4 g

17.112 $2\,Cl^-(aq) \rightarrow Cl_2(g) + 2\,e^-$

11 million tons = 11×10^6 tons; Cl_2, 70.91 amu

$$11 \times 10^6\,\text{tons} \times \frac{907,200\,\text{g}}{1\,\text{ton}} \times \frac{1\,\text{mol}\,Cl_2}{70.91\,\text{g}\,Cl_2} = 1.41 \times 10^{11}\,\text{mol}\,Cl_2$$

$$\text{Charge} = 1.41 \times 10^{11}\,\text{mol}\,Cl_2 \times \frac{2\,\text{mol}\,e^-}{1\,\text{mol}\,Cl_2} \times \frac{96,500\,C}{1\,\text{mol}\,e^-} = 2.72 \times 10^{16}\,C$$

$1\,J = 1\,C \times 1\,V$;   Energy $= (2.72 \times 10^{16}\,C)(4.5\,V) = 1.22 \times 10^{17}\,J$

$$kWh = (1.22 \times 10^{17}\,J)\left(\frac{1\,kWh}{3.6 \times 10^6\,J}\right) = 3.4 \times 10^{10}\,kWh$$

17.113 (a) From: $B + A^+ \rightarrow B^+ + A$, A^+ is reduced more easily than B^+
           From: $C + A^+ \rightarrow C^+ + A$, A^+ is reduced more easily than C^+
           From: $B + C^+ \rightarrow B^+ + C$, C^+ is reduced more easily than B^+
           $A^+ + e^- \rightarrow A$
           $C^+ + e^- \rightarrow C$
           $B^+ + e^- \rightarrow B$
      (b) A^+ is the strongest oxidizing agent;   B is the strongest reducing agent
      (c) $A^+ + B \rightarrow B^+ + A$

17.114 (a) oxidizing agents: PbO_2, H^+, $Cr_2O_7^{2-}$;   reducing agents: Al, Fe, Ag
      (b) PbO_2 is the strongest oxidizing agent. H^+ is the weakest oxidizing agent.
      (c) Al is the strongest reducing agent.   Ag is the weakest reducing agent.
      (d) oxidized by Cu^{2+}: Fe and Al;   reduced by H_2O_2: PbO_2 and $Cr_2O_7^{2-}$

17.115 From Appendix D:
      $AgBr(s) + e^- \rightarrow Ag(s) + Br^-(aq)$ $E° = 0.07\,V$
      (a) oxidation: $C_6H_4(OH)_2(aq) \rightarrow C_6H_4O_2(aq) + 2\,H^+(aq) + 2\,e^-$ $E° = -0.699\,V$
           reduction: $\underline{2[AgBr(s) + e^- \rightarrow Ag(s) + Br^-(aq)]}$ $\underline{E° = 0.07\,V}$
           overall: $2\,AgBr(s) + C_6H_4(OH)_2(aq) \rightarrow$
                                   $2\,Ag(s) + 2\,Br^-(aq) + C_6H_4O_2(aq) + 2\,H^+(aq)$
      overall $E° = -0.699\,V + 0.07\,V = -0.63\,V$
      Because the overall $E°$ is negative, the reaction is nonspontaneous when $[H^+] = 1.0\,M$.

(b) $E^o(\text{in } 1.0 \text{ M OH}^-) = E^o(\text{in } 1.0 \text{ M H}^+) - \dfrac{0.0592 \text{ V}}{n} \log \dfrac{[\text{Br}^-]^2[\text{H}^+]^2[\text{C}_6\text{H}_4\text{O}_2]}{[\text{C}_6\text{H}_4(\text{OH})_2]}$

$[\text{H}^+] = \dfrac{K_w}{[\text{OH}^-]} = \dfrac{1.0 \times 10^{-14}}{1.0} = 1.0 \times 10^{-14} \text{ M}$

$E^o(\text{in } 1.0 \text{ M OH}^-) = -0.63 \text{ V} - \dfrac{(0.0592 \text{ V})}{2} \log \dfrac{(1)^2(10^{-14})^2(1)}{(1)} = +0.20 \text{ V}$

17.116 (a) $3 \text{ CH}_3\text{CH}_2\text{OH}(aq) + 2 \text{ Cr}_2\text{O}_7^{2-}(aq) + 16 \text{ H}^+(aq) \rightarrow$
$\qquad\qquad\qquad\qquad 3 \text{ CH}_3\text{CO}_2\text{H}(aq) + 4 \text{ Cr}^{3+}(aq) + 11 \text{ H}_2\text{O}(l)$

oxidation:
$3 \text{ CH}_3\text{CH}_2\text{OH}(aq) + 3 \text{ H}_2\text{O}(l) \rightarrow 3 \text{ CH}_3\text{CO}_2\text{H}(aq) + 12 \text{ H}^+(aq) + 12 \text{ e}^- \qquad E^o = -0.058\text{V}$
reduction:
$2 \text{ Cr}_2\text{O}_7^{2-}(aq) + 28 \text{ H}^+(aq) + 12 \text{ e}^- + \rightarrow 4 \text{ Cr}^{3+}(aq) + 14 \text{ H}_2\text{O}(l) \qquad \underline{E^o = \;\;1.36 \text{ V}}$
$\qquad\qquad\qquad\qquad\qquad\qquad\qquad\qquad\qquad\qquad\qquad\qquad \text{overall } E^o = \;\;1.30 \text{ V}$

(b) $E = E^o - \dfrac{0.0592 \text{ V}}{n} \log \dfrac{[\text{CH}_3\text{CO}_2\text{H}]^3[\text{Cr}^{3+}]^4}{[\text{CH}_3\text{CH}_2\text{OH}]^3[\text{Cr}_2\text{O}_7^{2-}]^2[\text{H}^+]^{16}}$

$\text{pH} = 4.00, \;\; [\text{H}^+] = 0.000\ 10 \text{ M}$

$E = 1.30 \text{ V} - \dfrac{(0.0592 \text{ V})}{12} \log\left(\dfrac{(1.0)^3(1.0)^4}{(1.0)^3(1.0)^2(0.000\ 10)^{16}} \right)$

$E = 1.30 \text{ V} - \dfrac{(0.0592 \text{ V})}{12} \log \dfrac{1}{(0.000\ 10)^{16}} = 0.98 \text{ V}$

17.117 (a) $\Delta G^o = -nFE^o$
$\Delta G^o_3 = \Delta G^o_1 + \Delta G^o_2 \;\; \text{therefore} \;\; -n_3FE^o_3 = -n_1FE^o_1 + (-n_2FE^o_2)$
$n_3E^o_3 = n_1E^o_1 + n_2E^o_2$
$E^o_3 = \dfrac{n_1 E^o_1 + n_2 E^o_2}{n_3}$

(b) $E^o_3 = \dfrac{(3)(-0.04 \text{ V}) + (2)(0.45 \text{ V})}{1} = 0.78 \text{ V}$

(c) E^o values would be additive ($E^o_3 = E^o_1 + E^o_2$) if reaction (3) is an overall cell reaction because the electrons in the two half reactions, (1) and (2), cancel. That is, $n_1 = n_2 = n_3$ in the equation for E^o_3.

17.118 anode: $\qquad \text{Ag}(s) + \text{Cl}^-(aq) \rightarrow \text{AgCl}(s) + \text{e}^-$
cathode: $\qquad \underline{\text{Ag}^+(aq) + \text{e}^- \rightarrow \text{Ag}(s)}$
overall: $\qquad \text{Ag}^+(aq) + \text{Cl}^-(aq) \rightarrow \text{AgCl}(s) \qquad E^o = 0.578 \text{ V}$

For $\text{AgCl}(s) \rightleftharpoons \text{Ag}^+(aq) + \text{Cl}^-(aq) \qquad E^o = -0.578 \text{ V}$

$E^o = \dfrac{0.0592 \text{ V}}{n} \log K; \quad \log K = \dfrac{nE^o}{0.0592 \text{ V}} = \dfrac{(1)(-0.578 \text{ V})}{0.0592 \text{ V}} = -9.76$

$K = K_{sp} = 10^{-9.76} = 1.7 \times 10^{-10}$

17.119 (a)   anode: $Cu(s) \rightarrow Cu^{2+}(aq) + 2\,e^-$ $E° = -0.34$ V

         cathode: $\underline{2\,Ag^+(aq) + 2\,e^- \rightarrow 2\,Ag(s)}$ $\underline{E° = 0.80\ V}$

         overall: $2\,Ag^+(aq) + Cu(s) \rightarrow Cu^{2+}(aq) + 2\,Ag(s)$ $E° = 0.46$ V

$$E = E° - \frac{0.0592\,\text{V}}{n} \log \frac{[Cu^{2+}]}{[Ag^+]^2} = 0.46\,\text{V} - \frac{(0.0592\,\text{V})}{2} \log\left(\frac{1.0}{(0.050)^2}\right) = 0.38\,\text{V}$$

(b) $[Ag^+] = \dfrac{K_{sp}}{[Br^-]} = \dfrac{5.4 \times 10^{-13}}{1.0\,\text{M}} = 5.4 \times 10^{-13}\,\text{M}$

$$E = E° - \frac{0.0592\,\text{V}}{n} \log \frac{[Cu^{2+}]}{[Ag^+]^2} = 0.46\,\text{V} - \frac{(0.0592\,\text{V})}{2} \log\left(\frac{1.0}{(5.4 \times 10^{-13})^2}\right) = -0.27\,\text{V}$$

The cell potential for the spontaneous reaction is E = 0.27 V.

The spontaneous reaction is: $Cu^{2+}(aq) + 2\,Ag(s) + 2\,Br^-(aq) \rightarrow 2\,AgBr(s) + Cu(s)$

(c) $Cu^{2+}(aq) + 2\,e^- \rightarrow Cu(s)$ $E° = 0.34$ V

     $\underline{2\,Ag(s) + 2\,Br^-(aq) \rightarrow 2\,AgBr(s) + 2\,e^-}$ $\underline{E° = ?}$

     $Cu^{2+}(aq) + 2\,Ag(s) + 2\,Br^-(aq) \rightarrow 2\,AgBr(s) + Cu(s)$ $E° = 0.27$ V

$E° = ? = 0.27$ V $- 0.34$ V $= -0.07$ V

For: $AgBr(s) + e^- \rightarrow Ag(s) + Br^-(aq)$

the standard reduction potential is $E° = 0.07$ V

17.120 $4\,Fe^{2+}(aq) + O_2(g) + 4\,H^+(aq) \rightarrow 4\,Fe^{3+}(aq) + 2\,H_2O(l)$

oxidation $4\,Fe^{2+}(aq) \rightarrow 4\,Fe^{3+}(aq) + 4\,e^-$ $E° = -0.77$ V

reduction $O_2(g) + 4\,H^+(aq) + 4\,e^- \rightarrow 2\,H_2O(l)$ $\underline{E° = 1.23\ V}$

                                          overall $E° = 0.46$ V

$$P_{O_2} = 160\,\text{mm Hg} \times \frac{1.00\,\text{atm}}{760\,\text{mm Hg}} = 0.211\,\text{atm}$$

$$E = E° - \frac{0.0592\,\text{V}}{n} \log \frac{[Fe^{3+}]^4}{[Fe^{2+}]^4[H^+]^4(P_{O_2})}$$

$$E = 0.46\,\text{V} - \frac{0.0592\,\text{V}}{4} \log \frac{(1 \times 10^{-7})^4}{(1 \times 10^{-7})^4(1 \times 10^{-7})^4(0.211)}$$

$E = 0.46$ V $- 0.42$ V $= 0.04$ V

Because E is positive, the reaction is spontaneous.

17.121 $H_2MoO_4(aq) + As(s) \rightarrow Mo^{3+}(aq) + H_3AsO_4(aq)$

$H_2MoO_4(aq) \rightarrow Mo^{3+}(aq)$

$H_2MoO_4(aq) \rightarrow Mo^{3+}(aq) + 4\,H_2O(l)$

$6\,H^+(aq) + H_2MoO_4(aq) \rightarrow Mo^{3+}(aq) + 4\,H_2O(l)$

$[3\,e^- + 6\,H^+(aq) + H_2MoO_4(aq) \rightarrow Mo^{3+}(aq) + 4\,H_2O(l)] \times 5$

                                   (reduction half reaction)

$As(s) \rightarrow H_3AsO_4(aq)$

$As(s) + 4\,H_2O(l) \rightarrow H_3AsO_4(aq)$

$$As(s) + 4 H_2O(l) \rightarrow H_3AsO_4(aq) + 5 H^+(aq)$$
$$[As(s) + 4 H_2O(l) \rightarrow H_3AsO_4(aq) + 5 H^+(aq) + 5 e^-] \times 3 \quad \text{(oxidation half reaction)}$$

Combine the two half reactions.
$$30 H^+(aq) + 5 H_2MoO_4(aq) + 3 As(s) + 12 H_2O(l) \rightarrow$$
$$5 Mo^{3+}(aq) + 3 H_3AsO_4(aq) + 15 H^+(aq) + 20 H_2O(l)$$
$$15 H^+(aq) + 5 H_2MoO_4(aq) + 3 As(s) \rightarrow 5 Mo^{3+}(aq) + 3 H_3AsO_4(aq) + 8 H_2O(l)$$

$5 \times [H_2MoO_4(aq) + 2 H^+(aq) + 2 e^- \rightarrow MoO_2(s) + 2 H_2O(l)]$	$E^\circ = +0.646$ V
$5 \times [MoO_2(s) + 4 H^+(aq) + e^- \rightarrow Mo^{3+}(aq) + 2 H_2O(l)]$	$E^\circ = -0.008$ V
$3 \times [As(s) + 3 H_2O(l) \rightarrow H_3AsO_3(aq) + 3 H^+(aq) + 3 e^-]$	$E^\circ = -0.240$ V
$3 \times [H_3AsO_3(aq) + H_2O(l) \rightarrow H_3AsO_4(aq) + 2 H^+(aq) + 2 e^-]$	$E^\circ = -0.560$ V

$$15 H^+(aq) + 5 H_2MoO_4(aq) + 3 As(s) \rightarrow 5 Mo^{3+}(aq) + 3 H_3AsO_4(aq) + 8 H_2O(l)$$

$$\Delta G^\circ = -nFE^\circ = -(10 \text{ mol e}^-)\left(\frac{96,500 \text{ C}}{1 \text{ mol e}^-}\right)(0.646 \text{ V})\left(\frac{1 \text{ J}}{1 \text{ C} \cdot \text{V}}\right) = -623,390 \text{ J} = -623.4 \text{ kJ}$$

$$\Delta G^\circ = -nFE^\circ = -(5 \text{ mol e}^-)\left(\frac{96,500 \text{ C}}{1 \text{ mol e}^-}\right)(-0.008 \text{ V})\left(\frac{1 \text{ J}}{1 \text{ C} \cdot \text{V}}\right) = 3,860 \text{ J} = +3.9 \text{ kJ}$$

$$\Delta G^\circ = -nFE^\circ = -(9 \text{ mol e}^-)\left(\frac{96,500 \text{ C}}{1 \text{ mol e}^-}\right)(-0.240 \text{ V})\left(\frac{1 \text{ J}}{1 \text{ C} \cdot \text{V}}\right) = 208,440 \text{ J} = +208.4 \text{ kJ}$$

$$\Delta G^\circ = -nFE^\circ = -(6 \text{ mol e}^-)\left(\frac{96,500 \text{ C}}{1 \text{ mol e}^-}\right)(-0.560 \text{ V})\left(\frac{1 \text{ J}}{1 \text{ C} \cdot \text{V}}\right) = 324,240 \text{ J} = +324.2 \text{ kJ}$$

$$\Delta G^\circ(\text{total}) = -623.4 \text{ kJ} + 3.9 \text{ kJ} + 208.4 \text{ kJ} + 324.2 \text{ kJ} = -86.9 \text{ kJ} = -86,900 \text{ J}$$
$$1 \text{ V} = 1 \text{ J/C}$$

$$\Delta G^\circ = -nFE^\circ; \quad E^\circ = \frac{-\Delta G^\circ}{nF} = \frac{-(-86,900 \text{ J})}{(15 \text{ mol e}^-)\left(\dfrac{96,500 \text{ C}}{1 \text{ mol e}^-}\right)} = +0.060 \text{ J/C} = +0.060 \text{ V}$$

17.122 First calculate E° for the galvanic cell in order to determine E°_1.

anode:	$5 [2 Hg(l) + 2 Br^-(aq) \rightarrow Hg_2Br_2(s) + 2 e^-]$	$E^\circ_1 = ?$
cathode:	$2 [MnO_4^-(aq) + 8 H^+(aq) + 5 e^- \rightarrow Mn^{2+}(aq) + 4 H_2O(l)]$	$E^\circ_2 = 1.51$ V
overall:	$2 MnO_4^-(aq) + 10 Hg(l) + 10 Br^-(aq) + 16 H^+(aq) \rightarrow$	
	$2 Mn^{2+}(aq) + 5 Hg_2Br_2(s) + 8 H_2O(l)$	

$$n = 10 \text{ mol e}^-$$

$$E = E^\circ - \frac{0.0592 \text{ V}}{n} \log \frac{[Mn^{2+}]^2}{[Br^-]^{10}[MnO_4^-]^2[H^+]^{16}}$$

$$1.214 \text{ V} = E^\circ - \frac{(0.0592 \text{ V})}{10} \log \left(\frac{(0.10)^2}{(0.10)^{10}(0.10)^2(0.10)^{16}}\right)$$

$$1.214 \text{ V} = E^\circ - \frac{(0.0592 \text{ V})}{10} \log \frac{1}{(0.10)^{26}} = E^\circ - 0.154 \text{ V}$$

$E^\circ = 1.214 + 0.154 = 1.368 \text{ V}$

$E^\circ_1 + E^\circ_2 = 1.368 \text{ V}; \quad E^\circ_1 + 1.51 \text{ V} = 1.368 \text{ V}; \quad E^\circ_1 = 1.368 \text{ V} - 1.51 \text{ V} = -0.142 \text{ V}$

oxidation:	$2 \text{ Hg(l)} \rightarrow \text{Hg}_2^{2+}(aq) + 2 \text{ e}^-$	$E^\circ = -0.80 \text{ V}$ (Appendix D)
reduction:	$\underline{\text{Hg}_2\text{Br}_2(s) + 2 \text{ e}^- \rightarrow 2 \text{ Hg(l)} + 2 \text{ Br}^-(aq)}$	$\underline{E^\circ = +0.142 \text{ V}}$ (from E°_1)
overall:	$\text{Hg}_2\text{Br}_2(s) \rightarrow \text{Hg}_2^{2+}(aq) + 2 \text{ Br}^-(aq)$	$E^\circ = -0.658 \text{ V}$

$$E^\circ = \frac{0.0592 \text{ V}}{n} \log K; \quad \log K = \frac{nE^\circ}{0.0592 \text{ V}} = \frac{(2)(-0.658 \text{ V})}{0.0592 \text{ V}} = -22.2$$

$K = K_{sp} = 10^{-22.2} = 6 \times 10^{-23}$

17.123

oxidation:	$\text{Cu}^+(aq) \rightarrow \text{Cu}^{2+}(aq) + \text{e}^-$	$E^\circ = -0.15 \text{ V}$
reduction:	$\underline{\text{Cu}^{2+}(aq) + 2 \text{ CN}^-(aq) + \text{e}^- \rightarrow \text{Cu(CN)}_2^-(aq)}$	$\underline{E^\circ = 1.103 \text{ V}}$
overall:	$\text{Cu}^+(aq) + 2 \text{ CN}^-(aq) \rightarrow \text{Cu(CN)}_2^-(aq)$	$E^\circ = 0.953 \text{ V}$

$$E^\circ = \frac{0.0592 \text{ V}}{n} \log K; \quad \log K = \frac{nE^\circ}{0.0592 \text{ V}} = \frac{(1)(0.953 \text{ V})}{0.0592 \text{ V}} = 16.1$$

$K = K_f = 10^{16.1} = 1 \times 10^{16}$

17.124 (a)

anode:	$4[\text{Al(s)} \rightarrow \text{Al}^{3+}(aq) + 3 \text{ e}^-]$	$E^\circ = 1.66 \text{ V}$
cathode:	$\underline{3[\text{O}_2(g) + 4 \text{ H}^+(aq) + 4 \text{ e}^- \rightarrow 2 \text{ H}_2\text{O(l)}]}$	$\underline{E^\circ = 1.23 \text{ V}}$
overall:	$4 \text{ Al(s)} + 3 \text{ O}_2(g) + 12 \text{ H}^+(aq) \rightarrow 4 \text{ Al}^{3+}(aq) + 6 \text{ H}_2\text{O(l)}$	$E^\circ = 2.89 \text{ V}$

(b) & (c) $\quad E = E^\circ - \dfrac{2.303 \, RT}{nF} \log \dfrac{[\text{Al}^{3+}]^4}{(P_{\text{O}_2})^3 [\text{H}^+]^{12}}$

$$E = 2.89 \text{ V} - \frac{(2.303)\left(8.314 \, \frac{\text{J}}{\text{K} \cdot \text{mol}}\right)(310 \text{ K})}{(12 \text{ mol e}^-)(96{,}500 \text{ C/mol e}^-)} \log\left(\frac{(1.0 \times 10^{-9})^4}{(0.20)^3 (1.0 \times 10^{-7})^{12}}\right)$$

$E = 2.89 \text{ V} - 0.257 \text{ V} = 2.63 \text{ V}$

17.125 $\quad E_1 = E^\circ_1 - \dfrac{0.0592 \, \text{V}}{6} \log \dfrac{[\text{Cu}^{2+}]^3 (P_{\text{NO}})^2}{[\text{NO}_3^-]^2 [\text{H}^+]^8}$ and $E_2 = E^\circ_2 - \dfrac{0.0592 \, \text{V}}{2} \log \dfrac{[\text{Cu}^{2+}] (P_{\text{NO}_2})^2}{[\text{NO}_3^-]^2 [\text{H}^+]^4}$

(a) $\quad E_1 = 0.62 \text{ V} - \dfrac{0.0592 \text{ V}}{6} \log \dfrac{(0.10)^3 (1.0 \times 10^{-3})^2}{(1.0)^2 (1.0)^8} = 0.71 \text{ V}$

$\quad E_2 = 0.45 \text{ V} - \dfrac{0.0592 \text{ V}}{2} \log \dfrac{(0.10)(1.0 \times 10^{-3})^2}{(1.0)^2 (1.0)^4} = 0.66 \text{ V}$

Reaction (1) has the greater thermodynamic tendency to occur because of the larger positive potential.

(b) $E_1 = 0.62 \text{ V} - \dfrac{0.0592 \text{ V}}{6} \log \dfrac{(0.10)^3(1.0 \times 10^{-3})^2}{(10.0)^2(10.0)^8} = 0.81 \text{ V}$

$E_2 = 0.45 \text{ V} - \dfrac{0.0592 \text{ V}}{2} \log \dfrac{(0.10)(1.0 \times 10^{-3})^2}{(10.0)^2(10.0)^4} = 0.83 \text{ V}$

Reaction (2) has the greater thermodynamic tendency to occur because of the larger positive potential.

(c) Set the two equations equal to each other and solve for x.

$0.62 \text{ V} - \dfrac{0.0592 \text{ V}}{6} \log \dfrac{(0.10)^3(1.0 \times 10^{-3})^2}{(x)^2(x)^8} = 0.45 \text{ V} - \dfrac{0.0592 \text{ V}}{2} \log \dfrac{(0.10)(1.0 \times 10^{-3})^2}{(x)^2(x)^4}$

$0.17 - \dfrac{0.0592 \text{ V}}{6}[(-9) - 10 \log x] = -\dfrac{0.0592 \text{ V}}{2}[(-7) - 6 \log x]$

$0.0516 = 0.0789 \log x; \quad \dfrac{0.0516}{0.0789} = \log x; \quad 0.654 = \log x$

$[HNO_3] = x = 10^{0.654} = 4.5 \text{ M}$

Multiconcept Problems

17.126 (a) $4 \text{ CH}_2\text{=CHCN} + 2 \text{ H}_2\text{O} \rightarrow 2 \text{ NC(CH}_2)_4\text{CN} + \text{O}_2$

(b) $\text{mol e}^- = 3000 \text{ C/s} \times 10.0 \text{ h} \times \dfrac{3600 \text{ s}}{1 \text{ h}} \times \dfrac{1 \text{ mol e}^-}{96,500 \text{ C}} = 1119.2 \text{ mol e}^-$

mass adiponitrile =

$1119.2 \text{ mol e}^- \times \dfrac{1 \text{ mol adiponitrile}}{2 \text{ mol e}^-} \times \dfrac{108.14 \text{ g adiponitrile}}{1 \text{ mol adiponitrile}} \times \dfrac{1.0 \text{ kg}}{1000 \text{ g}} = 60.5 \text{ kg}$

(c) $1119.2 \text{ mol e}^- \times \dfrac{1 \text{ mol O}_2}{4 \text{ mol e}^-} = 279.8 \text{ mol O}_2$

$PV = nRT$

$V = \dfrac{nRT}{P} = \dfrac{(279.8 \text{ mol})\left(0.082\,06 \dfrac{\text{L·atm}}{\text{K·mol}}\right)(298 \text{ K})}{\left(740 \text{ mm Hg} \times \dfrac{1 \text{ atm}}{760 \text{ mm Hg}}\right)} = 7030 \text{ L O}_2$

17.127 (a) $2 \text{ MnO}_4^-(aq) + 5 \text{ H}_2\text{C}_2\text{O}_4(aq) + 6 \text{ H}^+(aq) \rightarrow$
$2 \text{ Mn}^{2+}(aq) + 10 \text{ CO}_2(g) + 8 \text{ H}_2\text{O}(l)$

(b) oxidation: $5[\text{H}_2\text{C}_2\text{O}_4(aq) \rightarrow 2 \text{ CO}_2(g) + 2 \text{ H}^+(aq) + 2 \text{ e}^-]$ $\qquad E^° = 0.49 \text{ V}$
reduction: $2[\text{MnO}_4^-(aq) + 8 \text{ H}^+(aq) + 5 \text{ e}^- \rightarrow \text{Mn}^{2+}(aq) + 4 \text{ H}_2\text{O}(l)]$ $\qquad \underline{E^° = 1.51 \text{ V}}$
$\qquad\qquad\qquad\qquad\qquad\qquad\qquad\qquad\qquad\qquad\qquad\qquad$ overall $E^° = 2.00 \text{ V}$

(c)

$\Delta G^° = -nFE^° = -(10 \text{ mol e}^-)\left(\dfrac{96,500 \text{ C}}{1 \text{ mol e}^-}\right)(2.00 \text{ V})\left(\dfrac{1 \text{ J}}{1 \text{ C·V}}\right) = -1,930,000 \text{ J} = -1,930 \text{ kJ}$

$$E^\circ = \frac{0.0592\ V}{n}\log K;\quad \log K = \frac{nE^\circ}{0.0592\ V} = \frac{(10)(2.00\ V)}{0.0592\ V} = 338;\quad K = 10^{338}$$

(d) $Na_2C_2O_4$, 134.0 amu

$$1.200\ g\ Na_2C_2O_4 \times \frac{1\ mol\ Na_2C_2O_4}{134.0\ g\ Na_2C_2O_4} \times \frac{2\ mol\ KMnO_4}{5\ mol\ Na_2C_2O_4} = 3.582 \times 10^{-3}\ mol\ KMnO_4$$

$$molarity = \frac{3.582 \times 10^{-3}\ mol}{0.032\ 50\ L} = 0.1102\ M$$

17.128 (a) $Cr_2O_7^{2-}(aq) + 6\ Fe^{2+}(aq) + 14\ H^+(aq) \rightarrow 2\ Cr^{3+}(aq) + 6\ Fe^{3+}(aq) + 7\ H_2O(l)$

(b) The two half reactions are:

oxidation: $\quad Fe^{2+}(aq) \rightarrow Fe^{3+}(aq) + e^-$ $\hphantom{reduction:xxxxxxxxxxxxxxxxxxxxxxxx}$ $E^\circ = -0.77\ V$

reduction: $\quad Cr_2O_7^{2-}(aq) + 14\ H^+(aq) + 6\ e^- \rightarrow 2\ Cr^{3+}(aq) + 7\ H_2O(l)$ $\quad E^\circ = 1.36\ V$

At the equivalence point the potential is given by either of the following expressions:

(1) $E = 1.36\ V - \dfrac{0.0592\ V}{6}\log\dfrac{[Cr^{3+}]^2}{[Cr_2O_7^{2-}][H^+]^{14}}$

(2) $E = 0.77\ V - \dfrac{0.0592\ V}{1}\log\dfrac{[Fe^{2+}]}{[Fe^{3+}]}$

where E is the same in both because equilibrium is reached and the solution can have only one potential. Multiplying (1) by 6, adding it to (2), and using some stoichiometric relationships at the equivalence point will simplify the log term.

$$7E = [(6 \times 1.36\ V) + 0.77\ V] - (0.0592\ V)\log\frac{[Fe^{2+}][Cr^{3+}]^2}{[Fe^{3+}][Cr_2O_7^{2-}][H^+]^{14}}$$

At the equivalence point, $[Fe^{2+}] = 6[Cr_2O_7^{2-}]$ and $[Fe^{3+}] = 3[Cr^{3+}]$. Substitute these equalities into the previous equation.

$$7E = [(6 \times 1.36\ V) + 0.77\ V] - (0.0592\ V)\log\frac{6[Cr_2O_7^{2-}][Cr^{3+}]^2}{3[Cr^{3+}][Cr_2O_7^{2-}][H^+]^{14}}$$

Cancel identical terms.

$$7E = [(6 \times 1.36\ V) + 0.77\ V] - (0.0592\ V)\log\frac{6[Cr^{3+}]}{3[H^+]^{14}}$$

mol $Fe^{2+} = (0.120\ L)(0.100\ mol/L) = 0.0120\ mol\ Fe^{2+}$

$$mol\ Cr_2O_7^{2-} = 0.0120\ mol\ Fe^{2+} \times \frac{1\ mol\ Cr_2O_7^{2-}}{6\ mol\ Fe^{2+}} = 0.002\ 00\ mol\ Cr_2O_7^{2-}$$

$$volume\ Cr_2O_7^{2-} = 0.002\ 00\ mol \times \frac{1\ L}{0.120\ mol} = 0.0167\ L$$

At the equivalence point assume mol Fe^{3+} = initial mol Fe^{2+} = 0.0120 mol
Total volume at the equivalence point is 0.120 L + 0.0167 L = 0.1367 L

$$[Fe^{3+}] = \frac{0.0120\ mol}{0.1367\ L} = 0.0878\ M;\quad [Cr^{3+}] = [Fe^{3+}]/3 = (0.0878\ M)/3 = 0.0293\ M$$

$[H^+] = 10^{-pH} = 10^{-2.00} = 0.010$ M

$7E = [(6 \times 1.36 \text{ V}) + 0.77 \text{ V}] - (0.0592 \text{ V})\log\dfrac{6(0.0293)}{3(0.010)^{14}} = 8.93 - 1.585 = 7.345$ V

$E = \dfrac{7.345 \text{ V}}{7} = 1.05$ V at the equivalence point.

17.129 $2 H_2(g) + O_2(g) \rightarrow 2 H_2O(l)$

(a) $\Delta H^\circ = 2 \Delta H^\circ_f(H_2O) = (2 \text{ mol})(-285.8 \text{ kJ/mol}) = -571.6$ kJ

$\Delta S^\circ = 2 S^\circ(H_2O) - [2 S^\circ(H_2) + S^\circ(O_2)]$

$\Delta S^\circ = (2 \text{ mol})(69.9 \text{ J/(K} \cdot \text{mol})) - [(2 \text{ mol})(130.6 \text{ J/(K} \cdot \text{mol})) + (1 \text{ mol})(205.0 \text{ J/(K} \cdot \text{mol}))]$

$\Delta S^\circ = -326.4$ J/K $= -0.3264$ kJ/K

$95 \,^\circ\text{C} = 368$ K

$\Delta G^\circ = \Delta H^\circ - T\Delta S^\circ = -571.6$ kJ $-(368 \text{ K})(-0.3264 \text{ kJ/K}) = -451.5$ kJ

1 V $= 1$ J/C

$\Delta G^\circ = -nFE^\circ; \quad E^\circ = -\dfrac{\Delta G^\circ}{nF} = -\dfrac{-451.5 \times 10^3 \text{ J}}{(4)(96,500 \text{ C})} = 1.17$ J/C $= 1.17$ V

(b) $E = E^\circ - \dfrac{2.303\,RT}{nF}\log\dfrac{1}{(P_{H_2})^2(P_{O_2})}$

$E = 1.17 \text{ V} - \dfrac{(2.303)\left(8.314\,\dfrac{\text{J}}{\text{mol}\cdot\text{K}}\right)(368 \text{ K})}{(4 \text{ mol e}^-)(96,500 \text{ C/mol e}^-)}\log\left(\dfrac{1}{(25)^2(25)}\right)$

$E = 1.17$ V $+ 0.077$ V $= 1.25$ V

17.130 (a) $\quad Zn(s) + 2 Ag^+(aq) + H_2O(l) \rightarrow ZnO(s) + 2 Ag(s) + 2 H^+(aq)$

$\Delta H^\circ_{rxn} = \Delta H^\circ_f(ZnO) - [2 \Delta H^\circ_f(Ag^+) + \Delta H^\circ_f(H_2O)]$

$\Delta H^\circ_{rxn} = [(1 \text{ mol})(-350.5 \text{ kJ/mol})] - [(2 \text{ mol})(105.6 \text{ kJ/mol}) + (1 \text{ mol})(-285.8 \text{ kJ/mol})]$

$\Delta H^\circ_{rxn} = -275.9$ kJ

$\Delta S^\circ = [S^\circ(ZnO) + 2 S^\circ(Ag)] - [S^\circ(Zn) + 2 S^\circ(Ag^+) + S^\circ(H_2O)]$

$\Delta S^\circ = [(1 \text{ mol})(43.7 \text{ J/(K}\cdot \text{mol})) + (2 \text{ mol})(42.6 \text{ J/(K}\cdot \text{mol}))]$

$\quad - [(1 \text{ mol})(41.6 \text{ J/(K}\cdot \text{mol})) + (2 \text{ mol})(72.7 \text{ J/(K}\cdot \text{mol})) + (1 \text{ mol})(69.9 \text{ J/(K}\cdot \text{mol}))$

$\Delta S^\circ = -128.0$ J/K

$\Delta G^\circ = \Delta H^\circ - T\Delta S^\circ = -275.9$ kJ $- (298 \text{ K})(-128.0 \times 10^{-3} \text{ kJ/K}) = -237.8$ kJ

(b) 1 V $= 1$ J/C

$\Delta G^\circ = -nFE^\circ \quad E^\circ = \dfrac{-\Delta G^\circ}{nF} = \dfrac{-(-237.8 \times 10^3 \text{ J})}{(2 \text{ mol e}^-)\left(\dfrac{96,500 \text{ C}}{1 \text{ mol e}^-}\right)} = 1.232$ J/C $= 1.232$ V

$E^\circ = \dfrac{0.0592 \text{ V}}{n}\log K; \quad \log K = \dfrac{nE^\circ}{0.0592 \text{ V}} = \dfrac{(2)(1.232 \text{ V})}{0.0592 \text{ V}} = 41.62$

$K = 10^{41.62} = 4 \times 10^{41}$

(c) $E = E° - \dfrac{0.0592\,V}{n} \log \dfrac{[H^+]^2}{[Ag^+]^2}$

The addition of NH_3 to the cathode compartment would result in the formation of the $Ag(NH_3)_2^+$ complex ion which results in a decrease in Ag^+ concentration. The log term in the Nernst equation becomes larger and the cell voltage decreases.

On mixing equal volumes of two solutions, the concentrations of both solutions are cut in half.

	Ag^+(aq)	+	$2\,NH_3$(aq)	⇌	$Ag(NH_3)_2^+$(aq)
before reaction (M)	0.0500		2.00		0
assume 100% reaction	−0.0500		−2(0.0500)		+0.0500
after reaction (M)	0		1.90		0.0500
assume small back rxn	+x		+2x		−x
equil (M)	x		1.90 + 2x		0.0500 − x

$K_f = 1.7 \times 10^7 = \dfrac{[Ag(NH_3)_2^+]}{[Ag^+][NH_3]^2} = \dfrac{(0.0500 - x)}{(x)(1.90 + 2x)^2} \approx \dfrac{0.0500}{(x)(1.90)^2}$

Solve for x. $x = [Ag^+] = 8.15 \times 10^{-10}$ M

$E = E° - \dfrac{0.0592\,V}{n} \log \dfrac{[H^+]^2}{[Ag^+]^2} = 1.232\,V - \dfrac{0.0592\,V}{2} \log \dfrac{(1.00\,M)^2}{(8.15 \times 10^{-10}\,M)^2} = 0.694\,V$

(d) Calculate new initial concentrations because of dilution to 110.0 mL.

$M_i \times V_i = M_f \times V_f; \quad M_f = [Cl^-] = \dfrac{M_i \times V_i}{V_f} = \dfrac{0.200\,M \times 10.0\,mL}{110.0\,mL} = 0.0182\,M$

$M_i \times V_i = M_f \times V_f; \quad M_f = [Ag^+] = \dfrac{M_i \times V_i}{V_f} = \dfrac{0.0500\,M \times 100.0\,mL}{110.0\,mL} = 0.0455\,M$

$M_i \times V_i = M_f \times V_f; \quad M_f = [NH_3] = \dfrac{M_i \times V_i}{V_f} = \dfrac{2.00\,M \times 100.0\,mL}{110.0\,mL} = 1.82\,M$

Now calculate the $[Ag^+]$ as a result of the following equilibrium:

	Ag^+(aq)	+	$2\,NH_3$(aq)	⇌	$Ag(NH_3)_2^+$(aq)
before reaction (M)	0.0455		1.82		0
assume 100% reaction	−0.0455		−2(0.0455)		+0.0455
after reaction (M)	0		1.73		0.0455
assume small back rxn	+x		+2x		−x
equil (M)	x		1.73 + 2x		0.0455 − x

$K_f = 1.7 \times 10^7 = \dfrac{[Ag(NH_3)_2^+]}{[Ag^+][NH_3]^2} = \dfrac{(0.0455 - x)}{(x)(1.73 + 2x)^2} \approx \dfrac{0.0455}{(x)(1.73)^2}$

Solve for x. $x = [Ag^+] = 8.94 \times 10^{-10}$ M

For AgCl, $K_{sp} = 1.8 \times 10^{-10}$

IP $= [Ag^+][Cl^-] = (8.94 \times 10^{-10}\,M)(0.0182\,M) = 1.6 \times 10^{-11}$

IP $< K_{sp}$, AgCl will not precipitate.

Now calculate new initial concentrations because of dilution to 120.0 mL.

$$M_i \times V_i = M_f \times V_f; \quad M_f = [Br^-] = \frac{M_i \times V_i}{V_f} = \frac{0.200 \text{ M} \times 10.0 \text{ mL}}{120.0 \text{ mL}} = 0.0167 \text{ M}$$

$$M_i \times V_i = M_f \times V_f; \quad M_f = [Ag^+] = \frac{M_i \times V_i}{V_f} = \frac{0.0500 \text{ M} \times 100.0 \text{ mL}}{120.0 \text{ mL}} = 0.0417 \text{ M}$$

$$M_i \times V_i = M_f \times V_f; \quad M_f = [NH_3] = \frac{M_i \times V_i}{V_f} = \frac{2.00 \text{ M} \times 100.0 \text{ mL}}{120.0 \text{ mL}} = 1.67 \text{ M}$$

Now calculate the $[Ag^+]$ as a result of the following equilibrium:

	$Ag^+(aq)$	$+$	$2 NH_3(aq)$	$\rightleftharpoons$	$Ag(NH_3)_2^+(aq)$
before reaction (M)	0.0417		1.67		0
assume 100% reaction	–0.0417		–2(0.0417)		+0.0417
after reaction (M)	0		1.59		0.0417
assume small back rxn	+x		+2x		–x
equil (M)	x		1.59 + 2x		0.0417 – x

$$K_f = 1.7 \times 10^7 = \frac{[Ag(NH_3)_2^+]}{[Ag^+][NH_3]^2} = \frac{(0.0417 - x)}{(x)(1.59 + 2x)^2} \approx \frac{0.0417}{(x)(1.59)^2}$$

Solve for x. $x = [Ag^+] = 9.70 \times 10^{-10}$ M
For AgBr, $K_{sp} = 5.4 \times 10^{-13}$
IP $= [Ag^+][Br^-] = (9.70 \times 10^{-10} \text{ M})(0.0167 \text{ M}) = 1.6 \times 10^{-11}$
IP $>$ K_{sp}, AgBr will precipitate.

17.131 (a) anode: $Fe(s) + 2 OH^-(aq) \rightarrow Fe(OH)_2(s) + 2 e^-$
cathode: $2 \times [NiO(OH)(s) + H_2O(l) + e^- \rightarrow Ni(OH)_2(s) + OH^-(aq)]$
overall: $Fe(s) + 2 NiO(OH)(s) + 2 H_2O(l) \rightarrow Fe(OH)_2(s) + 2 Ni(OH)_2(s)$

(b)

$$\Delta G^\circ = -nFE^\circ = -(2 \text{ mol } e^-)\left(\frac{96{,}500 \text{ C}}{1 \text{ mol } e^-}\right)(1.37 \text{ V})\left(\frac{1 \text{ J}}{1 \text{ C} \cdot \text{V}}\right) = -264{,}410 \text{ J} = -264 \text{ kJ}$$

$$E^\circ = \frac{0.0592 \text{ V}}{n} \log K; \quad \log K = \frac{nE^\circ}{0.0592 \text{ V}} = \frac{(2)(1.37 \text{ V})}{0.0592 \text{ V}} = 46.3$$

$K = 10^{46.3} = 2 \times 10^{46}$

(c) It would still be 1.37 V because OH^- does not appear in the overall cell reaction. The overall cell reaction contains only solids and one liquid, therefore the cell voltage does not change because there are no concentration changes.

(d) $Fe(OH)_2$, 89.86 amu; 1 A = 1 C/s

$$\text{mol } e^- = (0.250 \text{ C/s})(40.0 \text{ min})\left(\frac{60 \text{ s}}{1 \text{ min}}\right)\left(\frac{1 \text{ mol } e^-}{96{,}500 \text{ C}}\right) = 6.22 \times 10^{-3} \text{ mol } e^-$$

$$\text{mass } Fe(OH)_2 = (6.22 \times 10^{-3} \text{ mol } e^-) \times \frac{1 \text{ mol } Fe(OH)_2}{2 \text{ mol } e^-} \times \frac{89.86 \text{ g } Fe(OH)_2}{1 \text{ mol } Fe(OH)_2} = 0.279 \text{ g}$$

H$_2$O molecules consumed =

$$(6.22 \times 10^{-3} \text{ mol e}^-) \times \frac{2 \text{ mol H}_2\text{O}}{2 \text{ mol e}^-} \times \frac{6.022 \times 10^{23} \text{ H}_2\text{O molecules}}{1 \text{ mol H}_2\text{O}} = 3.75 \times 10^{21} \text{ H}_2\text{O molecules}$$

17.132 (a) Oxidation half reaction: 2 [C$_4$H$_{10}$(g) + 13 O^{2-}(s) → 4 CO$_2$(g) + 5 H$_2$O(l) + 26 e$^-$]

Reduction half reaction: 13 [O$_2$(g) + 4 e$^-$ → 2 O^{2-}(s)]

Cell reaction: 2 C$_4$H$_{10}$(g) + 13 O$_2$(g) → 8 CO$_2$(g) + 10 H$_2$O(l)

(b) ΔH° = [8 ΔH°$_f$(CO$_2$) + 10 ΔH°$_f$(H$_2$O)] − [2 ΔH°$_f$(C$_4$H$_{10}$)]

ΔH° = [(8 mol)(−393.5 kJ/mol) + (10 mol)(−285.8 kJ/mol)]

$$- [(2 \text{ mol})(-126 \text{ kJ/mol})] = -5754 \text{ kJ}$$

ΔS° = [8 S°(CO$_2$) + 10 S°(H$_2$O)] − [2 S°(C$_4$H$_{10}$) + 13 S°(O$_2$)]

ΔS° = [(8 mol)(213.6 J/(K · mol)) + (10 mol)(69.9 J/(K · mol))]

$$- [(2 \text{ mol})(310 \text{ J/(K} \cdot \text{mol)}) + (13 \text{ mol})(205 \text{ J/(K} \cdot \text{mol)})] = -877.2 \text{ J/K}$$

ΔG° = ΔH° − TΔS° = −5754 kJ − (298 K)(−877.2 × 10^{-3} kJ/K) = −5493 kJ

1 V = 1 J/C

$$\Delta G° = -nFE°; \quad E° = -\frac{\Delta G°}{nF} = -\frac{-5493 \times 10^3 \text{ J}}{(52)(96{,}500 \text{ C})} = 1.09 \text{ J/C} = 1.09 \text{ V}$$

ΔG° = −RT ln K

$$\ln K = \frac{-\Delta G°}{RT} = \frac{-(-5493 \text{ kJ})}{(8.314 \times 10^{-3} \text{ kJ/K})(298 \text{ K})} = 2217$$

K = e^{2217} = 7 × 10^{962}

On raising the temperature, both K and E° will decrease because the reaction is exothermic (ΔH° < 0).

(c) C$_4$H$_{10}$, 58.12 amu; 10.5 A = 10.5 C/s

$$\text{mass C}_4\text{H}_{10} = 10.5 \text{ C/s} \times 8 \text{ hr} \times \frac{60 \text{ min}}{1 \text{ hr}} \times \frac{60 \text{ s}}{1 \text{ min}} \times \frac{1 \text{ mol e}^-}{96{,}500 \text{ C}} \times \frac{2 \text{ mol C}_4\text{H}_{10}}{52 \text{ mol e}^-} \times$$

$$\frac{58.12 \text{ g C}_4\text{H}_{10}}{1 \text{ mol C}_4\text{H}_{10}} = 7.00 \text{ g C}_4\text{H}_{10}$$

$$n = 7.00 \text{ g C}_4\text{H}_{10} \times \frac{1 \text{ mol C}_4\text{H}_{10}}{58.12 \text{ g C}_4\text{H}_{10}} = 0.120 \text{ mol C}_4\text{H}_{10}$$

20 °C = 20 + 273 = 293 K

$$PV = nRT \quad V = \frac{nRT}{P} = \frac{(0.120 \text{ mol})\left(0.082\ 06 \dfrac{\text{L} \cdot \text{atm}}{\text{K} \cdot \text{mol}}\right)(293 \text{ K})}{\left(815 \text{ mm Hg} \times \dfrac{1.00 \text{ atm}}{760 \text{ mm Hg}}\right)} = 2.69 \text{ L}$$

17.133 (a) cathode:

(1) MnO$_2$(s) + 4 H$^+$(aq) + 2 e$^-$ → Mn^{2+}(aq) + 2 H$_2$O(l)　　　　E° = +1.22 V

(2) Mn(OH)$_2$(s) + OH$^-$(aq) → MnO(OH)(s) + H$_2$O(l) + e$^-$　　　　E° = +0.380 V

(3) Mn^{2+}(aq) + 2 OH$^-$(aq) → Mn(OH)$_2$(s)　　　K = 1/K$_{sp}$ = 1/(2.1 × 10^{-13}) = 4.8 × 10^{12}

(4) $\underline{4 \times [H_2O(l) \rightarrow H^+(aq) + OH^-(aq)]}$ $K = (K_w)^4 = (1.0 \times 10^{-14})^4 = 1.0 \times 10^{-56}$
     $MnO_2(s) + H_2O(l) + e^- \rightarrow MnO(OH)(s) + OH^-(aq)$

$\Delta G^o_1 = -nFE^o = -(2 \text{ mol } e^-)\left(\dfrac{96,500 \text{ C}}{1 \text{ mol } e^-}\right)(1.22 \text{ V})\left(\dfrac{1 \text{ J}}{1 \text{ C} \cdot \text{V}}\right) = -235,460 \text{ J} = -235.5 \text{ kJ}$

$\Delta G^o_2 = -nFE^o = -(1 \text{ mol } e^-)\left(\dfrac{96,500 \text{ C}}{1 \text{ mol } e^-}\right)(0.380 \text{ V})\left(\dfrac{1 \text{ J}}{1 \text{ C} \cdot \text{V}}\right) = -36,670 \text{ J} = -36.7 \text{ kJ}$

$\Delta G^o_3 = -RT \ln K = -(8.314 \times 10^{-3} \text{ kJ/K})(298 \text{ K}) \ln (4.8 \times 10^{12}) = -72.3 \text{ kJ}$

$\Delta G^o_4 = -RT \ln K = -(8.314 \times 10^{-3} \text{ kJ/K})(298 \text{ K}) \ln (1.0 \times 10^{-56}) = +319.5 \text{ kJ}$

$\Delta G^o(\text{total}) = -235.5 \text{ kJ} - 36.7 \text{ kJ} - 72.3 \text{ kJ} + 319.5 \text{ kJ} = -25.0 \text{ kJ} = -25,000 \text{ J}$

$1 \text{ V} = 1 \text{ J/C}$

$\Delta G^o = -nFE^o; \ E^o = \dfrac{-\Delta G^o}{nF} = \dfrac{-(-25,000 \text{ J})}{(1 \text{ mol } e^-)\left(\dfrac{96,500 \text{ C}}{1 \text{ mol } e^-}\right)} = +0.259 \text{ J/C} = +0.259 \text{ V}$

$E^o_{\text{cathode}} = +0.259 \text{ V}$

anode:
(1) $Zn(s) \rightarrow Zn^{2+}(aq) + 2 e^-$ $E^o = +0.76 \text{ V}$
(2) $\underline{Zn^{2+}(aq) + 2 OH^-(aq) \rightarrow Zn(OH)_2(s)}$ $K = 1/K_{sp} = 1/(4.1 \times 10^{-17}) = 2.4 \times 10^{16}$
     $Zn(s) + 2 OH^-(aq) \rightarrow Zn(OH)_2(s) + 2 e^-$

$\Delta G^o_1 = -nFE^o = -(2 \text{ mol } e^-)\left(\dfrac{96,500 \text{ C}}{1 \text{ mol } e^-}\right)(0.76 \text{ V})\left(\dfrac{1 \text{ J}}{1 \text{ C} \cdot \text{V}}\right) = -146,680 \text{ J} = -146.7 \text{ kJ}$

$\Delta G^o_2 = -RT \ln K = -(8.314 \times 10^{-3} \text{ kJ/K})(298 \text{ K}) \ln (2.4 \times 10^{16}) = -93.4 \text{ kJ}$

$\Delta G^o(\text{total}) = -146.7 \text{ kJ} - 93.4 \text{ kJ} = -240.1 \text{ kJ} = -240,100 \text{ J}$

$1 \text{ V} = 1 \text{ J/C}$

$\Delta G^o = -nFE^o; \ E^o = \dfrac{-\Delta G^o}{nF} = \dfrac{-(-240,100 \text{ J})}{(2 \text{ mol } e^-)\left(\dfrac{96,500 \text{ C}}{1 \text{ mol } e^-}\right)} = +1.24 \text{ J/C} = +1.24 \text{ V}$

$E^o_{\text{anode}} = +1.24 \text{ V}$
$E^o_{\text{cell}} = E^o_{\text{cathode}} + E^o_{\text{anode}} = 0.259 \text{ V} + 1.24 \text{ V} = 1.50 \text{ V}$

(b) $FeO_4^{2-}(aq) \rightarrow Fe(OH)_3(s)$
$FeO_4^{2-}(aq) \rightarrow Fe(OH)_3(s) + H_2O(l)$
$FeO_4^{2-}(aq) + 5 H^+(aq) \rightarrow Fe(OH)_3(s) + H_2O(l)$
$FeO_4^{2-}(aq) + 5 H^+(aq) + 3 e^- \rightarrow Fe(OH)_3(s) + H_2O(l)$
$FeO_4^{2-}(aq) + 5 H^+(aq) + 5 OH^-(aq) + 3 e^- \rightarrow Fe(OH)_3(s) + H_2O(l) + 5 OH^-(aq)$
$FeO_4^{2-}(aq) + 5 H_2O(l) + 3 e^- \rightarrow Fe(OH)_3(s) + H_2O(l) + 5 OH^-(aq)$
$FeO_4^{2-}(aq) + 4 H_2O(l) + 3 e^- \rightarrow Fe(OH)_3(s) + 5 OH^-(aq)$

(c) K_2FeO_4, 198.04 amu; MnO_2, 86.94 amu

$$\text{coulombs} = 10.00 \text{ g } K_2FeO_4 \times \frac{1 \text{ mol } K_2FeO_4}{198.04 \text{ g } K_2FeO_4} \times \frac{3 \text{ mol } e^-}{1 \text{ mol } K_2FeO_4} \times \frac{96,500 \text{ C}}{1 \text{ mol } e^-} =$$

1.46×10^4 C from 10.00 g K_2FeO_4

$$\text{coulombs} = 10.00 \text{ g } MnO_2 \times \frac{1 \text{ mol } MnO_2}{86.94 \text{ g } MnO_2} \times \frac{1 \text{ mol } e^-}{1 \text{ mol } MnO_2} \times \frac{96,500 \text{ C}}{1 \text{ mol } e^-} =$$

1.11×10^4 C from 10.00 g MnO_2

17.134 (a) $4 [Au(s) + 2 CN^-(aq) \rightarrow Au(CN)_2^-(aq) + e^-]$ (oxidation half reaction)

$O_2(g) \rightarrow 2 H_2O(l)$
$O_2(g) + 4 H^+(aq) \rightarrow 2 H_2O(l)$
$4 e^- + O_2(g) + 4 H^+(aq) \rightarrow 2 H_2O(l)$ (reduction half reaction)

Combine the two half reactions.
$4 Au(s) + 8 CN^-(aq) + O_2(g) + 4 H^+(aq) \rightarrow 4 Au(CN)_2^-(aq) + 2 H_2O(l)$
$4 Au(s) + 8 CN^-(aq) + O_2(g) + 4 H^+(aq) + 4 OH^-(aq)$
$\qquad\qquad\qquad \rightarrow 4 Au(CN)_2^-(aq) + 2 H_2O(l) + 4 OH^-(aq)$
$4 Au(s) + 8 CN^-(aq) + O_2(g) + 4 H_2O(l)$
$\qquad\qquad\qquad \rightarrow 4 Au(CN)_2^-(aq) + 2 H_2O(l) + 4 OH^-(aq)$
$4 Au(s) + 8 CN^-(aq) + O_2(g) + 2 H_2O(l) \rightarrow 4 Au(CN)_2^-(aq) + 4 OH^-(aq)$
(b) Add the following five reactions together. $\Delta G°$ is calculated below each reaction.
$4 [Au^+(aq) + 2 CN^-(aq) \rightarrow Au(CN)_2^-(aq)]$ $K = (K_f)^4$
$\Delta G° = -RT \ln K = -(8.314 \times 10^{-3} \text{ kJ/K})(298 \text{ K}) \ln (6.2 \times 10^{38})^4 = -885.2$ kJ

$O_2(g) + 4 H^+(aq) + 4 e^- \rightarrow 2 H_2O(l)$ $E° = 1.229$ V

$\Delta G° = -nFE° = -(4 \text{ mol } e^-)\left(\dfrac{96,500 \text{ C}}{1 \text{ mol } e^-}\right)(1.229 \text{ V})\left(\dfrac{1 \text{ J}}{1 \text{ C} \cdot \text{V}}\right) = -474,394 \text{ J} = -474.4$ kJ

$4 [H_2O(l) \rightleftharpoons H^+(aq) + OH^-(aq)]$ $K = (K_w)^4$
$\Delta G° = -RT \ln K = -(8.314 \times 10^{-3} \text{ kJ/K})(298 \text{ K}) \ln (1.0 \times 10^{-14})^4 = +319.5$ kJ

$4 [Au(s) \rightarrow Au^{3+}(aq) + 3 e^-]$ $E° = -1.498$ V

$\Delta G° = -nFE° = -(12 \text{ mol } e^-)\left(\dfrac{96,500 \text{ C}}{1 \text{ mol } e^-}\right)(-1.498 \text{ V})\left(\dfrac{1 \text{ J}}{1 \text{ C} \cdot \text{V}}\right) = +1,734,684 \text{ J} = +1,734.7$ kJ

$4 [Au^{3+}(aq) + 2 e^- \rightarrow Au^+(aq)]$ $E° = 1.401$ V

$\Delta G° = -nFE° = -(8 \text{ mol } e^-)\left(\dfrac{96,500 \text{ C}}{1 \text{ mol } e^-}\right)(1.401 \text{ V})\left(\dfrac{1 \text{ J}}{1 \text{ C} \cdot \text{V}}\right) = -1,081,572 \text{ J} = -1,081.6$ kJ

Overall reaction:
$4 Au(s) + 8 CN^-(aq) + O_2(g) + 2 H_2O(l) \rightarrow 4 Au(CN)_2^-(aq) + 4 OH^-(aq)$
$\Delta G° = -885.2 \text{ kJ} - 474.4 \text{ kJ} + 319.5 \text{ kJ} + 1,734.7 \text{ kJ} - 1,081.6 \text{ kJ} = -387.0$ kJ

17.135 The overall cell reaction is:

$$2\ Fe^{3+}(aq) + 2\ Hg(l) + 2\ Cl^-(aq) \rightarrow 2\ Fe^{2+}(aq) + Hg_2Cl_2(s)$$

The Nernst equation can be applied to separate half reactions.

One half reaction is for the calomel reference electrode.

$$2\ Hg(l) + 2\ Cl^-(aq) \rightarrow Hg_2Cl_2(s) + 2\ e^- \qquad E^\circ = -0.28\ V$$

When $[Cl^-] = 2.9$ M,

$$E_{calomel} = E^\circ - \frac{0.0592\ V}{n}\log\frac{1}{[Cl^-]^2} = -0.28\ V - \frac{0.0592\ V}{2}\log\frac{1}{(2.9)^2} = -0.25\ V$$

Balance the titration redox reaction: $\quad MnO_4^-(aq) + Fe^{2+}(aq) \rightarrow Mn^{2+}(aq) + Fe^{3+}(aq)$

$$[Fe^{2+}(aq) \rightarrow Fe^{3+}(aq) + e^-]\ \text{x}\ 5$$

$$MnO_4^-(aq) \rightarrow Mn^{2+}(aq)$$
$$MnO_4^-(aq) \rightarrow Mn^{2+}(aq) + 4\ H_2O(l)$$
$$MnO_4^-(aq) + 8\ H^+(aq) \rightarrow Mn^{2+}(aq) + 4\ H_2O(l)$$
$$MnO_4^-(aq) + 8\ H^+(aq) + 5\ e^- \rightarrow Mn^{2+}(aq) + 4\ H_2O(l)$$

Combine the two half reactions.

$$MnO_4^-(aq) + 5\ Fe^{2+}(aq) + 8\ H^+(aq) \rightarrow Mn^{2+}(aq) + 5\ Fe^{3+}(aq) + 4\ H_2O(l)$$

initial mol Fe^{2+} = (0.010 mol/L)(0.1000 L) = 0.0010 mol Fe^{2+}

mL MnO_4^- needed to reach endpoint =

$$0.0010\ \text{mol}\ Fe^{2+}\ \text{x}\ \frac{1\ \text{mol}\ MnO_4^-}{5\ \text{mol}\ Fe^{2+}}\ \text{x}\ \frac{1.00\ L}{0.010\ \text{mol}\ MnO_4^-}\ \text{x}\ \frac{1000\ mL}{1.00\ L} = 20.0\ mL$$

(a) initial mol Fe^{2+} = (0.010 mol/L)(0.1000 L) = 0.0010 mol Fe^{2+}

mol MnO_4^- in 5.0 mL = (0.010 mol/L)(0.0050 L) = 0.000 050 mol MnO_4^-

$$MnO_4^-(aq) + 5\ Fe^{2+}(aq) + 8\ H^+(aq) \rightarrow Mn^{2+}(aq) + 5\ Fe^{3+}(aq) + 4\ H_2O(l)$$

	MnO_4^-	Fe^{2+}		Fe^{3+}
before (mol)	0.000 050	0.0010		0
change (mol)	−0.000 050	−5(0.000 050)		(0.000 050)
after (mol)	0	0.000 75		0.000 25

Again, the Nernst equation can be applied to separate half reactions.

The other half reaction is: $\quad Fe^{3+}(aq) + e^- \rightarrow 2\ Fe^{2+}(aq) \qquad E^\circ = +0.77\ V$

E for the half reaction after adding 5.0 mL of MnO_4^- is

$$E_{Fe^{3+}/Fe^{2+}} = E^\circ - \frac{0.0592\ V}{n}\log\frac{[Fe^{2+}]}{[Fe^{3+}]} = 0.77\ V - \frac{0.0592\ V}{1}\log\frac{(0.000\ 75)}{(0.000\ 25)} = 0.74\ V$$

(Note in the Nernst equation above, we are taking a ratio of Fe^{2+} to Fe^{3+} so we can ignore volumes and just use moles instead of molarity.)

$$E_{cell} = E_{Fe^{3+}/Fe^{2+}} + E_{calomel} = 0.74\ V + (-0.25\ V) = 0.49\ V$$

(b) initial mol Fe^{2+} = (0.010 mol/L)(0.1000 L) = 0.0010 mol Fe^{2+}

mol MnO_4^- in 10.0 mL = (0.010 mol/L)(0.0100 L) = 0.000 10 mol MnO_4^-

$$MnO_4^-(aq) + 5\ Fe^{2+}(aq) + 8\ H^+(aq) \rightarrow Mn^{2+}(aq) + 5\ Fe^{3+}(aq) + 4\ H_2O(l)$$

before (mol)	0.000 10	0.0010	0
change (mol)	−0.000 10	−5(0.000 10)	+5(0.000 10)
after (mol)	0	0.000 50	0.000 50

E for the half reaction after adding 10.0 mL of MnO_4^- is

$$E_{Fe^{3+}/Fe^{2+}} = E° - \frac{0.0592\,V}{n} \log \frac{[Fe^{2+}]}{[Fe^{3+}]} = 0.77\,V - \frac{0.0592\,V}{1} \log \frac{(0.000\,50)}{(0.000\,50)} = 0.77\,V$$

$$E_{cell} = E_{Fe^{3+}/Fe^{2+}} + E_{calomel} = 0.77\,V + (-0.25\,V) = 0.52\,V$$

(c) initial mol Fe^{2+} = (0.010 mol/L)(0.1000 L) = 0.0010 mol Fe^{2+}

mol MnO_4^- in 19.0 mL = (0.010 mol/L)(0.0190 L) = 0.000 19 mol MnO_4^-

$$MnO_4^-(aq) + 5\ Fe^{2+}(aq) + 8\ H^+(aq) \rightarrow Mn^{2+}(aq) + 5\ Fe^{3+}(aq) + 4\ H_2O(l)$$

before (mol)	0.000 19	0.0010	0
change (mol)	−0.000 19	−5(0.000 19)	+5(0.000 19)
after (mol)	0	0.000 05	0.000 95

E for the half reaction after adding 19.0 mL of MnO_4^- is

$$E_{Fe^{3+}/Fe^{2+}} = E° - \frac{0.0592\,V}{n} \log \frac{[Fe^{2+}]}{[Fe^{3+}]} = 0.77\,V - \frac{0.0592\,V}{1} \log \frac{(0.000\,05)}{(0.000\,95)} = 0.85\,V$$

$$E_{cell} = E_{Fe^{3+}/Fe^{2+}} + E_{calomel} = 0.85\,V + (-0.25\,V) = 0.60\,V$$

(d) 21.0 mL is past the endpoint so the MnO_4^- is in excess and all of the Fe^{2+} is consumed.

initial mol Fe^{2+} = (0.010 mol/L)(0.1000 L) = 0.0010 mol Fe^{2+}

mol MnO_4^- in 21.0 mL = (0.010 mol/L)(0.0210 L) = 0.000 21 mol MnO_4^-

$$MnO_4^-(aq) + 5\ Fe^{2+}(aq) + 8\ H^+(aq) \rightarrow Mn^{2+}(aq) + 5\ Fe^{3+}(aq) + 4\ H_2O(l)$$

before (mol)	0.000 21	0.0010	0	0
change (mol)	−0.000 20	−5(0.000 20)		+5(0.000 20)
after (mol)	0.000 01	0	0.000 20	0.0010

Because the Fe^{2+} is totally consumed, there is a new half reaction:

$MnO_4^-(aq) + 8\ H^+(aq) + 5\ e^- \rightarrow Mn^{2+}(aq) + 4\ H_2O(l)$ $\qquad E° = 1.51\,V$

The total volume = 100.0 mL + 21.0 mL = 121.0 mL = 0.1210 L

$[MnO_4^-]$ = 0.000 01 mol/0.1210 L = 0.000 083 M

$[Mn^{2+}]$ = 0.000 20 mol/0.1210 L = 0.001 65 M

We need to determine $[H^+]$ in order to determine the half reaction potential.

$[H_2SO_4]_{dil} \cdot 121.0$ mL = $[H_2SO_4]_{conc} \cdot 100.0$ mL

$[H_2SO_4]_{dil}$ = [(1.50 M)(100.0 mL)]/121.0 mL = 1.24 M

We can ignore the small amount of H^+ consumed by the titration itself, because the H_2SO_4 concentration is so large.

Consider the dissociation of H_2SO_4. From the complete dissociation of the first proton, $[H^+] = [HSO_4^-] = 1.24$ M.

For the dissociation of the second proton, the following equilibrium must be considered:

	$HSO_4^-(aq)$	$\rightleftharpoons$	$H^+(aq)$	+	$SO_4^{2-}(aq)$
initial (M)	1.24		1.24		0
change (M)	$-x$		$+x$		$+x$
equil (M)	$1.24 - x$		$1.24 + x$		x

$$K_{a2} = \frac{[H^+][SO_4^{2-}]}{[HSO_4^-]} = 1.2 \times 10^{-2} = \frac{(1.24 + x)(x)}{1.24 - x}$$

$$x^2 + 1.252x - 0.0149 = 0$$

Use the quadratic formula to solve for x.

$$x = \frac{-(1.252) \pm \sqrt{(1.252)^2 - 4(1)(-0.0149)}}{2(1)} = \frac{-1.252 \pm 1.276}{2}$$

$x = -1.264$ and 0.012

Of the two solutions for x, only the positive value of x has physical meaning, since x is the $[SO_4^{2-}]$.

$[H^+] = 1.24 + x = 1.24 + 0.012 = 1.25$ M

$$E_{MnO_4^-/Mn^{2+}} = E° - \frac{0.0592\,V}{n} \log \frac{[Mn^{2+}]}{[MnO_4^-][H^+]^8}$$

$$= 1.51\ V - \frac{0.0592\,V}{5} \log \frac{(0.001\ 65)}{(0.000\ 083)(1.25)^8} = 1.50\ V$$

$$E_{cell} = E_{MnO_4^-/Mn^{2+}} + E_{calomel} = 1.50\ V + (-0.25\ V) = 1.25\ V$$

Notice that there is a dramatic change in the potential at the equivalence point.

Hydrogen, Oxygen, and Water

18.1 $PV = nRT$; $PV = \dfrac{g}{\text{molar mass}} RT$

$$d_{H_2} = \frac{g}{V} = \frac{P(\text{molar mass})}{RT} = \frac{(1.00\ \text{atm})(2.016\ \text{g/mol})}{\left(0.08206\ \dfrac{\text{L}\cdot\text{atm}}{\text{K}\cdot\text{mol}}\right)(298\ \text{K})} = 0.0824\ \text{g/L}$$

$1\ \text{L} = 1000\ \text{mL} = 1000\ \text{cm}^3$
$d_{H_2} = 0.0824\ \text{g}/1000\ \text{cm}^3 = 8.24 \times 10^{-5}\ \text{g/cm}^3$

$$\frac{d_{air}}{d_{H_2}} = \frac{1.185 \times 10^{-3}\ \text{g/cm}^3}{8.24 \times 10^{-5}\ \text{g/cm}^3} = 14.4; \text{ Air is 14 times more dense than } H_2.$$

18.2   For every 100.0 g, there are:
61.4 g O, 22.9 g C, 10.0 g H, 2.6 g N, and 3.1 g other

$$22.9\ \text{g C} \times \frac{1\ \text{mol C}}{12.011\ \text{g C}} = 1.907\ \text{mol C}$$

$$10.0\ \text{g H} \times \frac{1\ \text{mol H}}{1.008\ \text{g H}} = 9.921\ \text{mol H}$$

Assume the sample contains 1.907 mol ^{13}C and 9.921 mol D.

$$\text{mass } ^{13}C = 1.907\ \text{mol } ^{13}C \times \frac{13.0034\ \text{g } ^{13}C}{1\ \text{mol } ^{13}C} = 24.8\ \text{g } ^{13}C$$

$$\text{mass D} = 9.921\ \text{mol D} \times \frac{2.0141\ \text{g D}}{1\ \text{mol D}} = 20.0\ \text{g D}$$

(a) Total mass if all H is D is:
61.4 g O + 22.9 C + 20.0 g D + 2.6 g N + 3.1 g other = 110.0 g

$$\text{mass \% D} = \frac{20.0\ \text{g D}}{110.0\ \text{g}} \times 100\% = 18.2\%\ \text{D}$$

(b) Total mass if all C is ^{13}C is:
61.4 g O + 24.8 g ^{13}C + 10.0 g H + 2.6 g N + 3.1 g other = 101.9 g

$$\text{mass \% } ^{13}C = \frac{24.8\ \text{g } ^{13}C}{101.9\ \text{g}} \times 100\% = 24.3\%\ ^{13}C$$

(c) The isotope effect for H is larger than that for C because D is two times the mass of ^1H while ^{13}C is only about 8% heavier than ^{12}C.

(d) Because the mass % of H in the human body is 10.0%, the mass of H in a 150-pound person is (150 lbs x 0.100) 15.0 pounds. ^2H weighs twice as much as ^1H. If all the ^1H were replaced by ^2H, the mass of ^2H would be 30.0 pounds and the 150-pound person would have gained 15.0 pounds.

18.3   (a) $2\,Ga(s) + 6\,H^+(aq) \rightarrow 3\,H_2(g) + 2\,Ga^{3+}(aq)$

(b) $H_2O(g) + C(s) \xrightarrow{1000\,°C} CO(g) + H_2(g)$

18.4   (a) SiH_4, covalent   (b) KH, ionic   (c) H_2Se, covalent

18.5   (a) $SrH_2(s) + 2\,H_2O(l) \rightarrow 2\,H_2(g) + Sr^{2+}(aq) + 2\,OH^-(aq)$
        (b) $KH(s) + H_2O(l) \rightarrow H_2(g) + K^+(aq) + OH^-(aq)$

18.6 $CaH_2(s) + 2\,H_2O(l) \rightarrow 2\,H_2(g) + Ca^{2+}(aq) + 2\,OH^-(aq)$
        CaH_2, 42.09 amu; 25 °C = 298 K

$$PV = nRT; \qquad n_{H_2} = \frac{PV}{RT} = \frac{(1.00\ \text{atm})(2.0 \times 10^5\ \text{L})}{\left(0.082\ 06\ \dfrac{\text{L}\cdot\text{atm}}{\text{K}\cdot\text{mol}}\right)(298\ \text{K})} = 8.18 \times 10^3\ \text{mol}\ H_2$$

$$8.18 \times 10^3\ \text{mol}\ H_2 \times \frac{1\ \text{mol}\ CaH_2}{2\ \text{mol}\ H_2} \times \frac{42.09\ \text{g}\ CaH_2}{1\ \text{mol}\ CaH_2} \times \frac{1\ \text{kg}}{1000\ \text{g}} = 1.7 \times 10^2\ \text{kg}\ CaH_2$$

18.7   (a) (1) ZrH_x, interstitial   (2) PH_3, covalent   (3) HBr, covalent   (4) LiH, ionic
        (b) (1) and (4) are likely to be solids at 25 °C. (2) and (3) are likely to be gases at 25 °C.
        Covalent hydrides, like (2) and (3), form discrete molecules and have only relatively
        weak intermolecular forces, resulting in gases. (4) is an ionic metal hydride with strong
        ion-ion forces holding the 3-dimensional lattice together in the solid state. (1) is an
        interstitial hydride with the metal atoms in a solid crystal lattice and H's occupying
        holes.
        (c) $LiH(s) + H_2O(l) \rightarrow H_2(g) + Li^+(aq) + OH^-(aq)$

18.8   Assume 12.0 g of Pd with a volume of 1.0 cm^3.
        $V_{H_2} = 935\ cm^3 = 935\ mL = 0.935\ L$

$$PV = nRT; \qquad n_{H_2} = \frac{PV}{RT} = \frac{(1.00\ \text{atm})(0.935\ \text{L})}{\left(0.082\ 06\ \dfrac{\text{L}\cdot\text{atm}}{\text{K}\cdot\text{mol}}\right)(273\ \text{K})} = 0.0417\ \text{mol}\ H_2$$

$n_H = 2\,n_{H_2} = 0.0834\ \text{mol}\ H$

$$12.0\ \text{g Pd} \times \frac{1\ \text{mol Pd}}{106.42\ \text{g Pd}} = 0.113\ \text{mol Pd}$$

$Pd_{0.113}H_{0.0834}$
$Pd_{0.113/0.113}H_{0.0834/0.113}$
$PdH_{0.74}$
g H = (0.0834 mol H)(1.008 g/mol) = 0.0841 g H

$$d_H = 0.0841\ \text{g/cm}^3; \qquad M_H = \frac{0.0834\ \text{mol}}{0.001\ \text{L}} = 83.4\ M$$

18.9 $2\, KMnO_4(s) \rightarrow K_2MnO_4(s) + MnO_2(s) + O_2(g)$
$KMnO_4$, 158.03 amu; 25 °C = 298 K

$$mol\ O_2 = 0.200\ g\ KMnO_4 \times \frac{1\ mol\ KMnO_4}{158.03\ g\ KMnO_4} \times \frac{1\ mol\ O_2}{2\ mol\ KMnO_4} = 6.33 \times 10^{-4}\ mol\ O_2$$

$$PV = nRT;\quad V = \frac{nRT}{P} = \frac{(6.33 \times 10^{-4}\ mol)\left(0.082\ 06\ \dfrac{L \cdot atm}{K \cdot mol}\right)(298\ K)}{1.00\ atm} = 0.0155\ L$$

$$V = 0.0155\ L \times \frac{1000\ mL}{1\ L} = 15.5\ mL\ O_2$$

18.10   A is Li; B is Ga; C is C
(a) Li_2O, Ga_2O_3, CO_2
(b) Li_2O is the most ionic. CO_2 is the most covalent.
(c) CO_2 is the most acidic. Li_2O is the most basic.
(d) Ga_2O_3 is amphoteric and can react with both $H^+(aq)$ and $OH^-(aq)$.

18.11   (a) $Li_2O(s) + H_2O(l) \rightarrow 2\, Li^+(aq) + 2\, OH^-(aq)$
(b) $SO_3(l) + H_2O(l) \rightarrow H^+(aq) + HSO_4^-(aq)$
(c) $Cr_2O_3(s) + 6\, H^+(aq) \rightarrow 2\, Cr^{3+}(aq) + 3\, H_2O(l)$
(d) $Cr_2O_3(s) + 2\, OH^-(aq) + 3\, H_2O(l) \rightarrow 2\, Cr(OH)_4^-(aq)$

18.12   (a) Rb_2O_2 Rb +1, O −1, peroxide   (b) CaO Ca +2, O −2, oxide
(c) CsO_2 Cs +1, O −1/2, superoxide   (d) SrO_2 Sr +2, O −1, peroxide
(e) CO_2 C +4, O −2, oxide

18.13   (a) $Rb_2O_2(s) + H_2O(l) \rightarrow 2\, Rb^+(aq) + HO_2^-(aq) + OH^-(aq)$
(b) $CaO(s) + H_2O(l) \rightarrow Ca^{2+}(aq) + 2\, OH^-(aq)$
(c) $2\, CsO_2(s) + H_2O(l) \rightarrow O_2(g) + 2\, Cs^+(aq) + HO_2^-(aq) + OH^-(aq)$
(d) $SrO_2(s) + H_2O(l) \rightarrow Sr^{2+}(aq) + HO_2^-(aq) + OH^-(aq)$
(e) $CO_2(g) + H_2O(l) \rightleftharpoons H^+(aq) + HCO_3^-(aq)$

18.14 σ^*_{2p} ___

π^*_{2p} ↑↓   ↑

π_{2p} ↑↓   ↑↓

σ_{2p} ↑↓

σ^*_{2s} ↑↓

σ_{2s} ↑↓
        O_2^-

O_2^- is paramagnetic with one unpaired electron.

$$O_2^-\ bond\ order = \frac{\left(\begin{array}{c} number\ of \\ bonding\ electrons \end{array}\right) - \left(\begin{array}{c} number\ of \\ antibonding\ electrons \end{array}\right)}{2} = \frac{8 - 5}{2} = 1.5$$

18.15 H—$\overset{\cdot\cdot}{\underset{\cdot\cdot}{O}}$—$\overset{\cdot\cdot}{\underset{\cdot\cdot}{O}}$—H The electron dot structure indicates a single bond (see text Table 18.2) which is consistent with an O–O bond length of 148 pm.

18.16 $PbS(s) + 4 H_2O_2(aq) \rightarrow PbSO_4(s) + 4 H_2O(l)$

18.17 The white particles are most likely tiny particles of insoluble $CaCO_3$.
$Ca^{2+}(aq) + CO_3^{2-}(aq) \rightarrow CaCO_3(s)$

18.18 (a) $2 Li(s) + 2 H_2O(l) \rightarrow H_2(g) + 2 Li^+(aq) + 2 OH^-(aq)$
(b) $Sr(s) + 2 H_2O(l) \rightarrow H_2(g) + Sr^{2+}(aq) + 2 OH^-(aq)$
(c) $Br_2(l) + H_2O(l) \rightleftharpoons HOBr(aq) + H^+(aq) + Br^-(aq)$

18.19 mass of H_2O = 5.62 g – 3.10 g = 2.52 g H_2O

$$2.52 \text{ g } H_2O \times \frac{1 \text{ mol } H_2O}{18.02 \text{ g } H_2O} = 0.140 \text{ mol } H_2O$$

$$3.10 \text{ g } NiSO_4 \times \frac{1 \text{ mol } NiSO_4}{154.8 \text{ g } NiSO_4} = 0.0200 \text{ mol } NiSO_4$$

$$\text{number of } H_2O\text{'s in hydrate} = \frac{n_{H_2O}}{n_{NiSO_4}} = \frac{0.140 \text{ mol}}{0.0200 \text{ mol}} = 7$$

Hydrate formula is $NiSO_4 \cdot 7 H_2O$

18.20 Hydrogen can be stored as a solid in the form of solid interstitial hydrides or in the recently discovered tube-shaped molecules called carbon nanotubes.

18.21 $H_2(g) + 1/2 O_2(g) \rightarrow H_2O(g)$ $\Delta H° = -242 \text{ kJ}$

$$\text{mol } H_2 = 1.45 \times 10^6 \text{ L} \times \frac{0.088 \text{ kg}}{1 \text{ L}} \times \frac{1000 \text{ g}}{1 \text{ kg}} \times \frac{1 \text{ mol } H_2}{2.016 \text{ g } H_2} = 6.33 \times 10^7 \text{ mol } H_2$$

$$q = 6.33 \times 10^7 \text{ mol } H_2 \times \frac{242 \text{ kJ}}{1 \text{ mol } H_2} = 1.5 \times 10^{10} \text{ kJ}$$

$$\text{mass } O_2 = 6.33 \times 10^7 \text{ mol } H_2 \times \frac{0.5 \text{ mol } O_2}{1 \text{ mol } H_2} \times \frac{32.00 \text{ g } O_2}{1 \text{ mol } O_2} \times \frac{1 \text{ kg}}{1000 \text{ g}} = 1.0 \times 10^6 \text{ kg } O_2$$

Key Concept Problems

18.22 (a) (1) covalent   (2) ionic   (3) covalent   (4) interstitial
(b) (1) H, +1; other element, –3
(2) H, –1; other element, +1
(3) H, +1; other element, –2

18.23 (a) A, NaH; B, PdH_x; C, H_2S; D, HI
(b) NaH (ionic); PdH_x (interstitial); H_2S and HI (covalent)

(c) H_2S and HI (molecular); NaH and PdH_x (3-dimensional crystal)

(d) NaH: Na +1, H −1

H_2S: S −2, H +1

HI: I −1, H +1

18.24 (a) Because of the unpaired electron, the compound is a superoxide. The chemical formula is KO_2.

(b) Because of the unpaired electron, the compound is paramagnetic and attracted by a magnetic field.

(c) O_2 has a bond order of 2; O_2^- has a bond order of 1.5. The O−O bond length in O_2^- is longer and the bond energy is smaller than in O_2.

(d) The solution is basic because of the following reaction that produces OH^-:

$$2\ KO_2(s)\ +\ H_2O(l)\ \rightarrow\ O_2(g)\ +\ 2\ K^+(aq)\ +\ HO_2^-(aq)\ +\ OH^-(aq)$$

18.25 (a) (1) −2, +4; (2) −2, +6; (3) −2, +2

(b) (1) covalent; (2) covalent; (3) ionic

(c) (1) acidic; (2) acidic; (3) basic

(d) (1) carbon; (2) sulfur

18.26 (a) (1) −2, +2; (2) −2, +1; (3) −2, +5

(b) (1) three-dimensional; (2) molecular; (3) molecular

(c) (1) solid; (2) gas or liquid; (3) gas or liquid

(d) (2) hydrogen; (3) nitrogen

18.27 (a) A, CaO; B, Al_2O_3; C, SO_3; D, SeO_3

(b) CaO (basic); Al_2O_3 (amphoteric); SO_3 and SeO_3 (acidic)

(c) CaO (most ionic); SO_3 (most covalent)

(d) CaO and Al_2O_3 (3-dimensional crystal); SO_3 and SeO_3 (molecular)

(e) CaO (highest melting point); SO_3 (lowest melting point)

18.28 (a) The ionic hydride (4) has the highest melting point.

(b) (1), (2), and (3) are covalent hydrides. (1) and (2) can hydrogen bond, (3) cannot. Consequently, (3) has the lowest boiling point.

(c) (1), water, and (4), the ionic hydride react together to form $H_2(g)$.

18.29 (1) is CO_2. The molecule is linear because C has two charge clouds. There are two C=O double bonds.

(2) is SO_2. The molecule is bent because S has three charge clouds. There is one S–O single bond and one S=O double bond. SO_2 has two resonance structures so each S–O bond appears to be a bond and a half.

CO_2 has the stronger bonds.

18.30 (a) A, KH; B, MgH$_2$; C, H$_2$O; D, HCl
(b) HCl
(c) KH(s) + H$_2$O(l) → H$_2$(g) + K$^+$(aq) + OH$^-$(aq)
MgH$_2$(s) + 2 H$_2$O(l) → 2 H$_2$(g) + Mg^{2+}(aq) + 2 OH$^-$(aq)
(d) HCl reacts with water to give an acidic solution. KH and MgH$_2$ react with water to give a basic solution.

18.31 A is Rb and B is Ba
(a) oxides: Rb$_2$O and BaO; peroxides: Rb$_2$O$_2$ and BaO$_2$; superoxides: RbO$_2$ and Ba(O$_2$)$_2$
(b) The superoxides RbO$_2$ and Ba(O$_2$)$_2$ are paramagnetic.
(c) BaO(s) + H$_2$O(l) → Ba^{2+}(aq) + 2 OH$^-$(aq)
BaO$_2$(s) + H$_2$O(l) → Ba^{2+}(aq) + HO$_2^-$(aq) + OH$^-$(aq)
Ba(O$_2$)$_2$(s) + H$_2$O(l) → O$_2$(g) + Ba^{2+}(aq) + HO$_2^-$(aq) + OH$^-$(aq)

Section Problems
Chemistry of Hydrogen (Sections 18.1–18.5)

18.32 Quantitative differences in properties that arise from the differences in the masses of the isotopes are known as isotope effects.
Examples: H$_2$ and D$_2$ have different melting and boiling points.
H$_2$O and D$_2$O have different dissociation constants.

18.33 Deuterium can be separated from protium by passing an electric current through a solution of an inert electrolyte in ordinary water. H$_2$O is preferentially electrolyzed. Over time, the solution becomes enriched in D$_2$O.

18.34 $\frac{\text{mass }^2\text{D} - \text{mass }^1\text{H}}{\text{mass }^1\text{H}} \times 100\% = \frac{2.0141 \text{ amu} - 1.0078 \text{ amu}}{1.0078 \text{ amu}} \times 100\% = 99.85\%$

$\frac{\text{mass }^3\text{H} - \text{mass }^2\text{H}}{\text{mass }^2\text{H}} \times 100\% = \frac{3.0160 \text{ amu} - 2.0141 \text{ amu}}{2.0141 \text{ amu}} \times 100\% = 49.74\%$

The differences in properties will be larger for H$_2$O and D$_2$O rather than for D$_2$O and T$_2$O because of the larger relative difference in mass for H and D versus D and T. This is supported by the data in Table 18.1.

18.35 (a) ocean volume = (1.35 x 10^{18} m^3)(100 cm/m)^3 = 1.35 x 10^{24} cm^3
mass of H$_2$O = (1.35 x 10^{24} cm^3)(1.00 g/cm^3) = 1.35 x 10^{24} g

1.35 x 10^{24} g H$_2$O x $\frac{1 \text{ mol H}_2\text{O}}{18.02 \text{ g H}_2\text{O}}$ = 7.49 x 10^{22} mol H$_2$O

mol H = 7.49 x 10^{22} mol H$_2$O x $\frac{2 \text{ mol H}}{1 \text{ mol H}_2\text{O}}$ = 1.50 x 10^{23} mol H

mol D = (1.50 x 10^{23} mol H)(0.000 156 atom fraction D) = 2.34 x 10^{19} mol D
mass of D = (2.34 x 10^{19} mol D)(2.0141 g/mol)(1 kg/1000 g) = 4.71 x 10^{16} kg D
(b) mol T = (1.50 x 10^{23} mol H)(0.01 x 10^{-16} atom fraction T) = 1.50 x 10^5 mol T
mass of T = (1.50 x 10^5 mol T)(3.0160 g/mol)(1 kg/1000 g) = 5 x 10^2 kg T

18.36   There are 18 kinds of H_2O.

$H_2{}^{16}O$	$H_2{}^{17}O$	$H_2{}^{18}O$
$D_2{}^{16}O$	$D_2{}^{17}O$	$D_2{}^{18}O$
$T_2{}^{16}O$	$T_2{}^{17}O$	$T_2{}^{18}O$
$HD^{16}O$	$HD^{17}O$	$HD^{18}O$
$HT^{16}O$	$HT^{17}O$	$HT^{18}O$
$DT^{16}O$	$DT^{17}O$	$DT^{18}O$

18.37   There are 10 kinds of PH_3.

PH_3	PH_2D	PH_2T
PD_3	PD_2H	PD_2T
PT_3	PT_2H	PT_2D	PHDT

18.38   (a) $Zn(s) + 2 H^+(aq) \rightarrow H_2(g) + Zn^{2+}(aq)$
(b) at 1000 °C, $H_2O(g) + C(s) \rightarrow CO(g) + H_2(g)$
(c) at 1100 °C with a Ni catalyst, $H_2O(g) + CH_4(g) \rightarrow CO(g) + 3 H_2(g)$
(d) There are a number of possibilities. (b) and (c) above are two; electrolysis is another: $2 H_2O(l) \rightarrow 2 H_2(g) + O_2(g)$

18.39   (a) $Fe(s) + 2 H^+(aq) \rightarrow H_2(g) + Fe^{2+}(aq)$
(b) $Ca(s) + 2 H_2O(l) \rightarrow H_2(g) + Ca^{2+}(aq) + 2 OH^-(aq)$
(c) $2 Al(s) + 6 H^+(aq) \rightarrow 3 H_2(g) + 2 Al^{3+}(aq)$
(d) $C_2H_6(aq) + 2 H_2O(l) \xrightarrow[\text{Catalyst}]{\text{Heat}} 2 CO(g) + 5 H_2(g)$

18.40   The steam-hydrocarbon reforming process is the most important industirial preparation of hydrogen.

$$CH_4(g) + H_2O(g) \xrightarrow[\text{Ni catalyst}]{1100\,°C} CO(g) + 3 H_2(g)$$

$$CO(g) + H_2O(g) \xrightarrow{400\,°C} CO_2(g) + H_2(g)$$
$$CO_2(g) + 2 OH^-(aq) \rightarrow CO_3{}^{2-}(aq) + H_2O(l)$$

18.41   (a) $3 Fe(s) + 4 H_2O(g) \rightarrow Fe_3O_4(s) + 4 H_2(g)$
(b) $C_3H_8(g) + 3 H_2O(g) \rightarrow 3 CO(g) + 7 H_2(g)$
(c) $CO(g) + H_2O(g) \rightarrow CO_2(g) + H_2(g)$

18.42   (a) LiH, 7.95 amu; CaH_2, 42.09 amu
$LiH(s) + H_2O(l) \rightarrow H_2(g) + Li^+(aq) + OH^-(aq)$
$CaH_2(s) + 2 H_2O(l) \rightarrow 2 H_2(g) + Ca^{2+}(aq) + 2 OH^-(aq)$

You obtain $\dfrac{1 \text{ mol } H_2}{7.95 \text{ g LiH}} = 0.126$ mol H_2/g LiH and $\dfrac{2 \text{ mol } H_2}{42.09 \text{ g } CaH_2} = 0.0475 \text{ mol } H_2/\text{g } CaH_2;$ Therefore, LiH gives more H_2.

(b) $25 \text{ °C} = 298 \text{ K}$

$PV = nRT;$ $n_{H_2} = \dfrac{PV}{RT} = \dfrac{(150 \text{ atm})(100 \text{ L})}{\left(0.082\ 06 \dfrac{L \cdot atm}{K \cdot mol}\right)(298 K)} = 613.4 \text{ mol } H_2$

mass $CaH_2 = 613.4 \text{ mol } H_2 \times \dfrac{1 \text{ mol } CaH_2}{2 \text{ mol } H_2} \times \dfrac{42.09 \text{ g } CaH_2}{1 \text{ mol } CaH_2} \times \dfrac{1 \text{ kg}}{1000 \text{ g}} = 12.9 \text{ kg } CaH_2$

18.43 $PV = nRT$

$n_{H_2} = \dfrac{PV}{RT} = \dfrac{\left(740 \text{ mm Hg} \times \dfrac{1 \text{ atm}}{760 \text{ mm Hg}}\right)(1.99 \times 10^8 \text{ L})}{\left(0.082\ 06 \dfrac{L \cdot atm}{K \cdot mol}\right)(293 \text{ K})} = 8.059 \times 10^6 \text{ mol } H_2$

$n_C = n_{H_2} = 8.059 \times 10^6 \text{ mol C}$

mass $C = (8.059 \times 10^6 \text{ mol C})(12.011 \text{ g C/mol})(1 \text{ kg}/1000 \text{ g}) = 9.68 \times 10^4 \text{ kg C}$

18.44   (a) MgH_2, H^- (b) PH_3, covalent   (c) KH, H^- (d) HBr, covalent

18.45   (a) H_2Se, covalent   (b) RbH, H^- (c) CaH_2, H^- (d) GeH_4, covalent

18.46 H_2S covalent hydride, gas, weak acid in H_2O
NaH   ionic hydride, solid (salt like), reacts with H_2O to produce H_2
PdH_x metallic (interstitial) hydride, solid, stores hydrogen

18.47 $TiH_{1.7}$ – an interstitial metal hydride, nonstoichiometric, solid, probably high melting
HCl – a covalent hydride, molecular, gas at 25 °C
CaH_2 – an ionic hydride, saltlike, white solid, high melting

18.48   (a) CH_4, covalent bonding   (b) NaH, ionic bonding

18.49   (a) CaH_2, ionic bonding   (b) NH_3, covalent bonding

18.50   (a) H—S̈e—H , bent   (b) H—Äs—H, trigonal pyramidal
                                              |
                                              H

(c)   H   , tetrahedral
            |
       H—Si—H
            |
            H

18.51 (a)

$$H-Ge-H$$

, tetrahedral

(b) $H-\overset{\cdot\cdot}{\underset{\cdot\cdot}{S}}-H$, bent

(c) $H-\overset{\cdot\cdot}{N}-H$, trigonal pyramidal
with H below N

18.52 A nonstoichiometric compound is a compound whose atomic composition cannot be expressed as a ratio of small whole numbers. An example is PdH_x. The lack of stoichiometry results from the hydrogen occupying holes in the solid state structure.

18.53 Hydrogen atoms in interstitial hydrides are mobile because they occupy some of the holes between the larger metal atoms and can jump from the occupied holes to nearby unoccupied holes.

18.54 (a) TiH_2, 49.90 amu; Assume 1.0 cm^3 of TiH_2 which has a mass of 3.75 g.

$$3.75 \text{ g TiH}_2 \times \frac{1 \text{ mol TiH}_2}{49.90 \text{ g TiH}_2} = 0.0752 \text{ mol TiH}_2$$

$$0.0752 \text{ mol TiH}_2 \times \frac{2 \text{ mol H}}{1 \text{ mol TiH}_2} = 0.150 \text{ mol H}$$

$$0.150 \text{ mol H} \times \frac{1.008 \text{ g H}}{1 \text{ mol H}} = 0.151 \text{ g H}$$

$d_H = 0.15$ g/cm^3; the density of H in TiH_2 is about 2.1 times the density of liquid H_2.

(b)

$$PV = nRT; \quad V = \frac{nRT}{P} = \frac{\left(0.15 \text{ g} \times \frac{1 \text{ mol}}{2.016 \text{ g}}\right)\left(0.082\ 06 \frac{\text{L} \cdot \text{atm}}{\text{K} \cdot \text{mol}}\right)(273 \text{ K})}{1.00 \text{ atm}} = 1.7 \text{ L H}_2$$

$$1.7 \text{ L} = 1.7 \times 10^3 \text{ mL} = 1.7 \times 10^3 \text{ cm}^3$$

18.55 LiH, 7.949 amu
Assume 1.0 cm^3 of LiH which has a mass of 0.78 g.

$$0.78 \text{ g LiH} \times \frac{1 \text{ mol LiH}}{7.949 \text{ g LiH}} = 0.0981 \text{ mol LiH}$$

$$0.0981 \text{ mol LiH} \times \frac{1 \text{ mol H}}{1 \text{ mol LiH}} = 0.0981 \text{ mol H}$$

$$0.0981 \text{ mol H} \times \frac{1.008 \text{ g H}}{1 \text{ mol H}} = 0.0989 \text{ g H}; \quad d_H = 0.0989 \text{ g/cm}^3$$

$$(5.0 \text{ cm}^3) \times (0.0989 \text{ g H/cm}^3) \times \frac{1 \text{ mol H}}{1.008 \text{ g H}} \times \frac{6.022 \times 10^{23} \text{ H atoms}}{1 \text{ mol H}} = 3.0 \times 10^{23} \text{ H atoms}$$

$$(5.0 \text{ cm}^3) \times (0.0989 \text{ g H/cm}^3) \times \frac{1 \text{ mol H}}{1.008 \text{ g H}} \times \frac{1 \text{ mol H}_2}{2 \text{ mol H}} = 0.245 \text{ mol H}_2$$

$$PV = nRT; \quad V = \frac{nRT}{P} = \frac{(0.245 \text{ mol})\left(0.082 \ 06 \ \dfrac{L \cdot atm}{K \cdot mol}\right)(293 \text{ K})}{\left(740 \text{ mm Hg} \times \dfrac{1.00 \text{ atm}}{760 \text{ mm Hg}}\right)} = 6.0 \text{ L H}_2$$

Chemistry of Oxygen (Sections 18.6–18.12)

18.56 (a) O_2 is obtained in industry by the fractional distillation of liquid air.
(b) In the laboratory, O_2 is prepared by the thermal decomposition of $KClO_3(s)$.

$$2 \text{ KClO}_3(s) \ \overset{\text{heat}}{\underset{\text{MnO}_2}{\rightarrow}} \ 2 \text{ KCl}(s) + 3 \text{ O}_2(g)$$

18.57 In nature, oxygen is found as $O_2(g)$ in the atmosphere, in the form of H_2O in the hydrosphere, and combined with other elements in the form of silicates, carbonates, oxides, and other oxygen containing minerals in the lithosphere.

18.58 $2 \text{ H}_2\text{O}_2(aq) \ \overset{\text{catalyst}}{\rightarrow} \ 2 \text{ H}_2\text{O}(l) + \text{O}_2(g)$
H_2O_2, 34.01 amu; 25 °C = 298 K

$$\text{mol O}_2 = 20.4 \text{ g H}_2\text{O}_2 \times \frac{1 \text{ mol H}_2\text{O}_2}{34.01 \text{ g H}_2\text{O}_2} \times \frac{1 \text{ mol O}_2}{2 \text{ mol H}_2\text{O}_2} = 0.300 \text{ mol O}_2$$

$$PV = nRT; \quad V = \frac{nRT}{P} = \frac{(0.300 \text{ mol})\left(0.082 \ 06 \ \dfrac{L \cdot atm}{K \cdot mol}\right)(298 \text{ K})}{1.00 \text{ atm}} = 7.34 \text{ L O}_2$$

18.59 $2 \text{ C}_2\text{H}_2(g) + 5 \text{ O}_2(g) \rightarrow 4 \text{ CO}_2(g) + 2 \text{ H}_2\text{O}(g) \quad \Delta H° = -2511 \text{ kJ}$

C_2H_2, 26.04 amu

$$\text{mass C}_2\text{H}_2 = 1000 \text{ kJ} \times \frac{2 \text{ mol C}_2\text{H}_2}{2511 \text{ kJ}} \times \frac{26.04 \text{ g C}_2\text{H}_2}{1 \text{ mol C}_2\text{H}_2} = 20.74 \text{ g} = 20.7 \text{ g C}_2\text{H}_2$$

$PV = nRT$

$$V = \frac{nRT}{P} = \frac{\left(\dfrac{5}{2} \times 20.74 \text{ g} \times \dfrac{1 \text{ mol}}{26.04 \text{ g}}\right)\left(0.082 \ 06 \ \dfrac{L \cdot atm}{K \cdot mol}\right)(273 \text{ K})}{1.00 \text{ atm}} = 44.6 \text{ L O}_2$$

18.60 (a) $4 \text{ Li}(s) + \text{O}_2(g) \rightarrow 2 \text{ Li}_2\text{O}(s)$ (b) $P_4(s) + 5 \text{ O}_2(g) \rightarrow P_4\text{O}_{10}(s)$
(c) $4 \text{ Al}(s) + 3 \text{ O}_2(g) \rightarrow 2 \text{ Al}_2\text{O}_3(s)$ (d) $Si(s) + \text{O}_2(g) \rightarrow \text{SiO}_2(s)$

18.61 (a) $2 Ca(s) + O_2(g) \rightarrow 2 CaO(s)$ (b) $C(s) + O_2(g) \rightarrow CO_2(g)$
(c) $4 As(s) + 5 O_2(g) \rightarrow 2 As_2O_5(s)$ (d) $4 B(s) + 3 O_2(g) \rightarrow 2 B_2O_3(s)$

18.62   :Ö::Ö: The electron dot structure shows an O=O double bond. It also shows all electrons paired. This is not consistent with the fact that O_2 is paramagnetic.

18.63   The highest occupied molecular orbital in O_2 is the doubly degenerate π^*_{2p} orbital which contains 2 unpaired electrons. The bond order is 2 because O_2 has four π_{2p}, two σ_{2p} bonding electrons, and two π^*_{2p} antibonding electrons. (See text, Figure 7.20.)

18.64 $Li_2O < BeO < B_2O_3 < CO_2 < N_2O_5$ (see Figure 18.6)

18.65 $P_4O_{10} < SiO_2 < GeO_2 < Ga_2O_3 < K_2O$ (see Figure 18.6)

18.66 $N_2O_5 < Al_2O_3 < K_2O < Cs_2O$ (see Figure 18.6)

18.67 $BaO < SnO_2 < SO_3 < Cl_2O_7$ (see Figure 18.6)

18.68   (a) CrO_3 (higher Cr oxidation state)   (b) N_2O_5 (higher N oxidation state)
(c) SO_3 (higher S oxidation state)

18.69   (a) CrO (lower Cr oxidation state)   (b) SnO (lower Sn oxidation state)
(c) As_2O_3 (lower As oxidation state)

18.70   (a) $Cl_2O_7(l) + H_2O(l) \rightarrow 2 H^+(aq) + 2 ClO_4^-(aq)$
(b) $K_2O(s) + H_2O(l) \rightarrow 2 K^+(aq) + 2 OH^-(aq)$
(c) $SO_3(l) + H_2O(l) \rightarrow H^+(aq) + HSO_4^-(aq)$

18.71   (a) $BaO(s) + H_2O(l) \rightarrow Ba^{2+}(aq) + 2 OH^-(aq)$
(b) $Cs_2O(s) + H_2O(l) \rightarrow 2 Cs^+(aq) + 2 OH^-(aq)$
(c) $N_2O_5(s) + H_2O(l) \rightarrow 2 H^+(aq) + 2 NO_3^-(aq)$

18.72   (a) $ZnO(s) + 2 H^+(aq) \rightarrow Zn^{2+}(aq) + H_2O(l)$
(b) $ZnO(s) + 2 OH^-(aq) + H_2O(l) \rightarrow Zn(OH)_4^{2-}(aq)$

18.73   (a) $Ga_2O_3(s) + 3 H^+(aq) + 3 HSO_4^-(aq) \rightarrow 2 Ga^{3+}(aq) + 3 SO_4^{2-}(aq) + 3 H_2O(l)$
(b) $Ga_2O_3(s) + 2 OH^-(aq) + 3 H_2O(l) \rightarrow 2 Ga(OH)_4^-(aq)$

18.74   A peroxide has oxygen in the –1 oxidation state, for example, H_2O_2. A superoxide has oxygen in the –1/2 oxidation state, for example, KO_2.

18.75   (a) peroxide   (b) oxide   (c) superoxide   (d) peroxide

18.76   (a) BaO_2 (b) CaO (c) CsO_2 (d) Li_2O (e) Na_2O_2

18.77 (a) $BaO_2(s) + H_2O(l) \rightarrow Ba^{2+}(aq) + HO_2^-(aq) + OH^-(aq)$
(b) $2\,RbO_2(s) + H_2O(l) \rightarrow O_2(g) + 2\,Rb^+(aq) + HO_2^-(aq) + OH^-(aq)$

18.78

	O_2	O_2^-	O_2^{2-}
σ^*_{2p}			
π^*_{2p}	↑ ↑	↑↓ ↑	↑↓ ↑↓
π_{2p}	↑↓ ↑↓	↑↓ ↑↓	↑↓ ↑↓
σ_{2p}	↑↓	↑↓	↑↓
	Bond order = 2	Bond order = 1.5	Bond order = 1

(a) The O–O bond length increases because the bond order decreases. The bond order decreases because of the increased occupancy of antibonding orbitals.
(b) O_2^- has 1 unpaired electron and is paramagnetic. O_2^{2-} has no unpaired electrons and is diamagnetic.

18.79 O_2^+

σ^*_{2p}
π^*_{2p} ↑   —
π_{2p} ↑↓ ↑↓
σ_{2p} ↑↓

(a) O_2^+ bond order = 2½, so the O–O bond should be shorter than that in O_2.
(b) O_2^+ should be paramagnetic, with one unpaired electron.

18.80 (a) $H_2O_2(aq) + 2\,H^+(aq) + 2\,I^-(aq) \rightarrow I_2(aq) + 2\,H_2O(l)$
(b) $3\,H_2O_2(aq) + 8\,H^+(aq) + Cr_2O_7^{2-}(aq) \rightarrow 2\,Cr^{3+}(aq) + 3\,O_2(g) + 7\,H_2O(l)$

18.81 (a) $2\,Fe^{2+}(aq) + 2\,H^+(aq) + H_2O_2(aq) \rightarrow 2\,Fe^{3+}(aq) + 2\,H_2O(l)$
(b) $IO_4^-(aq) + H_2O_2(aq) \rightarrow IO_3^-(aq) + O_2(g) + H_2O(l)$

18.82

Ozone has two resonance structures, consistent with two equivalent O–O bond lengths.

18.83 To distinguish O_3 from O_2, bubble the gas into a basic solution containing KI and a starch indicator. If the solution turns blue, the gas is O_3.
$O_3(g) + 2\,I^-(aq) + H_2O(l) \rightarrow O_2(g) + I_2(aq) + 2\,OH^-(aq)$
The I_2 that is produced combines with the starch indicator, giving the blue color.
O_2 does not oxidize I^-.

18.84 $3\,O_2(g) \xrightarrow{\text{electric discharge}} 2\,O_3(g)$

18.85 $3\,O_2(g) \rightarrow 2\,O_3(g)$ $\Delta H° = 285\,kJ$
O_2, 32.00 amu;   energy required = $10.0\,g\,O_2 \times \dfrac{1\,mol\,O_2}{32.00\,g\,O_2} \times \dfrac{285\,kJ}{3\,mol\,O_2} = 29.7\,kJ$

Chemistry of Water (Sections 18.13–18.15)

18.86 (a) $2 F_2(g) + 2 H_2O(l) \rightarrow O_2(g) + 4 HF(aq)$

(b) $Cl_2(g) + H_2O(l) \rightleftharpoons HOCl(aq) + H^+(aq) + Cl^-(aq)$

(c) $I_2(s) + H_2O(l) \rightarrow HOI(aq) + H^+(aq) + I^-(aq)$

(d) $Ba(s) + 2 H_2O(l) \rightarrow H_2(g) + Ba^{2+}(aq) + 2 OH^-(aq)$

18.87 No, because in H_2O hydrogen is in its highest oxidation state and oxygen is in its lowest oxidation state. Therefore, neither atom can undergo a reaction in which it is both oxidized and reduced.

18.88 $AlCl_3 \cdot 6 H_2O$

$$\left[\begin{array}{c} \text{OH}_2 \\ \text{H}_2\text{O} \quad | \quad \text{OH}_2 \\ \text{Al} \\ \text{H}_2\text{O} \quad | \quad \text{OH}_2 \\ \text{OH}_2 \end{array}\right]^{3+}$$

18.89 The Fe^{3+} cation is bonded to the O atom of six H_2O molecules, which are located at the corners of a regular octahedron.

$$\left[\begin{array}{c} \text{OH}_2 \\ \text{H}_2\text{O} \quad | \quad \text{OH}_2 \\ \text{Fe} \\ \text{H}_2\text{O} \quad | \quad \text{OH}_2 \\ \text{OH}_2 \end{array}\right]^{3+}$$

18.90 $CaSO_4 \cdot \frac{1}{2} H_2O$, 145.15 amu; H_2O, 18.02 amu
Assume one mole of $CaSO_4 \cdot \frac{1}{2} H_2O$

$$\text{mass \% } H_2O = \frac{\text{mass } H_2O}{\text{mass hydrate}} \times 100\% = \frac{\frac{1}{2}(18.02 \text{ g})}{145.15 \text{ g}} \times 100\% = 6.21\%$$

18.91 $CuSO_4 \cdot 5 H_2O$, 249.69 amu; H_2O, 18.02 amu
Assume one mole of $CuSO_4 \cdot 5 H_2O$.

$$\text{mass \% } H_2O = \frac{\text{mass } H_2O}{\text{mass hydrate}} \times 100\% = \frac{5 \times 18.02 \text{ g}}{249.69 \text{ g}} \times 100\% = 36.08\%$$

18.92 $CaSO_4 \cdot \frac{1}{2} H_2O$, 145.15 amu; H_2O, 18.02 amu
mass of H_2O lost = 3.44 g – 2.90 g = 0.54 g H_2O

$$2.90 \text{ g } CaSO_4 \cdot \tfrac{1}{2} H_2O \times \frac{1 \text{ mol}}{145.15 \text{ g}} = 0.020 \text{ mol } CaSO_4 \cdot \tfrac{1}{2} H_2O$$

$$0.54 \text{ g } H_2O \times \frac{1 \text{ mol}}{18.02 \text{ g}} = 0.030 \text{ mol}$$

number of H₂O's lost = $\dfrac{0.030 \text{ mol}}{0.020 \text{ mol}}$ = 1.5 H₂O per CaSO₄ · ½ H₂O formed

The mineral gypsum is CaSO₄ · 2 H₂O; x = 2

18.93   Assume that you start with 1.00 g CoCl₂ sample. When the CoCl₂ is exposed to moist air, there is an 83.0% increase in mass to 1.83 g.

CoCl₂, 129.84 amu; H₂O, 18.02 amu

mass of CoCl₂ · xH₂O = 1.83 g;   mass CoCl₂ = 1.00 g

mass H₂O = 1.83 g − 1.00 g = 0.83 g

0.83 g H₂O x $\dfrac{1 \text{ mol H}_2\text{O}}{18.02 \text{ g H}_2\text{O}}$ = 0.046 mol H₂O

1.00 g CoCl₂ x $\dfrac{1 \text{ mol CoCl}_2}{129.84 \text{ g CoCl}_2}$ = 0.007 70 mol CoCl₂

x = $\dfrac{\text{mol H}_2\text{O}}{\text{mol CoCl}_2}$ = $\dfrac{0.046 \text{ mol}}{0.007 \ 70 \text{ mol}}$ = 6.0;   The hydrate formula is CoCl₂ · 6 H₂O

18.94   Convert mi³ to cm³; $(1.0 \text{ mi}^3)\left(\dfrac{1609 \text{ m}}{1 \text{ mi}}\right)^3\left(\dfrac{100 \text{ cm}}{1 \text{ m}}\right)^3$ = 4.2 x 10¹⁵ cm³

mass of sea water = volume x density = (4.2 x 10¹⁵ cm³)(1.025 g/cm³) = 4.3 x 10¹⁵ g

mass of salts = (0.035)(4.3 x 10¹⁵ g)$\left(\dfrac{1 \text{ kg}}{1000 \text{ g}}\right)$ = 1.5 x 10¹¹ kg

18.95   volume = (1.00 m³)(100 cm/m)³ = 1.00 x 10⁶ cm³

mass = (1.00 x 10⁶ cm³)(1.025 g/cm³) = 1.025 x 10⁶ g = 1025 kg

kg Mg²⁺ = (1025 kg)(1.35 g Mg²⁺/kg)(1 kg/1000 g) = 1.38 kg Mg²⁺

Chapter Problems

18.96   React H₂O with a reducing agent to produce H₂. Ca or Al could be used.

18.97   N₂(g) + 3 H₂(g) → 2 NH₃(g);   H₂, 2.016 amu; NH₃, 17.03 amu

(11.4 x 10⁶ tons NH₃)(2000 pounds/ton)(453.5 g/pound) = 1.03 x 10¹³ g NH₃

1.03 x 10¹³ g NH₃ x $\dfrac{1 \text{ mol NH}_3}{17.03 \text{ g NH}_3}$ = 6.05 x 10¹¹ mol NH₃

6.05 x 10¹¹ mol NH₃ x $\dfrac{3 \text{ mol H}_2}{2 \text{ mol NH}_3}$ = 9.07 x 10¹¹ mol H₂

9.07 x 10¹¹ mol H₂ x $\dfrac{2.016 \text{ g H}_2}{1 \text{ mol H}_2}$ x $\dfrac{1 \text{ pound}}{453.5 \text{ g}}$ x $\dfrac{1 \text{ ton}}{2000 \text{ pounds}}$ = 2.02 x 10⁶ tons

mass H₂ = 2.02 million tons H₂

18.98   Butadiene, C_4H_6, 54.09 amu; 2.7 kg = 2700 g

moles H_2 = 2700 g C_4H_6 x $\dfrac{1 \text{ mol } C_4H_6}{54.09 \text{ g } C_4H_6}$ x $\dfrac{2 \text{ mol } H_2}{1 \text{ mol } C_4H_6}$ = 99.8 mol H_2

At STP, P = 1.00 atm and T = 273 K

$PV = nRT$; $V = \dfrac{nRT}{P} = \dfrac{(99.8 \text{ mol})\left(0.082\ 06\ \dfrac{L \cdot atm}{K \cdot mol}\right)(273 \text{ K})}{1.00 \text{ atm}}$ = 2.2 x 10^3 L of H_2

18.99   (a) $Ca(OH)_2$ (b) Cr_2O_3 (c) RbO_2 (d) Na_2O_2 (e) BaH_2 (f) H_2Se

18.100 (a) B_2O_3, diboron trioxide   (b) H_2O_2, hydrogen peroxide
        (c) SrH_2, strontium hydride   (d) CsO_2, cesium superoxide
        (e) $HClO_4$, perchloric acid   (f) BaO_2, barium peroxide

18.101 (a), (b), and (d) are different kinds of water and have similar properties. (c) and (e) are
        different kinds of hydrogen peroxide and have similar properties. The properties of (a),
        (b), and (d) are quite different from those of (c) and (e).

18.102 (a) 6; $^{16}O_2$, $^{17}O_2$, $^{18}O_2$, $^{16}O^{17}O$, $^{16}O^{18}O$, $^{17}O^{18}O$

        (b) 18 $^{16}O_3$ $^{16}O_2{}^{17}O$ $^{18}O_3$
                      $^{17}O_2{}^{16}O$ $^{17}O_3$ $^{16}O_2{}^{18}O$
                      $^{18}O_2{}^{16}O$ $^{18}O_2{}^{17}O$ $^{17}O_2{}^{18}O$
                      $^{16}O^{17}O^{16}O$ $^{17}O^{18}O^{17}O$ $^{16}O^{17}O^{18}O$
                      $^{16}O^{18}O^{16}O$ $^{17}O^{16}O^{17}O$ $^{18}O^{16}O^{18}O$
                      $^{17}O^{18}O^{16}O$ $^{18}O^{16}O^{17}O$ $^{18}O^{17}O^{18}O$

18.103 (a) Al, +3   O, –2   (b) Sr, +2   O, –1
        (c) Sn, +4   O, –2   (d) Cs, +1   O, –1/2

18.104 (a) $2 H_2(g) + O_2(g) \rightarrow 2 H_2O(l)$
        (b) $O_3(g) + 2 I^-(aq) + H_2O(l) \rightarrow O_2(g) + I_2(aq) + 2 OH^-(aq)$
        (c) $H_2O_2(aq) + 2 H^+(aq) + 2 Br^-(aq) \rightarrow 2 H_2O(l) + Br_2(aq)$
        (d) $2 Na(l) + H_2(g) \rightarrow 2 NaH(s)$
        (e) $2 Na(s) + 2 H_2O(l) \rightarrow H_2(g) + 2 Na^+(aq) + 2 OH^-(aq)$

18.105 (a) $2 H_2(g) + O_2(g) \rightarrow 2 H_2O(l)$
        (b) $5 H_2O_2(aq) + 2 MnO_4^-(aq) + 6 H^+(aq) \rightarrow 5 O_2(g) + 2 Mn^{2+}(aq) + 8 H_2O(l)$
        (c) $2 F_2(g) + 2 H_2O(l) \rightarrow O_2(g) + 4 HF(aq)$

18.106 K is oxidized by water. F_2 is reduced by water.
        Cl_2 and Br_2 disproportionate when treated with water.

18.107 mass $= (2.0 \times 10^6$ kg bromine$)\left(\dfrac{1000 \text{ g}}{1 \text{ kg}}\right)\left(\dfrac{1 \text{ kg sea water}}{0.065 \text{ g Br}^-}\right) = 3.08 \times 10^{10}$ kg sea water

$$\text{volume} = \dfrac{\text{mass}}{\text{density}} = \dfrac{(3.08 \times 10^{10}\text{kg})\left(\dfrac{1000 \text{ g}}{1 \text{ kg}}\right)}{1.025 \text{ g/cm}^3} = 3.00 \times 10^{13} \text{ cm}^3 = 3.00 \times 10^{13} \text{ mL}$$

volume $= 3.00 \times 10^{13}$ mL $\times \dfrac{1 \text{ L}}{1000 \text{ mL}} = 3.0 \times 10^{10}$ L

If recovery is only 20%, you need to process five times this volume.
processed volume $= 5(3.0 \times 10^{10}$ L$) = 1.5 \times 10^{11}$ L

18.108 (a) $CO(g) + 2 H_2(g) \rightarrow CH_3OH(l)$
$\Delta H° = \Delta H°_f(CH_3OH) - \Delta H°_f(CO)$
$\Delta H° = (1 \text{ mol})(-239.2 \text{ kJ/mol}) - (1 \text{ mol})(-110.5 \text{ kJ/mol}) = -128.7$ kJ
(b) $CO(g) + H_2O(g) \rightarrow CO_2(g) + H_2(g)$
$\Delta H° = \Delta H°_f(CO_2) - [\Delta H°_f(CO) + \Delta H°_f(H_2O)]$
$\Delta H° = (1 \text{ mol})(-393.5 \text{ kJ/mol}) - [(1 \text{ mol})(-110.5 \text{ kJ/mol})$
$\qquad\qquad + (1 \text{ mol})(-241.8 \text{ kJ/mol})] = -41.2$ kJ
(c) $2 KClO_3(s) \rightarrow 2 KCl(s) + 3 O_2(g)$
$\Delta H° = 2 \Delta H°_f(KCl) - 2 \Delta H°_f(KClO_3)$
$\Delta H° = (2 \text{ mol})(-436.5 \text{ kJ/mol}) - (2 \text{ mol})(-397.7 \text{ kJ/mol}) = -77.6$ kJ
(d) $6 CO_2(g) + 6 H_2O(l) \rightarrow 6 O_2(g) + C_6H_{12}O_6(s)$
$\Delta H° = \Delta H°_f(C_6H_{12}O_6) - [6 \Delta H°_f(CO_2) + 6 \Delta H°_f(H_2O)]$
$\Delta H° = (1 \text{ mol})(-1273.3 \text{ kJ/mol}) - [(6 \text{ mol})(-393.5 \text{ kJ/mol})$
$\qquad\qquad + (6 \text{ mol})(-285.8 \text{ kJ/mol})] = 2802.5$ kJ

18.109 $SO_2(g) + H_2O(l) \rightleftharpoons H_2SO_3(aq)$

$H_2SO_3(aq) + H_2O(l) \rightleftharpoons H_3O^+(aq) + HSO_3^-(aq)$

$SO_3(g) + H_2O(l) \rightleftharpoons H_2SO_4(aq)$

$H_2SO_4(aq) + H_2O(l) \rightleftharpoons H_3O^+(aq) + HSO_4^-(aq)$

SO_3 gives a higher concentration of H_3O^+ ions because H_2SO_4 is a strong acid and H_2SO_3 is a weak acid.

Multiconcept Problems

18.110 $2 O_3(g) \rightarrow 3 O_2(g) \qquad \Delta H° = -258$ kJ
$PV = nRT$

$$n_{O_3} = \dfrac{PV}{RT} = \dfrac{\left(63.6 \text{ mm Hg} \times \dfrac{1.00 \text{ atm}}{760 \text{ mm Hg}}\right)(1.000 \text{ L})}{\left(0.082\ 06 \dfrac{\text{L} \cdot \text{atm}}{\text{K} \cdot \text{mol}}\right)(293 \text{ K})} = 3.48 \times 10^{-3} \text{ mol O}_3$$

$q = 3.48 \times 10^{-3}$ mol $O_3 \times \dfrac{285 \text{ kJ}}{2 \text{ mol O}_3} \times \dfrac{1000 \text{ J}}{1 \text{ kJ}} = 496$ J are liberated.

18.111 NaH, 24.00 amu; Because the unit cell is like NaCl, it contains 4 NaH's.

$$\text{unit cell mass} = \frac{24.00 \text{ g NaH}}{1 \text{ mol NaH}} \times \frac{1 \text{ mol NaH}}{6.022 \times 10^{23} \text{ NaH}} \times \frac{4 \text{ NaH}}{1 \text{ unit cell}} = 1.59 \times 10^{-22} \text{ g}$$

$$\text{unit cell volume} = 1.59 \times 10^{-22} \text{ g} \times \frac{1 \text{ cm}^3}{1.39 \text{ g}} = 1.14 \times 10^{-22} \text{ cm}^3$$

$$\text{unit cell edge} = \sqrt[3]{1.14 \times 10^{-22} \text{ cm}^3} = 4.85 \times 10^{-8} \text{ cm}$$

$$\text{unit cell edge} = 4.85 \times 10^{-8} \text{ cm} \times \frac{1 \times 10^{-2} \text{ m}}{1 \text{ cm}} \times \frac{1 \text{ pm}}{1 \times 10^{-12} \text{ m}} = 485 \text{ pm}$$

$$\text{unit cell edge} = 2r_{Na^+} + 2r_{H^-} = 485 \text{ pm}$$

$$r_{H^-} = \frac{485 \text{ pm} - (2r_{Na^+})}{2} = \frac{485 \text{ pm} - 2(102 \text{ pm})}{2} = 140 \text{ pm}$$

18.112 $MH_2(s) + 2 HCl(aq) \rightarrow 2 H_2(g) + M^{2+}(aq) + 2 Cl^-(aq)$
$PV = nRT$

$$n_{H_2} = \frac{PV}{RT} = \frac{\left(750 \text{ mm Hg} \times \dfrac{1.00 \text{ atm}}{760 \text{ mm Hg}}\right)(1.000 \text{ L})}{\left(0.082\ 06 \dfrac{\text{L} \cdot \text{atm}}{\text{K} \cdot \text{mol}}\right)(293 \text{ K})} = 0.0410 \text{ mol H}_2$$

$$0.0410 \text{ mol H}_2 \times \frac{1 \text{ mol MH}_2}{2 \text{ mol H}_2} = 0.0205 \text{ mol MH}_2$$

$$\text{molar mass} = \frac{1.84 \text{ g}}{0.0205 \text{ mol}} = 89.8 \text{ g/mol}$$

mass M = 89.8 – mass of 2 H = 89.8 – 2(1.008) = 87.7 g/mol;   M = Sr; SrH_2

18.113 (a) 77 °C = 350 K
The molar mass (mol. mass) of the sulfur oxide (SO_x) gas can be computed from its density.

$$PV = nRT; \quad PV = \frac{g}{\text{mol. mass}} RT: \quad d = \frac{g}{V} = \frac{P(\text{mol. mass})}{RT}$$

$$\text{mol. mass} = \frac{dRT}{P} = \frac{(2.64 \text{ g/L})\left(0.082\ 06 \dfrac{\text{L} \cdot \text{atm}}{\text{K} \cdot \text{mol}}\right)(350 \text{ K})}{\left(720 \text{ mm Hg} \times \dfrac{1.00 \text{ atm}}{760 \text{ mm Hg}}\right)} = 80.04 \text{ g/mol}$$

Subtract the molar mass of S from the SO_x molar mass and then divide by the molar mass of O. This will give the value of x.

$$x = \frac{80 - 32}{16} = 3; \quad \text{The formula for the sulfur oxide is } SO_3.$$

(b) SO_3, 80.06 amu
$SO_3(g) + H_2O(l) \rightarrow H_2SO_4(aq)$
$\Delta H^\circ_{rxn} = \Delta H^\circ_f(H_2SO_4) - [\Delta H^\circ_f(SO_3) + \Delta H^\circ_f(H_2O)]$

ΔH°_{rxn} = [(1 mol)(–909.3 kJ/mol)] – [(1 mol)(–395.7 kJ/mol) + (1 mol)(–285.8 kJ/mol)]

ΔH°_{rxn} = –227.8 kJ

mass SO_3 = (2.64 g/L)(0.2500 L) = 0.660 g SO_3

$$0.660 \text{ g } SO_3 \times \frac{1 \text{ mol } SO_3}{80.06 \text{ g } SO_3} = 8.24 \times 10^{-3} \text{ mol } SO_3$$

q = (8.24 x 10^{-3} mol)(–227.8 kJ/mol) = –1.88 kJ; 1.88 kJ of heat is released.

(c) mol H_2SO_4 = mol SO_3 = 8.24 x 10^{-3} mol

$H_2SO_4(aq)$ + 2 NaOH(aq) → 2 $H_2O(l)$ + $Na_2SO_4(aq)$

$$8.24 \times 10^{-3} \text{ mol } H_2SO_4 \times \frac{2 \text{ mol NaOH}}{1 \text{ mol } H_2SO_4} = 0.0165 \text{ mol NaOH}$$

$$0.0165 \text{ mol NaOH} \times \frac{1 \text{ L}}{0.160 \text{ mol}} = 0.103 \text{ L} = 103 \text{ mL}$$

18.114 N_2O_5, 108.01 amu

$N_2O_5(g)$ + $H_2O(l)$ → 2 $HNO_3(aq)$

2 $HNO_3(aq)$ + Zn(s) → $H_2(g)$ + $Zn(NO_3)_2(aq)$

$$5.4 \text{ g } N_2O_5 \times \frac{1 \text{ mol } N_2O_5}{108.01 \text{ g } N_2O_5} \times \frac{2 \text{ mol } HNO_3}{1 \text{ mol } N_2O_5} \times \frac{1 \text{ mol } H_2}{2 \text{ mol } HNO_3} = 0.050 \text{ mol } H_2$$

$$PV = nRT; \quad P_{H_2} = \frac{nRT}{V} = \frac{(0.050 \text{ mol})\left(0.082\ 06 \dfrac{L \cdot atm}{K \cdot mol}\right)(298 \text{ K})}{0.500 \text{ L}} = 2.45 \text{ atm}$$

$$P_{H_2} = 2.45 \text{ atm} \times \frac{760 \text{ mm Hg}}{1.00 \text{ atm}} = 1862 \text{ mm Hg}$$

(a) In H there is 0.0156 atom % D.

To get P_{HD} multiply P_{H_2} by the atom % D and then by 2 because H_2 is diatomic.

P_{H_2} = (1862 mm Hg)(0.000156)(2) = 0.58 mm Hg

$$(b) \quad n_{HD} = \frac{PV}{RT} = \frac{\left(0.58 \text{ mm Hg} \times \dfrac{1.00 \text{ atm}}{760 \text{ mm Hg}}\right)(0.5000 \text{ L})}{\left(0.082\ 06 \dfrac{L \cdot atm}{K \cdot mol}\right)(298 \text{ K})} = 1.56 \times 10^{-5} \text{ mol HD}$$

$$1.56 \times 10^{-5} \text{ mol HD} \times \frac{6.022 \times 10^{23} \text{ HD molecules}}{1 \text{ mol HD}} = 9.4 \times 10^{18} \text{ HD molecules}$$

(c) P_{D_2} = (1862 mm Hg)(0.000 156)^2 = 4.53 x 10^{-5} mm Hg

$$n_{D_2} = \frac{PV}{RT} = \frac{\left(4.53 \times 10^{-5} \text{ mm Hg} \times \dfrac{1.00 \text{ atm}}{760 \text{ mm Hg}}\right)(0.5000 \text{ L})}{\left(0.082\ 06 \dfrac{L \cdot atm}{K \cdot mol}\right)(298 \text{ K})} = 1.22 \times 10^{-9} \text{ mol } D_2$$

$$1.22 \times 10^{-9} \text{ mol } D_2 \times \frac{6.022 \times 10^{23} \text{ } D_2 \text{ molecules}}{1 \text{ mol } D_2} = 7.3 \times 10^{14} \text{ } D_2 \text{ molecules}$$

18.115 (a) $2 Na(s) + 2 HCl(aq) \rightarrow H_2(g) + 2 Na^+(aq) + 2 Cl^-(aq)$

$PV = nRT$

$$n_{H_2} = \frac{PV}{RT} = \frac{\left(434 \text{ mm Hg} \times \frac{1.00 \text{ atm}}{760 \text{ mm Hg}}\right)(0.500 \text{ L})}{\left(0.082\,06 \frac{\text{L} \cdot \text{atm}}{\text{K} \cdot \text{mol}}\right)(295 \text{ K})} = 0.0118 \text{ mol } H_2$$

$$\text{mol Na} = 0.0118 \text{ mol } H_2 \times \frac{2 \text{ mol Na}}{1 \text{ mol } H_2} = 0.0236 \text{ mol Na}$$

$$\text{mass Na} = 0.0236 \text{ mol Na} \times \frac{22.99 \text{ g Na}}{1 \text{ mol Na}} = 0.543 \text{ g Na}$$

$$\text{mass \% Na} = \frac{0.543 \text{ g Na}}{5.26 \text{ g amalgam}} \times 100\% = 10.3\%$$

(b) total mol HCl = $(0.2000 \text{ mol/L})(0.2500 \text{ L}) = 0.050\,00$ mol HCl

$$\text{mol HCl reacted} = 0.0236 \text{ mol Na} \times \frac{2 \text{ mol HCl}}{2 \text{ mol Na}} = 0.0236 \text{ mol HCl}$$

mol HCl remaining = $0.050\,00$ mol $- 0.0236$ mol $= 0.0264$ mol HCl

$$\text{in 50.00 mL, mol HCl} = \frac{0.0264 \text{ mol HCl}}{250.0 \text{ mL}} \times 50.00 \text{ mL} = 0.005\,28 \text{ mol HCl}$$

$$\text{volume NaOH} = 0.005\,28 \text{ mol HCl} \times \frac{1.000 \text{ L}}{0.1000 \text{ mol NaOH}} = 0.0528 \text{ L} = 52.8 \text{ mL}$$

18.116

$$\ln P_2 = \ln P_1 + \frac{\Delta H_{vap}}{R}\left(\frac{1}{T_1} - \frac{1}{T_2}\right)$$

$$\Delta H_{vap} = \frac{(\ln P_2 - \ln P_1)(R)}{\left(\frac{1}{T_1} - \frac{1}{T_2}\right)}$$

$P_1 = 75$ mm Hg; $T_1 = 89\,°C = 89 + 273 = 362$ K
$P_2 = 319.2$ mm Hg; $T_2 = 125\,°C = 125 + 273 = 398$ K

$$\Delta H_{vap} = \frac{[\ln(319.2) - \ln(75.0)](8.314 \times 10^{-3} \text{ kJ/(K} \cdot \text{mol)})}{\left(\frac{1}{362 \text{ K}} - \frac{1}{398 \text{ K}}\right)} = 48.2 \text{ kJ/mol}$$

Now calculate the normal boiling point.
$P_1 = 75$ mm Hg; $T_1 = 89\,°C = 89 + 273 = 362$ K
$P_2 = 760$ mm Hg; $T_2 = ?$

$$\ln P_2 = \ln P_1 + \frac{\Delta H_{vap}}{R}\left(\frac{1}{T_1} - \frac{1}{T_2}\right)$$

$$(\ln P_2 - \ln P_1)\left(\frac{R}{\Delta H_{vap}}\right) = \frac{1}{T_1} - \frac{1}{T_2}$$

Solve for T_2 (the boiling point for H_2O_2 at 760 mm Hg).

$$\frac{1}{T_1} - (\ln P_2 - \ln P_1)\left(\frac{R}{\Delta H_{vap}}\right) = \frac{1}{T_2}$$

$$\frac{1}{362\ K} - [\ln(760) - \ln(75.0)]\left(\frac{8.314 \times 10^{-3}\ kJ/(K \cdot mol)}{48.2\ kJ/mol}\right) = \frac{1}{T_2} = 0.002\ 363/K$$

$T_2 = 1/0.002\ 363/K = 423\ K = 423 - 273 = 150\ °C$ (boiling point for H_2O_2)

The calculated boiling point is the same as the one listed in section 18.11.

18.117 $MgBr_2$, 184.11 amu

Ocean volume = $1.35 \times 10^{18}\ m^3 \times \dfrac{1\ L}{10^{-3}\ m^3} \times \dfrac{1000\ mL}{1\ L} = 1.35 \times 10^{24}\ mL$

1.35 g Mg^{2+}/kg seawater

0.065 g Br^-/kg seawater

mass of seawater = $(1.35 \times 10^{24}\ mL)(1.025\ g/mL)(1\ kg/1000\ g) = 1.384 \times 10^{21}\ kg$

mass Mg^{2+} = $(1.35\ g\ Mg^{2+}/kg)(1.384 \times 10^{21}\ kg) = 1.87 \times 10^{21}\ g\ Mg^{2+}$

number of Mg^{2+} ions = $1.87 \times 10^{21}\ g\ Mg^{2+} \times \dfrac{6.022 \times 10^{23}\ Mg^{2+}\ ions}{24.305\ g\ Mg^{2+}} = 4.63 \times 10^{43}\ Mg^{2+}$ ions

mass Br^- = $(0.065\ g\ Br^-/kg)(1.384 \times 10^{21}\ kg) = 9.0 \times 10^{19}\ g\ Br^-$

number of Br^- ions = $9.0 \times 10^{19}\ g\ Br^- \times \dfrac{6.022 \times 10^{23}\ Br^-\ ions}{79.904\ g\ Br^-} = 6.8 \times 10^{41}\ Br^-$ ions

The number of Mg^{2+} ions is in excess, so it is the number of Br^- ions that will determine how much $MgBr_2$ can be made.

tons $MgBr_2$ = $6.8 \times 10^{41}\ Br^-$ ions $\times \dfrac{1\ MgBr_2}{2\ Br^-} \times \dfrac{184.11\ g\ MgBr_2}{6.022 \times 10^{23}\ MgBr_2} \times \dfrac{1\ kg}{1000\ g} \times$

$\dfrac{2.2046\ lb}{1\ kg} \times \dfrac{1\ ton}{2000\ lb} = 1.1 \times 10^{14}$ tons $MgBr_2$

19.1 (a) B is above Al in group 3A, and therefore B is more nonmetallic than Al.

(b) Ge and Br are in the same row of the periodic table, but Br (group 7A) is to the right of Ge (group 4A). Therefore, Br is more nonmetallic.

(c) Se (group 6A) is more nonmetallic than In because it is above and to the right of In (group 3A).

(d) Cl (group 7A) is more nonmetallic than Te because it is above and to the right of Te (group 6A).

19.2 (a) HNO_3 H_3PO_4

Nitrogen can form very strong $p\pi$ - $p\pi$ bonds. Phosphorus forms weaker $p\pi$ - $p\pi$ bonds, so it tends to form more single bonds.

(b) The larger S atom can accommodate six bond pairs in its valence shell, but the smaller O atom is limited to two bond pairs and two lone pairs.

19.3 Carbon forms strong π bonds with oxygen. Silicon does not form strong π bonds with oxygen, and what results are chains of alternating silicon and oxygen singly bonded to each other.

19.4 An ethane-like structure is unlikely for diborane because it would require 14 valence electrons and diborane only has 12. The result is two three-center, two-electron bonds between the borons and the bridging hydrogen atoms.

19.5 $H-C \equiv N\!:$ The carbon is sp hybridized.

19.6 $Hb-O_2 + CO \rightleftharpoons Hb-CO + O_2$

Mild cases of carbon monoxide poisoning can be treated with O_2. Le Châtelier's principle says that adding a product (O_2) will cause the reaction to proceed in the reverse direction, back to $Hb-O_2$.

19.7 (a) $Si_8O_{24}^{16-}$ (b) $Si_2O_5^{2-}$

19.8   Formal charge = $\left(\begin{array}{c}\text{Number of}\\\text{valence electrons}\\\text{in free atom}\end{array}\right) - \dfrac{1}{2}\left(\begin{array}{c}\text{Number of}\\\text{bonding}\\\text{electrons}\end{array}\right) - \left(\begin{array}{c}\text{Number of}\\\text{nonbonding}\\\text{electrons}\end{array}\right)$

Nitrous oxide, N_2O

$\underset{0\ \ +1\ \ -1}{:N\equiv N-\ddot{O}:} \longleftrightarrow \underset{-1\ +1\ \ \ 0}{:\ddot{N}=N=\ddot{O}:} \longleftrightarrow \underset{-2\ +1\ +1}{:\ddot{N}-N\equiv O:}$

The first structure is most important because it has a –1 formal charge on an electronegative oxygen.

Nitric oxide, NO

$\underset{0\ \ \ 0}{\cdot\ddot{N}=\ddot{O}:} \longleftrightarrow \underset{-1\ +1}{:\ddot{N}=\dot{O}:}$ paramagnetic

The first structure is more important because it has no formal charges.

Nitrous acid, HNO_2

$\underset{0\ \ \ 0\ \ \ 0}{H-\ddot{O}-\dot{N}=\ddot{O}:} \longleftrightarrow \underset{+1\ \ 0\ \ -1}{H-\ddot{O}=N-\ddot{O}:}$

The first structure is more important because it has no formal charges.

Nitrogen dioxide, NO_2

$\underset{-1\ +1\ \ 0}{:\ddot{O}-\dot{N}=\ddot{O}:} \longleftrightarrow \underset{0\ +1\ -1}{:\ddot{O}=\dot{N}-\ddot{O}:}$ paramagnetic

Both structures are of equal importance.

Nitric acid, HNO_3

$\underset{\substack{0\ +1\ \ 0}}{H-\ddot{O}-N=\ddot{O}:} \longleftrightarrow \underset{\substack{0\ +1\ -1}}{H-\ddot{O}-N-\ddot{O}:} \longleftrightarrow \underset{\substack{+1\ +1\ -1}}{H-\ddot{O}=N-\ddot{O}:}$

$\qquad\ \ \underset{-1}{:\ddot{O}:} \qquad\qquad\ \ \underset{0}{\overset{||}{O}:} \qquad\qquad\ \ \underset{-1}{:\ddot{O}:}$

The first two structures are of equal importance. Both are more important than structure three because it has a +1 formal charge on an electronegative oxygen.

19.9 H_3PO_4 $H_6P_4O_{13}$

$$\begin{array}{c}\qquad\quad O\\ \qquad\quad \|\\ H-O-P-O-H\\ \qquad\quad |\\ \qquad\quad O\\ \qquad\quad |\\ \qquad\quad H\end{array}$$

$$\begin{array}{c}\quad O\quad\ \ O\quad\ \ O\quad\ \ O\\ \quad \|\quad\ \ \|\quad\ \ \|\quad\ \ \|\\ H-O-P-O-P-O-P-O-P-O-H\\ \quad |\quad\ \ |\quad\ \ |\quad\ \ |\\ \quad O\quad\ \ O\quad\ \ O\quad\ \ O\\ \quad |\quad\ \ |\quad\ \ |\quad\ \ |\\ \quad H\quad\ \ H\quad\ \ H\quad\ \ H\end{array}$$

The chemical structures show the condensation of four phosphoric acid molecules:

H—O—P(=O)(—O—H)—O—H + H—O—P(=O)(—O—H)—O—H + H—O—P(=O)(—O—H)—O—H + H—O—P(=O)(—O—H)—O—H →

H—O—P(=O)(—O—H)—O—P(=O)(—O—H)—O—P(=O)(—O—H)—O—P(=O)(—O—H)—O—H + 3 H₂O

$$4\,H_3PO_4(aq) \rightarrow H_6P_4O_{13}(aq) + 3\,H_2O(l)$$

19.10 (a) SO_3^{2-}, HSO_3^{-}, SO_4^{2-}, HSO_4^{-} (b) HSO_4^{-} (c) SO_3^{2-} (d) HSO_4^{-}

19.11 (a) H—S̈—H , bent.

(b) :Ö—S̈=Ö: ⟷ :Ö=S̈—Ö: , bent, S is sp^2 hybridized.

(c)

:Ö—S̈=Ö: ⟷ :Ö=S̈—Ö: ⟷ :Ö—S(=Ö)—Ö:

trigonal planar, S is sp^2 hybridized.

19.12 Because copper is always conducting, a photocopier drum coated with copper would not hold any charge so no document image would adhere to the drum for copying.

Key Concept Problems

19.13 (a) 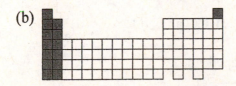 main-group elements

(b) 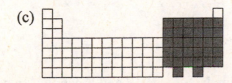 s-block elements

(c) p-block elements

(d) 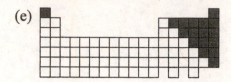 main-group metals

(e) 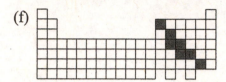 nonmetals

(f) 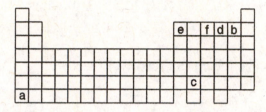 semimetals

19.14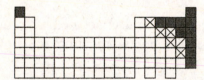

19.15 (a) gases, (b) liquid, and (c) solids

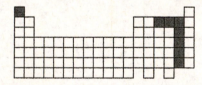

(a) ■
(b) ▨
(c) ⊠

(d) diatomic molecules

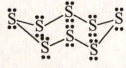

19.16 (a) PF_5 and SF_6 (b) CH_4 and NH_4^+ (c) CO and NO_2 (d) P_4O_{10}

19.17 (a) N_2, O_2, F_2, P_4 (tetrahedral), S_8 (crown-shaped ring), Cl_2

(b) :N≡N: :Ö=Ö: :F̈–F̈:   :C̈l—C̈l:

(c) The smaller N and O can form strong π bonds, whereas P and S cannot. In both F_2 and Cl_2, the atoms are joined by a single bond.

19.18 (a)

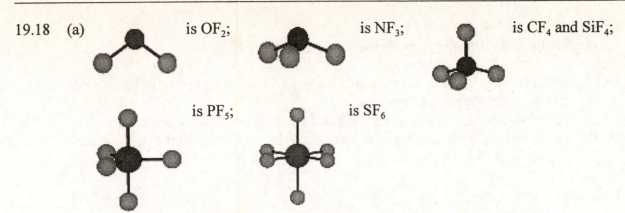

is OF$_2$;   is NF$_3$;   is CF$_4$ and SiF$_4$;

is PF$_5$;   is SF$_6$

(b)   The small N atom is limited to three nearest neighbors in NF$_3$, whereas the larger P atom can accommodate five nearest neighbors in PF$_5$. N uses its three unpaired electrons in bonding to three F atoms and has one lone pair, whereas P uses all five valence electrons in bonding to five F atoms. Both C and Si have four valence electrons and use sp^3 hybrid orbitals to bond to four F atoms.

19.19 (a) CO$_2$, Cl$_2$O$_7$, SO$_3$, N$_2$O$_5$

   (b)   :Ö=C=Ö:

:Ö   Ö   Ö:
  \ | /
   Cl   Cl
  /   \ / \
:Ö:Ö: :Ö:Ö:

Ö:
‖
S   (resonance structures are needed)
:Ö   Ö:

:Ö   Ö:
  \\   //
   N—Ö—N   (resonance structures are needed)
  /   \\
:Ö   Ö:

19.20 (a) H$_2$O, CH$_4$, HF, B$_2$H$_6$, NH$_3$

   (b) H—Ö—H   H   H—F̈:   H   H   H   H—N̈—H
                   | \ / \ / |
              H—C—H   B   B   H
                   | / \ / \
                   H   H H H

There is a problem in drawing an electron-dot structure for B$_2$H$_6$ because this molecule is electron deficient and has two three-center, two-electron bonds.

19.21 (a) SiO$_4^{4-}$ (b) Si$_3$O$_{10}^{8-}$ (c) Si$_4$O$_{12}^{8-}$

Section Problems
General Properties and Periodic Trends (Sections 19.1–19.2)

19.22   (a) Cl (group 7A) is to the right of S (group 6A) in the same row of the periodic table. Cl has the higher ionization energy.
(b) Si is above Ge in group 4A. Si has the higher ionization energy.
(c) O (group 6A) is above and to the right of In (group 3A) in the periodic table. O has the higher ionization energy.

19.23   K < Al < P < F

19.24   (a) Al is below B in group 3A. Al has the larger atomic radius.
(b) P (group 5A) is to the left of S (group 6A) in the same row of the periodic table. P has the larger atomic radius.
(c) Pb (group 4A) is below and to the left of Br (group 7A) in the periodic table. Pb has the larger atomic radius.

19.25   O < S < As < Sn

19.26   (a) I (group 7A) is to the right of Te (group 6A) in the same row of the periodic table. I has the higher electronegativity.
(b) N is above P in group 5A. N has the higher electronegativity.
(c) F (group 7A) is above and to the right of In (group 3A) in the periodic table. F has the higher electronegativity.

19.27   Ge < P < N < O

19.28   (a) Sn is below Si in group 4A. Sn has more metallic character.
(b) Ge (group 4A) is to the left of Se (group 6A) in the same row of the periodic table. Ge has more metallic character.
(c) Bi (group 5A) is below and to the left of I (group 7A) in the periodic table. Bi has more metallic character.

19.29   (a) S is is above Te in group 6A. S has more nonmetallic character.
(b) Cl (group 7A) is to the right of P (group 5A) in the same row of the periodic table. Cl has more nonmetallic character.
(c) Br (group 7A) is above and to the right of Bi (group 5A) in the periodic table. Br has more nonmetallic character.

19.30   In each case the more ionic compound is the one formed between a metal and nonmetal.
(a) CaH_2 (b) Ga_2O_3 (c) KCl   (d) $AlCl_3$

19.31   In each case the more covalent molecule is the one formed between two nonmetals.
(a) PCl_3 (b) NO   (c) NH_3 (d) SiO_2

19.32   Molecular   (a) B_2H_6 (c) SO_3 (d) $GeCl_4$
Extended three-dimensional structure   (b) $KAlSi_3O_8$

19.33   Molecular   (b) P_4O_{10} (c) $SiCl_4$
Extended three-dimensional structure   (a) KF   (d) $CaMgSi_2O_6$

19.34   Nonmetal and semimetal oxides are acidic. Metal oxides are basic or amphoteric.
(a) P_4O_{10} (b) B_2O_3 (c) SO_2
(d) N_2O_3 N_2O_3 is in the same group but above As_2O_3 in the periodic table.

19.35   The more metallic the element in the oxide, the more basic is the oxide.
(a) SnO_2 (b) In_2O_3 (c) Al_2O_3 (d) BaO

19.36   (a) Sn   (b) Cl   (c) Sn   (d) Se   (e) B

19.37   (a) F   (b) Al   (c) N   (d) Si   (e) S

19.38   The smaller B atom can bond to a maximum of four nearest neighbors, whereas the larger Al atom can accommodate more than four nearest neighbors.

19.39   The smaller C atom can bond to a maximum of four nearest neighbors, whereas the larger Ge atom can accommodate more than four nearest neighbors.

19.40   In O_2 a π bond is formed by 2p orbitals on each O. S does not form strong π bonds with its 3p orbitals, which leads to the S_8 ring structure with single bonds.

19.41   In N_2 π bonds are formed by 2p orbitals on each N. P does not form strong π bonds with its 3p orbitals, which leads to the P_4 tetrahedral structure with single bonds.

The Group 3A Elements (Sections 19.3–19.5)

19.42   +3 for B, Al, Ga and In;   +1 for Tl

19.43   (a) Na +1, B +3, F –1   (b) Ga +3, Cl –1
(c) Tl +1, Cl –1   (d) B +3, H –1

19.44   Boron is a hard semiconductor with a high melting point. Boron forms only molecular compounds and does not form an aqueous B^{3+} ion. $B(OH)_3$ is an acid.

19.45   Boron is a semimetal, whereas all the other 3A elements are metals. Boron has a much smaller atomic radius and a higher electronegativity than the other group 3A elements.

19.46 $2\ BBr_3(g) + 3\ H_2(g) \xrightarrow[1200\ °C]{Ta\ wire} 2\ B(s) + 6\ HBr(g)$

This reaction is used to produce crystalline boron.

19.47 $B_2O_3(l) + 3 Mg(s) \rightarrow 2 B(s) + 3 MgO(s)$

19.48   (a) An electron deficient molecule is a molecule that doesn't have enough electrons to form a two-center, two-electron bond between each pair of bonded atoms. B_2H_6 is an electron deficient molecule.
(b) A three-center, two-electron bond has three atoms bonded together using just two electrons. The B-H-B bridging bond in B_2H_6 is a three-center, two-electron bond.

19.49

The terminal B-H bonds in diborane are ordinary two-center, two-electron bonds. The bridging B–H–B bonds in diborane are three-center, two-electron bonds. A three-center, two-electron bond is longer than a two-center, two-electron bond because it has less electron density between each pair of adjacent atoms.

19.50   (a) Al   (b) Tl   (c) B   (d) B

19.51   (a) Ga   (b) B   (c) Tl   (d) B

The Group 4A Elements (Sections 19.6–19.9)

19.52   (a) Pb   (b) C   (c) Si   (d) C

19.53   (a) Pb   (b) Si   (c) C   (d) C

19.54   (a) $GeBr_4$, tetrahedral; Ge is sp^3 hybridized.
(b) CO_2, linear; C is sp hybridized.
(c) CO_3^{2-}, trigonal planar; C is sp^2 hybridized.
(d) $SnCl_3^-$, trigonal pyramidal; Sn is sp^3 hybridized.

19.55   (a) SiO_4^{4-}, tetrahedral; Si is sp^3 hybridized.
(b) NH_4^+, tetrahedral; N is sp^3 hybridized.
(c) $SnCl_2$, bent; Sn is sp^2 hybridized.
(d) N_2O, linear; N is sp hybridized.

19.56   Diamond is a very hard, high melting solid. It is an electrical insulator.
Diamond has a covalent network structure in which each C atom uses sp^3 hybrid orbitals to form a tetrahedral array of σ bonds. The interlocking, three-dimensional network of strong bonds makes diamond the hardest known substance with the highest melting point for an element. Because the valence electrons are localized in the σ bonds, diamond is an electrical insulator.

19.57 Graphite has a two-dimensional sheetlike structure in which each C atom uses sp^2 hybrid orbitals to form trigonal planar σ bonds to three neighboring C atoms. In addition, each C atom uses its remaining p orbital, perpendicular to the plane of the sheet, to form a π bond. Because each C atom must share its π bond with its three neighbors, the π electrons are delocalized and are free to move in the plane of the sheet. As a result, the electrical conductivity of graphite in a direction parallel to the sheets is about 10^{20} times greater than the conductivity of diamond. The conductivity of graphite perpendicular to the sheets of C atoms is lower because electrons must hop from one sheet to the next. The carbon sheets in graphite are separated by a distance of 335 pm and are held together by weak London dispersion forces. Consequently, the sheets can easily slide over one another, thus accounting for the slippery feel of graphite and its use as a lubricant.

19.58 (a) carbon tetrachloride, CCl_4 (b) carbon monoxide, CO   (c) methane, CH_4

19.59 CaC_2, calcium carbide;   C, –1

19.60 Some uses for CO_2 are:
   (1) To provide the "bite" in soft drinks; $CO_2(aq) + H_2O(l) \rightleftharpoons H_2CO_3(aq)$
   (2) CO_2 fire extinguishers; CO_2 is nonflammable and 1.5 times more dense than air.
   (3) Refrigerant; dry ice, sublimes at –78 °C.

19.61 CO bonds to hemoglobin and prevents it from carrying O_2. CN^- bonds to cytochrome oxidase and interferes with the electron transfer associated with oxidative phosphorylation.

19.62 $SiO_2(l) + 2\,C(s) \rightarrow Si(l) + 2\,CO(g)$
   (sand)

   Purification of silicon for semiconductor devices:
   $Si(s) + 2\,Cl_2(g) \rightarrow SiCl_4(l)$; $SiCl_4$ is purified by distillation.

   $SiCl_4(g) + 2\,H_2(g) \xrightarrow{heat} Si(s) + 4\,HCl(g)$; Si is purified by zone refining.

19.63 (a) silica sand (SiO_2)

   $SiO_2(l) + 2\,C(s) \xrightarrow{heat} Si(l) + 2\,CO(g)$

   (b) cassiterite (SnO_2)

   $SnO_2(s) + 2\,C(s) \xrightarrow{heat} Sn(l) + 2\,CO(g)$

   (c) galena (PbS)
   $2\,PbS(s) + 3\,O_2(g) \rightarrow 2\,PbO(s) + 2\,SO_2(g)$
   $PbO(s) + CO(g) \rightarrow Pb(l) + CO_2(g)$

19.64   (a) SiO_4^{4-} (b) $Si_4O_{13}^{10-}$

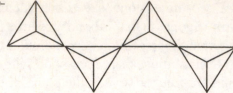

The charge on the anion is equal to the number of terminal O atoms.

19.65

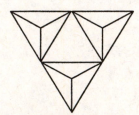

$Si_4O_{12}^{8-}$

19.66   (a) $Si_3O_{10}^{8-}$
(b) The charge on the anion is 8–. Because the Ca^{2+} to Cu^{2+} ratio is 1:1, there must be 2 Ca^{2+} and 2 Cu^{2+} ions in the formula for the mineral. There are also 2 waters. The formula of the mineral is: $Ca_2Cu_2Si_3O_{10} \cdot 2\,H_2O$

19.67   (a) spodumene, $LiAlSi_2O_6$

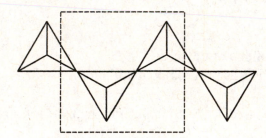

Repeating unit, $Si_2O_6^{4-}$

(b) wollastonite, $Ca_3Si_3O_9$

$Si_3O_9^{6-}$

(c) thortveitite, $Sc_2Si_2O_7$

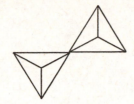

$Si_2O_7{}^{6-}$

(d) albite, $NaAlSi_3O_8$ - three dimensional SiO_2 structure with Al^{3+} in place of one-fourth of the Si^{4+} ions. (An equal number of Na^+ ions balance the charge.)

The Group 5A Elements (Sections 19.10–19.12)

19.68   (a) P   (b) Sb and Bi   (c) N   (d) Bi

19.69   (a) N   (b) Bi   (c) P   (d) N

19.70   (a) N_2O, +1   (b) N_2H_4, −2   (c) Ca_3P_2, −3
       (d) H_3PO_3, +3   (e) H_3AsO_4, +5

19.71   (a) NO, +2   (b) HNO_2, +3   (c) PH_3, −3
       (d) P_4O_{10}, +5   (e) $H_5P_3O_{10}$, +5

19.72   :N≡N:
       N_2 is unreactive because of the large amount of energy necessary to break the N≡N triple bond.

19.73   (a) N_2O,   linear
       :N≡N—Ö: ⟷ :N̈=N=Ö: ⟷ :N̈—N≡O:
       (b) NO,   linear, paramagnetic   :Ṅ=Ö:

       (c) NO_2,   bent, paramagnetic

       :Ö—N̈=Ö: ⟷ :Ö=N̈—Ö:

19.74   (a) NO_2^-, bent
       (b) PH_3, trigonal pyramidal
       (c) PF_5, trigonal bipyramidal
       (d) PCl_4^+, tetrahedral

19.75   (a) PCl_6^-, octahedral
       (b) N_2O, linear
       (c) H_3PO_3, tetrahedral
       (d) NO_3^-, trigonal planar

19.76   White phosphorus consists of tetrahedral P_4 molecules with 60° bond angles.

Red phosphorus is polymeric.
White phosphorus is reactive due to the considerable strain in the P_4 molecule.

19.77   (a) phosphorus(III) oxide, P_4O_6 (b) phosphorus(V) oxide, P_4O_{10}

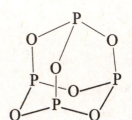   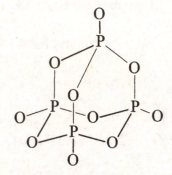

(c) phosphorous acid   (d) orthophosphoric acid

          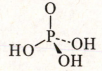

(e) polymetaphosphoric acid

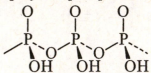

19.78   (a) The structure for phosphorous acid is

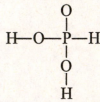

Only the two hydrogens bonded to oxygen are acidic.
(b) Nitrogen forms strong π bonds, and in N_2 the nitrogen atoms are triple bonded to each other. Phosphorus does not form strong pπ - pπ bonds, and so the P atoms are single bonded to each other in P_4.

584

19.79 (a) Nitric acid is a strong oxidizing agent, but phosphoric acid is not because the nitrogen atom is smaller and more electronegative than the phosphorus atom. This favors its reduction.
(b) P, As, and Sb can use d orbitals to become five-coordinate. Nitrogen cannot.

19.80 (a) $2 NO(g) + O_2(g) \rightarrow 2 NO_2(g)$
(b) $4 HNO_3(aq) \rightarrow 4 NO_2(aq) + O_2(g) + 2 H_2O(l)$
(c) $3 Ag(s) + 4 H^+(aq) + NO_3^-(aq) \rightarrow 3 Ag^+(aq) + NO(g) + 2 H_2O(l)$
(d) $N_2H_4(aq) + 2 I_2(aq) \rightarrow N_2(g) + 4 H^+(aq) + 4 I^-(aq)$

19.81 (a) N_2 is separated from liquid air by fractional distillation.
(b) NH_3, Haber process $N_2(g) + 3 H_2(g) \rightarrow 2 NH_3(g)$
(c) HNO_3, Ostwald process

$$4 NH_3(g) + 5 O_2(g) \xrightarrow[\text{Pt/Rh catalyst}]{850\,°C} 4 NO(g) + 6 H_2O(g)$$

$2 NO(g) + O_2(g) \rightarrow 2 NO_2(g)$
$3 NO_2(g) + H_2O(l) \rightarrow 2 HNO_3(aq) + NO(g)$
(d) H_3PO_4
For food additive:
$P_4(l) + 5 O_2(g) \rightarrow P_4O_{10}(s)$
$P_4O_{10}(s) + 6 H_2O(l) \rightarrow 4 H_3PO_4(aq)$
For fertilizers:
$Ca_3(PO_4)_2(s) + 3 H_2SO_4(aq) \rightarrow 2 H_3PO_4(aq) + 3 CaSO_4(s)$

The Group 6A Elements (Sections 19.13–19.14)

19.82 (a) O   (b) Te   (c) Po   (d) O

19.83 (a) Po   (b) O   (c) O   (d) S

19.84 (a) rhombic sulfur -- yellow crystalline solid (mp 113 °C) that contains crown-shaped S_8 rings.
(b) monoclinic sulfur -- an allotrope of sulfur in which the S_8 rings pack differently in the crystal.
(c) plastic sulfur -- when sulfur is cooled rapidly, the sulfur forms disordered, tangled chains, yielding an amorphous, rubbery material called plastic sulfur.
(d) Liquid sulfur between 160 and 195 °C becomes dark reddish-brown and very viscous forming long polymer chains (S_n, n > 200,000).

19.85 The dramatic increase in the viscosity of molten sulfur at 160-195 °C is due to opening of the S_8 rings, yielding S_8 chains that form long polymer chains, which become entangled. Above 200 °C, the polymer chains begin to fragment into smaller pieces, with a decrease in viscosity.

19.86　(a) hydrogen sulfide, H_2S　(b) sulfur dioxide, SO_2　(c) sulfur trioxide, SO_3

　　　　　lead(II) sulfide, PbS　　　　sulfurous acid, H_2SO_3　　　sulfur hexafluoride, SF_6

19.87　(a) HgS, -2　(b) $Ca(HSO_4)_2$, $+6$　(c) H_2SO_3, $+4$　(d) FeS_2, -1　(e) SF_4, $+4$

19.88　Sulfuric acid (H_2SO_4) is manufactured by the contact process, a three-step reaction sequence in which (1) sulfur burns in air to give SO_2, (2) SO_2 is oxidized to SO_3 in the presence of a vanadium(V) oxide catalyst, and (3) SO_3 reacts with water to give H_2SO_4.

　　　　(1) $S(s) + O_2(g) \rightarrow SO_2(g)$

$$\text{(2) } 2\,SO_2(g) + O_2(g) \xrightarrow[\text{V}_2\text{O}_5 \text{ catalyst}]{\text{heat}} 2\,SO_3(g)$$

　　　　(3) $SO_3(g) + H_2O \text{ (in conc } H_2SO_4) \rightarrow H_2SO_4(l)$

19.89　(a) $Na_2SO_3(s) + 2\,H^+(aq) \rightarrow SO_2(g) + H_2O(l) + 2\,Na^+(aq)$

　　　　(b) $CH_3C(S)NH_2(aq) + H_2O(l) \rightarrow CH_3C(O)NH_2(aq) + H_2S(aq)$

　　　　(c) $NaOH(aq) + H_2SO_4(aq) \rightarrow NaHSO_4(aq) + H_2O(l)$

19.90　(a) $Zn(s) + 2\,H_3O^+(aq) \rightarrow Zn^{2+}(aq) + H_2(g) + 2\,H_2O(l)$

　　　　(b) $BaSO_3(s) + 2\,H_3O^+(aq) \rightarrow H_2SO_3(aq) + Ba^{2+}(aq) + 2\,H_2O(l)$

　　　　(c) $Cu(s) + 2\,H_2SO_4(l) \rightarrow Cu^{2+}(aq) + SO_4^{2-}(aq) + SO_2(g) + 2\,H_2O(l)$

　　　　(d) $H_2S(aq) + I_2(aq) \rightarrow S(s) + 2\,H^+(aq) + 2\,I^-(aq)$

19.91　(a) $ZnS(s) + 2\,H_3O^+(aq) \rightarrow Zn^{2+}(aq) + H_2S(g) + 2\,H_2O(l)$

　　　　(b) $H_2S(aq) + 2\,Fe^{3+}(aq) \rightarrow S(s) + 2\,Fe^{2+}(aq) + 2\,H^+(aq)$

　　　　(c) $Fe(s) + 2\,H_3O^+(aq) \rightarrow Fe^{2+}(aq) + H_2(g) + 2\,H_2O(l)$

　　　　(d) $BaO(s) + H_3O^+(aq) + HSO_4^-(aq) \rightarrow BaSO_4(s) + 2\,H_2O(l)$

19.92　(a) Acid strength increases as the number of O atoms increases.

　　　　(b) In comparison with S, O is much too electronegative to form compounds of O in the +4 oxidation state.　Also, an S atom is large enough to accommodate four bond pairs and a lone pair in its valence shell, but an O atom is too small to do so.

　　　　(c) Each S is sp^3 hybridized with two lone pairs of electrons.　The bond angles are therefore 109.5°.　A planar ring would require bond angles of 135°.

19.93　(a) O is more electronegative than S because of its smaller size.

　　　　(b) S forms long S_n chains by forming single bonds because it does not effectively π-bond with itself.　In contrast, O can effectively π-bond with itself.

　　　　(c) SO_3 has three bond pairs and no lone pairs, therefore the geometry is trigonal planar. SO_3^{2-} has three bond pairs and one lone pair, therefore the geometry is trigonal pyramidal.

Halogen Oxoacids and Oxoacid Salts (Section 19.15)

19.94 (a) $HBrO_3$, +5 (b) HIO, +1 (c) $NaClO_2$, +3 (d) $NaIO_4$, +7

19.95 (a) KBrO, +1 (b) H_5IO_6, +7 (c) $NaBrO_3$, +5 (d) $HClO_2$, +3

19.96 (a) iodic acid (b) chlorous acid (c) sodium hypobromite
(d) lithium perchlorate

19.97 (a) potassium chlorite (b) metaperiodic acid (c) hypobromous acid
(d) sodium bromate

19.98 (a) HIO_3 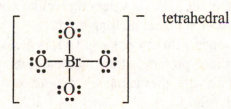 trigonal pyramidal

(b) ClO_2^- bent

(c) HOCl bent

(d) IO_6^{5-} octahedral

19.99 (a) BrO_4^- tetrahedral

(b) ClO_3^- 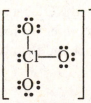 trigonal pyramidal

(c) HIO_4 tetrahedral

(d) HOBr 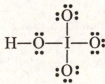 bent

19.100 Oxygen atoms are highly electronegative. Increasing the number of oxygen atoms increases the polarity of the O–H bond and increases the acid strength.

19.101 Acid strength increases in the order HIO < HBrO < HClO because electronegativity increases in the order I < Br < Cl. The higher the electronegativity the more polarized is the O–H bond.

19.102 (a) $Br_2(l) + 2 OH^-(aq) \rightarrow OBr^-(aq) + Br^-(aq) + H_2O(l)$
(b) $Cl_2(g) + H_2O(l) \rightarrow HOCl(aq) + H^+(aq) + Cl^-(aq)$
(c) $3 Cl_2(g) + 6 OH^-(aq) \rightarrow ClO_3^-(aq) + 5 Cl^-(aq) + 3 H_2O(l)$

19.103 $8 KClO_3(s) + C_{12}H_{22}O_{11}(s) \rightarrow 8 KCl(s) + 12 CO_2(g) + 11 H_2O(g)$

Chapter Problems

19.104 (a) $Na_2B_4O_7 \cdot 10 H_2O$
(b) $Ca_3(PO_4)_2$
(c) elemental sulfur, FeS_2, PbS, HgS, $CaSO_4 \cdot 2 H_2O$

19.105 (a) H_2SO_4 is used in the manufacturing of soluble phosphate and ammonium sulfate fertilizers.
(b) N_2 is used to produce NH_3 by the Haber process. NH_3 is used for fertilizers. N_2 gas is used for inert atmospheres.
(c) NH_3 is the starting material for the industrial synthesis of other important N-containing compounds, such as HNO_3. NH_3 itself is used as a fertilizer.
(d) HNO_3 is used for fertilizers, explosives, plastics, and dyes.
(e) H_3PO_4 is used for fertilizers and as a food additive.

19.106 (a) LiCl is an ionic compound. PCl_3 is a covalent molecular compound.
The ionic compound, LiCl, has the higher melting point.
(b) Carbon forms strong π bonds with oxygen, and CO_2 is a covalent molecular compound with a low melting point. Silicon prefers to form single bonds with oxygen. SiO_2 is three dimensional extended structure with alternating silicon and oxygen singly bonded to each other. SiO_2 is a high melting solid.
(c) Nitrogen forms strong π bonds with oxygen and NO_2 is a covalent molecular compound with a low melting point. Phosphorus prefers to form single bonds with oxygen. P_4O_{10} is a larger covalent molecular compound than NO_2, with a higher melting point.

19.107 (a) Ga   (b) In   (c) Pb
Metals are better electrical conductors than nonmetals (S and P) or semimetals (B).

19.108 In silicate and phosphate anions, both Si and P are surrounded by tetrahedra of O atoms, which can link together to form chains and rings.

19.109 Both NH_3 and PH_3 are colorless gases at room temperature. Both have a trigonal pyramidal geometry. NH_3 can hydrogen bond, PH_3 cannot. Aqueous solutions of NH_3 are basic. Aqueous solutions of PH_3 are neutral.

19.110 Earth's crust: O, Si, Al, Fe;   Human body: O, C, H, N

19.111 H_2SO_4, N_2, O_2, $CH_2=CH_2$, CaO, NH_3, H_3PO_4, $CH_3CH=CH_2$, Cl_2, and NaOH.

19.112 C, Si, Ge and Sn have allotropes with the diamond structure.
Sn and Pb have metallic allotropes.
C (nonmetal), Si (semimetal), Ge (semimetal), Sn (semimetal and metal), Pb (metal)

19.113 (a) $NH_4^+(aq) + OH^-(aq) \rightarrow NH_3(g) + H_2O(l)$
(b) $CO_3^{2-}(aq) + 2\,H^+(aq) \rightarrow CO_2(g) + H_2O(l)$
(c) $2\,NaBH_4 + I_2 \rightarrow B_2H_6(g) + H_2(g) + 2\,NaI$
(d) $CaC_2(s) + 2\,H_2O(l) \rightarrow C_2H_2(g) + Ca(OH)_2(aq)$
(e) $NH_4NO_3(l) \rightarrow N_2O(g) + 2\,H_2O(g)$
(f) $Cu(s) + 2\,NO_3^-(aq) + 4\,H^+(aq) \rightarrow Cu^{2+}(aq) + 2\,NO_2(g) + 2\,H_2O(l)$

19.114 (a) $H_3PO_4(aq) + H_2O(l) \rightleftharpoons H_3O^+(aq) + H_2PO_4^-(aq)$
H_3PO_4 is a Brønsted-Lowry acid.
(b) $B(OH)_3(aq) + 2\,H_2O(l) \rightleftharpoons B(OH)_4^-(aq) + H_3O^+(aq)$
$B(OH)_3$ is a Lewis acid.

19.115 H_2SO_4 is present in acid rain.
$S(s) + O_2(g) \rightarrow SO_2(g)$ (burning of fossil fuels)
$2\,SO_2(g) + O_2(g) \rightarrow 2\,SO_3(g)$ (in the atmosphere)
$SO_3(g) + H_2O(l) \rightarrow H_2SO_4(aq)$ (rain)

19.116 (a) In diamond each C is covalently bonded to four additional C atoms in a rigid three-dimensional network solid. Graphite is a two-dimensional covalent network solid of carbon sheets that can slide over each other. Both are high melting because melting requires the breaking of C–C bonds.
(b) Chlorine does not form perhalic acids of the type H_5XO_6 because its smaller size favors a tetrahedral structure over an octahedral one.

19.117 (a) C as diamond   (b) $Cl_2(g) + H_2O(l) \rightarrow HOCl(aq) + H^+(aq) + Cl^-(aq)$
(c) NO   (d) NO_2 (e) BF_3 (f) Al_2O_3 (g) Si   (h) HNO_3
(i) C as diamond, graphite, and fullerene.

19.118 The angle required by P_4 is 60°. The strain would not be reduced by using sp^3 hybrid orbitals because their angle is ~109°.

19.119 Carbon is a versatile element that can form millions of very stable compounds with elements such as N, O, and H. Biomolecules contain chains and rings with many C–C bonds. Si–Si bonds are much less stable and chains of Si atoms are uncommon. In addition, carbon can form very stable pπ-pπ multiple bonds. On the other hand, the chemistry of silicon (which cannot form stable pπ-pπ bonds) is dominated by structures based on the SiO_4^{4-} anion.

Multiconcept Problems

19.120 (a) $\cdot\ddot{N}=\ddot{O}:$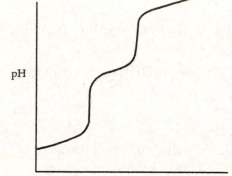

The O–N–O bond angle should be ~120°.

(b)

σ^*_{2p}	—
π^*_{2p}	↑ —
σ_{2p}	↑↓
π_{2p}	↑↓ ↑↓
σ^*_{2s}	↑↓
σ_{2s}	↑↓

The bond order is 2½ with one unpaired electron.

19.121 (a)

HO—P—H with O double bonded and OH

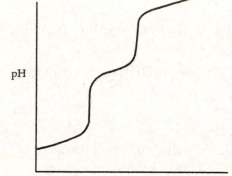

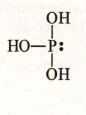

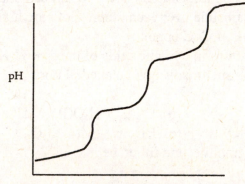

(b) $K_{a1} = 1.0 \times 10^{-2}$; $pK_{a1} = -\log(1.0 \times 10^{-2}) = 2.00$

$K_{a2} = 2.6 \times 10^{-7}$; $pK_{a2} = -\log(2.6 \times 10^{-7}) = 6.59$

At the first equivalence point, $pH = \dfrac{pK_{a1} + pK_{a2}}{2} = \dfrac{2.00 + 6.59}{2} = 4.29$

mmol HPO_3^{2-} = (30.00 mL)(0.1240 mmol/mL) = 3.72 mmol HPO_3^{2-}

volume NaOH to reach second equivalence point

$$= 3.72 \text{ mmol } HPO_3^{2-} \times \frac{2 \text{ mmol NaOH}}{1 \text{ mmol } HPO_3^{2-}} \times \frac{1.00 \text{ mL}}{0.1000 \text{ mmol NaOH}} = 74.40 \text{ mL}$$

At the second equivalence point only Na_2HPO_3, a basic salt, is in solution.

$$[HPO_3^{2-}] = \frac{3.72 \text{ mmol}}{30.00 \text{ mL} + 74.40 \text{ mL}} = 0.0356 \text{ mmol/mL} = 0.0356 \text{ M}$$

$$K_b = \frac{K_w}{K_{a2}} = \frac{1.0 \times 10^{-14}}{2.6 \times 10^{-7}} = 3.8 \times 10^{-8}$$

$$HPO_3^{2-}(aq) + H_2O(l) \rightleftharpoons H_2PO_3^{2-}(aq) + OH^-(aq)$$

initial (M)	0.0356	0	~0
change (M)	–x	+x	+x
equil (M)	0.0356 – x	x	x

$$K_b = \frac{[H_2PO_3^-][OH^-]}{[HPO_3^{2-}]} = 3.8 \times 10^{-8} = \frac{(x)(x)}{0.0356 - x} \approx \frac{x^2}{0.0356}$$

Solve for x.

$$x = [OH^-] = \sqrt{(3.8 \times 10^{-8})(0.0356)} = 3.69 \times 10^{-5} \text{ M}$$

$$[H_3O^+] = \frac{K_w}{[OH^-]} = \frac{1.0 \times 10^{-14}}{3.69 \times 10^{-5}} = 2.71 \times 10^{-10} \text{ M}$$

$$pH = -\log[H_3O^+] = -\log(2.71 \times 10^{-10}) = 9.57$$

19.122

$2 \text{ In}^+(aq) + 2 e^- \rightarrow 2 \text{ In}(s)$	$E^\circ = -0.14 \text{ V}$
$\text{In}^+(aq) \rightarrow \text{In}^{3+}(aq) + 2 e^-$	$E^\circ = 0.44 \text{ V}$
$3 \text{ In}^+(aq) \rightarrow \text{In}^{3+}(aq) + 2 \text{ In}(s)$	$E^\circ = 0.30 \text{ V}$

$1 \text{ J} = 1 \text{ V} \cdot \text{C}$

$\Delta G^\circ = -nFE^\circ = -(2)(96,500 \text{ C})(0.30 \text{ V}) = -5.8 \times 10^4 \text{ J} = -58 \text{ kJ}$

Because $\Delta G^\circ < 0$, the disproportionation is spontaneous.

$2 \text{ Tl}^+(aq) + 2 e^- \rightarrow 2 \text{ Tl}(s)$	$E^\circ = -0.34 \text{ V}$
$\text{Tl}^+(aq) \rightarrow \text{Tl}^{3+}(aq) + 2 e^-$	$E^\circ = -1.25 \text{ V}$
$3 \text{ Tl}^+(aq) \rightarrow \text{Tl}^{3+}(aq) + 2 \text{ Tl}(s)$	$E^\circ = -1.59 \text{ V}$

$\Delta G^\circ = -nFE^\circ = -(2)(96,500 \text{ C})(-1.59 \text{ V}) = +3.07 \times 10^5 \text{ J} = +307 \text{ kJ}$

Because $\Delta G^\circ > 0$, the disproportionation is nonspontaneous.

19.123 (a) $2 NH_4NO_3(s) \rightarrow 2 N_2(g) + O_2(g) + 4 H_2O(g)$; NH_4NO_3, 80.04 amu

(b) volume $= 1.80 \text{ m}^3 \times \left(\frac{100 \text{ cm}}{1 \text{ m}} \right)^3 = 1.80 \times 10^6 \text{ cm}^3$

mass $NH_4NO_3 = (1.80 \times 10^6 \text{ cm}^3)(1.725 \text{ g/cm}^3) = 3.105 \times 10^6 \text{ g}$

total mols of gas =

$$3.105 \times 10^6 \text{ g } NH_4NO_3 \times \frac{1 \text{ mol } NH_4NO_3}{80.04 \text{ g } NH_4NO_3} \times \frac{7 \text{ mol gas}}{2 \text{ mol } NH_4NO_3} = 1.36 \times 10^5 \text{ mol gas}$$

$500\ ^{\circ}C = 773\ K; \quad PV = nRT$

$$V = \frac{nRT}{P} = \frac{(1.36 \times 10^5\ \text{mol})\left(0.082\ 06\ \dfrac{L \cdot atm}{K \cdot mol}\right)(773\ K)}{1.00\ atm} = 8.62 \times 10^6\ L$$

(c) $\Delta H^{\circ} = 4\ \Delta H^{\circ}_f(H_2O) - 2\ \Delta H^{\circ}_f(NH_4NO_3)$

$\Delta H^{\circ} = [(4\ \text{mol})(-241.8\ \text{kJ/mol})] - [(2\ \text{mol})(-365.6\ \text{kJ/mol})] = -236.0\ \text{kJ/2 mol } NH_4NO_3$

$$q = 3.105 \times 10^6\ g\ NH_4NO_3 \times \frac{1\ \text{mol } NH_4NO_3}{80.04\ g\ NH_4NO_3} \times \frac{-236.0\ \text{kJ}}{2\ \text{mol } NH_4NO_3} = -4.58 \times 10^6\ \text{kJ}$$

4.58×10^6 kJ of heat are released in this reaction.

19.124 NH_4NO_3, 80.04 amu; $\quad (NH_4)_2HPO_4$, 132.06 amu

(a) % N in $NH_4NO_3 = \dfrac{2 \times (14.007\ \text{amu N})}{80.04\ \text{amu } NH_4NO_3} \times 100\% = 35.0\%\ N$

% N in $(NH_4)_2HPO_4 = \dfrac{2 \times (14.007\ \text{amu N})}{132.06\ \text{amu } (NH_4)_2HPO_4} \times 100\% = 21.2\%\ N$

Let x equal the fraction of NH_4NO_3 and $(1 - x)$ equal the fraction of $(NH_4)_2HPO_4$ in the mixture.

$0.3381 = x(0.350) + (1 - x)(0.212) = 0.138x + 0.212$

$0.1261 = 0.138x \qquad x = \dfrac{0.1261}{0.138} = 0.9138$

$(1 - x) = (1 - 0.9138) = 0.0862$

There is 8.62 % $(NH_4)_2HPO_4$ and 91.38% NH_4NO_3 in the mixture.

(b) sample mass = 0.965 g

mass $NH_4NO_3 = (0.965\ g)(0.9138) = 0.8818\ g$

mass $(NH_4)_2HPO_4 = 0.965\ g - 0.8818\ g = 0.0832\ g$

mol $NH_4NO_3 = 0.8818\ g \times \dfrac{1\ \text{mol } NH_4NO_3}{80.04\ g\ NH_4NO_3} = 0.0110\ \text{mol } NH_4NO_3$

mol $(NH_4)_2HPO_4 = 0.0832\ g \times \dfrac{1\ \text{mol } (NH_4)_2HPO_4}{132.06\ g\ (NH_4)_2HPO_4} = 6.30 \times 10^{-4}\ \text{mol } (NH_4)_2HPO_4$

$[NH_4^+] = \dfrac{0.0110\ \text{mol} + 2(6.30 \times 10^{-4}\ \text{mol})}{0.0500\ L} = 0.245\ M$

$[HPO_4^{2-}] = \dfrac{6.30 \times 10^{-4}\ \text{mol}}{0.0500\ L} = 0.0126\ M$

The equilibrium constant for the transfer of a proton from the cation of a salt to the anion of the salt is equal to $\dfrac{K_a K_b}{K_w}$.

$K_a(NH_4^+) = 5.56 \times 10^{-10}$ and $K_b(HPO_4^{2-}) = 1.61 \times 10^{-7}$

$K = \dfrac{K_a K_b}{K_w} = \dfrac{(5.56 \times 10^{-10})(1.61 \times 10^{-7})}{1.0 \times 10^{-14}} = 8.95 \times 10^{-3}$

$$NH_4^+(aq) + HPO_4^{2-}(aq) \rightleftharpoons H_2PO_4^-(aq) + NH_3(aq)$$

initial (M)	0.245	0.0126	0	0
change (M)	$-x$	$-x$	$+x$	$+x$
equil (M)	$0.245 - x$	$0.0126 - x$	x	x

$$K = \frac{[H_2PO_4^-][NH_3]}{[NH_4^+][HPO_4^{2-}]} = 8.95 \times 10^{-3} = \frac{x^2}{(0.245 - x)(0.0126 - x)}$$

$0.991x^2 + 0.002\,31x - (2.76 \times 10^{-5}) = 0$

Use the quadratic formula to solve for x.

$$x = \frac{(-0.002\,31) \pm \sqrt{(0.002\,31)^2 - (4)(0.991)(-2.76 \times 10^{-5})}}{2(0.991)} = \frac{(-0.002\,31) \pm 0.010\,71}{2(0.991)}$$

$x = 0.004\,24$ and $-0.006\,65$

Of the two solutions for x, only the positive value of x has physical meaning because x is equal to both the $[H_2PO_4^-]$ and $[NH_3]$.

$$K_a(NH_4^+) = \frac{[H_3O^+][NH_3]}{[NH_4^+]} = 5.56 \times 10^{-10} = \frac{[H_3O^+](0.004\,24)}{(0.245 - 0.004\,24)}$$

$$[H_3O^+] = \frac{(5.56 \times 10^{-10})(0.245 - 0.004\,24)}{0.004\,24} = 3.157 \times 10^{-8} \text{ M}$$

$pH = -\log[H_3O^+] = -\log(3.157 \times 10^{-8}) = 7.50$

19.125 (a) $P_4(s) + 5\,O_2(g) \rightarrow P_4O_{10}(s)$

$\qquad P_4O_{10}(s) + 6\,H_2O(l) \rightarrow 4\,H_3PO_4(aq)$

(b) P_4, 123.90 amu

$$\text{mol } H_3PO_4 = 5.00 \text{ g } P_4 \times \frac{1 \text{ mol } P_4}{123.90 \text{ g } P_4} \times \frac{1 \text{ mol } P_4O_{10}}{1 \text{ mol } P_4} \times \frac{4 \text{ mol } H_3PO_4}{1 \text{ mol } P_4O_{10}} = 0.1614 \text{ mol}$$

$$[H_3PO_4] = \frac{0.1614 \text{ mol}}{0.2500 \text{ L}} = 0.646 \text{ M}$$

For the dissociation of the first proton, the following equilibrium must be considered:

$$H_3PO_4(aq) + H_2O(l) \rightleftharpoons H_3O^+(aq) + H_2PO_4^-(aq)$$

initial (M)	0.646	~0	0
change (M)	$-x$	$+x$	$+x$
equil (M)	$0.646 - x$	x	x

$$K_{a1} = \frac{[H_3O^+][H_2PO_4^-]}{[H_3PO_4]} = 7.5 \times 10^{-3} = \frac{x^2}{0.646 - x}$$

$x^2 + (7.5 \times 10^{-3})x - (4.84 \times 10^{-3}) = 0$

Solve for x using the quadratic formula.

$$x = \frac{-(7.5 \times 10^{-3}) \pm \sqrt{(7.5 \times 10^{-3})^2 - (4)(1)(-4.84 \times 10^{-3})}}{2(1)} = \frac{(-7.5 \times 10^{-3}) \pm 0.139}{2}$$

$x = 0.0657$ and -0.0733;   Of the two solutions for x, only the positive value of x has physical meaning, because x is the $[H_3O^+]$.

$x = 0.0657 \text{ M} = [H_2PO_4^-] = [H_3O^+]$

Only the dissociation of the first proton contributes a significant amount of H_3O^+.

$pH = -\log[H_3O^+] = -\log(0.0657) = 1.18$

(c) $3 \text{ Ca}^{2+}(aq) + 2 \text{ H}_3PO_4(aq) \rightarrow \text{Ca}_3(PO_4)_2(s) + 6 \text{ H}^+(aq)$

$\text{Ca}_3(PO_4)_2$, 310.18 amu

$$\text{mass Ca}_3(PO_4)_2 = 0.1614 \text{ mol H}_3PO_4 \times \frac{1 \text{ mol Ca}_3(PO_4)_2}{2 \text{ mol H}_3PO_4} \times \frac{310.18 \text{ g Ca}_3(PO_4)_2}{1 \text{ mol Ca}_3(PO_4)_2} = 25.0 \text{ g}$$

(d) $\text{Zn}(s) + 2 \text{ H}^+(aq) \rightarrow \text{H}_2(g) + \text{Zn}^{2+}(aq)$; the gas is H_2.

$$\text{mol H}_2 = 0.1614 \text{ mol H}_3PO_4 \times \frac{6 \text{ mol H}^+}{2 \text{ mol H}_3PO_4} \times \frac{1 \text{ mol H}_2}{2 \text{ mol H}^+} = 0.242 \text{ mol H}_2$$

$PV = nRT$; $20 \text{ °C} = 293 \text{ K}$

$$V = \frac{nRT}{P} = \frac{(0.242 \text{ mol})\left(0.082\,06 \dfrac{\text{L} \cdot \text{atm}}{\text{K} \cdot \text{mol}}\right)(293 \text{ K})}{\left(742 \text{ mm Hg} \times \dfrac{1.00 \text{ atm}}{760 \text{ mm Hg}}\right)} = 5.96 \text{ L}$$

19.126 $\text{N}_2O_4(g) \rightleftharpoons 2 \text{ NO}_2(g)$

$$P_{Total} = 753 \text{ mm Hg} \times \frac{1.00 \text{ atm}}{760 \text{ mm Hg}} = 0.991 \text{ atm}$$

$P_{Total} = P_{N_2O_4} + P_{NO_2} = 0.991 \text{ atm}$

$P_{N_2O_4} = 0.991 \text{ atm} - P_{NO_2}$

$$K_p = \frac{(P_{NO_2})^2}{P_{N_2O_4}} = 0.113; \qquad K_p = \frac{(P_{NO_2})^2}{(0.991 \text{ atm} - P_{NO_2})} = 0.113$$

$(P_{NO_2})^2 + 0.113(P_{NO_2}) - 0.112 = 0$

Use the quadratic formula to solve for P_{NO_2}.

$$P_{NO_2} = \frac{-(0.113) \pm \sqrt{(0.113)^2 - (4)(1)(-0.112)}}{2(1)} = \frac{-0.113 \pm 0.679}{2}$$

$P_{NO_2} = -0.0396$ and 0.283

Of the two solutions for P_{NO_2}, only 0.283 has physical meaning because NO_2 can't have a negative partial pressure.

$P_{NO_2} = 0.283 \text{ atm}$ and $P_{N_2O_4} = 0.991 \text{ atm} - P_{NO_2} = 0.991 - 0.283 = 0.708 \text{ atm}$

	$\text{N}_2O_4(g)$	$\rightarrow$	$2 \text{ NO}_2(g)$
before reaction (atm)	0.708		0.283
change (atm)	−0.708		+2(0.708)
after reaction (atm)	0		$0.283 + 2(0.708) = 1.70 \text{ atm}$

$PV = nRT; \quad 25\ °C = 298\ K$

$$n_{NO_2} = \frac{PV}{RT} = \frac{(1.70\ atm)(0.5000\ L)}{\left(0.082\ 06\ \frac{L\cdot atm}{K\cdot mol}\right)(298\ K)} = 0.0348\ mol\ NO_2$$

(a)

	$2\ NO_2(aq)$	$+\ 2\ H_2O(l)\ \rightarrow$	$HNO_2(aq)$	$+\ H_3O^+(aq)$	$+\ NO_3^-(aq)$
before reaction (mol)	0.0348		0	~0	0
change (mol)	−0.0348		+0.0348/2	+0.0348/2	+0.0348/2
after reaction (mol)	0		0.0174	0.0174	0.0174

(b) $[HNO_2] = [H_3O^+] = 0.0174\ mol/0.250\ L = 0.0696\ M$

	$HNO_2(aq)$	$+\ H_2O(l)\ \rightleftharpoons$	$H_3O^+(aq)$	$+\ NO_2^-(aq)$
initial (M)	0.0696		0.0696	0
change (M)	−x		+x	+x
equil (M)	0.0696 − x		0.0696 + x	x

$$K_a = \frac{[H_3O^+][NO_2^-]}{[HNO_2]} = 4.5 \times 10^{-4} = \frac{(0.0696\ +\ x)x}{(0.0696\ -\ x)}$$

$x^2 + 0.070\ 05 - (3.132 \times 10^{-5}) = 0$

Use the quadratic formula to solve for x.

$$x = \frac{-(0.070\ 05) \pm \sqrt{(0.070\ 05)^2 - (4)(1)(-3.132 \times 10^{-5})}}{2(1)} = \frac{-0.070\ 05 \pm 0.070\ 94}{2}$$

$x = -0.0705$ and 4.45×10^{-4}

Of the two solutions for x only 4.45×10^{-4} has physical meaning because −0.0705 leads to negative concentrations.

$[NO_2^-] = x = 4.45 \times 10^{-4}\ M = 4.4 \times 10^{-4}\ M$

$[H_3O^+] = 0.0696 + x = 0.0696\ +\ 4.45 \times 10^{-4} = 0.0700\ M$

$pH = -\log[H_3O^+] = -\log(0.0700) = 1.15$

(c) Total solution molarity $= [NO_3^-] + [NO_2^-] + [H_3O^+] + [HNO_2]$

$= 0.0696\ M + (4.45 \times 10^{-4}) + 0.0700\ M + 0.0692\ M = 0.2092\ M$

$$\Pi = MRT = (0.2092\ M)\left(0.082\ 06\ \frac{L\cdot atm}{K\cdot mol}\right)(298\ K) = 5.12\ atm$$

(d) $mol\ H_3O^+ = (0.0700\ mol/L)(0.250\ L) = 0.0175\ mol\ H_3O^+$

$mol\ HNO_2 = (0.0692\ mol/L)(0.250\ L) = 0.0173\ mol\ HNO_2$

Total mol of acid to neutralize $= 0.0175\ +\ 0.0173 = 0.0348\ mol$

$$mass\ CaO = 0.0348\ mol\ acid \times \frac{1\ mol\ CaO}{2\ mol\ acid} \times \frac{56.08\ g\ CaO}{1\ mol\ CaO} = 0.976\ g\ CaO$$

20 Transition Elements and Coordination Chemistry

20.1 (a) V, [Ar] $3d^3 4s^2$ (b) Co^{2+}, [Ar] $3d^7$
(c) Mn^{4+} in MnO_2, [Ar] $3d^3$ (d) Cu^{2+} in $CuCl_4^{2-}$, [Ar] $3d^9$

20.2

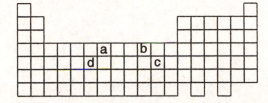

      (a) Mn   (b) Ni^{2+}

      (c) Ag   (d) Mo^{3+}

20.3 Z_{eff} increases from left to right across the first transition series.
(a) The transition metal with the lowest Z_{eff} (Ti) should be the strongest reducing agent because it is easier for Ti to lose its valence electrons. The transition metal with the highest Z_{eff} (Zn) should be the weakest reducing agent because it is more difficult for Zn to lose its valence electrons.
(b) The oxoanion with the highest Z_{eff} (FeO_4^{2-}) should be the strongest oxidizing agent because of the greater attraction for electrons. The oxoanion with the lowest Z_{eff} (VO_4^{3-}) should be the weakest oxidizing agent because of the lower attraction for electrons.

20.4 (a) $Cr_2O_7^{2-}$ (b) Cr^{3+} (c) Cr^{2+} (d) Fe^{2+} (e) Cu^{2+}

20.5 (a) $Cr(OH)_2$ (b) $Cr(OH)_4^-$ (c) CrO_4^{2-} (d) $Fe(OH)_2$ (e) $Fe(OH)_3$

20.6 $[Cr(NH_3)_2(SCN)_4]^-$

20.7 In $Na_4[Fe(CN)_6]$ each sodium is in the +1 oxidation state (+4 total); each cyanide (CN^-) has a −1 charge (−6 total). The compound is neutral; therefore, the oxidation state of the iron is +2.

20.8 (a)

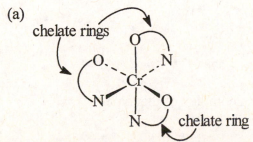

(b) Cr^{3+} is the Lewis acid. The glycinate ligand is the Lewis base. Nitrogen and oxygen are the ligand donor atoms. The chelate rings are identified in the drawing.
(c) The coordination number is 6. The coordination geometry is octahedral. The chromium is in the +3 oxidation state.

20.9   (a) tetraamminecopper(II) sulfate   (b) sodium tetrahydroxochromate(III)

      (c) triglycinatocobalt(III)   (d) pentaaquaisothiocyanatoiron(III) ion

20.10 (a) $[Zn(NH_3)_4](NO_3)_2$ (b) $Ni(CO)_4$ (c) $K[Pt(NH_3)Cl_3]$ (d) $[Au(CN)_2]^-$

20.11 (a) Two diastereoisomers are possible.

            cis   trans

     (b) No isomers are possible for a tetrahedral complex of the type MA_2B_2.

     (c) Two diastereoisomers are possible.

     (d) No isomers are possible for a complex of this type.

     (e) Two diastereoisomers are possible.

           trans   cis

     (f) No diastereoisomers are possible for a complex of this type.

20.12 (1) and (2) are the same. (3) and (4) are the same. (1) and (2) are different from (3) and (4).

20.13 (a) chair, no   (b) foot, yes   (c) pencil, no
      (d) corkscrew, yes   (e) banana, no   (f) football, no

20.14 (a) (2) and (3) are chiral and (1) and (4) are achiral.
      (b) enantiomer of (2)   enantiomer of (3)

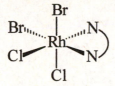 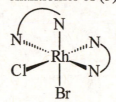

20.15 (a) $[Fe(C_2O_4)_3]^{3-}$ can exist as enantiomers.

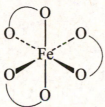 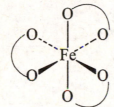

      (b) $[Co(NH_3)_4en]^{3+}$ cannot exist as enantiomers.
      (c) $[Co(NH_3)_2(en)_2]^{3+}$ can exist as enantiomers.

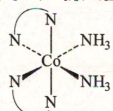

      (d) $[Cr(H_2O)_4Cl_2]^+$ cannot exist as enantiomers.

20.16 (a) The ion is absorbing in the red (625 nm), so the most likely color for the ion is blue.
      (b) 625 nm = 625 x 10^{-9} m

$$E = h\frac{c}{\lambda} = (6.626 \times 10^{-34}\,J \cdot s)\left(\frac{3.00 \times 10^8\,m/s}{625 \times 10^{-9}\,m}\right) = 3.18 \times 10^{-19}\,J$$

20.17 (a) Fe^{3+} [Ar]   ↑ ↑ ↑ ↑ ↑
                           3d   4s   4p

          $[Fe(CN)_6]^{3-}$ [Ar]   ↑↓ ↑↓ ↑ | ↑↓ ↑↓   ↑↓   ↑↓ ↑↓ ↑↓
                                     3d   4s   4p

                     d^2sp^3 1 unpaired e^-

(b)   Co²⁺   [Ar]   ↓↑ ↓↑ ↑ ↑   ___   ___ ___ ___
                          3d   4s   4p

[Co(H₂O)₆]²⁺   [Ar]   ↓↑ ↓↑ ↑ ↑ ↑ | ↓↑   ↓↑ ↓↑ ↓↑   ↓↑ ↓↑ | ___ ___ ___
                          3d   4s   4p   4d

                                   sp³d²   3 unpaired e⁻

(c)   V³⁺   [Ar]   ↑ ↑ ___ ___ ___   ___   ___ ___ ___
                          3d   4s   4p

[VCl₄]⁻   [Ar]   ↑ ↑ ___ ___ ___ | ↓↑   ↓↑ ↓↑ ↓↑ |
                          3d   4s   4p

                                   sp³   2 unpaired e⁻

(d)   Pt²⁺   [Xe]   ↓↑ ↓↑ ↓↑ ↑ ↑   ___   ___ ___ ___
                          5d   6s   6p

[PtCl₄]²⁻   [Xe]   ↓↑ ↓↑ ↓↑ ↓↑ | ↓↑   ↓↑   ↓↑ ↓↑ | ___
                          5d   6s   6p

                                   dsp²   no unpaired e⁻

20.18

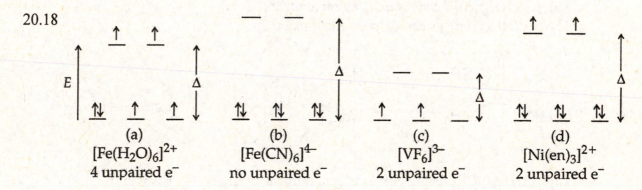

          (a)   (b)   (c)   (d)

    [Fe(H₂O)₆]²⁺   [Fe(CN)₆]⁴⁻   [VF₆]³⁻   [Ni(en)₃]²⁺

   4 unpaired e⁻   no unpaired e⁻   2 unpaired e⁻   2 unpaired e⁻

20.19   Both [NiCl₄]²⁻ and [Ni(CN)₄]²⁻ contain Ni²⁺ with a [Ar] 3d⁸ electron configuration.

(a)   [NiCl₄]²⁻ (tetrahedral)   (b)   [Ni(CN)₄]²⁻ (square planar)

        ↓↑ ↑ ↑   ___
        xy xz yz   x²–y²

                                      ↓↑
                                     xy

                                      ↓↑
       ↓↑ ↓↑   z²
      z² x²–y²

  2 unpaired electrons   ↓↑ ↓↑
                                  xz yz

                           no unpaired electrons

20.20   Although titanium has a large positive E° for oxidation, the bulk metal is remarkably immune to corrosion because its surface becomes coated with a thin, protective oxide film.

20.21   Actual yield = 1.00×10^5 tons Ti $\times \dfrac{2000 \text{ lbs}}{1 \text{ ton}} \times \dfrac{453.6 \text{ g}}{1 \text{ lb}} = 9.07 \times 10^{10}$ g Ti

Theoretical yield = $\dfrac{9.07 \times 10^{10} \text{ g Ti}}{0.935} = 9.70 \times 10^{10}$ g Ti

mol Ti = 9.70×10^{10} g Ti $\times \dfrac{1 \text{ mol Ti}}{47.88 \text{ g Ti}} = 2.03 \times 10^9$ mol Ti

mol Cl_2 = 2.03×10^9 mol Ti $\times \dfrac{2 \text{ mol } Cl_2}{1 \text{ mol Ti}} = 4.06 \times 10^9$ mol Cl_2

$20 \,^{\circ}C = 293$ K;   PV = nRT

$$V = \frac{nRT}{P} = \frac{(4.06 \times 10^9 \text{ mol})\left(0.082\ 06 \ \dfrac{L \cdot atm}{K \cdot mol}\right)(293 \text{ K})}{\left(740 \text{ mm Hg } \times \dfrac{1.00 \text{ atm}}{760 \text{ mm Hg}}\right)} = 1.00 \times 10^{11} \text{ L of } Cl_2$$

Key Concept Problems

20.22

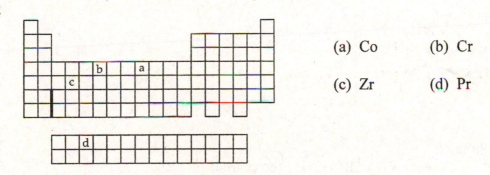

(a) Co   (b) Cr

(c) Zr   (d) Pr

20.23   (a) A^{2+}, [Ar] $3d^8$ (b) B^+, [Kr] $4d^{10}$ (c) C^{3+}, [Kr] $4d^3$ (d) DO_4^{2-}, [Ar] $3d^2$

20.24   (a) The atomic radii decrease, at first markedly and then more gradually. Toward the end of the series, the radii increase again. The decrease in atomic radii is a result of an increase in Z_{eff}. The increase is due to electron-electron repulsions in doubly occupied d orbitals.
(b) The densities of the transition metals are inversely related to their atomic radii. The densities initially increase from left to right and then decrease toward the end of the series.
(c) Ionization energies generally increase from left to right across the series. The general trend correlates with an increase in Z_{eff} and a decrease in atomic radii.
(d) The standard oxidation potentials generally decrease from left to right across the first transition series. This correlates with the general trend in ionization energies.

20.25 (a) $NH_2–CH_2–CH_2–NH_2$ is a bidentate ligand. It can form a chelate ring using the atoms indicated in bold.

(b) $CH_3–CH_2–CH_2–NH_2$ is a monodentate ligand.

(c) $NH_2–CH_2–CH_2–NH–CH_2–CO_2^-$ is a tridentate ligand. It can form chelate rings using the atoms indicated in bold.

(d) $NH_2–CH_2–CH_2–NH_3^+$ is a monodentate ligand. The first N can coordinate to a metal.

20.26 (1) dichloroethylenediamineplatinum(II)

(2) trans-diammineaquachloroplatinate(II) ion

(3) amminepentachloroplatinate(IV) ion

(4) cis-diaquabis(ethylenediamine)platinum(IV) ion

20.27 (a) $Na[Au(CN)_2]$

1 Na^+ 2 CN^-

The oxidation state of the Au is +1.   Coordination number = 2; Linear

$$\left[CN-Au-NC\right]^-$$

(b) $[Co(NH_3)_5Br]SO_4$

1 Br^- 1 SO_4^{2-} 5 NH_3 (no charge)

The oxidation state of the Co is +3.   Coordination number = 6; Octahedral

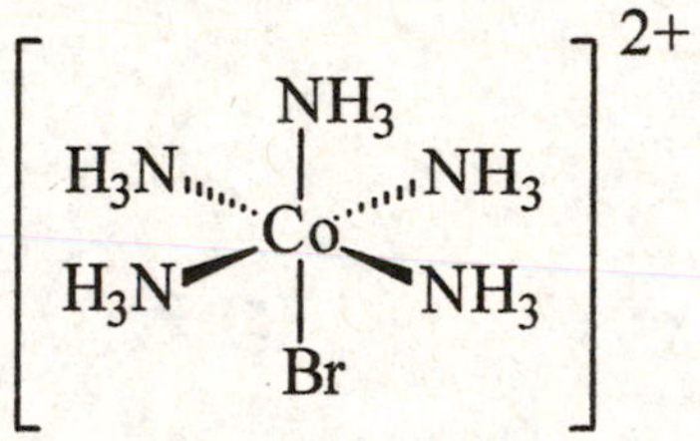

(c) $Pt(en)Cl_2$

2 Cl^- en = $NH_2CH_2CH_2NH_2$ (no charge)

The oxidation state of the Pt is +2.   Coordination number = 4; Square planar

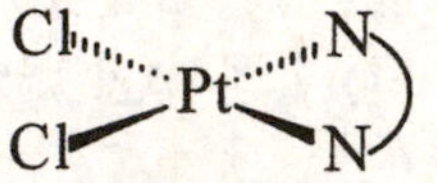

(d) $(NH_4)_2[PtCl_2(C_2O_4)_2]$

2 NH_4^+ 2 Cl^- 2 $C_2O_4^{2-}$

The oxidation state of the Pt is +4.   Coordination number = 6; Octahedral

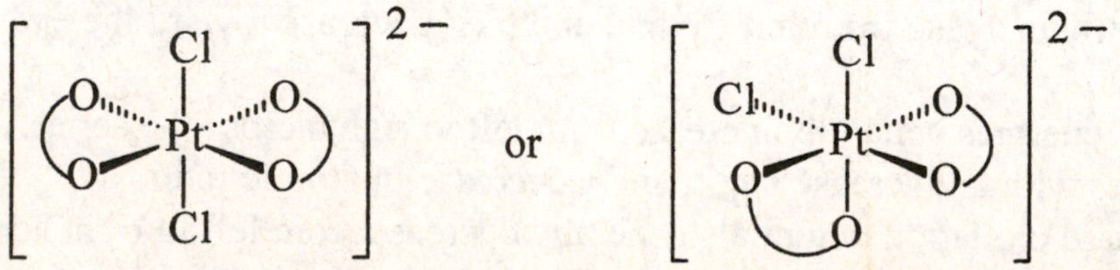

20.28. (a) (1) cis;   (2) trans;   (3) trans;   (4) cis
(b) (1) and (4) are the same.   (2) and (3) are the same.
(c) None of the isomers exist as enantiomers because their mirror images are identical.

20.29 (a) (1) chiral;   (2) achiral;   (3) chiral;   (4) chiral
(b)

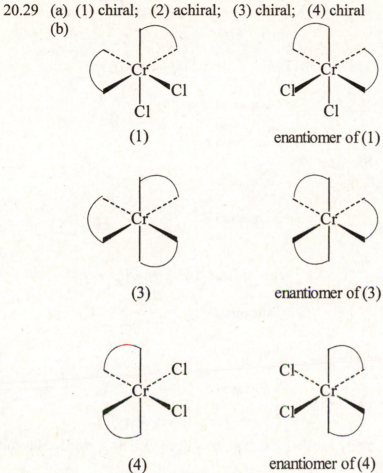

(1)   enantiomer of (1)

(3)   enantiomer of (3)

(4)   enantiomer of (4)

(c) (1) and (4) are enantiomers.

20.30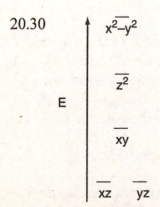

20.31 The tetrahedral complex is chiral because it is not identical to its mirror image. The square planar complex is achiral because it has a symmetry plane (the plane of the molecule).

Section Problems
Electron Configurations and Properties of Transition Elements (Sections 20.1–20.2)

20.32 (a) Cr, [Ar] $3d^5 4s^1$ (b) Zr, [Kr] $4d^2 5s^2$ (c) Co^{2+}, [Ar] $3d^7$
        (d) Fe^{3+}, [Ar] $3d^5$ (e) Mo^{3+}, [Kr] $4d^3$ (f) Cr(VI), [Ar] $3d^0$

20.33 (a) Mn, [Ar] $3d^5 4s^2$ (b) Cd, [Kr] $4d^{10} 5s^2$ (c) Cr^{3+}, [Ar] $3d^3$
        (d) Ag^+, [Kr] $4d^{10}$ (e) Rh^{3+}, [Kr] $4d^6$ (f) Mn(VI), [Ar] $3d^1$

20.34 (a) Cu^{2+},   [Ar] $3d^9$ ⇅ ⇅ ⇅ ⇅ ↑   1 unpaired e^-
                           3d

       (b) Ti^{2+},   [Ar] $3d^2$ ↑ ↑ _ _ _   2 unpaired e^-
                           3d

       (c) Zn^{2+},   [Ar] $3d^{10}$ ⇅ ⇅ ⇅ ⇅ ⇅   no unpaired e^-
                           3d

       (d) Cr^{3+},   [Ar] $3d^3$ ↑ ↑ ↑ _ _   3 unpaired e^-
                           3d

20.35 (a) Sc^{3+},   [Ar] $3d^0$ _ _ _ _ _   no unpaired e^-
                           3d

       (b) Co^{2+},   [Ar] $3d^7$ ⇅ ⇅ ↑ ↑ ↑   3 unpaired e^-
                           3d

       (c) Mn^{3+},   [Ar] $3d^4$ ↑ ↑ ↑ ↑ _   4 unpaired e^-
                           3d

       (d) Cr^{2+},   [Ar] $3d^4$ ↑ ↑ ↑ ↑ _   4 unpaired e^-
                           3d

20.36 Ti is harder than K and Ca largely because the sharing of d, as well as s, electrons results in stronger metallic bonding.

20.37 Mo has a higher melting point than Y or Cd. Melting points increase as the number of unpaired d electrons available for metallic bonding increases, and then decrease as the d electrons pair up and become less available for bonding.

20.38 (a) The decrease in radii with increasing atomic number is expected because the added d electrons only partially shield the added nuclear charge. As a result, Z_{eff} increases. With increasing Z_{eff}, the electrons are more strongly attracted to the nucleus, and atomic size decreases.
       (b) The densities of the transition metals are inversely related to their atomic radii.

20.39 Ti > V > Cr > Mn   Atomic radius decreases with increasing Z_{eff}.

20.40 The smaller than expected sizes of the third-transition series atoms are associated with what is called the lanthanide contraction, the general decrease in atomic radii of the f-block lanthanide elements. The lanthanide contraction is due to the increase in Z_{eff} as the 4f subshell is filled.

20.41   Zr and Hf are about the same size. The lanthanide contraction for Hf is due to the increase in Z_{eff} as the 4f subshell is filled.

20.42
Sc	$(631 + 1235) = 1866$ kJ/mol
Ti	$(659 + 1310) = 1969$ kJ/mol
V	$(651 + 1410) = 2061$ kJ/mol
Cr	$(653 + 1591) = 2224$ kJ/mol
Mn	$(717 + 1509) = 2226$ kJ/mol
Fe	$(762 + 1562) = 2324$ kJ/mol
Co	$(760 + 1648) = 2408$ kJ/mol
Ni	$(737 + 1753) = 2490$ kJ/mol
Cu	$(745 + 1958) = 2703$ kJ/mol
Zn	$(906 + 1733) = 2639$ kJ/mol

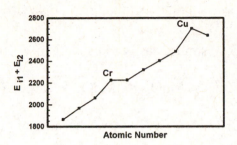

Across the first transition element series, Z_{eff} increases and there is an almost linear increase in the sum of the first two ionization energies. This is what is expected if the two electrons are removed from the 4s orbital. Higher than expected values for the sum of the first two ionization energies are observed for Cr and Cu because of their anomalous electron configurations (Cr $3d^5\ 4s^1$; Cu $3d^{10}\ 4s^1$). An increasing Z_{eff} affects 3d orbitals more than the 4s orbital and the second ionization energy for an electron from the 3d orbital is higher than expected.

20.43   Oxidation potentials generally decrease from left to right across the transition series. The general trend correlates with an increase in ionization energy, which is due in turn to an increase in Z_{eff} and a decrease in atomic radius.

20.44   (a) $Cr(s) + 2\,H^+(aq) \rightarrow Cr^{2+}(aq) + H_2(g)$ (b) $Zn(s) + 2\,H^+(aq) \rightarrow Zn^{2+}(aq) + H_2(g)$
        (c) N.R.   (d) $Fe(s) + 2\,H^+(aq) \rightarrow Fe^{2+}(aq) + H_2(g)$

20.45   (a) $Mn(s) + 2\,H^+(aq) \rightarrow Mn^{2+}(aq) + H_2(g)$ (b) N.R.
        (c) $2\,Sc(s) + 6\,H^+(aq) \rightarrow 2\,Sc^{3+}(aq) + 3\,H_2(g)$
        (d) $Ni(s) + 2\,H^+(aq) \rightarrow Ni^{2+}(aq) + H_2(g)$

Oxidation States (Section 20.3)

20.46   (b) Mn   (d) Cu

20.47   (b) Al   (d) Sc

20.48   Sc(III), Ti(IV), V(V), Cr(VI), Mn(VII), Fe(VI), Co(III), Ni(II), Cu(II), Zn(II)

20.49   The highest oxidation state for the group 3B-7B metals is the group number, corresponding to the loss of all valence s and d electrons. For the later transition metals, loss of all valence electrons is energetically prohibitive because of the increasing Z_{eff}. Therefore, only lower oxidation states are accessible for the later transition metals.

20.50 Cu^{2+} is a stronger oxidizing agent than Cr^{2+} because of a higher Z_{eff}.

20.51 Ti^{2+} is a stronger reducing agent than Ni^{2+} because of a smaller Z_{eff}.

20.52 Cr^{2+} is more easily oxidized than Ni^{2+} because of a smaller Z_{eff}.

20.53 Fe^{3+} is more easily reduced than V^{3+} because of a higher Z_{eff}.

20.54 Mn^{2+} < MnO_2 < MnO_4^- because of increasing oxidation state of Mn.

20.55 $Cr_2O_7^{2-}$ < Cr^{3+} < Cr^{2+} because of decreasing oxidation state of the Cr.

Chemistry of Selected Transition Elements (Section 20.4)

20.56 (a) $Cr_2O_3(s) + 2 Al(s) \rightarrow 2 Cr(s) + Al_2O_3(s)$
       (b) $Cu_2S(l) + O_2(g) \rightarrow 2 Cu(l) + SO_2(g)$

20.57 (a) $Fe(s) + NO_3^-(aq) + 4 H^+(aq) \rightarrow Fe^{3+}(aq) + NO(g) + 2 H_2O(l)$
       (b) $3 Cu(s) + 2 NO_3^-(aq) + 8 H^+(aq) \rightarrow 3 Cu^{2+}(aq) + 2 NO(g) + 4 H_2O(l)$
       (c) $Cr(s) + NO_3^-(aq) + 4 H^+(aq) \rightarrow Cr^{3+}(aq) + NO(g) + 2 H_2O(l)$

20.58 $Cr(OH)_2$ < $Cr(OH)_3$ < $CrO_2(OH)_2$
       Acid strength increases with polarity of the O–H bond, which increases in turn with the oxidation state of Cr.

20.59 acid: $Cr(OH)_3(s) + OH^-(aq) \rightarrow Cr(OH)_4^-(aq)$
       base: $Cr(OH)_3(s) + 3 H_3O^+(aq) \rightarrow Cr^{3+}(aq) + 6 H_2O(l)$

20.60 (c) $Cr(OH)_3$

20.61 (c) Cu^+ $2 Cu^+(aq) \rightarrow Cu(s) + Cu^{2+}(aq)$

20.62 (a) Add excess KOH(aq) and Fe^{3+} will precipitate as $Fe(OH)_3(s)$. $Na^+(aq)$ will remain in solution.
       (b) Add excess NaOH(aq) and Fe^{3+} will precipitate as $Fe(OH)_3(s)$. $Cr(OH)_4^-(aq)$ will remain in solution.
       (c) Add excess NH_3(aq) and Fe^{3+} will precipitate as $Fe(OH)_3(s)$. $Cu(NH_3)_4^{2+}(aq)$ will remain in solution.

20.63 (a) Add H_2S to precipitate CuS, or add NaOH(aq) to precipitate $Cu(OH)_2$. $K^+(aq)$ will remain in solution.
       (b) Add excess NaOH(aq) to precipitate $Cu(OH)_2$. $Cr(OH)_4^-(aq)$ will remain in solution.
       (c) Add excess NaOH(aq) to precipitate $Fe(OH)_3$. $Al(OH)_4^-(aq)$ will remain in solution.

20.64 (a) $Cr_2O_7^{2-}(aq) + 6 Fe^{2+}(aq) + 14 H^+(aq) \rightarrow 2 Cr^{3+}(aq) + 6 Fe^{3+}(aq) + 7 H_2O(l)$
       (b) $4 Fe^{2+}(aq) + O_2(g) + 4 H^+(aq) \rightarrow 4 Fe^{3+}(aq) + 2 H_2O(l)$

(c) $Cu_2O(s) + 2 H^+(aq) \rightarrow Cu(s) + Cu^{2+}(aq) + H_2O(l)$

(d) $Fe(s) + 2 H^+(aq) \rightarrow Fe^{2+}(aq) + H_2(g)$

20.65 (a) $6 Cr^{2+}(aq) + Cr_2O_7^{2-}(aq) + 14 H^+(aq) \rightarrow 8 Cr^{3+}(aq) + 7 H_2O(l)$

(b) $3 Cu(s) + 2 NO_3^-(aq) + 8 H^+(aq) \rightarrow 3 Cu^{2+}(aq) + 2 NO(g) + 4 H_2O(l)$

(c) $Cu^{2+}(aq) + 4 NH_3(aq) \rightarrow Cu(NH_3)_4^{2+}(aq)$

(d) $Cr(OH)_4^-(aq) + 4 H^+(aq) \rightarrow Cr^{3+}(aq) + 4 H_2O(l)$

20.66 (a) $2 CrO_4^{2-}(aq) + 2 H_3O^+(aq) \rightarrow Cr_2O_7^{2-}(aq) + 3 H_2O(l)$
(yellow)   (orange)

(b) $[Fe(H_2O)_6]^{3+}(aq) + SCN^-(aq) \rightarrow [Fe(H_2O)_5(SCN)]^{2+}(aq) + H_2O(l)$
(red)

(c) $3 Cu(s) + 2 NO_3^-(aq) + 8 H^+(aq) \rightarrow 3 Cu^{2+}(aq) + 2 NO(g) + 4 H_2O(l)$
(blue)

(d) $Cr(OH)_3(s) + OH^-(aq) \rightarrow Cr(OH)_4^-(aq)$
$2 Cr(OH)_4^-(aq) + 3 HO_2^-(aq) \rightarrow 2 CrO_4^{2-}(aq) + 5 H_2O(l) + OH^-(aq)$
(yellow)

20.67 (a) $Cu^{2+}(aq) + 4 NH_3(aq) \rightarrow [Cu(NH_3)_4]^{2+}(aq)$
(dark blue)

(b) $Cr_2O_7^{2-}(aq) + 2 OH^-(aq) \rightarrow 2 CrO_4^{2-}(aq) + H_2O(l)$
(orange)   (yellow)

(c) $Fe^{3+}(aq) + 3 OH^-(aq) \rightarrow Fe(OH)_3(s)$
(red-brown)

(d) $3 CuS(s) + 8 H^+(aq) + 2 NO_3^-(aq) \rightarrow 3 Cu^{2+}(aq) + 3 S(s) + 2 NO(g) + 4 H_2O(l)$
(blue)   (yellow)

Coordination Compounds; Ligands (Sections 20.5–20.6)

20.68 Ni^{2+} accepts six pairs of electrons, two each from the three ethylenediamine ligands. Ni^{2+} is an electron pair acceptor, a Lewis acid. The two nitrogens in each ethylenediamine donate a pair of electrons to the Ni^{2+}. The ethylenediamine is an electron pair donor, a Lewis base. The formation of $[Ni(en)_3]^{2+}$ is a Lewis acid-base reaction.

20.69 Lewis acid (electron pair acceptor) is Fe^{3+}
Lewis base (electron pair donor) is $C_2O_4^{2-}$

20.70 (a) $[Ag(NH_3)_2]^+$ (b) $[Ni(CN)_4]^{2-}$ (c) $[Cr(H_2O)_6]^{3+}$

20.71

	Coordination Number
(a) $[AgCl_2]^-$	2
(b) $[Cr(H_2O)_5Cl]^{2+}$	6
(c) $[Co(NCS)_4]^{2-}$	4
(d) $[ZrF_8]^{4-}$	8
(e) $Co(NH_3)_3(NO_2)_3$	6
(f) $[Fe(EDTA)(H_2O)]^-$	7

20.72 (a) $AgCl_2^-$
2 Cl^-
The oxidation state of the Ag is +1.
(b) $[Cr(H_2O)_5Cl]^{2+}$
4 H_2O (no charge)   1 Cl^-
The oxidation state of the Cr is +3.
(c) $[Co(NCS)_4]^{2-}$
4 NCS^-
The oxidation state of the Co is +2.
(d) $[ZrF_8]^{4-}$
8 F^-
The oxidation state of the Zr is +4.
(e) $Co(NH_3)_3(NO_2)_3$
3 NH_3 (no charge)   3 NO_2^-
The oxidation state of the Co is +3.

20.73 (a) $[Ni(CN)_5]^{3-}$
5 CN^-
The oxidation state of the Ni is +2.
(b) $Ni(CO)_4$
4 CO (no charge)
The oxidation state of the Ni is 0.
(c) $[Co(en)_2(H_2O)Br]^{2+}$
2 en ($NH_2CH_2CH_2NH_2$, no charge)   1 H_2O (no charge)   1 Br^-
The oxidation state of the Co is +3.
(d) $[Cu(H_2O)_2(C_2O_4)_2]^{2-}$
2 H_2O (no charge)   2 $C_2O_4^{2-}$
The oxidation state of the Cu is +2.

20.74 (a) $Ni(CO)_4$ (b) $[Ag(NH_3)_2]^+$ (c) $[Fe(CN)_6]^{3-}$ (d) $[Ni(CN)_4]^{2-}$

20.75 (a) $Ir(NH_3)_3Cl_3$ (b) $[Cr(H_2O)_2(C_2O_4)_2]^-$ (c) $[Pt(en)_2(SCN)_2]^{2+}$

20.76

The iron is in the +3 oxidation state, and the coordination number is six. The geometry about the Fe is octahedral. The oxalate ligand is behaving as a bidentate chelating ligand. There are three chelate rings, one formed by each oxalate ligand.

20.77

$[Pt(en)_2]^{2+}$ is square planar. Ethylenediamine is a neutral bidentate chelating ligand. The coordination number of the Pt is four, and the oxidation number of the Pt is +2.

20.78 (a) $Co(NH_3)_3(NO_2)_3$
3 NH_3 (no charge)   3 NO_2^-
The oxidation state of the Co is +3.
(b) $[Ag(NH_3)_2]NO_3$
2 NH_3 (no charge)   1 NO_3^-
The oxidation state of the Ag is +1.
(c) $K_3[Cr(C_2O_4)_2Cl_2]$
3 K^+ 2 $C_2O_4^{2-}$ 2 Cl^-
The oxidation state of the Cr is +3.
(d) $Cs[CuCl_2]$
1 Cs^+ 2 Cl^-
The oxidation state of the Cu is +1.

20.79 (a) $(NH_4)_3[RhCl_6]$
3 NH_4^+ 6 Cl^-
The oxidation state of the Rh is +3.
(b) $[Cr(NH_3)_4(SCN)_2]Br$
4 NH_3 (no charge)   2 SCN^- 1 Br^-
The oxidation state of the Cr is +3.
(c) $[Cu(en)_2]SO_4$
2 en ($NH_2CH_2CH_2NH_2$, no charge)   1 SO_4^{2-}
The oxidation state of the Cu is +2.
(d) $Na_2[Mn(EDTA)]$
2 Na^+ 1 $EDTA^{4-}$
The oxidation state of the Mn is +2.

Naming Coordination Compounds (Section 20.7)

20.80 (a) tetrachloromanganate(II)   (b) hexaamminenickel(II)
(c) tricarbonatocobaltate(III)   (d) bis(ethylenediamine)dithiocyanatoplatinum(IV)

20.81 (a) tetrachloroaurate(III)   (b) hexacyanoferrate(II)
(c) pentaaquaisothiocyanatoiron(III)   (d) diamminedioxalatochromate(III)

20.82 (a) cesium tetrachloroferrate(III)   (b) hexaaquavanadium(III) nitrate
(c) tetraamminedibromocobalt(III) bromide   (d) diglycinatocopper(II)

20.83 (a) tetraamminecopper(II) sulfate   (b) hexacarbonylchromium(0)
(c) potassium trioxalatoferrate(III)
(d) amminecyanatobis(ethylenediamine)cobalt(III) chloride

20.84   (a) $[Pt(NH_3)_4]Cl_2$ (b) $Na_3[Fe(CN)_6]$ (c) $[Pt(en)_3](SO_4)_2$ (d) $Rh(NH_3)_3(SCN)_3$

20.85   (a) $[Ag(NH_3)_2]NO_3$ (b) $K[Co(H_2O)_2(C_2O_4)_2]$
(c) $Mo(CO)_6$ (d) $[Cr(NH_3)_2(en)_2]Cl_3$

Isomers (Sections 20.8–20.9)

20.86

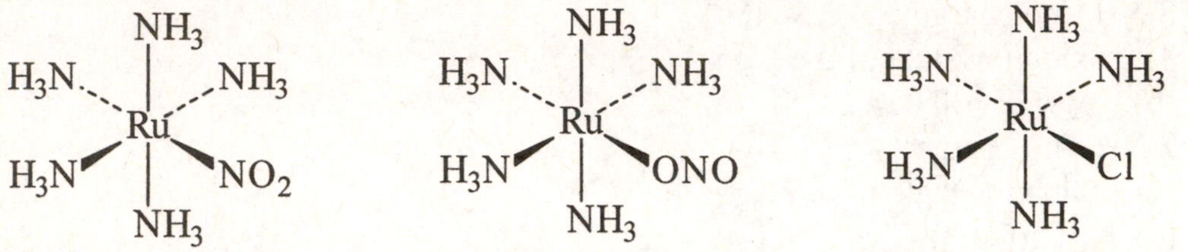

$[Ru(NH_3)_5(NO_2)]Cl$ $[Ru(NH_3)_5(ONO)]Cl$ $[Ru(NH_3)_5Cl]NO_2$

$[Ru(NH_3)_5(NO_2)]Cl$ and $[Ru(NH_3)_5(ONO)]Cl$ are linkage isomers.
$[Ru(NH_3)_5Cl]NO_2$ is an ionization isomer of both $[Ru(NH_3)_5(NO_2)]Cl$ and
$[Ru(NH_3)_5(ONO)]Cl$.

20.87

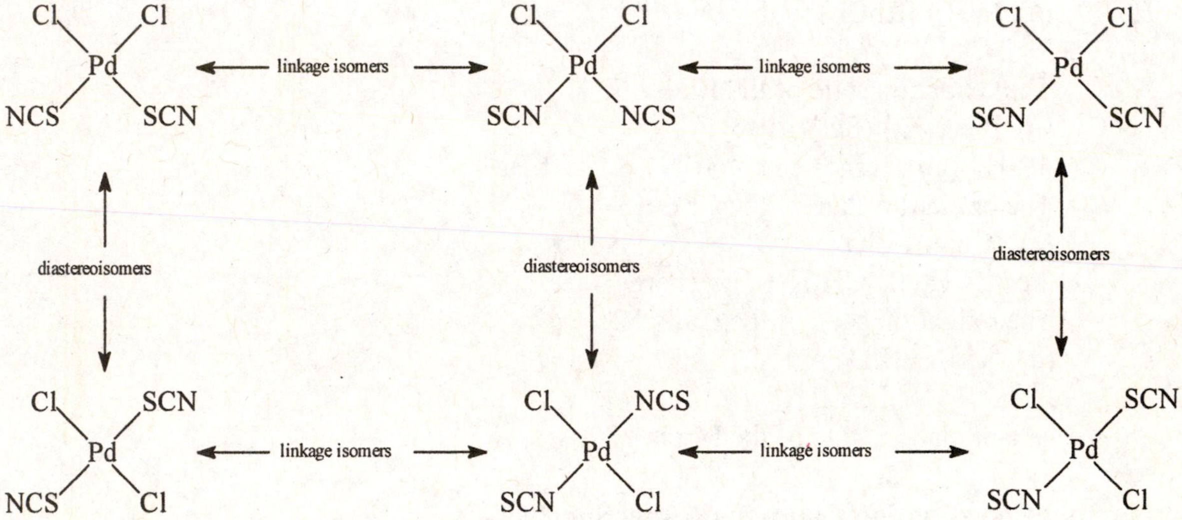

20.88   (a) $[Cr(NH_3)_2Cl_4]^-$ can exist as cis and trans diastereoisomers.

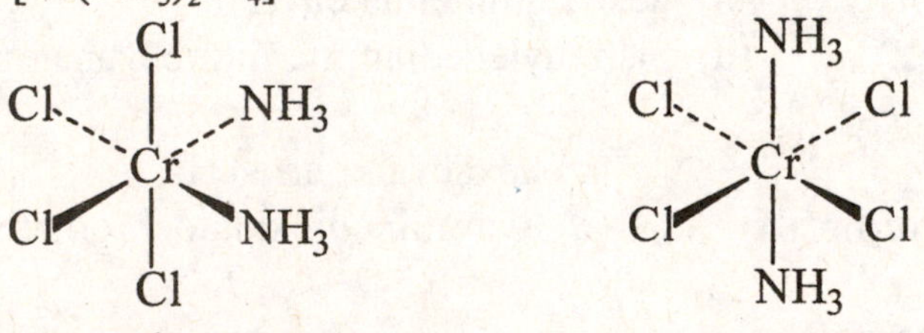

(b) $[Co(NH_3)_5Br]^{2+}$ cannot exist as diastereoisomers.
(c) $[FeCl_2(NCS)_2]^{2-}$ (tetrahedral) cannot exist as diastereoisomers.

(d) $[PtCl_2Br_2]^{2-}$ (square planar) can exist as cis and trans diastereoisomers.

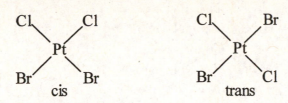

20.89 (a) $Pt(NH_3)_2(CN)_2$ can exist as two (cis and trans) diastereoisomers.

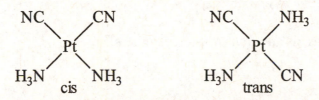

(b) $[Co(en)(SCN)_4]^-$ cannot exist as diastereoisomers.

(c) $[Cr(H_2O)_4Cl_2]^+$ can exist as two (cis and trans) diastereoisomers.

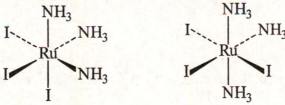

(d) $Ru(NH_3)_3I_3$ can exist as two diastereoisomers.

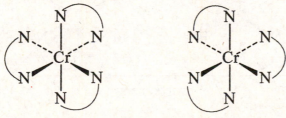

20.90 (c) cis-$[Cr(en)_2(H_2O)_2]^{3+}$ (d) $[Cr(C_2O_4)_3]^{3-}$

20.91 (a) $[Cr(en)_3]^{3+}$ can exist as enantiomers.

(b) cis-$[Co(en)_2(NH_3)Cl]^{2+}$ can exist as enantiomers.

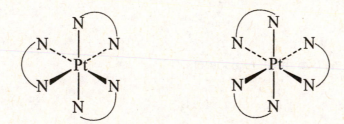

(c) trans-$[Co(en)_2(NH_3)Cl]^{2+}$ cannot exist as enantiomers.
(d) $[Pt(NH_3)_3Cl_3]^+$ cannot exist as enantiomers.

20.92 (a) $Ru(NH_3)_4Cl_2$ can exist as cis and trans diastereoisomers.

cis   trans

(b) $[Pt(en)_3]^{4+}$ can exist as enantiomers.

(c) $[Pt(en)_2ClBr]^{2+}$ can exist as both diastereoisomers and enantiomers.

diastereoisomers

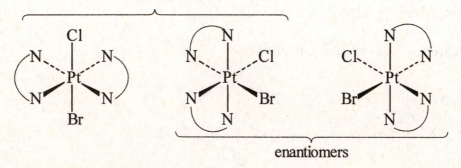

enantiomers

20.93 (a) $[Rh(C_2O_4)_2I_2]^{3-}$ can exist as both diastereoisomers and enantiomers.

diastereoisomers

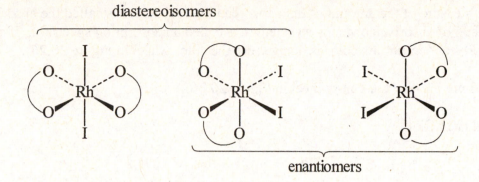

enantiomers

(b) $[Cr(NH_3)_2Cl_4]^-$ can exist as cis and trans diastereoisomers.

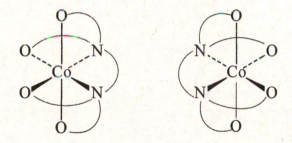

cis   trans

(c) $[Co(EDTA)]^-$ can exist as enantiomers.

20.94 Plane-polarized light is light in which the electric vibrations of the light wave are restricted to a single plane. The following chromium complex can rotate the plane of plane-polarized light.

$[Cr(en)_3]^{3+}$

20.95 A racemic mixture is a mixture that contains equal amounts of two enantiomers. A racemic mixture does not affect plane-polarized light because the effect of one enantiomer is canceled by that of the other enantiomer.

Color of Complexes; Valence Bond and Crystal Field Theories (Sections 20.10–20.12)

20.96 The measure of the amount of light absorbed by a substance is called the absorbance, and a graph of absorbance versus wavelength is called an absorption spectrum. If a complex absorbs at 455 nm, its color is orange (use the color wheel in Figure 20.26).

20.97 ~525 nm   (use the color wheel in Figure 20.26)

20.98 (a) $[Ti(H_2O)_6]^{3+}$

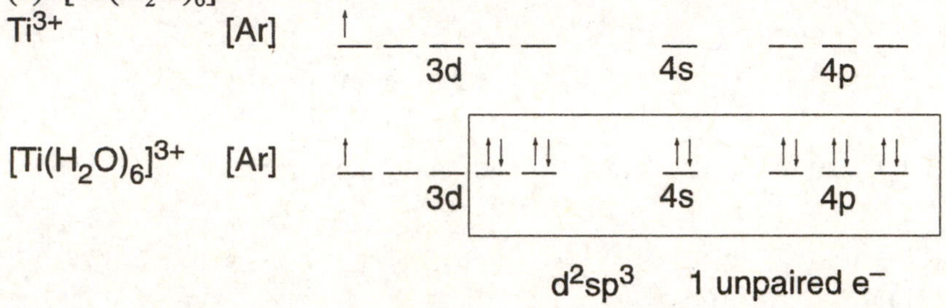

Ti^{3+} [Ar]   ↑ _ _ _ _   _   _ _ _
                  3d   4s   4p

$[Ti(H_2O)_6]^{3+}$ [Ar]   ↑ _ _ | ↑↓ ↑↓ | ↑↓ | ↑↓ ↑↓ ↑↓ |
                          3d   4s   4p

d^2sp^3 1 unpaired e^-

(b) $[NiBr_4]^{2-}$

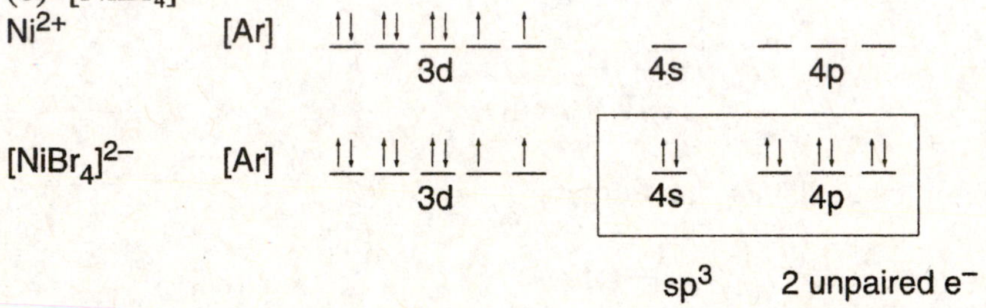

Ni^{2+} [Ar]   ↑↓ ↑↓ ↑↓ ↑ ↑   _   _ _ _
                  3d   4s   4p

$[NiBr_4]^{2-}$ [Ar]   ↑↓ ↑↓ ↑↓ ↑ ↑ | ↑↓ | ↑↓ ↑↓ ↑↓ |
                      3d   4s   4p

sp^3 2 unpaired e^-

(c) $[Fe(CN)_6]^{3-}$ (low-spin)

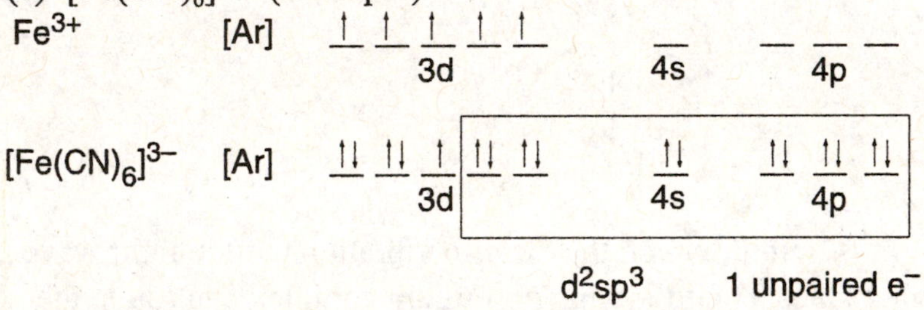

Fe^{3+} [Ar]   ↑ ↑ ↑ ↑ ↑   _   _ _ _
                  3d   4s   4p

$[Fe(CN)_6]^{3-}$ [Ar]   ↑↓ ↑↓ ↑ | ↑↓ ↑↓ | ↑↓ | ↑↓ ↑↓ ↑↓ |
                        3d   4s   4p

d^2sp^3 1 unpaired e^-

(d) $[MnCl_6]^{3-}$ (high-spin)

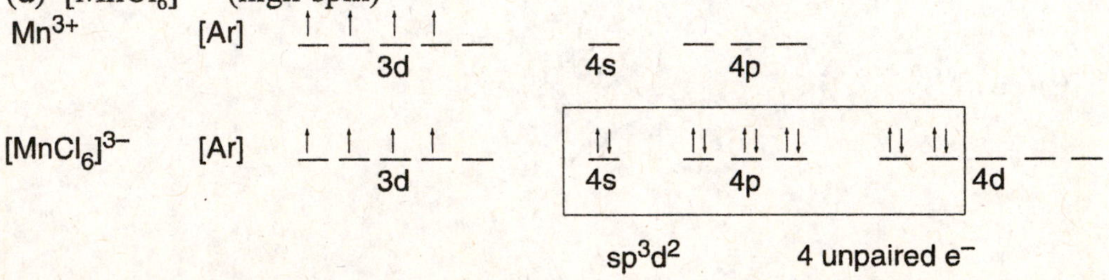

Mn^{3+} [Ar]   ↑ ↑ ↑ ↑ _   _   _ _ _
                  3d   4s   4p

$[MnCl_6]^{3-}$ [Ar]   ↑ ↑ ↑ ↑ _ | ↑↓ | ↑↓ ↑↓ ↑↓ | ↑↓ ↑↓ | _ _ _
                      3d   4s   4p   4d

sp^3d^2 4 unpaired e^-

20.99 (a) $[AuCl_4]^-$

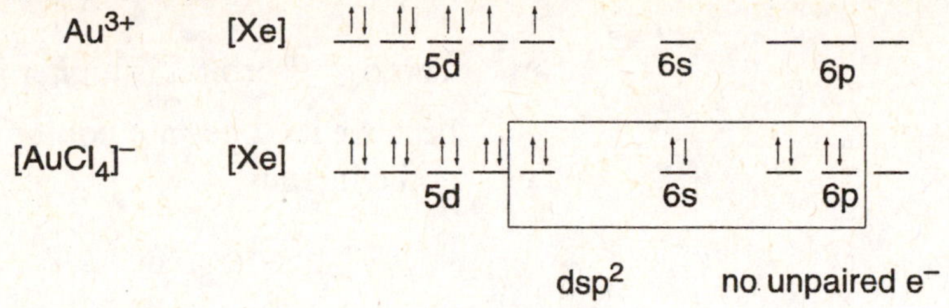

Au^{3+} [Xe] ⇅ ⇅ ⇅ ↑ ↑   ___   __ __ __
                   5d   6s   6p

$[AuCl_4]^-$ [Xe] ⇅ ⇅ ⇅ ⇅ ⟦⇅   ⇅   ⇅ ⇅⟧ __
                   5d   6s   6p

dsp^2 no unpaired e$^-$

(b) $[Ag(NH_3)_2]^+$

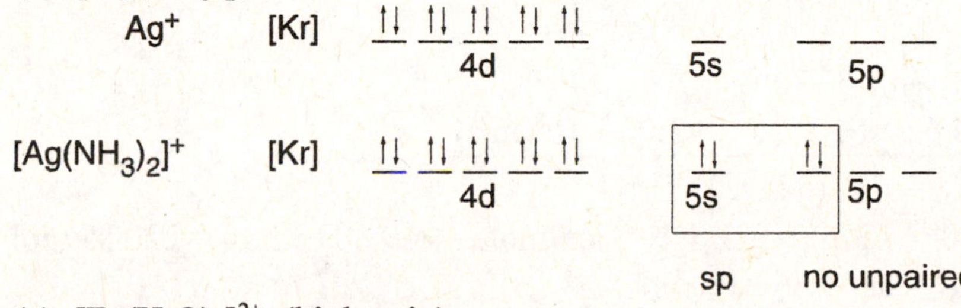

Ag$^+$ [Kr] ⇅ ⇅ ⇅ ⇅ ⇅   ___   __ __ __
                4d   5s   5p

$[Ag(NH_3)_2]^+$ [Kr] ⇅ ⇅ ⇅ ⇅ ⇅   ⟦⇅   ⇅⟧ __
                        4d   5s   5p

sp   no unpaired e$^-$

(c) $[Fe(H_2O)_6]^{2+}$ (high-spin)

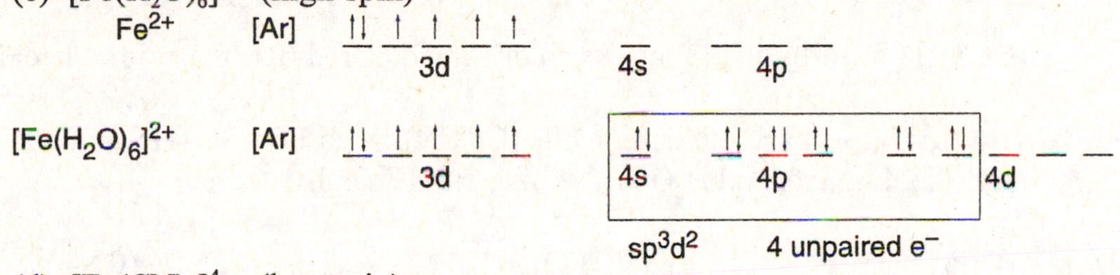

Fe^{2+} [Ar] ⇅ ↑ ↑ ↑ ↑   ___   __ __ __
                   3d   4s   4p

$[Fe(H_2O)_6]^{2+}$ [Ar] ⇅ ↑ ↑ ↑ ↑   ⟦⇅   ⇅ ⇅ ⇅   ⇅ ⇅⟧ __ __ __
                           3d   4s   4p   4d

sp^{3}d^2 4 unpaired e$^-$

(d) $[Fe(CN)_6]^{4-}$ (low-spin)

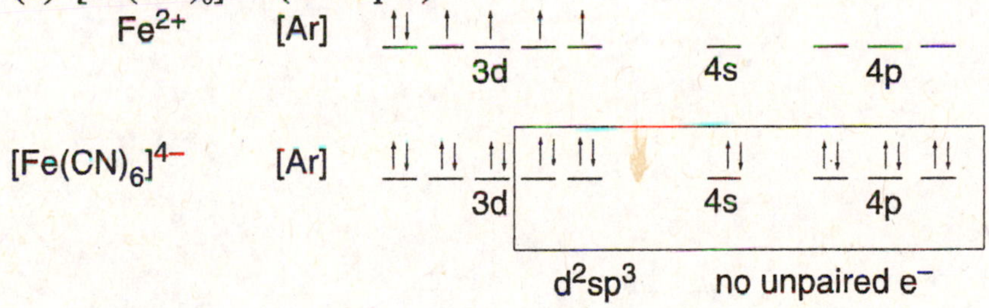

Fe^{2+} [Ar] ⇅ ↑ ↑ ↑ ↑   ___   __ __ __
                   3d   4s   4p

$[Fe(CN)_6]^{4-}$ [Ar] ⇅ ⇅ ⇅ ⟦⇅ ⇅   ⇅   ⇅ ⇅ ⇅⟧
                         3d   4s   4p

d^{2}sp^3 no unpaired e$^-$

20.100 $[Ti(H_2O)_6]^{3+}$ Ti^{3+} 3d^1

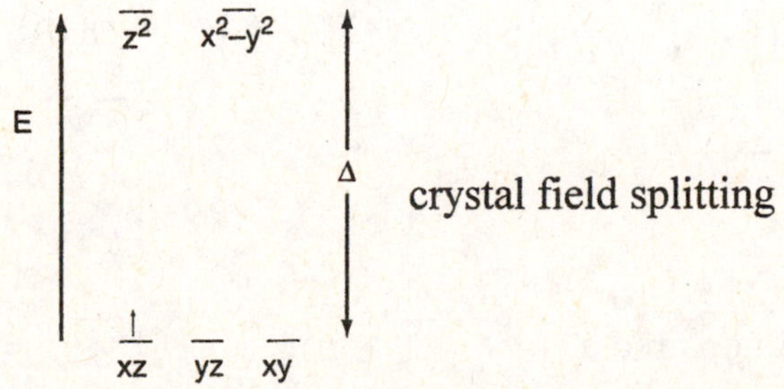

crystal field splitting

$[Ti(H_2O)_6]^{3+}$ is colored because it can absorb light in the visible region, exciting the electron to the higher-energy set of orbitals.

20.101

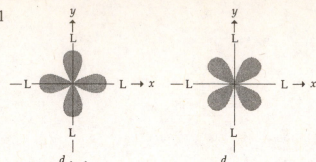

$d_{x^2-y^2}$ d_{xy}

The $d_{x^2-y^2}$ orbital is higher in energy because its lobes are pointing directly at the ligands.

20.102 $\lambda = 544$ nm $= 544 \times 10^{-9}$ m

$$\Delta = \frac{hc}{\lambda} = \frac{(6.626 \times 10^{-34} \text{ J} \cdot \text{s})(3.00 \times 10^8 \text{ m/s})}{(544 \times 10^{-9} \text{ m})} = 3.65 \times 10^{-19} \text{ J}$$

$\Delta = (3.65 \times 10^{-19} \text{ J/ion})(6.022 \times 10^{23} \text{ ion/mol}) = 219{,}803 \text{ J/mol} = 220 \text{ kJ/mol}$

For $[Ti(H_2O)_6]^{3+}$, $\Delta = 240$ kJ/mol

Because $\Delta_{NCS^-} < \Delta_{H_2O}$ for the Ti complex, NCS^- is a weaker-field ligand than H_2O. If $[Ti(NCS)_6]^{3-}$ absorbs at 544 nm, its color should be red (use the color wheel in Figure 20.26).

20.103 $[Cr(H_2O)_6]^{3+}$ absorbs at ~580 nm while $[Cr(CN)_6]^{3-}$ absorbs at ~415 nm. CN^- is a stronger-field ligand, and H_2O is a weaker-field ligand.

20.104   (a) $[CrF_6]^{3-}$ (b) $[V(H_2O)_6]^{3+}$ (c) $[Fe(CN)_6]^{3-}$

     — —   — —   — —

↑ ↑ ↑
3 unpaired e⁻

         ↑ ↑ —
         2 unpaired e⁻

                       ↑↓ ↑↓ ↑
                       1 unpaired e⁻

20.105   (a) $[Cu(en)_3]^{2+}$ (b) $[FeF_6]^{3-}$ (c) $[Co(en)_3]^{3+}$

     ↑↓ ↑   ↑ ↑   — —

                  ↑ ↑ ↑
↑↓ ↑↓ ↑↓   5 unpaired e⁻   ↑↓ ↑↓ ↑↓
1 unpaired e⁻   no unpaired e⁻

20.106 $Ni^{2+}(aq)$ $Zn^{2+}(aq)$

   ↑ ↑   ↑↓ ↑↓

   ↑↓ ↑↓ ↑↓   ↑↓ ↑↓ ↑↓

$Ni^{2+}(aq)$ is green because the Ni^{2+} ion can absorb light, which promotes electrons from the filled d orbitals to the higher energy half-filled d orbitals. $Zn^{2+}(aq)$ is colorless because the d orbitals are completely filled and no electrons can be promoted, so no light is absorbed.

20.107 Cr^{3+} is $3d^3$ and should be colored because absorption of light can promote electrons from the lower-energy to the higher-energy d orbitals. Y^{3+} is $4d^0$ and should be colorless because it can't exhibit d-d transitions.

20.108   Weak-field ligands produce a small Δ. Strong-field ligands produce a large Δ. For a metal complex with weak-field ligands, $\Delta < P$, where P is the pairing energy, and it is easier to place an electron in either d_{z^2} or $d_{x^2-y^2}$ than to pair up electrons; high-spin complexes result. For a metal complex with strong-field ligands, $\Delta > P$ and it is easier to pair up electrons than to place them in either d_{z^2} or $d_{x^2-y^2}$; low-spin complexes result.

20.109   Ligands do not point directly at any d-orbitals, and the crystal field splitting is small. Because $\Delta < P$, high-spin complexes result.

20.110   __ x^2-y^2

    ↑↓ xy

    ↑↓ z^2

    ↑↓ ↑↓ xz, yz

Square planar geometry is most common for metal ions with d^8 configurations because this configuration favors low-spin complexes in which all four lower energy d orbitals are filled, and the higher energy $d_{x^2-y^2}$ orbital is vacant.

20.111   (a) $[Pt(NH_3)_4]^{2+}$ (square planar)

    __ x^2-y^2

     ↑↓ xy

     ↑↓ z^2

    ↑↓ ↑↓ xz, yz
    no unpaired e^-

(b) $[MnCl_4]^{2-}$ (tetrahedral)

$\underline{\uparrow}$ $\underline{\uparrow}$ $\underline{\uparrow}$
xy xz yz

$\underline{\uparrow}$ $\underline{\uparrow}$
x^2-y^2 z^2
5 unpaired e^-

(c) $[Co(NCS)_4]^{2-}$ (tetrahedral)

$\underline{\uparrow}$ $\underline{\uparrow}$ $\underline{\uparrow}$
xy xz yz

$\underline{\uparrow\downarrow}$ $\underline{\uparrow\downarrow}$
x^2-y^2 z^2
3 unpaired e^-

(d) $[Cu(en)_2]^{2+}$ (square planar)

$\underline{\uparrow}$ x^2-y^2

$\underline{\uparrow\downarrow}$ xy

$\underline{\uparrow\downarrow}$ z^2

$\underline{\uparrow\downarrow}$ $\underline{\uparrow\downarrow}$ xz, yz
1 unpaired e^-

Chapter Problems

20.112 (a) $[Mn(CN)_6]^{3-}$ Mn^{3+} [Ar] $3d^4$
CN$^-$ is a strong-field ligand. The Mn^{3+} complex is low-spin.

$\underline{}$ $\underline{}$

$\underline{\uparrow\downarrow}$ $\underline{\uparrow}$ $\underline{\uparrow}$ 2 unpaired e^-, paramagnetic

(b) $[Zn(NH_3)_4]^{2+}$ Zn^{2+} [Ar] $3d^{10}$
$[Zn(NH_3)_4]^{2+}$ is tetrahedral.

$\underline{\uparrow\downarrow}$ $\underline{\uparrow\downarrow}$ $\underline{\uparrow\downarrow}$

$\underline{\uparrow\downarrow}$ $\underline{\uparrow\downarrow}$ no unpaired e^-, diamagnetic

(c) $[Fe(CN)_6]^{4-}$ Fe^{2+} [Ar] $3d^6$

CN^- is a strong-field ligand. The Fe^{2+} complex is low-spin.

— —

⇅ ⇅ ⇅   no unpaired e^-, diamagnetic

(d) $[FeF_6]^{4-}$ Fe^{2+} [Ar] $3d^6$

F^- is a weak-field ligand. The Fe^{2+} complex is high-spin.

↑ ↑

⇅ ↑ ↑   4 unpaired e^-, paramagnetic

20.113   (a) $[Ni(H_2O)_6]^{2+}$ Ni^{2+} [Ar] $3d^8$

H_2O is a weak-field ligand.

↑ ↑

⇅ ⇅ ⇅   2 unpaired e^-, paramagnetic

(b) $[Co(CN)_6]^{3-}$ Co^{2+} [Ar] $3d^6$

CN^- is a strong-field ligand. The Co^{3+} complex is low-spin.

— —

⇅ ⇅ ⇅   no unpaired e^-, diamagnetic

(c) $[HgI_4]^{2-}$ Hg^{2+} [Xe] $5d^{10}$

$[HgI_4]^{2-}$ is tetrahedral.

⇅ ⇅ ⇅

⇅ ⇅   no unpaired e^-, diamagnetic

(d) $[Cu(NH_3)_4]^{2+}$ Cu^{2+} [Ar] $3d^9$

$[Cu(NH_3)_4]^{2+}$ is square-planar.

↑

⇅

⇅

⇅ ⇅   1 unpaired e^-, paramagnetic

20.114   (a) $4 [Co^{3+}(aq) + e^- \rightarrow Co^{2+}(aq)]$

$\underline{2 H_2O(l) \rightarrow O_2(g) + 4 H^+(aq) + 4 e^-}$

$4 Co^{3+}(aq) + 2 H_2O(l) \rightarrow 4 Co^{2+}(aq) + O_2(g) + 4 H^+(aq)$

(b) $4 Cr^{2+}(aq) + O_2(g) + 4 H^+(aq) \rightarrow 4 Cr^{3+}(aq) + 2 H_2O(l)$

(c) $3 [Cu(s) \rightarrow Cu^{2+}(aq) + 2 e^-]$
$\underline{Cr_2O_7{}^{2-}(aq) + 14 H^+(aq) + 6 e^- \rightarrow 2 Cr^{3+}(aq) + 7 H_2O(l)}$
$3 Cu(s) + Cr_2O_7{}^{2-}(aq) + 14 H^+(aq) \rightarrow 3 Cu^{2+}(aq) + 2 Cr^{3+}(aq) + 7 H_2O(l)$

(d) $2 CrO_4{}^{2-}(aq) + 2 H^+(aq) \rightarrow Cr_2O_7{}^{2-}(aq) + H_2O(l)$

20.115 (a) $2 [5 e^- + 8 H^+(aq) + MnO_4{}^-(aq) \rightarrow Mn^{2+}(aq) + 4 H_2O(l)]$
$\underline{5 [C_2O_4{}^{2-}(aq) \rightarrow 2 CO_2(g) + 2 e^-]}$
$16 H^+(aq) + 2 MnO_4{}^-(aq) + 5 C_2O_4{}^{2-}(aq) \rightarrow 2 Mn^{2+}(aq) + 8 H_2O(l) + 10 CO_2(g)$

(b) $6 [Ti^{3+}(aq) + H_2O(l) \rightarrow TiO^{2+}(aq) + 2 H^+(aq) + e^-]$
$\underline{6 e^- + Cr_2O_7{}^{2-}(aq) + 14 H^+(aq) \rightarrow 2 Cr^{3+}(aq) + 7 H_2O(l)}$
$6 Ti^{3+}(aq) + Cr_2O_7{}^{2-}(aq) + 2 H^+(aq) \rightarrow 6 TiO^{2+}(aq) + 2 Cr^{3+}(aq) + H_2O(l)$

(c) $2 [MnO_4{}^-(aq) + e^- \rightarrow MnO_4{}^{2-}(aq)]$
$\underline{H_2O(l) + SO_3{}^{2-}(aq) + 2 OH^-(aq) \rightarrow SO_4{}^{2-}(aq) + 2 H_2O(l) + 2 e^-}$
$2 MnO_4{}^-(aq) + SO_3{}^{2-}(aq) + 2 OH^-(aq) \rightarrow 2 MnO_4{}^{2-}(aq) + SO_4{}^{2-}(aq) + H_2O(l)$

(d) $4 [H_2O(l) + Fe(OH)_2(s) \rightarrow Fe(OH)_3(s) + H^+(aq) + e^-]$
$\underline{4 H^+(aq) + 4 e^- + O_2(g) \rightarrow 2 H_2O(l)}$
$2 H_2O(l) + 4 Fe(OH)_2(s) + O_2(g) \rightarrow 4 Fe(OH)_3(s)$

20.116 $mol\ Fe^{2+} = (0.1000\ L)(0.400\ mol/L) = 0.0400\ mol\ Fe^{2+}$
$mol\ Cr_2O_7{}^{2-} = (0.1000\ L)(0.100\ mol/L) = 0.0100\ mol\ Cr_2O_7{}^{2-}$
$6 [Fe^{2+}(aq) \rightarrow Fe^{3+}(aq) + e^-]$
$\underline{Cr_2O_7{}^{2-}(aq) + 14 H^+(aq) + 6 e^- \rightarrow 2 Cr^{3+}(aq) + 7 H_2O(l)}$
$6 Fe^{2+}(aq) + Cr_2O_7{}^{2-}(aq) + 14 H^+(aq) \rightarrow 6 Fe^{3+}(aq) + 2 Cr^{3+}(aq) + 7 H_2O(l)$

	Fe^{2+}	$Cr_2O_7{}^{2-}$	Fe^{3+}	Cr^{3+}
initial (mol)	0.0400	0.0100	0	0
change (mol)	−0.0400	$-\dfrac{0.0400}{6}$	+0.0400	$+\dfrac{0.0400}{3}$
after (mol)	0	0.00333	0.0400	0.0133

$[Fe^{3+}] = \dfrac{0.0400\ mol}{0.2000\ L} = 0.200\ M$

$[Cr^{3+}] = \dfrac{0.0133\ mol}{0.200\ L} = 0.0665\ M$

$[Cr_2O_7{}^{2-}] = \dfrac{0.00333\ mol}{0.200\ L} = 0.0166\ M$

20.117   mol $Cr_2(SO_4)_3$ = (0.0400 L)(0.030 mol/L) = 0.0012 mol
mol $Cr(OH)_4^-$ = 2(mol $Cr_2(SO_4)_3$) = 2(0.0012 mol) = 0.0024 mol
2 [$Cr(OH)_4^-(aq) \rightarrow CrO_4^{2-}(aq) + 4\ H^+(aq) + 3\ e^-$]
3 [$2\ e^- + 2\ H^+(aq) + H_2O_2(aq) \rightarrow 2\ H_2O(l)$]
_____
2 $Cr(OH)_4^-(aq)$ + 3 $H_2O_2(aq) \rightarrow$ 2 $CrO_4^{2-}(aq)$ + 2 $H^+(aq)$ + 6 $H_2O(l)$
   add 2 OH^- to both sides of the reaction

2 $Cr(OH)_4^-(aq)$ + 3 $H_2O_2(aq)$ + 2 $OH^-(aq) \rightarrow$ 2 $CrO_4^{2-}(aq)$ + 8 $H_2O(l)$

mol H_2O_2 = (0.0100 L)(0.20 mol/L) = 0.0020 mol
For 0.0024 mol $Cr(OH)_4^-$, you need (3/2)(0.0024 mol) = 0.0036 mol H_2O_2.
Because you have only 0.0020 mol H_2O_2, H_2O_2 is the limiting reactant.

$$\text{mol } CrO_4^{2-} = 0.0020 \text{ mol } H_2O_2 \times \frac{2 \text{ mol } CrO_4^{2-}}{3 \text{ mol } H_2O_2} = 0.001\ 33 \text{ mol } CrO_4^{2-}$$

$$[CrO_4^{2-}] = \frac{0.001\ 33 \text{ mol}}{0.100 \text{ L}} = 0.013 \text{ M}$$

20.118   (a) sodium aquabromodioxalatoplatinate(IV)
(b) hexaamminechromium(III) trioxalatocobaltate(III)
(c) hexaamminecobalt(III) trioxalatochromate(III)
(d) diamminebis(ethylenediamine)rhodium(III) sulfate

20.119 $[Mn(CN)_6]^{3-}$ Mn^{3+} $3d^4$

valence bond theory

[Ar] ↿⇂ ↿ ↿ | ↿⇂ ↿⇂   ↿⇂   ↿⇂ ↿⇂ ↿⇂ | d^2sp^3
       3d   4s   4p

crystal field theory

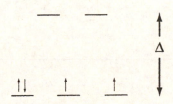

Crystal field model predicts 2 unpaired electrons.

20.120 Cl⁻ is a weak-field ligand, whereas CN⁻ is a strong-field ligand. Δ for $[Fe(CN)_6]^{3-}$ is larger than the pairing energy P; Δ for $[FeCl_6]^{3-}$ is smaller than P. Fe^{3+} has a $3d^5$ electron configuration.

$[FeCl_6]^{3-}$
(high spin)

$[Fe(CN)_6]^{3-}$
(low spin)

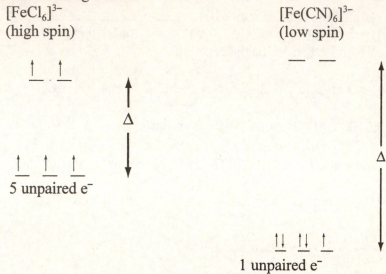

Because of the difference in Δ, $[FeCl_6]^{3-}$ is high-spin with five unpaired electrons, whereas $[Fe(CN)_6]^{3-}$ is low-spin with only one unpaired electron.

20.121 Cr^{3+} is a $3d^3$ ion. Regardless of the crystal field splitting energy, the three electrons singly occupy the three lower energy d orbitals.

20.122 A choice between high-spin and low-spin electron configurations arises only for complexes of metal ions with four to seven d electrons, the so-called d^4-d^7 complexes. For d^1-d^3 and d^8-d^{10} complexes, only one ground-state electron configuration is possible. In d^1-d^3 complexes, all the electrons occupy the lower-energy d orbitals, independent of the value of Δ. In d^8-d^{10} complexes, the lower-energy set of d orbitals is filled with three pairs of electrons, while the higher-energy set contains two, three, or four electrons, again independent of the value of Δ.

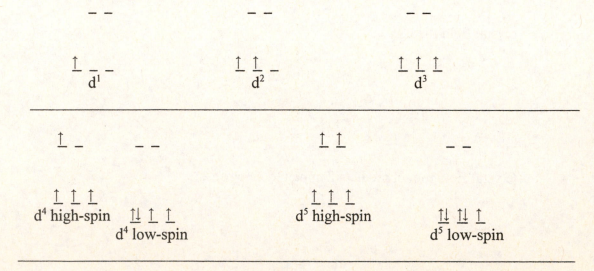

↿ ↿   – –   ↿ ↿   ↿ –

⇅ ↿ ↿
d^6 high-spin   ⇅ ⇅ ⇅   ⇅ ⇅ ↿   ⇅ ⇅ ⇅
           d^6 low-spin d^7 high-spin d^7 low-spin

↿ ↿   ⇅ ↿   ⇅ ⇅

⇅ ⇅ ⇅   ⇅ ⇅ ⇅   ⇅ ⇅ ⇅
d^8 d^9 d^{10}

20.123 (a) $[Mn(H_2O)_6]^{2+}$ high-spin Mn^{2+}, $3d^5$

↿ ↿

↿ ↿ ↿
5 unpaired e⁻

(b) $Pt(NH_3)_2Cl_2$ square-planar Pt^{2+}, $5d^8$

–

⇅

⇅

⇅ ⇅
no unpaired e⁻

(c) $[FeO_4]^{2-}$ tetrahedral Fe(VI), $3d^2$

– – –

↿ ↿
2 unpaired e⁻

(d) $[Ru(NH_3)_6]^{2+}$ low-spin Ru^{2+}, $4d^6$

– –

⇅ ⇅ ⇅
no unpaired e⁻

20.124 $[CoCl_4]^{2-}$ is tetrahedral. $[Co(H_2O)_6]^{2+}$ is octahedral. Because $\Delta_{tet} < \Delta_{oct}$, these complexes have different colors. $[CoCl_4]^{2-}$ has absorption bands at longer wavelengths.

20.125 $[Rh(en)_2(NO_2)(SCN)]^+$

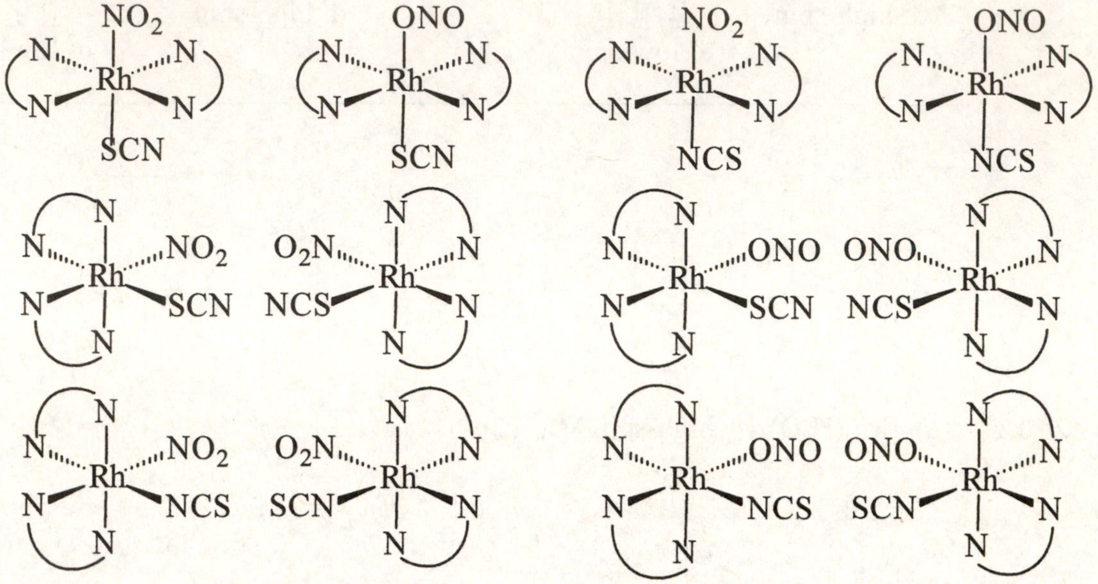

20.126 $Co(gly)_3$

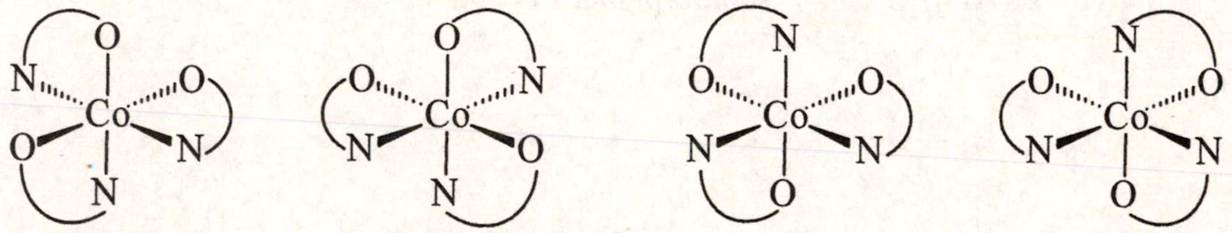

20.127 $Pt(gly)_2Cl_2$

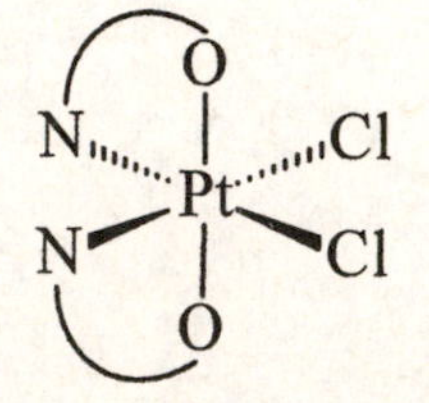

1 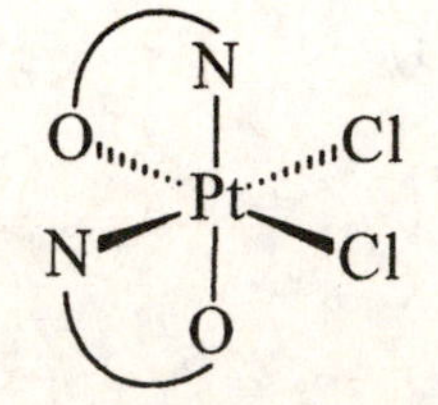   **2** 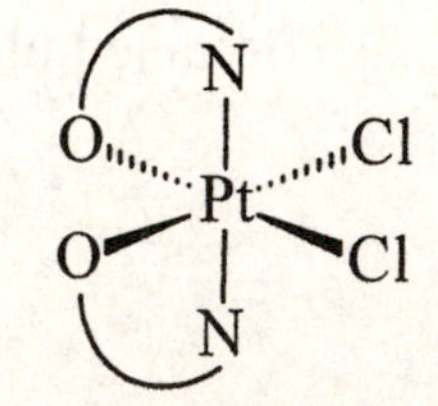   **3**

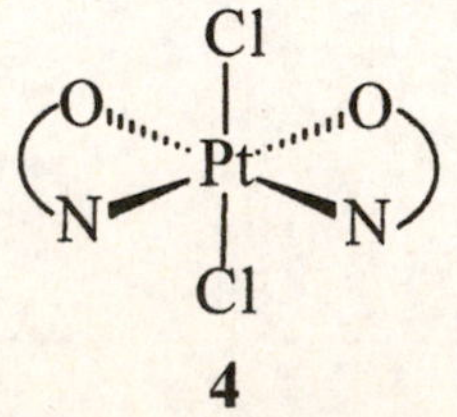

4 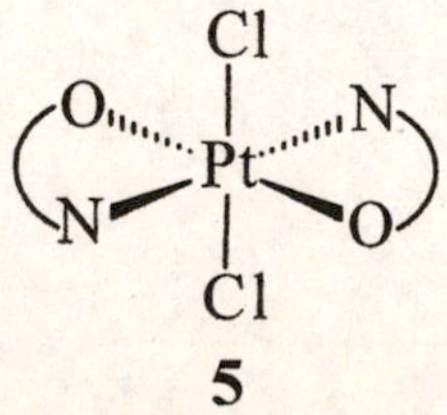   **5**

1, 2, 3, and **4** have a dipole moment. **1, 2,** and **3** can exist as enantiomers.

20.128

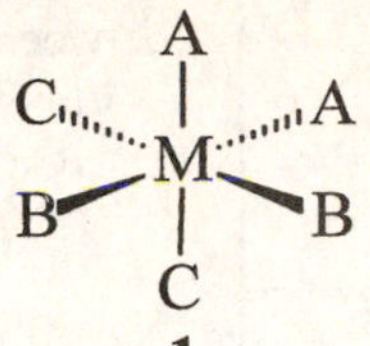

1 can exist as enantiomers.

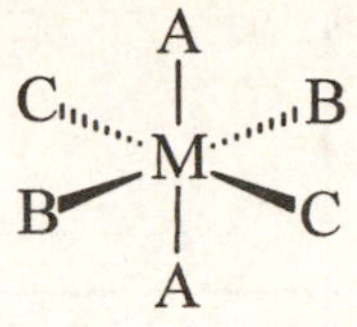

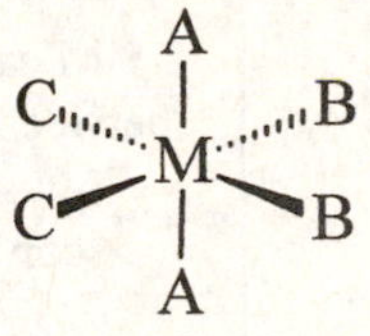

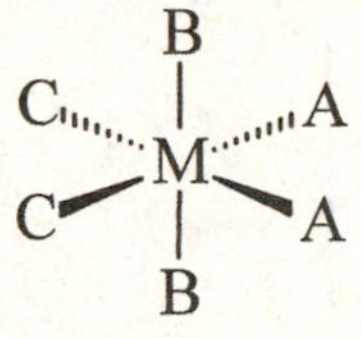

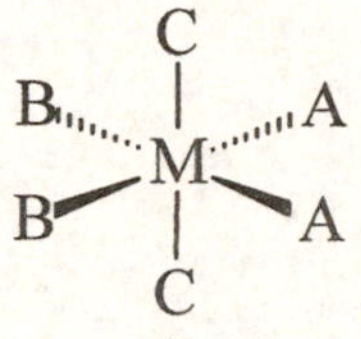

20.129 The nitro ($-NO_2$) complex is orange which means it absorbs in the blue region (see color wheel) of the visible spectrum. The nitrito ($-ONO$) is red which means it absorbs in the green region. The energy of the absorbed light is related to ligand field strength. Blue is higher energy than green, therefore nitro ($-NO_2$) is the stronger field ligand.

20.130

	weak-field ligands	strong-field ligands	
Ti^{2+} [Ar] $3d^2$	— — ↑ ↑ —	— — ↑ ↑ —	BM cannot distinguish between high-spin and low-spin electron configurations
	$BM = \sqrt{2(2+2)} = 2.83$	$BM = \sqrt{2(2+2)} = 2.83$	
V^{2+} [Ar] $3d^3$	— — ↑ ↑ ↑	— — ↑ ↑ ↑	BM cannot distinguish between high-spin and low-spin electron configurations
	$BM = \sqrt{3(3+2)} = 3.87$	$BM = \sqrt{3(3+2)} = 3.87$	
Cr^{2+} [Ar] $3d^4$	↑ — ↑ ↑ ↑	— — ↑↓ ↑ ↑	BM can distinguish between high-spin and low-spin electron configurations
	$BM = \sqrt{4(4+2)} = 4.90$	$BM = \sqrt{2(2+2)} = 2.83$	

Mn^{2+} [Ar] $3d^5$	↑ ↑ ↑ ↑ ↑	— — ↑↓ ↑↓ ↑	BM can distinguish between high-spin and low-spin electron configurations
	$BM = \sqrt{5(5+2)} = 5.92$	$BM = \sqrt{1(1+2)} = 1.73$	
Fe^{2+} [Ar] $3d^6$	↑ ↑ ↑↓ ↑ ↑	— — ↑↓ ↑↓ ↑↓	BM can distinguish between high-spin and low-spin electron configurations
	$BM = \sqrt{4(4+2)} = 4.90$	$BM = 0$	
Co^{2+} [Ar] $3d^7$	↑ ↑ ↑↓ ↑↓ ↑	↑ — ↑↓ ↑↓ ↑↓	BM can distinguish between high-spin and low-spin electron configurations
	$BM = \sqrt{3(3+2)} = 3.87$	$BM = \sqrt{1(1+2)} = 1.73$	
Ni^{2+} [Ar] $3d^8$	↑ ↑ ↑↓ ↑↓ ↑↓	↑ ↑ ↑↓ ↑↓ ↑↓	BM cannot distinguish between high-spin and low-spin electron configurations
	$BM = \sqrt{2(2+2)} = 2.83$	$BM = \sqrt{2(2+2)} = 2.83$	
Cu^{2+} [Ar] $3d^9$	↑↓ ↑ ↑↓ ↑↓ ↑↓	↑↓ ↑ ↑↓ ↑↓ ↑↓	BM cannot distinguish between high-spin and low-spin electron configurations
	$BM = \sqrt{1(1+2)} = 1.73$	$BM = \sqrt{1(1+2)} = 1.73$	
Zn^{2+} [Ar] $3d^{10}$	↑↓ ↑↓ ↑↓ ↑↓ ↑↓	↑↓ ↑↓ ↑↓ ↑↓ ↑↓	BM cannot distinguish between high-spin and low-spin electron configurations
	$BM = 0$	$BM = 0$	

20.131 ML_2 __ z^2

          __ __ xz, yz

          __ __ x^2-y^2, xy

20.132   (a)

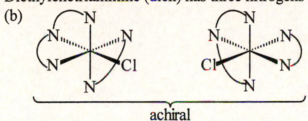

1   **2**

(b) Isomer **2** would give rise to the desired product because it has two trans NO_2 groups.

20.133   (a) A tridentate ligand bonds to a metal using electron pairs on three donor atoms. Diethylenetriammine (dien) has three nitrogens which it can use to bond to a metal.

(b)

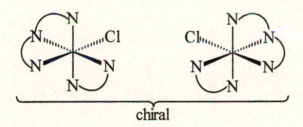

achiral

chiral

20.134   (a) $(NH_4)[Cr(H_2O)_6](SO_4)_2$, ammonium hexaaquachromium(III) sulfate

Cr^{3+} __ __

     $\uparrow$ $\uparrow$ $\uparrow$
     3 unpaired e^-

(b) $Mo(CO)_6$, hexacarbonylmolybdenum(0)

Mo^0 __ __

     $\uparrow\downarrow$ $\uparrow\downarrow$ $\uparrow\downarrow$
   low-spin, no unpaired e^-

627

(c) $[Ni(NH_3)_4(H_2O)_2](NO_3)_2$, tetraamminediaquanickel(II) nitrate

Ni^{2+} ↑ ↑

↑↓ ↑↓ ↑↓
2 unpaired e^-

(d) $K_4[Os(CN)_6]$, potassium hexacyanoosmate(II)

Os^{2+} __ __

↑↓ ↑↓ ↑↓
low-spin, no unpaired e^-

(e) $[Pt(NH_3)_4](ClO_4)_2$, tetraammineplatinum(II) perchlorate

Pt^{2+} __
↑↓
↑↓
↑↓ ↑↓
low spin, no unpaired e^-

(f) $Na_2[Fe(CO)_4]$, sodium tetracarbonylferrate(–II)

Fe^{2-} ↑↓ ↑↓ ↑↓

↑↓ ↑↓
no unpaired e^-

20.135 (a) Fe^{2+}, sodium pentacyanonitrosylferrate(II)

(b) __ __

↑↓ ↑↓ ↑↓
low-spin, no unpaired e^-

20.136 For transition metal complexes, observed colors and absorbed colors are generally complementary. Using the color wheel (Figure 20.26), the absorbed colors in the table are complementary colors to those observed.

	Observed Color	Absorbed Color	Approximate λ (nm)
$Cr(acac)_3$	red	green	530
$[Cr(H_2O)_6]^{3+}$	violet	yellow	580
$[CrCl_2(H_2O)_4]^+$	green	red	700
$[Cr(urea)_6]^{3+}$	green	red	700
$[Cr(NH_3)_6]^{3+}$	yellow	violet	420
$Cr(acetate)_3(H_2O)_3$	blue-violet	orange-yellow	600

The magnitude of Δ is comparable to the energy of the absorbed light from the low energy red end to the high energy violet end (ROYGBIV). The red of $[CrCl_2(H_2O)_4]^+$ is lower energy than the yellow of $[Cr(H_2O)_6]^{3+}$, so $Cl^- < H_2O$. Because $[CrCl_2(H_2O)_4]^+$ and $[Cr(urea)_6]^{3+}$ are both red, Δ for 6 urea's is approximately equal to Δ for 2 Cl^-'s and 4 H_2O's. Therefore, urea is between Cl^- and H_2O.

The spectrochemical series is: $Cl^- <$ urea $<$ acetate $< H_2O <$ acac $< NH_3$

Multiconcept Problems

20.137 (a) $Fe^{3+}(aq) + 3\,C_2O_4^{2-}(aq) \rightleftharpoons [Fe(C_2O_4)_3]^{3-}(aq) \qquad K_f = 3.3 \times 10^{20}$

$$[Fe(C_2O_4)_3]^{3-}(aq) \rightleftharpoons Fe^{3+}(aq) + 3\,C_2O_4^{2-}(aq) \qquad K = 1/K_f = 3.0 \times 10^{-21}$$

	$[Fe(C_2O_4)_3]^{3-}(aq) \rightleftharpoons$	$Fe^{3+}(aq) +$	$3\,C_2O_4^{2-}(aq)$
initial (M)	0.100	0	0
change (M)	$-x$	$+x$	$+3x$
equil (M)	$0.100 - x$	x	$3x$

$$K = \frac{[Fe^{3+}][C_2O_4^{2-}]^3}{[[Fe(C_2O_4)_3]^{3-}]} = 3.0 \times 10^{-21} = \frac{x(3x)^3}{0.100 - x} \approx \frac{x(3x)^3}{0.100}$$

$$3.0 \times 10^{-22} = 27x^4$$

$$x = [Fe^{3+}] = \sqrt[4]{3.0 \times 10^{-22}/27} = 1.8 \times 10^{-6}\ M$$

(b) $[Fe(C_2O_4)_3]^{3-}(aq) \rightleftharpoons Fe^{3+}(aq) + 3\,C_2O_4^{2-}(aq) \qquad K_1 = 3.0 \times 10^{-21}$

$3\,H_3O^+(aq) + 3\,C_2O_4^{2-}(aq) \rightleftharpoons 3\,HC_2O_4^-(aq) + 3\,H_2O(l) \qquad K_2 = (1/K_{a2})^3$

$\underline{3\,H_3O^+(aq) + 3\,HC_2O_4^-(aq) \rightleftharpoons 3\,H_2C_2O_4(aq) + 3\,H_2O(l) \qquad K_3 = (1/K_{a1})^3}$

$[Fe(C_2O_4)_3]^{3-}(aq) + 6\,H_3O^+(aq) \rightleftharpoons Fe^{3+}(aq) + 3\,H_2C_2O_4(aq) + 6\,H_2O(l)$

$$K_{overall} = K_1 K_2 K_3 = (3.0 \times 10^{-21})\left(\frac{1}{6.4 \times 10^{-5}}\right)^3\left(\frac{1}{5.9 \times 10^{-2}}\right)^3 = 5.6 \times 10^{-5}$$

$\Delta G = -RT\ln K = -[8.314 \times 10^{-3}\ kJ/(K \cdot mol)](298\ K)\ln(5.6 \times 10^{-5}) = 24.3\ kJ/mol$

The reaction is nonspontaneous because ΔG is positive.

(c) ↑ ↑

↑ ↑ ↑
5 unpaired e^-

(d)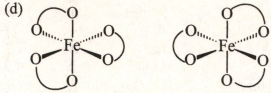

The complex is chiral. Enantiomers are shown.

20.138 (1) $Ni(H_2O)_6^{2+}(aq) + 6\,NH_3(aq) \rightleftharpoons Ni(NH_3)_6^{2+}(aq) + 6\,H_2O(l)$ $K_f = 2.0 \times 10^8$
(2) $Ni(H_2O)_6^{2+}(aq) + 3\,en(aq) \rightleftharpoons Ni(en)_3^{2+}(aq) + 6\,H_2O(l)$ $K_f = 4 \times 10^{17}$
(a) Reaction (2) should have the larger entropy change because three bidentate en ligands displace six water molecules.
(b) $\Delta G^\circ = \Delta H^\circ - T\Delta S^\circ$
Because ΔH°_1 and ΔH°_2 are almost the same, the difference in ΔG° is determined by the difference in ΔS°. Because ΔS°_2 is larger than ΔS°_1, ΔG°_2 is more negative than ΔG°_1 which is consistent with the greater stability of $Ni(en)_3^{2+}$.
(c) $\Delta H^\circ - T\Delta S^\circ = \Delta G^\circ = -RT \ln K_f$
$\Delta H^\circ_1 - T\Delta S^\circ_1 - (\Delta H^\circ_2 - T\Delta S^\circ_2) = -RT \ln K_f(1) - [-RT \ln K_f(2)]$

$T\Delta S^\circ_2 - T\Delta S^\circ_1 = RT \ln K_f(2) - RT \ln K_f(1) = RT \ln \dfrac{K_f(2)}{K_f(1)}$

$\Delta S^\circ_2 - \Delta S^\circ_1 = R \ln \dfrac{K_f(2)}{K_f(1)} = [8.314\ \text{J/(K} \cdot \text{mol)}] \ln \dfrac{4 \times 10^{17}}{2.0 \times 10^8}$

$\Delta S^\circ_2 - \Delta S^\circ_1 = 178\ \text{J/(K} \cdot \text{mol)}$ or $180\ \text{J/(K} \cdot \text{mol)}$

20.139 (a) $[Fe^{2+}(aq) \rightarrow Fe^{3+}(aq) + e^-] \times 5$ $E^\circ = -0.77$ V
     (oxidation half reaction)

$MnO_4^-(aq) \rightarrow Mn^{2+}(aq)$
$MnO_4^-(aq) \rightarrow Mn^{2+}(aq) + 4\,H_2O(l)$
$8\,H^+(aq) + MnO_4^-(aq) \rightarrow Mn^{2+}(aq) + 4\,H_2O(l)$
$8\,H^+(aq) + MnO_4^-(aq) + 5\,e^- \rightarrow Mn^{2+}(aq) + 4\,H_2O(l)$ $E^\circ = 1.51$ V
     (reduction half reaction)
Combine the two half reactions.
$5\,Fe^{2+}(aq) + MnO_4^-(aq) + 8\,H^+(aq) \rightarrow 5\,Fe^{3+}(aq) + Mn^{2+}(aq) + 4\,H_2O(l)$

(b) $E^\circ = -0.77$ V $+ 1.51$ V $= 0.74$ V
$\Delta G^\circ = -nFE^\circ = -(5\ \text{mol}\ e^-)\left(\dfrac{96{,}500\ \text{C}}{\text{mol}\ e^-} \right)(0.74\ \text{V})\left(\dfrac{1\ \text{J}}{1\ \text{C} \cdot \text{V}} \right)\left(\dfrac{1\ \text{kJ}}{1000\ \text{J}} \right) = -3.6 \times 10^2\ \text{kJ}$

$$E^\circ = \frac{0.0592 \text{ V}}{n} \log K$$

$$\log K = \frac{nE^\circ}{0.0592 \text{ V}} = \frac{(5)(0.74 \text{ V})}{0.0592 \text{ V}} = 62.5; \quad K = 10^{62.5} = 3 \times 10^{62}$$

(c)

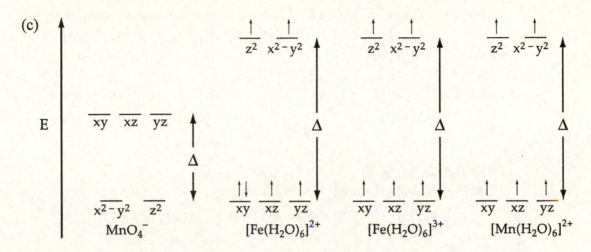

(d) The paramagnetism of the solution increases as the reaction proceeds. When Fe^{2+}(aq) is oxidized to Fe^{3+}(aq), the number of unpaired electrons goes from 4 to 5. In addition, the Mn^{2+}(aq) produced has 5 unpaired electrons.

(e) 34.83 mL = 0.03483 L

mol MnO_4^-(aq) = (0.051 32 mol/L)(0.034 83 L) = 1.787×10^{-3} mol MnO_4^-

$$\text{mass Fe} = 1.787 \times 10^{-3} \text{ mol } MnO_4^- \times \frac{5 \text{ mol } Fe^{2+}}{1 \text{ mol } MnO_4^-} \times \frac{55.847 \text{ g Fe}}{1 \text{ mol } Fe^{2+}} = 0.4990 \text{ g Fe}$$

$$\text{mass \% Fe} = \frac{0.4990 \text{ g Fe}}{1.265 \text{ g sample}} \times 100\% = 39.45 \% \text{ Fe}$$

20.140 (a) $Cr(s) + 2 H^+(aq) \rightarrow Cr^{2+}(aq) + H_2(g)$

(b) mol Cr = 2.60 g Cr x $\dfrac{1 \text{ mol Cr}}{52.00 \text{ g Cr}}$ = 0.0500 mol Cr

mol H_2SO_4 = (0.050 00 L)(1.200 mol/L) = 0.060 00 mol H_2SO_4
The stoichiometry between Cr and H_2SO_4 is one to one, therefore Cr is the limiting reagent because of the smaller number of moles.

mol H_2 = 0.0500 mol Cr x $\dfrac{1 \text{ mol } H_2}{1 \text{ mol Cr}}$ = 0.0500 mol H_2

25 °C = 298 K
PV = nRT

$$V = \frac{nRT}{P} = \frac{(0.0500 \text{ mol})\left(0.082\ 06\ \dfrac{\text{L} \cdot \text{atm}}{\text{K} \cdot \text{mol}}\right)(298 \text{ K})}{\left(735 \text{ mm Hg} \times \dfrac{1.00 \text{ atm}}{760 \text{ mm Hg}}\right)} = 1.26 \text{ L of } H_2$$

(c) 0.060 00 mol H_2SO_4 can provide 0.1200 mol H^+. 0.0500 mol Cr reacts with 2 x (0.0500 mol H^+) = 0.100 mol H^+. This leaves 0.0200 mol H^+ and 0.0600 mol SO_4^{2-}, which will give, after neutralization, 0.0200 mol HSO_4^- and 0.0400 mol SO_4^{2-}.

$[HSO_4^-]$ = 0.0200 mol/0.050 00 L = 0.400 M

$[SO_4^{2-}]$ = 0.0400 mol/0.050 00 L = 0.800 M

The pH of this solution can be determined from the following equilibrium:

$$HSO_4^-(aq) + H_2O(l) \rightleftharpoons H_3O^+(aq) + SO_4^{2-}(aq)$$

initial (M)	0.400	0	0.800
change (M)	–x	+x	+x
equil (M)	0.400 – x	x	0.800 + x

$$K_{a2} = \frac{[H_3O^+][SO_4^{2-}]}{[HSO_4^-]} = 1.2 \times 10^{-2} = \frac{(x)(0.800 + x)}{0.400 - x}$$

$x^2 + 0.812x - 0.0048 = 0$

Use the quadratic formula to solve for x.

$$x = \frac{-(0.812) \pm \sqrt{(0.812)^2 - 4(1)(-0.0048)}}{2(1)} = \frac{-0.812 \pm 0.8237}{2}$$

x = 0.005 85 and –0.818

Of the two solutions for x, only the positive value of x has physical meaning, because x is the $[H_3O^+]$.

$[H_3O^+]$ = x = 0.005 85 M

pH = $-\log[H_3O^+]$ = $-\log(0.005 85)$ = 2.23

(d) Crystal field d-orbital energy level diagram

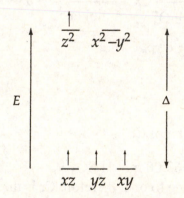

Valence bond orbital diagram

$Cr(H_2O)_6^{2+}$ [Ar] 3d 4s 4p 4d

sp^3d^2 4 unpaired e^-

(e) The addition of excess KCN converts $Cr(H_2O)_6^{2+}$(aq) to $Cr(CN)_6^{4-}$(aq). CN^- is a strong field ligand and increases Δ changing the chromium complex from high spin, with 4 unpaired electrons, to low spin, with only 2 unpaired electrons.

20.141 (a) $t_{1/2} = \dfrac{0.693}{k} = \dfrac{0.693}{3.2 \times 10^{-5} \text{ s}^{-1}} = 2.16 \times 10^4 \text{ s}$

$t_{1/2} = (2.16 \times 10^4 \text{ s})(1 \text{ h}/3600 \text{ s}) = 6.0 \text{ h}$

(b) Let A = trans-$[Co(en)_2Cl_2]^+$

$\ln\dfrac{[A]_t}{[A]_o} = -kt;$ $\ln[A]_t - \ln[A]_o = -kt;$ $\ln[A]_t = \ln[A]_o - kt$

$\ln[A]_t = \ln(0.138) - (3.2 \times 10^{-5} \text{ s}^{-1})(16.5 \text{ h})(3600 \text{ s/h}) = -3.88$

$[A]_t = e^{-3.88} = 0.021 \text{ M}$

(c) trans-$[Co(en)_2Cl_2]^+$(aq) → trans-$[Co(en)_2Cl]^{2+}$(aq) + Cl^-(aq)   (slow)

   trans-$[Co(en)_2Cl]^{2+}$(aq) + H_2O(l) → trans-$[Co(en)_2(H_2O)Cl]^{2+}$(aq)   (fast)

(d)

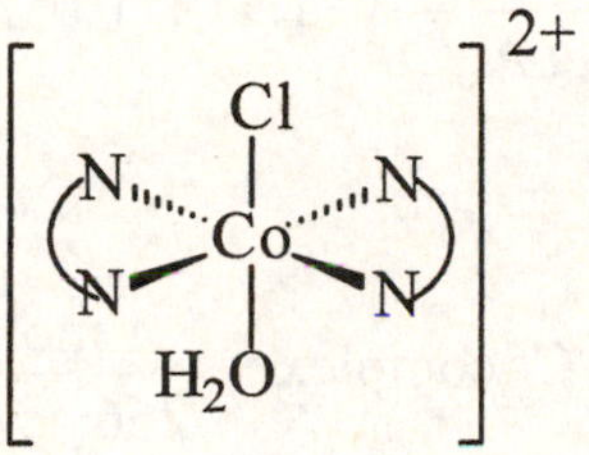

The reaction product is achiral because it has several mirror planes.

(e)

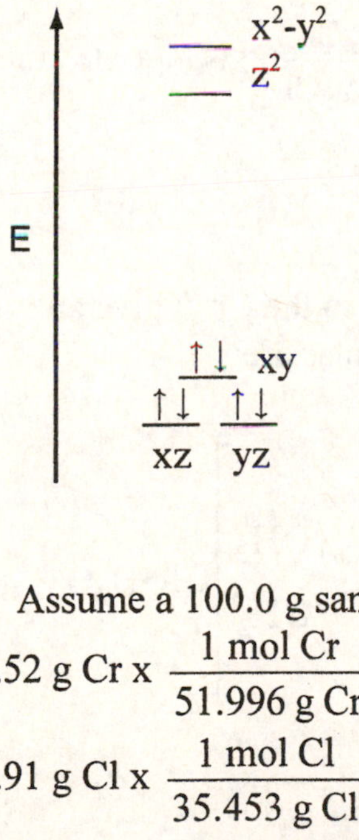

20.142 (a) Assume a 100.0 g sample of the chromium compound.

$19.52 \text{ g Cr} \times \dfrac{1 \text{ mol Cr}}{51.996 \text{ g Cr}} = 0.3754 \text{ mol Cr}$

$39.91 \text{ g Cl} \times \dfrac{1 \text{ mol Cl}}{35.453 \text{ g Cl}} = 1.126 \text{ mol Cl}$

$40.57 \text{ g H}_2\text{O} \times \dfrac{1 \text{ mol H}_2\text{O}}{18.015 \text{ g H}_2\text{O}} = 2.252 \text{ mol H}_2\text{O}$

$Cr_{0.3754}Cl_{1.126}(H_2O)_{2.252}$, divide each subscript by the smallest, 0.3754.

$Cr_{0.3754/0.3754}Cl_{1.126/0.3754}(H_2O)_{2.252/0.3754}$

$CrCl_3(H_2O)_6$

(b) $Cr(H_2O)_6Cl_3$, 266.45 amu; AgCl, 143.32 amu

For **A**:   mol Cr complex = mol Cr = 0.225 g Cr complex $\times \dfrac{1 \text{ mol Cr complex}}{266.45 \text{ g Cr complex}} =$

8.44×10^{-4} mol Cr

mol Cl = mol AgCl = 0.363 g AgCl $\times \dfrac{1 \text{ mol AgCl}}{143.32 \text{ g AgCl}} = 2.53 \times 10^{-3}$ mol Cl

$\dfrac{\text{mol Cl}}{\text{mol Cr}} = \dfrac{2.53 \times 10^{-3} \text{ mol Cl}}{8.44 \times 10^{-4} \text{ mol Cr}} = 3 \text{ Cl/Cr}$

For **B**:   mol Cr complex = mol Cr = 0.263 g Cr complex $\times \dfrac{1 \text{ mol Cr complex}}{266.45 \text{ g Cr complex}} =$

9.87×10^{-4} mol Cr

mol Cl = mol AgCl = 0.283 g AgCl $\times \dfrac{1 \text{ mol AgCl}}{143.32 \text{ g AgCl}} = 1.97 \times 10^{-3}$ mol Cl

$\dfrac{\text{mol Cl}}{\text{mol Cr}} = \dfrac{1.97 \times 10^{-3} \text{ mol Cl}}{9.87 \times 10^{-4} \text{ mol Cr}} = 2 \text{ Cl/Cr}$

For **C**:   mol Cr complex = mol Cr = 0.358 g Cr complex $\times \dfrac{1 \text{ mol Cr complex}}{266.45 \text{ g Cr complex}} =$

1.34×10^{-3} mol Cr

mol Cl = mol AgCl = 0.193 g AgCl $\times \dfrac{1 \text{ mol AgCl}}{143.32 \text{ g AgCl}} = 1.34 \times 10^{-3}$ mol Cl

$\dfrac{\text{mol Cl}}{\text{mol Cr}} = \dfrac{1.34 \times 10^{-3} \text{ mol Cl}}{1.34 \times 10^{-3} \text{ mol Cr}} = 1 \text{ Cl/Cr}$

Because only the free Cl^- ions (those not bonded to the Cr^{3+}) give an immediate precipitate of AgCl, the probable structural formulas are:

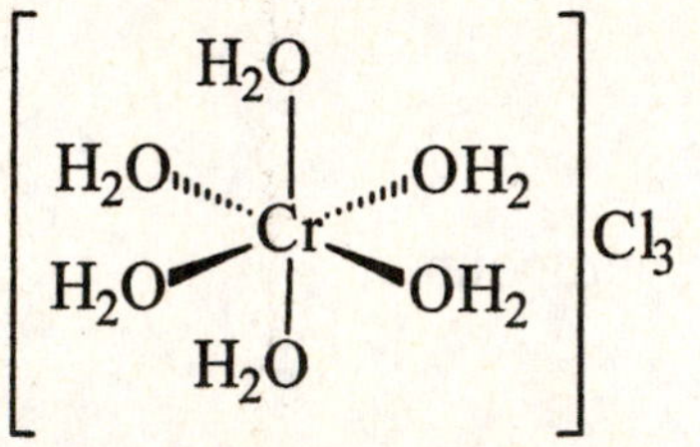

A

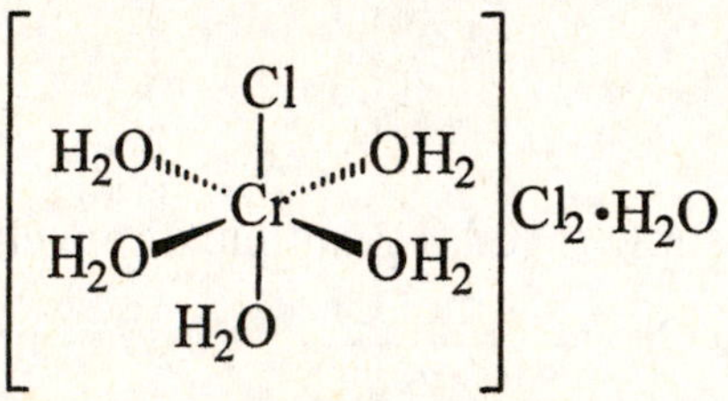

B

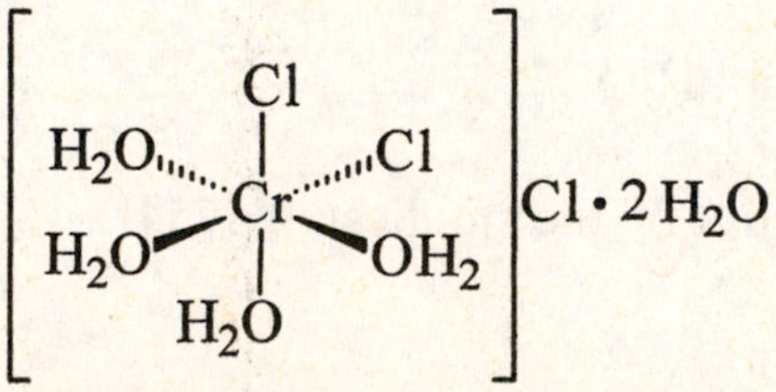

C

Structure **C** can exist as either cis or trans diastereoisomers.

(c) H_2O is a stronger field ligand than Cl^-. Compound **A** is likely to be violet absorbing in the yellow. Compounds **B** and **C** have weaker field ligands and would appear blue or green absorbing in the orange or red, respectively.

(d) $\Delta T = K_f \cdot m \cdot i$
For **A**, $i = 4$; for **B**, $i = 3$; and for **C**, $i = 2$.
For **A**, $\Delta T = K_f \cdot m \cdot i = (1.86\ °C/m)(0.25\ m)(4) = 1.86\ °C$
freezing point $= 0\ °C - \Delta T = 0\ °C - 1.86\ °C = -1.86\ °C$
For **B**, $\Delta T = K_f \cdot m \cdot i = (1.86\ °C/m)(0.25\ m)(3) = 1.39\ °C$
freezing point $= 0\ °C - \Delta T = 0\ °C - 1.39\ °C = -1.39\ °C$
For **C**, $\Delta T = K_f \cdot m \cdot i = (1.86\ °C/m)(0.25\ m)(2) = 0.93\ °C$
freezing point $= 0\ °C - \Delta T = 0\ °C - 0.93\ °C = -0.93\ °C$

20.143   (a) & (b) Let $\overbrace{X \quad Y}$ represent the tfac$^-$ ligand.

$$A =$$

$$B =$$

enantiomers

(c) $k_1 = 0.0889\ h^{-1}$, $T_1 = 66.1\ °C = 66.1 + 273.15 = 339.2\ K$
$k_2 = (0.0870\ min^{-1})(60\ min/1\ h) = 5.22\ h^{-1}$, $T_2 = 99.2\ °C = 99.2 + 273.15 = 372.3\ K$

$$\ln\left(\frac{k_2}{k_1}\right) = \left(\frac{-E_a}{R}\right)\left(\frac{1}{T_2} - \frac{1}{T_1}\right)$$

$$E_a = -\frac{[\ln k_2 - \ln k_1](R)}{\left(\dfrac{1}{T_2} - \dfrac{1}{T_1}\right)}$$

$$E_a = -\frac{[\ln(5.22) - \ln(0.0889)][8.314 \times 10^{-3}\ kJ/(K \cdot mol)]}{\left(\dfrac{1}{372.3\ K} - \dfrac{1}{339.2\ K}\right)} = 129\ kJ/mol$$

(d) $\underline{}$ $\underline{}$

$\uparrow\downarrow$ $\quad$ $\uparrow\downarrow$ $\quad$ $\uparrow\downarrow$

no unpaired e^-, diamagnetic

20.144 (a) $K = \dfrac{[Cr_2O_7^{2-}]}{[CrO_4^{2-}]^2[H^+]^2} = 1.00 \times 10^{14}$

$[Cr_2O_7^{2-}]/[CrO_4^{2-}]^2 = 1.00 \times 10^{14} [H^+]^2$

In neutral solution, $[H^+] = 1.0 \times 10^{-7}$ and $[Cr_2O_7^{2-}]/[CrO_4^{2-}]^2 = 1$, so $[Cr_2O_7^{2-}]$ and $[CrO_4^{2-}]$ are comparable.

In basic solution, $[H^+] < 1.0 \times 10^{-7}$ and $[Cr_2O_7^{2-}]/[CrO_4^{2-}]^2 < 1$, so $[CrO_4^{2-}]$ predominates.

In acidic solution, $[H^+] > 1.0 \times 10^{-7}$ and $[Cr_2O_7^{2-}]/[CrO_4^{2-}]^2 > 1$, so $[Cr_2O_7^{2-}]$ predominates.

(b) At pH = 4.000, the $[H^+] = 1.00 \times 10^{-4}$ M

Let $x = [Cr_2O_7^{2-}]$ and $y = [CrO_4^{2-}]$

$\dfrac{[Cr_2O_7^{2-}]}{[CrO_4^{2-}]^2} = [H^+]^2(1.00 \times 10^{14})$

$\dfrac{[Cr_2O_7^{2-}]}{[CrO_4^{2-}]^2} = (1.00 \times 10^{-4})^2(1.00 \times 10^{14})$

$\dfrac{[Cr_2O_7^{2-}]}{[CrO_4^{2-}]^2} = 1.00 \times 10^6 = \dfrac{x}{y^2}$

Because there are 2 Cr atoms per $Cr_2O_7^{2-}$, the total Cr concentration is $2[Cr_2O_7^{2-}] + [CrO_4^{2-}]$, and therefore $2x + y = 0.100$.

$\dfrac{x}{y^2} = 1.00 \times 10^6$ and $2x + y = 0.100$ M; solve these simultaneous equations.

$x = (1.00 \times 10^6)y^2$ and $x = (0.100 - y)/2$; substitute $(0.100 - y)/2$ for x

$(0.100 - y)/2 = (1.00 \times 10^6)y^2$

$(2.00 \times 10^6)y^2 + y - 0.100 = 0$

Use the quadratic formula to solve for y.

$y = \dfrac{-(1) \pm \sqrt{(1)^2 - 4(2.00 \times 10^6)(-0.100)}}{2(2.00 \times 10^6)} = \dfrac{(-1) \pm (894.2)}{4.00 \times 10^6}$

$y = -2.24 \times 10^{-4}$ and $2.233 \times 10^{-4} = 2.23 \times 10^{-4}$

Of the two solutions for y, only the positive value of y has physical meaning because y is the $[CrO_4^{2-}]$.

$[CrO_4^{2-}] = 2.23 \times 10^{-4}$ M

$[Cr_2O_7^{2-}] = x = (1.00 \times 10^6)y^2 = (1.00 \times 10^6)(2.233 \times 10^{-4}$ M$)^2 = 4.99 \times 10^{-2}$ M

(c) At pH = 2.000, the $[H^+] = 1.00 \times 10^{-2}$ M

Let x = $[Cr_2O_7^{2-}]$ and y = $[CrO_4^{2-}]$

$$\frac{[Cr_2O_7^{2-}]}{[CrO_4^{2-}]^2} = [H^+]^2(1.00 \times 10^{14})$$

$$\frac{[Cr_2O_7^{2-}]}{[CrO_4^{2-}]^2} = (1.00 \times 10^{-2})^2(1.00 \times 10^{14})$$

$$\frac{[Cr_2O_7^{2-}]}{[CrO_4^{2-}]^2} = 1.00 \times 10^{10} = \frac{x}{y^2}$$

Because there are 2 Cr atoms per $Cr_2O_7^{2-}$, the total Cr concentration is $2[Cr_2O_7^{2-}]$ + $[CrO_4^{2-}]$, and therefore 2x + y = 0.100.

$\frac{x}{y^2} = 1.00 \times 10^{10}$ and 2x + y = 0.100 M; solve these simultaneous equations.

x = $(1.00 \times 10^{10})y^2$ and x = (0.100 – y)/2; substitute (0.100 – y)/2 for x

(0.100 – y)/2 = $(1.00 \times 10^{10})y^2$

$(2.00 \times 10^{10})y^2 + y - 0.100 = 0$

Use the quadratic formula to solve for y.

$$y = \frac{-(1) \pm \sqrt{(1)^2 - 4(2.00 \times 10^{10})(-0.100)}}{2(2.00 \times 10^{10})} = \frac{(-1) \pm (8.944 \times 10^4)}{4.00 \times 10^{10}}$$

y = -2.24×10^{-6} and $2.236 \times 10^{-6} = 2.24 \times 10^{-6}$

Of the two solutions for y, only the positive value of y has physical meaning because y is the $[CrO_4^{2-}]$.

$[CrO_4^{2-}] = 2.24 \times 10^{-6}$ M

$[Cr_2O_7^{2-}] = x = (1.00 \times 10^{10})y^2 = (1.00 \times 10^{10})(2.236 \times 10^{-6}$ M$)^2 = 5.00 \times 10^{-2}$ M

Metals and Solid-State Materials

21.1 The most common oxidation state for the 3B transition metals (Sc, Y, and La) is 3+. The 3+ oxidation state of the cations conveniently matches the 3– charge of the phosphate ion resulting in a large lattice energy and corresponding insolubility for MPO_4 compounds.

21.2 (a) $Cr_2O_3(s) + 2 Al(s) \rightarrow 2 Cr(s) + Al_2O_3(s)$
(b) $Cu_2S(s) + O_2(g) \rightarrow 2 Cu(s) + SO_2(g)$
(c) $PbO(s) + C(s) \rightarrow Pb(s) + CO(g)$
(d) $2 K^+(l) + 2 Cl^-(l) \xrightarrow{electrolysis} 2 K(l) + Cl_2(g)$

21.3 $CaO(s) + SiO_2(s) \rightarrow CaSiO_3(l)$ (slag)
The O^{2-} in CaO behaves as a Lewis base and SiO_2 is the Lewis acid. They react with each other in a Lewis acid-base reaction to yield $CaSiO_3$ (Ca^{2+} and SiO_3^{2-}).

21.4 The electron configuration for Hg is $[Xe] 4f^{14} 5d^{10} 6s^2$. Assuming the 5d and 6s bands overlap, the composite band can accomodate 12 valence electrons per metal atom. Weak bonding and a low melting point are expected for Hg because both the bonding and antibonding MOs are occupied.

21.5 (a) The composite s-d band can accomodate 12 valence electrons per metal atom.
Hf $[Xe] 6s^2 4f^{14} 5d^2$, 4 valence electrons (4 bonding, 0 antibonding)
The s-d band is 1/4 full, so Hf is picture (1).
Pt $[Xe] 6s^2 4f^{14} 5d^8$, 10 valence electrons (6 bonding, 4 antibonding)
The s-d band is 5/6 full, so Pt is picture (2).
Re $[Xe] 6s^2 4f^{14} 5d^5$, 7 valence electrons (6 bonding, 1 antibonding)
The s-d band is 7/12 full, so Re is picture (3).
(b) Re has an excess of 5 bonding electrons and it has the highest melting point and is the hardest of the three.
(c) Pt has an excess of only 2 bonding electrons and it has the lowest melting point and is the softest of the three.

21.6 Ge doped with As is an n-type semiconductor because As has an additional valence electron. The extra electrons are in the conduction band. The number of electrons in the conduction band of the doped Ge is much higher than for pure Ge, and the conductivity of the doped semiconductor is higher.

21.7 (a) (1), silicon; (2), white tin; (3), diamond; (4), silicon doped with aluminum
(b) (3) < (1) < (4) < (2)
Diamond (3) is an insulator with a large band gap. Silicon (1) is a semiconductor with a band gap smaller than diamond. The conduction band is partially occupied with a few

electrons and the valence band is partially empty. Silicon doped with aluminum (4) is a p-type semiconductor that has fewer electrons than needed for bonding and has vacancies (positive holes) in the valence band. White tin (2) has a partially filled s-p composite band and is a metallic conductor.

21.8 $E = 222 \text{ kJ/mol} \times \dfrac{1000 \text{ J}}{1 \text{ kJ}} \times \dfrac{1 \text{ mol}}{6.02 \times 10^{23}} = 3.69 \times 10^{-19} \text{ J}$

$\nu = \dfrac{E}{h} = \dfrac{3.69 \times 10^{-19} \text{ J}}{6.626 \times 10^{-34} \text{ J} \cdot \text{s}} = 5.57 \times 10^{14} \text{ s}^{-1}$

$\lambda = \dfrac{c}{\nu} = \dfrac{3.00 \times 10^{8} \text{ m/s}}{5.57 \times 10^{14} \text{ s}^{-1}} = 5.39 \times 10^{-7} \text{ m} = 539 \times 10^{-9} \text{ m} = 539 \text{ nm}$

21.9   8 Cu at corners $8 \times 1/8 = 1$ Cu
       8 Cu on edges $8 \times 1/4 = \underline{2 \text{ Cu}}$
                            Total $= 3$ Cu

       12 O on edges $12 \times 1/4 = 3$ O
       8 O on faces $8 \times 1/2 = \underline{4 \text{ O}}$
                            Total $= 7$ O

21.10 $Si(OCH_3)_4 + 4 \text{ H}_2O \rightarrow Si(OH)_4 + 4 \text{ HOCH}_3$

21.11 $Ba[OCH(CH_3)_2]_2 + Ti[OCH(CH_3)_2]_4 + 6 \text{ H}_2O \rightarrow BaTi(OH)_6(s) + 6 \text{ HOCH}(CH_3)_2$

                 heat
$BaTi(OH)_6(s) \rightarrow BaTiO_3(s) + 3 \text{ H}_2O(g)$

21.12   (a) cobalt/tungsten carbide is a ceramic-metal composite.
       (b) silicon carbide/zirconia is a ceramic-ceramic composite.
       (c) boron nitride/epoxy is a ceramic-polymer composite.
       (d) boron carbide/titanium is a ceramic-metal composite.

21.13   The smaller the particle, the larger the band gap and the greater the shift in the color of the emitted light from the red to the violet. The yellow quantum dot is larger because yellow is closer to the red than is the blue.

21.14   atom diameter $= 250 \text{ pm} = 250 \times 10^{-12} \text{ m} = 0.25 \times 10^{-9} \text{ m} = 0.25 \text{ nm}$
       (a) 5.0 nm/0.25 nm = 20 atoms on an edge of nanoparticle
       total atoms in nanoparticle $= 20^3 = 8,000$
       interior atoms in nanoparticle $= (20 - 2)^3 = 18^3 = 5,832$
       surface atoms in nanoparticle = total atoms – interior atoms = 8,000 – 5,832 = 2,168
       % of atoms on nanoparticle surface = (2,168/8,000) x 100% = 27%

       (b) 10 nm/0.25 nm = 40 atoms on an edge of nanoparticle
       total atoms in nanoparticle $= 40^3 = 64,000$

interior atoms in nanoparticle = $(40 - 2)^3 = 38^3 = 54,872$
surface atoms in nanoparticle = total atoms – interior atoms = $64,000 - 54,872 = 9,128$
% of atoms on nanoparticle surface = $(9,128/64,000) \times 100\% = 14\%$

Key Concept Problems

21.15 A – metal oxide;   B – metal sulfide;   C – metal carbonate;   D – free metal

21.16 (a) electrolysis   (b) roasting a metal sulfide
  (c)   A = Li, electrolysis.
           B = Hg, roasting of the metal sulfide.
           C = Mn, reduction of the metal oxide.
           D = Ca, reduction of the metal oxide.

21.17 (a) (1) and (4) are semiconductors; (2) is a metal; (3) is an insulator
  (b) (3) < (1) < (4) < (2). The conductivity increases with decreasing band gap.
  (c) (1) and (4) increases; (2) decreases; (3) not much change.

21.18 (a) (2), bonding MO's are filled.
  (b) (3), bonding and antibonding MO's are filled.
  (c) (3) < (1) < (2). Hardness increases with increasing MO bond order.

21.19

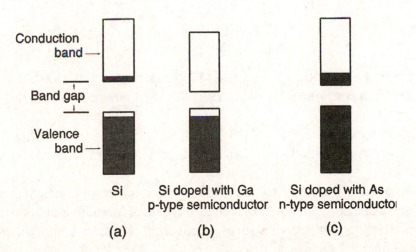

(d) The electrical conductivity of the doped silicon in both cases is higher than for pure silicon. Si doped with Ga is a p-type semiconductor with many more positive holes in the conduction band than in pure Si. This results in a higher conductivity. Si doped with As is an n-type semiconductor with many more electrons in the conduction band than in pure Si. This also results in a higher conductivity.

21.20

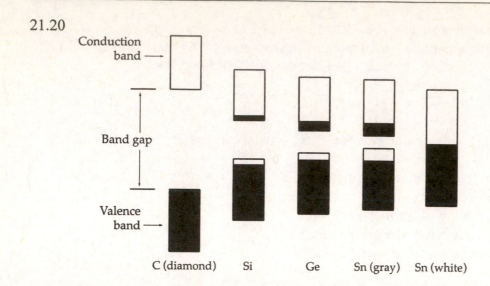

C (diamond)   Si   Ge   Sn (gray)   Sn (white)

21.21   For LEDs the energy of the light emitted is roughly equal to the band gap energy, the higher the energy the shorter the wavelength, and the smaller the energy the longer the wavelength. LED (1) has the larger band gap energy so it emits blue light; LED (2) has the smaller band gap energy so it emits red light.

Section Problems
Sources of the Metallic Elements (Section 21.1)

21.22 TiO_2, MnO_2, and Fe_2O_3

21.23   Pd, Pt, and Au

21.24   (a) Cu is found in nature as a sulfide.   (b) Zr is found in nature as an oxide.
        (c) Pd is found in nature uncombined.   (d) Bi is found in nature as a sulfide.

21.25   (a) V is found in nature as an oxide.   (b) Ag is found in nature as a sulfide.
        (c) Rh is found in nature uncombined.   (d) Hf is found in nature as an oxide.

21.26   The less electronegative early transition metals tend to form ionic compounds by losing electrons to highly electronegative nonmetals such as oxygen. The more electronegative late transition metals tend to form compounds with more covalent character by bonding to the less electronegative nonmetals such as sulfur.

21.27   The early transition metals, which include Fe, generally occur as oxides because these less electronegative metals tend to form ionic compounds by losing electrons to the very electronegative oxygen. On the other hand, oxides of the s-block metals are strongly basic and far too reactive to exist in an environment that contains acidic oxides such as CO_2 and SiO_2. Consequently, s-block metals, including Ca, are found in nature as carbonates, as silicates, and in the case of Na and K, as chlorides.

21.28 (a) Fe_2O_3, hematite   (b) PbS, galena
       (c) TiO_2, rutile   (d) $CuFeS_2$, chalcopyrite

21.29 (a) cinnabar, HgS   (b) bauxite, $Al_2O_3 \cdot x\ H_2O$
       (c) sphalerite, ZnS   (d) chromite, $FeCr_2O_4$

Metallurgy (Section 21.2)

21.30 The flotation process exploits the differences in the ability of water and oil to wet the surfaces of the mineral and the gangue. The gangue, which contains ionic silicates, is moistened by the polar water molecules and sinks to the bottom of the tank. The mineral particles, which contain the less polar metal sulfide, are coated by the oil and become attached to the soapy air bubbles created by the detergent. The metal sulfide particles are carried to the surface in the soapy froth, which is skimmed off at the top of the tank. This process would not work well for a metal oxide because it is too polar and will be wet by the water and sink with the gangue.

21.31 Sometimes, ores are concentrated by chemical treatment. In the Bayer process, Ga_2O_3 is separated from Fe_2O_3 by treating the ore with hot aqueous NaOH. The amphoteric Ga_2O_3 dissolves as $Ga(OH)_4^-$, but the basic Fe_2O_3 does not.
$Ga_2O_3(s) + 2\ OH^-(aq) + 3\ H_2O(l) \rightarrow 2\ Ga(OH)_4^-(aq)$

21.32 Because $E° < 0$ for Zn^{2+}, the reduction of Zn^{2+} is not favored.
Because $E° > 0$ for Hg^{2+}, the reduction of Hg^{2+} is favored.
The roasting of CdS should yield CdO because, like Zn^{2+}, $E° < 0$ for the reduction of Cd^{2+}.

21.33 $Mg^{2+}(aq) + 2\ e^- \rightarrow Mg(s)$ $E° = -2.37\ V$
      $Zn^{2+}(aq) + 2\ e^- \rightarrow Zn(s)$ $E° = -0.76\ V$

      The very negative reduction potential for Mg^{2+} indicates that Mg^{2+} is much more difficult to reduce than the Zn^{2+}.

21.34 (a) $V_2O_5(s) + 5\ Ca(s) \rightarrow 2\ V(s) + 5\ CaO(s)$
       (b) $2\ PbS(s) + 3\ O_2(g) \rightarrow 2\ PbO(s) + 2\ SO_2(g)$
       (c) $MoO_3(s) + 3\ H_2(g) \rightarrow Mo(s) + 3\ H_2O(g)$
       (d) $3\ MnO_2(s) + 4\ Al(s) \rightarrow 3\ Mn(s) + 2\ Al_2O_3(s)$

               electrolysis
       (e) $MgCl_2(l) \quad \rightarrow \quad Mg(l) + Cl_2(g)$

21.35 (a) $Fe_2O_3(s) + 3\ H_2(g) \rightarrow 2\ Fe(s) + 3\ H_2O(g)$
       (b) $Cr_2O_3(s) + 2\ Al(s) \rightarrow 2\ Cr(s) + Al_2O_3(s)$
       (c) $Ag_2S(s) + O_2(g) \rightarrow 2\ Ag(s) + SO_2(g)$
       (d) $TiCl_4(g) + 2\ Mg(l) \rightarrow Ti(l) + 2\ MgCl_2(l)$

               electrolysis
       (e) $2\ LiCl(l) \quad \rightarrow \quad 2\ Li(l) + Cl_2(g)$

21.36 $2 ZnS(s) + 3 O_2(g) \rightarrow 2 ZnO(s) + 2 SO_2(g)$

$\Delta H° = [2 \Delta H°_f(ZnO) + 2 \Delta H°_f(SO_2)] - [2 \Delta H°_f(ZnS)]$

$\Delta H° = [(2 \text{ mol})(-350.5 \text{ kJ/mol}) + (2 \text{ mol})(-296.8 \text{ kJ/mol})]$

$- (2 \text{ mol})(-206.0 \text{ kJ/mol}) = -882.6 \text{ kJ}$

$\Delta G° = [2 \Delta G°_f(ZnO) + 2 \Delta G°_f(SO_2)] - [2 \Delta G°_f(ZnS)]$

$\Delta G° = [(2 \text{ mol})(-320.5 \text{ kJ/mol}) + (2 \text{ mol})(-300.2 \text{ kJ/mol})]$

$- (2 \text{ mol})(-201.3 \text{ kJ/mol}) = -838.8 \text{ kJ}$

$\Delta H°$ and $\Delta G°$ are different because of the entropy change associated with the reaction. The minus sign for $(\Delta H° - \Delta G°)$ indicates that the entropy is negative, which is consistent with a decrease in the number of moles of gas from 3 mol to 2 mol.

21.37 $HgS(s) + O_2(g) \rightarrow Hg(l) + SO_2(g)$

$\Delta H° = \Delta H°_f(SO_2) - \Delta H°_f(HgS)$

$\Delta H° = (1 \text{ mol})(-296.8 \text{ kJ/mol}) - (1 \text{ mol})(-58.2 \text{ kJ/mol}) = -238.6 \text{ kJ}$

$\Delta G° = \Delta G°_f(SO_2) - \Delta G°_f(HgS)$

$\Delta G° = (1 \text{ mol})(-300.2 \text{ kJ/mol}) - (1 \text{ mol})(-50.6 \text{ kJ/mol}) = -249.6 \text{ kJ}$

$\Delta H°$ and $\Delta G°$ are different because of the entropy change associated with the reaction. The plus sign for $(\Delta H° - \Delta G°)$ indicates that the entropy is positive, which is consistent with a solid going to liquid and no change in the number of moles of gas.

21.38 $FeCr_2O_4(s) + 4 C(s) \rightarrow Fe(s) + 2 Cr(s) + 4 CO(g)$

ferrochrome

(a) $FeCr_2O_4$, 223.84 amu;   Cr, 52.00 amu;   236 kg = 236 x 10^3 g

$$\text{mass Cr} = 236 \times 10^3 \text{ g} \times \frac{1 \text{ mol FeCr}_2\text{O}_4}{223.84 \text{ g}} \times \frac{2 \text{ mol Cr}}{1 \text{ mol FeCr}_2\text{O}_4} \times \frac{52.00 \text{ g Cr}}{1 \text{ mol Cr}} \times \frac{1.00 \text{ kg}}{1000 \text{ g}} = 110 \text{ kg Cr}$$

(b) $$\text{mol CO} = 236 \times 10^3 \text{ g} \times \frac{1 \text{ mol FeCr}_2\text{O}_4}{223.84 \text{ g}} \times \frac{4 \text{ mol CO}}{1 \text{ mol FeCr}_2\text{O}_4} = 4217.3 \text{ mol CO}$$

$$PV = nRT; \quad V = \frac{nRT}{P} = \frac{(4217.3 \text{ mol})\left(0.082\,06\, \frac{L \cdot atm}{K \cdot mol}\right)(298 \text{ K})}{\left(740 \text{ mm Hg} \times \frac{1.00 \text{ atm}}{760 \text{ mm Hg}}\right)} = 1.06 \times 10^5 \text{ L CO}$$

21.39 $Cu_2S(l) + O_2(g) \rightarrow 2 Cu(l) + SO_2(g)$

Cu_2S, 159.16 amu; Cu, 63.546 amu; SO_2, 64.06 amu

(a) mass of Cu = 1.4×10^{10} kg Cu $\times \dfrac{1000 \text{ g}}{1 \text{ kg}} = 1.4 \times 10^{13}$ g Cu

$$\text{mass Cu}_2\text{S} = 1.4 \times 10^{13} \text{ g Cu} \times \frac{1 \text{ mol Cu}}{63.546 \text{ g Cu}} \times \frac{1 \text{ mol Cu}_2\text{S}}{2 \text{ mol Cu}} \times \frac{159.16 \text{ g Cu}_2\text{S}}{1 \text{ mol Cu}_2\text{S}} \times \frac{1 \text{ kg}}{1000 \text{ g}}$$

mass Cu_2S = 1.8×10^{10} kg Cu_2S

(b) $mol\ SO_2 = 1.8 \times 10^{10}\ kg\ Cu_2S \times \dfrac{1000\ g}{1\ kg} \times \dfrac{1\ mol\ Cu_2S}{159.16\ g\ Cu_2S} \times \dfrac{1\ mol\ SO_2}{1\ mol\ Cu_2S}$

$mol\ SO_2 = 1.1 \times 10^{11}\ mol\ SO_2$
$PV = nRT$

$V = \dfrac{nRT}{P} = \dfrac{(1.1 \times 10^{11}\ mol)\left(0.082\ 06\ \dfrac{L \cdot atm}{K \cdot mol}\right)(273\ K)}{1.0\ atm} = 2.5 \times 10^{12}\ L\ SO_2$

(c) H_2SO_4, 98.08 amu

$mass\ H_2SO_4 = 1.1 \times 10^{11}\ mol\ SO_2 \times \dfrac{1\ mol\ H_2SO_4}{1\ mol\ SO_2} \times \dfrac{98.08\ g\ H_2SO_4}{1\ mol\ H_2SO_4} \times \dfrac{1\ kg}{1000\ g}$

$mass\ H_2SO_4 = 1.1 \times 10^{10}\ kg\ H_2SO_4$

21.40 $Ni^{2+}(aq) + 2\ e^- \rightarrow Ni(s);\qquad 1\ A = 1\ C/s$

$mass\ Ni = 52.5\ \dfrac{C}{s} \times 8\ h \times \dfrac{3600\ s}{1\ h} \times \dfrac{1\ mol\ e^-}{96{,}500\ C} \times \dfrac{1\ mol\ Ni}{2\ mol\ e^-} \times \dfrac{58.69\ g\ Ni}{1\ mol\ Ni} \times \dfrac{1.00\ kg}{1000\ g}$

$mass\ Ni = 0.460\ kg\ Ni$

21.41 $Cu^{2+}(aq) + 2\ e^- \rightarrow Cu(s)$

$Charge = 7.50\ kg \times \dfrac{1000\ g}{1\ kg} \times \dfrac{1\ mol\ Cu}{63.546\ g} \times \dfrac{2\ mol\ e^-}{1\ mol\ Cu} \times \dfrac{96{,}500\ C}{1\ mol\ e^-} = 2.28 \times 10^7\ C$

$Time = \dfrac{2.28 \times 10^7\ C}{40.0\ C/s} \times \dfrac{1\ h}{3600\ s} = 158\ h$

Iron and Steel (Section 21.3)

21.42 $Fe_2O_3(s) + 3\ CO(g) \rightarrow 2\ Fe(l) + 3\ CO_2(g)$
Fe_2O_3 is the oxidizing agent. CO is the reducing agent.

21.43 Coke (C) is converted to CO, the reducing agent used in the production of Fe.
$C(s) + CO_2(g) \rightarrow 2\ CO(g)$
$2\ C(s) + O_2(g) \rightarrow 2\ CO(g)$

21.44 Slag is a byproduct of iron production, consisting mainly of $CaSiO_3$. It is produced from the gangue in iron ore.

21.45 Limestone is added to the blast furnace in the commercial process for producing iron to remove the gangue from the iron ore. At the high temperatures of the blast furnace, the limestone decomposes to lime (CaO), a basic oxide that reacts with SiO_2 and other acidic oxides present in the gangue. The product, called slag, is a molten material consisting mainly of calcium silicate.
$CaCO_3(s) \rightarrow CaO(s) + CO_2(g)$
$CaO(s) + SiO_2(s) \rightarrow CaSiO_3(l)$ (slag)

21.46 Molten iron from a blast furnace is exposed to a jet of pure oxygen gas for about 20 minutes. The impurities are oxidized to yield a molten slag that can be poured off.
$$P_4(l) + 5\,O_2(g) \rightarrow P_4O_{10}(l)$$
$$6\,CaO(s) + P_4O_{10}(l) \rightarrow 2\,Ca_3(PO_4)_2(l)\ (slag)$$

$$2\,Mn(l) + O_2(g) \rightarrow 2\,MnO(s)$$
$$MnO(s) + SiO_2(s) \rightarrow MnSiO_3(l)\ (slag)$$

21.47 Slag forms in the basic oxygen process because the impurities are oxidized, and the acidic oxides that form react with the basic CaO to yield a molten slag that can be poured off. Phosphorus, for example, is oxidized to P_4O_{10}, which then reacts with CaO to give molten calcium phosphate:
$$P_4(l) + 5\,O_2(g) \rightarrow P_4O_{10}(l)$$
$$6\,CaO(s) + P_4O_{10}(l) \rightarrow 2\,Ca_3(PO_4)_2(l)\ (slag)$$

Manganese also passes into the slag because its oxide is basic and reacts with added SiO_2, yielding molten manganese silicate.
$$2\,Mn(l) + O_2(g) \rightarrow 2\,MnO(s)$$
$$MnO(s) + SiO_2(s) \rightarrow MnSiO_3(l)\ (slag)$$

21.48 $SiO_2(s) + 2\,C(s) \rightarrow Si(s) + 2\,CO(g)$
$Si(s) + O_2(g) \rightarrow SiO_2(s)$
$CaO(s) + SiO_2(s) \rightarrow CaSiO_3(l)\ (slag)$

21.49 $CaSO_4(s) + 3\,C(s) \rightarrow CaO(s) + S(l) + 3\,CO(g)$
$2\,S(l) + 3\,O_2(g) \rightarrow 2\,SO_3(g)$
$CaO(s) + SO_3(g) \rightarrow CaSO_4(l)\ (slag)$

Bonding in Metals (Section 21.4)

21.50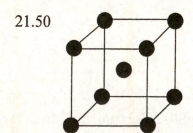
Each K has a single valence electron and has eight nearest neighbor K atoms. The valence electrons can't be localized in an electron-pair bond between any particular pair of K atoms.

21.51 Cesium has just one valence electron per Cs atom and crystallizes in a body-centered cubic structure in which each Cs atom is surrounded by eight other Cs atoms. Consequently, the valence electrons can't be localized in a bond between any particular pair of Cs atoms. Instead, the electrons are delocalized and belong to the crystal as a whole. In the electron-sea model, we visualize the crystal as a three-dimensional array of metal cations immersed in a sea of delocalized electrons that are free to move throughout the crystal. The continuum of delocalized, mobile valence electrons acts as an electrostatic glue that holds the metal cations together.

21.52   Malleability and ductility of metals follow from the fact that the delocalized bonding extends in all directions. When a metallic crystal is deformed, no localized bonds are broken. Instead, the electron sea simply adjusts to the new distribution of cations, and the energy of the deformed structure is similar to that of the original. Thus, the energy required to deform a metal is relatively small.

21.53   The electron-sea model affords a simple qualitative explanation for the electrical and thermal conductivity of metals. Because the electrons are mobile, they are free to move away from a negative electrode and toward a positive electrode when a metal is subjected to an electrical potential. The mobile electrons can also conduct heat by carrying kinetic energy from one part of the crystal to another.

21.54   The energy required to deform a transition metal like W is greater than that for Cs because W has more valence electrons and hence more electrostatic "glue".

21.55   Na has one valence electron. Mg has two valence electrons. There are more electrons per cation in the electron sea for Mg than for Na. There is more electrostatic glue for Mg, and hence the higher melting point.

21.56   The difference in energy between successive MOs in a metal decreases as the number of metal atoms increases so that the MOs merge into an almost continuous band of energy levels. Consequently, MO theory for metals is often called band theory.

21.57

21.58   The energy levels within a band occur in degenerate pairs; one set of energy levels applies to electrons moving to the right, and the other set applies to electrons moving to the left. In the absence of an electrical potential, the two sets of levels are equally populated. As a result there is no net electric current. In the presence of an electrical potential those electrons moving to the right are accelerated, those moving to the left are slowed down, and some change direction. Thus, the two sets of energy levels are now unequally populated. The number of electrons moving to the right is now greater than the number moving to the left, and so there is a net electric current.

21.59   An electrical potential can shift electrons from one set of energy levels to the other only if the band is partially filled. If the band is completely filled, there are no available vacant energy levels to which electrons can be excited, and therefore the two sets of levels must remain equally populated, even in the presence of an electrical potential. This means that an electrical potential can't accelerate the electrons in a completely filled band. Materials that have only completely filled bands are therefore electrical insulators. By contrast, materials that have partially filled bands are metals.

21.60 (a)   (b)

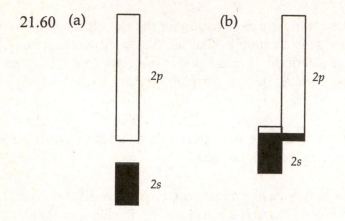

Diagram (b) shows the 2s and 2p bands overlapping in energy and the resulting composite band is only partially filled. Thus, Be is a good electrical conductor.

21.61 (a)   (b)

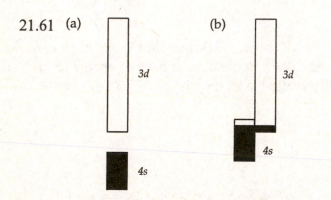

Diagram (b) shows the 4s and 3d bands overlapping in energy and the resulting composite band is only partially filled. Thus, diagram (b) agrees with the fact that Ca is a good electrical conductor.

21.62 Transition metals have a d band that can overlap the s band to give a composite band consisting of six MOs per metal atom. Half of the MOs are bonding and half are antibonding, and thus one expects maximum bonding for metals that have six valence electrons per metal atom. Accordingly, the melting points of the transition metals go through a maximum at or near group 6B.

21.63 Transition metals have a d band that can overlap the s band to give a composite band consisting of six MOs per metal atom. Half of the MOs are bonding and half are antibonding, and thus we might expect maximum bonding and hardness for metals that have six valence electrons per metal atom. The hardness will decrease as the number of valence electrons (beyond six) increases because they go into antibonding MOs. Cu has three more valence electrons than Fe, and Cu is softer.

Semiconductors and Semiconductor Applications (Sections 21.5–21.6)

21.64 A semiconductor is a material that has an electrical conductivity intermediate between that of a metal and that of an insulator. Si, Ge, and Sn (gray) are semiconductors.

21.65 (a) The valence band is the band of bonding molecular orbitals in a semiconductor.
(b) The conduction band is the band of antibonding molecular orbitals in a semiconductor.
(c) The band gap is the energy difference between the valence band and the conduction band in a semiconductor.
(d) Doping is the addition of a small amount of impurities to increase the conductivity of a semiconductor.

21.66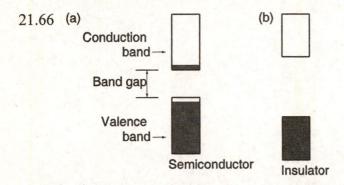

The MOs of a semiconductor are similar to those of an insulator, but the band gap in a semiconductor is smaller. As a result, a few electrons have enough energy to jump the gap and occupy the higher-energy, conduction band. The conduction band is thus partially filled, and the valence band is partially empty. When an electrical potential is applied to a semiconductor, it conducts a small amount of current because the potential can accelerate the electrons in the partially filled bands.

21.67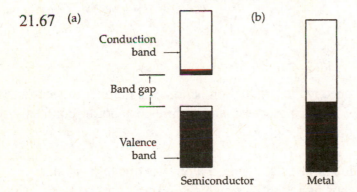

Metals have partially filled bands with no band gap between the highest occupied and lowest unoccupied MOs, which makes them conductors. Semiconductors have a moderate band gap between the valence and conduction bands. A few electrons have enough energy to jump the gap and occupy the conduction band. The result is a partially filled valence and conduction band. When an electrical potential is applied to a semiconductor, it conducts a small amount of current.

21.68 As the band gap increases, the number of electrons able to jump the gap and occupy the higher-energy conduction band decreases, and thus the conductivity decreases.

21.69 The electrical conductivity of a semiconductor increases with increasing temperature because the number of electrons with sufficient energy to occupy the conduction band increases as the temperature rises. At higher temperatures, there are more charge carriers (electrons) in the conduction band and more vacancies in the valence band.

21.70 An n-type semiconductor is a semiconductor doped with a substance with more valence electrons than the semiconductor itself. Si doped with P is an example.

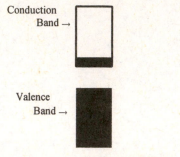

Conduction Band →

Valence Band →

n-Type semiconductor

21.71 A p-type semiconductor is a semiconductor doped with a substance with fewer valence electrons than the semiconductor itself. Si doped with B is an example.

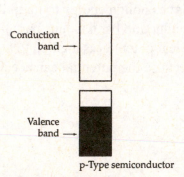

Conduction band →

Valence band →

p-Type semiconductor

21.72 In the MO picture, the extra electrons occupy the conduction band. The number of electrons in the conduction band of the doped Ge is much greater than for pure Ge, and the conductivity of the doped semiconductor is correspondingly higher.

21.73 Si doped with Ga is a p-type semiconductor because each Ga atom has one less valence electron than Si. The valence band is thus partially filled, which accounts for the electrical conductivity. The conductivity is greater than that of pure Si because the doped Si has many more positive holes in the valence band. That is, doped Si has more vacant MOs available to which electrons can be excited by an electrical potential.

21.74 (a) p-type (In is electron deficient with respect to Si)
(b) n-type (Sb is electron rich with respect to Ge)
(c) n-type (As is electron rich with respect to gray Sn)

21.75 (a) n-type (As is electron rich with respect to Ge)
(b) p-type (B is electron deficient with respect to Ge)
(c) n-type (Sb is electron rich with respect to Si)

21.76 Al_2O_3 < Ge < Ge doped with In < Fe < Cu

21.77 NaCl < Si < gray Sn < gray Sn doped with Sb < Ag

21.78 In a diode, current flows only when the junction is under a forward bias (negative battery terminal on the n-type side). A p-n junction that is part of a circuit and subjected to an alternating potential acts as a rectifier, allowing current to flow in only one direction, thereby converting alternating current to direct current.

21.79 LEDs produce light as a result of the combination of electrons and holes in the region of a p-n junction. This only occurs under a forward bias (negative battery terminal on the n-type side). Under a reverse bias (negative battery terminal on the p-type side), almost no current flows because the electrons and holes move away from one another.

21.80 An LED and a photovoltaic cell are both p-n junctions, but the two devices involve opposite processes. An LED converts electrical energy to light; a photovoltaic, or solar, cell converts light to electricity.

21.81 Both LEDs and diode lasers produce light as a result of the combination of electrons and holes in the region of a p-n junction. Compared to an LED, however, the light from a diode laser is more intense, more highly directional, and all of the same frequency and phase.

21.82 $E = 193 \text{ kJ/mol} \times \dfrac{1000 \text{ J}}{1 \text{ kJ}} \times \dfrac{1 \text{ mol}}{6.02 \times 10^{23}} = 3.21 \times 10^{-19} \text{ J}$

$\nu = \dfrac{E}{h} = \dfrac{3.21 \times 10^{-19} \text{ J}}{6.626 \times 10^{-34} \text{ J·s}} = 4.84 \times 10^{14} \text{ s}^{-1}$

$\lambda = \dfrac{c}{\nu} = \dfrac{3.00 \times 10^8 \text{ m/s}}{4.84 \times 10^{14} \text{ s}^{-1}} = 6.20 \times 10^{-7} \text{ m} = 620 \times 10^{-9} \text{ m} = 620 \text{ nm, orange light}$

21.83 $\lambda = 470 \text{ nm} = 470 \times 10^{-9} \text{ m}$

$E = h\dfrac{c}{\lambda} = (6.626 \times 10^{-34} \text{ J·s})\left(\dfrac{3.00 \times 10^8 \text{ m/s}}{470 \times 10^{-9} \text{ m}}\right)(6.02 \times 10^{23}/\text{mol}) = 2.55 \times 10^5 \text{ J/mol}$

$E = 2.55 \times 10^5 \text{ J/mol} \times \dfrac{1 \text{ kJ}}{1000 \text{ J}} = 255 \text{ kJ/mol}$

Superconductors (Section 21.7)

21.84 (1) A superconductor is able to levitate a magnet.
(2) In a superconductor, once an electric current is started, it flows indefinitely without loss of energy. A superconductor has no electrical resistance.

21.85

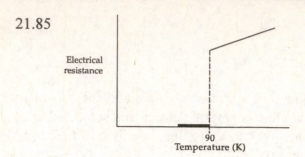

21.86   Some K^+ ions are surrounded octahedrally by six C_{60}^{3-} ions; others are surrounded tetrahedrally by four C_{60}^{3-} ions.

21.87   In $YBa_2Cu_3O_7$;

atom	coordination numbers
Cu	4 and 5
Y	8
Ba	10

Ceramics and Composites (Sections 21.8–21.9)

21.88   Ceramics are inorganic, nonmetallic, nonmolecular solids, including both crystalline and amorphous materials. Ceramics have higher melting points, and they are stiffer, harder, and more resistant to wear and corrosion than are metals.

21.89   The bonding in ceramics consists of either directed covalent bonds or ionic bonds. This is very different from the electron-sea model of bonding for metals.

21.90   Ceramics have higher melting points, and they are stiffer, harder, and more wear resistant than metals because they have stronger bonding. They maintain much of their strength at high temperatures, where metals either melt or corrode because of oxidation.

21.91   Oxide ceramics don't react with oxygen because they are already fully oxidized.

21.92   The brittleness of ceramics is due to strong chemical bonding. In silicon nitride each Si atom is bonded to four N atoms and each N atom is bonded to three Si atoms. The strong, highly directional covalent bonds prevent the planes of atoms from sliding over one another when the solid is subjected to a stress. As a result, the solid can't deform to relieve the stress. It maintains its shape up to a point, but then the bonds give way suddenly and the material fails catastrophically when the stress exceeds a certain threshold value. By contrast, metals are able to deform under stress because their planes of metal cations can slide easily in the electron sea.

21.93

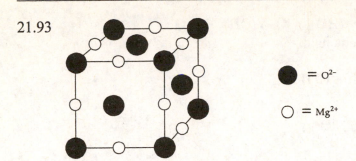

$\bullet$ = O^{2-}

$\circ$ = Mg^{2+}

The brittleness of oxide ceramics, as well as their hardness, stiffness, and high melting points, is due to strong ionic bonding. The strong ionic bonding prevents planes of ions from sliding over one another when the solid is subjected to a stress. As a result, the solid can't deform to relieve the stress. It maintains its shape up to a point, but then the bonds give way suddenly and the material fails catastrophically. For example, if the stress displaces a plane of ions by one-half a unit-cell edge length, the resulting Mg^{2+}–Mg^{2+} and O^{2-}–O^{2-} repulsions cause the crystal to shatter. By contrast, metals are able to deform under stress because their planes of cations can slide easily in the electron sea.

21.94 Ceramic processing is the series of steps that leads from raw material to the finished ceramic object.

21.95 Sintering, which occurs below the melting point, is a process in which the particles of the powder are "welded" together without completely melting. During sintering, the crystal grains grow larger and the density of the material increases as the void spaces between particles disappear.

21.96 A sol is a colloidal dispersion of tiny particles. A gel is a more rigid gelatin-like material consisting of larger particles.

21.97 As an example, pure $Ti(OCH_2CH_3)_4$ is dissolved in an appropriate organic solvent, and water is added to bring about a hydrolysis reaction:
$Ti(OCH_2CH_3)_4 + 4\ H_2O \rightarrow Ti(OH)_4(s) + 4\ HOCH_2CH_3$
The $Ti(OH)_4$ forms in the liquid as a colloidal dispersion called a sol. Subsequent reactions eliminate water and form oxygen bridges between Ti atoms:
$(HO)_3Ti–O–H + H–O–Ti(OH)_3 \rightarrow (HO)_3Ti–O–Ti(OH)_3 + H_2O$
Because all of the OH groups can undergo the above reaction, the particles of the sol link together through a three-dimensional network of oxygen bridges, and the sol is converted to a more rigid, gelatin-like material called a gel.

21.98 $Zr[OCH(CH_3)_2]_4 + 4\ H_2O \rightarrow Zr(OH)_4 + 4\ HOCH(CH_3)_2$

21.99 $Zn(OCH_2CH_3)_2 + 2\ H_2O \rightarrow Zn(OH)_2 + 2\ HOCH_2CH_3$

21.100 $(HO)_3Si–O–H + H–O–Si(OH)_3 \rightarrow (HO)_3Si–O–Si(OH)_3 + H_2O$
Further reactions of this sort give a three-dimensional network of Si–O–Si bridges. On heating, SiO_2 is obtained.

21.101 $(HO)_2Y–O–H + H–O–Y(OH)_2 \rightarrow (HO)_2Y–O–Y(OH)_2 + H_2O$
On drying and sintering, Y_2O_3 is obtained.

21.102 $3\ SiCl_4(g) + 4\ NH_3(g) \rightarrow Si_3N_4(s) + 12\ HCl(g)$

21.103 $2\ BCl_3(g) + 3\ H_2(g) \rightarrow 2\ B(s) + 6\ HCl(g)$

21.104   Graphite/epoxy composites are good materials for making tennis rackets and golf clubs because of their high strength-to-weight ratios.

21.105   Silicon carbide-reinforced alumina is stronger and tougher than pure alumina. The silicon carbide whiskers have great strength along their axis because most of the chemical bonds are aligned in that direction. In addition the silicon carbide whiskers can prevent microscopic cracks from propagating.

Chapter Problems

21.106   The chemical composition of the alkaline earth minerals is that of metal sulfates and sulfites, MSO_4 and MSO_3.

21.107   77 K is the boiling point of the readily available liquid N_2.

21.108   Band theory better explains how the number of valence electrons affects properties such as melting point and hardness.

21.109   W   [Xe] $4f^{14}\ 5d^4\ 6s^2$ Au   [Xe] $4f^{14}\ 5d^{10}\ 6s^1$
These facts are better explained with the MO band model. Transition metals have a d band that can overlap the s band to give a composite band consisting of six MOs per metal atom. Half of the MOs are bonding and half are antibonding, and thus one expects maximum bonding for metals that have six valence electrons per metal atom. Accordingly, the hardness and melting points of the transition metals go through a maximum at or near group 6B.

21.110   V   [Ar] $3d^3\ 4s^2$ Zn   [Ar] $3d^{10}\ 4s^2$
Transition metals have a d band that can overlap the s band to give a composite band consisting of six MOs per metal atom. Half of the MOs are bonding and half are antibonding. Strong bonding and a high enthalpy of vaporization are expected for V because almost all of the bonding MOs are occupied and all of the antibonding MOs are empty. Weak bonding and a low enthalpy of vaporization are expected for Zn because both the bonding and the antibonding MOs are occupied.

21.111   (a) MgO, insulator
(b) Si doped with Sb (Sb is electron rich with respect to Si), n-type semiconductor
(c) white tin, metallic conductor
(d) Ge doped with Ga (Ga is electron deficient with respect to Ge), p-type semiconductor
(e) stainless steel, metallic conductor

21.112 With a band gap of 130 kJ/mol, GaAs is a semiconductor. Because Ge lies between Ga and As in the periodic table, GaAs is isoelectronic with Ge.

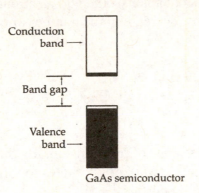

GaAs semiconductor

21.113 (a) $P_4(l) + 5\ O_2(g) \rightarrow P_4O_{10}(l)$
$6\ CaO(s) + P_4O_{10}(l) \rightarrow 2\ Ca_3(PO_4)_2(l)$ (slag)
(b) mass $P_4 = (0.0045)(3.4 \times 10^3\ kg) = 15.3\ kg\ P_4$

$$mol\ P_4O_{10} = 15.3 \times 10^3\ g \times \frac{1\ mol\ P_4}{123.9\ g} \times \frac{1\ mol\ P_4O_{10}}{1\ mol\ P_4} = 123.5\ mol\ P_4O_{10}$$

$$mass\ CaO = 123.5\ mol\ P_4O_{10} \times \frac{6\ mol\ CaO}{1\ mol\ P_4O_{10}} \times \frac{56.08\ g\ CaO}{1\ mol\ CaO} \times \frac{1\ kg}{1000\ g} = 42\ kg\ CaO$$

21.114 $YBa_2Cu_3O_7$, 666.20 amu; $Cu(OCH_2CH_3)_2$, 153.67 amu
$Y(OCH_2CH_3)_3$, 224.09 amu; $Ba(OCH_2CH_3)_2$, 227.45 amu

$$mol\ Cu(OCH_2CH_3)_2 = 75.4\ g \times \frac{1\ mol}{153.67\ g} = 0.4907\ mol\ Cu(OCH_2CH_3)_2$$

mass $Y(OCH_2CH_3)_3 = 0.4907\ mol\ Cu(OCH_2CH_3)_2 \times$

$$\frac{1\ mol\ Y(OCH_2CH_3)_3}{3\ mol\ Cu(OCH_2CH_3)_2} \times \frac{224.09\ g\ Y(OCH_2CH_3)_3}{1\ mol\ Y(OCH_2CH_3)_3} = 36.7\ g\ Y(OCH_2CH_3)_3$$

mass $Ba(OCH_2CH_3)_2 = 0.4907\ mol\ Cu(OCH_2CH_3)_2 \times$

$$\frac{2\ mol\ Ba(OCH_2CH_3)_2}{3\ mol\ Cu(OCH_2CH_3)_2} \times \frac{227.45\ g\ Ba(OCH_2CH_3)_2}{1\ mol\ Ba(OCH_2CH_3)_2} = 74.4\ g\ Ba(OCH_2CH_3)_2$$

mass $YBa_2Cu_3O_7 = 0.4907\ mol\ Cu(OCH_2CH_3)_2 \times$

$$\frac{1\ mol\ YBa_2Cu_3O_7}{3\ mol\ Cu(OCH_2CH_3)_2} \times \frac{666.20\ g\ YBa_2Cu_3O_7}{1\ mol\ YBa_2Cu_3O_7} = 109\ g\ YBa_2Cu_3O_7$$

21.115 (a) In an n-type InP semiconductor the valence band is completely filled and the conduction band is partially full. Cd has only two 5s and no 5p electrons. Adding Cd to the n-type InP semiconductor would add positive holes that would combine with free electrons. This results in a decrease in the number of electrons in the conduction band and a decrease in the conductivity.

(b) In a p-type InP semiconductor the valence band is partially filled and the conduction band is empty. The charge carriers are positive holes. Se has 6 valence electrons. Adding Se to the p-type InP semiconductor would add electrons that would combine with positive holes. This results in a decrease in the number of positive holes and a decrease in the conductivity.

21.116   (a) $6 Al(OCH_2CH_3)_3 + 2 Si(OCH_2CH_3)_4 + 26 H_2O \rightarrow$

$$6 Al(OH)_3(s) + 2 Si(OH)_4(s) + 26 HOCH_2CH_3$$
$$\text{sol}$$

(b) H_2O is eliminated from the sol through a series of reactions linking the sol particles together through a three-dimensional network of O bridges to form the gel.
$(HO)_2Al–O–H + H–O–Si(OH)_3 \rightarrow (HO)_2Al–O–Si(OH)_3 + H_2O$

(c) The remaining H_2O and solvent are removed from the gel by heating to produce the ceramic, $3 Al_2O_3 \cdot 2 SiO_2$.

21.117

$$\text{sunlight}$$
$$AgCl \qquad \rightleftarrows \qquad Ag + Cl$$
$$\text{(transparent)} \quad \text{dark} \quad \text{(opaque)}$$

21.118   (a)

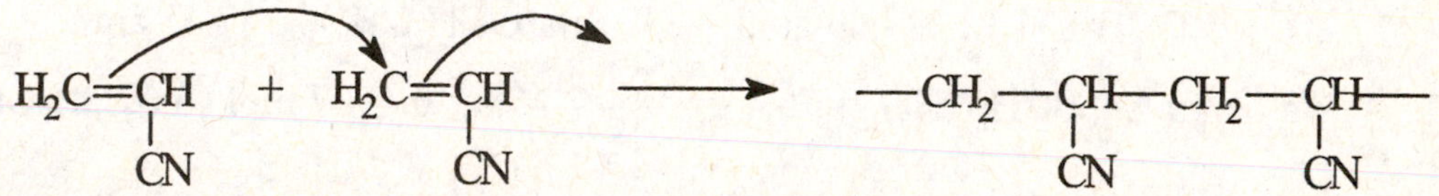

(b) $\Delta H° = D_{C=C} - 2 D_{C–C} = 611 \text{ kJ} - 2(350) \text{ kJ} = -89 \text{ kJ/unit};$ exothermic

21.119

Zn has 2 valence electrons. Ga has 3 valence electrons. When ZnSe is doped with Ga, there are extra electrons in the semiconductor. Extra electrons lead to n-type semiconductors.

21.120　(a)

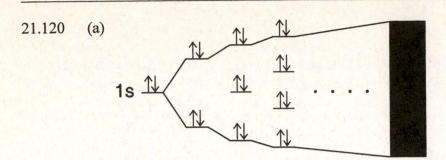

1s

This material is an insulator because all MOs are filled, preventing the movement of electrons.

(b)

Neutral hydrogen atoms have only 1 valence electron, compared with 2 in H^-.
Partially empty antibonding MOs will allow the movement of electrons, so the doped material will be a conductor.
(c)　The missing electrons in the doped material create "holes" which are positive charge carriers.　This type of doped material is a p-type conductor.

21.121　(a)

(b)　The electrical conductivity of a semiconductor increases with increasing temperature because the number of electrons with sufficient energy to occupy the conduction band increases as the temperature rises.
(c) (i)　The conductivity of GaAs would increase when doped with Zn because it would produce positive holes in the valence band.　Zn doped GaAs would be a p-type semiconductor.
(ii)　The conductivity of GaAs would increase when doped with S because it would put extra electrons in the conduction band.　S doped GaAs would be an n-type semiconductor.

Multiconcept Problems

21.122 $Cr_2O_7^{2-}(aq) + 6 Fe^{2+}(aq) + 14 H^+(aq) \rightarrow 6 Fe^{3+}(aq) + 2 Cr^{3+}(aq) + 7 H_2O(l)$

mol $Cr_2O_7^{2-}$ = (0.038 89 L)(0.018 54 mol/L) = 7.210 x 10^{-4} mol $Cr_2O_7^{2-}$

mass Fe = 7.210 x 10^{-4} mol $Cr_2O_7^{2-}$ x $\dfrac{6 \text{ mol Fe}^{2+}}{1 \text{ mol Cr}_2O_7^{2-}}$ x $\dfrac{55.847 \text{ g Fe}^{2+}}{1 \text{ mol Fe}^{2+}}$ = 0.2416 g Fe^{2+}

mass % Fe = $\dfrac{0.2416 \text{ g}}{0.3249 \text{ g}}$ x 100% = 74.36% Fe

21.123 660 nm = 660 x 10^{-9} m and 3.0 mW = 3.0 x 10^{-3} W = 3.0 x 10^{-3} J/s

$E = h\dfrac{c}{\lambda} = (6.626 \text{ x } 10^{-34} \text{ J·s})\left(\dfrac{3.00 \text{ x } 10^8 \text{ m/s}}{660 \text{ x } 10^{-9} \text{ m}}\right) = 3.0 \text{ x } 10^{-19}$ J/photon

# of photons/s = $\dfrac{3.0 \text{ x } 10^{-3} \text{ J/s}}{3.0 \text{ x } 10^{-19} \text{ J/photon}}$ = 1.0 x 10^{16} photons/s

# of electrons/s = # of photons/s = 1.0 x 10^{16} electrons/s

# of moles of electrons/s = 1.0 x 10^{16} electrons/s x $\dfrac{1 \text{ mol e}^-}{6.02 \text{ x } 10^{23} \text{ e}^-}$ = 1.7 x 10^{-8} mol e$^-$/s

A = 1.7 x 10^{-8} mol e$^-$/s x $\dfrac{96,500 \text{ C}}{1 \text{ mol e}^-}$ = 0.0016 C/s = 0.0016 A = 1.6 x10^{-3} A = 1.6 mA

21.124 (a) 431 pm = 431 x 10^{-12} m
There are 4 oxygen atoms in the face-centered cubic unit cell.
mass of unit cell = (5.75 g/cm^3)(431 x 10^{-12} m)^3(100 cm/1 m)^3 = 4.604 x 10^{-22} g

mass of Fe in unit cell = (4.604 x 10^{-22} g) – 4 O atoms x $\dfrac{15.9994 \text{ g O}}{6.022 \text{ x } 10^{23} \text{ O atoms}}$

$= 3.541 \text{ x } 10^{-22}$ g Fe

number of Fe atoms in unit cell = 3.541 x 10^{-22} g Fe x $\dfrac{6.022 \text{ x } 10^{23} \text{ Fe atoms}}{55.847 \text{ g Fe}}$

$= 3.818$ Fe atoms

For Fe$_x$O, x = $\dfrac{3.818 \text{ Fe atoms}}{4 \text{ O atoms}}$ = 0.954

(b) The average oxidation state of Fe = $\dfrac{+2}{0.954}$ = 2.096

(c) Let X equal the fraction of Fe^{3+} and Y equal the fraction of Fe^{2+} in wustite.
So, X + Y = 1 and 3X + 2Y = 2.096
Y = 1 – X
3X + 2(1 –X) = 2.096
3X + 2 – 2X = 2.096
X + 2 = 2.096
X = 2.096 – 2 = 0.096
9.6% of the Fe in wustite is Fe^{3+}.

(d) $d = 431$ pm

$$d = \frac{n\lambda}{2\sin\theta} = \frac{3 \cdot 70.93 \, pm}{2\sin\theta} = 431 \, pm$$

$$\sin\theta = \frac{3 \cdot 70.93 \, pm}{2 \cdot 431 \, pm} = 0.247 \text{ and } \theta = 14.3 \, °$$

(e) The presence of Fe^{3+} in the semiconductor leads to missing electrons that create "holes" which are positive charge carriers. This type of doped material is a p-type semiconductor.

21.125 $SiO_2(s) + 2 \, C(s) \rightarrow Si(s) + 2 \, CO(g)$
(a) $\Delta H° = 2 \, \Delta H°_f(CO) - \Delta H°_f(SiO_2)$
$\Delta H° = (2 \, mol)(-110.5 \, kJ/mol) - (1 \, mol)(-910.7 \, kJ/mol) = 689.7 \, kJ$
$\Delta S° = [S°(Si) + 2 \, S°(CO)] - [S°(SiO_2) + 2 \, S°(C)]$
$\Delta S° = [(1 \, mol)(18.8 \, J/(K \cdot mol)) + (2 \, mol)(197.6 \, J/(K \cdot mol))]$
$\qquad\qquad - [(1 \, mol)(41.5 \, J/(K \cdot mol)) + (2 \, mol)(5.7 \, J/(K \cdot mol))]$
$\Delta S° = 361.1 \, J/K = 361.1 \times 10^{-3} \, kJ/K$
$\Delta G° = \Delta H° - T\Delta S° = 689.7 \, kJ - (298.15 \, K)(361.1 \times 10^{-3} \, kJ/K) = 582.1 \, kJ$
(b) The reaction is endothermic because $\Delta H° > 0$.
(c) The number of moles of gas increases from 0 to 2 mol, therefore, $\Delta S° > 0$.
(d) Because $\Delta G° > 0$, the reaction is nonspontaneous at 25 °C and 1 atm pressure of CO.
(e) To determine the crossover temperature, set $\Delta G° = 0$ and solve for T.
$\Delta G° = 0 = \Delta H° - T\Delta S°$

$\Delta H° = T\Delta S°$; $T = \dfrac{\Delta H°}{\Delta S°} = \dfrac{689.7 \, kJ}{361.1 \times 10^{-3} \, kJ/K} = 1910 \, K = 1637 \, °C$

21.126 $Ni(s) + 4 \, CO(g) \rightleftharpoons Ni(CO)_4(g)$
$\Delta H° = -160.8 \, kJ; \, \Delta S° = -410 \, J/K = -410 \times 10^{-3} \, kJ/K$
(a) 150 °C = 423 K
$\Delta G° = \Delta H° - T\Delta S° = -160.8 \, kJ - (423 \, K)(-410 \times 10^{-3} \, kJ/K) = +12.6 \, kJ$
$\Delta G° = -RT \ln K$

$\ln K = \dfrac{-\Delta G°}{RT} = \dfrac{-12.6 \, kJ/mol}{[8.314 \times 10^{-3} \, kJ/(K \cdot mol)](423 \, K)} = -3.58$

$K = K_p = e^{-3.58} = 0.028$
(b) 230 °C = 503 K
$\Delta G° = \Delta H° - T\Delta S° = -160.8 \, kJ - (503 \, K)(-410 \times 10^{-3} \, kJ/K) = +45.4 \, kJ$
$\Delta G° = -RT \ln K$

$\ln K = \dfrac{-\Delta G°}{RT} = \dfrac{-45.4 \, kJ/mol}{[8.314 \times 10^{-3} \, kJ/(K \cdot mol)](503 \, K)} = -10.86$

$K = K_p = e^{-10.86} = 1.9 \times 10^{-5}$

(c) $\Delta S°$ is large and negative because as the reaction proceeds in the forward direction, the number of moles of gas decrease from four to one.
Because $\Delta S°$ is negative, $-T\Delta S°$ is positive, and as T increases, $\Delta G°$ becomes more positive because $\Delta G° = \Delta H° - T\Delta S°$.

(d) The reaction is exothermic because ΔH° is negative.

$$Ni(s) + 4\ CO(g) \rightleftharpoons Ni(CO)_4(g) + heat$$

Heat is added as the temperature is raised and the reaction proceeds in the reverse direction to relieve this stress, as predicted by Le Châtelier's principle. As the reverse reaction proceeds, the partial pressure of CO increases and the partial pressure of $Ni(CO)_4$ decreases. K_p decreases as calculated because $K_p = \dfrac{P_{Ni(CO)_4}}{(P_{CO})^4}$.

21.127 $C(s) + CO_2(g) \rightarrow 2\ CO(g)$

(a) CO_2, 44.01 amu

$$mol\ CO_2 = 100.0\ g\ CO_2 \times \frac{1\ mol\ CO_2}{44.01\ g\ CO_2} = 2.272\ mol\ CO_2$$

$\Delta H^\circ = [2\ \Delta H^\circ_f(CO)] - \Delta H^\circ_f(CO_2)$
$\Delta H^\circ = (2\ mol)(-110.5\ kJ/mol) - (1\ mol)(-393.5\ kJ/mol) = 172.5\ kJ$
$\Delta S^\circ = [2\ S^\circ(CO)] - [S^\circ(C) + S^\circ(CO_2)]$
$\Delta S^\circ = (2\ mol)(197.6\ J/(K \cdot mol)) - [(1\ mol)(5.7\ J/(K \cdot mol)) + (1\ mol)(213.6\ J/(K \cdot mol))]$
$\Delta S^\circ = 175.9\ J/K = 175.9 \times 10^{-3}\ kJ/K$
at 500 °C (773 K):
$\Delta G^\circ = \Delta H^\circ - T\Delta S^\circ = 172.5\ kJ - (773\ K)(175.9 \times 10^{-3}\ kJ/K) = 36.5\ kJ$

$$\ln K_p = \frac{-\Delta G^\circ}{RT} = \frac{-36.5\ kJ/mol}{[8.314 \times 10^{-3}\ kJ/(K \cdot mol)](773\ K)} = -5.68$$

$K_p = e^{-5.68} = 3.4 \times 10^{-3}$

$$P_{CO_2} = \frac{nRT}{V} = \frac{(2.272\ mol)\left(0.082\ 06\ \dfrac{L \cdot atm}{K \cdot mol}\right)(773\ K)}{50.00\ L} = 2.88\ atm$$

	$C(s)$ +	$CO_2(g)$	$\rightleftharpoons$	$2\ CO(g)$
initial (atm)		2.88		0
change (atm)		–x		+2x
equil (atm)		2.88 – x		2x

$$K_p = \frac{(P_{CO})^2}{P_{CO_2}} = 3.4 \times 10^{-3} = \frac{(2x)^2}{2.88 - x}$$

$4x^2 + (3.4 \times 10^{-3})x - 9.79 \times 10^{-3} = 0$
Use the quadratic formula to solve for x.

$$x = \frac{-(3.4 \times 10^{-3}) \pm \sqrt{(3.4 \times 10^{-3})^2 - (4)(4)(-9.79 \times 10^{-3})}}{2(4)} = \frac{(-3.4 \times 10^{-3}) \pm (0.396)}{8}$$

x = 0.049 07 and –0.049 93
Of the two solutions for x, only the positive value of x has physical meaning because 2x is the partial pressure of CO.

$P_{CO_2} = 2.88 - x = 2.88 - 0.049\ 07 = 2.831\ atm$

$P_{CO} = 2x = 2(0.049\ 07) = 0.0981\ atm$

$P_{total} = P_{CO_2} + P_{CO} = 2.831 + 0.0981 = 2.93\ atm$

$$[CO] = \frac{n}{V} = \frac{P}{RT} = \frac{(0.0981 \text{ atm})}{\left(0.082\ 06\ \dfrac{\text{L} \cdot \text{atm}}{\text{K} \cdot \text{mol}}\right)(773\ \text{K})} = 1.6 \times 10^{-3}\ \text{M}$$

$$[CO_2] = \frac{n}{V} = \frac{P}{RT} = \frac{(2.831 \text{ atm})}{\left(0.082\ 06\ \dfrac{\text{L} \cdot \text{atm}}{\text{K} \cdot \text{mol}}\right)(773\ \text{K})} = 4.46 \times 10^{-2}\ \text{M}$$

(b)   at 1000 °C (1273 K):

$$\Delta G^\circ = \Delta H^\circ - T\Delta S^\circ = 172.5 \text{ kJ} - (1273 \text{ K})(175.9 \times 10^{-3} \text{ kJ/K}) = -51.4 \text{ kJ}$$

$$\ln K_p = \frac{-\Delta G^\circ}{RT} = \frac{-(-51.4 \text{ kJ/mol})}{[8.314 \times 10^{-3} \text{ kJ/(K} \cdot \text{mol})](1273 \text{ K})} = 4.86$$

$$K_p = e^{4.86} = 1.3 \times 10^2$$

$$P_{CO_2} = \frac{nRT}{V} = \frac{(2.272 \text{ mol})\left(0.082\ 06\ \dfrac{\text{L} \cdot \text{atm}}{\text{K} \cdot \text{mol}}\right)(1273 \text{ K})}{50.00 \text{ L}} = 4.75 \text{ atm}$$

$$\begin{array}{lccc}
 & C(s) & + \quad CO_2(g) & \rightleftharpoons \quad 2\ CO(g) \\
\text{initial (atm)} & & 4.75 & 0 \\
\text{change (atm)} & & -x & +2x \\
\text{equil (atm)} & & 4.75 - x & 2x
\end{array}$$

$$K_p = \frac{(P_{CO})^2}{P_{CO_2}} = 1.3 \times 10^2 = \frac{(2x)^2}{4.75 - x}$$

$$4x^2 + (1.3 \times 10^2)x - 617.5 = 0$$

Use the quadratic formula to solve for x.

$$x = \frac{-(1.3 \times 10^2) \pm \sqrt{(1.3 \times 10^2)^2 - (4)(4)(-617.5)}}{2(4)} = \frac{(-1.3 \times 10^2) \pm (163.6)}{8}$$

x = 4.200 and −36.70

Of the two solutions for x, only the positive value of x has physical meaning because 2x is the partial pressure of CO.

$$P_{CO_2} = 4.75 - x = 4.75 - 4.200 = 0.55 \text{ atm}$$

$$P_{CO} = 2x = 2(4.200) = 8.40 \text{ atm}$$

$$P_{total} = P_{CO_2} + P_{CO} = 0.55 + 8.40 = 8.95 \text{ atm}$$

$$[CO] = \frac{n}{V} = \frac{P}{RT} = \frac{(8.40 \text{ atm})}{\left(0.082\ 06\ \dfrac{\text{L} \cdot \text{atm}}{\text{K} \cdot \text{mol}}\right)(1273 \text{ K})} = 8.04 \times 10^{-2}\ \text{M}$$

$$[CO_2] = \frac{n}{V} = \frac{P}{RT} = \frac{(0.55 \text{ atm})}{\left(0.082\ 06\ \dfrac{\text{L} \cdot \text{atm}}{\text{K} \cdot \text{mol}}\right)(1273 \text{ K})} = 5.3 \times 10^{-3}\ \text{M}$$

(c) $\Delta G^\circ = \Delta H^\circ - T\Delta S^\circ$;   The equilibrium shifts to the right with increasing temperature because ΔH° is positive (endothermic reaction) and ΔS° is positive. Therefore, ΔG° is more negative at higher temperatures.

21.128   (a) $(NH_4)_2Zn(CrO_4)_2(s) \rightarrow ZnCr_2O_4(s) + N_2(g) + 4 H_2O(g)$

(b) mol $(NH_4)_2Zn(CrO_4)_2 = 10.36$ g $(NH_4)_2Zn(CrO_4)_2$ x $\dfrac{1 \text{ mol } (NH_4)_2Zn(CrO_4)_2}{333.45 \text{ g } (NH_4)_2Zn(CrO_4)_2}$

$= 0.03107$ mol $(NH_4)_2Zn(CrO_4)_2$

mass $ZnCr_2O_4 = 0.03107$ mol $(NH_4)_2Zn(CrO_4)_2$ x $\dfrac{1 \text{ mol } ZnCr_2O_4}{1 \text{ mol } (NH_4)_2Zn(CrO_4)_2}$

x $\dfrac{233.38 \text{ g } ZnCr_2O_4}{1 \text{ mol } ZnCr_2O_4} = 7.251$ g $ZnCr_2O_4$

(c) mol N_2 + mol $H_2O = 0.03107$ mol $(NH_4)_2Zn(CrO_4)_2$ x $\dfrac{5 \text{ mol gas}}{1 \text{ mol } (NH_4)_2Zn(CrO_4)_2}$

$= 0.1554$ mol gaseous by-products

292 °C = 565 K; $PV = nRT$

$V = \dfrac{nRT}{P} = \dfrac{(0.1554 \text{ mol})\left(0.082\ 06 \dfrac{L \cdot atm}{K \cdot mol}\right)(565 \text{ K})}{\left(745 \text{ mm Hg x } \dfrac{1.00 \text{ atm}}{760 \text{ mm Hg}}\right)} = 7.35$ L

(d) A face-centered cubic unit cell has four octahedral holes and eight tetrahedral holes. This unit cell contains one Zn^{2+} ion in a tetrahedral hole and two Cr^{3+} ions in octahedral holes, therefore 1/8 of the tetrahedral holes and 1/2 of the octahedral holes are filled.

(e)

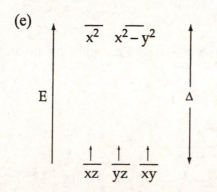

Octahedral Cr^{3+} has three unpaired electrons in the lower-energy d orbitals (xy, xz, yz). Cr^{3+} can absorb visible light to promote one of these d electrons to one of the higher-energy d orbitals making this compound colored. All of the d orbitals in Zn^{2+} are filled and no d electrons can be promoted, consequently the Zn^{2+} ion does not contribute to the color.

21.129   (a)

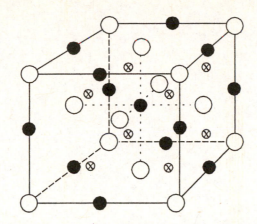

$\bigcirc$ = C_{60}^{3-}

$\bullet$ = M^+ in an octahedral hole

$\otimes$ = M^+ in a tetrahedral hole

(b) There are 4 C_{60}^{3-} ions, 4 octahedral holes, and 8 tetrahedral holes per unit cell.

(c) Octahedral holes: (1/2,1/2,1/2), (1/2,0,0), (0,1/2,0), (0,0,1/2)
Tetrahedral holes: (1/4,1/4,1/4), (3/4,1/4,1/4), (1/4,3/4,1/4), (3/4,3/4,1/4), (1/4,1/4,3/4), (3/4,1/4,3/4), (1/4,3/4,3/4), (3/4,3/4,3/4)

(d) Let the unit cell edge = a.
The face diagonal is equal to 4R = 4(500 pm) = 2000 pm

$a^2 + a^2 = (2000)^2$; $2a^2 = 4 \times 10^6$; $a^2 = 2 \times 10^6$; $a = \sqrt{2 \times 10^6} = 1414$ pm
$a = 2R(C_{60}^{3-}) + 2R(\text{octahedral hole}) = 1414$ pm

$$R(\text{octahedral hole}) = \frac{1414 \text{ pm} - 2R(C_{60}^{3-})}{2} = \frac{1414 \text{ pm} - 2(500 \text{ pm})}{2} = 207 \text{ pm}$$

The tetrahedron that defines the tetrahedral hole can be thought of as being found inside a cube with edge = a/2 = 707 pm. This cube is located in one corner of the unit cell. The face diagonal of this cube = $2R(C_{60}^{3-})$ = 1000 pm.

The body diagonal of this cube = $\sqrt{707^2 + 1000^2} = 1225$ pm
Body diagonal = $2R(C_{60}^{3-}) + 2R(\text{tetrahedral hole}) = 1225$ pm

$$R(\text{tetrahedral hole}) = \frac{1225 \text{ pm} - 2R(C_{60}^{3-})}{2} = \frac{1225 \text{ pm} - 2(500 \text{ pm})}{2} = 112 \text{ pm}$$

(e) Na^+ will fit into the octahedral and tetrahedral holes without expanding the C_{60}^{3-} framework. K^+ and Rb^+ will fit into the octahedral holes without expanding the C_{60}^{3-} framework but will fit into the tetrahedral holes only if the C_{60}^{3-} framework is expanded.

21.130

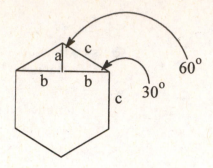

$c = 141.5$ pm $= 141.5 \times 10^{-12}$ m

$\cos(60) = a/c$ and $\sin(60) = b/c$

$a = \cos(60) \cdot c = (0.5)(141.5 \times 10^{-12}$ m$) = 7.075 \times 10^{-11}$ m

$b = \sin(60) \cdot c = (0.866)(141.5 \times 10^{-12}$ m$) = 1.225 \times 10^{-10}$ m

diameter $= 1.08$ nm $= 1.08 \times 10^{-9}$ m $= 1080 \times 10^{-12}$ m $= 1080$ pm

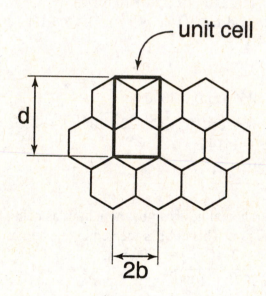

$d = 2a + 2c = 2(7.075 \times 10^{-11}$ m$) + 2(141.5 \times 10^{-12}$ m$) = 4.245 \times 10^{-10}$ m

$d = 424.5 \times 10^{-12}$ m $= 424.5$ pm

$2b = 2(1.225 \times 10^{-10}$ m$) = 2.450 \times 10^{-10}$ m $= 245.0 \times 10^{-12}$ m $= 245.0$ pm

Area of unit cell $= (424.5$ pm$)(245.0$ pm$) = 1.04 \times 10^5$ pm^2/cell

No. of C atoms/cell $= (4)(1/4) + (2)(1/2) + 2 = 4$ C atoms/cell

Surface area of nanotube $= \pi d l = \pi(1080$ pm$)(1.0 \times 10^9$ pm$) = 3.39 \times 10^{12}$ pm^2

No. of C atoms $= (3.39 \times 10^{12}$ pm$^2)\left(\dfrac{1 \text{ cell}}{1.04 \times 10^5 \text{ pm}^2}\right)\left(\dfrac{4 \text{ C atoms}}{\text{cell}}\right) = 1.3 \times 10^8$ C atoms

22 Organic Chemistry

22.1

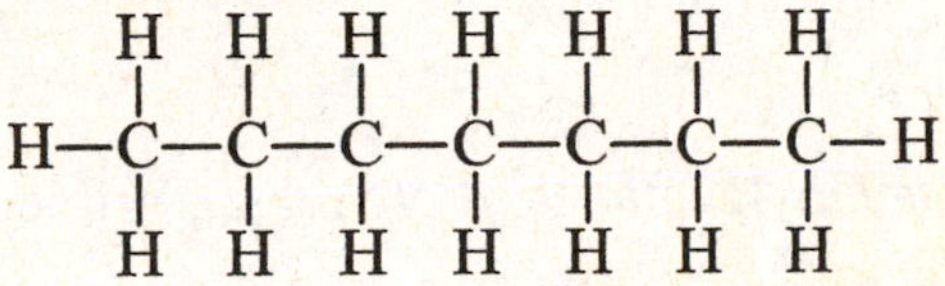

$$\begin{array}{cccccccccccccc}
& H & & H & & H & & H & & H & & H & & H \\
& | & & | & & | & & | & & | & & | & & | \\
H- & C & - & C & - & C & - & C & - & C & - & C & - & C & -H \\
& | & & | & & | & & | & & | & & | & & | \\
& H & & H & & H & & H & & H & & H & & H
\end{array}$$

22.2

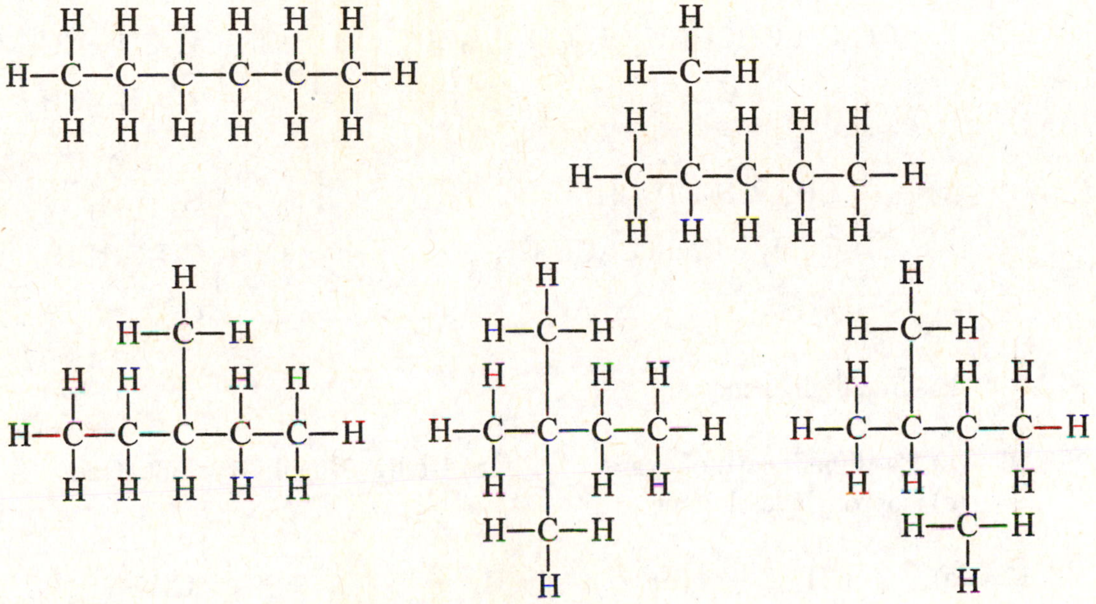

$$\begin{array}{cccccccccccc}
& H & & H & & H & & H & & H & & H \\
& | & & | & & | & & | & & | & & | \\
H- & C & - & C & - & C & - & C & - & C & - & C & -H \\
& | & & | & & | & & | & & | & & | \\
& H & & H & & H & & H & & H & & H
\end{array}$$

$$\begin{array}{c}
H \\ | \\ H-C-H \\
\end{array}$$
$$\begin{array}{ccccccccc}
& H & & & H & & H & & H \\
& | & & & | & & | & & | \\
H- & C & - & C & - & C & - & C & - & C & -H \\
& | & & | & & | & & | & & | \\
& H & & H & & H & & H & & H
\end{array}$$

22.3

$$CH_3CH_2CH_2CH_2CH_3 \qquad CH_3CH_2\overset{\overset{\displaystyle CH_3}{|}}{C}HCH_3 \qquad CH_3\overset{\overset{\displaystyle CH_3}{|}}{\underset{\underset{\displaystyle CH_3}{|}}{C}}CH_3$$

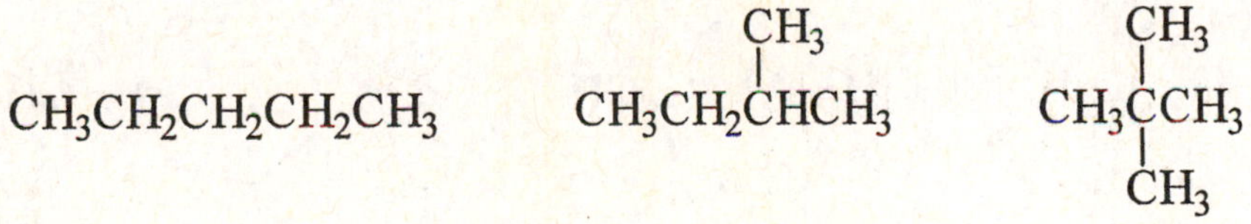

22.4 C_7H_{16}

$$CH_3CH_2CH_2\overset{\overset{\displaystyle CH_3}{|}}{\underset{\underset{\displaystyle CH_3}{|}}{C}}CH_3$$

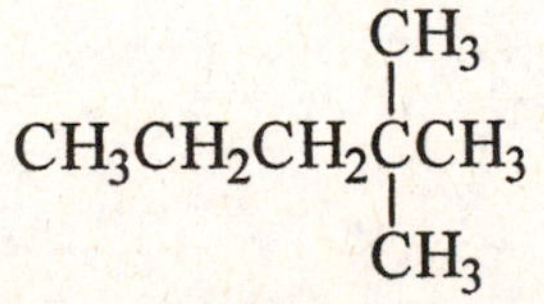

22.5 Structures (a) and (c) are identical. They both contain a chain of six carbons with two –CH$_3$ branches at the fourth carbon and one –CH$_3$ branch at the second carbon. Structure (b) is different, having a chain of seven carbons.

22.6 The two structures are identical. The compound is

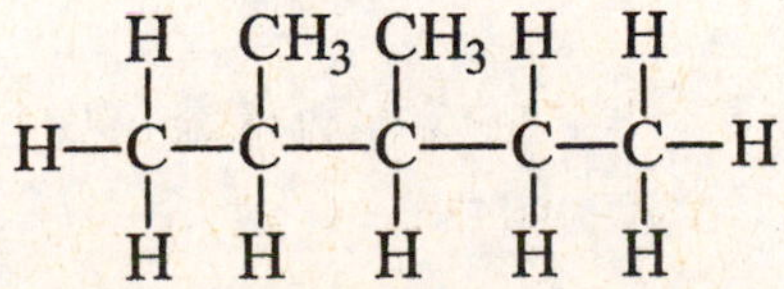

$$\begin{array}{ccccccccc}
& H & & CH_3 & & CH_3 & & H & & H \\
& | & & | & & | & & | & & | \\
H- & C & - & C & - & C & - & C & - & C & -H \\
& | & & | & & | & & | & & | \\
& H & & H & & H & & H & & H
\end{array}$$

22.7   (a) $CH_3CH_2CH_2CH_2CH_3$ pentane

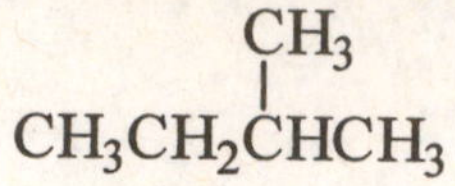

$$CH_3CH_2\underset{\underset{CH_3}{|}}{CH}CH_3$$ 2-methylbutane

$$CH_3\underset{\underset{CH_3}{|}}{\overset{\overset{CH_3}{|}}{C}}CH_3$$ 2,2-dimethylpropane

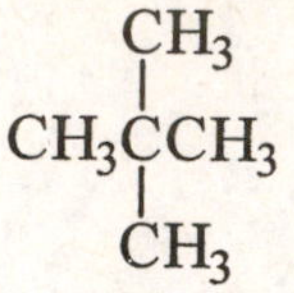

(b) 3,4-dimethylhexane   (c) 2,4-dimethylpentane   (d) 2,2,5-trimethylheptane

22.8   (a)

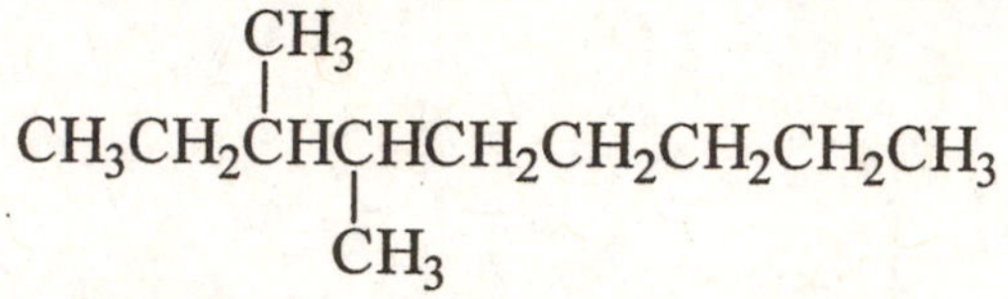

$$CH_3CH_2\underset{\underset{CH_3}{|}}{CH}\overset{\overset{CH_3}{|}}{CH}CH_2CH_2CH_2CH_2CH_3$$

(b)

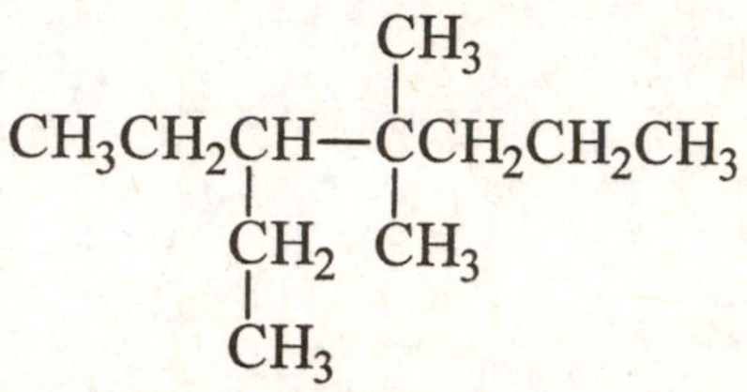

$$CH_3CH_2CH{-}\underset{\underset{\underset{CH_3}{|}}{CH_2}}{\overset{\overset{CH_3}{|}}{C}}CH_2CH_2CH_3$$

(c)

$$CH_3\underset{\underset{CH_3}{|}}{\overset{\overset{CH_3}{|}}{C}}CH_2\overset{\overset{CH_2CH_2CH_3}{|}}{CH}CH_2CH_2CH_2CH_3$$

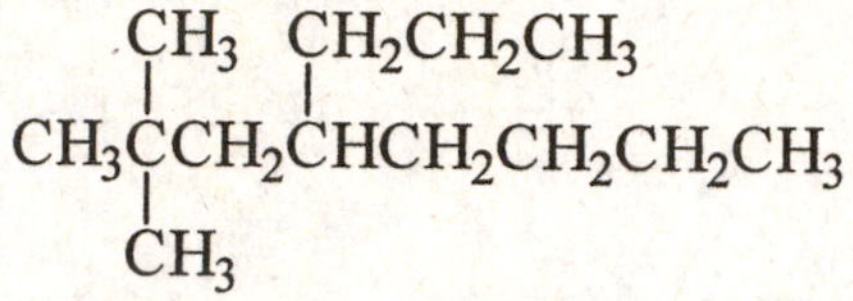

(d)

$$CH_3\underset{\underset{CH_3}{|}}{\overset{\overset{CH_3}{|}}{C}}CH_2\overset{\overset{CH_3}{|}}{CH}CH_3$$

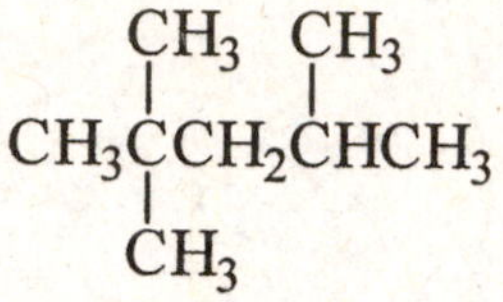

22.9   2,3-dimethylhexane

22.10 (a) 1,4-dimethylcyclohexane   (b) 1-ethyl-3-methylcyclopentane
(c) isopropylcyclobutane

22.11 (a)   (b) 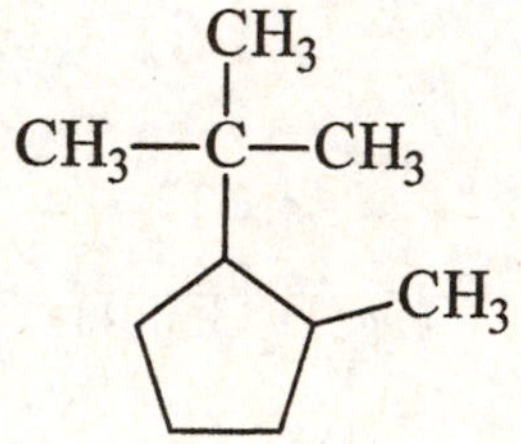   (c)

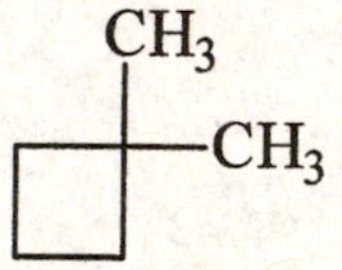

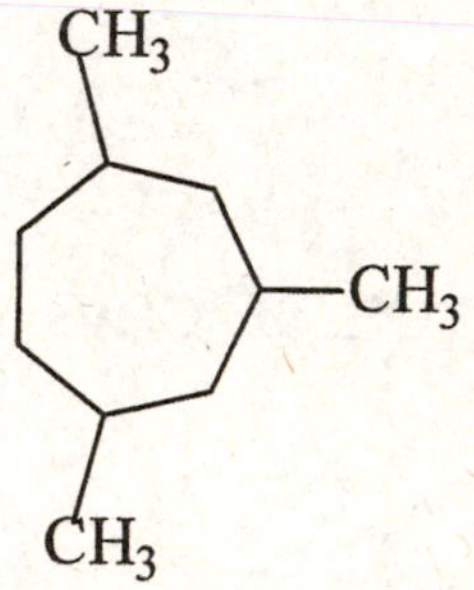

(d)

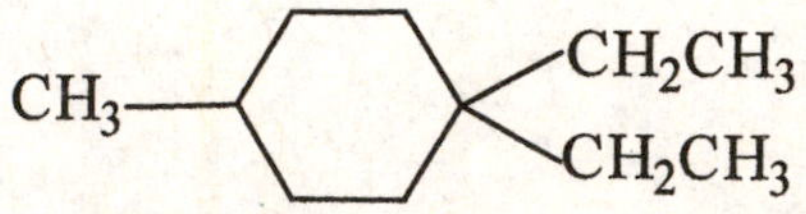

22.12

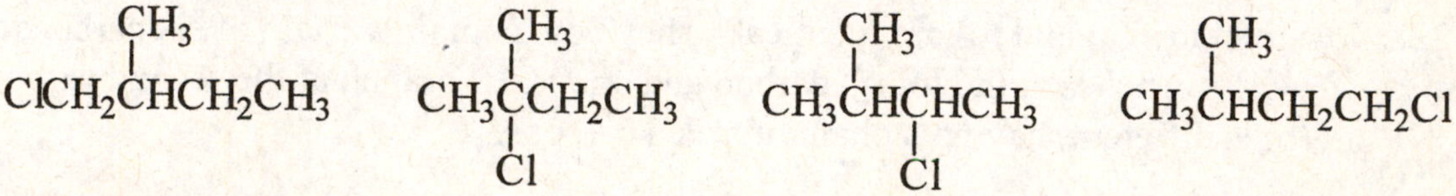

$$ClCH_2\underset{\underset{CH_3}{|}}{CH}CH_2CH_3 \qquad CH_3\underset{\underset{Cl}{|}}{\overset{\overset{CH_3}{|}}{C}}CH_2CH_3 \qquad CH_3\overset{\overset{CH_3}{|}}{CH}\underset{\underset{Cl}{|}}{CH}CH_3 \qquad CH_3\overset{\overset{CH_3}{|}}{CH}CH_2CH_2Cl$$

22.13 (a) 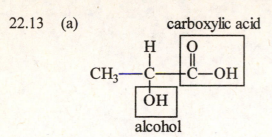 (b)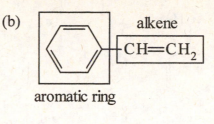

22.14 (a)

$$CH_3\overset{\overset{\displaystyle O}{\|}}{C}H$$

(b)

$$CH_3CH_2\overset{\overset{\displaystyle O}{\|}}{C}OH$$

22.15 (a) 3-methyl-1-butene   (b) 4-methyl-3-heptene   (c) 3-ethyl-1-hexyne

22.16 (a)

$$\underset{\overset{\displaystyle |}{CH_3}}{\overset{\overset{\displaystyle CH_3}{|}}{CH_3CCH}}=CHCH_2CH_3$$

(b)

$$CH_3C\equiv CC\underset{\overset{\displaystyle |}{CHCH_3}}{\overset{\overset{\displaystyle CH_3}{|}}{}}HCH_2CH_2CH_3$$

(c)

$$\underset{H}{\overset{CH_3CH_2}{}}C=C\underset{CH_2CH_2CH_3}{\overset{H}{}}$$

(d)

$$\underset{\overset{\displaystyle |}{CH_3CH}}{\overset{\overset{\displaystyle CH_3}{|}}{}}\underset{H}{\overset{}{}}C=C\underset{H}{\overset{CH_2CH_3}{}}$$

22.17 (a) $CH_3CH_2CH_2CH_3$ (b)

$$CH_3\underset{\overset{\displaystyle |}{}}{\overset{\overset{\displaystyle Br}{|}}{C}}H\underset{\overset{\displaystyle |}{}}{\overset{\overset{\displaystyle Br}{|}}{C}}HCH_3$$

(c)

$$CH_3CH_2\overset{\overset{\displaystyle OH}{|}}{C}HCH_3$$

22.18

$$CH_3\underset{\overset{\displaystyle |}{H}}{\overset{\overset{\displaystyle H}{|}}{C}}-\underset{\overset{\displaystyle |}{H}}{\overset{\overset{\displaystyle OH}{|}}{C}}CH_2CH_3 \qquad CH_3\underset{\overset{\displaystyle |}{H}}{\overset{\overset{\displaystyle OH}{|}}{C}}-\underset{\overset{\displaystyle |}{H}}{\overset{\overset{\displaystyle H}{|}}{C}}CH_2CH_3$$

22.19

22.20 (a)   Br ... Br   (b) NO_2 ... Cl   (c) CH_2CH_3 ... CH_2CH_3

22.21 (a)

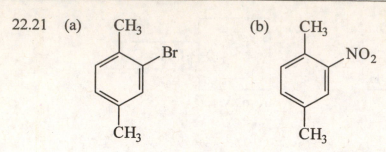

(b)

22.22

22.23 (a)

$\overset{+}{NH_2CH_3}$

Cl^-

(b) $CH_3CH_2CH_2\overset{+}{NH_3}$ Cl^-

22.24 (a) $\underset{\underset{CH_3}{|}}{CH_3CHCH_2CH_2}\overset{O}{\overset{||}{C}}-OH$

(b) $\overset{O}{\overset{||}{C}}-O-\underset{\underset{CH_3}{|}}{CHCH_3}$

(c) $CH_3CH_2\overset{O}{\overset{||}{C}}-NHCH_2CH_3$

22.25 (a) $\overset{O}{\overset{||}{C}}NH_2$ CH_3

(b) $\underset{\underset{Cl}{|}}{CH_3CHCH_2}\overset{O}{\overset{||}{C}}O\underset{\underset{CH_3}{|}}{CHCH_2CH_3}$

22.26 $\underset{\underset{CH_3}{|}}{CH_3CHCHCH_2}\overset{O}{\overset{||}{C}}O\underset{\underset{CH_3}{|}}{CHCH_3}$
      $\underset{CH_3}{}$

22.27 (a) CO_2CH_3
          $CH=CH_2$

(b) $HOCH_2CH_2CH_2OH$ $+$ $\overset{O}{\overset{||}{HOC}}CH_2CH_2\overset{O}{\overset{||}{C}}OH$

22.28

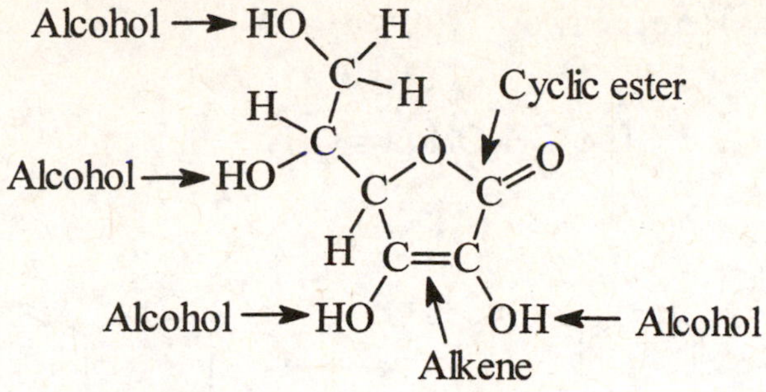

Alcohol → HO   H
           \ /
            C
           / \
          H   H
      H |
       \   C — HO ← Alcohol
        C   / \   Cyclic ester
       / \ /   O   C=O
  HO C   C   /
       \   \   /
        H   C = C
           /   \
      HO   OH ← Alcohol
       ↑ Alcohol ↑ Alkene

Key Concept Problems

22.29

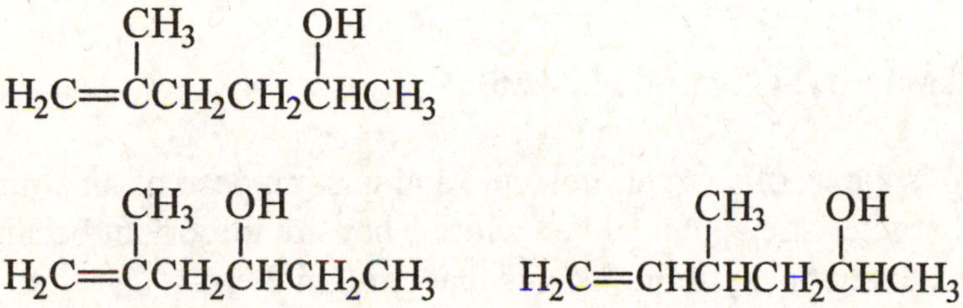

$$\underset{CH_3}{H_2C=\overset{|}{C}CH_2CH_2\overset{OH}{\overset{|}{C}H}CH_3}$$

$$H_2C=\overset{\overset{CH_3}{|}}{C}CH_2\overset{\overset{OH}{|}}{C}HCH_2CH_3 \qquad H_2C=CH\overset{\overset{CH_3}{|}}{C}HCH_2\overset{\overset{OH}{|}}{C}HCH_3$$

22.30 (a)

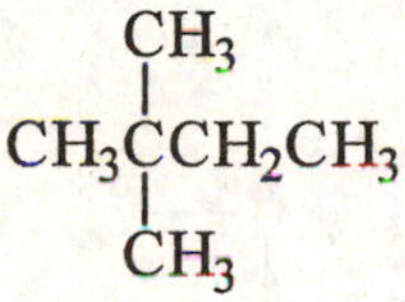

$$CH_3\overset{\overset{CH_3}{|}}{\underset{\underset{CH_3}{|}}{C}}CH_2CH_3$$

(b)

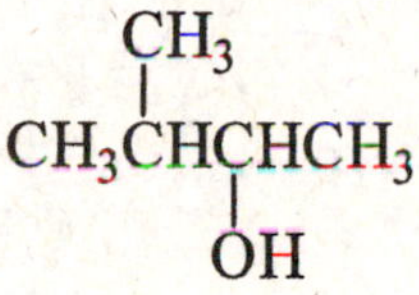

$$CH_3\overset{\overset{CH_3}{|}}{C}H\overset{\underset{OH}{|}}{C}HCH_3$$

22.31 (a) 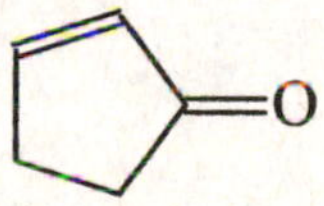 [cyclopentanone structure with =O]   (b) 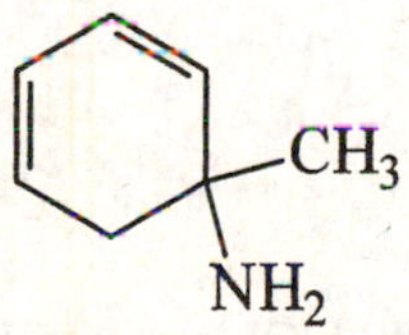 [cyclohexadiene structure with CH_3 and NH_2]

22.32 (a) alkene, ketone, ether   (b) alkene, amine, carboxylic acid

22.33 (a) 2,3-dimethylpentane   (b) 2-methyl-2-hexene

22.34 $CH_2=CCl_2$

22.35

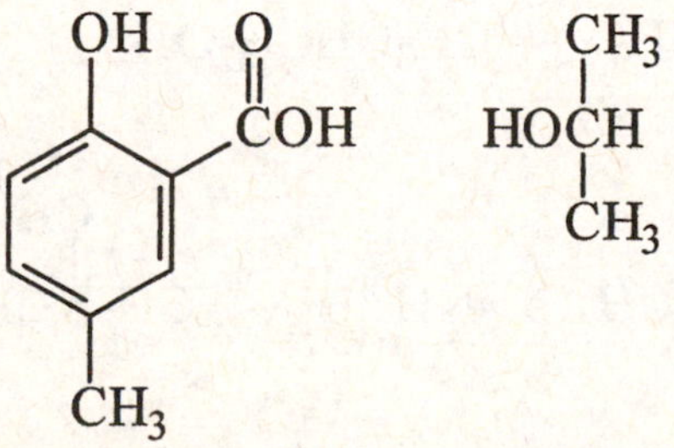

[aromatic ring with OH and COOH substituents and CH_3] $HO\overset{\overset{CH_3}{|}}{\underset{\underset{CH_3}{|}}{C}}H$

669

22.36 There are many possibilities. Here are two:

CH₃CH₂CCH=CCH₃ with O (double bond) on carbon and CH₃

CH₃CCH₂CH₂C=CH₂ with O and CH₃

22.37

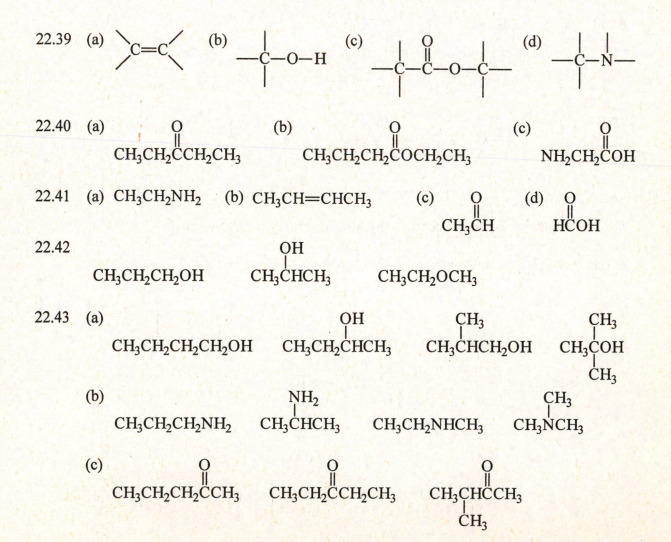

Section Problems
Functional Groups and Isomers (Sections 22.2, 22.8)

22.38 A functional group is a part of a larger molecule and is composed of an atom or group of atoms that has a characteristic chemical behavior. They are important because their chemistry controls the chemistry in molecules that contain them.

22.39 (a) C=C (b) $-\text{C}-\text{O}-\text{H}$ (c) $-\text{C}-\text{C}-\text{O}-\text{C}-$ (with O) (d) $-\text{C}-\text{N}-$

22.40 (a) CH₃CH₂CCH₂CH₃ (with O) (b) CH₃CH₂CH₂COCH₂CH₃ (with O) (c) NH₂CH₂COH (with O)

22.41 (a) CH₃CH₂NH₂ (b) CH₃CH=CHCH₃ (c) CH₃CH (with O) (d) HCOH (with O)

22.42 CH₃CH₂CH₂OH   CH₃CHCH₃ (with OH)   CH₃CH₂OCH₃

22.43 (a) CH₃CH₂CH₂CH₂OH   CH₃CH₂CHCH₃ (with OH)   CH₃CHCH₂OH (with CH₃)   CH₃COH (with CH₃, CH₃)

(b) CH₃CH₂CH₂NH₂   CH₃CHCH₃ (with NH₂)   CH₃CH₂NHCH₃   CH₃NCH₃ (with CH₃)

(c) CH₃CH₂CH₂CCH₃ (with O)   CH₃CH₂CCH₂CH₃ (with O)   CH₃CHCCH₃ (with O, CH₃)

(d)

$$\text{CH}_3\text{CH}_2\text{CH}_2\text{CH}_2\overset{\displaystyle O}{\overset{\|}{\text{C}}}\text{H} \qquad \text{CH}_3\overset{\text{CH}_3}{\underset{|}{\text{CH}}}\text{CH}_2\overset{\displaystyle O}{\overset{\|}{\text{C}}}\text{H} \qquad \text{CH}_3\text{CH}_2\underset{\underset{\text{CH}_3}{|}}{\text{CH}}\overset{\displaystyle O}{\overset{\|}{\text{C}}}\text{H} \qquad \text{CH}_3\underset{\underset{\text{CH}_3}{|}}{\text{C}}\!\!-\!\!\overset{\text{CH}_3}{\underset{}{\text{CH}}}\overset{\displaystyle O}{\overset{\|}{\text{C}}}\text{H}$$

22.44 (a) alkene and aldehyde   (b) aromatic ring, alcohol, and ketone

22.45 ester, aromatic ring, and amine

Alkanes (Sections 22.2–22.7)

22.46 In a straight-chain alkane, all the carbons are connected in a row. In a branched-chain alkane, there are branching connections of carbons along the carbon chain.

22.47 An alkane is a compound that contains only carbon and hydrogen and has only single bonds. An alkyl group is the part of an alkane that remains when a hydrogen is removed.

22.48 In forming alkanes, carbon uses sp^3 hybrid orbitals.

22.49 Because each carbon is bonded to its maximum number of atoms and cannot bond to additional atoms, an alkane is said to be saturated.

22.50 C_3H_9 contains one more H than needed for an alkane.

22.51 (a) $\underline{\text{C}}\text{H}_3\!\!=\!\!\text{CHCH}_2\text{CH}_2\text{OH}$ Underlined carbon has five bonds.

(b) $\text{CH}_3\text{CH}_2\text{CH}\!\!=\!\!\overset{\displaystyle O}{\overset{\|}{\underline{\text{C}}}}\text{CH}_3$ Underlined carbon has five bonds.

(c) $\text{CH}_3\text{CH}_2\text{C}\!\!\equiv\!\!\underline{\text{C}}\text{H}_2\text{CH}_3$ Underlined carbon has six bonds.

22.52 (a) 4-ethyl-3-methyloctane   (b) 4-isopropyl-2-methylheptane
(c) 2,2,6-trimethylheptane   (d) 4-ethyl-4-methyloctane

22.53 2,2,4-trimethylpentane

22.54 (a)

$$\text{CH}_3\text{CH}_2\underset{\underset{\text{CH}_2\text{CH}_3}{|}}{\text{CH}}\text{CH}_2\text{CH}_2\text{CH}_3$$

(b)

$$\text{CH}_3\underset{\underset{\text{CH}_3}{|}}{\overset{\overset{\text{CH}_3}{|}}{\text{C}}}\!\!-\!\!\overset{\overset{\text{CH}_3}{|}}{\text{CH}}\text{CH}_2\text{CH}_3$$

(c)

$$\text{CH}_3\text{CH}_2\underset{\underset{\text{CH}_3}{|}}{\overset{\overset{\text{CH}_2\text{CH}_3}{|}}{\text{C}}}\!\!-\!\!\underset{\underset{\text{CH}_3}{|}}{\text{CH}}\text{CH}_2\text{CH}_2\text{CH}_3$$

(d)

$$\text{CH}_3\underset{\underset{\text{CH}_3}{|}}{\text{CH}}\text{CH}_2\text{CH}_2\underset{\underset{\overset{\text{CH}_3}{\underset{|}{\text{CH}}}}{|}}{\text{CH}}\text{CH}_2\text{CH}_2\text{CH}_3$$

22.55 (a) 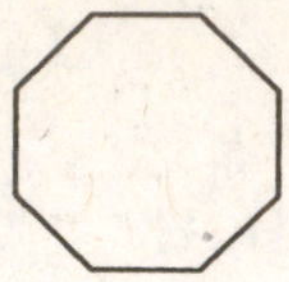 (b) 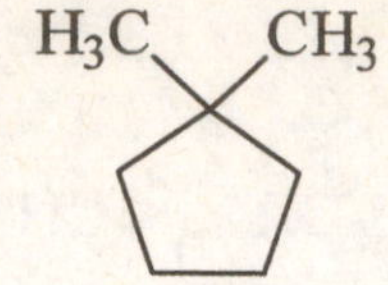 (c) 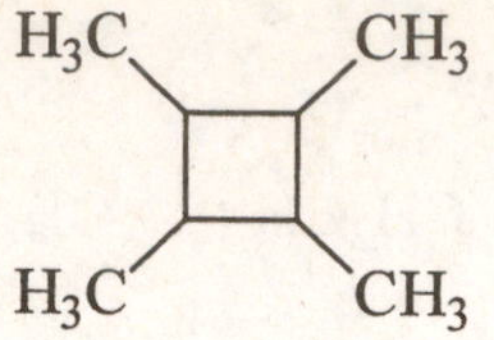 (d) 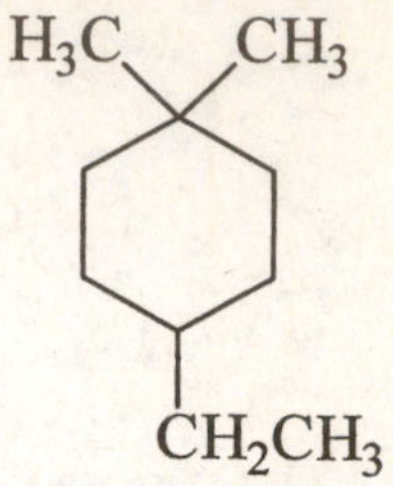

22.56 (a) 1,1-dimethylcyclopentane
(b) 1-isopropyl-2-methylcyclohexane
(c) 1,2,4-trimethylcyclooctane

22.57 (a) The longest chain contains six carbons and the molecule should be named from a hexane root; the correct name is 3,3-dimethylhexane.
(b) The longest chain contains seven carbons and the molecule should be named from a heptane root; the correct name is 3,5-dimethylheptane.
(c) The ring is a cycloheptane ring and the methyl groups are in the 1 and 3 position; the correct name is 1,3-dimethylcycloheptane.

22.58 The structures are shown in Problem 22.2.
hexane, 2-methylpentane, 3-methylpentane, 2,2-dimethylbutane, and 2,3-dimethylbutane

22.59 $CH_3CH_2CH_2CH_2CH_2CH_2CH_3$
heptane

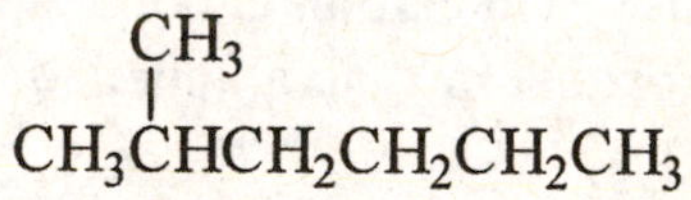
2-methylhexane

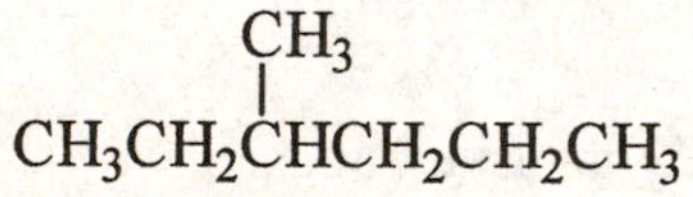
3-methyhexane

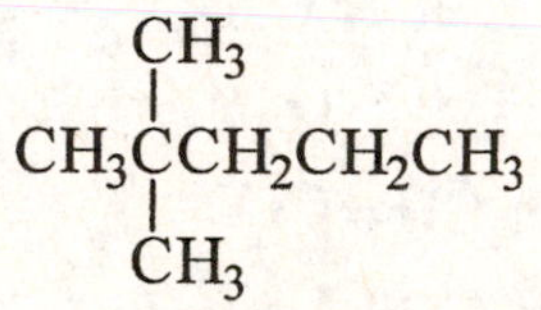
2,2-dimethylpentane

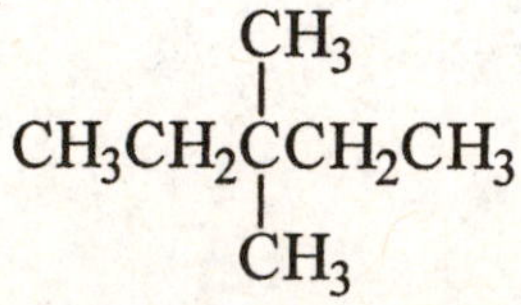
3,3-dimethylpentane

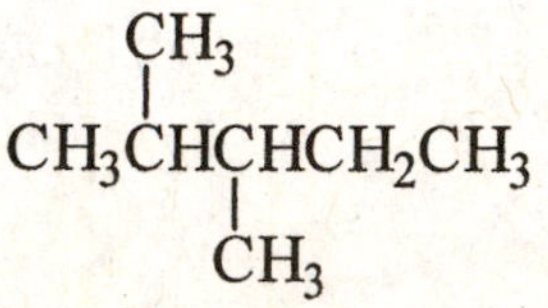
2,3-dimethylpentane

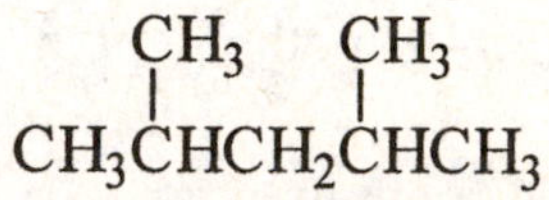
2,4-dimethypentane

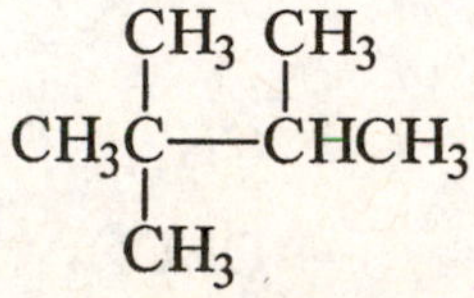
2,2,3-trimethylbutane

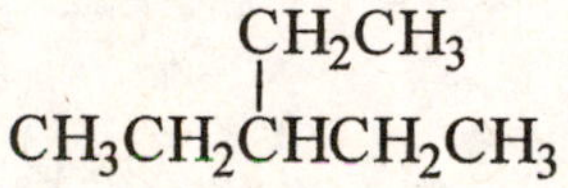
3-ethylpentane

22.60　(a)

$$Cl$$
$$ClCH_2CH_2CH_2CH_2CH_2CH_3 \quad CH_3\overset{\displaystyle |}{C}HCH_2CH_2CH_2CH_3 \quad CH_3CH_2\overset{\displaystyle |}{C}HCH_2CH_2CH_3$$

(b)

$$\underset{ClCH_2CH_2\overset{\displaystyle |}{C}HCH_2CH_3}{\overset{\displaystyle CH_3}{}} \quad \underset{CH_3\overset{\displaystyle |}{C}H\overset{\displaystyle |}{C}HCH_2CH_3}{\overset{\displaystyle Cl\ CH_3}{}} \quad \underset{CH_3CH_2\overset{\displaystyle |}{\underset{\displaystyle |}{C}}CH_2CH_3}{\overset{\displaystyle CH_3}{} \atop \overset{\displaystyle Cl}{}} \quad \underset{CH_3CH_2\overset{\displaystyle |}{C}HCH_2CH_3}{\overset{\displaystyle CH_2Cl}{}}$$

(c)

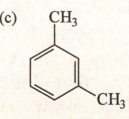

22.61　Reaction (a) is likely to have a higher yield because there is only one possible monochlorinated substitution product. Reaction (b) has four possible monochlorinated substitution products, which would result in a lower yield of the one product shown.

Alkenes, Alkynes, and Aromatic Compounds (Sections 22.9–22.11)

22.62　(a) sp^2　　　(b) sp　　　(c) sp^2

22.63　Alkenes, alkynes, and aromatic compounds are said to be unsaturated because they do not contain as many hydrogens as their alkane analogs.

22.64　Today the term "aromatic" refers to the class of compounds containing a six-membered ring with three double bonds, not to the fragrance of a compound.

22.65　An addition reaction is the reaction of an XY molecule with an alkene or alkyne.

$$\ce{C=C} + XY \longrightarrow \underset{\displaystyle |}{\overset{\displaystyle X\ \ Y}{-C-C-}}$$

22.66　(a) $CH_3CH=CHCH_2CH_3$　(b) $HC\equiv CCH_2CH_3$　(c)

673

22.67 $CH_2=C=CHCH_2CH_3$ $CH_2=CHCH=CHCH_3$ $CH_2=CHCH_2CH=CH_2$

$CH_3CH=C=CHCH_3$ $\overset{\overset{\displaystyle CH_3}{|}}{CH_2=CCH=CH_2}$ $\overset{\overset{\displaystyle CH_3}{|}}{CH_2=C=CCH_3}$

22.68 (a) 4-methyl-2-pentene   (b) 3-methyl-1-pentene
     (c) 1,2-dichlorobenzene, or o-dichlorobenzene
     (d) 2-methyl-2-butene   (e) 7-methyl-3-octyne

22.69 (a) $\overset{\overset{\displaystyle H\ \ H}{|\ \ |}}{CH_3C=CCH_2CH_2CH_3}$ (b) $\overset{\overset{\displaystyle CH_3}{|}}{CH_3CHCH=CHCH_2CH_3}$ (c) $\overset{\overset{\displaystyle CH_3}{|}}{CH_2=CCH=CH_2}$

22.70 $CH_2=CHCH_2CH_2CH_3$ $CH_3CH=CHCH_2CH_3$ $\overset{\overset{\displaystyle CH_3}{|}}{CH_2=CCH_2CH_3}$
     1-pentene   2-pentene

2-methyl-1-butene

$\overset{\overset{\displaystyle CH_3}{|}}{CH_3C=CHCH_3}$ $\overset{\overset{\displaystyle CH_3}{|}}{CH_2=CHCHCH_3}$ Only 2-pentene can exist as cis-trans isomers.

2-methyl-2-butene   3-methyl-1-butene

22.71 $HC\equiv CCH_2CH_2CH_3$ $CH_3C\equiv CCH_2CH_3$ $\overset{\overset{\displaystyle CH_3}{|}}{HC\equiv CCHCH_3}$
     1-pentyne   2-pentyne

3-methyl-1-butyne

22.72 (a) $CH_2=CHCH_2CH_2CH_2CH_3$ This compound cannot form cis-trans isomers.
     (b) $CH_3CH=CHCH_2CH_2CH_3$ This compound can form cis-trans isomers because of the different groups on each double bond C.
     (c) $CH_3CH_2CH=CHCH_2CH_3$ This compound can form cis-trans isomers because of the different groups on each double bond C.

22.73 (a) $\overset{\overset{\displaystyle CH_3}{|}}{CH_3CHCH=CHCH_3}$
     This compound can form cis-trans isomers because of the different groups on each double bond C.
     (b) $\overset{\overset{\displaystyle CH=CH_2}{|}}{CH_3CH_2CHCH_3}$ This compound cannot form cis-trans isomers.
     (c) $\overset{\overset{\displaystyle Cl}{|}}{CH_3CH=CHCHCH_2CH_3}$
     This compound can form cis-trans isomers because of the different groups on each double bond C.

22.74 (a)

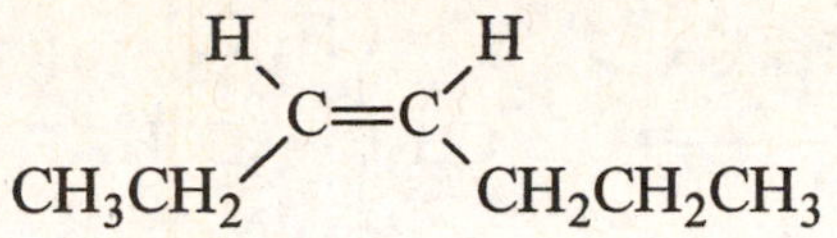

$$CH_3CH_2 \overset{H}{\underset{}{C}}=\overset{H}{\underset{CH_2CH_2CH_3}{C}}$$

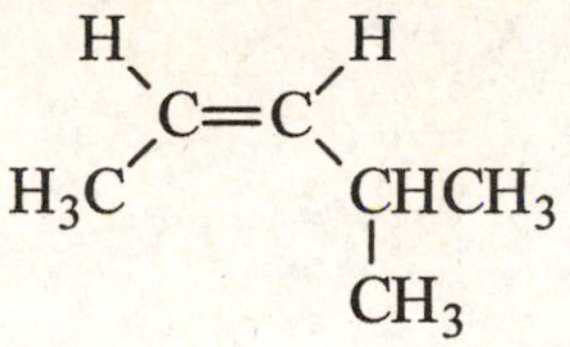

(b)

$$H_3C \overset{H}{\underset{}{C}}=\overset{H}{\underset{\underset{CH_3}{|}}{C}} \overset{}{CHCH_3}$$

(c)

$$\underset{H}{\overset{CH_3}{\underset{}{CH_3CH}}} \overset{}{\underset{}{C}}=\overset{H}{\underset{\underset{CH_3}{|}}{C}} \overset{}{CHCH_3}$$

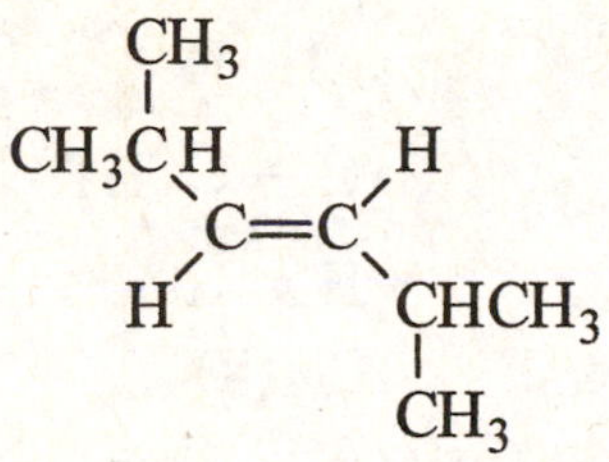

22.75 (a)

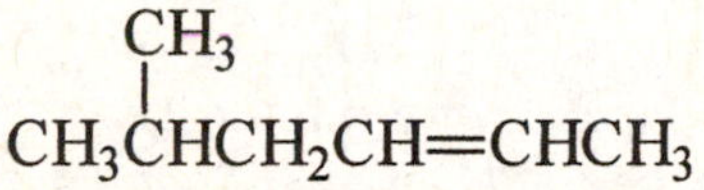

$$\underset{}{\overset{CH_3}{\underset{|}{CH_3CHCH_2CH=CHCH_3}}}$$
5-methyl-2-hexene

(b)

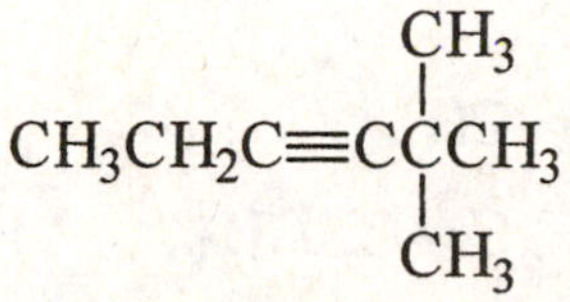

$$CH_3CH_2C{\equiv}C\underset{\underset{CH_3}{|}}{\overset{\overset{CH_3}{|}}{C}}CH_3$$
2,2-dimethyl-3-hexyne

(c)

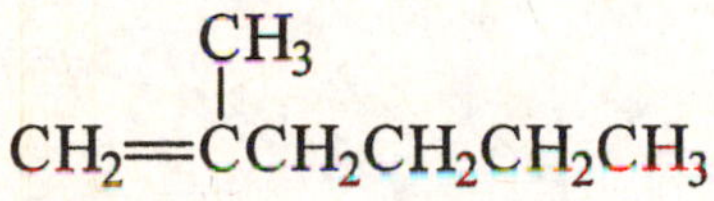

$$\underset{}{\overset{CH_3}{\underset{|}{CH_2=CCH_2CH_2CH_2CH_3}}}$$
2-methyl-1-hexene

(d)

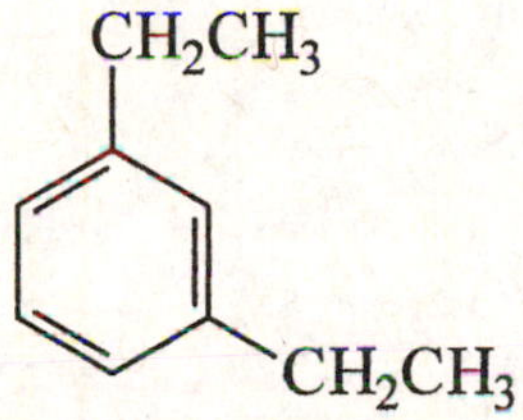

CH₂CH₃ ... CH₂CH₃
1,3-diethylbenzene

22.76   Cis-trans isomers are possible for substituted alkenes because of the lack of rotation about the carbon-carbon double bond. Alkanes and alkynes cannot form cis-trans isomers because alkanes have free rotation about carbon-carbon single bonds and alkynes are linear about the carbon-carbon triple bond.

22.77   Small-ring cycloalkenes don't exist as cis-trans isomers because the trans isomer could not close the carbon-carbon chain back on itself to form a ring.

22.78 (a)

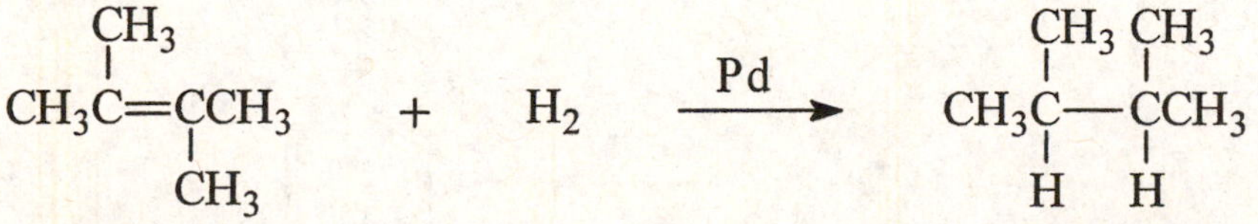

$$\underset{\underset{CH_3}{|}}{\overset{\overset{CH_3}{|}}{CH_3C}}=CCH_3 \quad + \quad H_2 \quad \overset{Pd}{\longrightarrow} \quad \underset{\underset{H}{|} \; \underset{H}{|}}{\overset{\overset{CH_3}{|} \; \overset{CH_3}{|}}{CH_3C-CCH_3}}$$

(b)

$$\underset{\underset{CH_3}{|}}{\overset{\overset{CH_3}{|}}{CH_3C}}=CCH_3 \quad + \quad Br_2 \quad \longrightarrow \quad \underset{\underset{Br}{|} \; \underset{Br}{|}}{\overset{\overset{CH_3}{|} \; \overset{CH_3}{|}}{CH_3C-CCH_3}}$$

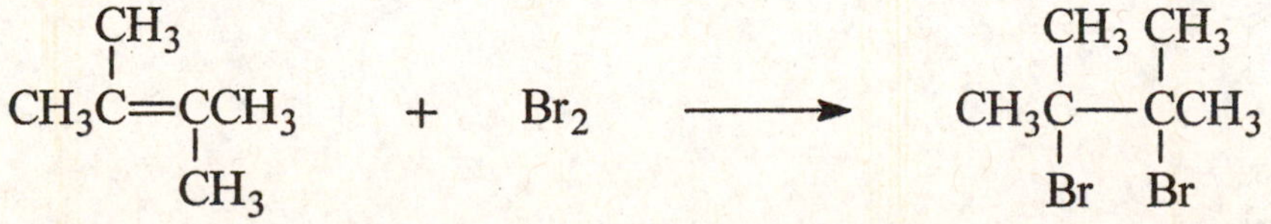

(c)

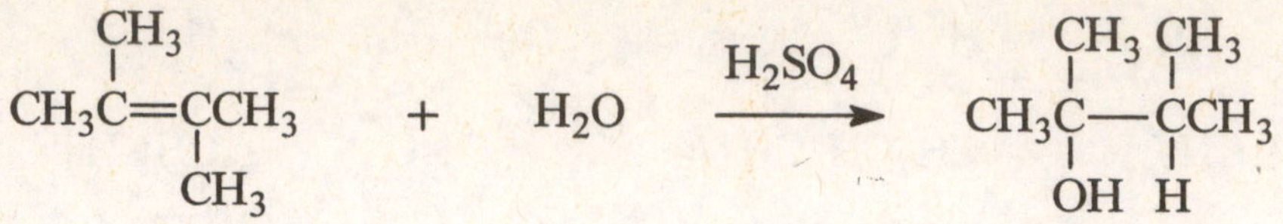

$$CH_3C{=}CCH_3 \ (\text{with } CH_3 \text{ groups}) \ + \ H_2O \ \xrightarrow{H_2SO_4} \ CH_3\underset{OH}{C}{-}\underset{H}{C}CH_3 \ (\text{with } CH_3, CH_3 \text{ groups})$$

22.79 (a)

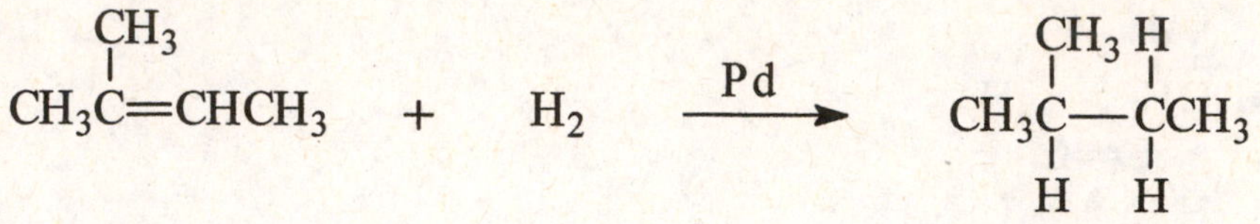

$$CH_3\underset{|}{\overset{CH_3}{C}}{=}CHCH_3 \ + \ H_2 \ \xrightarrow{Pd} \ CH_3\underset{H}{\overset{CH_3}{C}}{-}\underset{H}{\overset{H}{C}}CH_3$$

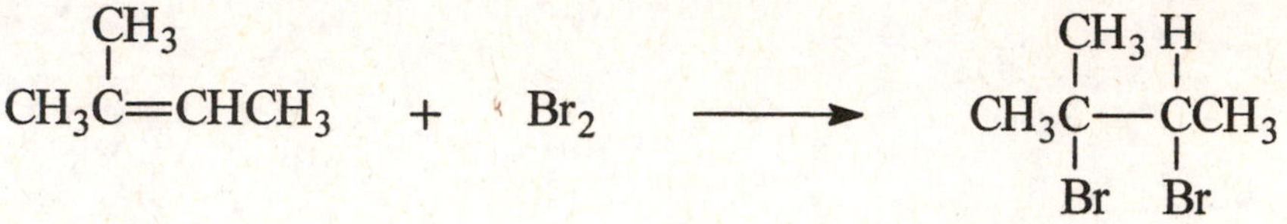

(b)

$$CH_3\underset{|}{\overset{CH_3}{C}}{=}CHCH_3 \ + \ Br_2 \ \longrightarrow \ CH_3\underset{Br}{\overset{CH_3}{C}}{-}\underset{Br}{\overset{H}{C}}CH_3$$

(c)

$$CH_3\underset{|}{\overset{CH_3}{C}}{=}CHCH_3 \ + \ H_2O \ \xrightarrow{H_2SO_4} \ CH_3\underset{OH}{\overset{CH_3}{C}}{-}\underset{H}{\overset{H}{C}}CH_3 \ + \ CH_3\underset{H}{\overset{CH_3}{C}}{-}\underset{OH}{\overset{H}{C}}CH_3$$

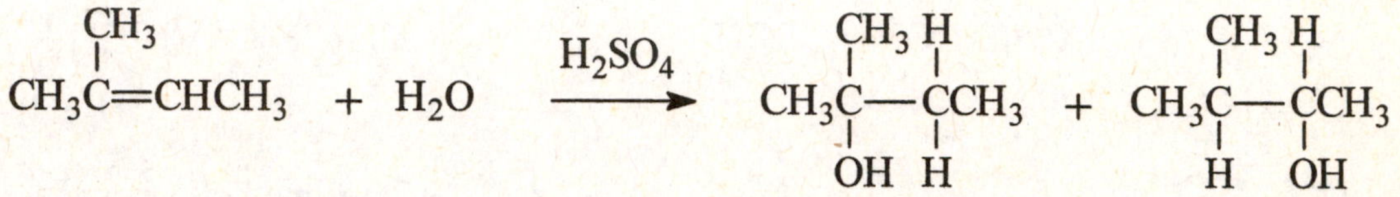

22.80 (a)

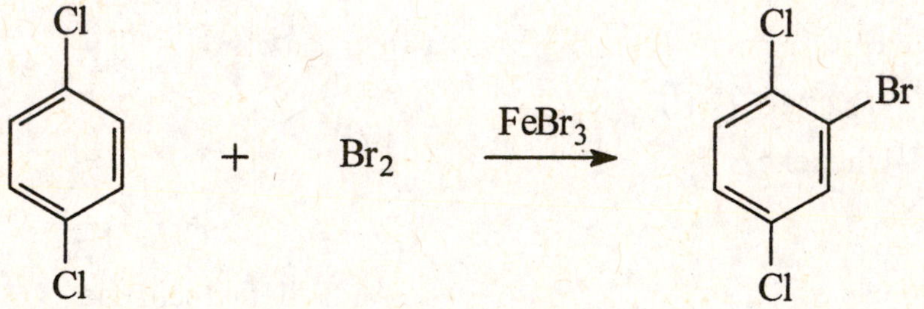

1,4-dichlorobenzene $+ \ Br_2 \ \xrightarrow{FeBr_3}$ 1,4-dichloro-2-bromobenzene

(b)

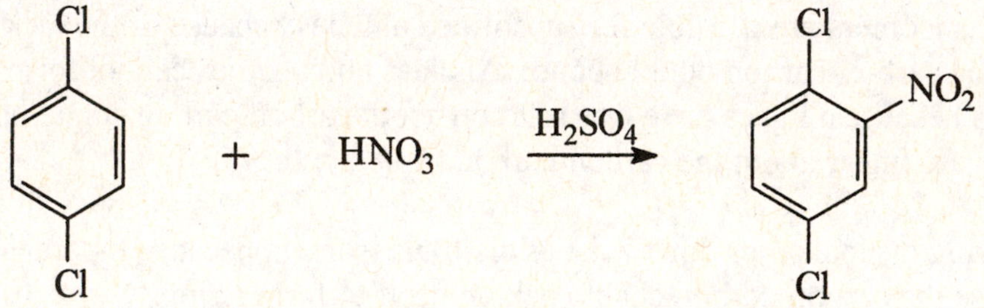

1,4-dichlorobenzene $+ \ HNO_3 \ \xrightarrow{H_2SO_4}$ 1,4-dichloro-2-nitrobenzene

(c)

1,4-dichlorobenzene $+ \ Cl_2 \ \xrightarrow{FeCl_3}$ 1,2,4-trichlorobenzene

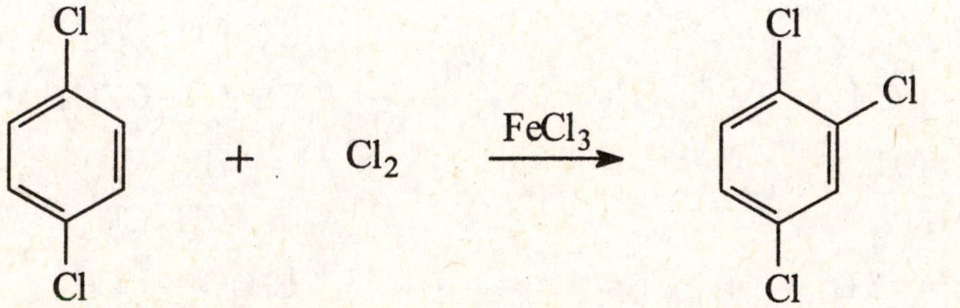

22.81

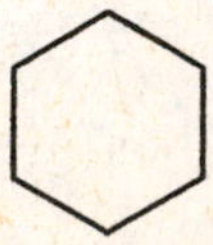

cyclohexane

Alcohols, Amines, and Carbonyl Compounds (Sections 22.12–22.14)

22.82 (a)

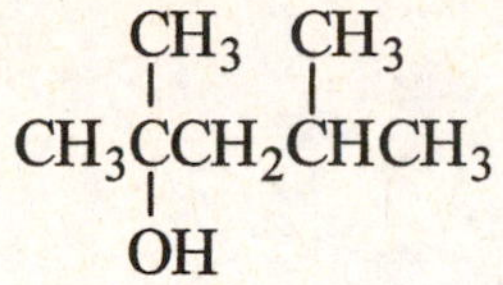

$$CH_3CCH_2CHCH_3$$

with CH_3 and CH_3 above, and OH below.

(b) 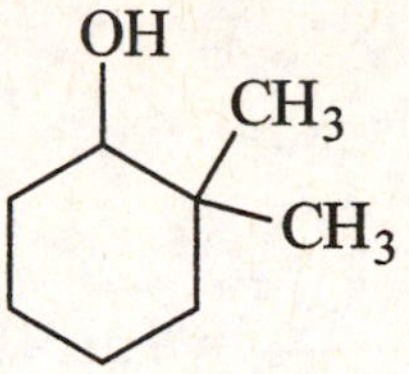 cyclohexane ring with OH, CH_3, CH_3 substituents

(c)

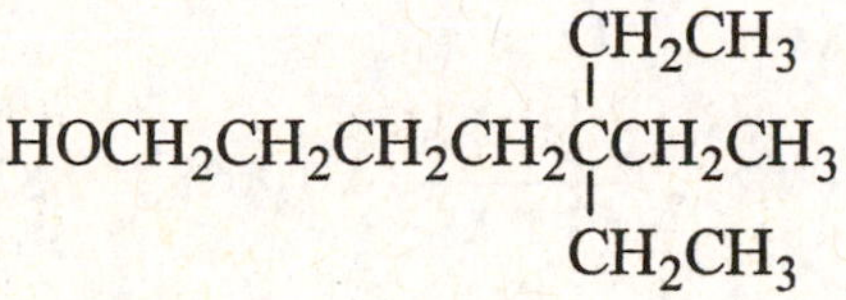

$$HOCH_2CH_2CH_2CH_2CCH_2CH_3$$

with CH_2CH_3 above and CH_2CH_3 below.

(d)

$$CH_3CH_2CCH_2CH_2CH_3$$

with CH_2CH_3 above and OH below.

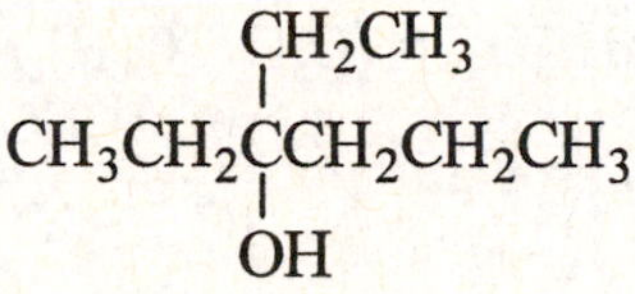

22.83 (a) $CH_3CH_2CH_2NH_2$ (b) $(CH_3CH_2)_2NH$ (c) $CH_3CH_2CH_2NHCH_3$

22.84 Quinine; a base will dissolve in aqueous acid, but menthol is insoluble.

22.85 Pentanoic acid will react with aqueous $NaHCO_3$ to yield CO_2, but methyl butanoate will not.

22.86 An aldehyde has a terminal carbonyl group. A ketone has the carbonyl group located between two carbon atoms.

22.87 In aldehydes and ketones, the carbonyl-group carbon is bonded to atoms (H and C) that don't attract electrons strongly. In carboxylic acids, esters, and amides, the carbonyl-group carbon is bonded to an atom (O or N) that does attract electrons strongly.

22.88 The industrial preparation of ketones and aldehydes involves the oxidation of the related alcohol.

22.89 Carboxylic acids, esters, and amides undergo carbonyl-group substitution reactions, in which a group –Y substitutes for the –OH, –OC, or –N group of the starting material.

22.90 (a) ketone (b) aldehyde (c) ketone (d) amide (e) ester

22.91 (a)

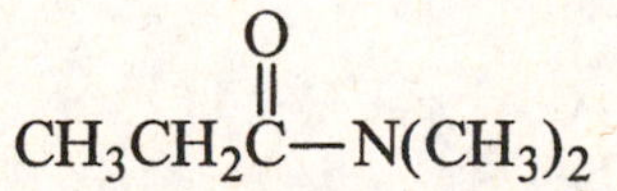

$$CH_3CH_2\overset{O}{\overset{\|}{C}}{-}N(CH_3)_2$$
N,N-dimethylpropanamide

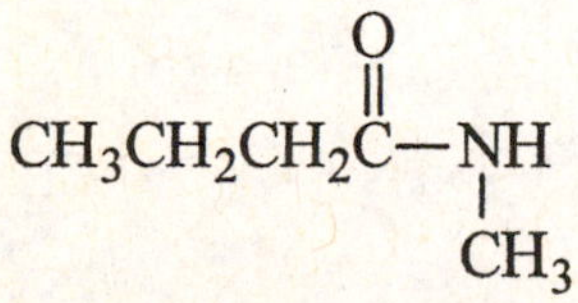

$$CH_3CH_2CH_2\overset{O}{\overset{\|}{C}}{-}NH$$ with CH_3 below
N-methylbutanamide

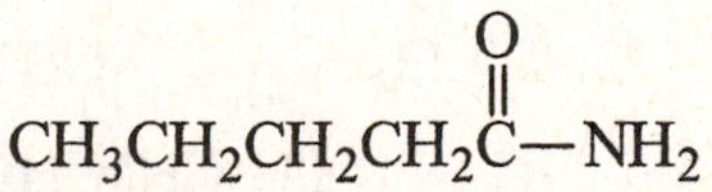

$$CH_3CH_2CH_2CH_2\overset{O}{\overset{\|}{C}}{-}NH_2$$
pentanamide

(b)

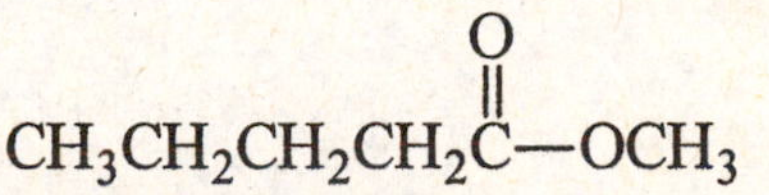

$$CH_3CH_2CH_2CH_2\overset{O}{\overset{\|}{C}}{-}OCH_3$$
methyl pentanoate

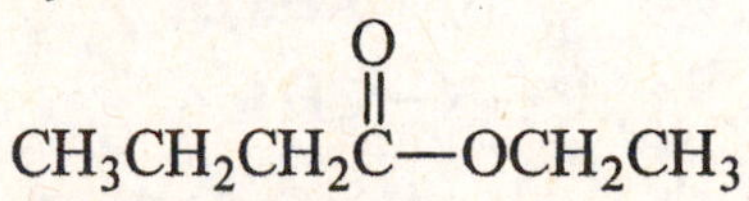

$$CH_3CH_2CH_2\overset{O}{\overset{\|}{C}}{-}OCH_2CH_3$$
ethyl butanoate

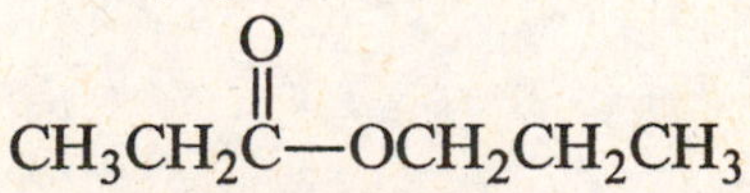

$$CH_3CH_2\overset{O}{\overset{\|}{C}}{-}OCH_2CH_2CH_3$$
propyl propanoate

22.92

$$C_6H_5CO_2H(aq) + H_2O(l) \rightleftharpoons H_3O^+(aq) + C_6H_5CO_2^-(aq)$$

initial (M)	1.0	~0	0
change (M)	$-x$	$+x$	$+x$
equil (M)	$1.0 - x$	x	x

$$K_a = \frac{[H_3O^+][C_6H_5CO_2^-]}{[C_6H_5CO_2H]} = 6.5 \times 10^{-5} = \frac{x^2}{1.0-x} \approx \frac{x^2}{1.0}$$

$$x = [H_3O^+] = [C_6H_5CO_2H]_{diss} = 0.0081 \text{ M}$$

$$\% \text{ dissociation} = \frac{[C_6H_5CO_2H]_{diss}}{[C_6H_5CO_2H]_{initial}} \times 100\% = \frac{0.0081 \text{ M}}{1.0 \text{ M}} \times 100\% = 0.81\%$$

22.93 (a)

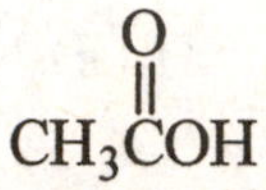

$$\underset{\underset{}{\overset{\overset{O}{\parallel}}{CH_3C}}OH}{}$$

(b)

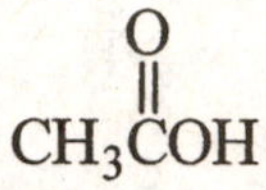

$$CH_3COH$$

(c)

$$CH_3C-O^- \quad Na^+$$

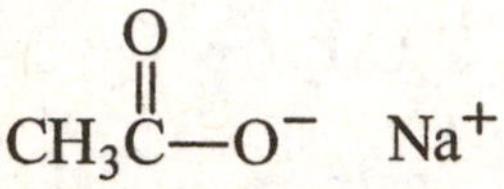

22.94 (a) methyl 4-methylpentanoate
(b) 4,4-dimethylpentanoic acid
(c) 2-methylpentanamide

22.95 (a) N,N-dimethyl-4-methylhexanamide
(b) isopropyl 2-methylpropanoate
(c) N-ethyl-p-chlorobenzamide

22.96 (a)

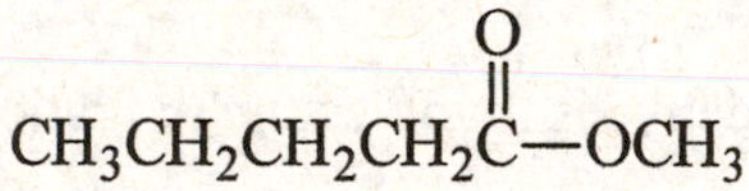

$$CH_3CH_2CH_2CH_2C-OCH_3$$

(b)

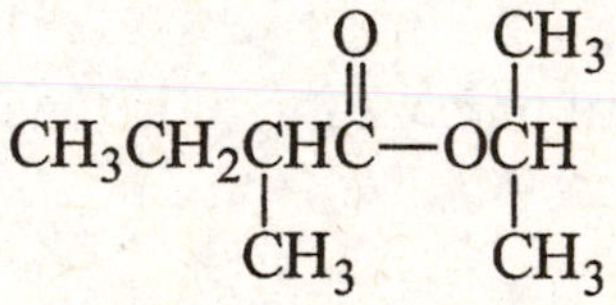

$$CH_3CH_2CHC-OCH$$

(c)

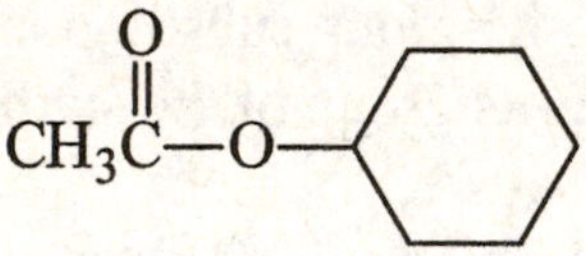

$$CH_3C-O-$$

22.97 (a)

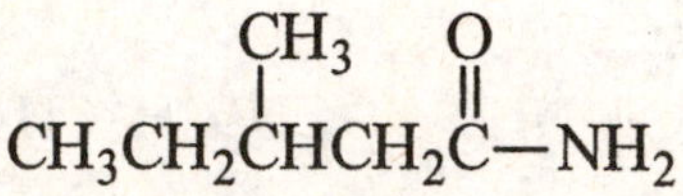

$$CH_3CH_2CHCH_2C-NH_2$$

(b)

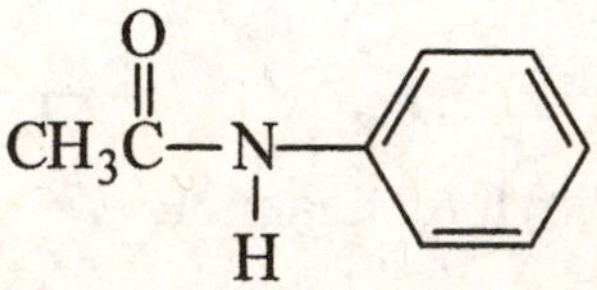

$$CH_3C-N-$$

(c)

$$C-N-CH_3$$

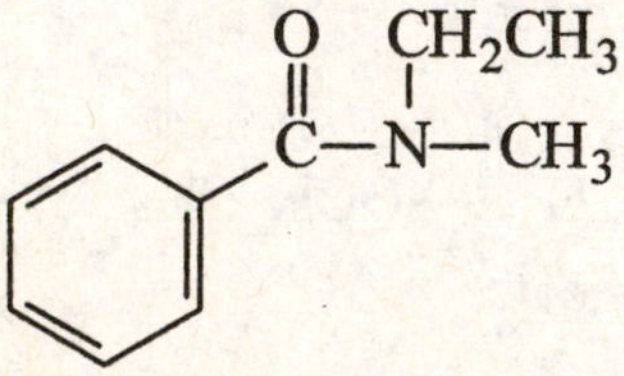

22.98 (a)

$$CH_3CH_2CH_2CH_2\overset{\overset{\displaystyle O}{\|}}{C}OH \;+\; CH_3OH \;\xrightarrow{\;H^+\;}\; CH_3CH_2CH_2CH_2\overset{\overset{\displaystyle O}{\|}}{C}OCH_3 \;+\; H_2O$$

(b)

$$CH_3CH_2\underset{\underset{\displaystyle CH_3}{|}}{\overset{\overset{\displaystyle O}{\|}}{C}}HCOH \;+\; \underset{\underset{\displaystyle CH_3}{|}}{H}\overset{\overset{\displaystyle CH_3}{|}}{C}OH \;\xrightarrow{\;H^+\;}\; CH_3CH_2\underset{\underset{\displaystyle CH_3}{|}}{C}H\overset{\overset{\displaystyle O}{\|}}{C}O\underset{\underset{\displaystyle CH_3}{|}}{C}H \;+\; H_2O$$

(c)

$$CH_3\overset{\overset{\displaystyle O}{\|}}{C}OH \;+\; HO-\text{(cyclohexyl)} \;\xrightarrow{\;H^+\;}\; CH_3\overset{\overset{\displaystyle O}{\|}}{C}O-\text{(cyclohexyl)} \;+\; H_2O$$

22.99 (a)

$$CH_3CH_2\overset{\overset{\displaystyle CH_3}{|}}{C}HCH_2\overset{\overset{\displaystyle O}{\|}}{C}OH \;+\; NH_3 \;\xrightarrow{\;Heat\;}\; CH_3CH_2\overset{\overset{\displaystyle CH_3}{|}}{C}HCH_2\overset{\overset{\displaystyle O}{\|}}{C}NH_2 \;+\; H_2O$$

(b)

$$CH_3\overset{\overset{\displaystyle O}{\|}}{C}OH \;+\; NH_2-\text{(phenyl)} \;\xrightarrow{\;Heat\;}\; CH_3\overset{\overset{\displaystyle O}{\|}}{C}-\underset{\underset{\displaystyle H}{|}}{N}-\text{(phenyl)} \;+\; H_2O$$

(c)

$$\text{(phenyl)}-\overset{\overset{\displaystyle O}{\|}}{C}OH \;+\; H\underset{\underset{\displaystyle CH_2CH_3}{|}}{N}CH_2CH_3 \;\xrightarrow{\;Heat\;}\; \text{(phenyl)}-\overset{\overset{\displaystyle O}{\|}}{C}-\underset{\underset{\displaystyle CH_2CH_3}{|}}{N}\overset{\overset{\displaystyle CH_3}{}}{}CH_2CH_3 \;+\; H_2O$$

22.100 amine, aromatic ring, and ester

$$H_2N-\text{(phenyl)}-\overset{\overset{\displaystyle O}{\|}}{C}OH$$

carboxylic acid

$$HOCH_2CH_2\underset{\underset{\displaystyle CH_2CH_3}{|}}{N}CH_2CH_3$$

alcohol

22.101

$$CH_3(CH_2)_{14}\overset{\overset{\displaystyle O}{\|}}{C}OK \qquad CH_3(CH_2)_7CH=CH(CH_2)_7\overset{\overset{\displaystyle O}{\|}}{C}OK$$

$$CH_3(CH_2)_{16}\overset{\overset{\displaystyle O}{\|}}{C}OK \qquad HOCH_2\underset{\underset{\displaystyle OH}{|}}{C}HCH_2OH$$

Polymers (Section 22.15)

22.102 Polymers are large molecules formed by the repetitive bonding together of many smaller molecules, called monomers.

22.103 Polyethylene results from the polymerization of a simple alkene by an addition reaction to the double bond. Nylon results from the sequential reaction of two difunctional molecules.

22.104

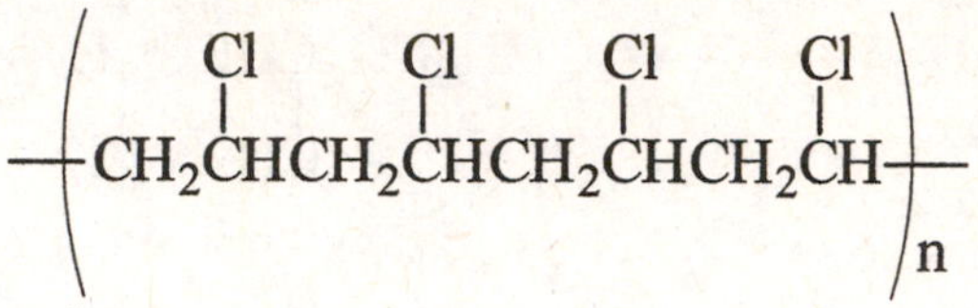

$$\left(\!CH_2\overset{Cl}{\underset{|}{C}}HCH_2\overset{Cl}{\underset{|}{C}}HCH_2\overset{Cl}{\underset{|}{C}}HCH_2\overset{Cl}{\underset{|}{C}}H\!\right)_{\!n}$$

22.105 (a)

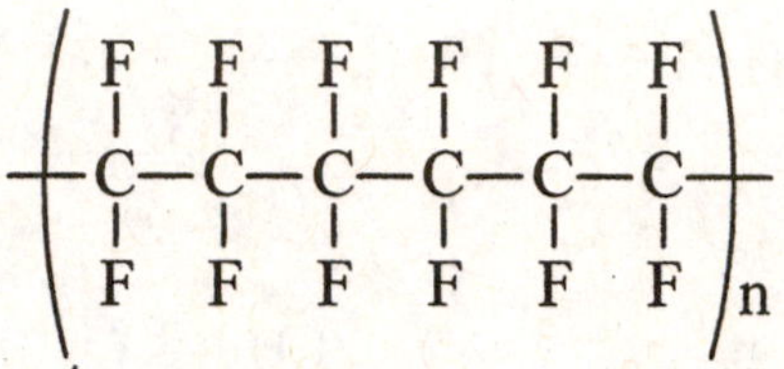

$$\left(\!\begin{matrix}F&F&F&F&F&F\\ |&|&|&|&|&|\\ C&-C&-C&-C&-C&-C\\ |&|&|&|&|&|\\ F&F&F&F&F&F\end{matrix}\!\right)_{\!n}$$

(b)

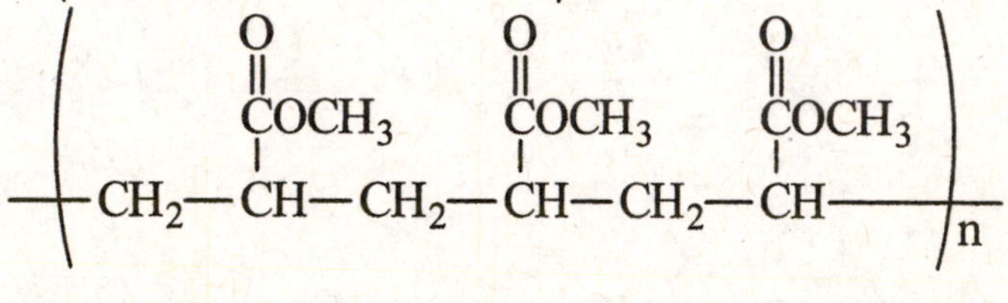

$$\left(\!CH_2-\overset{\overset{O}{\|}}{\underset{|}{C}OCH_3}CH-CH_2-\overset{\overset{O}{\|}}{\underset{|}{C}OCH_3}CH-CH_2-\overset{\overset{O}{\|}}{\underset{|}{C}OCH_3}CH\!\right)_{\!n}$$

22.106 (a) $CH_2\!=\!\underset{\underset{CN}{|}}{CH}$ (b) $CH_2\!=\!\underset{\underset{CH_3}{|}}{CH}$ (c) $CH_2\!=\!CCl_2$

22.107 $H_2C\!=\!\underset{\underset{CO_2CH_3}{}}{\overset{\overset{CN}{|}}{C}}$

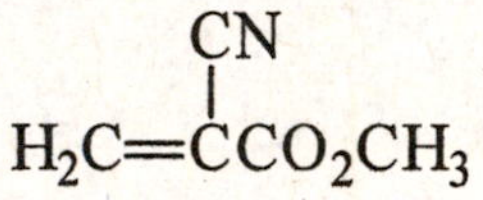

22.108

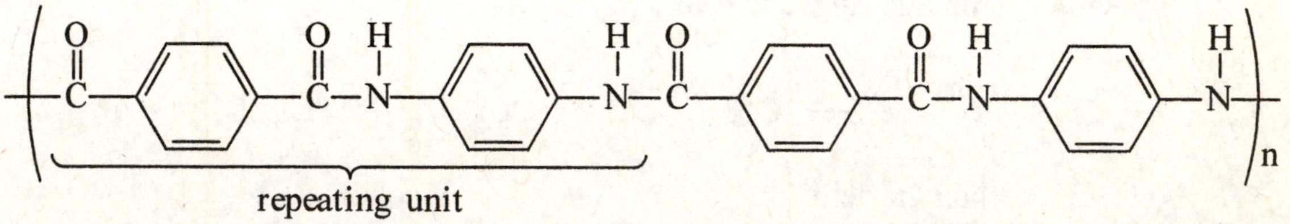

$$\left(\!\overset{\overset{O}{\|}}{C}-\!\!\!\bigcirc\!\!\!-\overset{\overset{O}{\|}}{C}-\overset{\overset{H}{|}}{N}-\!\!\!\bigcirc\!\!\!-\overset{\overset{H}{|}}{N}-\overset{\overset{O}{\|}}{C}-\!\!\!\bigcirc\!\!\!-\overset{\overset{O}{\|}}{C}-\overset{\overset{H}{|}}{N}-\!\!\!\bigcirc\!\!\!-\overset{\overset{H}{|}}{N}\!\right)_{\!n}$$

repeating unit

22.109

$$\left(\!OCH_2CH_2O\overset{\overset{O}{\|}}{C}CH_2CH_2\overset{\overset{O}{\|}}{C}OCH_2CH_2O\overset{\overset{O}{\|}}{C}CH_2CH_2\overset{\overset{O}{\|}}{C}\!\right)_{\!n}$$

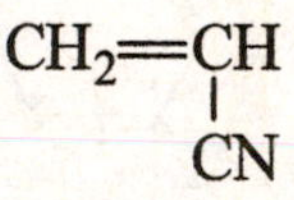

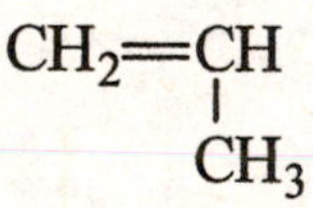

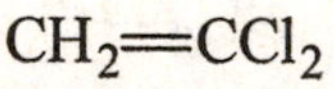

repeating unit

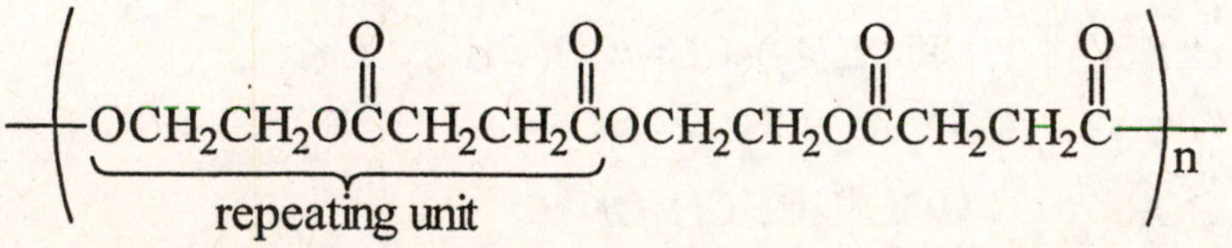

Chapter Problems

22.110 (a)
$$CH_3CHCH_2CH_2CH_2CH_2CH_3$$
with CH_3 substituent on the second carbon

(b)
$$CH_3CHCH_2CHCH_2CH_3$$
with CH_3 and CH_2CH_3 substituents

(c)
$$CH_3CH_2CH-CCH_2CH_2CH_2CH_3$$
with CH_3, CH_2CH_3, and CH_3 substituents

(d)
$$CH_3CHCH_2CCH_2CH_2CH_3$$
with CH_3 and CH_3 substituents

(e) cyclopentane with two CH_3 groups

(f)
$$CH_3CH_2CHCHCH_2CH_2CH_3$$
with CH_3 and $CHCH_3$ / CH_3 substituents

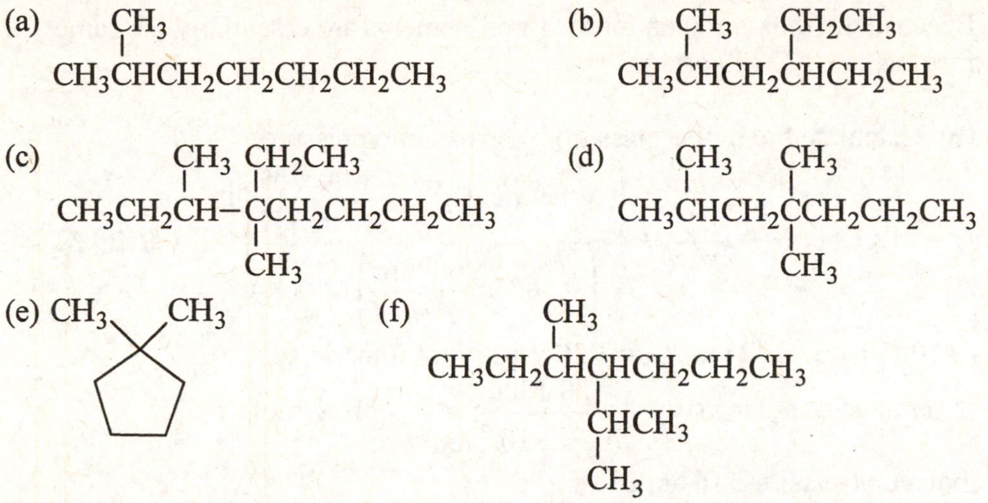

22.111 (a) 2,3-dimethylhexane   (b) 4-isopropyloctane
       (c) 4-ethyl-2,4-dimethylhexane (d) 3,3-diethylpentane

22.112 Cyclohexene will react with Br_2 and decolorize it. Cyclohexane will not react.

22.113 Cyclohexene will react with Br_2 and decolorize it. Benzene will not react with Br_2 without a catalyst.

22.114 (a)
$$CH_3CH_2CH_2CH_2CHCH_3$$
with CH_3 substituent

(b) benzene ring with Br, NO_2, and Br substituents

(c) cyclohexane ring with CH_3, CH_3, and OH substituents

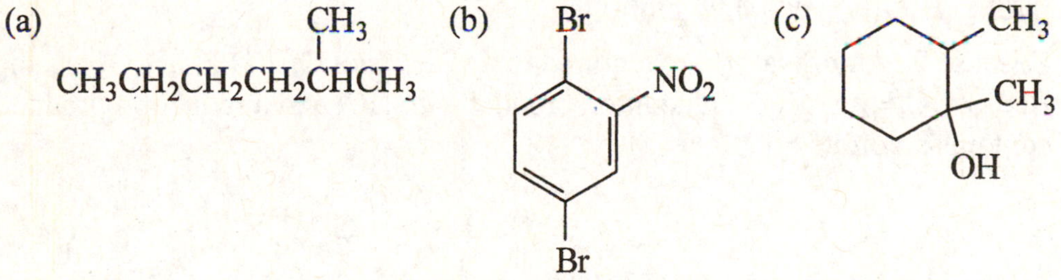

22.115
$$\left(-NHCH_2CH_2CH_2CH_2CH_2\overset{O}{\overset{\|}{C}}NHCH_2CH_2CH_2CH_2CH_2\overset{O}{\overset{\|}{C}}-\right)_n$$
repeating unit

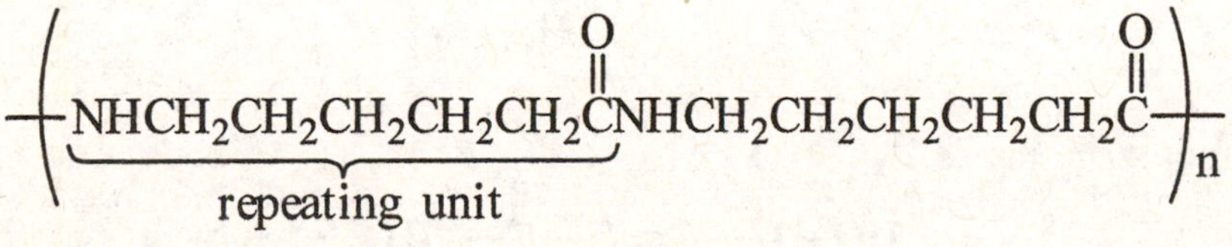

Multiconcept Problems

22.116 (a) Calculate the empirical formula. Assume a 100.0 g sample of fumaric acid.
$$41.4 \text{ g C} \times \frac{1 \text{ mol C}}{12.01 \text{ g C}} = 3.45 \text{ mol C}$$

$$3.5 \text{ g H} \times \frac{1 \text{ mol H}}{1.008 \text{ g H}} = 3.47 \text{ mol H}$$

681

Chapter 22 – Organic Chemistry

$$55.1 \text{ g O} \times \frac{1 \text{ mol O}}{16.00 \text{ g O}} = 3.44 \text{ mol O}$$

Because the mol amounts for the three elements are essentially the same, the empirical formula is CHO (29 amu).

(b) Calculate the molar mass from the osmotic pressure.

$$\Pi = MRT; \quad M = \frac{\Pi}{RT} = \frac{\left(240.3 \text{ mm Hg} \times \dfrac{1.00 \text{ atm}}{760 \text{ mm Hg}}\right)}{\left(0.082\,06 \dfrac{L \cdot atm}{K \cdot mol}\right)(298 \text{ K})} = 0.0129 \text{ M}$$

$$(0.1000 \text{ L})(0.0129 \text{ mol/L}) = 1.29 \times 10^{-3} \text{ mol fumaric acid}$$

$$\text{fumaric acid molar mass} = \frac{0.1500 \text{ g}}{1.29 \times 10^{-3} \text{ mol}} = 116 \text{ g/mol}$$

molecular mass = 116 amu

(c) Determine the molecular formula. $\quad \dfrac{\text{molar mass}}{\text{empirical formula mass}} = \dfrac{116}{29} = 4$

molecular formula = $C_{(1 \times 4)}H_{(1 \times 4)}O_{(1 \times 4)} = C_4H_4O_4$

From the titration, the number of carboxylic acid groups can be determined.

$$\text{mol } C_4H_4O_4 = 0.573 \text{ g} \times \frac{1 \text{ mol } C_4H_4O_4}{116 \text{ g}} = 0.004\,94 \text{ mol } C_4H_4O_4$$

mol NaOH used = (0.0941 L)(0.105 mol/L) = 0.0099 mol NaOH

$$\frac{\text{mol NaOH}}{\text{mol } C_4H_4O_4} = \frac{0.0099 \text{ mol}}{0.004\,94 \text{ mol}} = 2$$

Because 2 mol of NaOH are required to titrate 1 mol $C_4H_4O_4$, $C_4H_4O_4$ is a diprotic acid. Because $C_4H_4O_4$ gives an addition product with HCl and a reduction product with H_2, it contains a double bond.

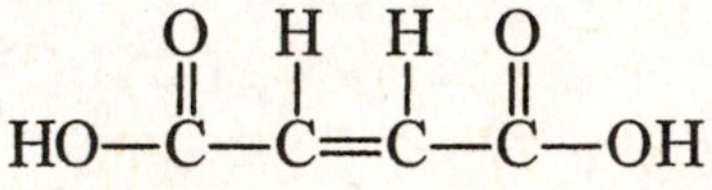

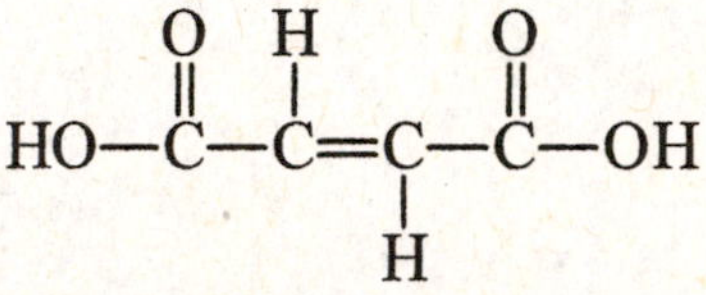

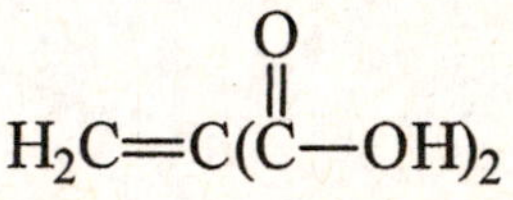

(d) The correct structure is

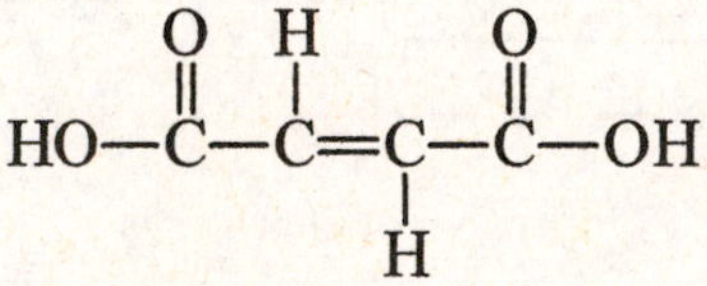

22.117 (a) CO_2, 44.01 amu; H_2O, 18.02 amu

$$\text{mol } CO_2 = 0.1213 \text{ g } CO_2 \times \frac{1 \text{ mol } CO_2}{44.01 \text{ g } CO_2} = 0.00276 \text{ mol } CO_2$$

$$\text{mol } H_2O = 0.0661 \text{ g } H_2O \times \frac{1 \text{ mol } H_2O}{18.02 \text{ g } H_2O} = 0.00367 \text{ mol } H_2O$$

$$\text{mass C} = 0.00276 \text{ mol } CO_2 \times \frac{1 \text{ mol C}}{1 \text{ mol } CO_2} \times \frac{12.011 \text{ g C}}{1 \text{ mol C}} = 0.0332 \text{ g C}$$

$$\text{mass H} = 0.00367 \text{ mol } H_2O \times \frac{2 \text{ mol H}}{1 \text{ mol } H_2O} \times \frac{1.008 \text{ g H}}{1 \text{ mol H}} = 0.00740 \text{ g H}$$

$$\text{mass O} = 0.0552 \text{ g sample} - 0.0332 \text{ g C} - 0.00740 \text{ g H} = 0.0146 \text{ g O}$$

$$\text{mol C} = 0.00276 \text{ mol } CO_2 \times \frac{1 \text{ mol C}}{1 \text{ mol } CO_2} = 0.00276 \text{ mol C}$$

$$\text{mol H} = 0.00367 \text{ mol } H_2O \times \frac{2 \text{ mol H}}{1 \text{ mol } H_2O} = 0.00734 \text{ mol H}$$

$$\text{mol O} = 0.0146 \text{ g O} \times \frac{1 \text{ mol O}}{16.00 \text{ g O}} = 0.000913 \text{ mol O}$$

$C_{0.00276} H_{0.00734} O_{0.000913}$ (divide each subscript by the smallest)

$C_{0.00276 / 0.000913} H_{0.00734 / 0.000913} O_{0.000913 / 0.000913}$

$C_{3.023} H_{8.039} O$

C_3H_8O

$$2\, C_3H_8O(l) + 9\, O_2(g) \rightarrow 6\, CO_2(g) + 8\, H_2O(l)$$

(b) C_3H_8O is a molecular formula because a multiple such as $C_6H_{16}O_2$ is not possible.

(c)

Alcohol   Ether

$CH_3CH_2CH_2OH$ $CH_3\overset{\overset{\displaystyle OH}{|}}{C}HCH_3$ $CH_3CH_2OCH_3$

(d) Acetone is $CH_3\overset{\overset{\displaystyle O}{\|}}{C}CH_3$. The most likely structure for C_3H_8O is $CH_3\overset{\overset{\displaystyle OH}{|}}{C}HCH_3$.

(e) C_3H_8O, 60.10 amu

$$\text{mol } C_3H_8O = 5.000 \text{ g } C_3H_8O \times \frac{1 \text{ mol } C_3H_8O}{60.10 \text{ g } C_3H_8O} = 0.08320 \text{ mol } C_3H_8O$$

$$\Delta H^\circ_{combustion} = \frac{-166.9 \text{ kJ}}{0.08320 \text{ mol}} = -2006 \text{ kJ/mol} = -2006 \text{ kJ}$$

$$C_3H_8O(l) + 9/2\, O_2(g) \rightarrow 3\, CO_2(g) + 4\, H_2O(l)$$

$$\Delta H^\circ_{combustion} = [\, 3\, \Delta H^\circ_f(CO_2) + 4\, \Delta H^\circ_f(H_2O)] - \Delta H^\circ_f(C_3H_8O)$$

$$\Delta H^\circ_f(C_3H_8O) = [\, 3\, \Delta H^\circ_f(CO_2) + 4\, \Delta H^\circ_f(H_2O)] - \Delta H^\circ_{combustion}$$

$$= [(3 \text{ mol})(-393.5 \text{ kJ/mol}) + (4 \text{ mol})(-285.8 \text{ kJ/mol})] - (-2006 \text{ kJ})$$

$$= -317.7 \text{ kJ}$$

$$\Delta H^\circ_f = -317.7 \text{ kJ/mol}$$

22.118 (a) 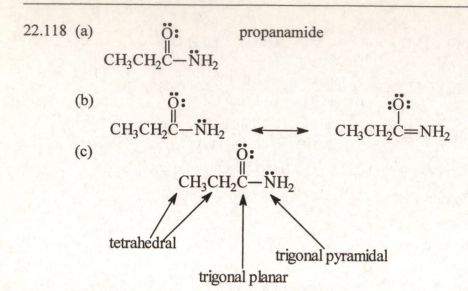 propanamide

(b)

(c)

tetrahedral

trigonal planar

trigonal pyramidal

(d) An observed trigonal planar N does not agree with the VSEPR prediction. The second resonance structure is consistent with a trigonal planar N.